CONIFERS OF THE WORLD

CONIFERS OF THE WORLD

THE COMPLETE REFERENCE

JAMES E. ECKENWALDER

TIMBER PRESS
PORTLAND • LONDON

Frontispiece: Port Orford cedars (*Chamaecyparis lawsoniana*).

Black-and-white photographs of trees in nature by Zsolt Debreczy and István Rácz. Drawings of cones, seed scales, and seeds by Janet Mockler. Other drawings in the introductory matter by Takashi Sawa. Maps by University of Toronto Cartography Office. Photomicrographs by Ronald Dengler. Black-and-white foliage images and color photographs by the author unless otherwise indicated.

Published in 2009 by Timber Press, Inc.

The Haseltine Building
133 S.W. Second Avenue, Suite 450
Portland, Oregon 97204-3527
www.timberpress.com

2 The Quadrant
135 Salusbury Road
London NW6 6RJ
www.timberpress.co.uk

ISBN-13: 978-0-88192-974-4

Designed by Christi Payne
Printed in China

Library of Congress Cataloging-in-Publication Data
Eckenwalder, James E., 1949-
Conifers of the world : the complete reference / James E. Eckenwalder. -- 1st ed.
p. cm.
Includes bibliographical references and index.
ISBN 978-0-88192-974-4
1. Conifers. 2. Conifers--Classification. I. Title.
QK494.E34 2009
585--dc22

2009016604

A catalog record for this book is also available from the British Library.

CONTENTS

Color photographs follow page 64

PREFACE

Successive waves of new interpretation since the 1960s have altered our understanding of conifer relationships. These revisions arise from closer investigation of previously obscure or overlooked groups (primarily those of the tropics and southern hemisphere), from new lines of evidence (most recently from DNA data), from discovery of new conifers (like *Wollemia*), and from changing conceptual and theoretical perspectives, including a grouping method called cladistic analysis and its accompanying phylogenetic approach to classification.

Changing views of relationships result in new classifications that, in turn, lead to name changes for conifers. Many of these changes have little effect because they involve unfamiliar plants, but some affect trees well known to gardeners, foresters, and botanists. Because there are often uncertainties in and legitimate differences of opinion over the classification of various conifers, a given conifer may be referred to by several different names, the correct one varying with the classification accepted by an author. One example familiar to many readers living in a cool temperate region is Alaska yellow cedar (or Nootka cypress), a handsome, narrowly conical tree with drooping branches described in this book under the probably unfamiliar scientific name *Cupressus nootkatensis.* Other contemporary accounts, adopting different classifications of the cypresses as a whole, refer to it by the even more unfamiliar names *Callitropsis nootkatensis* or *Xanthocyparis nootkatensis.* Ironically, the most familiar botanical name for this tree, *Chamaecyparis nootkatensis,* the only one used during most of the last century, and the one still commonly used in gardens and in print, is not correct in any classification acceptable today. Alaska yellow cedar is not actually related to the five tree species originally placed in the genus *Chamaecyparis,* including hinoki and sawara cypresses and Port Orford cedar (also called Lawson cypress).

As an undergraduate student, I completed a senior thesis investigating the relationships of these and other cypresses with redwood trees and their relatives. As a result, I proposed a classification of these trees that placed them all in a single family rather than the two separate families that were then almost universally accepted (though some authors assigned them to more than two). The testing and subsequent general adoption of this proposal was one of many factors stimulating the cascade of work that led to so many recent name changes in conifers.

Presenting and explaining these and other changes in classification and the revisions of names that followed was one of my primary goals in writing this book. Since the taxonomic evidence and decisions leading to such changes are so often open to interpretation, I discuss the varying views of relationship that remain possible in many cases and highlight other situations in which more study is needed using additional fieldwork, more material, and newly examined characters. Surprising to many people unfamiliar with conifer studies are some instances of genera that look like they should be related to one another but are decidedly not. For instance, firs (*Abies*) and spruces (*Picea*), which seem to have so much in common, were confused with one another by early professional taxonomists, including Linnaeus, but are really rather distant, as evidenced by numerous traits if not the most obvious ones. In contrast, there are genera that seem rather dissimilar, such as Douglas firs (*Pseudotsuga*) and larches (*Larix*), that are, nonetheless, closely related, again with support from many characteristics, using both traditional and contemporary taxonomic methods.

A second purpose here is to present descriptions and identification guides for all of the world's living conifers as they occur in nature, without the historical bias in favor of temperate-climate ("hardy") genera and species characteristic of most books of this kind. This comprehensive coverage of wild conifers necessitates, in the inevitable trade-offs of bookmaking, a corresponding de-emphasis on descriptions of the thousands of individual cultivars that grace our gardens, though I provide an indication of the kinds of features that one can expect to find in the cultivars within each genus. As part of this endeavor, I hope to draw greater attention to the full spectrum of diversity in conifers, which are too often perceived as primitive poor cousins and also-rans to the flowering plants. The diversity of form, features, and ecological adaptations can be astonishing to those only familiar with members of the dozen most widespread and abundant genera of the pine (Pinaceae), yew (Taxaceae), and cypress (Cupressaceae) families in the northern hemisphere. These familiar genera, including the pines (*Pinus*), firs (*Abies*), spruces (*Picea*), larches (*Larix*), yews (*Taxus*), junipers (*Juniperus*), and arborvitaes (*Thuja*), include less than a fifth of the almost 70 living genera of conifers and belong to only half of the six conifer families. While these dozen familiar northern genera admittedly contain over half the living species of conifers worldwide, much of the diversity in conifers springs from the contrasting characteristics of the genera rather than from the species within a genus, which more or less represent variations on a theme. For this reason, the descriptions of genera are an important part of the presentation in this book, with certain kinds of information, like general cone structure, microscopic leaf structure, wood structure, pollen form, fossil record, chromosome numbers, cultivation notes, and broad cultivar characteristics, appearing mostly at this level. Therefore, it is important to read the description and discussion of the genus as well as the species account for a particular plant because key characteristics that can help with confirming an identification may be found only in the text relating to the genus.

My third concern is to simplify the language of the plant descriptions and identification guides as much as possible in order to make them accessible to a wider audience than just the botanical cognoscenti. This becomes a difficult balancing act at times and requires a certain measure of compromise because I do not want to sacrifice accuracy to achieve accessibility. While plain English wording may sometimes seem to lack the precision implicit in technical botanical descriptive terminology, botanical terms are not always used in a consistent manner in the literature (and are sometimes, though rarely, even used incorrectly), so their meanings are less precise than one might naturally expect.

My overriding motivation behind all of these considerations, however, is to share my fascination and enthusiasm for these wonderful plants. Inspired by researchers and thinkers in such diverse fields as taxonomy, dendrology, embryology, evolution, forest genetics, horticulture, morphology, paleobotany, phytochemistry, and phytogeography, I have been directly engaged with the classification and biology of conifers (and other gymnosperms) for more than 35 years. It is astonishing to me that, with all the workers in the field of conifer studies, the enormous economic importance of conifers and their conspicuous presence in or dominance of so many cultivated and natural landscapes, conifers are still usually perceived as poor cousins to the flowering plants, outmoded or well on their way to an inevitable demise. This perspective does not reflect the marvelous trees and shrubs that I know and admire. Even considering those that are relicts from the Tertiary period, like redwood (*Sequoia sempervirens*), giant sequoia (*Sequoiadendron giganteum*), bald cypress (*Taxodium distichum*), dawn redwood (*Metasequoia glyptostroboides*), and golden larch (*Pseudolarix amabilis*), all once widely distributed across the northern hemisphere and now restricted to small corners of single continents, one finds them often as thriving, commonly dominant, members of their present habitats, holding great stores of genetic variation that can serve as the basis for their continued success and responsiveness to environmental changes. Many other conifers, notably junipers (*Juniperus*), cypress pines (*Callitris*), araucarian pines (*Araucaria*), and pines (*Pinus*), are clearly actively generating new species even today and are quite capable of competing with flowering plants in many environments that they already occupy or into which they have been introduced. I hope here, therefore, to give a summary of some of the things we know about conifers, to encourage further research, and most of all to celebrate conifer diversity.

When I started this book, I naively thought that it would be a relatively easy matter to bring together and present the information I wished to convey. I guessed this might take perhaps as long as 2 or 3 years. More than a dozen years later, it appears that my time expectations were rather optimistic. Most conifer genera lack current worldwide formal taxonomic revisions, and some monographs and reviews that exist are inconsistent with one another, so there were few genera that did not need at least some thought to present properly. In summarizing and refining current knowledge on the classification, identification, and characteristics of the world's conifers, there were three major tasks that proved time consuming. Firstly, numerous taxonomic decisions needed to be made as I went along. For each genus, I considered all generally accepted taxa (species, subspecies, and varieties) along with more recently described ones that had not yet gained general acceptance and decided which ones I thought merited recognition and which should be relegated to the synonymy of others (and if so, with which accepted taxa should they be merged). Each decision required a review of evidence from morphology, distribution, ecology, chemistry, and molecular studies taken both from the literature and as I saw them in living trees and shrubs and in preserved herbarium specimens. Then, in describing each accepted species, I needed to integrate and evaluate previously published descriptions with what I saw in the plants themselves, filling in gaps in existing portrayals where I could. A few conifer species, however, are still so poorly documented that their complete description cannot yet be compiled. Some characteristics are included for species of all genera, while others, mostly microscopic, are included only for those genera in which they are helpful in distinguishing the species. Lastly, in the search for plain-language descriptions, I found myself going back and forth among various wordings before I settled on a descriptive terminology that met my need for combining accuracy with clarity and accessibility.

While new discoveries of and about conifers are made all the time, it is time to bring this work to a close. I hope it will allow users to identify any conifer they encounter in the wild and many of those found in gardens, except for those whose cultivar traits lie too far outside of what may usually be found in nature. Throughout the text, I point out many groups in which future research in the garden, field, herbarium, and laboratory would be helpful in solving taxonomic questions and difficulties. Let us hope that these uncertainties are soon resolved and that the next overview of the world's conifers can present these successes without, at the same time, having to report the extinctions of any conifers due to global climate change and the pressures caused by continuing human population growth.

ACKNOWLEDGMENTS

Every author of a factual book is beholden to numerous predecessors. For an author dealing with the classification of plants, the debt can extend concretely back to Linnaeus' *Species Plantarum,* published in 1753. With regards to this book specifically, Krüssmann's *Manual of Cultivated Conifers* was precedent setting. In the course of making taxonomic evaluations, compiling descriptions, and framing my discussion, I consulted thousands of references, large and small. The bibliography favors the newer literature; older works, even when consulted, are often only included through their citation in more recent articles and books. I am more directly indebted to Aljos Farjon, Chris Quinn, Charles Miller, Aaron Liston, John Silba, Zsolt Debreczy, and István Rácz for sending me copies of some of their own works. To all researchers who continue to make significant conifer observations and whose work may not be reflected in this book, my sincere apologies. With great reluctance, I found it necessary at arbitrary points, varying by species, to put the literature aside in the interest of finishing the book.

Descriptive data in the literature were corroborated and extended by observations in the field in Europe, Siberia, and North, Central, and South America, as well as by examining preserved herbarium specimens of wild-growing and cultivated trees collected for and preserved at numerous herbaria, including those of the Fairchild Tropical Garden; Missouri Botanical Garden; New York Botanical Garden; Oregon State University; Rancho Santa Ana Botanic Garden; Royal Botanic Garden, Edinburgh; Royal Botanic Gardens, Kew; and University of California, Berkeley, all of whose curators and officials I thank for access and for numerous other courtesies. I also thank these same people and institutions as well as those at the Bedgeberry Pinetum; Hoyt Arboretum; Humber Nurseries; Jardin Botanique de Montréal; Royal Botanical Garden, Hamilton; San Francisco Botanical Garden (formerly Strybing Arboretum), Toronto Botanical Garden; University of California, Berkeley, Botanical Garden; University of Guelph Arboretum; and University of Washington Botanic Gardens for permission to collect specimens from living trees in their collections, and all those homeowners who unwittingly allowed me to nick snippets of twigs from their front yards. Data on largest individuals in the United States are derived from the American Forestry Association's Social Register of Big Trees. A special thanks to Tim Dickinson, Deb Metsger, and Jenny Bull of my home herbarium, the Vascular Plant Herbarium of the Royal Ontario Museum.

My views on conifers have benefited from discussions with many friends and colleagues, including Bob Adams, Jim Basinger, Bert Brehm, Ib Christensen, Bill Critchfield, Aljos Farjon, Bob Fincham, Michael Frankis, Al Gordon, Sean Graham, Bob Hill, Ken Hill, Dean Kelch, David de Laubenfels, Ben LePage, Aaron Liston, Beth McIver, Steve Manchester, Jack Maze, Robert Mill, Chris Page, Bob Price, Chris Quinn, Keith Rushforth, Howard Schorn, Saša Stefanović, Dennis Stevenson, Ruth Stockey, John Strother, Steven Strauss, Barry Tomlinson, and Tom Zanoni.

All illustrations in this book are original and newly published here. I thank Janet Mockler for the drawings of cones, seeds, and seed scales and Takashi Sawa for additional drawings and diagrams. Thanks also to Byron Moldofsky, Jane Davie, and Mariange Beaudry of the Cartography Office at the University of Toronto. Thanks to István Rácz and Zsolt Debreczy for their black-and-white photographs of plant habits in the field. Further examples of the breadth and depth of their tree photography can be found in their forthcoming book, *Conifers Around the World.* I am grateful to Ron Dengler and Spencer Barrett for the use of their slides and to Taylor Feild, Jim Lounsberry, and Rowan Sage, who offered me theirs.

Nancy Dengler provided facilities, supplies, and funding for examining cross-sectional anatomy of needles, Sasha Terry did much of the sectioning, and Ron Dengler oversaw this project, preserved the material, and took the photomicrographs used here, as well as many others. My thanks to all three and to Nancy, also, for reading and providing helpful comments on parts of the manuscript.

I am especially indebted to a grant from the Stanley Smith Horticultural Trust, facilitated initially by Robert Ornduff. Not only did the trust fund the major portion of all the illustrations, it gave me a much needed vote of confidence at the beginning of this lengthy project and strengthened my desire to make the wording as jargon free as possible for all those interested in conifers.

Lastly, I acknowledge my enormous debt to two individuals who were central to this project from its inception through to its completion and without whom, it would not have been done. Dale Johnson (and all the staff at Timber Press) have been truly patient and helpful during the long gestation of this project. And, of course, there would not have been a book without the organizational abilities, typing and word-processing skills, tenacity, encouragement, and love of my wife, Susan Eckenwalder, who bore the postponement of so much with astonishing patience and grace, and to whom I dedicate this volume.

CHAPTER 1

CONIFER CLASSIFICATION

Not all conifers are as instantly recognizable members of the group as are pines, spruces, or junipers for northern-hemisphere people. Understanding the taxonomic limits of the group, and its classification into families, genera, species, subspecies, and varieties, and the relationships among those included, are helpful in making full use of this book to identify wild and cultivated conifers.

WHAT ARE CONIFERS?

Conifers are a group of seed plants (taxonomically an order, subclass, class, or division), all of which are descended from a common ancestor in the late Paleozoic, more than 300 million years ago, that they do not share with any of the other four living groups of seed plants. The nearly 550 species of conifers are found all around the world (although in differential abundance and prominence), on every continent (except Antarctica), and on many islands. Many conifers are familiar plants, especially those belonging to the most widespread genera: pines (*Pinus*), firs (*Abies*), spruces (*Picea*), and junipers (*Juniperus*) in the northern hemisphere (Plate 1), and yellowwoods (*Podocarpus*) in the southern. Taken together, these five genera contain about 300 species, more than half the living conifer species, and occur in almost all the places where any conifers are found.

All but about 15 species of conifers are evergreen, even in temperate and colder climates. Most flowering plants are also evergreen (especially those of the tropics and warm temperate regions) but typically are referred to as broad-leaved evergreens to distinguish them from the needle- and scale-leaved conifers. While the majority of conifers have needle-, scale-, or clawlike leaves, a few species have broader leaves that are a far cry from pine needles or juniper scales. Despite some variations, however, their distinctive leaf forms are among the most obvious characteristics uniting the conifers, since most of these forms are shared across the different families.

The name conifer means cone-bearer, but this characteristic, inclusive of the seeds in a pine cone, is neither found in all conifers nor confined to them. Still, the particular structure of the seed cones, when present, is unique to conifers. This structure basically involves seed scales in the axils of (and often more or less united with) bracts, with both kinds of organs attached to the axis of the cone. This is a compound cone because it is a highly condensed, branched, reproductive shoot consisting of a cone axis clothed with modified leaves (the bracts), each bearing a small branch in its axil (the seed scale). Seed scales do not look like branches, but their equivalence to shoots has been demonstrated by studies of the development of modern conifer seed cones and of the structures of ancient fossil conifers. Divergent interpretations of the specifics of each of these lines of evidence have not changed the basic concept that the seed scales of conifer seed cones consist of a number of ancestrally separate seed-bearing structures (and often non-seed-bearing ones as well) of an axillary dwarf shoot that became united phylogenetically and developmentally into a single structure. None of the other seed plant groups has seed-bearing organs with this structure. Gnetophytes are the most similar because they too have compound seed cones, but theirs lack seed scales and instead have reproductive dwarf shoots in the

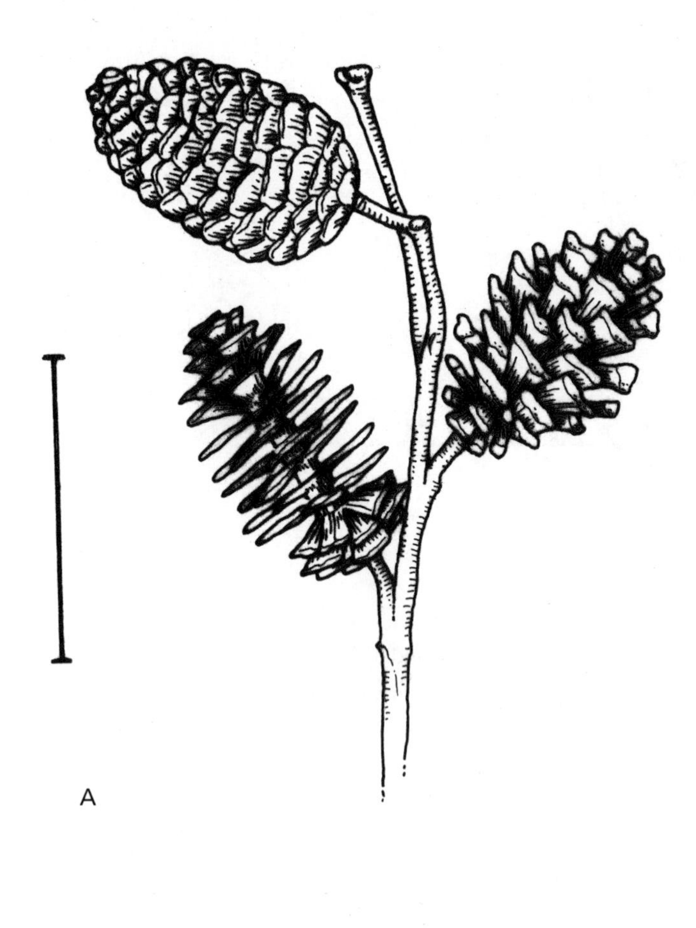

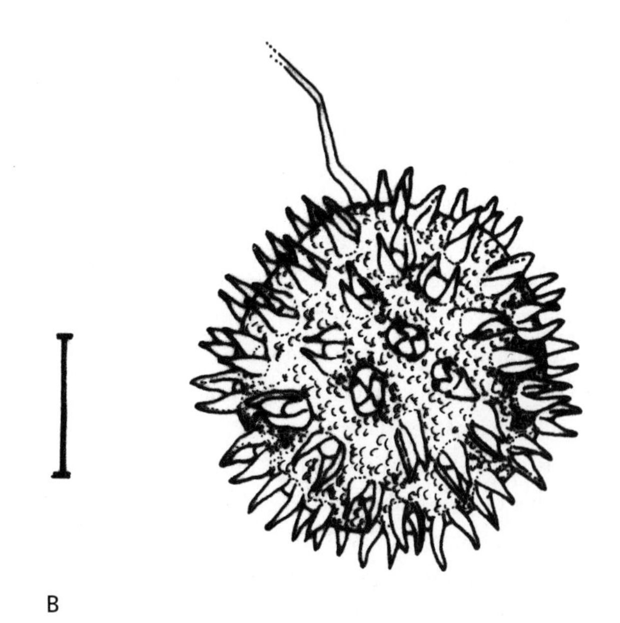

Conelike seed-bearing structures of two species that are flowering plants, not conifers: European alder (*Alnus glutinosa,* A; scale, 2 cm) and sweet gum (*Liquidambar styraciflua,* B; scale, 1 cm).

axils of the bracts that consist of separate rather than united parts. Most of the other seed plant groups have simple seed cones or none at all.

Besides the compound seed cones and needle (or scale) leaves, conifers also have distinctive wood that differs from the wood of cycads in being more compact, and from that of gnetophytes and most flowering plants in having water-conducting tissue consisting solely of tracheids, without the larger-diameter vessels that have greater water-conduction capacity. Similar wood is, however, found in *Ginkgo,* which differs from conifers in its fan-shaped leaf form and seed-bearing structures: large seeds are paired at the tips of slender stalks and lack any associated bracts or scales.

Another feature that loosely unites the conifers and separates them from the other seed plants is the structure of pollen grains. Conifers mostly have one of two basic types of pollen grains. The most distinctive forms, at least among living seed plants, are the ones with two or three air bladders, the forms found in most Pinaceae, including the common pines, firs, and spruces, and in most Podocarpaceae, including the widespread yellowwoods. The other predominant form is nearly spherical, with a bumpy to smooth surface but few other obvious features. Both forms have more than the two or three included nuclei found in flowering plant pollen grains. The latter also have different, more complicated wall structures as well as myriad variations in shape, germination regions, and sculpturing of the surface. Cycads and *Ginkgo* have more or less boat-shaped grains, while many gnetophytes have pollen grains with longitudinal furrows.

Thus even though there is no single feature that unites all conifers and sets them apart from the other groups of living seed plants, there is ample evidence for their unity and distinctness. This unity reflects their common heritage from an ancestral conifer that is uniquely their ancestor. At the same time, their obvious similarities to the other seed plants reflect an even more ancient common ancestor, shared by all the seed plants, that lived more than 350 million years ago.

TAXONOMY OF CONIFERS

Taxonomic concepts about the relationships of conifers to each other and to other plants are continuously reviewed in light of the ever expanding base of botanical and horticultural knowledge. This leads to changing classifications and sometimes to name changes as a result of new understanding of relationships. Although never really dormant, taxonomic work on conifers returned to a high activity level beginning in 1969 after two decades of more modest interest. New conifers have been discovered at a fairly steady pace since Linnaeus's publication of *Species Plantarum* in 1753, even when there was no particular interest in conifer classifications. Discovery of new conifers, especially new genera, has sometimes been a catalyst for renewed interest in conifer classification, frequently leading to a reevaluation of generic and familial concepts. Some aspects of conifer classification at all taxonomic levels saw revision and reinterpretation during this period of increased activity. Much progress has come about as a result of the ability to read sequences of DNA for particular genes, such as the large submit of the photosynthetic enzyme rubisco (ribulose bis-phosphate carboxylase-oxygenase), the most abundant protein in the world. DNA studies settle some old uncertainties about the interrelationships among conifers but at the same time introduce new ones, partly as a result of the sample of species used in any given study and partly as a result of actual inconsistencies in the relationships implied by the DNA sequences of different genes. In addition to molecular studies, progress in understanding conifer relationships also comes about as a result of development of other new lines of evidence, including chemical analyses of leaf and wood oils, ultrastuctural studies of pollen grain walls, developmental studies of seed cones, and investigation of chromosome complements as well as through new methods of taxonomic interpretation and more ready access to conifers throughout the world. The following paragraphs review several shifts in taxonomic concepts relating to conifers at different taxonomic levels.

ARE GYMNOSPERMS A REAL TAXONOMIC GROUP?

Gymnosperms ("naked seeds") consist of 546 conifer species plus three other seed plants groups: 250 or so species of cycads, the maidenhair tree (*Ginkgo biloba*), and about 50 species of the three gnetophyte or chlamydosperm genera—*Ephedra* (Mormon teas or joint firs), *Gnetum*, and the bizarre and unique *Welwitschia bainesii*. This aggregate of fewer than a thousand gymnosperm species contrasts with the remaining group of seed plants, the flowering plants or angiosperms ("boxed seeds"), which is the largest group of all with about 250,000 species. Separation of seed plants into these two traditional major groups has more to do with the overwhelming numerical dominance of angiosperms than with the coherence of meaningful features in gymnosperms. Many alternative arrangements of the five living seed plant groups have been proposed. Evolutionarily, conifers have been variously linked individually to *Ginkgo*, gnetophytes, and even some angiosperms, but not to cycads (to the exclusion of the other groups). Molecular studies are ambiguous about gnetophytes and *Ginkgo*, each of which has sometimes been linked to conifers. A few studies even suggest that conifers are more closely linked to flowering plants than are cycads, which, along with gnetophytes, are often considered the closest living relatives of angiosperms.

The status of gymnosperms as a real taxonomic group depends on the pattern of relationships of the four gymnosperm groups. If any one is more closely related to the angiosperms than to the other gymnosperm groups, then gymnosperms fail the phylogenetic test of monophyly for recognition as a taxon. It would mean that the most recent common ancestor of the gymnosperms was also the ancestor of the angiosperms. If you take all fossil seed plants into account, it is clear that gymnosperms are paraphyletic rather than monophyletic because the most ancient seed plants, the ancestors of all living seed plants including flowering plants, would be considered gymnosperms, and the traditional gymnosperms exclude at least one group, the angiosperms, descended from that ancestor. However, that leaves open the question of whether those gymnosperms alive today have a common ancestor exclusive of the angiosperms. This question is as yet unanswered because different genes give phylogenies that are either consistent or inconsistent with the monophyly of living gymnosperms, including conifers.

WHICH PLANTS SHOULD BE INCLUDED AMONG THE CONIFERS?

Two major uncertainties about the composition of conifers merit review, one historical and one contemporary. Traditionally, conifers included the families included in this book (with varyingly detailed family arrangements and assignments). As a result of his studies of fossil conifers in the 1930s, Rudolf Florin proposed that the yew family, Taxaceae (excluding *Cephalotaxus*), is so different from the other conifer families that they are essentially unrelated and should be placed in their own order Taxales separate from the Coniferales. This suggestion was fairly generally accepted because, despite many similarities in morphology to other conifers, the yews (or taxads) seemed to completely lack seed cones and Florin's studies implied that they had a long fossil history separate from other conifers. A substantial minority of conifer taxonomists did not accept this separation because of the numerous characteristics they share with other conifers. Even these taxonomists, however, often put the yews in a distinct suborder. DNA sequences unambiguously show that yews belong squarely within the conifers and are closely related to Cupressaceae, thus effectively quashing the idea of a separate order or suborder.

An especially close relationship between gnetophytes and conifers was never suspected on morphological grounds, despite the inclusion of gnetophytes in many books on conifers. Instead, gnetophytes were usually thought to be more closely related to flowering plants. Unexpectedly, some molecular studies imply that gnetophytes may be the sister group of Pinaceae *within* the conifers. Using the genes that give this result, the relationship is very strongly supported. Many other genes, however, reveal no connection whatever between gnetophytes and conifers. The apparent close relationship between gnetophytes and Pinaceae may simply be an artifact: a result of so-called long branch attraction in which highly divergent taxa are linked together because neither is closely related to anything else. While it seems very unlikely that gnetophytes will ultimately be shown really to belong within the conifers, it would be reasonably consistent with their morphology to think that they might be the sister group to the conifers. Both groups might very well have arisen from an ancestor that would be assigned to an extinct Paleozoic group of seed plants called the cordaites, which have compound pollen and seed cones and large, strap-shaped leaves unlike those of either of the two modern groups.

FAMILY PROBLEMS

In the 19th and early 20th centuries, most taxonomists recognized just two families of conifers: Pinaceae for those conifers with dry seed cones and predominantly wind-dispersed seeds, and Taxaceae for those with fleshy ones dispersed by fruit-eating birds and mammals. Then in 1926, Robert Pilger divided each of these into three or four families based on the structure of the seed cones, giving seven families: Pinaceae sensu stricto (in the narrow sense), Araucariaceae, Cupressaceae, and Taxodiaceae from the old Pinaceae sensu lato (in the broad sense), and Taxaceae sensu stricto, Cephalotaxaceae, and Podocarpaceae from the old Taxaceae sensu lato. This scheme of families was generally accepted for the next 50 years.

Two of the families, however, Cupressaceae and Taxodiaceae, in contrast to all the others, were distinguished primarily by their leaf arrangement: opposite or whorled in Cupressaceae and spiral in Taxodiaceae. All other features supposedly characterizing Cupressaceae could be found in one or another genus of Taxodiaceae. And then in 1948, when the living *Metasequoia glyptostroboides* (dawn redwood) was described, it proved to be an obvious close relative of redwood (*Sequoia sempervirens*) and giant sequoia (*Sequoiadendron giganteum*), both typical members of Taxodiaceae, and yet it had an opposite leaf arrangement. Separation of Cupressaceae and Taxodiaceae then became untenable, as suggested by David de Laubenfels in 1965. General acceptance of a revised family arrangement only followed after I presented a strong statement of the evidence in 1976. Fifteen years after that, molecular studies began to confirm my view that the traditional Cupressaceae are embedded within the traditional Taxodiaceae, being more closely related to *Glyptostrobus, Taxodium,* and *Cryptomeria* than these genera are to other genera of the traditional Taxodiaceae. The combined family is known as Cupressaceae, the earlier name of the two.

These same molecular studies show that Japanese umbrella pine (*Sciadopitys*), traditionally considered an anomalous and distinctive member of the Taxodiaceae, actually deserves a family of its own, as sometimes suggested in the past, because all members of the Cupressaceae sensu lato share a common ancestor with all of the Taxaceae that is not shared with *Sciadopitys.* Sciadopityaceae is then the sister family of the Cupressaceae-Taxaceae pair.

Taxaceae in this book includes plum yews (*Cephalotaxus*), which Pilger put in their own family, Cephalotaxaceae, because they have an obvious seed cone that is lacking in the other yews. *Cephalotaxus* resembles the other taxads in all other respects, even in the structure of the seeds. Molecular studies show that there are three groups of taxads rather than two, splitting Pilger's Taxaceae sensu stricto into two groups that are not more closely related to each other than either is to *Cephalotaxus.* There are two possible treatments of this situation: either recognize three separate families (Taxaceae, Cephalotaxaceae, and Amentotaxaceae) or three subfamilies of a single family (Taxaceae). Because of the obvious unity among these groups, which is at least as strong as among genera included in each of the other conifer families, the taxads are all included in a single family here.

As a result of various studies of morphology, development, and DNA sequences, there have thus been some shifts in the family structure proposed by Pilger. His Pinaceae, Araucariaceae, and Podocarpaceae remain very much as he defined them (except that two new genera have been discovered since his work: *Cathaya* in the Pinaceae and *Wollemia* in the Araucariaceae). His other four families, however, were reorganized into three, merging Taxodiaceae into Cupressaceae and Cephalotaxaceae into Taxaceae while separating out Sciadopityaceae.

In contrast to this relative conservatism or even modest consolidation, a few workers advocate further subdivision of conifer families. The most extreme example of this family splitting comes from studies of seed structure in Taxaceae and Podocarpaceae. The 24 genera in these two families are broken up into 33 genera and assigned to 20 families distributed among seven orders in a new class Podocarpopsida. The relationships implied by these proposals are completely at variance with the preponderance of morphological data and all molecular data. Furthermore, assigning almost every genus (as recognized here) to its own family is hardly taxonomically informative. This shows the kinds of difficulties that can arise when taxonomists rely entirely on just a single line of evidence: so-called one-character taxonomy.

HOW MANY GENERA OF CONIFERS ARE THERE?

In contrast to the relative conservatism of the families accepted by conifer taxonomists, subdivision of genera (as opposed to discovery of previously undescribed species belonging to new genera) has gone on at a fairly steady pace going back to the time of Carl Linnaeus and the beginning of scientific conifer classification and nomenclature. Linnaeus assigned the 26 conifer species he named in his 1753 *Species Plantarum* to seven genera, of which *Cupressus, Juniperus, Pinus, Taxus,* and *Thuja* each had two or more species. By 1841, these 26 species had been distributed among 13 genera (the ones in which these species are still included today) with only *Juniperus* and *Pinus* retaining more than one of the original species. *Juniperus* is the only Linnaean genus that is not currently subdivided, although a proposal to separate the scale-leaved and needle-leaved junipers as separate genera was advanced in 1754 but has generally not been accepted since, although it was adopted in Russian and Chinese literature during the 1930s to as recently as the 1970s, and even sometimes today. By 1841, also, three eastern Asian genera and the first seven from the southern hemisphere were added. By the centennial of *Species Plantarum,* the total number of genera of conifers accepted had reached 34, half the 67 recognized in this book. Discoveries of new genera have continued at a modest pace ever since. In recent times, a bigger source of increase in the number of generally recognized genera has been through splitting of old genera. This has tended to concentrate on different groups of conifers at different times. The preponderance of modern splitting has involved the southern hemisphere conifers.

Before 1960 there was a flurry of work on Cupressaceae by different authors, leading to breaking up of the incense cedar genus (*Libocedrus* sensu lato) and assignment of its 11 species to five genera (four of which are accepted here: *Austrocedrus, Calocedrus, Libocedrus* sensu stricto, and *Papuacedrus*). *Libocedrus* itself was split from *Thuja* in 1847 based on a species that had been described 5 years earlier. Before the splitting, the species assigned to *Libocedrus* were rather unusual in having distinct distribution areas in the northern and southern hemispheres. The series of studies that subdivided *Libocedrus* kept the northern segregate (*Calocedrus*) as a close relative of the southern ones, but more recent DNA studies show that it is not even related to them but rather to other northern genera. The generic segregation began as an attempt to emphasize the distinctiveness of northern and southern Cupressaceae. This viewpoint has since strengthened. Molecular studies show that the southern and northern genera of the cypresses (Cupressaceae sensu stricto) represent two entirely separate lineages although they are sister to each other. All 10 northern genera are most closely related to each other, and all nine southern genera are each other's closest relatives with no cross-hemisphere generic relationships.

One group within the Cupressaceae that may also see changes in generic circumscription as molecular data accumulate is the pair *Cupressus* and *Juniperus.* They include predominantly northern hemisphere species with a single juniper extending into the southern hemisphere in East Africa. The genus *Xanthocyparis* was described in this group but is here retained in *Cupressus* because of the basic agreement of the two contained species with others in the genus. However, some work, based on the arrangement of species in the molecular phylogeny, hints that *C. nootkatensis* and *C. vietnamensis* might each belong to a different genus and that the New and Old World species of *Cupressus* might have to be treated as separate genera. This would be because *Juniperus* appears to fall between the cypresses of the two hemispheres, with *C. nootkatensis* and *C. vietnamensis* as successive sister species to the rest of *Cupressus* and *Juniperus.* If these preliminary results hold up, they would represent a major shift in the arrangement of the cypress and juniper species into genera.

The family that has seen the most modern generic splitting is Podocarpaceae. The two largest traditional genera in this family, *Podocarpus* (the second largest conifer genus after *Pinus*) and *Dacrydium,* began to be divided in 1969, less than a decade after the (near) completion of a monograph of *Podocarpus* that retained it in the traditional sense. Previous taxonomic and morphological work recognized the heterogeneity of both genera and the difficulty of clearly distinguishing them from one another. These complexities were accommodated by formal botanical sections in *Podocarpus* and by informal groupings in *Dacrydium.* Recognizing the fundamental heterogeneity of these genera compared to accepted genera in other families, especially Pinaceae and Cupressaceae, de Laubenfels (1969, 1972) redistributed the eight previously recognized sections of *Podocarpus* sensu lato among five segregate genera (*Dacrycarpus, Nageia, Parasitaxus, Podocarpus* sensu stricto, and *Prumnopitys*), retaining only section *Podocarpus* in the pruned but still large genus *Podocarpus.* He also split *Dacrydium* into two genera: *Dacrydium* and *Falcatifolium.* Then Christopher Quinn showed that most of the New Zealand species of *Dacrydium* have morphologies and patterns of embryo development so unlike each other and those of *D. cupressinum* (rimu, type species of the genus) and its related tropical species that they merited assignment to an additional three genera: *Halocarpus, Lagarostrobos,* and *Lepidothamnus.* Later, Christopher Page raised the three sections of *Nageia* and the two sections of *Prumnopitys* to generic rank, as *Afrocarpus, Nageia,* and *Retrophyllum,* and as *Prumnopitys* and *Sundacarpus,* respectively. Finally (if there is such a thing as finality in taxonomy), Brian Molloy assigned the Tasmanian and

New Zealand species of *Lagarostrobos* to two separate genera, as *L. franklinii* and *Manoao colensoi*, respectively. Several DNA studies show that almost all these changes are amply justified and that many of the segregate genera are not even related to the traditional genus in which they were so long included. The only proposed segregate genus among this group that clearly cannot be sustained is *Sundacarpus*. The single species (*Prumnopitys amara*) is deeply embedded among the species of *Prumnopitys* in molecular phylogenies to such an extent that it is not even justifiable to place it in a separate section within the genus (as it was when all these species were still included in *Podocarpus*). Separation of *Manoao* from *Lagarostrobos* is also somewhat questionable, but if they were put together again, the recombined genus might also have to include *Parasitaxus ustus* from New Caledonia and be called *Parasitaxus*. Unlike the example of *Sundacarpus*, this is a case where either treatment is justifiable, neither being in conflict with the available evidence on relationships. The further massive generic splitting proposed subsequently by Alexey Bobrov and Alexander Melikian is not accepted here since it is in conflict with the preponderance of morphological and molecular evidence now available.

As a result of assessing these reevaluations of generic relationships, 67 genera are accepted in this book. This is a substantial increase over the 53 accepted in a standard work in English on conifers from a previous generation, Dallimore, Jackson, and Harrison's *Handbook of Coniferae and Taxaceae*, ed. 4 (1966), despite the fact that just a single entirely new genus (*Wollemia*) has been discovered since that work. It represents an even greater increase from the 46 genera recognized in the pivotal work of Pilger (1926), from the preceding generation, before the discovery of just another three entirely new genera (*Cathaya*, *Metasequoia*, and *Pseudotaxus*). Most generic relationships in known conifers have now been examined with varying degrees of thoroughness using DNA sequences. Based on these results, it would seem that the number of genera of conifers recognized is now likely to become fairly stable. Certain disputes will no doubt continue where there is no clearly correct answer, perhaps some entirely new genera remain to be discovered, and a few workers, focusing on their own very narrow research interests without reference to the larger picture, may well go off the deep end in giving generic recognition to small differences. Nonetheless, most of the previously understudied conifer groups have now been examined, and there is increasing consensus about which of them to recognize as genera.

This is not to say that there is still not much work to be done on classification within genera. All the larger genera (*Abies*, *Juniperus*, *Picea*, *Pinus*, and *Podocarpus*), for instance, would benefit from revised classifications of their subgenera and sections. Spanning several continents and each with a number or rare and localized species as well as commoner and more widespread ones, taxonomic revision of these five genera and their species groupings presents many logistical and practical problems that will not easily be overcome. Completely satisfying infrageneric classifications of these genera are among the most important remaining difficulties in conifer taxonomy.

THE LIMITS OF SPECIES

In contrast to the somewhat measured increases and relative stability of generic limits in conifers, species limits and the numbers of species recognized have been much more varied. This book accepts and describes 546 species of conifers, at the lower end of the number conventionally recognized. At the other extreme, some authors would accept 600 or more species. The differences are due largely to how a number of widespread phylads (groups of mutually most closely related populations) are treated, whether as species complexes (favored by "splitters") or as complex species (favored by "lumpers," like this author). In such complexes, populations from different parts of the geographic range of the phylad are recognizably distinct from one another, usually in overlapping quantitative traits.

There may be just two geographic variants with adjoining ranges, like pitch pine (*Pinus rigida* subsp. *rigida*) and pond pine (*P. rigida* subsp. *serotina*), which are often treated as separate species. In such cases, the behavior of the variants where they come together is important in deciding whether to treat them as species or as infraspecific variants. If they completely intergrade at their points of contact, taxonomists are inclined to place them within the same species. When they are reproductively isolated from one another at these meeting zones, producing only few, sterile or nonviable hybrids or none at all (a fairly rare situation among the species in question), then they would normally be considered separate species. If, as most commonly occurs, there are some hybrids between them but not a complete blending together, then treating them as separate species or as varieties or subspecies of a single species becomes more a matter of preference and tends to be strongly influenced by tradition in the group in question.

In other examples, such as Pillars of Hercules fir (*Abies pinsapo*) or Japanese white pine (*Pinus parviflora*), there may be several variants (four in *A. pinsapo*, five in *P. parviflora*) separated by geographic gaps in their ranges. In these cases, the geographic variants never come together naturally. Therefore, how they would behave in contact is unknown, even though they can almost always be crossed with one another artificially to produce viable offspring. Here again, there is no clearly correct taxonomy, and choosing between the alternative treatments is somewhat arbitrary and often follows tradition.

The most complicated and difficult cases involve hierarchies of variation, in which populations across a region differ modestly from one another but less than they do from the populations of other regions. For example, not only are the two subspecies of Aleppo pine (*Pinus halepensis* subsp. *brutia* and subsp. *halepensis*) often treated as separate species, but each contains geographic variation that is sometimes recognized taxonomically at either the infraspecific level or as further segregate species. The two subspecies originally had relatively limited regions of contact in the eastern Mediterranean. Plantation culture has brought them together in many places, and spontaneous populations of seedlings adjacent to plantations show a great deal of interbreeding. However, choosing to treat them as subspecies of a single species means that the variation within each of them is not given formal taxonomic recognition although the provenance variation has practical implications for forestry. Similar considerations apply to European black pine (*P. nigra*), with a similar distribution to Aleppo pine (though not found in North Africa) and similar variation patterns. Black pine is very rarely divided into separate species the way Aleppo pine commonly is, showing the inconsistency of traditional treatments.

As a general rule, I tend to treat the types of complexes just described, as well as many others, as variants of a single species rather than as separate species. This is to prevent their relationships to one another from being obscured by treating their status as equivalent to that of other species in the same genus. If they are simply listed as species, how do you know that *Pinus halepensis* and *P. brutia* (as it is known when treated as a species) are much more closely related to each other than either is to *P. nigra*, for example?

These kinds of problems arise primarily in the five large, widespread genera (*Abies, Juniperus, Picea, Pinus,* and *Podocarpus*), but there are also examples in intermediate-sized ones, like *Agathis, Callitris, Cupressus,* and *Dacrydium.* I am relatively conservative in the lumping accepted in this book, mostly using infraspecific variants previously proposed. There are a number of additional complexes that should probably be treated in a similar fashion but that need further study. The firs of the northern side of the Mediterranean, from *Abies alba* to *A. nordmanniana,* and many species complexes in *Podocarpus* are obvious examples.

It should go without saying that lumping should not extend to taxa that are not closely related, but some authors propose just that. Finding a suitable balance between lumping and splitting is always a challenge and is often inherently subjective, a result of emphasizing different aspects of agreed-upon relationships. We can look forward to continuing debates about how many conifer species there are for a long time to come.

SUBSPECIES VERSUS VARIETIES

Botanical taxonomy and nomenclature provide for two main ranks of taxa within species: subspecies and varieties. A given species may be divided into either subspecies or varieties, or both ranks may be recognized, in which case the varieties are nested within subspecies. In most plant classifications, only one or the other of these ranks is used. For conifers, as for many other plant groups, species are traditionally divided into varieties, the only infraspecific rank used by Linnaeus for plants. Use of subspecies became popular in the middle of the 20th century with the rise of biosystematic approaches to taxonomy that tried to assess the relationships among populations based on their relative potential for interbreeding. In general in plant classification, the two ranks are used for essentially equivalent infraspecific variants in different genera. Individual taxonomists typically use one or the other of these ranks based on their training. I come from a taxonomic tradition that favors subspecies, but since conifer species are traditionally divided into varieties, that is the infraspecific rank used predominantly in this book. Subspecies are used only where there are names available at that rank from previous authors, and the two ranks are never mixed in a single species. There seems to be no taxonomic value in wholesale conversion of varieties to subspecies or vice versa in a group as some authors have done, simply because they favor one rank or the other.

In a very few cases, there might be merit in using both ranks to express variation in some of the more complex species. This would be tempting for the cases of Aleppo pine (*Pinus halepensis*) and European black pine (*P. nigra*) already mentioned. In looking at these cases and others like them, I finally decided that the variation within the recognized subspecies did not really need formal taxonomic recognition. The differences between populations within subspecies seem to be fairly minor and inconsistent. From a practical standpoint for forestry and horticulture, it is probably more useful to establish a classification of provenances or cultivars based on characters meaningful to foresters or gardeners than to try to use taxonomic ranks for this purpose. Thus, in these and in other cases, use of two infraspecific ranks would need to be justified based on established taxonomic criteria rather than on a desire to give some specific variants taxonomic recognition. This sets a fairly high bar that is not likely to be reached very often, and so we are likely to continue to see conifer classifications with species divided into only varieties or subspecies but not both. Any attempt to use both should also keep in mind the cumbersomeness of quadrinomials (of genus, species, subspecies, and variety) and the potential for imprecision or confusion, or both, when one of the infraspecific ranks is omitted when referring to a particular variant, as allowed by the *International Code of Botanical Nomenclature* (which governs botanical names).

CHAPTER 2

CONIFER NAMES

Conifers are given different kinds of names by botanists, horticulturists, and other people. Scientific or botanical names, cultivar names, and common names have different purposes and origins, and are governed by different rules.

SCIENTIFIC NAMES

The conifers are listed here in alphabetical order by their scientific names. As every botanist knows, these names are used to identify species in all scientific literature, even if used only once in a paper on nontaxonomic subjects like physiology or ecology. Scientific names are governed by rules described in the *International Code of Botanical Nomenclature.* The *Code* and its rules are reviewed and revised, if necessary, every 6 years at an International Botanical Congress when botanists from around the world gather for this and other purposes. Any proposed changes must be adopted by participants in an open vote at the nomenclature session so that the nomenclatural rules of the *Code* truly represent the wishes of the taxonomists who are governed by them. The goal of these well agreed upon rules is to promote stability of names, so that a given species will be known by the same name throughout the scientific literature. The core of the rules is the principle of priority: that insofar as it agrees with other rules, the earliest name for a species published on or after 1 May 1753 (the date of Linnaeus's pivotal *Species Plantarum*) is the one by which it should be known.

The form of scientific names is tightly regulated by the *Code,* up to a point. The name of a species, in the form of a binomial, consists of a single-word, capitalized generic name followed by a single-word specific epithet beginning with a lowercase letter (for example, *Larix decidua* or *Pinus ponderosa*). The *Code* allows capitalization of specific epithets derived from generic names or from people's names, but in such cases a lowercase initial is always correct as well and is preferred under the present *Code* (for example, *Picea abies* and *P. wilsonii* are preferred to *P. Abies* and *P. Wilsonii*). The generic name is often abbreviated to a single letter after its first appearance as long as the reference is unambiguous. Amazingly enough, there are no examples among conifers of species with the same specific epithet belonging to genera with the same first letter, a phenomenon that is fairly common among flowering plants. Sharing of specific epithets like *alba* or *virginiana* is one reason that species names must always include the generic name or its abbreviation. The epithet alone is not enough. The binomials are italicized because they are treated as if they were in Latin, whether or not they are actually derived from Latin. And, while the *Code* specifies that scientific names are to be considered as Latin, it is completely open about the possible sources of names. They can be taken from any language or even be composed arbitrarily. The *Code* asks only that they not be too awkward in Latin and that Latin terminations be used for specific epithets.

In practice, a large majority of generic names of conifers are classical Latin and Greek names or are compounded straightforwardly from Latin and Greek roots. Among the few exceptions are *Manoao* (from Maori), *Sequoia* (from Cherokee), *Tsuga* (from Japanese, further combined in the half-Greek compounds *Nothotsuga* and *Pseudotsuga*), and *Wollemia* (from an unspecified Australian language, possibly Darkinyung). Specific epithets are much more often derived from languages other than Latin and Greek (although these are still abundant sources), and latinizations of surnames and place names in a variety of languages are commonplace. Some specific epithets are latinizations of common names in other languages, like *Cedrus deodara,* from Sanskrit through Hindi ("tree of the gods," reflecting the ritual use of aromatic cedar woods in many cultures).

I often include brief explanations of scientific names wherever they are not as obvious as a familiar geographic name (like *Tsuga canadensis, Juniperus chinensis,* or *Taxus floridana*). These explanations may be found in different parts of the genus and species accounts. If the epithet is descriptive, the explanation will usually be found associated with a corresponding part of the species description by words like "hence the scientific name." These explanations refer to Latin unless otherwise specified. On the other hand, if the epithet commemorates a person, that person will be identified in a separate sentence within the discussion that follows the description. Less obvious geographic names might be found in the discussion or in the statement describing the range of the species.

Names of subspecies or varieties, when these occur, consist of a species name followed by the abbreviation subsp. or var. and an additional epithet. These subspecies and varietal (collectively, infraspecific) epithets are essentially just like specific epithets, being governed by the same rules for their formation. In fact, one subspecies or variety will always repeat the epithet of the species to which it belongs (for instance, *Taxodium distichum* var. *distichum*). This is the subspecies or variety that includes the type of the whole species, a specimen that fixes the name. This type subspecies or variety is not necessarily the typical one: it may be rarer than another subspecies or variety in the species. For instance, lodgepole pine (*Pinus contorta* var. *latifolia*) has a much wider distribution than the type-containing shore pine (*P. contorta* var. *contorta*). Any other infraspecific epithet(s) must agree grammatically (at least in Latin) with the specific epithet and generic name. Thus, for pond cypress (*Taxodium distichum* var. *imbricarium*), the varietal epithet *imbricarium* is of neuter grammatical gender like the binomial it follows. Each variety or subspecies within a species, if there are any, has its own type, a different specimen for each named.

A final component of scientific names consists of their authorities, which directly follow the name and are not italicized. The authorities include the original author(s) of the name or epithet and the author(s) responsible for the exact form it takes today (called the combination). Use of authorities is required by the *Code* as a bibliographic tracking device and to avoid confusion between inadvertent duplicate uses of the same name by different authors. Such duplicate names are called homonyms. They were much more prevalent in the past when slower and more erratic communication across the world meant that many taxonomists worked in isolation from one another. Even in nontaxonomic botanical articles, editors insist on use of authorities the first time a name appears. The way the authorities are presented provides a bit of a clue about the history of the name. If the name appears in exactly the same form as when it was originally described, then the name of the original author(s) follows the species name directly: *Pinus sylvestris* Linnaeus or *Abies homolepis* P. Siebold & Zuccarini. If the name is a different combination (an epithet associated with a different genus or species) than that proposed by the original author(s), then the original author appears in parentheses, followed by the author who proposed the combination: *Tsuga canadensis* (Linnaeus) Carrière or *Picea jezoensis* (P. Siebold & Zuccarini) Carrière. In these examples, Linnaeus included eastern hemlock in his broad concept of *Pinus* in 1763, and Siebold and Zuccarini first described Yezo spruce as a species of *Abies* in 1842. Carrière then transferred both species in 1855 to the genera in which they are still placed. Authors' surnames are often abbreviated, but here they are written out in full. Furthermore, I use the minimal initials needed to distinguish authors of conifer names and combinations from all other authors of the same surname who made contributions to plant nomenclature, including mushrooms and seaweeds, as these organisms are all covered by the botanical *Code.*

The same considerations relating to original and parenthetical authors also apply to citation of authorities for varieties and subspecies, but these may seem further confusing because some of them do not have authorities after the epithet and others do. For example, in slash pine we have *Pinus elliottii* Engelmann var. *elliottii,* contrasting with *P. elliottii* Engelmann var. *densa* Little & K. Dorman. In this and all other divided species, an authority is not cited for the type subspecies or variety that repeats the species epithet. That is because such names are considered not to have been coined directly by any authority. Instead, they are regarded as automatically created names (autonyms) that arose when the species was first subdivided into subspecies or varieties. In the example of slash pine, when Engelmann described *P. elliottii* in 1880, he did not divide it into varieties. The type variety (var. *elliottii*) was only created (nomenclaturally speaking, of course) when Little and Dorman described southern Florida slash pine (variety *densa*) as a separate variety within the species in 1952. We cannot ascribe variety *elliottii* to Engelmann because he never used this name. Nor can we write either "Little & K. Dorman" or "(Engelmann) Little & K. Dorman" because it is neither a new taxon (Engelmann had already described it as a species) nor a new combination because it was not moved from one species to another. Therefore, these type subspecies and varieties have no author, unlike all the other names in this book, including all the species and genera. There are about 200 subspecies and varieties of conifer species listed and described in this book, and about 80 of these are autonyms, listed without citation of authorities.

There is also an autonym rule that applies to subgenera and sections within genera so that the subgenus or section that contains the type species of the genus is given without an authority while all others have authorities. Subdivisions of genera are used very sparingly in this book because very few genera are large enough to have been divided into subgenera or sections and because some proposed schemes do little to improve understanding of relationships in their genera. The two subgenera of *Pinus* and the two sections of *Juniperus* both seem well supported. They would be cited as *Pinus* Linnaeus subg. *Pinus* (containing the type of the genus, *P. sylvestris* Linnaeus, Scots pine) and *Pinus* Linnaeus subg. *Strobus* J. Lemmon (with its type *P. strobus* Linnaeus, eastern white pine), and *Juniperus* Linnaeus sect. *Juniperus* (containing the type of the genus, *J. communis* Linnaeus, common juniper) and *Juniperus* Linnaeus sect. *Sabina* Spach (with its type, *J. sabina* Linnaeus, savin).

Full names of all authorities associated with the scientific names in this book, along with their dates of birth and death (where known), are given in Appendix 2. That list also includes the most common abbreviation for these names that might be found in other conifer literature.

Those Pesky Name Changes

If the goal of the nomenclatural rules is stability, why then do so many names seem to change over time, creating frustration among many gardeners and other nontaxonomists, including many ecologists and plant physiologists? There are two most common causes for these name changes, one driven by the rules of the *Code* and the other by increasing scientific knowledge. In the first case, a familiar name can be displaced by a newly discovered older name. This can happen either because of a closer examination of obscure scientific literature or because of a changed interpretation of what plant an author actually had in mind when an older name was coined. There have been attempts to prevent these kinds of changes by adding a provision to the *Code* that would simply prevent obscure old names from being adopted by compiling a list of "names in common use" that would be closed to additions from older literature once it was finalized. These attempts were repudiated by the International Botanical Congress held in St. Louis in 1999. Most participants felt that these provisions are not really needed because overlooked older names are really only a small part of name changes today, having had their heyday in previous decades. Still, the correct name for a conifer can turn on small points of interpretation.

A good example resides in two shrubby Australian podocarps from Tasmania and New South Wales: Tasman dwarf pine and cascade pine. Until 1951, these two species were known, respectively, as *Pherosphaera hookeriana* and *P. fitzgeraldii*. Then, in 1951, J. Garden and L. A. Johnson argued, successfully as far as the majority of conifer taxonomists were concerned, that the genus name *Pherosphaera* and the species name *P. archeri* are actually synonyms of *Microcachrys* and *M. tetragona*. They then proposed the new generic name *Microstrobos* and the species names *M. niphophilus* (Tasman dwarf pine) and *M. fitzgeraldii* (cascade pine). These names were quickly adopted and used for the next 50 years, but in 2004 R. K. Brummitt, R. R. Mill, and A. Farjon revived the issue and came to a different conclusion. Garden and Johnson's original argument turned on the idea that when J. D. Hooker used the name *P. hookeriana* for the Tasman dwarf pine, he was describing a different species than W. Archer had when he first described the species. They thought that Archer was inadvertently renaming strawberry pine (*Microcachrys tetragona*), a related and similar podocarp. So they interpreted Hooker's *P. hookeriana* as a name independent of Archer's and thus wrong under the *Code*. Brummitt and coauthors argued, to the contrary, that Archer had also described Tasman dwarf pine when he originally introduced the name *Pherosphaera hookeriana*, and, therefore, this is the correct name for the species that had been known essentially exclusively as *Microstrobos niphophilus* for the preceding 50 years. But this latter conclusion also results from one possible interpretation of certain unclear wording in the original publications as well as in Brummitt, Mill, and Farjon's selection of a type specimen (technically lectotypification, a process that establishes a type specimen for a name when the original author of the name failed to designate one) in such a way as to have this outcome. As a result, it is unclear what scientific name will be used for Tasman dwarf pine after all the arguments have been presented and discussed. So I continue to use *Microstrobos* in this book. This kind of case is fairly uncomfortable for taxonomists as well as for other kinds of botanists, gardeners, and even conservation policy makers and enforcers. Luckily, such cases are rare in conifers, but this one shows that reinterpretations of issues that seemed to be settled can upset stability in nomenclature.

The other prominent reason for name changes is much more common and goes beyond reinterpretation of nomenclatural niceties. These changes arise because of changing taxonomic concepts, which then affect nomenclature. In conifers, the largest source of such changes involves differences of opinion about how inclusive to make individual species. As noted in Chapter 1, splitters, who emphasize differences between taxa, and lumpers, who emphasize similarities, can produce very different classifications of the same plants. More complex classifications that recognize few species but subdivide them into varieties or subspecies are different yet again.

Each of these different classifications will be accompanied by at least some different names for the same plants.

Many of the conifers treated as varieties and subspecies in this book are considered separate species by some other taxonomists. Examples include the four varieties of *Abies pinsapo* or *Pinus culminicola* and the three subspecies of *Cedrus libani.* Most of these taxa retain the same epithet whether they are treated at the rank of species or below, and only the combination in which the epithet is presented varies. One of them, however, border pinyon, bears a different epithet when it is treated as a species (*P. discolor*) rather than a variety (*P. culminicola* var. *bicolor*). Such differences can arise because of various different rules in the *Code* but principally for two reasons. On the one hand, epithets have different dates for purposes of priority at the different ranks at which they were used. Hence, one may be the earliest epithet available at the rank of variety, for example, but not of species. This is especially true if the equivalent variety and species have somewhat different circumscriptions, that is if they do not include exactly the same sets of populations. An epithet can also be precluded from use at one rank versus another by issues of homonymy, when the name was used independently earlier for another plant altogether.

In essence, the *Code* does not legislate for or against any particular classification, it only governs the names that must be applied as the result of adopting a particular classification. The kinds of name shifts that result will always be present, because even with genetic markers, there are no rules about how much variation should be encompassed within a species and when, instead, it should be apportioned among segregate species. This is not to say that anything goes in taxonomy. Some proposed classifications are in conflict with one or more lines of evidence about relationships. Such classifications are not likely to be adopted by taxonomists other than the ones who propose them. Nonetheless, legitimate differences of taxonomic opinion, consequent differences in classifications, and their attendant name changes will be with us for the foreseeable future. As a result, while most conifers have a single correct scientific name, some others have two or more equally correct names, depending on the classification adopted. It is important to note that taxonomists are not compelled to accept the most recently proposed classification for a group of plants. They will do so and abandon a previous classification only if they are convinced by the arguments for the changes presented by the author(s) of the revision. Despite what nontaxonomists might think about the apparent prevalence of name changes, taxonomists are, on the whole, a conservative group and prefer not to alter classifications and names unless there seem to be compelling reasons to do so.

Synonyms and Cross-References

The legacy of name changes due to changing taxonomic concepts, to differences between splitters and lumpers, and to just plain mistakes means that many of the conifer names accepted in this book accumulated synonyms over the years. Carefully combing the botanical literature, as Aljos Farjon has done for his *World Checklist and Bibliography of Conifers* (1998), which should be consulted for essentially complete lists of conifer synonyms, shows that there are, on average, about four synonyms for every accepted name for a species, subspecies, or variety of conifer. Many of these are obsolete names long abandoned in the botanical literature. Some were only ever used once, in the papers in which they were coined, and were never adopted in botanical literature generally. A few were even soon abandoned by their original authors. These kinds of names are unlikely to be encountered by anyone but specialists and thus are not included here. I am selective in my choice of synonyms to list, emphasizing those that are in common use or that represent relatively recently described taxa that I do not accept here. I also list those names accepted in major contemporary floras but that I do not accept in this book. I do not, however, generally list synonymous combinations that represent mechanical switching between subspecies and varieties simply because an author prefers one rank over the other, without any independent taxonomic study. Occasionally, I list synonyms as subspecies for conifers accepted here as varieties or vice versa, but this is only when both versions are in more or less common use. Likewise, I typically do not list many more recent synonyms based on somewhat haphazard switching of varieties from one species to another or the reduction of some species to varieties of others to which they do not seem particularly closely related. Even though I do not list these synonyms, it should still be possible to see what names I accept for the taxa involved. Most of these taxa have also been treated as species more recently, and so, in many cases, it is possible to look up those epithets in the Index and find a cross-reference to the accepted name. Names in the Index are presented without their authors' names, which are given in the synonymies. The selected synonyms that I chose to list will be found as the last sentence in each description of family, genus, species, subspecies, or variety, and preceding the paragraphs of the discussion.

CULTIVAR NAMES

Just as scientific names are governed by the *International Code of Botanical Nomenclature* (*ICBN*), cultivar names are governed by the *International Code of Nomenclature of Cultivated Plants* (*ICNCP* or the cultivated plant *Code*), which is coordinated with the *ICBN.* Cultivar names follow the Latin name of the species or just the generic name. As with species names, cultivar names are supposed to be used only once within a genus, but many repeated names (like 'Nana' or 'Aureo variegata') were used in the past for cultivars in different species of the same genus (as in *Chamaecyparis* or *Picea*), and not all of these have yet been brought into conformity with the cultivated plant *Code.* The two examples I gave of old names also conflict with the present rules because they are in Latin. Cultivar names proposed since the beginning of 1959 have to be in a modern language rather than in Latin, as they were earlier when no nomenclatural distinction was made between cultivars and botanical varieties, both being called "varieties." Thus cultivar names, even those actually in Latin, are not italicized the way scientific names are. Further graphical distinction from specific, varietal, and subspecific epithets is also provided by starting cultivar names with a capital letter and enclosing them in single quotation marks. Before the 1995 edition of the *ICNCP,* use of the abbreviation "cv." in front of the cultivar name was an accepted alternative to the use of quotation marks. This practice is no longer considered correct, but it has not yet disappeared. The cultivated plant *Code* includes many rules about exactly what kinds of words can be used to establish new cultivar names. Without the same kind of international participation by its stakeholders as the botanical *Code,* the cultivated plant *Code* and the Registration Authorities for the different genera established by it are often overlooked by the nursery trade. Recognizing this reality, the *ICNCP* acknowledges that trade names are often used instead of the officially registered cultivar names. The same cultivar is frequently sold under different trade names in different countries or by competing companies. Likewise, it is common to find different clones under a single cultivar or trade name. So nomenclatural difficulties can be found in the garden as well as in the wild.

COMMON NAMES

Common names for each species are listed after the scientific name. English common names are listed first, with the one I prefer at the beginning. The majority of conifer species have no established English common names. Only those species, about 200 of them, native to regions where English is a national language have what could be termed genuine English common names. A great many of these names, like a large proportion of those in other languages, are not really exclusive to single species. Only roughly half of all conifer species are cultivated at any distance outside of their native range, including in English-speaking regions, and many of these are found only in botanical gardens and large estate gardens or specialist collections where use of scientific names is the norm. Nonetheless, authors of conifer books find it useful to propose English names for these cultivated species and for some others that are confined to their native ranges. These names are typically either translations of the scientific name or commemorate some part of the native geographic range or some person either directly or even rather peripherally associated with the species. Where I suggest additional English names in this book, I usually follow these traditions or adapt a common name from another language. This latter has already been the practice with some Japanese species and especially in New Zealand, where Maori names for most native conifers are used by New Zealanders generally. For most of the conifers commonly encountered by English speakers, I follow precedent and choose an existing English name as the preferred name. Whenever a conifer is native to an English-speaking country, I choose an English name used in that country in preference to names given to it elsewhere in cultivation. As a result, I try not to restrict application of the "generic" names cedar, cypress, and pine to the single genus from which each was derived (*Cedrus, Cupressus,* and *Pinus,* respectively) by using made-up designators like "false" as my preferred English name. However, I also list such names if they are in fairly common use.

Additional common names in languages other than English are listed after the English name(s) when the species to which they apply occur in areas where languages other than English are the main languages. If a species occurs only in an area with a single national language, then a common name in that language is given. Most conifer species cross linguistic boundaries and cover areas with two or more predominant languages. Species growing in linguistically complex areas such as New Guinea or Europe may have common names in many languages. As with traditional English common names, these names may be as imprecise as pine or cedar, especially when several species of a genus or of related genera are found in the area and distinguishing among them is not of any particular cultural or economic significance to the local people. In Europe, on the other hand, there are so few native species of conifers that distinctive names are often found. Some of these names, of course, especially classical Greek and Latin ones, are the source of the scientific names of widespread genera indigenous to Europe and the Mediterranean region: *Abies, Cedrus, Cupressus, Juniperus, Picea, Pinus,* and *Taxus.* Other classical folk generic names became specific epithets, such as pinaster (*Pinus pinaster*), pinea (*P. pinea*), and sabina (*J. sabina*), while still others simply vanished or were later assigned as scientific generic names to plants other than conifers. I try to give a selection of common names for European conifers that emphasizes national languages of the more populous countries where the species occurs, without being entirely consistent about this.

A broadly similar policy has been used for the linguistically complex Malesian region, embracing Indonesia, Malaysia, the Philippines, and Papua New Guinea. Only a fraction of the possible common names are listed, especially in New Guinea, where hundreds of languages are mostly spoken by a few thousand or fewer speakers each. I select names from the largest linguistic groups where possible. Some of these I use as the basis for proposed English names, trying to use names that apply only to the species in question or that apply to it more often than to other species.

For each common name listed besides the English names, I give the language from which it is derived in parentheses immediately following it or after two or more names in the same language. Many languages, like English, have more than one name for a given species, replacing each other geographically or with different interest groups in the society. The names I list are mostly derived from the floristic and monographic works noted in the bibliography. Some of the names, like many of the Chinese, North American French, and English names are clearly "official" common names rather than what people call the plant in conversation. Since it is essentially impossible to tell for sure if this is the case and since, as with birds, officially imposed common names can actually come to be used by people, such names are given here unless there is some compelling alternative.

Despite all the difficulties cited with common names, including regional and individual variation and a common lack of complete specificity, with the same name being used for different species, in any one area there are likely to be relatively few confusions in the application of common names among people who have a genuine interest in the plants, for whatever reason. In many cases, a common name elicited from a local informant may be the best entry point for getting at the identity of a tall forest tree. Likewise, there are many literary references to conifers that use common names exclusively, including some scientific works. It is a fact of human life that widespread and economically important species (like *Nageia wallichiana* in Malesia, *Picea abies* in Europe, or *Abies balsamea* in North America) will have many common names throughout their range. By including some of these names in this work, I hope to reflect a little of the cultural richness that conifers provide throughout the world.

DISCOVERY OF CONIFER GENERA AND SPECIES

The work of the early conifer taxonomists, like that of all subsequent botanists in the field, depended on the bravery, resourcefulness, dedication, and skill of plant explorers of many nations as well as on the knowledge and good grace of the local people they so often relied upon. In every part of the world, and beginning at different times in each, these hardy individuals collected the plants, seeds, and specimens that brought the world's conifers into scientific consciousness and into gardens. From early explorers like the Bartrams, Michaux, Thunberg, and Wallich, through 19th century giants like W. J. Hooker, Douglas, and Jeffrey and their 20th century counterparts like Meyer, Wilson, Forrest, and Farrer, right down to the present, hundreds of collectors made major and minor contributions to our knowledge of conifers. The stories of the more famous individuals are well known, both through their own writings and through those of admiring contemporaries and later biographers and garden and botanical historians. Their legacy, and also that of many less familiar collectors, is always kept before us because many of the conifers they brought into the larger world were named for them by the plant taxonomists who were only too pleased to recognize their essential role in the scientific enterprise. The realms of discovery and description also frequently coincided in single individuals engaged in both activities, often at different times of their lives and careers. And we cannot forget the substantial minority of collectors who died in the field of disease, accident, or misadventure. Even in this era of ready world travel and widespread amenities, plant collecting is still often hard, uncomfortable, and sometimes quite dangerous. We owe a continuing debt to all these individuals and those yet to come.

Back in the garden, herbarium, and laboratory, and since the publication of Linnaeus's *Species Plantarum* in 1753, the starting point for botanical nomenclature, few decades passed without the description of a currently accepted genus, and none missed out entirely on at least one of the species included here. While Linnaeus and his contemporaries in the 1750s (particularly Philip Miller) got things off to a good start, with eight genera and 29 species, there was a real falloff in activity in the following few decades. In fact, the 1770s, with no new genera that stuck and only two species, represents the slowest period in the history of conifer discovery until the 2000s.

It was not until the middle of the 19th century, with European and American colonial and commercial expansion and the first major books devoted to conifers, that description of new conifers took off. In fact, from the 1820s through 1860s, 26 genera (39% of the presently recognized ones) and 229 species (42% of the species accepted here) were described. This was the most fruitful half-century in the discovery of conifer species and genera. Not only had a total of 41 of the 67 genera (61%) been named by 1870, but species belonging to an additional 15 genera, that were only later recognized as being distinct, had also been described. Thus only 11 totally new genera remained to be discovered after

1870 (of course, there may still be a few unknowns lurking in the bushes). At the same time, 59% of all the species recognized today had been described by 1870.

Following these decades of discovery, there were subsequent secondary periods of heightened activity, like around 1900 and following the Second World War in the 1940s, with a gradual diminution through into the 1980s. As far as novel discoveries go, this activity was largely reflected in newly described species while the genera proposed concurrently were mostly split from ones previously recognized. Even among the 11 new genera described after 1870, and for which no species had been described by 1870, five were based on species initially described in preexisting genera (two in the 1880s, one in 1908, and two in the 1930s). *Taiwania* (1906), *Austrotaxus* (1922), *Microbiota* (1923), *Cathaya* (1962), and *Wollemia* (1995) are the only genera published after 1870 based solely on new species described with them.

Metasequoia (1948) falls into this group only because the genus was reestablished for extant plants based on the then newly discovered *M. glyptostroboides.* It had actually first been described a few years earlier, in 1941, for some fossils that had originally been assigned to *Sequoia* in the 19th century. While this exact nomenclatural situation is unique among conifers, there are other cases where fossil species of a genus were already known before it was described based on living plants and were only recognized as members of the genus in retrospect. This is the case for *Cathaya* and probably for *Wollemia,* for instance.

The patterns of discovery and naming so far described apply only to those taxa actually accepted here. Many additional genera, species, and varieties were proposed to the scientific and horticultural communities over the years that are presently considered synonymous with previously published taxa. Thus, while only two genera and six species newly described in the 1990s and 2000s are accepted and included here, many other names were proposed at all ranks, from variety and subspecies to order and class. These names fall primarily into three groups.

First are genera and species with recent dates but that really represent deliberate renamings required by nomenclatural difficulties with the original name and any other older names. For these names, such as *Phyllocladus toatoa* Molloy, there is no implication of a new discovery, even in those cases where the newly renamed species also has to be fully described as if new for technical reasons. Such renamings are largely accepted here, but the species they apply to are considered to have been discovered at the time of their first description, even though the original name is not valid.

A second set of new names of genera and species comes from the further splitting of previously recognized taxa that does not seem justified by presently available evidence. The continued subdivision of the reduced cores of *Dacrydium* and *Podocarpus* that survived the justified separation of genera recognized by de Laubenfels, Quinn, and Page in the 1960s, 1970s, and 1980s is an example of taxonomic innovation with little support among conifer taxonomists.

A third group consists of genera and species believed to be newly known by their authors. Most seed plants have considerable variation among populations, so that many of the more recently proposed highly localized new species or varieties described from China and Mexico, for example, just seem like populations of previously described species with more or less extreme traits compared to various better-known populations. In some other cases, the proposed new taxa were actually described before, often in obscure literature, and so the recent naming is superfluous even if the taxa are truly distinct. Thus, even if *Cupressus nootkatensis* and the more recently described *C. vietnamensis* are ultimately removed from *Cupressus* to their own segregate genus, it would have to be called *Callitropsis* Oersted (1864) rather than *Xanthocyparis* Farjon & T. H. Nguyên (2002).

Taxonomists will continue to describe new conifers for the foreseeable future. Some will represent unimpeachable new discoveries, while others, like the rejected categories just discussed, will be the result of judgment calls that may not be embraced by their authors' taxonomic colleagues. All taxonomic treatments are provisional, however, and some of the more recently described varieties, species, and genera that are not accepted here may gain acceptance if additional data and experience show that they are worthy of recognition. Exploration and innovation continue apace. The world has not yet yielded up all of its conifer denizens for our instruction and delight.

CHAPTER 3

CONIFERS IN NATURE AND IN THE GARDEN

Conifers are a conspicuous and economically important part of both natural and cultivated landscapes throughout much of the world. Understanding the factors that shape their distribution and prevalence may help protect them in the environments that we are changing.

GEOGRAPHIC DISTRIBUTION OF CONIFERS

Range maps are presented for each genus that show the aggregate distribution of all of its species in nature. These maps are generalized and cannot be used to infer detailed distributional data due to the lack of precision inherent in the scale of the maps. The maps are solely to convey a general impression of the regions of the world in which each genus may be found and not whether they can be found in any particular place, an increasingly difficult proposition given global warming and large-scale habitat destruction in some parts of the world. Furthermore, there is no attempt here to map the distributions of individual species, a task that has never been completed for all conifers, although there are good maps for a few particular regions in the world and for a few taxonomic groups worldwide.

LATITUDINAL AND ELEVATIONAL RANGES

Conifers are found on all six continents that support trees (but not in Antarctica), on most large continental islands (except those of the high arctic), and on a surprising number of oceanic islands. They extend virtually to the southern tips of the three southern continents: to 34°S in the Cape Region of South Africa, to 43°30′S in Tasmania (Australia) and 48°S in Stewart Island (New Zealand), and to 55°S in Tierra del Fuego (Chile). In the northern continents, land extends well north of the arctic tree line, where the conifers reach their northern limit, but this is still much farther north than the southern limit in the southern hemisphere. In North America, conifers reach their northern limit at about 69°N at various places in Alaska, western Canada, and Greenland. Conifers are found above 70°N at various points across Eurasia from Norway across the Russian arctic, reaching their northernmost stands in the Lena Valley of Siberia at close to 73°N.

Only one or a few species is found at each of these extremes of northerly or southerly distribution. Different species are found at the southern extremes of conifer distribution on each continent, although *Libocedrus* and *Podocarpus* are both genera with species at the southern limits on two of the three continents. Species are also mostly different between Eurasia and North America in the north, but common juniper (*Juniperus communis*) is shared at the northern limit of conifers across both continents as well as on mid-Atlantic Iceland, where it reaches near 67°N. Additional northern-extreme species in North America include creeping juniper (*J. horizontalis*), tamarack (*Larix laricina*), and white and black spruces (*Picea glauca* and *P. mariana*). In Eurasia, common juniper is joined at the tree line in different places by Dahurian larch (*Larix gmelinii*), Siberian spruce (*Picea abies* subsp. *obovata*), Scots and Siberian dwarf pines (*Pinus sylvestris* and *P. pumila*), and even by an introduced species, mugo pine (*P. mugo*), which is naturalized on dunes of Arctic Ocean beaches at the northern tip of Norway. In

South America, only a single conifer species, Fuegian incense cedar (*Libocedrus uvifera*), reaches the southern limit of conifers here. At the southern tip of Africa, a continent poor in conifer species, true yellowwood (*Podocarpus latifolius*) is joined by berg cypress (*Widdringtonia nodiflora*) and often by naturalized stands of maritime pine (*Pinus pinaster*) and other pine species. In Australasia, nine native conifers (about equally distributed between Cupressaceae and Podocarpaceae) and the introduced and naturalized Monterey pine (*P. radiata*) are found in southern Tasmania, while half of all New Zealand native conifers (10 of 20 species, all Podocarpaceae) are found on the southernmost Stewart Island, which has an area of just 1,750 square kilometers. Thus the northern and southern limits of the conifers as a whole are different in character. The arctic limit is climatically determined at the extreme of the ecological tolerances of this group of woody plants. In contrast, the antarctic limit is reached where the southern landmasses come to an end. Only in South America does the land seem to come close to the limits of ecological tolerance of the conifers.

The elevational limits of conifers extend from sea level to about 4,800 m. None grow in the hot, dry, desert depressions that reach below sea level on land, like Death Valley, the Dead Sea, and the Caspian Depression. Still, over half of all conifer genera (36) and almost a third of the species (177) sometimes grow within 100 m of sea level. Ten species (but no genera) extend no higher, although *Actinostrobus* tops out at 350 m, the only genus not reaching to at least 500 m. At the other elevational extreme, over 4,000 m separate the highest conifers from the peak of Mount Everest at 8,848 m. Far fewer conifers reach these upper limits in subalpine regions than the lower one at sea level. Only seven species in three genera (*Abies* and *Larix* with one species each and *Juniperus* with five) extend above 4,500 m, and another 13 species and three genera (*Dacrycarpus* and *Picea* each with just a single species and *Pinus* with two) exceed 4,000 m. Single species of *Cupressus* and *Phyllocladus* and two species of *Podocarpus* reach 4,000 m along with additional *Abies, Dacrycarpus, Juniperus,* and *Picea* species.

Despite the wide total overall elevational range of conifers, from salty seasides and lowland swamps and plains to steep and rocky subalpine slopes, the majority of them are plants of modest elevations in the lower mountains and foothills. More than 60% of conifer species and about 90% of the genera may be found between 500 and 1,500 m elevation (although most not exclusively within this zone). This is the band in which the global mean elevation of the land surface (840 m) lies. Roughly equal numbers of species and genera reach the top of their elevational range within each 500 m between 1,000 and 3,000 m, for about 60% of all conifers (324 species in 41 genera) over this 2,000-m interval. Of the remainder, almost equal numbers of species reach their upper limits in the 1,800 m above 3,000 m (110 species) as in the interval between sea level and 1,000 m (102 species). This evenness is not found among the genera, for three times as many top out above 3,000 m (20 genera) as below 1,000 m (six genera). This difference, of course, is partly an automatic consequence of the fact that each of the 39 genera (about 60%) with two or more species is composed of species with different elevational ranges. Thus the maximum elevation reached by the genera is higher, on average, than that of the species they contain.

The elevational band of greatest diversity for conifer genera and species lies between 500 and 1,500 m, and the number of both falls off with elevation above this level. Although this decline in diversity with altitude could be due to any number of ecological factors, including climatic tolerances and interactions with pests and competitors, a simpler explanation could be at play. There is a strong relationship between the number of conifer species and genera present at a given elevation and the total available land surface of the Earth at that elevation, but it only holds for land above 1,000 m. For each of the 500-m intervals from 1,000 to 5,000 m, species numbers have a statistical correlation of 95% with land surface, while for genera it is 97%, very high correlations for an ecological relationship. When these numbers are squared (to yield the coefficient of determination, symbolized as r^2), we find that 90% of the variation in species numbers and 94% of the variation in number of genera with elevation is directly related to the total land surface area at each elevation, without invoking any other environmental factor. This is not to say that environmental tolerances are not important in conifer distribution. Clearly, the elevational range of individual species is set by their tolerances of temperature and moisture, in concert with other environmental factors, intensity of attacks by pests and diseases, and competition with hardwoods and other conifers. What it does say, however, is that, despite the effects of environment on any one species, the number of species expected to tolerate the conditions of a given elevational band needs no other explanation than its available space. In fact, the relationship is very close to one conifer species per 1,000 square kilometers for all lands above 1,000 m in elevation.

These relationships break down below 1,000 m, where over 70% of the land surface of the Earth occurs. While there are more conifer genera and species between 500 and 1,000 m than there are between 1,000 and 1,500 m, the increase is essentially inconsequential (adding four genera and three species) compared to the approximate doubling of the land surface area. The effect is even more extreme in passing to the interval from sea level to 500 m, where a further tripling of the land surface area is accompanied by an actual decline in the numbers of conifer genera and species present. Clearly, there is much more competition from angiosperms for ecological space in the lowlands below 1,000 m than there is in the hills and mountains above this level. Only in the cold boreal region and a few cool temperate regions are conifers dominant in sizable chunks of the lowlands. Since the boreal forests are regions of relatively low biological diversity in general, they do not contribute much to the species counts of conifers in the lowlands.

The relationship of land surface area to the number of species and genera reaching their maximum elevation in each band is also strong (correlations of 99% and 89%, respectively) only for the top half of the elevational gradient, the land above 2,500 m. Extending the comparison down to 1,000 m weakens the relationship, lowering the correlations for species and genera to 79% and 70%, respectively, while adding that last 1,000 m down to sea level essentially eliminates it. This highlights the simple fact that very few conifer species and genera are entirely restricted to the lowlands where most of the Earth's land lies.

Species composition also affects the total elevational range of each genus, though to a lesser extent than one might think. The five largest genera (*Pinus, Podocarpus, Juniperus, Abies,* and *Picea*) all have elevational ranges spanning 4,000 m or more, while no genus with fewer than five species has an elevational range as large as 4,000 m. The other four genera with distributions spanning 4,000 m or more are, not surprisingly, the ones that join the largest five in the subalpine zones: *Phyllocladus* (five species), *Dacrycarpus* (nine), *Larix* (10), and *Cupressus* (16). Looking at the lower end of the elevational ranges, no genus with four or more species (76 of them) has a span of less than 1,500 m, and only two of them are distributed over less than 2,000 m. The biggest surprise surrounds the 28 monotypic genera, each with just a single species contributing to its elevational range. These genera (or the equivalent individual species) display wide variation in the height of their elevational spans, from *Wollemia nobilis,* with 150 m from top to bottom of the only known population, to the more widely distributed *Papuacedrus papuana* and *Platycladus orientalis,* each spanning about 3,300 m. The median elevational range for the monotypic genera (14 above and 14 below) is about 1,100 m, well below the minimum for genera with four or more species. The median for the 13 genera with two or three species is 2,000 m, and for the next 16 genera, with 4–10 species each, is 3,500 m, while the 10 largest genera have a median span of 4,000 m. Not surprisingly, the more species a genus has in general, the wider its elevational range is likely to be.

RICHNESS OF CONIFER FLORAS

Conifer species and genera are very unevenly distributed around the world. Asia is, by far, the most richly endowed continent, with 39 of the 67 genera (58%) and about 200 of the 546 species (37%), while Africa, Europe, and South America are poorest, each with eight or nine genera (12–13%) and about 25–30 species (4–5%). Australasia and the South Pacific Islands, with 27 genera (40%) and 143 species (about 26%), and North and Central America and the West Indies, with 17 genera (25%) and more than 150 species (27%), fall in between those extremes. As has often been noted, the diversity is concentrated in and around the Pacific Ocean, where far more genera and species of conifers occur than in the lands surrounding the Atlantic (with the Mediterranean Sea) or Indian Oceans. More than 90% (61 of 67) of all living conifer genera are found in the lands in and around the Pacific compared to 30% for the Atlantic (20 genera) and 34% for the Indian Ocean (23 genera). There are comparable disparities for species, with three times as many species around the Pacific Ocean (400, almost three-quarters of the total) as around the Atlantic Ocean (138, one-quarter), which has twice as many as surround the Indian Ocean (73, a little more than an eighth).

Given this broad pattern of distribution, which individual country or region has the richest conifer flora in the world? This is not an easy question to answer because there are different plausible measures of diversity depending on whether you simply count up all the species in a country, irrespective of size, or whether you calculate on a per square kilometer basis. Likewise, the answers are somewhat different if genera are considered rather than species. We might expect larger countries to have richer conifer floras than smaller countries, both because they have more land surface area and because they can be expected to have a greater diversity of environments, including greater elevational range and more climates and vegetation types. Where the country is located, particularly with respect to latitudinal position and range, has an obvious effect, as does association with ocean basins of varying total diversity. Thus the two largest countries in the world, Russia (37 species in seven genera) and Canada (34 species in nine genera), have much smaller conifer floras (ranking eighth and ninth in species numbers, respectively) than the next two largest countries, their neighbors to the south, China and the United States.

On simple counts, China has the richest conifer flora, with 107 species, about 10% more (on about 2% more land) than the 96 in the United States. There is a much greater disparity in the number of genera since China, which has by far the greatest number of conifer genera among the world's nations, has almost twice as many (29) as the United States (16). This latter number is equaled or barely exceeded only by Taiwan (16), Japan (17), and Vietnam (18), which each have just one-third to two-fifths the number of species found in the United States. Most of the conifers in Vietnam are spillovers from China and occur only near the border with that country, and the conifers of Taiwan are also mostly shared with mainland China, but a much smaller proportion of those of Japan are found also in its neighbors.

While areas of high conifer diversity may be found across various regions of China, there is only one such region in the United States—California—which has all but two genera (*Larix* and *Taxodium*) and more than half of all the conifer species (51 of 96) in the entire country. If California were a separate country, it would have the fourth richest conifer flora in numbers of species. California also contributes to the diversity of the third largest conifer flora, that of Mexico (76 species in 10 genera but, like California and the United States as a whole, richest in pines), since 12 species in the Mexican flora are found only in northern Baja California as well as in California (and beyond in some cases).

The actual fourth largest national conifer flora is that of Indonesia, with 49 species in 12 genera, half the species number found in the United States but with two-thirds as many genera. Indonesia spans two biogeographic regions, the Australasian and Indo-Malayan, and consists of thousands of islands of all sizes. Two of the islands, New Guinea and Borneo, respectively the second and third largest islands in the world (after Greenland), are each shared with another country. If each of these islands were a country by itself, both would rank among the top 10 in conifer diversity, each being comparable to all of Canada. New Guinea has 34 species in 10 genera and Borneo has 32 species in 8 genera, thus bracketing Canada's 34 species in 9 genera. On the other hand, if one considers the whole Malesian region, including the Philippines, Malaysia, Papua New Guinea, Singapore, and Brunei in addition to Indonesia, the total conifer count, 69 species in 12 genera, still falls below that of Mexico.

The fifth largest conifer flora belongs not to a large continental nation like China, the United States, or Mexico, nor to a large, complex, and varied archipelago like Indonesia, but to a modest-sized, single island, much of which is covered with toxic serpentine soils. New Caledonia, with its 44 conifer species in 13 genera, may not be an independent nation (it is an overseas department of France), but it is a self-contained administrative unit that can be treated as a "conifer country." This is the richest conifer flora in the world on an area basis, with an amazing 23 species per 10,000 square kilometers. By way of contrast, China, the United States, India, Venezuela, and Chile each have about 0.1 (to 0.15) species per 10,000 square kilometers while Russia and Canada only have 0.02 and 0.03, respectively, the latter the same as for the whole of Europe, which has just 30 species in eight genera. Unlike those of most mainland and many island nations, the New Caledonian conifers are all endemic, restricted entirely to the main island, except for a single species, *Araucaria columnaris,* commonly substituting for Norfolk Island pine (*A. excelsa*) in cultivation, which also extends to the adjoining dependent Isle of Pines and Society Islands. Based on DNA studies, the numerous species of *Araucaria* and handful of *Agathis* and *Podocarpus* species seem to have evolved very recently rather than being relicts of an ancient conifer radiation. Furthermore, three of the genera (*Austrotaxus, Neocallitropsis,* and *Parasitaxus*), belonging to three different families, are also endemic, and a fourth (*Acmopyle*) occurs elsewhere only as a single rare species in Fiji.

In comparison, among the large continental nations, Russia has just a single endemic genus (*Microbiota,* which may someday be found in adjacent northeastern China or North Korea), Canada, Brazil, and India have none, and the United States has two (*Sequoia* and *Sequoiadendron*), but the champions in conifer endemic genera are China and Australia. Mainland China has five endemic genera (*Cathaya, Metasequoia, Nothotsuga, Pseudolarix,* and *Pseudotaxus*) and a sixth that is shared with Taiwan (*Taiwania*). Australia has seven endemic genera (*Actinostrobus, Microstrobos, Wollemia, Athrotaxis, Diselma, Lagarostrobos,* and *Microcachrys*), of which the last four are found today only in Tasmania (but were more widespread in the geologic past). Other nationally endemic genera include *Halocarpus* and *Manoao* (New Zealand) and *Sciadopitys* and *Thujopsis* (Japan), while *Papuacedrus* is endemic to the island of New Guinea although found in both Papuan nations. Thus New Caledonia is third in endemic genera only to China and Australia, each with over 400 times its area. Of course, many genera that are national endemics today were more widely distributed in the geologic past, based on clear fossil evidence. These paleoendemics include *Cathaya, Metasequoia, Pseudolarix, Sequoia,* and *Sequoiadendron* in the northern hemisphere and *Athrotaxis, Fitzroya,* and *Wollemia* in the southern. Very few nationally endemic conifer genera are likely to be neoendemics that evolved relatively recently and remained in their original home. *Neocallitropsis* is certainly one of these, based on DNA studies that show it is nested within *Callitris,* but no others have been demonstrated to date (though the exceedingly specialized *Parasitaxus* could be another one). Instead, the remaining nationally endemic conifer genera are probably also paleoendemics, even though there is not yet the fossil evidence to confirm this supposition.

Australia, as the sixth largest country in the world, provides an interesting contrast in conifer diversity to the fifth largest, Brazil. Australia, with 41 species in 13 genera, follows New Caledonia as the sixth most diverse conifer flora, but Brazil, about 10% larger, falls way down on the list, with only eight species in three genera (with *Araucaria angustifolia* only in the far south). This is fairly typical for northern South America, which hosts only three genera of conifers (*Podocarpus, Prumnopitys,* and *Retrophyllum*). The largest national conifer flora in the region is that of Venezuela, the fifth largest country there (and just a ninth as large as Brazil), with 14 species in all three genera, including several endemic species. Its relative richness for northern South America is probably due to its possession of three separate mountain systems, the cordilleran ranges, coastal mountains, and Guiana Highlands, each with somewhat different species of conifers. Even Chile and Argentina in southern South America, with far more genera (eight) than in the north, have fewer species than Venezuela (nine and 11, respectively). The additional genera found in southern South America occur in temperate habitats, where most of the species and two of the genera (*Fitzroya* and *Saxegothaea*) would be endemic to Chile if they did not creep just over the border into tiny portions of Argentina.

Falling between Australia and Argentina in size, India, with 30 species in 13 genera, also falls between them in the size of conifer flora, which is found almost entirely in the Himalaya and in the mountainous regions adjacent to China in the northeast, where genera like *Amentotaxus* barely make an appearance. The rest of the country is almost devoid of native conifers, but India still manages to support all the conifers found in the Indian subcontinent. India has the same number of species and five more genera than all of Europe, which has three times its area.

Taken together, the two Koreas, with 20 species and 12 genera, are the only other continental nation, besides those already mentioned, to reach a total of 20 species or more. Only a few island nations reach this total. Besides Indonesia (and the individual shared islands, New Guinea and Borneo) and New Caledonia, already discussed, only Japan, Taiwan, the Philippines, and New Zealand have at least 20 species each. Japan is the richest of these, with 40 species in 17 genera (more genera but fewer species than in New Caledonia), and it is also the largest of the four in the aggregate area of its five main islands and many smaller islands. It stretches from boreal spruce-fir forests in northerly Hokkaidō to tropical shoreline pines and junipers in the southerly Ryukyu Islands near Taiwan. The latter, only a tenth the size of Japan, have almost as many genera (16) but many fewer species (25) than Japan. Fewer species and genera still are found in the Philippines and New Zealand, which have almost identically sized conifer floras of 10 genera with 21 and 20 species, respectively, on very similar land areas. The distribution and location of those areas is very different, however, with New Zealand concentrated in three main islands mostly in the temperate zone, while the Philippines consists of hundreds of tropical islands. There is also a difference in the nature of their conifer floras, with that of New Zealand all endemic, including two endemic genera, while that of the Philippines is mostly shared with its island neighbors to the south, Indonesia and Malaysia.

At the bottom of conifer richness, many islands have just a single genus of conifers with one or a handful of species. These include the largest island in the world, icy Greenland, with one species of *Juniperus,* and the fourth largest, tropical Madagascar, with four species of *Podocarpus.* Near the bottom of island nations bearing conifers are Jamaica and Puerto Rico with two and one species of *Podocarpus,* respectively. Similar is tiny Tonga of only 700 square kilometers, again with a single species of *Podocarpus,* the last outpost of conifers heading eastward into the western Pacific. As one heads back toward Asia, the conifer flora gets progressively richer as one passes from Tonga through larger Vanuatu (four species in four genera) and Fiji (eight species in six genera) back to New Caledonia, which is only about 5% larger than Fiji, which is about 25% larger than Vanuatu, which is about 20 times larger than Tonga. This whole chain of islands lies at about the same latitude, near 20°S, so their disparity in species richness would seem to be accurately portrayed as an attenuation out into the Pacific from Asian and Australasian source regions.

Assessment of all the disparities and similarities in amounts of conifer diversity just discussed would suggest that numbers of genera and species cannot simply be predicted from such geographic features as size of a land or island mass, its latitudinal position and range, maximum elevation, and distance to neighbors. Instead, there would appear to be additional historical factors that strongly influence conifer richness. These include natural factors, like climatic and tectonic fluctuations through geologic time, but also the arbitrariness of national boundaries with respect to basic geographic and biogeographic divisions. We have a lot more information about the ecological correlates controlling the distributions of individual conifer species than we do about how these individual distributions aggregate into conifer floras. We are very far from understanding (and predicting) how many conifer species and genera we might expect to find in a given location or larger area on the basis of its physical environment. There is much left to be done in understanding and accurately modeling conifer biogeography, distribution, abundance, and diversity. To answer the question posed at the beginning of this section, however, it would appear that China has the richest conifer flora if you only consider the total number of genera and species within a nation, but, if you take geographic area into account, New Caledonia clearly has the largest number of genera and species for its area.

CONIFERS IN HORTICULTURE

Conifers are cultivated to varying extents throughout the world, both as unselected forms that are commonly referred to as "the species" and in the form of an enormous number of cultivars selected over the past 200 years or more. With species growing naturally from the equator to the arctic timberline and to the southern tips of Africa, Australia, and South America, from sea level to the alpine timberline, and from the world's deserts to the rain forests, there is some conifer that can be planted in any garden or near almost any home in the world. Within the limits of their climatic tolerances, most species are not terribly demanding with respect to soil composition and moisture, although some species from the wettest places, like coastal southern Chile, grow poorly without abundant atmospheric humidity. Of course, no one species will grow throughout the combined natural range of all conifers, the biggest limitation (in cultivation) being winter temperatures on both the low and high sides.

There is much more experience with winter cold hardiness than with the potential success of cold-climate species grown in warm climates. Cold hardiness is usually expressed in terms of zones numbered 1–11 according to the USDA (US Department of Agriculture) plant hardiness zone scale, but there are a variety of competing schemes. The USDA hardiness zones are used here (listed at the end of each species description immediately following the distribution and habitat information) because of their familiarity to gardeners. The zones are based on the coldest temperature a plant can expect to encounter during the winter in an average year (the average annual minimum temperature). So the hardiness zones do not reflect the absolute minimum temperatures that might be encountered within a zone. Each zone represents an interval of 10° Fahrenheit (about 5.6° Celsius), starting with less than −50°F (−45°C) for Zone 1 and extending to above 40°F (4.5°C) for Zone 11. The maximum cold hardiness (lowest zone) for most species of conifers is not well known. For one thing, only about half of all conifer species are cultivated away from home. The majority of the remainder have not been tested very extensively. Only about 100–125 species are fairly widely grown, and not all of these have really been pushed to their limits. As a result, the hardiness zones listed here may well not express the maximum potential cold hardiness, even for some fairly commonly cultivated species. For the many species that are not yet cultivated, the zones have been guessed from the climates they experience in their native habitats, tempered where possible by experience with any neighboring species that may happen to be cultivated.

Since the vast majority of species not yet cultivated come from in or near the tropics, zones presented for temperate species are likely to be a little more accurate than those for tropical species. Overall, more that half of all conifer species (51.6%) are tender, assigned to Zones 9 and 10, with an average annual minimum temperature no less than 20°F (6.6°C); another 41.7% are temperate (Zones 5–8), with an average minimum winter temperature no less than −20°F (−28.8°C); and only about one in 15 (6.7%) are very cold hardy, assigned to Zones 1–4, with winter temperatures plummeting (in Zone 1) to below −50°F (−45°C). Thus, while we think of conifers as boreal, cold-climate plants, because that is where they seem to attain their greatest ecological dominance (not to mention their association with wintry Christmas trees), most conifer species are actually tropical to warm temperate. Nonetheless, the great majority of conifer species experience some degree of frost at sometime during their long lives, which, on average, are much longer than those of the hardwoods with which they grow. Thus they are more likely to experience extremes of weather fluctuation.

When considering the minimum winter temperatures experienced by plants, it is important to recognize that there is a very strong correlation between the depth of the lowest temperatures experienced in a place and the duration of these cold temperatures. So, while parts of Zone 10 may experience freezing temperatures every year, this may consist of just a single frosty night, while plants in Zone 1 will experience months of bone-chilling weather. This places some limitation on the likelihood that a conifer will be hardy in a zone much colder than its native range. Nonetheless, there are still probably a few conifers assigned to Zones 9 and 10 that might be hardy in warmer temperate zones, and a higher proportion of the temperate species (Zones 5–8) that might grow in climates of Zone 4 or even 3. One of the reasons for considering the latter is that the numbers of people, large cities, and botanical gardens are much smaller in Zones 4 and colder than they are in each of the warmer zones. This means that there has simply been less experience with conifer species in the cold zones than in Zone 5 and warmer. As a rule of thumb, it may always be worth attempting to grow a conifer in an area one zone colder than the one to which it is assigned here. Many conifer species occur naturally over a range of hardiness zones, while the original seed sources for individuals in cultivation are usually much more restricted. If you add the variability in hardiness of different individuals growing in any one place, there would seem to be ample opportunity for pushing the known temperature boundaries by new introductions of species already in cultivation.

It is much less clear how much warmer a climate you can move a conifer to than that of its native habitat. Some of the factors that may be involved include prolonged heat during sum-

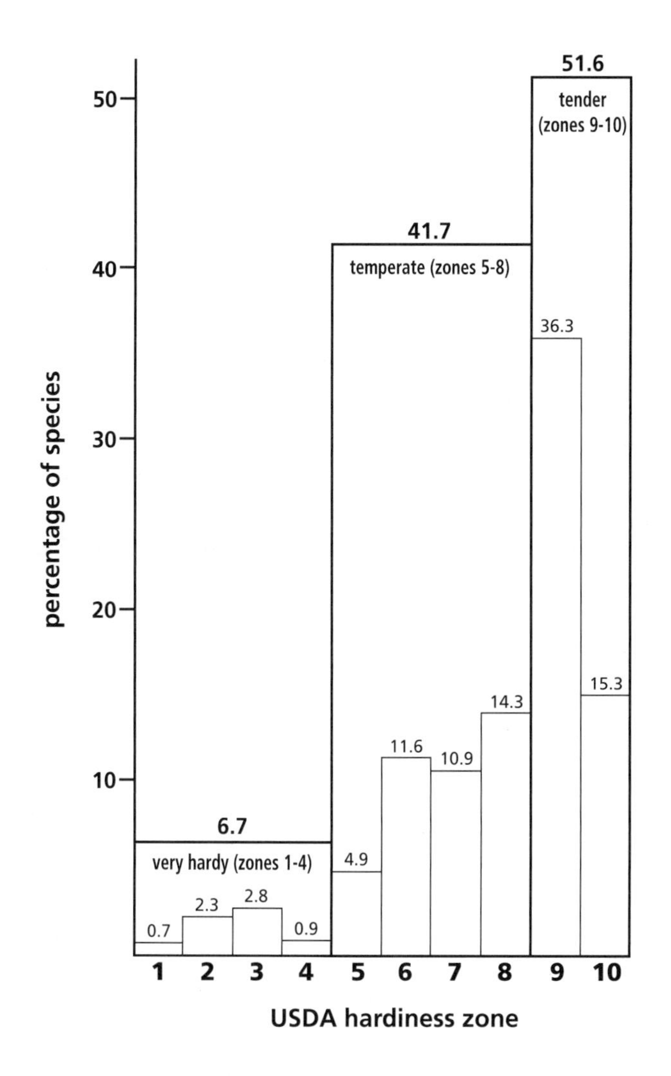

Average Annual Minimum Temperature

Temperature (°C)			Zone	Temperature (°F)		
−45.6	and	below	1		Below	−50
−42.8	to	−45.5	2a	−45	to	−50
−40.0	to	−42.7	2b	−40	to	−45
−37.3	to	−40.0	3a	−35	to	−40
−34.5	to	−37.2	3b	−30	to	−35
−31.7	to	−34.4	4a	−25	to	−30
−28.9	to	−31.6	4b	−20	to	−25
−26.2	to	−28.8	5a	−15	to	−20
−23.4	to	−26.1	5b	−10	to	−15
−20.6	to	−23.3	6a	−5	to	−10
−17.8	to	−20.5	6b	0	to	−5
−15.0	to	−17.7	7a	5	to	0
−12.3	to	−15.0	7b	10	to	5
−9.5	to	−12.2	8a	15	to	10
−6.7	to	−9.4	8b	20	to	15
−3.9	to	−6.6	9a	25	to	20
−1.2	to	−3.8	9b	30	to	25
1.6	to	−1.1	10a	35	to	30
4.4	to	1.7	10b	40	to	35
4.5	and	above	11	40	and	above

Maximum hardiness of the 546 species of conifers by tested or expected individual US Department of Agriculture hardiness zones and grouped into three broad hardiness categories. No conifers are known to require a mean annual minimum temperature of 40°F (4.5°C) or above (Zone 11).

mers (or lack thereof) and excess warmth during the winter that might interfere with dormancy. Both factors are associated with the balance between photosynthesis and respiration, which determines whether a plant is actually able to grow through net accumulation of the carbon fixed in photosynthesis (including the structural cellulose of wood) and lost in respiration. These factors interact with water and nutrient requirements to make some conifer species more frost tender than might be expected from the temperatures they experience in their native habitat.

Because so little is known of the physiology of most conifers, it is presently hard to predict how a species will perform in zones warmer or colder than its native region. For instance, Himalayan cypress (*Cupressus torulosa*) grows up to 3,300 m in the western Himalaya, becoming a large tree exceptionally to 50 m tall in the face of prolonged periods of freezing weather. Yet the species barely survives the modest winter frosts it encounters when planted outdoors in western Europe. On the other hand, Oriental arborvitae (*Platycladus orientalis*), while hardy to Zone 6 (down to −9°F, −23°C) and quite capable of sustaining prolonged frozen winters, also thrives at sea level in southern Nicaragua, where the lowest recorded temperature is 59°F (15°C) and the average high of every month exceeds 86°F (30°C). So, at the moment, there is no substitute for actual experience in the ground with conifers rather than predictions based purely on the native habitat. One reason for the uncertainty in extrapolating hardiness potential from native habitat is that many conifers are presently found in smaller areas than they were in the recent and more distant geologic past. *Cathaya, Fokienia, Glyptostrobus, Metasequoia, Pseudolarix,* and *Taxodium* are all examples of genera with just a single (or two) species that occurred widely across the northern hemisphere during the Tertiary but that are now restricted to limited corners of just a single continent. Their climatic tolerances might well be (and clearly are in some cases, like *Metasequoia* and *Taxodium*) much broader than you would predict from their present distributions. There is clearly a need for more work on climatic tolerances of conifers, particularly on physiological studies that might shed light on ways to test for the limits of tolerance without requiring massive field trials. Since half of all conifer species are not yet cultivated outside their homelands, there is tremendous scope for expansion of the repertoire of conifers available to grace parks and gardens, particularly in the tropics and subtropics.

Besides climatic tolerances, a major consideration relating to conifer species in cultivation is how big the plant will get. The stature of conifers in the wild ranges from dwarf shrubs less than 30 cm tall, such as some species of *Juniperus*, *Lepidothamnus* (*L. laxifolius* can produce cones on shoots just 5 cm high), and *Microbiota*, to forest giants from the northwestern coast of North America, pushing or exceeding 100 m tall, notably noble fir (*Abies procera*), Douglas fir (*Pseudotsuga menziesii*), redwood (*Sequoia sempervirens*), and giant sequoia (*Sequoiadendron giganteum*). These giants, of course, are hundreds to thousands of years old, so individuals of these species will not soon reach such heights in cultivation. Nonetheless, the tallest individuals of these giant species in Britain (where good records are kept), planted 100–150 years ago, exceed 40–50 m in height. They are not trees for a small suburban lot, dwarfing such a site in as little as 20–25 years. It is a common mistake to underestimate the growth potential of the most frequently cultivated conifers, often planting them too close to buildings or other structures. So, even though the sizes given in the text for each of the species reflect mature individuals in their native habitats, it is still a good idea to give some attention to these sizes when considering conifer species to plant in a given site. There are many more modest-sized species for small yards that, given careful selection for growth form, bark patterns, foliage form and color, and persistent seed cones, can create a very big impression.

CULTIVARS OF CONIFERS

Over the past 200 years, and at a varying but generally accelerating pace, thousands of cultivars, representing about a quarter of the world's conifer species, have been selected both from the wild and in cultivation. In more recent times, perhaps 3,000 cultivars have been more or less available, more than five times as many as the total number of conifer species and averaging about 20 for each of the species from which they were selected. This average is made up of many species with just one or a handful of available cultivars and a few species with hundreds each, notably *Juniperus chinensis* (with its hybrid *J. ×pfitzeriana*), *Picea abies*, *Taxus baccata* (with its hybrid *T. ×media*), *Thuja occidentalis*, *Tsuga canadensis*, and three of the species of *Chamaecyparis* (*C. lawsoniana*, *C. obtusa*, and *C. pisifera*).

Because of its complete worldwide coverage of conifer species, this book had no room left for extensive listing and descriptions of cultivars. Under each appropriate genus description, however, are comments on the general types of variations selected for in the cultivars derived from its species or hybrids, or both. Among the conifers, virtually all cultivars are vegetatively propagated selections that therefore maintain the distinctive characteristics for which they were selected. Overall, cultivar selection all across the conifers has favored primarily variations in habit (especially dwarfing, narrowly upright spires, weepers, and spreaders), foliage color (especially intense blues and yellows and variegated foliage), and, to a somewhat lesser extent, needle length (usually shorter) and foliage form (especially persistent juvenility and threadlike foliage). Because of the large number of slow-growing and dwarf conifer cultivars in commerce, cultivars will be, on average, smaller than plants of the same species in the wild at a comparable age. Cultivars selected for reproductive features, like early or prolific seed cone production and striking colors of immature seed cones, make up a minute fraction of all available cultivars.

The amazing array of cultivars available from local garden centers or from specialist nurseries were selected both from natural populations and from trees in cultivation. Although far from the most common source of novelties, individual wild trees displaying distinctive characteristics have yielded cultivars in several genera. Far more common is propagation from the witches'-brooms (densely bunched twigs) that appear as dark balls of unusual growth in the crowns of wild trees of many species. Some enthusiasts have made a career out of collecting and naming these (often diseased) dwarfs. Most dwarf and slow-growing conifer cultivars, however, were derived by careful inspection of the seedlings grown from large seed lots. Many foliage color variants were also spotted in seed lots, but such color variants have also been derived from sports (or somatic mutations) at the branch tips of other cultivars. Finally, a number of spreading cultivars were derived from cuttings of horizontal branches, which in some species fail to make a leader with upright growth. It is important to note that some branches of these cultivars, as well as ones derived by the preceding processes, such as dwarf Alberta spruce (*Picea glauca* 'Conica') or many variegated cultivars, may revert to the normal growth form or foliage color characteristic of the species in the wild. Such reversions should be cut out in order to preserve the distinctive characteristics of the cultivar.

In broad general terms, a cultivar is likely to be a little more fussy about garden conditions than unselected individuals of the same species. This is especially true for color variants, which may, depending on the cultivar, either require full sun to bring out the richest color or, contrarily, require protection from sun so that they do not get sunburned and produce browned and drying foliage. Growth rates are also important since even the slowest-growing cultivars still grow throughout their life, as plants must. People will often buy a dwarf conifer and place it in the garden without full attention to how large it will become in 5, 10, or 20 years. While most conifers readily tolerate light pruning during shoot growth, only a few—yews (*Taxus*) and northern white cedar (*Thuja occidentalis*), for instance, both of which make good hedge plants for this reason—tolerate the heavy pruning into old wood that would be required to shrink back a dwarf that has outgrown its site. For this and other reasons, it is more important to inquire about the site requirements of a potential cultivar purchase than of a species in its unselected form, for which the size and ecological information in this book can serve as some guide to appropriate conditions in the garden.

GARDEN USES OF CONIFERS

Conifers have a large number of garden uses, but three stand out as most common in practice. Use as specimen trees predominates in parks and larger gardens, but even small gardens frequently sport a conifer on the front lawn in temperate climates. Many conifers have a regular, rhythmic pattern of growth that yields a pleasant symmetrical, conical shape while young and for many additional years. Typically slower growing than hardwoods, they adopt the more irregular, character-laden habits of maturity much more slowly as well. Their enormous potential longevity, much greater on average than that of hardwoods, and their great potential stature give specimen conifers a long-lasting impact in the landscape. Such trees require little pruning, either in youth or old age, and develop their characteristic forms with minimal encouragement. There are a variety of statures, mature shapes, bark patterns, foliage textures, and colors to choose from when selecting conifers as specimen trees. This use of conifers employs the greatest proportion of unselected, seed-grown individuals, although there are certainly many cultivars available as well.

Use of dwarf conifers as accents in beds, borders, and rock gardens is similar to their use as specimen trees in showcasing them as individual plants. This use is where cultivars play their greatest role, there being few unselected conifers compact enough for these sites. The dwarf conifers used in these ways provide a variety of informal and more geometric shapes to choose from, including spheres, half spheres, broad to narrow cones, slender spires, broad and flat-topped cylinders, low disks, and rock huggers. The full range of foliage colors is found here as well, from the golds and ice blues that sparkle from a distance or glow at close range, to bright or somber greens that harmonize with many other colors, to bronzes and purples (along with reassuring greens) that emerge among the winter snows when perennials are hidden out of sight. The panorama of effects that can be achieved with dwarf conifers has made them a major component of garden design in general and a focal point in many gardens. For some enthusiasts, they are the whole point. Indeed, the broad scope of available cultivars make dwarf conifers one of the more understandable entries in the rarified category of "collectors' plants."

The remaining common garden uses are much more prosaic and treat the plants as a mass rather than as individual specimens. This, of course, refers to their use in hedges, shelterbelts, foundation plantings, and sidewalk planters near urban centers. In these applications, in contrast to specimen plantings, pruning is commonplace and often essential for maintaining the desired effect and mass. This means that only those conifers that can tolerate heavy pruning can be used, so these applications are dominated by yews (especially *Taxus baccata, T. cuspidata,* and their hybrid, *T.* ×*media*), by various cupressaceous cedars (like *Thuja occidentalis, Chamaecyparis lawsoniana,* and *Cupressus* ×*leylandii*), by mugo pine (*Pinus mugo* subsp. *mugo*), and by a wealth of junipers (especially the Pfitzer junipers, *Juniperus* ×*pfitzeriana*). The same plants are the chief conifer contributors to living, green topiary statuary, garden features that combine the emphasis on individuality of specimen planting with the methodical pruning maintenance regimen of hedges. In all these applications, deep green is the predominant color, although it is possible to find some hedges made up of yellow or blue (or sometimes variegated) cultivars, or even by a mixture of different foliage hues. Here, as with specimen dwarf conifers and unlike landscape trees, foliage is most of the story. Bark of trunks and branching structure are essentially hidden in these applications and play little part in the landscape effect, so they are not ordinarily a consideration.

In addition to focusing on shape, color, and other aspects of the appearance of the plant when selecting conifers for the garden, it is essential to consider site factors. These include potential neighbors, available space, proximity to structures, amount and timing of sunlight, drainage, and soil chemistry (with emphasis on acidity or predominance of calcium). With due attention to these selection criteria, conifers can be found that will be suitable for any site in the garden.

CONSERVATION OF CONIFERS

The breadth of distribution, natural abundance, and histories of direct exploitation and habitat degradation of the conifers are all so varied that their conservation status ranges from "you could not exterminate it, even if you wanted to," to species that are hanging on by a thread and could become extinct within a decade. Local species of modest elevations in regions of gentle topography are generally most threatened, but even widespread species can be at risk if their habitat is in demand by people for other uses or if they are directly exploited beyond their capacity for regeneration.

About a third of all conifer species (186 species) are threatened or vulnerable throughout their geographic ranges. Many of these have at least one population in a national park, nature or forest reserve, or other governmentally protected area. Unfortunately, such protection is often nominal, and even in developed countries, conservation areas may be open to mineral exploration and to historic grazing rights, while in developing countries they may be subject to illegal tree cutting or even to clearing for agriculture by impoverished peasants.

Even ecotourism can damage sensitive environments through shear numbers of people loving trees to death. A famous example is compaction of soil around giant sequoia trees (*Sequoiadendron giganteum*) in California and kauris (*Agathis australis*) in New Zealand, from trampling by hundreds of thousands of visitors, leading to death of surface roots and reduced water and nutrient uptake, with corresponding stresses. A few conifers suffer simply by being in spots that people want to visit for their scenic value rather than for the species itself. Examples of this include cascade pine (*Microstrobos fitzgeraldii*), growing around waterfalls in the Blue Mountains near Sydney, Australia, and the interesting rheophytic (growing in streams) New Caledonian corkwood (*Retrophyllum minor*) in the Plaine des Lacs, with its waterfalls and lakes, just 50 km from Nouméa, capital of New Caledonia and arrival point of virtually all tourists to the island.

These kinds of cases are rare, however. Tourism is far less of a threat to conifers than the direct and indirect pressures exerted by a continuously rising population almost everywhere that conifers grow. The threats basically boil down to three categories: (1) conversion of tracts of land once supporting conifers to other uses: agricultural (including plantations of nonnative conifers), residential, commercial, and industrial; (2) cutting of trees on the remaining tracts to provide lumber, fiber (for paper and structural composites, like fiberboard), fuel, and other resources; and (3) human-induced environmental changes.

Threats from Land Conversion

The single largest threat to conifers is the clearance of forest land containing them for other purposes so that, unlike with logging, not even the possibility of regeneration is left. While steadily expanding housing tracts in North America and the suburban aggregations of shopping malls and automobile infrastructure to serve them mostly replace the agricultural and industrial lands that reflected previous conversions from natural vegetation, they still keep the land alienated from its original cover and, in some instances, further adversely affect conifer populations. For example, stands of southern Florida slash pine (*Pinus elliottii* var. *densa*) occupy the highest and least flood-prone lands near Miami that are coveted for the best building sites. The pines that are left for scenery in a new housing development typically begin to decline and succumb within a few years to the injuries they sustained, both above and below ground, during construction. A similar fate can befall stands of ponderosa pine (*P. ponderosa*) in the western United States, but this species has a vast range compared to that of southern Florida slash pine, or Monterey and bishop pines (*P. radiata* and *P. muricata*) in desirable coastal California, so the potential impact is smaller, though the increased human density does increase the frequency and severity of fires that, in turn, destroy more houses built among fire-prone trees.

Felling and fires are tools of choice in clearing tropical forest for homesteads. These forests are not dominated by conifers, but, as rural populations increase and more accessible farmland is taken over by larger agricultural operations, the old slash and burn systems, which used very long rotation times that allowed significant forest regeneration, are breaking down. Conifer populations are affected because rotation times are reduced or even eliminated, hampering regeneration and leading to depletion of local resources so that subsistence farmers then travel into the montane forests, which house more conifers, to search for wood for fuel and construction. Under these circumstances, conifers tend to persist only on the steepest and most inaccessible sites, which certainly are not suitable for all species. Haiti is a country where almost everything has been cut and what has not has been burned repeatedly. Its native species of *Juniperus*, *Pinus*, and *Podocarpus* are critically endangered because there is no place left for them to grow since the whole landscape is occupied by people.

Even when conditions are not so dire, conversion of forest land to forest plantations of exotic trees is also problematic for native conifer species. Usually, conversion takes place on the best sites for forest growth, and so native forests, including conifers, remain on poorer land that may not be so favorable for regeneration and maintenance of viable populations in the long term. Governments of many countries, faced with crises in national timber production, aggressively searched the world for the conifer species with the fastest growth rates under their conditions and supplemented their finds with additional breeding programs. New Zealand is notable for enormously larger stands of Monterey pine (*Pinus radiata*) than those found in its native California, and the trees have a much finer form (from a forestry point of view) than can be found in the wild stands today, which were among the first victims of heavy exploitation on the western coast of the United States over 200 years ago. South Africa and East African countries concentrated on tropical pines from Mexico and elsewhere, and several of the introductions became invasive species threatening native vegetation, including some species of *Widdringtonia*. Plantation forestry is sometimes a matter of afforestation, planting trees on previously unforested land, as it is in Scotland, where Sitka spruce (*Picea sitchensis*) covers former heathlands, at least some of which were forested at an even earlier period. More often, however, it is a matter of reforestation, replanting forest on recently forested land, as it is in New Zealand where the Monterey pine plantations were once mixed podocarp and broad-leaved forest. Even attempts to reforest with native tree species usually result in lower species diversity (and certainly a different species distribution) than was found in the original forest being restored.

Threats from Direct Exploitation

Even when plantations are present, they rarely provide all the forest resources required, and so natural forests are also exploited and, in most countries, provide the great majority of such resources. This is especially true in countries like Canada, Russia, and the United States, which have large areas of boreal forest and (at least in their boreal regions) low population densities. Because of their wood structure, wherever conifers grow, even when they are not in the volumes found in the boreal forest, they are the best and most heavily exploited sources of wood for large and small construction and for paper pulp. The latter, of course, is exploited almost entirely industrially, but conifer wood for construction is obtained both industrially and by local people, while the third major worldwide use of conifers, for fuel, is largely a matter of local consumption. All of these can severely tax thinly distributed rain-forest conifers. As one example, when first discovered by botanists in 1999, Vietnamese yellow cedar (*Cupressus vietnamensis*), as one of the very few conifers in the vicinity, was already largely restricted to the most inaccessible sites within its very narrow range because of local exploitation for the fragrant, highly decay- and water-damage-resistant wood found typically in *Cupressus* and other Cupressaceae. A fourth important use for wild conifers, extraction of resins, like naval stores in the southeastern United States or Manila copal and other dammars in Malesia, is usually of less conservation concern than the other three because the trees tend to be tapped rather than felled and so may still reproduce, even if their growth rate is reduced.

Threats from Environmental Change

Human-induced environmental changes include a wide range of different factors. One that is very common and often has a very negative effect on conifer regeneration is the increased frequency or severity, or both, of fires in natural conifer forests near people. Conifers, with resin canals permeating their tissues, often including their trunks, are flammable trees, and fire is an important component of regeneration for many of them, in part through reducing competition with less-fire-tolerant flowering plants. With a very few exceptions, however, conifers cannot persist when the frequency of severe fires is so great that their return intervals are shorter than the time it takes for an individual to reach reproductive maturity. The critically endangered Mulanje cedar (*Widdringtonia whytei*) now persists in natural stands on Mount Mulanje, Malawi, only in pockets that fire cannot reach. Likewise, the endemic *Juniperus* species of Haiti barely hold on in the face of frequent anthropogenic fires. Other organisms, of course, are also threatened by an excessive fire regimen within conifer forests, for example, the monarch butterflies (*Danaus plexipus*) overwintering in groves of oyamel fir (*Abies religiosa*) in southern Mexico, where the rate of exploitation is far greater than it was when these wintering grounds were first discovered in 1976.

Ironically, human-enforced fire suppression can be as much of a risk to conifer populations as too many fires. A notable example is the giant sequoia (*Sequoiadendron giganteum*), which, during decades of fire suppression in national parks in the Sierra Nevada of California, showed drastically reduced regeneration. The open, parklike groves of these enormous trees began to close in with more shade-tolerant species, like California red fir (*Abies magnifica*), which crowded out the seedling and sapling giant sequoias. Another common effect of fire suppression is the accumulation of fuel, which leads to far more intense and damaging fires when they finally occur, as they inevitably do. This has the greatest consequence for conifers, perhaps, when residential communities nestled among pines (like ponderosa pine, *Pinus ponderosa*) or spruces (such as Engelmann spruce, *Picea engelmannii*) at the urban–wildland interface are burned out and rebuilt to the exclusion of trees in the immediate vicinity and with a broader buffer zone added. This would never be a threat to widespread species, but for very local ones in popular areas, like southern Florida slash pine (*Pinus elliottii* var. *densa*) near Miami or bishop pine (*P. muricata*) and Monterey pine (*P. radiata*) in coastal California, it can have a significant effect.

At least as important an anthropogenic environmental change for conifers as alternations in fire frequency and intensity is pollution, primarily air pollution. During the 1970s and 1980s, especially, large areas of Norway spruce (*Picea abies*) forest in central Europe, particularly in regions like the Black Forest of Germany, were damaged by acid precipitation resulting from the burning of high-sulfur coal in power plants. Trees were killed outright or the photosynthetic tissue in their needles was so damaged that their growth rates were drastically reduced. At the same time, conifers in the mountains surrounding the Los Angeles basin in California, such as white fir (*Abies concolor*), Coulter, Jeffrey, ponderosa, and sugar pines (*Pinus coulteri, P. jeffreyi, P. ponderosa,* and *P. lambertiana*), and bigcone Douglas fir (*Pseudotsuga macrocarpa*), were severely damaged by photochemical smog (including ozone and other reactive compounds) arising from rush-hour automobile exhaust. In a third location, the southern Appalachians of North Carolina and Virginia, subalpine stands of red spruce (*Picea rubens*) showed enormous decline and mortality, but here the causes were less obvious. With huge grants going into research projects dedicated to uncovering the causes of decline, many researchers demonstrated that red spruce foliage is highly susceptible to condensation of acidic fog and to effects of photochemical smog containing volatile organic compounds arising from emissions from cars and from the trees themselves. The latter natural emissions give rise to the hazes that historically gave the Great Smoky Mountains their name. Other researchers contended that stand age structure was sufficient to explain the decline, suggesting that the trees were simply growing old. While both kinds of factors no doubt contributed to the decline, damage here as well as in Los Angeles, central Europe, and in other parts of the developed world have since receded somewhat because of much stronger air pollution rules. Unfortunately, the scenario is playing itself out again in the developing world, particularly in China, where many rare conifers, already stressed by the impact of large human populations, now face the pollution attending explosive economic growth with little or no environmental controls.

Another form of pollution associated with human activity is biological invasion by aggressive pests and disease. Many nonnative insects have been introduced around the world and cause substantial economic losses to crop plants and forests, including many conifers. A few of them are so aggressive and damaging that they can actually threaten species that never evolved defenses against them. The hemlock woolly adelgid (*Adelges tsugae*), an aphid from Asia, is expanding through populations of both the widespread eastern hemlock (*Tsuga canadensis*) and the rare Carolina hemlock (*T. caroliniana*) in eastern North America, with extremely high mortality among infested trees. It has little effect on Asian hemlock species or on western hemlock (*T. heterophylla*), which was the first North American hemlock the adelgid encountered when it arrived on the West Coast in 1924. It reached the northeastern United States in the 1950s, perhaps on contaminated nursery stock, arguably the most common long-distance expansion mode for such pests. It had a lag period of decades before the true severity of the infestation manifested itself, and only since about 2000 have attempted biological control measures shown promise.

The similar balsam woolly adelgid (*Adelges piceae*) has a comparable record of differential impact on different host species and delayed expansion on the one that is most susceptible. It reached the northeastern United States and southeastern Canada from Europe around 1900 and spread through populations of balsam fir (*Abies balsamea*) without causing excessive damage. It reached the western coast of North America in 1929 and attacked most of the native firs of the region but, again, without catastrophic effects. When it reached the southern Appalachians in 1956, however, it attacked subalpine stands of Fraser fir (*A. fraseri*) with an intensity not seen on other species, leading to mortality rates of 90–99%. It is possible that, as with the red spruce (*Picea rubens*), with which it shares these high-elevation subalpine forests, Fraser fir was especially susceptible to balsam woolly adelgid because the trees were already stressed by air pollution.

Far to the north, red spruce trees in Prospect Point Park in Halifax, Nova Scotia, gave the city and province a scare in 1998 when they were found to be infested and in the process of being killed by brown spruce longhorn beetles (*Tetropium fuscum*), a European pest not previously encountered in North America. Most longhorn beetles attack weakened, dying, or recently dead trees, but the trees in Prospect Point Park had been mature and healthy. In order to save the important timber industry of Nova Scotia, the drastic step of cutting and burning all infested and potentially infested trees was taken, despite the fact that the spruces are the pride and joy of the park.

Similar examples of actual and potential devastation of conifer species by alien organisms attend the introduction of pathogenic fungi and bacteria. A present example is the increasing and increasingly widespread mortality experienced by Port Orford cedar (*Chamaecyparis lawsoniana*), a very local species in the Coast Ranges of southwestern Oregon and northwestern California, as a result of the spread of the blight-causing water mold *Phytophthora lateralis,* a fungus of unknown origin. Once the fungus reached the range of Port Orford cedar, around 1952, it began to expand throughout the range of the species, even though it can only spread naturally downstream of an infection site because it spreads by waterborne spores. Its colonization of ever more drainage systems within the range of Port Orford cedar must be credited (if that is the correct word) to the movement of infected mud on the tires of logging trucks and other vehicles entering the forests for recreation (about 75% of new locations and all long-distance ones) and on the shoes and boots of hikers and paws of animals (about 25% of new locations and mostly close to older ones). Many invasive species benefit from human activities and disturbances, which often speed their expansions and carry them much farther than they might otherwise reach.

Human impacts can also alter the relationship between conifer species and native potential pests and pathogens. Stinking cedar (*Torreya taxifolia*), a very narrow endemic of the Apalachicola River valley in the panhandle region of northern Florida, has been brought to the verge of elimination from the wild in the state park named for it (Torreya State Park) by a disease whose cause remains somewhat elusive. The immediate cause may be a fungus in the genus *Scytalidium* that was probably introduced to the region from nearby areas in the 1950s with establishment of slash pine (*Pinus elliottii*) plantations. Stinking cedar was common on the slopes surrounding the valley, but the disease, which spread rapidly during the 1960s and 1970s, has reduced the population to fewer than 1,500 stunted trees, each surviving by repeated suckering before the shoots are killed, never reaching sexual maturity. The few living individuals are being lost at a rate of 5% per year. The trees may be more susceptible to the disease following clearance of the uplands surrounding the river for plantations and agriculture, which may have drastically altered the water relations of the slopes, stressing the trees and allowing the fungus to become pathogenic in their roots.

With increased and increasingly rapid international trade that spans the globe, the kinds of catastrophic invasions just sampled are playing themselves out in many places. Highly localized species, whether rare or common within their limited ranges, are most susceptible to these introduced pests and diseases. The ability of these invaders to hitchhike on nursery stock is just one of many avenues open to them that make it extremely difficult to limit their spread once they are established in an area. Fragmentation of the landscape that, on the surface, might provide barriers between patches of suitable habitat, unfortunately does not, because we travel the intervening areas inadvertently carrying these pests and polluting pristine areas with them.

Response to Climate Change

The last environmental change to discuss, and perhaps the greatest threat to conifer diversity in the future, is that in climate. We are already feeling the effects of change, but even if we drastically reduce our production of carbon dioxide and other greenhouse gases, Earth's climate will continue to warm for the foreseeable future. The consequences for conifers are difficult to predict accurately, but they are not likely to be beneficial. On average, wherever conifers grow, they favor cooler sites than nearby flowering plants so that warming trends may tip the balance toward angiosperms, keeping in mind that low nutrient status of soils and a landscape-level fire regimen are other environmental factors favorable to conifers. Of course, conifers, like most plants, are adaptable and resilient and have survived much climatic change in the past, especially during the rapid glacial–interglacial fluctuations of the past 1.6 million years. One thing about the present warming, however, that differs from conditions during the repeated abandonment and reoccupation of Canada, the northern United States, and northern Eurasia is that human land use covers much of the surface needed for migration and establishment. Even if today's conifers are capable of establishing new populations by northward dispersal of propagules generation after generation, there may be few available places for the seeds to germinate.

That consideration, of course, does not apply to the presently treeless arctic tundra, where human influences (besides climate change) are very low. Even if temperatures become warm enough to support trees north of the arctic circle, any conifers establishing there would still face increasingly long seasons of 24-hour nights northward into the Canadian Arctic Archipelago. This was true the last time the high arctic was clothed with forests, during the very warm Eocene, about 50 million years ago. Axel Heiberg Island (and presumably other arctic islands) then supported forests of *Metasequoia* (dawn redwood), *Glyptostrobus* (Chinese swamp cypress), *Picea* (spruce), *Pseudolarix* (golden larch), and *Larix* (larch), among others, in an environment often referred to as the tropical arctic. This is obviously an exaggeration, and most of the trees endured the dark, if not bitterly cold, winters never experienced in the tropics by being or becoming deciduous. It is not clear how spruces survived these Eocene arctic winters nor how today's evergreen conifers of the boreal and temperate zones would fare under similar circumstances.

Invasion of newly climatically suitable areas, when possible, is now and will continue to be more obvious than retreat from the southerly portions of their range. Very few species extend to the geographic limits of their full climatic tolerances either to the north or south (except for species at the tree line) because, as they approach these limits, they are more limited by competition with other species. So, as climates warm, the southern edge of the range (or northern edge in the southern hemisphere) is not immediately thrust into unfavorable climate since it will take a while for the northern edge (or southern edge in the southern hemisphere) of unfavorable climate to actually reach the present southern range limit. Add to this the fact that many conifers are very long-lived trees and it becomes clear that they are likely to persist at their southern limits through several human lifetimes. It may thus seem, for a time, as if they are actually expanding their ranges with global warming, even if they are in trouble in both the south and the north. Collapse of populations and vegetational turnover at the warmer edge of the range may then be either gradual or sudden as older trees begin to die off and are not replaced by seedlings and saplings, which may have ceased to become established decades before. This kind of resistance to change followed by sudden conversion to new forest types has been documented many times in the fossil record and is particularly clear in the record of Carboniferous coal swamps in the eastern United States and elsewhere.

Conservation in Situ

While all these threats to conifers should be of real concern to everyone, and if human population growth continues unabated not even the most abundant and widespread conifer species may remain safe from the changes of population decline and extinction, at the moment only about 20 conifer species (4% of the total) are critically endangered. A great many of the rare species around the world have at least some populations within national parks, nature reserves, and other more or less protected areas. The vast majority of these refuges were established around scenic wonders or threatened animal populations rather than the plants, including conifers, that sustain them. Nonetheless, the conifers that happen to be in these parks and preserves are accorded a measure of protection that is of real consequence for many of them. A few parks and reserves were established explicitly to preserve populations or tracts of conifers. Torreya State Park in Florida, mentioned earlier, houses the rare and local Florida yew (*Taxus floridana*) as well as stinking cedar (*Torreya taxifolia*). It was established before disease virtually eliminated the latter from its slopes. California has many state and national parks and other protected areas centered on conifers. These include the redwood (*Sequoia sempervirens*) and giant sequoia (*Sequoiadendron giganteum*) parks, the Schulman Grove in the White Mountains, housing the world's oldest trees (Methuselah and other Great Basin bristlecone pines, *Pinus longaeva*), and the Torrey Pines State Reserve, housing the tiny mainland population of *P. torreyana.* Numerous botanical reserves, some of which were established to protect conifer species or communities, are found in places like New Caledonia and China. While these reserves do not ensure the continued survival of endangered conifer species, they certainly help, now and at least into the near future.

Conservation ex Situ

Conservation away from natural stands (ex situ conservation) is also extremely important. While dedicated conservation programs in botanical gardens are useful, having a species in general cultivation is fundamentally better insurance. There are only a few cases as yet of the benefits of being in cultivation for conifer conservation. The most obvious is stinking cedar (*Torreya taxifolia*), already mentioned. The only individuals of this species free from disease and capable of producing seed are found in cultivation far from the natural stands. There has also been a program of vegetative propagation by rooted cuttings of genetically diverse individuals from the natural range (the disease affects only the roots). Maintenance of genetic diversity is the essential foundation for all conservation programs. Luckily, most conifers start out at the higher end of the genetic diversity spectrum, at least among the relative handful studied. The distribution of that diversity is such that a good sampling of a single population will capture much of the allelic diversity in a whole species (although additional variation will be captured with more populations). This applies primarily to allozymes and other molecular markers while morphological and physiological diversity will often have a strong geographic component (related to adaptation to differing environments) and will require more extensive sampling to preserve it effectively.

Even species like red pine (*Pinus resinosa*), Torrey pine (*P. torreyana*), and western red cedar (*Thuja plicata*) that buck the conifer trend by having essentially no electrophoretic (allozyme) variation (indicating that they went through "recent" bottlenecks of drastically reduced population size) usually show variation (albeit harder to measure) among populations in ecologically significant characteristics. One exception may be Wollemi pine (*Wollemia nobilis*), whose wild individuals are, to all intents and purposes, a single clone in a single population. Thousands of propagules of this species were produced by seeds, cuttings, and micropropagation (tissue culture, which is widely used in orchid propagation). As this species spreads across the globe in cultivation, it will be unlikely to become extinct, even if a disease emerges that is lethal to all individuals. It will be interesting to monitor these far-flung individuals to see at what rate new genetic diversity may arise in their offspring, whether as sports (somatic mutations) or as seedling variation.

The goal of all conservation programs must be the maintenance of viable, genetically diverse populations in the wild. This will become increasingly challenging as land available for natural vegetation continues to shrink and global change alters the climates of those areas that remain. Let us hope that we can meet these challenges and that future generations will continue to learn from and be inspired by these fascinating and often magnificent trees and shrubs.

CHAPTER 4

CONIFER MORPHOLOGY

This book describes the morphology (external structures) and, to a lesser extent, the anatomy (internal structures) of conifers as an aid to identifying them and distinguishing them from related species. Following established convention, the various parts of the plant body are described in a fairly straightforward order, passing from appearance at a distance (the habit and crown), through the support structures (trunk, bark, branching, twigs, and buds) and foliage (including the internal structure of the leaves), to the position, appearance, and parts of the reproductive structures (pollen cones, pollen grains, seed cones, and seeds). Although the classification of conifers into families and genera is based primarily on their reproductive structures, these groups can also often be identified by other aspects of their plant bodies: the leaves, stems, and roots that constitute their vegetative morphology. Species within a genus are often distinguished primarily by vegetative features. The identification guides in this book at all levels use vegetative characters wherever possible because reproductive characters are transient in most cases compared to leaves and twigs or are not produced in large numbers in every year following sexual maturity.

Using the identification guides and descriptions requires some knowledge of the terminology applied to both vegetative and reproductive organs of conifers. I avoid technical terminology throughout the text in cases where plainer language could be substituted. Other conifer texts are somewhat inconsistent in their use of botanical terms, and a few use them incorrectly, so I incorporate some additional appropriate technical terms into this discussion even though they are not used in the text. The Glossary contains and explains a fuller range of terminology from the descriptions than I could accommodate in this brief introduction to conifer structure. I hope the use of English terms makes the descriptions more readily meaningful without sacrificing scientific accuracy.

ROOTS

The root systems of most conifers are predominantly shallow, consisting of a mass of fibrous roots in the upper layers of the soil arising from just a few major structural roots. These shallow root systems make many conifers particularly susceptible to windthrow. This applies to specimen trees as well as to trees in the wild. There is not a great deal of variation in the root systems of conifers, but a few features are potentially useful in identification.

All conifers, like most seed plants, have their feeding roots associated with symbiotic fungi that contribute to their ability to absorb nutrients from the soil. These mycorrhizal associations are of two fundamentally different kinds, involving different major groups of fungi. All members of the Pinaceae, and only members of this family among conifers, have ectomycorrhizae, in which the fungal mycelium envelops the feeding roots on the outside and penetrates the interior between the cells to form what is known as a Hartig net. Infected roots are readily recognizable as being much thicker than uninfected roots, more sparsely branched, and somewhat clublike or coral-like. The ectomycorrhizal associations often have basidiomycetes as their fungal partners, most of which reproduce by means of typical mushrooms. Unlike the mushrooms on the forest floor and the alterations to root morphology that mark ectomycorrhizae, there are usually no visible external signs of endomycorrhizae. These associations, which characterize all families of conifers except Pinaceae, have the fungal mycelium, including the reproductive spores, confined to the cortex of the roots that surround their vascular tissue. Here, the fungal mycelium actually penetrates cells and produces microscopic bladders (vesicles) and densely branched clumps (arbuscules) that give endomycorrhizae the alternate name vesicular-arbuscular mychorrizae. The endomycorrhizal fungi are all members of the order Glomales, all of which are endomycorrhizal associates and which belong to a completely different group of fungi than the mushrooms, that is, the zygomycetes. Unlike most endomycorrhizal conifers, the roots of many Podocarpaceae also have swollen growths, called nodules, that resemble those associated with nitrogen-fixing bacteria in plants such as legumes. In those Podocarpaceae investigated, however, these nodules contain endomycorrhizal fungi.

A relatively few conifers spread vegetatively through their root systems. Port Orford cedar (*Chamaecyparis lawsoniana*), for example, has spreading lower branches that strike roots in contact with the soil and ultimately become separate trunks that support their own crowns. The same thing happens with black spruce (*Picea mariana*) when its lower branches are overgrown by sphagnum moss in boreal bogs. Two other bog species, Chilean pygmy cedar (*Lepidothamnus fonkii*) and pygmy pine (*L. laxifolius*), produce special horizontal branches that creep along under the bog surface and produce dwarf upright shoots at intervals. Some shrubby junipers (*Juniperus*), yews (*Taxus*), and arborvitaes (*Thuja*), among others, show more compact vegetative increase by aboveground branches that root into the substrate as they spread.

Another uncommon underground phenomenon in conifers is root grafting, by which the roots of genetically distinct neighbors become physically connected to one another. This enables them to communicate nutrients, chemical signals, and pathogens with each other via their vasculature. Because it is so cryptic, we have no idea how common this phenomenon is, but among the few species already known to do this are eastern hemlock (*Tsuga canadensis*) and red pine (*Pinus resinosa*). It was discovered in them because they sometimes produce living stumps, individual cut trees that may, nonetheless, continue to live for decades, as evidenced by the continuing growth of bark over the cut surface.

Equally remarkable are the cypress knees of bald cypress (*Taxodium distichum*) and Chinese swamp cypress (*Glyptostrobus pensilis*). These conical or rounded outgrowths of the root system emerge from sodden soils or through flooding water to a height of up to 1–2 m. They are presumed to promote gas exchange for the waterlogged roots, which otherwise might drown.

BRANCHES AND CROWN

Most young conifers have a very regular growth habit, with a straight central trunk bearing regular circles of horizontal branches separated by branch-free intervals. This pattern arises from the typical flushing growth rhythm of conifers in which the terminal bud is surrounded by a ring of lateral buds and both types of bud burst simultaneously and grow out as shoots concurrently. The side branches from these flushes of growth may be simple and unbranched in their first year, like those of blue spruce (*Picea pungens*) or Wollemi pine (*Wollemia nobilis*), or increasingly complexly branched, up to the elaborate systems of Norfolk Island pine (*Araucaria excelsa*) or hinoki cedar (*Chamaecyparis obtusa*). The relative uniformity of young trees is varyingly gradually replaced with age by widely different characteristic branching patterns. Typically, large forest-tree species retain a single main trunk while shrubs and trees of more open habitats tend to divide into several trunks or main branches near the ground. As the tree grows and lower portions become shaded, many of the original rings of branches are shed, or single branches of some rings may be retained to become part of the framework of the crown. The form of such major limbs can vary from something resembling a fork or with a sharp upward angle in many Pinaceae to distinctly L-shaped, with a short horizontal segment turning vertically through a sharp elbow.

The branching pattern of young conifers means that almost all of them have a characteristic conical shape when growing without interference. Maturation often brings changes in crown form for two main reasons: inherent changes in relative growth of different parts and the hazards of life. Chances of having encountered the latter increase with age, so long-lived conifers often have broken, irregular crowns. The several-thousand-year-old Great Basin bristlecone pines (*Pinus longaeva*) of the White Mountains of California, with a few spiky living branches emerging from a craggy trunk more dead than alive, are an extreme example of this creeping breakup of the crown. With fewer environmental hazards or less time for them to accumulate, different conifers still develop distinct crown forms that depart dramatically from their youthful cones. While some species retain a conical form for quite a few years (100 or more in the case of giant sequoias, *Sequoiadendron giganteum*, for example), few but those with relatively short life spans, like Serbian spruce (*Picea omorika*), balsam and subalpine firs (*Abies balsamea* and *A. lasiocarpa*), or the upright form of Mediterranean cypress (*Cupressus sempervirens*), typically retain such a form for life. Most begin to round off the tops of their crown as they mature, and a few, like dwarf kauri (*Agathis ovata*), cedar of Lebanon (*Cedrus libani*), or Monterrey cypress (*Cupres-

sus macrocarpa), can become quite flat-topped. A few species mature with the lollipop shape of children's tree drawings, notably the Mediterranean stone pine (*Pinus pinea*). Most conifers, however, have some variant of a dome-shaped canopy at maturity, varying from fairly narrow, as in many firs (for example, noble fir, *A. procera*), through moderate, as in Arizona cypress (*Cupressus arizonica*), to quite broad, as in giant sequoia. While the outline of the crown may be generally dome-shaped, it is often interrupted by gaps due to broken or diseased limbs so that perfect domes are in the minority. Perhaps the species most likely to express perfect domes are shrubby species, such as plateau pinyon (*Pinus culminicola*) or mugo pine (*P. mugo*). Shrubs also add creeping or ground-hugging crown shapes, such as those of creeping juniper (*Juniperus horizontalis*), pygmy pine (*Lepidothamnus laxifolius*), and cascade pine (*Microstrobos fitzgeraldii*). A crown shape falling between conical and dome-shaped is an egg-shaped form, such as found in northern white cedar (*Thuja occidentalis*). Whatever the overall shape, there is considerable variation in the total depth of the crown, which often varies with the density of surrounding vegetation, and in where the widest portion of the crown occurs, whether at or near the bottom, the middle, or, less commonly, the top. An important factor influencing the crown shape is the relative angle to the trunk and length of a branch at different positions within the crown.

TRUNK AND BARK

The trunk itself may take a few different forms. Some species, like bald cypress (*Taxodium distichum*), have the base of the trunk greatly swollen, while others, like some podocarps (for instance, totara, *Podocarpus totara*) and dawn redwood (*Metasequoia glyptostroboides*), may be fluted and buttressed. Above the base, a few massive species, like kauri (*Agathis australis*), Sitka spruce (*Picea sitchensis*), and giant sequoia (*Sequoiadendron giganteum*), have almost cylindrical trunks with very little taper for 50 m or more. Most other arborescent conifers show obvious taper from the ground to the crown, and some species, particularly more open-grown ones or smaller trees, taper strikingly.

Perhaps the most conspicuous differences among the trunks of mature conifers lie in their barks. Even young trunks, in which the bark is basically smooth in most species, vary in features such as color and various irregularities of the surface, like resin pockets in fir (*Abies*) species, low horizontal ribs in some *Araucaria* species, and especially prominent lenticels in Nikko fir (*Abies homolepis*) and Wollemi pine (*Wollemia nobilis*). Barks of these and about 100 other conifers are pictured in the book by Vaucher (2003) while 78 are portrayed in the one by Debreczy and Rácz (2000). The color of smooth, young bark is typically greenish, grayish, reddish, or various shades of brown and is sometimes mottled, especially in tropical species, which may be colonized rather quickly by lichens, algae, and mosses. With increasing age, more and more differences become apparent among conifers belonging to different families and genera and even among species within some of the larger genera. The changes that come with age are often also mirrored by changes along the shoot system, from limbs through the upper trunk and then passing down to the base, with the peripheral parts showing more juvenile-like characteristics. The most obvious distinction is between those species that retain fairly smooth, thin barks on large trunks and those that become quite thick and rough, with many kinds of intermediate conditions expressed by at least some species.

Smooth-barked species, which are present everywhere but more frequent in the tropics, usually remain so by continuous peeling or flaking. Most species of spruce (*Picea*) shed thin flakes, as do flaky fir (*Abies squamata*) and Tecate cypress (*Cupressus guadalupensis* var. *forbesii*). If thicker scales are shed, as in matai (*Prumnopitys taxifolia*) or yellow silver pine (*Lepidothamnus intermedius*), the result is a pockmarked bark often described as "hammered." Most spectacularly among the flakers, lacebark pine (*Pinus bungeana*) sheds irregular patches that are thin in young trees and thicker in adults. As in the familiar flowering plant genus *Platanus* (the sycamores or plane trees), this produces a spectacularly colorful mottling that in very large old trees, like many near Beijing, becomes predominantly gleaming white, reflected in the Chinese name *bai pi song* (white skin pine). In contrast to these patchy species, other thin-barked trees shed long, vertical strips, like some trees of totara (*Podocarpus totara*) or common yew (*Taxus baccata*), while others are somewhat intermediate, like plum yews (for example, longleaf plum yew, *Cephalotaxus fortunei*).

The distinction between patch- and strip-shedding species seems to be related to a major distinction in thick-barked conifers, that between blocky barks and fibrous ones, a reference to texture and structure. Fibrous barks maintain coherence over long vertical stretches. It is relatively easy to shred fibrous bark lengthwise but difficult to snap it across. In contrast, blocky barks fracture readily in both directions. Both types of bark may become vertically ridged and furrowed. Fibrous barks are almost all variants of this form, but blocky barks develop many other patterns.

The primary differences among the various ridged and furrowed fibrous barks lie in the thickness of the bark and hence the prominence of the ridges and furrows. At the extremes, species like Mediterranean cypress (*Cupressus sempervirens*) and Montezuma bald cypress (*Taxodium mucronatum*) have very shallow furrows separating low, narrow ridges, while redwood (*Sequoia sempervirens*) and giant sequoia (*Sequoiadendron giganteum*) have highly fire-resistant bark up to 30 cm thick or more with massive, flat-topped ridges plunging into deep, V-shaped furrows. Mature fibrous barks are most often some shade of reddish brown and commonly weather to gray, though sometimes with a reddish cast or revealing brighter color in the bottom of the furrows. A few species, like redwood, giant sequoia, and incense cedar (*Calocedrus decurrens*), remain red, even at great sizes and ages. Fibrous bark (and related smooth forms) characterizes virtually all Cupressaceae, most Taxaceae, and many Podocarpaceae, as well as Japanese umbrella pine (*Sciadopitys verticillata*), the sole species of Sciadopityaceae.

Ridged and furrowed blocky barks also vary greatly in thickness (though without the extreme thickness represented by redwood and giant sequoia bark) and show a little broader range in color than corresponding fibrous barks. They are somewhat more likely to show strongly interlaced ridges (as in European larch, *Larix decidua*), interrupted ridges (as in deodar cedar, *Cedrus deodara*), and strong color contrasts in the furrows (as in Douglas fir, *Pseudotsuga menziesii*). In true firs, like European silver fir (*Abies alba*) and red fir (*A. magnifica*), the tops of the ridges often retain the lenticels and resin pockets that covered the young trunk. There are often intermediate forms between ridged and furrowed barks and the two other most common forms of blocky barks, checkerboard (or alligator) barks and platy barks. The most familiar checkerboard bark is that of alligator juniper (*Juniperus deppeana*), somewhat anomalous here because fibrous bark is the norm in Cupressaceae. Cedar of Lebanon (*C. libani*) has a bark form that is somewhat intermediate between ridges and furrows and a checkerboard. Many Pinaceae (especially pines) have platy barks, with plates varying in size and regularity of shape. Smaller plates are found in shortleaf pine (*Pinus echinata*), Parry pinyon (*P. quadrifolia*), and golden larch (*Pseudolarix amabilis*), for example, and the largest ones cover large trunks of ponderosa pine (*Pinus ponderosa*). Maritime pine (*P. pinaster*) is a good example of a bark intermediate between plates and ridges, and lodgepole pine (*P. contorta*) has bark intermediate between plates and flaking scales. The full range of blocky bark forms also has a wider range of colors than do fibrous barks, ranging from very pale cream and yellow, through orange, red, purple, and brown, to hues of gray that bump into black. Blocky barks and related smooth forms characterize all Pinaceae and Araucariaceae and many Podocarpaceae.

BRANCHLETS AND BUDS

First-year twigs (branchlets) of most conifers are not tremendously variable but do show some features of interest. In most conifers, the branchlets are apparently completely clothed by attached (decurrent) leaf bases (also called leaf cushions). This produces the familiar pattern of grooves of varying prominence that mark the boundaries between adjacent leaf bases. In situations of extreme crowding of leaves, as with many scale-leaved species, the twigs may be further hidden by overlapping leaf blades, including free tips or tightly pressed (appressed) portions, or both. With time and thickening (through secondary growth), the attached leaf bases of needle-leaved as well as scale-leaved conifers become separated from one another by strips of bark and usually are finally shed. How long this takes, and how long the leaf bases remain green, vary among species and genera. There is some relationship to the number of years that the leaf blades live, but this just provides a lower limit to the persistence of the leaf bases. Trunks of monkey puzzle tree (*Araucaria araucana*), for instance, retain leaf blades and bases many years, long after the blades have turned brown and ceased to photosynthesize, and the two are finally shed together. In spruce (*Picea*) and hemlock (*Tsuga*) species, the needles are shed from a woody peg (sterigma) at the summit of the leaf base. These sterigmata are retained and make the twigs rough for several years after leaf fall. Among pines, some, like ponderosa pine (*Pinus ponderosa*), retain their leaf bases for several years, even after leaf fall, while others, like eastern white pine (*P. strobus*), whose leaves fall in their second or third year, quickly shed the leaf bases also to produce smooth twigs within 2 or 3 years. In both these pines, as well as in all other pine species, the leaf bases attached to the twigs belong to the scale leaves that shelter the bundles of needlelike leaves, which do not themselves have conspicuous leaf bases. In many species of *Abies*, as in most North American species, for example, white fir (*A. concolor*), there is little or no evidence of the decurrent leaf bases, and the leaves seem to attach directly to a naked twig, leaving behind a smooth, round scar when they fall. In other species, especially some Asian species, like momi fir (*A. firma*), the leaf bases are marked by shallow grooves. There are many other cases where the depths of the grooves and hence the prominence of the leaf bases may be useful for distinguishing species within a genus.

The tips of the shoots may be protected by specialized buds while the shoot apex is in a resting phase, or, in other cases, such buds may be scarcely developed. Conifers with scale leaves, like many Cupressaceae and Podocarpaceae, typically lack distinguishable buds with special bud scales so that the shoot apex is surrounded only by small, immature ordinary foliage leaves that will later mature and reach full size when the shoot resumes growth. Examples include cypresses (*Cupressus*), junipers (*Juniperus*), and arborvitaes (*Thuja*) in the Cupressaceae and dwarf pines (*Lepidothamnus*) and some rimus (*Dacrydium*) in the Podocarpaceae. Buds in some species with needlelike leaves are also surrounded simply by arrested foliage leaves so that the shoot seems to come to a point without any swelling at the tip. This happens, for example, in China firs (*Cunninghamia*), juvenile *Taiwania* (both Cupressaceae), and many araucarias, like monkey puzzle tree (*Araucaria araucana*). More often, however, there are at least a few modified leaves associated with the resting shoot tips, the bud scales. In many yellowwoods (*Podocarpus*) and kauris (*Agathis*), among others, the buds are relatively unspecialized and the bud scales remain green. The overall shape of such buds and the shapes of the individual bud scales are very useful for identification of species in some of these genera. The most strongly developed buds are found in Pinaceae, in which the bud scales (or cataphylls) are highly specialized for protecting the winter buds. These bud scales are hard and closely overlap around the buds. Variation among taxa is found again in the shape of the buds and bud scales but also in the texture and color of the scales, in whether they are hairy or not and in the amount and color of any resin that might coat them. Buds may not be the richest source of characters for conifer classification and identification, but they are often useful adjuncts when taken in combination with other features.

LEAVES AND LEAF ARRANGEMENTS

Although leaves, whether scales leaves, needles, or broader forms, take second place to cones in formal aspects of conifer classification, they still play a prominent role there and in identification of genera and species. Leaf arrangement or phyllotaxis, the pattern of leaf attachment to the shoots, contributes a great deal to the appearance of individual conifer genera and species. There are four main patterns found in conifers: spiral, decussate, tricussate, and bijugate. Spiral (or helical or alternate) phyllotaxis, in which successive leaves are displaced to varying extents around and along the length of the twig, tracing a spiral, is far the most common. It has many variants, depending on how tight the spirals are and whether production of foliage leaves is intermittent. If the leaves are very evenly spaced and well separated from one another vertically but fairly crowded around the twig, an appearance of a twig bristling with leaves all around may be produced, as in spruces (*Picea*). This sort of appearance is usually more prominent in vertical shoots than in horizontal ones, on which the leaves of the lower side may be displaced a little outward to receive more light. Short side shoots with minimal leaf separation along their length will produce tufts of leaves like those found in larches (*Larix*) and true cedars (*Cedrus*). The most extreme form of such spur shoots is found in pines (*Pinus*), in which the bud in the axil of a non-photosynthetic scale leaf on the shoot produces just two, three, or five (or in a very few species only one or as many as eight) needles (the fascicle) surrounded at the base by a sheath of papery scale leaves. The spur shoot tissue is so undeveloped that the needles are effectively all in a single ring or whorl. A false superficial appearance of whorls (or pseudowhorls) is found in taxa such as Japanese umbrella pine (*Sciadopitys verticillata*) and some species of yellowwood (*Podocarpus*), in which large foliage leaves are concentrated near the tip of each growth increment and long stretches in between bear mostly less conspicuous scale leaves, which are often brown, thus making this intervening region between pseudowhorls look bare.

In decussate, tricussate, and bijugate phyllotaxis, there are two (decussate and bijugate) or three (tricussate) leaves attached at the same level (or node) on and at equal distances around the twig. In decussate phyllotaxis (predominant in many Cupressaceae and also found in *Agathis* and a few genera of Podocarpaceae, like *Nageia* and *Microcachrys*), successive opposite pairs are shifted sideways at right angles to each other so that there are four rows of leaves down the twigs, made up of crisscross pairs. A comparable arrangement is found in tricussate phyllotaxis, characteristic of some Cupressaceae, notably *Actinostrobus, Callitris,* and *Juniperus* subg. *Juniperus,* among others, so that successive trios are centered in the gaps between leaves of the preceding trio, and there are thus six rows of leaves, with a shift of 60° from node to node. Successive pairs are only slightly displaced in bijugate phyllotaxis, found in *Torreya* and *Cephalotaxus,* producing a double spiral, analogous to a DNA molecule with the spokes pointing outward. There is another minor phyllotactic variant involving whorled leaves. In *Neocallitropsis* and (sometimes) in stringybark pine (*Callitris macleayana*), the phyllotaxis is tetracussate, with successive whorls of four leaves offset by 45° from each other and producing eight rows (leading to an abandoned segregate generic name for stringybark pine: *Octoclinis*). One of the reasons why *Calocedrus* and *Tetraclinis* are interpreted as having opposite leaves even though they appear to be whorled, with the base of both facial and lateral leaves (see below) seemingly at the same level, is because the leaves are lined up in only four rows rather than the eight expected with whorls of four. The shift in position of successive pairs or whorls along a shoot, as well as of the single leaves in spiral phyllotaxis, is a mechanical outcome of finding space on the dome of the shoot apex during leaf initiation.

Underlying spiral, decussate, and bijugate leaf arrangements may all be modified in horizontal branchlets and branches of some conifers into a distichous presentation, with all or most leaves secondarily reoriented into a single plane on either side of the twig parallel to the ground. Most conifers with this arrangement achieve it by twisting of the petiole, but some also have varying degrees of twisting of the twig internodes between the leaf attachments. There is also considerable variation in the tightness of the distichy achieved, from very strict horizontal presentation to rather ragged rows that sweep out a substantial arc above and below the horizontal plane. Species of true firs (*Abies*) run the gamut from entirely distichous arrangements in lower branches and juveniles of such species as grand fir (*A. grandis*) and white fir (*A. concolor*) to complete occupancy of the semicylinder of space above the horizontal plane in upper, coning branches of most species or throughout the whole crown of such species as noble fir (*A. procera*) and red fir (*A. magnifica*). Some other conifers with spiral phyllotaxis that have a predominantly distichous leaf arrangement include bald cypress (*Taxodium distichum*), redwood (*Sequoia sempervirens*), eastern hemlock (*Tsuga canadensis*), Pacific yew (*Taxus brevifolia*), and miro (*Prumnopitys ferruginea*), representing four of the six conifer families. The two genera with bijugate phyllotaxis, *Amentotaxus* and *Cephalotaxus*, both Taxaceae, typically have a distichous presentation of the leaves, but this can be quite ragged in some plum yews, or the two rows may be raised to meet in a V rather than at 180°, as is often the case in Japanese plum yew (*C. harringtonii*). Distichous presentation of fundamentally decussate needles is found, among others, in *Nageia*, kauris (*Agathis*), and dawn redwood (*Metasequoia*). In the latter, it is achieved by alternate clockwise and counterclockwise 90° twisting of the internodes. Many Cupressaceae, the ones known as cedars with frondlike branch systems, achieve planation, a kind of distichy, by branching only from the lateral leaves and (almost) never from the members of the facial pairs. Wollemi pine (*Wollemia nobilis*), which has decussate leaves on upright shoots, has a unique variant of distichy. One member of each leaf pair is horizontal and the other projects above the twig at an angle. Since succeeding pairs reverse which needle is horizontal and which is raised, the result is a shallow V of needles sitting above a pair of nearly horizontal rows. Because of the alternate twisting of the pairs, all the horizontal needles on one side of the twig arise on the same relative position of the shoot apex, just as they would if the arrangement remained decussate. While most conifers have one of a few common leaf arrangements, distinctive arrangements, like those of *Wollemia* or *Sciadopitys*, can be very helpful in identification, particularly in combination with details of leaf form and structure.

Leaves have three distinct portions: the base by which they attach to the twig and connect with its axial vasculature; the blade, which (except in various nonphotosynthetic specializations, like bracts and bud and pollen cone scales) contains most of the chloroplasts and hence is the main photosynthetic region; and the petiole (leaf stalk) that links base and blade. Since the leaf bases (also called leaf cushions) are flattened against and intimately united with the twigs in conifers and completely cover them, they are described under Branchlets and Buds. Petioles are variably developed in conifers but are never as prominent as they are in many broad-leaved trees, for example, in quaking aspen (*Populus tremuloides*), whose long, flattened petioles keep the round blades fluttering incessantly, even in the lightest breeze. No conifer leaves flutter, and their longest petioles are no more than a few millimeters long. Many conifers, in fact, have no discernable petiole, their blades attaching directly to the leaf base without the intervention of a contracted portion. Scale leaves rarely, if ever, have a petiole, and some needlelike leaves also lack them, though less commonly. This is one of the ways that you can distinguish spruces with flattened needles (like Serbian spruce, *Picea omorika*), which lack petioles, from most hemlocks (*Tsuga*), which have a short but very slender and sharply delineated one. Most conifers with needlelike leaves, however, have at least some trace of a petiole, with a constriction of the tissue below the blade. The length and cross-sectional shape of the petiole are useful characters in some genera, as in the true firs (*Abies*).

The leaf blade displays many features useful for identification, but there is much convergence in leaf characteristics among unrelated conifers because of the strong association between leaf traits and environmental conditions. This does not mean that all the different conifers growing in a particular habitat have similar leaves, only that certain leaf forms are more common in some environments than in others. Many (perhaps most) conifers of hot, dry, or seasonally dry habitats have scalelike leaves with tips pressed tightly against the twigs (or more accurately, against the bases of longitudinally adjoining leaves), but the detailed form of the leaves is quite varied. There is even less constraint in all more permissive environments, and tropical rain forests support conifers with the largest leaves and a full range of sizes and forms down from there. Conifer leaves range in size from tiny scales 1 mm or less long to broad, oval needles as long as 30 cm. There is a complete continuum in length between these extremes. Juvenile leaves of vigorously growing young trees are often much larger (as much as by two or three times) than those of fully mature individuals, which tend to have lower overall growth rates. There is also a near continuum in leaf shapes of conifers. The most extensive work on conifer leaf shapes was done by David de Laubenfels, who divided them into four types, with most conifer species bearing two or three types during their lifetime or even at the same time along a single shoot. Although it is convenient to assign leaves to predefined types, it is important to keep in mind that these types are somewhat arbitrary segments of a continuum of forms. Thus the terms scalelike and needlelike (or scale leaves and needles) are used throughout this book, but there are numer-

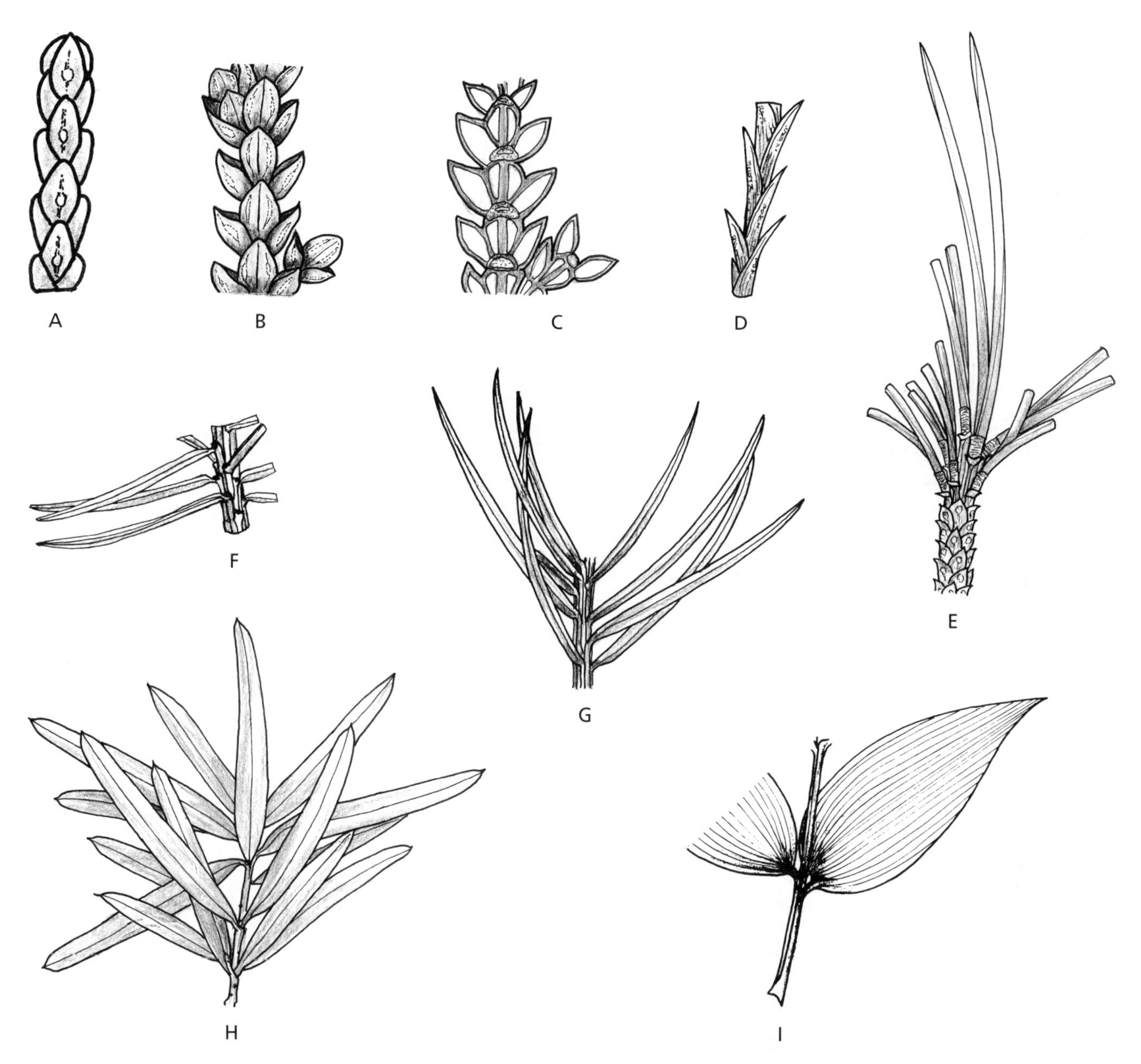

Examples of conifer leaf types. (A) Alaska yellow cedar (*Cupressus nootkatensis*), scalelike leaves in crisscross (decussate) pairs without differentiation into facial and lateral pairs and with (dorsal) glands, ×4. (B, C) Hiba arborvitae (*Thujopsis dolabrata*), scalelike leaves in crisscross pairs on a flattened branchlet spray with strong differentiation into facial and lateral pairs and equally strong differentiation of branchlet upper side (B), with small, weak stomatal patches, and underside (C), with intensely white, waxy stomatal zones occupying most of the surface between dark raised rims and midribs, ×2. (D) Giant sequoia (*Sequoiadendron giganteum*), awl-like leaves in a spiral arrangement and bridging the gap between scale- and needlelike leaf forms, ×2. (E) Bosnian pine (*Pinus heldreichii*), needlelike leaves in bundles (fascicles) of two surrounded by a sheath of papery scale leaves and attached in the axils of thicker, tougher scale leaves clothing the twig in a spiral arrangement, the individual needles of a pair long and slender with a semicircular cross section, ×1. (F) Blue spruce (*Picea pungens*), needlelike leaves attached singly in a spiral arrangement to a peglike woody base (sterigma) that runs down onto and clothes the twigs, the needles long and slender and more or less diamond-shaped in cross section, ×1. (G) Longleaf plum yew (*Cephalotaxus fortunei*), needlelike leaves attached in spiraling pairs (bijugate) and partially rearranged into a single plane on either side of the twig (distichous), with (decurrent) bases running down onto and clothing the twig, each needle somewhat sickle-shaped and flattened top to bottom (dorsiventrally), with a strong midrib and sharply distinguished upper and undersides, ×0.5. (H) Oleander podocarp (*Podocarpus neriifolius*), needlelike leaves attached singly in a spiral arrangement but concentrated near the end of each growth increment, each needle flattened top to bottom, broad, with an evident midrib and clearly demarcated upper and under sides, ×0.5. (I) Smooth-bark kauri (*Agathis robusta*), needlelike leaves attached in (opposite) pairs in either a decussate or distichous arrangement, each needle flattened top to bottom, narrowly egg-shaped, with numerous parallel veins rather than a single midvein, ×0.5.

ous transitions between them, and many species go back and forth between them with relative ease. Bog pine of New Zealand (*Halocarpus bidwillii*) and Chinese juniper (*Juniperus chinensis*) are good examples of conifers with abrupt switches in both directions between scale leaves and needles, while redwood (*Sequoia sempervirens*) and taiwania (*Taiwania cryptomerioides*) show gradual transitions between scales and two different types of needles. The needles of redwood are straight and flattened top to bottom while those of taiwania are clawlike and somewhat flattened side to side. Because of such transitions, it is perhaps better to think in terms of describing particular features of the leaves rather than trying to pigeonhole them into a few necessarily arbitrarily delimited categories or a larger number of more narrowly defined types that are harder to remember.

There are a number of different features by which the external form of conifer leaves may be described, no one of which is singularly more important than any others. The seemingly most obvious and straightforward features are the absolute and relative length and width of the blade as well as the position of the greatest width along the length. The difficulty here, especially with scale leaves, is in deciding where the leaf base (or petiole, or both) ends and the leaf blade begins. With scale leaves, there is no petiole. The widest part of the blade (or the free portion of the leaf) is right at its attachment to the leaf base, and the leaves appear to run right down onto the twig without interruption. Because of this, measurements of scale leaves, such as those found in northern white cedar (*Thuja occidentalis*) or eastern red cedar (*Juniperus virginiana*), generally encompass the whole visible portion of the leaf, including the leaf base, unless the description explicitly refers to the free portion of the leaf.

Having the blade (a term that hardly seems to apply to scale leaves) be widest at the point of attachment to the leaf base would appear to encompass almost all scale leaves and the clawlike subset of needle leaves, such as are found in Japanese cedar (*Cryptomeria japonica*) and many species of rimus (*Dacrydium*). Additional features of scale leaves include whether the free tip is appressed forward against the twig or stands straight out from it (or something in between), whether there is a gland on the outer (lower or abaxial) face, whether this face is keeled, and if so, whether the keel is rounded or sharply angled. Also important for taxa with scalelike leaves, especially Cupressaceae with decussate (successively crisscross pairs) leaf arrangement, is whether all the leaf pairs are essentially similar or whether pairs that lie to the sides of the branchlets (lateral pairs) are distinct from pairs (facial pairs) occupying the upper and lower surfaces of the twigs, which in such cases are always arranged in frondlike sprays. The lateral pairs, when differentiated, are often somewhat ax-head-like compared to the tightly appressed facial leaves and may also be either longer, the same, or shorter than the facial leaves.

Everything other than scalelike leaves may be construed as needlelike, although this is quite a heterogeneous assemblage. Needlelike leaves may be straight or bent in simple or, more often, compound curves. They may be cylindrical (singleleaf pinyon, *Pinus monophylla*), semicircular (Scots pine, *P. sylvestris*), triangular (ponderosa pine, *P. ponderosa*), or square (blue spruce, *Picea pungens*) in cross section or may bear flattened blades beside the midrib that vary from the narrowest of fringes (white spruce, *P. glauca*) to broad extensions almost as wide as the leaf is long (tumu dammar, *Agathis orbicula,* which lacks a single midrib). Needles with flattened blades may be flattened top to bottom, the common condition found for example, in yellowwoods (*Podocarpus*), redwood (*Sequoia sempervirens*), yews (*Taxus*), and hemlocks (*Tsuga*), or they may be flattened side to side, as in *Acmopyle,* Chihuahua spruce (*Picea chihuahuana*), or *Retrophyllum.* Besides curvature, relative width, and the position of the widest point on the blade, the shape of a needle is also influenced by the straightness of the margin and by the shapes of the base and apex of the blade. All these features interact with the greatest width and its position along the blade. Given two species with leaf blades of a similar relative width, one that has nearly parallel sides for an extended distance, like many *Podocarpus* species, for instance, will have a steeper curvature to the base and apex of the blade (and hence these will be relatively blunter) than one, like some *Agathis* species, that narrow continuously from the widest point to both the base and apex. If the widest point is nearer to one end of the leaf than the other, it will affect the relative bluntness of the general outline of the blade tip versus its base. The exact shapes of the blade apex and base are not simply determined by these factors because they also respond to other design requirements, like overlap with other leaves, shedding of excess rain or snow, protection against herbivores, and developmental constraints. Since all these different kinds of factors interact with one another, as well as operating partly independently of one another, it seems much more effective to describe the individual components that contribute to the appearance of the leaves than to establish a refined series of types that need to be memorized.

The embryonic leaves of conifers, the seed leaves or cotyledons, have been studied fairly intensively, though not yet exhaustively, and are sometimes distinctive among conifer taxa. A very few species, like bunya-bunya (*Araucaria bidwillii*), have hypogeal germination in which the cotyledons remain within the seed coat underground while the root and shoot emerge but remain attached. The vast majority of conifers, however, have epigeal germination, in which the cotyledons emerge above the ground and shed the seed coat. These cotyledons are all needlelike and usually straight, and have one or two midveins, depending on the family or genus, although occasional individual cotyledons may have extra vascular strands. What varies most among conifers is the number of cotyledons. Two cotyledons, like the condition in the numerically predominant dicotyledonous flowering plants, are found in at least some members of all conifer families except Pinaceae. It is the only number regularly found in Taxaceae and Sciadopityaceae and the predominant number in Podocarpaceae and Cupressaceae, which also have species with four cotyledons. Araucariaceae species are more or less evenly split between two and four. Cupressaceae add three as a common number (especially in some species of *Juniperus*), and a few genera have more than four. Five or more is the norm in the Pinaceae, which has the greatest variation in number of the conifer families, and this variation may even be useful in species identification, as it can be in true firs (*Abies*) and pines (*Pinus*).

MICROSCOPIC STRUCTURE OF LEAVES

The internal structure of conifer leaves reflects both the evolutionary history and the present environments and challenges of the species that bear them. I describe the general arrangement of the leaf tissues as seen in a magnified cross section of the blade for each of the genera. Within a few genera, like the pines (*Pinus*), spruces (*Picea*), true firs (*Abies*), yellowwoods (*Podocarpus*), and catkin yews (*Amentotaxus*), some of these minute features can also be helpful in species identifications, particularly the size, number, and relative positions of resin canals. These resin canals, basically hollow pipes with walls made up of specialized wall cells, are part of the defensive machinery of the leaves and contain the sticky terpenoid resins whose chemical identities have also been analyzed by taxonomists. The position of the resin canals, whether alternating with the vascular bundles or tucked underneath them, is one of the easiest reliable ways to distinguish vegetative specimens of the confusingly similar *Nageia* (Podocarpaceae) and *Agathis* (Araucariaceae).

The midvein, which may be made up of a single vascular strand or a pair (in the hard pines, *Pinus* subg. *Pinus,* and the Japanese umbrella pine, *Sciadopitys verticillata*), is continuous with the vasculature of the supporting twig. The original connections may later be severed and replaced with new ones if the leaf lives at least several years, a feature that has only begun to be investigated for its taxonomic and ecological significance. Species of *Nageia* and many Araucariaceae have several to many parallel veins running the length of the leaf rather than a single (or double) midvein, as in all other conifers.

Vascular bundles may be accompanied by additional transport cells (transfusion tissue) that helps link the vascular tissue to the photosynthetic tissue, and the arrangement of these cells can prove useful in identification of some groups, like *Podocarpus.* The vascular bundles may also have varying amounts of support tissue nearby, made up to a large extent of thick-walled cells called sclereids. Where there are masses of these above and below the midvein, it becomes obvious on the leaf surface through a prominent raised midrib. Sclereids and other support cells that help to stiffen the leaf and make it less attractive to herbivores may also be found just beneath the epidermis (the details of distribution of this hypodermis is useful in distinguishing the needles of different pines) or variously scattered through the photosynthetic tissue of the blade (the mesophyll), as in some species of *Podocarpus* and *Amentotaxus* (where they can sometimes be seen as shadows when the needles are held to the light). Such support tissues are not generally found in scalelike leaves and are best developed in relatively large needles or those, as in pines, that are very long for their diameter so that they would droop without the extra support (which some do, like many white pines, including *Pinus strobus,* and a few hard pines, like Mexican weeping pine, *P. patula*).

The photosynthetic tissue itself, responsible for what is, after all, the main function of foliage leaves, often consists of two distinct zones: the palisade layer and the spongy mesophyll. The palisade layer receives most of the light, typically lies under the upper surface of the leaf, and consists of one or more solid or interrupted layers of tightly packed cells conspicuously elongated perpendicular to the leaf surface. In needles, like those of *Acmopyle,* that are flattened side to side, palisade layers may underlie both sides. Most conifer leaves that are flattened top to bottom, however, have spongy mesophyll in the lower portion of the leaf, beneath the midvein. This tissue is much more open than the palisade layer and consists of more rounded or irregular cells that touch each other only at some points, leaving extensive air spaces among them that open up to the atmosphere via stomates in the adjacent epidermis. The spongy mesophyll (via the epidermis) is thus the immediate conduit for transpiration, the evaporation of water from the leaf into the atmosphere that drives the flow of water from the soil into the plant and is essential for the light-harvesting reactions of photosynthesis. During these reactions, which take place in the chloroplast, water is split to yield an electron that gradually gives up its energy to the organic molecules in the electron transport chain of photosystems II and I. The spongy mesophyll also provides extensive surfaces for the absorption of carbon dioxide and the release of oxygen, the other gases of photosynthesis, also exchanged with the atmosphere through the stomates.

The stomates (or stomata) are one of the more obvious features of the epidermis in conifers, the outer skin of the leaf. Although individually tiny, the stomates are mostly arranged in distinct lines or patches. These are often aggregated into bands that are very obvious on the surface of the leaf because they are marked by a pale coating of externally secreted wax distinct from any that might be borne on the rest of the leaf surface. There are well-studied differences among the stomates of conifers in their structure and that of their surrounding subsidiary cells, but most of these require special preparations that are beyond the means of most people who want to identify conifers. One fine detail of the epidermis around the stomate that is routinely mentioned in the descriptions of genera is the presence (or not) of a Florin ring, a set of low ridges on the cuticle (the outer wall layer that makes the epidermis waterproof) surrounding the guard cells (the cells that make the lips of the stomate and open and close to allow gas exchange or prevent water loss).

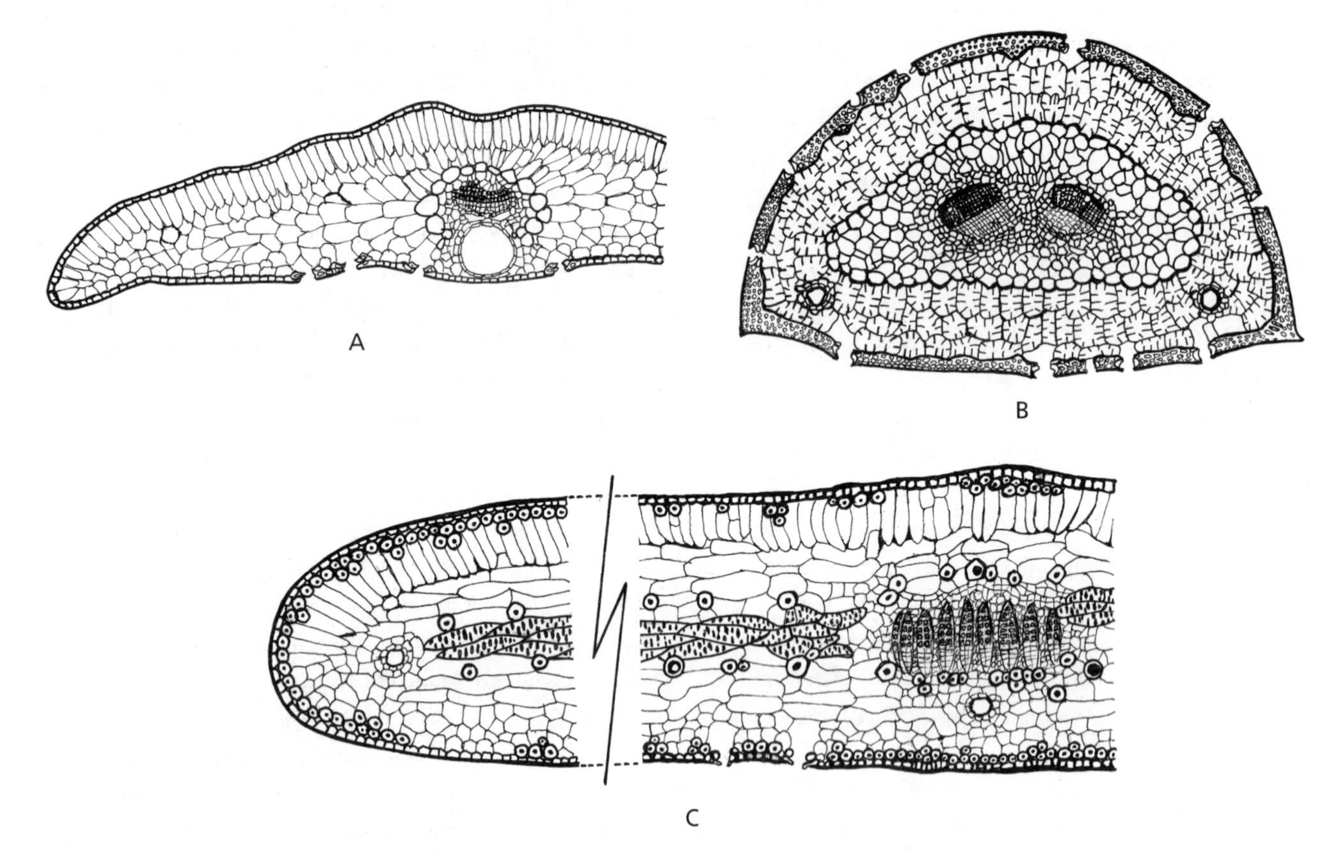

Diagrammatic, microscopically enlarged cross sections of conifer leaves to show the general distribution of tissues. (A) Eastern hemlock (*Tsuga canadensis*), epidermis surrounding leaf without underlying hypodermis is interrupted on the underside by stomates that open into the loosely packed spongy mesophyll while the upper side is lined by the tightly packed tall cells of the palisade parenchyma, and the midvein lies right in the center of the needle with the single strand of vascular tissue consisting of thicker-walled xylem at the top above the thinner-walled phloem, both embedded in nonphotosynthetic parenchyma, which in turn is surrounded by the single cell layer of the endodermis except where this is interrupted beneath by a single, large, cell-lined resin canal, about ×60. (B) Maritime pine (*Pinus pinaster*), epidermis is solidly underlain by two or more layers of thick-walled hypodermis, both tissues pierced all around by stomates that open into loose spongy mesophyll that completely surrounds the very large midvein, which contains two distinct vascular strands (as in all hard pines of subgenus *Pinus*), embedded in nonphotosynthetic tissue, including extra transport (transfusion) tissue at the sides, all inside the single cell layer of endodermis with its thick, waterproofed outer walls, while a pair of resin canals surrounded by several cell layers lie well outside this at the outer corners of the needle, about ×40. (C) São Tome yellowwood (*Afrocarpus mannii*), epidermis only partially underlain by thick-walled hypodermis one or two cells deep (where present), both interrupted on the underside by stomates in a well-defined stomatal band on either side of the midrib, which has a single midvein or partially separated wedges of xylem and phloem with little sign of a surrounding endodermis, and flanked by bands of extra transport (transfusion and accessory transfusion) tissue supported by longitudinal strands of thick-walled fibers seen here in cross section, the needle also traversed by three resin canals, one beneath the midrib and one out at each edge near the junction between the palisade parenchyma layer that lines the upper side of the needle and the spongy mesophyll lining the underside, about ×75.

Otherwise, the arrangement of stomates on the leaves is most generally useful in identification. The most obvious aspect of arrangement, and one that has long been used in describing and distinguishing conifers, relates to which side(s) of the leaf bear stomates. If stomates are found only on the upper side of the leaf, the leaves are epistomatic. This condition is fairly common in scale-leaved species from arid habitats, like some junipers (*Juniperus* subg. *Sabina*) and cypresses (*Cupressus*). The upper surface of the leaf in these species is pressed against the twig, and confining stomates to this surface helps reduce water loss. Epistomaty is somewhat less common in needle-bearing species, at least at the level of genus. One of the two subgenera of pines, the soft or white pines (*Pinus* subg. *Strobus*), has epistomatic needles, and so does one of the two subgenera of junipers, the gin junipers (*Juniperus* subg. *Juniperus*).

More common is the presence of stomates on both the upper and lower surfaces of the leaf, then referred to as amphistomatic. The other subgenus of pines, the hard or yellow pines *Pinus* subg. *Pinus*), is amphistomatic. This is one good way (among many) of distinguishing an atypical five-needled hard pine, like Arizona pine (*P. arizonica*), from a typical soft pine, or of distinguishing a two- or three-needled pinyon pine (such as Colorado pinyon, *P. edulis*, or Mexican pinyon, *P. cembroides*, both soft pines) from a typical hard pine. Many scale-leaved species of Cupressaceae, Podocarpaceae, and Araucariaceae are also amphistomatic. So are many other needle-leaved species besides pines, including most spruces (*Picea*), a few true firs (*Abies*), and half the species of *Nageia*. Amphistomatic leaves often have a very uneven distribution of stomates, typically with fewer on the upper surface than on the lower. This is the case with the amphistomatic firs, which sometimes have stomates on the upper side nearly confined to the tip of the needle. Even the spruces, with their four-sided needles, giving a roughly square cross section, usually have one or two fewer rows of stomates on the two upper faces than they do on the lower pair.

This favoring of the lower surface culminates in hypostomatic leaves, in which the stomates are confined to that surface. The great majority of hypostomatic leaves are needlelike (in the broad sense that includes all elongate leaves, whether they are narrow or broad). Because the lower surface of scale leaves is the one that faces outward, it would be very unusual to have stomates on it but not on the more protected inwardly facing upper side. Many hypostomatic needles have the stomates in two well-defined bands on either side of the midvein. The bands may be marked by white, gray, yellowish, or brown wax or hairs (trichomes) and are often slightly (or even deeply) sunken below the rest of the underside. Within the bands, the stomates many be haphazardly scattered over the surface or may be collected in straight lines that vary from interrupted or irregular, or with the guard cells pointed randomly, to continuous and tight with the long axis of all the stomates set right along the line. The number of lines of stomates is often a useful character, as it is in spruces and true firs.

The different types of stomatal distribution are under strong selection by the environment with respect to water availability. As a result, closely related species may display different stomatal arrangements. Thus, among the silver firs, balsam fir (*Abies balsamea*) is hypostomatic while its relative subalpine fir (*A. lasiocarpa*) is amphistomatic, and the same contrast separates grand fir (*A. grandis*) from white fir (*A. concolor*). Of all the microscopic features of conifers that are useful in classification and identification, the stomates (although not their fine details, like the number and arrangement of subsidiary cells surrounding the guard cells) are the easiest to observe since one can often see details of their arrangements with a simple 10× magnifying glass.

Such a hand lens is also useful for observing glands on the leaves. When initially formed, the glands, which contain fragrant terpenoid resins, are covered over by the cuticle, but they may rupture, leading to dots of white, dried resin on the face of the leaves. Resin glands are almost confined to scalelike leaves, especially those of the arborvitaes (*Thuja*), junipers (*Juniperus*), and cypresses (*Cupressus*). There are differences between species in shapes of the glands (from round to greatly elongated) and in their positions on the backs of the scales. A few species of juniper (for instance one-seed juniper, *J. monosperma*) and cypresses (like Sargent cypress, *C. sargentii*) have glands that typically rupture, and so these species have leaves marked by sticky white spots as the resin dries and crystallizes.

The glands contribute to the distinctive fragrances of these genera. Fresh foliage of each species of *Thuja* smells different from all the others, and this can be helpful in identifying these otherwise quite similar trees. The internal resin canals described earlier also contribute to the distinctive fragrances of different species and can be checked by crushing or cutting the foliage. Fragrances are notoriously difficult to express, so they are not generally described here. Nonetheless, as one learns to identify conifer species, it is useful to try to remember their fragrances as an additional tool in separating them from their relatives. Thus one of the common tricks for separating Jeffrey pine (*Pinus jeffreyi*) from young ponderosa pine (*P. ponderosa*) without the mature platelike bark is to sniff the bark furrows for the distinctive smell of vanilla that signals the presence of hexane in the bark resin of Jeffrey pine.

POLLEN CONES

Pollen-bearing cones of conifers are more uniform in structure than the seed cones and have relatively few structural features to distinguish them. Thus they tend to differ between families and some genera but often are not useful in separating species. When they differ between species, it is usually a matter of size or arrangement and not of any real difference in structure. Pollen cones are shed more or less quickly after release of the pollen, so they are short-lived and not available for investigation as mature organs during most of the year. This, coupled with the comparative lack of distinctive features, has left pollen cones relatively neglected in conifer classification and identification.

All conifer pollen cones are built on much the same plan. There is a central axis bearing pollen scales (microsporophylls) in a spiral, decussate (crisscross), or whorled arrangement that is usually the same as the phyllotaxis of the leaves. This is referred to as a simple cone, contrasting with the compound seed cones, because the pollen scales are equivalent to leaves attached directly to the axis. Pollen scales consist of a slender stalk in the interior of the cone and a blade of varying shape making up the surface, with two (rarely one) or more (as many as 20 in some species of *Araucaria,* such as chandelier araucarian, *A. muelleri*) pollen sacs (microsporangia) attached to the lower edge or inner face at the bottom. The pollen sacs may be round to greatly elongated and attached to the stalk (in Pinaceae, with only two sacs) or free, as in Araucariaceae with their more numerous sacs.

The two types of pollen grains found in these sacs, the simple spherical ones and those bearing two or three air bladders, are described in Chapter 1, under What Is a Conifer?, detailing characteristic features of conifers as a taxonomic group distinct from other seed plants. The pollen grains with bladders are a little more varied than those without and are sometimes identifiable to species though more commonly to genus, and the spherical ones are as often identifiable only to a family or to a group of genera as they are to genus.

Below the lowest pollen scales there may be sterile bracts of various shapes, as in kauris (*Agathis*) and some yellowwoods (*Podocarpus*), and sometimes bud scales or their remnants, if individual pollen cones emerge from buds. The cones may be

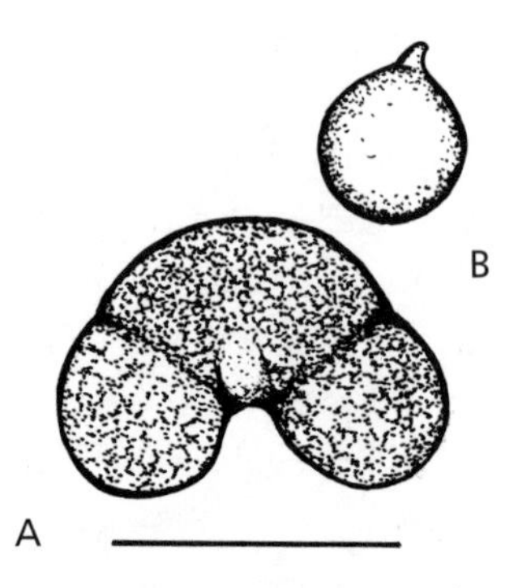

The two main general types of pollen grains found in the conifers, those with two air bladders (bisaccate), represented here by cathaya (*Cathaya argyrophylla,* A), and the smooth, spherical ones, represented here by sugi cedar (*Cryptomeria japonica,* B), which has a germination papilla absent in pollen grains of most species with this type.

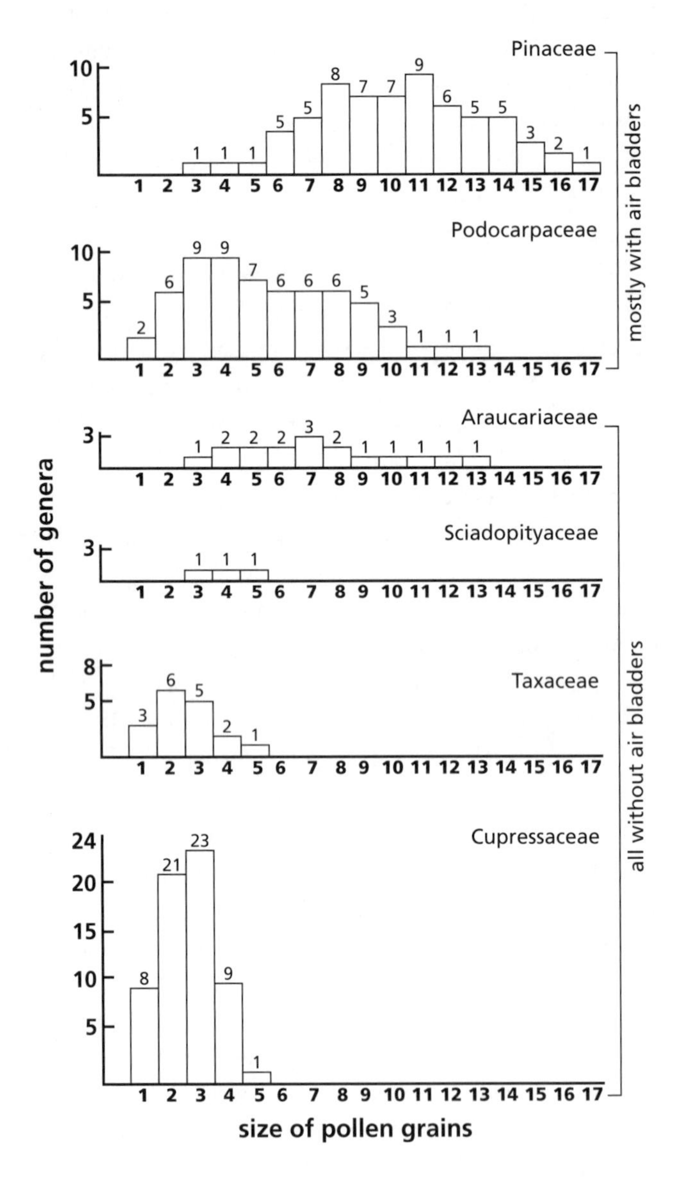

Distribution of pollen grain size among the genera within each family of conifers. Although there is much overlap among families, medium sizes can be quite different for families with similar morphological types of grains. Each size class has a range of 5 μm, starting with 20–25 μm for column 1 and going to 120–125 μm for column 16, with three exceptions: columns 9 (60.1–70 μm), 14 (100.1–115 μm), and 17 (126–145 μm). Individual genera have species spanning more than one size class, so the totals of the genera noted on the columns exceed the total number of genera for each family.

stalked or not (sessile), and the stalk may be naked or have additional minute bracts along it. They may be single or clustered and may be positioned terminally at the tips of branchlets, attached in the axils (armpits) of ordinary foliage leaves on twigs that are otherwise unspecialized, or attached in the axils of bracts on specialized reproductive shoots. The latter are found in several Podocarpaceae, including some yellowwoods and *Prumnopitys* species, and Cupressaceae, including bald cypress (*Taxodium distichum*) and dawn redwood (*Metasequoia glyptostroboides*), among others.

These reproductive branch systems bearing pollen cones have been interpreted as corresponding to the compound pollen cones of another living group of seed plants, the gnetophytes (*Gnetum, Welwitschia,* and *Ephedra* or Mormon tea), and to those of Paleozoic conifer ancestors, the long extinct Cordaitales. The compound pollen cones of cordaites, like their similar seed cones, were much more compact structures than the corresponding modern branch systems, with bracts of the primary axis bearing axillary fertile dwarf shoots that were smaller or less conelike themselves than the modern axillary simple cones. On the whole, it seems somewhat more likely that the modern compound branch systems are a newer development rather than a direct retention of the ancestral condition.

The true situation is especially uncertain with respect to members of the yew family (Taxaceae). Many of these, including yews (*Taxus*), have rather different pollen scales than those of the other conifers, with (as many as seven) pollen sacs all around the periphery of a head at the summit of a central stalk. The most plausible proposed interpretation of these so-called peltate sporangiophores is that each one actually represents a simple pollen cone by itself in which the original pollen scales became fused side to side in evolutionary time. Thus the whole cone, which contains several of these structures, is actually a compound pollen cone, unlike the simple ones it superficially resembles. *Amentotaxus, Cephalotaxus,* and *Pseudotaxus* all seem to show intermediate features that would support this interpretation. Of course, as with relevant Cupressaceae and Podocarpaceae, the compound pollen cones need not be remnants from the most ancient conifers but a secondary return to compound structures from a simple ancestry via de novo specialization of reproductive branches. Paleobotanical, developmental, and molecular studies may settle the issue some day.

SEED CONES

Most of the formal classification of conifers at the family and genus level is based on the structure of the seed-bearing organs. In most conifers, these are compound cones or structures that can easily be seen as variations on a compound cone. In contrast to the pollen scales of pollen cones, the seed-bearing (ovuliferous) scales are not attached to the central axis in the position they would have if they were modified leaves. Instead, the positions of leaves are occupied by bracts, and the seed scales are in the axils of these bracts. As described in Chapter 1, this led to the hypothesis that what appears to be an unexpected leaf in the axil of a leaf is actually a transformation of the expected axillary branch that is common to all seed plants. Rudolf Florin's work with cordaites and Paleozoic ancestral conifers confirmed this interpretation by showing that these ancient plants actually had a dwarf fertile shoot with seed-bearing and sterile appendages (leaves) in the bract axil rather than the modern seed scale. Mature conifer seed cones rarely conform to the unmodified model of a cone axis bearing seed scales free in the axils of the bracts. The most familiar example that does conform to this model is Douglas fir (*Pseudotsuga menziesii*), with its three-lobed bracts looking like the hind legs and tail of a mouse diving between the seed scales. The bracts of bristlecone fir (*Abies bracteata*) are somewhat similar, but the middle "tail" is about 3 cm long and sticks straight out from the cone, giving the species its scientific and common names. Other forms of such exserted bracts are found also in some true firs (*Abies*) and larches (*Larix*), for example. In most Pinaceae, however, as in pines (*Pinus*) and spruces (*Picea*), the bracts enlarge much less than the seed scales during development so that they never show externally in the mature cone and make an inconspicuous protrusion at the base of each scale when a seed cone is ripped apart.

In the other families, while the bracts may enlarge during cone development, they are fused to varying degrees with the seed scales they subtend. This concept of fusion is an evolutionary interpretation. What actually happens developmentally is intercalary (internal) growth of common tissue below the original free tips of the bracts and seed scales. Thus the bract and scale do not literally become stuck together during development but rather become fused through the shared growth. Fusion of the bract to the seed scale is so complete in most Cupressaceae that all that can be seen are cone scales (sometimes referred to as bract-scale complexes) with a faint (in junipers, *Juniperus*) to prominent (in redwood, *Sequoia*) line across the exposed face representing the boundary between bract and seed scale. In some Cupressaceae, such as *Athrotaxis* and China fir (*Cunninghamia*), the seed scale is reduced to just a tiny, seed-bearing pad near the base of the bract, and the part of the cone scale that you see externally is basically just the bract. Other families are somewhat intermediate. For example, Araucariaceae exhibit less extensive fusion so that there is a deep notch across the cone scale between the bract and seed scale portions. In many conifers, the bracts play their most important functional role early in the development of the seed cone, at the time of pollination, when their arrangement and shape often strongly influence wind currents around the cone and hence how pollen grains are captured from the air.

In addition to varying degrees of fusion to the bract, seed scales show many useful features for classification and identification. The number of seeds borne on each scale is fixed (with anomalies) for most families, but highly varied in Cupressaceae. Seed scales uniformly bear just one seed in Araucariaceae and Podocarpaceae (though rarely, individual scales may have more) and two in Pinaceae. In these families, too, the seeds are embedded in the scales, although the manner of embedding is quite different in each. Species of *Araucaria* have the seeds sunken into a pit in the seed scale, and the whole scale, which is winged, acts as the dispersal unit. Pinaceae also generally have wind-dispersed seeds, but here the seeds are covered only by a thin skin from the top layer of the seed scale. This detaches from the scale as a terminal wing that comes out from one end of and clings to the seed. The two main groups of genera in the family differ in whether the seed lies entirely on the surface of the scale and hence has a naked lower surface (as in *Pinus* and *Picea*) or whether one end projects beyond the inner edge of the scale and is thus partially wrapped by a continuation of the skin beneath (as in *Abies* and *Cedrus*). These differences are also related to differences in details of pollination between the groups of genera. Podocarpaceae are largely animal dispersed, with various parts of the seed cones fleshy in different genera. The portion of the seed scales that enfolds the seed, called the epimatium in this family, is usually fleshy, though not in all genera. In the largest genus in the family, the yellowwoods (*Podocarpus*), the two or three potentially fertile bracts (those subtending functional seed scales) typically fuse and become a bright red berrylike mass (the podocarpium) beneath the darker and less fleshy seed within its epimatium (or, less often, two or three such structures).

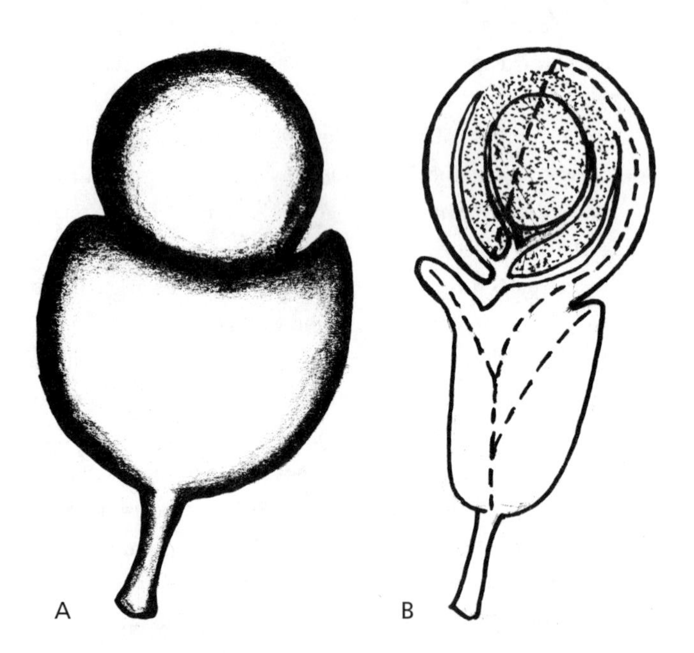

Seed cone of plum pine (*Podocarpus elatus*) in surface (A) and cross section (B). The lower portion is the podocarpium, consisting of two united, fleshy bracts (the dashed lines representing their midveins), one of which has an axillary seed scale (the epimatium) wrapping around the seed (stippled), which includes a single, partially hardened, seed coat (integument) surrounding the megasporangium in which the embryo resides (not shown), about ×2.

Seeds are not embedded in the seed scale in the other three families, unless the fleshy, cuplike, or entirely embracing aril that surrounds the seeds in Taxaceae is actually a highly modified seed scale, an interpretation for which there is no evidence. I return to Taxaceae later since their seed-bearing structures mostly bear no obvious resemblance to the bract and seed scale structure of other conifers. The single species of Sciadopityaceae, Japanese umbrella pine (*Sciadopitys verticillata*), has five to nine seeds per scale while Cupressaceae has the most varied conditions, with from one seed per scale in subalpine arborvitae (*Microbiota decussata*) and some junipers (*Juniperus*) to as many as 12 in some species of cypress pine (*Callitris*) and 20 in some species of cypress (*Cupressus*). Both families have small seeds with two or three small, nearly equal to highly unequal wings (none in most junipers, subalpine arborvitae, and Oriental arborvitae, *Platycladus orientalis*) along the sides of the seed body that are outgrowths of the seed coat. These wings have nothing to do with the seed scale except that they may be shaped by how the seeds are packed into the growing and maturing cone.

The total number of bracts and scales as well as the number of these that are fertile varies greatly among conifer families, genera, and species. Many Podocarpaceae, especially *Podocarpus* and *Dacrydium*, have two or three bract-scale complexes, of which only one may be fertile. Likewise, many Cupressaceae with decussate or tricussate phyllotaxis have seed cones with just two or three pairs or trios of bract-scale complexes, of which only the middle pair (or trio) may be fertile. Examples of the latter condition include the various genera of incense cedars (*Austrocedrus, Calocedrus, Libocedrus,* and *Papuacedrus*), which have a much enlarged, paddlelike middle pair of scales. Even when all scales are fertile, they may still be unequal, as in *Callitris,* in which the inner trio of large scales carries more seeds than the smaller outer trio. The same is true for the inner and outer pairs of Alaska yellow cedar (*Cupressus nootkatensis*). Most of the northern hemisphere decussate-leaved Cupressaceae (the subfamily Cupressoideae) have more than two or three pairs of scales, with the greatest numbers in some cypresses (*Cupressus*) and Fujian cedar (*Fokienia hodginsii*), with as many as seven or eight pairs.

The alternate-leaved species of Cupressaceae, formerly treated as a separate redwood family, have even more scales, with as many as about 50 in giant sequoia (*Sequoiadendron giganteum*). The only member of the redwood group with opposite leaves and cone scales, dawn redwood (*Metasequoia glyptostroboides*), has as many as 14 pairs of scales, falling between the Cupressoideae and the other redwoods in this feature. Still larger numbers are found in some Araucariaceae and most Pinaceae. The champs are probably species like bunya-bunya (*Araucaria bidwillii*), sugar pine (*Pinus lambertiana*), Norway spruce (*Picea abies*), and noble fir (*Abies procera*) with as many as 200 or more scales (so many that people are not enthusiastic about counting them). The largest number of cone scales in the Podocarpaceae is only 20–30, in Prince Albert yew (*Saxegothaea conspicua*) and in strawberry pine (*Microcachrys tetragona*), which bears its scales in offset whorls of four. Japanese umbrella pine (*Sciadopitys verticillata*, Sciadopityaceae) has about 70 scales while the only Taxaceae to exceed one functional seed scale in a cone (at a time) are plum yews (*Cephalotaxus*), with as many as eight pairs.

Corresponding only partly to these differences in numbers of scales are differences in the size and shape of the whole cone. Cones range in size from 2–3 mm in diameter in *Microbiota* and some species of junipers (*Juniperus*, for example, eastern redcedar, *J. virginiana*), to (exceptionally) 60 cm long in sugar pine (*Pinus lambertiana*) and 20 cm or more in diameter in bunya-bunya (*Araucaria bidwillii*) and Coulter pine (*P. coulteri*). The smallest seed cones are generally nearly spherical, as are many much larger ones, the largest spherical cones being some species of kauri (*Agathis*) with cones up to about 12 cm in diameter, while bald cypress (*Taxodium distichum*) lies in between, at up to 4 cm.

Much more common than spherical seed cones are egg-shaped and oblong ones. Some are nearly cylindrical in their middle portions, like Norway spruce (*Picea abies*) and European silver fir (*Abies alba*), but most taper strongly on either side of their widest point. The great majority of conifers have cones that are longer than they are wide. This is the case in the largest conifer cones just mentioned, those of sugar and Coulter pines and of bunya-bunya. Very few species have cones like those of some individuals of eastern dammar (*Agathis dammara*) that are oblate, like squished spheres, and hence wider than they are long. Just as with leaves, there is a near continuum of species linking all the extremes in seed cone shape. On top of that, the shape for any single cone often changes between the time of maturity and the release of seeds. This is due to the spreading of the seed scales, broadening the cone without usually changing the length compared to the unopened, green, but mature state. Many members of Pinaceae, including most pines, spruces, larches (*Larix*), and hemlocks (*Tsuga*) as well as some Cupressaceae, like China firs (*Cunninghamia*) and arborvitaes (*Thuja*), display this kind of change. Others in the same families, including firs and true cedars (*Cedrus*) in the Pinaceae, and cypresses (*Cupressus*) and redwood (*Sequoia sempervirens*) in the Cupressaceae, show little change in shape because the scales already stick straight out from the cone axis and separate not by curling back from the axis, but by shattering (the mentioned Pinaceae) or by separating vertically, mostly by simple shrinkage while drying (the mentioned Cupressaceae). Shattering at maturity is universal in Araucariaceae, while *Sciadopitys* (Sciadopityaceae) has scales that separate vertically, so neither of these families has seed cones that change shape dramatically after maturity. The remaining two families, Podocarpaceae and Taxaceae, generally have fleshy cones and so display little shape change except a little wrinkling, perhaps, as the fleshy parts dry. The foregoing applies only to fully mature cones. All these different cone types may well undergo dramatic changes in shape as they grow and reach maturity.

One of the factors contributing to overall cone shape is the shape of the external face of individual seed scales. For those conifers with dry or woody cones bearing wind-dispersed seeds, three features are particularly prominent: the relative width compared to the height of the face, the presence or not of some horizontal demarcation between the bract and ovuliferous scale portions of a fused cone scale, and the appearance of any protuberances that might be present. The latter, which is often referred to as an umbo if there is only one, can be quite dramatic. The large, heavy cones of Coulter pine (*Pinus coulteri*) and gray pine (*P. sabiniana*) have thick scales armed with great, ferocious, curved spines. In general, the external faces of the seed scales of Araucariaceae, Cupressaceae, Pinaceae, and Sciadopityaceae range down from there to completely smooth. In between the extremes may be found a variety of prickles, from fragile ones like those of Colorado bristlecone pine (*P. aristata*) to rather sharp ones like those of Jeffrey pine (*P. jeffreyi*). Other common features on the face of the cone scale include warts (or tubercles) in some species of *Cupressus* and *Callitris* and a tonguelike projection of various lengths representing the free tip of the bract portion of the scale in many Cupressaceae and Araucariaceae.

Pine species in general are unique among conifers in having a double raised area on the scale face (or at the tip in subsection *Strobus*, the typical white or soft pines). This double raised area, with an apophysis arising within the umbo, is due to extended rest periods during the 2-year period of cone maturation, although other genera with 2-year maturation, like true cedars (*Cedrus*) or cypresses (*Cupressus*), show no sign of comparable organization. Another seed scale feature with a restricted taxonomic distribution (in *Cryptomeria, Glyptostrobus,* and *Taxodium*) is the dissolution of the end of the scale into a series of teeth corresponding to the potential positions of seeds on the inner face. These teeth appear early in seed cone development and seem to correspond to the separate seed-bearing leaves of the axillary fertile dwarf shoots of ancestral conifers before their evolutionary fusion into a single, united seed scale. Some other Cupressaceae also start out with such teeth, but these are swamped by later growth. Given the rather advanced phylogenetic position of *Cryptomeria* and its close relatives, it is unlikely that its conspicuous teeth represent an uninterrupted retention direct from ancestral conifers in the now extinct late Paleozoic and early Mesozoic family Voltziaceae. Instead, there has probably been some developmental switch that has allowed the reemergence of the trait.

Most of these features of bracts and seed scales reach their full expression only with maturation of the cones. This can take as long as three growing seasons in a few species of pines. Other conifers complete the process within either one or two growing seasons, the exact period varying among species and genera. Even those conifers with the fastest development have seed cones that usually do not reach maturity until the autumn (in temperate climate species), so conifer reproduction is generally slower than it is in most flowering plants. The great majority of conifers complete seed cone maturation within a year of pollination, but a few genera require two, including true cedars (*Cedrus*), cypresses (*Cupressus*), and all the pines except those that take 3 years. This is one of many features that distinguish species of cypress with small seed cones from the superficially similar false-cypresses (*Chamaecyparis*), whose cones mature within a single season. The same difference separates giant sequoia (*Sequoiadendron*) from coast redwood (*Sequoia*).

Once cones and the seeds they contain mature, they may shatter, releasing both seed scales and seeds; their scales may gape, allowing seeds to pass between them; or they may persist unopened for varying lengths of time, as long as decades in some. Most cones that shatter at maturity are large cylinders held upright over the foliage, usually near the top of the crown. Such cones are borne by all Araucariaceae, and by true firs (*Abies*) and true cedars (*Cedrus*) in the Pinaceae. All these are typically wind dispersed and have seeds with large wings (or the whole cone scale acts as a wing in *Araucaria*). One exception to this pattern is bald cypresses (*Taxodium*), whose cones are near spheres only about 2–3 cm in diameter that are borne in no particular orientation and contain plump, angular, wingless seeds. Bald cypresses grow in swamps, and the seeds, which float, are released directly into (flood)waters upon which they are dispersed. Another exception is *Keteleeria,* the closest relative of the true firs, which, like *Abies,* has large, cylindrical, upright cones, but these remain intact at maturity and release their seeds by spreading of the scales. The many other members of Pinaceae and Cupressaceae whose cones remain intact at maturity and spread their scales to release seeds retain those cones for varying lengths of time. Some are shed soon after the seeds while others may be retained for several years, gradually weathering and decaying, but a good proportion lose them piecemeal over the course of the winter or nearly all at once in the spring with or soon after the emergence of new growth. For example, the conspicuous, large, brown cones that decorate Norway spruce (*Picea abies*) through the winter months, progressively releasing their seeds, are shed fairly abruptly with the onset of summer and the growth of new cones.

Conifers whose cones neither shatter nor open at maturity evince at least three completely different seed-dispersal strategies with different cone longevities. The shortest persistence after maturity are found in those species with fleshy cones, cone parts, or seeds: essentially all Taxaceae, nearly all Podocarpaceae, and junipers (*Juniperus*) in the Cupressaceae. Typically, when the cones reach maturity, as marked by their softening and becoming brightly colored (most often red but also many other colors), they are more or less promptly picked off by fruit-eating birds or mammals. It does not always happen this way in practice; for example, the cones may drop and rot in piles under large yellowwoods (*Podocarpus* and *Afrocarpus*) in South Africa. Dispersal by frugivores is most common in the tropics and in warm temperate lands while wind dispersal predominates in cold temperate and boreal climates. Exceptions to this generalization include the wind-dispersed pines and araucarians that may be found in the tropics and bird-dispersed yews (*Taxus*) in north temperate realms, among other inconsistencies.

A second type of animal dispersal, based on seed caching, results in cones that last longer on the tree than fleshy ones. Cones of these species, mostly pines, are not at all fleshy and are generally similar to those of wind-dispersed species except that the scales may remain closed or gape just a little, but not enough for the seeds to be released. These seeds are extracted by squirrels and corvids (crows, jays, and nutcrackers), which then cache them in scattered seed stores, some of which escape winter consumption and subsequently germinate. This dispersal mode has evolved repeatedly among completely unrelated pines, such as arolla pine (*Pinus cembra,* a soft pine) and Mediterranean stone pine (*P. pinea,* a hard pine) in Europe, whitebark pine (*P. albicaulis,* a soft pine) and gray pine (*P. sabiniana,* a hard pine) in North America, and chilgoza pine (*P. gerardiana*) and Chinese white pine (*P. armandii*), belonging to unrelated white pine groups, in Asia. All these have very large seeds in deep pits in the seed scales that either lack wings entirely or have short, nonfunctional ones. Anomalous within this ecological group are the numerous pinyon pines (*Pinus* subsection *Cembroides*) of Mexico and the southwestern United States, which have the typical seeds of the group and are cached by scrub jays, but their cones open widely, plainly exposing the seeds. The habitats occupied by squirrel- and corvid-dispersed species are extremely varied, including subalpine barrens, the fringes of the Mediterranean, grassy foothills, and mountain slopes bearing the first trees immediately above deserts, but all are rather open woodlands or shrublands.

In contrast to this environmental diversity, the third dispersal group, while taxonomically varied, consists of species living in environments that experience frequent fires. These are the closed-cone conifers that retain viable seed within their unopened cones for years or even decades (delayed opening is referred to as serotiny) until a forest fire, often killing the parent tree, causes the cones to open, sometimes with an audible snap, and release their seeds onto the clean ash bed from which potentially competing shrubs have been burned out. Once again, a number of unrelated pine groups display this trait, though all are hard pines, including jack pine (*Pinus banksiana*) in the North American boreal forest, knobcone pine (*P. attenuata*) in the shrubby chaparral of California, and cluster pine (*P. pinaster*) of Mediterranean woodlands. Besides pines, the closed-cone trait associated with postfire dispersal is common in cypresses (*Cupressus*) of the northern hemisphere and cypress pines (*Callitris*) of the southern, both members of the Cupressaceae. Many of these fire-associated conifers also have very thick, woody cone scales compared to their less fire-prone relatives, and these presumably provide some thermal insulation.

Among all the conifers, the seed-bearing structures of the yews (Taxaceae) are least conelike. Indeed, their single seeds at the tips of short, bracted shoots led Florin and others to regard them as belonging to a taxonomic order separate from the Coniferales, although this view is not generally accepted today. Within the family, only plum yews (*Cephalotaxus*) have an obvious though highly reduced seed cone, with decussate seed scales each bearing a pair of ovules. Many plum yew cones have just a single ovule reaching maturity, and this seed dwarfs the cone from which it emerges. The seed coat of plum yews has a complex structure, and its outer layer corresponds to the more clearly separated aril of the other genera of the family. Seeds of catkin yews (*Amentotaxus*) and nutmeg yews (*Torreya*) are most similar to those of plum yews in size and in the nearly complete enclosure of the remainder of the seed by the fairly fibrous aril. The arils of yews (*Taxus*) and white-berry yew (*Pseudotaxus chienii*), in contrast, are very fleshy, open cups in which the seeds sit, while those of New Caledonian yew (*Austrotaxus spicata*) are somewhat intermediate between these two extremes. An aril is also described for some Podocarpaceae, such as celery pines (*Phyllocladus*), but in the latter instance, at least, the supposed aril might actually be a highly modified epimatium (an enclosing seed scale) rather than an outgrowth of the stalk immediately beneath the seed as with true arils. In some yews, a leaf axil that has once borne and dispersed a seed continues to do so for several years via a persistent stubby shoot. This shoot has been interpreted as a transformed seed cone that matures its unrecognizably reduced seed scales at a rate of one per year. This is the most promising attempt to date to relate the single terminal seeds of yews to ordinary conifer cones.

CONIFER TRAITS AS ADAPTATIONS

The kinds of morphological features just described and many others that appear throughout the genus and species descriptions have been honed by natural selection in response to present or past environmental conditions. Some features are obviously presently adaptive, like the serotinous cones of many conifers in fire-prone habitats. It has even been shown that the frequency of serotiny in populations of pitch pine (*Pinus rigida*) in the New Jersey Pine Barrens increases with the frequency at which they are exposed to fires, so this feature is under active selection. We should not, however, expect all traits of each species to represent adaptations to current conditions. Instead, many, perhaps most, conspicuous morphological differences between species and genera represent inheritances from ancestors, recent or remote. Each trait has a different time depth and arose under different environmental conditions. Thus each species displays a mosaic of characteristics with varying degrees of tightness of fit to its habitat. Few if any of the traits are strongly disadvantageous because they would be subject to strong negative selection, but many may be relatively neutral and thus not confer any particular fitness advantage or disadvantage. As well, the environments experienced by different populations and individuals are never uniform, even for a species like Wollemi pine (*Wollemia nobilis*), which consists of what is essentially a single population confined to just one river canyon. Selection acts differently on the members of a species scattered through its present geographic range (which is itself dynamic, as evidenced by much range change, even in the geologically recent past), so different genotypes are favored in different locations. If different portions of a species range are physically separated from one another or show consistent environmental differences, morphologically different races may evolve. If these races are reasonably conspicuous and consistent in their differences, they may be recognized formally as subspecies or varieties within a species (and some taxonomists may separate them as species in their own right). More often, however, the great store of morphological variation within a species shows no consistent geographic pattern, and so it simply contributes to the variation in shapes and sizes of organs encompassed by the descriptions. The dynamic nature of conifer variation also means that the descriptions here are unlikely to include the true extremes of measurements and shapes for a species, though the majority of individuals and organs should fit comfortably within the stated ranges.

CHAPTER 5

PALEOBOTANY AND EVOLUTION

Fossil conifers are found all over the Earth, from the arctic to the antarctic, in many places that house no conifers today as well as in their present regions of abundance and diversity. The record consists of a wide range of different kinds of fossils, including pollen grains, wood, pollen and seed cones, and foliage preserved in three rather different manners. Impressions show shapes of leaves and various surface patterns but preserve no original material. Compressions preserve a carbonized film, and this frequently includes the resistant cuticle and thus reveals the detailed structure of the epidermis, which has been shown to differ considerably among different conifers. The cuticles are separated from the rock matrix and mounted on microscope slides for examination with light microscopy or on stubs for scanning electron microscopy. Finally, petrified material can preserve the internal structure of the fossilized organs, sometimes including fine cellular details. These can be revealed either by grinding slices of the fossil so thinly that they transmit light or by etching the rock with acid and trapping the resistant cell walls that stick up in a thin film of transparent plastic. Unfortunately, the different organs and types of preservation require different environmental and geological conditions, so it is extremely rare to have a complete picture of the external morphology and internal anatomy of all the organs of a fossil conifer. The relatively few whole plant reconstructions, however, are gradually increasing in number as paleobotanists painstakingly piece together the scattered bits. Even without these most completely known fossils, however, numerous individual organs are available and open to interpretation so that a fairly coherent picture of the origin and evolution of conifers has emerged. New fossils are discovered and described all the time, continuing to refine and enrich this picture.

The earliest known conifers date from the late Carboniferous period, some 300 million years ago, the period of the great coal age swamp forests, but they were not part of these forests. Growing on uplands surrounding the swamps, their remains rarely entered the fossil record, and when they did, they were often in the form of charcoalized fragments that had washed down after forest fires. Nothing is known of their reproductive structures, but their foliage, like that of most later ancient conifers, was reminiscent of shoots of Norfolk Island pine (*Araucaria heterophylla*) and other similar araucarians. Conifers became much more abundant in the drier Permian period that followed the Carboniferous at the end of the Paleozoic era. The great coal swamps, with their forests of tree-sized horsetails and club mosses, were replaced by dry-ground forests, and these were dominated, in part, by conifers. None of these belonged to any of the six still extant families but to extinct families, like Majonicaceae, Emporiaceae, and Utrechtiaceae, in which the dwarf reproductive shoots in the axils of the seed cone bracts still consisted of separate individual sterile and seed-bearing leaves and were not yet organized into definite, unified seed scales. Following the catastrophic Permian extinctions and the beginning of the Mesozoic era of reptiles, conifers continued to diversify during the Triassic period, showing increased variety in foliage and reproductive structures. By about 225 million years ago, late in the Triassic, fossils ascribed by some authors to several extant families—Araucariaceae, Cupressaceae, and Podocarpaceae—are present. Others are considered to be precursors to Pinaceae, if not members of that family. It is not at all certain that any of these early attributions to extant families are correct, but during the course of the following Jurassic period, fossils belonging to Araucariaceae and Taxaceae become part of the conifer floras that also included representatives of now extinct families, such as Cheirolepidiaceae and Miroviaceae. Conifers, in fact, were dominants of these early and mid-Mesozoic forests, but the first incontrovertible evidence of fully characteristic members (referred to as the crown groups) of the remaining extant families—Cupressaceae, Pinaceae, Podocarpaceae, and Sciadopityaceae—are only recorded in the last period of the Mesozoic, the Cretaceous, which began about 145 million years ago. Even then, very few extant genera are recorded from the Cretaceous, and most of those that are, like *Pinus,* first appeared late in that period. The vast majority of the living genera for which we have a fossil record are first identified in sediments dating from the early or mid-Tertiary period, after the catastrophic close of the Mesozoic 65 million years ago. Despite the much greater time depth for conifers as a whole than for flowering plants, the extant conifer genera are thus comparable in age to many familiar genera of broad-leaved trees, like poplars (*Populus*), oaks (*Quercus*), maples (*Acer*), and birches (*Betula*). The records of individual conifer genera are briefly reviewed in the discussion for each genus, but a few general remarks are appropriate here. There is no particular association between the age of a genus and the number of species it has today. The oldest extant genera, for example, those dating from the Cretaceous, run the gamut from presently monotypic (single species) ones, like *Sciadopitys* and *Sequoia,* to the biggest of all, *Pinus,* with about 100 species. On the other hand, relatively recent genera, while they might be expected to be small, like the monotypic, narrowly endemic (very localized) *Microbiota,* can also be quite large, like *Juniperus,* with its more that 50 species.

Much more captivating to the imagination is the fact that some genera that today are narrowly endemic monotypes, or even on the verge of extinction, were much more widely distributed in the past, and some were dominants over vast areas. Throughout the early and mid- Tertiary, volcanic activity and other tectonic processes dammed streams, creating vast swamps over many areas of the northern hemisphere. These swamps, in contrast to those of the Paleozoic coal age, were dominated by conifers, particularly three genera of Cupressaceae with much more restricted ranges today. There are now two species of bald cypress (*Taxodium*) in southern North America, from the southeastern United States to northern Guatemala, while the closely related Chinese swamp cypress (*Glyptostrobus*) and the more distant dawn redwood (*Metasequoia*) are monotypic genera with even more limited natural distributions in historic times, the former in swampy parts of southern China and adjacent Vietnam, particularly near the coast, and the latter in a local region of central China, where it grows in ravines and along the floodplains of broader stream valleys. A few other genera that were widespread during the Tertiary but are much more restricted today include *Cathaya* and *Pseudolarix* of the Pinaceae, *Tetraclinis* of the Cupressaceae, and notably, *Wollemia* of the Araucariaceae, all of which are monotypic today. Even some genera with a number of species today were still more widely distributed in the past, including *Araucaria,* which extended into the northern hemisphere as well as more widely in the southern during the Mesozoic. A fairly common pattern among northern hemisphere genera with species in North America and eastern Asia today was to have been present in Europe as well during the Tertiary and to have been eliminated from there as the climate got colder late in the period or during the upheavals of the Pleistocene epoch ice ages. Arborvitaes (*Thuja*) and nutmeg yews (*Torreya*) are examples of this kind of pattern, while other genera also disappeared from North America during the Tertiary, like plum yews (*Cephalotaxus*).

The four geographic components just considered, eastern and western North America, Europe, and eastern Asia, together make up a holarctic distribution today or an arctotertiary distribution in the past. Most of the conifer genera that had a full arctotertiary distribution in the past became extinct in one or more of the four regions so that only a few genera retain a holarctic distribution today, including true firs (*Abies*), larches (*Larix*), spruces (*Picea*), and pines (*Pinus*), all of the Pinaceae, yews (*Taxus*) of the Taxaceae, and junipers (*Juniperus*) of the Cupressaceae. Note that four of these are among the five largest conifer genera. There appears to have been some safety in species numbers during Tertiary and Quaternary extinctions as those genera that experienced loss of one or more regions were all monotypic or had just a handful of species. A pattern similar to the remnant arctotertiary distributions of the northern hemisphere may be found with such southern temperate genera as *Athrotaxis, Austrocedrus, Dacrycarpus, Fitzroya, Lepidothamnus,* and *Libocedrus,* each of which was more widespread in the past, but the corresponding patterns there have been much less extensively worked out than those in the north. There are many reasons for this lower level of knowledge, including fewer paleobotanists having started work later on the fewer relevant deposits in the smaller southern landmasses. A little bit of it may also be due to a greater focus on an earlier southern hemisphere distribution pattern that resulted from the concentration of all the southern lands, including Antarctica, into the ancient continent of Gondwana, which housed forests during the Permian dominated by another extinct gymnosperm group unrelated to conifers, the glossopterids, with large, tongue-shaped leaves. In between the north and south temperate regions, the fossil record of conifers in the modern tropical zones is so poorly known that we cannot really discern anything useful about historical patterns of geographic distribution there during the Tertiary.

This history of regional extinctions in both the northern and southern hemispheres during the Tertiary and Quaternary periods might give an impression of decline in the conifer flora through the course of the Cenozoic era, but this would be somewhat misleading. While it is true that most regions now have lower conifer generic diversity than they had during the Tertiary, global diversity remained essentially unchanged since genera that disappeared from one place usually persisted elsewhere. There are only a few known cases of global extinction of a widespread genus during the Tertiary, including *Cunninghamiostrobus,* a genus related to *Cunninghamia,* and *Pityostrobus,* a "garbage can" morphogenus containing species not assignable to any of the extant genera of Pinaceae. In fact, taken at face value, the Tertiary record of conifers is actually one of steady progressive generic diversification during the earlier Paleogene and slowing down during the succeeding Neogene. About a dozen of each of the extant genera are first known from each of the Cretaceous, Paleocene, Eocene, and Oligocene epochs (the latter three making up the Paleogene), but only two in the Miocene and one in the Pliocene (together making up the Neogene). Fifteen of the extant genera have no known fossil record, but they are likely to be spread across the same time distribution, the way undecided voters often are, rather than being clumped in any one epoch. There are also many genera described from Paleogene deposits of each epoch that are now extinct, but these are almost all based on very little material from just a few places and hence of uncertain status. A number of these are members of the Podocarpaceae that are based on scraps of foliage of restricted distribution in space and time so that one cannot be sure that they are not unusual species of still extant genera. None of these poorly known genera was common or widely distributed. Likewise, the widely distributed genera that became extirpated in most regions were essentially either all monotypic or with few species. No extant genus that is known to have had numerous species during the Tertiary is now a locally restricted monotypic endemic.

Conifers were an important component of temperate ecosystems throughout the Tertiary, and they remain so today. The boreal forest, which is dominated by a few species of conifers, is the single largest forest formation on Earth today. The evolutionary roots of the conifers may lie deep in the past, but they are hardly a spent force. While suffering significant losses in global dominance after the rise of angiosperms during the Cretaceous, they have since held their own during the last 60 million years, and they continued to evolve new genera during that interval, just as have the flowering plants. The overall pattern can be viewed in an evolutionary context in terms of adaptation to the new environmental conditions that are always emerging in the world. Because reproductive structures (except pollen grains) are more transient and less numerous than vegetative structures, the history of the evolutionary diversification of pollination and seed-dispersal mechanisms is less clear than that of photosynthetic specializations. The forms of conifer foliage in various fossil floras often helps in interpreting the environments in which those conifers grew, especially in concert with clues from the rocks themselves, including evidence on the nature of the soils (paleosols) that supported them. While the range of known fossil conifer foliage is vast, understanding how the different foliage forms interrelate evolutionarily lags far behind. In particular, the steps leading to the more extreme foliage forms, such as the broad multiveined leaves of *Nageia* or the leaflike branching systems (phylloclades) of *Phyllocladus,* is worthy of further investigation using both fossils and the most powerful molecular tools available to developmental biologists. There is a great deal to learn still about conifers of the past and their evolution.

Plate 1. Trees of three of the five most widely distributed northern hemisphere conifer genera (*Larix, Picea,* and *Pinus*), growing side by side.

Plate 2. Green mature seed cone of balsam fir (*Abies balsamea*). Photo Ronald Dengler.

Plate 3. Shattering seed cones of balsam fir (*Abies balsamea*),

Plate 4. White fir (*Abies concolor*), like all *Abies* species, has seed cones concentrated at the top of the crown.

Plate 5. Resin-coated mature seed cones of Fraser fir (*Abies fraseri*) with protruding (exserted) bracts.

Plate 6. Subalpine forest of subalpine fir (*Abies lasiocarpa*) and Engelmann spruce (*Picea engelmannii*).

Plate 7. Mature seed cones and foliage of Outeniqua yellowwood (*Afrocarpus falcatus*).

Plate 8. Fine specimen of serpentine forest kauri (*Agathis lanceolata*). Photo Spencer Barrett.

Plate 9. Fresh young leaves of Melanesian kauri (*Agathis macrophylla*) showing opposite leaf arrangement. Photo Spencer Barrett.

Plate 10. Shrubby maquis minière dominated by conifers on serpentine soil in New Caledonia. Photo Spencer Barrett.

Plate 11. Low forest of dwarf kauri (*Agathis ovata*) overtopping maquis minière on a ridge. Photo Spencer Barrett.

Plate 12. Pock marked bark of smooth-bark kauri (*Agathis robusta*). Photo Spencer Barrett.

Plate 13. New growth of two Cook pines (*Araucaria columnaris*). Photo Spencer Barrett.

Plate 14. Seaside stand of Cook pine (*Araucaria columnaris*) in nature. Photo Spencer Barrett.

Plate 15. Juvenile foliage of chandelier araucaria (*Araucaria muelleri*). Photo Spencer Barrett.

Plate 16. Pollen cones and foliage of Chilean cedar (*Austrocedrus chilensis*), showing white stomatal zones.

Plate 17. Threadlike foliage of winged chandelier cypress (*Callitris neocaledonica*). Photo Spencer Barrett.

Plate 18. Montane forest of incense cedar (*Calocedrus decurrens*), white fir (*Abies concolor*), and red fir (*A. magnifica*) in the Sierra Nevada of California.

Plate 19. Seed cones of Atlas cedar (*Cedrus libani* subsp. *atlantica*).

Plate 20. Tiny seed cones with protruding seeds of Japanese plum yew (*Cephalotaxus harringtonii*).

Plate 21. Open mature seed cones and foliage of Port Orford cedar (*Chamaecyparis lawsoniana*).

Plate 22. Fresh foliage of a silvery variegated cultivar of sawara cypress (*Chamaecyparis pisifera*).

Plate 23. Green mature seed cones of sawara cypress (*Chamaecyparis pisifera* 'Filifera').

Plate 24. Forest of sugi cedar (*Cryptomeria japonica*) in eastern China. Photo Spencer Barrett.

Plate 25. Pollen and seed cones of China fir (*Cunninghamia lanceolata*).

Plate 26. Foliage of candelabra rimu (*Dacrydium araucarioides*). Photo Spencer Barrett.

Plate 27. Mature seed cones and foliage of Chilean alerce (*Fitzroya cupressoides*).

Plate 28. Juxtaposed juvenile and adult foliage with mature seed cones of Chinese juniper (*Juniperus chinensis*).

Plate 29. The aptly named creeping juniper (*Juniperus horizontalis*), one of the lowest-growing conifers. Photo Ronald Dengler.

Plate 30. Mature seed cones of European larch (*Larix decidua*) after shedding their seeds.

Plate 31. Characteristic fluted trunk of dawn redwood (*Metasequoia glyptostroboides*).

Plate 32. Bright green deciduous foliage of dawn redwood (*Metasequoia glyptostroboides*).

Plate 33. Immature seed cones and young foliage of dawn redwood (*Metasequoia glyptostroboides*).

Plate 35. Pollen cones and foliage of chandelier cypress (*Neocallitropsis pancheri*). Photo Spencer Barrett.

Plate 34. Trees of chandelier cypress (*Neocallitropsis pancheri*) in nature. Photo Spencer Barrett.

Plate 36. Leaflike twigs (phylloclades) of tanekaha (*Phyllocladus trichomanoides*). Photo Ronald Dengler.

Plate 38. Subalpine forest of Engelmann spruce (*Picea engelmannii*), showing the tree line.

Plate 37. Mature seed cone of Norway spruce (*Picea abies*). Photo Ronald Dengler.

Plate 39. Bursting buds of white spruce (*Picea glauca*) still covering the emerging new shoots.

Plate 40. Mature seed cones of white spruce (*Picea glauca*) after seed shed persisting into winter. Photo Ronald Dengler.

Plate 41. Green mature seed cones and foliage of black spruce (*Picea mariana*). Photo Ronald Dengler.

Plate 42. Seedlings of Jeffrey pine (*Pinus jeffreyi*).

Plate 45. Fresh and decayed seed cones of singleleaf pinyon (*Pinus monophylla*). Photo Ronald Dengler.

Plate 44. Twig of singleleaf pinyon (*Pinus monophylla*), showing the bundles with just one needle. Photo Ronald Dengler.

Plate 43. Sparse subalpine stand of Great Basin bristlecone pine (*Pinus longaeva*) in the very arid White Mountains of eastern California. Photo Ronald Dengler.

Plate 46. Grass stage of longleaf pine (*Pinus palustris*). Photo Ronald Dengler.

Plate 47. Older saplings of longleaf pine (*Pinus palustris*).

Plate 49. Pollen cones of red pine (*Pinus resinosa*) at the base of a new shoot before it elongates. Photo Ronald Dengler.

Plate 50. Characteristic bark of chir pine (*Pinus roxburghii*).

Plate 48. Mixed montane conifer forest with ponderosa pine (*Pinus ponderosa*).

Plate 51. Seedling of eastern white pine (*Pinus strobus*) showing the transition from singly attached seedling needles to bundles of five adult needles. Photo Ronald Dengler.

Plate 52. Needle bundles of eastern white pine (*Pinus strobus*) as the scale leaves of the sheath are shed. Photo Ronald Dengler.

Plate 53. First-year seed cone of Scots pine (*Pinus sylvestris*) shortly after pollination.

Plate 54. Mature seed cones of Scots pine (*Pinus sylvestris*) after opening and seed release.

Plate 55. Confused sexuality of Japanese black pine (*Pinus thunbergii*) at the time of pollination, with some seed cones scattered among the pollen cones at the base of the shoots as well as in their normal position at the tip.

Plate 56. Pollen cones and foliage of Buddhist pine (*Podocarpus macrophyllus*).

Plate 57. Immature seed cones of Buddhist pine (*Podocarpus macrophyllus*) shortly after pollination.

Plate 58. Ripe seed cones and foliage of miro (*Prumnopitys ferruginea*). Photo Ronald Dengler.

Plate 59. Mature and immature seed cones, pollen cones, and foliage of bigcone Douglas fir (*Pseudotsuga macrocarpa*). Photo Ronald Dengler.

Plate 62. Mature giant sequoia (*Sequoiadendron giganteum*). Photo Ronald Dengler.

Plate 60. Green mature seed cones of Douglas fir (*Pseudotsuga menziesii*).

Plate 61. Waxy new growth of redwood (*Sequoia sempervirens*). Photo Ronald Dengler.

Plate 65. Seeds of foundation yew (*Taxus ×media*) with mature and immature arils. Photo Ronald Dengler.

Plate 63. Pollen cones, green mature seed cones, and foliage of bald cypress (*Taxodium distichum*). Photo Ronald Dengler.

Plate 64. Pollen cones of foundation yew (*Taxus ×media*). Photo Ronald Dengler.

Plate 66. Mature seeds of California nutmeg (*Torreya californica*). Photo Ronald Dengler.

Plate 67. Foliage of sapling Wollemi pine (*Wollemia nobilis*). Photo Spencer Barrett.

CHAPTER

6

CONIFER IDENTIFICATION

Conifer species within many genera, and even some belonging to different genera, can be difficult to distinguish, especially during the months when pollen and seed cones are absent, immature, or inaccessible. Here, I provide some hints to aid identification and introduce the identification guides used in the book.

EVERGREEN OR DECIDUOUS?

The life span of leaves of conifers ranges from less than 6 months (for high-latitude or subalpine *Larix* species) to 15 years or more (for *Pinus longaeva* and several species of *Araucaria*). Most species are evergreen and have leaves that last for 2–5 years before they lose their chlorophyll and fall off either individually or as whole discarded branchlets. Only a handful of species belonging to five genera (*Glyptostrobus, Larix, Metasequoia, Pseudolarix,* and *Taxodium*) are deciduous, with leaves that are shed annually with the onset of the unfavorable season before new leaves expand at the beginning of the following growing season. A few evergreen species, such as the North American *Pinus strobus,* also shed all their leaves from one whole growing season together during a brief seasonal leaf fall, but this occurs only after one or more recent annual leaf crops have grown out so that the tree is never bare. Thus only deciduous species lack leaves during their dormant season. During the growing season, however, deciduous and evergreen species alike are covered with foliage, so how can you tell which species are deciduous?

There are two basic strategies. One is to recognize foliage characteristics common to all the deciduous species that can be used to distinguish them from evergreen species, without reference to their taxonomic identity. The other is to know specific traits of the five deciduous genera and how they may be distinguished from the most similar evergreens.

In following the first strategy, there are two basic clues. Most reliable is to recognize that all leaves present are from the current growing season. This can be done if growth increments on the twigs can be identified, whether by their separation by remnant rings of bud scales at the base of each, by branching at its tip, or by shorter needles at its beginning, end, or both. Since all the deciduous species have more or less needlelike leaves, the fact that conifers with scalelike leaves generally lack all three of these features, so that the growth increments are extremely difficult to recognize, is irrelevant, as scale-leaved conifers are all evergreen. This first clue, possession of just a single-age population of needles, can be supplemented by the second, concerning the texture of the needles. Leaves of deciduous species are much softer and more delicate than those of the temperate evergreen species with which they grow. That is because they lack the various tough, strengthening tissues found in leaves of neighboring evergreens, which must, after all, survive and remain functional through freezing temperatures during at least one winter, and usually more. A practical manifestation of this softness is that when twigs are cut and left out of water, the needles on those of deciduous trees shrivel and curl up very quickly, within a day, while those of evergreen species retain their shape for many days, if not indefinitely, as they dry. Taken together, these two clues provide considerable confidence that a given conifer is deciduous.

The other strategy, that of learning to recognize the five deciduous conifer genera, is helped by the fact that they fall into two groups. Larches (*Larix*) and the golden larch (*Pseudolarix amabilis*), both members of the pine family (Pinaceae), differ from almost all other members of their family and, indeed, from all other conifers in having their needles crowded all around the tips of long-lived spur shoots (shoots with essentially no stem showing in the leafy portion because of a lack of elongation between successive leaves). The only other genus with this characteristic is the evergreen *Cedrus* (true cedars), which, as noted, has harder needles that do not shrivel up after twigs are cut. Some features of the seed and pollen cones that distinguish these three genera may be found in the identification guide to genera of Pinaceae, Group A. The other three deciduous genera all belong to the Cupressaceae and typically have needles attached distichously along either side of deciduous short shoots (with internodal elongation, thus not spur shoots) with which they fall. New short shoots are produced in the same positions on the permanent branchlets each year for several to many years, depending on the degree of shading by the developing crown. Dawn redwood (*Metasequoia glyptostroboides*) is unique among all conifers in having decussate (crisscross pairs of) needles secondarily brought into a distichous arrangement by twisting of the branchlets between successive leaf pairs, alternately clockwise and counterclockwise. Chinese swamp cypress (*Glyptostrobus pensilis*) and American bald cypresses (*Taxodium*) both have spirally attached needles, which in at least some twigs also stick out all around the short shoots rather than being distichously arranged. These two genera are closely related and can only be fully reliably distinguished by aspects of their seed cones, as detailed in the identification guide to Cupressaceae, Groups A and B. Their foliage is reminiscent of that of many other conifers, including other Cupressaceae (redwood, *Sequoia sempervirens*) and some Taxaceae (especially yews, *Taxus*) and Podocarpaceae (such as species of *Dacrycarpus* and *Falcatifolium* as well as some of *Podocarpus*), and even to a certain extent some Pinaceae (especially some true firs, *Abies*). They differ, of course, in being deciduous, with much softer, weaker needles than any of the other mentioned conifers, as well as differing completely in their seed cones and in the arrangement of their pollen cones on catkinlike, hanging, specialized reproductive shoots. Montezuma bald cypress (*Taxodium mucronatum*), from Mexico, can be barely evergreen, with 1-year-old shoots cast off with the emergence of the new ones, but this never happens with the northern *T. distichum*. All told, the five deciduous conifer genera are among the easiest ones to recognize, in part because of their deciduousness.

DO SEED CONES MATURE IN ONE, TWO, OR THREE GROWING SEASONS?

Rates of seed maturation in conifers are slower than those of many flowering plants. Still, the majority of conifer genera have seed cones that mature within a year of pollination. About a third of the genera, including *Agathis, Cedrus, Cephalotaxus, Cupressus,* and *Microcachrys,* have species that require two growing seasons for seed and cone maturation. A few species of *Araucaria, Pinus,* and a few other genera even require three growing seasons. Identifying species that require 2 or 3 years for cone maturation can be simple: if two or three distinct size categories of mature seed cones are present on a single tree during the summer to near the end of the growing season, then they take 2 or 3 years to reach maturity, respectively. The different-size cones in *Pinus* are also obviously attached to growth increments of the present year and of the previous year or two, but this is not at all obvious in genera like *Cedrus* and *Cupressus,* so the only real clue is usually cone size classes. Even recognizing this may be difficult sometimes because an individual tree may not produce a crop of seed cones every year. In these circumstances, you can only hope that there are at least a few cones present, even in off years. On the other hand, difficulties can occur when species retain their mature seed cones for many years after maturation, as often happens in species of *Callitris, Cupressus,* or *Pinus* from fire-prone habitats. Here, the serotinous cones of these trees only open and disperse seeds onto a clean, ashy seed bed after fires burn off potential competitors and the heat stimulates their cone opening. Nonetheless, this is really not much of a difficulty in these cases because once the seed cones mature they do not subsequently enlarge much or at all, so they do not contribute confusing additional size categories, despite representing additional generations of years of origin. As a result of serotinous retention, trees like jack pine (*P. banksiana*) or knobcone pine (*P. attenuata*) can have a dozen or more successive whorls of seed cones, looking progressively grayer and more weathered as you go back along branches from the growth increment of the current year. For the most part, recognition of distinct size classes of seed cones is relatively straightforward and can be useful sometimes in distinguishing members of different genera. For instance, maturation in 2 years is one of many features that links Alaska yellow cedar (*Cupressus nootkatensis*) with other species of *Cupressus* rather than with *Chamaecyparis* in which it has long been placed by most authors and whose species it superficially more closely resembles.

SEX DISTRIBUTION IN CONIFERS

No conifers regularly have the hermaphroditic occurrence of both pollen- and seed-bearing organs in a single cone the way that both stamens and pistils are present in most angiosperm flowers. Instead, pollen and seed scales are parts of separate cones. Pollen and seed cones co-occur in single individuals of monoecious species, usually in different parts of the crown. In dioecious species, by contrast, the pollen and seed cones are borne on different unisexual individuals. About two-thirds of the genera are monoecious and one-third dioecious.

Sex distributions are rarely completely fixed developmentally. Anomalous conditions may result from unusual weather conditions, wounding, or sometimes just as an infrequent but normal variation. Thus individuals belonging to genera or species described as dioecious will sometimes have a few pollen cones on female individuals and, more rarely, a few seed cones on male individuals. Apparent unisexual individuals in monoecious species are more common because seed cones are often not produced in noticeable abundance every year, and pollen cones are transient compared to seed cones. It is difficult, then, to know for sure whether individuals of a normally monoecious species with just one type of cone present in a given season are truly genetically unisexual or just responding to particular environmental circumstances. Such environmental perturbations may well be responsible for two other, even rarer, anomalies in sex distribution. A familiar example of the first type of anomaly, the occurrence of cones of one sex in positions normally associated with the other sex, may be found in some cultivated individuals of Japanese black pine (*Pinus thunbergii*). In these individuals, baby seed cones may be found in crowded rings at the base of the new growth increment where pollen cones usually occur rather than in a sparser ring of three or four at its tip. The other rare anomaly involves the occurrence of pollen and seed organs in the same cone. This is much rarer than the corresponding phenomenon in flowering plants because of a fundamental difference in the structure of pollen and seed cones: pollen scales are equivalent to leaves while seed scales are highly compressed whole shoots. This means that one type of organ cannot simply take the place of the other. A seed scale corresponds to a whole pollen cone. Thus, in examples of bisexual cones that I have seen in China fir (*Cunninghamia lanceolata*), for instance, the rather normal-looking seed scales of the upper portion of the cone are replaced at the very base by similar bracts with complete pollen cones sticking out between them.

Despite the occurrence of such anomalies in sex distributions, the vast majority of conifer species are clearly normally either monoecious or dioecious. There are strong genetic and ecological consequences to these two different types of sex distributions. The primary genetic consequence is that self-pollination is impossible in dioecious species, where each individual is exclusively either male or female. Although monoecious species typically have a physical separation of pollen and seed cones, with seed cones in the upper reaches of the crown and pollen cones beneath, wind can still carry pollen up within the crown and effect self-pollination. Often there is strong inbreeding depression in these monoecious species, which has been measured in Douglas fir (*Pseudotsuga menziesii*) for instance, that leads to a lower rate of survival and growth in the selfed offspring versus outcrossed ones from the same mother tree. Those selfed offspring that are fit enough to survive to reproductive maturity will have many more homozygous genes than outcrossed members of the population, which can have some effect on future generations and the genetic structure of the population in comparison to dioecious species. Genetic tests in Douglas fir and other economically important monoecious conifers show that the proportion of homozygosity in a population, indicative of selfing, decreases from seeds to saplings to mature trees, reflecting the fitness advantage of heterozygotes over homozygote.

Considering reproductive ecology, there is a very strong association in conifers between the type of sex distribution and the mode of seed dispersal. Those conifers (all Taxaceae, all Podocarpaceae except *Saxegothaea,* as well as *Juniperus* in the Cupressaceae) whose seeds are accompanied by fleshy structures and are dispersed by frugivorous (fruit-eating) birds and mammals are overwhelmingly dioecious. There are exceptions, of course, but most of those exceptions are of the type just described in which a few individuals may have some cones of the other sex. There are only a handful of frugivore-dispersed species that are normally monoecious. This includes ground hemlock (*Taxus canadensis*), three species of *Phyllocladus* (*P. alpinus, P. aspleniifolius,* and *P. trichomanoides*), and seven species of *Juniperus* (with four others having a significant fraction of bisexual individuals). In contrast, those conifers with dry seeds and cones lacking any fleshy structures are overwhelmingly monoecious. This is true whether the seeds are dispersed by wind, by water, or even by seed-eating animals (not frugivores), like squirrels or nutcrackers (relatives of jays). As with frugivore-dispersed genera, there are exceptions to the association, but again most of these involve anomalous individuals. A few species buck the trend, however, and oddly enough, most of these occur in southern Chile, where monkey puzzle tree (*Araucaria araucana*), Chilean cedar (*Austrocedrus chilensis*), Chilean alerce (*Fitzroya cupressoides*), and Fuegian incense cedar (*Libocedrus uvifera*) are all predominantly or exclusively dioecious, as are the frugivore-dispersed members of the Podocarpaceae with which they grow. This regional anomaly is unexplained, but the prevailing explanation for the otherwise tight association between mode of dispersal and sex distribution relates to trade-offs between the energy apportioned to seed dispersal and the ability to maintain both sexes in a single individual. This theory has not been tested experimentally but remains the most convincing explanation for the association, despite the strong family connection to dispersal modes. This is because all three groups with fleshy disseminules are most closely related to different dry-coned and dry-seeded predominantly monoecious groups. Podocarpaceae are closest to Araucariaceae, Taxaceae to Cupressaceae and Sciadopityaceae, and *Juniperus* is deeply embedded in the Cupressaceae near *Cupressus*. Thus they have all acquired their frugivore dispersal mode independently of one another and represent separate instances of dioecy accompanying fleshy disseminules. Nonetheless, a more experimental approach to this interesting ecological and genetic association would be welcome.

HOW TO USE THE IDENTIFICATION GUIDES

The numerous identification guides in this book focus on mature individuals. Even when reproductive features are not used, the guides depend on contrasts of character measurements that may not be applicable to juvenile foliage. Nor are they appropriate for many named cultivars, which typically have foliage features selected for their departures from their species' norms. Granted these caveats, the guides should allow readers to put a name on any reproductively mature conifer in the wild or as unselected individuals in cultivation. Identifications should be carefully checked with any relevant illustrations and a detailed reading of the appropriate description(s), which give full ranges of values for the characters included in the guide as well as for many other characters that may clinch the assignment.

Included are identification guides that pick out conifers among the other groups of seed-bearing plants, assign an unknown conifer to one of the six extant plant families of conifers, distinguish the 67 individual genera within families, and identify the species within each genus that has two or more species. Thus a conifer may be identified starting with the finest taxonomic resolution that can apply to it a priori: knowing it is a seed plant but not being sure if it is a conifer, knowing that it is a conifer but not to which family it belongs, knowing its family but not the genus, or recognizing it as belonging to a familiar genus but not a familiar species.

Identification guides at these different levels are all constructed the same way. They present, for groups of as many as about 10 species (or other taxa) at a time (but usually fewer), a condensed tabulation with more or less the fewest characters that definitively distinguish that group of species.

The distinguishing characters are described first, then their values for all the species are listed in the same order. Rather than accurately describing the actual values for each character found in each species, the overall range for all of the species is divided into segments defined by particular breakpoints. Assignment of the species to particular segments for the character usually does not take extreme values into account. Instead, these values usually represent the average leaf (or other organ) or the values found in the majority of leaves, although sometimes the guides call for the largest organs to be found in the specimen. In most guides, characters of leaves or twigs are given first, followed by characters of pollen or seed cones, or both, if necessary, as they often are, to distinguish similar species. The species in the guide are ordered by their values for the first character and then, insofar as possible, by their values on each of the remaining characters, in order. This makes direct comparison of the species in the guide much easier.

For the 10 genera with more then 10 species, the genus is first divided into arbitrary groups on the basis of one or two characters, then the species in each are distinguished by whatever characters and break points are most appropriate for the species in that group. The characters used will thus vary from group to group within the larger genera. These groups are generally just ones of convenience and carry no implication that the species within them are more closely related to each other than they are to species in other groups. In fact, some variable species will be found in more than one group because different individuals span the break points for group assignment.

CHAPTER

7

SEED PLANTS AND CONIFER FAMILIES

In addition to the true conifers treated in this book, living seed plants include the flowering plants, cycads, gnetophytes and *Ginkgo biloba.*

Identification Guide to the Major Groups of Seed Plants

The five living groups of seed plants may be distinguished by whether the seeds are sealed within a carpel or borne naked in a cone or at the tip of a shoot, with at least their tip exposed at the time of pollination; by whether the pollen-producing structures are simple (the main axis bearing pollen scales directly) or compound (the main axis bearing bracts with axillary dwarf reproductive shoots); by whether the seed-producing structures are simple or compound; and by whether the leaves are simple or compound:

cycads (10–11 genera, 150–300 species), seeds in cone (or open whorl of sporophylls), pollen-producing structures simple, seed-producing structures simple, leaves pinnately compound

ginkgo (one genus, one species), seeds at tip of a shoot, pollen-producing structures simple, seed-producing structures simple, leaves simple

conifers (67 genera, 546 species), seeds in cone or at tip of a shoot, pollen-producing structures simple, seed-producing structures compound (or appearing simple), leaves simple

gnetophytes (three or four genera, 90–105 species), seeds in cone, pollen-producing structures compound, seed-producing structures compound, leaves simple

flowering plants (ca. 13,000 genera, ca. 250,000 species), seeds in carpels, pollen-producing structures simple, seed-producing structures simple, leaves simple or variously compound. In addition to these features, most flowering plants (and only they) have both pollen- and seed-producing organs as part of the same reproductive structures: their flowers. Although this hermaphroditism occurs in the majority of flowering plants, there are still far more flowering plant species with unisexual flowers than the total number of species of the other four seed plant groups combined.

Identification Guide to Conifer Families

The six families of conifers each vary so much in leaf form and other vegetative characteristics that they are most readily recognized by their seed-bearing structures, although these are quite varied also. Thus it is difficult to devise a simple guide that works for all genera. The six living conifer families may be distinguished by whether the seeds are embedded in the cone scale or attached to its surface at one end, by the number of seeds per fertile cone scale, by the types of fleshy structures associated with mature cones, if any, and for only those members of the Araucariaceae, Sciadopityaceae, and Cupressaceae seemingly identical with respect to these features (and definitely not for all members of these families), by the ordinary type of foliage leaf:

Araucariaceae, seeds embedded or free, one per cone scale, mature cones without fleshy structures, leaves needlelike to bladelike with several to many separate veins

Podocarpaceae, seeds embedded, one per cone scale, mature cones with bracts (the podocarpium), seed scales (the epimatium), an aril or a fleshy outer seed coat layer

Pinaceae, seeds embedded (recognizable as the wing wrapped around one side of the seed), two per cone scale, mature cones without fleshy structures

Taxaceae, seeds free, one (and cone apparently absent) or two (sticking out of a tiny, soft cone in *Cephalotaxus*) per cone scale, mature cones with an aril or a fleshy outer seed-coat layer

Sciadopityaceae, seeds free, five to nine per cone scale, mature cones without fleshy structures, leaves double needles with two separate veins

Cupressaceae, seeds free, 1–20 per cone scale, mature cones without fleshy structures (or the whole cone in *Juniperus*), leaves scalelike or needles with a single vein

Each family description is followed by an identification guide to its genera. Please turn to the genera, which are arranged alphabetically in the next chapter, for further information on them and their species.

Araucariaceae J. B. Henkel & W. Hochstetter

MONKEY PUZZLE TREES AND OTHER ARAUCARIANS

Evergreen trees with scalelike, clawlike, needlelike, or broad-bladed, often multiveined leaves attached individually in a spiral arrangement or in distichous or crisscross (decussate) pairs. Branching usually very regular and horizontal, with most branches in widely separated alternating pairs or periodic near whorls of three or more (usually four to seven), producing a tiered appearance. Pollen cones usually comparatively large (largest among the conifers, on average), single in the axils of leaves or at the tips of branches, often with distinctive bracts at the base. Pollen scales very numerous, attached spirally or in whorls by a long, slender stalk, with several elongate pollen sacs attached just at one end to the inside of the outer, expanded portion of the pollen scale. Pollen grains spherical, without air bladders. Seed cones single and upright at the tips of branches, generally shattering at maturity, fairly woody or leathery, without any fleshy structures. Seed scales very numerous, the upper, ovuliferous (seed-bearing) scale portion and the lower, bract portion nearly equal in length and either united with one another for almost their entire length or free for up to half of it. Seeds almost always single on the seed scale, relatively large, some the largest among those of all conifers, wind dispersed, either via the whole, detached, winglike seed scale in which the seed is embedded or through a highly asymmetrical pair of wings growing out from the seed coat. Cotyledons four with a single midvein or two with a pair of veins. Chromosome base number $x = 13$. Three genera with 35 species in southern South America and in the western Pacific from Southeast Asia to New Zealand.

Identification Guide to Araucariaceae

The three genera of Araucariaceae are readily distinguished by their leaf arrangement on horizontal branches:

Araucaria, leaves spiral

Agathis, leaves opposite to subopposite and in two horizontal rows (distichous)

Wollemia, leaves opposite and in four rows, two horizontal and two forming a broad V above the horizontal plane

Cupressaceae S. F. Gray

CYPRESSES AND REDWOODS

Evergreen or deciduous trees or shrubs with scale-, claw-, or flattened needlelike leaves with a single, undivided midvein. Leaves attached individually in a spiral arrangement, in crisscross pairs, or in alternating trios or fours. In many species with leaves attached in spirals or crisscross pairs, foliage spreads distichously to the right and left of the twigs in a plane parallel to the ground rather than sticking out all around them. Branching typically fairly evenly spread out, without concentration into regular tiers. Branches often turning up at their ends with age like arms bent at the elbow. Pollen cones small to medium in size, single at the tips of otherwise ordinary branchlets or in the axils of ordinary or modified foliage leaves and then often grouped in various ways along specialized twigs, rarely cupped by bracts at the base. Pollen scales few to many, attached spirally, in crisscross pairs, or in alternating trios or fours, with a broad, upturned blade bearing 2–10 round pollen sacs along the inside lower edge near the attachment to the short, slender stalk. Pollen grains spherical, sometimes with a prominent papilla, without air bladders. Seed cones single and usually horizontal or hanging at the tips of otherwise ordinary branches, remaining intact (rarely shattering) at maturity, more or less woody and lacking any fleshy structures (except the whole cone berrylike in *Juniperus*, in which it is called a galbulus). Seed scales few to many, the upper, ovuliferous (seed-bearing) scale portion and the lower, bract portion either nearly equal in length and united with one another for almost their entire length or the ovuliferous scale reduced to an insignificant pad of tissue on the upper side of the bract, or seemingly lost entirely. Seeds 1–30 on the upper surface of each seed scale in one to five rows, small, released from the seed cone at maturity (retained within it in *Juniperus*), wind, bird, or passively dispersed, commonly with two or three nearly equal (to very unequal), narrow to broad seed wings derived from the seed coat and extending the whole length of the seed body. Cotyledons usually two (to six), each with one or two veins. Chromosome base number $x = 11$ (33 in *Sequoia*). Twenty-eight genera with 136 species in both the northern and southern hemispheres but absent from many regions, like most of Africa and all of South America except central and southern Chile and adjacent Argentina. Synonym: Taxodiaceae Saporta.

Identification Guide to Cupressaceae

Among those genera of Cupressaceae with leaves in pairs, *Metasequoia* is unique in having needlelike leaves that are positioned strictly distichously, in two flat rows (by twisting of the branchlets between successive leaf pairs) rather than remaining in crisscross (decussate) pairs lined up in four ranks. For convenience in identification, the remaining 27 genera of the family may be divided into six artificial groups by whether the leaves are borne alternately, in twos (or fours or eights), or in threes, by whether branchlets emerge three-dimensionally or distichously in two ranks and, for Groups D and E only, by whether lateral leaves (those facing in the direction of the plane of distichous branching) are similar to facial leaves (those facing at right angles to the plane of branching) or conspicuously larger than them:

Group A, leaves alternate, branchlets three-dimensional

Group B, leaves alternate, branchlets distichous

Group C, leaves in twos, fours, or eights, branchlets three-dimensional

Group D, leaves in twos, fours, or eights, branchlets distichous, lateral leaves similar to facial leaves

Group E, leaves in twos, fours, or eights, branchlets distichous, lateral leaves larger than facial leaves

Group F, leaves in threes, branchlets three-dimensional

Group A. These genera overlap considerably in obvious foliage features so they are most reliably distinguished by characteristics of their seeds and seed cones. The six genera in this group may be distinguished by whether the leaves are soft or hard in texture, by how many seeds are borne on each cone scale, by how many wings edge the seeds, and by whether the cone scales end in a series of pointed teeth or not:

Taxodium, leaves soft, seeds two, wings none, cone scales toothed or not

Glyptostrobus, leaves soft, seeds two, wing one, cone scales toothed or not

Taiwania, leaves hard, seeds two, wings two, cone scales not toothed

Athrotaxis, leaves hard, seeds three to six, wings two, cone scales not toothed

Cryptomeria, leaves hard, seeds three to six, wings two, cone scales toothed

Sequoiadendron, leaves hard, seeds more than six, wings two, cone scales not toothed

Group B. These four genera may be distinguished by whether the foliage is deciduous or evergreen, by how many seeds are borne on each cone scale, and by how many wings edge the seeds:

Taxodium, foliage deciduous or partially evergreen, seeds two, wings none

Glyptostrobus, foliage deciduous or partially evergreen, seeds two, wing one

Cunninghamia, foliage evergreen, seeds three, wings two

Sequoia, foliage evergreen, seeds more than three, wings two

Group C. Among these genera, *Juniperus* is unique in having a berrylike seed cone that remains closed at maturity, while all the rest have more or less woody seed cones that usually spread open at maturity (though sometimes only after fires). Besides *Juniperus,* the remaining seven genera may be distinguished by whether leaves of adults are needle- or scalelike, by the length of most mature seed cones, and by the number of fertile seed scales in most cones:

Neocallitropsis, leaves needlelike, seed cones 7–10 mm, seed scales two whorls of four

Callitris, leaves needle- or scalelike, seed cones more than 10 mm, seed scales two whorls of four

Widdringtonia, leaves needle- and scalelike, seed cones at least 10 mm, seed scales two pairs

Microbiota, leaves scalelike, seed cones less than 7 mm, seed scales one pair

Diselma, leaves scalelike, seed cones less than 7 mm, seed scales two pairs

Libocedrus, leaves scalelike, seed cones 7–10 mm, seed scales one pair

Cupressus, leaves scalelike, seed cones more than 10 mm, seed scales three to seven pairs

Group D. Among these genera, only the seed cones of *Cupressus* require two growing seasons to reach maturity (so that two sizes of cones are present), those of the other genera maturing in a single growing season. Besides *Cupressus,* the remaining four genera in this group may be distinguished by whether most of the visible portion of the leaves on the branchlets is the attached leaf base or the free (although often pressed against the twig) blade tip, by the number of fertile scales in the seed cones, and by the number and size of the seed wings:

Calocedrus, leaf base conspicuous, fertile scales one pair, one broad and one narrow seed wing

Tetraclinis, leaf base conspicuous, fertile scales two pairs, two broad seed wings

Platycladus, free blade conspicuous, fertile scales two pairs, seed wings absent

Chamaecyparis, free blade conspicuous, fertile scales more than two pairs, two narrow seed wings

Group E. These six genera may be distinguished by the relative length of facial leaves compared to lateral leaves and by the length and position of the prickle on the exposed face of the seed scale that marks the free tip of the fused bract:

Austrocedrus, facial leaves much shorter than the lateral, prickle short and at the tip

Papuacedrus, facial leaves much shorter than the lateral, prickle short and near the middle

Libocedrus, facial leaves much shorter than the lateral, prickle very long and near the middle

Fokienia, facial leaves about as long as the lateral leaves, prickle short and near the middle

Thuja, facial leaves about as long as the lateral leaves, prickle short and near the tip

Thujopsis, facial leaves about as long as the lateral leaves, prickle short to long, very stout, and near the tip

Group F. These four genera may be distinguished by whether the adult leaves are scalelike and tight to the twigs (sometimes with a sharp, spreading, short tip) or more or less needlelike and standing out from them, by whether the seed cones are dry or berrylike at maturity, and by whether each seed scale has a vertical row of colorful and progressively enlarged bracts at its base:

Actinostrobus, leaves scalelike, seed cones dry, seed scale with colorful bracts

Callitris, leaves scalelike, seed cones dry, seed scale without enlarged bracts

Fitzroya, leaves needlelike, seed cones dry, seed scale without enlarged bracts

Juniperus, leaves needlelike, seed cones berrylike, seed scale without enlarged bracts

Pinaceae Adanson

PINES, SPRUCES, AND FIRS

Evergreen or deciduous trees or shrubs with narrow, needlelike, single veined (the midrib sometimes divided into two bundles) leaves attached, individually or in tufts, in a spiral arrangement. Branching usually with regular tiers of three to six horizontal branches and with a few, smaller and weaker branches scattered along the trunk in between the tiers. Pollen cones small or large, single in the axils of leaves or seemingly clustered at the tips of twigs, commonly with some bud scales or bracts around the base. Pollen scales numerous, attached spirally, with two elongate pollen sacs attached to the base of the upturned blade of the scale and, as they extend inward, attached also on either side of the long, slender stalk. Pollen grains mostly with two air bladders to the sides of the roughly equal-sized body (without bladders in *Pseudotsuga* and *Larix* and with a single sac encircling the grain in most *Tsuga* species). Seed cones single and upright to dangling at the tips of (usually) short branches, either shattering or remaining intact at maturity, more or less woody, without any fleshy structures. Seed scales moderately to very numerous, the upper, ovuliferous scale portion free for most of its length from the lower bract portion and usually a little to much longer than it, but sometimes the bract sticking out beyond the ovuliferous scale at maturity, as in *Nothotsuga, Pseudotsuga,* and some species of *Larix* and *Abies.* Seeds in pairs on the seed scales, fairly small to moderate in size, wind (or rarely animal) dispersed as individual seeds released from the seed scales in which they were embedded and which contribute the asymmetrical, terminal seed wing found in almost all species. Cotyledons (two to) four to eight (or more), each with one or two veins. Chromosome base number *x* almost always 12 (13 in *Pseudotsuga menziesii*, 22 in *Pseudolarix*). Eleven genera with 195 species distributed almost throughout the northern hemisphere south of about 70°N, with *Pinus* extending south to about 13°N in Central America, 28°N in the Canary Islands and northeastern Africa, and just south of the equator in Sumatra (Indonesia). Synonym: Abietaceae Berchtold & J. Presl.

Identification Guide to Pinaceae

Pinus differs from all other genera in the family (and from all other living conifers) in having the green needles in tight bundles (fascicles) of two to eight (a single species with one) surrounded at the base (at least initially) by a closely fitting sheath of papery scale leaves (which later spread or fall off in subgenus *Strobus*). The remaining 10 genera, which all have their needles attached individually to the twigs, may be divided into two arbitrary groups for convenience in identification based on whether most of the needles are attached in tufts at the tips of short shoots or are spread out along the length of long shoots:

Group A, short shoots

Group B, long shoots

Group A. These three genera may be distinguished by whether they are evergreen or deciduous, by whether the seed cones are upright and shattering at maturity or hanging and shed as a whole without shattering, and by whether the pollen cones are single or clustered at the tips of the short shoots:

Cedrus, evergreen, seed cones upright and shattering, pollen cones single

Pseudolarix, deciduous, seed cones upright and shattering, pollen cones clustered

Larix, deciduous, seed cones hanging or upright and intact, pollen cones single

Group B. These seven genera may be distinguished by whether the seed cones are erect or pendent at maturity, by whether they shatter or remain intact with seed release, by whether the bracts stick out from between the scales or are hidden by them, by whether the seeds have resin pockets, by whether the leaves are attached to woody pegs (sterigmata) that persist on the twigs after the leaves have fallen, by whether the leaves have hairs (or scars of hairs) along their margins, and by whether the pollen cones are single in the axils of the leaves or clustered at the tips of the twigs:

Abies, seed cones upright, shattering, bracts sticking out or hidden, seeds with resin pockets, leaves without sterigmata or marginal hairs, pollen cones single

Keteleeria, seed cones upright, remaining intact, bracts hidden or rarely visible, seeds with resin pockets, leaves without sterigmata or marginal hairs, pollen cones clustered

Nothotsuga, seed cones upright, remaining intact, bracts sticking out, seeds with resin pockets, leaves with sterigmata but not marginal hairs, pollen cones clustered

Pseudotsuga, seed cones pendent, remaining intact, bracts sticking out, seeds without resin pockets, leaves without sterigmata or marginal hairs, pollen cones single

Tsuga, seed cones pendent, remaining intact, bracts hidden, seeds with resin pockets, leaves with sterigmata but not marginal hairs, pollen cones single

Picea, seed cones pendent, remaining intact, bracts hidden, seeds without resin pockets, leaves with sterigmata but not marginal hairs, pollen cones single

Cathaya, seed cones pendent, remaining intact, bracts hidden, seeds without resin pockets, leaves with sterigmata and marginal hairs, pollen cones single

Podocarpaceae Endlicher

YELLOWWOODS AND OTHER PODOCARPS

Evergreen trees or shrubs with scale-, claw-, or flattened, needlelike or broader leaves, usually with a single, undivided midvein (multiple veins in *Nageia*). Leaves attached individually in a spiral arrangement or in pairs radiating around the twig or arranged distichously in more or less flat rows on either side of predominantly horizontal branchlets. Branching typically with extra, weaker branches along the trunk between the main tiers of three to five major branches. Pollen cones small to moderate in size, single at the tips of branchlets or single to clustered in axils of foliage leaves or bracts on otherwise ordinary branchlets or on specialized reproductive shoots, often with one or a few rows of distinctive bracts at the base. Pollen scales usually numerous to very numerous, attached spirally or apparently so by a (usually) short stalk, with two (or four) round to slightly elongated pollen sacs attached to the stalk and to the lower, outer edge of the upturned more or less triangular blade. Pollen grains usually with two (or three) variously large to small air bladders (or spherical and lacking bladders in *Saxegothaea*). Seed cones single or clustered and often upright on short or long, specialized reproductive branchlets in the axils of leaves or at the tips of otherwise ordinary shoots, almost always with some kind of fleshy structure(s), which vary greatly among the genera. Seed scales typically few, often just a single fertile one, with only a little fusion between the sometimes fleshy lower bract portion (when called a podocarpium) and the more frequently fleshy (or leathery) upper ovuliferous scale portion (called an epimatium in this family). Seeds almost always single and folded within the epimatium, small to large, plump and wingless, usually animal dispersed by themselves or as part of the intact seed cone, with a fleshy seed coat, aril, epimatium, or podocarpium, or any combination of these. Cotyledons normally two, each with two veins. Chromosome base numbers $x = 9–19$. Eighteen genera with 156 species throughout the southern hemisphere and north to about 25°N in Mexico and the West Indies, about 15°N in East Africa, and 35°N in Japan. Synonym: Phyllocladaceae (Pilger) C. Bessey.

Identification Guide to Podocarpaceae

Among the genera of Podocarpaceae, *Parasitaxus* is unique in being entirely reddish purple, including the leaves, which are scalelike and not photosynthetic. *Phyllocladus* is unique in having photosynthesis carried out not by normal foliage leaves but by green, flattened branch systems (phylloclades) with toothed edges corresponding to tiny scale leaves. For convenience in identification, the remaining 16 genera may be divided into three artificial groups by whether the leaves are opposite (or nearly so) or alternate and, for the latter, by whether the foliage leaves of adults are broadest right where they run down onto the twigs (are scale-, claw-, or awl-like) and spiral all around the twigs, or instead, narrow to a distinct petiole (are needlelike of various straight or curved shapes) and spread to the sides of the twigs more or less in two distichous rows:

Group A, leaves opposite

Group B, leaves alternate, broad-based and spiral

Group C, leaves alternate, petioled and distichous

Group A. These four genera may be distinguished by whether the leaves are scalelike or expanded, by how many veins they have, and by whether the leaf arrangement of horizontal shoots is substantially distichous or more three-dimensional:

Microcachrys, leaves scalelike, vein one, horizontal shoots three-dimensional

Afrocarpus, leaves expanded, vein one, horizontal shoots three-dimensional or distichous

Retrophyllum, leaves expanded, vein one, horizontal shoots distichous

Nageia, leaves expanded, veins several to many, horizontal shoots distichous

Group B. These seven genera cannot be reliably identified by vegetative features alone (other than microscopic ones) but may be distinguished by whether transitions between juvenile and adult foliage are abrupt (with juvenile leaf types conspicuous on some shoots of mature plants) or gradual, by whether juvenile leaves (when present) stick out all around the twigs or are arranged in two distichous rows, by whether the adult leaves are tight to the twigs or stick out from them, by how many fertile scales are found in most seed cones, by how long most mature seeds are, and by the shape of the seeds in cross section:

Dacrycarpus, foliage transition abrupt, juvenile leaves distichous, adult leaves tight, fertile scales one to five, seeds at least 3 mm, circular

Halocarpus, foliage transition abrupt, juvenile leaves all around, adult leaves tight, fertile scales two to five, seeds at least 3 mm, somewhat flattened

Manoao, foliage transition gradual, juvenile leaves all around, adult leaves tight, fertile scales two to five, seeds at least 3 mm, circular

Lagarostrobos, foliage transition gradual, juvenile leaves all around, adult leaves tight, fertile scales five to eight, seeds 2–3 mm, somewhat flattened

Microstrobos, foliage transition gradual, juvenile leaves all around, adult leaves tight or sticking out, fertile scales two to eight, seeds less than 2 mm, somewhat flattened or angular

Lepidothamnus, foliage transition gradual, juvenile leaves all around, adult leaves sticking out, fertile scale one, seeds at least 3 mm, circular

Dacrydium, foliage transition gradual, juvenile leaves all around, adult leaves sticking out, fertile scale one, seeds at least 3 mm, somewhat flattened

Group C. These six genera require seed cones for confident identification and may be distinguished by whether the leaves are flattened side to side or top to bottom, by whether they are mostly complexly S-curved (falcate) or mostly fairly straight or simply curved, by the number of fertile scales in the seed cones, by whether these are tightly packed together or widely separated, by whether a podocarpium (a grown-together group of juicy, swollen, colored bracts) is present, and by the length of most mature seeds:

Dacrycarpus, leaves flattened side to side, falcate, fertile scales one to three, tight, podocarpium present, seeds 3–7.5 mm

Acmopyle, leaves flattened side to side, falcate, fertile scales one or two, tight, podocarpium present, seeds at least 7.5 mm

Falcatifolium, leaves flattened top to bottom, falcate, fertile scales one or two, tight, podocarpium present or weakly developed, seeds 5–7.5 mm

Prumnopitys, leaves flattened top to bottom, falcate to straight, fertile scales as many as 12, widely separated, podocarpium absent, seeds more than 7.5 mm

Saxegothaea, leaves flattened top to bottom, straight or somewhat curved, fertile scales 15–20, tight and resembling a typical conifer cone, podocarpium absent, seeds 3–5 mm

Podocarpus, leaves flattened top to bottom, mostly straight, fertile scales one to three, tight, podocarpium present, seeds at least 5 mm

Sciadopityaceae Luerssen

UMBRELLA PINE

Evergreen tree with needlelike, two-veined cladodes (leaflike shoots) inserted in dense umbrella-spoke-like rings at intervals along the shoots, each in the axil of a brown, nonphotosynthetic scale leaf. Scale leaves (the true leaves) attached in a spiral arrangement, very tight at the end of each growth increment, where the whorls of cladodes occur, and much more open in between the whorls. Branching mostly in periodic near whorls of three to five but with a scattering of weaker single branches between the whorls. Pollen cones small, individually nearly spherical, attached in dense globular clusters at the tips of otherwise ordinary foliage shoots, without distinctive bracts at the base of each cone but the whole cluster with remnant bud scales at the base. Pollen scales numerous, attached spirally by a long, slender stalk, with two elongate pollen sacs attached along the stalk and to the base of the roundly triangular, upturned blade. Pollen grains spherical, without air bladders. Seed cones single and upright at the tips of stubby shoots, remaining intact at maturity and gaping between the scales to release the seeds, without any fleshy structures. Seed scales numerous, the upper, ovuliferous scale portion extending well beyond the lower, bract portion and united with it for roughly half their length. Seeds as many as 12 in a single row attached to the upper surface of the seed scale, relatively small, wind dispersed via a pair of nearly symmetrical seed wings running along either side as outgrowths from the seed coat. Cotyledons two, each with a single midvein. Chromosome base number $x = 10$. One genus, *Sciadopitys,* with a single species in southern Japan.

Taxaceae S. F. Gray

YEWS AND PLUM YEWS

Evergreen trees and shrubs with flattened, needlelike leaves with a single midvein and attached individually or in pairs in a spiral (or double spiral) arrangement and almost always distichously extending out horizontally on either side of the twigs. Branching with many extra branches in between the regular whorls of three to five. Pollen cones small, attached singly, either in the axils of ordinary foliage leaves or along specialized reproductive shoots, with bud scales around the base and sometimes with other bracts internally that suggest that these cones have an underlying compound structure, like that of ancestral conifers and a few extant genera of other families. Pollen scales few, attached spirally or in pairs by a long, slender stalk that is usually surrounded by a circle (or half circle) of elongate pollen sacs attached to a more or less umbrella- or shield-shaped head. Pollen grains spherical, without air bladders. Seed cones usually reduced beyond recognition, single and without particular orientation, at the tip of specialized reproductive shoots in the axils of foliage leaves, always with fleshy structures. Seed scales usually apparently absent, except in *Cephalotaxus,* in which there are a few crisscross pairs of scales consisting almost entirely of the bract portion, with a highly reduced ovuliferous scale portion. Seeds single (two on each scale in *Cephalotaxus* but only one usually maturing), medium-sized to large, plump, animal dispersed, often with a fleshy outer seed coat layer and also surrounded in whole or in part by a cuplike or larger, fleshy aril with a few (rows of) bracts around the base, without wings and free from the seed scale (when present). Cotyledons two, each with a single midvein. Chromosome base numbers $x = 11$ and 12. Six genera with 23 species throughout much of Eurasia and North America and in New Caledonia, with a concentration of genera in eastern Asia. Synonyms: Amentotaxaceae Y. Kudô & Yo. Yamamoto, Cephalotaxaceae F. Neger.

Identification Guide to Taxaceae

The six genera of Taxaceae may be readily identified by their leaf arrangement, by whether the seeds are solitary at the tips of branches or borne on small cones, by the size of the mature seeds, by whether the aril is an open cup, almost encloses the seed except for the tip, or is absent, and by the color of the fully mature aril (or seed coat):

Amentotaxus, leaves in opposite spiraling pairs (bijugate), seeds solitary, more than 1.5 cm, aril almost enclosing, red or purple

Cephalotaxus, leaves in opposite spiraling pairs, seeds in cones, 1.5 cm or more, aril absent, seed coat red, purple, or brown

Pseudotaxus, leaves spiral, seeds solitary, less than 1 cm, aril an open cup, white

Taxus, leaves spiral, seeds solitary, less than 1 cm, aril an open cup, red, orange, or yellow

Austrotaxus, leaves spiral, seeds solitary, 1–1.5 cm, aril almost enclosing, deep purple

Torreya, leaves spiral, seeds solitary, 1.5 cm or more, aril almost enclosing, pale to dark green with purple markings or reddish yellow

CHAPTER

8

CONIFER GENERA AND SPECIES

Identification guides to the species are provided after the description of each genus with two or more species, along with geographic and taxonomic guides for the largest genera.

Abies P. Miller

FIR, TRUE FIR, SILVER FIR

Pinaceae

Evergreen, modest-sized to large trees, with a conical crown composed of regular tiers of relatively short, horizontal branches. Trunk usually single, straight, with smooth, nonfibrous bark bearing rows of conspicuous resin pockets when young, remaining smooth with age or becoming variously furrowed or scaly. Twigs all elongate, without distinction into long and short shoots, smooth or grooved. Winter buds well developed, scaly. Leaves spirally arranged but sometimes two-ranked or nearly so, needlelike, usually more or less flattened, straight or curved upward, the margin smooth, the tip pointed, rounded, or notched, the base leaving a flat, round scar on the twig when detached.

Plants monoecious. Pollen cones single, emerging from scaly buds in the axils of leaves of year-old twigs, hanging in rows beneath twigs in the lower portion of the crown, elongate. Pollen scales numerous, each scale with two pollen sacs. Pollen grains very large (body 75–145 μm long, slightly larger than those of *Picea* and much larger than those of *Pinus*), with two round air bladders about half the size of the oblong body and diverging at a little more than a right angle, the body and bladders variously minutely wrinkled or warty but differently from each other. Seed cones roughly cylindrical or barrel-shaped, single on upright stalks in the axils of leaves on shoots at the top of the crown, upright at pollination and at maturity, maturing in a single season. Cone shattering at maturity to release the seeds and numerous, spirally arranged, tightly packed, flat seed scales from a long persistent central axis of variable shape, from very narrowly cylindrical to quite swollen in the middle. Seed scales wedge- to kidney-shaped, with a thin stalk. Bracts attached to the seed scales only at their base, leaflike in shape with a central point, shorter than the scales, or longer than them and sticking out between and beyond them to varying extents. Seeds two per scale, wedge-shaped, the wing derived from the seed scale, extending about the length of the body beyond it and much wider, also covering the body above and wrapping about halfway around it. Cotyledons 4–10, each with one vein. Chromosome base number $x = 12$.

Wood soft, usually not fragrant, off-white to light tan, with little difference between sap- and heartwood. Grain fairly even, with a gradual transition from early- to latewood. Resin canals ordinarily absent but often with crystals in the ray cells.

Stomates forming two prominent stomatal bands beneath and often with additional single lines of stomates above. Each stomate shallowly to deeply sunken beneath and largely hidden by the one to three cycles of four (to six) subsidiary cells, which may have very crinkly cell walls but lack a Florin ring. Leaf cross section almost always with a single resin canal (rarely more) on each side of the double-stranded midvein near the edge or middle and with transfusion tissue lying beneath and sometimes lapping up a little around the outer sides of the midvein within the endodermis. Photosynthetic tissue with a dense, multilayered palisade of relatively short cells beneath the upper epidermis and adjoining hypodermis, which is typically thickest at the outer edges of the leaf and sometimes present only there. Spongy mesophyll filling up the lower two-thirds of the leaf, with especially extensive air spaces above the stomatal bands and without extra sclereids scattered through the tissue.

Forty species across the northern hemisphere, primarily in mountain regions. References: Chater 1993, Christensen 1997, Coode and Cullen 1965, Farjon 1990, Farjon and Rushforth 1989, Fu et al. 1999b, Hunt 1993, Komarov 1986, Lee 1979, Liu 1971, Martínez 1963, K. C. Sahni 1990, Yamazaki 1995.

The name *Abies* is the classical Latin name for European silver fir (*A. alba*), which grows in mountain ranges throughout the Italian peninsula. The resin pockets in the bark are sources of natural adhesives and mounting media for microscope slides (especially Canada balsam), useful because they have the same refractive index as glass and so do not cause distortion. The soft wood is not particularly good for construction or cabinetry, but the lack of odor and staining heartwood extractives makes it useful for pulp and (progressively less frequently today) for food boxes and crates. The perfectly conical shape and fragrant, dark green foliage of short, long-lasting needles make firs favorites as Christmas trees in North America and Europe, where they are raised for the purpose. Their shapeliness also makes them favorites in the open garden, particularly those that have long needles and some tolerance to drier conditions (like *A. concolor*) or colorful seed cones on young trees (like *A. koreana*). Although some species have been in gardens for centuries, cultivar selection has been fairly modest, emphasizing variations in cone size and early maturity, needle length and color (such as highly glaucous blue-greens, chlorotic yellows, and variegated foliage), and growth habits (including narrow columns, flattened spreaders, and many dwarfs).

The various firs are closely related and rather uniform in appearance, so classification of *Abies* is difficult, and there is little agreement among different authors on the number of species, subspecies, and varieties or their arrangement into subgenera, sections, subsections, and series. Even the best attempts to arrange the genus have points of weakness. There is a strong geographic component to the taxonomy of the firs, and the main distributional regions—North America, western Eurasia, and eastern Eurasia—are largely occupied by different sections. There may be a few sections in both eastern Asia and North America, but this has yet to be convincingly demonstrated. Even in the well-botanized parts of the world, as in western Eurasia and northern North America, much more work needs to be done to settle some of the difficulties, but things are even more confused in China, Mexico, and Guatemala, where collections are much sparser and the trees are threatened by uncontrolled exploitation. Comparison of DNA sequences may help to settle the arrangement of sections but will not, by itself, solve the problems of species limits without extensive new field study.

There are many hybrids, both in nature and in cultivation, but these are between species of the same section or of closely related sections. Most crosses between species of different sections fail. As a result of intercrossability, many sections consist of a geographic sequence of species (or subspecies or varieties) linked by more or

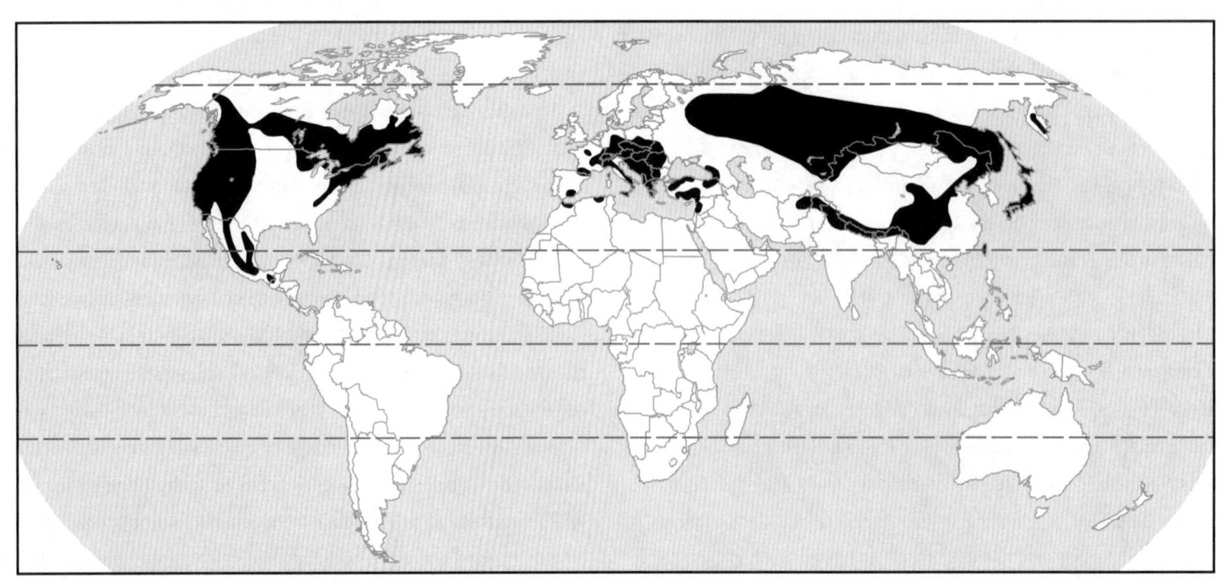

Distribution of *Abies* across the northern continents, scale 1 : 200 million

less extensive areas of intermediate populations. Different classifications treat the intervening populations as hybrids, subspecies, or varieties of one or the other species on either side of the cross, or as species in their own right. This is the situation in western Eurasia and North Africa, where all species are intercrossable, and even species that are geographically separated today might have been in contact and hybridized as they were forced together at various times during the ice ages.

The true firs are predominantly trees of cool, humid environments, ranging from sea level in boreal and maritime forests to the alpine tree line in the mountains. Often, a single mountain range will support two or more species with different but overlapping elevational distributions. They generally do not reach the arctic tree line and, on average, have a more southerly overall range than do the similarly distributed spruces (*Picea*) and larches (*Larix*). They appear to have occupied similar environments since their first convincing appearance in the fossil record in Washington state during the mid-Eocene, about 47 million years ago. Subsequent fossils are widespread but uncommon across middle latitudes of the northern hemisphere, constituting a continuous record throughout the Tertiary of pollen, foliage, cone scales, and seeds. There is even an extensive pollen record in Iceland, where the genus is no longer found.

Geographic Guide to *Abies*

Europe, North Africa, and western Asia
- *Abies alba*
- *A. cephalonica*
- *A. cilicica*
- *A. nebrodensis*
- *A. nordmanniana*
- *A. pinsapo*
- *A. sibirica*

Siberia, Russian Far East, northeastern China, and Korea
- *Abies holophylla*
- *A. koreana*
- *A. nephrolepis*
- *A. sibirica*

Japan, Sakhalin, and Kurile Islands
- *Abies firma*
- *A. homolepis*
- *A. mariesii*
- *A. sachalinensis*
- *A. veitchii*

Himalaya
- *Abies delavayi*
- *A. densa*
- *A. forrestii*
- *A. pindrow*
- *A. spectabilis*

Southwestern and central China
- *Abies beshanzuensis*
- *A. chensiensis*
- *A. delavayi*
- *A. fargesii*
- *A. recurvata*
- *A. squamata*

Taiwan
- *Abies kawakamii*

Eastern North America
- *Abies balsamea*
- *A. fraseri*

Western North America
- *Abies amabilis*
- *A. bracteata*
- *A. concolor*
- *A. grandis*
- *A. lasiocarpa*
- *A. magnifica*
- *A. procera*

Mexico and Guatemala
- *Abies durangensis*
- *A. guatemalensis*
- *A. hickelii*
- *A. religiosa*
- *A. vejarii*

Taxonomic Guide to *Abies*

The classifications of *Abies* by Liu (1971) and Farjon and Rushforth (1989) cannot be adopted without modifications suggested by DNA studies. These are still quite incomplete, having sampled only a little more than half the taxa (species and varieties) in the genus. Nonetheless, it is clear that many fir species are much less differentiated genetically than they appear to be morphologically. Thus the classification adopted here, through based on those of Liu, and Farjon and Rushforth, is simpler than either of their classifications, using only the rank of section (they collectively used subgenera, sections, and subsections) and fewer of these than they accepted. In addition, both classifications tended to overemphasize the length of the bract. As a result, closely related species were sometimes placed in distinct taxonomic categories, like *Abies balsamea* and *A. fraseri* in eastern North America or the long- and short-bracted Mediterranean firs. The simplified classification adopted here brings such taxa together into larger groups.

Abies sect. *Grandis* Engelmann (synonyms: sect. *Oiamel* Franco, sect. *Vejarianae* Tang S. Liu)
Abies concolor
A. durangensis
A. grandis
A. guatemalensis
A. hickelii
A. religiosa
A. vejarii
Abies sect. *Bracteata* Engelmann
Abies bracteata
Abies sect. *Balsamea* Engelmann (synonyms: sect. *Elate* Hickel, sect. *Pichta* Mayr)
Abies balsamea
A. fraseri
A. koreana
A. lasiocarpa
A. nephrolepis
A. sachalinensis
A. sibirica
A. veitchii
Abies sect. *Amabilis* (Matzenko) Farjon & Rushforth
Abies amabilis
A. mariesii
Abies sect. *Nobilis* Engelmann
Abies magnifica
A. procera
Abies sect. *Abies* (synonym: sect. *Piceaster* Spach)
Abies alba
A. cephalonica
A. cilicica
A. nebrodensis
A. nordmanniana
A. pinsapo
Abies sect. *Momi* Mayr (synonyms: sect. *Chensienses* (Matzenko) Tang S. Liu, sect. *Homolepides* (Franco) Tang S. Liu)
Abies beshanzuensis
A. chensiensis
A. fargesii
A. firma
A. holophylla
A. homolepis
A. kawakamii
A. recurvata
Abies sect. *Pseudopicea* Hickel (synonyms: sect. *Elateopsis* Matzenko ex Tang S. Liu, sect. *Pindrau* Mayr)
Abies delavayi
A. densa
A. forrestii
A. pindrow
A. spectabilis
A. squamata

Identification Guide to *Abies*

The species of *Abies,* especially as young trees, are difficult to identify with conviction. Mature seed cones as well as foliage from upper and lower branches is usually required, and even then it is difficult to separate some species. A few species have distinctive features that make them stand out within this fairly uniform genus. When dealing with trees in nature, there are rarely more than three or four species in a given area, so identification is simplified, although hybridization may introduce some complications. In cultivation, many more species are possible in any given locality, the trees are often immature, and hybrids become an even more likely source of confusion. This identification guide cannot include hybrids because so many are possible and so few have been characterized. Because there is currently no entirely satisfactory system of botanical sections for *Abies,* this identification guide uses groups of convenience, established simply for ease of identification. These groups should not be construed as indicating relationships among the species they contain. Although related species will often be found in the same group, they need not be. Because of variation among individuals, a single species may be found in more than one group. For ease of identification, the species of *Abies* may be divided into six groups based on the color of the mature young branchlets and the color of the hairs on them, if any:

Group A, branchlets with a distinct reddish or purplish cast, red-haired

Group B, branchlets reddish or purplish, dark-haired

Group C, branchlets reddish or purplish, pale-haired or hairless

Group D, branchlets with a yellowish or grayish cast, red-haired

Group E, branchlets yellowish or grayish, dark-haired

Group F, branchlets yellowish or grayish, pale-haired or hairless

Group A. These five species may be distinguished by whether there are continuous lines of stomates on the upper surface of the needles, how many lines of stomates usually make up each stomatal band beneath, the position of the resin canals—deep inside the needle (median) or near the lower epidermis (marginal)—and the appearance of the bracts in the intact seed cones:

Abies sachalinensis, no stomatal lines above, fewer than 10 stomatal lines beneath, resin canals median, bracts hidden

A. fargesii, no stomatal lines above, 10 or more stomatal lines beneath, resin canals median, bracts sticking up

A. forrestii, no stomatal lines above, 10 or more stomatal lines beneath, resin canals marginal, bracts sticking up

A. magnifica, with stomatal lines above, fewer than 10 stomatal lines beneath, resin canals marginal, bracts usually hidden

A. procera, with stomatal lines above, fewer than 10 stomatal lines beneath, resin canals marginal, bracts turned down

Group B. These four species may be distinguished by the length of most buds, the number of lines of stomates in the stomatal bands beneath, and the position of the resin canals (see Group A for explanation):

Abies grandis, most buds less than 5 mm, no more than seven stomatal lines on each side beneath, resin canals marginal

A. guatemalensis, most buds no more than 5 mm, more than seven stomatal lines on each side beneath, resin canals marginal

A. spectabilis, most buds at least 5 mm, at least seven stomatal lines on each side beneath, resin canals marginal

A. squamata, most buds at least 5 mm, at least seven stomatal lines on each side beneath, resin canals median

Group C. Among these 13 species, *Abies delavayi* is unique in having the needle margins strongly curled under to hide the stomatal bands, while all the others are either flat or slightly curled. *Abies hickelii* stands out in having at least two pairs of resin canals in the needles and usually three or more, while the others have only one pair. *Abies bracteata* differs from all the others in its narrow, pointed buds up to 2.5 cm long and in bristle tips on the seed cone bracts up to 3 cm long, while in the other species the buds are plumper and no more than half as long, and the bract tips are no more than 1 cm long and usually much shorter. The remaining 10 species may be distinguished by the length of most buds, whether the leaf tip is sharply pointed or not, the position of the resin canals in the needles (see Group A for explanation), the length of most seed cones, whether the unripe mature seed cones are dominantly purple or some shade of green or brown, and the appearance of the bracts:

Abies vejarii, most buds no more than 5 mm, leaf tips pointed, resin canals marginal, most seed cones less than 11 cm, purple, bracts hidden to sticking up

A. pinsapo, most buds no more than 5 mm, leaf tip pointed, resin canals marginal or median, most seed cones more than 11 cm, brown, bracts hidden

A. durangensis, most buds no more than 5 mm, leaf tip blunt, resin canals marginal, most seed cones less than 11 cm, green, bracts hidden to sticking up

A. guatemalensis, most buds no more than 5 mm, leaf tip notched, resin canals marginal, most seed cones less than 11 cm, purple, bracts hidden to sticking up

A. veitchii, most buds less than 5 mm, leaf tip notched, resin canals nearly marginal, most seed cones less than 11 cm, purple or green, bracts bent down

A. pindrow, most buds less than 5 mm, leaf tip notched, resin canals nearly marginal, most seed cones at least 11 cm, purple, bracts hidden

A. religiosa, most buds less than 5 mm, leaf tip notched or pointed, resin canals marginal, most seed cones more than 11 cm, purple, bracts bent down

A. cephalonica, most buds at least 5 mm, leaf tip pointed, resin canals marginal, most seed cones more than 11 cm, brown, bracts bent down

A. forrestii, most buds more than 5 mm, leaf tip notched, resin canals marginal, most seed cones no more than 11 cm, purple, bracts sticking up

A. fargesii, most buds at least 5 mm, leaf tip notched, resin canals median, most seed cones less than 11 cm, purple, bracts sticking up

Group D. These five species may be distinguished by the length of the buds, the number of lines of stomates on top of the needles, the position of the resin canals in the needles (see Group A for explanation), the color of the unripe mature seed cones, and the appearance of the bracts:

Abies mariesii, buds no more than 3 mm, no stomatal lines above, resin canals marginal, unripe mature seed cones purple, bracts hidden

A. kawakamii, buds more than 3 mm, fewer than six stomatal lines above, resin canals marginal, unripe mature seed cones purple, bracts hidden

A. lasiocarpa, buds at least 3 mm, no more than six stomatal lines above, resin canals median, unripe mature seed cones purple, bracts hidden

A. fraseri, buds at least 3 mm, no more than six stomatal lines above, resin canals median, unripe mature seed cones purple, bracts bent down

A. concolor, buds about 3 mm, at least six stomatal lines above, resin canals marginal, unripe mature seed cones green or purple, bracts hidden

Group E. Among these nine species, *Abies nebrodensis* stands out because its needles are rarely more than 1.5 cm long, while in the other species many or all of the needles exceed 1.5 cm. The remaining eight species may be distinguished by the length of most buds, the number of lines of stomates in each stomatal band beneath, the position of the resin canals (see Group A for explanation), the length and color of the unripe mature seed cones, and the appearance of the bracts:

Abies amabilis, most buds up to 4 mm, fewer than eight stomatal lines beneath, resin canals marginal, unripe mature seed cones at least 8 cm, purple or green, bracts hidden

A. sibirica, most buds up to 4 mm, fewer than eight stomatal lines beneath, resin canals marginal or median, unripe mature seed cones up to 8 cm, purple, bracts hidden

A. balsamea, most buds more than 4 mm, up to eight stomatal lines beneath, resin canals median, unripe mature seed cones less than 8 cm, purple, bracts hidden or the tips sticking out

A. nephrolepis, most buds more than 4 mm, at least eight stomatal lines beneath, resin canals median, unripe mature seed cones up to 8 cm, purple, bracts sticking up

A. koreana, most buds around 4 mm, at least eight stomatal lines beneath, resin canals median, unripe mature seed cones less than 8 cm, purple, bracts bent down

A. recurvata, most buds at least 4 mm, more than eight stomatal lines beneath, resin canals marginal or median, unripe mature seed cones more than 8 cm, purple, bracts hidden or sticking up

A. firma, most buds at least 4 mm, more than eight stomatal lines beneath, resin canals median (sometimes with a second marginal pair), unripe mature seed cones at least 8 cm, green, bracts sticking up

A. nordmanniana, most buds at least 4 mm, at least eight stomatal lines beneath, resin canals marginal, unripe mature seed cones more than 8 cm, green, bracts bent down

Group F. Among these 12 species, *Abies concolor* is the only one with (up to 18) lines of stomates along the whole length of the upper surface of the needles. The other species entirely lack stomates above or have a few broken rows near the tips of the needles. The remaining 10 species may be distinguished by the length of most buds, the length of the longest ordinary needles, the number of stomatal lines in each stomatal band on the lower side of the needles, the position of the resin canals (see Group A for explanation), the length and color of the unripe mature seed cones, the appearance of the bracts, and (for *A. alba* and *A. nordmanniana* only) the position of the needles on lower branches and the width of the seed cones:

Abies nebrodensis, most buds no more than 5 mm, longest needles less than 3 cm, no more than 11 stomatal lines beneath, resin canals marginal, most seed cones no more than 12 cm long, unripe mature seed cones green, bracts bent down

A. alba, most buds no more than 5 mm, longest needles no more than 3 cm, fewer than 11 stomatal lines beneath, resin canals marginal, most seed cones at least 12 cm long, no more than 4 cm wide, unripe mature seed cones green, bracts bent down, needles on lower branches spreading out to the sides

A. nordmanniana, most buds no more than 5 mm, longest needles no more than 3 cm, fewer than 11 stomatal lines beneath, resin canals marginal, most seed cones at least 12 cm long, at least 4 cm wide, unripe mature seed cones green, bracts bent down, needles on lower branches angled forward above to cover the twig

A. cilicica, most buds less than 5 mm, longest needles no more than 3 cm, fewer than 11 stomatal lines beneath, resin canals marginal, most seed cones more than 12 cm long, unripe mature seed cones green, bracts hidden

A. homolepis, most buds no more than 5 mm, longest needles no more than 3 cm, at least 11 stomatal lines beneath, resin canals median, most seed cones no more than 12 cm long, unripe mature seed cones purple, bracts hidden

A. firma, most buds no more than 5 mm, longest needles more than 3 cm, more than 11 stomatal lines beneath, resin canals median (sometimes with a second marginal pair), most seed cones about 12 cm long, unripe mature seed cones green, bracts sticking up

A. recurvata, most buds at least 5 mm, longest needles no more than 3 cm, about 11 stomatal lines beneath, resin canals nearly marginal, most seed cones less than 12 cm long, unripe mature seed cones purple, bracts hidden or sticking up

A. beshanzuensis, most buds at least 5 mm, longest needles at least 3 cm, about 11 stomatal lines beneath, resin canals marginal, most seed cones no more than 12 cm long, unripe mature seed cones green, bracts hidden or bent down

A. densa, most buds more than 5 mm, longest needles more than 3 cm, about 11 stomatal lines beneath, resin canals marginal, most seed cones no more than 12 cm long, unripe mature seed cones purple, bracts hidden to sticking up

A. holophylla, most buds at least 5 mm, longest needles more than 3 cm, no more than 11 stomatal lines beneath, resin canals median, most seed cones no more than 12 cm long, usually green, bracts hidden

A. chensiensis, most buds more than 5 mm, longest needles more than 3 cm, more than 11 stomatal lines beneath, resin canals marginal or median, most seed cones no more than 12 cm long, green, bracts hidden

Abies alba P. Miller

EUROPEAN (OR COMMON) SILVER FIR, SAPIN PECTINÉ (FRENCH), WEISSTANNE (GERMAN), ABETE BIANCO (ITALIAN), ABEITE (SPANISH), SWIERK (POLISH), SMEREKA (UKRAINIAN)

Abies sect. *Abies*

Tree to 45(–60) m tall, with trunk to 1.5(–3) m in diameter. Bark light gray, becoming checkered with age. Branchlets densely hairy, not grooved. Buds 3.5–4(–6) mm long, not resinous. Needles arranged to the sides and above the twigs, 1.5–3 cm long, dark green above, the tip notched (or pointed). Individual needles plump in cross section and with a resin canal on either side near the edge just inside the lower epidermis, without stomates above and with seven to nine rows of stomates in each silvery stomatal band beneath. Pollen cones about 2 cm long, greenish yellow. Seed cones roughly cylindrical, 10–20 cm long, 3–4 cm across, green or purple-tinged when young, maturing reddish brown. Bracts emerging from between the scales and turned down. Persistent cone axis narrowly conical. Seed body 7–11 mm long, the wing a little longer. Cotyledons mostly five. Central and southern Europe from the Pyrenees to Bulgaria and the Poland-Belarus frontier. Forming pure stands or mixed with other species in montane forests; (300–)500–1,950 m. Zone 5.

European silver fir (*Abies alba*) in nature.

Upper (A) and undersides (B) of twig of European silver fir (*Abies alba* 'Pyramidalis') with 4 years of growth, ×0.5.

Upper (A) and undersides (B) of twigs of Pacific silver fir (*Abies amabilis*) with 2 years of growth, ×0.5.

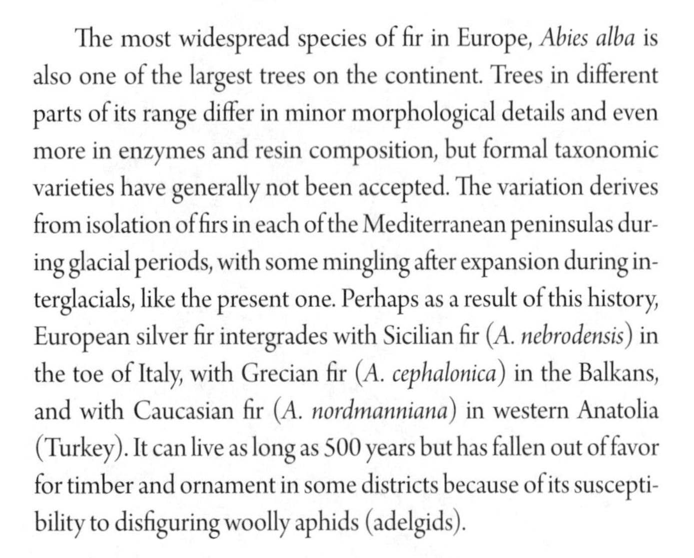

The most widespread species of fir in Europe, *Abies alba* is also one of the largest trees on the continent. Trees in different parts of its range differ in minor morphological details and even more in enzymes and resin composition, but formal taxonomic varieties have generally not been accepted. The variation derives from isolation of firs in each of the Mediterranean peninsulas during glacial periods, with some mingling after expansion during interglacials, like the present one. Perhaps as a result of this history, European silver fir intergrades with Sicilian fir (*A. nebrodensis*) in the toe of Italy, with Grecian fir (*A. cephalonica*) in the Balkans, and with Caucasian fir (*A. nordmanniana*) in western Anatolia (Turkey). It can live as long as 500 years but has fallen out of favor for timber and ornament in some districts because of its susceptibility to disfiguring woolly aphids (adelgids).

Abies amabilis D. Douglas ex Jas. Forbes

PACIFIC SILVER FIR

Abies sect. *Amabilis*

Tree to 75(–82) m tall, with trunk to 2.5 m in diameter. Bark silvery gray, finally breaking up somewhat with age. Branchlets with light brown hairs, not grooved. Buds 3–4(–5) mm long, resinous at the tips or overall. Needles arranged to the sides on lower branches and also angled forward above and covering the twigs on upper branches, (0.7–)1.5–2.5(–3) cm long, bright green above, the tips notched on lower branches, pointed on upper ones. Individual needles plump in cross section and with a resin canal on either side near the edge just inside the lower epidermis, without stomates above and with five or six rows of stomates in each white stomatal band beneath. Pollen cones 10–15 mm long, bright red. Seed cones roughly barrel-shaped, 8–10(–15) cm long, 3.5–5 (–6.5) cm across, purple (or green) when young, maturing purplish brown. Bracts much shorter than the seed scales and hidden by them. Persistent cone axis narrowly conical. Seed body 10–12 mm long, the wing up to 1.5 times as long. Cotyledons four to seven. Pacific Coast of North America from southeasternmost Alaska, through British Columbia, to northeasternmost California. Forming pure stands or more commonly mixed with one or more of a dozen other conifers on slopes in the coastal and montane mixed conifer forests; 0–1,850 m. Zone 6.

Abies amabilis is not closely related to other North American firs and does not hybridize with any of them in nature or in experiments. It lives up to its scientific name (Latin for "lovely") when young but becomes gaunt with age. It retains a narrow growth form throughout life, which may exceed 400 years, after reaching maturity at 20–30 years. Within its narrow distributional limit (no more than about 200 km from the sea), Pacific silver fir has a wide ecological tolerance, growing in both open and densely shady sites and often attaining dominance over some of its associates. The largest known individual in the United States in combined dimensions when measured in 1999 was 66.5 m tall, 3.6 m in diameter, with a spread of 8.2 m. The wood is similar to that of the associated western hemlock (*Tsuga heterophylla*), although not as strong, and is used (somewhat inappropriately) for a range of construction uses alongside its forest associates as well as (more appropriately) for container plywood. *Abies amabilis* often grows with grand fir (*A. grandis*) without hybridizing with this supposed relative. It is morphologically similar to the Japanese Maries fir (*A. mariesii*), but the extent of their reproductive compatibility is unknown.

Abies ×arnoldiana Nitzelius

Tree to 15 m tall. A garden hybrid (*Abies koreana* × *A. veitchii*) first arising in cultivation in the Göteborg Botanic Garden (Sweden) from seed sent by the Arnold Arboretum (hence the scientific name) near Boston. The two parents are closely related species that do not overlap in nature but occupy similar subalpine habitats in Japan (Veitch fir, *A. veitchii*) and Korea (Korean fir, *A. koreana*), respectively. The hybrids combine the early cone production of Korean fir with the more rapid growth and slightly longer needles of Veitch fir. The needles widen toward the tip like those of Korean fir. Zone 5.

Abies balsamea (Linnaeus) P. Miller

BALSAM FIR, SAPIN BAUMIER (FRENCH)
Plates 2 and 3
Abies sect. *Balsamea*

Tree to 25(–30) m tall, with trunk to 1(–1.5) m in diameter. Bark gray, finally splitting up irregularly into shallow blocks with age. Branchlets with sparse, short gray hairs, not grooved. Buds 4.5–5.5 mm long, resinous. Needles arranged to the sides on lower branches, curved upward on the top side of the twigs in upper branches, 1.5–2.5 cm long, dark green above, the tips rounded to pointed. Individual needles plump in cross section and with a large resin canal on either side in the center, near the midvein, with up to three rows of stomates above, particularly near the tip, and with four to eight rows in each white stomatal band beneath. Pollen cones 4–6 mm long, of various colors from greenish through reddish tinges to bluish purple. Seed cones oblong, 4–7(–8) cm long, (1.5–)2–3 cm across, dull purple when young, maturing brown. Bracts shorter than the seed scales and hidden by them or the main blade just about as long as, and the point then sticking straight out beyond, the scales. Persistent cone axis narrowly cylindrical. Seed body 3–6 mm long, the wing up to twice as long. Cotyledons mostly four. Northeastern North America from northern Labrador to northern Virginia and West Virginia west to central Alberta and northeastern Iowa and central Minnesota. Forming pure stands or mixed in with other trees in boreal and mixed forests, and to the alpine tree line in montane forests of the central and northern Appalachians and related mountains; 0–1,500(–1,900) m. Zone 2. Synonyms: *A. balsamea* var. *phanerolepis* Fernald, *A.* [×]*phanerolepis* (Fernald) Tang S. Liu.

The scientific and common names refer to the fragrant resin (Canada balsam) that has historically been an important mounting medium for microscope slides. It is obtained in essentially pure form directly from the resin pockets on the bark. The tree is used today primarily as a pulp species and as one of the three most popular Christmas trees in North America. Balsam fir is extremely shade-tolerant and commonly establishes under the canopy of other species. It is short-lived, however, typically less

Natural stand of young balsam firs (*Abies balsamea*).

Underside of twig of balsam fir (*Abies balsamea*) with 4 years of growth, ×1.

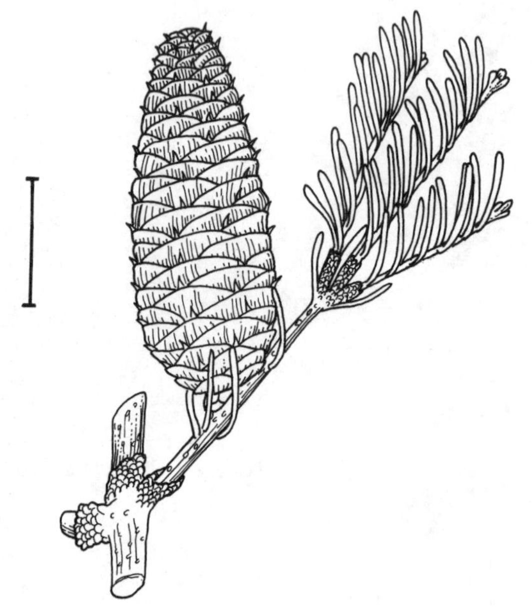

Mature seed cone of balsam fir (*Abies balsamea*) before shattering; scale, 2 cm.

than 150 years, so stands begin to decline soon after attaining dominance. In the United States, the largest known individual in combined dimensions when measured in 1993 was only 30.5 m tall, 1.2 m in diameter, with a spread of 13.7 m, much smaller than many other fir species. In the mountains of New England, stands of balsam fir and red spruce (*Picea rubens*) are subject to progressive windthrow, creating waves of regeneration. Mortality is also caused by many fungal diseases and insects, especially spruce budworm, which has killed millions of trees during each of a dozen severe outbreaks since 1704.

Balsam fir is very closely related to Fraser fir (*Abies fraseri*) of the high southern Appalachians. The two species are isolated today but may well have been in contact during the ice ages. The southernmost populations of *A. balsamea* in northern Virginia and West Virginia are somewhat intermediate between balsam and Fraser firs and have sometimes been considered hybrids between them. A great deal of morphological, biochemical, and genetic data suggests that this is not the case and that balsam firs with emergent bracts (the most obvious trait distinguishing Fraser fir from balsam fir) occur in the wettest parts of the range of the species, near the Atlantic coast in the north and on high mountains southward. Throughout this area, the comparative length of the bracts increases fairly smoothly with elevation, so there is little justification for recognizing trees with emergent bracts as a separate variety. At the other end of its distribution, in Alberta, balsam fir hybridizes with subalpine fir (*A. lasiocarpa*), which is also a close relative but much more distinct than Fraser fir. These three species are fully interfertile with one another in artificial crosses but are not very crossable with other firs that have been tried.

Abies beshanzuensis M. H. Wu

BAISHAN FIR, BAISHANZU LENGSHAN (CHINESE)

Abies sect. *Momi*

Tree to 20(–30) m tall, with trunk to 0.6(–1) m in diameter. Bark pale gray, breaking up somewhat and becoming dark reddish brown with age. Branchlets hairless or minutely hairy, especially in the deep grooves between the leaf bases. Buds 4–7 mm long, with a thin film of white resin. Needles arranged to the sides of the twigs in several rows on lower branches, also or exclusively turned upward on upper branches with seed cones, (1–)1.5–3.5(–5) cm long, glossy dark green above, the tip usually notched. Individual needles flat in cross section and with a resin canal on either side usually touching the lower epidermis near the margin, without stomates above and with 10–12 rows of stomates in each silvery to greenish white stomatal band beneath. Pollen cones (10–)15–25 mm long, red. Seed cones elongate egg-shaped to cylindrical, (6–)8–12 cm long, (3–)3.5–4.5(–5) cm across, yellowish to dark green when young, maturing light to dark brown. Bracts about as long as the scarcely hairy seed scales and bent back slightly oven them, sometimes hidden. Persistent cone axis narrowly conical. Seed body (6–)8–12 mm long, the wing about as long. Cotyledons four to six. Southern China, in southern Zhejiang, western Jiangxi, southwestern Hunan, and northern Guangxi. Scattered emergents above lower montane mixed hardwood forest with few conifers; 1,400–1,850(–2,100) m. Zone 8. Synonyms: *A. beshanzuensis* var. *ziyuanensis* (L. K. Fu & S. L. Mo) L. K. Fu & N. Li, *A. dayuanensis* Q. X. Liu, *A. fabri* var. *beshanzuensis* (M. H. Wu) Silba, *A. fabri* var. *ziyuanensis* (L. K. Fu & S. L. Mo) Silba, *A. yuanbaoshanensis* Y. J. Lu & L. K. Fu, *A. ziyuanensis* L. K. Fu & S. L. Mo.

Abies beshanzuensis is only known from a few populations on widely separated mountains scattered across southern China, one of which, Baishanzu, gave its name to the species. These populations differ a little from one to the next and were originally described as separate species. Taken together, however, they are no more varied than many other silver fir species, and even separating them as varieties seems unnecessary. This species is related to Momi fir (*A. firma*) and Manchurian fir (*A. holophylla*) to the northeast and to Shaanxi fir (*A. chensiensis*) to the north but the details of these relationships are unclear and require further study. Although some populations are partly protected within nature reserves, *A. beshanzuensis* is so rare that it is of considerable conservation concern. Its growth at much lower elevations than most southern Chinese firs makes it of potential interest for ecological studies and for warm temperate forestry, either by itself or as a contributor to tree breeding programs. The stands in the Baishan range are the only population of firs in mainland eastern China.

Abies ×*borisii-regis* Mattfeld

BALKAN FIR

Tree to 30 m tall, resembling its parents (*Abies alba* × *A. cephalonica*) in features common to both and variably combining the characteristics by which they differ. Branchlets are usually hairy, like those of European silver fir (*A. alba*), while the buds are usually resinous, like those of Grecian fir (*A. cephalonica*). Needles are usually straight, like those of *A. alba,* but are pointed upward (as well as to the sides) rather than somewhat forward on the top of the twig, as in *A. cephalonica.* Tips of needles are particularly variable, even on a single tree, from a little less notched than in *A. alba,* through rounded, to even more pointed than in *A. cephalonica.* While the resin canals in the leaves are often near the lower epidermis, as they usually are in both parents, they are sometimes in the middle of each side of the leaf, a condition occasionally found also in *A. cephalonica.* Cylindrical seed cones are 10–13 cm long and 4–5 cm across, overlapping with both parents but generally shorter than those of *A. cephalonica* and wider than those of *A. alba.* Bracts are a little shorter than those of either parent, and little more than the tip sticks out between the scales. It usually sticks straight out, unlike either parent, but may bend downward as it does in them. The trees commonly lack cones. Balkan Peninsula from southern Albania and Bulgaria south through Greece. Mountain forests, usually with one of its parents; (400–)600–1,700(–2,000) m. Zone 6. Synonyms: *A. alba* var. *acutifolia* Turrill, *A. cilicica* var. *borisii-regis* (Mattfeld) Silba.

Firs of the central Balkans on either side of 40°N are exceptionally variable and combine the characteristics of European silver fir (*Abies alba*) and Grecian fir (*A. cephalonica*), which meet here. Trees north of 40°N have more features of *A. alba* and sometimes grow with apparently pure *A. alba.* South of 40°N the balance shifts toward *A. cephalonica,* and the intermediates usually grow among more numerous trees of pure *A. cephalonica.* Intergradation probably dates back to glacial periods, when *A. alba,* retreating before the advancing glaciers, invaded the territory of

Natural stand of hybrid Balkan firs (*Abies* ×*borisii-regis*).

A. cephalonica, along with the other Mediterranean peninsulas. These hybrids are named for Boris III, king of Bulgaria from the end of the First World War to the end of the Second World War, the period when Johannes Mattfeld found them in the Rhodope Mountains of Bulgaria and described them as an independent species.

Abies bracteata (D. Don) P. Poiteau

BRISTLECONE FIR, SANTA LUCIA FIR

Abies sect. *Bracteata*

Tree to 35(–55) m tall, with trunk to 1.3 m in diameter. Bark grayish brown, breaking up slightly with age. Branchlets hairless, thinly waxy at first, not or weakly grooved. Buds (12–)15–20(–25) mm long, sharp-pointed, not resinous. Needles arranged all around the twigs or mostly to the sides, but rather densely so, (2.5–)3–6 cm long, very stiff, dark green above, the tips sharply pointed. Individual needles slightly plump in cross section and with a small resin canal on either side at the far edge, just inside the lower epidermis, without stomates above and with 8–10 rows of stomates in each greenish white stomatal band beneath. Pollen cones about 3 cm long, yellowish brown. Seed cones egg-shaped, (4.5–)7–10 cm long, 4–6 cm across, reddish green when young, maturing pale purplish brown. Bract blades about as long as the seed scales and emerging beyond them, the stiff, narrow tips extending out another 1–3 cm (hence the scientific and common names). Persistent cone axis broadly conical, widest below the midpoint. Seed body 6–10 mm long, the wing about as long. Cotyledons mostly seven. Santa Lucia Mountains of coastal Monterey County and

Natural stand of young bristlecone firs (*Abies bracteata*).

Old bristlecone fir (*Abies bracteata*) in nature.

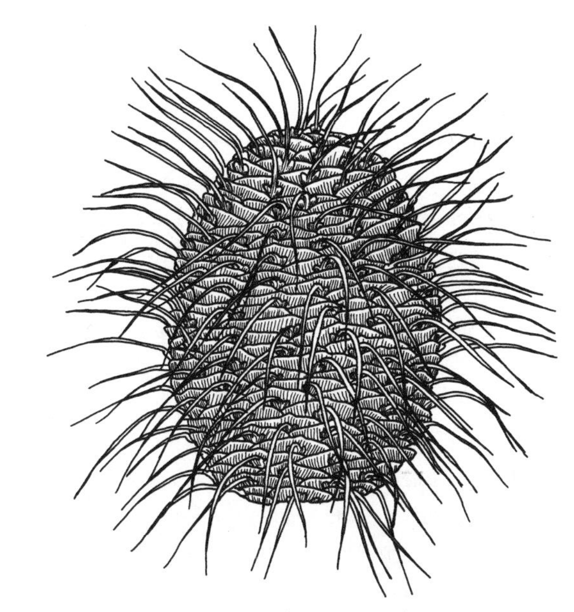

Mature seed cone of bristlecone fir (*Abies bracteata*) before shattering, ×0.5.

Underside of twig of bristlecone fir (*Abies bracteata*) with 2 years of growth, ×0.5.

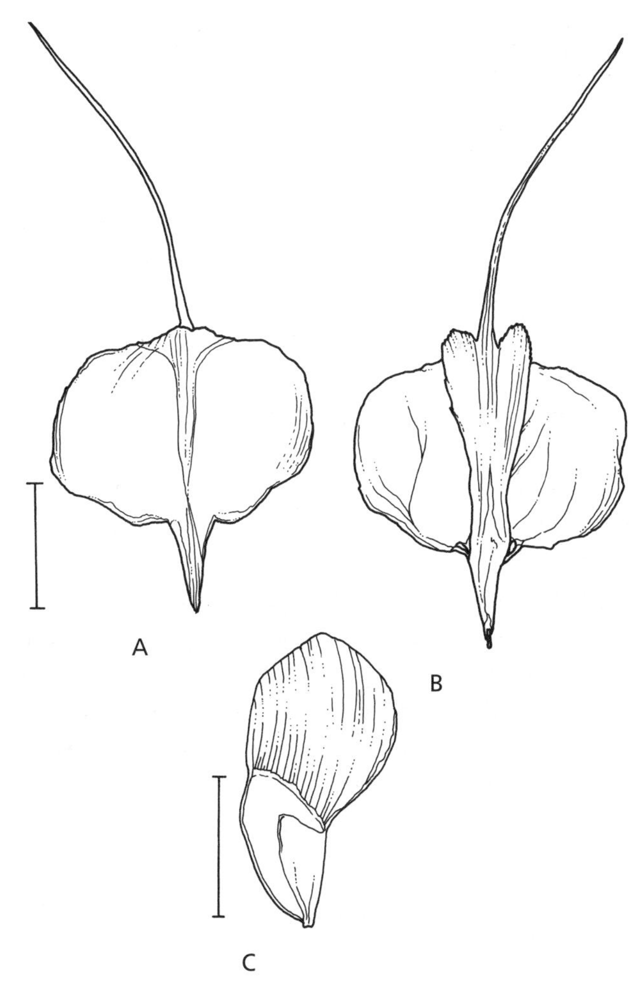

Upper (A) and undersides (B) of shed seed scale with bract and seed (C) of bristle cone fir (*Abies bracteata*); scale, 1 cm.

northern San Luis Obispo County, California. Scattered among evergreen hardwoods on steep slopes above the redwood forest; (200–)600–900(–1,600) m. Zone 8.

Bristlecone fir is so set off from other firs that it is frequently placed in a subgenus apart from all others. The long, stiff, sharp needles, long pointy buds, and long, stiff bract tips, often speckled with resin, are unique among the firs. Fossils similar to bristlecone fir are found in Miocene sediments in Nevada that are about 15 million years old, so the species has been distinct for some time. *Abies bracteata* is a very local and uncommon species with an unusual ecology. For most of its elevational range it only grows with hardwoods, not other conifers. At its lower limit it approaches the redwood forest while at the top it reaches into the mixed conifer forest of pines (*Pinus lambertiana, P. coulteri,* and *P. ponderosa*), Douglas fir (*Pseudotsuga menziesii*), and incense cedar (*Calocedrus decurrens*). Because of its rarity, it has no economic uses except as a distinctive ornamental.

Abies cephalonica J. C. Loudon

GRECIAN FIR, KUKUNARIA, ELATE (GREEK)

Abies sect. *Abies*

Tree to 30 m tall, with trunk to 1(–2) m in diameter. Bark grayish brown, becoming furrowed with age. Branchlets not hairy, prominently grooved between the leaf bases. Buds 4.5–7 mm long, not resinous. Needles arranged straight out all around the twigs or twisting to point upward, (1.5–)1.8–3(–3.5) cm long, shiny dark green above, the tips variably pointed, often prickly. Individual needles flat or plumpish in cross section and with a resin canal on either side near the edge just inside the lower epidermis or well away from it, with only a few stomates near the tip or with up to seven lines of stomates in the groove above and with 6–8(–10) lines in each white stomatal band beneath. Pollen cones 12–18 mm long, purplish red. Seed cones cylindrical, (10–)12–16 cm long, (3.5–)4–5 cm across, brownish green when young, maturing reddish brown. Bracts longer than the seed scales and bent back over them, sometimes almost covering them. Persistent cone axis narrowly conical. Seed body 6.5–9 mm long, the wing up to 1.5 times as long. Cotyledons five to seven. Mountains of Greece. Forming pure stands or mixed with black pine (*Pinus nigra*) at higher elevations and with other species near its lower limit; (600–)900–1,600(–2,000) m. Zone 6. Synonyms: *A. cephalonica* var. *apollonis* (J. Link) Beissner, *A. cephalonica* var. *graeca* (Fraas) Tang S. Liu.

The exact northern limit of Grecian fir is difficult to identify because by 40°N most or all of the trees are intermediates with European silver fir (*Abies alba*), known as Balkan fir (*A.* ×*borisii-regis*). The influence of these hybrids extends down into the Peloponnesus because *A. alba* occurred this far south during glaciations. The firs on Mount Athos on the Ionian Island of Kefallinia (or Cephalonia as it was called during the British occupation when seeds were sent back and the species was described) are, however, a relatively pure expression of the species. It is the main forest tree of the mountains of the Peloponnesus and an important timber tree here and elsewhere, including places like Italy, where it is grown in plantations. In addition to *A. alba,* there is also influence from Caucasian fir (*A. nordmanniana*) of Turkey in northeastern Greece, particularly on Mount Athos.

Abies chensiensis Tieghem

SHAANXI FIR, QINLING LENGSHAN (CHINESE)

Abies sect. *Momi*

Tree to 40(–50) m tall, with trunk to 2 m in diameter. Bark dark gray, splitting into long ridges with age. Branchlets hairless or with a few hairs at first in the grooves between the leaf bases. Buds 7–10 mm long, with a thin coating of resin. Needles arranged to the sides in several rows on lower branches, the upper rows of leaves shorter and angled forward, those of branches with seed cones curving upward above the twigs, 1.5–5(–7) cm long, shiny bright green above, the tip forked on young trees, becoming notched on lower branches, and finally pointed on branches with seed cones. Individual needles widest beyond the middle, plump in cross section with a resin canal on either side near the margins touching the lower epidermis on lower branches, well away from the epidermis on branches bearing seed cones, without stomates in the groove above and with 14–20 lines of stomates in each grayish green stomatal band beneath. Pollen cones 5–10 mm long. Seed cones cylindrical or elongate egg-shaped, (7–)8–11(–14) cm long, 3–5 cm across, green when young, maturing reddish brown. Bracts three-quarters or less as long as the minutely woolly seed scales and hidden by them. Persistent cone axis narrowly conical. Seed body 8–10 mm long, the wing about as long. Cotyledons four to six. Central and southwestern China. In mixed forests with other conifers and a few hardwoods on deep rich soils of mountain valleys and protected slopes; 2,300–3,000(–3,500) m. Zone 6. Two varieties.

Over the years, different authorities presented a broader or narrower picture of this distinctive silver fir, particularly with regard to the relationships between central and southwestern Chinese populations. The maturation-related shift in the position of the resin canals of the needles from contact with the epidermis to well away from it is an unusual feature of central Chinese populations. Shaanxi fir is related to the west-central and southwestern

Chinese Min fir (*Abies recurvata*), and it is with some hesitation that long-needled trees from the southwest are included as a variety of the geographically relatively remote *A. chensiensis* rather than of the nearer *A. recurvata*. The classification of these species, which may not be closely related to the more common southwestern Delavay fir (*A. delavayi*) and its relatives, deserves further attention. Shaanxi fir is a rare tree occurring in scattered populations on widely separated mountains. While some stands are in nature reserves, others are still harvested for construction timber.

Abies chensiensis Tieghem **var. *chensiensis***
SHAANXI FIR, QINLING LENGSHAN (CHINESE)
Needles to 5 cm long, with resin canals well away from the epidermis on branches bearing seed cones, and with deeply forked tips on juvenile trees. Seed cones up to 11 cm long. Central China, from southern Gansu through southern Shaanxi, formerly called Shensi (hence the scientific name), to southwestern Henan and northwestern Hubei.

Abies chensiensis var. salouenensis (Bordères & H. Gaussen) Silba
SALWYN FIR, DAGUO LENGSHAN (CHINESE)
Needles to 7 cm long, with resin canals touching the epidermis on branches bearing seed cones, and with merely notched tips on juvenile trees. Seed cones up to 14 cm long. Southwestern China—southeastern Xizang (Tibet) and northwestern Yunnan—and adjacent northeastern India (Arunachal Pradesh). Synonyms: *Abies chensiensis* subsp. *salouenensis* (Bordères & H. Gaussen) K. Rushforth, *A. chensiensis* subsp. *yulongxueshanensis* K. Rushforth, *A. chensiensis* var. *yulongxueshanensis* (K. Rushforth) Silba, *A. ernestii* var. *salouenensis* (Bordères & H. Gaussen) W. C. Cheng & L. K. Fu, *A. recurvata* var. *salouenensis* (Bordères & H. Gaussen) C. T. Kuan.

Abies cilicica (F. Antoine & Kotschy) Carrière

CILICIAN FIR, TAURUS FIR, ILLEDEN, TOROS GÖKNAN (TURKISH),
Abies sect. *Abies*
Tree to 30 m tall, with trunk to 0.7(–1) m in diameter. Bark gray, becoming flaky and then furrowed with age. Branchlets with scattered short brown hairs or none, grooved between the leaf bases. Buds 3–4 mm long, not resinous or with sparse resin. Needles straight, angled forward, arranged to the sides and above the twigs, 2.5–4 cm long, bright green above, the tips bluntly pointed, rounded, or notched. Individual needles flat in cross section and with a resin canal on either side near the edge usually just inside the lower epidermis but sometimes deep within, usually without stomates in the groove above and with six or seven rows of stomates in each greenish white stomatal band beneath. Pollen cones 10–15 mm long, red. Seed cones cylindrical 15–20(–30) cm long, 4–6 cm across, green when young, maturing reddish brown. Bracts usually much shorter than the scales and hidden by them, occasionally just poking straight out between them. Persistent cone axis narrowly conical. Seed body 9–12 mm long, the wing a little longer. Cotyledons seven or eight (or nine). Mountains of Lebanon and south-central Turkey, in the western and central Taurus Ranges (ancient Cilicia, hence the species name). Forming pure stands or mixed with cedar of Lebanon (*Cedrus libani*) and other conifers and also with evergreen hardwoods at lower elevations; 1,000–1,800(–2,100) m. Zone 6. Synonym: *A. cilicica* subsp. *issaurica* Coode & Cullen.

Cilician fir is relatively isolated geographically from other Mediterranean firs. It is a tree of thin rocky soils, often on limestone, in summer-dry mountain forests. There is no direct evidence of past contact and hybridization with other species of *Abies*. However, it shares with Caucasian fir (*A. nordmanniana*) the tendency for western populations to have hairless twigs and a bit of resin on the buds, while eastern populations are slightly hairy and lack resin, but there is no sharp discontinuity between these groups that would warrant formal taxonomic separation. *Abies cilicica* is closely related to *A. nordmanniana*, differing primarily in its hidden bracts. This has led some taxonomists to place them in separate sections of the genus, sections *Piceaster* and *Abies*, respectively, but these sections are doubtfully distinct so all of the Mediterranean species are best placed together in a single section.

Abies concolor (G. Gordon & Glendinning) F. G. Hildebrand

WHITE FIR
Plates 4 and 18
Abies sect. *Grandis*
Tree to 60 m tall, with trunk to 2 m in diameter. Bark light gray, breaking up into dark gray ridges and reddish furrows with age. Branchlets hairless or with yellowish hairs at first, not grooved. Buds 3–5 mm long, resinous. Needles arranged to the sides on lower branches and also turned upward on branches with seed cones, (1.5–)2–6(–7) cm long, light green or bluish green with wax both above and beneath (hence the scientific name, "same color"), the tips rounded or shallowly notched. Individual needles fairly flat to almost round in cross section and with a resin canal on either side near the edge just inside the lower epidermis, with (5–)7–12(–18) rows of stomates above (fewer toward the tip)

and four to eight rows in each stomatal band beneath. Pollen cones (6–)12–20 mm long, dark red. Seed cones roughly cylindrical, 6–13 cm long, 3–4.5 cm across, green, grayish green, or dull purple when young, maturing light brown. Bracts less than half as long as the seed scales and hidden by them. Persistent cone axis narrowly conical. Seed body 6–12 mm long, the wing up to 1.5 times as long. Cotyledons (five or) six or seven (to nine). Widespread through the mountains of southwestern North America, from central Oregon, southeastern Idaho, and central Colorado, south to northeastern Sonora and northern Baja California (Mexico). Growing with various other conifers in different parts of its range; (600–)900–3,000(–3,500) m. Zone 5. Two varieties.

White fir (named for the pale needles) varies a great deal in morphological and biochemical features throughout its range. Most distinctive are the trees of California since they lack the waxy coating on the leaves found elsewhere, and these are distinguished as variety *lowiana*. They are a major timber tree in the Sierra Nevada, where they grow at generally lower elevations than red fir (*Abies magnifica*). These two species do not hybridize where they grow intermixed, and they are not crossable in artificial hybridization experiments. Nor does white fir hybridize with subalpine fir (*A. lasiocarpa*) in their zones of overlap in the Rocky Mountain region. It does, however, hybridize and intergrade with the closely related grand fir (*A. grandis*) in northwestern California and southwestern Oregon. These hybrids can be recognized at an early age by the dark green needles with a few lines of stomates above. More work is needed on the taxonomic variation of the many scattered populations of this species. The largest known individual in combined dimensions when measured in 1997 was 66.2 m tall, 2.2 m in diameter, with a spread of 11.9 m.

Upper side of twig of white fir (*Abies concolor*) with 2 years of growth, ×0.5.

Abies concolor (G. Gordon & Glendinning) F. G. Hildebrand **var. *concolor***

ROCKY MOUNTAIN WHITE FIR

Needles mostly 4 cm or more long, with a waxy coating and about 12 rows of stomates above, and with a rounded tip. Eastern part of the range of the species, eastern Nevada to Colorado, south to Sonora. Synonyms: *Abies concolor* var. *martinezii* Silba, *A. lowiana* var. *viridula* Debreczy & I. Rácz.

Abies concolor* var. *lowiana (G. Gordon) J. Lemmon

CALIFORNIA WHITE FIR

Needles mostly 4 cm or less long, without wax and with about seven rows of stomates above, the tip slightly notched. Western part of the range of the species, from Oregon to Baja California. Synonyms: *Abies concolor* var. *baja-californica* Silba, *A. lowiana* (G. Gordon) A. Murray bis.

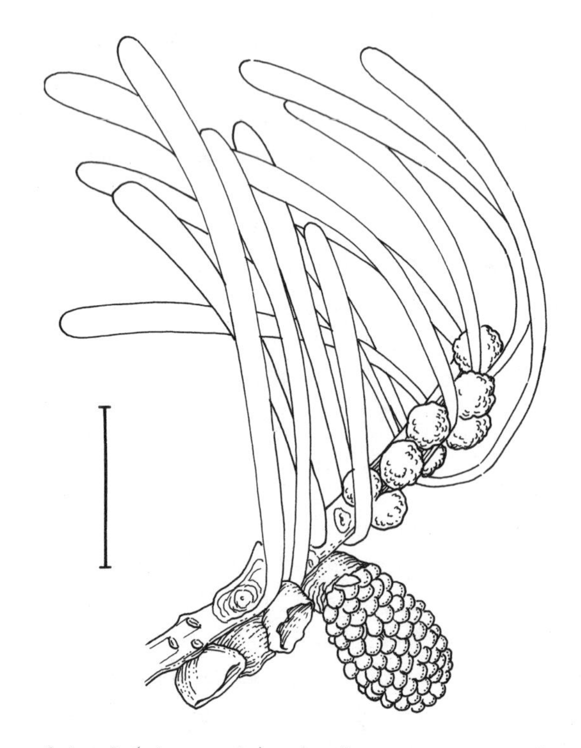

Twig of white fir (*Abies concolor*) with pollen cone, remnant pollen cone bud scales, and buds of future pollen cones; scale, 1 cm.

Abies delavayi Franchet

DELAVAY FIR, YUNNAN FIR, CANGSHAN LENGSHAN, YUAN BIAN ZHONG (CHINESE)

Abies sect. *Pseudopicea*

Tree to 40 m tall, with trunk to 1(–1.5) m in diameter. Bark light gray, darkening, browning, flaking, and then becoming deeply ridged and furrowed with age. Branchlets hairless or transiently minutely hairy in the shallow grooves between the leaf bases. Buds 4–8(–12) mm long, coated with reddish brown resin. Needles arranged all around the twigs or, more commonly, concentrated to their sides and angled forward above them, (1–)1.5–3(–4.5) cm long, shiny dark green above, the tips blunt or notched. Individual needles flat in cross section but with the margins rolled under, often covering the stomatal bands, with a small resin canal on either side touching the lower epidermis near the leaf margins, without stomates above and with 9–11 lines of stomates in each white stomatal band beneath. Pollen cones 20–35(–40) mm long, reddish purple. Seed cones oblong or barrel-shaped, 6–11(–14) cm long, 3–4.5(–5.5) cm across, violet black with a thin waxy coating when young, maturing blackish brown. Bracts about as long as the slightly hairy seed scales and with an added bristle tip, usually sticking up, out, or down a little between the scales, sometimes hidden by them at the top of the seed cone or throughout. Persistent cone axis swollen in the middle. Seed body 5–8 mm long, the wing a little shorter. Cotyledons four to six. Southwestern China—western Sichuan, southeastern Xizang (Tibet), northwestern Yunnan—and adjacent Myanmar. Forming pure stands in the subalpine belt or mixed with other conifers and hardwoods there and in the montane forest below; (1,500–)2,500–4,000(–4,300) m. Zone 8.

Delavay fir is the most widespread of the confusing fir species on the high mountains of southwestern China. It stands out among them by virtue of its rolled leaf margins. There is considerable variation in the length of the needles and the length of the bracts, and this variation has been used to define segregate varieties and species. However, this variation follows a common pattern among silver firs in which trees of a species at low elevations have hidden bracts, while members of the same species higher up have emergent bracts. Because of its relative abundance, *Abies delavayi* is an important commercial species, but like other species of *Abies*, its wood is weak and not very durable. In addition to paper pulp and inside uses, the wood is used in coffin making for those who cannot afford the preferred coffin timbers, like the fragrant and highly durable woods of cypresses (*Cupressus*) and China fir (*Cunninghamia lanceolata*). The species name honors Père Pierre Jean Marie Delavay (1834–1895), a French missionary and botanist who collected the type specimen and sent seed back to Europe during his stay in Yunnan, China, from 1884 until his death there on New Year's Eve in 1895.

Abies delavayi Franchet **var. *delavayi***

DELAVAY FIR, CANG SHAN LENGSHAN (CHINESE)

Branchlets reddish or purplish brown. Rolled leaf margins often completely hiding the stomatal bands. A bit of the blade of the bracts as well as the point usually emerging between the seed scales. Southwestern portion of the range of the species, in Xizang (Tibet) and Yunnan (China) and in Myanmar. Synonyms: *Abies delavayi* var. *motuoensis* W. C. Cheng & L. K. Fu, *A. delavayi* var. *nukiangensis* (W. C. Cheng & L. K. Fu) Farjon & Silba, *A. nukiangensis* W. C. Cheng & L. K. Fu.

Abies delavayi **var. *fabri*** (M. T. Masters) D. Hunt

FABER FIR, LENGSHAN (CHINESE)

Branchlets yellowish brown. Rolled leaf margins usually revealing part of the stomatal bands. Only the point of the bract usually emerging between the seed scales. Northeastern part of the range of the species, in Sichuan (China). Synonym: *Abies fabri* (M. T. Masters) Craib.

Abies densa W. Griffith

HIMALAYAN RED FIR, SIKKIM FIR, DUNSHING (DZONGKHA OF BHUTAN)

Abies sect. *Pseudopicea*

Tree to 50(–60) m tall, with trunk to 2.5 m in diameter. Bark gray, becoming flaky and then ridged and furrowed with age. Branchlets reddish brown (hence the common name), distinctly grooved between the leaf bases, with or without some dark brown hairs in the grooves. Buds about 8 mm long, a little resinous. Needles arranged to the sides and above the twigs, (1.3–)2.5–5 cm long, light green above, the tips notched. Individual needles flat in cross section with the margins rolled under and with a small resin canal on either side near the edge just inside the lower epidermis, without stomates above and with 8–12 lines of stomates in each white stomatal band beneath. Pollen cones 20–40(–50) mm long, violet. Seed cones cylindrical, 8–12(–15) cm long. 3.5–5(–6) cm across, dark violet when young, maturing dark purplish brown. Bracts as long as or a little shorter than the seed scales and often just peeking out between them, especially near the bottom of the cone. Persistent cone axis narrowly conical. Seed body 5–8 mm long, the wing up to 1.5 times as long. Cotyledons four to six. Eastern Himalaya from eastern Nepal to Arunachal Pradesh (India) and adjacent China in Xizang (Tibet). Forming pure stands or mixed with other conifers and hardwoods in a distinct belt within the montane forest (2,450–)2,750–3,750(–4,200) m. Zone 8. Synonym: *A. spectabilis* var. *densa* (W. Griffith) Silba.

The taxonomic status of the firs of the eastern Himalaya remains controversial. Both the number of species and their affinities are disputed. Some authors see the eastern firs as simply extensions of the western Himalayan pindrow fir (*Abies pindrow*) and Webb fir (*A. spectabilis*). Others see an extension of one of the western species, usually Webb fir, joined by a Chinese species, like Delavay fir (*A. delavayi*). The course adopted here is to recognize a single eastern Himalayan species different from either of the western species and from any primarily Chinese species. Himalayan fir differs from both western Himalayan species in its rolled leaf margins and slightly emergent bracts of the seed cone. The needles of Delavay fir have the margins even more strongly rolled under and the bracts stick out a littler farther over the whole cone. In these features, the Himalayan red fir is intermediate between Webb fir and Delavay fir, but a hybrid origin remains nothing more than an intriguing possibility in the absence of further study.

Abies durangensis M. Martínez

DURANGO FIR, GUALLAMÉ BLANCO, PINABETE (SPANISH)

Abies sect. *Grandis*

Tree to 40 m tall, with trunk to 1(–1.5) m in diameter. Bark gray, darkening, reddening, and becoming deeply furrowed with age. Branchlets hairless or with short hairs in the deep grooves between the leaf bases. Buds 4–5 mm long, resin-coated. Needles arranged to the sides of and angled forward above the twigs, (1.5–)2.5–5 (–6.5) cm long, shiny light green above, the tip rounded or bluntly pointed. Individual needles flat or a little plump in cross section and with a resin canal on either side touching the lower epidermis near the edge, without stomates above or with (one to) three to five incomplete lines of stomates in the groove near the tip and with 4–10 lines in each white stomatal band beneath. Pollen cones 10–20 mm long, purplish red. Seed cones cylindrical, 5–9(–11) cm long, 3–4(–4.5) cm across, green when young, maturing yellowish brown. Bracts from half as long as the hairy seed scales and hidden by them to about as long and peeking up between them. Persistent cone axis narrowly cylindrical. Seed body 6–8(–10) mm long, the wing 1.5–2 times as long. Cotyledons five to eight. Northwestern and north-central Mexico from Chihuahua and Coahuila to northwestern Michoacán. Mixed with pines (*Pinus*), oaks (*Quercus*), and other conifers and hardwoods on deep, moist, rich soils of mountain ravines and canyons; (1,450–)1,800–2,500(–2,900) m. Zone 9. Synonyms: *A. coahuilensis* I. M. Johnston, *A. durangensis* var. *coahuilensis* (I. M. Johnston) M. Martínez, *A. neodurangensis* Debreczy, I. Rácz & R. Salazar.

Like the other Mexican and Central American silver firs, *Abies durangensis* occurs as a series of separated populations in pockets of favorable habitat in the mountains. The largest populations are in the Sierra Madre Occidental of the states of Chihuahua and Durango (hence the scientific name). Even though it typically grows at lower elevations than other Mesoamerican species, its habitat is still highly discontinuous. Populations in different locations vary in needle length, number of stomates on the upper side of the leaf, size of the seed cones, and relative prominence of the bracts, and some of these populations were described as varieties or species. Durango fir is closely related to white fir (*A. concolor*) toward the north and Guatemalan fir (*A. guatemalensis*) toward the south. Its distinctions from these species need to be clarified in the context of a thorough study of all three species at the population level. The leaf oils and enzymes of *A. durangensis* are the same as those of southern populations of *A. concolor*, and even those of the more distinctive Véjar fir (*A. vejarii*) of northeastern Mexico, so a population study might well lead to revision of presently accepted taxonomic and geographic boundaries.

Abies* ×*equi-trojani (P. Ascherson & Sintenis ex Boissier) Mattfeld

TROJAN FIR

Tree to 30 m tall, with trunk to 0.7 m in diameter. Similar to both presumptive parents (*Abies cephalonica* × *A. nordmanniana*) in their many shared features and differing from each in a few additional characters. In contrast to Caucasian fir (*A. nordmanniana*), *A.* ×*equi-trojani* combines young branchlets lacking hairs with buds lacking resin. Leaf tips are pointed, like those of Grecian fir (*A. cephalonica*), but seed cones are more like those of *A. nordmanniana*. The overall appearance is fairly intermediate between these two closely related species. Western Turkey, confined to Kazdag (Ida Mountain), near Troy (hence the species name), and Chataldag, a little farther inland. Forming pure stands at high elevation but mixed with oaks (*Quercus*) and beech (*Fagus orientalis*) below; (400–)800–1,700 m. Zone 6. Synonyms: *A. nordmanniana* subsp. *equi-trojani* (P. Ascherson & Sintenis ex Boissier) Coode & Cullen, *A.* ×*olcayana* Ata & Merev.

Trojan fir is confined to two isolated mountain ranges in northwestern Turkey and is not now in contact with either of the presumed parent species. It has been treated as a species in its own right or as a variety or subspecies of both *Abies cephalonica* and *A. nordmanniana*. Its morphological intermediacy and apparent hybrid vigor in growth rate, coupled with extensive pollen sterility, argue in favor of a hybrid origin. It yields an excellent timber, highly prized for construction, so the limited populations are subjected to excessive exploitation and are threatened by removal of the large, reproductive individuals. Further investigation of its seed production, in light of its pollen sterility, seems warranted.

Abies fargesii Franchet

FARGES FIR, BASHAN LENGSHAN (CHINESE)

Abies sect. *Momi*

Tree to 40 m tall, with trunk to 2 m in diameter. Bark gray, flaking, darkening, browning and finally breaking into deep ridges and furrows with age. Branchlets hairless to shaggy with reddish hairs, especially in the grooves between the leaf bases. Buds 3–6(–8) mm long, variably resinous. Needles arranged to the sides of the twigs in several rows, the upper rows shorter and bent up above the twigs on branches bearing seed cones, 1–2.5(–4.5) cm long, shiny bright green to dark green above, the tips usually notched, but sometimes blunt or pointed. Individual needles flat in cross section and with a modest to large resin canal on either side (usually) away from the lower epidermis and also well in from the leaf margins, the margins sometimes slightly rolled under, sometimes with a few short lines of stomates in the groove above near the tip and with 8–12 lines of stomates in each greenish white to silvery stomatal band beneath. Pollen cones 10–15 mm long, red. Seed cones elongate egg-shaped to cylindrical, (3–)5–9(–10) cm long, 3–4.5 cm across, violet when young, maturing purplish or reddish brown. Bracts about as long as the fuzzy seed scales or sometimes a little shorter, the narrow tip usually a little longer and peeking up between the scales to bending down a little over them, or sometimes hidden. Persistent cone axis swollen in the middle. Seed body 5–8 mm long the wing about as long. Cotyledons four to six. Central China, from southern Gansu to western Henan, south to northern Guizhou and northwestern Yunnan. Forming pure stands or mixed with other conifers and hardwoods on mountain slopes in the subalpine zone and below; (1,500–)2,100–3,700(–4,000) m. Zone 6. Three varieties.

Farges fir is an important tree of the dense montane conifer forests of central China, occurring in a wider range of forests over a broader elevational range than either Shaanxi fir (*Abies chensiensis*) or Min fir (*A. recurvata*), which have similar overall distributions. Although it does not grow as large and the wood is not as dense as that of *A. recurvata, A. fargesii* is common enough to be an important commercial species for construction, pulpwood, and carpentry, including coffin making. The recognized varieties differ in hairiness of the shoots and the length of the bracts in the seed cones. The species as a whole is related to Delavay fir (*A. delavayi*) and Forrest fir (*A. forrestii*), which replace it toward the west. The species name honors Père Paul Guillaume Farges (1844–1912), a French missionary and naturalist who collected the type specimen during botanical collecting in western China.

Abies fargesii var. *fanjingshanensis* (W. L. Huang, Y. L. Tu & S. Z. Fang) Silba

FANJING MOUNTAIN FIR, FANJINGSHAN LENGSHAN (CHINESE)

Branchlets hairless or slightly hairy in the grooves. Bracts hidden by the seed scales. Northern Guizhou; 2,100–2,350 m. Synonym: *Abies fanjingshanensis* W. L. Huang, Y. L. Tu & S. Z. Fang.

Natural stand of wind-trimmed Fanjing Mountain firs (*Abies fargesii* var. *fanjingshanensis*).

Young mature Farges fir (*Abies fargesii* var. *fargesii*) in nature.

Abies fargesii Franchet **var. *fargesii***

FARGES FIR, BASHAN LENGSHAN (CHINESE)

Branchlets hairless or slightly hairy in the grooves. Bracts sticking out slightly between the seed scales. Southern Gansu and northeastern Sichuan to western Henan and northwestern Hubei (China); (1,500–)2,000–3,800 m. Synonyms: *Abies fargesii* var. *hupehensis* Silba, *A. fargesii* var. *sutchuensis* Franchet.

Abies fargesii* var. *faxoniana (Rehder & E. H. Wilson) Tang S. Liu

FAXON FIR, MINJIANG LENGSHAN (CHINESE)

Branchlets covered with reddish hairs except on main branches. Bracts sticking out slightly between the seed scales. Southern Gansu, Sichuan, northwestern Yunnan (China), and southeastern Xizang (Tibet); 2,700–4,000 m. Synonyms: *Abies chayuensis* W. C. Cheng & L. K. Fu, *A. delavayi* var. *faxoniana* (Rehder & Wilson) A. B. Jackson, *A. fabri* subsp. *minensis* (Bordères & H. Gaussen) K. Rushforth, *A. fabri* var. *minensis* (Bordères & H. Gaussen) Silba, *A. ferreana* Bordères & H. Gaussen, *A. ferreana* var. *longibracteata* L. K. Fu & N. Li, *A. forrestii* var. *chayuensis* (W. C. Cheng & L. K. Fu) Silba, *A. forrestii* var. *ferreana* (Bordères & H. Gaussen) Farjon & Silba.

Abies firma P. Siebold & Zuccarini

MOMI FIR, MOMI (JAPANESE)

Abies sect. *Momi*

Tree to 40(–50) m tall, with trunk to 1.5(–2) m in diameter. Bark gray, becoming browner, flaky and then furrowed with age. Young branchlets with tiny, dark hairs in the shallow grooves between the leaf bases. Buds 3–5(–10) mm long, sparingly or not resinous. Needles arranged to the sides and angled upward in several rows, (1.5–)2–3.5 cm long, shiny dark green above, the tips sharply forked in young trees, simply notched or even bluntly pointed on old ones. Individual needles flat in cross section and with a resin canal on either side midway between the midvein and the edge and deep inside the leaf tissue, sometimes with an extra small resin canal near the margin or each side, usually without stomates above and with 14–16 lines of stomates in each gray stomatal band beneath. Pollen cones 10–20(–30) mm long, yellowish green. Seed cones elongate egg-shaped, (8–)10–13(–15) cm long, (3–)4–5 cm across, yellowish green when young, maturing grayish green. Bract blades about as long as the seed scales and with longer tips that point upward between the scales. Persistent cone axis narrowly conical. Seed body 8–10 mm long, the wing less than 1.5 times as long. Cotyledons four (or five). Central and southern Honshū (primarily on the Pacific Ocean side of central Honshū), Shikoku, and Kyūshū (Japan). Forming pure stands or more commonly mixed with numerous other conifers in lower montane forests; (50–)300–1,000(–1,600) m. Zone 6.

Momi fir is the most widespread and common silver fir of Japan, in part because it is the lowest-elevation species on all the main islands except Hokkaidō, where it is replaced by Sakhalin fir (*Abies sachalinensis*). It is genetically very closely related to Nikko fir (*A. homolepis*), which is the next fir species up in the elevational sequence, and these two can hybridize where they overlap to produce Mitsumine fir (*A. ×umbellata*). Although active hybridization seems localized today, being restricted to central Honshū, populations of *A. firma* in each part of its range share DNA traits with adjoining populations of *A. homolepis* that are not found in momi firs elsewhere. This implies a longer and more widespread history of hybridization between the two species than one would suspect from the rarity of obviously intermediate trees. There is also DNA evidence, but no obvious morphological traces, of limited past hybridization with Veitch fir (*A. veitchii*), with which it is not generally in contact today. The scientific name of momi fir emphasizes its stiff needles.

Abies forrestii Coltman-Rogers

FORREST FIR, CHUANDIAN LENGSHAN (CHINESE)

Abies sect. *Pseudopicea*

Tree to 30(–40) m tall, with trunk to 1(–1.5) m in diameter. Bark gray, reddening and becoming fissured with age. Branchlets hairless to very fuzzy, especially in the shallow grooves between the leaf bases. Buds 4–10 mm long, thickly coated with white to yellowish resin. Needles arranged to the sides in several rows and also rising above the twigs and angled forward to cover them, 1.5–3(–4) cm long, shiny dark green to bluish green above, the tip usually notched but sometimes bluntly to sharply pointed. Individual needles flat or a little plump in cross section and with a resin canal on either side touching the lower epidermis, or sometimes away from it, a little way in from the margin, which may be straight or a little curled under, without stomates in the groove above and with 9–11 lines of stomates in each white stomatal band beneath. Pollen cones 25–45 mm long, purple. Seed cones elongate egg-shaped to cylindrical, (6–)7–11(–14) cm long, (3.5–)4–5.5(–6) cm across, dark purple when young, maturing purplish or blackish brown. Bract body about as long as the minutely fuzzy seed scales, with a projecting tip and sticking straight up between the scales, or sometimes curled back over them. Persistent cone axis broadly conical or swollen in the middle. Seed body 7–10 mm long, the wing about as long. Cotyledons four to six. Southwestern China, in southwestern Sichuan, northwestern Yunnan, and southeastern Xizang (Tibet). Forming pure stands near the alpine tree line or mixed with other conifers and progressively more hardwoods below this; (2,400–)2,900–4,000(–4,500) m. Zone 7. Two varieties.

The firs of southwestern China—Forrest fir, Delavay fir (*Abies delavayi*), and Farges fir (*A. fargesii*)—present a confusing assemblage of forms, both in the wild and in cultivation. Some authorities accept as many as 13 species and varieties in this rugged region. A more conservative stance is taken here, recognizing four varieties among the three species just mentioned, with other varieties elsewhere and, conversely, other unrelated species in the area. The pattern of variation found in *A. forrestii* and its relatives strongly suggests hybridization as one cause of the taxonomic confusion they present. Solving the problems posed by these species will require both experiments and new fieldwork. Forrest fir once covered large areas of the subalpine region in southwestern China, but it has been heavily exploited for construction and cabinetry timber, paper pulp, and tannins from the bark. On the other had, it has been used to a limited extent in reforestation projects. Both varieties of this species, which are scattered across the whole geographic range, are named for George Forrest (1873–1932), a British plant collector who made seven expeditions to Yunnan between 1904 and 1932, dying on the last one.

Abies forrestii Coltman-Rogers **var. *forrestii***

FORREST FIR, CHUANDIAN LENGSHAN (CHINESE)

Branchlets hairless or a little hairy at first. Leaves mostly toward the upper end of the range of the species in length. Bracts of the seed cones with a narrow, prickle tip. Synonyms: *Abies chengii* K. Rushforth, *A. delavayi* var. *forrestii* (Coltman-Rogers) A. B. Jackson, *A. forrestii* var. *chengii* (K. Rushforth) Silba.

Abies forrestii* var. *georgei (M. Orr) Farjon

GEORGE FIR, CHANGBAO LENGSHAN (CHINESE)

Branchlets densely (often reddish) hairy. Leaves mostly toward the lower end of the range of the species in length. Bracts of the seed cones with a broad, triangular tip. Synonyms: *Abies delavayi* var. *georgei* (M. Orr) Melville, *A. delavayi* var. *smithii* (Viguié & H. Gaussen) Tang S. Liu, *A. forrestii* var. *smithii* Viguié & H. Gaussen, *A. georgei* M. Orr, *A. georgei* var. *smithii* (Viguié & H. Gaussen) W. C. Cheng & L. K. Fu.

Abies fraseri (Pursh) Poiret

FRASER FIR, SHE BALSAM

Plate 5

Abies sect. *Balsamea*

Tree to 15(–30) m tall, with trunk to 1 m in diameter. Bark silvery gray, becoming slightly flaky with age. Branchlets with dense reddish hairs, not grooved. Buds 2–4 mm long, very resinous. Needles arranged to the sides on lower branches and also angled upward on higher branches, (1–)1.5–2(–2.5) cm long, dark green above, the tips slightly notched or rounded (to bluntly pointed). Individual needles flat in cross section and with a large resin canal on either side midway from the edges, surfaces, and midrib, with up to three rows of stomates above, particularly near the tip, and with 8–12 rows in each greenish white stomatal band beneath. Pollen cones 8–10 mm long, greenish through reddish yellow. Seed cones oblong, 3.5–6(–8) cm long, 2.5–4 cm across, dark purple when young, maturing purplish brown. Paler bracts longer than the seed scales, sticking well out beyond them and bending down to cover them. Persistent cone axis narrowly conical. Seed body 4–6 mm long, the wing about as long. Cotyledons mostly five. High southern Appalachian Mountains of southwestern Virginia, western North Carolina, and southeastern Tennessee. Forming pure stands or mixed with red spruce (*Picea rubens*) or other conifers and hardwoods in the wet subalpine zone; (1,200–)1,500–2,050 m. Zone 5. Synonyms: *A. balsamea* subsp. *fraseri* (Pursh) E. Murray, *A. balsamea* var. *fraseri* (Pursh) Spach.

Natural stand of Fraser fir (*Abies fraseri*) ravaged by acid rain.

Fraser fir is named after Scottish plant explorer John Fraser (1750–1811), who collected the type specimen and introduced seeds to Britain. It is abundant within its limited natural range but not found on every mountain high enough to support it. Like the associated red spruce, trees of Fraser fir experienced extensive mortality and reduced regeneration during the last decades of the 20th century, probably linked to acidic air pollution arising at lower elevations and concentrated in the droplets of mist that envelop them much of the time. The largest known individual in combined dimensions when measured in 1990 was 27.1 m tall, 1.2 m in diameter, with a spread of 15.9 m. Trees are often short-lived in cultivation, but dark green, dense, and symmetrical when young, making them popular Christmas trees in North America. In a region rich with valuable pine trees, these small, frequently misshapen conifers of high elevations are little exploited, and many are protected by government agencies. *Abies fraseri* is very

Upper side of twig of Fraser fir (*Abies fraseri*) with 3 years of growth, ×1.

closely related to balsam fir (*A. balsamea*), with few differences separating them. They are maintained as separate species rather than as varieties of a single species largely by tradition because of the strikingly different appearance provided by the seed cone bracts and, to a lesser extent, by the number of stomatal rows. They are scarcely separable in other features of morphology, biochemistry, or genetics.

Abies grandis (D. Douglas ex D. Don in A. Lambert) Lindley

GRAND FIR

Abies sect. *Grandis*

Tree to 75(–90) m tall (hence the scientific and common names), with trunk to 1.5(–2) m in diameter. Bark gray when young, becoming brown and shallowly ridged and furrowed with age. Branchlets minutely hairy at first, not grooved. Buds 1.5–3 mm long, resinous. Needles predominantly arranged to the sides of the twigs on both upper and lower branches, (1–)2–5(–6) cm long, dark green above, the tips usually notched. Individual needles flat to slightly plump in cross section and with a small resin canal on either side near the edges just inside the lower epidermis, without stomates above and with five to seven rows of stomates in each white stomatal band beneath. Pollen cones 12–18 mm long,

Young grand fir (*Abies grandis*) in nature.

Young mature Guatemalan fir (*Abies guatemalensis*) in nature.

various colors from greenish through reddish tinges to purple. Seed cones more or less cylindrical, (5–)6–10(–12) cm long, 3–3.5(–4) cm across, various colors from dark green through gray to purple when young, maturing reddish brown. Bracts much shorter than the densely hairy seed scales and hidden by them. Persistent cone axis narrowly conical. Seed body 6–9 mm long, the wing about as long or a little longer. Cotyledons mostly five or six. Northwestern United States and adjacent Canada from southern British Columbia to coastal northern California and central Idaho. Mixed conifer forests of Douglas fir (*Pseudotsuga menziesii*), Sitka spruce (*Picea sitchensis*), western hemlock (*Tsuga heterophylla*), western red cedar (*Thuja plicata*), and redwood (*Sequoia sempervirens*) on moist soils; 0–1,500(–1,850) m. Zone 6. Synonym: *A. grandis* var. *idahoensis* Silba.

Grand fir is a characteristic tree of the giant lowland coastal rain forest of the Pacific Northwest and of moderate elevations in the Cascade Range and northern Rocky Mountains where many other Coast Range–Cascade species have disjunct occurrences. The eastern populations in the Rocky Mountains of Idaho and Montana were segregated as a distinct variety, but this is hardly tenable. In this lack of significant differentiation between the coastal and inland populations, grand fir resembles most of its associates with similar distributions, like western hemlock (*Tsuga heterophylla*) and western red cedar (*Thuja plicata*). Grand fir frequently grows with Pacific silver fir (*Abies amabilis*) without hybridizing, casting doubt on the assignment of both species to the same section of the genus, as advocated by some taxonomists. It overlaps, hybridizes, and intergrades with the closely related white fir (*A. concolor*) in the southern part of its range. Many of the hybrids are readily recognizable by having leaves that are both dark green above and also bear stomatal lines. This is the tallest silver fir. It is cut along with other conifers for timber, plywood, and pulp. In the United States, the largest known individual in combined dimensions when measured in 1987 was 78.4 m tall, 2.0 m in diameter, with a spread of 11.0 m.

Abies guatemalensis Rehder

GUATEMALAN FIR, PINABETE (SPANISH)

Abies sect. *Grandis*

Tree to 35(–45) m tall, with trunk to 1(–1.5) m in diameter. Bark grayish brown, breaking up somewhat with age. Branchlets somewhat hairy, especially near the tips on lower branches, grooved between the leaf bases. Buds 4–5 mm long, resinous. Needles arranged mostly to the sides in several ranks and with a few shorter needles angled forward above the twigs, (1–)2.5–5.5 cm long, shiny dark or light green above, the tip usually notched but sometimes bluntly pointed. Individual needles flat in cross section and with a resin canal on either side near the outer edge and touching the lower epidermis, without or with three or four discontinuous rows of stomates in the groove above near the tip and with 8–12 rows in each silvery white stomatal band beneath. Pollen cones 15–25 mm long, yellow. Seed cones oblong, (6–)8–10(–12) cm long, (2.5–)4–5.5 cm across, purple when young, maturing dark brown. Bracts from half as long as the minutely hairy seed scales and hidden by them to about as long and peaking out between them. Persistent cone axis narrowly conical. Seed body 8–10 mm long, the wing a little longer. Cotyledons five or six. Pacific slope of southern Mesoamerica, from Coahuila, San Luis Potosí, and eastern Guerrero (southern Mexico) to Santa Barbara (western Honduras). Forming pure stands or, more commonly, mixed with other conifers and hardwoods on moist soils in the high mountains; (1,500–)3,200–3,800(–4,100) m. Zone 9. Synonyms: *A. flinckii* K. Rushforth, *A. guatemalensis* var. *jaliscana* M. Martínez, *A. guatemalensis* var. *longibracteata* Debreczy & I. Rácz, *A. guatemalensis* var. *tacanensis* (C. Lundell) M. Martínez, *A. religiosa* var. *emarginata* Loock & M. Martínez, *A. tacanensis* C. Lundell.

Mature Hickel fir (*Abies hickelii*) in nature.

Guatemalan fir is the most southerly of all world species of silver firs, reaching below 15°N in the high mountains near Lake Atitlán in Guatemala (from which it was first described and received its species name). It is also the most southerly of the western North American group of species that includes Durango fir (*Abies durangensis*), white fir (*A. concolor*), and grand fir (*A. grandis*), the group as a whole extending north of 50°N. The boundaries among these species are poorly defined, and future research may well subdivide them differently. For example, the populations of Colima, Jalisco, and Michoacán (Mexico) at the northwestern limit of the species' range were originally described as a widely separated variety of *A. guatemalensis,* one of oyamel fir (*A. religiosa*), and a separate species including these two varieties. The length of the leaves and seed cones bracts are so variable that recognition of varieties based on these seems hard to justify. The northwestern populations are most distinct, with narrower seed cones and earlier flushing than usual for the species, and the most strongly differentiated isozyme frequencies (although still embedded among the southern and eastern populations). These differences are weak, and so until more convincing evidence for recognition is put forward, variety *jaliscana* is not accepted as distinct here. Although producing excellent lumber, *A. guatemalensis* is too rare to be a major commercial species, and many of the best stands are in government forest reserves. Like *A. religiosa,* the boughs are cut as decorations on religious holidays and are also used to construct overnight shelters, just as those of balsam fir (*A. balsamea*) are favored for the same purpose in the boreal forest of Canada.

Abies hickelii Flous & H. Gaussen

HICKEL FIR, PINABETE (SPANISH)

Abies sect. *Grandis*

Tree to 30 m tall, with trunk to 1 m in diameter. Bark gray, darkening and breaking up into blocks with age. Branchlets with transient long hairs in the deep grooves between the leaf bases. Buds 4–5 mm long, covered with thick, yellow resin. Needles thinly or thickly arranged to the sides of the twigs, (1–)2–3.5(–6) cm long, shiny bright green or yellowish green above, the tip notched to bluntly pointed. Individual needles flat in cross section and with (two or) three or four (to six) resin canals scattered around the periphery on either side, touching the lower epidermis but a little inside of the upper epidermis, without or with a few broken lines of stomates near the tip above and with seven to nine lines of stomates in each grayish green stomatal band beneath. Pollen cones 10–15 mm long, red. Seed cones elongate egg-shaped, 6–8(–12) cm long, (2.5–)3.5–4(–5) cm across, purple or green when young, maturing brown. Bracts usually a little longer than the hairy seed scales, and the broad, triangular tip sticking out and up between them. Persistent cone axis narrowly conical. Seed body 6–8(–11) mm long, the wing a little shorter to a little longer. Cotyledons five or six. Southern Mexico, primarily in Oaxaca and Chiapas, with outliers in Guerrero, Hidalgo, and Veracruz. Forming pure stands near the top of its elevational range or mixed with pine (*Pinus*), oak (*Quercus*), and other conifers and hardwoods on mountain slopes; (1,650?–)2,500–3,000 m. Zone 9. Synonyms: *A. hickelii* var. *oaxacana* (M. Martínez) Farjon & Silba, *A. hidalgensis* Debreczy, I. Rácz & Guízar, *A. oaxacana* M. Martínez, *A. zapotekensis* Debreczy, I. Rácz & G. Ramírez.

No other true fir has as many resin canals in the needles as this rare species of the high mountains of southern Mexico. Only one or two others ever have more than one canal on either side of the leaf. In *Abies hickelii* the number of canals may be higher in cone-bearing than in lower branches, and there may also be geographic or elevational variation in resin canal number. Variations in resin canals, as well as in size of seed cones, were used to subdivide Hickel fir into varieties or even distinct species. No one, however, has undertaken the careful range-wide comparisons of population samples required to evaluate the proposed additional species or varieties. In the absence of such studies, *A. hickelii* is maintained here in a broad sense that encompasses a great deal of variation in leaf arrangement, size, and anatomy, and in seed cone size and bract length. All species of *Abies* show variation in these features with age of the tree and with position on mature trees. The species name honors Paul Robert Hickel (1865–1935), a French dendrologist who published on *Abies,* if not on the Mexican species.

Abies holophylla Maximowicz

MANCHURIAN FIR, CHOS NAMU (KOREAN), SHAN SONG (CHINESE)

Abies sect. *Momi*

Tree to 40(–50) m tall, with trunk to 1.5 m in diameter. Bark gray, browning and becoming ridged and furrowed with age. Young branchlets hairy in the deep grooves. Buds 4–8 mm long, sparsely resinous or not. Needles arranged straight out to the sides on lower branches but directed upward on high branches bearing seed cones, 2–4.5 cm long, glossy dark green above, the tips usually sharp-pointed (not notched like its close relatives, hence the scientific name, "entire leaf"). Individual needles straight and stiff, flat or a little plump in cross section with a large resin canal near the center of either side, without or with a few lines of stomates near the tip above and with 9–11 lines in each grayish green stomatal band beneath. Pollen cones 10–15 mm long, reddish yellow. Seed cones cylindrical, (6–)10–12(–14) cm long, (3–)3.5–4.5(–5) cm across, green (or purplish green) when young, maturing light brown. Bracts less than half as long as the densely hairy seed scales and hidden by them. Persistent cone axis narrowly conical or slightly swollen near the base. Seed body 6–8(–10) mm long, the wing up to 1.5 times as long. Cotyledons five or six. Cheju (Quelpart) Island, Korean Peninsula and nearby northeastern China, and extreme southeastern Russian Far East. Forming pure stands or more commonly mixed with other conifers in boreal and mixed forests in dryish mountain valleys; 0–800(–1,500) m. Zone 6.

Manchurian fir is ecologically comparable to its close relative momi fir (*Abies firma*), which is the corresponding low-elevation fir in Japan. The two species share stiff needles, but those of Manchurian fir are longer and narrower than those of *A. firma,* with a sharp-pointed, not notched tip. When seed cones are present, they lack the emerging bract tips of momi fir but are otherwise similar. Despite its large size, *A. holophylla,* like other species of this group, has soft, weak wood that is not much exploited other than for pulp. It is less shade-tolerant and somewhat more drought resistant than most of the other Asian maritime species.

Abies homolepis P. Siebold & Zuccarini

NIKKO FIR, URAJIRO-MOMI (JAPANESE)

Abies sect. *Momi*

Tree to 30(–40) m tall, with trunk to 1(–1.5) m in diameter. Bark gray, becoming browner and flaky with age. Branchlets hairless and shiny, deeply grooved between the leaf bases. Buds 3–5 mm long, moderately to heavily resinous. Needles arranged to the sides of the twigs in several rows, 1–2.5(–3.5) cm long, shiny dark green above, the tips sharply forked on young trees, simply notched or blunt on older ones. Individual needles flat in cross section and with a large resin canal in the center of each side of the leaf, without stomates above and with 11–13 lines of stomates in each white stomatal band beneath. Pollen cones 10–20 mm long, yellowish brown. Seed cones (7–)9–12 cm long, 3–4 cm across, dark purple when young, ripening purplish brown. Bracts about half as long as the seed scales and hidden by them (hence the scientific name, "uniform scales," in contrast to *A. firma*). Persistent

Vigorous mature Nikko fir (*Abies homolepis*) in nature.

cone axis narrowly cylindrical. Seed body 6–9 mm long, the wing about as long. Cotyledons four (or five). Central Honshū, Kii Peninsula, and Shikoku (Japan). Occasionally forming pure stands but more often mixed with numerous other montane forest conifers and fewer hardwoods in the elevational belt between *A. firma* and *A. veitchii*; (700–)1,000–1,800(–2,200) m. Zone 5.

Nikko fir is the next species up from momi fir (*Abies firma*) in the vertical zonation of Japanese firs. It is closely related to momi fir but much less widespread, as one would expect from its higher elevation. It hybridizes with momi fir at its lower limit to form Mitsumine fir (*A. ×umbellata*), and with Veitch fir (*A. veitchii*) at its upper limit, but the latter hybrid has not been named.

Abies ×insignis Carrière ex E. Bailly

Tree to 30 m or more tall. A garden plant arising repeatedly by spontaneous and controlled hybridization of its parents (*Abies nordmanniana* × *A. pinsapo*) growing in cultivation. It has the black hairs on the twigs of Caucasian fir (*A. nordmanniana*) but differs in also having resinous buds (western races of *A. nordmanniana* usually have resinous buds but hairless twigs), resin canals well inside the leaf tissue, and slightly larger cones on average, with the bracts hidden or sticking out or up a little bit and not bent down over the scales. It differs from Pillars of Hercules fir (*A. pinsapo*) in its hairy twigs, frequently emerging bracts, and leaves concentrated to the sides and above, with quite variable tips. It was the first of the fir hybrids to arise and be distributed in cultivation, beginning in 1850. Different cultivars vary in their particular combinations of parental characters, and several were named independently. Zone 6. Synonyms: *A. alba* var. *pardei* (H. Gaussen) Silba, *A. pardei* H. Gaussen.

Abies kawakamii (Hayata) Keis. Ito

TAIWAN FIR, TAIWAN LENGSHAN (CHINESE)

Abies sect. *Momi*

Tree to 20(–35) m tall, with trunk to 1 m in diameter. Bark pale gray, soon becoming scaly and later darkening and becoming furrowed. Branchlets densely hairy, grooved between the leaf bases. Buds 4–5 mm long, very resinous. Needles arranged densely to the sides and above the twigs, where shorter, (0.5–)1–2(–3) cm long, glossy bright deep green above, the tips rounded or slightly notched. Individual needles flat or slightly plump in cross section and with a resin canal on either side touching the lower epidermis near the edges, with a few lines of stomates above and with 8–12 lines of stomates in each white stomatal band beneath. Pollen cones 9–13 mm long, greenish yellow. Seed cones oblong, 5–7.5(–9) cm long, 3–4.5 cm across, reddish purple when young, maturing purplish brown. Bracts half to two-thirds as long as the fuzzy seed scales and hidden by them. Persistent cone axis conical or swelling slightly below the middle. Seed body (6–)7–9 mm long, the wing about as long. Cotyledons four or five. Central mountain ranges of Taiwan. Forming pure stands and mixed with other conifers and a few hardwoods in the subalpine zone, particularly on the northern and northeastern side of the peaks; (2,400–)2,800–3,500(–3,800) m. Zone 6.

The affinities of this sole silver fir species of Taiwan are not entirely clear. It was originally described as a variety of Maries fir (*Abies mariesii*) of Japan, which it resembles somewhat in morphology, resin composition, and subalpine habitat. However, these resemblances are not particularly convincing, and some classifications ally it with Siberian balsam fir (*A. sibirica*), Korean fir (*A. koreana*), and Veitch fir (*A. veitchii*) while others place it among the central Chinese species. New data must be accumulated before the relationships of *A. kawakamii* can be recognized with confidence, but the few molecular data available favor placement in section *Momi*, a group of central Chinese and Japanese species, within which its geographically nearest neighbor, Baishan fir (*A.*

Natural pure stand of Taiwan fir (*Abies kawakamii*).

beshanzuensis), also belongs. The specific epithet honors Takiwa Kawakami (1871–1915), the Japanese naturalist who collected the type specimen on Yü Shan (Mount Morrison), the highest peak in Taiwan. With its relatively small stature and remote habitat, this species has little commercial importance. At the top of its elevational distribution, it mixes with and gives way to scaly juniper (*Juniperus squamata*), while below it is replaced by several conifers, with Chinese hemlock (*Tsuga chinensis*) dominating.

Mature Korean fir (*Abies koreana*) in nature.

Upper side (mostly) of twig of Korean fir (*Abies koreana*) with 3 years of growth, ×0.5.

Abies koreana E. H. Wilson

KOREAN FIR, KUSANG NAMU (KOREAN)

Abies sect. *Balsamea*

Tree to 18 m tall, with trunk to 0.8 m in diameter. Bark pale gray, darkening and breaking up into blocks with age. Young branchlets with a few hairs in the shallow grooves between the leaf bases, but these soon lost. Buds 3–5 mm long, lightly resinous. Needles standing out all around the twigs, although shorter, denser, and curved above, mostly 1–2 cm long, shining dark green above, widest near the pointed (on cone-bearing branches) or notched tip. Individual needles flat or plump in cross section, the edges slightly rolled down, and with a resin canal near the center of either side, without stomates on the upper side and with 8–10 lines of stomates in each gleaming white stomatal band beneath. Pollen cones 8–15 mm long, red. Seed cones cylindrical, 4–6(–7) cm long, 2–3 cm across, rich purple when young, maturing purplish brown. Bracts about as long as the minutely hairy seed scales and sticking out a little and back over them. Persistent cone axis narrowly conical. Seed body 5–6 mm long, the wing only about half as long. Cotyledons four or five. Cheju (Quelpart) Island and nearby mainland South Korea. Mostly mixed with other conifers and hardwoods in the subalpine forest of mountain summits; 1,000–1,850 m. Zone 6.

Korean fir is an elegant small tree popular in cultivation for its intensely silvery stomatal bands contrasting with the deep green upper leaf surfaces and its production of intensely purple seed cones from a young age. Numerous cultivars emphasize these traits. In nature, it is confined to the summits of the highest peaks in southern Korea, a habitat corresponding to its close relative Veitch fir (*Abies veitchii*) of Japan. Northward, it is replaced by another close relative, Amur fir (*A. nephrolepis*), which has a much wider geographic and elevational range, in a pattern mirroring those of the related species pairs in section *Balsamea,* Sakhalin fir (*A. sachalinensis*)–Veitch fir (*A. veitchii*) and balsam fir (*A. balsamea*)–Fraser fir (*A. fraseri*).

Abies lasiocarpa (W. J. Hooker) T. Nuttall

SUBALPINE FIR

Plate 6

Abies sect. *Balsamea*

Tree to 30(–40) m tall, with trunk to 1.2(–2) m in diameter. Bark reddish to dark gray when young, soon breaking up into scales or

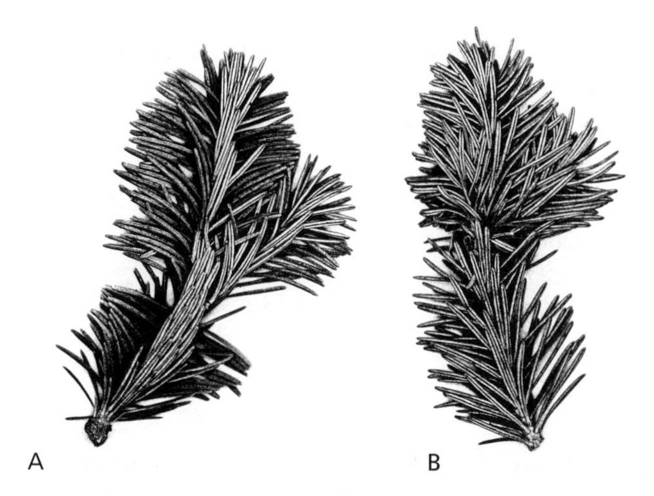

Upper (A) and undersides (B) of twigs of subalpine fir (*Abies lasiocarpa*) with 2 years of growth, ×0.5.

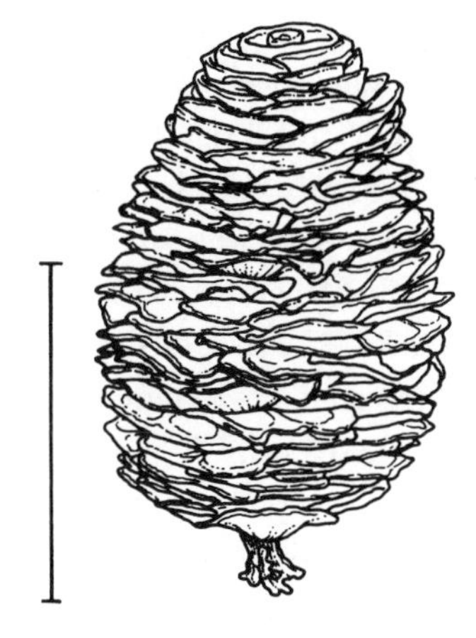

Mature seed cone of subalpine fir (*Abies lasiocarpa*) at point of shattering; scale, 5 cm.

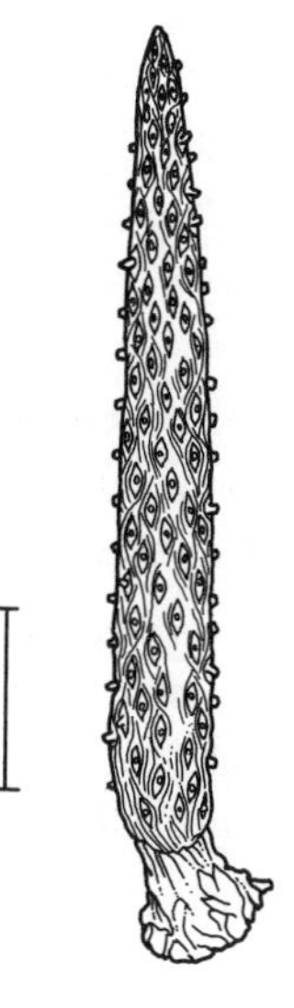

Central axis of seed cone of subalpine fir (*Abies lasiocarpa*) after shattering and shedding the seeds and seed scales; scale, 2 cm.

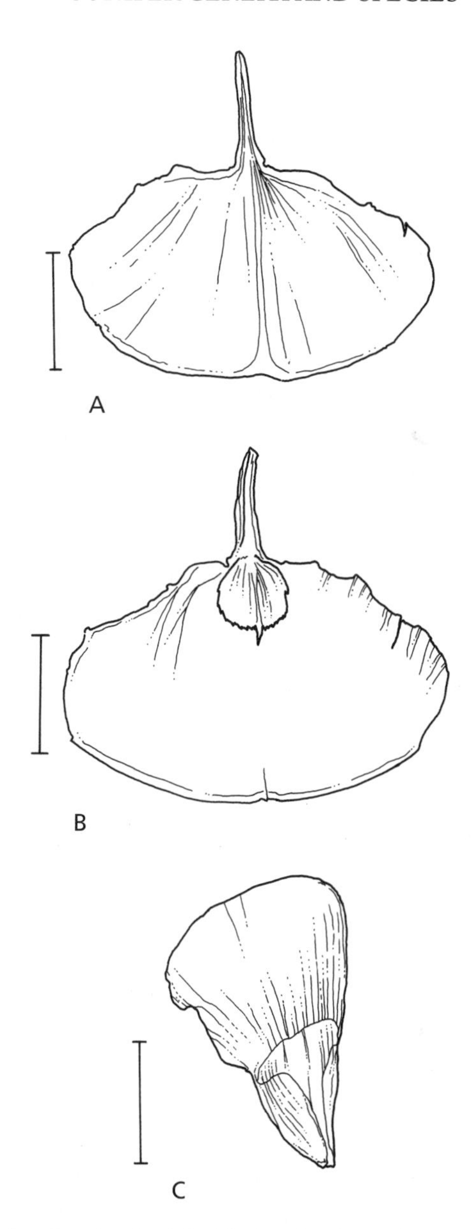

Upper (A) and undersides (B) of shed seed scale with bract and seed (C) of subalpine fir (*Abies lasiocarpa*); scale, 1 cm.

becoming thick and corky with age in the southern Rocky Mountains. Branchlets reddish brown, hairy, finely grooved between the leaf attachments. Buds 3–6 mm long, resinous at the tips or overall. Needles arranged all around the twigs but turned to the sides and above on both upper and lower branches, (1–)1.5–2.5(–3) cm long, light green or bluish green with wax above, the tips notched or rounded. Individual needles flat in cross section and with a large resin canal in the center of either side, with three to six rows of stomates in a groove above and three to five rows in each stomatal band beneath. Pollen cones 13–20 mm long, greenish purple. Seed cones cylindrical, (5–)7–10(–12) cm long, (2–)3–4 cm across, dull purple when young, maturing purplish brown. Bracts much shorter than the densely short-hairy scales (hence the scientific name, "hairy fruit") and hidden by them (except when stunted by cone insects). Persistent cone axis narrowly cylindrical. Seed body 5–7 mm long, the wing up to about twice as long. Cotyledons three to six. Widespread in western North America, from eastern Alaska and the southern Yukon to northwestern California, southeastern Arizona, and southwestern New Mexico. Growing with other conifers such as Engelmann spruce (*Picea engelmannii*) and lodgepole pine (*Pinus contorta*), in subalpine forests; (600–)1,500–2,700(–3,650) m. Zone 2. Synonym: *A. balsamea* subsp. *lasiocarpa* (W. J. Hooker) Boivin. Three varieties.

Subalpine fir is the smallest of the western North America firs though larger than its close relatives balsam fir (*Abies balsamea*) and Fraser fir (*A. fraseri*) in eastern North America. The largest known individual in the United States in combined measurements when measured in 1992 was 38.1 m tall, 2.0 m in diameter, with a spread of 7.9 m. The trees are usually much smaller than this in their harsh subalpine forests, sometimes reduced to shrubs right at the alpine timberline. Because of its small size and high mountain habitat, subalpine fir is not an important commercial timber species. It hybridizes with *A. balsamea* where they come in contact in central Alberta but does not seem to hybridize with the unrelated Pacific silver fir (*A. amabilis*) and noble fir (*A. procera*) where it overlaps with them at its lower elevational limit in the Cascade Range of Oregon and Washington. The species exhibits extensive geographic variation, with differences in resin composition and morphology among trees of the Pacific slope, southern Rocky Mountains, and northern and central Rocky Mountains. These led some authors to recognize two or three separate species, but because the morphological differences are few, they are here considered varieties. Pacific subalpine fir (variety *lasiocarpa*) and Rocky Mountain subalpine fir (variety *bifolia*) intergrade across the Continental Divide in the northern Rocky Mountains.

***Abies lasiocarpa* var. *arizonica* (Merriam) J. Lemmon**

CORKBARK FIR

Bark thick and corky. Fresh leaf scars tan around the edges. Southern Colorado through Arizona and New Mexico.

***Abies lasiocarpa* var. *bifolia* (A. Murray bis) Eckenwalder**

ROCKY MOUNTAIN SUBALPINE FIR

Bark thin. Fresh leaf scars tan around the edges. East of the Continental Divide from the Yukon to Utah and central Colorado. Synonym: *Abies bifolia* A. Murray bis.

Abies lasiocarpa* (W. J. Hooker) T. Nuttall var. *lasiocarpa

PACIFIC SUBALPINE FIR

Bark thin. Fresh leaf scars red around the edges. Pacific slope from eastern Alaska and the Yukon to northwestern California.

***Abies magnifica* A. Murray bis**

RED FIR

Plate 18

Abies sect. *Nobilis*

Tree to 55(–60?) m tall, with trunk to 2.6 m in diameter. Bark grayish when young, soon becoming reddish brown and strongly ridged and furrowed. Branchlets with dense reddish hairs at first, grooved between the leaf bases. Buds 3–5 mm long, not resinous, sparsely hairy. Needles all pointing upward on the twig from an abruptly bent leaf base, 2–3.5 cm long, bluish or grayish green above, the tips rounded on lower branches to pointed on cone-bearing branches. Individual needles flat to very plump in cross section and with a resin canal on either side near the edge just inside the lower epidermis, with 8–13 rows of stomates above and four or five rows in each waxy stomatal band beneath. Pollen cones 15–20 mm long, reddish brown. Seed cones elongate egg-shaped, (10–)14–20 cm long, (3–)7–10 cm across, purple when young, maturing yellowish brown. Bracts a little shorter than the seed scales and hidden by them or (in variety *shastensis*) slightly longer and bent back over them without fully covering them. Seed body 12–15 mm long, the wing about as long. Cotyledons seven or eight. Southeastern Oregon to northern Coast Ranges and southern Sierra Nevada of California. Forming pure stands or mixed with other conifers, including white fir (*A. concolor*), incense cedar (*Calocedrus decurrens*), and giant sequoia (*Sequoiadendron giganteum*), in midelevation montane coniferous forest; 1,400–2,700(–2,900) m. Zone 6. Two varieties.

Red fir is a major timber tree of exceptionally high productivity in the Sierra Nevada of California. Its seedlings have very high shade tolerance and a high tolerance for crowding and so replace even the giant sequoia in the absence of fires. The largest known individual in combined dimensions when measured in 1996 was 52.5 m tall, 3.0 m in diameter, with a spread of 12.5 m. It is closely related to noble fir (*Abies procera*), which replaces it northward via a zone of overlap and intergradation in southwestern Oregon and northern California. The two species are artificially intercrossable but only fully so with red fir as the seed parent. The Shasta red fir (variety *shastensis*), with its protruding bracts, is often considered

Natural stand of red fir (*Abies magnifica*).

a product of introgression in the overlap zone, but the only characteristic in which it resembles noble fir is the emergence of the bracts. In all other features which differentiate red and noble firs, it is like red fir or somewhat distinct from either. Like analogous bracted variants of balsam fir (*A. balsamea*), then, it is probably not a result of hybridization but simply of variation within red fir. Red fir also commonly grows with the generally lower-elevation white fir (*A. concolor*) wherever the elevational ranges of these two common species overlap. They are unrelated and never hybridize with one another.

Abies magnifica A. Murray bis **var. *magnifica***
CALIFORNIA RED FIR
Bracts entirely hidden by the seed scales. Southern portion of range in California from Mount Lassen southward.

Abies magnifica* var. *shastensis J. Lemmon
SHASTA RED FIR
Bracts bent back over seed scales but not completely covering them. Northern portion of range in California and Oregon from Mount Lassen northward (including Mount Shasta). Synonym: *Abies* [×]*shastensis* (J. Lemmon) J. Lemmon.

Abies mariesii M. T. Masters
MARIES FIR, AOMORI-TODOMATSU, Ō-SHIRABISO (JAPANESE)
Abies sect. *Amabilis*
Tree to 25(–30) m tall, or just 5 m or less at the alpine timberline, with trunk to 1 m in diameter. Bark purple-tinged very pale gray, finally breaking up somewhat and darkening with age on the lower trunk. Branchlets densely covered with fine reddish brown hairs, deeply but narrowly grooved. Buds 2–3 mm long, resinous and hairy. Needles arranged to the sides and angling forward, but shorter and even more densely curved upward and covering the twigs, (0.6–)1.5–2.5 cm long, dark green above, the tips rounded or notched. Individual needles flat in cross section, with a resin canal on either side touching the lower epidermis near the margin of the leaf, without stomates above and with (10–)12(–14) lines of stomates in each white stomatal band beneath. Pollen cones 15–20 mm long, purplish brown. Seed cones egg-shaped, (4–)6–9 cm long, (1.5–)2–4(–4.5) cm across, very dark purple when young, maturing dark purplish brown. Bracts up to about half as long as the internally densely hairy seed scales and hidden by them. Persistent cone axis spindle-shaped. Seed body (4–)6–8 mm long, the wing up to 1.5 times as long. Cotyledons four to six. Mountains of central and northern Honshū (Japan). Forming pure stands near timberline or mixed with several other conifers and a few hardwoods on moist soils in the subalpine zone; (600–)1,000–2,600(–2,900) m. Zone 6.

Maries fir forms the top band in the elevational zonation of Japanese fir species and is one of the most important trees of the subalpine forest. It is not closely related to any of the other Japanese species, and genetic data from DNA and enzymes provides no evidence of hybridization with Veitch fir (*Abies veitchii*), its nearest neighbor. There is evidence, however, of postglacial migration from Pleistocene refuges, with loss of genetic diversity northward from central Honshū. Although not yet confirmed by chemical or DNA data, the closest relationship of Maries fir appears to be with the American Pacific silver fir (*A. amabilis*), a species in its turn unrelated to the other North American firs. A possible relationship to the Taiwan fir (*A. kawakamii*), as reflected in early taxonomic treatments, has little support from either molecules or morphology. Charles Maries (ca. 1851–1902) brought this species to the attention of European botanists after finding it on Mount Hakkoda in 1878, while collecting for the Veitch nursery firm, and it was described from his collections there.

Abies nebrodensis (Lojacano) G. Mattei

SICILIAN FIR, ABETE DELLE NEBRODI (ITALIAN)

Abies sect. *Abies*

Tree to 15 m tall, with trunk to 0.6 m in diameter. Bark light gray, becoming scaly with age and ultimately slightly furrowed at the base. Branchlets without hairs, slightly grooved between the leaf bases. Buds 4–5(–9) mm long, not prominently resinous. Needles stiff, arranged straight out to the sides and above the twigs but with a gap beneath them, 8–15(–20) mm long, bright, shiny green above, the tip blunt to prickly. Individual needles plump in cross section and with a small resin canal on either side near the edge, just inside the lower epidermis, without or with a few broken lines of stomates near the tip above and with 6–11 rows of stomates in each broad stomatal band beneath. Pollen cones 15–20 mm long, purple. Seed cones cylindrical, 7–12(–20) cm long, 3–4(–5) cm across, green when young, maturing yellowish brown. Bracts about as long as the minutely hairy seed scales, sticking out between them, and bent down over them. Persistent cone axis narrowly conical. Seed body (6–)8–10(–12) mm long, the wing a little longer. Mountains of northern Sicily and perhaps the toe of Italy. Once forming pure forests or mixed with black pine (*Pinus nigra*) and other trees and shrubs at its lower limit; 1,400–1,600(–2,000) m. Zone 6. Synonym: *A. alba* var. *nebrodensis* (Lojacano) Svoboda.

One of the last mature wild Sicilian firs (*Abies nebrodensis*).

As the premier timber tree of Sicily, *Abies nebrodensis* was brought to the verge of extinction over the centuries, with just one mature tree remaining among a score of saplings in its last natural stand on Monte Cervo. Reforestation efforts prevented its complete demise, but the genetic base of the species cannot help but have been affected by the population bottleneck. Sicilian fir is close to European silver fir (*A. alba*) and has been treated as a variety or subspecies of it or as a hybrid between *A. alba* and Algerian fir (*A. pinsapo* var. *numidica*) or Grecian fir (*A. cephalonica*). There is no substantive evidence to back up any of these interpretations. Populations of *A. alba* in southernmost mainland Italy show some evidence of past hybridization with *A. nebrodensis,* presumably dating from the late glacial when the two were packed together in the southern Italian refugium.

Abies nephrolepis (Trautvetter) Maximowicz

AMUR FIR, PIKHTA AMURSKAYA (RUSSIAN), CHOU LENGSHAN (CHINESE), PUNBI NAMU (KOREAN)

Abies sect. *Balsamea*

Tree to 25(–35) m tall, with trunk to 0.8(–1.2) m in diameter. Bark pale gray, eventually becoming shallowly fissured. Branchlets with a few to many dark hairs in the shallow grooves between the leaf bases. Buds 4–5 mm long, thinly resinous mostly at the tip. Needles angled forward, arranged to the sides and above the twigs, (1–)1.5–2(–3) cm long, shiny bright green above, the tips pointed on cone-bearing shoots, otherwise shallowly notched. Individual needles flat in cross section with the edges rolled down and with a resin canal closer to the top than the bottom of either side, often with two or three short lines of stomates in the groove above near the tip and with 8–10 lines in each grayish white stomatal band beneath. Pollen cones 35–45 mm long, reddish purple. Seed cones elongate egg-shaped to cylindrical, (4–)5–7.5(–9.5) cm long, (1.5–)2–3(–3.5) cm in diameter, reddish purple (or green) when young, maturing purplish brown. Bracts a little shorter than the fuzzy, kidney-shaped seed scales (hence the scientific name, "kidney scale") but with a long, slender tip that sticks out between the scales without bending back down over them. Persistent cone axis narrowly conical. Seed body 4–6 mm long, the wing almost as long. Cotyledons four or five. Southeastern Russian Far East east of the Uda River, south to Shanxi province (China) and North Chŏlla province (South Korea), very discontinuous south of Russia. Forming pure stands or scattered among other conifers and deciduous hardwoods in wet areas on well-drained mountain soils; (300–)700–2,100 m. Zone 4.

Amur fir is scarcely seen in cultivation, being replaced almost entirely by its rarer close relative, Korean fir (*Abies koreana*), which is essentially similar but more elegant. Amur fir overlaps in the northwestern part of its range with another relative, Siberian balsam fir (*A. sibirica*). The two are somewhat segregated elevationally since Siberian balsam fir, here in the southern part of its distribution, occurs at somewhat higher elevations than Amur fir at the northern part of its range. The two species cross occasionally to form a sterile hybrid, *A.* ×*sibirico-nephrolepis*. As a tree of relatively small stature, Amur fir is not much exploited in forestry, but is used for pulp and to a lesser extent in construction and carpentry.

Abies nordmanniana (Steven) Spach

CAUCASIAN FIR, DOĞU KARADENIZ GÖKNAN (TURKISH), SOTCHI (GEORGIAN), EGEVIN (ARMENIAN), AMZA (ABKHAZIAN), PIKHTA KAVKAZSKAYA (RUSSIAN)

Abies sect. *Abies*

Tree to 50(–70) m tall, with trunk to 1.5(–2) m in diameter. Bark gray, becoming shallowly furrowed with age. Branchlets with scattered to dense reddish or blackish hairs or none, grooved between the leaf bases. Buds 3.5–5(–6) mm long, not resinous or slightly (to heavily) resinous. Needles arranged to the sides and angled forward above and covering the twigs on lower branches, curved upward on coning branches, (1.5–)2–3(–4) cm long, bright dark green above, the tips prominently notched or rounded. Individual needles flat in cross section and with a resin canal on either side near the edge just inside the lower epidermis or deeper, without stomates or with a few discontinuous lines of stomates near the tip above and with eight or nine lines in each white stomatal band beneath. Pollen cones 8–12(–20) mm long, reddish purple or yellow. Seed cones cylindrical, (10.5–)12–16.5(–20) cm long, 4–6 cm across, green when young, maturing reddish brown. Bract blades about as long as the seed scales (longer with the tips), extending between them and bending down over them and then up at the tip. Persistent cone axis narrowly conical. Seed body (8–)10–12 mm long, the wing about as long to 1.5 times as long. Cotyledons six or seven. Mountains surrounding the Black Sea on the northeast, east, and south, from southernmost Russia through Georgia and Armenia to Turkey west to Ulu Dăg [Mount Olympus]. Forming pure stands or mixed with Caucasian spruce (*Picea orientalis*) and other trees; (900–)1,300–1,900(–2,200) m. Zone 5. Synonyms: *A.* [×]*bornmuelleriana* Mattfeld, *A. nordmanniana* subsp. *bornmuelleriana* (Mattfeld) Coode & Cullen.

Caucasian fir is a handsome tree of the cool, humid Black Sea mountains, including the western Caucasus and Pontic Mountains, where the trees can reach an age of 500 years. It is named for Alexander Nordmann (1803–1866), a Finnish botanist who collected the type specimen in 1836 near the headwaters of the Kura River in Georgia and introduced it into general cultivation. The hairiness of the twigs and the amount of resin on the buds vary tremendously throughout the range of the species. Trees in

Young mature Caucasian fir (*Abies nordmanniana*) in nature.

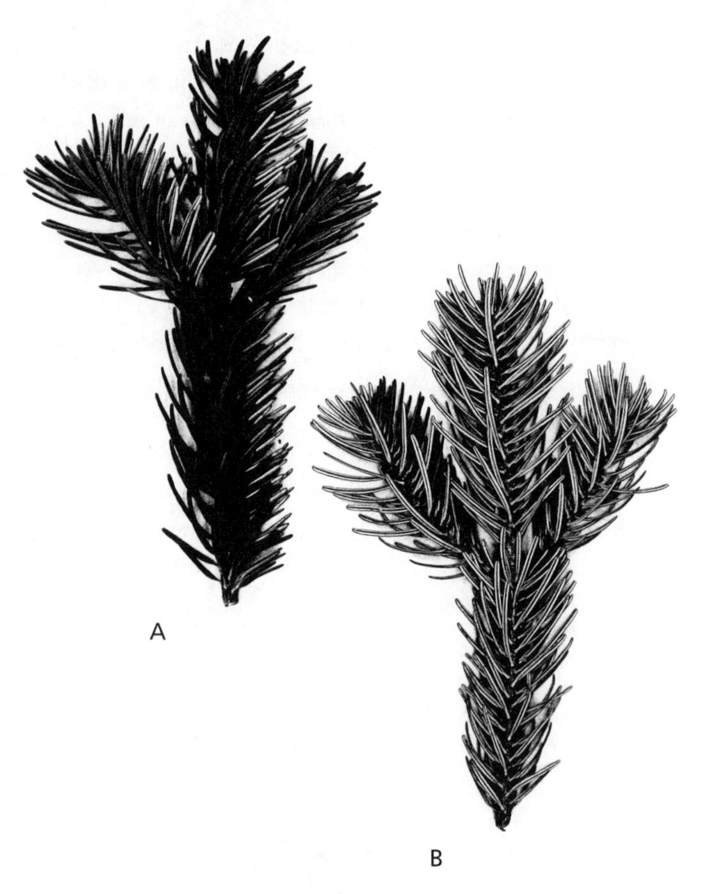

Upper (A) and undersides (B) of twigs of Caucasian fir (*Abies nordmanniana*) with 2 years of growth, ×0.5.

western Turkey usually have glabrous twigs and resinous buds, while those in the Pontic Mountains usually have hairy twigs and dry buds, but the distinction is not sharp and breaks down further in the Caucasus. There thus seems to be little justification for separating the western populations as a separate species (*Abies bornmuelleriana*) or subspecies. Though they may, in part, owe their characteristics to past hybridization with Grecian fir (*A. cephalonica*), they do not seem to represent direct hybrids either, unlike Trojan fir (*A. ×equi-trojani*). Because of the continuous nature of variation in the Caucasian fir, it is not divided here it into two or more varieties. The size of the tree and the quality of its wood make it an important timber tree throughout its native range, and rapid growth has led to its use as a plantation tree elsewhere.

Abies pindrow (Royle ex D. Don) Royle

PINDROW FIR, HIMALAYAN SILVER FIR, PINDRAU (HINDI)

Abies sect. *Pseudopicea*

Tree to 60 m tall, with trunk to 2.5(–3) m in diameter. Bark silvery gray, darkening and becoming strongly ridged and furrowed with age. Branchlets hairless, shallowly grooved between the leaf bases. Buds 3–4(–8) mm long, covered with white resin. Needles arranged mostly to the sides of the twigs in several rows of different lengths, (2–)2.5–6(–9) cm long, shiny bright green above, the tips sharply forked to just notched. Individual needles flat in cross section and with a resin canal on either side near the edge a little separated from the lower epidermis, usually without stomates above and with six to eight lines of stomates in each dull light gray stomatal band beneath. Pollen cones 10–15 mm long, brown. Seed cones cylindrical, 10–14(–18) cm long, (4–)5–6.5(–7.5) cm across, dark violet when young, maturing dark reddish brown. Bracts up to half as long as the seed scales and hidden by them. Persistent cone axis narrowly conical. Seed body 10–12 mm long, the wing about 1.5 times as long. Cotyledons four to six. Himalaya, from northeastern Afghanistan to Nepal and adjacent China in Xizang (Tibet). Forming pure stands or mixed with other conifers and hardwoods on cool, moist, snowy sites in the mountains; (300–)2,100–2,700(–3,700) m. Zone 8. Synonyms: *A. gamblei* P. Hickel, *A. pindrow* var. *brevifolia* Dallimore & A. B. Jackson.

Pindrow fir (from the common name pindrau in the Simla Hills) has an extensive distribution at moderate elevations in the western Himalaya and is heavily exploited for timber. It generally grows at lower elevations than the related Webb fir (*Abies spectabilis*), but the two species have a broad zone of overlap and frequently hybridize to produce a variety of intermediates that are more similar to pindrow fir at lower elevations and to Webb fir higher up. Some of the variants described in each species relate to these hybrids. Much like deodar cedar (*Cedrus deodara*), this Himalayan pair is far separated from the nearest Mediterranean silver firs and scarcely contacts the Chinese species toward the east. The exact degree of relationship between them is controversial so that some authors put them in their own taxonomic section together, while others place each in a separate section with other species. Here, they are placed together with other Himalayan and western Chinese species in section *Pseudopicea.*

Abies pinsapo Boissier

PILLARS OF HERCULES FIR, PINSAPO (SPANISH)

Abies sect. *Abies*

Tree to 30(–50) m tall, with trunk to 0.8(–1.5) m in diameter. Bark grayish brown, becoming deeply ridged and furrowed with age. Branchlets hairless, often shiny, grooved between the leaf bases. Buds 3.5–5 mm long, resinous or not. Needles arranged straight out all around the twigs or bent upward or denser toward the sides, (0.6–)1–2(–2.5) cm long, shiny dark green above, the tip sharp to blunt, rounded, or even slightly notched. Individual needles stiff, thick in cross section and with a resin canal on either

Upper (A) and undersides (B) of twigs of Spanish fir (*Abies pinsapo*) with 2 years of growth, ×0.5.

Battered old Spanish fir (*Abies pinsapo*) in nature.

side near the edge either just inside the lower epidermis or deep within the leaf tissue, with 1–14 continuous or discontinuous lines of stomates in a variably expressed central groove above, especially near the apex, and 6–7(–10) lines in each greenish white stomatal band beneath. Pollen cones 5–7(–15) mm long, reddish purple. Seed cones cylindrical, (9–)12–18(–25) cm long, 3.5–5(–6) cm across, greenish, yellowish, or reddish brown when young, maturing yellowish to purplish brown. Bracts no more than half as long as the seed scales and hidden by them. Persistent cone axis narrowly conical. Seed body 7–10 mm long, the wing up to 1.5 times as long. Cotyledons five to seven. Mountains of the southwestern Mediterranean in southeastern Spain, northern Morocco, and northeastern Algeria. Forming pure stands or mixed with Atlas cedar (*Cedrus libani* subsp. *atlantica*) in rocky soils at higher elevations and with other conifers and evergreen hardwoods below; (1,000–)1,200–1,800(–2,100) m. Zone 6. Four varieties.

Populations of trees in the four widely scattered regions of this species differ from one another and were described as separate species. They are closely related and are here treated as varieties of a single species, but their relationships have never been thoroughly assessed at a population level. Because there are so few populations of each of these firs, however, even the most detailed analyses are likely to leave a subjective element in deciding their taxonomic status. These are among the most drought tolerant of firs and, while they can benefit from abundant moisture, they are not tolerant of waterlogged soils. The twigs can be mashed in water to produce a soap, thus explaining the species and Spanish common name ("soap pine").

Abies pinsapo var. marocana (Trabut) Ceballos & Bolaños

RIF FIR, MOROCCAN FIR, SAPIN DE MAROC (FRENCH)

Bud scales loose, smooth-edged, resinous. Needles crowded above the twigs, with shallow resin canals and 1–4(–11) interrupted lines of stomates above. Rif Mountains near Chechaouèn, Morocco. Synonym: *Abies marocana* Trabut.

Abies pinsapo var. numidica (de Lannoy ex Carrière) Salomon

ALGERIAN FIR, SAPIN DES BABORS (FRENCH)

Bud scales loose, minutely fringed, nonresinous. Needles crowded above the twigs, sometimes pointed somewhat backward, with shallow resin canals and three to eight interrupted or continuous lines of stomates above. Mounts Babor and Thababor, northeastern Algeria. Synonym: *Abies numidica* de Lannoy ex Carrière.

Abies pinsapo Boissier **var. *pinsapo***
SPANISH FIR, PINSAPO (SPANISH)
Bud scales tight, smooth-edged, resinous. Needles spreading all around the twigs, with deep resin canals (except shallow in juvenile leaves) and 6–14 continuous lines of stomates above. Mountains of Granada and Malaga provinces in southern Spain.

Abies pinsapo* var. *tazaotana (Côzar ex E. Villar) Pourtet
TAZAOTAN FIR, SAPIN DE TAZAOT (FRENCH)
Bud scales loose, smooth-edged, nonresinous. Needles crowded above the twigs, with shallow resin canals and one or two inconspicuous lines of stomates above. Tazaot massif, northern Morocco. Synonym: *Abies tazaotana* Côzar ex E. Villar.

Abies procera Rehder
NOBLE FIR
Abies sect. *Nobilis*
Tree to 70(–85) m tall, with trunk to 2.5(–3) m in diameter. Bark dark gray, soon becoming reddish gray, narrowly ridged and furrowed. Branchlets finely hairy, shallowly grooved between the leaf bases. Buds 2.5–3.5 mm, not resinous, hairy on the basal scales, hidden among leaves. Needles all pointing upward on the twig from an abruptly bent leaf base, (1–)2–3(–3.5) cm long, bluish green with wax above before becoming dull dark green, the tips notched to rounded, occasionally pointed. Individual needles flat to slightly plump in cross section and with a resin canal on either side just inside the lower epidermis somewhere from the midpoint to near the edge, with (or without) up to 14 rows of stomates in the groove above, sometimes divided between two stomatal bands, and (four to) six or seven rows in each stomatal band beneath. Pollen cones 15–25 mm long, reddish purple. Seed cones barrel-shaped, 10–15(–30) cm long, 5–6.5(–8) cm across, green or reddish purple when young, maturing light brown. Bracts yellow-green, longer than the seed scales and bent back over to cover them. Seed body 12–13 mm long, the wing a little longer. Cotyledons mostly five or six. Cascade and Coast Ranges of Washington, Oregon, and northwestern California. Forming pure stands or, more commonly, mixed with almost any conifers of the region; (60–)900–2,200(–2,700) m. Zone 6.

Noble fir is the largest American fir, explaining its common and scientific names ("tall"). The largest known individuals, in combined measurements, when measured in 1989 were (1) 83.0 m tall, 2.5 m in diameter, with a spread of 14.9 m and (2) 69.2 m tall, 2.9 m in diameter, with a spread of 12.5 m. It is an important

Upper (A) and undersides (B) of twigs of noble fir (*Abies procera*) with 2 years of growth, ×0.5.

commercial species but not ordinarily singled out among its associated conifers. The timber was historically called "larch," leading to the name Larch Mountain in the Coast Ranges of Oregon where *Larix* does not occur. In the south it overlaps and hybridizes with the closely related red fir (*Abies magnifica*) but does not hybridize with any of the other *Abies* species with which it grows, all being rather distantly related.

Abies recurvata M. T. Masters
MIN FIR, ZIGUO LENGSHAN (CHINESE)
Abies sect. *Momi*
Tree to 40(–60) m tall, with trunk to 1.5(–2.5) m in diameter. Bark dark gray, browning, flaking, and becoming ridged and furrowed with age. Branchlets hairless or with short-lived, small, dark hairs in the deep grooves between the leaf bases. Buds 4–8 mm long, variably resinous. Needles arranged to the sides and above the twigs, angled backward on the leader and other exceptionally vigorous shoots, 1–2.5(–3.5) cm long, sometimes widest near the tip, shiny bright green or dulled with wax above, the tip bluntly to sharply pointed, even prickly on young trees, or occasionally notched. Individual needles flat in cross section and with a resin canal on either side touching the lower epider-

Young mature Min fir (*Abies recurvata*) in nature.

mis or away from it near the margin, without or with five to seven (to eight) incomplete lines of stomates near the tip in the groove above and with 9–13 lines in each greenish white stomatal band beneath. Pollen cones 10–15 mm long, red. Seed cones elongate egg-shaped to cylindrical, (4–)5–8(–10) cm long, 2.5–3.5(–5) cm across, violet-purple when young, maturing grayish brown. Bracts usually less than three-quarters as long as the fuzzy seed scales and hidden by them but sometimes the tips just sticking out in the lower half of the cone. Persistent cone axis thick and swollen in the middle. Seed body (5–)7–10 mm long, the wing a little shorter. Cotyledons four to six. West-central China, from southern Gansu to southeastern Xizang (Tibet). Often forming pure stands or mixed with numerous other conifers and some hardwoods in a high montane forest in valleys and on gentle slopes; 2,300–3,800 m. Zone 6. Two varieties.

The recurved needles that give Min fir (from the Min River valley in Sichuan) its scientific name are one of the most distinctive features of the species but are not its most common leaf arrangement, being found only on the most vigorous non-cone-bearing shoots. This is the largest silver fir in China, and the fact that its wood is among the hardest in the genus has led to heavy exploitation for construction timber. The wider affinities of the Min fir are uncertain, and different authorities ally it with the western Chinese species related to Delavay fir (*Abies delavayi*) on the basis of the thick cone axis, or with the northeastern Chinese and Japanese species related to Manchurian fir (*A. holophylla*) and Nikko fir (*A. homolepis*). It is included with the latter group here, section *Momi.* Few features separate the two recognized varieties of the species, and little is known of their behavior in their overlap zone in Sichuan.

Abies recurvata var. *ernestii* (Rehder) K. Rushforth

WILSON FIR, HUANGGUO LENGSHAN, YUAN BIAN ZHONG (CHINESE)

Branchlets sometimes lightly hairy, the grooves very deep. Buds 6–8 mm long. Needles generally less strongly recurved, often 2–3.5 cm long, the resin canals touching the epidermis. Cone axis more strongly swollen. Western Sichuan (China) and eastern Xizang (Tibet). Synonyms: *Abies chensiensis* var. *ernestii* (Rehder) Tang S. Liu, *A. ernestii* Rehder.

Abies recurvata M. T. Masters var. *recurvata*

MIN FIR, ZIGUO LENGSHAN (CHINESE)

Branchlets hairless, less deeply grooved. Buds 4–6 mm long. Needles sometimes strongly recurved, often 1–2.5 cm long, the resin canals away from the epidermis. Cone axis more nearly thickly conical. Southern Gansu to northwestern Sichuan (China).

Abies religiosa (Humboldt, Bonpland & Kunth) D. F. L. Schlechtendal & Chamisso

OYAMEL FIR, SACRED FIR, OYAMEL, PINABETE (SPANISH)

Abies sect. *Grandis*

Tree to 45(–60) m tall, trunk to 1.5(–2) m in diameter. Bark grayish white, darkening or browning and breaking up into irregular plates with age. Branchlets at first sparsely to densely hairy in the prominent grooves between the leaf bases, the hairs disappearing with age. Buds 2–4 mm, thickly covered with white to yellowish resin. Needles arranged to the sides in several rows and also forward over the twig on lower branches and upward on higher branches with seed cones (1–)3–5(–9) cm long, shiny bright green to waxy bluish green above, the tip usually pointed or rounded but sometimes notched. Individual needles flat in cross section and with a resin canal on each side touching the lower epidermis near the outer margin, without stomates or with about five interrupted lines of stomates in the groove above, especially near the tip, and with 8–12 lines in each bluish white waxy stomatal band beneath. Pollen cones (12–)20–40 mm long, red. Seed cones cylindrical to oblong, (8–)12–16 cm long, 4–6.5(–8) cm across, purple when young, maturing purplish brown. Bracts about as long as the minutely fuzzy seed scales or somewhat longer, emerging between the scales and bent down over them. Persistent cone axis narrowly conical. Seed body (8–)9–12(–14) mm long, the wing 1–1.5 times as long. Cotyledons five or six. Central Mexico, from Jalisco and Guerrero to Hidalgo and Veracruz. Forming pure stands or mixed with pines (*Pinus*), oaks (*Quercus*), or other hardwoods on mountainsides and in mountain valleys; (1,800–)2,500–3,500(–4,100) m. Zone 9. Synonyms: *A. colimensis* Rushforth & Narave, *A. religiosa* var. *glaucescens* (G. Gordon) Carrière.

Oyamel fir was the first of the Mesoamerican fir species to be described. It occurs at high elevations across the central Mexican volcanic belt on numerous separated mountains. It is highly variable across this range but even more so with elevation. In at least some locations, it exhibits the common fir pattern of greater emergence of the bracts with elevation. Needle length and shape of the needle tip also vary confusingly. There seems little justification for subdividing the species without a thorough study across its whole distribution. Throughout central Mexico the branches are cut to decorate churches on feast days, explaining the scientific name. The common name oyamel is the Mexican Spanish continuation of the classical Aztec name for the tree, *oyametl.* The dense, windless groves of oyamel fir, protected from frost, are the overwintering grounds of the monarch butterflies (*Danaus plexippus*) that migrate south from eastern North America each autumn.

Abies sachalinensis (Friedr. Schmidt) M. T. Masters

SAKHALIN FIR, HOKKAIDŌ FIR, TODOMATSU (JAPANESE), PIKHTA SAKHALINSKAYA (RUSSIAN), YAYUP (AINU)

Abies sect. *Balsamea*

Tree to 30(–40) m tall, or dwarfed at 2–3 m in exposed habitats, with trunk to 1 m in diameter. Bark grayish white, remaining smooth or breaking up into grayish brown scaly ridges with age. Branchlets densely hairy in the prominent shallow grooves between the leaf bases or overall. Buds 3–5 mm long, hairy and moderately to heavily resinous. Needles arranged evenly and densely out to the sides and above the twigs and hiding them, (1–)1.5–3(–3.5) cm long, intense bright green above, the tips blunt, rounded, or slightly notched. Individual needles flat to slightly plump in cross section and with a resin canal near the center of each side (rarely near the lower epidermis), without stomates above and with seven to nine lines of stomates in each white stomatal band beneath. Pollen cones 10–15 mm long, red. Seed cones tapering strongly in both directions from the midpoint, (2.5–)5–7(–8) cm long, 2–2.5(–3) cm across, variously greenish brown through purple to black when young, maturing purplish brown. Bracts yellowish green to reddish brown, a little shorter to a little longer than the densely minutely hairy scales and hidden by them to emerging from and bent back over them. Persistent cone axis narrowly cylindrical to conical. Seed body (4–)5–6 mm long, the wing half to two-thirds as long. Cotyledons four or five. Sakhalin (hence the scientific name) and southeastern Kamchatka (Russia) to Hokkaidō (Japan). Forming pure stands or mixed with other conifers and hardwoods in subalpine, boreal, and mixed forests; (0–)100–1,650 m. Zone 6. Two varieties.

Sakhalin fir is the only silver fir species occurring in its northern island range, where it grows with Sakhalin spruce (*Picea glehnii*), Yezo spruce (*P. jezoensis*), and Dahurian larch (*Larix gmelinii*) in a mixture similar to the North American boreal forest. At its upper limit, it overlaps with the lower range of the subalpine dwarf Siberian pine (*Pinus pumila*), which has no equivalent in the flat expanses of the eastern and central North American boreal forest where balsam fir (*Abies balsamea*) occurs. The Sakhalin fir is very closely related to Veitch fir (*A. veitchii*) of the mountains farther south in Japan, with which it shares many DNA variants. In this respect, Sakhalin fir also shows a correspondence to balsam fir, with its isolated high-elevation vicariant (geographically separated sister species) Fraser fir (*A. fraseri*) in the southern Appalachians. In some classifications, as in the one adopted here, Sakhalin fir is treated as belonging to the same botanical section *Balsamea* as balsam fir. Sakhalin fir is also closely related to Amur fir (*A. nephrolepis*) and Korean fir (*A. koreana*), which replace it in the maritime forests of the Amur region and Korea. Together with subalpine fir (*A. lasiocarpa*) in western North America and Siberian balsam fir (*A. sibirica*) across Russia, these species constitute a replacement series that occupies most of the boreal region of the northern hemisphere. The two varieties of Sakhalin fir are unusual in overlapping so much with each other in geographic distribution and their relationship should be given careful study.

Abies sachalinensis* var. *mayriana K. Miyabe & Y. Kudô

Bark remaining smooth in age. Bracts yellowish green when young, sticking far out between the seed scales. Seed wings brown. Southern Sakhalin (Russia) and Hokkaidō (Japan), especially southward and at lower elevations.

Abies sachalinensis (Friedr. Schmidt) M. T. Masters **var. *sachalinensis***

Bark roughening with age. Bracts light brown when young, hidden by the seed scales or sticking out a little between them. Seed wings purple. Throughout the range of the species but more common northward and at higher elevations on Hokkaidō (Japan). Synonyms: *Abies sachalinensis* var. *gracilis* (V. Komarov) Farjon, *A. sachalinensis* var. *nemorensis* H. Mayr.

Abies sibirica Ledebour

SIBERIAN BALSAM FIR, PICHTA [SIBIRSKAYA] (RUSSIAN)

Abies sect. *Balsamea*

Tree to 30(–40) m tall, with slender trunk to 0.6(–1) m in diameter. Bark very gray, cracking into flakes and then plates with age. Branchlets thinly short hairy at first, scarcely to strongly grooved. Buds 2–4(–4.5) mm long, heavily white-resinous. Needles arranged to the sides and above the twigs, directed exclusively upward on high branches, (1–)1.5–2.5(–5) cm long, shiny bright green above, the tips shallowly notched or rounded on lower branches to pointed on those bearing seed cones. Individual needles flattened to plump in cross section and with a resin canal on either side just inside the lower epidermis or deep within the leaf, without stomates above or with a few short lines of stomates near the tip and with four to seven lines in each dull light green stomatal band beneath. Pollen cones about 15 mm long, red. Seed cones elongate egg-shaped, 5–8(–10) cm long, 2–4 cm across, deep purple when young, maturing light reddish brown. Bracts only about a third as long as the seed scales and hidden by them. Persistent cone axis narrowly conical. Seed body 5–6 mm long, the wing up to twice as long. Cotyledons four or five. Northern and central Eurasia from northeastern European Russia and the southern Ural Mountains across Siberia (hence the species name) north of the Chinese and Mongolian border region to the Uda and Amur Rivers and central Heilongjiang province (China), with an outlier in the Tian Shan (Kyrgyzstan). Forming pure stands or mixed with other boreal (taiga) and montane forest conifers and hardwoods; 0–2,000(–2,800) m. Zone 2. Two varieties.

Siberian balsam fir is the most widely distributed true fir, a prominent member of the vast taiga forests of Siberia as well as the conifer forests of the mountains of central Asia. This distribution

Narrow spire of young mature Siberian fir (*Abies sibirica*) in nature.

corresponds ecologically to that of balsam fir (*Abies balsamea*) and subalpine fir (*A. lasiocarpa*) combined in North America, and these species are closely related, sharing small seed cones and notched needles with median resin canals. Despite its huge geographic and elevational range, Siberian balsam fir has relatively little recorded variation. The isolated populations in the Tian Shan differ from the rest of the species in a few characters and are recognized as a distinct variety, but the continuous range has not been taxonomically subdivided. Toward the east, Siberian balsam fir is replaced by the related Amur fir (*A. nephrolepis*) and Sakhalin fir (*A. sachalinensis*), both with bracts emerging slightly between more kidney-shaped seed scales. It may hybridize with Amur fir in their zone of overlap west of the Amur River, but the possible hybrids (*A. ×sibirico-nephrolepis*) have also been interpreted as part of the variation within Amur fir.

Abies sibirica var. _semenovii_ (B. Fedtschenko) Tang S. Liu

Branchlets deeply grooved. Buds thinly covered with resin. Needles with marginal resin canals and five to seven rows of stomates in each stomatal band. Blades of seed cone bracts wider than long. Tian Shan.

Upper (A) and undersides (B) of twigs of Siberian fir (*Abies sibirica*) with 1 (A) or 2 (B) years of growth, ×0.5.

Abies sibirica Ledebour **var. _sibirica_**

Branchlets shallowly grooved. Buds thickly covered with resin. Needles with median resin canals and four or five rows of stomates in each stomatal band. Blades of seed cone bracts about as wide as long. Range of the species except the Tian Shan.

Abies ×sibirico-nephrolepis Takenouchi & S. S. Chien

Tree to 25 m tall. Resembling its parents (*Abies nephrolepis* × *A. sibirica*) in features common to both and variably combining or falling between them in features by which they differ. Perhaps more similar to Amur fir (*A. nephrolepis*) but differs in its shorter needles, about 1.5 cm long, and shorter bracts, less than three-quarters as long as the seed scales, between which they may just barely peek out, but are usually hidden. Hidden bracts are reminiscent of *A. sibirica*, but Siberian balsam fir has more pentagonal seed scales. Seeds are also more like those of Siberian balsam fir, with their much longer wings than in *A. nephrolepis*, but they are sterile. Seed cones are at the upper end of size for both species.

Found with its parents and other conifers and hardwoods in bottoms of mountain valleys where both occur.

The status of these trees is in some doubt. The *Flora of China* (Fu et al. 1999b) treats them simply as part of the variation of Amur fir, but their combination of features, and especially the sterile seeds, suggest a hybrid origin. They have only been reported from northern Heilongjiang province (China), but if they really are hybrids, they should also be found in the much larger area of overlap in the Amur region of Russia.

Abies spectabilis (D. Don) Spach

WEBB FIR, HIMALAYAN SUBALPINE FIR, BADAR (KASHMIRI), GOBRASALLA (NEPALESE)

Abies sect. *Pseudopicea*

Tree to 40(–60) m tall, with trunk to 2(–3) m in diameter, but often much smaller at high elevations. Bark silvery gray, darkening and becoming ridged and furrowed with age. Young branchlets densely dark-hairy in the deep grooves between the leaf bases but soon shedding the hairs. Buds 4.5–6(–10) mm long, covered with brownish resin. Needles arranged in several rows straight out to the sides and others forward above the twigs, the different rows of different lengths, 1.5–4(–6.5) cm long, shiny deep green above, the tips generally strongly notched to forked. Individual needles flat in cross section and with a resin canal on either side near the edge, touching the lower epidermis, without stomates above and with 7–15 lines of stomates in each silvery white stomatal band beneath. Pollen cones 30–50 mm long, purple. Seed cones barrel-shaped, 7–15(–20) cm long, 4–7 cm across, dark violet when young, maturing purplish brown. Bracts half as long as the seed sales and hidden by them. Persistent cone axis narrowly conical. Seed body 9–11 mm long, the wing about as long. Cotyledons four to six. Western Himalaya from northeastern Afghanistan to Nepal and adjacent China in Xizang (Tibet). Forming pure stands or mixed with other conifers, including pindrow fir (*A. pindrow*), and hardwoods in the subalpine forest; (1,650–)2,100–3,800(–4,300[–5,350?]) m. Zone 8. Synonym: *A. spectabilis* var. *brevifolia* (A. Henry) Rehder.

Webb fir (from an older scientific name for the species) replaces pindrow fir (*Abies pindrow*) at higher elevations. It is often stunted in exposed sites at high elevations and, even when grown well, has a broad, flat-topped crown unusual among silver firs (explaining the scientific name, "remarkable"). It differs from pindrow fir in a number of characteristics, including its hairy, deeply grooved twigs, and shorter needles with resin canals in a different position. These characteristics are variously combined in their hybrids, and some of the hybrids even have two pairs of resin canals, a pair in the position found in each parent. It is unclear what happens at the contact of Webb fir with Himalayan red fir (*A. densa*) in Nepal, and many authors include the eastern Himalayan *A. densa* as a synonym of Webb fir. The more lightly grooved twigs, rolled needle margins, and emerging bract tips of the Himalayan red fir separate the two species fairly easily, but their relationship deserves further study. Because of its high elevational distribution and shorter stature, Webb fir is somewhat less exploited for timber than pindrow fir but has much the same uses.

Abies squamata M. T. Masters

FLAKY FIR, LINPI LENGSHAN (CHINESE)

Abies sect. *Pseudopicea*

Tree to 40 m tall, trunk to 1(–1.6) m in diameter. Bark purplish brown, soon peeling in thin, papery flakes (hence the common and scientific names, "scaly") and ultimately becoming blocky. Branchlets densely hairy, prominently grooved between the leaf bases. Buds 4–6 mm long, covered with a thin, white resin that soon wears off. Needles on lower branches arranged predominantly to the sides, those on higher branches with seed cones pointing upward, (1–)1.5–3 cm long, green or bluish green above, the tips blunt or sharp or even prickle-tipped in young trees. Individual needles flat or a little plump in cross section and with a resin canal on either side of the midrib halfway out to the edge and well away from the lower epidermis but even farther from the upper one, with 3–7(–15) broken lines of stomates in the groove above near the tip and seven to nine lines in each white to greenish white stomatal band beneath. Pollen cones 2–3 cm long, purple. Seed cones oblong, 5–7(–8) cm long, 2–3(–4) cm across, violet when young, maturing violet brown to almost black. Bracts about as long as the slightly hairy seed scales and with their tips sticking up or down between them. Persistent cone axis swollen below the middle. Seed body 5–6 mm long, the wing a little longer. Cotyledons four or five. South-central China from southern Gansu and southern Qinghai through Sichuan to southeastern Xizang (Tibet) and northern Yunnan. Forming pure stands or mixed with other conifers and hardwoods, predominantly in subalpine forest; (3,000–)3,500–4,200(–4,700) m. Zone 6.

With its papery bark, flaky fir is the most distinctive of the difficult group of central Chinese silver firs. It is also the highest elevation *Abies* species in the region, and scaly juniper (*Juniperus squamata*) is the only conifer growing above it in these mountains. Throughout its main elevational range it grows together with Lijiang spruce (*Picea likiangensis*) while other conifers are mixed in below, including additional species of *Abies*, like Min fir (*A. recurvata*) and Farges fir (*A. fargesii*). Its wood is used for carpentry, construction, and pulp.

Abies ×umbellata H. Mayr

MITSUMINE FIR, MITSUMINE MOMI (JAPANESE)

Tree to 35 m tall, with trunk to 1.5 m in diameter. More similar to Nikko fir (*Abies homolepis*) in appearance than to momi fir (*A. firma*), but the few differences from the former parent tend toward the latter one. Most notable differences from Nikko fir lie in the seed cones, which are broader in the hybrid (4–5 cm across), green rather than purple, and with longer bracts that just stick out between the seed scales, especially toward the base of the cone. Needles have more lines of stomates (13–15) in each stomatal band. Some hairs in the deep grooves on the twigs might be expected in their first season but are generally not found. Mount Mitsumine and other peaks of central Honshū (Japan). Growing with *A. homolepis* in the vicinity of *A. firma*; 1,000–1,200 m. Zone 5. Synonym: *A. homolepis* var. *umbellata* (H. Mayr) E. H. Wilson.

This hybrid has a restricted distribution in nature today, but past hybridization between its parent species has left an enduring legacy in the sharing of certain DNA patterns between them. The scientific name describes the broad, spreading crown that is unusual in silver firs but certainly not restricted to this hybrid.

Abies ×vasconcellosiana Franco

Tree to 30 m or more tall. Garden plant resulting from spontaneous hybridization of its geographically widely separated parents (*Abies pindrow* × *A. pinsapo*) growing in cultivation. Like both parents, it has hairless twigs and seed cones with hidden bracts. It differs from pindrow fir (*A. pindrow*) in less prominently grooved twigs, shorter, stiffer leaves, 1–3 cm long, with blunt tips and deeply embedded resin canals, and narrower seed cones 4–5 cm across. It differs from Pillars of Hercules fir (*A. pinsapo*) in more heavily resinous buds and longer, blunter, needles of variable length that are directed to the sides and above the twigs, those above shorter than the ones to the sides and arching forward. The circumstances of origin are unknown, but the trees were first noticed at about 20 years of age in a park in Portugal where both parents were growing and were named for Professor J. de Carvalho e Vasconcellos. Zone 6.

Abies veitchii Lindley

VEITCH FIR, SHIRABISO (JAPANESE)

Abies sect. *Balsamea*

Tree to 25(–30) m tall or dwarfed at the alpine timberline, with trunk to 0.8(–1) m in diameter. Bark grayish white, usually remaining so with age. Branchlets densely covered with short brown hairs that are soon lost, not or shallowly grooved between the leaf bases. Buds 3–4.5 mm long, moderately to heavily resinous. Needles arranged to the sides and above the twigs, angled gently forward, (0.5–)1–2.5 cm long, shiny dark green above, the tips flat or slightly notched. Individual needles flat to a little plump in cross section and with a resin canal on either side toward the outer edge nearer the lower epidermis than the upper but not touching it, without or with a few scattered stomates in the groove above near the tip and with 13–15 lines of stomates in each white stomatal band beneath. Pollen cones 8–15 mm long, purplish brown. Seed cones cylindrical, (3–)4–6(–8) cm long, (1.5–)2–2.5 cm across, blackish purple (rarely green) when young, maturing dark purplish brown. Bracts a little shorter than or about as long as the externally hairy seed scales and usually peeking out a little between and bent down over them. Persistent cone axis narrowly cylindrical. Seed body 4.5–6 mm long, the wing half as long or a little less. Cotyledons four or five. Central Honshū, Kii Peninsula, and central Shikoku (Japan). Forming pure stands or mixed with other conifers in the subalpine forest; (800–)1,200–1,900(–2,800) m. Zone 3. Synonyms: *A. veitchii* var. *nikkoensis* H. Mayr; *A. veitchii* var. *olivacea* Shirasawa, *A. veitchii* var. *sikokiana* (T. Nakai) Kusaka.

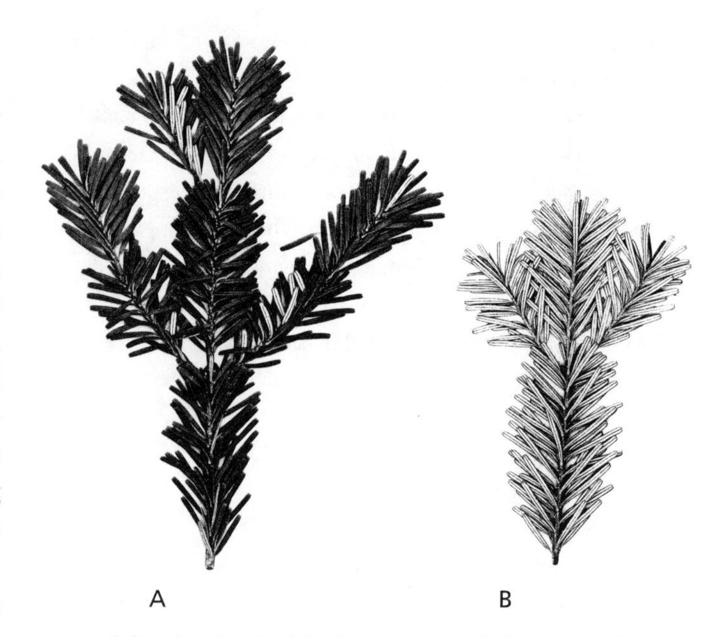

Upper (A) and undersides (B) of twigs of Veitch fir (*Abies veitchii*) with 3 (A) or 2 (B) years of growth, ×0.5.

Veitch fir shares the subalpine zone of central Honshū with the unrelated Maries fir (*Abies mariesii*), generally at a lower elevation, but is the only high subalpine fir southward in its range. These two species are the shortest of the Japanese silver firs, and this, along with their subalpine habitat, makes them of little economic importance. The specific epithet honors John Gould Veitch (1839–1870), who collected specimens and seeds of the species on Mount Fuji in 1860 for his family's nursery firm in England. Veitch fir is closely related to Sakhalin fir (*A. sachalinensis*), which replaces it north of Honshū with a much broader elevational distribution. It is also close to Korean fir (*A. koreana*), and some individuals on Shikoku resemble this neighboring species in their somewhat wedge-shaped needles. These needles are shorter than those of most Veitch firs, and their cones are also smaller, while trees of Korean fir resemble most Veitch firs in these respects. Despite these differences, there seems little justification for giving these variants formal taxonomic recognition. Other proposed varieties based on seed cone color or size are equally untenable.

Abies vejarii M. Martínez

VÉJAR FIR, HAYARÍN (SPANISH)

Abies sect. *Grandis*

Tree to 40 m tall, with trunk to 0.5(–1) m in diameter. Bark gray, becoming browner and breaking up into scales and then ridges and furrows with age. Branchlets hairless or slightly hairy in the shallow grooves between the leaf bases. Buds 3–5 mm long, thickly coated with resin. Needles arranged to the sides of and angled forward above the twigs, or radiating all around on branches bearing seed cones (1–)1.5–2.5(–3) cm long, dark green to waxy grayish green above, the tip bluntly to sharply pointed. Individual needles flat in cross section and with a resin canal on either side touching the lower epidermis near the margin, usually with 7–10 broken lines of stomates over the surface above and with 5–10 lines in each white stomatal band beneath. Pollen cones 5–10 mm long, red. Seed cones oblong to almost spherical, 6–10(–15) cm long, 4–5(–7) cm across, dark purple when young, maturing blackish brown. Bracts a little shorter to a little longer than the minutely fuzzy seed scales and hidden by them or sticking up between them. Persistent cone axis narrowly conical. Seed body 8–12 mm long, the wing a little shorter to a little longer. Cotyledons four to six. Sierra Madre Oriental of northeastern Mexico from southwestern Coahuila to southwestern Tamaulipas. Mixed with pines (*Pinus*), oaks (*Quercus*), and other conifers and hardwoods on mountain slopes and in high canyons; 2,000–3,000(–3,300) m. Zone 7. Synonyms: *A. mexicana* M. Martínez, *A. vejarii* var. *macrocarpa* M. Martínez, *A. vejarii* subsp. *mexicana* (M. Martínez) Farjon, *A. vejarii* var. *mexicana* (M. Martínez) Tang S. Liu.

Véjar fir typically grows at higher elevations than its neighbor (and possible relative) to the west, Durango fir (*Abies durangensis*). In the region near Monterrey (Nuevo León) and Saltillo (Coahuila) it grows as low as 2,000 m. The stands at lower elevations here have smaller seed cones with hidden bracts, while those growing higher up in the mountains have bracts sticking out between the scales of larger seed cones. These two forms have been treated as distinct varieties or even separate species, but the same pattern of increasing bract protrusion with elevation is found in many silver firs, including balsam fir (*A. balsamea*) in northeastern North America and Sakhalin fir (*A. sachalinensis*) in northeastern Asia. The leaf structure, if not the arrangement, is very similar to that of white fir (*A. concolor*) to the north, and Véjar fir may ultimately prove to belong to that species. The species name honors Octavio Véjar Vázquez, who was Mexico's minister of public education, responsible for promotion of the arts and sciences, during the presidency of Manuel Avila Camacho (1940–1946), when it was discovered and described.

Abies ×*vilmorinii* M. T. Masters

Tree to 25 m or more tall. Garden plant resulting from controlled pollination or spontaneous hybridization of its parents (*Abies cephalonica* × *A. pinsapo*) growing in cultivation. Like both parents, it has shining, hairless twigs bearing pointed needles which may have deeply seated resin canals. It differs from Grecian fir (*A. cephalonica*) in having prominent rows of stomates on the leaves above and seed cones with a nipplelike tip and bracts barely emerging between the seed scales or not at all. It differs from Pillars of Hercules fir (*A. pinsapo*) in its long needles, 1.5–3 cm long, concentrated to the sides and above the twigs and its seed cones with visible bracts. It was first raised in 1868 by Henry de Vilmorin (1843–1899), head of the famed Vilmorin firm from 1866 until his death, in his nursery at Verrières after deliberately crossing the parents in 1867. Zone 6.

Acmopyle Pilger

ACMOPYLE

Podocarpaceae

Small to medium, evergreen trees. Trunk usually single. Thin, fibrous bark, initially smooth but flaking in scales over time. Crown conical at first, becoming rather open with age. Branchlets weakly differentiated into long and short shoots, the former constituting the main axes of growth and bearing scale leaves, the latter bearing foliage leaves, green for at least the first year, hairless, prominently grooved between the elongate, attached leaf bases. Resting buds ill-defined, consisting of a loose aggregate of green scale leaves. Leaves spirally inserted, predominantly of two distinct forms, scale leaves and foliage leaves. Scale leaves found on long shoots, reproductive axes, resting buds, and at the base of short shoots, triangular, flattened top to bottom in the ordinary way. Foliage leaves found on short shoots needlelike, longest near the middle and much shorter near the beginning and end of each shoot, extending in a single plane on either side of the twig through bending of the leaf at its point of attachment, then running down onto the twig without a petiole. Needles standing out from the twig in a forwardly directed, shallow S shape, flattened side to side rather than top to bottom.

Plants dioecious. Pollen cones on simple or sparingly branched reproductive branchlets emerging from the tips of short shoots or along the length of long shoots that are densely clothed with scale leaves. Simple shoots bearing one or two pollen cones at their tips and the branched ones bearing a varied number of such single or paired pollen cones. Each cone cylindrical, with a few small bracts at the base and numerous, densely spirally arranged pollen scales, each scale bearing two pollen sacs. Pollen grains medium (body 40–50 μm long), with two (rarely three) somewhat kidney-shaped air bladders that are smaller than the ovoid body and diverge at about 100° away from the germination furrow, fairly smooth but with coarse internal sculpturing, the cap of the body thicker and more coarsely sculptured than the furrow side. Seed cones single or two or three from the ends of short shoots, each at the tip of a separate densely scaly stalk, highly modified and reduced, without free bracts at the base and with (four or) five to eight bracts united with the cone axis and becoming swollen and fleshy and covered with warts (as a structure called a podocarpium), maturing in (1–)2 years. All of the bracts except the uppermost one with at least a remnant of a seed scale, one or two of the bracts each with a single, plump, unwinged seed that is united for about half of its length with the fleshy seed scale (a structure called an epimatium), the edge of the epimatium marked by a distinct ridge around the seed. Seed with the fleshy layer of the seed coat and epimatium surrounding an extremely hard and thick inner shell, the ovule opening upward or slightly to the side (hence the scientific name, Greek for "tip gate"). Cotyledons two, each with two veins. Chromosome base number $x = 10$.

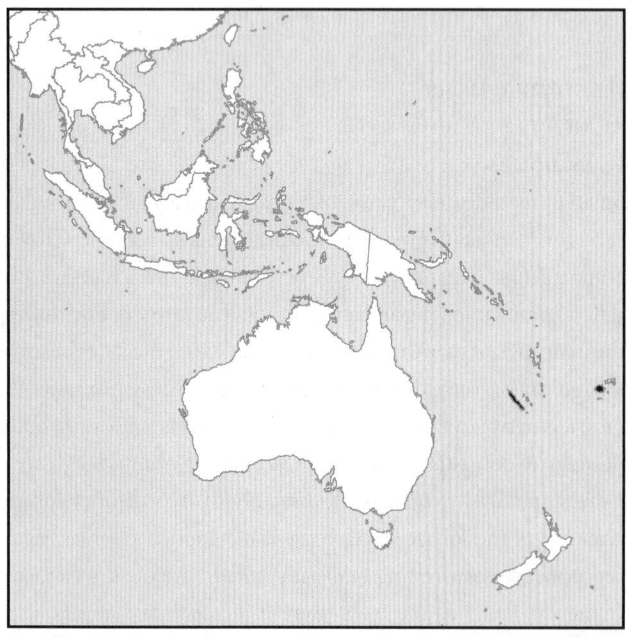

Distribution of *Acmopyle* in New Caledonia and Fiji, scale 1 : 119 million

Wood soft, light, not fragrant, yellowish brown to light brown, with little distinction between sap- and heartwood. Grain fine and even, with well-marked growth rings because of the sharp distinction of narrow layers of latewood. Resin canals absent and individual resin parenchyma cells few.

Stomates in a waxy stomatal band on either side of the midrib on the underside and in individual lines on the upper side. Each stomate sunken beneath the four (to six) surrounding subsidiary cells, which may be shared between adjacent stomates in a single line, without a Florin ring, plugged with wax. Photosynthetic tissue on just one side, the side facing up, corresponding either to the right or left side of ordinary foliage leaves in other conifers depending on whether the leaf is positioned to the left or to the right of the twig. The single resin canal then lying either to the right or left of the midvein rather than below it (although in strict morphological terms it is still on the abaxial side of the midvein, the side away from the twig, just as it is in *Podocarpus,* for example).

Two species in Fiji and New Caledonia. References: Bush and Doyle 1997, de Laubenfels 1969, 1972, Florin 1940a, Hill and Brodribb 1999, Hill and Carpenter 1991, Mill et al. 2001, Sahni 1920.

The species of *Acmopyle,* although found in a very few botanical gardens, are not in general cultivation, and no cultivar selection has been undertaken.

Acmopyle has a distinctive combination of features, some of which are shared with several other podocarp genera while others are unique to it. The seed cone, with its well-developed podocarpium, is somewhat similar to that of *Podocarpus* and *Dacrycarpus* but consists of more bracts. The warty surface of the bracts and their preservation of rudimentary seed scales even in those without seeds is also distinctive. The upright seed leads to a variation on the pollination method found in *Podocarpus* and many other podocarps in which pollen is scavenged from particular areas of the young seed cone by a very large pollination drop. Rather than having the whole ovule pointed back down toward the cone axis as in most other podocarps, *Acmopyle* has a special down-curved beak that directs the pollination drop downward onto the correct surfaces, including the uppermost bract, the only one that is completely devoid of any trace of a seed scale.

The type of foliage leaf found on the short shoots, flattened side to side, is very unusual in conifers and, among podocarp genera, is otherwise found only in *Falcatifolium* and in juvenile foliage of *Dacrycarpus*. Similar fossilized foliage that can be assigned with some confidence to *Acmopyle* because it preserves the distinctive structure of the epidermis is known as early as the late Paleocene more than 55 million years ago, making it one of the older living genera of conifers. Much older fossils resembling *Acmopyle* and dating from the Jurassic, more than 140 million years ago, are not established as belonging to this genus or even as being closely related to it. The verified fossils have been found in Antarctica, Argentina, Tasmania, New South Wales, and Western Australia, all regions outside its present range of distribution but conforming well to a typical southern hemisphere distribution that reflects the late Mesozoic breakup of the southern hemisphere supercontinent Gondwanaland. No fossils of *Acmopyle* are known from sediments younger than the early Oligocene, more than 30 million years ago, so there is a long, uncharted path from the fossil record to the present-day distribution.

The persistence of the genus may be related, in part, to the presence of nitrogen-fixing nodules on the roots, structures present in some other podocarps as well. Its age might suggest that *Acmopyle* is not particularly closely related to any other podocarps despite its overlapping features with other genera. Not surprisingly, then, DNA studies do not provide any convincing account of its relationships, but it may be closest to a group consisting of *Podocarpus* and it near relatives and *Dacrydium* and its near relatives. This would be in accordance with its resemblance to some of these genera in such things as possession of a podocarpium, while retaining other features that are more primitive within the family, like the greater number of bracts making up the seed cone.

Identification Guide to *Acmopyle*

The two species of *Acmopyle* may be distinguished by the width of most foliage leaves, by whether the edges of the foliage leaves and scale leaves have tiny hairs along their margins (are ciliate or not), by the length and width of the pollen cones, and by the length of the scaly stalks supporting the seed cones, the length of most of the podocarpiums, and the length of the seed:

> *Acmopyle pancheri*, most foliage leaves up to 2.5 mm wide, not ciliate, pollen cones at least 1 cm by 2 mm, scaly stalks supporting the seed cones at least 1 cm, most podocarpiums at least 1 cm, seed at least 1 cm
>
> *A. sahniana*, most foliage leaves more than 2.5 mm wide, ciliate, pollen cones less than 1 cm, scaly stalks supporting the seed cones up to 2 mm, most podocarpiums less than 1 cm, seed less than 1 cm

Acmopyle pancheri (Ad. Brongniart & Gris) Pilger

NEW CALEDONIAN ACMOPYLE

Tree to 15(–25) m tall, with trunk to 0.4 m in diameter. Bark light to dark brown, weathering gray and flaking to reveal brown patches with age. Crown shallow and irregular. Short shoots typically (3–)5–10 cm long. Foliage leaves hairless, waxy bluish green at first, becoming dark green above with maturity, varying in size depending on whether the tree is growing in the shade or in more open, sunny sites, 1–1.5 cm long and 1.8–2.5 mm wide in the sun, 1.5–2 cm long and 2.5–3 mm wide when shade grown. Pollen cones 1–2 cm long and 2–3 mm wide. Seed cones on a scaly stalk 9–22 mm long, waxy light gray over a dark green to yel-

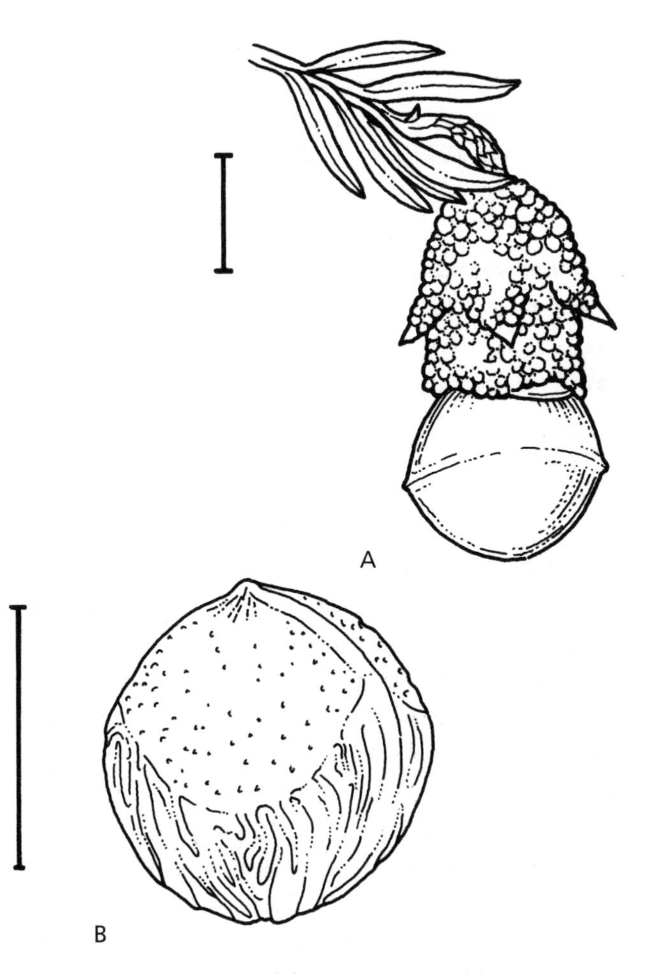

Twig with dangling seed cone (A) and cleaned seed (B) of New Caledonian acmopyle (*Acmopyle pancheri*); scale, 1 cm.

lowish green surface, the podocarpium (united fleshy bracts and cone axis) 8–18 mm long. Combined seed coat and epimatium spherical, 10–11 mm in diameter. Throughout most of the length of New Caledonia but most abundant in the southern fifth. Scattered as a canopy tree in drier portions of rain forests and in the understory of mossy forests, usually on well-weathered serpentine soils; (250–)750–1,250 m. Zone 10. Synonym: *Acmopyle alba* J. Buchholz.

New Caledonian acmopyle is a rare species but nowhere near as restricted as many other New Caledonian conifers. Because of its small stature, in part, it is not particularly threatened by direct exploitation for timber. It is, however, threatened by general habitat destruction associated with the mining industry because it grows on the serpentine soils that are indicators of exploitable nickel and other mineral deposits. It is part of ex situ conservation efforts in botanical gardens in the northern hemisphere, and in relation to these efforts, its pollination biology has been studied under glasshouse conditions. Similar work also needs to be done in the field in relation to the efficiency and aerodynamics of pollen release and transport as well as ovule receptivity. The species name honors Jean Pancher (1814–1877), who collected plants (including the type specimen of this species) for the French colonial administration between 1857 and 1869 and who later died on the island after returning from 5 years back in France.

Acmopyle sahniana J. Buchholz

FIJIAN ACMOPYLE

Tree to 8(–12) m tall, with trunk to 0.2 m in diameter. Bark grayish brown, generally smooth, dimpled at first and later sparingly minutely warty. Crown openly dome-shaped, with widely spaced, upwardly angled to horizontal branches. Short shoots typically 4–11 cm long. Foliage leaves sparingly hairy along the edge, shiny dark green above, with two narrow, waxy whitish green stomatal bands beneath, (0.2–)1–2.5 cm long and (0.6–) 2–3.5(–4.8) mm wide. Pollen cones 1.5–7.5 mm long and 1–2 mm wide. Seed cones on a scaly stalk about 5–6 mm long, the podocarpium (united fleshy bracts and cone axis) dark green to dark purple with traces of grayish wax, (5–)7–9 mm long by 7–8 mm thick. Combined seed coat and epimatium grayish purple with a whitish waxy coating, 7–9 mm long by 5–6 mm thick. Viti Levu, Fiji. As scattered trees or small groves in the understory of lower montane rain forest; 800–1,050 m. Zone 10.

This extremely rare species is known only from a handful of stands on two widely separated mountains on the main island of Fiji. Its extreme rarity makes it of conservation concern even if the stands are protected. There is anecdotal evidence that individual trees may change sex or bear both pollen and seed cones simultaneously, which would be quite unusual in Podocarpaceae and is not known in New Caledonian acmopyle (*Acmopyle pancheri*). The species name honors Birbal Sahni (1891–1949), an Indian paleobotanist who did his doctoral research on the anatomy of New Caledonian acmopyle soon after it was placed in this then new genus and who recognized that some specimens from Fiji probably belonged to the same genus.

Actinostrobus Miquel

SANDPLAIN-CYPRESS

Cupressaceae

Evergreen shrubs or small trees with fibrous furrowed bark peeling in thin flakes and strips. Densely branched from the base with stiff horizontal or ascending branches to form an irregularly conical crown. Branchlets cylindrical or three-angled, branching three-dimensionally. Without specialized resting buds. Seedling leaves in alternating quartets at first, followed by alternating trios, needlelike, well separated on the twig by the long leaf bases running down onto it and standing out from it at an angle of about 60°, the free tips progressively shorter in juvenile and adult foliage. Adult leaves in alternating trios, scalelike, keeled, with elongate leaf bases running down on the twigs and prominent, triangular free tips with thickened margins angling outward to give a slightly bristly texture to the foliage.

Plants monoecious. Pollen cones numerous, single at the tips of short lateral branchlets, oblong, with five to seven alternating trios of pollen scales, each scale with two to four pollen sacs. Pollen grains small (25–35 µm in diameter), spherical, minutely but conspicuously bumpy, without a germination pore. Seed cones numerous, single at the tips of short lateral branches close to the trunk, maturing in a single season but remaining closed on the tree after maturity. Cones spherical to oval at maturity, apparently with a single whorl of six triangular fertile seed scales attached at the base around a central column and closing side to side so that tips of all six meet at the top, actually of two alternating trios, one with slightly wider scales than the other. Fertile scales each with one to three sterile scales beneath grading into the foliage leaves of the branchlet and with one or two seeds. Seeds oval, with (two or) three unequal to nearly equal triangular wings derived from the seed and extending along the whole length of the seed body or beyond. Cotyledons two, each with one vein. Chromosome base number $x = 11$.

Wood odorless, soft, light, with pale sapwood weakly contrasting with the small core of slightly darker heartwood. Grain somewhat uneven and moderately fine, with poorly defined growth rings marked by one or two rows of smaller latewood tracheids with wall thickness similar to that of those in earlywood. Resin canals absent but sometimes with traumatic resin pockets and with fairly numerous, scattered individual resin parenchyma cells often more concentrated near the ring boundaries and thus helping to delimit them.

Stomates in a single narrow band on either side of the midline both above and beneath. Each stomate deeply sunken beneath the four or five (to six) surrounding subsidiary cells, which are topped by a steep-sided Florin ring inside an encircling channel. Leaf cross section with a single midvein linked also by transfusion tissue to the branchlet core. Outer (lower) face of leaf with a thin layer of hypodermis inside the epidermis adjacent to the palisade photosynthetic tissue, which lies outside a very open spongy mesophyll accompanied by resin cavities.

Three species in southwestern Australia. References: Bowman and Harris 1995, Clifford and Constantine 1980, Farjon 2005b, Gardner 1964.

Actinostrobus is closely related to *Callitris,* as long recognized by their morphology and ecology and now seen also in their DNA sequences. Both genera inhabit the dry forests and woodlands of Australia, corresponding ecologically to the more distantly related *Widdringtonia* in southern Africa and *Cupressus, Juniperus,* and *Tetraclinis* in northern Africa and elsewhere in the northern hemisphere. The two genera also share leaves in trios with conspicuous elongate leaf bases, giving a jointed appearance to the branchlets.

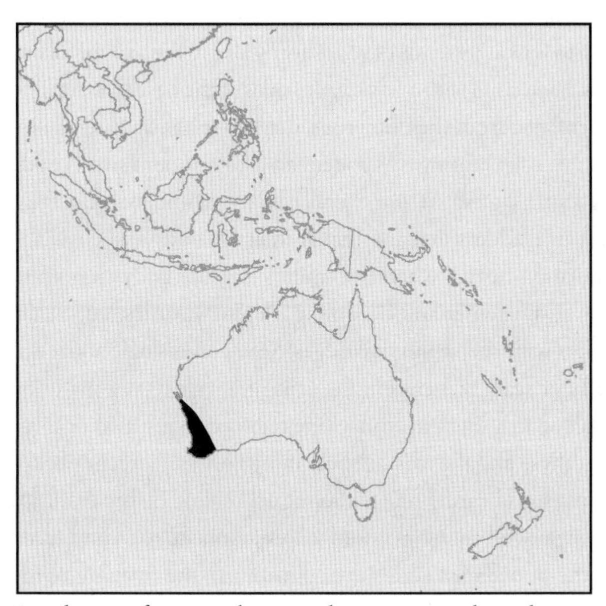

Distribution of *Actinostrobus* in southwestern Australia, scale 1 : 119 million

Actinostrobus, however, has longer free leaf tips than *Callitris,* leading to a slight bristliness. The seed cones of both genera have an apparent whorl of six triangular fertile scales, but the outer trio in *Callitris* is much smaller than the inner trio, and their tips do not reach to the tip of the cones, unlike the nearly equal, colorful trios of *Actinostrobus* (named from the Greek for "star cone"). The scales of *Callitris* usually have more than one or two seeds and do not have neat rows of sterile scales beneath them. The shrubs are too small to be of much economic significance for wood products. Grown to a limited extent in botanical gardens and Australian home gardens, there has been no cultivar selection in this genus. There is no known fossil record for *Actinostrobus,* which is not surprising given the dryland habitats of the species.

Identification Guide to *Actinostrobus*

The three species of *Actinostrobus* can be distinguished by the diameter of the branchlets, the length of the free adult leaf tip, the number of sterile scales beneath each seed scale and by the shape of the tip of the seed cones:

> *Actinostrobus acuminatus,* branchlet diameter 0.5–1 mm, free adult leaf tip length 1.5–2.5 mm, sterile scale one, seed cone tip pointed
>
> *A. arenarius,* branchlet diameter 0.5–1 mm, free adult leaf tip length 0.5–1.5 mm, sterile scales two or three, seed cone tip pointed
>
> *A. pyramidalis,* branchlet diameter 1–1.5 mm, free adult leaf tip length 0.5–2 mm, sterile scales two or three, seed cone tip rounded

Actinostrobus acuminatus Parlatore

DWARF CYPRESS, CREEPING PINE, MOORE CYPRESS PINE

Shrub to 1 m. Branchlets 0.5–1 mm in diameter. Free tips of adult scale leaves 1.5–2.5 mm long. Pollen cones 3–5 mm long. Seed cones oval, to 1.5 cm long, about two-thirds as wide, with a narrow prolonged tip. Seed scales with just a single sterile scale beneath. Seeds dark brown, the body 7–12 mm long, the wings to 6 mm wide. Southwestern Western Australia. Arid scrub of sand plains, salt sands, and along dry river courses in the interior; 50–200 m. Zone 9.

Dwarf cypress is the shortest species of *Actinostrobus,* but it spreads widely by thin underground stems to form extensive clonal patches. Seed production is limited and may respond favorably to fires as does the vegetative spread. Although small of stature, their mode of surviving fires makes them long-lived, with maximum ages estimated at 300 years.

Actinostrobus arenarius C. A. Gardner

SANDPLAIN-CYPRESS

Tree to 6 m tall. Branchlets 0.5–1 mm in diameter. Free tips of adult scale leaves 0.5–1.5 mm long, sometimes ending in a prickle. Pollen cones 3–4 mm long. Seed cones broadly oval, 1.2–1.6 cm long, about as wide, with a short pointed tip. Seed scales with two or three sterile scales beneath. Seeds light brown, the body 3–6 mm long, the wings to 5 mm wide. Southwestern Western Australia. Coastal Mediterranean climate scrubland on limestone; 50–300 m. Zone 9. Synonym: *Actinostrobus pyramidalis* var. *arenarius* (C. A. Gardner) Silba.

Sandplain-cypress is the tallest species of *Actinostrobus.* The seed cones remain closed on the tree many years until opened by the heat of a fire. Then seeds from cones up to 6 years old are released in abundance and germinate on the newly cleared soil. The species typically lives only about 20 years between the frequent fires but may reach 100 years in protected areas.

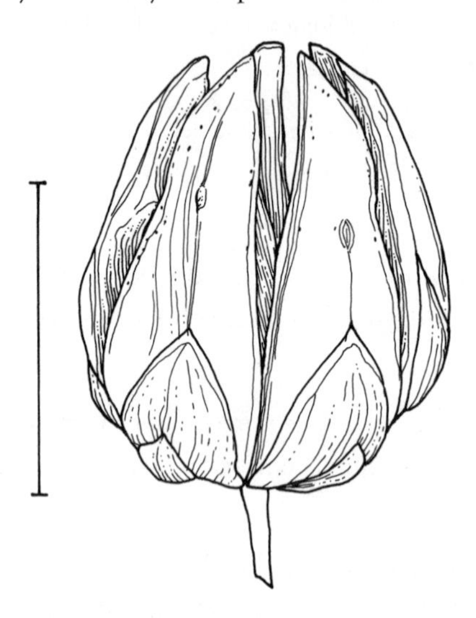

Seed cone of Swan River cypress (*Actinostrobus pyramidalis*); scale, 1 cm.

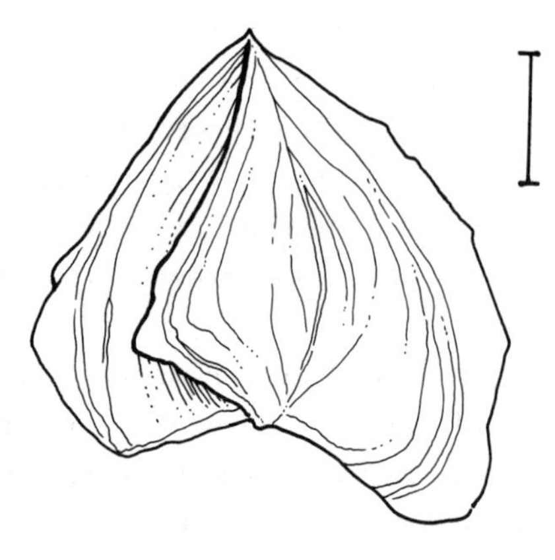

Three-winged seed of Swan River cypress (*Actinostrobus pyramidalis*); scale, 2 mm.

Actinostrobus pyramidalis Miquel

SWAN RIVER CYPRESS, [WESTERN AUSTRALIAN] SWAMP CYPRESS, KING GEORGE'S CYPRESS PINE

Shrub, or tree to 5 m tall. Branchlets 1–1.5 mm in diameter. Free tips of adult scale leaves 0.5–2 mm long. Pollen cones 5–6 mm long. Seed cones spherical, 1–1.2 cm long, about as wide, the tip rounded. Seed scales with two or three sterile scales beneath. Seeds light brown, the body 2.5–4 mm long, the wings to 3 mm wide. Southwestern Western Australia. Along rivers or on salt sands; 50–350 m. Zone 9. Swan River cypress makes an upright, densely pyramidal small tree in cultivation, with bright green leaves. It is successful on a range of well-drained soils.

Afrocarpus (J. Buchholz & N. Gray) C. Page

YELLOWWOOD

Podocarpaceae

Tall forest trees with a straight, clear trunk and a shallow, broad crown. Bark scaly and peeling in irregular flakes. Branchlets all elongate, without distinction into short and long shoots, hairless, remaining green for the first year, often square in cross section and prominently grooved between the attached leaf bases. Resting buds well developed, the bud scales tightly arranged in crisscross pairs. Leaves variously arranged on a single individual, primarily in crisscross pairs, but members of a pair often displaced longitudinally, and some shoots fully spirally arranged, radiating around the twigs or brought into a plane by twisted stalks, roughly sword-shaped and usually obviously curved to one side.

Plants dioecious. Pollen cones cylindrical, single or in clusters of two to four at the ends of very short stalks in the axils of foliage leaves. Each cone with a few sterile scales at the base that are little differentiated from the numerous, densely spirally arranged pollen scales, each bearing two pollen sacs. Pollen grains small (body 25–40 μm long, 50–70 μm overall), with two obliquely directed, round air bladders that are smaller than the body and cover the germination furrow when the grain is dry, the body smooth to grainy at the center and becoming rougher near the edges, the bladders with coarse internal ridges. Seed cones single at the tips of leafy or scaly branches in the axils of foliage leaves, maturing and falling in a single season. Cone reduced to just two or three bracts, these and the axis not becoming fleshy, all except the uppermost seed-bearing one withering and falling before maturity. The single seed round, without a crest, embedded in and fused with the upper seed scale (the epimatium), the opening of the ovule pointing down back to the cone axis. Cotyledons two, each with two veins. Chromosome base number $x = 12$.

Wood soft to moderately hard, light to moderately heavy, not fragrant, light yellowish brown to light brown, with little distinction between sapwood and heartwood. Grain fine and even, the transition from early- to latewood gradual, the variably expressed growth rings sometimes marked by a noticeably darker ring of latewood at the ring boundary. Resin canals absent but with discontinuous vertical lines of resin cells in the wood parenchyma.

Leaf surfaces with nearly equal numbers of stomates in numerous parallel lines both above and beneath. Each stomate sunken beneath and partially hidden by the four or five (to six) surrounding subsidiary cells, with just a hint of a Florin ring. Leaf cross section with a single, prominent midrib that is slightly raised above and below, with one resin canal beneath the midrib and wings of transfusion tissue and accessory transfusion tissue extending out sideways from the midrib nearly to the leaf margin. Photosynthetic tissue forming a prominent palisade layer beneath the epidermis and underlying, nearly continuous (except beneath the stomates), hypodermis on either or both faces. Numerous strengthening fibers lie in the looser spongy mesophyll, particularly at the boundary with the midrib and transfusion tissue.

Two species, one on São Tomé Island and the other discontinuous throughout the eastern side of Africa from northern Ethiopia to the Cape region of South Africa. References: Buchholz and Gray 1948a, de Laubenfels 1969, 1987, Gray 1953a, Hill and Pole 1992, Leistner 1966, Melville 1955, Page 1988. Synonyms: *Decussocarpus* sect. *Afrocarpus* (J. Buchholz & N. Gray) de Laubenfels, *Nageia* sect. *Afrocarpus* (J. Buchholz & N. Gray) de Laubenfels, *Podocarpus* sect. *Afrocarpus* J. Buchholz & N. Gray.

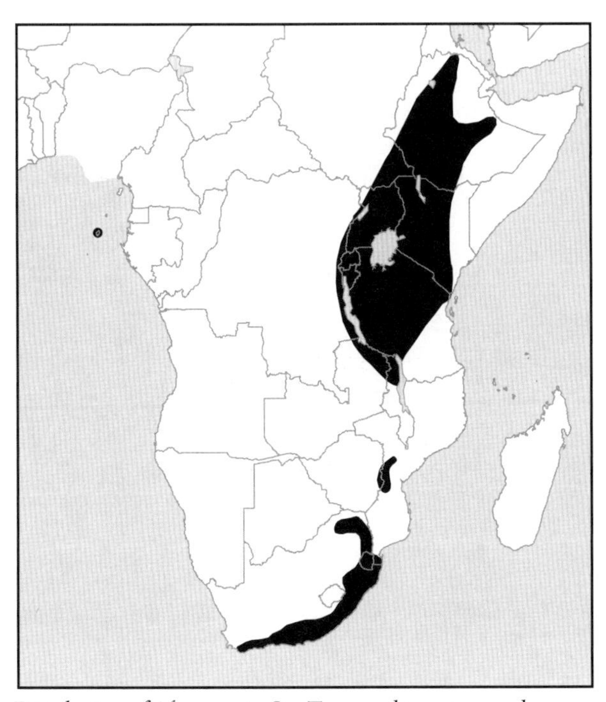

Distribution of *Afrocarpus* in São Tome and in eastern and southern Africa, scale 1 : 70 million

Both species are cultivated outdoors in mild climates and also for indoor landscaping but no cultivar selection has taken place. Common yellowwood (*Afrocarpus falcatus*) is the most important native softwood timber tree in southern Africa and most of East Africa as well but has been largely replaced commercially by faster-growing plantation-grown pines from North and Central America.

These two yellowwood species have been subjected to more than their fair share of taxonomic change, both with respect to species boundaries and to their relationships to other podocarps. As many as five species were carved out of the geographically separate portions of the range of *Afrocarpus falcatus*, based on small and inconsistent differences in leaf size and in the thickness of the hard shell surrounding the seed within the epimatium. When taxonomists came to see that recognition of so many species in this group was untenable, some authors mistakenly merged some of them with the West African *A. mannii* rather than with the South African *A. falcatus*. Although acceptance of just two species in the group with the limits followed here is probably the prevailing view, a few authors still recognize one or two of the segregate species. Oddly enough, perhaps, the two species of *Afrocarpus* display nearly the extremes in size of geographic range found among all the podocarps. While *A. falcatus* is one of the two or three most widely distributed species in the family, *A. mannii* is one of the most local, being completely confined in nature to the peak of the small volcanic island of São Tomé in the Gulf of Guinea, presumably as what is called a relictual endemic, a species now restricted to a single limited region but that was once more widely distributed. There is no evidence for this idea because there is no known fossil record for *Afrocarpus*, in part because of the other source of taxonomic confusion surrounding these species, their relationships to other podocarps. These two species are one of the more controversial segregants from *Podocarpus* because of their apparent similarity to the African species of that genus, especially those of section *Scytopodium*. Even now, many African botanists are reluctant to exclude them from *Podocarpus*, where they were long placed. Nonetheless, they have long been recognized as a distinctive element within *Podocarpus*.

Although Pilger (1903), in the only complete (for its time) fully descriptive monograph of Podocarpaceae, placed the species of *Afrocarpus* in the section of *Podocarpus* that is now recognized as the genus *Prumnopitys*, Buchholz and Gray (1948a), in their revision of *Podocarpus* in the inclusive sense now generally abandoned, put them in their own section and provided the name *Afrocarpus* for it. The name was a reference to their place of origin and their status as species of *Podocarpus* (it means "African fruit"). When de Laubenfels (1969) began the modern breakup of *Podocarpus* in the traditional sense, he joined the species of *Afrocarpus* with members of two other sections characterized by opposite leaf pairs in a separate genus to which he mistakenly gave a new name. Later (1987), he transferred all the species to *Nageia,* the correct name for this combined genus, but kept the three original sections of Buchholz and Gray. Finally, Page (1988), recognizing the morphological heterogeneity of this grouping, treated each former section as a genus in its own right, found here under the names *Afrocarpus, Nageia,* and *Retrophyllum,* Some contemporary studies, including DNA studies, point to a close relationship among the three genera, potentially validating their inclusion as sections of a single genus. Other studies, however, suggest that they are not each other's closest relatives. Although they have rather similar seed cones, they are here maintained as separate genera because of their substantial differences in leaf structure at both the macroscopic and microscopic level and because of the absence of any definitive evidence that they are most closely related to each other to the exclusion of other podocarps.

Identification Guide to *Afrocarpus*

The two species of *Afrocarpus* are easily distinguished by the length and width of the leaves of mature trees and by the length of the seeds:

Afrocarpus falcatus, leaves up to 5 cm long, mostly less than 5 mm wide, seeds less than 3 cm

A. mannii, leaves at least 7.5 cm long, at least 5 mm wide, seeds more than 3 cm

Afrocarpus falcatus (Thunberg) C. Page

OUTENIQUA YELLOWWOOD, COMMON YELLOWWOOD, OUTENIEKWAGEELHOUT (AFRIKAANS), UMGEYA (XHOSA AND ZULU), MSE MAWE (BANTU), MU SENGERA (KIKUYU)
Plate 7

Tree to 45(–60) m tall under optimum moist conditions, though shorter, sometimes no more than 10 m, in dry habitats. Trunk to 1.5(–2.4) m in diameter. Bark relatively smooth, dark brown to grayish, flaking in irregular, curled scales to reveal brighter purplish brown patches beneath. Crown dome-shaped, the relatively few large, crooked branches radiating from the trunk and rapidly proliferating to support the dense foliage. Branchlets green, squarish in cross section, deeply grooved between the elongated attached leaf bases. Resting buds about 2.5–3 mm long and 0.7–2 mm wide, with tightly overlapping bud scales with hard, short-pointed tips. Leaves usually in crisscross pairs, although the members of a pair are often well offset from one another, fairly evenly distributed along the branchlets, twisted at the base to lie flat. Individual needles waxy bluish green to yellowish green, (1–)2–4(–5) cm long (to 15 cm in juvenile trees), (1.5–)2–5(–6) mm wide, often distinctly curved to one side (hence the scientific name, Latin for "sickle-shaped"), widest in the lower half and from there gradually and then more abruptly tapering to the short, sharp tip, tapering more abruptly to the rounded base and very short stalk. Midrib only slightly raised above and beneath, with one resin canal beneath the midrib and 14–20 lines of stomates on each side. Pollen cones 0.4–1 cm long (to 1.5 cm after the pollen is shed) and 2–3.5 mm wide, one to four directly in the axils of the leaves or at the tip of a very short, leafless stalk less than 5 mm long. Pollen scales with a bluntly triangular, irregularly toothed tip 0.5–1 mm long and 0.7–1.5 mm wide. Seed cones on a leafy to scaly stalk to 2(–3) cm long, the reproductive part with two or three unequal bracts, these not at all fleshy and the lower ones drying and falling before maturity, this region about 3 mm long by 2 mm thick. Fertile seed scale one, the combined seed coat and epimatium yellowish to light reddish brown, nearly spherical, (0.6–)1.5–2(–2.5) cm in diameter, forming a resinous, thin fleshy coating over a bumpy, bony shell 1–4(–6) mm thick, grooved along one side. Discontinuous through eastern and southern Africa, from Eritrea to the Cape Region of South Africa with a gap from northern Malawi to the northern border of South Africa with Zimbabwe. Usually mixed with other high forest trees in coastal and montane forests but occasionally forming pure stands in the vicinity of permanent streams; (100–)1,000–2,000(–3,000) m. Zone 10. Synonyms: *Afrocarpus dawei* (Stapf) C. Page, *A. gaussenii* (Woltz) C. Page, *A. gracilior* (Pilger) C. Page, *A. usambarensis* (Pilger) C. Page, *Decussocarpus falcatus* (Thunberg) de Laubenfels, *D. gracilior* (Pilger)

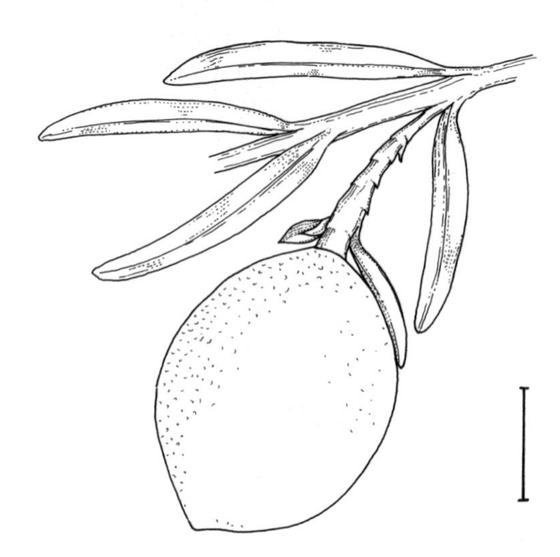

Seed cone of Outeniqua yellowwood (*Afrocarpus falcatus*); scale, 1 cm.

de Laubenfels, *Nageia falcata* (Thunberg) Carrière, *Podocarpus dawei* Stapf, *P. falcatus* (Thunberg) Endlicher, *P. gaussenii* Woltz, *P. gracilior* Pilger, *P. usambarensis* Pilger.

Outeniqua yellowwood is less widely distributed and less abundant than true yellowwood (*Podocarpus latifolius*), with which it often grows. It is easily distinguished from true yellowwood and from all of the other African yellowwoods in the genus *Podocarpus* by its flaky, not furrowed bark, by its abundant stomates on both sides of the leaves, by the leafy or scaly stalk of the seed cones, and by the complete lack of fleshiness in the bracts and axis of the seed cone. Despite these differences, there has historically been confusion with Cape yellowwood (*P. elongatus*), and older literature often used the latter name for Outeniqua yellowwood. It is still a common tree, despite more than a century of heavy exploitation for its excellent, durable timber. The great individuals that made it the tallest tree in South Africa are almost all gone now, but abundant reproduction and rapid growth ensure its continued role in the African high forests.

It sets abundant crops of seed every 2–4 years, and even with destruction of many seeds by seed insects and dispersal of many others by monkeys and fruit-eating birds, there may still be heaps of seeds beneath the parent trees in a bumper-crop seed year. The major seed dispersers, however, may be fruit-eating bats, the resinous epimatium and seed coat discouraging other potential dispersers. Germination may extend over several years, so this species does not suffer from the short seed viability of many tropical conifers and other trees. Although it does best in regions of high rainfall, it is much more tolerant of drought than true yellowwood, compensating for the lack of moisture with a lower stature of just 10 m but still carrying a large crown on an impressive bole. In both wet and drier forest, growth is episodic, producing many incomplete growth rings that make it impossible to age the tree using a simple increment core and requiring examination of the whole cross section of a felled tree for an accurate tally of the growth rings. Such complete studies show that the species can live 600 years. The leaves and twigs contain the anticancer agent taxol (or paclitaxel), first discovered in the unrelated (other than being conifers also) yews (*Taxus*).

The populations of Outeniqua yellowwood in eastern Africa were often distinguished from *Afrocarpus falcatus* as one to three separate species, but the variation across the group appears to be essentially continuous. Still, one can acknowledge the possible justification for dividing the continuum into segments, as species or varieties, based on the sizes of the pollen and seed cones and on the thickness of the stony layer in the seed. On the other hand, there seems to be no justification for merging the larger-seeded variants with the other species of the genus accepted here, São Tomé yellowwood (*A. mannii*), otherwise entirely restricted to that island. There are real discontinuities between the West African species and the eastern and southern African one that are not found among the scattered populations of the latter. One other species was described from Madagascar, but this was simply based on a cultivated specimen of Outeniqua yellowwood, and no species of *Afrocarpus* is known to occur naturally on the island. Outeniqua yellowwood is, indeed, widely cultivated in suitable climates throughout the world, as in eastern Australia, where it has even become naturalized.

Afrocarpus mannii (J. Hooker) C. Page

SÃO TOMÉ YELLOWWOOD, PINHEIRO DA TERRA (PORTUGUESE)

Tree to 17 m tall, with trunk to 0.7 m in diameter. Bark smooth, grayish brown to dark gray, flaking in irregular patches. Crown conical at first, becoming rounded with slender, rising branches bearing drooping branchlets. Twigs grouped in whorls, bright green at first but soon becoming brown, deeply grooved between the long, attached leaf bases. Resting buds 2.5–3 mm long by about 1.5 mm wide, the tightly overlapping scales with slightly spreading, pointed

Twig of São Tome yellowwood (*Afrocarpus mannii*) with both opposite and alternate leaf arrangements on two episodes of growth, ×0.5.

tips. Leaves widely spaced, slightly drooping, lasting 4 years. Individual needles shiny bright green, (2–)7.5–15(–16) cm long (to 18 cm in seedlings), (4–)7–11(–12) mm wide, nearly straight or with a slight, simple S-shaped curve, widest point variable, from below to beyond the middle, tapering from there gradually or more abruptly to the drawn-out, pointed tip and fairly abruptly to the wedge-shaped base, inserted on the stem essentially without a stalk. Midrib hardly raised either above or beneath, with a single resin canal beneath the midrib, and lines of stomates forming bands on either side both above and beneath. Pollen cones 1.2–2.5 cm long and 2–3.5 mm in diameter, one or two directly in the axils of leaves. Pollen scales with an upturned, broadly triangular, finely toothed tip about 1 mm long and 1.5 mm wide. Seed cones on a leafy stalk to 3 cm long, the reproductive part with two or three bracts, the lowest of which dry up and fall away, the axis hardening somewhat but not becoming at all fleshy or leathery. Fertile seed scale one, the combined seed coat and epimatium shiny reddish brown, pear-shaped, 3–4 cm long by 2–2.5 cm thick, fleshy and resinous surrounding a warty stony layer 4–7 mm thick, slightly grooved along one side. Restricted to the peak of São Tomé Island in the Gulf of Guinea in West Africa. Scattered in montane rain forest on the slopes of the peak; (500–) 1,000–2,000 m. Zone 10. Synonyms: *Decussocarpus mannii* (J. Hooker) de Laubenfels, *Nageia mannii* (J. Hooker) O. Kuntze, *Podocarpus mannii* J. Hooker.

São Tomé yellowwood has one of the most restricted natural distributions of all conifers, being entirely confined to the peak of the tiny volcanic island of São Tomé, with only 855 square kilometers of surface area. It is not known whether the tree originally occurred as well across the whole island since the lowlands were entirely cleared for agriculture long before any botanical exploration had been undertaken. Although a tree of modest size, it is fast growing and handsome and has been cultivated on the mainland of West Africa and elsewhere. Trees growing at the summit of the mountain near 2,024 m are stunted, but they attain their full stature wherever they are less exposed. The species name honors Gustav Mann (1836–1916), a German-born gardener and forester who collected the type material in 1861, during the British Niger Expedition of 1859–1862, and then spent the remainder of his career with the Indian Forest Service.

Agathis R. A. Salisbury

KAURI, DAMMAR

Araucariaceae

Evergreen, often giant trees, or rarely shrubs, most commonly becoming emergent over surrounding species with age. Trunk cylindrical to slightly fluted, sometimes forked, often straight and with little taper throughout the massive bole, which may be free of branches for as little as 1–2 m or as much as 15 m or more. Bark fairly thin, remaining relatively smooth in most species, flaking continuously to produce varied patterns of pockmarks, rarely becoming ridged and furrowed with age (*Agathis ovata*), exuding when wounded a sparse to copious white, yellow, or reddish resin that dries brown. Crown dense, conical and regular in youth, with closely spaced crisscross pairs of horizontal to upwardly angled branches (or apparent whorls of four to six), becoming more open and irregular, and egg-shaped, dome-shaped, or flat-topped with age, with slender to stout, horizontal to upwardly angled branches curved up at the ends and bearing crisscross pairs of horizontal branchlets. Branchlets all elongate, without distinction into long and short shoots, remaining (various shades of) green for several years, hairless, scarcely marked by to prominently grooved between attached leaf bases. Resting buds well developed, more or less spherical, tightly wrapped by a few crisscross pairs of specialized but green and soft-textured bud scales. Leaves attached in crisscross pairs but arranged in two (sometimes untidy) flat rows on horizontal shoots by bending and twisting of the petioles and of the branchlets between the leaves, the two leaves in a pair sometimes somewhat displaced in position from each other along the twig. Leaves of seedlings and juvenile trees generally similar to those of adults in shape, structure, and arrangement. Leaf shapes quite variable within trees and species, not needlelike, broad, flattened top to bottom, and variously egg-shaped or elliptical, with pointed to rounded tips and rounded to wedge-shaped bases on a wide and thick, short, flattened petiole.

Plants monoecious. Pollen cones single on short stalks in the axils of otherwise ordinary foliage leaves. Each pollen cone more or less cylindrical, with (none or) two to six (to eight) crisscross pairs of spoon-shaped bracts at the base (the lowest pair sometimes leaflike) and with numerous (often hundreds of), spirally arranged pollen scales. Each scale with 2–8(–14) elongate pollen sacs attached to the lower edge of the rounded or faceted external face of the scale and extending inward parallel to the scale stalk. Pollen grains medium to large (40–65 µm in diameter), nearly spherical, without air bladders or apparent germination structures, minutely and irregularly roughened over the whole surface. Seed cones single and upright at the tips of specialized, thick, leafy reproductive shoots in the axils of foliage leaves, sometimes two or three of these in a near whorl around the supporting shoot. Each seed cone nearly spherical (hence the scientific name, Greek for "a ball of string") to noticeably longer than wide or a little wider than long, with many (often a few hundred), tightly packed seed scales intimately united with their accompanying bracts, maturing in two seasons and shattering to release the seeds. Each seed scale (and bract) fan-shaped, thin, the exposed face somewhat thickened, horizontally elongately diamond-shaped, sometimes with a more or less well developed projecting triangular tip, the base with a notch or indentation on one or both sides, the whole scale shed at maturity. Seed one per scale, attached to a bump on, but not apparently embedded in, the upper surface of the scale near its middle, the micropyle pointed back toward the cone axis. Seed body roughly a flattened egg shape, with one very large wing (larger than the body) attached on one side in the upper half and a much reduced, rudimentary wing on the other. Resin canals absent but sometimes with individual resin parenchyma cells that may look like resin-filled tracheids. Cotyledons two, each with several parallel veins and resembling the foliage leaves. Chromosome base number $x = 13$.

Wood soft, light to moderately heavy, with nearly white to light brown sapwood more or less sharply contrasting with creamy, yellowish, or pinkish brown to rich reddish brown heartwood, commonly shiny or speckled, or both. Grain very even and fine to moderately coarse, with somewhat cryptic growth rings marked by a gradual transition to smaller, thicker-walled latewood tracheids.

Stomates confined to and almost completely covering the lower surface of the leaf, with various orientations within their many interrupted and somewhat irregular lines. Each stomate tucked beneath the four or five (to six) inner subsidiary cells and surrounded by a prominent, steep, nearly complete, (usually) sunken Florin ring. Without a single well-marked midvein and midrib but with numerous, closely spaced, parallel veins produced by repeated branching near the leaf base from two veins in the petiole. Veins in the leaf blade alternating with single resin canals (or vertical pairs in *Agathis borneensis* and *A. ovata*, at least), not noticeably raised above or beneath and not disrupting the nearly continuous single (or partially double or triple) palisade layer beneath the upper epidermis. Leaf tissue commonly with a variable density and distribution of compact and branched sclereids and with small strands of transfusion tissue paralleling and near the upper side of the veins.

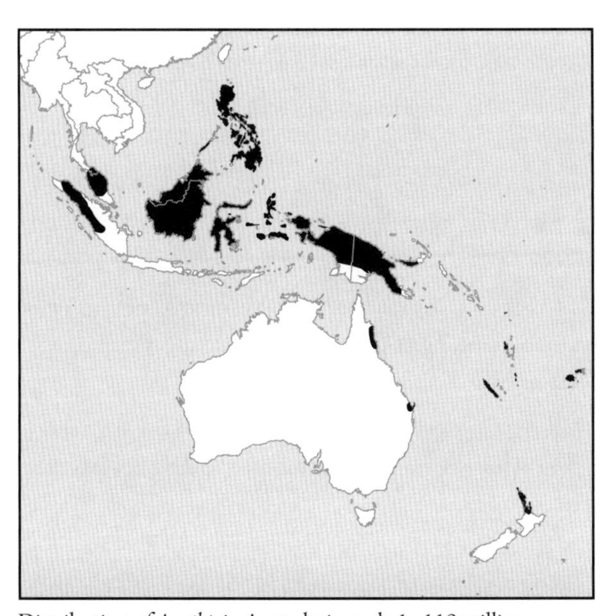

Distribution of *Agathis* in Australasia, scale 1 : 119 million

Fifteen species distributed discontinuously from Sumatra (Indonesia), Malaya (Malaysia) and Luzon (Philippines) across northern and central Malesia and through Melanesia to southern Queensland (Australia), northernmost New Zealand, and Fiji. References: de Laubenfels 1972, 1988, K. Hill 1998, R. Hill and Brodribb 1999, Hyland 1978, Whitmore 1977, 1980.

Like some of the giant conifers of the northern hemisphere, for instance, Douglas fir (*Pseudotsuga menziesii*), redwood (*Sequoia sempervirens*), and giant sequoia (*Sequoiadendron giganteum*), kauris are often relatively fast growing. Combined with their tolerance of growing in single-species stands and the considerable value of the timber and resins that they yield, this has led to serious efforts at plantation culture. Natural stands in many regions were severely depleted of old-growth kauris. As well, the commercial exploitation of dammar resins by wounding of the thin bark damages the trees over time and leads to deformation and decline. The rapid growth of kauri trees allows them to keep up with hardwoods in young stands, but their great dominance is achieved primarily through longevity. They outlast their original hardwood neighbors and one or more successor generations, growing all the while. Despite the superior quality of such old-growth trees for timber production, plantations can yield commercially viable harvest within reasonable rotation times, especially in those tropical regions where there are few other conifers suitable for plantation culture. Some species are cultivated for ornament in the tropics and warm temperate regions, but there has been no cultivar selection for horticultural purposes in the genus.

Considering the substantial economic importance of kauri trees, both as sources of timber and of commercial resins, there is surprisingly little consensus on an appropriate classification of the species and of their relationships within the genus. The two most recent overall treatments of *Agathis* differ one-and-a-half-fold in the number of species recognized (13 versus 21), and their groupings of species display only a little overlap. Two less recent formal treatments for only a portion of the range of the genus recognize additional species and accept conflicting assignments to the same number (three) of groups as the more recent treatments. On the other had, two more recent studies of epidermal structures delineate yet another grouping of species or none at all. The few DNA studies conducted as of 2007 include only a fraction of the known morphological and geographic variation and do not identify any convincing taxonomic structure within the genus. One of the problems is that most of the species, no matter how many one cares to recognize, form an intergrading mass while those with any really distinctive features, like the faceted bosses of the pollen scales in *A. labillardieri* and *A. microstachya* or the prominent triangular projections of the seed scales of *A. australis, A. endertii,* and *A. ovata,* form groupings of too few species to be of much practical significance (other than facilitating identification).

Taking up the issue of how many species to recognize, one problem is that, while the upper montane populations in western Malesia differ collectively from the surrounding lowland and lower montane populations, they also differ to varying but lesser degrees from each other and from the upper montane populations of Sulawesi, the Philippines, and the Moluccas. Most are treated here as belonging to the same species as the north-central Malesian dammars, but other taxonomists separate several into their own, more local species. Likewise, the number of species occurring in Melanesia and the relationships to one another are unsettled. For example, the five species accepted for New Caledonia in most contemporary literature are usually considered to be endemic, and relationships beyond the island have not been entertained. Yet the trees usually called *Agathis corbassonii* are very similar to those of *A. macrophylla* in Vanuatu, little more than 250 km away, just about the length of New Caledonia itself. Since they also have a similar ecology, rather different from that of other New Caledonian species of *Agathis,* the New Caledonian trees could well be treated as being part of *A. macrophylla,* the earliest name for the combined species.

The whole genus is rife with such uncertainties, in large part because of both complications due to political fragmentation of the region and the difficulties of obtaining good specimens from these towering trees with both pollen and seed cones linked to substantial foliage samples. The same difficulties impede the assembly of comprehensive samples for DNA studies. The known fossil record of *Agathis* is too sparse to be of much help and is confined to Australia and New Zealand. While a few Mesozoic fossils were assigned to the genus, these lack fully diagnostic features that would confirm their identity. Instead, the earliest secure specimens date from the mid-Eocene of southern Australia, about 40 million years ago. The pollen is not readily separable from that of some araucarias and has contributed little to understanding the history and distribution of the genus, except during the last 12,000 years, for which period the progressive restriction of New Zealand kauri (*A. australis*) to the northern tip of North Island has been documented.

Identification Guide to *Agathis*

Leaves show so much overlap among the 15 species of *Agathis* that it is impossible to identify them securely using only those features of the leaves that are visible with the naked eye or with the modest magnification of a hand lens. Identification is still difficult when measurements of the pollen and seed cones are included, as they are in this guide, so careful reading of the descriptions after a tentative identification is made is especially important here. For convenience of identification, the species may be divided into two groups by the length of most leaves of mature trees:

Group A, most leaves up to 7 cm

Group B, most leaves at least 7 cm

Group A. In addition to the characters given next, *Agathis microstachya* differs from all other species in this group by having the exterior faces of the pollen scales distinctly faceted, while they are rounded in the rest. The nine species in Group A may be distinguished by the width of most leaves of mature trees, the length and the width of the mature pollen cones, and the diameter of most mature seed cones:

Agathis atropurpurea, most leaves less than 1.5 cm wide, pollen cones less than 2 cm long, up to 0.7 cm wide, most seed cones less than 7 cm

A. microstachya, most leaves less than 1.5 cm wide, pollen cones less than 2 cm long, up to 0.7 cm wide, most seed cones at least 7 cm

A. moorei, most leaves less than 1.5 cm wide, pollen cones 2–3.5 cm long, 0.7–1 cm wide, most seed cones more than 7 cm

A. corbassonii, most leaves less than 1.5 cm wide, pollen cones 3.5–5 cm long, 0.7–1 cm wide, most seed cones more than 7 cm

A. australis, most leaves less than 1.5 cm wide, pollen cones 3.5–5 cm long, more than 1.5 cm wide, most seed cones up to 7 cm

A. orbicula, most leaves more than 1.5 cm wide, pollen cones less than 2 cm long, less than 0.7 cm wide, most seed cones less than 7 cm

A. endertii, most leaves more than 1.5 cm wide, pollen cones 2–3.5 cm long, up to 0.7 cm wide, most seed cones less than 7 cm

A. dammara, most leaves more than 1.5 cm wide, pollen cones 2–5 cm long, 0.7–1.5 cm wide, most seed cones at least 7 cm

A. ovata, most leaves more than 1.5 cm wide, pollen cones 3.5–5 cm long, 1–1.5 cm wide, most seed cones less than 7 cm

Group B. In addition to the characters given next, *Agathis labillardieri* differs from the other species in this group by having the exterior faces of the pollen scales distinctly faceted rather than rounded. The seven species in Group B have too much overlap in leaf size to be distinguished by it. Instead, they may be distinguished by the length and width of most mature pollen cones, by whether the underside of the leaves is whitened by a thin film of wax or not, and by the length and width of larger mature seed cones:

Agathis lanceolata, most pollen cones 2–3.5 cm long, 0.7–1 cm wide, underside of leaves not waxy, seed cones at least 10.5 cm long, more than 7.5 cm wide

A. labillardieri, most pollen cones 2–3.5 cm long, 1–1.5 cm wide, underside of leaves not waxy, seed cones less than 10.5 cm long, more than 7.5 cm wide

A. dammara, most pollen cones 2–5 cm long, 0.7–1.5 cm wide, underside of leaves usually not waxy, seed cones up to 10.5 cm long, more than 7.5 cm wide

A. macrophylla, most pollen cones 2–5 cm long, at least 1.5 cm wide, underside of leaves waxy, seed cones at least 10.5 cm long, more than 7.5 cm wide

A. montana, most pollen cones 3.5–5 cm long, 0.7–1 cm wide, underside of leaves not waxy, seed cones less than 10.5 cm long, less than 7.5 cm wide

A. robusta, most pollen cones more than 5 cm long, 0.7–1 cm wide, underside of leaves not waxy, seed cones at least 10.5 cm long, more than 7.5 cm wide

A. borneensis, most pollen cones more than 5 cm long wide, more than 1.5 cm wide, underside of leaves not waxy, seed cones less than 10.5 cm long, less than 7.5 cm wide

Agathis atropurpurea B. Hyland

BLACK KAURI, BLUE KAURI

Tree to 50 m tall, with smoothly cylindrical, unbuttressed trunk to 1.5(–2) m in diameter. Bark smooth and reddish purple to purplish black at first (hence the scientific name, Latin for "blackish purple"), weathering brown to grayish brown and peeling in large scales to produce a coarsely mottled pattern with age. Crown relatively narrow, egg-shaped to cylindrical or dome-shaped with slender, steeply upwardly arched to nearly horizontal branches bearing tufted, upwardly angled branchlets. Branchlets often bluish green with wax when young, fairly densely clothed with foliage. Leaves yellowish green to bright green, without a waxy film beneath at maturity, 3–5.5(–7) cm long, 0.5–2 cm wide, widest near the middle, tapering smoothly and quickly to the rounded or roundly triangular tip and a little more gradually to the narrowly wedge-shaped base on a very short petiole 1–2 mm long. Pollen cones 0.9–1.5(–2) cm long, (4–)5–7.5 mm thick, with three to five pairs of larger, clasping sterile scales at the base on a stalk about 3–6 mm long. Each pollen scale with two to five pollen sacs and a rounded external face. Seed cones nearly spherical, 3.5–5.5 cm long, 3.5–5 cm thick. Seed scales without a large, tonguelike projection. Seed body about 10 mm long and 6 mm wide, the larger wing about as large, the smaller one roundly triangular, projecting about 2 mm. Limited to a stretch of about 100 km south of Cairns in coastal ranges and the Atherton Tableland, Queensland (Australia). Scattered in the canopy of or emergent above lower montane rain forest on soils derived from granite substrates; 900–1,500 m. Zone 9.

This rare species has smaller seed cones than do the other two kauri species that grow with it. However, its pollen cones, while smaller than those of smooth-bark kauri (*Agathis robusta*), are similar in size to those of bull kauri (*A. microstachya*). They are readily distinguished by their pollen cones, nonetheless, because the exposed face of the pollen scale in black kauri is rounded rather than faceted as it is in bull kauri. Although black kauri attains a similar stature to the other two Australian species, the wood is often inferior because of a greater prevalence of spiral grain. Individual black kauri trees may live more than 1,000 years.

Agathis australis (D. Don) R. A. Salisbury

KAURI (MAORI AND ENGLISH), NEW ZEALAND KAURI

Tree to 40(–60) m tall, with a massively cylindrical (or somewhat squared off), unbuttressed trunk to 4.5(–7) m in diameter and free of branches for up to 20(–30) m in fully mature trees. Bark very resinous, smooth and light brown at first, thickening, weathering very light gray and flaking continuously to produce fine mottling on a background of intersecting, shallow and narrow spiral ridges with great size and age. Crown conical at first and then passing through a prolonged, narrowly cylindrical "ricker" phase before shedding the lower branches and becoming broadly dome- or vase-shaped as the tree enters the canopy at maturity, with a ring of very heavy, upwardly angled, widely spreading limbs at the summit of the unbranched trunk bearing crowded tufts of branchlets. Branchlet bluish green with wax, this sometimes wearing off and then the twigs becoming green before turning brown with age, densely clothed with foliage. Leaves yellowish green to bright green (to coppery green in juveniles) above, variably bluish green with wax beneath, (1.5–)2–3.5(–5) cm long (to 10 cm in

Crowns of two emergent old kauris (*Agathis australis*) in nature.

juveniles), (0.3–)0.5–1(–1.5) cm wide (to 2 cm in juveniles), widest near or before (or a little beyond) the middle, tapering gradually or abruptly to the rounded or bluntly triangular tip (sharply triangular in juveniles) and less gradually to the wedge-shaped to rounded base attached directly to the twig or on a very short, broad petiole 1–3(–5) mm long. Pollen cones (2–)2.5–5(–6) cm long, (0.7–)1.2–2 cm thick, with four to six pairs of larger, sterile scales at the base, the lowest (none or) one or two pairs leaflike, on a stalk (0–)2–20 mm long. Each pollen scale with two to five pollen sacs and a rounded outer face. Seed cones with remnants of a waxy film at maturity, often spotted with resin, roughly spherical, often a little wider than long, (4.5–)5.5–7.5(–9) cm long and thick. Seed scales with a conspicuous, tonguelike projection. Seed body about 7–10 mm long and 3.5–5 mm wide, the larger wing about 7–15 mm by 5–10 mm, the smaller one bluntly triangular, barely projecting 0–2 mm. Northwestern tip of North Island, New Zealand, north of 38°06′S. Forming pure stands in youth, later scattered singly or in small groves or dominating as an emergent over mixed rain forest with various hardwoods and other conifers, with best development on slightly drier sites with limited mineral nutrients; (0–)80–350(–600) m. Zone 8.

While it was traditionally second in importance to totara (*Podocarpus totara*) for the Maori of North Island, kauri was the premier softwood of colonial New Zealand. Nearly 1.5 million hectares of virgin kauri forest were reduced to a remnant of just 6,200 hectares by 1975. Not all of the losses were directly due to logging, however. Many beautiful forests succumbed to both inadvertently and deliberately set fires during the late 19th and early 20th centuries. Since 1985, all remaining old-growth stands on state land (less than 150 hectares!) are protected, accessible for logging only during a national emergency. Maturing second-growth stands now occupy another 60,000 hectares, and some regeneration and plantation culture with kauri is underway, although it pales into insignificance in comparison to the planting of Monterey pine (*Pinus radiata*) throughout the country.

The uses of kauri wood have been quite varied, from rough construction and boatbuilding (including Maori seagoing dugout canoes) to turnery and finely detailed carving. Craft uses emphasize relatively uncommon wavy and mottled figured grains. The most unusual of these are found in swamp kauris, logs dug up from bogs after being submerged for centuries or millennia (the oldest are radiocarbon dated at 40,000 years old or more). Other present sources of old-growth timber include the tops of trees felled during the logging era (the wood is durable in contact with the ground) and reuse of timber from old buildings. In addition to the wood, the resin of kauri was a major export commodity as a primary constituent of varnishes in the days before synthetic resins and petroleum-based products became commonplace. The majority of this kauri gum was obtained not from living or recently felled trees but from enormous accumulated masses of material mined from the soil beneath present and prehistoric forests, including the same wetlands that yield swamp kauri.

Fossil evidence from swamp kauris and pollen profiles suggests that kauri forests never extended beyond their present limits during the current glacial cycle and reached their current southern limit just 3,000 years ago in an expansion out of a refuge in the northern half of the Northland district that began just 7,000 years ago. The extreme geographic restriction of kauri is a little peculiar considering that it grows well in cultivation throughout New Zealand at least as far south as Dunedin in the Otago district of South Island, more than 850 km south of its natural southern limit. Even with its limited range in New Zealand, kauri is the most southerly species in the genus (hence the scientific name, Latin for "southern"). It is the hardiest species of *Agathis* and is widely, if sparsely, cultivated throughout moist warm temperate regions of the world. Growth is relatively rapid, both in cultivation and in native stands, so the forest giants of old-growth stands are not enormously old. Their age is respectable, however. Trees often reach 600 years, with a maximum recorded age near 1,700 years. This is quite long enough, in combination with climatic sensitivity, to make kauri useful in dendroclimatological studies. Its growth variations track the weather changes resulting from the Southern Oscillation, the southern hemisphere climatic cycle linked to the northern El Niño (together referred to as ENSO). Using correlations of kauri growth rings with the modern record of Southern Oscillations, a reconstruction of the dates and intensities of these events has been pushed back into the 1700s. Such extended chronologies can help in understanding the ENSO cycle and thus lead to more accurate and detailed forecasts of these climatically powerful phenomena.

The tonguelike projection on the seed scales of kauri is only found in two other species in the genus, dwarf kauri (*Abies ovata*) of New Caledonia and bulok dammar (*A. endertii*) of Borneo. All three also share the waxy underside of the leaf found in about half the *Agathis* species. It is not clear whether these three species are especially closely related, however, although dwarf kauri does superficially resemble a miniature version of kauri. Each of the three differs from the others in some features in which they more closely resemble other species in the genus, and DNA studies do not shed much light on their affinities. Nevertheless, kauri is by far the best-studied species of the genus with respect to its ecology, reproductive biology, and resin chemistry, among other aspects of its biology. It is just that other species in the genus are so poorly studied, and the genus as a whole is so homogeneous, that there are few points of comparison for working out their relationships to one another.

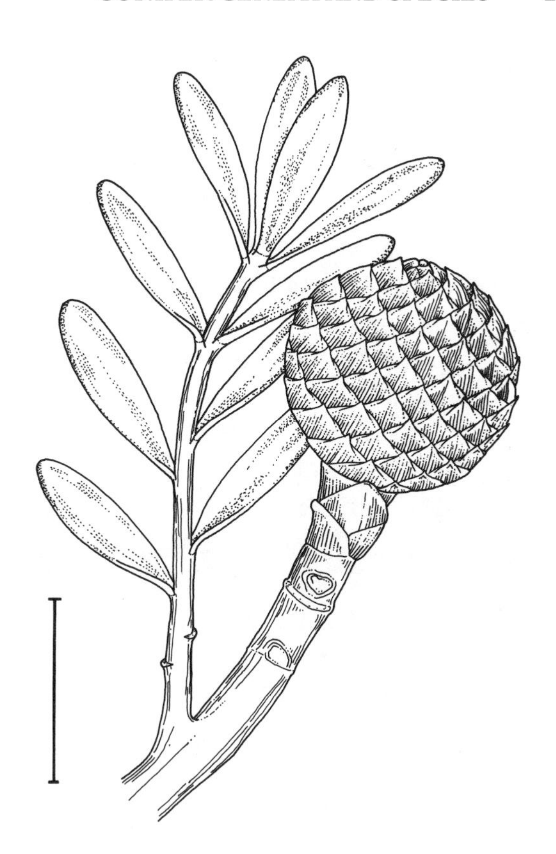

Slightly immature seed cone and replacement foliage shoot of kauri (*Agathis australis*); scale, 5 cm.

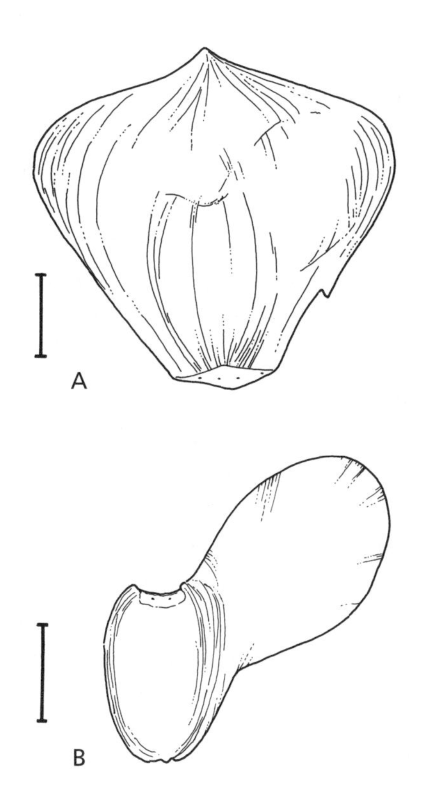

Upper side of shed seed cone scale (A) and seed (B) of kauri (*Agathis australis*); scale, 5 mm.

Agathis borneensis O. Warburg

WESTERN DAMMAR, DAMMAR MINYAK, BINDANG, SANUM (MALAY), NUJU, TULONG (DAYAK)

Tree to 35(–60) m tall, with cylindrical, unbuttressed trunk to 1.6(–3) m in diameter. Bark smooth and light brown to light gray at first, darkening with age and weathering, sometimes becoming almost black, and flaking in small, thin or thick scales to become densely and deeply pockmarked, often spotted with resin. Crown narrowly dome-shaped, with irregularly spaced, large, stiffly spreading to upwardly angled branches bearing tufts of branchlets. Branchlets green from the beginning, without a waxy film, densely clothed with foliage. Leaves light green to yellowish green above and beneath, (3–)6–9(–12) cm long (to 14 cm in juveniles), (1–)2–3.5 cm wide (to 4 cm in juveniles), widest near the middle (a little before the middle in juveniles), tapering gradually to the triangular tip, usually prolonged in a narrower sharp to blunt point, and more abruptly to the concavely wedge-shaped base on a short petiole about 5–10 mm long. Pollen cones distinctly widest near the middle, 3–7(–9) cm long, (14–)20–30(–40) mm thick, with two or three pairs of small, clasping sterile scales at the base, on a stalk 1–10 mm long. Each pollen scale with two to seven pollen sacs and a rounded external face. Seed cones often flushed with purple just before maturity, not waxy, distinctly longer than wide, (6–)7–10(–15) cm long, 5.5–8(–13) cm thick. Seed scales without a large, tonguelike projection. Seed body 10–12 mm long and (5–)8–9 mm wide, the larger wing 15–20 mm by 9–16 mm, the smaller one bluntly triangular, projecting about 1–2 mm. Southern Malay Peninsula, northern and central Sumatra, and Borneo (Malaysia and Indonesia). Scattered as an emergent over various kinds of lowland and lower montane rain forests but forming pure stands on sterile white sands; (2–)300–1,000(–1,400) m. Zone 9. Synonyms: *Agathis alba* (Blume) Foxworthy (misapplied), *A. rhomboidalis* O. Warburg.

Throughout its region of occurrence, western dammar is a common dominant on especially sterile, peaty substrates but still manages to reach enormous sizes. It is much larger than the only other *Agathis* species in the region, mountain dammar (*A. dammara* subsp. *flavescens*), which is only found on scattered mountains in the upper montane zone here, west of its continuous region in the Philippines and Sulawesi. These two dammars were much confused with one another in the past, but western dammar differs from the eastern dammar (*A. dammara*) most conspicuously in its much thicker, bulging pollen cone. Somewhat more obscure but always present is a difference in the leaf cross sections. Western dammar is unique in the genus in having at least some of the resin canals between the veins stacked up in vertical pairs. (They are single in all other species.) With its occurrence at a variety of elevations and on soils of greatly varying fertility, this species exhibits considerable variation, particularly in the size, shape, and thickness of adult leaves. As a result, it has been divided into as many as five species, but all have similar pollen and seed cones, and all share the paired resin canals in the leaves, besides which, the varying leaf forms all intergrade with one another. Only one of these segregates is listed in the synonymy above because the others were abandoned long ago and are not recognized in contemporary conifer literature. It is a highly prized timber tree and has been subjected to excessive exploitation for its wood. The resin has been much less sought after than that of mountain dammar. There is limited plantation culture of the species even though under favorable conditions it makes excellent growth, while on poor sites it is one of the few trees that can thrive. It makes a handsome specimen tree in the humid tropics but is rarely cultivated for ornament.

Agathis corbassonii de Laubenfels

RED KAURI, KAORI ROUGE (FRENCH)

Tree to 25(–35) m tall, with tapering, unbuttressed trunk to 1(–1.5) m in diameter. Bark dark gray and smooth at first, weathering light gray to reddish brown, flaking in broad, thin scales, and finally becoming very rough in large trees and dripping with abundant resin. Crown sparse and open, spreading, dome-shaped, with well-spaced, horizontal or gently upswept branches bearing broad tufts of branchlets. Branchlets waxy bluish green at first, densely clothed with foliage. Leaves dark green above, paler and bluish green with a thin, white, waxy film beneath, 4.5–7.5 cm long (to 11 cm in juveniles), 0.6–1(–1.5) cm wide (to 2 cm in juveniles), widest at or near the middle (below the middle in juveniles), tapering very gradually to the rounded tip (triangular in juveniles) and to the wedge-shaped base on a very short petiole 1–3 mm long. Pollen cones 2.5–5 cm long, 6–9 mm thick, with four or five pairs of larger, somewhat spreading sterile scales at the base on a stalk 3–6 mm long. Each pollen scale with four pollen sacs and a rounded external face. Seed cones without a conspicuous waxy film at maturity, round but a little longer than wide, about 10 cm long and 9 cm thick. Seed scales with a broad, short, upturned lip but without a large, tonguelike projection. Seed body 10–12(–15) mm long and 6–8(–10) mm wide, the larger wing about 20 mm by 10–15 mm, the smaller one broadly triangular, projecting about 2–3 mm. Discontinuously distributed in low mountains of the northern two-thirds of New Caledonia, from near Pouébo to near Bouloupari. Scattered as an emergent above rain forests along ridge tops and on slopes where soils are not derived from serpentine; 300–700 m. Zone 10.

Red kauri was long confused with white kauri (*Agathis moorei*), each named for its wood color. Many botanists thought that this difference, while important for forestry, was ecological in origin. One forester who thought otherwise was Michel Corbasson, director of the tropical forestry center in Nouméa, New Caledonia, after whom the species was named and who helped establish a dry forest reserve on the outskirts of Nouméa in 1972 that later expanded to become the city's zoo and botanical garden, which also bears his name. While red kauri and white kauri share similar ranges and neither grows over ultrabasic substrates in serpentine-derived soils, red kauri grows, perhaps, on drier sites. Its narrower leaves, proportionately the narrowest in the genus, with a waxy film beneath not found in either white kauri or serpentine forest kauri (*A. lanceolata*), the other two large kauris of the forests of New Caledonia, are consistent with a relatively dry habitat. The affinities of the species are not all that clear, although a relationship with the very similar Melanesian kauri (*A. macrophylla*) has been suggested. Resolution of this difficulty awaits further study of both botanical features and DNA sequences. Published DNA sequences do not contain enough differences between species to clarify the relationships among them. Red kauri has not yet been as heavily exploited for timber as has serpentine forest kauri. The majority of its range lies outside of the mining district, where forest exploitation in general is most intense, and some stands are in parks and biological reserves. There is still some cause for concern, however, considering the ever increasing impact of human activities in its range.

Agathis dammara (A. Lambert) L. Richard

EASTERN DAMMAR

Tree to 40(–70) m tall, much smaller on poor sites, with unbuttressed (or irregularly swollen, lobed, and distorted when subjected to tapping for resin), cylindrical, straight or less often crooked trunk to 1.5(–3) m in diameter. Bark smooth, with horizontal lines of resin pockets, and reddish brown to grayish brown at first, weathering gray and shedding thick scales or larger patches at varying rates to yield a scaly to densely and deeply pockmarked, mottled surface. Crown conical to narrowly cylindrical at first, opening up and becoming more irregular and dome- to vase-shaped at maturity, with relatively slender, scattered, upwardly angled branches bearing irregular, well-spaced tufts of branchlets. Branchlets not waxy, dark green to very yellowish green, densely to more sparsely clothed with foliage. Leaves shiny light green to dark green or yellowish green above, similar beneath or sometimes paler and with a thin film of wax, (2.5–)4–8(–10) cm long (to 15 cm in juveniles), (0.8–)1.5–3.5(–4) cm wide (to 5 cm in juveniles), widest near the middle (before the middle in juveniles), tapering fairly quickly to a rounded to roundly triangular tip, sometimes with a narrowed, prolonged point (especially in juveniles), and with a similar taper to the wedge-shaped or roundly wedge-shaped base narrowing to a short petiole 3–10 mm long. Pollen cones (1–)2.5–5(–6) cm long, (6–)8–15 mm thick, with four (or five) pairs of slightly larger, clasping, sterile scales (or the lowest pair sometimes somewhat spreading and leaflike) on a stalk (0.1–)3–5(–10) mm long. Each pollen scale with (3–) 5–10(–12) pollen sacs and a rounded external face. Seed cones without a waxy film at maturity, round but generally a little longer than wide, (6–)6.5–8.5(–10.5 or perhaps 12) cm long, 6–7.5 (–10) cm thick. Seed scales without a large, tonguelike projection. Seed body 9–14(–16) mm long and 6–8(–10) mm wide, the larger wing 14–25 mm by 8–17 mm, the smaller one triangular, projecting about 1–3 mm (rarely expanded up to about 10 mm). Distributed more or less continuously from Calayan Island, north of Luzon through the Philippines, Sulawesi, and the Moluccas (both Indonesia) to Buru, Ambon, and Seram, and discontinuously westward on scattered mountains in northwestern Borneo (Sabah and Sarawak) and the Gunong Tahan massif of southern Malaya (both Malaysia). Scattered as an emergent above various wet-forest types, ranging from lowland rain forests to upper montane mossy forests; 0–2,220 m. Zone 10. Two subspecies.

Eastern dammar, like its neighboring counterpart western dammar (*Agathis borneensis*), is a very important timber tree, at least locally. The wood has physical properties similar to those of the unrelated northern hemisphere Sitka spruce (*Picea sitchensis*) and so is used for some of the same purposes requiring a white wood with good dimensional stability as well as strength and toughness. In many places, particularly in the Philippines and on Java (where the species is cultivated in plantations), the resin, known as Manila copal (or white dammar, hence the scientific name) in international commerce, is the most important product. This material accumulates in high branches of the trees, but the difficulty in collecting it has led to repeated wounding of the boles as an easier means of obtaining the resin. This practice disfigures the trees and makes them essentially useless for timber production. The existing limited level of plantation culture of eastern dammar within the region comes nowhere near meeting the losses due to exploitation of old-growth trees. The species is also cultivated for timber as well as resin, along with some other dammars and kauris, on Java, where it does not occur naturally.

Eastern dammar is very closely related to western dammar and differs from that species most obviously in its skinnier pollen cones and fatter seed cones. While the two species have largely separate ranges, eastern dammar also occurs above western dammar on a number of mountains within the range of that species in Malaysia. These isolated occurrences are sometimes treated as separate species but differ little from trees in the relatively continuous eastern portion of the range.

The eastern trees, in turn, are sometimes split between two species, with overlapping more northerly versus more southerly ranges. There are, indeed, minor differences between these groups within the species, and they are here treated as subspecies. The more southerly plants were the first to be formally named and so they retain the name subspecies *dammara.* Compared to the more northerly plants, they grow at lower elevations, tend to have larger leaves on average, and have thicker pollen cones with more pollen sacs on each pollen scale. There is little difference in the seed cones and seeds except that the expanded wing may be a little larger in subspecies *dammara.* Although the northern trees grow at progressively lower elevations in the northern portion of their range, they typically grow above subspecies *dammara* where they overlap. This difference in ecology, along with the shared minor differences in morphology, place the disjunct trees of the mountains of Malaya and Borneo into the northern subspecies. This group, therefore, takes the name of subspecies *flavescens,* a name that was first applied just to the trees on Gunong Tahan in peninsular Malaya, because this is the only name available at the rank of subspecies.

With its complicated variation and somewhat fragmented range across a linguistically fragmented region, eastern dammar has many additional synonyms in older literature that are not recognized as distinct in more recent taxonomic and floristic works and so are not listed here. It also has many additional common names than the small selection of more widely used ones given here.

Agathis dammara (A. Lambert) L. Richard subsp. *dammara*

LOWLAND DAMMAR, DAMMAR RADJA (MALAY), HULONTUU, KAWO (DAYAK)

Twigs and leaves green. Adult leaves not waxy beneath, (5–)6–8(–10) cm long, (1.5–)2–3.5(–4) cm wide. Pollen cones (2–)2.5–5(–6) cm long, 10–15 mm thick. Pollen scales with (5–)7–10(–12) pollen sacs. Larger seed wing 20–24 mm by 16–17 mm. Southeastern portion of the range of the species throughout Sulawesi and the Moluccas, with outliers in Palawan and Samar. Lowland and lower montane rain forests; 0–1,200 m. Synonyms: *Agathis alba* (Blume) Foxworthy, *A. celebica* (Koorders) O. Warburg, *A. loranthifolia* R. A. Salisbury.

Agathis dammara subsp. *flavescens* (H. Ridley) T. Whitmore

MOUNTAIN DAMMAR, ALMACIGA (SPANISH), GALAGALA (TAGALOG), TENGILAN (DAYAK)

Twigs and leaves often strongly yellowish green, especially on exposed sites over poor soils. Adult leaves sometimes waxy beneath, (2.5–)4–7 cm long, (0.8–)1.5–3(–3.5) cm wide. Pollen cones (1–)2.5–3.5(–4.5) cm long, (6–)8–10(–11) mm thick. Pollen scales with (three to) five or six pollen sacs. Larger seed wing 14–20 mm by 8–14 mm. Northern and western portion of the range of the species, throughout the Philippines west to the Malay Peninsula and south to central Sulawesi and the northern Moluccas. Montane rain forests and mossy forests; (250–)1,200–2,000(–2,200) m. Synonyms: *Agathis celebica* subsp. *flavescens* (H. Ridley) T. Whitmore, *A. flavescens* H. Ridley, *A. kinabaluensis* de Laubenfels, *A. lenticula* de Laubenfels, *A. philippinensis* O. Warburg.

Agathis endertii Meijer Drees

BULOK DAMMAR

Tree to 45(–65) m tall, with cylindrical, unbuttressed trunk to 1 m in diameter or more. Bark smooth and brown at first, weathering gray and flaking in scales to produce a mottled patchwork of colors. Crown regularly conical at first, becoming broadly vase-shaped and irregular with age, with a few, heavy, upwardly angled branches at the top of the bole bearing tufts of branchlets. Branchlets green from the beginning or with a thin film of wax initially, densely clothed with foliage. Leaves dark green above, bluish green beneath with a thin waxy film, (3.5–)5–7(–9) cm long (to 11 cm in juveniles), (1.5–)2–3(–3.5) cm wide (to 4.5 cm in juveniles), widest near or a little before the middle, tapering smoothly and gradually to the roundly triangular or rounded tip (or sharply triangular in juveniles) and more abruptly to the roundly wedge-shaped base (narrowly wedge-shaped in juveniles) on a short petiole 3–6 mm long. Pollen cones 2.5–4 cm long, 7–9 mm thick, with two or three pairs of larger, clasping sterile scales at the base, stalkless (or on a very short stalk to 2 mm). Each pollen scale with three or four pollen sacs and a rounded external face. Seed cones without a waxy film at maturity, rounded but distinctly longer than wide, 6–7 cm long, 4.5–6 cm thick. Seed scales with a conspicuous, roundly triangular, projecting tip. Seed body 10–11 mm long and 7–8 mm wide, the larger wing 15–20 mm by 14–15 mm, the smaller one roundly triangular, projecting about 2 mm or effectively absent. Discontinuously distributed in northern and central Borneo, in both Malaysia (Sabah and Sarawak) and Indonesia (Kalimantan). Scattered as an emergent in lowland and lower montane forests, often over sandstone substrates; 200–1,400(–1,600) m. Zone 10.

Although bulok dammar is widespread in Borneo, it is poorly known and seemingly often confused with western dammar (*Agathis borneensis*), which has very similar foliage. Bulok dammar never has the paired resin canals between the veins that are characteristic of western dammar, and both the pollen and seed cones are smaller than those of the more common and widespread species. In addition, the triangular projection of the seed scales readily distinguishes bulok dammar from all other species in Borneo, even when only the scattered scales from a shattered mature seed cone are present on the ground. These free tips are shared only with kauri (*A. australis*) of New Zealand and dwarf kauri (*A. ovata*) of New Caledonia, both of which also have persistently waxy undersides of the leaves, among other features, so these three species may be especially closely related within the genus. This proposal has been formalized by creating a section of the genus for them, but confirmation of their relationship to the exclusion of other species by DNA studies would be welcome. The species name honors Frederik Endert (1891–1953), a Dutch forester born in Java and long associated with the Forest Research Institute in Bogor, who collected in Borneo and Sulawesi, although he did not collect the types of this species.

Agathis labillardieri O. Warburg

PEN DAMMAR, AGLO, FUKO, IDJIR, KOBA, LEGATULUS, OSIER, WARKAI (PAPUAN LANGUAGES)

Tree to 40(–60) m tall, with smoothly cylindrical, unbuttressed trunk to 1(–2) m in diameter. Bark brown and smooth at first, weathering gray and flaking in scales to produce a mottled patchwork. Crown densely conical at first, becoming more open and broadly dome-shaped with age, with upwardly angled branches bearing tufts of branchlets. Branchlets green from the beginning, openly clothed with foliage. Leaves shiny light to dark green above, a little paler but not waxy beneath, (3–)6–9(–10) cm long (to 12.5 cm in juveniles), (1–)1.5–2.5 cm wide (to 3.5 cm in juveniles), widest before the middle, tapering very gradually in a shallow S curve to the narrowly to roundly triangular tip (sharply triangular in juveniles) and more abruptly to the rounded to roundly wedge-shaped base on a short petiole 3–7 mm long. Pollen cones (1.8–)2.5–3.5 cm long, 10–15 mm thick, with two or three pairs of clasping, sterile scales at the base on a stalk (0.1–)2–6(–10) mm long. Each pollen scale with (3–)5–12 pollen sacs and a faceted external face. Seed cones not waxy at maturity, round but a little longer than wide, 8.5–10 cm long, 7.5–9 cm thick. Seed scales without a large, tonguelike projection. Seed body 12–13 mm long and 7–8 mm wide, the larger wing about 20 mm by 15 mm, the smaller one triangular, projecting about 2–3 mm. Discontinuous and locally common throughout the hilly lowlands and modest elevations of Irian Jaya (western New Guinea, Indonesia), including the islands at the western tip and those in Cenderawasih Bay, extending east in the Sepik River valley of northwestern Papua New Guinea to about 143°30′E. Scattered or forming groves as an emergent over rain forests on a variety of soil types, including those derived from serpentine; (0–)250–1,200(–1,800) m. Zone 10.

Only bull kauri (*Agathis microstachya*) of Queensland (Australia) shares the faceted pollen scales found in pen dammar, and these two species appear to be closely related. Pen dammar has larger pollen cones than does bull kauri, and its leaves average a little larger, as do its seeds and seed cones. It contrasts sharply in morphology and habitat with the only other *Agathis* species in New Guinea, Papuan kauri (*A. robusta* subsp. *nesophila*), of the eastern highlands. Despite sharing the island, these two taxa do not appear to be closely related and have separated ranges so that they never occur together in nature. Oddly, their counterparts in Australia, bull kauri, and kauri pine (*A. robusta* subsp. *robusta*), reverse the elevational relationships found in New Guinea, with kauri pine at lower elevations than bull kauri. Within its native range, pen dammar is an important (and locally overexploited) source of timber and resin and is even grown in plantation culture to a certain extent. The species name honors Jaques-Julien de La Billardière (1755–1834), a well-traveled French botanist who collected the type material on Waigeo Island, at the northwestern tip of New Guinea, during extensive botanical explorations in Australasia as a naturalist on the French La Pérouse expedition to the region in 1791–1792.

Agathis lanceolata O. Warburg

SERPENTINE FOREST KAURI, KOGHIS KAURI, KAORI DE FORÊT (FRENCH)

Plate 8

Tree to 30(–40) m tall, with smoothly cylindrical, unbuttressed trunk to 2(–3) m in diameter and often unbranched for up to 18 m or more. Bark smooth and light gray at first, weathering darker gray, peeling in thin, irregular flakes to reveal reddish brown patches, and finally becoming shallowly and narrowly furrowed. Crown dense, conical to cylindrical in youth, becoming broadly vase-shaped and spreading with age, with a few, upwardly angled major limbs at the top of the unbranched bole, forking repeatedly and bearing numerous branchlets near their tips. Branchlets not conspicuously waxy at maturity, densely clothed with foliage. Leaves light green at first, maturing shiny dark green above, paler and duller beneath, 6–8 cm long (to 13 cm in juveniles), 1.5–2 cm wide (to 4.5 cm in juveniles), widest near the middle (before the middle in juveniles), tapering steadily to the rounded to roundly triangular tip and slightly more gradually to the wedge-shaped base on a short petiole 3–8 mm long. Pollen cones (1.5–)2–2.5 cm long, 7–10 mm thick, with three or four pairs of larger, clasping, sterile scales at the base (of which the lowest pair may be somewhat expanded and leaflike) on a short stalk 2–5 mm long. Each pollen scale with three to five pollen sacs and a humped external face. Seed cones not waxy-coated at maturity, round but a little longer than wide, 8–12 cm long, 7–10 cm thick. Seed scales briefly upturned at the edge but without a large, tonguelike projection. Seed body 12–15 mm long and 6–7 mm wide, the larger wing 15–20 mm by 11–13 mm, the smaller one triangular to somewhat bladelike, projecting about 3–4 mm. Fairly well distributed in the southern quarter of New Caledonia and much more discontinuous northward through the southern two-thirds of the island north to Tonine Mountain near Touho. Scattered as an emergent above the canopy of rain forests, mostly on serpentine-derived soils; 150–500(–1,100) m. Zone 10.

Unlike the other two large forest kauris of New Caledonia, red kauri (*Agathis corbassonii*) and white kauri (*A. moorei*), serpentine forest kauri is almost always found on ultramafic soils while the other two never are. Once a characteristic species of the southern rain forests, this species has been much more heavily exploited than its northern congeners because of its occurrence in the mining districts and proximity to population centers. It was first exploited for its resin and later, more heavily, for timber. Few accessible commercial stands remain today, but some of the best of these are protected in botanical reserves. Among the handful of species with which it has been compared, black kauri (*A. atropurpurea*) of Australia has the most similar chemical composition of diterpene (20-carbon) acids in its resin. Since comparable studies for other New Caledonian kauris have not been published, this data is not particularly helpful for understanding the relationships of this species. The species name, Latin for "shaped like a little lance," applies equally well in describing the leaves of several other species of *Agathis* as well, including red and white kauris.

Agathis macrophylla (Lindley) M. T. Masters

MELANESIAN KAURI, PACIFIC KAURI, KAURI (MELANESIAN PIDGIN), DURE (SANTA CRUZIAN), NEIJEV, NENDU (NI-VANUATU), NDAKUA (FIJIAN)

Plate 9

Tree to 30(–45) m tall, with cylindrical, straight, unbuttressed to slightly basally swollen trunk to 1.5(–3) m in diameter, typically unbranched for no more than 10(–20) m. Bark light brown and smooth at first, weathering light gray, flaking in round scales to reveal pinkish brown to reddish brown fresh bark, and finally becoming heavily pockmarked and mottled. Crown fairly open, conical at first, passing through cylindrical to broadly vase-shaped with a domed top in age, with fairly slender to heavy, sharply upwardly angled to practically horizontal branches bearing branchlets along their length. Branchlets bluish green with a persistent, thin waxy film, openly to densely clothed with foliage. Leaves shiny dark green above, duller dark green to bluish green beneath and with or without a thin film of wax, (3.5–)6–9(–12) cm long (to 15 cm in juveniles, hence the scientific name, Latin for "big leaf"), (1–)2–3 cm wide (to 4 cm in juveniles), widest before the middle, tapering steadily or gradually and then more abruptly to the rounded or roundly triangular tip (sharply triangular in juveniles) and quite abruptly to the roundly wedge-shaped to rounded base on a short petiole 1–8 mm long. Pollen cones (2–)4–6(–8) cm long (to 12 cm after the pollen is shed), (8–)15–20(–25) mm thick, with three or four pairs of larger, loosely cupping, sterile scales at the base on a short stalk (0–)3–5(–7) mm long. Each pollen scale with 6–14 pollen sacs and a humped external face. Seed cones with remnants of a waxy film at maturity, nearly spherical to a little longer than wide, 10–13 cm long, 8–10 cm thick. Seed scales without a large, tonguelike projection. Seed body 10–12 mm long and 7–10 mm wide, the larger wing 15–17 mm by 12–15 mm, the smaller one narrowly triangular, projecting up to about 10 mm. Sporadically distributed among the islands of Melanesia, from the Santa Cruz group (Solomon Islands), through Espiritu Santo, Aneityum, and Erromango in Vanuatu, to Vanua Levu, Viti Levu, and two smaller islands in Fiji. Scattered or forming small groves in, or emergent above, the canopy of rain forests on gentle to steep mountain slopes; (0–)50–700(–900) m. Zone 10. Synonyms: *Agathis obtusa* (Lindley) A. Morrison, *A. silbae* de Laubenfels, *A. vitiensis* (B. Seemann) E. Drake.

Melanesian kauri is the largest native tree in much of its range. While the wood is not durable and not as suitable for construction as other local trees, it was historically much appreciated in boatbuilding both by native Melanesians and by export markets in Australia. Just as important was the resin (called *makandre* in Fijian), which is used in caulking and glazing and is burned on torches. The timber has been and still is an important export commodity on several islands, providing both solid lumber and veneer. Because the islands are fairly small, these industries are often short-lived. It took only 40 years, from 1924 to 1964, to send all of the old-growth kauri on Vanikoro in the Santa Cruz group (the island from which the tree was first described) to Australian markets. This kind of depredation has been repeated elsewhere, but luckily, regeneration is generally good so that Melanesian kauri is not being eliminated, even if large trees are. Growth, which occurs primarily during the warm wet season, is fairly rapid for established trees so second growth stands are now maturing. Melanesian kauris do not appear to live as long as some of the other species in the genus, but a maximum known age of more than 600 years is still pretty respectable. In addition to natural stands, there are timber plantations and seed orchards both within its native range and in far-flung sites such as Java and Hawaii. With its handsome, large, glossy leaves, it is also planted to a limited extent in botanical gardens but is not in general horticulture in the tropics.

There is considerable variation among populations in stature, resin production, and botanical characteristics, and thus trees from the different island groups were originally described as separate species. However, they are all more similar to one another than they are to any of the other species of *Agathis* (except the quite similar and possibly conspecific red kauri, *A. corbassonii*, of New Caledonia), and so are kept together here. For instance, while the chemical composition of the resin varies quite a bit from species to species among kauris generally, resins from trees in Fiji and Vanikoro are quite similar. Perhaps the most distinctive kauris in this region are trees from Espiritu Santo in northern Vanuatu that were named *A. silbae*. They may, indeed, prove to be distinct from Melanesian kauri, but available material is still incomplete and inconclusive, and thus the case for separation is not yet fully convincing. There are many other groups of populations in *Agathis* whose species status requires further investigation in the field, herbarium, and laboratory.

Agathis microstachya J. F. Bailey & C. White

BULL KAURI

Tree to 50 m tall, with smoothly cylindrical, unbuttressed trunk to 1.5(–2.8) m in diameter. Bark smooth and light brown at first, darkening, weathering grayish brown, and becoming coarsely scaly and roughened by persistent, relatively small flakes with age. Crown fairly open, cylindrical to broadly dome- or vase-shaped, with a few, heavy, upwardly arched limbs at the summit of the unbranched trunk and bearing open tufts of branchlets. Branchlets green from the beginning, relatively sparsely clothed with foliage. Leaves shiny bright to dark green above, paler beneath, (2–)4–7(–10) cm long (to 15 cm in juveniles), (0.5–)1.5–2.5(–3) cm wide (to 4 cm in juveniles), widest near or before the middle, tapering gradually and then more abruptly to the rounded or roundly triangular tip (narrowly triangular in juveniles) and more abruptly to the rounded or wedge-shaped base on a very short petiole 1–2 mm long. Pollen cones (1–)1.5–2 cm long, 6–10 mm thick, with four or five pairs of larger, clasping sterile scales at the base on a stalk (0–)1–3 mm long. Each pollen scale with two to five pollen sacs and a faceted external face. Seed cones with remnants of a waxy film at maturity, round but a little longer than wide, (7.5–)8.5–10(–11.5) cm long, 6.5–9(–10) cm thick. Seed scales without a large, tonguelike projection. Seed body about 10 mm long and 5 mm wide, the larger wing about 25 mm by 15 mm, the smaller one bladelike, projecting about 8 mm. Limited to a stretch of less than 100 km on the Atherton Tableland and vicinity southwest of Cairns, Queensland (Australia). Scattered as an emergent above the canopy of lowland rain forests on a variety of soil types; 400–900 m. Zone 9.

Although rare, bull kauri has been exploited along with smooth-bark kauri (*Agathis robusta*) for its large volume of excellent construction timber, as long as it is kept from contact with soil and weather. It grows at lower elevations than the rather similar black kauri (*A. atropurpurea*), from which it differs in its paler bark, larger seed cones, and generally wider pollen cones with faceted pollen scales. Based on intermediacy in morphology and resin chemistry, hybrids have been reported where these two species grow together. Faceted pollen scales link bull kauri to pen dammar (*A. labillardieri*) of western New Guinea, the only other kauri species with this feature. Pen dammar typically has wider leaves and pollen cones that are about twice the size of those of bull kauri. Along with black kauri and tumu dammar (*A. orbicula*), bull kauri has the smallest pollen cones in the genus, just up to 2 cm long (hence the scientific name, Greek for "small spike of grain"). Juvenile leaves are larger and distinctly more pointed than those of mature trees. Individuals can live more than 1,000 years.

Agathis montana de Laubenfels

MOUNT PANIÉ KAURI

Tree to 15(–20) m tall, with relatively short trunk to 1 m in diameter. Bark smooth and light brown at first, weathering grayish brown and flaking in thin sheets and later in small, irregular scales to reveal reddish brown fresh patches, often streaked or blotched with fresh or dried resin. Crown dense to open, broadly vase-shaped with a flattened top made up of stout, contorted branches bearing tufts of branchlets. Branchlets green from the beginning, densely clothed with foliage. Leaves dark green above, paler but without a waxy film beneath, 6–8 cm long (to 10 cm in juveniles), 1.5–2.5 cm wide (to 3 cm in juveniles), widest near the middle (or sometimes a little before the middle in juveniles), tapering fairly rapidly to the roundly triangular to rounded tip and to the roundly wedge-shaped base on a short petiole 1–6 mm long. Pollen cones 4–5 cm long, 8–10 mm thick, with five or six pairs of larger, clasping, sterile scales at the base and attached directly in the leaf axils without a stalk or on a very short stalk 1–2 mm long. Each pollen scale with five to eight pollen sacs and a rounded external face. Seed cones not conspicuously waxy at maturity, round but noticeably longer than wide, about 8–9 cm long and 6–7 cm thick. Seed scales turned up at the outer rim but without a large, tonguelike projection. Seed body about 10 mm long and 7 mm wide, the larger wing 18–20 mm by 14–15 mm, the smaller one roundly triangular, projecting about 4–5 mm. Restricted to and common along the summits of the Mount Panié range (hence the names), from Mount Ignambi to Mount Panié in northern New Caledonia, with a few stands in the Roches d'Ouaïème just across the Ouaïème River to the south. Forming the canopy of the summit cloud forest, alone or in combination with other species, on soils derived from sedimentary rocks; 900–1,630 m. Zone 9.

Mount Panié kauri has the highest elevational range of all the New Caledonian *Agathis* species. It was long confused with serpentine forest kauri (*A. lanceolata*) but appears to be more closely related to its closer neighbor, white kauri (*A. moorei*), which grows at lower elevations nearby, also on soils derived from sedimentary rocks rather than from serpentine. It differs from both those species in its much longer pollen cones and generally proportionately a little broader leaves. When first described in 1969 it was thought to be restricted to the summit plateau of Mount Panié itself, but subsequent surveys show that it occupies the whole Massif du Panié above about 1,000 m, lending further credence, perhaps, to its recognition as a distinct species. Thanks to its relatively low stature and out-of-the-way habitat. Mount Panié kauri has essentially not been exploited the way its lowland forest relatives have. Furthermore, the summit forest on Mount Panié is protected in the Mount Panié Special Botanical Reserve, so this species is probably the least threatened of the kauris of New Caledonia.

Agathis moorei (Lindley) M T. Masters

WHITE KAURI, KAORI BLANC, KAORI DU NORD (FRENCH), DICOU (KANAK)

Tree to 25(–30) m tall, with cylindrical, unbuttressed trunk to 2(–3) m in diameter and unbranched for up to about 18 m in old trees. Bark smooth and light gray at first, weathering darker grayish brown, flaking in thin scales, and becoming progressively rougher and both pockmarked and persistently scaly with age. Crown dense, conical at first, passing through cylindrical to broadly dome-shaped with age, with relatively slender horizontal branches turned up at the ends and bearing branchlets along their length. Branchlets yellowish green from the beginning, densely clothed with foliage. Leaves shiny bright green above, paler and duller but not waxy beneath, 5–7 cm long (to 20 cm in juveniles), 0.8–1.5 cm wide (to 3.5 cm in juveniles), widest near the middle (before the middle in juveniles), tapering smoothly to the roundly triangular tip (sharply triangular in juveniles) and somewhat to dramatically more abruptly to the wedge-shaped to roundly wedge-shaped base on a very short petiole 1–3 mm long. Pollen cones 2.5–3(–4) cm long, 8–10 mm thick, with seven or eight pairs of larger, tightly clasping sterile scales (of which the lowest pair may be somewhat leaflike) on a stalk 8–12(–25) mm long. Each pollen scale with about 7–10 pollen sacs and a rounded external face. Seed cones with some remnants of a waxy film at maturity, almost spherical to a little longer than wide, 10–15 cm long and 9–12 cm thick. Seed scales slightly turned up at the rim but without a large, tonguelike projection. Seed body skinnier than in other species, 20–22 mm long and 7–8 mm wide, the larger wing about 30 mm by 20 mm, the smaller one narrowly triangular, projecting about 4–5 mm. Widely but discontinuously distributed through the northern two-thirds of New Caledonia, from the Diahot River valley near Ouégoa in the north southward to the Thio River valley near Thio. Scattered or forming small groves in, or as an emergent above, the canopy of lowland rain forests on soils derived from sedimentary rocks; (30–)200–700(–1,000) m. Zone 10.

Although white kauri is often cited from serpentine-derived soils at Prony at the southern tip of the island, the trees there appear to have been planted long ago. The species is essentially completely replaced by serpentine forest kauri (*Agathis lanceolata*) on ultrabasic soils of the mining districts in the south. These two big lowland kauris are similar in general appearance, but white kauri has more numerous sterile scales at the base of the pollen cones. It is a smaller tree than serpentine forest kauri, but with the depletion of the latter in the more accessible regions, it has become the primary timber kauri on the island. The wood has a wide variety of uses, including as paper pulp and in the manufacture of matchsticks, two uses that do not seem to be the most fitting end for a 500-year-old forest giant. Luckily, some fine stands are protected in parks and reserves, and there is also at least a limited amount of plantation culture. The species name honors Charles Moore (1820–1905), Scottish-born director of the Royal Botanic Garden, Sydney, who collected the type material of this species as well as that of Melanesian kauri (*A. macrophylla*) and who also named dwarf kauri (*A. ovata*) and some other species now considered synonyms of earlier names.

Agathis orbicula de Laubenfels

TUMU DAMMAR

Tree to 40 m tall, with trunk to 1 m in diameter. Bark reddish brown and smooth at first, weathering dark brown and flaking in irregular, thick patches. Crown dense, compact, with contorted branches bearing tufts of branchlets. Branchlets waxy bluish green at first, densely clothed with foliage. Leaves dark green above, bluish green with a waxy film beneath, (2–)2.4–4 cm long (to 6.5 cm in juveniles), (0.8–)1.2–2.4 cm wide (to 2.8 cm in juveniles), proportionately the widest and roundest in the genus (hence the scientific name), widest near the middle, tapering rapidly to the rounded or roundly triangular tip and even more abruptly to the rounded to broadly wedge-shaped base on a short petiole 3–7 mm long. Pollen cones 0.8–1.5 cm long, 4–6 mm thick, with two or three pairs of larger, clasping, sterile scales at the base on a stalk

2–6 mm long. Each pollen scale with two or three pollen sacs and a humped external face. Seed cones much longer than wide, about 7 cm long and 4.5 cm thick. Seed scales without a large, tonguelike projection. Seed not seen or otherwise described. Discontinuously distributed in northern Borneo from southern Sabah to central Sarawak (Malaysia). Scattered in lower montane rain forests and in scrub of sandstone plateau barrens (kerangas), where sometimes dominant; 450–1,050 m. Zone 10.

Were it not for its extremely small pollen cones, tumu dammar might be accommodated in mountain dammar (*Agathis dammara* subsp. *flavescens*), where the short, broad leaves with a waxy film beneath and skinny seed cones, as well as the low-elevation habitat, would represent extremes within the variation of that taxon. The pollen cones, however, fall completely outside of the range of variation within mountain dammar, including the populations found in Borneo, which are sometimes treated as two additional endemic species. The relationships of tumu dammar might be clarified by well-designed DNA studies involving multiple populations of tumu dammar and of eastern and western dammars (*A. dammara* and *A. borneensis*) from Borneo and beyond.

Agathis ovata (C. Moore ex Vieillard) O. Warburg

DWARF KAURI, SCRUB KAURI, KAORI NAIN, KAORI DE MONTAGNE (FRENCH)

Plate 11

Shrub, maturing at as little as 1 m high, to small tree to 8(–13) m (or exceptionally to 25 m on favorable and protected sites), with cylindrical or knobbly, unbuttressed or slightly swollen trunk to 0.7(–1.2) m in diameter, branched from near the base or free of branches for up to 10(–15) m. Bark light brown and smooth at first, weathering dark to very pale gray, flaking in thick, irregular scales, and becoming very thick and deeply and irregularly ridged and furrowed with age. Crown dense and conical at first, becoming more open, very broadly dome-shaped, and flat-topped, with a cluster of thick, upwardly angled, contorted branches, subdividing rapidly and bearing tufts of upright branchlets at their tips. Branchlets with a thin waxy film, densely clothed with foliage. Leaves dark green to yellowish green above, paler and with a grayish waxy film beneath, (3–)4–8 cm long, (1–)1.5–5 cm wide, more or less egg-shaped (hence the scientific name), widest near or beyond the middle (before the middle in juveniles), tapering abruptly to the rounded to roundly triangular tip and equally abruptly or somewhat more gradually to the roundly wedge-shaped base on a short petiole 1–5 mm long. Pollen cones with a waxy film, 3–5 cm long, 10–15 mm thick, with five or six pairs of larger, tightly clasping, sterile scales at the base (of which the lowest pair is leaflike, extending 12–15 mm) on a stalk 3–15 mm long. Each pollen scale with six to eight pollen sacs and a rounded external face. Seed cones with remnants of a waxy film at maturity, round but a little longer than wide, about 6 cm long and 5 cm thick. Seed scales with a prominent, triangular, projecting tip about 3–4 mm long. Seed body 9–12 mm long and 8–9 mm wide, the larger wing 10–15 mm by 9–14 mm, the smaller one earlike, projecting about 3 mm. Widely distributed and common in the southern third of New Caledonia south from the Ouenghi River valley, with an outlier on Mont Ménazi near Kouaoua, some 65 km north of the more continuous distribution. Most commonly scattered in the canopy of or as widely spaced emergents above maquis minière shrublands on serpentine-derived ultrabasic soils (Plate 10) but also a component of adjacent rain-forest canopies; 150–800(–1,150) m. Zone 10.

As one of the larger and more conspicuous of the scattered trees emergent above the scrubby maquis minière that is so ubiquitous in the more populous southern end of New Caledonia, dwarf kauri is the *Agathis* species most likely to be noticed by visitors to the island. It has the thickest and roughest bark of any kauri species, an asset in the fire-ravaged maquis landscape it most frequently inhabits. Even with this protection, dwarf kauri is much shorter lived in the maquis vegetation than when it is growing in the wetter, less fire-prone rain forests. The maximum age of maquis-grown trees has been estimated to reach 400 years, a quite respectable age and one implying survival of many fires, while forest-grown trees may live another century.

The projecting tip of the seed scales is a feature not found in other New Caledonian kauris but one shared with the New Zealand kauri (*Agathis australis*). In many ways, the growth habit of dwarf kauri does resemble a miniature version of its giant New Zealand cousin, and some authors thought that this was its closest relationship. Some very preliminary DNA studies, however, suggest that all the Melanesian kauris, including the five New Caledonian species and Melanesian kauri (*A. macrophylla*) itself, are more closely related than any of them (dwarf kauri included) is to other species in the genus. Resolution of its relationships await a more comprehensive investigation of the whole genus. Early botanical understanding of dwarf kauri was marred by confusion with serpentine forest kauri (*A. lanceolata*), which is the more common species in the southern rain forests on ultrabasic soils and is a much larger tree, even in those places where the two species grow together in the same forest stands. These two species are not particularly similar to one another, but early material was sparse and early literature descriptions were largely based on leaf shapes, with little opportunity to assess the considerable variability within species. The names now used for these two species are the earliest ones that can be associated with type specimens, hence their application is unambiguous.

Agathis robusta (C. Moore ex F. J. Mueller) F. M. Bailey

SMOOTH-BARK KAURI

Plate 12

Tree to 40(–60) m tall, with smoothly cylindrical, unbuttressed or basally flared trunk to 1.5(–2.5) m in diameter. Bark smooth and orange-brown to brown at first, weathering grayish brown to gray and flaking continuously in small, closely spaced scales to produce fine mottling with age. Crown dense to open, broadly dome- to broadly vase-shaped, with a few, widely spreading, upwardly angled to nearly horizontal, sometimes contorted, heavy limbs at the summit of the unbranched trunk and bearing open, upwardly angled or horizontal tufts of branchlets. Branchlets green from the beginning, moderately densely clothed with foliage. Leaves shiny bright to dark green above, a little paler beneath, (3–)7–10 cm long (to 13 cm in juveniles), (1–)2–3 cm wide (to 4.5 cm in juveniles), widest near or a little before the middle, tapering smoothly or gradually and then more abruptly to the sharply to bluntly triangular tip and a little more abruptly to the roundly wedge-shaped base on a short petiole (1–)3–10 mm long. Pollen cones 4–9(–11) cm long, 7–14 mm thick, with (three or) four to seven pairs of larger, loosely clasping, sterile scales at the base on a stalk 2–9 mm long. Each pollen scale with two to eight pollen sacs and an irregularly rounded external face. Seed cones not waxy at maturity, plump but noticeably longer than wide, 8.5–15 cm long, 6.5–10.5 cm thick. Seed scales without a large tonguelike projection. Seed body 10–15 mm long and 6–7 mm wide, the larger wing 24–26 mm by 10–16 mm, the smaller one sharply triangular, projecting about 3–4 mm. Scattered in discontinuous regions along the coast of Queensland (Australia), on New Britain and in eastern New Guinea (Papua New Guinea). Scattered as an emergent over various types of lowland and montane rain forest or forming closed canopy stands at drier rain-forest margins on a variety of substrates but with best development on soils derived from granite; 0–2,000 m. Zone 9. Two subspecies.

Smooth-bark kauri is the most widely distributed of the three co-occurring kauri species in Queensland and has one of the widest ranges in the genus. Like the other widespread species, the range of smooth-bark kauri consists of discontinuous segments often treated as belonging to different species. Here they are treated as a single species with two subspecies. Although the lowland plants in New Guinea and New Britain are here assigned to the same subspecies as trees in the eastern highlands, they have also been assigned to the same taxon as the lowland populations in Australia. The affinities of these northern lowland populations are very much in need of review with complete material and field observations. The two regions occupied in northern and southern Queensland are almost the same as those inhabited by bunya-bunya (*Araucaria bidwillii*), another giant member of the Araucariaceae.

Underside of twig of smooth-bark kauri (*Agathis robusta*) with one growth increment, ×0.5.

Smooth-bark kauri was once an important timber tree in Australia, and the resin exuded from the wounded trunks was shown to have the best potential for producing turpentine among all the Australian conifers. However, commercial stands were depleted before an industry surrounding resin collection could develop, as it did with the dammars (*Agathis dammara*) of Malesia. Resin production by wounded smooth-bark kauri is relatively sparse and, unlike some other species in the genus, the raw resin blackens on exposure to air due to its iron and manganese content. Although natural stands of emergent trees have either been depleted or are protected, the species still has some presence as a commercial timber tree because of limited plantation culture. It is also widely grown as an ornamental within its native range and elsewhere in the humid tropics, without any apparent cultivar selection. Wood from plantation trees is of high quality, even through the dimensions are not those of primary stands.

Smooth-bark kauri appears to be less long lived in old-growth stands than the other two Australian species, but determining their age is complicated by the fact that growth rings are not formed every year. Nonetheless, the oldest known individuals are reported to be a little more than 600 years old, compared to more than 1,000 for black kauri (*Agathis atropurpurea*) and bull kauri (*A. microstachya*). Smooth-bark kauri is also a tree of generally lower elevations than these other two species, at least in Australia.

Agathis robusta subsp. *nesophila* T. Whitmore

PAPUAN KAURI, ASONG, MUWAKA, OGAPA (CHIMBU LANGUAGES)

Pollen cones 9–13 mm in diameter. Seed cones 8.5–10 cm long, 6.5–8 cm in diameter. Chimbu and Eastern Highland provinces to West New Britain and Central provinces (Papua New Guinea); (200–)900–2,000 m. Synonym: *Agathis spathulata* de Laubenfels.

Agathis robusta (C. Moore ex F. J. Mueller) F. M. Bailey subsp. *robusta*

KAURI PINE, QUEENSLAND KAURI

Pollen cones 7–9(–14) mm in diameter. Seed cones 9–15 cm long, 8–10.5 cm in diameter. Fraser Island and nearby mainland near Gympie, southern Queensland, and vicinity of Cairns, from Cookstown to Ingham, northern Queensland (Australia); 0–900 m. Synonym: *Agathis palmerstonii* (F. J. Mueller) F. M. Bailey.

Amentotaxus Pilger

CATKIN YEW

Taxaceae

Evergreen small trees or shrubs. Trunk(s) one or more, cylindrical, straight and erect, usually slender. Bark obscurely fibrous, thin, smooth at first, flaking in small scales, and becoming shallowly and irregularly ridged and furrowed with age. Crown cylindrical to narrowly dome-shaped, rather open to moderately dense, with well-separated horizontal to slightly rising branches in pairs successively at right angles to their predecessors or in near whorls of (three or) four. Branchlets all elongate, without distinction into long and short shoots, hairless, remaining green 1–2 years or more, usually a little squared off in cross section, completely clothed by and prominently grooved between the attached leaf bases up to about 1.5 cm long. Resting buds well developed, squarish in cross section, pointedly egg-shaped, with 5–10 pairs of glossy brown, keeled, triangular bud scales, most of which are shed with bud burst and shoot elongation. Leaves attached in pairs, each pair at right angles to the preceding and following pairs, presented in two flat rows to the sides of the twigs by bending of the petioles and twisting of the internodes between the pairs. Individual leaves broadly sword-shaped, straight or slightly curved forward or S-shaped, flattened top to bottom, leathery. Midrib prominent, sharply or more broadly raised within an ill-defined groove above, slightly raised to slightly grooved beneath within a broadly depressed or raised green band flanked by conspicuous, broad, waxy stomatal bands, these in turn flanked by green marginal bands, uniformly dark green above, the leaf edges often narrowly turned down.

Plants dioecious. Pollen cones arranged in pairs along and in a congested clump at the end of a specialized, leafless, drooping spike (hence the scientific name, Latin for "catkin yew") with three or four (usually) spikes emerging from buds near or at the tip of the previous year's growth and surrounding a vegetative bud. Each pollen cone spherical or slightly elongated, with about four or five alternating pairs of pollen scales. Each pollen scale with a central stalk, two to eight pollen sacs on the lower side (sometimes on the upper as well), and usually with a triangular free tip above. Pollen grains small to medium (20–45 µm in diameter), spherical or slightly squashed, without air bladders, very finely (to almost undetectably) bumpy but otherwise almost featureless. Seed cones single at the tips of leafless, drooping stalks in the axils of ordinary foliage leaves near the tip of a growth increment, highly reduced, not at all conelike, consisting solely of 4–6(–10) alternating pairs of thin, rounded bracts tightly packed around the base of a single large seed with a fleshy outer seed coat surrounded by and united with a fleshy aril for most of its length. Seeds maturing in a single season. Cotyledons two, each with one vein. Chromosome base number $x = 7$.

Wood relatively hard and heavy, fragrant, smooth in texture, yellowish brown. Grain medium and even, with somewhat obscure growth rings marked by a gradual transition to a narrow band of latewood. Resin canals absent but with numerous, conspicuous, individual resin parenchyma cells and tracheids (wood cells) with double spiral thickening.

Stomates densely crowded within the stomatal bands so that they share many of their surrounding subsidiary cells, aligned with the long axis of the leaf but not arranged in distinct lines. Each stomate sunken beneath and surrounded by the (4–)8–10 radiating subsidiary cells, which are flat and not topped by a Florin ring. Midvein single, completely surrounded by a discontinuous cylinder of small sclereids, with one small to large resin canal immediately beneath it, and flanked by small bands of transfusion tissue. Photosynthetic tissue with a single or partially double palisade layer lining the whole upper surface beneath the epidermis and with horizontal spongy mesophyll filling the blade on either side of the midvein and usually accompanied by fibers that produce wrinkles perpendicular to the midrib on drying.

Two species in southeastern Asia from northeastern India (Arunachal Pradesh) through central and southern China to Taiwan, south to southern Vietnam. References: Ferguson 1992, Fu et al. 1999e, Keng 1969, Li 1952, T. H. Nguyên and Vidal 1996, K. C. Sahni 1990, Tomlinson and Zacharias 2001.

While the catkin yews are not confined to primary forests and can tolerate some disturbance, they are rare plants. They are threatened throughout their range by direct cutting for their wood and even more so by general forest clearance. A few reserves were established to protect them or a broader sample of plants and animals, but these reserves cover only a small part of their morphological variation. Partly as a consequence of their rarity, the catkin yews have scarcely been investigated for the kinds of cancer-fighting chemicals discovered in so many other members of the yew family. Neither of the catkin yew species is in general cultivation (they are rarely found even in botanical gardens), and there has been no cultivar selection.

Although the members of this genus share many characteristics with other genera of Taxaceae (for instance, spikes of pollen cones are also found in the New Caledonian *Austrotaxus*), they still have a distinctive combination of features that makes them easy to recognize. Even the leaves, with their conspicuous, broad stomatal bands bearing densely packed stomates sharing many subsidiary cells, are readily identified as *Amentotaxus* (although they were confused with *Podocarpus* when the latter was less well understood). There have been considerable taxonomic uncertainties surrounding the genus. While it was sometimes included with plum yews (*Cephalotaxus*) in Cephalotaxaceae or with nutmeg yews (*Torreya*) or by itself in Amentotaxaceae, DNA studies strongly support the traditional view that all these genera belong with *Taxus, Pseudotaxus,* and *Austrotaxus* in the Taxaceae. Within the Taxaceae, it joins *Torreya* to form one of three groups of genera (botanical tribe or subfamily). The structure of the seed and aril are very similar to those of *Torreya,* while the paired leaves are also found in *Cephalotaxus.* (The leaf arrangement in *Torreya* and *Cephalotaxus,* with spiraling pairs, is actually somewhat different from that of *Amentotaxus,* which has crisscross pairs.)

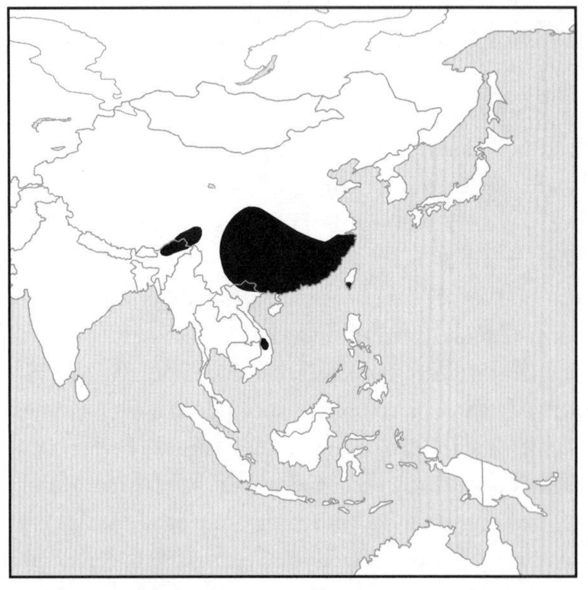

Distribution of *Amentotaxus* in southeastern Asia, scale 1 : 120 million

More controversial is the number of species in the genus. The genus was generally considered to be monotypic, consisting only of *Amentotaxus argotaenia,* until H. L. Li divided it into four species in 1952. It has been common to recognize three or four species in the genus ever since, with some authors accepting as many as six or seven. The genus is divided into two species here because, even with large geographic gaps in the range, there appear to be few real character discontinuities and some proposed distinguishing features are of uncertain value. Much more needs to be known about the genetic and evolutionary relatedness of the different populations. There simply is not enough information about variation within the genus to accept many of the proposed segregates as separate species. Unfortunately, the continuing depletion of these intriguing plants with habitat loss makes obtaining a coherent range-wide understanding ever more difficult.

Because of the distinctiveness of the leaves, even in the absence of seed and pollen cones, there is a well-established fossil record of *Amentotaxus* extending from the Paleocene to the upper Miocene (about 60 million to 10 million years ago) in Europe and from the mid-Cretaceous to the Miocene (about 100 million to 10 million years ago) in North America, but without a known corresponding record in Asia except for a possible pollen record from the Miocene of Taiwan.

Identification Guide to *Amentotaxus*

The two species of *Amentotaxus* may be distinguished (with difficulty) by whether the stomatal bands are white or yellow to reddish brown, whether they are generally narrower or wider than the green marginal band, how many pollen sacs there are on each pollen scale of one of the lower pollen cones, and the usual color of the skin of the mature seed and surrounding aril:

Amentotaxus argotaenia, stomatal bands white, narrower than marginal ones, up to four pollen sacs, seed bright red to purplish red

A. yunnanensis, stomatal bands white or colored, wider than marginal ones, more than four pollen sacs, seed reddish purple to purple

Amentotaxus argotaenia (Hance) Pilger

NORTHERN CATKIN YEW, SUI HUA SHAN (CHINESE), SAM HOA BÔNG (VIETNAMESE)

Shrub, or tree to 7(–20) m tall, with trunk to 0.2(–0.5) m in diameter. Bark reddish brown at first, weathering light to dark gray. Leaves (2–)3–11(–15) cm long, (3.5–)6–10(–12.5) mm wide, glossy yellowish green to dark green above, straight or slightly curved, widest near or before the middle, tapering gradually and then more abruptly to the sharply to bluntly triangular tip, and more abruptly to the roundly wedge-shaped or rounded base on a very short petiole to 3(–4) mm long. Stomatal bands (0.6–)1–2(–3) mm wide, grayish white to greenish white, flanking a green midrib band 1–1.5(–2) mm wide and flanked by green marginal bands (0.5–)1.5–3 mm wide. Pollen spikes (1–)2–4(–10) in a cluster, each (1–)4–6.5 cm long. Pollen scales of lower cones on spikes with two to four (or five) pollen sacs. Seed stalks 1–1.5(–2.5) cm long. Seeds 2–2.5(–3.5) cm long, 1–1.5(–2.5) cm in diameter, bright red to purplish red. Throughout eastern and central China (from Jiangsu to Guangdong west to southern Gansu, eastern Sichuan, Guizhou, and Guangxi) and in adjacent northern Vietnam (Hoang Liên Son to Cao Lang), with an outlier in northeastern India (Arunachal Pradesh). Scattered in the understory or in thickets at the edges of wet warm temperate and subtropical, montane broad-leaved evergreen forests, especially in valleys and along streamsides on a variety of substrates; (300–)500–1,500(–1,900) m. Zone 8. Three varieties.

Northern catkin yew is much more widely distributed than is its more tropical relative, southern catkin yew (*Amentotaxus yunnanensis*). As a result, it occurs in many different specific forest associations over a wide range of climatic conditions. There is corresponding variation in morphological characteristics across the scattered occurrences of the species. This variation is here recognized in the form of three varieties, an arrangement that may well require revision as more genetic data on these plants becomes available.

Amentotaxus argotaenia (Hance) Pilger **var. *argotaenia***

Leaves (3–)6–11 cm long, 6–8(–11) mm wide, crossed by fibers that dry as ridges perpendicular to the midrib. Pollen cone spikes 3.5–6.5 cm long, in groups of as many as four. Seeds 2–2.5 cm long and 1–1.5 cm in diameter on stalks 1–1.5 cm long. Throughout the range of the species except in northeastern India. Synonyms: *Amentotaxus argotaenia* var. *cathayensis* (H. L. Li) P. C. Keng, *A. cathayensis* H. L. Li.

Amentotaxus argotaenia **var. *assamica*** (D. K. Ferguson) Eckenwalder

Leaves (2.5–)7–15 cm long, (3.5–)7–12.5 mm wide, without fibers extending out from the midrib. Pollen cone spikes 4–5.5 cm long, in groups of as many as four. Seeds 2–3.5 cm long and 1.5–2.5 cm in diameter on stalks 1.5–2.5 cm long. Known only from the Dafla Hills, Arunachal Pradesh, northeastern India. Synonym: *Amentotaxus assamica* D. K. Ferguson.

Amentotaxus argotaenia **var. *brevifolia*** K. M. Lan & F. H. Zhang

DUAN YE SUI HUA SHAN (CHINESE)

Leaves 2–3.7 cm long, 5–7 mm wide, with fibers extending out from the midrib. Pollen cone spikes 1.5–5.5 cm long, in groups of as many as 10. Seeds 2–2.5 cm long and 1–1.5 cm in diameter on stalks 1–1.5 cm long. Known only from southern Guizhou (China).

Amentotaxus yunnanensis H. L. Li

SOUTHERN CATKIN YEW, YUN NAN SUI HUA SHAN, TAI WAN SUI HUA SHAN (CHINESE), DE TÙNG VÂN NAM, DINH TÙNG (VIETNAMESE)

Shrub, or small tree to 10(–20) m tall, with trunk to 0.3(–1) m in diameter. Bark grayish brown to reddish brown, weathering light gray. Leaves (3–)5–10(–15) cm long, (4.5–)7–12(–15) mm wide, glossy dark green to bluish green above, straight to distinctly S-curved, widest near or before the middle to nearly parallel-sided, tapering very gradually to the sharply to bluntly triangular tip and more abruptly to the roundly wedge-shaped

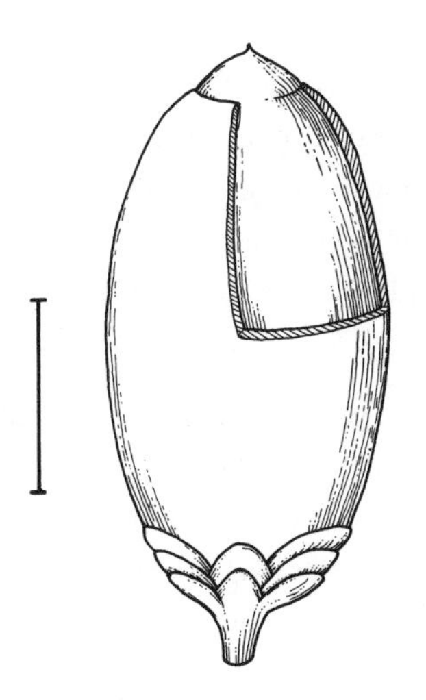

Seed cone of northern catkin yew (*Amentotaxus argotaenia*) with aril partially cut away to show seed; scale, 1 cm.

base on a very short petiole to 2(–5) mm long. Stomatal bands (1–)2–4(–5) mm wide, white to rusty brown, flanking a green midrib band (0.8–)1.5–2(–2.5) mm wide and flanked by green marginal bands 0.5–1.5(–2) mm wide. Pollen spikes (1–)3–4(–6) in a cluster, each (2.5–)5–10(–15) cm long. Pollen scales of lower cones on spikes with four to eight pollen sacs. Seed stalks 1.5–2.5(–3) cm long. Seeds 2–3 cm long, 1–1.5 cm in diameter, purple to reddish purple. Discontinuous in southern Taiwan and from southwestern Guizhou (China) south to southern Vietnam. Scattered or gregarious in the understory or canopy of wet or seasonally dry lower montane broad-leaved evergreen or mixed forests or bamboo thickets, especially on slopes, in valleys, or along streamsides, often on limestone-derived soils; (500–)1,000–1,600(–2,300) m. Zone 9. Two varieties.

Southern catkin yew is known from very few populations in a variety of habitats, with corresponding variation in morphology, especially in the size and shape of the leaves and of the stomatal bands and flanking regions. While these populations are sometimes assigned to several different species, they are here treated as belonging to two weakly differentiated varieties of a single species. Many populations are represented in herbarium collections only by one sex or the other, or even just by sterile specimens, so increasing knowledge could lead to changes in the varieties accepted here.

Treelet of Taiwan catkin yew (*Amentotaxus yunnanensis* var. *formosana*) in nature.

Amentotaxus yunnanensis **var. *formosana*** (H. L. Li) Silba

TAIWAN CATKIN YEW, TAI WAN SUI HUA SHAN (CHINESE), ĐINH TÙNG (VIETNAMESE)

Leaves mostly in the shorter portion of the range for the species, usually less than 8.5 mm wide. Stomatal bands mostly up to 2.5 mm wide, white. Pollen spikes and seeds generally in the lower portion of the range for the species. Southern Taiwan (Taitung and Pingtung) and southern Vietnam (Gia Lai-Công Tum and Lâm Dông). Synonyms: *Amentotaxus formosana* H. L. Li, *A. poilanei* (de Ferré & Rouane) D. K. Ferguson, *A. yunnanensis* var. *poilanei* de Ferré & Rouane.

Amentotaxus yunnanensis H. L. Li **var. *yunnanensis***

YUNNAN CATKIN YEW, YUN NAN SUI HUA SHAN (CHINESE), DE TÙNG VÂN NAM (VIETNAMESE)

Leaves mostly in the longer portion of the range for the species, usually more than 8.5 mm wide. Stomatal bands mostly 3 mm or more wide, yellowish white or pale to rusty brown. Pollen spikes and seeds generally in the upper portion of the range for the species. Southwesternmost corner of Guizhou through southeastern Yunnan (China) to

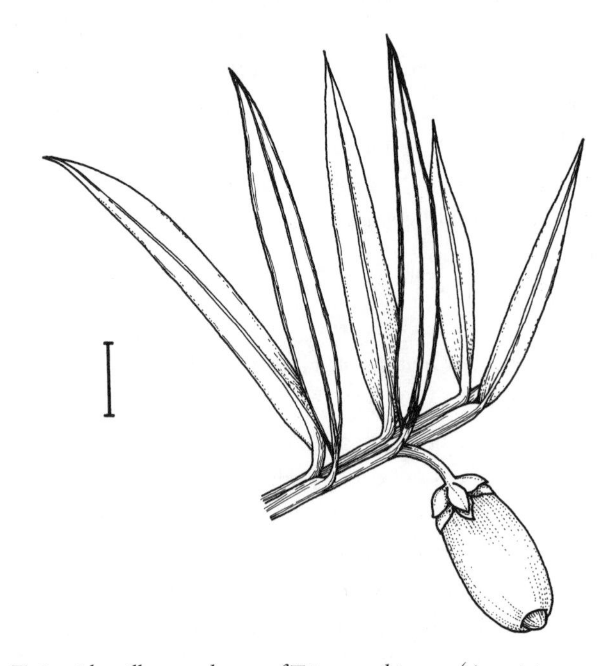

Twig with axillary seed cone of Taiwan catkin yew (*Amentotaxus yunnanensis* var. *formosana*); scale, 1 cm.

Male treelet of Yunnan catkin yew (*Amentotaxus yunnanensis* var. *yunnanensis*) in nature, with larger leaves than Taiwan catkin yew (variety *formosana*).

adjacent northern Vietnam (Hoang Liên Song to Cao Lang). Synonyms: *Amentotaxus argotaenia* var. *yunnanensis* (H. L. Li) P. C. Keng, *A. hatuyenensis* T. H. Nguyên.

Araucaria A. L. Jussieu

ARAUCARIAN PINE

Araucariaceae

Short to tall evergreen trees. Trunk usually single, straight, and slender to massive, with little taper beneath the crown. Bark thin to irregularly corky, often with horizontal lines of lenticels and enlarging persistent leaf bases when young, remaining smooth with age or peeling and flaking and becoming scaly and roughened and sometimes even moderately ridged and furrowed. Resin canals absent, but sometimes with scattered, individual resin-filled tracheids. Crown generally open, often becoming strikingly flat-topped with age, either moderately shallow because of the shedding of lower branches or extending practically to the ground because such shed branches are replaced from adjoining buds on the trunk, such crowns (as in *Araucaria columnaris*) being narrowly cylindrical. Branches arranged in regular pseudowhorls, typically turned up at the ends, the lower ones sometimes pendulous. Branchlets all elongate, without distinction into long and short shoots, attached directly to the primary branches, often just out to the sides, but sometimes all around, elongating but usually not themselves branching further, coarse to very thick, remaining green for several years until shed by themselves or with the supporting branch. Resting buds unspecialized, without distinct bud scales. Leaves crowded, spirally attached, rarely (*A. bidwillii*) somewhat two-ranked, stiff, clawlike to scalelike and slightly flattened top to bottom or side to side (in the 15 species of section *Eutacta*) or spearheadlike (in sharpness as well as in shape) and flattened top to bottom (in the four species in the remaining three sections).

Plants monoecious or occasionally dioecious. Pollen cones single (or in small clusters) at the tips of branchlets or in leaf axils or at the end of short side shoots, cylindrical, to 22 cm long. Each pollen cone with a few sterile scales at the base of numerous, densely spirally arranged pollen scales. Pollen scales with a more or less triangular exposed blade at the end of a very slender stalk that is surrounded by some 4–20 elongate cylindrical pollen sacs attached only at one end to the base of the scale head. Pollen grains large to very large (50–100 μm in diameter), nearly spherical and generally featureless. Seed cones solitary and upright at the ends of branches, large, often nearly spherical, to 25 cm wide, some among the most massive of conifer seed cones, ripening in 2–3 years, then disintegrating. Seed scales fused to the bract for half or more of their length, leaving a slot between the two in the outer portion, very numerous (often in the hundreds), densely overlapping, thickened at the tip, often winged at the sides, the bract prolonged as a narrowly triangular point. Each seed scale with a single, large seed (some the largest among conifers) without wings and embedded in a depression in the seed scale, in which they are retained and fall with the scales from the shattered cones. Cotyledons two, each with two veins, or four cotyledons, each with one vein. Chromosome base number $x = 13$.

Wood soft and weak, of medium weight, with nearly white to yellowish sapwood not very distinct from the similar or slightly darker white to light brown heartwood that is sometimes streaked with darker brown or red. Grain very fine and even, usually with little or no evidence of growth rings, but sometimes with a few slightly flatter, thicker-walled latewood tracheids.

Stomates in a few to many regular (or somewhat irregular and interrupted) longitudinal lines (which may be arranged in bands or patches) on both the upper and lower surfaces (often nearly restricted to the lower surface in *Araucaria bidwillii*). Individual stomates highly varied among different species, sunken beneath and partially hidden by the one to three circles of four to six often crinkly walled subsidiary cells, the inner circle lacking a Florin ring but plugged with wax. Leaf cross section with a single midvein (clawlike and scalelike leaves) or with additional smaller veins paralleling it on either side (broader leaves), at least the midvein flanked by some transfusion and sclerenchyma tissue. Epidermis underlain by an often multilayered hypodermis everywhere except next to the lines of stomates. Photosynthetic tissue consisting of a more or less well defined palisade layer on the upper side (and sometimes on the lower as well) and looser spongy mesophyll beneath the palisade. Resin canal usually single, centered under the midvein or sometimes with others off to the sides of the leaf and above the midvein.

Nineteen species in New Guinea, Northeastern Australia, Norfolk Island, New Caledonia, and southern Brazil to Chile. References: de Laubenfels 1972, 1988, Setoguchi et al. 1998, Wieland 1935, Wilde and Eames 1952.

The majority of species of *Araucaria* are tall rain-forest emergents, with massive columnar trunks carrying their canopies to 30–60 m (or even 89 m in *A. hunsteinii*) above the forest floor. Many are important timber trees, or were before their stands were depleted. Species with the largest seeds, like *A. araucana, A. bidwillii,* and *A. brasiliensis,* have also been important seasonal food resources for local people living in the vicinity of stands. In fact, the name *Araucaria* is the common name of monkey puzzle tree (*A. araucana*) used by one of these peoples, the Araucanians of Chile. Two-thirds of the species are restricted to New Caledonia, where they occur primarily on serpentine soils. Few of these are in cultivation and all species on the island belong to section

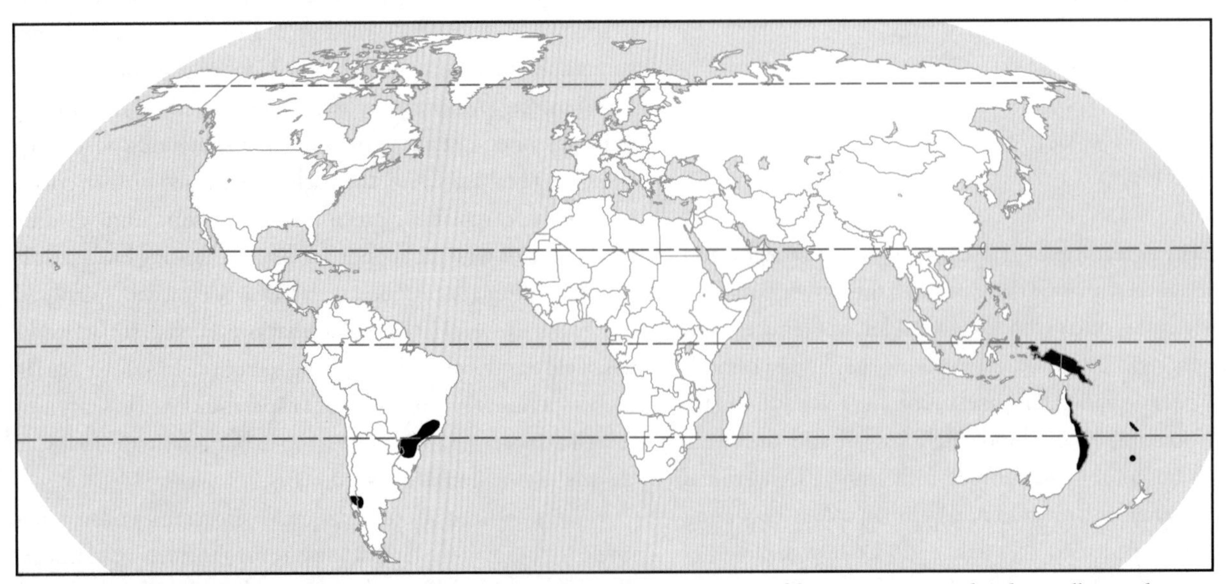

Distribution of *Araucaria* in Australasia and South America with nearly three-quarters of the species restricted to the small Australasian islands of New Caledonia and Norfolk, scale 1 : 200 million

Eutacta, the largest of the four sections. A handful of species are widely and extensively cultivated, with *A. heterophylla* and *A. bidwillii* (bunya-bunya) in the tropics and *A. araucana* extending into cooler but still mild regions, like southern coastal British Columbia in Canada, and with *A. heterophylla* also produced in millions as a house plant. Despite this there has been virtually no cultivar selection in the genus.

The living sections form an almost continuous series in their distinguishing characteristics, even with respect to microscopic features (Stockey and Ko 1986). Nonetheless, there is reasonable paleobotanical and phylogenetic justification for recognizing the sections, and they are retained here also as a convenient first division in the identification guide. Were it not for the fossil record, the three smaller sections (sections *Araucaria, Bunya,* and *Intermedia*), containing only four extant species among them, might well be merged into a single section that would be called section *Araucaria* because it contains the type of the genus, monkey puzzle tree (*A. araucana*).

DNA studies reveal little genetic divergence among the New Caledonian species, implying that the taxonomic and ecological diversity of this group has been the result of a relatively recent radiation rather than relictual retention of ancient species that might have been anticipated from the long fossil record of the genus.

It is uncertain when *Araucaria* first evolved, but Araucarian seed scales that could belong to the genus first appeared in Eurasia, North America, and the then Gondwanan India in the late Triassic, more than 200 million years ago. Even earlier, in the Paleozoic, the earliest known conifers had foliage and growth habits remarkably similar to those of some extant araucarias, such as Norfolk Island pine (*A. heterophylla*) and other species of section *Eutacta,* but their cone structures are not araucarian and they were presumably the ancestors of all extant conifers, not just *Araucaria.* Undoubted members of the genus had an essentially worldwide tropical distribution throughout the remainder of the Mesozoic, after which they became extinct in the northern hemisphere and progressively restricted to their present distribution in the southern hemisphere during the Tertiary. The fossil record contains some morphological features not found among the extant species so that additional taxonomic sections were named just for extinct species.

Taxonomic Guide to *Araucaria*

Araucaria sect. *Araucaria* (synonym: sect. *Columbea* Endlicher)
- *Araucaria angustifolia*
- *A. araucana*

Araucaria sect. *Bunya* Wilde & Eames
- *Araucaria bidwillii*

Araucaria sect. *Intermedia* C. T. White
- *Araucaria hunsteinii*

Araucaria sect. *Eutacta* Endlicher
- *Araucaria bernieri*
- *A. biramulata*
- *A. columnaris*
- *A. cunninghamii*
- *A. heterophylla*
- *A. humboldtensis*
- *A. laubenfelsii*
- *A. luxurians*
- *A. montana*
- *A. muelleri*
- *A. nemorosa*
- *A. rulei*
- *A. schmidii*
- *A. scopulorum*
- *A. subulata*

Identification Guide to *Araucaria*

For convenience of identification, the 19 species of *Araucaria* may be divided into two groups based on whether the leaves are predominantly flat and straight and stand out from the twig, or are scale-, claw-, or needlelike, and curl forward to varying degrees:

Group A, leaves flat and straight

Group B, leaves scale-, claw-, or needlelike and curled

Group A. These four species may be distinguished by the length and width of the adult foliage leaves and whether they are fairly uniform in length along the twigs or quite variable from one leaf to the next:

Araucaria angustifolia, leaves less than 7 cm long, less than 1 cm wide, uniform along the twigs

A. bidwillii, leaves less than 7 cm long, less than 1 cm wide, variable along the twigs

A. araucana, leaves less than 7 cm long, more than 1 cm wide, uniform along the twigs

A. hunsteinii, leaves more than 7 cm long, more than 1 cm wide, uniform along the twigs

Group B. These 15 species may be further divided into four subgroups based on the length of most adult leaves and whether the juvenile leaves are scalelike or needlelike:

Group B1, most adult leaves less than 7 mm long, juvenile leaves scalelike

Group B2, most adult leaves less than 7 mm long, juvenile leaves needlelike

Group B3, most adult leaves more than 7 mm long, juvenile leaves scalelike

Group B4, most adult leaves more than 7 mm long, juvenile leaves needlelike

Group B1. These four species may be distinguished by the length and width of the adult foliage leaves:

Araucaria bernieri, leaves less than 4 mm (down to 2 mm) long, up to 2.5 mm wide

A. scopulorum, leaves up to 4 mm (down to 3 mm) long, at least 2.5 mm wide

A. subulata, leaves at least 4 mm long, up to 2.5 mm wide

A. humboldtensis, leaves more than 4 mm long, at least 4 mm wide

Group B2. These four species may be distinguished by the width and length of the larger ordinary adult foliage leaves:

Araucaria cunninghamii, leaves up to 2 mm wide, more or less than 6 mm long

A. columnaris, leaves 4–5 mm wide, up to 6 mm long

A. luxurians, leaves 4–5 mm wide, at least 6 mm long

A. heterophylla, leaves at least 5 mm wide, up to 6 mm long

Group B3. These three species may be distinguished by the width of most adult foliage leaves:

Araucaria schmidii, leaves less than 3 mm wide

A. montana, leaves 7–8 mm wide

A. rulei, leaves more than 10 mm wide

Group B4. These four species may be distinguished by the width of most adult foliage leaves:

Araucaria nemorosa, leaves less than 4 mm wide

A. biramulata, leaves 5–6 mm wide

A. laubenfelsii, leaves 8–10 mm wide

A. muelleri, leaves more than 14 mm wide

Branchlet of Parana pine (*Araucaria angustifolia*) without distinct growth increments, ×0.5.

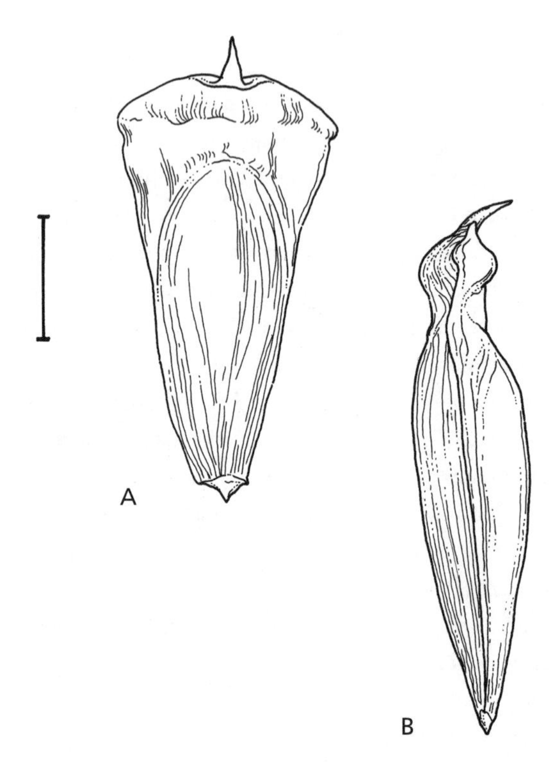

Upper-face (A) and side (B) views of shed, unwinged seed cone scale of Parana pine (*Araucaria angustifolia*) with embedded seed; scale, 1 cm.

Araucaria angustifolia (A. Bertoloni) O. Kuntze

PARANA PINE, PINHO BRASILEIRO, CURIY (PORTUGUESE), PINO PARANÁ, PINO MÍSIONERO (SPANISH)

Araucaria sect. *Araucaria*

Tree to 35(–50) m tall, with trunk to 1.5(–2) m in diameter, clear of branches for most of its height at maturity or with a few, scattered, shorter replacement branches below. Bark reddish brown to grayish brown, remaining thin and smooth, with horizontally expanded leaf attachment scars and shallow, irregular, discontinuous, vertical furrows with ragged edges. Crown passing from conical through deeply to shallowly dome-shaped and ultimately to tabletop flat, like an umbrella, with a few tiers of four to eight closely spaced, thin, horizontal branches turning up abruptly near the ends and bearing large, ball-like tufts of branchlets. Branchlets green for the first year or two, turning tan, smooth and clearly visible between the leaves. Leaves densely spirally arranged, standing nearly straight out or angled a little forward along the twigs, sword- to spearhead-shaped, flat or a little bowed outward around the midline. Each needle 1.7–4.5(–6) cm long, 5–8 mm wide, green or blue-green, wider and shorter on branches with seed cones. Stomates mostly parallel to the long axis of the leaf, in closely spaced, discontinuous lines on both faces. Pollen cones (7–)10–20 cm long, 1.5–2.5(–3) cm thick, each pollen scale with about 10 pollen sacs in two rows. Seed cones nearly spherical to broadly egg-shaped, often wider than high, 10–16 cm in diameter, green or brown at maturity externally, revealing a bright red interior upon shattering, with a brown, club-shaped remnant cone axis. Seed scales 7.5–9 cm long, 2.5–4 cm wide, conspicuously potbellied around the embedded seed, abruptly contracted beyond it just beneath the outside face, with a slender, down-curved free tip 6–8(–10) mm long. Seeds light brown, about 5 cm long, 2 cm wide. Cotyledons remaining underground with the seed coat at germination. Southern Brazil and northeastern Argentina. Forming dense to open pure stands or mixed with evergreen hardwoods in wet to seasonally dry uplands; 500–2,000 m. Zone 9. Synonym: *A. brasiliana* A. Richard.

As the only large conifer in its native range, Parana pine is an important timber tree. Seed production is abundant, so there is often good reproduction in exploited stands. It is also grown in plantations within and outside of its native region and as an ornamental in many tropical areas. The large, nutritious seeds have been an important food source for local people and wildlife. As with many subalpine pines in Eurasian and western North America, these seeds are dispersed, in part, through seed caching by a jay, in this case the plush-crested jay (*Cyanocorax chrysops*).

Because of its economic importance there has been a fair amount of study of genetic variation in Parana pine. These studies accord with the relatively low genetic variation found in other araucarians compared to most conifers. Furthermore, there is somewhat more genetic differentiation between populations than is commonly found in conifers, with their wind pollinated, highly outcrossed breeding system.

Parana pine and monkey puzzle tree (*Araucaria araucana*) are each other's closest relatives, the only extant members of section *Araucaria*, and are quite similar to each other. The most obvious difference between the two is that the later described Parana pine has narrower leaves (hence the scientific name). The two species also differ in their pollen and seed cones, but these are not always present. In an attempt to extend the ecological range of the South American araucarias, Parana pine has been crossed successfully with monkey puzzle tree. These experiments, carried out in the mid-20th century, never led to commercially viable forest plantations of hybrid araucarias. It would be interesting to look at the horticultural potential of these hybrids since both parent species are attractive ornamentals in suitable climates.

Araucaria araucana (G. Molina) K. Koch

MONKEY PUZZLE TREE, CHILE PINE, PINO ARAUCARIA, PEHUÉN, PINO DE NEUQUÉN (SPANISH)

Araucaria sect. *Araucaria*

Tree to 30(–50) m tall, with trunk to 1.5(2.5) m in diameter, smaller in cultivation, becoming free of branches for much of its height. Bark grayish brown, either smooth with persistent whorls of eyelike branch scars and numerous narrow rings originating from leaf scars

Young monkey puzzle tree (*Araucaria araucana*) in nature.

or roughened with short, corky horizontal blocky bumps aligned in straight to irregular vertical ridges. Crown passing from conical to dome-shaped, with closely spaced tiers of widely spreading, more or less horizontal, slender branches with well spaced, regular coarse branchlets. Branchlets extending out to the sides of the branches, sparsely branched themselves, remaining green for several years, densely clothed with and largely hidden by the foliage. Leaves densely, spirally arranged, angled a little forward along the twigs, broadly spearhead- to pointedly egg-shaped, generally flat but a little bowed outward around the midline, slightly narrowed at their points of attachment. Each needle 2.5–3.5(–5) cm long, 15–20(–25) mm wide, with sharp, prickly tip, dark green, persisting about 10–15 years. Stomates parallel to the long axis of the leaf, in numerous, closely spaced, discontinuous lines on both faces. Pollen cones 8–12 cm long, 4–5 cm thick, each pollen scale with 12–20 pollen sacs in two rows. Seed cones standing upright, nearly spherical, (10–)15–20 cm in diameter, brown at maturity, the surface hidden by the free scale tips. Seed scales densely spirally arranged, 4–5 cm long, 1.5–2 cm wide, wedge-shaped, with a narrow neck between the outer face and the swollen, seed-bearing portion and with a triangular, upturned free tip

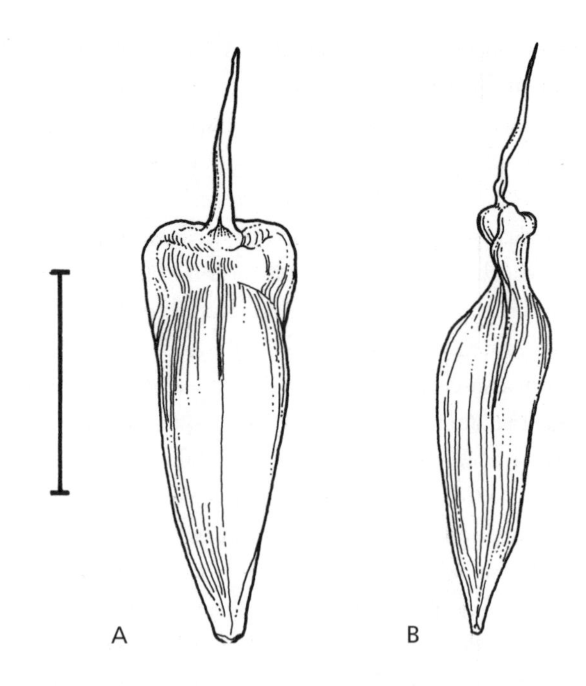

Upper-face (A) and side (B) views of shed, unwinged seed cone scale of monkey puzzle tree (*Araucaria araucana*) with embedded seed; scale, 2 cm.

Characteristic open natural stand of monkey puzzle trees (*Araucaria araucana*).

1.8–2.5(–4) cm long. Seeds 2.5–3.5(–4.5) cm long, narrowly ovoid, red-brown, edible (*piñones*). Cotyledons remaining underground within the seed shell during germination. South-central Chile and adjacent Argentina, on either side of the Andes. Commonly forming open pure stands on fertile, well-drained, evenly moist soil with moist air; tolerant of coastal winds and salt; (600–)900–1,500(–1,800) m. Zone 8. Synonym: *A. imbricata* Pavon.

Monkey puzzle tree is by far the best-known species of *Araucaria* in cultivation in temperate regions, prized for its dramatic architectural form. As it was the first species of *Araucaria* introduced to botanists and is the formal type of the genus, it would keep the name *Araucaria* if the genus were ever split up. In its homeland it was an important timber tree until it was depleted by overexploitation. Its large, nutritious seeds have been an important seasonal food for the Araucanian people of Chile (hence the scientific name), and their interests in the tree highlight the interactions between indigenous resource rights and biological conservation efforts. Like its closest relative, Parana pine (*A. angustifolia*), the only other extant member of section *Araucaria,* monkey puzzle tree is probably dispersed by seed-caching vertebrates, although jays, which disperse Parana pine, are now absent within its range. In keeping with this dispersal mode, both species are predominantly dioecious, with separate pollen- and seed-producing individuals, a habit more frequently associated with conspicuously fleshy seeds or seed cones in conifers. Monkey puzzle tree has another regeneration strategy that is unusual among conifers. It sprouts readily from cut stumps or fallen trees, and even from underground roots. Many araucarians have dormant buds along their stems that allow them to replace shed or damaged branches, but few of these produce upright-growing replacement stems the way monkey puzzle tree does. This common name reflects a fanciful envisioning of a monkey climbing a tree of *A. araucana* (none live within its native range) and stuck out at the tips of the branchlets by the viciously prickly, forwardly directed, very stiff needles.

Mature monkey puzzle tree (*Araucaria araucana*) in nature.

Araucaria bernieri J. Buchholz
BERNIER COLUMNAR ARAUCARIA
Araucaria sect. *Eutacta*

Tree to 50(–55) m tall in the south or as little as 4 m at maturity in the north, with trunk usually to 1 m in diameter. Bark gray, smooth, ringed through peeling in narrow bands. Crown narrowly cylindrical, with closely spaced tiers of five to seven slender, short, horizontal branches frequently shed and renewed with replacements bearing regular combs of branchlets. Branchlets slender, upwardly angled, making a V above the supporting branch, densely clothed with and largely hidden by the foliage, often shed intact after a few years. Juvenile leaves scalelike, spreading from the twigs, with a gentle forward curve, to 7 mm long, slightly flattened from side to side. Adult leaves scalelike, spreading from the twigs, strongly curled forward, overlapping, strongly keeled, pointed, 2–3.5 mm long. Stomates slanted away from the main axis of the leaf, in two dense bands on the inner face, in patches at the base of the outer faces on either side of the keel and with thin discontinuous lines extending out to the leaf tips. Pollen cones 4–9 cm long, 8–16 mm thick, with a bluish white waxy coating, each pollen scale with four to six pollen sacs in a single row or partially doubled. Seed cones egg-shaped, 10–11 cm long, 7.5–8 cm wide, with a bluish white waxy coating. Seed scales 2.5–3 cm long and wide, nearly round in outline, with two thin wings each occupying about a third of the width on either side of the embedded seed and with a narrowly triangular free tip 4.5–6 mm long sticking straight out. Seeds 2–2.5 cm long, 8–10 mm wide. Cotyledons rising above ground during germination. New Caledonia, mostly in the southern half with outliers at the northern tip. Lowland rain forests; 100–600(–800) m. Zone 9.

Bernier columnar araucaria is most closely related to upland columnar araucaria (*Araucaria subulata*), with which it was long confused but differs in the smaller, less flattened needles and the bluish white waxy coating on both the pollen and seed cones. It shares its unusual columnar growth habit with the more interior

A. subulata and with the more coastal Cook pine (*A. columnaris*). Like those other species, it was an important timber tree but only in the south, since trees in the northern stands on serpentine soils are too dwarfed for effective exploitation. As loss and replacement of limbs proceeds during the life of a tree, the branch architecture becomes increasingly complex, and a tree may appear to have several superposed crowns within the overall cylindrical outline.

Araucaria bidwillii W. J. Hooker

BUNYA-BUNYA, BUNYA PINE

Araucaria sect. *Bunya*

Tree to 40(–50) m tall, with trunk to 1.5 m in diameter, free from branches for most of its height in dense stands or more or less clothed with replacement branches in the open. Bark brown to nearly black, smooth, thick, peeling in thin layers but some flakes accumulating as short, horizontal lines of warts. Crown conical at first, becoming broadly dome-shaped above and sometimes with additional, irregular dome-shaped crowns contributing to an overall cylindrical outline beneath the main crown with tiers of four to seven (or more) very slender, elongate branches bearing tufts of branchlets only at and near their tips. Branchlets remaining green for several years, widely spaced out to the sides of the supporting branches and densely clothed with but not hidden by the foliage. Leaves sword-shaped on sterile branches, standing straight out from the twigs or angled a little forward, more or less all gathered to the sides by twisting of the petioles, 2–5 cm long, 5–10 mm wide, very unequal in length along the branchlets, stiff, very dark green above, paler beneath, ending in a long, stiff point. Leaves more densely arranged on the uppermost and fruiting branches, shorter, often only 2 cm long and curved inward. Stomates aligned with the long axis of the leaf, in more numerous discontinuous lines beneath than above. Pollen cones to (4–)10–15(–20) cm long, 1.2–1.5 cm thick, each pollen scale with 10–12 pollen sacs in two rows. Seed cones very large, nearly spherical to a little elongate, 25–30 cm high, 20–25 cm wide, weighing up to nearly 5 kg, with as many as 150 seeds. Seed scales to 6.5 cm long and 7 cm wide, broadly winged on either side of the embedded seed, markedly tapered at the apex, with an upturned, triangular free tip about 6 mm long. Seeds nearly pear-shaped, 5–6 cm long and 2.5 cm wide, edible. Cotyledons remaining in the seed coat underground during germination. Australia, along the coast of Queensland. Forming pure stands or mixed with hoop pine (*A. cunninghamii*) and evergreen hardwoods as an emergent above coastal montane rain forest; (150–)1,000–1,200 m. Zone 9.

Young bunya-bunya (*Araucaria bidwillii*) in nature.

Although of very limited natural distribution, being confined to two small areas in coastal Queensland, in both the south and the north, bunya-bunya was one of the premier timber trees of early Australia and was heavily exploited. Today, native stands are largely confined to national parks, such as the fine ones in Bunya Mountains National Park near Brisbane. The tree is widely cultivated in suitable climates, however, both in Australia and throughout the subtropics and the Mediterranean region. Large trees in cultivation tend to be felled after a time because of the hazard posed as the enormous seed cones fall intact at maturity, to shatter upon impact, releasing the seed scales and seeds. These seeds were an important seasonal food for those aboriginal groups lucky enough to include stands of bunya-bunya within their lands, just as seeds of the related South American species, Parana pine (*Araucaria angustifolia*) and monkey puzzle tree (*A. araucana*), were exploited. The cones were collected directly from the crowns, and large old trees can still be seen with the climbing ladders cut into their trunks. The species name honors John Bidwill (1815–1853), a Queensland government botanist who carried the type specimen, actually collected by Andrew Petrie in 1838, to William Hooker at Kew, who credited them to Bidwill in describing the species.

Twig of bunya-bunya (*Araucaria bidwillii*) with a single growth increment, ×0.5.

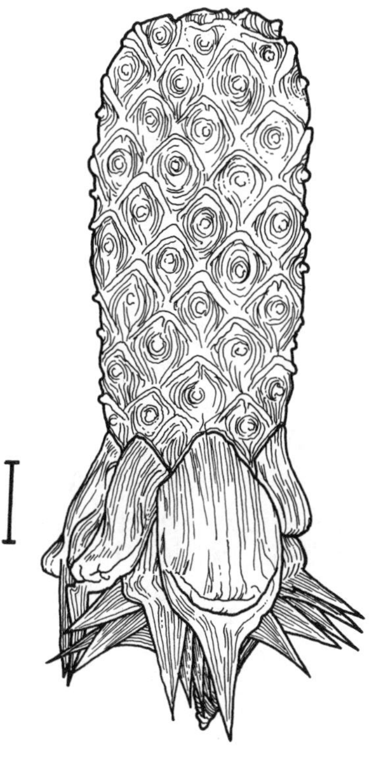

Remnant seed cone axis of bunya-bunya (*Araucaria bidwillii*) after shedding the seed scales; scale, 1 cm.

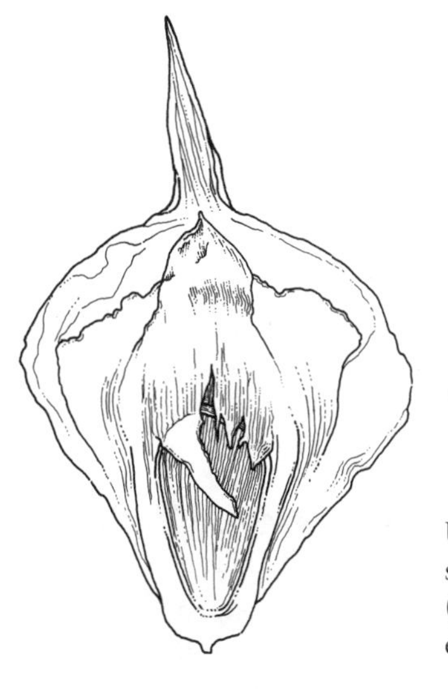

Upper face of shed, unwinged seed cone scale of bunya-bunya (*Araucaria bidwillii*) with embedded seed; scale, 2 cm.

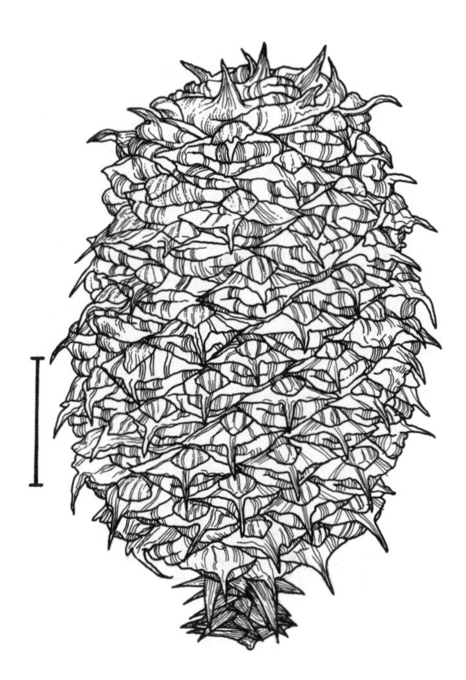

Mature seed cone of bunya-bunya (*Araucaria bidwillii*) before shattering; scale, 5 cm.

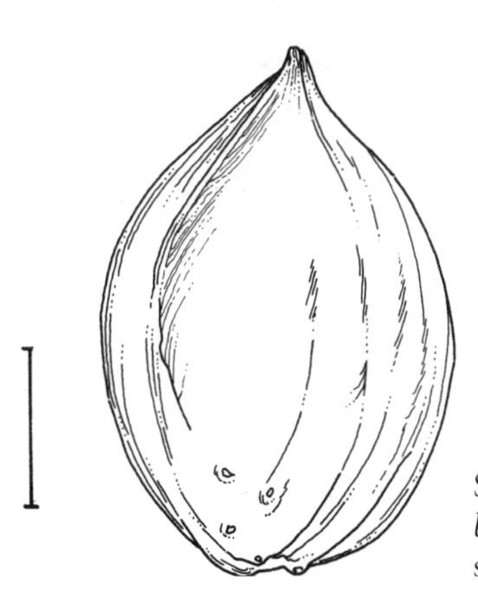

Seed of bunya-bunya (*Araucaria bidwillii*) removed from seed scale; scale, 1 cm.

Araucaria biramulata J. Buchholz

PIGGYBACK ARAUCARIA

Araucaria sect. *Eutacta*

Tree to 15(–30) m tall, with trunk to 1 m in diameter. Bark gray to brownish, peeling in narrow horizontal strips or later in thick flakes. Crown flattened at the top, broadly cylindrical, with closely spaced tiers of (four or) five (or six) slender, upswept original and replacement branches of varying lengths all the way up and down the trunk, though fairly consistent within the whorls. Branches often bearing secondary branches near the trunk (hence the scientific name, Latin for "twice-branched"), both types ending in upright, elongate tufts of radiating branchlets. Branchlets fairly coarse, about 1 cm in diameter, often slightly contorted, remaining green 1–2 years but hidden by the leaves in adult foliage. Needles broadly awl-shaped in juvenile and transitional foliage, scalelike and egg-shaped in adults, flattened and more or less keeled both above and below, (5–)7–9 mm long, 4–6 mm wide at the base. Stomates at right angles to the long axis of the leaf, in irregularly spaced, discontinuous rows on either side of the keel both above and below. Pollen cones 6–7 cm long, 1.5–2 cm wide, each pollen scale with seven or eight pollen sacs in a full and a partial row. Seed cones 9–10 cm long and nearly as wide, light brown at maturity. Seed scales about 3 cm long and 2.5 cm wide including the papery wings extending out to either side of and wider than the embedded seed, with a narrowly triangular free tip 8–10 mm long generally sticking straight out. Seed sword-shaped, about 2.5 cm long and 1 cm wide. Cotyledons emerging above ground during germination. New Caledonia, mostly in the interior well away from the sea. Emerging above thickets on steep rocky slopes; 250–1,000(–1,130) m. Zone 9.

Piggyback araucaria is not especially distinctive and is often overlooked among similar species. Lush araucaria (*Araucaria luxurians*), Port Boise araucaria (*A. nemorosa*), and Cook pine (*A. columnaris*) all live exclusively on the coast while piggyback araucaria reaches it only in a couple of locations. Its foliage is most similar to that of lush araucaria, but its needles are fairly uniform in length along the branchlets while those of lush araucaria are very uneven, with alternating zones of longer and shorter needles. It differs from all three species in having the free tip of the seed scale sticking straight out rather than directed up or down over the cones. This difference also distinguished it from another species with similar foliage, Mount Humboldt araucaria (*A. humboldtensis*), like piggyback araucaria a tree of the interior but much smaller and with a candelabra-like crown mostly lacking the lower branches typically found in piggyback araucaria.

Araucaria columnaris (J. R. Forster) W. J. Hooker

COOK PINE, PIN COLONNAIRE (FRENCH)

Plates 13 and 14

Araucaria sect. *Eutacta*

Tree to 35(–60) m tall, with trunk to 0.5(–1) m in diameter. Bark thin, gray, peeling in horizontal strips. Crown very narrowly cylindrical, with lower branches usually falling and replaced by new short branches from buds on the trunk, this portion of the stem then appearing like a green column (hence the scientific name). Branches in short horizontal tiers of (four or) five to seven, with long, slender, whiplike side branches bearing tightly spaced combs of branchlets. Branchlets remaining green 2–3 years, largely hidden by both juvenile and adult foliage, shed intact after a few years along with their supporting branch. Juvenile leaves more or less clawlike, curled forward and inward along the branchlets, pointed, 4–7 mm long. Adult leaves scalelike, egg-shaped, with a broad, low, keel or crest on the outer (lower) face, 5–7 mm long, 4–5 mm wide, stiff, bent forward and inward, densely and tightly overlapping, resembling a braided rope. Stomates at right angles to the long axis of the leaf, in irregularly spaced, short lines making up two bands around the midrib on both the inner and outer faces. Pollen cones 5–10 cm long, 13–22 mm thick, each pollen scale with 8–10 pollen sacs in a single or partially doubled row. Seed cones a little longer than spherical, 10–15 cm long, 7–11 cm thick, dark green at maturity. Seed scales 3–4 cm long, 4–5 cm wide including the papery wings extending out on either side of the narrower, embedded seed, with a sharp, narrowly triangular fee tip 7–10 mm long turned up along the surface of the cone. Southernmost New Caledonia, including the Isle of Pines and the nearby Loyalty Islands. Forming open pure stands as the only

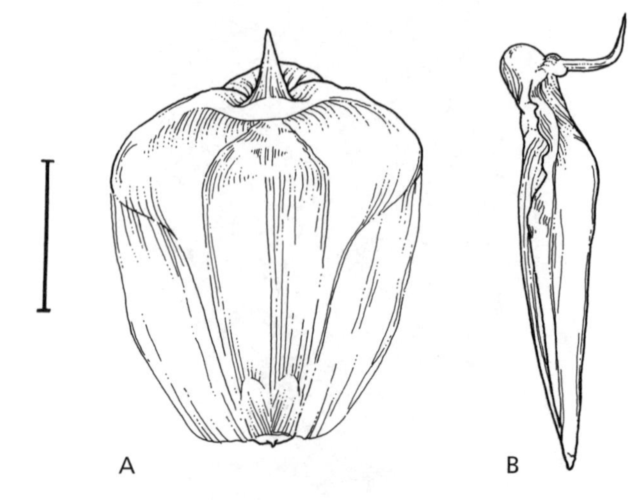

Upper-face (A) and side (B) views of shed, winged seed cone scale of Cook pine (*Araucaria columnaris*) with embedded seed; scale, 1 cm.

trees on coastal coral soils; 0–100 m. Zone 9. Synonyms: *A. cookii* R. Brown ex Endlicher, *A. excelsa* (A. Lambert) R. Brown.

Cook pine is very similar to Norfolk Island pine (*Araucaria heterophylla*) and often mistakenly distributed as that species, especially in nursery stock exported from Hawaii, where Cook pine has been long enough established to produce abundant seed. It differs in the narrower branch spread, shorter leaves, larger pollen and seed cones, and broader cotyledons, 3–5 mm wide versus 1.5 mm. The narrow, columnar growth habit so characteristic of Cook pine in its native stands is usually absent in cultivation, where the trees retain their original branches for much longer and hence develop a broadly conical crown. The columnar form is found, however, in localities struck by hurricanes or other strong winds. Cook pine is the only New Caledonian conifer found on calcareous substrates.

Araucaria cunninghamii D. Don

HOOP PINE

Araucaria sect. *Eutacta*

Tree to 30–60(–70) m tall, with trunk to 1.8(–3) m in diameter. Bark rough, reddish brown to grayish brown, flaking in patches and peeling in horizontal strips (or "hoops," hence the common name). Crown dense, deep, egg-shaped, with a persistent framework of numerous original tiers of four to seven sturdy upwardly angled to horizontal branches bearing many, closely spaced, radiating branchlets densely clothed with foliage. Branchlets remaining green for several years before being shed intact or becoming incorporated into the branch framework, only partially hidden by the leaves. Juvenile leaves needlelike, sword-shaped, sticking straight out from the twigs, dense but with space between them, (8–)10–20(–27) mm long, sometimes bluish green with wax, straight, sharply pointed. Adult leaves clawlike, more densely packed and overlapping, 4–5(–10) mm long, strongly keeled on both sides, short-pointed and curved inward, green to blue-green. Stomates mostly tilted away from the long axis of the leaf in either direction, in two sparse bands on the upper (inner) face and two patches at the base of the outer face. Pollen cones (2–)4–10 cm long, 8–10 mm thick, each pollen scale with five to eight pollen sacs in a single row. Seed cones visibly longer than spherical, 6–12 cm long, 5–8 cm wide, grayish brown at maturity. Seed scales blackish brown with reddish brown wings, 2.3–3 cm long, 3–3.5 cm wide, including the papery, fragile wings, each 10–12 mm wide and wider than the 10-mm-wide seed-bearing portion, with a triangular, upturned free tip 2–7 mm long. Seeds almond-shaped, 1.2–2 cm long, 4–7 mm wide. Cotyledons emerging above ground during germination. Northeastern Australia, New Guinea. Canopy emergent above rain forests of mixed and varied compositions; 0–1,500(–2,745) m. Zone 9. Two varieties.

Hoop pine was one of the premier native timber trees of Australia and was severely depleted during a century of over-exploitation, despite (or because of) its wide distribution in the accessible coastal regions. While most araucarias have a sparse, even gaunt canopy, such as that found in monkey puzzle tree (*Araucaria araucana*), hoop pine retains its primary tiers of branches and even its branchlets, typically short-lived in other species, for far longer than its congeners, and thus has the densest canopy in the genus, sometimes appearing remarkably cypress-like or cypress pinelike from a distance for open-grown trees. It is moderately common in cultivation, both in Australia and in other warm, frost-free climates. While it belongs to the same section of Araucaria (section *Eutacta*) as Norfolk Island pine (*A. heterophylla*) and all the New Caledonian species of the genus, it is the most taxonomically isolated species in the section, sister to all the other species in DNA-based phylogenies. Australian and New Guinean populations are slightly differentiated from each other genetically but are far more closely related to each other than to any of the other species. The species name honors early Anglo-Australian botanist Allan Cunningham (1791–1839), who collected the type specimen.

Araucaria cunninghamii D. Don **var.** ***cunninghamii***

HOOP PINE, MORETON BAY PINE

Tree to 30–60 m tall. Juvenile leaves 10–15 mm long. Adult leaves about 4–5 mm long. Pollen cones (2–)4–8 cm long. Seed cones 6–10 cm long, 5–7.5 cm thick. Eastern coast of Australia, from the northern Cape York Peninsula (Queensland) to the Clarence River (northern New South Wales). Often mixed with bunya-bunya (*Araucaria bidwillii*) where the ranges of the two species overlap in montane and coastal rain forests, commonly in their drier portions.

Araucaria cunninghamii **var.** ***papuana*** Lauterbach

PAPUAN HOOP PINE, PIEN

Tree to 50–70 m tall. Very similar to variety *cunninghamii* but not as symmetrically branched and without the candelabra-type growth. Juvenile leaves to 27 mm long. Adult needles awl-shaped, about 1 cm long, sharp, acuminate, and curved forward. Pollen cones 9–10 cm long. Seed cones 7–12 cm long, 6–8 cm thick. Seed scales longer, base narrower, more acuminate at the apex. Throughout the spine of New Guinea and along some coasts in both Irian Jaya and Papua New Guinea. Emergent above various rain-forest formations; 60–2,745 m. Synonym: *Araucaria beccarii* O. Warburg.

Araucaria heterophylla (R. A. Salisbury) Franco

NORFOLK ISLAND PINE

Araucaria sect. *Eutacta*

Tree to 50–60(–70) m tall, with trunk to 2(–3) m in diameter. Bark light brown, peeling in thin scales and layers. Crown long remaining conical, becoming somewhat egg-shaped with age, with very regular tiers of four to seven slender, horizontal branches bearing two evenly spaced combs of branchlets for much of their length. Branchlets slender, extending horizontally or shallowly V-shaped in young trees, extending upward in a narrower V near the top of mature trees, remaining green many years before being shed intact, largely hidden by the foliage. Juvenile leaves soft, awl-shaped, sticking straight out or bent inward and not overlapping, 8–12(–20) mm long. Adult leaves stiff and hard, scalelike or transitional between claw- and scalelike, slightly keeled, tightly overlapping like a braided cord or whip, 4–6(–10) mm long. Stomates ranging from parallel to perpendicular to the long axis of the leaf, in incomplete, irregular lines occupying large patches on the inner (upper) face and confined to the base of the outer face where

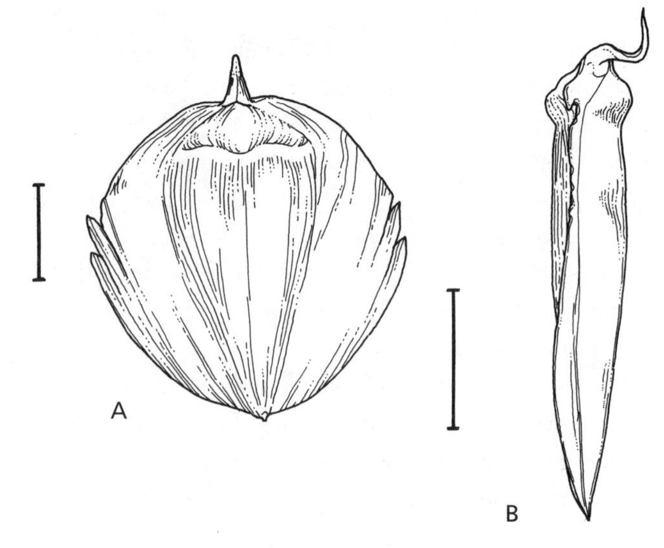

Upper face (A) and side (B) views of shed, winged seed scale of Norfolk Island pine (*Araucaria heterophylla*); scale, 1 cm.

Young mature Norfolk Island pine (*Araucaria heterophylla*) in nature.

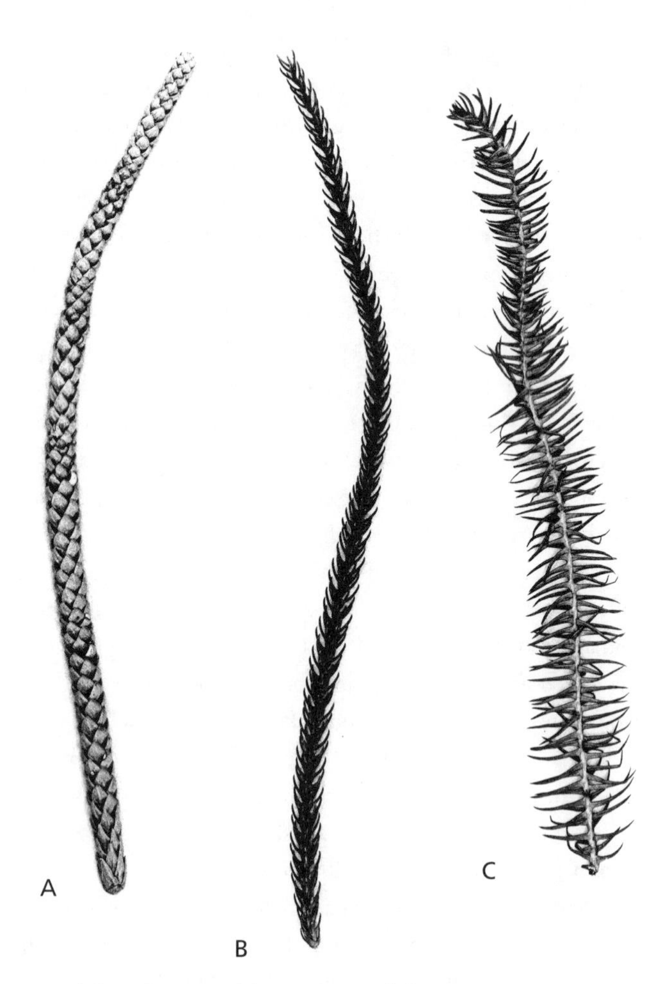

Adult (A), intermediate (B), and juvenile (C) twigs of Norfolk Island pine (*Araucaria heterophylla*) with two weakly distinguishable growth increments, ×0.5.

they are hidden by the overlapping leaves. Pollen cones 3.5–5 cm long, 10–13 mm thick, each pollen scale with 6–10 pollen sacs in a single or partially doubled row. Seed cones approximately spherical (7.5–)9–12 cm long and wide, green at maturity. Seed scales 2.5–4.5 cm long and 1.8–3.8 cm wide, including the papery wings, which are narrower than the 1.2–2 cm wide seed-bearing portion, with a sharp, narrowly triangular upturned tip 5–12 mm long. Seeds almond-shaped, about 3 cm long and 1.5 cm wide. Cotyledons emerging above ground during germination. Norfolk Island. Forming pure stands or towering above mixed rain forests; 0–250(–315) m. Zone 9.

Although the natural range of Norfolk Island pine is confined to that tiny speck of 34 square kilometers of land in the South Pacific midway along a line between New Caledonia and New Zealand, this species and monkey puzzle tree (*Araucaria araucana*) are the two most familiar araucarias in cultivation outdoors in warm and moderate climates, contributing some of the most architectural tree forms to the cultivated landscape. Norfolk Island pine is also a common greenhouse subject and indoor accent plant in cooler climates. It is as well one of the most characteristic landscape features of its native Norfolk Island. Despite the narrowness of this natural range, it is a common tree on the island, and there is no evidence of restricted genetic variation (at least no more so than in other species of *Araucaria*). In fact, there are more cultivars of Norfolk Island pine, varying in needle length, shape, and color, than there are for all of the other species combined (except for monkey puzzle tree, which has a similar number). While mature foliage is rare in cultivation and effectively absent indoors, the striking differences between juvenile and adult foliage are responsible for the species name (Latin for "different leaves"), especially in comparison to the much more uniform monkey puzzle tree, the only other *Araucaria* familiar to botanists at the time that *A. heterophylla* was described. Because of earlier confusion with the similar Cook pine (*A. columnaris*), this species has been mistakenly called *A. excelsa*, a synonym of that New Caledonian species. Cultivated plants sold as Norfolk Island pine are often Cook pines, especially those originating as nursery stock from Hawaii. Norfolk Island pine has a much wider branch spread, longer leaves, smaller pollen and seed cones, and narrower cotyledons (only 1.5 mm wide versus 3.5 mm) than Cook pine. Despite their strong similarity, the two species are not particularly closely related. DNA studies show that Cook pine is closest to the other New Caledonian species, whose relationships are too messy to be resolved as yet, while Norfolk Island pine is sister to the whole New Caledonian group and thus no more closely related to Cook pine than to any of the other species.

Araucaria humboldtensis J. Buchholz

MOUNT HUMBOLDT ARAUCARIA

Araucaria sect. *Eutacta*

Tree to (6–)10–15(–20) m tall, with trunk to 0.3(–0.5) m in diameter. Bark light gray, smooth, peeling in scales or horizontal strips. Crown very sparse, with a few scattered, short branches along the trunk overreached by a shallow, flat-topped umbrella at the very summit composed of a few tiers of four to six upwardly angled, thin branches with tufts of branchlets at their tips densely clothed with foliage. Branchlets coarse, nearly upright, remaining green 2–3 years before being shed intact, essentially hidden by the leaves. Leaves scalelike, broadly pointed, keeled, 5–6 mm long, 3–5 mm wide, dark green with yellow-green tips. Stomates with no preferred orientation, in two bands without obvious discrete lines on the inner (upper) face and two basal patches on the outer face. Pollen cones 5–8 cm long, 15 mm thick, each pollen scale with five to eight pollen sacs in a single row. Seed cones nearly spherical or a little longer, to 9 cm long, 8 cm wide, waxy bluish green. Seed scales about 3 cm long by 3 cm wide, including the two papery wings about as wide as the seed-bearing portion, with an upturned, narrowly triangular free tip 4–7 mm long. Seeds almond-shaped, 2.5 cm long and 9 mm wide. Cotyledons emerging above ground during germination. Mountains in the southern fifth of New Caledonia. Scattered as a common emergent above short lower montane forest on serpentine soils often mixed with other species of *Araucaria*; 750–1,550 m. Zone 9.

Mount Humboldt araucaria is one of the characteristic species of both shrubby maquis minière and low rain forests within its limited range within easy reach of the capital, Nouméa. Its absolutely flat crown and very sparse lower canopy distinguish it from associated species and give the region one of its most familiar plant monuments, all along the southern side of the lower slopes of Mount Humboldt.

Araucaria hunsteinii K. Schumann

KLINKI PINE, PA'A, YOMEJO, KARINA, ND'UK (PAPUAN LANGUAGES)

Araucaria sect. *Intermedia*

Tree to 60(–90) m tall, often clear of branches for two-thirds of its height, with trunk to 1.2(–2) m in diameter. Bark dark reddish brown, rough, flaking and peeling in rings, secreting abundant clear resin. Crown dome-shaped, thin, with widely spaced tiers of five or six branches, turned up at their ends and bearing open tufts of branchlets densely clothed with foliage. Branchlets green at first, coarse, completely covered by the attached leaf base and partially obscured by the leaf blades, except in the gaps between them, falling intact after a few years. Juvenile needles arranged horizontally out to the sides of and well spaced along the branchlets by twisting of the leaf bases, sword-shaped, widest near the middle, flat, (0.2–) 2.5–7.5 cm long, 1–10 mm wide, largest near the middle of each growth increment. Adult needles in several rows all around and closely spaced along the branchlets, angled forward, sword-shaped, widest near the base, flat but bowed outward around the midline, hard and sharp-pointed, waxy when young, (2.5–)5–10(–15) cm long, 1–2 cm wide, grading continuously from shortest at the base to largest near the tip of each growth increment. Pollen cones (4–) 15–22 cm long, 1.8–2.5 cm thick, each pollen scale with 6–10 pollen sacs in a single row. Seed cones nearly spherical to more elongated, 18–25 cm long, 12.5–16 cm thick, often waxy when young. Seed scales (3–)4–5(–6) cm long, 7–10 cm wide, including the huge, papery, triangular wings, extending out 3–4 cm on either side of the swollen, central seed-bearing portion, with an upturned, fragile, spinelike free tip up to 2 cm long but usually broken off in the mature cones. Seeds narrowly almond-shaped, 2.5–4 cm long, 8–12 mm wide. Cotyledons emerging above ground during germination. Papua New Guinea. Emergent over oaks and other evergreen hardwoods in lower montane rain forests and some drier forests; (520–)700–1,500(–2,100) m. Zone 9. Synonyms: *A. klinkii* Lauterbach, *A. schumanniana* O. Warburg.

Klinki pine is the tallest tree in New Guinea and the whole Malesian region. It stands so far above the hardwood forests it towers over and has such a thin crown that it appears to have no competitive interactions with these associates, at least once it reaches the fully emergent state. As a consequence, the cross sectional area of the trunks of all the hardwoods in forests on closely matched (even adjacent) sites is the same whether Klinki pine is present or not. This effect, which has been called additive basal area by ecologists, is rare in emergents, which usually have more interaction with the canopy layer, shading it and reducing its productivity. This is the case with the other *Araucaria* of New Guinea, the more widespread Papuan hoop pine (*A. cunninghamii*), which has a denser crown that nestles above the canopy of the embedding hardwood forest rather than standing entirely free of it. Although hoop and Klinki pines may grown together, hoop pine is generally a tree of higher elevations, the dividing line for their preferred habitats lying at around 1,000 m. Klinki pine does not regenerate well in shade or deep organic soils so large disturbances seem to be required for establishment of stands. As the postdisturbance stands age, the early hardwoods gradually die out over the centuries while the Klinki pines continue growing, leaving successive generations of hardwoods far below them. Klinki pines can live 500 years, not as long as some conifers but longer than most hardwoods.

A softwood of this size could hardly be ignored by foresters, and Klinki pine has been heavily exploited. Because of its initial ecological requirements, it tends to be eliminated from stands if the surrounding hardwoods are not removed at the same time. The old-growth stands exploited during the 20th century may well have resulted from agricultural clearance by native Papuans centuries before, just as the emergent stands of white pine (*Pinus strobus*) that attracted lumbermen to the Great Lakes region of North America in the 18th and 19th centuries were probably initiated following forest clearance for agriculture by native Americans in the 16th and 17th centuries or earlier. Klinki pine is not a tree likely to be overlooked by forest-dwelling people, and it has many names in eastern Papuan languages, only a few of which are listed above.

Klinki pine is taxonomically somewhat isolated within the genus *Araucaria*, being placed in a section by itself, section *Intermedia*, so named because it is intermediate in some characteristics between the bulk of the genus that makes up section *Eutacta*, like hoop pine and Norfolk Island pine (*A. heterophylla*), and the three other species in the genus, bunya-bunya (*A. bidwillii*) in section *Bunya* and Parana pine (*A. angustifolia*) and monkey puzzle tree (*A. araucana*) in section *Araucaria*. Molecular evidence shows that it groups with the latter three species against the 15 species in section *Eutacta*, and its foliage is certainly more similar to theirs, but its seed scales are a little more like those in section *Eutacta*, and the cotyledons emerge above ground rather than staying inside the seed coat the way they do in bunya-bunya, Parana pine, and monkey puzzle tree. Actually, the seed scales, with their enormously expanded wings, are way beyond what is found in section *Eutacta*.

Araucaria laubenfelsii Corbasson

MOUNT MOU ARAUCARIA

Araucaria sect. *Eutacta*

Tree to 10–30(–50) m tall, with trunk to 0.5(–1) m in diameter. Bark gray, smooth, peeling in horizontal strips at first, later in irregular scales. Crown narrowly cylindrical, with a flattened or dome-shaped, umbrella-like cap composed of a few tiers of four (or five) slender, upwardly angled branches above many tiers of short replacement branches all the way down the trunk, both old and new branches bearing numerous radiating branchlets densely clothed with foliage. Branchlets coarse, 2–3 cm thick, partially or fully hidden by the leaves, remaining green 2–3 years before being shed intact. Juvenile leaves awl-like but slightly curved forward, standing out from the branchlets, 10–15 mm long. Adult leaves scalelike, overlapping, bent forward over the twig, broadly egg-shaped, thick, keeled, bowed outward around the midline, bluntly pointed, varying in length along the twig, 1–2 cm long. Stomates mostly at right angles to the long axis of the leaf, in numerous interrupted but dense rows on both faces. Pollen cones 12–15 cm long, 2–3 cm thick, each pollen scale with 10–12 pollen sacs in two rows. Seed cones longer than wide, 10–12 cm long, 8–9 cm wide, green at maturity. Seed scales about 3 cm long and 3 cm wide, including the thin papery wings on either side of and each equal in width to the swollen, central seed-bearing portion, with a sharply triangular, upturned tip 8–10 mm long. Seeds about 15 mm long, 10 mm wide. Cotyledons appearing above ground during germination. Southern New Caledonia, especially on Mount Mou and nearby peaks. Emergent above montane rain forest on serpentine soils; (400–)900–1,300 m. Zone 10?

Mount Mou araucaria was long confused with Rule araucaria (*Araucaria rulei*) and only described as a distinct species in 1968. In almost all respects, however, it is more similar to lush araucaria (*A. luxurians*), having similar pollen cones, seed scales, and leaves that also vary conspicuously along the length of the twigs. These leaves are larger than those of lush araucaria, more or less intermediate in size between those of lush araucaria and those of chandelier araucaria (*A. muelleri*). The species name honors David de Laubenfels (b. 1925), whose work on the previously understudied tropical rain-forest conifers initiated a new understanding of generic boundaries among Podocarpaceae and who has also described or reinterpreted many species of tropical conifers, including several species of *Araucaria* and the related *Agathis* from New Caledonia.

Araucaria luxurians (Ad. Brongniart & Gris) de Laubenfels

LUSH ARAUCARIA, COAST ARAUCARIA, SAPIN DE NOËL (FRENCH)

Araucaria sect. *Eutacta*

Tree to 20–30 m or more tall but often much smaller, with trunk to 0.5(–1) m in diameter. Bark reddish brown at first, weathering gray, peeling in narrow horizontal bands interrupted by occasional vertical furrows and overlain by irregular flakes. Crown narrowly conical to narrowly dome-shaped with numerous tiers of five to seven thin branches, the original tiers long retained and also supplemented with additional new branches in the older portions, all ending in tufts of upright, radiating branchlets densely clothed with foliage. Branchlets with mature foliage 1–2 cm in diameter, hidden by the leaves, remaining green 2–3 years before being shed intact. Juvenile needles clawlike, erect, 6–12 mm long. Adult leaves scalelike, egg-shaped, bowed out around the midline, shallowly crested, varying in length along the shoots, 5–7 mm long. Stomates mostly tilted away from the long axis of the leaves, in numerous, closely spaced, interrupted lines gathered in loose bands on both surfaces. Pollen cones 12–17 cm long, 2.5–3 cm wide, each pollen scale with 12–15 pollen sacs in two rows. Seed cones roughly spherical, 10–12 cm long, 8–10 cm wide, yellowish green at maturity. Seed scales 3–3.5 cm long, 3.5–4.5 cm wide, including the thin papery wings about as wide as the central, seed-bearing portion, with a sharply triangular, upturned tip 6–10 mm long. Seeds narrowly egg-shaped, 2.5–3 cm long, 0.8–1.2 cm wide. Cotyledons appearing above ground during germination. New Caledonia, mostly in the southern half. On serpentine cliffs by the sea; 0–200 m. Zone 9.

Lush araucaria was originally described as a variety of Cook pine (*Araucaria columnaris*) and shares some morphological characteristics and its seashore habitat with that species. It differs, among other features, in a broader, often more conical rather than cylindrical growth habit and in leaves that vary in length conspicuously along a branchlets rather than remaining fairly uniform. Its occurrence on serpentine soils (like the majority of New Caledonian conifers), rather than the coral-based calcareous substrates favored by Cook pine (although it is not restricted to them), is also helpful in distinguishing the two in the field if not in cultivation. The species name refers to the fuller crown found in lush araucaria compared to Cook pine, although it may actually be more closely related to Mount Mou araucaria (*A. laubenfelsii*), which it more strongly resembles. As one of the fuller and more symmetrically crowned New Caledonian *Araucaria* species, lush araucaria is appreciated as a garden ornamental in its homeland as well as one of the best local Christmas trees. It could be more widely grown elsewhere.

Araucaria montana Ad. Brongniart & Gris

SUMMIT ARAUCARIA

Araucaria sect. *Eutacta*

Columnar tree to (10–)20–30(–40) m tall, with trunk to 0.6 (–1.2) m in diameter. Bark brown, weathering gray, peeling in horizontal strips at first, later in irregular patches. Crown dense to thin, regularly conical below, capped by a rounded to flat top, with numerous upwardly angled tiers of four original and replacement branches both at the original levels and in between, reaching about 2.5 m before being shed and bearing persistent tufts of long, spirally arranged branchlets densely clothed with foliage. Branchlets coarse, 15–30 mm wide (including the leaves), partially visible between the leaves, remaining green 3–4 years before being shed intact. Juvenile leaves scalelike, spearhead-shaped, curled forward and tightly overlapping, 10 mm long, 4–5 mm wide near the base. Adult leaves scalelike, egg-shaped, spreading away from the twig at a forward angle, 11–14(–20) mm long, 7–8(–10) mm wide, keeled on the outer face and bowed outward around the midline. Stomates tilted away from the long axis of the leaf, sometimes attaining a right angle, tightly packed in closely spaced, short or interrupted lines making up two broad bands on the inner (upper) face and with patches at the base of the outer face extending in a few lines to near the tip. Pollen cones 8–13.5 cm long, 2–3 cm wide, each pollen scale with 10–12 pollen sacs in two rows. Seed cones broadly egg-shaped, 8–10 cm long, 6–8 cm thick, dark green at maturity. Seed scales 3–3.5 cm long, 2–2.5 cm wide, including the papery wings that are commonly incomplete and narrower than the central seed-bearing portion, with an upturned, broadly triangular tip 5–10 mm long. Seeds almond-shaped, about 2 cm long by 1 cm wide. Cotyledons appearing above ground during germination. Throughout the length of New Caledonia. Forming thin, pure stands of emergents above maquis minière shrublands or low cloud forest, mostly on serpentine-derived soils along ridges and high plateaus of the interior mountains (hence the scientific name); (250–)600–1,200(–1,330) m. Zone 9.

Summit araucaria may be the most widespread and most common *Araucaria* of New Caledonia. It is certainly a characteristic feature of unforested mountain landscapes in many parts of the main island. When well grown it can be an attractive ornamental, with a full crown of large, dark green leaves. Under less favorable conditions it can appear rather ratty, so it is not much grown. Although it is, overall, the most common New Caledonian species of *Araucaria,* reaches a fair size, and has good-quality wood, its scattered stands are not suitable for commercial exploitation.

Araucaria muelleri (Carrière) Ad. Brongniart & Gris

CHANDELIER ARAUCARIA, MUELLER ARAUCARIA, PIN CANDÉLABRE (FRENCH)

Plate 15

Araucaria sect. *Eutacta*

Tree to (10–)20–25(–30) m tall, with trunk to 0.3(–0.5) m in diameter. Bark light gray, peeling in flakes or horizontal strips and further roughened by branch scars. Crown flat-topped or broadly rounded to dome-shaped, with a few to many widely spaced tiers of four slender branches extending out to up to 4 m and turning up abruptly at the ends with spirally arranged tufts of very long, upright branchlets densely clothed with foliage. Branchlets up to 60 cm long, coarse, 3–5 cm in diameter, mostly hidden by the leaves, remaining green 2–3 years and then falling intact when shed. Juvenile leaves needlelike, sword-shaped, flat, standing out from the branchlets at a forward angle, 20–25 mm long, continuing to enlarge in succeeding years until their death. Adult leaves thick, leathery, scalelike, pointedly egg-shaped, conspicuously keeled, dark green to olive green, 30–35 mm long, 15–20 mm wide, densely overlapping. Stomates mostly at right angles to the long axis of the leaf, in widely spaced, regular but occasionally interrupted rows on both faces. Pollen cones 13–25 cm long, 3–4.5 cm wide, each pollen scale with 18–20 pollen sacs in two or three rows. Seed cones broadly egg-shaped, 11–15 cm long, 8–10 cm wide, dark green with yellow bract tips at maturity. Seed scales 3–3.5 cm long, 3–4 cm wide including the fragile, papery, often incomplete wings about as wide as the central, seed-bearing portion. Seeds narrowly almond-shaped, about 2.5 cm long and 1 cm wide. Cotyledons appearing above ground during germination. Southern tip of New Caledonia in the vicinity of the Plaine des Lacs. Forming sparse groves above thickets on serpentine; (150–)250–1,100 m. Zone 9.

Chandelier araucaria has the largest leaves among all the New Caledonian *Araucaria* species, and largest within the whole section *Eutacta.* The leaves are tied for widest in the genus with those of monkey puzzle tree (*A. araucana*) and are exceeded in length only by monkey puzzle tree, Klinki pine (*A. hunsteinii*), and bunya-bunya (*A. bidwillii*), each belonging to a different one of the other three sections of the genus. Their large size contributes to the candlelike appearance of the upright, straight branchlets, making the whole crown look like an antique chandelier. It is certainly one of the characteristic trees of the Plaine des Lacs and is planted in the restoration of the natural vegetation around the Chûtes de la Madeleine, one of the most important scenic tourist spots in New Caledonia. Reduction in abundance of the species has resulted more from unrestrained wildfires and habitat loss than from timber harvest because chandelier araucaria is usually too small and scattered for commercial exploitation.

Chandelier araucaria is most closely related to the very similar Rule araucaria (*Araucaria rulei*), which displays a more delicate version of the same growth habit. These are the only two species of *Araucaria* in New Caledonia linked to each other to the exclusion of others in DNA studies. Even if future DNA studies begin to discern more of the relationships among the species in this more recently evolved flock, these two are likely to remain the most closely related pair within the group. The species name honors Ferdinand von Mueller (1825–1896), state botanist of Victoria, Australia, who did some work on New Caledonian plants, but not, in fact, on this species.

Araucaria nemorosa de Laubenfels

PORT BOISE ARAUCARIA

Araucaria sect. *Eutacta*

Tree to 15–30 m tall (at different sites), with trunk to 0.5 m in diameter. Bark reddish brown, weathering gray, curling back and peeling along narrow horizontal strips or flaking in irregular scales. Crown narrowly conical to narrowly egg-shaped, with numerous, unequally spaced original and replacement tiers of (four or) five to seven slender, upwardly angled branches up to 4 m long bearing long combs of persistent branchlets rising on either side to form a V. Branchlets slender, 8–12 mm in diameter, including the leaves, partially obscured by them, remaining green for several years before being shed intact or with the whole branch. Juvenile and adult leaves both clawlike, keeled on both faces, standing out from the branchlets at a slight forward angle, curled inward slightly, and overlapping loosely. Juvenile leaves 4–8 mm long, 0.8–1.2 mm thick. Adult leaves varying in length along the branchlets, 6–10 mm long, 1.5–3 mm thick. Stomates mostly nearly at right angles to the long axis of the leaf, in variably spaced, interrupted lines making up continuous bands on either side of the keel on both faces. Pollen cones 8–10 cm long, 12–15 mm thick, each pollen scale with 6–10 pollen sacs in a single row. Seed cones spherical or a little longer, 10–12 cm long, 8.5–10 cm thick, green at maturity. Seed scales about 3 cm long, only slightly narrower, including the papery wings that are slightly wider than the central, seed-bearing portion, with a narrow, bristlelike, downturned tip 1–2 cm long. Seeds narrowly almond-shaped, 1.8–2.5 cm long, about 1 cm thick. Cotyledons appearing above ground during germination. Southernmost New Caledonia in the vicinity of Port Boisé. In a few small, dense groves (hence the scientific name, Latin for "of the groves") near stands of Cook pine (*A. columnaris*) on serpentine-derived soils at the shore; 0–35 m. Zone 10?

Port Boise araucaria is one of the rarest living conifers and is threatened with extinction in its native groves by strip mining that feeds on the metal-laden serpentine rocks that underlie so much of New Caledonia. Even if the native stands are extinguished, a few plantations may keep the species alive, though it is also a rare species in cultivation, even in its homeland. It grows quite well as a specimen tree or in plantations and is an attractive tree in cultivation but too similar in general appearance to Cook pine and Norfolk Island pine (*Araucaria heterophylla*) to have garnered much horticultural interest. It was long unrecognized among the more numerous stands of Cook pine that surround it but differs in its much broader growth habit with longer combs of branchlets bearing longer, narrower needles. Its coastal groves occupy one of the wettest regions in New Caledonia, with an annual rainfall of 2 m. It is not clear to which species it is most closely related, although similarities in the pollen scales and the variability in needle lengths along the twigs make piggyback araucaria (*A. biramulata*) and lush araucaria (*A. luxurians*) candidates, while the confusion with Cook pine should not be discounted. One feature of Port Boise araucaria not found in any other species of *Araucaria* relates to the pollen cones. The bracts at their base are the narrowest and most awl-like in the genus, contrasting very strongly with the broad blade of the pollen scales.

Araucaria rulei F. J. Mueller ex Lindley

RULE ARAUCARIA

Araucaria sect. *Eutacta*

Tree to 20–25(–30) m tall under favorable conditions but often shorter, with trunk to 0.3(–0.6) m in diameter. Bark dark brown at first, weathering stark white, peeling in horizontal strips or later in irregular patches. Crown deeply and broadly dome-shaped, with numerous, widely spaced, original and replacement tiers of four (or five) slender branches up to 4 m long, bearing additional upright secondary branches on their upper sides near the trunk, abruptly turned up at their tips, and bearing compact tufts of radiating, upright branchlets densely clothed with foliage. Branchlets up to 50 cm long, ropelike, coarse, 2.5–3 cm thick, including the leaves, by which they are largely hidden, remaining green 2–3 years before being shed intact. Juvenile leaves clawlike, narrowly triangular, standing out from the twig at a forward angle, with leaves of varying lengths intermixed rather than in graded arrangements, 12–15 mm long, 6–8 mm wide. Adult leaves glossy dark green, scalelike, pointedly egg-shaped, keeled on the outer (lower) face and bowed outward around the midline, curved inward, densely overlapping, stiff, leathery to hard, quite variable in size and differing from plant to plant, (15–)20–25 mm long, 11–14 mm wide. Stomates in line with, tilted away from or at right angles to the long axis of the leaf, in variably spaced, discontinuous rows making up a broad band on the inner (upper) face and two patches on either side of the keel at the base of the outer face where hidden by overlapping leaves. Pollen cones (5–)12–15 cm long, 3–3.5 cm wide, each pollen scale with 12–15 pollen sacs in two or three rows. Seed cones a little longer than spherical, 10–12 cm long by 8–9 cm thick, dark green at maturity with red free bract tips. Seed scales 3–3.5 cm long, 2.5–3 cm wide including the narrow, papery wings 4–7 mm wide, with a thornlike, flat, upturned tip 1.5–2 cm long. Seeds 1.5–2 cm long and 7–10 mm wide. Cotyledons appearing above ground during germination. Throughout the main island of New Caledonia, with localities more concentrated in the south. Forming open woodlands on mountain slopes above a shrubby hardwood understory on rocky thin red soils derived from serpentine; (150–)400–1,200 m. Zone 9.

Although Rule araucaria is one of the most widespread araucarias of New Caledonia, individual populations are infrequent and small so the species was always rare. Since it is also restricted to mineral-rich substrates, it is severely threatened by mining activities even though exploitation for timber has not been a real problem. The sites where the species grows are dry, not only because of the coarse, rocky soils but also because the rainfall is only about 1 m per year, not enough to equal annual evapotranspiration in this hot climate and thus leading to drought stress outside of the rainy season. Rule araucaria is closely related to chandelier araucaria (*Araucaria muelleri*), resembling a more delicate version of that dramatic species. It has a similar arrangement of candlelike branchlets, but the crown is much deeper and fuller and the top is rounded rather than tabletop flat. It is occasionally cultivated in botanical gardens in warm climates, like Australia, or in large greenhouses and conservatories elsewhere. There are even a few named cultivars. The species name honors John Rule, a Melbourne plantsman whose botanical collector, W. Duncan, found the species in New Caledonia and thus brought it to the attention of plant taxonomists.

Araucaria schmidii de Laubenfels

MOUNT PANIÉ ARAUCARIA

Araucaria sect. *Eutacta*

Tree to 20–30 m tall, with trunk to 0.3(–0.6) m in diameter. Bark mottled brown and light to dark gray, peeling in narrow horizontal strips. Crown a narrow cylinder of many tiers of four or more replacement branches overtopped by a broader, flat-topped umbrella of a few tiers of four slender original branches extending out to up to 2 m and bearing lengthy combs of upwardly angled branchlets densely clothed with foliage. Branchlets closely spaced, forming a V above the branches, fairly coarse, 1–2.5 cm thick, including the leaves, which partially to completely hide them, remaining green 3–4 years before being shed intact. Juvenile leaves yellowish green, clawlike, standing out from the twigs with a forward angle and curled in at the tips, strongly keeled, diamond-shaped in cross section, about 18 mm long and 2 mm wide. Adult leaves dark green, between claw- and scalelike, spearhead-shaped, curling forward along and tightly overlapping around the branchlets, keeled, triangular in cross section, 7–10 mm long and 1.5–2 mm wide. Stomates mostly at or near right angles to the long axis of the leaf, in closely spaced, relatively continuous lines making up two bands on the inner (upper) face and two patches at the overlapped base of the outer face. Pollen cones 5–10 cm long, 10–15 mm thick, each pollen scale with 6–10 pollen sacs in a single row. Seed cones nearly spherical, 8.5–10 cm in diameter, dark green at maturity with pale green to light reddish brown free bract tips. Seed scales 2.5–3.5 cm long, 2–3 cm wide, including the papery wings narrower than the central seed-containing portion, with a narrowly triangular, upturned tip 10–20 mm long. Seeds narrowly almond-shaped, 2–2.5 cm long, about 1 cm wide. Cotyledons appearing above ground during germination. Mount Panié-Colnett range, northeastern New Caledonia. In dense to sparse pure stands emergent above wet montane shrublands and dwarf cloud forest on schistose soils, with an annual rainfall of more than 4 m; 1,300–1,650 m. Zone 9?

When Mount Panié araucaria was first described it was known only from slopes around the summit plateau of that peak, at 1,500–1,600 m. It has subsequently been collected at a number of locations above 1,300 m, extending from Mount Panié northward along the continuous ranges to Mount Colnett. The juvenile leaves are unusual among the New Caledonian *Araucaria* species in being so much longer than the adult needles, which are also relatively more similar in general form compared to the norm for these species. The whole effect of the juvenile foliage is rather like that of northern hemisphere spruce trees, such as Norway spruce (*Picea abies*) or blue spruce (*P. pungens*), although there is no detailed correspondence, of course. The species name honors Maurice Schmid, the French-born New Caledonian botanist from Centre ORSTOM in Nouméa who collected the type specimen during work on Mount Panié in the 1960s.

Araucaria scopulorum de Laubenfels

CLIFF ARAUCARIA

Araucaria sect. *Eutacta*

Tree to (4–)15–20 m tall, with trunk to 0.3(–0.6) m in diameter. Bark smooth, reddish brown, weathering through gray to white, peeling in thin horizontal strips and irregular flakes. Crown broadly and shallowly dome-shape above an extremely irregular cylindrical lower crown, with numerous tiers of five to seven slender, horizontal or upwardly angled original and replacement branches up to 2.5 m long, turned up at the ends and bearing short to long upwardly angled combs of branchlets densely clothed with foliage. Branchlets very closely spaced along and in two rows forming a V above the supporting branch, fine, about 6–8 mm in diameter, including the leaves, which completely hide them, remaining green 2–3 years and reaching about 50 cm long before being shed intact. Juvenile needles clawlike, standing out from the twigs at a slight forward angle, strongly keeled on the outer (under) face and a little on the inner, 6–7 mm long, 1.5 mm wide. Adult leaves scalelike, egg-shaped, thick and hard, keeled, standing out from the branchlets and curled forward at the tips, loosely to tightly overlapping, yellowish green to dark green, 3–4 mm long, 2.5–3 mm wide. Stomates mostly tilted away from the long axis of the leaf, in irregular, interrupted rows forming two broad bands on the inner (upper) face and scattered thinly across the lower face. Pollen cones 3–5 cm long, 7–11 mm thick, each pollen scale with six to eight pollen sacs in a single row. Seed cones pineapple-like, 8–10 cm long, 7–8 cm thick, yellowish green at maturity with dark green, free bract tips. Seed scales 3–3.5 cm long, 2.5–3 cm wide, including the fragile papery wings slightly wider than the central seed-bearing portion but often torn, with drooping, triangular, free tip 4.5–6 mm long. Seeds lozenge-shaped, about

1.5 cm long, 9 mm wide. Cotyledons appearing above ground during germination. Central eastern coast of New Caledonia with an outlier at the northwestern tip. Very sparse emergents above coastal thickets on serpentine cliffs (hence the scientific name, Latin for "of the cliffs"); 0–200(–500) m. Zone 10?

Cliff araucaria is a rare species confined to immediate coastal bluffs and vicinity, much like Cook pine (*Araucaria columnaris*), lush araucaria (*A. luxurians*), and Port Boise araucaria (*A. nemorosa*). Among these coastal species with a general similarity in variably columnar growth habit, it overlaps in distribution only with lush araucaria. The two species are easily distinguished by two readily observable characteristics. The branchlets of cliff araucaria are arranged in a pair of combs held in a broad V above the supporting branch while those of lush araucaria spiral around the branches in a shaving-brush-like tuft. As well, the leaves of cliff araucaria are fairly uniform in size along the branchlet while those of lush araucaria vary conspicuously. Cliff araucaria also has more branches in each tier (five to seven or more) than is typical for lush araucaria (four), and there can be differences in growth habit, but do not count on either of these features to distinguish them. In details of foliage and cones, cliff araucaria more closely resembles Bernier columnar araucaria (*A. bernieri*), a rain-forest giant found farther inland for the most part, with smaller leaves and larger pollen and seed cones on average.

Araucaria subulata Vieillard

VIEILLARD COLUMNAR ARAUCARIA

Araucaria sect. *Eutacta*

Tree to 50 m tall, with trunk to 0.5(–1) m in diameter. Bark light gray to grayish white, thin, peeling in narrow horizontal strips and marked by the buildup of numerous dark gray corky lenticels aligned in rings. Crown conical when young, becoming narrowly columnar with a flat top by frequent shedding and replacement of numerous, closely space tiers of five to seven short, slender horizontal branches, extending to about 2 m before falling and bearing paired combs of branchlets rising on either side and densely clothed with foliage. Branchlets slender, 5–10 mm thick, including the leaves, which hide them, remaining green 2–4 years while elongating to 30–50 cm before being shed intact. Juvenile leaves claw-like, strongly keeled, densely overlapping, 4–5 mm long, about 1 mm wide. Adult leaves between claw- and scalelike, spearhead-shaped, keeled, standing out from the twig at a forward angle and curved forward and inward, bright to dark green, 4–6 mm long, 1.5–2.5 mm wide. Stomates ranging from in line with to at right angles to the long axis of the leaf, in irregular, discontinuous lines making up bands on the inner (upper) face and irregular patches at the base of the outer face, with irregular extensions upward. Pollen cones 5–10 cm long, 12–13 mm thick, each pollen scale with 8–10 pollen sacs in a single row. Seed cones a little longer than spherical, (7.5–)10–12 cm long, (6–)7–10 cm wide, bright green with yellowish green free bract tips. Seed scales about 3 cm long, 3–3.5 cm wide, including the papery wings about as wide as the central seed-bearing portion, with an upwardly angled, narrowly triangular free tip 6–8 mm long. Cotyledons appearing above ground during germination. Southern half of New Caledonia, in the mountains of the interior. Scattered as an emergent above rain forests, usually on serpentine-derived soils; (300–)500–900(–1,900) m. Zone 9. Synonym: *A. balansae* Ad. Brongniart & Gris.

Vieillard columnar araucaria resembles Bernier columnar araucaria (*Araucaria bernieri*) and Cook pine (*A. columnaris*) in its exaggeratedly columnar, maypolelike growth habit. These three species replace each other at generally progressively lower elevations from the cloudy montane haunts of Vieillard columnar araucaria to the shoreline stands of Cook pine. Although Vieillard columnar araucaria has a fairly broad distribution in southern New Caledonia, individual populations are small and widely scattered so the species is fairly rare. It is threatened by both mining development on its mineral-containing, serpentine substrates and by timber harvesting since the wood is of excellent quality and many stands are accessible. It has been recorded in error from Mount Ignambi in northern New Caledonia on the basis of a single specimen of doubtful origin, but repeated searches have not subsequently found it there. The species name, Latin for "awl-like," may refer to the juvenile leaves or to the free tips of the seed cone bracts, neither of which is particularly distinctive in the genus but do differ from most of the species that had been described by 1862, the year that this one was published.

Athrotaxis D. Don

TASMANIAN CEDAR

Cupressaceae

Small to moderately large evergreen trees with thin, shallowly furrowed, fibrous bark peeling in long, thin strips on a straight or contorted trunk. Densely branched with stiff, ascending branches. Crown roundly conical at first, remaining so (*Athrotaxis cupressoides*) or becoming broadly dome- to ball-shaped with age (*A. selaginoides*) and remaining dense where not broken up in the vicissitudes of time. Branchlets cylindrical, upright, branching from all sides or roughly two-ranked. Resting buds unspecialized, consisting solely of arrested immature foliage leaves. Leaves in a dense spiral, forwardly directed, overlapping, scalelike to broadly awl-shaped or clawlike, with a more or less well developed keel along the lower midline, tight against the twig or spreading.

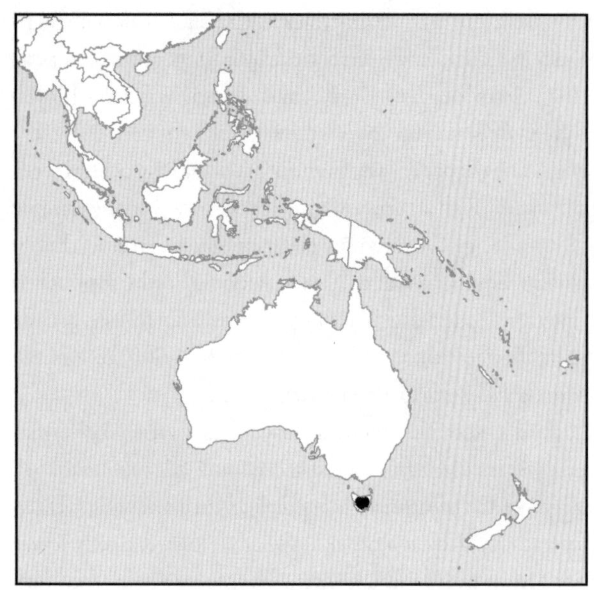
Distribution of *Athrotaxis* in Tasmania, scale 1 : 119 million

Plants monoecious. Pollen cones numerous, single at the ends of ordinary branchlets, with 25–35 tightly spirally arranged pollen scales, each scale with two pollen sacs. Pollen grains small (25–30 μm in diameter) spherical, minutely bumpy and with a small germination papilla. Seed cones numerous, single at the ends of ordinary branchlets, maturing in a single season, with 15–25 spirally arranged seed scales. Bract intimately fused with and more massive than the seed-bearing portion, each scale with three to six inverted seeds. Seeds small, oval, with two wings derived from the seed coat extending the whole length of and as wide as or slightly wider than the body and giving an overall heart-shaped outline. Cotyledons two, each with one vein. Chromosome base number $x = 11$.

Wood soft and weak, fairly light, with yellowish sapwood sharply contrasting with pink to reddish brown heartwood. Grain very evenly and moderately fine, with strongly delimited growth rings marked by a gradual transition to flatter, thick-walled latewood tracheids. Resin canals absent but with sparse to moderate scattered individual resin-filled parenchyma cells, especially within or near the latewood.

Stomates irregularly oriented, confined to small patches at the base of the leaf underside and collected in two broad bands extending the length of the upper side. Each stomate sunken beneath the (four or) five or six (or seven) subsidiary cells, which are often shared between adjacent stomates and topped by a low Florin ring. Leaf cross section with a central midvein flanked by transfusion tissue extending out to the sides. Resin canal prominent, following the midvein beneath and sometimes flanked by additional, discontinuous canals or cavities. Photosynthetic tissue with a thin palisade layer on the lower (outer) surface and a mass of spongy mesophyll, both surrounded by a thin, single-layered hypodermis (except thicker at the leaf corners and absent beneath the stomates) directly beneath the epidermis.

Two species in Tasmania, Australia. References: Cullen and Kirkpatrick 1988, Enright and Hill 1995, Farjon 2005b; Hill et al. 1993.

Neither of the living species is of particular economic importance, but both are greatly reduced in range within their nonagricultural montane habitats because they do not regenerate well after the fires that became more prevalent with European colonization since the 1800s. Both species, however, can produce new plants from root sprouts, an uncommon trait among conifers, and both have the potential for a long life span and are therefore useful in dendrochronological studies. While both species are cultivated to a limited extent in botanical gardens in wet, cool, temperate regions, they are not in general cultivation, and no cultivar selection has taken place. The name *Athrotaxis,* Greek for "crowded arrangement," refers to the organs of the seed cones, but these are no more crowded than those of other conifer genera, including *Cryptomeria,* which David Don described at the same time as he described this genus.

Athrotaxis is the only living genus of southern hemisphere Cupressaceae with a spiral adult leaf arrangement. Based on DNA sequences it appears to be most closely related to the northern hemisphere *Taiwania.* These two genera fall between the basal *Cunninghamia* and all other genera of the family in molecular phylogenies. Locally common hybrids between the two species occur where their ranges overlap ecologically. These used to be treated as a third, intermediate species, but DNA studies confirm their hybrid origin.

Although restricted to Tasmania today, Tertiary fossils are also found in mainland eastern Australia, New Zealand, and perhaps in South America, providing a Tertiary distribution similar to that of *Fitzroya* or *Libocedrus.* The oldest fossils of the genus date from the Cretaceous. There are also numerous fossils in the northern hemisphere that are similar to *Athrotaxis,* particularly in the Cretaceous. These species belong to related extinct genera and link *Athrotaxis* to a *Cunninghamia*-like ancestor common to all extant Cupressaceae.

Identification Guide to *Athrotaxis*

The two species of *Athrotaxis* and their hybrid may be distinguished by the form of the leaves, their length, the diameter of the seed cones, and the shape of the seed scale tip (exclusive of the bract tip):

Athrotaxis cupressoides, leaves scalelike, 2–4 mm long, seed cones 8–12 mm, seed scale tip rounded

A. ×laxifolia, leaves scalelike, 4–6 mm long, seed cones 10–15 mm, seed scale tip pointed

A. selaginoides, leaves clawlike, 6–12 mm long, seed cones 12–18 mm, seed scale tip pointed

Athrotaxis cupressoides D. Don

PENCIL PINE

Tree to 15(–30) m tall, with trunk to 1(–1.5) m in diameter in sheltered habitats, but more commonly less than 5 m and often dwarfed near the alpine timberline. Bark light brown, scaly on the branches. Crown conical, dense, extending to the ground except in larger individuals. Forming clonal patches by root suckers. Branchlets completely hidden by the leaves. Adult scale leaves tight to the twig, dark green, the margin translucent, 2–4 mm long, with a blunt or slightly pointed tip. Pollen cones 4–6 mm long. Seed cones nearly spherical, 8–12(–15) mm in diameter, the scales with a distinct rounded tip beyond the pointed tip of the bract portion. Seeds about 3 mm long. Central and southern Tasmania. Forming pure stands or mixed with other trees in montane rain forests and subalpine scrublands; (600–)900–1,250 (–1,360) m. Zone 8.

Pencil pine occurs under harsher conditions than does King Billy pine (*Athrotaxis selaginoides*), at higher elevations and in more exposed or waterlogged sites. It is the more frost tolerant of the two species and thus more prevalent in subalpine vegetation. Where the two occur together, often by lakes and streamsides, they may hybridize to form *A. ×laxifolia.* Pencil pine is not shade tolerant and requires full sun for germination. However, it has no fire tolerance and is slow to reinvade large burns. Once established, individual trees may live to 1,300 years, but the total longevity of clones is unknown. The species name (Latin for "like *Cupressus*") reflects the general similarity of the foliage to that of cypress trees, but only in the form of the leaves, because in pencil pine they are spirally arranged instead of in opposite pairs, as they are in all species of *Cupressus.*

Athrotaxis ×laxifolia W. J. Hooker

SUMMIT CEDAR

A hybrid between pencil pine (*Athrotaxis cupressoides*) and King Billy pine (*A. selaginoides*), growing with them and intermediate in foliage characteristics, bearing scale leaves 4–6 mm long with pointed tips that lie loosely against the twig. Seed cones nearly spherical, 10–15 mm in diameter, the seed scale tip scarcely evident beyond the point of the bract portion. Central and southern Tasmania. Growing with its parents; 900–1,200 m. Zone 8.

Athrotaxis ×laxifolia was originally described as a rare third species of the genus, intermediate between the other two. DNA fingerprints confirm its long-suspected hybridity. It is fully fertile, and its offspring are highly varied, spanning the range of leaf and cone form from one parent to the other.

Young trees of summit cedar (*Athrotaxis ×laxifolia*) on the left and King Billy pine (*A. selaginoides*) on the right in nature, with Hungarian botanist Zsolt Debreczy.

Athrotaxis selaginoides D. Don

KING BILLY PINE, KING WILLIAM PINE

Tree to 15(–40) m tall, with trunk to 2.2(–3) m in diameter. Bark reddish brown weathering gray, soft and corky on branches. Crown deeply dome-shaped, dense, above a long, bare trunk. Forming clonal patches by root suckering. Adult leaves 6–8 (–12) mm long, broadly clawlike, dark green with lighter stomatal bands along the outside, the margin dark, standing out from the twig at the base but curved forward and overlapping one another, the tip pointed and tucked in toward the twig. Pollen cones 6–8 mm long. Seed cones nearly spherical, 12–18 mm in diameter, the seed-bearing portion very inconspicuous beyond the pointed tip of the bract portion. Seeds about 2 mm long. Central and southern Tasmania. Forming pure stands or mixed with myrtle beech (*Nothofagus cunninghamii*) and celery top pine (*Phyllocladus aspleniifolius*) in montane rain forests on infertile soils; (20–) 750–1,200(–1,300) m. Zone 9.

King Billy pine is less frost hardy than pencil pine (*Athrotaxis cupressoides*), but it is more shade tolerant, so it often occurs in denser forests at lower elevations than its relative. They often occur in proximity, however, crossing sparingly to produce *A.* ×*laxifolia.* Like pencil pine, King Billy pine is also capable of root suckering but is intolerant of fires. It is also very slow growing, as one would expect from its cool, nutrient-poor habitat, but it is very long-lived, often exceeding 800 years to a maximum of about 1,300. The wood is prized for cabinetry and boatbuilding, but supplies are declining due to exploitation and to fires. At least a third of the stands of King Billy pine in Tasmania were completely destroyed by fires, most accidentally started by campers, loggers, or other visitors to the mountain forests. The species name (Latin for "like *Selaginella*") reflects a general resemblance of the foliage

Old mature King Billy pine (*Athrotaxis selaginoides*) in nature.

Natural stand of King Billy pine (*Athrotaxis selaginoides*) with trees of all ages.

Open mature seed cone of King Billy pine (*Athrotaxis selaginoides*) after seed dispersal; scale, 1 cm.

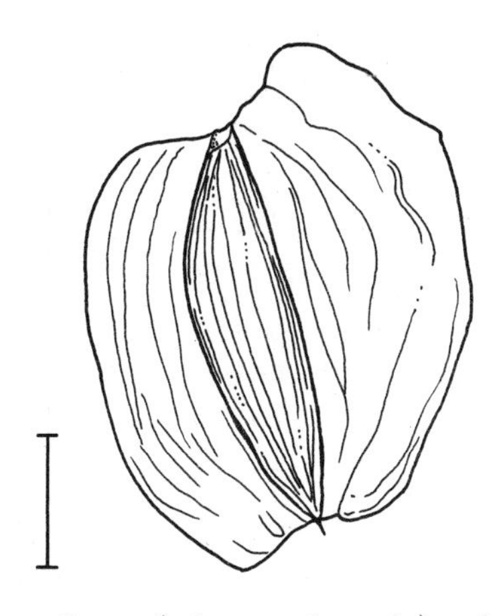

Seed of King Billy pine (*Athrotaxis selaginoides*); scale, 1 mm.

to that of many tropical spike mosses, which, like the related club mosses (*Lycopodium*), are vascular plants, not true mosses.

Austrocedrus Florin & Boutelje

CHILEAN INCENSE CEDAR

Cupressaceae

Evergreen trees and shrubs with a single or multiple trunks bearing fibrous, deeply furrowed bark peeling in strips. Densely branched from the base when young with short horizontal or upswept branches to form a dense, conical crown that broadens and flattens with age above a clear trunk two-thirds the height of the tree. Branchlets arising only from lateral leaves, in pairs or alternately, to form flattened, fernlike sprays with distinct upper and lower sides, mostly hidden by the leaves and attached leaf base. Without definite winter buds. Seedling leaves in alternating quartets, needlelike, standing out from and well spaced on the stem, soon giving way to juvenile lateral sprays. Juvenile and adult leaves in alternating pairs, scalelike, densely clothing the branchlets, differentiated into facial and lateral pairs; the lateral pairs much larger than the facial ones, like large saw teeth in juveniles but the free tip much reduced in adults. Successive facial pairs widely separated by the touching bases of the lateral pairs, some with a round resin gland. Margins of lateral leaves thickened and the whole area inside these margins on the lower side covered with a whitish, waxy stomatal area.

Plants monoecious or dioecious. Pollen cones numerous, single at the ends of lateral branchlets, oblong, with four or five alternating pairs of pollen scales, each scale with three or four round pollen sacs. Pollen grains small (20–30 µm in diameter), spherical, minutely bumpy and with a very faint, small germination pore. Seed cones single at the ends of short branchlets, maturing in a single season, oblong, with two alternating pairs of seed scales, only the inner, larger, paddle-shaped pair fertile. Each scale apparently with intimate fusion of the bract to the seed part, the bract extending almost the whole length of the scale and coming to a small prickle just below its tip, the fertile pair each with two seeds. Seeds oval, with two unequal wings derived from the seed coat, the inner, larger wing expanding across the seed scale, the outer wing a mere fringe. Cotyledons two, each with one vein. Chromosome base number $x = 11$.

Wood fragrant, soft, light, decay resistant, notably knotty, the yellowish white sapwood moderately distinct from the pale yellowish brown heartwood. Grain even and somewhat coarse, with well-defined growth rings marked by a gradual transition to much smaller and thicker-walled latewood tracheids. Resin canals absent but with widely scattered individual resinous parenchyma cells sometimes concentrated within the latewood.

Stomates in more or less waxy patches on both sides of the lateral leaves (but more prominent on the side facing down). Individual stomates tucked just under four to six subsidiary cells, which are frequently shared by adjacent stomates both within and between rows, and which are topped by a lobed but continuous Florin ring of varying height but do not bear additional papillae. Leaf cross section with a single-stranded midvein accompanied below by a single large resin canal with transfusion tissue in between. Photosynthetic tissue accompanied by scattered sclereids, all loose and spongy without a well-organized palisade layer beneath the epidermis and adjacent discontinuous hypodermis.

One species in southern South America. References: Biloni 1990, Dodd and Rafii 1995, Farjon 2005b, Florin and Boutelje 1954, Hill and Carpenter 1989, McIver and Aulenback 1994, Rodríguez R. et al. 1983, Veblen et al. 1995.

Austrocedrus is one of the southern hemisphere incense cedars once included within *Libocedrus* (hence the scientific name, Latin for "southern cedar"). It differs from species of *Libocedrus* and *Papuacedrus* in its wood, the thickened margin of the leaves and their arrangement of stomates, and in having the bract tip at the tip of the seed scale rather than in the middle (*Libocedrus*) or toward the base (*Papuacedrus*). DNA sequences support its relationship to, but distinction from, the other incense cedars. Although *A. chilensis* is fairly commonly cultivated as a specimen tree in moist, moderate climates, primarily in larger landscapes like parks and botanical gardens, there has been no cultivar selection. Fossils have been found in Miocene deposits of Argentina, where it is found today, but also earlier, in Oligocene sediments in Tasmania across the South Pacific from its present range.

Old mature Chilean cedar (*Austrocedrus chilensis*) in nature.

Austrocedrus chilensis (D. Don) Pichi Sermolli & Bizzarri

CHILEAN CEDAR, CIPRÉS [DE LA CORDILLERA] (SPANISH)

Plate 16

Tree to 20(–37) m tall, with trunk to 1.5(–2.5) in diameter or multistemmed tall shrub in arid steppes. Bark grayish tan. Branchlet sprays fanlike. Adult lateral leaves 2–4 mm long, facial leaves 0.5–1 mm long. Pollen cones 4–5 mm long. Seed cones 1–2 cm long, the outer pair of scales less than half as long as the fertile pair. Seeds 3–5 mm long, the larger wing extending the same distance beyond the tip of the seed. Chile and adjacent Argentina, from 32°S to 44°S. Varied forests and woodlands, from Mediterranean climate woodland near Santiago (Chile), to moist montane forests at moderate elevations on either side of the Andes, to sparse dry woodlands bordering Patagonian steppe in Argentina; 250–1,800 m. Zone 8. Synonym: *Libocedrus chilensis* (D. Don) Endlicher.

Chilean cedar thrives under an unusually broad range of moisture conditions. It would probably extend into wetter situations were it not excluded from them by the species of southern beech (*Nothofagus*). It is largest in moist montane forests and usually shrubby at the borders of the Patagonian steppe. Despite the differences between these two environments, there is not much genetic difference between the cedars growing in them. On the other hand, the isolated northern groves in the Mediterranean woodlands near Santiago are genetically distinct from those of the continuous range farther south. The trees can reach 1,000 years old but rarely exceed 500 years, often succumbing to fires in the drier parts of their range.

Underside of branchlet spray of Chilean cedar (*Austrocedrus chilensis*), ×1.

Open mature seed cone of Chilean cedar (*Austrocedrus chilensis*) in side view after seed dispersal; scale, 1 cm.

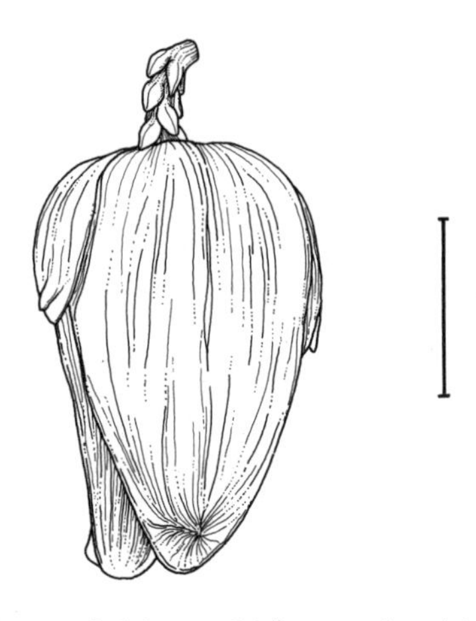

Open mature seed cone of Chilean cedar (*Austrocedrus chilensis*) in face view after seed dispersal; scale, 5 mm.

Distribution of *Austrocedrus chilensis* in Chile and adjacent Argentina, scale 1 : 35 million

Austrotaxus R. Compton

NEW CALEDONIAN YEW

Taxaceae

Evergreen trees or shrubs. Trunk cylindrical or irregular, often branched from near the base. Bark fibrous, smooth at first, flaking in scales or short strips and then becoming shallowly furrowed between flat ridges peeling in strips. Crown dense, cylindrical to dome-shaped, with numerous thick, crooked branches bearing alternate or paired branchlets. Branchlets all elongate, without distinction into long and short shoots, hairless, remaining green for the first year or two, smooth and without apparent attached leaf bases. Resting buds small, distinct but somewhat loosely constructed of hard, triangular bud scales that fall with shoot expansion leaving conspicuous rings of scars. Leaves spirally attached and crowded toward the end of the growth increment. Each leaf needlelike, sword-shaped, straight to moderately sickle-shaped, flattened top to bottom.

Plants dioecious. Pollen cones attached singly in the axils of about 15 spirally arranged bracts along crowded, short reproductive spikes, the spikes in the axils of fallen bud scales ringing the base of new flushes of foliage. Each pollen cone united with its bract, much reduced, consisting of four or five crowded pollen scales with little or no free tip and one or two pollen sacs, the whole individual pollen cone giving the appearance of a single pollen scale with about seven pollen sacs. Pollen grains small (25–30 μm in diameter), nearly spherical, without any apparent germination aperture, minutely but conspicuously warty, otherwise featureless. Seed cones single or two or three around the base of a new flush of foliage, each at the tip of a very short, entirely scaly stalk in the axil of a fallen bud scale. Each seed cone without any trace of seed scales, consisting of a single seed and aril seated in a cup of bracts at the tip of the stalk. Seeds plump, closely covered except at the tip by but not united with the aril, only the aril fleshy. Outer seed coat hard, opening straight opposite and pointing away from the stalk, maturing in a single season and falling with the withering of the aril if the cones are not earlier removed by fruit-eating birds or mammals. Cotyledons two, each with one vein. Chromosome base number unknown.

Wood moderately hard and heavy, light brown. Grain fine and very even, with indistinct growth rings marked by a gradual transition to a few rows of smaller-celled but not darker latewood. Resin canals absent but with a few, scattered individual resin parenchyma cells of the same size as the tracheids (wood cells), which lack spiral thickening.

Without stomates above and with a broad band of stomates on either side of the midrib beneath covering most of the lower surface, each band consisting of many interrupted, and occasionally irregular lines of stomates. Each stomate tucked beneath and largely hidden by the four to six subsidiary cells, which are often shared by adjacent stomates in a line and topped by a steep, incomplete Florin ring. Midvein single, prominent, grooved above and raised beneath, without resin canals, flanked by cylinders of transfusion tissue, and with broad wings of accessory transfusion tissue extending out to the margins. Photosynthetic tissue forming a single palisade layer covering the upper side of the leaves directly beneath the epidermis without an intervening hypodermis. One species in New Caledonia. References: Cope 1998, de Laubenfels 1972; Jaffré 1995.

Austrotaxus is the only genus of Taxaceae native in the southern hemisphere (hence the scientific name, Latin for "southern yew"). It is most similar to catkin yew (*Amentotaxus*) in overall appearance but is actually more closely related to yew (*Taxus*) and white-cup yew (*Pseudotaxus*). It belongs with the latter genera in one of three lineages in the family, while catkin yew belongs in another one. All three genera in this group lack resin canals in the leaves, a very unusual feature in conifers, but *Austrotaxus* differs from the other two genera in many features, including its larger leaves and seeds, with a thinner, tight-fitting aril, and its spiky pollen cones. While it seems to share these spikes with *Amentotaxus,* they are also structurally rather similar to the more condensed compound pollen cones of *Pseudotaxus.* Alone among genera of Taxaceae, *Austrotaxus* lacks spiral thickenings in the walls of its wood cells (tracheids). *Austrotaxus spicata* is not in general cultivation, although it might be in a very few botanical gardens, and there has been no cultivar selection. There is no known fossil record for the genus, although if there were pollen grains in the published record, they would be very difficult to distinguish from those of other genera.

Austrotaxus spicata R. Compton

NEW CALEDONIAN YEW

Shrub, or tree to 15(–25) m tall, with trunk to 0.3(–0.7) m in diameter. Bark reddish brown, richer and darker where scales flake away, weathering grayish brown. Leaves 4.5–11(–18) cm long, (2.5–)4–6(–8) mm wide, shiny dark green above, paler beneath, widest near the middle, tapering very gradually to the sharply and usually narrowly triangular tip and to the narrowly wedge-shaped base on a petiole (2–)4–8(–10) mm long, the edges turned down. Pollen cone spikes (7–)10–15(–18) mm long, 2–3 mm in diameter. Seeds with a deep purple aril, 10–12 mm long, 6–8 mm in diameter, on a stalk 2–4 mm long. Northern two-thirds of New Caledonia, from Mount Ignambi near Pouébo south to Mount Nakada near Canala. Scattered in the understory of montane rain forests or in the canopy of lower-stature cloud forests on exposed ridge tops on various substrates but generally not on serpentine-derived soils; 500–1,000(–1,350) m. Zone 10.

Vegetatively, New Caledonian yew resembles species of the unrelated yellowwood genus (*Podocarpus*) more than it does its own closest relatives, the northern hemisphere yews (*Taxus* and

Distribution of *Austrotaxus spicata* in northern and central New Caledonia, scale 1 : 119 million

Male twig of New Caledonian yew (*Austrotaxus spicata*) with both sides of leaves and a cluster of pollen cones, ×= 0.75.

Pseudotaxus). The resemblance even extends to the microscopic structure of the wood, which lacks the spiral thickening of the cell walls of the tracheids found in all other Taxaceae, prompting at least one wood anatomist to consider it as a member of the Podocarpaceae. One look at the pollen cones and the seeds with their aril, however, is enough to show its true affinities and distinguish it from neighboring New Caledonian yellowwoods. It is one of only four species of conifer in New Caledonia that do not grow on ultramafic soils derived from serpentine substrates. It is fairly common and is not much subject to direct exploitation. Furthermore, its absence from mining districts, which are restricted to the mineral-rich serpentine areas, reduce the effects of habitat destruction, which are so devastating for many of New Caledonia's conifers. The species is not in general cultivation but would make a picturesque specimen of moderate growth in moist to wet tropical regions. The species name, from Latin, refers to the pollen spikes.

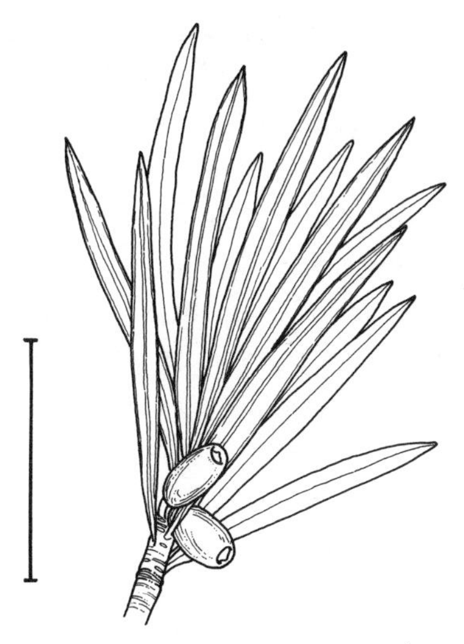

Female twig with seed cones of New Caledonian yew (*Austrotaxus spicata*); scale, 5 cm.

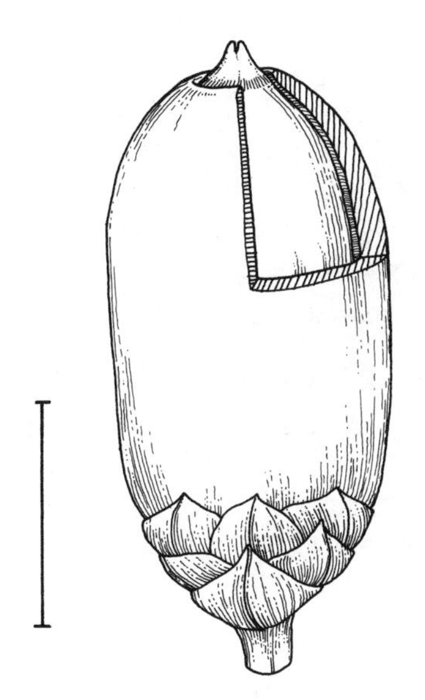

Seed cone of New Caledonian yew (*Austrotaxus spicata*) with aril partially cut away to show seed; scale, 5 mm.

Callitris Ventenat

CYPRESS PINE

Cupressaceae

Evergreen trees and shrubs. Bark fibrous, vertically furrowed, persisting and thickening with age. Densely, often irregularly branched from the base, with horizontal or gently ascending slender branches. Branchlets cylindrical or three-angled, branching three-dimensionally but with just one branch at a node. Without definite winter buds. Seedling and juvenile leaves in alternating quartets, crowded, sword-shaped, the free blades much longer than the attached leaf bases that run down onto the twig, standing straight out from the twig or at angles down to about 60°. Juvenile leaves usually relatively short-lived but recurring regularly on adult trees of some species, often giving way abruptly to adult foliage within a few successive attachments. Adult leaves in alternating trios (hence the scientific name, Latin for "beautiful threes"), scalelike, keeled or rounded, the free tips triangular, pressed forward against the branchlets, and tiny in comparison to the usually elongate attached leaf bases, giving the branchlets a jointed appearance.

Plants monoecious. Pollen cones numerous, single or clustered at or near the tips of the branchlets, oblong, with 3–10 alternating trios (rarely quartets) of pollen scales, each scale with two to six round pollen sacs. Pollen grains small (20–30 µm in diameter), spherical, minutely but conspicuously bumpy and with a tiny, obscure germination pore. Seed cones numerous, single or in clusters at the tips of short, stout branchlets, maturing in two seasons, then often persisting and remaining closed after maturity. Cones spherical to oval, with two alternating trios (or rarely quartets) of woody, triangular seed scales attached together at the base around a central, pyramidal or three-lobed column. Seed scales closing side to side, the inner trio usually longer than the outer one and touching at the tip of the cone. The scales with a variably expressed point below their tips, each outer scale with (none to) three to nine seeds in (none or) one to three rows, each inner scale with (2–)6–12 seeds in (one or) two to four rows, or fewer by abortion. Seeds oval and somewhat flattened, with (one or) two (or three) asymmetrical wings derived from the seed coat, the wings narrower to wider than the seed body and extending its whole length or beyond. Cotyledons two, each with one vein. Chromosome base number $x = 11$.

Wood unusually dense and hard, usually without pronounced heartwood colors, distinct growth rings, or resin canals, but often with distinctive, thin, microscopic bands across the bordered pits of the rounded tracheids.

Stomates lining the grooves between the attached leaf bases and sometimes also extending out onto the inner faces of the free tips at their bases. Each stomate sunken in a narrow pit beneath the four to six (or seven) subsidiary cells, which may be shared between adjacent stomates within and between rows and are topped by a high, steep, interrupted Florin ring. Leaf cross section with a single midvein hugging the twig side of the attached leaf bases and scarcely extending into the free tip. Midvein accompanied toward the outside by a single large resin canal and flanked by varying quantities of transfusion tissue. Photosynthetic tissue forming a prominent dense palisade inside the epidermis and hypodermis, giving way inward to a spongy mesophyll that includes patches of sclereids, often in inverse proportion to the amount of transfusion tissue. Seventeen species in Australia and New Caledonia. References: Baker and Smith 1910, de Laubenfels 1972, Farjon 2005b, Garden 1957, Thompson and Johnson 1986.

The strong, durable, resistant wood of the larger species of *Callitris* is highly prized for timber. A few species of the genus are in general cultivation in Australia, but elsewhere they are nearly restricted to botanical gardens and tree collections, primarily in Mediterranean climate regions. Cultivar selection has been minimal but a few foliage color variants are grown.

Callitris is the largest genus of Cupressaceae in the southern hemisphere. It is the characteristic conifer of dry forests and woodlands throughout Australia (though joined by its close relative *Actinostrobus* in the southwest), including the arid interior, where it is the only conifer. With their seed cones often remaining closed until scorched by fires, the Australian species are roughly equivalent ecologically to species of *Cupressus* in the northern hemisphere. The ecology in New Caledonia is rather different, the two species being found in both rain forests and dense shrublands, just like their close relative *Neocallitropsis.*

Callitris is easily distinguished from *Actinostrobus* by its lack of conspicuous rows of bracts at the base of the seed cone, the inequality of the two trios of seed scales, and the shorter free tips of the adult leaves. *Neocallitropsis* never develops scalelike leaves and has narrower seed cones scales always in quartets. Natural hybridization occurs in only a few of the possible combinations of overlapping species. The fossil record of *Callitris* is quite sparse, as might be expected for a dryland genus, but it apparently extends back to the mid-Cretaceous.

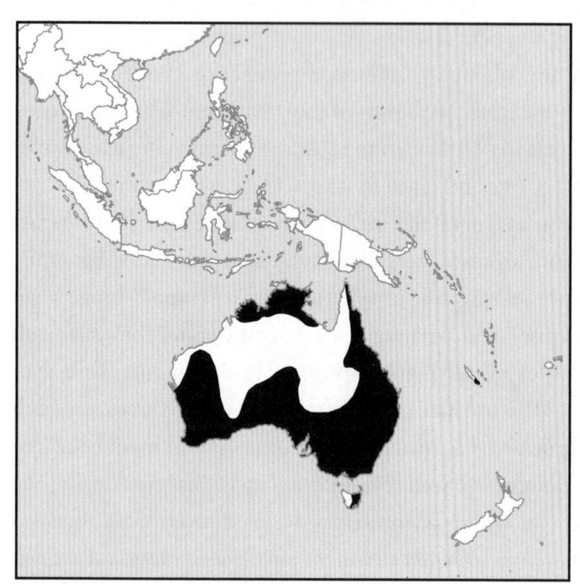

Distribution of *Callitris* in Australia and southern New Caledonia, scale 1 : 119 million

Identification Guide to *Callitris*

For ease of identification, the 17 species of *Callitris* may be divided into three informal groups based on whether the branchlets with their attached leaf bases are rounded or obtusely or sharply keeled, whether the seeds are numerous or just two on the seed scales, and whether the central column of the seed cones is simple or three-lobed:

Group A, branchlets rounded, seeds numerous, seed cone column simple

Group B, branchlets obtusely keeled, seeds numerous, seed cone column three-lobed

Group C, branchlets sharply keeled, seeds two, seed cone column simple or three-lobed

Group A. These six species may be distinguished by the predominant length of the adult leaves, the diameter of most unopened mature seed cones, whether the foliage is green or waxy, and the density of warts on the seed scales:

Callitris intratropica, leaves 2.5 mm or less, closed seed cones less than 18 mm, foliage dark green, warts absent

C. columellaris, leaves 2.5 mm or less, closed seed cones 18 mm or more, foliage dark green, warts absent

C. glaucophylla, leaves 2.5 mm or less, closed seed cones 18 mm or more, foliage waxy, warts absent

C. canescens, leaves more than 2.5 mm, closed seed cones less than 18 mm, foliage light green, warts absent

C. preissii, leaves more than 2.5 mm, closed seed cones 18 mm or more, foliage usually dark green, warts sparse

C. verrucosa, leaves more than 2.5 mm, closed seed cones 18 mm or more, foliage dark green or waxy, warts dense

Group B. These five species may be distinguished by the adult leaf length, the diameter of the unopened mature seed cones, and the size of the point at the tip of the seed scales:

Callitris rhomboidea, leaves less than 4 mm, unopened seed cones 20 mm or less, seed scale point stout

C. endlicheri, leaves less than 4 mm, closed seed cones 20 mm or less, seed scale point small

C. monticola, leaves less than 4 mm, closed seed cones up to 25 mm, seed scale point small

C. oblonga, leaves 4 mm or more, closed seed cones 20 mm or less, seed scale point small

C. muelleri, leaves 4 mm or more, closed seed cones more than 20 mm, seed scale point small

Group C. These six species may be distinguished by the adult leaf length, the diameter of the seed cone stalk, and the diameter of the unopened mature seed cone:

Callitris neocaledonica, leaves down to 2 mm, seed cone stalk 4 mm or less, closed seed cone less than 10 mm

C. baileyi, leaves down to 2 mm, seed cone stalk 4 mm or less, closed seed cone 10–15 mm

C. macleayana, leaves down to 2 mm, seed cone stalk more than 4 mm, closed seed cone more than 15 mm

C. sulcata, leaves 3 mm or more, seed cone stalk 4 mm or less, closed seed cone 10 mm or less

C. drummondii, leaves 3 mm or more, seed cone stalk more than 4 mm, closed seed cone 10–15 mm

C. roei, leaves 3 mm or more, seed cone stalk more than 4 mm, closed seed cone more than 15 mm

Callitris baileyi C. White

BLACKBUTT CYPRESS PINE

Tree to 15(–20) m tall, often shorter, with trunk to 0.5 m in diameter. Bark tan, weathering gray, thin, flaking in small scales, building up into low ridges between shallow furrows. Crown narrow, upright. Branchlets triangular. Juvenile foliage not persistent in mature plants. Adult leaves, including bases, 2–5 mm long, usually just 2 mm on the smallest branchlets, sharply keeled. Pollen cones single or in clusters of two or three, 2–3 mm long, with three to five (to seven) trios of pollen scales, each scale with two or three pollen sacs. Seed cones single on slender stalks 2 mm or more in diameter, not persisting after maturity, more or less spherical, 10–13 mm in diameter, with a simple, weakly three-angled, stout central column about 3 mm long. Each scale with a small, sharp point below the tip and grooved below it to the base outside. Seeds two to four on each scale, light brown, the body 3–4 mm long, 1.5–2 mm wide, with two very unequal wings, the smaller a narrow fringe about as long as the seed body, the larger 5–9 mm long, 3–4 mm wide. Southeastern Queensland and northeastern New South Wales, Australia. Open *Eucalyptus* forest of drier ranges inland from the Pacific Coast; 100–500(–1,000) m. Zone 9.

Blackbutt cypress pine is one of the more local species in the Queensland–New South Wales border region where seven of Australia's 15 species of *Callitris* grow. It is named after Frederick Bailey (1827–1915), Colonial Botanist of Queensland but not the collector of the type specimen of this species, however many others he may have brought to light.

Callitris canescens (Parlatore) S. T. Blake

SCRUBBY CYPRESS PINE, MORRISON CYPRESS PINE

Shrub, or tree to 5(–6) m tall, with trunk to 0.3 m in diameter, usually dividing into several major limbs at or near the ground. Bark reddish brown, weathering gray, thin, flaking in small scales. Crown dense, widely spreading, puffy and irregular. Branchlets rounded. Juvenile foliage not persistent in mature plants. Adult leaves, including bases, to 4 mm long, much shorter on the branchlets, light green or waxy (canescent, hence the species name), not keeled, the outer face flat. Pollen cones single, 1.5–2 mm long, with three or four (to six) trios of pollen scales, each scale with three or four pollen sacs. Seed cones single or in clusters of two to four on thick stalks 2–7 mm in diameter, long persistent, nearly spherical, but slightly wider than long, 13–19 mm in diameter, with a short, thick central column 2–3 mm long. Scales smooth when closed, wrinkling somewhat when opened, without any obvious point near the tip. Seeds three to seven on each scale, black, the body 2.5–5 mm long, 1–2 mm wide, with two nearly equal wings as long as the body and (0–)1–3 mm wide, a third wing to 1 mm wide sometimes present. Southern quarter of Australia in Western Australia and South Australia. Open *Eucalyptus* scrubland, primarily on calcareous soils; (0–)100–300(–400) m. Zone 9. Synonym: *Callitris morrisonii* R. T. Baker.

Scrubby cypress pine has been widely confused with other species, including dwarf cypress pine (*Callitris drummondii*), Rottnest Island pine (*C. preissii*), and mallee pine (*C. verrucosa*).

These errors are partly responsible for reports of smooth rather than warty cones in the latter two species, although both do have smooth cones upon occasion, especially when hybridizing with white cypress pine (*C. glaucophylla*).

Callitris columellaris F. J. Mueller

COAST CYPRESS PINE, RICHMOND CYPRESS PINE

Tree to 18(–35) m tall, with trunk to 1.5 m in diameter or forking near the base. Bark reddish brown, weathering grayish brown, becoming strongly ridged and furrowed with maturity. Crown dense, narrow, with short horizontal or upturned branches. Branchlets rounded. Juvenile foliage not persistent in mature plants. Adult leaves, including bases, 1–4 mm long, rich green, not keeled. Pollen cones in clusters of (one to) three to five, 3–4(–5) mm long, with three to five trios of pollen scales, each with three or four pollen sacs. Seed cones single on slender stalks, not persisting after maturity, egg-shaped to almost spherical, 1.5–2 cm in diameter with a thick, single, angled central column (called a columella, hence the species name) 5–7 mm long. Scales mostly smooth but with a small prickle near the tip, separating to near the base. Seeds 4–10 on each scale, brown or black, the body 5–7 mm long, 2.5–3.5 mm wide, one transparent wing somewhat larger than the other, both longer than the seed body, 6–9 mm long, (0–)2–6 mm wide, sometimes with a third wing to 1 mm wide. Southeastern Queensland and northeastern New South Wales, Australia. Forming dense stands on sandy soils of the immediate coastal region; 0–50(–100) m. Zone 9. Synonym: *Callitris arenosa* A. Cunningham.

Mature coast cypress pine (*Callitris columellaris*) in nature.

With its dense, billowing, bright green, columnar crown, coast cypress pine may be the most ornamental species in the genus. It is closely related to white cypress pine (*Callitris glaucophylla*) and northern cypress pine (*C. intratropica*), with which it is often merged without formal recognition. At other times, the three were treated as varieties of a single species under the name *C. columellaris,* the oldest name. White cypress pine is ecologically distinct from the other two, in addition to its combination of larger cones with dark green foliage. Although the wood has an attractive figure and takes a good polish, it is not much exploited for timber. The oldest trees reach about 200 years.

Callitris drummondii (Parlatore) F. J. Mueller

DWARF CYPRESS PINE, DRUMMOND CYPRESS PINE

Shrub or sometimes a tree to 3 m tall, with one or more trunks to 0.2(–0.3) m in diameter. Bark reddish brown, weathering gray, flaking in scales and becoming ridged and furrowed at the base. Crown compact, rounded. Branchlets triangular. Juvenile foliage not persistent in mature plants. Adult leaves, including bases, (2–)3–4 mm long, bright green, sharply keeled. Pollen cones single or in pairs, 3–6 mm long, with four to seven (to nine) trios or pollen scales, each scale with three pollen sacs. Seed cones single or clustered, on stout stalks, long persistent, nearly spherical, but slightly wider than long, 13–15 mm in diameter, with a three-angled central column 3 mm tall and 3 mm broad at the base. Scales with a small or inconspicuous prickle, wrinkling with age, the inner trio each with two seeds. Seeds dark brown, the body 2–3 mm long, 1–2 mm wide, with two very unequal light brown wings, the smaller as long as the body and 1–1.5 mm wide, the larger twice as long and 2–3.5 mm wide. Southern coast of Western Australia. Coastal thickets; 0–100 m. Zone 9.

Dwarf cypress pine is a local species closely related to Roe cypress pine (*Callitris roei*) and replaces it along the coast of Western Australia in much the same way that coast cypress pine (*C. columellaris*) replaces white cypress pine (*C. glaucophylla*) along the northern coast of New South Wales. The species name honors James Drummond (ca. 1786–1863), a Scottish-born Australian who was Government Botanist and Superintendent of the Government Gardens in Perth, Western Australia, when he collected the type specimen.

Callitris endlicheri (Parlatore) F. M. Bailey
BLACK CYPRESS PINE, RED CYPRESS PINE
Tree to 15(–25) m tall, with trunk to 1 m in diameter. Bark dark brown, deeply furrowed, with narrow ridges. Crown conical, open, with horizontal or gently upraised branches. Branchlets three-winged. Juvenile leaves gradually giving way to adult foliage through transitional leaf forms. Adult leaves, including bases, 1–3(–4) mm long, sometimes with a bristly tip, yellow-green to dark green or occasionally somewhat waxy, with a prominent, rounded keel. Pollen cones mostly single, 1.5–3 mm long, with three to six trios of pollen scales, each scale with two to four pollen sacs. Seed cones single, or in small clusters, on slender stalks, persisting after maturity, roughly spherical, 15–20 mm in diameter, with a three-lobed central column, the lobes sometimes separate, spreading 2–8 mm across. Scales with a small but distinct triangular swelling near the tip, otherwise smooth, separating to near the base, Seeds five to nine on each scale, brown to black, sometimes waxy, the body 3–6 mm long, 2–3 mm wide, with two slightly unequal wings about as long as the body, (0–)2–5 mm wide. Southeastern Australia from southeastern Queensland through New South Wales to northeastern Victoria. Open woodlands and forests with *Eucalyptus* and *Acacia* on stony hills and ridges; (100–)250–750(–1,200) m. Zone 9. Synonym: *Callitris calcarata* (A. Cunningham ex Mirbel) F. J. Mueller.

Black cypress pine is the second most widespread species of the interior of eastern Australia after white cypress pine (*Callitris glaucophylla*) and often grows with the latter where their ranges overlap. These two species are not closely related and there is no evidence of hybridization between them. The durable wood has an attractive figure and is commercially important, particularly within its natural range. The species name commemorates Stephan Endlicher (1804–1849), an Austro-Hungarian botanist who explored southeastern Australia and described many of its plants.

Callitris glaucophylla Joy Thompson & L. A. Johnson
WHITE CYPRESS PINE
Tree to 20(–30) m tall, or stunted in the driest portions of interior Australia, with trunk to 1 m in diameter. Bark grayish brown, strongly ridged and furrowed. Crown conical, open, broadening with age, formed from bottom to top by drooping, horizontal, and rising branches,. Branchlets cylindrical, with three grooves. Juvenile foliage not persistent in mature plants. Adult leaves, including bases, 1–3 mm long, usually bluish gray from a waxy coating (hence the species name, Greek for "gray leaf"), not keeled. Pollen cones single or in clusters of two or three, 3–6 mm long, with four or five trios of pollen scales, each scale with three or four pollen sacs. Seed cones single or in small clusters on slender stalks 1–4 mm in diameter, not persisting after maturity, egg-shaped to almost spherical, 1.5–2.5 cm in diameter, with a simple, usually slender central column to 5(–7) mm long. Scales with an inconspicuous point below the tip, smooth when closed, wrinkling upon opening, opening almost to the base. Seeds 4–12 per scale, chestnut to dark brown, the body 4–7 mm long, 3–4 mm wide, with two roughly equal wings extending the length of the body or a little beyond, 2–5 mm wide, sometimes with a third wing to 2 mm wide. Central and southern Australia in all mainland states. Open woodlands and forests with *Eucalyptus* and *Acacia* on well-drained soils, primarily in the interior; (50–)100–500(–1,000) m. Zone 9. Synonyms: *Callitris columellaris* var. *campestris* Silba, *C. glauca* R. Brown ex R. T. Baker & H. G. Smith, *C. hugelii* (Carrière) Franco (misapplied).

White cypress pine is by far the most widely distributed species of *Callitris*, occurring discontinuously in the interior throughout the southern two-thirds of Australia. As the only conifer of most of this region, its attractive, fragrant, dense, termite-resistant wood gains added importance. It is sympatric with several other species throughout its extensive range but appears to hybridize only with the related Rottnest Island pine (*C. preissii*). It is nowhere sympatric with the two most closely related species, coast cypress pine (*C. columellaris*) and northern cypress pine (*C. intra-*

Mature seed cone of white cypress pine (*Callitris glaucophylla*) after shedding seeds; scale, 1 cm.

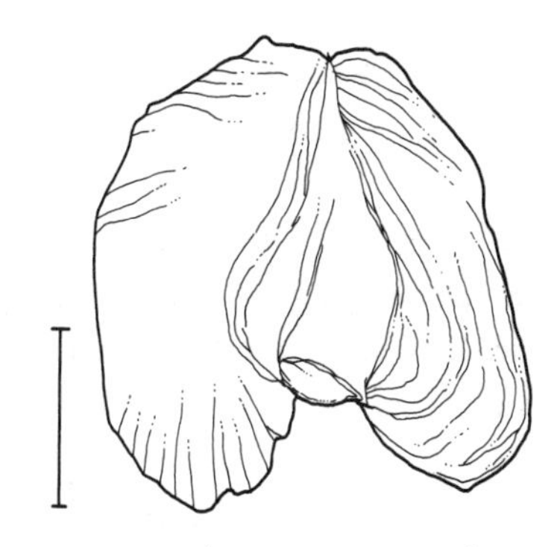

Seed of white cypress pine (*Callitris glaucophylla*); scale, 4 mm.

tropica). It is not always separated from these two species in floristic and taxonomic treatments. Outside of Australia, white cypress pine has become naturalized in central Florida.

Callitris intratropica R. T. Baker & H. G. Smith

NORTHERN CYPRESS PINE

Tree to 20 m tall, with trunk to 0.8 m in diameter, or shrubby on drier sites. Bark reddish brown, weathering gray, soon becoming deeply ridged and furrowed. Crown rounded, broadly spreading. Branchlets cylindrical, with three grooves. Juvenile foliage not persistent in mature plants. Adult leaves, including bases, mostly 2 mm long, gray green from a waxy coating, not keeled. Pollen cones mostly in clusters of two or three, 2–4 mm long, with three or four trios of pollen scales, each scale with three pollen sacs. Seed cones single on slender stalks, not persisting after maturity, nearly spherical, but wider than long, 12–18 mm in diameter, with a short, simple, three-angled central column 3–5 mm long. Scales with a small but distinct point below the tip, smooth, opening to near the base. Seeds three to seven on each scale, chestnut brown, the body 3–6 mm long, 2–3 mm wide, with two nearly equal wings with translucent tips, 1–4 mm wide. Northernmost Australia from northeastern Western Australia to northern Queensland. Dry open or closed woodlands, forests, and scrublands on a variety of soils, from sands to clays and loams; (0–)50–300(–500) m. Zone 10. Synonym: *Callitris columellaris* var. *intratropica* (R. T. Baker & H. G. Smith) Silba.

Northern cypress pine is the most northerly species of *Callitris,* occurring only north of 18°S (hence "within the tropics"), and it does not overlap with any other species. It is closely related to coast cypress pine (*C. columellaris*) and white cypress pine (*C. glaucophylla*) but differs from the former in its gray foliage and from the latter in its smaller cones. The oldest individuals are recorded as reaching 250 years.

Callitris macleayana (F. J. Mueller) F. J. Mueller

STRINGYBARK PINE, PORT MACQUARIE PINE, BRUSH CYPRESS PINE

Tree to 25(–45) m tall, with trunk to 0.6(–1.3) m in diameter, unbranched for much of its height. Bark reddish gray, furrowed, retaining stringy peels. Crown dense, upright, with short branches, or thinner in closed forests. Branchlets triangular. Juvenile foliage with leaves in quartets, the blades 5–8 mm long, gradually shortening through transition foliage but coexisting even on mature trees with adult foliage and then sometimes bearing pollen and seed cones. Adult leaves in trios, 2–3.5 mm long, including the bases, bright green, sharply keeled. Pollen cones 4–8 mm long, with four to seven trios (or quartets on juvenile foliage) of pollen scales, each scale with three or four pollen sacs. Seed cones single on stout stalks, persisting long after maturity, pointed-ovoid, 2–3 cm in diameter, with a very short, three- (or four-) lobed or parted central column 0.1–3 mm long. Scales in quartets when born on juvenile foliage, the inner and outer whorls similar in length and width, divided to near the base, each scale broadly channeled, with a small but distinct point near the tip. Seeds two to four per scale, light brown, the body 3–6 mm long, 1.5–3 mm wide, with one wing to 6 mm wide and the other 1 mm. Southeastern Queensland and northeastern New South Wales, Australia. Various forest types but especially rain-forest margins of the Coast Ranges, primarily on sandy soils; (10–)200–800(–1,000) m. Zone 9. Synonym: *Callitris parlatorei* F. J. Mueller.

Stringybark pine is one of the most distinctive species of the genus, with its rain-forest-margin habitat, long, unbranched bole, persistent juvenile foliage, and pointy seed cones with scales sometimes in quartets. The seed cones are reminiscent of those of the related genus *Actinostrobus* but lack its attractive sterile scales at the base. Because of the scales in quartets on juvenile foliage, stringybark pine was originally described as a separate genus, *Octoclinis* F. Mueller, and even after it was recognized as a *Callitris* the adult phase was still often placed in a separate species, *C. parlatorei.* The confusions were finally laid to rest in 1910 with the publication of Baker and Smith's *A Research on the Pines of Australia,* but the affinities of the species are still puzzling. The wood lacks the figure characteristic of other species and is less fragrant than most. Despite the more humid, tropical environment of this species compared to its relatives, it develops annual growth rings because of the alternation of dry and wet seasons, and the size of these rings increases and decreases with the changing length of different wet seasons. An uncommon tree and much more sensitive to fires than other *Callitris* species, stringybark pine may be an example of a species with improved regeneration following fire suppression and increased mechanical disturbance during the 20th century. The species name commemorates William S. Macleay (1792–1865), London-born Australian plantsman who collected the type specimen and is not to be confused with his younger cousin William J. Macleay (1820–1891), also a noted naturalist and first president of the Linnean Society of New South Wales.

Callitris monticola J. Garden

STEELHEAD

Shrub to 3–4 m, with one or more trunks to 0.1 m in diameter. Bark brown, weathering gray, remaining smooth by flaking in small scales. Crown narrow with upwardly angled branches.

Branchlets bluntly triangular. Juvenile foliage not persistent in mature plants. Adult leaves, including bases, 2–4 mm long, blue-green with a waxy coating, with a prominent, but low, rounded keel. Pollen cones 2.5–4 mm long, with three to five trios of pollen scales, each scale with two or three pollen sacs. Seed cones single or in clusters on stout stalks, long persisting, flattened-spherical, 1.5–2.5 cm in diameter, with a central column of three spreading lobes to 5 mm long. Scales with a small point below the tip, smooth and grayish waxy when young, becoming wrinkled and the wax persisting only around the edges when open, the opening scales leaving a large, solid base. Seeds three to five on each scale, dark brown to black, the body 3.5–5 mm long, 1.5–2.5 mm wide, with two nearly equal wings a little longer than the body, 1.5–3 mm wide. Border ranges between southeastern Queensland and northeastern New South Wales, Australia. Forest openings on rocky ridges and slopes (hence the scientific name, Latin for "mountain dweller"); (300–)500–1,000(–1,200) m. Zone 9.

Steelhead replaces the similar Illawarra pine (*Callitris muelleri*), which is found much farther south in New South Wales. Both are shrubs of rocky outcrops, but *C. monticola* is found more often on granitic rocks and *C. muelleri* on sandstones. The small size and rarity of steelhead preclude any significant economic uses for the species, but it is threatened by the increasing frequency of anthropogenic fires.

Callitris muelleri (Parlatore) F. J. Mueller

ILLAWARRA PINE, MUELLER CYPRESS PINE

Shrub, or multistemmed tree to 3(–8) m tall, with trunks to 0.2 m in diameter. Bark brown, weathering gray, remaining smooth by flaking in small scales. Crown dense, narrowly cylindrical, with upright branches. Branchlets rounded-triangular. Juvenile foliage sometimes persisting on older plants, with blades 5–12 mm long. Adult leaves, including bases, 4–7(–12) mm long, dark green, with a prominent, rounded to sharp keel. Pollen cones solitary or in clusters of two or three, 2–4 mm long, with three to six trios of pollen scales, each scale with two or three pollen sacs. Seed cones single or in clusters on stout stalks, persisting after maturity, flattened-spherical, 2–3 cm in diameter, with a three-lobed central column 1–4 mm long. Scales with a small sharp point near the tip, wrinkled when open, the opening slits leaving a large, solid base. Seeds 5–10 on each scale, brown to black, the body 3.5–5 mm long, 2.5–4 mm wide, with two slightly unequal wings a little longer than the body, (0–)1–4.5 mm wide, sometimes with a third wing to 1.5 mm wide. Coast ranges of southeastern New South Wales, Australia. Forest openings on rocky sandstone ridges; (0–) 100–500(–750) m. Zone 9.

Illawarra pine is one of the most ornamental small species of the genus, forming a dark green upright accent. Although the wood has an attractive figure, it has little commercial value because of the plant's small size and the awkward locations in which it grows. The adult leaves are the longest in the genus, and the persistence of juvenile foliage on adult individuals, while less pronounced than in stringybark pine (*Callitris macleayana*), is still unusual. The species name commemorates Ferdinand von Mueller (1825–1896), German-born state botanist of Victoria, Australia, and Director of the Botanical and Zoological Gardens in Melbourne.

Callitris neocaledonica Dummer

WINGED CHANDELIER CYPRESS,
CÈDRE CANDELABRA (FRENCH)
Plate 17

Tree to 15 m tall, shorter at higher elevations, with trunk to 1 m in diameter. Bark thin, shredding, dark brown, turning gray with age. Crown broad, umbrella-shaped, formed by long branches turned up at their tips. Branchlets triangular. Juvenile leaf blades 5–9 mm long, gradually shortening to the adult form, the juvenile foliage commonly persistent into adulthood. Adult leaves, including leaf bases 2–4(–5) mm long, dark green, sharply keeled. Pollen cones single, 2–3 mm long, with three to five trios of pollen scales, each scale with three to five pollen sacs. Seed cones single on slender stalks, not persisting after maturity, ovoid, 7–9 mm in diameter, with a three-angled central column 5–6 mm long. Scales with a strong conical point near the tip, smooth, opening to near the base, the inner and outer trios of seed scales nearly equal in length and both narrow, but the outer trio slightly longer and forming the apex of the cone. Seeds two on each inner scale, brown, with two unequal wings about 1 mm wide, the larger one extending above and below the seed body. Southern New Caledonia. A subcanopy tree in the rain forest or emergent above scrublands on serpentine soils; (700–)900–1,350(–1,500) m. Zone 9? Synonym: *Callitris sulcata* var. *alpina* R. Compton.

Winged chandelier cypress is the only species of *Callitris* in which the outer trio of scales form the tip of the seed scale. It is one of the highest-elevation species in the genus. It grows far above threadleaf chandelier cypress (*C. sulcata*) but sometimes grows with the much rarer, related, chandelier cypress (*Neocallitropsis pancheri*), which it resembles in growth habit, cone form, and to a lesser extent in the juvenile foliage. Chandelier cypress, however, never develops adult foliage and so always has its leaves and cones scales in quartets rather than trios.

Callitris oblonga A. Richard & L. Richard

RIVER PINE, PIGMY CYPRESS PINE, TASMANIAN CYPRESS PINE

Tall shrub, or tree to 2–4(–10) m tall, often branching near the base, with trunk to 25 cm in diameter. Bark brown, weathering grayish brown, flaking in scales and becoming shallowly furrowed. Crown dense, broadly cylindrical, with upturned branches. Branchlets rounded-triangular. Juvenile foliage not persistent in mature plants. Adult leaves, including bases, 4–5 mm long, bluish green with wax, with a rounded keel. Pollen cones usually in clusters of three to five, 1–2 mm long, with three or four trios of pollen scales, each scale with two or three pollen sacs. Seed cones solitary or clustered on stout, short branchlets, persisting long after maturity, egg-shaped (close enough to the oblong of the species name), 1.2–2.0 cm in diameter, with a three-lobed, short central column 2–3 mm long. Scales with a short but distinct point below the tip, wrinkled when open, the outer trio about half as long as the inner. Seeds 5–15 on each scale, dark brown to black, the body 2.5–4 mm long, 1.5–2.5 mm wide, with two nearly equal wings (0–)1–2(–3) mm wide, sometimes with a third wing to 1 mm wide. Northeastern Tasmania and southeastern and northeastern New South Wales, Australia. Sandy riversides; (0–)100–1,000(–1,300) m. Zone 8.

The riparian habitat of river pine is unusual within the genus. In Tasmania it is largely confined to the South Esk River drainage, where it can dominate sandy streamsides, in part, perhaps, by vegetative propagation. The trees from New South Wales occupy similar habitats, but it is not certain whether they are the same species as the Tasmanian populations or represent a distinct, closely related species. In neither case is the wood of any commercial value because of the limited distribution and small size of the trees. There is competitive displacement between river pine and Oyster Bay pine (*Callitris rhomboidea*) in their limited range of overlap in Tasmania. They never hybridize in nature because their pollen release is displaced by 6 months from each other, river pine in winter and Oyster Bay pine in summer.

Callitris preissii Miquel in J. G. Lehman

ROTTNEST ISLAND PINE, MURRAY CYPRESS PINE, SLENDER CYPRESS PINE

Tree to 15(–30) m tall, with trunk to 70 cm in diameter. Bark dark brown, furrowed. Crown broadly conical, open, with spreading branches. Branchlets cylindrical. Juvenile foliage not persisting in mature plants. Adult leaves, including bases, 1–4 mm long, dark green, or sometimes bluish with wax, not keeled. Pollen cones single or in clusters, 3–6 mm long, with 3–7(–12) trios of pollen scales, each scale with two to four pollen sacs. Seed cones single or clustered on stout stalks thickening to 12 mm in diameter, persisting long after maturity, spherical to somewhat egg-shaped, (1.5–)2.5–3(–4) cm in diameter, with a short, broadly pyramidal central column to 5 mm long. Scales with an inconspicuous point near the tip, wrinkled when open and often with few to many large warts reaching 3 mm in diameter. Seeds 6–12(–16) per seed scale, dark brown to black, the body 3–5 mm long, 1.5–2.5 mm wide, with two nearly equal wings slightly longer than the body, (1–)3–6 mm wide. Widespread across southern mainland Australia from southwestern Australia to New South Wales. Open woodlands and scrublands on various soils, from limestone pavements to deep sands; 0–300(–500) m. Zone 9. Synonyms: *Callitris gracilis* R. T. Baker, *C. gracilis* subsp. *murrayensis* (J. Garden) K. Hill, *C. preissii* subsp. *murrayensis* J. Garden, *C. propinqua* R. Brown ex R. T. Baker & H. G. Smith, *C. robusta* R. Brown ex R. T. Baker & H. G. Smith, *C. tuberculata* R. Brown ex R. T. Baker & H. G. Smith.

Rottnest Island pine has the most variable seed cones in the genus, and this is reflected in the extensive synonymy. The seed cones are usually large but vary considerably in the prevalence of warts. Some of the variation, and the presence of wax on the leaves, may be due to hybridization with both white cypress pine (*Callitris glaucophylla*) and mallee pine (*C. verrucosa*). Hybridization with the latter is so extensive that mallee pine has been treated as a subspecies of Rottnest Island pine, but they are quite distinct in the chemistry of their leaf oils. In Western Australia the species is confined to the coast, where it was first described, and here it reaches its largest size. Unfortunately, it has been much depleted there through overexploitation of its useful timber. The species name commemorates Johann Preiss (1811–1883), a German botanist who explored western Australia.

Callitris rhomboidea R. Brown ex A. Richard & L. Richard

OYSTER BAY PINE, PORT JACKSON PINE

Tree to 12(–30) m tall, or sometimes shrubby, with a single or forking trunk to 0.4 m in diameter. Bark brown, weathering grayish brown, hard, flaking in scales and becoming shallowly furrowed. Crown variable, with upright or spreading branches. Branchlets rounded-triangular. Juvenile foliage somewhat delayed in giving way to adult foliage but not persisting to adulthood. Adult leaves, including bases 1.5–4 mm long, dark green or silvered with wax, with a strong, rounded keel. Pollen cones mostly single, 1–2 mm long, with three to five trios of pollen scales, each scale with three or four pollen sacs. Seed cones usually tightly clustered on stout stalks thickening to 8 mm in diameter, long persistent after maturity, roughly spherical to

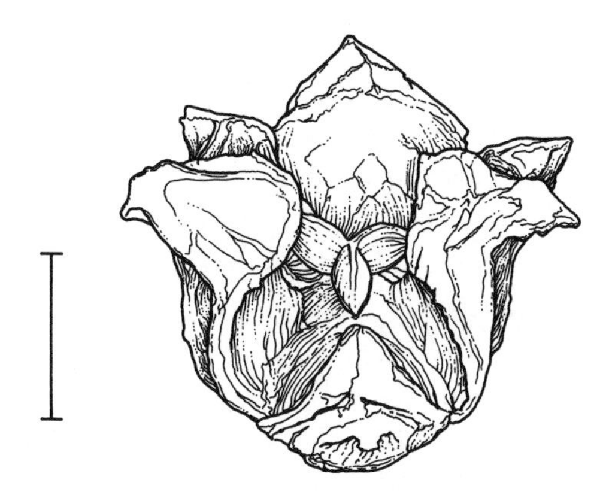

Mature seed cone with prominent columella of Oyster Bay pine (*Callitris rhomboidea*) after shedding seeds; scale, 1 cm.

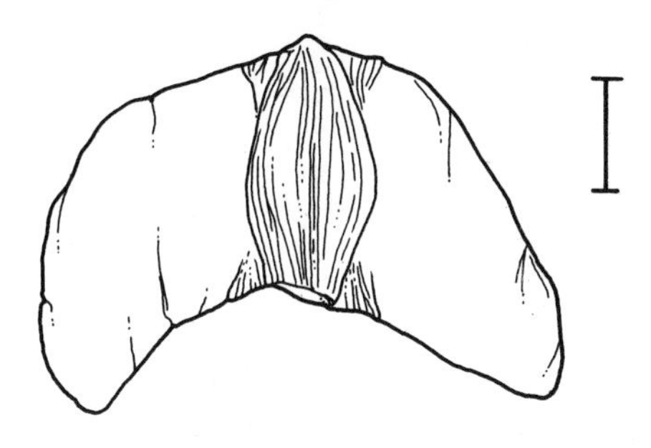

Seed of Oyster Bay pine (*Callitris rhomboidea*); scale, 2 mm.

somewhat shorter, 1.5–2 cm in diameter, with a three-lobed or three-parted central column 3 mm high or less. Scales with a prominent pyramidal point below the tip (thus the scales have a "rhomboid" shape), deeply wrinkled when mature, opening to leave a solid base. Seeds 4–10 on each scale, dark brown to blackish, the body 2.5–4 mm long, 1.5–2 mm wide, with two nearly equal wings slightly longer than the body, 1–4 mm wide. Eastern Australia from southeastern Queensland to eastern Tasmania, west to Kangaroo Island, South Australia. Various open forests, woodlands, and scrublands on a wide variety of substrates, primarily near the coast; 0–300(–750) m. Zone 9. Synonym: *Callitris tasmanica* (Bentham) R. T. Baker & H. G. Smith.

This widespread but uncommon species has sometimes been subdivided based on habit, the broader, spreading trees being separated as *Callitris tasmanica.* With its exceptionally thick scales on persistent, closed seed cones, Oyster Bay pine regenerates well after fires but is also capable of maintaining itself in mature stands. A highly ornamental small tree, Oyster Bay pine is widespread but uncommon throughout most of its range. It is most closely related to black cypress pine (*C. endlicheri*) and basically replaces it near the coast, as coast cypress pine (*C. columellaris*) replaces white cypress pine (*C. glaucophylla*). The trees are too small to provide commercial timber, but are used in woodworking.

Callitris roei (Endlicher) F. J. Mueller

ROE CYPRESS PINE

Shrub, or tree to 3 m tall, with one or more trunks to 0.2 m in diameter. Bark reddish brown, weathering gray, long remaining smooth by flaking, finally becoming shallowly furrowed near the ground. Crown open, wide, with thin, wavy, outstretched branches. Branchlets sharply triangular. Juvenile foliage not persisting in mature plants. Adult leaves, including bases, 2–3.5(–5) mm long, bluish green with wax, with a sharp triangular keel. Pollen cones single, 3–6 mm long, with 3–6(–10) trios of pollen scales, each scale with three or four pollen sacs. Seed cones single or clustered on stout branchlets, flattened spherical, to 2 cm in diameter, persisting long after maturity, with a three-angled, stout central column 4 mm long. Scales with a strong conical point below the tip, shallowly grooved to the base, the inner trio each with two seeds. Seeds chestnut brown, the body 3–4 mm long, 2–2.5 mm wide, with two unequal wings, the larger one 3–6 mm wide, 5–7 mm long, the smaller 1–2 mm wide. Southwestern Western Australia. Open woodlands; 100–300(–500) m. Zone 9.

Roe cypress pine has the strongest leaf keels of those species with triangular keels and has some of the coarsest branchlets in the genus. Like the other two endemic southwestern Australian species, its close relative dwarf cypress pine (*Callitris drummondii*) and the unrelated scrubby cypress pine (*C. canescens*), it is much less well known than the eastern species. The species name honors John Roe (1797–1878), the English-born Australian naturalist who collected the original material in 1847 during his long tenure as Surveyor-General of Western Australia.

Callitris sulcata (Parlatore) Schlechter

THREADLEAF CHANDELIER CYPRESS, NIÉ,
SAPIN DE COMBOUI (FRENCH)

Tree to 5–12 m tall, with trunk to 0.4 m in diameter. Bark light brown, weathering gray, hard, smooth at first, flaking in scales and becoming deeply furrowed at the base. Crown dense, with tufted branches. Branchlets sharply triangular. Juvenile foliage giving way abruptly to adult forms, moderately persistent, with blades 18–26 mm long. Adult leaves, including bases, 4–6 mm long, with a sharp keel. Pollen cones single, 3–5 mm long, with three to five trios of pollen scales, each scale with four to six pollen sacs. Seed cones single on slender stalks, not persistent, nearly spherical, about 1 cm in diameter, with a three-angled, slender central column 3 mm long. Scales with a short, but strong, conical point near the tip, with a groove (or sulcus) down the center when open (hence the scientific name), separating to near the base, the inner trio of scales each with two seeds. Seeds 4–5 mm long, with two unequal wings, each about 2 mm wide but the larger one longer than the seed body and the smaller one shorter. Southern New Caledonia. Scattered or forming groves in rain forests and scrublands in river valleys; (15–)50–200(–300) m. Zone 10?

Threadleaf chandelier cypress grows at much lower elevations than winged chandelier cypress (*Callitris neocaledonica*) and on deeper soils. It is also less divergent in appearance from most Australian species than is winged chandelier cypress. Although the wood is fragrant (with an odor of camphor), it is less so than in chandelier cypress (*Neocallitropsis pancheri*) and has not been as heavily exploited for either the wood or its distilled oil.

Callitris verrucosa (A. Cunningham ex Endlicher) F. J. Mueller

MALLEE (CYPRESS) PINE

Shrub, or tree to 8(–15) m tall, usually dividing from the base, with trunk(s) to 0.6 m in diameter. Bark light brown, weathering gray, flaking in scales and becoming thick, stringy, and deeply ridged and furrowed. Crown rounded, spreading, open, with upwardly angled branches. Branchlets cylindrical. Juvenile foliage not persisting in mature plants. Adult leaves, including bases, 2–4 mm long, dark green, yellow-green, or waxy, not keeled. Pollen cones in clusters of (two) or three (to five), 1.5–3.5 mm long, with four to nine trios of pollen scales, each scale with two or three pollen sacs. Seed cones usually single on stout stalks thickening to 10 mm, persisting long after maturity, nearly spherical, 1.5–2.5(–3) cm in diameter, with a simple, short, pyramidal central column 2–3 mm high. Scales with an inconspicuous point below the tip lost among the profusion of small warts 1–2 mm wide that cover the surface (hence the scientific name, Latin for "warty"). Seeds 6–12(–16) on each scale, chestnut to dark brown, the body 2.5–6 mm long, 1.5–3 mm wide, with two nearly equal wings (1–)2–4(–6) mm wide, slightly longer than the body, sometimes with a third wing to 2 mm wide. Southern third of mainland Australia from Western Australia to New South Wales. Fire-prone scrublands, primarily on nutrient-poor, sandy soils; (0–)50–200 (–300) m. Zone 9? Synonym: *Callitris preissii* subsp. *verrucosa* (A. Cunningham ex Endlicher) J. Garden.

Mallee pine is the common cypress pine of the vast mallee scrublands of southern Australia. These fire-washed communities are named for the mallee eucalypts that resprout from lignotubers (underground storage stems). Mallee pine is somewhat fire resistant, with its thick bark, but its main response to a killing fire is the subsequent release of the numerous seeds accumulated in years of unopened cones retained on the branches.

Calocedrus W. Kurz

INCENSE CEDAR

Cupressaceae

Evergreen trees with a single straight trunk bearing smooth, scaly bark when young, becoming fibrous, deeply furrowed, and peeling in thin strips with age. Densely branched from the base with short, horizontal or gently ascending branches. Crown remaining narrowly columnar through life or broadening to a dome with age. Branchlets in flattened fernlike sprays held vertically. Without definite winter buds. Seedling leaves in alternating quartets, needlelike, standing out from and well spaced on the stem, seedling phase generally short-lived, giving rise to adult branchlets by the second year. Adult leaves in alternating pairs that resemble aligned quartets because their bases are so close, scalelike, dense,

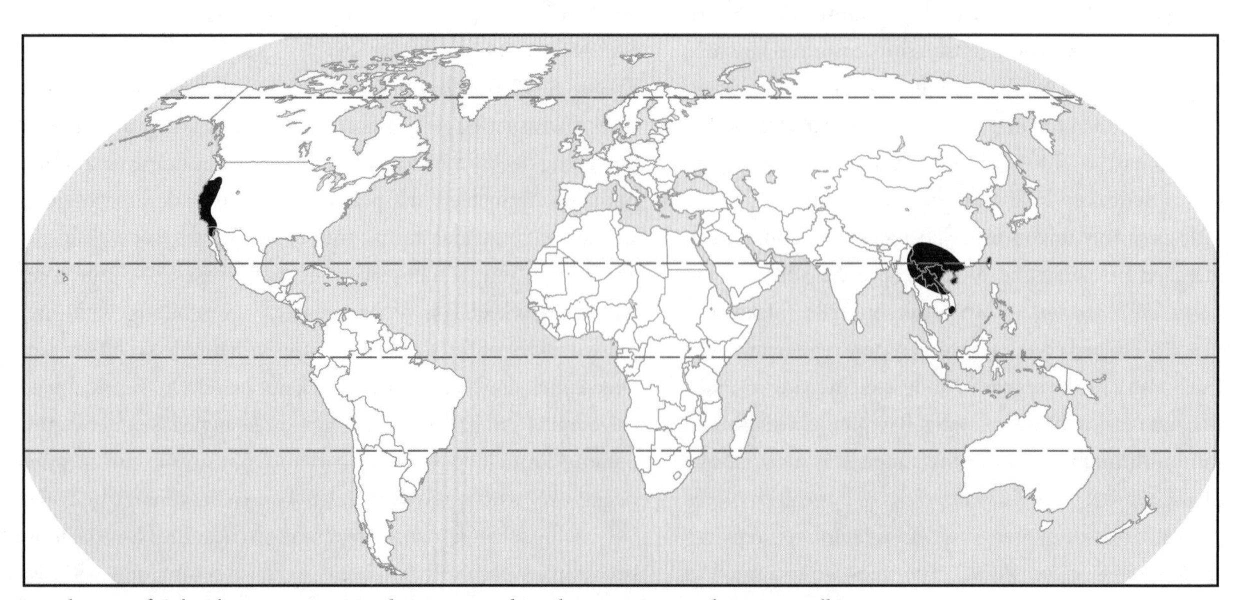

Distribution of *Calocedrus* in western North America and southeastern Asia, scale 1 : 200 million

the bases running down onto the branchlets, the lateral and facial pairs dissimilar, without glands or with inconspicuous glands near the tips of facial leaves.

Plants monoecious. Pollen cones numerous, single at the ends of short lateral branchlets, oblong, with 6–10 alternating pairs of pollen scales. Each scale with three or four round pollen sacs. Pollen grains small (25–40 μm in diameter), nearly spherical, with a minute germination bump, otherwise almost featureless. Seed cones relatively sparse, single at the ends of short side branchlets that may be as long as the cones. Leaves of cone-bearing branchlets tightly crowded, without the elongation of the leaf bases characterizing foliage leaves. Cones maturing in a single season, oblong, with three alternating pairs of scales. Each seed scale bearing a bump just below the tip corresponding to the fused bract portion, only the middle pair fertile, with two ovules on each scale, the inner pair as long as the fertile pair but united as a flat plate between them, the outer pair much smaller. Seeds oval, with two very unequal wings derived by outgrowth of the seed coat at the top, the larger, inner wing spreading out to the tip of the seed scale. Cotyledons two, each with one vein. Chromosome base number $x = 11$.

Wood somewhat fragrant, light, soft and weak but highly decay resistant, with very pale yellowish white sapwood sharply contrasting with pinkish heartwood that browns with age. Grain moderately to very fine and even, with weakly defined to prominent growth rings marked by an abrupt transition to a few rows of much smaller and somewhat thicker walled latewood tracheids. Resin canals absent but with abundant individual resin parenchyma cells scattered throughout the growth increment or concentrated in and near the latewood.

Stomates in elongate bands in the grooves between leaves and on the more protected side of the branchlets and fewer and scattered on the more exposed side. Each stomate sunken beneath the four to six (or seven) subsidiary cells and surrounded by a nearly continuous, bumpy Florin ring. Subsidiary cells and other epidermal cells in the stomatal regions also often bearing additional large, round to elongate papillae. Leaf cross section with a single midvein near the twig in the attached leaf base and scarcely entering the free tip. Midvein accompanied toward the outside of the lateral leaves only or of both types by a very large resin canal and flanked by small wings of transfusion tissue. Photosynthetic tissue forming a complete but shallow palisade layer all along the exposed portion of the leaves beneath the epidermis and the adjoining thin hypodermis. Scattered sclereids lie just beneath the palisade layer, and the remainder of the leaf is occupied by the looser photosynthetic spongy mesophyll.

Three species in western North America and eastern Asia. References: Averyanov et al. 2008, Brunsfeld et al. 1994, Farjon 2005b, Florin 1930, Florin and Boutelje 1953, Fu et al. 1999g, Li 1953. Synonym: *Heyderia* K. Koch.

Although originally placed with southern hemisphere incense cedars in *Libocedrus,* DNA evidence shows that the northern hemisphere incense cedars are more closely related to *Platycladus* and *Microbiota* on the one hand, and to *Cupressus* and *Juniperus* on the other, placing them squarely among the northern hemisphere genera. The paddle-shaped seed scales must have evolved independently in the northern and southern incense cedars. The southern hemisphere genera lack the flat plate between the fertile scales that is so prominent in *Calocedrus.* The arrangement of the leaves in apparent whorls of four leaves of equal length, and the consistently alternate branching are also clear points of distinction for *Calocedrus.* The fragrant wood that gives them their common name is found much more widely among Cupressaceae than in just the genera called incense cedars. Nor are the three species of *Calocedrus* the only ones that merit the translation of the Greek- and Latin-derived name "beautiful cedars." The North American *C. decurrens* is in general cultivation while the Asian species are largely confined to botanical gardens. A modest number of cultivar selections emphasize dwarf forms and foliage color variants. The fossil record for *Calocedrus* is modest but extends back to the Oligocene in western North America. This record is based on foliage and cones since pollen is not reliably distinguishable from that of related genera.

Identification Guide to *Calocedrus*

The three species of *Calocedrus* may be distinguished by the length of the adult lateral leaves, whether they have prominent stomatal areas beneath, the number of scale pairs on the pollen cones, and the length of the mature seed cones:

> *Calocedrus formosana,* lateral leaves 1.5–2.5 mm, stomatal zones prominent, pollen cone scales six to eight pairs, seed cones 1–1.5 cm
>
> *C. decurrens,* lateral leaves 2–3 mm, stomatal zones obscure, pollen cone scales six to eight pairs, seed cones 2–3 cm
>
> *C. macrolepis,* lateral leaves 3–4 mm, stomatal zones prominent, pollen cone scales 7–10 pairs, seed cones 1.5–2 cm

Calocedrus decurrens (J. Torrey) Florin

INCENSE CEDAR

Plate 18

Tree to 46 m tall, with trunk to 3.7 m in diameter. Bark rich reddish brown. Crown cylindrical to narrowly conical, dense at first but becoming more ragged with age. Branchlet sprays yellowish green to dark green on both sides, without conspicuous stomatal

Upper side of foliage spray of incense cedar (*Calocedrus decurrens*) with pollen cones, ×0.5.

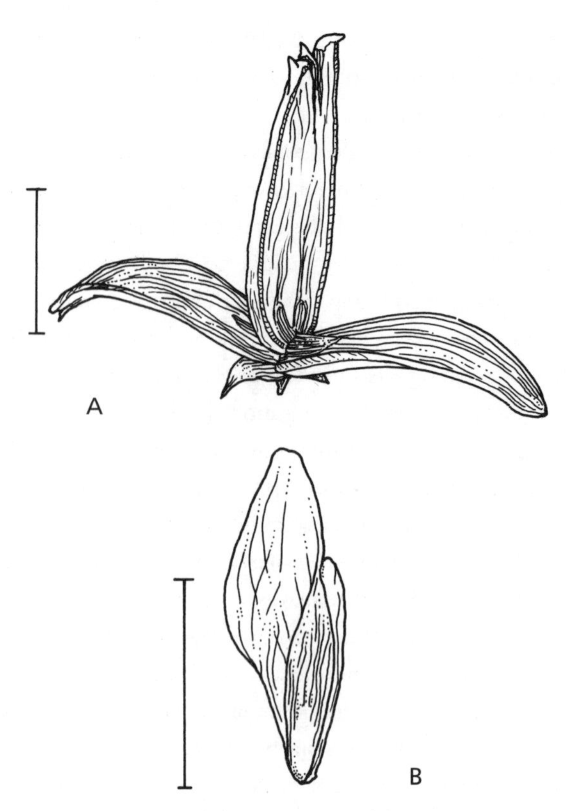

Open mature seed cone (A) and shed seed (B) of incense cedar (*Calocedrus decurrens*); scale, 1 cm.

Mature incense cedar (*Calocedrus decurrens*) in nature.

zones. Scale leaves of branchlets 2–3 mm long (to 14 mm on main twigs), the short, bluntly triangular free tips directed forward but not tight to the twig, usually with inconspicuous glands. Pollen cones 4–6 mm long, with six to eight pairs of pollen scales. Seed cones 2–3 cm long, hanging down and light tan or yellowish brown at maturity. Seeds 1.5–2.5 cm long, including the wing. Pacific North America from Oregon to Baja California (Mexico). Montane conifer forests; 300–2,800 m. Zone 7. Synonyms: *Heyderia decurrens* (J. Torrey) K. Koch, *Libocedrus decurrens* J. Torrey.

Incense cedar is one of the characteristic species of dry conifer forests in the mountains of California and Oregon. The fragrant wood is resistant to decay and weathering in outdoor use as shingles or fence posts but is subject to discoloring heart rot in the living tree. It replaced eastern red cedar (*Juniperus virginiana*) as the main domestic source of pencil wood in North America before itself being displaced by tropical hardwoods. It grows in drier forests than the somewhat similar but more northerly western red cedar (*Thuja plicata*), and its larger, hanging seed cones with fewer, thicker scales and apparently whorled leaves with proportionately much longer attached leaf bases easily distinguish it from other local Cupressaceae. The largest known individual in combined dimensions when measured in 1997 was 50.3 m tall, 3.8 m in diameter, with a spread of 14.9 m.

Calocedrus formosana (Florin) Florin

TAIWAN INCENSE CEDAR, TAI WAN CUI BAI (CHINESE)

Tree to 23 m tall, with trunk to 3 m in diameter, trunk often crooked. Bark purplish brown, long remaining smooth and flaking. Crown narrowly conical, broadening with age. Branchlet sprays dark green above, lighter green beneath with prominent stomatal zones. Scale leaves of branchlets 1.5–2.5 mm long, the free tips very short and rounded, usually without glands. Pollen cones 4–6 mm long, with six to eight pairs of pollen scales. Seed cones 1–1.5 cm long, at the tips of short branchlets to 3 mm long. Seeds 8–12 mm long, including the wing. Northern and central Taiwan (also known as Formosa, hence the scientific name). Montane conifer forests; 300–1,900 m. Zone 9. Synonyms: *Calocedrus macrolepis* var. *formosana* (Florin) W. C. Cheng & L. K. Fu, *Heyderia formosana* (Florin) H. L. Li, *Libocedrus formosana* Florin.

Unlike the two Taiwanese species of the related genus *Chamaecyparis,* this uncommon tree apparently has little economic importance and is rarely cultivated outside of its native range. It is sometimes treated as a variety of Chinese incense cedar (*Calocedrus macrolepis*) but has shorter leaves and pollen and seed cones and lives in moister habitats than that species.

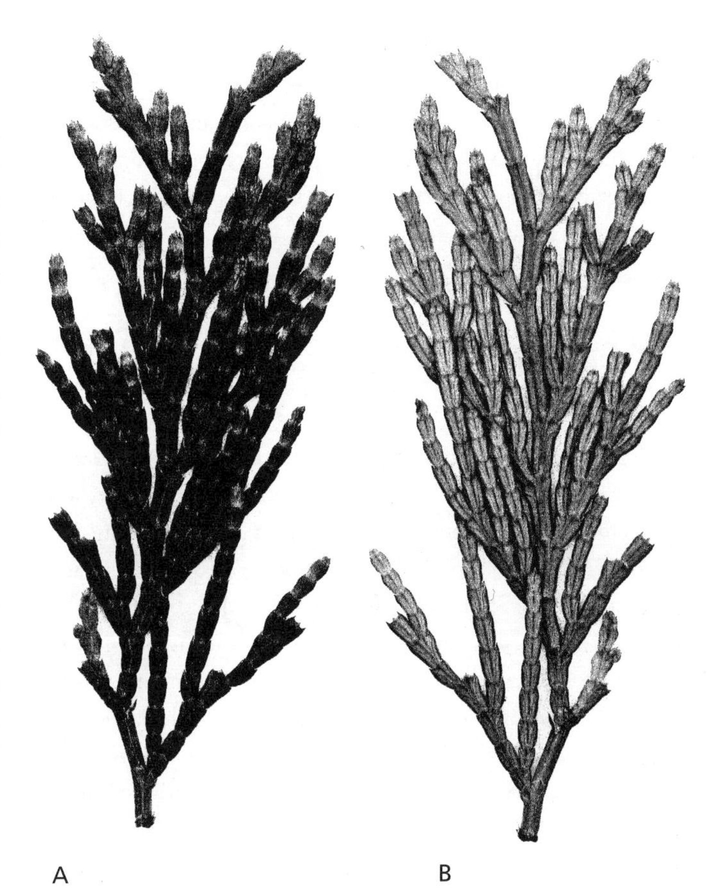

Upper (A) and undersides (B) of foliage spray of Taiwan incense cedar (*Calocedrus formosana*), ×1.

Calocedrus macrolepis W. Kurz

CHINESE INCENSE CEDAR, CUI BAI (CHINESE)

Tree to 30(–35) m tall, with trunk to 1.5 m in diameter. Bark grayish brown. Crown conical, greatly broadening and rounding with age but remaining fairly dense. Branchlet sprays dark green above, paler beneath with greenish white, waxy stomatal zones. Scale leaves of adult branchlets 3–4 mm long (to 8 mm on main twigs), the free tips rounded, giving a beadlike appearance to the branchlets, without glands. Juvenile leaves broader and flatter, with spreading, pointed tips, reminiscent of Fujian cedar (*Fokienia hodginsii*). Pollen cones 6–8 mm long, with 7–10 pairs of pollen scales. Seed cones (1–)1.5–2 cm long, at the tips of cylindrical or four-angled branchlets 3–17 mm long. Seeds 10–18 mm long, including the wing. Southwestern China and adjacent Myanmar and Vietnam. Canopy tree in dry, open, montane mixed forests; 1,000–2,000 m. Zone 9. Synonyms: *Calocedrus rupestris* Averyanov, T. H. Nguyên & K. L. Phan, *Heyderia macrolepis* (W. Kurz) H. L. Li, *Libocedrus macrolepis* (W. Kurz) Bentham & J. Hooker.

Because Chinese incense cedar is the main conifer of the upper slopes and reaches of dry, hot river valleys in Yunnan (China), it is heavily exploited for both timber and firewood. It is becoming depleted as a result of this pressure and because it may be replaced after harvest by stands of Yunnan pine (*Pinus yunnanensis*) and tan oak (*Lithocarpus* species) that do better in cutover conditions. It is rarely cultivated outside of its native range but should do well

Old mature Chinese incense cedar (*Calocedrus macrolepis*) cultivated within its native range.

in drier warm temperate regions like some parts of Australia and South Africa. Some authors separate populations in northern Vietnam as a distinct species, with populations in southern Vietnam retained as a disjuncts of *Calocedrus macrolepis.* While this segregate of Chinese incense cedar is not accepted here, it merits further study with mature material.

Cathaya W. Y. Chun & Kuang

CATHAYA

Pinaceae

Evergreen trees with a straight, single trunk bearing horizontal branches. Bark nonfibrous, smooth at first, becoming flaky and then furrowed. Crown conical in youth, gradually flattening and becoming dome-shaped with age. Shoots weakly differentiated into two kinds. Sterile side shoots shortened and with needles concentrated at the tips, but not long-lived like those of *Larix* and *Cedrus.* Long main shoots with seasonally varying spacing between leaves, the more condensed sections with the needles in pseudowhorls like those of the short shoots. Twigs shallowly ridged and furrowed, completely clothed by the woody, persistent, leaf bases. Leaf attachment scars round or squarish, flush with the ridges or raised on a low peg. Winter buds well developed, with thin scales falling cleanly in spring, not resinous. Leaves closely or openly spiraled, needlelike, flat, straight or slightly curved to one side, gradually tapering to a short stalk.

Plants monoecious. Pollen cones single, with several sticking out from separate buds in the axils of needles at the tips of the previous year's twigs. Pollen scales numerous, densely spiraled, each with two elongated pollen sacs attached to its stalk. Pollen grains medium to large (body 45–60 µm long, 55–75 µm overall), with two round air bladders a little smaller than the oblong body and covered with a more open, squiggly, minute sculpturing than the body. Seed cones single from buds in the axils of needles on twigs of the previous year, unstalked, upright at pollination but sticking straight out from the twigs at maturity, maturing in a single season but persisting several years. Seed scales round, spirally arranged, spreading at maturity to release the seeds. Bract attached to the seed scale only at the base, much smaller, triangular with a prolonged, pointed tip. Seeds two per scale, the body egg-shaped, the asymmetrical wing similar and larger in both length and width but much smaller than the seed scale and derived from it, cupping the seed body from below. Cotyledons three or four, each with one vein. Chromosome base number $x = 12$.

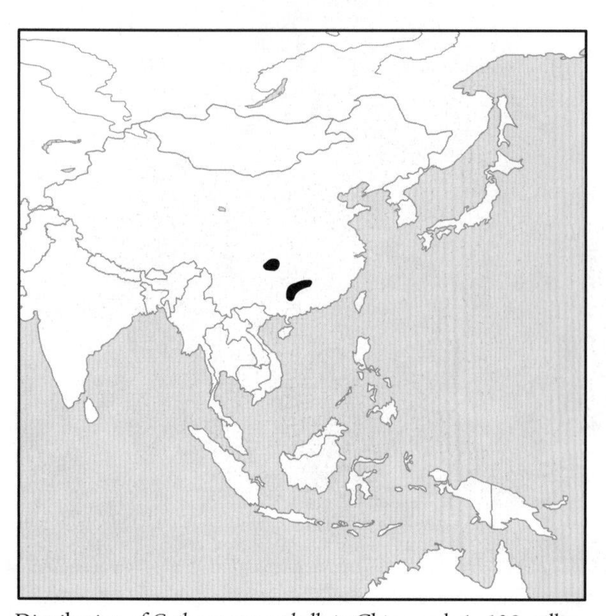

Distribution of *Cathaya argyrophylla* in China, scale 1 : 120 million

Wood similar to that of *Pseudotsuga* in the distinctive, spirally thickened water-conducting cells (tracheids) but differing in the more prominent growth rings formed by a more gradual transition from early- to latewood. Vertical resin ducts confined to the latewood and earlywood–latewood transition.

Stomates arranged in two broad bands beneath, one on either side of the midrib, each band consisting of 10–20 lines of stomates. Individual stomates sunken beneath and tucked in under the four to six subsidiary cells, which are topped by a stout Florin ring. Leaf cross section with transfusion tissue right beneath the single-stranded midvein and two resin canals, one on each side out toward the leaf margin just inside the lower epidermis at the outer edge of the stomatal band. Photosynthetic tissue without a palisade layer but with a somewhat radial organization of the part of the spongy mesophyll that lies beneath the upper leaf epidermis and adjoining discontinuous to continuous thin hypodermis.

One species in China. References: Fu et al. 1999c, Page 1988, Sivak 1976, Wang et al. 1998, T. S. Ying et al. 1993, Yu and Zeng 1992.

The affinities of *Cathaya* (the name honors its country of origin, since Cathay is a historical name for China) are controversial and appear to vary with the kind of evidence examined. Most features favor an alliance with *Pseudotsuga* and *Larix,* but DNA studies suggest a closer, albeit still distant, relationship with *Pinus* and *Picea.* Despite placement of the sole species in *Tsuga* by some authors, there are no important similarities with this genus or any other genus in the abietoid subfamily of the Pinaceae. Although cultivated in botanical gardens in China, *C. argyrophylla* is scarcely known outside of its homeland, and no cultivar selection has taken place. *Cathaya* has a sparse fossil record across Eurasia, consisting mostly of pollen grains as early as the late Oligocene, about 25 million years ago, and it has also been reported from shoots, cones, and seeds in Miocene sediments of Idaho.

Cathaya argyrophylla W. Y. Chun & Kuang

CATHAYA, YIN SHAN (CHINESE)

Tree to 24 m tall, with trunk to 40(–85) cm in diameter. Bark gray, irregularly and narrowly fissured. New shoots yellowish brown, ribbed with attached leaf bases, densely short hairy, the hairs wearing off by the second year. Buds light yellowish brown, 6–8 mm long. Needles sticking straight out all around the branch, 4–6 cm long and 2.5–3 mm wide, up to 3 cm long on short shoots, dark green above, with two white, stomatal bands beneath, each band with 11–17 lines of stomates, the tip rounded to bluntly pointed, the edges hairy at first, curled slightly under. Seed cones

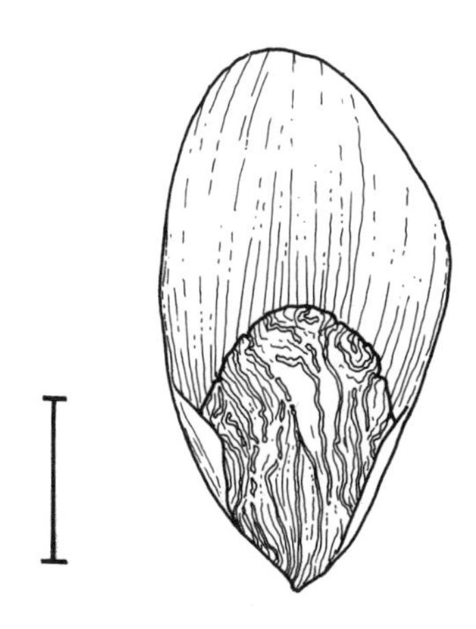

Seed of cathaya (*Cathaya argyrophylla*); scale, 2 mm.

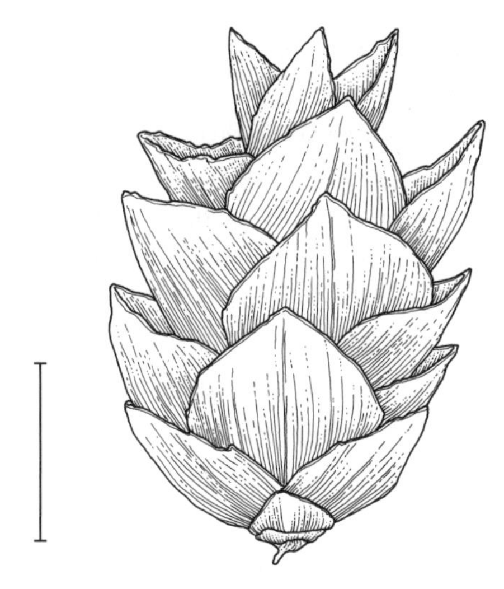

Mature seed cone of cathaya (*Cathaya argyrophylla*) after seed dispersal; scale, 1 cm.

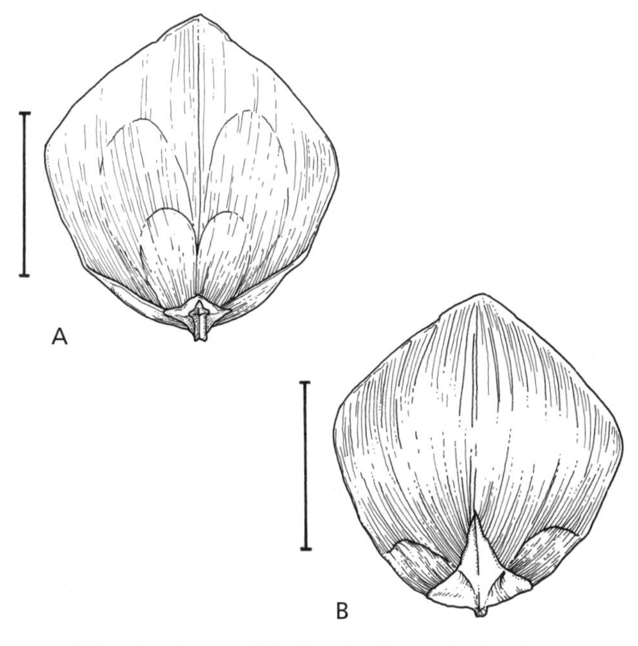

Upper (A) and undersides (B) of artificially detached seed cone scale of cathaya (*Cathaya argyrophylla*); scale, 5 mm.

Old mature cathaya (*Cathaya argyrophylla*) in nature.

green before maturity, ripening dark brown, 3–5 cm long, 1.5–2 cm wide, with 13–16 seed scales, these circular to oval, 15–25 mm long and 10–25 mm wide. Seed body 5–6 mm long, 3–4 mm thick, dark green with lighter mottling, the wing 5–9 mm longer. South-central China, from southeastern Sichuan to eastern Hunan, south to central Guangxi. Scattered or forming small groves in mixed forests on rocky ridges and cliffs; 900–1,900 m. Zone 6. Synonyms: *Cathaya nanchuensis* W. Y. Chun & Kuang, *Pseudotsuga argyrophylla* (W. Y. Chun & Kuang) Greguss, *Tsuga argyrophylla* (W. Y. Chun & Kuang) de Laubenfels & Silba.

Cathaya is a rare tree with perhaps no more than 4,000 individuals among about 30 known populations. These locations are so remote that the species is in little immediate danger from human disturbance. There is concern for the tree, however, because regeneration is often poor as the species has a high light requirement for establishment. This, no doubt, is why it is largely confined to rocky, open sites. The small size and wide separation of the populations led to higher genetic differences between populations than is usual in conifers and to lower genetic variability within populations. Unlike the even more restricted dawn redwood (*Metasequoia glyptostroboides*), discovered a decade earlier, *Cathaya* is rare in cultivation in China and almost unknown outside. The species name, Greek for "silver leaf," refers to the silvery white stomatal bands on the under side.

Cedrus Trew

CEDAR

Pinaceae

Evergreen trees with a straight single trunk bearing numerous, spirally arranged horizontal branches. Lower branches sometimes turning up and becoming massive and trunklike. Bark smooth at first, nonfibrous, flaking in scales, cracking to pass through a more or less checkerboard pattern, and then becoming deeply ridged and furrowed with age. Crown conical when young, broadening and flattening markedly with age, often spreading wider than high. Branchlets of two kinds: outstretched long shoots and permanently condensed short shoots. Long shoots extending the growth of the tree, minutely hairy at first, slightly ribbed from the attached, peglike leaf bases. Short shoots bearing most of the foliage and the cones, ringed with alternating groups of leaf scars and persistent bud scales. Winter buds well developed but small, scaly. Leaves spirally arranged and radiating all around the twigs, widely spaced on long shoots, crowded in a ring on short shoots. Leaf blade needlelike, three- or four-sided in cross section, wider than thick, straight, stiff, the tip pointed, the base narrowed to its joint with the raised peg on the twig.

Plants monoecious or partially dioecious. Pollen cones single at the tips of short shoots, upright, with numerous, densely spirally arranged pollen scales. Each scale bearing two somewhat egg-shaped pollen sacs attached along the stalk. Pollen grains large (body 45–80 μm long), with two round, wrinkly air bladders that blend smoothly into the unusually thick cap of the oblong, minutely bumpy body. Seed cones barrel-shaped or cylindrical, single at the tips of short shoots, upright from pollination through to maturity, maturing in two or three seasons. Cones breaking up at maturity to shed both the seeds and the tightly packed, densely spirally arranged, woody seed scales, leaving behind a narrowly conical, long-persistent cone stalk. Seed scales wedge-shaped beyond a narrower stalk, the tiny, hidden bract attached only at the base. Seeds two per scale, the body and the much larger, asymmetrical wing each roughly triangular. Seed wing derived from the skin of the seed scale and covering it, cupping the seed body on one side and wrapping around it a little on the other side. Cotyledons 6–10, each with one vein. Chromosome base number $x = 12$.

Wood soft, medium weight, light brown, the heartwood sweetly and sharply fragrant, decay resistant. Growth rings well defined, marked by inconspicuous latewood. Vertical resin canals absent but developing resin canals in the rays in response to wounding.

Each face of the needles with a narrow stomatal band. Individual stomates slightly sunken in a shallow pit, the four to six surrounding subsidiary cells, smooth, often shared between adjacent stomates in (but not between) a row, sometimes with a partial second circle along the sides only, and without a Florin ring. Leaf cross section with one resin canal on each side near the lower epidermis more than halfway out from the two-stranded midvein to the outer corner. Midvein accompanied on its lower side by transfusion and sclerenchyma tissue, the whole surrounded by a conspicuous cylinder of endodermis. Photosynthetic tissue with cells more or less radially aligned out from the midrib, the outermost layer not always forming a definite, dense palisade inside the epidermis and adjoining, nearly continuous hypodermis.

Two species in the Mediterranean Sea basin and the western Himalaya. References: Coode and Cullen 1965, Farjon 1990, Frankis 1988, Meikle 1977, K. C. Sahni 1990.

The aromatic wood of the true cedars has given these trees their social importance since ancient times and has also contributed to the ambiguities surrounding them. Although they are referred to here as the "true" cedars, the Greek name *kedros* (from which the Latin common name for the trees, *Cedrus,* is derived) was also applied to eastern Mediterranean junipers with equally fragrant wood, just as the name cedar is applied today to a number of other members of the family Cupressaceae unrelated to *Cedrus* and the Pinaceae. In fact, the name *Cedrus* was first formally applied in botanical nomenclature in 1755 to a genus of Cupressaceae, 2 years before Trew used it to give Linnaeus's *Pinus cedrus* its own genus. However, the present usage was preserved by formally conserving the name, with the cedar of Lebanon (*C. libani*) as the type, against the competing usage by Duhamel.

Members of the genus have been appreciated and cultivated in their native countries for centuries and are now grown worldwide in suitable climates. Cultivar selection has been fairly extensive, favoring foliage color (especially blue and golden variants) and growth habit (including many dwarf and weeping cultivars).

Both the number of species of true cedar and the position of the genus within the family Pinaceae are controversial. Many features of structure, leaf pigments, and immunological reactions place *Cedrus* among the abietoid genera, with a particularly close resemblance to the true firs (*Abies*), which have the most similar seed cones to those of *Cedrus.* DNA studies, in contrast, are inconsistent in their placement of *Cedrus* and its relationship to *Abies.* These studies often suggest that *Cedrus* is an early branch of the abietoid group, or even an early branch of the whole family, and not closely related to any other genus. The isolated position of the cedars among the Pinaceae is reinforced by a number of unusual characteristics, including maturation of pollen cones and pollination in the autumn rather than in the spring as in other Pinaceae. The chemistry of the fragrant wood is also distinctive within the family in lacking the resin acids found in many other Pinaceae and replacing them with unusual oil compounds not found elsewhere.

The geologically recently separated regional populations of true cedars are variously treated by different botanists as distinct species, subspecies, or varieties. The close relationship among these populations is emphasized by strongly overlapping or nearly identical characteristics for traits that usually vary among species within other genera of Pinaceae. For instance, while most genera of Pinaceae have species with distinctive microscopic sculpturing on the inner side of the protective cuticle layer covering the epidermis, the cedars cannot be told apart on this basis. The Himalayan cedar (*Cedrus deodara*) is the most distinct population, and almost all botanists accept it as an independent species. This is not the case with the populations in the Mediterranean region, traditionally assigned to either two or three species, primarily by where the trees come from. These morphologically intergrading, if geographically separated, regional populations are treated here as subspecies of a single species. Given the lack of clear morphological distinctions among them, treating these cedars as separate species would only be justifiable if they were reproductively isolated from one another, and there is presently no available systematic information about this.

The fossil record is also ambiguous about the antiquity and position of the true cedars. Pollen grains resembling those of *Cedrus* are found as early as the Jurassic, some 150 million years ago, and are widespread in the northern hemisphere, including North America. It is not clear that these fossils really belong to *Cedrus,* however, and identifiable needles, seeds, and cone scales are much later and much more restricted, being confined to western Eurasia from about the Oligocene, some 25 million years ago. True cedars were common in Europe throughout the Tertiary from then on but became restricted to southern Europe early in the Pleistocene ice ages and were finally eliminated from the continent. At some times during the Pleistocene, *Cedrus* spread

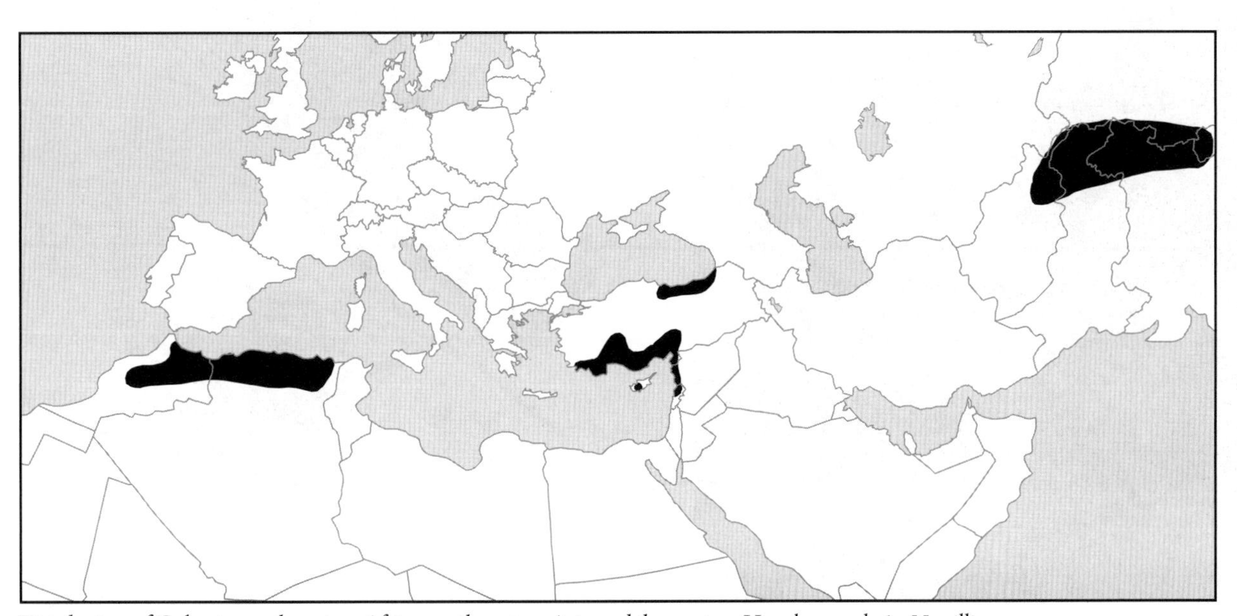

Distribution of *Cedrus* in northwestern Africa, southwestern Asia, and the western Himalaya, scale 1 : 55 million

across the Sahara in response to cooler, moister conditions. The western Himalayan portion of the range of the genus dates back at least to the late Pliocene, about 2 million years ago. Thus the present fragmented distribution of *Cedrus* is geologically very recent, a product of the Pleistocene glacial cycles.

Identification Guide to *Cedrus*

The two species of true cedar may be distinguished by the length of the needles on short shoots, the length of the winter buds, the length of the pollen cones, and the diameter of the largest seed cones:

Cedrus libani, short-shoot needles up to 2.5 cm, winter buds 2–3 mm, pollen cones up to 5 cm, seed cones up to 6 cm

C. deodara, short-shoot needles more than 2.5 cm, winter buds 1–2 mm, pollen cones more than 5 cm, seed cones more than 6 cm

Cedrus deodara (D. Don) G. Don

HIMALAYAN CEDAR, DEODAR CEDAR, DEODAR (HINDI)

Tree to 50(–75) m tall, with trunk to 3(–4.5) m in diameter. Bark grayish brown, narrowly and irregularly ridged. Crown remaining conical in forest-grown trees but breaking up and flattening in age on exposed sites, with long horizontal branches bearing drooping side branches. Young long shoots densely hairy with short, light brown hairs. Winter buds 1–2 mm long, not conspicuously resinous. Needles in tufts of 15–25(–30) on short shoots, dark green, 2.5–4(–6) cm long, with a long prickly point. Pollen cones (4–) 6–8 cm long, reddish brown. Seed cones 7.5–12.5 cm long, (5–) 6–8(–9) cm across, grayish green with reddish highlights before maturity, ripening dark reddish brown, broadly rounded, flattened, or indented at the tip. Seed scales 3–4 cm long, 3–4(–4.5) cm wide, with tan hairs on the hidden lower surface. Seed body 10–15 mm long, the wing 10–20 mm longer. Western Himalaya from eastern

Natural stand of young mature Himalayan cedars (*Cedrus deodara*).

Twig of Himalayan cedar *Cedrus deodara*) with 7 or more years of growth, ×0.4.

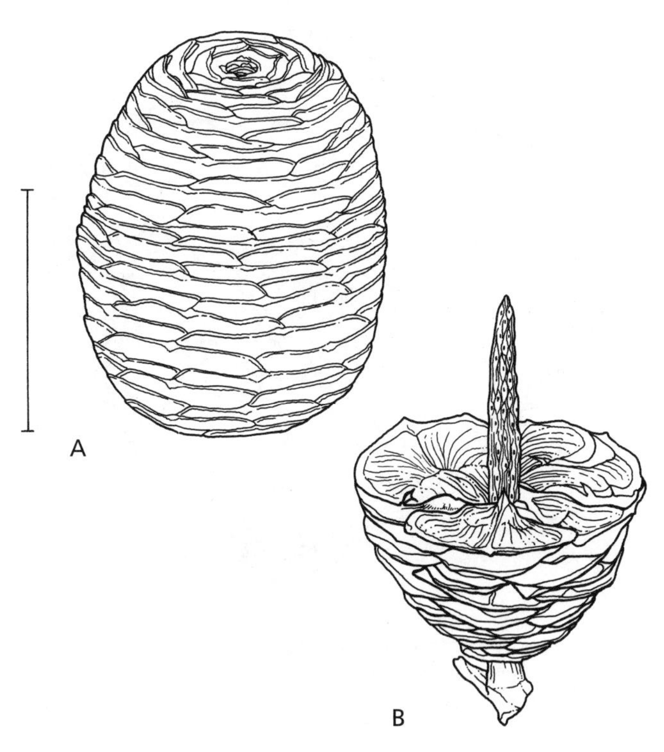

Mature seed cones of Himalayan cedar (*Cedrus deodara*) before (A) and after (B) beginning to shed the seed scales and seeds; scale, 5 cm.

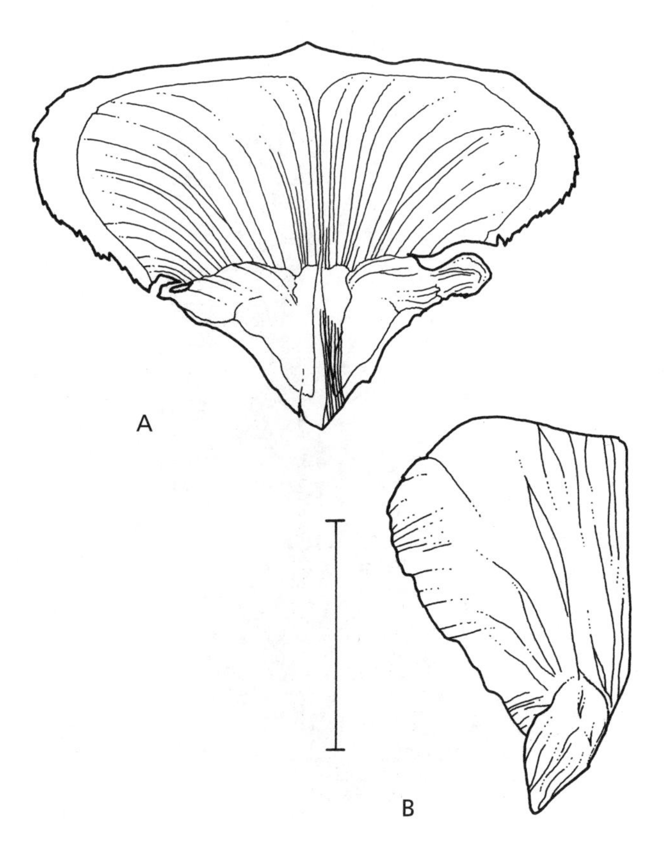

Upper side of shed seed scale (A) and seed (B) of Himalayan cedar (*Cedrus deodara*); scale, 1 cm.

Afghanistan to western Nepal. Forming extensive pure forests or mixed with other conifers and hardwoods in montane and subalpine forests; (1,200–)2,000–2,500(–3,300) m. Zone 7.

The Himalayan cedar is the grandest tree of the western Himalaya, giving credence to the scientific and Hindi common names, both derived from the Sanskrit *devadaru,* "tree of the gods." The celestial connection was reinforced, no doubt, by the qualities of the wood, the strongest softwood in the region and the one most resistant to decay and termite attack due to the characteristic oils that also give it its incenselike fragrance. Because of these qualities, the deodar cedar is cultivated to a limited extent as a plantation tree within its native range and beyond, as in China. On less favorable sites these plantations are subject to windthrow caused by the fungus *Heterobasidion annosum,* which kills and rots the roots. *Cedrus deodara* is also commonly planted as a handsome specimen tree wherever it is hardy, and numerous cultivars have been raised. Despite this indication of natural genetic variability, a study of enzymes showed no variation at all in five enzymes that were variable in *C. libani.* Perhaps *C. deodara,* like some species of pines (for example, *Pinus maximartinezii, P. resinosa,* and *P. torreyana*), has undergone an extreme reduction in population size at some time during or just before the present interglacial period.

Cedrus libani A. Richard

MEDITERRANEAN CEDAR

Tree to 30(–40) m tall or dwarfed at the alpine timberline, with trunk to 1.5(–2.5) m in diameter. Bark dark grayish brown, breaking up into vertically aligned, scaly blocks. Crown remaining conical in forest-grown trees but often broadening and flattening markedly with age in isolation, with long horizontal branches bearing horizontal or rising side branches. Young long shoots hairless or densely hairy with short, dark hairs. Winter buds 2–3 mm long, usually not conspicuously resinous. Needles in tufts of (15–)20–35(–45) on short shoots, dark green or grayish green with wax, 0.5–2.5 cm long (to 4 cm on long shoots), with a short point. Pollen cones 3–5 cm long, red. Seed cones 5–10(–12) cm long, 3–6 cm across, light green to grayish green with purplish highlights before maturity, ripening brown, broadly rounded, flat, indented, or with a central bump at the tip. Seed scales 2–3.5 cm long, 2.5–4 cm wide, with rusty hairs on the hidden lower surface. Seed body (8–)10–15 mm long, the wing 10–20 mm longer. Mediterranean region, in northwestern Africa and the northeastern Mediterranean. Forming pure stands or mixed with other conifers and hardwoods in montane and subalpine forests; (900–)1,300–2,500(–3,000) m. Zone 6. Three subspecies.

The range of the Mediterranean cedar is fragmented, and cedars from the different areas are often treated as three separate species. This is hardly justified since the trees in the different regions overlap substantially in their characteristics and are best treated as subspecies of a single species. Most distinctive is the population in Cyprus, which grows at lower elevations than the North African or adjacent mainland populations and is comparatively dwarfed in stature and in the number and length of the needles (explaining the subspecies name *brevifolia*). It also stands a little apart in its enzymes (although not as much as the Himalayan cedar) from the other two subspecies, which are essentially indistinguishable from each other in this feature. The Atlas cedar (subspecies *atlantica,* from the Atlas Mountains of northwestern Africa) is the most commonly cultivated subspecies of Mediterranean cedar and the one with the largest remaining natural populations. The cedar of Lebanon (subspecies *libani*), in contrast, has been subjected to ravages of overexploitation for thousands of years and is severely depleted, its regeneration further hampered by grazing by goats. The fleets of Egyptian pharaohs and of the Phoenicians, as well as the Temple of Solomon, were all built from the cedar forests that formerly clothed the Lebanon Mountains. The trees in southern Turkey are separated by some authors as a distinct subspecies or variety but are distinguished only by their narrower growth habit, and this has a lot to due with their growth in denser, more intact forests than may now be found in Syria and Lebanon. Numerous cultivars were raised as the Mediterranean cedar was carried around the warm temperate world, most derived from the African populations.

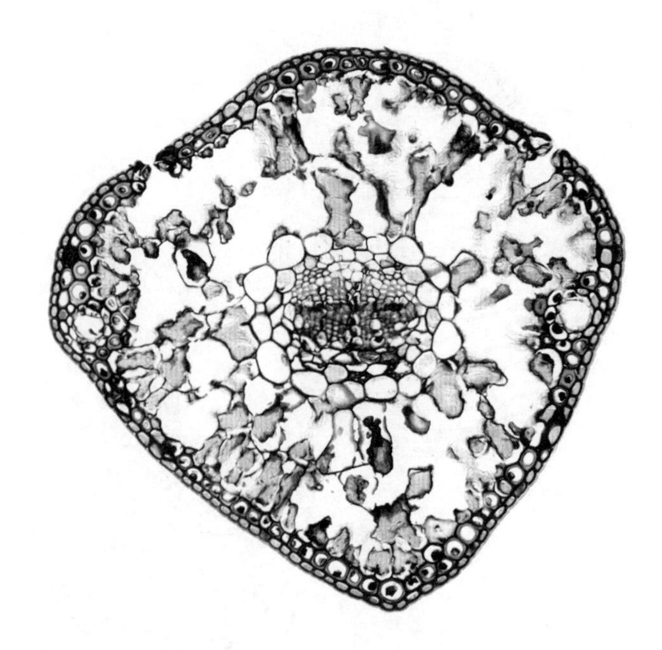

Cross section of preserved, stained, and sectioned needle of Atlas cedar (*Cedrus libani* subsp. *atlantica*) with fairly homogeneous photosynthetic tissue and a pair of resin canals.

Cedrus libani* subsp. *atlantica (Endlicher) Battandier & Trabut
ATLAS CEDAR
Plate 19
Short shoots with 20–35(–45) needles (1–)1.5–2.5 cm long. Pollen cones 3–4 cm long. Seed cones 5–8 cm long. Atlas and Rif Mountains of Morocco and Algeria. Synonyms: *Cedrus atlantica* (Endlicher) Carrière, *C. libani* var. *atlantica* (Endlicher) J. Hooker.

Cedrus libani* subsp. *brevifolia (J. Hooker) Meikle
CYPRUS CEDAR, KEDROS (GREEK),
KIBRIS SEDIRI (TURKISH)
Short shoots with 15–20 needles 0.5–2 cm long. Pollen cones 3–4 cm long. Seed cones (5–)7–10 cm long. Troodos Mountains of western Cyprus. Synonyms: *Cedrus brevifolia* (J. Hooker) Dode, *C. libani* var. *brevifolia* J. Hooker.

Underside of shoot of Atlas cedar (*Cedrus libani* subsp. *atlantica*) with 3 years of growth, ×1.

Upper side of shoot of cedar of Lebanon (*Cedrus libani* subsp. *libani*) with 5 or more years of growth, ×0.5.

Natural stand of mature cedars of Lebanon (*Cedrus libani* subsp. *libani*).

Cedrus libani A. Richard **subsp. *libani***

CEDAR OF LEBANON, ARZAT (ARABIC), KOKA KATRAN, SEDIR, LÜBNAN SEDIRI (TURKISH)

Short shoots with 20–35 needles (1–)2–2.5 cm long. Pollen cones 4–5 cm long. Seed cones 8–10(–12) cm long. Mountains adjacent to the northeastern Mediterranean coast, from the western Taurus Mountains of southwestern Turkey to the Mountains of Lebanon, with an outlier in Tokat province (Turkey) near the Black Sea. Synonyms: *Cedrus libani* subsp. *stenocoma* (O. Schwarz) Greuter & Burdet, *C. libani* var. *stenocoma* (O. Schwarz) P. H. Davis.

Cephalotaxus P. Siebold & Zuccarini ex Endlicher

PLUM YEW, COW'S TAIL PINE

Taxaceae

Evergreen trees and shrubs, usually with multiple stems from at or near the ground. Trunks cylindrical, eccentric, or sometimes fluted or buttressed. Bark obscurely or evidently fibrous, smooth and flaking in scales or peeling in narrow strips and becoming coarsely furrowed. Crown dense and dome-shaped in shrubs and some trees, open and irregular in others, with slender, gently upwardly angled branches initially in near whorls. Branchlets all elongate, without distinction into long and short shoots, hairless, remaining green for at least the first year, completely clothed by and deeply grooved between the elongate attached leaf bases. Resting buds well developed, with numerous hard brown bud scales persisting on the stem after shoot expansion. Leaves attached in opposite pairs with successive pairs spiraling around the twigs but presented in two flat, drooping, or V-shaped rows by bending and twisting of the petioles, Leaves sword-shaped, straight or slightly curved forward, with nearly parallel sides or tapering gradually toward the tip, flattened top to bottom, sometimes arching around the midrib.

Plants dioecious. Pollen cones in tight, spherical, scaly-stalked, spirally arranged clusters of 5–11 in the axils of leaves all along both sides of and underneath 1-year-old otherwise ordinary foliage shoots. Each pollen cone united with the supporting bract (except the terminal cone of the cluster) and with 1–10(–16) highly reduced pollen scales. Pollen scales each bearing 2–3(–7) by fusion of adjacent scales) egg-shaped, hanging pollen sacs. Pollen grains somewhat irregularly spherical with a low papilla at one end, small (25–35 µm in diameter), without air bladders, minutely and unevenly bumpy over the whole surface. Seed cones single or in groups of two to eight on separate scaly stalks in the axils of scale leaves or foliage leaves at or near the tip of otherwise ordinary 1-year-old shoots. Individual cones very compact, remaining fleshy, with four to eight pairs of fertile bracts, the successive pairs at right angles to one another. Each bract bearing two ovules on a highly reduced, attached seed scale, usually with only one (or two) seeds maturing in a cone. Seeds large, oval, the fleshy outer layer of the seed coat intimately united with a surrounding fleshy aril that extends almost all the way to the tip, maturing in two seasons. Cotyledons two, each with one vein. Chromosome base number $x = 12$.

Wood hard and heavy, the white to light brown sapwood moderately distinct from the brown heartwood. Grain very fine and even, with almost undetectable to moderately evident growth rings marked by very narrow bands of slightly denser latewood. Resin canals absent but usually with scattered or grouped individual resin cells and tracheids with spiral thickening.

Stomates confined to pale green or white stomatal bands beneath on either side of and at least as wide as the midrib region, each band consisting of 10–25 lines of stomates. Each stomate deeply sunken beneath and tucked under the four to six subsidiary cells, which are sometimes shared between adjacent stomates in a row (but not between rows), sometimes with a partial second circle of subsidiary cells along the sides, and flanked on both sides by a low ridge of cuticle, not quite forming a Florin ring. Midvein single, prominent, raised both above and beneath, with one small resin canal surrounded by two or more layers of wall cells immediately beneath it and wings of transfusion tissue on either side. Photosynthetic tissue with an irregular palisade layer directly beneath the upper epidermis without an intervening hypodermis, the palisade opening up beneath into the looser spongy mesophyll that extends down to the lower epidermis and the stomatal bands.

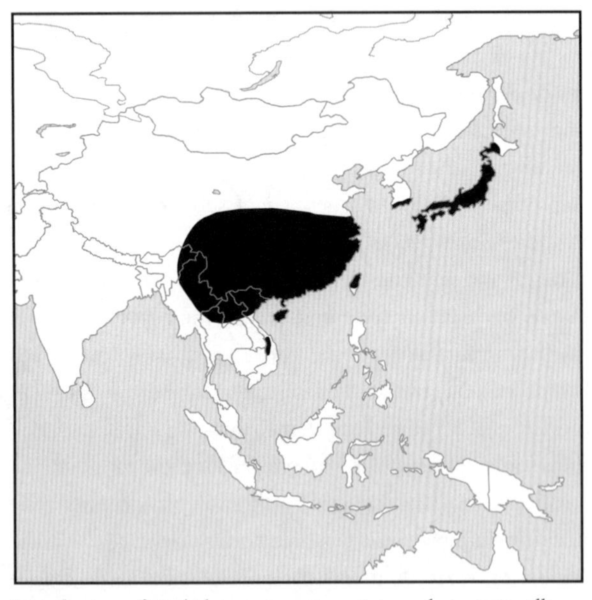

Distribution of *Cephalotaxus* in eastern Asia, scale 1 : 120 million

Five species in eastern and southern Asia from Japan and Korea to Indochina and eastern India. References: Cheng et al. 2000, Fu et al. 1999d, Gregor 1979, T. H. Nguyên and Vidal 1996, Lee 1979, Mai 1987, K. C. Sahni 1990, Schorn and Wehr 1996, Tripp 1995, Wilde 1975, Yamazaki 1995.

The scientific name (Latin for "head yew") refers to both the seed cones and the pollen cones, both of which are nearly spherical and very compact, constituting a head in botanical terminology. Members of the genus have been cultivated for centuries in eastern Asia and are now in modest general cultivation worldwide in suitable climates. Cultivar selection initially focused on the edible seeds but expanded subsequently to include variations in growth habit (especially fastigiate or spreading forms) and in size (longer or shorter than normal), attitude, and color of foliage (like golden or dark green).

Cephalotaxus is the only genus in the Taxaceae with seed cones that are readily identifiable as such. Because of this, Pilger in 1926 reversed his 1903 inclusion of the genus in Taxaceae and assigned it to a separate family all by itself. This separation was further emphasized by Florin in 1948, who kept *Cephalotaxus* among the conifers when he separated Taxaceae from the other conifers as a distinct order. Placement of *Cephalotaxus* in a monotypic family away from Taxaceae has been almost universally accepted since then, despite its very strong morphological similarities to other Taxaceae. Except for the supporting (and rather small) cones, the large plumlike seeds are rather similar to those of *Amentotaxus* and *Torreya,* as are the leaves. The distinctive leaf arrangement (technically referred to as bijugate) is like that of *Amentotaxus,* while the arrangement of the pollen cones is like that of *Torreya* though more complex. The pollen resembles that of *Pseudotaxus,* and pollen scales similar to those of *Taxus, Pseudotaxus,* and *Amentotaxus,* with five or more pollen sacs, are found at the tip of the terminal cone in the pollen cone clusters of *Cephalotaxus.* Spiral thickening of the wood cells (tracheids) is found in all these genera but is rare elsewhere in the conifers. Confirming these and other morphological, anatomical, and biochemical similarities, DNA studies show quite convincingly that the family Taxaceae, including *Cephalotaxus,* is a monophyletic group. These studies also show that *Cephalotaxus* constitutes one of three groups of genera (subfamilies) within the family, the other two being *Amentotaxus* and *Torreya* on the one hand and *Austrotaxus, Pseudotaxus,* and *Taxus* on the other. Which two of these three subfamilies are most closely related is somewhat ambiguous, although *Cephalotaxus* may be slightly more distantly related to the other two groups than they are to each other.

The division of *Cephalotaxus* into species has also been controversial. There has been no modern botanical monograph of the genus since Pilger's in 1903, and many floras and horticultural accounts recognize as many as 11 species instead of the five accepted here. Some of the additional proposed species are based on features like the color of the stomatal bands, which varies with the age of the foliage, or the presentation of the leaves, which varies with the amount of shade. Nonetheless, it is possible that some of the extra taxa could be recognized as additional botanical varieties of the five accepted species. On the other hand, *C. harringtonii, C. mannii,* and *C. sinensis* are similar enough to one another that they might be combined as a single species with further evidence. Clearly, more study of this group is needed. Unfortunately, this is becoming increasingly difficult as plum yews become rarer in nature due to direct exploitation of the larger species for wood and indirect damage through habitat destruction.

While *Cephalotaxus* is now confined to eastern and southern Asia, it was once much more widely distributed. Fossil foliage or seeds, or both, are known from the Eocene to the Pliocene of western North America (roughly 40 million to 4 million years ago) and from the Oligocene to the Pliocene of Europe (about 30 million to 4 million years ago) but are only known to extend back to the Miocene in eastern Asia (about 20 million years ago). However, some much older fossils, dating back to the Jurassic, more than 150 million years ago, share some similarities with *Cephalotaxus* and may be related to it, although this relationship is not firmly established.

Identification Guide to *Cephalotaxus*

The five species of *Cephalotaxus* may be distinguished by the length of most needles, their ratio of length to width, whether they overlap (or nearly so) side to side or are well separated from one another, and by the number of lines of stomates in each stomatal band:

Cephalotaxus oliveri, most needles less than 4 cm, length : width less than 10:1, overlapping, stomatal lines fewer than 18

C. sinensis, most needles less than 4 cm, length : width up to 10:1, separated, stomatal lines fewer than 18

C. mannii, most needles less than 4 cm, length : width up to 10:1, separated, stomatal lines more than 18

C. harringtonii, most needles less than 4 cm, length : width at least 10:1, separated, stomatal lines fewer than 18

C. fortunei, most needles at least 4 cm, length : width more than 10:1, separated, stomatal lines at least 18

Cephalotaxus fortunei W. J. Hooker

LONGLEAF PLUM YEW, FORTUNE PLUM YEW, SAN JIAN SHAN (CHINESE)

Tree or shrub to 20(–26) m tall, with trunk to 0.4 m in diameter. Bark reddish brown to purple, weathering gray, smooth at first, flaking in scales and then peeling in long shreds. Crown thin, open, and irregular. Branchlets turning greenish yellow in the second year. Leaves needlelike, well separated from one another, spreading flat or drooping at the tips, glossy dark green above, dull beneath, the light green to white stomatal bands much wider than the midrib region, each band with (13–)17–24 lines of stomates. Needles (1.5–)4–9(–12.5) cm long, (1.5–)2.5–5(–7) mm wide, sword-shaped, straight or slightly curved forward, widest before the middle, tapering very gradually to the narrowly triangular tip with a soft point up to 2 mm long, and more abruptly to the

Upper side of twig of longleaf plum yew (*Cephalotaxus fortunei*) with 2 years of growth, young pollen cone buds in the current year, and one leaf flipped over to the underside, ×0.5.

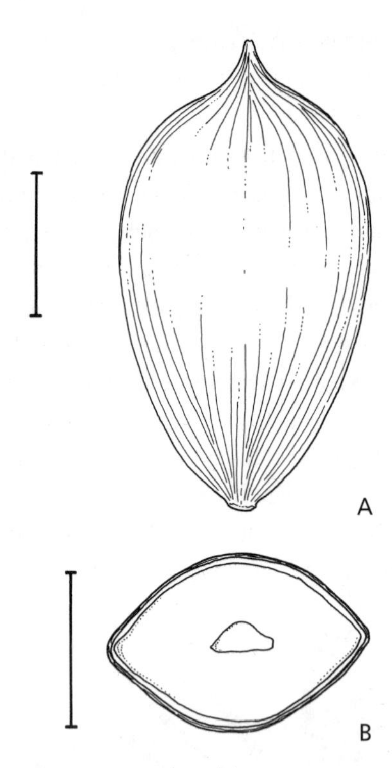

Face view (A) and cross section (B) of seed of longleaf plum yew (*Cephalotaxus fortunei*) with fleshy layer(s) removed; scale, 5 mm.

roundly wedge-shaped base on a very short petiole, flat around the midrib, the edges flat or narrowly turned down. Pollen cone clusters 5–10 mm in diameter with 6–14 pollen cones on a stalk (0–)2–5 mm long. Seed cones in groups of three to six on stalks 3–20 mm long. Seeds 1.5–4.5 cm long, the skin passing from yellowish green to purple or greenish brown with maturity, obscurely to prominently ribbed. Southern and central China and adjacent northern Myanmar, from southern Gansu and northern Shaanxi to Zhejiang, south to Yunnan, Guangxi, and Guangdong. Scattered in the canopy or understory of moist broad-leaved and mixed forest, from lowlands to subalpine forests, especially near streams, and in secondary forests and thickets; (200–)350–2,000(–3,700) m. Zone 7. Three varieties.

Longleaf plum yew is the most widely distributed species of *Cephalotaxus* in nature and, after Japanese plum yew (*C. harringtonii*), the second most common species of the genus in cultivation, where it is usually a spreading shrub with upswept branches to 2 m high. Old individuals may become much more massive and taller but usually remain multistemmed. Longleaf plum yew is one of the conifers commonly planted on temple grounds in China. Various authors treat each of the three botanical varieties as separate species, but they all seem to be each other's closest relatives and are readily distinguished as a group from the other species of the genus. The species name honors Robert Fortune (1812–1880), the Scottish plant explorer who smuggled living tea plants out of China for the East India Company and who collected the type specimen of this species in eastern China in 1848.

Cephalotaxus fortunei* var. *alpina H. L. Li

GAO SHAN SAN JIAN SHAN (CHINESE)

Shrub to 5(–13) m. Leaves (1.5–)3.5–7(–8) cm long, 1.5–3 (–3.5) mm wide. Pollen cones on stalks up to 2 mm long. Seeds reddish purple, 1.5–2.5 cm long, prominently ribbed. Western fringe of the range of the species, from northern Shaanxi to northern Yunnan (China). Understory of upper montane and subalpine coniferous and mixed forests; (1,100–)1,800–3,700 m. Synonym: *Cephalotaxus alpina* (H. L. Li) L. K. Fu.

Cephalotaxus fortunei W. J. Hooker **var. *fortunei***

SAN JIAN SHAN (CHINESE)

Tree or shrub to 20(–26) m. Leaves (3.5–)5–9(–12.5) cm long, (2–)3.5–5 mm wide. Pollen cones on stalks 2–5 mm long. Seeds reddish purple, 1.5–2.5 cm long, faintly striped. Throughout the range of the species except northern Shaanxi and northwestern Sichuan (China). Understory and canopy of lowland and lower montane mixed and broad-leaved forests and disturbed habitats; (200–)350–1,300 m. Synonyms: *Cephalotaxus fortunei* var. *concolor* Franchet, *C. kaempferi* K. Koch.

Cephalotaxus fortunei* var. *lanceolata (K. M. Feng) Silba

GONG SHAN SAN JIAN SHAN (CHINESE)

Tree or shrub to 20 m. Leaves 4.5–10 cm long, 4–7 mm wide. Pollen cones on stalks to 5 mm. Seeds greenish brown, 3.5–4.5 cm long, not obviously ribbed. Northwestern Yunnan (China) and adjacent northern Myanmar only. Scattered in the canopy and understory of moist montane hardwood forests along streams; around 1,900 m. Synonym: *Cephalotaxus lanceolata* K. M. Feng.

Cephalotaxus harringtonii (J. Knight ex Jas. Forbes) K. Koch

JAPANESE PLUM YEW, COW'S TAIL PINE, INUGAYA (JAPANESE), KAEBI CHA NAMU (KOREAN)

Plate 20

Tree or shrub to 8(–12) m, with trunk to 0.3(–0.6) m in diameter. Bark reddish brown, weathering grayish brown, peeling in elongate scales and becoming furrowed with age. Crown dense and fairly regular. Branchlets generally remaining green in the second year, sometimes becoming shiny brown. Leaves needlelike, well separated from one another, spreading flat or raised in a V, shiny

Upper (and under-) sides of twig of Japanese plum yew (*Cephalotaxus harringtonii*) with 2–5 years of growth, the actual number obscured by growth slowdowns, damage to shoot tips, and replacement branchlets, ×0.5.

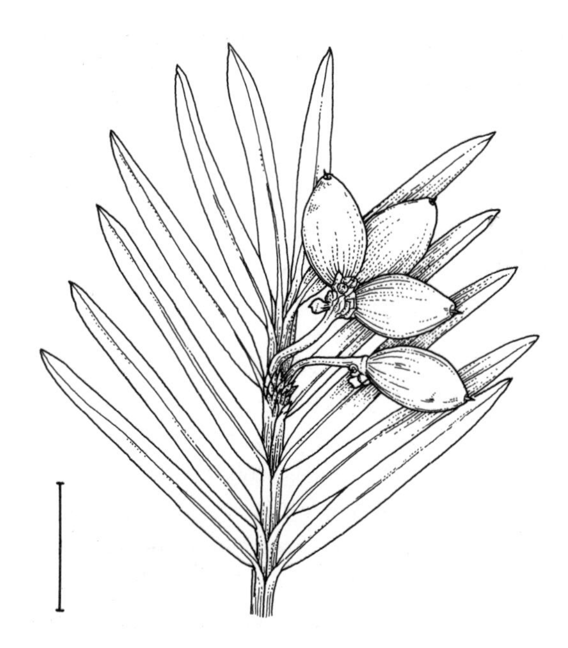

Twig of Japanese plum yew (*Cephalotaxus harringtonii*) with axillary, stalked seed cones; scale, 2 cm.

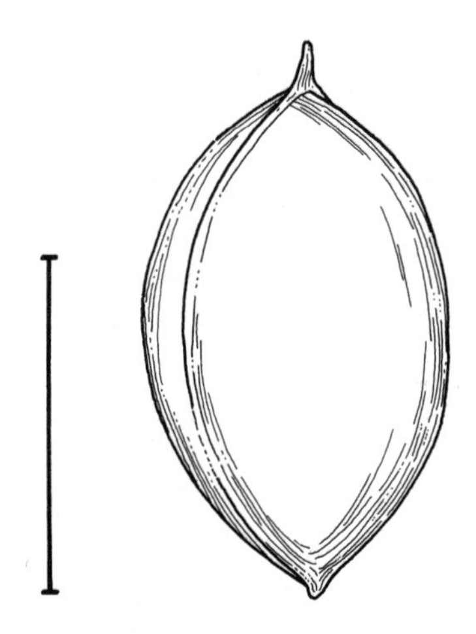

Seed of Japanese plum yew (*Cephalotaxus harringtonii*) with fleshy layer(s) removed; scale, 1 cm.

to dull dark green to yellowish green or bluish green above, duller beneath, the pale green to grayish green stomatal bands about twice as wide as the midrib region, each band with 12–15 lines of stomates. Needles (1–)2–3.5(–5) cm long, 2.5–4 mm wide, sword-shaped, straight or slightly curved forward, nearly parallel-sided, tapering abruptly to the roundly triangular tip, with a soft point less than 1 mm long, and to the broadly and roundly wedge-shaped base on a very short, thick petiole, flat around the midrib, the edges flat or narrowly turned down. Pollen cone clusters 4–8(–10) mm in diameter, with 6–10 pollen cones on a short stalk 3–4 mm long. Seed cones in groups of one or two (to six) on stalks 6–8(–12) mm long. Seeds (1.5–)2–2.5(–3) cm long, the skin passing through coppery yellow to red or purplish with maturity, obscurely ribbed. Common throughout Korea and the main islands of Japan. Scattered in the understory of deciduous hardwood forests; 0–1,000 m. Zone 7. Three varieties.

Japanese plum yew is the most common species of *Cephalotaxus* in cultivation in Europe and North America and is a little hardier than longleaf plum yew (*C. fortunei*). A number of cultivars have been selected, varying in habit, leaf color, and seed size and color, all of which must be propagated vegetatively. Seeds of Japanese plum yew have had a minor importance as human food and, contrastingly, as the source of an oil used in lamps. Although the wood is of good quality, the trees are usually so small that it is used primarily for small handmade objects like kitchen utensils. It was also used historically for making traditional bathtubs, taking advantage of the watertight qualities it shares with Japanese nutmeg yew (*Torreya nucifera*), which was the preferred wood for this purpose. Since the mid-1960s, the Japanese plum yew, more than any other species of *Cephalotaxus*, has been investigated intensively for its rich content of anticancer chemicals belonging to several groups of alkaloids. Those that are potent enough to be undergoing clinical trials against leukemia are called harringtonines after this species, one subset of a larger group of poisons named cephalotaxines after the genus. These compounds have some biosynthetic connection to the anticancer taxane alkaloids of yews (*Taxus*) and thus provide another link with other genera in the Taxaceae. The species name honors Charles Stanhope, the fourth Earl of Harrington, (1780–1851), whose famous garden of hedges and topiaries included examples of Japanese plum yew.

Cephalotaxus harringtonii (J. Knight ex Jas. Forbes) K. Koch var. *harringtonii*

JAPANESE PLUM YEW, INUGAYA (JAPANESE)

Tree to 8(–12) m tall, not suckering from horizontal branches or roots. Seeds dull purplish red, the flesh with an unpleasant resinous smell and taste. Pacific Ocean side of Honshū south through Shikoku and Kyūshū; 600–1,000 m. Synonyms: *Cephalotaxus drupacea* P. Siebold & Zuccarini, *C. drupacea* var. *pedunculata* (P. Siebold & Zuccarini) Miquel, *C. harringtonii* var. *drupacea* (P. Siebold & Zuccarini) G. Koidzumi, *C. pedunculata* P. Siebold & Zuccarini.

Cephalotaxus harringtonii var. *koreana* (T. Nakai) Rehder

KOREAN PLUM YEW, KAEBI CHA NAMU (KOREAN)

Low, tufted, upright shrub to 1.5(–3) m, not suckering from horizontal branches or roots. Seeds bright red to dull purplish red, the flesh sweet and tasty. Korea; low elevations. Synonym: *Cephalotaxus koreana* T. Nakai.

Cephalotaxus harringtonii var. *nana* (T. Nakai) Rehder

HAI-INUGAYA (JAPANESE), ANATNI (AINU)

Low, spreading shrub to 2 m, suckering from horizontal branches and roots. Seeds red, the flesh edible but not especially sweet. Hokkaidō and Japan Sea side of Honshū; 0–600 m. Synonym: *Cephalotaxus nana* T. Nakai.

Cephalotaxus mannii J. Hooker

KHASI PLUM YEW, HAINAN CUFEI (CHINESE), ĐI'NH TÙNG, PHI LƯỢC BÉ (VIETNAMESE)

Tree to 20(–25) m tall, with trunk to 0.6(–1.1) m in diameter. Bark light brown to reddish purple, smooth, with scattered bumpy lenticels, flaking away in scales. Crown dense at first, thinning with age and becoming broadly and shallowly dome-shaped, with horizontal branches turning up at the ends. Branchlets remaining green in the second year. Leaves needlelike, well separated from one another, spreading flat or raised in a shallow V, shiny dark green to yellowish green above, duller beneath, the pale green to grayish white or white stomatal bands more than three times as wide as the midrib region, each band with (10–)20–25 lines of stomates. Needles (1.5–)2.5–3.5(–5) cm long, (1.5–)2.5–4 mm wide, sword-shaped, straight or slightly curved forward near the tip, nearly parallel-sided or widest near or before the middle, tapering very gradually to the narrowly or roundly triangular tip with a soft point to 0.2 mm long, and more abruptly to the roundly wedge-shaped or broadly rounded base on a very short petiole, flat around the midrib, the edges narrowly turned down. Pollen cone clusters 4–5(–9) mm in diameter, with 6–10 pollen cones on a short stalk 1.5–4 mm long. Seed cones in groups of one to three (to five) on stalks 6–10(–15) mm long. Seeds (2–)2.2–3(–3.8) cm long, the skin passing through green to red with maturity. Very discontinuously distributed in hill country from northeastern India (eastward from the Bhutan border in Arunachal Pradesh and from the Khasi Hills in Meghalaya) and northern Myanmar, through southern China to southwestern Guangdong and Hainan and south in Vietnam to Lâm Dông province. Scattered in the understory or canopy of seasonally wet mixed or broad-leaved lower montane forests in valleys and on moist slopes on a variety of substrates; (500–)700–2,000 (–2,750) m. Zone 9. Synonyms: *Cephalotaxus griffithii* J. Hooker, *C. hainanensis* H. L. Li.

Khasi plum yew is the most tropical of the *Cephalotaxus* species. It occurs south and west of the range of its close relative Chinese plum yew (*C. sinensis*). These two species and Japanese plum yew (*C. harringtonii*) overlap in many characters, and their relationships to one another merit further study. Khasi plum yew yields a locally important timber, and extracts of the bark and foliage are used in traditional and modern medicine in China and Vietnam, presumably exploiting the rich array of alkaloids found in members of the genus, including anticancer agents. Stripping the bark kills the tree, as it does with Pacific yew (*Taxus brevifolia*) in North America. As a result of direct exploitation for these resources, and indirectly through habitat destruction, Khasi plum yew has been severely depleted in some regions. This is especially true on the island of Hainan, where the populations are sometimes treated as a separate species. Here the trees reach their greatest size and were once common but have been so heavily exploited that reproduction has been adversely affected in these dioecious plants. The species is also threatened in the eastern Himalaya, where there is enough variation that the scattered populations are sometimes assigned to two different species, even though there is little direct exploitation of the rather smaller trees. The species name honors Gustav Mann (1836–1916), a German-born plantsman and forester who collected the type specimen in the Khasi Hills of Meghalaya state during his long career with the Indian Forest Service.

Cephalotaxus oliveri M. T. Masters

OLIVER PLUM YEW, BI ZHI SAN JIAN SHAN (CHINESE)

Shrub, or tree to 4 m tall, with trunk, when present, to 0.2 m in diameter. Bark yellowish brown to grayish brown, smooth and peeling in scales. Crown thin and irregular. Branchlets remaining green in the second year or turning brown. Leaves needlelike, closely spaced so that adjoining leaves often touch or slightly overlap with one another, spreading flat or raised in a shallow V and drooping at the tips, shiny or dull dark green above, duller beneath, the white to whitish green stomatal bands about as wide as the midrib region, each band with 13–17 lines of stomates. Needles 1.5–2.5(–3.5) cm long, (2.3–)3–4(–4.5) mm wide, sword-shaped, straight or gently curved forward, nearly parallel-sided, tapering abruptly to the roundly triangular tip with a stiff, brittle point about 1 mm long, abruptly giving way to the squared-off or heart-shaped base on a very short, scarcely evident petiole to 0.5 mm long, arched around the midrib, the edges flat. Pollen cone clusters 5–7 mm in diameter with five to seven (to nine) pollen cones on a short stalk 2–3 mm long. Seed cones single or in groups of two or three on stalks 6–12 mm long. Seeds 2–3 cm long, the skin passing from pale reddish brown to red with maturity. Endemic to south-central China, from the western border of Yunnan to southwestern Hubei, eastern Jiangxi, and northwestern Guangdong. Scattered in the understory of moist warm temperate evergreen and mixed lower montane hardwood forests, especially in valleys near streams; 300–1,000(–1,800) m. Zone 8.

Oliver plum yew is the most morphologically distinctive species in the genus, especially in the shape and overlapping arrangement of the leaves. It cannot easily be confused with any of the other *Cephalotaxus* species. Like the other species, it is rich in cephalotaxine alkaloids, but its rarity and continuing decline are due more to general habitat destruction than to overexploitation for these cancer-fighting chemicals. It is, perhaps, the species in the genus most dependent on the moisture and shade of the mature forest understory and declines with forest disturbance and clearance. Its best-known stands are on Emei Shan (Mount Omei), one of the three sacred mountains of Sichuan, where it is relatively protected. The species name honors Daniel Oliver (1830–1916), keeper of the Royal Botanic Gardens, Kew, 1864–1890, who published an illustration of the species under a mistaken name while he was editor of *Icones Plantarum.*

Cephalotaxus sinensis (Rehder & E. H. Wilson) H. L. Li

CHINESE PLUM YEW, CUFEI (CHINESE)

Shrub, or tree to 8(–15) m tall, with trunk to 0.4(–1.2) m in diameter. Bark reddish brown or grayish brown, weathering gray, smooth at first, flaking in scales and becoming furrowed with age. Crown dense and symmetrical when young, becoming more irregular and open with age, with horizontal to noticeably drooping branches. Branchlets remaining green or turning light reddish brown in the second year. Leaves needlelike, slightly to well separated from one another along the twig, spreading flat to raised in a prominent V, shiny bright green to olive green above, duller beneath, the whitish green to brilliantly white or yellowish green stomatal bands about two to three (to four) times as wide as the midrib region, each band with (11–)13–15(–18) lines of stomates. Needles (1–)1.5–4(–7) cm long, 2–4(–6) mm wide, sword-shaped, straight or curved forward, nearly parallel-sided to evidently widest before or near the middle, tapering abruptly or gradually to the roundly or narrowly triangular tip with a soft point (0.2–)0.5–1(–1.5) mm long, and abruptly to the roundly wedge-shaped or rounded base on a short petiole to 1(–2.5) mm long, flat around the midrib, the edges flat or narrowly turned down. Pollen cone clusters (3–)4–7(–10) mm in diameter, with six or seven (to nine) pollen cones on a short stalk 1.5–3 mm long. Seed cones in groups of (one or) two to five (to eight) on stalks 2–8 mm long. Seeds (1.5–)1.8–2.5 cm long, the skin passing from greenish yellow to red or brownish purple with maturity and sometimes covered by a thin waxy film. Widely but discontinuously distributed in the southeastern third of China, from southern Gansu east to southern Jiangsu and south to the southern tier of mainland provinces from Yunnan to Guangdong and in Taiwan. Rare and widely scattered to locally common in the understory of various humid upper and lower montane forest types and thickets, especially in valleys and by streamsides on a wide range of substrates; (600–)900–2,500(–3,200) m. Zone 7. Two varieties.

Although Chinese plum yew is almost as widely distributed as longleaf plum yew (*Cephalotaxus fortunei*) and grows as a handsome, compact shrub, it is much less commonly cultivated outside of China than is the latter species. Likewise, it is less commonly cultivated than the much less naturally widely distributed Japanese plum yew (*C. harringtonii*), with which it was long confused. Chinese plum yew, Japanese plum yew, and Khasi plum yew (*C. mannii*) are all very similar to one another morphologically and replace each other geographically. The limited molecular studies available also suggest a close relationship. For one gene, called *mat*K, DNA studies found only one base pair difference between Chinese plum yew and Khasi plum yew, while there are five between Chinese plum yew and longleaf plum yew for the same gene. Thus the relationships of these three species to one another merit further detailed comparison involving populations from throughout the range of all three species. Plants with two subtly different leaf forms, accompanied by differences in the scaliness of the pollen cone stalks, are both found throughout the central portion of the range of Chinese plum yew. One of these forms is recognized as a separate species by some authors, but the differences appear minor, and the two agree in the majority of their features. A much more detailed understanding of the distribution and ecology of these forms is necessary to clarify their status. They could be unusually overlapping botanical varieties or ecotypes elicited by some environmental conditions that are not yet understood in the absence of a thorough investigation in the field and laboratory. It also seems possible that occasional hybridization with the co-occurring longleaf plum yew might be contributing to the complexity since some plants of Chinese plum yew have much longer needles than usual for the species. In contrast, the plants from Taiwan, also traditionally treated as a separate species can be comfortably considered a separate botanical variety, even though they are only weakly differentiated from the mainland plants, because of their geographic separation. Like the other species of *Cephalotaxus,* Chinese plum yew is a source of powerful anticancer cephalotaxine alkaloids, found throughout the plant, including the seeds. Uses of the attractive wood vary with the size of the plants. While furniture can be made from large individuals, shrubby specimens are used for small household items and carvings. Both forms of exploitation are cause for some conservation concern, especially in Taiwan, where the species is rather rare.

Cephalotaxus sinensis (Rehder & E. H. Wilson) H. L. *Li* var. *sinensis*

Usually a shrub to 4 m. Needles often raised in a V, commonly nearly parallel-sided, often less than 10 times as long as wide. Stomatal bands often very white with a thick layer of wax. Throughout the whole range of the species except Taiwan. 600–3,200 m. Synonyms: *Cephalotaxus drupacea* var. *sinensis* Rehder & E. H. Wilson, *C. harringtonii* var. *sinensis* (Rehder & E. H. Wilson) Rehder, *C. latifolia* W. C. Cheng & L. K. Fu ex L. K. Fu & R. Mill, *C. sinensis* var. *latifolia* W. C. Cheng & L. K. Fu (name invalid).

Cephalotaxus sinensis var. *wilsoniana* (Hayata) L. K. Fu & N. Li

Usually a tree to 10 m tall. Needles usually held flat, tapering from near the middle, usually a little more than 10 times as long as wide. Stomatal bands whitish green with a thin layer of wax. Taiwan only. 1,400–2,000 m. Synonyms: *Cephalotaxus harringtonii* var. *wilsoniana* (Hayata) Kitamura, *C. wilsoniana* Hayata.

Chamaecyparis Spach

CEDAR, CYPRESS, FALSE CYPRESS

Cupressaceae

Evergreen trees with a single to multiple straight trunks clothed in fibrous, furrowed bark that peels in long, thin strips. Open-grown trees densely branched from the base with upwardly arcing branches drooping at the tips. Crown conical to cylindrical, even when raised far above the ground on a bare trunk in mature forest-grown trees. Branchlets emerging alternately from lateral leaves on either side or just along the far side of a central shoot to form flattened, fernlike, horizontal or drooping sprays. Without definite winter buds. Seedling leaves in alternating quartets giving way to alternating pairs, needlelike, standing out from and well spaced on the stem, soon replaced by adult branchlets (except on permanently juvenile cultivars, the retinosporas). Adult leaves in alternating pairs, scalelike, overlapping, the facial and lateral leaves similar in size but the lateral leaves folded around the twig, keeled or not and with or without a resin gland.

Plants monoecious. Pollen cones numerous, each single at the tip of a branchlet, oblong. Pollen scales in three to eight alternating pairs, each scale with two to four pollen sacs. Pollen grains small (25–35 µm in diameter), spherical, minutely bumpy, with or without a small, ill-defined, circular germination pore. Seed cones numerous, single at the tips of short side branchlets, maturing in a single season, spherical to oblong, with three to seven alternating pairs of centrally attached seed scales. Each seed scale interpreted as derived from intimate fusion of the bract with the seed portion, the external face broadly hexagonal with a central horizontal line dividing the upper seed-bearing portion from the lower bract portion. Bract with a free, pointed, triangular tip. Uppermost pair of scales often fused into a single rodlike structure, the middle pairs fertile with (one or) two to four seeds per scale. Seeds oval, with two equal wings derived from the seed coat running the length of the body and about as wide. Cotyledons two, each with one vein. Chromosome base number $x = 11$.

Wood slightly to strongly fragrant, light, soft to moderately hard and strong, resistant to decay, with white to pale yellow sapwood sharply contrasting with the light yellowish brown to pinkish brown heartwood. Grain fairly even and very fine to moderately coarse, with well-defined growth rings marked by an abrupt transition to a few to several rows of smaller, thicker-walled latewood tracheids. Resin canals absent but with a fair number of individual resin parenchyma cells partially scattered through the growth increment and often concentrated in narrow bands.

Stomates arranged in patches or bands of varying position among the species, at least some patches conspicuous with deposits of whitish wax. Each stomate sunken beneath and largely hidden by the four to six surrounding subsidiary cells, which are often shared between adjacent stomates within and between rows, and which are topped by a steep, high, nearly continuous Florin ring. Sometimes also with a partial, irregular outer circle of subsidiary cells, and these and other epidermal cells often bearing large, round to greatly elongated papillae. Leaf cross section with a single large resin canal (which does not extend the whole length of the leaf) tucked under the single-stranded midvein, which is flanked by patches of transfusion tissue. Photosynthetic tissue forming a well-developed palisade of several cell layers all over the exposed portion of the leaf beneath the epidermis and thin, nearly continuous hypodermis but most of the photosynthetic volume inside consisting of spongy mesophyll laced with conspicuous large air spaces.

Five species in North America, Japan, and Taiwan. References: Farjon 2005b, Fu et al. 1999g, Li and Keng 1994, McIver 1994, Michener 1993, Yamazaki 1995, Zobel 1998.

Species of *Chamaecyparis* have a long history of cultivation. Because so many cultivars have been selected, each species constitutes a separate cultivar group for naming purposes under the *International Code of Nomenclature for Cultivated Plants.* As a consequence, you can find the same cultivar name (especially among older, Latinized ones) applied to two or more species unlike in most other conifer genera. Cultivar selection has embraced almost every vegetative aspect of the plants. There are permanently juvenile cultivars (called retinosporas and formerly sometimes treated as a separate genus), ones with sparsely branched foliage (the threadleaf or filifera types) or more densely branched, and with sprays that may be unusually elongated or foreshortened and exceptionally flat or intricately twisted and three-dimensional. Solid foliage colors embrace an enormous spectrum of greens, blues, and golds (and purples in winter), and variegated cultivars are legion. Growth habits range from enormous trees to compact dwarfs, with stiffly up-thrust to dolorously weeping branching, so much so in some cultivars that the plants never pick themselves up off the ground but spread from a central pile like a fountain. It is no exaggeration to say that there is a *Chamaecyparis* cultivar (or a few dozen) for virtually every position and design requirement in any garden (in an appropriate climatic zone, of course), from the smallest patch of green to the most monumental estate.

Even the smaller species of the genus, *Chamaecyparis pisifera* and *C. thyoides,* at up to 30 m tall or more, belie the Greek etymology of the generic name as "dwarf cypresses," and *C. formosensis* and *C. lawsoniana* at up to 65 m or more are truly gigantic. The genus has long been considered close to *Cupressus,* and many authors into the middle of the 20th century included these species in that genus. Since then, certain species, like *Cupressus funebris,* with its flattened sprays of branchlets and relatively small cones, have been assigned fairly indiscriminately to either genus. The apparent close relationship was also emphasized by hybridization between various species of *Cupressus* and the species usually known as *Chamaecyparis nootkatensis,* the Alaska yellow cedar, to form the hybrid (notho-) genus ×*Cupressocyparis.* However, the Alaska yellow cedar has

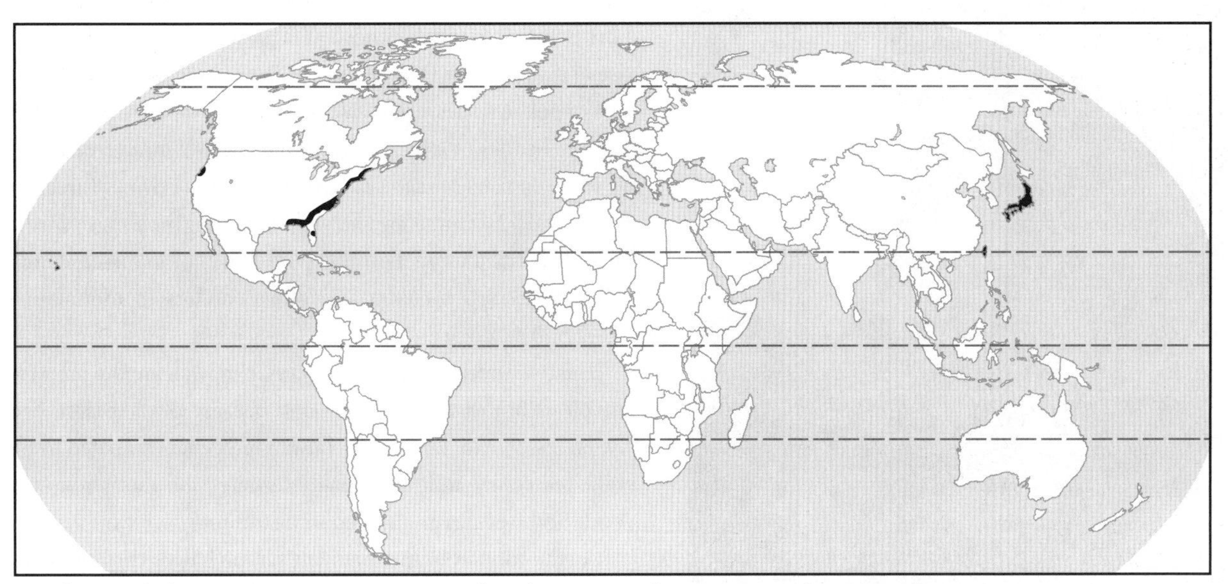

Distribution of *Chamaecyparis* in maritime regions of North America and eastern Asia, scale 1 : 200 million

long been recognized as anomalous within *Chamaecyparis,* with its four-scaled seed cones that take 2 years to mature and a leaf and heartwood chemistry that is closer to *Cupressus* than to the other relatively uniform species that are assigned to *Chamaecyparis.* DNA sequences strongly support a close alliance of the Alaska yellow cedar with *Cupressus,* as *Cupressus nootkatensis,* but show that the other *Chamaecyparis* species are not particularly closely related to *Cupressus.* Instead, *Chamaecyparis* is very close to *Fokienia,* which has similar cones but more flattened foliage sprays with broader lateral leaves.

The earliest known fossils of *Chamaecyparis* were recorded from the late Cretaceous of Vancouver Island, Canada, and antedate the Paleocene occurrences of the earliest known *Fokienia.* These fossils often have branchlets arising from both members of a pair of lateral leaves, a trait no longer found in *Chamaecyparis* and now confined to the southern hemisphere incense cedars (*Libocedrus* and related genera).

Old Taiwan red cedar (*Chamaecyparis formosensis*) in nature.

Identification Guide to *Chamaecyparis*

The five species of *Chamaecyparis* may be distinguished by whether the tips of the facial leaves touch the base of the next pair, how long the seed cones are, how many pairs of scales they have, and what color the pollen sacs are:

Chamaecyparis pisifera, tips of facial leaves touching, seed cones less than 8 mm, with up to six pairs of scales, pollen sacs yellow

C. thyoides, tips of facial leaves not touching, seed cones up to 8 mm, with fewer than five pairs of scales, pollen sacs yellow

C. lawsoniana, tips of facial leaves not touching, seed cones at least 8 mm, with up to five pairs of scales, pollen sacs red

C. obtusa, tips of facial leaves not touching, seed cones more than 8 mm, with up to five pairs of scales, pollen sacs yellow

C. formosana, tips of facial leaves not touching, seed cones more than 8 mm, with at least five pairs of scales, pollen sacs yellow

Chamaecyparis formosensis J. Matsumura

TAIWAN RED CEDAR, HONG KUAI (CHINESE)

Tree to 50(–65) m tall, with trunk to 5(–7) m in diameter. Bark reddish brown. Crown deeply dome-shaped. Branchlet sprays stiff or slightly drooping, with waxy white stomatal patches beneath. Scale leaves 1.5–3 mm long, without glands. Leaf tips short-pointed, triangular, those of facial leaves not touching the base of the next pair. Pollen cones 2–3 mm long, with six to eight pairs of pollen scales, each scale usually with three yellow pollen sacs. Seed cones oblong, 10–12 mm long, 8–9 mm across, with five to seven pairs of seed scales. Seeds (one or) two or three (to six) per scale, 2–3 mm long and wide, including the wings. Northern and central Taiwan. Forming pure groves or in mixed stands with *Chamaecyparis obtusa* and other conifers in mountains; 850–2,600 m. Zone 8.

The two species of *Chamaecyparis* in Taiwan are the premier coniferous timber trees of the island because of their large size (the largest tree in Taiwan is a Taiwan red cedar estimated to be 3,000 years old) and high quality termite- and decay-resistant wood. Both species are limited in distribution, with only about 7.5% of the coniferous forest land in Taiwan having their required combination of cool temperatures and high moisture content. A combination of shade intolerance, leading to minimal regeneration in established stands, and poor harvesting practices led to declines of both species. Although they often grow together, they actually occupy distinct but overlapping elevational bands, with Taiwan red cedar at lower elevations and Taiwan yellow cedar (*C. obtusa*) above it. Taiwan red cedar is the only species of the genus with oblong seed cones, the others having spherical ones. Taiwan was called Formosa by the occupying Japanese when the species was described so many Taiwanese species, like this one, bear either the epithet *formosensis* or *formosana*.

Upper side of branchlet spray of Port Orford cedar (*Chamaecyparis lawsoniana*) with bracts remaining after shedding pollen cones on growth during a single season, ×1.

Chamaecyparis lawsoniana (A. Murray bis) Parlatore
PORT ORFORD CEDAR, LAWSON CYPRESS, GINGER PINE
Plate 21

Tree to 50(–75) m tall, with trunk to 2(–5) m in diameter. Bark reddish brown, rather smooth, peeling in flakes at first and later in thin strips. Crown a relatively narrow spire. Lowest branches often layering to form groves around the parent tree. Branchlet sprays drooping, with waxy white X's beneath outlining the leaf

Clump of older Port Orford cedars (*Chamaecyparis lawsoniana*) in nature.

Natural stand of young mature Port Orford cedars (*Chamaecyparis lawsoniana*).

contacts. Scale leaves 2–3 mm long, with an elongate gland. Leaf tips triangular or protracted, those of the facial leaves not touching the base of the next pair. Pollen cones 2–4 mm long, with six to eight pairs of pollen scales, each scale with two or three red pollen sacs. Seed cones spherical, 8–12 mm in diameter, with three to five pairs of seed scales, these with a bluish green waxy bloom on the faces before maturity. Seeds two to four (to six) per scale, 2–5 mm long, the wings as broad as or broader than the body. Northwestern California and southwestern Oregon. Forming small groves or scattered in mixed conifer forests on a variety of substrates, including serpentine, within the coastal fog belt; 0–1,500 m. Zone 7.

The red pollen sacs of Port Orford cedar are one of its most distinctive features. Despite a lack of botanical varieties, it has proven to be extraordinarily variable in cultivation, producing numerous cultivars of distinctive size, growth habit, and foliage density, form, and color. Unfortunately, it has also proven to be susceptible to root-destroying fungal diseases, particularly when grown in inland localities away from the maritime climate characteristic of its limited natural range. Many of its named cultivars have thus been lost to gardens and from commerce. In some ways, the cultivars of the unrelated Leyland cypress (*Cupressus* ×*leylandii*) are more reminiscent of those of Port Orford cedar than they are of their own parents, Alaska yellow cedar (*C. nootkatensis*) and Monterey cypress (*C. macrocarpa*). The largest known individual in combined dimensions when measured in 1997 was 69.8 m tall, 3.6 m in diameter, with a spread of 11.9 m.

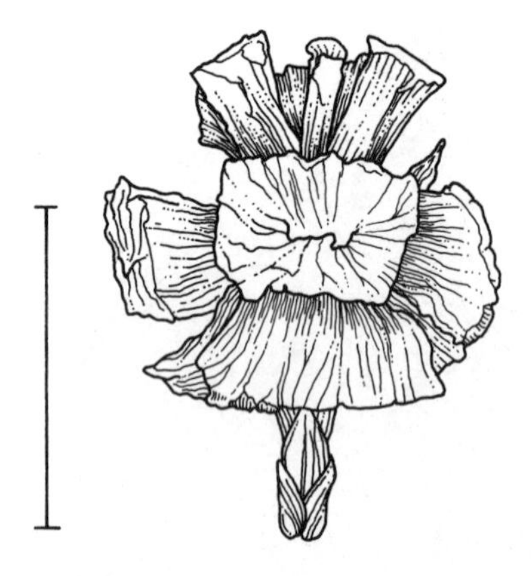

Open mature seed cone of Port Orford cedar (*Chamaecyparis lawsoniana*) after releasing seeds; scale, 1 cm.

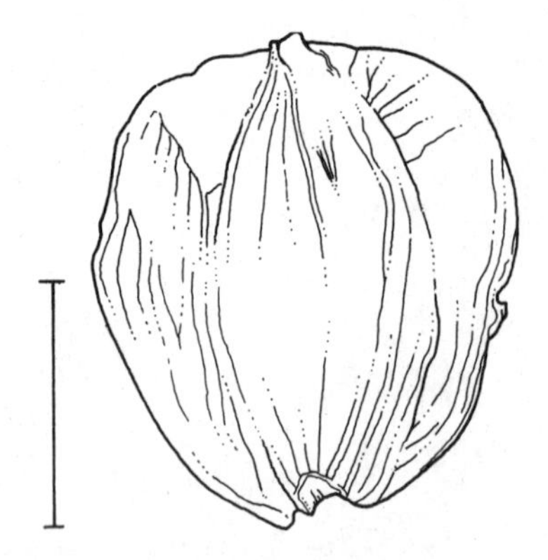

Face view of dispersed seed of Port Orford cedar (*Chamaecyparis lawsoniana*); scale, 2 mm.

Chamaecyparis obtusa (P. Siebold & Zuccarini) Endlicher

HINOKI CYPRESS, HINOKI (JAPANESE), TAIWAN BIAN BAI (CHINESE)

Tree to 30(–52) m tall, with trunk to 1.5(–3) m in diameter. Bark reddish brown, rather smooth, peeling in wide, thin strips. Crown broadly conical. Branchlet sprays drooping, often fan-shaped, with waxy white Y's beneath outlining the leaf contacts. Scale leaves 1–2.5 mm long, dark green, with a round gland. Leaf tips blunt (hence the scientific name), those of the facial leaves not touching the base of the next pair. Pollen cones about 4 mm long, with three to five pairs of pollen scales, each scale with three or four yellow pollen sacs. Seed cones spherical, 9–12 mm in diameter, with four or five pairs of seed scales. Seeds two to four (to six) per scale, 2–4 mm long, the wings no wider than the body. Japan and Taiwan; (10–)1,000–2,500(–2,800) m. Two varieties.

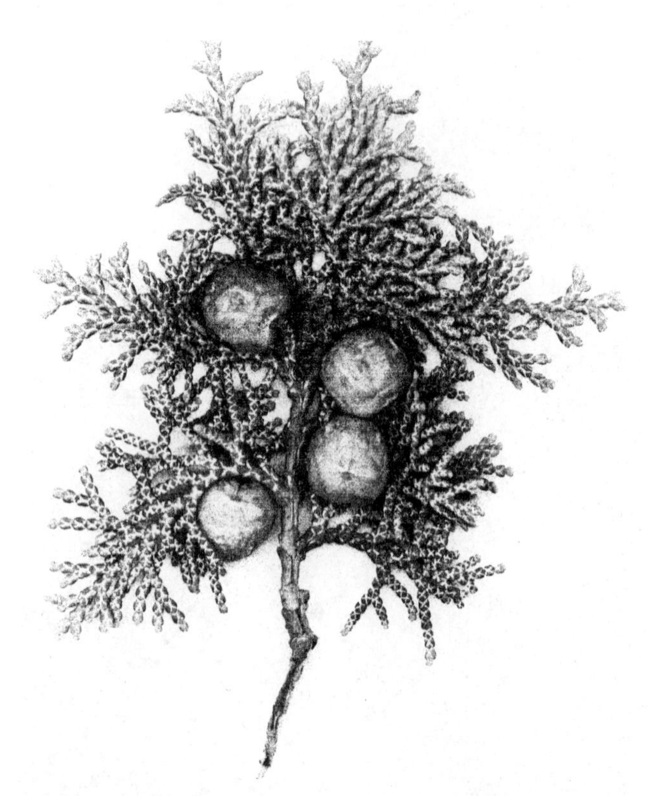

Underside of foliage spray of hinoki cypress (*Chamaecyparis obtusa* 'Gracilis') with green mature seed cones, ×1.

Like other temperate species of *Chamaecyparis,* hinoki cypress has produced many distinctive cultivars, all from the Japanese variety *obtusa.* In both of its native regions, it is a prized timber tree that has been severely exploited. In Japan it is also used in afforestation. Trees from the two regions are sometimes recognized as separate species, but they are very similar, differing primarily in stature and leaf and seed size.

Chamaecyparis obtusa* var. *formosana (Hayata) Rehder

TAIWAN YELLOW CEDAR, TAIWAN BIAN BAI (CHINESE)

Tree to 40 m tall, with trunk to 3 m in diameter. Scale leaves 0.8–2 mm long. Seeds 3–4 mm long. Northern and central Taiwan. Forming pure groves or mixed stands with Taiwan red cedar (*Chamaecyparis formosensis*) and other conifers, generally higher up than Taiwan red cedar; 1,200–2,800 m. Zone 8. Synonym: *C. taiwanensis* Masamune & Sig. Suzuki.

Chamaecyparis obtusa (P. Siebold & Zuccarini) Endlicher **var. *obtusa***

HINOKI CYPRESS, HINOKI (JAPANESE)

Tree to 52 m tall, with trunk to 2 m in diameter. Scale leaves 1–2.5 mm long. Seeds 2–3 mm long. Central and southern Japan, from central Honshū to Yaku Shima south of Kyūshū. Mixed conifer forests on steep mountain slopes and ridges; 10–2,200 m. Zone 3. Synonym: *Chamaecyparis obtusa* var. *breviramea* (Maximowicz) E. Regel.

Threadlike foliage spray of sawara cypress (*Chamaecyparis pisifera*), found only in cultivars such as this 'Filifera Aurea' but preserving the pointed leaves found in wild trees as well, ×1.

Chamaecyparis pisifera (P. Siebold & Zuccarini) Endlicher

SAWARA CYPRESS, RETINOSPORA, SAWARA (JAPANESE)

Plates 22 and 23

Tree to 30(–50) m tall, with trunk to 1(–2) m in diameter. Bark reddish brown, moderately ridged, peeling in thin strips and graying with age. Crown conical. Branchlet sprays stiff, with waxy white stomatal patches beneath. Scale leaves shiny dark green, 1.5–3 mm long, with an obscure gland. Leaf tips long-pointed and pressed against the twig or spreading, those of the facial leaves touching the base of the next pair. Pollen cones about 2 mm long, with four or six pairs of pollen scales, each scale with three or four yellow pollen sacs. Seed cones spherical, 5–6(–8) mm in diameter, with four to six pairs of seed scales. Seeds one or two per scale, 2–3 mm long, the wings distinctly broader than the body. Central Japan, mostly central Honshū with outliers to northern Honshū and Kyūshū. Mixed conifer forests in rocky, wet places on gentle slopes and in valleys; (110–)700–1,700(–2,400) m. Zone 3.

Sawara cypress has a more limited natural distribution in Japan than does hinoki cypress (*Chamaecyparis obtusa*), with which it commonly occurs. Both species are extensively planted for forestry and horticulture. They yield some of the premier coniferous timbers of Japan, and declining supplies led to importation of Alaska yellow cedar (*Cupressus nootkatensis*) from Alaska and British Columbia. Sawara cypress has much smaller seed cones than does hinoki cypress (*pisifera* means "pea-bearing"), and the pointed leaf tips make it the prickliest species. The tendency toward prickly leaves is exaggerated in juvenile forms (called retinosporas and once thought to be a separate genus). Many of the numerous cultivars have permanently juvenile leaves but also bear seed cones, though less abundantly than plants with fully adult foliage.

Chamaecyparis thyoides (Linnaeus) N. Britton, Sterns, and Poggenburg

ATLANTIC WHITE CEDAR

Tree to 20(–36) m tall, with trunk to 1(–2) m in diameter. Bark grayish brown, forming interconnected shallow ridges and peeling in thin strips. Crown narrowly spire-shaped. Branchlet sprays spreading, somewhat fan-shaped, with waxy white X's beneath outlining the leaf contacts. Scale leaves 2 mm long, yellowish green to bluish green, with a round gland. Leaf tips sharply triangular but pressed against the twig, those of the facial leaves not touching the base of the next pair. Pollen cones 2–4 mm long, with four to six pairs of pollen scales, each scale with (two or) three or four yellow pollen sacs. Seed cones spherical, 3.5–6(–8) mm in diameter, with three or four pairs of seed scales. Seeds one or two per scale, 2–2.5 mm long, the wings about half as wide as the body. Atlantic and

Gulf coastal regions of the United States, from southern Maine to Mississippi. Forming solid stands in bogs and swamps on peat or sterile acid sands, or mixed with other trees away from the wettest sites; 0–450 m. Zone 3. Synonyms: *Chamaecyparis henryae* H. L. Li, *C. thyoides* var. *henryae* (H. L. Li) E. Little.

The first species of *Chamaecyparis* to be cultivated in Europe and North America, Atlantic white cedar is generally considered less suitable as an ornamental than other species of the genus and has far fewer cultivars than Port Orford cedar (*C. lawsoniana*), hinoki cypress (*C. obtusa*), and sawara cypress (*C. pisifera*). Stands in Maine and New Hampshire are the most northerly occurrences of any species of *Chamaecyparis* but are not detectably different from other Atlantic stands. On the other hand, there has been a slight divergence of the disjunct populations of the Gulf Coast, which led H. L. Li to separate them as a distinct species, a distinction not generally supported today. Atlantic white cedar also has the lowest maximum elevation of any species in the genus and is the only one not found in mountains. The largest known individual in combined dimensions when measured in 1985 was 26.8 m tall, 1.5 m in diameter, with a spread of 12.8 m.

Cryptomeria D. Don

JAPANESE CEDAR

Cupressaceae

Evergreen trees with a single, straight, often massive trunk or forming clonal clusters. Bark fibrous, shallowly furrowed and peeling in vertical strips on narrow ridges. Moderately densely branched, the lower branches, if persistent, capable of rooting and turning up to form subsidiary trunks, although the stumps cannot resprout. Without specialized winter buds. Leaves densely spirally arranged, sticking out all around the twig, claw-shaped (to scalelike) in adults but longer and straighter in young trees, the bases running down on and completely clothing the twigs. Free portion flattened side to side and more or less diamond-shaped in cross section, tapering gradually to a soft, slightly incurved point.

Plants monoecious. Pollen cones clustered near the tips of the twigs but each single in the axil of a foliage leaf, oblong, with about 20 spirally arranged pollen scales. Each scale with a thin stalk and a rounded, triangular blade bearing three to five pollen sacs at its base. Pollen grains small (25–35 µm in diameter), nearly spherical, with a small, nearly straight, smooth germination papilla, the rest of the surface covered with minute bumps. Seed cones single at the tips of twigs, maturing in a single season, woody, nearly spherical but slightly pointed. Cone with 20–30 spirally arranged seed scales, the upper- and lowermost of which are sterile. Seed scales more or less wedge-shaped, the bract portion a little shorter than the fertile portion and fused to it for most of its length, the exposed part triangular and coming to a sharp point. Fertile portion consisting of four or five pointed lobes fused side to side at their bases (hence the scientific name, Greek for "hidden parts"), with three to five (or six) seeds. Seeds oblong, slightly flattened, with two very narrow wings derived from the seed coat. Cotyledons (two or) three, each with one vein. Chromosome base number $x = 11$.

Wood faintly fragrant, very soft, weak, and light but moderately decay resistant, with pale yellowish brown to pinkish brown sapwood gradually changing to a dark reddish brown heartwood. Grain moderately even and variably fine to fairly coarse, with sharply defined growth rings marked by a gradual transition to a broad band of much smaller and thicker-walled latewood tracheids. Resin canals absent but with individual resin parenchyma cells scattered through the growth increment and often somewhat concentrated into relatively narrow bands, especially in the latewood.

Stomates arranged in broad but relatively inconspicuous bands on each face. Each stomate oblique to the long axis of the bands, sunken slightly beneath the four to six subsidiary cells and not surrounded by a Florin ring, although there may be an inconspicuous trough around the subsidiary cells. Leaf cross section with a nearly central single-stranded midvein flanked by patches of transfusion tissue and paralleled beneath by a slender resin canal. Epidermis underlain all the way around (except under the stomatal lines) by a hypodermis. Photosynthetic tissue with a multilayered palisade occupying most of the upper portion above the midvein and a looser spongy mesophyll throughout the remainder.

One species in eastern Asia. References: Farjon 2005b, Ferguson 1967, Florin 1963, Fu et al. 1999h, Ohsawa 1994, Schlarbaum and Tsuchiya 1984a, Takaso and Tomlinson 1989a, Tsukada 1982, Wilson 1916, Yamazaki 1995.

Cryptomeria japonica is most extensively and intensively cultivated in Japan although it is widespread in botanical gardens and specialty collections elsewhere and enjoys modest general popularity. There has been extensive cultivar selection in Japan, involving growth habit (with many dwarfs of varying shape) and foliage characteristics, such as needle length and color and branching habit, including a dramatic cockscomb-bearing cristate cultivar. The most popular cultivar outside of Japan, however, is probably the permanently juvenile-foliaged *C. japonica* 'Elegans,' which also has a striking reddish bronze winter coloration.

Cryptomeria occupies an ambiguous position within the family Cupressaceae. One of the genera formerly separated into the family Taxodiaceae (or even its own monotypic family Cryptomeriaceae), it has shoot and cone structures reminiscent of those of some of the most ancient modern conifers, the so-called transition conifers. It has been compared especially to the late Permian genus *Pseudovoltzia,* which is one of the key genera in theoretical interpretations of the compound nature of conifer seed cones.

Most evidence suggests a closest relationship to *Glyptostrobus* and *Taxodium,* with *Glyptostrobus* showing many transitional characteristics between the two genera, negating the suggestion that *Cryptomeria* should be given its own family. The side-to-side seed-bearing teeth that make up the seed scale, so reminiscent of those of the extinct *Pseudovoltzia,* seem unique when examining mature cones of extant conifers. However, development studies of seed cones from their inception show that this kind of structure also underlies the more fused seed scales of the closely related *Glyptostrobus* and *Taxodium* as well as the less closely related *Cunninghamia* and even the more distant *Sciadopitys,* the only living member of the related family Sciadopityaceae. Thus seed cone structure, which seems anomalous at first blush, reinforces the close relationship of *Cryptomeria* to *Glyptostrobus* and *Taxodium,* uniting them into their own botanical tribe (Taxodieae) within the Cupressaceae.

Unlike these closest relatives and the more distantly related *Metasequoia* and *Sequoia,* however, *Cryptomeria* does not seem to have been a widespread, dominant member of northern hemisphere forests during the Tertiary. In fact, there are very few secure records of fossil *Cryptomeria* outside of Japan. Despite its similarities to the transition conifers, its confirmed fossil record extends back no earlier than the beginning of the Tertiary, far less than that of *Cunninghamia*-like conifers, the oldest modern members of the family.

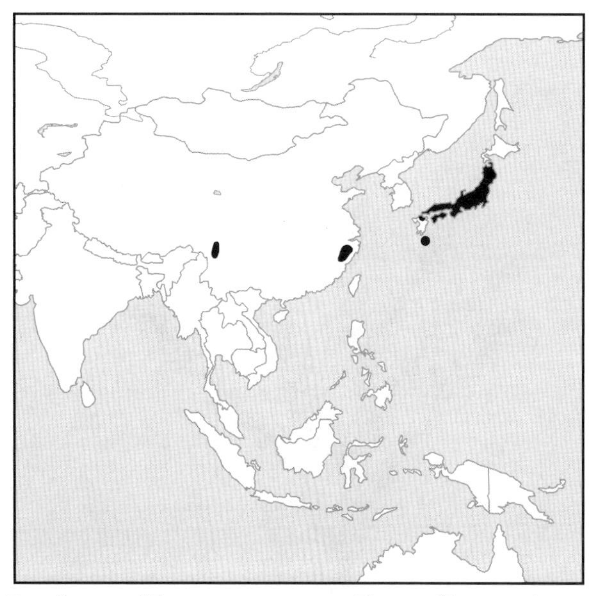

Distribution of *Cryptomeria japonica* in China and Japan, scale 1 : 120 million

Cryptomeria japonica (Linnaeus fil.) D. Don

SUGI CEDAR, JAPANESE CEDAR, CRYPTOMERIA, SUGI (JAPANESE), LIU SHAN (CHINESE)
Plate 24

Tree to 60 m tall, with trunk to 6 m in diameter. Bark rich reddish brown, up to 2.5 cm thick. Crown conical when young, becoming cylindrical with age. Needles shortest at the beginning and end of each growth increment, 3–12(–20) mm long. Pollen cones in clusters of (3–)15–30, 5–7 mm long, 2–3 mm thick. Seed cones (1–)1.5–2.5 cm long. Seeds three to five per scale, 4–6 mm long, 2–3 mm wide, including the narrow wings. Japan (Honshū to Yaku Shima) and central and southern China. Mixed in many forest types, sometimes forming pure stands; 50–1,600(–2,900) m. Zone 7. Synonyms: *Cryptomeria fortunei* Hooibrenk ex F. Otto & A. Dietrich, *C. japonica* var. *radicans* T. Nakai, *C. japonica* var. *sinensis* P. Siebold.

Emergent mature sugi cedar (*Cryptomeria japonica*) in nature.

Twig of sugi cedar (*Cryptomeria japonica*) with 5 or more years of growth weakly set off by shorter needles in between, ×0.5.

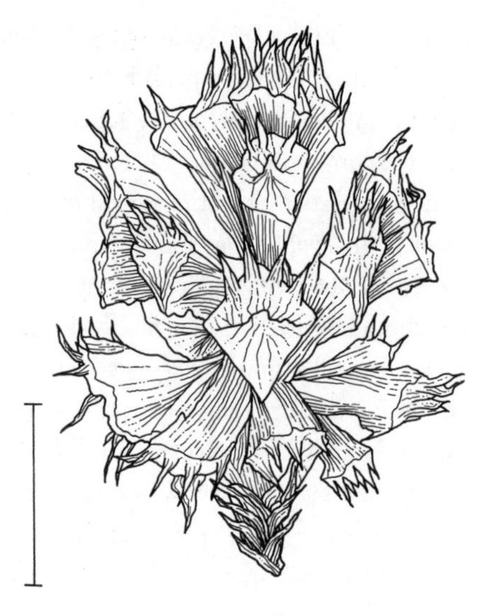

Open mature seed cone of sugi cedar (*Cryptomeria japonica*) after seed dispersal; scale, 1 cm.

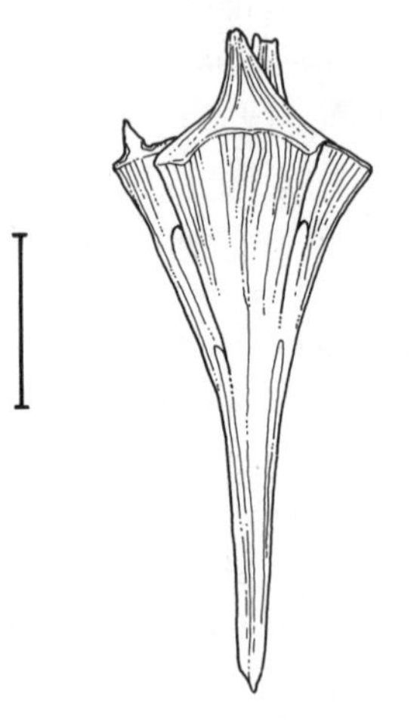

Artificially detached seed cone scale of sugi cedar (*Cryptomeria japonica*) with seed attachment scars; scale, 2 mm.

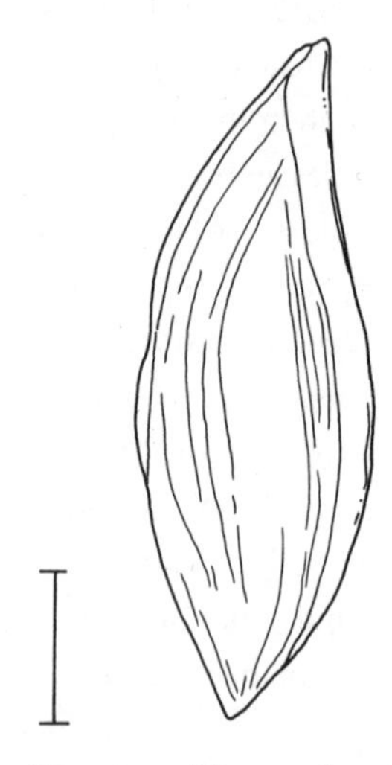

Face view of dispersed seed of sugi cedar (*Cryptomeria japonica*); scale, 1 mm.

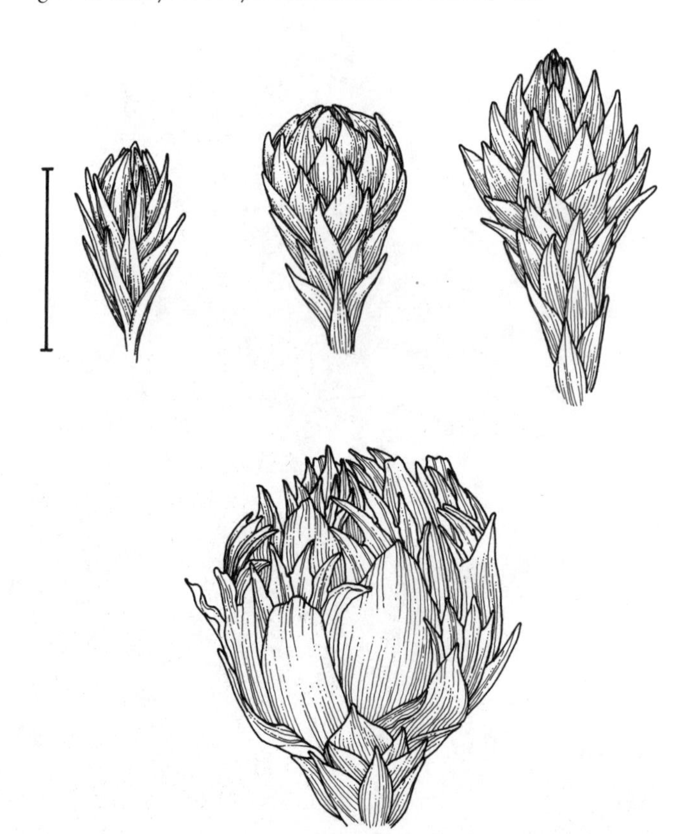

Development of seed cones of sugi cedar (*Cryptomeria japonica*). In the three smaller cones, only the bracts are visible, while the toothlike elements that coalesce into the seed scale portion are evident in the fourth cone; scale, 5 mm.

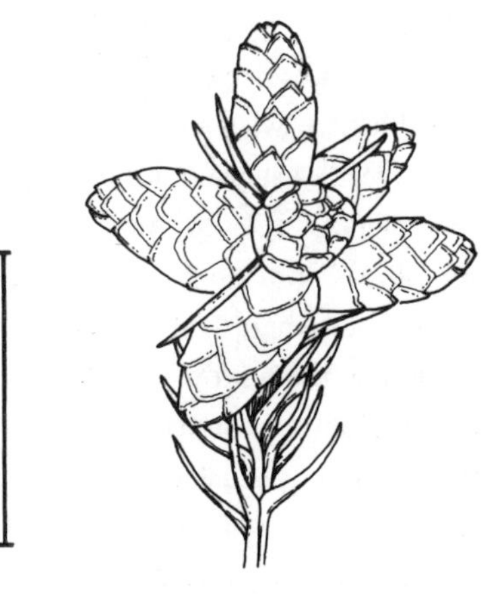

Terminal cluster of slightly immature pollen cones of sugi cedar (*Cryptomeria japonica*); scale, 1 cm.

Windblown sugi cedar (*Cryptomeria japonica*) at timberline in nature.

Sugi cedar is a quite variable plant in nature and has given rise to numerous divergent forms in cultivation, including some with one or two extra sets of chromosomes. In the light of this background variation, the smaller cones and foliage that are sometimes used to distinguish the Chinese plants as a separate variety or species do not seem worthy of formal recognition. Although widespread in China, there are few areas of natural stands, and early botanical visitors doubted whether it was truly native there, a doubt still held by some authors. Plantations are common in Japan as well, where this species is the largest and most popular native timber tree, though the wood is not as highly prized as that of hinoki cypress (*Chamaecyparis obtusa*). The extensive forest plantations of sugi cedar established since the 1940s had an unintended consequence in a great increase in hay fever symptoms during its pollen release in March and April. Although found throughout central Japan today, it was restricted to very narrow coastal refuges during the last ice age and thus barely escaped extinction in its main homeland.

Cunninghamia R. Brown

CHINA FIR

Cupressaceae

Evergreen trees with thin, fibrous bark peeling in irregular flakes on single, forked, or multiple trunks, regenerating by stump sprouting. Crown narrow, oval, with well-spaced whorls of short, horizontal branches. Shoot system weakly differentiated into permanent long shoots and short shoots that are shed intact after a few years. Resting buds roughly spherical, wrapped loosely by narrowly triangular, flat, leaflike bud scales. Leaves arranged in dense spirals but brought into two loose rows on either side of the stem on horizontal branches by twisting at the point of attachment with the twig, the bases running down the twigs and completely clothing them. Leaves needlelike, sword-shaped, flat, often bowed up in the middle to form an arch, minutely toothed along the margin, gradually tapering to the sharp, stiff tip, and slightly narrowed at the attachment to the twig.

Plants monoecious. Pollen cones cylindrical, in a starburst cluster of 15–30 from a common bud at the tip of an ordinary leafy twig. Each cone consisting of 30–100 spirally arranged pollen scales. Scales with a thin stalk and three pollen sacs at the base of the rounded-triangular, toothed, blade. Pollen grains small to medium (30–45 µm in diameter), more or less spherical, with an obscure, round germination pore and minute bumps over the whole surface. Seed cones single or in clusters of up to four at the tips of foliage shoots, often later displaced from the tip by new shoots growing out beyond them, broadly egg-shaped, maturing in their first season. Seed scales numerous, spirally arranged, thin, woody, triangular, minutely toothed, and dominated by the bract portion, the fused fertile portion reduced to a small, lobed pad of tissue supporting the (two or) three seeds. Seeds oblong, somewhat flattened, with two narrow wings derived from the seed coat extending the whole length of the body. Cotyledons two, each with one vein. Chromosome base number $x = 11$.

Wood strongly fragrant, soft, light and weak but extremely decay resistant, with a broad white sapwood sharply contrasting with the dark yellow to reddish brown heartwood. Grain moderately even and fine, with well-defined growth rings marked by an abrupt transition to a narrow band of denser latewood with much smaller, thicker-walled tracheids. Resin canals absent but with fairly abundant individual resin parenchyma cells scattered through the growth increment, especially in the outer half, and often in an open ring just inside the latewood.

Stomates in two prominent broad stomatal bands underneath and with more or less conspicuous ones on the upper side. Each stomate tucked under and partly hidden by the

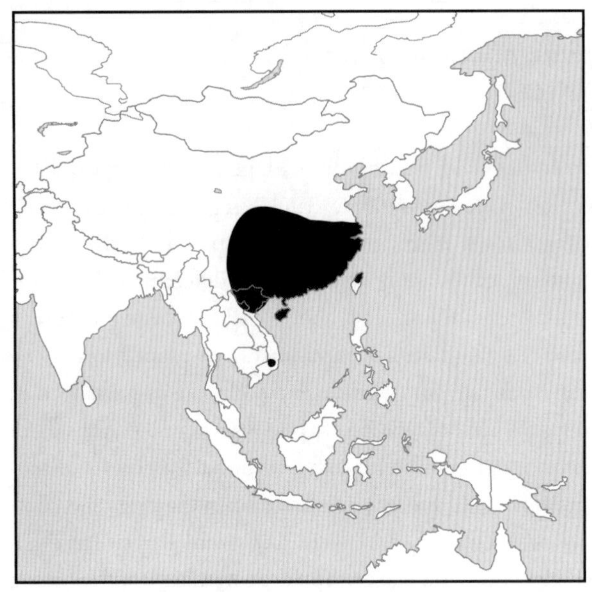
Distribution of *Cunninghamia* in southeastern Asia, scale 1 : 120 million

four to six irregularly shaped subsidiary cells, which are not surmounted by a Florin ring. Leaf cross section with a single-stranded midvein accompanied beneath by a large resin canal and to the sides by short wings of transfusion tissue. Often with an additional resin canal on either side about two thirds of the way out toward the margin from the midrib. Photosynthetic tissue forming a weakly developed palisade layer beneath the upper epidermis and adjoining nearly continuous (except beneath stomates) hypodermis. Spongy mesophyll with scattered clusters of sclereids, some of which mark the boundary between the spongy and palisade tissues.

Two species in China. References: Farjon 2005b, L. K. Fu 1992, Fu et al. 1999h, Fung 1994, Li 1963, Ohsawa 1994, T. S. Ying et al. 1993.

The wood of China fir, like that of many other Cupressaceae, is highly decay resistant and is a premier timber tree in China, being used as a coffin wood and for many other purposes. As a result, the trees are grown in China in plantations and forests managed for timber production. Elsewhere, they are moderately common in horticulture in warm temperate regions such as the western coast of North America or the southeastern United States. They are not always the most handsome trees, tending to have somewhat ragged crowns, and they are messier than the average conifer, with their masses of large shed shoots. Perhaps this explains why there has been so little cultivar selection, with only a handful of dwarfs and foliage color variants. The genus honors James Cunningham (d. ca. 1709), who was a physician with the East India Company in China when he acquired a specimen of *Cunninghamia lanceolata* for his garden in 1702 and sent specimens of it and other Chinese plants back to correspondents in England.

Because of the resemblance of the foliage to that of *Araucaria bidwillii,* some early taxonomists associated *Cunninghamia* with the Araucariaceae, but most authors, then as later, placed it with redwoods in the Taxodiaceae, now included in the Cupressaceae. Foliage and cones of *Cunninghamia,* nonetheless, are distinctive in the family, and it appears to have no particularly close relationships, although it is closest to its geographic neighbor, *Taiwania.* Instead, DNA studies suggest that its ancestors were the first members of the family to become distinct. This is borne out by the fossil record, which includes many well-preserved seed cones. Although fossils attributable to *Cunninghamia* itself only go back to the beginning of the Tertiary, structurally similar cones (*Cunninghamiostrobus* and *Elatides*) extend back to the late Jurassic, earlier than any other extant lineage of the family.

Despite differences in mature appearance, early development of cones in *Cunninghamia* is similar to that in *Cryptomeria* and other fairly distantly related genera of Cupressaceae, with each seed developing on a separate seed scale lobe (often rather toothlike early on). Later, development diverges as the bract undergoes massive enlargement in *Cunninghamia* unmatched by the seed scale lobes, while the latter also enlarge and become more or less united side to side in the remaining genera. *Cunninghamia* also shows a clear structural similarity (homology) between the seed cones and the clusters of male cones. The whole cluster has the same position at the end of a branch as the seed cones and also consists of a similarly shaped spiral of similarly shaped bracts. Of course, in the pollen-producing structure, each bract has a whole simple pollen cone in its axil consisting of dozens of pollen scales, each bearing three pollen sacs, rather than the three lobes, each bearing a single ovule (morphologically equivalent to a pollen sac as both are [based on] sporangia) that make up the seed scale portion of the cone scale. Still, in their early stages, before pollination, when they are still closed up like large buds, the two structures are remarkably similar. Occasionally, mixed cones occur in which the lowermost bracts subtend pollen cones while all the rest bear ovules in the normal fashion of seed cones.

Identification Guide to *Cunninghamia*

The two species of *Cunninghamia* may easily be distinguished by the length and width of the larger needles and by the length of the seed cones:

Cunninghamia konishii, needles 1.5–2 cm by 1.5–2.5 mm, seed cones 2–2.5 cm

C. lanceolata, needles 2–6 cm by 3–5 mm, seed cones 2.2–5 cm

Cunninghamia konishii Hayata

TAIWAN COFFIN FIR, LUAN DA SHAN (CHINESE)

Tree to 50 m tall, with trunk to 2.5 m in diameter. Bark light reddish brown, scaly. Needles often with a thin waxy coating lending a bluish green cast to the foliage, 1.5–2 cm long, 1.5–2.5 mm wide. Pollen cones in clusters of 14–16, 2–2.5 cm long, 4–5 mm thick. Seed cones (1.5–)2–2.5 cm long, 1.5–2 cm in diameter. Seeds 4–6 mm long. Northern and central Taiwan, with outliers in eastern mainland China, Laos, and Vietnam. Scattered in montane coniferous forests or in pure stands; (600–)1,300–2,000(–2,200) m. Zone 9. Synonym: *Cunninghamia kawakamii* Hayata.

Although Taiwan coffin fir was long considered endemic to Taiwan, specimens from nearby Fujian province (China) and from the borderland of northern Laos and Vietnam, while skimpy, appear to belong to it. Such specimens contribute to a feeling among taxonomists, including the authors of the treatment of *Cunninghamia* in the *Flora of China,* that Taiwan coffin fir is simply an extreme variant of China fir (*C. lanceolata*). A very limited set of genetic evidence points in the same direction but is hardly conclusive. There is little overlap between the two species in traditional foliage and cone characters, so it is easy to tell them apart. Conclusive evidence must await thorough study of both from throughout their ranges. The species name honors Nariaki Konishi, the Japanese botanist who collected the type specimen on Mount Luanta in central Taiwan in 1906.

Cunninghamia lanceolata (A. Lambert) W. J. Hooker

CHINA FIR, CUNNINGHAMIA, SHAN MU (CHINESE)

Plate 25

Tree to 25 m tall, with trunk to 3 m in diameter. Bark reddish brown, furrowed, peeling in long vertical strips. Needles light green when young, turning dark green above with maturity, (0.8–)3–6(–7) cm long, (2–)3–5(–7) mm wide, often with two narrow stomatal grooves above as well as the wider bands beneath. Pollen cones in clusters of 16–30, 1.5–3 cm long, 6–8 mm thick. Seed cones 2.5–5 cm long, 2.5–4 cm in diameter. Seeds 5–7 mm long. Central and southern China with outliers in northern Vietnam. Widespread on slopes in various forest communities and in plantations; (140–)250–2,500(–3,000) m. Zone 8. Synonym: *Cunninghamia unicanaliculata* D. Y. Wang & H. L. Liu.

Mature Taiwan coffin fir (*Cunninghamia konishii*) in nature.

Ancient Taiwan coffin fir (*Cunninghamia konishii*) in nature.

Mature China fir (*Cunninghamia lanceolata*) in nature.

Open mature seed cone of China fir (*Cunninghamia lanceolata*) after seed dispersal; scale, 2 cm.

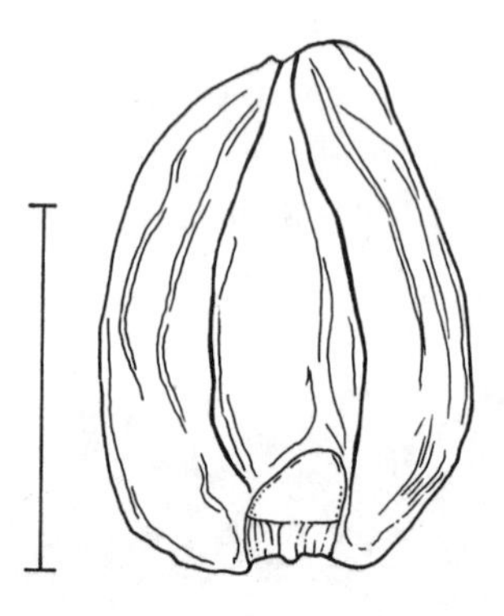

Face view of dispersed seed of China fir (*Cunninghamia lanceolata*); scale, 5 mm.

This is one of the premier timber trees of China, the soft, fragrant, durable wood serving many purposes, including coffin making. It has been severely depleted throughout its wide range but is also one of the most important plantation trees in China. Despite some difficulty with soil fungi, growth is typically very rapid, as much as 1 m per year, and stump sprouting after harvest allows for ready regrowth. Nonetheless, as with many redwood relatives, old-growth timber is favored, and, in the early 20th century, giant trees that had been buried in earthquakes two or three centuries before and which exceeded the dimensions of any available standing timber, were especially prized, being still sound and retaining their fragrance. A common abnormality is to have proliferous seed cones, in which the tip grows out as a leafy shoot. These dry up and die with maturation of the cone. Trees from southwestern Sichuan province differ from most others in some details of leaf structure and were described as a separate species, *Cunninghamia unicanaliculata,* here treated as synonymous with China fir. They might best be treated as a botanical variety of the species, but their status remains to be determined. The word *shan,* found in many Chinese conifer names and usually translated as "fir," applies first to this tree, all the others, including the true firs of the genus *Abies,* being linguistically derivative. The species name refers to the sword-shaped leaves.

Cupressus Linnaeus

CYPRESS

Cupressaceae

Evergreen trees or tall shrubs with a single or multiple trunks. Bark fibrous, usually furrowed and peeling in vertical strips but sometimes flaking to leave a smooth, mottled surface. Crown dense, mostly conical to egg-shaped, but ranging from stiffly narrowly columnar to picturesquely broadly flat-topped or weeping. Shoot system weakly differentiated into main axes and side shoots (branchlets), usually branching in three dimensions, sometimes emerging alternately from lateral leaves on either side of the main twigs to form flattened, fernlike sprays, the branchlets cylindrical or four-angled. Resting buds unspecialized, consisting solely of arrested immature foliage leaves. Seedling leaves in alternating quartets, needlelike, standing out from and well spaced on the stem, with one or two bands of stomates above and beneath, soon replaced by adult branchlets. Adult leaves in alternating, similar

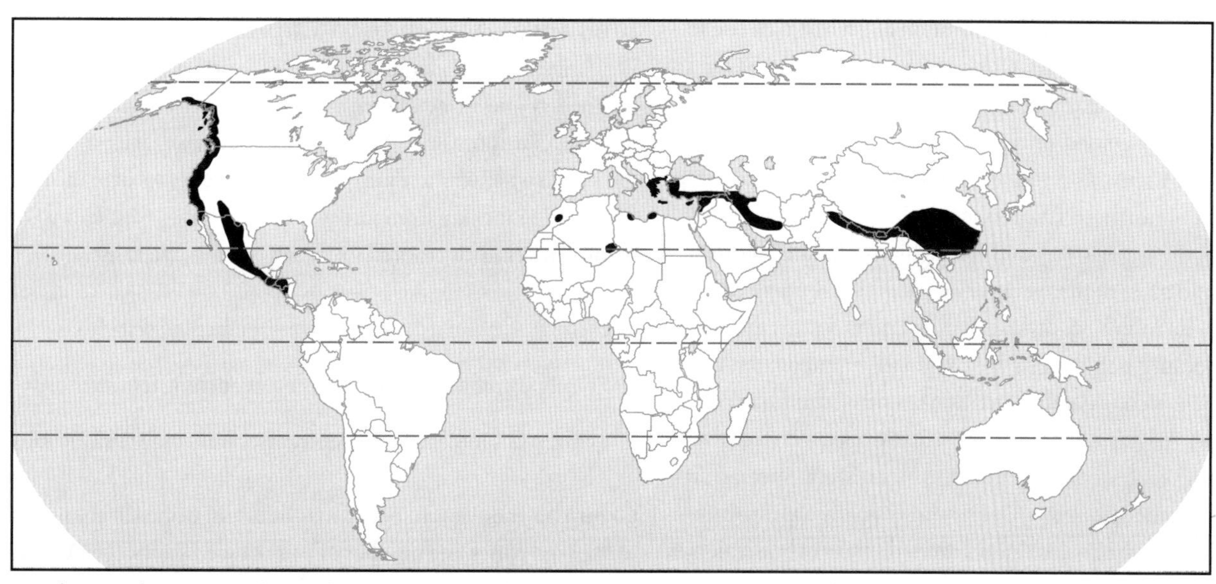

Distribution of *Cupressus* in the northern continents, scale 1 : 200 million

pairs (not as distinguishable lateral and facial pairs). Leaves of each pair scalelike, closely enclosing the twig, with the tip either pressed against it or pointing outward, sometimes with a prominent resin gland on the surface.

Plants monoecious. Pollen cones numerous, single at the tips of short branchlets, spherical to oblong. Pollen scales in 3–8(–12) alternating pairs, each scale with two to six pollen sacs. Pollen grains small (20–35 μm in diameter), spherical, minutely and irregularly bumpy, with a small, circular germination pore. Seed cones single or clustered at the tips of short side branchlets, spherical or oblong, woody, maturing in two seasons and then opening or remaining closed many years or until opened by fire. Cones with three to seven alternating pairs of centrally stalked, roughly hexagonal seed scales, or with two basally attached pairs, the uppermost pair sometimes united. Seeds 2–20 per seed scale, irregularly angular, often with a prominent attachment scar, variably narrowly two-winged, the wings derived from the seed coat. Cotyledons two (*Cupressus nootkatensis* and Old World species except *C. torulosa*) or three to six (*C. torulosa* and New World species except *C. nootkatensis*), each with one vein. Chromosome base number $x = 11$.

Wood faintly to moderately fragrant, light to medium weight, soft to fairly hard, often brittle, sometimes extremely decay resistant. Sapwood pale, generally whitish and giving way gradually to darker, bright yellow, yellowish brown, pinkish brown or reddish brown heartwood, sometimes speckled or streaked. Grain generally even and fine, with distinct but sometimes inconspicuous growth rings marked by a gradual transition to much smaller and slightly thicker walled latewood tracheids. Without resin canals but with fairly numerous individual resin parenchyma cells scattered through the growth increment or somewhat concentrated in variously positioned open rings.

Stomates in patches, primarily on hidden and protected surfaces of the leaves. Each stomate tucked in beneath and partly hidden by the four to six surrounding subsidiary cells, which usually carry a steep Florin ring as well as additional round to elongate papillae, which may also be found on other cells of the epidermis. Leaf cross section with a single-stranded midvein flanked by short wings of transfusion tissue. Resin canals up to three in number, lying just inside the epidermis in the attached leaf base but often not extending into the free portion. Photosynthetic tissue with a well-developed palisade layer beneath the epidermis and the very thin hypodermis on the outer face (developmentally the underside) of the leaf. Looser spongy mesophyll with very large air spaces making up the remaining photosynthetic tissue and most of the leaf volume, which usually lacks additional strengthening cells.

Seventeen species in western North America from Alaska south to Guatemala, the Mediterranean region, and the Himalayan region to southern China. References: Camus 1914, Eckenwalder 1993, Farjon 2005b, Fu et al. 1999g, Gadek and Quinn 1999, Silba 1983, Wolf 1948. Synonyms: *Callitropsis* Oersted, ×*Cupressocyparis* Dallimore, ×*Cuprocyparis* Farjon, *Xanthocyparis* Farjon & T. H. Nguyên.

Cupressus is the classical Latin name for Mediterranean cypress (*C. sempervirens*) and similar also to the ancient Greek equivalent (*kyparissos*), which underlies the derivation of *Chamaecyparis.* The island of Cyprus may be named for its population of *C. sempervirens* (or for its copper deposits), and this species has been cultivated for millennia in the Mediterranean Sea basin. Several other species are also in general cultivation and have given rise to a broad spectrum of cultivars. Cultivar selection has emphasized primarily habit (compact, columnar, or weeping) and foliage color (blue-green, golden, or variegated).

Although *Cupressus* is the second most widespread and diverse genus of northern Cupressaceae, there are far fewer species than in the vegetatively and (partially) ecologically similar *Juniperus,* and most of these are very local in distribution. Many are emergents in fire-prone shrublands and most are drought tolerant. Because populations are often well isolated from one another, there is considerable variation from population to population, and species concepts have usually been very narrow, often treating each isolated population as a distinct species. At the other extreme, *Cupressus funebris* and *C. nootkatensis,* species with smaller cones than usual and flattened sprays of branchlets, were mistakenly assigned to *Chamaecyparis* in the past, a genus traditionally considered closely related to *Cupressus.* In many 19th century works, species of *Chamaecyparis* were simply included as a section of *Cupressus.* The weight of evidence today, particularly from leaf chemistry and DNA sequences, shows conclusively that these two genera are not particularly closely related and that *C. funebris* and *C. nootkatensis* belong in *Cupressus* in the broad sense used here. Relationships are actually a little more complicated, and some authors separate *C. nootkatensis* and *C. vietnamensis* into their own genus while others suggest that New and Old World species might have to be placed in different genera to avoid merging *Juniperus* with *Cupressus.* Evidence for these more unconventional taxonomic arrangements are not presently particularly compelling, so a more traditional approach is taken here.

With the assignment of *C. nootkatensis* to *Cupressus,* the hybrid (notho-) genus ×*Cupressocyparis* (and its synonym ×*Cuprocyparis*), founded to include hybrids between *C. nootkatensis* and other species of *Cupressus,* such as the important Leyland cypress, simply becomes a synonym of *Cupressus.* The reduction of ×*Cupressocyparis* to *Cupressus* and the implausibility of *Tsuga mertensiana* as a proposed hybrid between *Tsuga* and *Picea* leaves only the redwood, *Sequoia sempervirens,* as a possible example of a hybrid between conifer genera.

As might be expected from a generally dryland genus, the fossil record of *Cupressus* is sparse. There is a single cone from Miocene deposits in Germany that is safely attributable to the genus as well as many more recent specimens from Pleistocene sediments in coastal California. In addition, the Cretaceous and early Tertiary genus *Mesocyparis* has some similarity to *Cupressus nootkatensis* and might be close to *Cupressus* genetically.

Identification Guide to *Cupressus*

For convenience of identification, the 17 species of *Cupressus* may be divided into three informal groups by whether the ultimate branchlets are fine (up to 1.5 mm thick) or coarse (at least 1.5 mm thick) and whether there are obvious fernlike sprays of branchlets arising by branching from just two rows of leaves (distichous) or branching is mostly from all four rows of leaves (tetrastichous):

Group A, ultimate branchlets fine, branching distichous

Group B, ultimate branchlets fine, branching tetrastichous

Group C, ultimate branchlets coarse, branching tetrastichous

Group A. These seven species may be distinguished by the length of the seed cones, the number of pairs of seed scales, the number of seeds per scale, the length of the pollen cones, and the presence or absence of drops of dried resin on the leaves (the three hybrid cypresses derived from *Cupressus nootkatensis* in cultivation—*C.* ×*leylandii, C.* ×*notabilis, C.* ×*ovensii*—also fall into Group A, and their descriptions should be checked when identifying cultivated cypresses with fernlike foliage sprays):

Cupressus vietnamensis, seed cones less than 1.5 cm, seed scale pairs two or three, seeds per scale up to three, pollen cones up to 3.5 mm, dried resin absent

C. nootkatensis, seed cones up to 1.5 cm, seed scale pairs two or three, seeds per scale fewer than eight, pollen cones at least 3 mm, dried resin absent

C. funebris, seed cones up to 1.5 cm, seed scale pairs at least three, seeds per scale fewer than eight, pollen cones at least 3 mm, dried resin absent

C. lusitanica, seed cones up to 1.5 cm, seed scale pairs at least three, seeds per scale at least eight, pollen cones at least 3 mm, dried resin absent

C. chengiana, seed cones at least 1.5 cm, seed scale pairs at least three, seeds per scale at least eight, pollen cones up to 3 mm, dried resin absent

C. macnabiana, seed cones at least 1.5 cm, seed scale pairs at least three, seeds per scale at least eight, pollen cones up to 3 mm, dried resin present

C. cashmeriana, seed cones at least 1.5 cm, seed scale pairs at least three, seeds per scale at least eight, pollen cones at least 3 mm, dried resin absent

Group B. These six species may be distinguished by the length and shape of the seed cones, the number of seeds per scale, the length of the pollen cones, and the presence or absence of drops of dried resin on the leaves:

Cupressus torulosa, seed cones up to 2.5 cm, oblong, seeds per scale up to 8, pollen cones more than 4 mm, dried resin absent

C. bakeri, seed cones up to 2.5 cm, oblong, seeds per scale at least 8, pollen cones less than 4 mm, dried resin present

C. goveniana, seed cones up to 2.5 cm, spherical, seeds per scale at least 8, pollen cones up to 4 mm, dried resin absent

C. duclouxiana, seed cones up to 2.5 cm, spherical, seeds per scale at least 8, pollen cones at least 4 mm, dried resin absent

C. guadalupensis, seed cones at least 2.5 cm, spherical, seeds per scale at least 8, pollen cones at least 4 mm, dried resin absent

C. sempervirens, seed cones at least 2.5 cm, oblong, seeds per scale at least 8, pollen cones at least 4 mm, dried resin absent

Group C. These four species may be distinguished by the length and shape of the seed cones and the presence or absence of drops of dried resin on the leaves:

Cupressus gigantea, seed cones up to 2 cm, oblong, dried resin absent

C. arizonica, seed cones 2–3 cm, spherical, dried resin present

C. sargentii, seed cones 2–3 cm, spherical, dried resin absent

C. macrocarpa, seed cones at least 3 cm, oblong, dried resin absent

Cupressus arizonica E. Greene

ARIZONA CYPRESS

Tree to 28 m tall, with trunk to 2 m in diameter. Bark ridged and furrowed or smooth and peeling in flakes. Crown dense, conical, broadening with age. Branchlets four-sided,1.3–2.3 mm in diameter, branching from all four rows of leaves. Scale leaves on branchlets 1–2 mm long, dark green or gray-green with wax, the edges minutely toothed, the back usually with a conspicuous drop of dried resin in an open gland. Pollen cones 2–5 mm long, about 2 mm wide, with (4–)5–8(–10) pairs of pollen scales, each with (three or) four to six pollen sacs. Seed cones spherical or a little elongated, (1.5–)2–3 cm long, gray or brown at maturity, often waxy before this, with three or four (to six) pairs of seed scales, each usually with a strong conical point on the face, especially on the upper scales, the surface otherwise smooth or warty. Seeds (5–)8–15(–20) per scale, (3–)4–6(–8) mm long, light to dark brown, sometimes with a thin to dense waxy coating or with resin pockets, or both. Cotyledons three to five (or six). Southwestern North America from south-central California to northern Zacatecas (Mexico). Dry to moist woodlands and scrublands on slopes and in valleys; 750–2,400(–2,825) m. Zone 7. Synonym: *Callitropsis arizonica* (E. Greene) D. Little. Five varieties.

Arizona cypress (named for the state, at that time a territory, from which it was first described in 1882) is the most drought resistant of the four commonly cultivated New World cypresses. It also has the widest distribution among the species in the southwestern United States, although its largest area of distribution is in northern Mexico. The range, however, is highly fragmented, and populations in different areas have different combinations of bark texture, leaf color (due to wax), prevalence of resin dots on leaves, and minor seed cone variations. As a result, several species were described and also later treated as varieties (or subspecies) of a single species. These segregates are not very convincing because they display most possible combinations of the varying characteristics, which also vary to a certain extent within populations. Some of the segregates were synonymized with one another in different ways, leading to recognition of fewer varieties. On the other hand, some DNA evidence supports species recognition for some of the varieties here, but it is not based on multiple population samples for each taxon. More attention must be given to thorough morphological, chemical, and molecular sampling of populations of these trees throughout their range, including Mexico, before the validity of discontinuities in their variation patterns from place to place can be assessed. The distinguishing characteristics of the regional varieties are presented for those who wish to recognize them. The largest known individual in the United States in combined dimensions, when measured in 1993, was 28.4 m tall and 1.9 m in diameter, with a spread of 14.6 m.

Cupressus arizonica E. Greene **var. *arizonica***

ROUGH ARIZONA CYPRESS, CEDRO BLANCO (SPANISH)

Bark of trunk fibrous and furrowed, dark brown or gray. Resin glands, on as few as half of the leaves, often inactive. Seed cones (1.2–)2–3 cm in diameter, remaining closed at maturity. Seeds (3–)4–5(–6) mm long, with wings 1 mm wide, usually not waxy. Southeastern Arizona to northern Zacatecas (Mexico).

Cupressus arizonica var. *glabra* (Sudworth) E. Little

SMOOTH ARIZONA CYPRESS

Bark of trunk smooth and flaking, red or reddish brown. Resin glands on nearly all leaves, usually active. Seed cones 2–2.5 (–3.2) cm in diameter, remaining closed at maturity. Seeds (3–)4–5(–8) mm long, with minute wings, usually waxy. Central Arizona. Synonyms: *Callitropsis glabra* (Sudworth) D. Little, *Cupressus glabra* Sudworth.

Cupressus arizonica var. *montana* (I. Wiggins) E. Little

SIERRA SAN PEDRO MARTÍR CYPRESS

Bark of trunk thin, fibrous and narrowly furrowed, red to brown. Resin glands frequent on leaves and active. Seed cones 1.5–2.5 cm in diameter, opening at maturity. Seeds 3–4(–5) mm long, with wings 1–5 mm wide, usually not waxy. Sierra San Pedro Martír, northern Baja California (Mexico). Synonyms: *Callitropsis montana* (I. Wiggins) D. Little, *Cupressus montana* I. Wiggins.

Young mature Piute cypress (*Cupressus arizonica* var. *nevadensis*) in nature.

Cupressus arizonica var. *nevadensis* (Abrams) E. Little

PIUTE CYPRESS

Bark of trunk fibrous and furrowed, grayish brown. Most leaves with an active resin gland. Seed cones (1.5–)2–3 (–3.5) cm in diameter, remaining closed at maturity. Seeds 3–6 mm long, with wings to 2 mm wide, occasionally slightly waxy. Kern County, south-central California. Synonyms: *Callitropsis nevadensis* (Abrams) D. Little, *Cupressus nevadensis* Abrams.

Cupressus arizonica var. *stephensonii* (C. Wolf) E. Little

CUYAMACA CYPRESS

Bark of trunk smooth and flaking, red or reddish brown. Resin glands often present, active or inactive. Seed cones (1–)2–2.5 cm in diameter, remaining closed at maturity. Seeds (4–)5–8 mm long, with wings 1–3 mm wide, usually waxy. Cuyamaca Mountains, San Diego County, California, and Sierra de Juárez, northern Baja California (Mexico). Synonyms: *Callitropsis stephensonii* (C. Wolf) D. Little, *Cupressus arizonica* subsp. *stephensonii* (C. Wolf) R. M. Beauchamp, *C. arizonica* var. *revealiana* Silba, *C. stephensonii* C. Wolf.

Regularly branching twig of Cuyamaca cypress (*Cupressus arizonica* var. *stephensonii*) with several orders of branchlets on about three weakly distinguishable years of growth, ×0.5.

Cupressus bakeri Jepson

MODOC CYPRESS

Tree to 30(–40) m tall, with trunk to 1 m in diameter. Bark smooth, reddish brown, initially flaking in patches and later building up in layers. Crown sparse, broadly columnar. Branchlets cylindrical or slightly four-sided, 0.5–1.3 mm in diameter, branching from all four rows of leaves. Scale leaves on branchlets about 2 mm long, grayish green with wax deposits, not toothed, usually with a conspicuous and active gland accumulating a drop of dried resin. Pollen cones produced from a young age (6–7 years), 2–3 mm long, 2–2.5 mm wide, with (three or) four (or five) pairs of pollen scales, each with (three or) four or five pollen sacs. Seed cones a little oblong, 1–2.3 cm long, silvery gray at maturity or turning brownish after opening, not waxy, with three or four pairs of seed scales, each with a conical point on the face, the surface very warty with resin pockets. Seeds (2–)7–11 per scale, mostly 3–4 mm long, light tan to medium brown, sometimes with a thin waxy coating or scattered resin pockets. Cotyledons three or four. Northern California and southwestern Oregon. Forming large or small stands or mixed with other conifers on volcanic or serpentine soils; 1,050–2,100 m. Zone 6. Synonyms: *Callitropsis bakeri* (Jepson) D. Little, *Cupressus bakeri* subsp. *matthewsii* C. Wolf.

Irregularly branching twig of Modoc cypress (*Cupressus bakeri*) with both adult and transitional juvenile foliage, ×1.

Modoc cypress is the most northerly of the New World dryland cypresses. It has the smallest seed cones among the California species, often with just three pairs of seed scales, so it forms a bridge to the atypical Alaska yellow cedar (*Cupressus nootkatensis*) of the Pacific Northwest coastal rain forests. The largest known individual is 39.3 m tall with a trunk diameter of 104 cm and a crown spread of 8.8 m. The trees can reach an age of 250 years or more, and individuals with diameters between 40 and 45 cm can be anywhere from 50 to 250 years old. Wolf (1948), in his monograph of New World cypresses, treated widely separated northwestern and southeastern groves of *C. bakeri* as distinct subspecies. Subsequent work has filled the gap between these areas, both geographically and morphologically, and there is no justification today for accepting Wolf's two subspecies. There is significant variation among the different groves, but this seems to be associated with soil types. Modoc cypress is most closely related to Arizona cypress (*C. arizonica*), especially the northern populations sometimes distinguished as Piute cypress (variety *nevadensis*). It is named after Milo Baker (1868–1961), a Californian plant explorer who collected specimens of it in 1898.

Cupressus cashmeriana Royle ex Carrière

HIMALAYAN WEEPING CYPRESS, CHENDEY (BHUTANESE)

Tree to 30(–45) m tall, with trunk to 2.5 m in diameter. Bark brown, fibrous, narrowly ridged and furrowed. Crown conical, weeping, with long, dangling branchlet sprays on horizontal to rising branches. Branchlets cylindrical, 1–1.5 mm in diameter, branching mostly from just two rows of leaves. Scale leaves on branchlets 1–2(–3) mm long, bright green to bluish green with wax, the edges minutely toothed, the back often with a shallow, inactive gland. Pollen cones 3–6 mm long, 1.5–2.5 mm wide, with (five to) seven or eight (or nine) pairs of pollen scales, each with (three or) four pollen sacs. Seed cones nearly spherical, 1–2 cm long, dark brown at maturity, often waxy before this, with four or five pairs of seed scales, each usually with a low central point on the face and fine lines radiating from there after opening. Seeds 10–12 per scale, 4–6 mm long, the wings 1–2 mm wide, brown, usually not waxy, smooth. Cotyledons two. Bhutan and neighboring states of India and Xizang (Tibet). Mixed, monsoon-influenced conifer forests; 1,500–3,000 m. Zone 9. Synonyms: *Cupressus assamica* Silba, *C. corneyana* Carrière (misapplied), *C. darjeelingensis* Silba, *C. himalaica* Silba, *C. himalaica* var. *darjeelingensis* (Silba) Silba, *C. pseudohimalaica* Silba.

This elegant cypress of the eastern Himalaya is not found wild in Kashmir, as the name would imply. It is, however, widely cultivated for its weeping habit but is often confused with Chinese weeping cypress (*Cupressus funebris*), which differs in having

Drooping twig of Min cypress (*Cupressus chengiana*) with branchlets arranged in flattened sprays, ×1.

Multiaged grove of Min cypress (*Cupressus chengiana*) in nature.

smaller cones and flattened branchlets, with some differentiation between facial and lateral leaf pairs. It is difficult to determine the exact native range of Himalayan weeping cypress because the trees are also frequently planted in their native region around villages and temples. The correct name for the species has also been uncertain because most names were based on trees in cultivation, and deciding on types, and whether they belong here, is difficult. The species has often gone under the earlier name *C. corneyana,* but the correct application of that name is unknown.

Cupressus chengiana S. Y. Hu

MIN CYPRESS, MIN JIANG BAI MU (CHINESE)

Tree to 30 m tall, with trunk to 1.7 m in diameter. Bark grayish brown, ridged and furrowed, peeling in thin, narrow strips. Crown dense, conical, with horizontal branches, broadening and thinning with age. Branchlets cylindrical or slightly flattened, 1–1.5(–2) mm in diameter, branching mostly from just two rows of leaves, but not particularly giving an impression of fernlike sprays. Scale leaves on branchlets 1–1.5 mm long, bright green, the back with a more or less conspicuous but relatively inactive gland. Pollen cones 2–3 (–4) mm long, with four to eight pairs of pollen scales, each with three or four pollen sacs. Seed cones spherical or a little elongated, (0.5–) 1–2 cm long, brown to reddish brown at maturity, sometimes waxy before this, with (three or) four or five (to seven) pairs of seed scales, these with a low, rounded point on the face, the surface otherwise smooth or slightly warty. Seeds four to eight per scale, 3–5 mm long, brown, not waxy, the wings to 1 mm wide. Cotyledons two. West-central China, in southernmost Gansu and western Sichuan. Mixed conifer forests of dry valleys and slopes of the headwaters of the Changjiang (Yangtze River) drainage; 900–3,000 m. Zone 8. Synonyms: *Cupressus chengiana* var. *jiangeensis* (N. Chao) Silba, *C. chengiana* var. *kansuensis* Silba, *C. chengiana* var. *wenchuanhsiensis* Silba, *C. fallax* Franco, *C. jiangeensis* N. Chao.

One of the best timber trees of its arid homeland, Min cypress has almost been eliminated from valley bottoms by overexploitation and is now mostly restricted to steep slopes. It was named for Chinese botanist Wan Chun Cheng (1904–1983), codescriber of dawn redwood (*Metasequoia glyptostroboides*), who collected the type specimen in the Min River valley (hence the Chinese name, "Min River cypress tree").

Cupressus duclouxiana P. Hickel ex A. Camus

YUNNAN CYPRESS, GAN XIANG BAI (CHINESE)

Tree to 25(–45) m tall, with trunk to 1(–3.5) m in diameter. Bark reddish brown, peeling in thin strips, narrowly ridged and furrowed. Crown dense, conical, with spreading or rising branches, broadening, rounding, and thinning with age. Branchlets cylindrical to squarish, 0.8–1.2(–1.4) mm in diameter, branching from all four rows of leaves, upright or occasionally somewhat drooping. Scale leaves on branchlets 1–2 mm long, dark green, or more commonly slightly bluish green with wax, the back usually with a conspicuous but inactive gland. Pollen cones (3–)4–7 mm long, with (6–)7–10 pairs of pollen scales, each with three or four pollen sacs. Seed cones spherical or slightly elongated, (1.2–)1.5–3 cm long, dark brown or somewhat grayish, often with a thin waxy coating that persists at maturity, with (three or) four or five (or six) pairs of seed scales, each with an inconspicuous to prominent central point on the face, the surface wrinkled after opening. Seeds (2–)5–20(–30) per scale, 3–5 mm long, brown, usually not waxy, the wings very narrow to 1 mm wide. Cotyledons two. West-central China from southeastern Xizang (Tibet) and southwestern Sichuan to central Yunnan. Mixed conifer forests in moist valleys; 1,400–3,300 m. Zone 9. Synonym: *Cupressus tonkinensis* Silba.

Yunnan cypress is less well known in the wild among the mountain valleys of southwestern China than as a tree commonly cultivated about temples and villages at lower elevations downstream. This is the sort of place where the French missionary Père François Ducloux, after whom it was named, found it while collecting plants in Yunnan province in 1904–1905. It was the first of the western Chinese cypresses to be described as distinct from Himalayan cypress (*Cupressus torulosa*), or even the western Eurasian Mediterranean cypress (*C. sempervirens*). There is still some confusion with Himalayan cypress, which is probably confined to the western Himalaya. Eastern Tibetan records of Himalayan cypress are better assigned either to Yunnan cypress or to the much coarser-twigged Tibetan cypress (*C. gigantea*). The taxonomy of the Himalayan and western Chinese cypresses merits further study.

Cupressus funebris Endlicher

CHINESE WEEPING CYPRESS, BAI MU (CHINESE)

Tree to 25(–35) m tall, with trunk to 2 m in diameter. Bark reddish brown, fibrous, smooth to shallowly furrowed. Crown dense, with short branches drooping at the ends with age and supporting weeping branchlets. Branchlets flattened, 1–1.5 mm wide, branching from just two rows of leaves (the lateral ones). Scale leaves on branchlets 1–1.5 mm long, light green to grayish green with wax, differentiated into facial and lateral pairs of similar size, the facial leaves with a shallow, elongate, inactive gland, the lateral leaves folded around the branchlet and glandless. Pollen cones 2.5–5 mm long, with four or five (to seven) pairs of pollen scales. Seed cones spherical (0.8–)1.2–1.5(–1.8) cm long, dark brown at maturity, with three or four (to six) pairs of seed scales, each with a prominent central point on the face, the surface otherwise smooth. Seeds (one to) five to nine per scale, 2.5–3.5 mm long, light brown, the wings narrow. Cotyledons two. Widespread across the provinces of central and southern China from Anhui, Zhejiang, and Fujian to Gansu, Sichuan, and Yunnan. Forming pure stands or scattered in mixed forests on moist soils; 200–2,000 m. Zone 9. Synonym: *Chamaecyparis funebris* (Endlicher) Franco.

Chinese weeping cypress is a favored tree in China for planting around temples, shrines, and cemeteries. The fragrant wood is highly prized for many uses, including cabinetry, coffins, construction, and boatbuilding. Although many authors include Chinese weeping cypress in *Chamaecyparis* because of its flattened branchlets and small seed cones with few seeds per scale, there is far better reason to retain it as an odd species of *Cupressus.* Although the seed cones are small, their form is that of *Cupressus* rather than *Chamaecyparis,* and they mature in their second season rather than in their first. The flattened foliage is an exaggeration of tendencies found in several other species of *Cupressus,* including Mexican cypress (*C. lusitanica*), Himalayan weeping cypress (*C. cashmeriana*), and Alaska yellow cedar (*C. nootkatensis*), all with small seed cones. *Cupressus* and *Chamaecyparis* are not closely related by their DNA structures, and many chemical compounds of wood and foliage in Chinese weeping cypress are typical of *Cupressus* rather than *Chamaecyparis.*

Cupressus gigantea W. C. Cheng & L. K. Fu

TIBETAN CYPRESS, JU BAI (CHINESE)

Tree to 45 m tall, with trunk to 3(–6) m in diameter. Bark bright reddish brown to grayish brown, fibrous, peeling in long thin strips to form furrows and ridges. Crown dense, conical, with short horizontal branches, flattening and thinning with age. Branchlets roundly four-angled, (1–)1.5–2 mm in diameter, branching from all four rows of leaves. Scale leaves on branchlets about 1 mm long, dark green or often bluish green with wax, the back often with an inconspicuous, inactive gland. Pollen cones 3–6(–8) mm long, about 2.5 mm wide, with (4–)6–8(–10) pairs of pollen scales, each scale with three to five pollen sacs. Seed cones slightly oblong, 1.5–2 cm long, dark brown with a grayish waxy coating, usually with six pairs of seed scales, each with a well-developed central, conical point on the face, the surface otherwise warty. Seeds four to eight per scale, 4–6 mm long, light brown, the wings very narrow. Cotyledons two. Southeastern Xizang (Tibet), along the Brahmaputra (Yarlung Zangbo) River and its tributaries. Forming groves of large trees in a general matrix of shrubby vegetation along streamsides and on gentle slopes; 3,000–3,400(–3,650) m. Zone 8? Synonyms: *Cupressus austrotibetica* Silba, *C. tongmaiensis* Silba, *C. tongmaiensis* var. *ludlowii* Silba, *C. torulosa* var. *gigantea* (W. C. Cheng & L. K. Fu) Farjon.

Tibetan cypress earns its scientific name *gigantea* as one of the tallest species in the genus as well as the largest tree of the arid valleys of the Tibetan Plateau. It is a very local species that resembles a coarser, larger version of Himalayan cypress (*Cupressus torulosa*) of the western Himalaya and has been treated as a variety of that species. The aromatic oils of its leaves are distinct from those of other species examined, including all other Chinese cypresses. The largest trees are reported to exceed 1,000 years in age, like other slow-growing conifers of arid regions.

Cupressus goveniana G. Gordon

GOWEN CYPRESS

Shrubby tree to 10 m tall, or rarely a large tree to 50 m, with trunk to 2 m in diameter, or with several small trunks from the base, or even just 1–2 cm in diameter when growing on hardpan soils. Bark reddish to grayish brown, thin, smooth to furrowed, fibrous. Crown sparse to dense, narrow to broad, with horizontal or upright branches. Branchlets cylindrical to slightly four-angled, 1–1.5 mm in diameter, branching from all four rows of leaves. Scale leaves of branchlets 1–2 mm long, bright or dark green to yellow-green, not conspicuously waxy, without a gland on the back, or with an inconspicuous, inactive one. Pollen cones 3–4 mm long, 1.5–2 mm wide, with (five or) six or seven (or eight) pairs of pollen scales, each with (three to) five or six pollen sacs. Seed cones spherical or slightly oblong, 1–2.5(–3) cm long, grayish brown at maturity, not waxy, with (three or) four or five pairs of seed scales, each usually fairly flat on the face, without prominent warts or wrinkles. Seeds 8–15 per scale, (2–)3–4(–5) mm long, brown to black, sometimes thinly waxy, occasionally with resin pockets, the wings thick and narrow. Cotyledons three or four (or five). Coastal central California, discontinuously from southern Mendocino County to northern Monterey County. Forming groves or scattered in closed-cone pine forest or margins of redwood forest, often on sterile, sandy or hardpan soils; 60–800 m. Zone 8. Synonym: *Callitropsis goveniana* (G. Gordon) D. Little. Three varieties.

There is variation among populations of Gowen cypress in foliage and seed characters, and these populations are divided geographically into three species, subspecies, or varieties. The four central populations in the Santa Cruz Mountains are most distinctive in morphology and chemistry, while varying among themselves in foliage and cones. Some of this variation might be due to past hybridization with Sargent cypress (*Cupressus sargentii*). The two populations near Carmel Bay in Monterey County are most similar to the numerous populations of Mendocino County. Most of the latter grow on shallow, impoverished, hardpan soils of ancient coastal terraces. They join shore pine (*Pinus contorta* var. *contorta*) to form the famous open pygmy forests of natural bonsai trees with an extraordinarily slow growth rate and reaching maturity with full sized cones at only a few decimeters tall. The dwarfism is not genetic and individuals respond rapidly if the hardpan is penetrated. Trees growing on deep soils nearby are among the tallest of all cypresses, the largest known individual being 43.3 m tall and 2.2 m in diameter, with a spread of 12.2 m. Preliminary DNA studies of single samples of each variety group them together relatively strongly compared to other groupings of California cypresses. Like other cypresses, however, the taxonomy of this species can only be resolved by thorough comparisons of populations. In the absence of such studies, the slight distinguishing features of the three proposed varieties are presented. There are abundant Quaternary fossils of this species along the coast of California that show that, like Monterey and bishop pines (*P. radiata* and *P. muricata*), it once had a much more continuous distribution throughout its present range and beyond to southern California. Recent fragmentation might explain some of the difficulties in resolving the relationships among populations of this species. It is named for James Gowen (d. 1862), a noted British rhododendron breeder who was secretary to the Horticultural Society of London when this species was described as new in its journal.

Natural stand of Santa Cruz cypress (*Cupressus goveniana* var. *abramsiana*).

Cupressus goveniana var. *abramsiana* (C. Wolf) E. Little

SANTA CRUZ CYPRESS

Foliage bright green. Seed cones often slightly oblong, 1.5–3 cm long. Seeds 3–5 mm long, brown, sometimes waxy, the attachment scar usually prominent. Santa Cruz Mountains, California. Synonyms: *Callitropsis abramsiana* (C. Wolf) D. Little, *Cupressus abramsiana* C. Wolf.

Cupressus goveniana G. Gordon var. *goveniana*

GOWEN CYPRESS

Foliage light green to yellowish green. Seed cones spherical, mostly 1–1.5 cm long. Seeds 2–4 mm long, dark brown to nearly black, usually not waxy, the attachment scar not prominent. Huckleberry Hill and Gibson Creek, around Carmel Bay, Monterey County, California.

Cupressus goveniana var. *pigmaea* J. Lemmon

MENDOCINO CYPRESS, PYGMY CYPRESS

Foliage dark green. Seed cones spherical, 1–2 cm long. Seeds 2–4 mm long, usually black but sometimes brown, usually not waxy, the attachment scar not prominent. Above the coast in Mendocino County from Fort Bragg to Albion, with scattered trees between Point Arena and Anchor Bay, California. Synonyms: *Callitropsis pigmaea* (J. Lemmon) D. Little, *Cupressus goveniana* subsp. *pigmaea* (J. Lemmon) Bartel, *C. pigmaea* (J. Lemmon) C. Sargent.

Cupressus guadalupensis S. Watson

GUADALUPE CYPRESS, CEDRO (SPANISH)

Tree to 15(–20) m tall, with single trunk to 1.2(–2.4) m in diameter or dividing near the base. Bark rich reddish brown, mottled with green and gray, smooth and flaking, becoming grayish brown, fibrous and furrowed at the base of large trunks. Crown widely spreading, open, rounded, with rising branches. Branchlets cylindrical to slightly four-sided, 1–1.5(–2) mm in diameter, branching from all four rows of leaves. Scale leaves on branchlets 1–1.5(–2) mm long, light to dark green, sometimes slightly waxy, the back often with an inconspicuous, inactive gland. Pollen cones 3–6 mm long, 2–5 mm wide, with (4–)6–8(–10) pairs of pollen scales, each with (three to) five pollen sacs. Seed cones nearly spherical or occasionally slightly oblong, (2–)2.5–4(–5) cm long, brown to grayish brown at maturity, not waxy, with (three or) four or five pairs of seed scales, each with a strong, or sometimes inconspicuous conical point on the face, otherwise smooth. Seeds 10–15 per scale, 5–6 mm long, dark brown, waxy or not, sometimes with resin pockets. Cotyledons three to six. Southwestern California to northeastern Baja California (Mexico). Forming groves standing out above scrublands on slopes; 450–1,000 m. Zone 9. Synonym: *Callitropsis guadalupensis* (S. Watson) D. Little. Two varieties.

The two commonly accepted varieties (sometimes considered species) of Guadalupe cypress are among the least distinguishable in a genus notorious for small distinctions. Some DNA studies show the two varieties to be closely related, if not to the exclusion of other cypresses, but these studies are based on single samples and suggest inconsistent relationships. The trees on Guadalupe Island (after which the species is named), off the coast of Baja California were decimated by the island's bountiful feral goats, which ate almost all regeneration for decades. The remaining picturesque groves contain mature, large trees that can exceed 250 years in age. This species may have given its name to Isla Cedros off Baja California, but California juniper (*Juniperus californica*), also called *cedro* but usually shrubbier than Guadalupe cypress, is also found on both Isla Cedros and Isla Guadalupe. Populations on the mainland grow in the highly fire-prone scrubland known as chaparral and are subject to frequent fires. Their cones remain closed many years after maturity, opening after fires to release the seeds. The resulting stands can be so crowded that mature, coning trees 20–30 years old may be only 1 m tall and 2–3 cm in diameter. Because of the frequent fires, the mainland trees are typically smaller than those of the islands. The largest known individual in California when measured in 2001 was 21.7 m tall, 0.8 m in diameter, with a spread of 14.6 m.

Cupressus guadalupensis var. *forbesii* (Jepson) E. Little

TECATE CYPRESS

New foliage bright light green, darkening and dulling with age. Pollen cones 3–4 mm long, with four to seven pairs of pollen scales. Seed cones mostly not more than 3 cm long. Orange and

Flaky smooth bark of Tecate cypress (*Cupressus guadalupensis* var. *forbesii*).

San Diego Counties, California, south in the coastal mountains of Baja California, Mexico, to San Quintín. Synonyms: *Callitropsis forbesii* (Jepson) D. Little, *Cupressus forbesii* Jepson.

Cupressus guadalupensis S. Watson **var. *guadalupensis***

GUADALUPE CYPRESS, CEDRO (SPANISH)

New foliage waxy bright blue-green, becoming green with age. Pollen cones 4–6 mm long, with (5–)8–10 pairs of pollen scales. Seed cones usually more than 3 cm long. Guadalupe and Cedros Islands, Baja California, Mexico.

Cupressus ×leylandii A. B. Jackson & Dallimore

LEYLAND CYPRESS

Tree (*Cupressus macrocarpa* × *C. nootkatensis*) to 36 m tall or more, with trunk to 1.2 m in diameter or more. Bark grayish, fibrous, furrowed, peeling in thin strips. Crown dense, usually narrowly conical, with short rising branches, gently nodding at the tips. Branchlets generally flattened, 1–2 mm wide, branching mostly just from the two rows of lateral leaves to form fernlike sprays but sometimes more irregular. Scale leaves on branchlets 1.5–2.5 mm long, the free tip at least half the length, gray-green to golden, the back not conspicuously glandular. Some cultivars sterile. Pollen cones 3–5 mm long, 2–3 mm wide, with five to seven pairs of pollen scales, each with three to six pollen sacs. Seed cones spherical, 1.5–2 cm long, dark brown at maturity, sometimes slightly waxy, with four pairs of seed scales, each usually with a distinct, flattened-conical point on the face, and lines radiating from there. Seeds about five on each scale, about 5 mm long, the wings about 1 mm wide, brown, dotted with resin pockets. Cotyledons two to four? Known only in cultivation. Zone 7. Synonyms: *Callitropsis ×leylandii* (A. B. Jackson & Dallimore) D. Little, ×*Cupressocyparis leylandii* (A. B. Jackson & Dallimore) Dallimore, ×*Cuprocyparis leylandii* (A. B. Jackson & Dallimore) Farjon.

Leyland cypress has arisen repeatedly from open-pollinated seeds of both parents since the original six seedlings were sown in 1882 by C. J. Leyland at Leighton Hall in Britain, which he inherited in 1892 along with Haggerston Hall, to which he moved the seedlings. Leighton Hall was later occupied by his nephew, J. M. Naylor, who corresponded with William Dallimore at Kew after finding two new seedlings. For most of the 20th century the trees were considered an intergeneric hybrid, since Alaska yellow cedar (*Cupressus nootkatensis*) was thought to be a species of *Chamaecyparis* until DNA sequences conclusively confirmed what many traditional features suggested, that it belonged with other New World species of *Cupressus.* The trees are even faster growing than either of their parents, which are the fastest-growing North American species. Since few trees are older than a century,

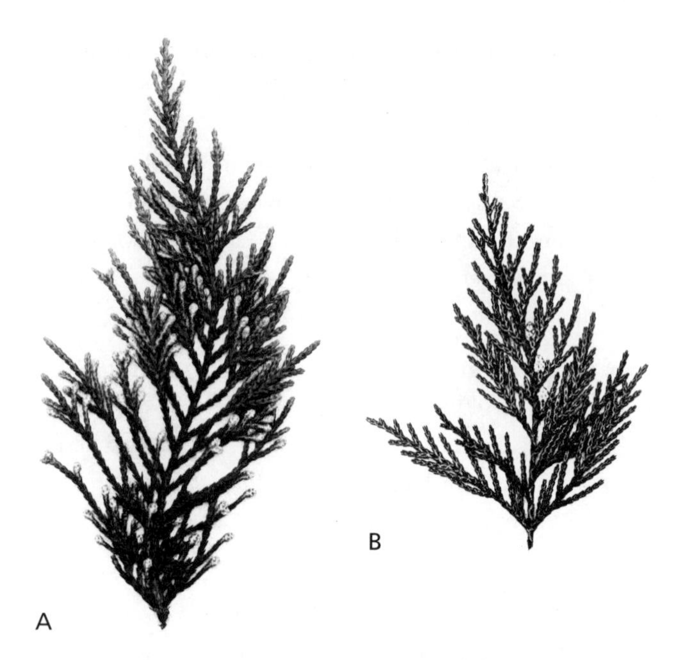

Upper side of branchlet spray with pollen cones (A) and underside with waxy white X's (B) of Leyland cypress (*Cupressus ×leylandii*), ×0.5.

Natural stand of old mature Mexican cypress (*Cupressus lusitanica*).

the maximum size they can reach is as yet unknown. They are important as large "green screens" in Britain and elsewhere where they are hardy. They are so widely planted in Britain, continental Europe, and warm temperate North America, in fact, and often in such inappropriate places considering their stature, that they border on pest status and have sparked many neighborly disputes.

Cupressus lusitanica P. Miller

MEXICAN CYPRESS, CEDAR OF GOA, CIPRÉS DE MÉXICO, TLASCAL (SPANISH)

Tree to 30 m tall, with trunk to 1.5(–3.5) m in diameter. Bark reddish brown, thin, fibrous, peeling in narrow strips. Crown variable, broadly conical or cylindrical, with horizontal branches. Branchlets four-sided, sometimes slightly flattened, 1–1.5(–2) mm in diameter, branching from all four rows of leaves or only from two, thus producing flattened sprays. Scale leaves on branchlets 1–2.5 mm long (to 10 mm with free tips to 4 mm on main shoots), dark green to bluish green with wax, rarely with an inactive resin gland, the edges usually smooth but sometimes minutely toothed. Pollen cones 3–4(–6) mm long, 2.5–3 mm wide, with six to eight pairs of pollen scales, each with (three or) four pollen sacs. Seed cones spherical, 1–1.5(–2) cm long, reddish or grayish brown at maturity, often waxy before this, with (two or) three or four pairs of seed scales, each usually with a strong conical point 5–6 mm high on the face and lines radiating from there. Seeds 8–12 per scale, 3–5 mm long, the wing to 1.5 mm wide, yellow-brown, usually not waxy and lacking resin pockets. Cotyledons (three or) four. Northern Mexico (southern Chihuahua and southern Coahuila) south through southern Guatemala to western Honduras (Santa Bárbara Mountain). Forming pure stands or usually mixed with other conifers in the mountains; (1,700–)2,200–2,900(–4,000) m. Zone 8. Synonym: *Callitropsis lusitanica* (P. Miller) D. Little. Two varieties.

Dense natural stand of young mature Mexican cypress (*Cupressus lusitanica*).

Mexican cypress is a highly ornamental species that has been widely cultivated in its homeland and montane districts throughout the tropics since at least the middle of the 17th century. That is when it was first cultivated by Carmelite friars in the Serra do Bussaco, near Coimbra in Portugal. This stand, which still exists today with trees over 300 years old, was the source of the material that Miller described, explaining the species name (Lusitania is the Roman province corresponding to Portugal). There has been much dispute about whether these trees are really the same as the Mexican cypress, and so some authors use a different name for this species, usually *Cupressus lindleyi.* It is now often difficult to determine whether stands in Central America are native or the result of centuries of planting, and many wild stands were severely depleted for their fine timber. As with other widespread species of *Cupressus* with fragmented distributions, Mexican cypress has a great many variants throughout its range, some of which were brought into cultivation and then served as the bases for additional described species, now largely forgotten. The natural variation in this species has not yet been well studied, but trees of the Huasteca region of east-central Mexico northeast of Mexico City were recognized as a separate species, *C. benthamii,* and are here treated as a weakly distinguished variety of *C. lusitanica.*

Cupressus lusitanica var. *benthamii* (Endlicher) Carrière

Branchlets in flattened sprays, arising from only two rows of leaves. Seed cones mostly with three pairs of scales. Mexico, border region where the states of Hidalgo, Puebla, and Veracruz come together. Synonyms: *Callitropsis benthamii* (Endlicher) D. Little, *Cupressus benthamii* Endlicher.

Cupressus lusitanica P. Miller var. *lusitanica*

Branchlets in three-dimensional sprays, arising from all four rows of leaves. Seed cones mostly with four pairs of scales. Throughout the range of the species. Synonyms: *Cupressus benthamii* var. *lindleyi* (Klotzsch ex Endlicher) M. T. Masters, *C. lindleyi* Klotzsch ex Endlicher, *C. lindleyi* var. *hondurensis* (Silba) Silba, *C. lusitanica* var. *hondurensis* Silba, *C. lusitanica* var. *lindleyi* (Klotzsch ex Endlicher) Carrière.

Spreading mature MacNab cypress (*Cupressus macnabiana*) in nature.

Cupressus macnabiana A. Murray bis

MACNAB CYPRESS

Shrubby tree to 10(–18) m tall, with trunk to 1.3 m in diameter, usually dividing near the base into three or more major diverging branches. Bark gray, rough, fibrous, to 2 cm thick, peeling in narrow, thin strips. Crown broad, irregular, with upwardly angled branches. Branchlets slightly four-sided, 0.5–1 mm in diameter, branching primarily from just two rows of leaves to form three-dimensional arrays of small frondlike, flattened sprays. Scale leaves on branchlets 1–1.5 mm long, grayish green with wax that becomes more prominent with age, the back with a conspicuous drop of dried resin in an open gland. Pollen cones 2–3 mm long, about 2 mm wide, with four to seven pairs of pollen scales, each with (three or) four or five pollen sacs. Seed cones nearly spherical or a little oblong, 1.5–2.5 cm long, grayish brown at maturity, generally not waxy, with three or four pairs of seed scales, each, particularly the upper pair, often with a strong, upcurved, conical point on the face that weathers away with age, the surface heavily wrinkled. Seeds 9–12 per scale, 2–5 mm long, the wings much narrower than the body, brown, sometimes slightly waxy, sometimes warty with resin pockets. Cotyledons (three or) four (or five). Northern California. Scrublands and open pine-oak woodlands on dry slopes and ridges, often on serpentine soils; 300–850 m. Zone 8. Synonym: *Callitropsis macnabiana* (A. Murray bis) D. Little.

MacNab cypress is a characteristic bush of the chaparral and woodlands on the foothills surrounding the Sacramento Valley of northern California. These communities are very fire prone, and the seed cones of MacNab cypress remain closed on the branches many years with seeds remaining viable until their release by fire. The accumulation of resin on the leaves makes the foliage among the most fragrant in the genus, and also contributes to its flammability. This is not one of the most ornamental species, and it is rarely cultivated outside of collections. It hybridizes naturally with *Cupressus sargentii* where they grow together. The largest known individual is 16.8 m tall, with a trunk 1.3 m in diameter and a crown spread of 13.7 m. The species is named for James MacNab (1810–1878), curator of the Royal Botanic Garden, Edinburgh, which supported the expedition on which William Murray, brother of its describer, collected it.

Cupressus macrocarpa K. Hartweg ex G. Gordon

MONTEREY CYPRESS

Tree to 25(–32) m tall, with trunk to 1.5(–4.5) m in diameter, often dividing near the base into three or more main branches. Bark light brown to gray, thick, irregularly ridged and furrowed, peeling in thin, fibrous strips. Crown variable, spirelike to broadly conical when young, becoming very broad, open, and flat-topped with age on exposed sites. Branchlets cylindrical, 1.5–2 mm in diameter, branching from all four rows of leaves. Scale leaves on branchlets 1.5–2 mm long, bright green, not waxy, usually without a gland on the back, but this inconspicuous and inactive if present. Pollen cones 4–6 mm long, 2.5–3 mm wide, with six or seven pairs of pollen scales, each with 6–8(–10) pollen sacs. Seed cones slightly oblong, 2.5–3.5(–4) cm long, grayish brown at maturity, not waxy, with four to six pairs of seed scales, each wrinkled on the face, with a low, inconspicuous central point. Seeds 10–16 per scale, 5–6 mm long, the wings to 2 mm wide, dark brown, usually not waxy, warty with resin pockets. Cotyledons (three or) four (or five). Narrowly restricted to the headlands at the mouth of Carmel Bay, Monterey County, California. Forming stands in coastal thickets on and behind sea bluffs; 5–35 m. Zone 8. Synonym: *Callitropsis macrocarpa* (K. Hartweg ex G. Gordon) D. Little.

Monterey cypress is an extremely narrow endemic, the two extant populations historically occupying less than 200 hectares

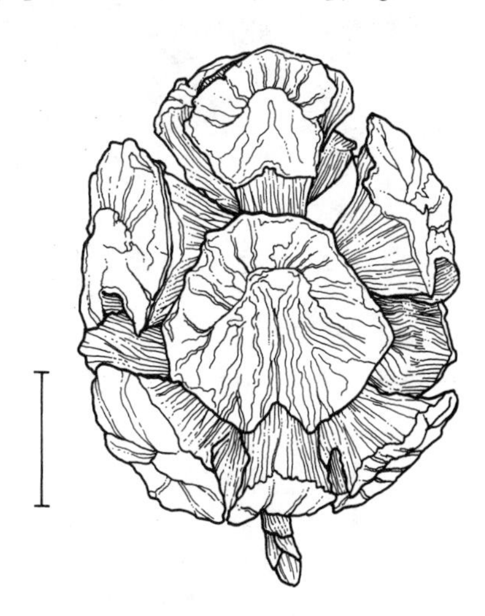

Open mature seed cone of Monterey cypress (*Cupressus macrocarpa*) after shedding seeds; scale, 1 cm.

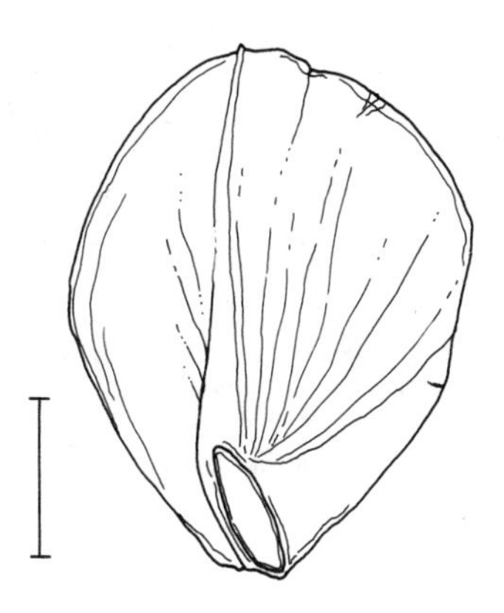

Attachment face view of dispersed seed of Monterey cypress (*Cupressus macrocarpa*); scale, 2 mm.

Mixed age natural stand of Monterey cypress (*Cupressus macrocarpa*).

of bluffs on the seaward side of much more extensive Monterey pine (*Pinus radiata*) stands. Late glacial fossils are uncommon but extend the range to Los Angeles County, California, some 375 km south of its present range. The species is one of the most massive of the cypresses, the largest known individual when measured in 1999 being 31.1 m tall, with a trunk diameter of 4.3 m and a spread of 35.4 m. Large, windblown individuals of the coastal bluffs are extremely picturesque, with a strong general resemblance to the cedars of Lebanon (*Cedrus libani* subsp. *libani*), a point noticed by Karl Hartweg, who collected the type specimen of the tree in 1846 and named it for its large cones. Considering the narrow range and limited gene pool of Monterey cypress, with fewer than 20,000 trees in nature, it has spawned a large number of cultivars, not to mention its hybrid offspring, Leyland cypress (*Cupressus* ×*leylandii*). It is one of the fastest-growing species and quickly forms a tree with presence, especially near coastal regions with moderate climates. Inland, it is highly susceptible to fungal pathogens, windburn, and other ailments that limit its attractiveness and life span.

Cupressus nootkatensis D. Don in A. Lambert

ALASKA YELLOW CEDAR, NOOTKA CYPRESS

Tree to 40(–50) m tall or dwarfed near the alpine tree line, with trunk to 2(–3.7) m in diameter. Bark gray-brown, raggedly ridged. Crown narrowly conical. Branchlets slightly flattened, 1–1.5 mm wide, branching from two rows of leaves, the branchlet sprays markedly dangling. Scale leaves on branchlets 1.5–2.5 mm long, yellow-green, scarcely differentiated into facial and lateral pairs since the latter are not keeled, sometimes glandular, with a fetid odor when bruised. Leaf tips triangular, tight to the branchlet, those of the facial leaves touching the base of the next pair. Pollen cones 2–5 mm long, 1.5–2 mm wide, with five or six pairs of pollen scales, each with two (or three) pollen sacs. Seed cones spherical, 6–13 mm long, dark reddish or purplish brown at maturity, often waxy before this, with two (or three) pairs of seed scales, each usually with a prominent, narrowly triangular point on the face and wrinkles radiating from there. Seeds (one or) two or three (to five) per scale, 2–5 mm long, light brown, the wings about as broad as the body. Cotyledons two. Northern

Open mature seed cone of Alaska yellow cedar (*Cupressus nootkatensis*) with central columella after shedding seeds; scale, 1 cm.

Face view of dispersed seed of Alaska yellow cedar (*Cupressus nootkatensis*); scale, 5 mm.

Pacific North America from Prince William Sound, Alaska, through British Columbia, to the Siskiyou Mountains of extreme northwestern California. Forming pure stands or usually mixed with other conifers, particularly western hemlock (*Tsuga heterophylla*) or mountain hemlock (*T. mertensiana*), on a variety of substrates, from rocky outcrops to waterlogged peats; 0–2,300 m. Zone 6. Synonyms: *Callitropsis nootkatensis* (D. Don in A. Lambert) Florin, *Chamaecyparis nootkatensis* (D. Don in A. Lambert) Sudworth, *Xanthocyparis nootkatensis* (D. Don in A. Lambert) Farjon & D. Harder.

The narrowly weeping habit of Alaska yellow cedar has made it a favorite in contemporary landscaping. It has hybridized spontaneously in cultivation with various other species of *Cupressus*, including Monterey cypress (*C. macrocarpa*), which yielded the vigorous Leyland cypress (*C. ×leylandii*). It is by far the most northerly and wet-climate species of *Cupressus* and has long been associated uncomfortably with the ecologically and geographically more similar *Chamaecyparis*. DNA sequences confirm that it is, instead, an odd species of *Cupressus*. Regeneration is limited within its native range, and the tree has experienced an unexplained high level of mortality among all age classes, leading to considerable conservation concern. The largest known individual in the United States in combined dimensions when measured in 1996 was 39.3 m tall, 3.7 m in diameter, with a spread of 8.2 m.

Underside of flattened branchlet spray of Alaska yellow cedar (*Cupressus nootkatensis*) with scars of pollen cone attachments on 2 years of growth distinguishable primarily by the vividness of the white waxy markings, ×1.

Cupressus ×notabilis (A. F. Mitchell) Eckenwalder

ALICE HOLT CYPRESS

Tree (*Cupressus arizonica* × *C. nootkatensis*) exceeding 18 m tall, with trunk exceeding 45 cm in diameter. Bark thin, reddish to purplish brown, peeling in vertical lines or flakes. Crown dense but tattered-appearing, conical, with gently arching branches. Branchlets slightly flattened, 1–1.5 mm wide, sparsely branching from just two rows of leaves. Scale leaves on branchlets about 2 mm long, light gray-green, the thin wax most evident at the base, not prominently glandular. Pollen cones 2.5–3.5 mm long, 1.5–2 mm wide, with five to seven pairs of pollen scales, each with three to five pollen sacs. Seed cones spherical, about 1.2 cm long, purple-brown at maturity with a bluish white waxy coating, with three (or four) pairs of seed scales, each with a flattened, triangular point on the face and lines radiating from there. Seeds three to five per scale, 4–5 mm long. Cotyledons two to four? Known only in cultivation. Zone 7. Synonyms: ×*Cupressocyparis notabilis* A. F. Mitchell, ×*Cuprocyparis notabilis* (A. F. Mitchell) Farjon.

Like Leyland cypress (*Cupressus ×leylandii*), the seed that gave rise to Alice Holt cypress also came from Leighton Hall in Britain, this time from an open-pollinated tree of smooth Arizona cypress (*C. arizonica* var. *glabra*). Two of the seedlings raised from this tree in 1957 at the British Forestry Commission's Alice Holt Station were hybrids, and the few trees in collections were propagated from one of these. They are vigorous, like the Leyland cypress, and could be marginally hardier, but they have a more ragged growth habit. The trees are too young to foretell how large they can become.

Cupressus ×ovensii (A. F. Mitchell) Eckenwalder

WESTONBIRT CYPRESS

Tree (*Cupressus lusitanica × C. nootkatensis*) of unknown potential height but reaching 17.4 m at 35 years, with trunk 30 cm in diameter at 35 years. Bark thin, of long, fibrous strips. Crown ragged, conical, with upraised branches with nodding tips and weeping branchlet sprays. Branchlets slightly flattened, 1–1.5 mm wide, branching regularly from the two rows of lateral leaves to form fernlike sprays. Scale leaves on branchlets 1.5–2 mm long, with a long free tip, deep bluish green, not conspicuously glandular. Pollen cones 3–5 mm long, 2–2.5 mm wide, with six or seven pairs of pollen scales, each with (two or) three (or four) pollen sacs. Seed cones spherical, about 1 cm long, purple-brown at maturity, with a whitish waxy coating before this, with three or four pairs of seed scales, each with a low, triangular point on the face, otherwise fairly smooth. Seeds three to five per scale, 3–5 mm long. Cotyledons two to four? Known only in cultivation. Zone 8. Synonyms: ×*Cupressocyparis ovensii* A. F. Mitchell, ×*Cuprocyparis ovensii* (A. F. Mitchell) Farjon.

Howard Ovens raised this hybrid in 1961 from a lone tree of Mexican cypress (*Cupressus lusitanica*) surrounded and sheltered by Alaska yellow cedar (*C. nootkatensis*) at the Westonbirt arboretum in Britain. The trees are vigorous but much broader and more open than Leyland cypress (*C. ×leylandii*). They are too young to be anywhere near their potential eventual size. Like many New World cypresses, the crushed foliage has a faintly lemony scent derived from the monoterpene limonene.

Grove of mature and seedling Sargent cypress (*Cupressus sargentii*) in nature.

Cupressus sargentii Jepson

SARGENT CYPRESS

Tree to 25(–45) m tall or shrubby and less than 10 m, with trunk to 1 m or more in diameter. Bark grayish brown to almost black, thick, fibrous, and furrowed, peeling in long, narrow strips. Crown variable, columnar to broadly conical, open to dense. Branchlets cylindrical or slightly four-sided, 1.5–2 mm thick, branching from all four rows of leaves. Scale leaves on branchlets 1.5–2 mm long, dull green, often with some waxiness, the margins smooth, usually with an inconspicuous, inactive, resin gland. Pollen cones 3–4(–5) mm long, about 2 mm wide, with (three to) five or six (to eight) pairs of pollen scales, each with three or four pollen sacs. Seed cones spherical, (1.5–)2–2.5(–3) cm long, brown or grayish brown at maturity, not waxy, with (two or) three or four (or five) pairs of seed scales, each with an inconspicuous to prominent conical point on the face, the surface also warty with resin pockets. Seeds 10–15 per scale, 4–6 mm long, with wings to 1(–2) mm wide, dark brown, often waxy. Cotyledons three or four (or five). Coast ranges of California. Scrublands, open pine-oak woodlands, and mixed conifer forests, usually on serpentine-derived soils; 200–1,100 m. Zone 8. Synonym: *Callitropsis sargentii* (Jepson) D. Little.

Sargent cypress is the most widely distributed cypress species in California, extending for almost 1,000 km along the inner Coast Ranges, away from the coast, from northern Mendocino County to central Santa Barbara County, but with a large gap south of San Francisco Bay. Although it varies considerably across this range, it has been relatively free of the confusing taxonomic subdivision that has been applied to other species. It is closely related to Gowen cypress (*Cupressus goveniana*) and was originally included within that species until Jepson distinguished it and named it for Charles Sargent (1841–1927), the founding director of the Arnold Arboretum of Harvard University in Massachusetts, who badgered Jepson and other correspondents for specimens and information while preparing his 14-volume *Silva of North America* (1890) and *Manual of the Trees of North America* (1905, 1922). It is one of the few cypresses to grow in contact with another species, and it hybridizes with MacNab cypress (*C. macnabiana*) in some of the places where they grow together. Although it is not sympatric with Gowen cypress today, it might have been responsible for the formation of Santa Cruz cypress (*C. goveniana* var. *abramsiana*) through hybridization. It has a high light requirement and is also sensitive to soil moisture, so it has not been successful in horticulture. The largest known individual when measured in 2000 was 17.7 m tall, 2.0 m in diameter, with a spread of 19.5 m. The champion that it replaced was uncharacteristically tall, at over 44 m, with a trunk 1.2 m in diameter and a crown spread of over 12 m.

Cupressus sempervirens Linnaeus

MEDITERRANEAN CYPRESS, FUNERAL CYPRESS, CIPRÉS (SPANISH), NEKOSTA ARRUNTA (BASQUE), CYPRÈS COMMUN (FRENCH), CIPRESSO (ITALIAN), SELVIA (ALBANIAN), KIPARIS (BULGARIAN), KYPARISSOS (GREEK), SERVI (TURKISH), SARV (ARABIC, FARSI)

Tree to 25(–40) tall, with trunk to 2(–4) m in diameter. Bark grayish brown, fibrous, longitudinally shallowly furrowed and ridged. Crown variable, from extremely spire-shaped to about as broad as tall, with upright or spreading branches, uniformly dense. Branchlets cylindrical or slightly four-sided, (0.5–)1–1.5 mm in diameter, branching from all four rows of leaves or favoring two of them. Scale leaves on branchlets 0.5–1(–1.5) mm long, dark green or sometimes bluish green with wax, the edges minutely toothed, the back often with an inconspicuous (sometimes evident), usually inactive resin gland. Pollen cones (2.5–)4–6(–8) mm long, 1.5–2.5(–3) mm wide, with 5–8 (–12) pairs of pollen scales, each with (three or) four or five pollen sacs. Seed cones usually oblong, (1.5–)2–3(–4) cm long, brown at maturity, not waxy, with (three to) five or six (or seven) pairs of seed scales, each with a short, sometimes inconspicuous conical point on the face, and ridges radiating from there when open. Seeds (1–)8–20(–25) per scale, 4–7 mm long, the wings less than 1 mm wide, reddish brown, not waxy, without resin pockets. Cotyledons two (or three). North Africa and northeastern Mediterranean to Iran, widely planted and sometimes naturalized in Mediterranean Europe, exact native distribution uncertain. Forming open woodlands on slopes, sometimes with pines or junipers, primarily on calcareous soils, predominantly near the coast in the Mediterranean portion of its range, 150–1,200(–2,000) m. Zone 8. Three varieties.

The narrow spires of cultivated *Cupressus sempervirens* are one of the most characteristic features of cemeteries, parks, and windbreaks in the Mediterranean region, as many iconic paintings attest. These fastigiate forms, far more than the spreading trees that make up wild stands, are also grown as landscape accents worldwide, especially in areas of Mediterranean climate. Although all species of *Cupressus* (as defined today) are evergreen, Linnaeus named this one *C. sempervirens* to contrast with the deciduous bald cypress (*Taxodium distichum*), which he included with *Cupressus* in his *Species Plantarum* of 1753. Louis Richard removed the deciduous species to his newly described genus *Taxodium* in 1810. Individuals can be long-lived and commonly exceed 500 years, while some are reported to reach 2,000 years. Trees from isolated populations in the Sahara and the Atlas Mountains of western North Africa are moderately distinctive and are treated here as varieties, although they were first describe as separate species. Regeneration of these populations is threatened by cutting for firewood and by tree-climbing goats. Experiments have been undertaken using ordinary Mediterranean cypress trees as surrogate mothers for Saharan cypress embryos. Trees at the eastern limits in Iran were also segregated, with less justification, and are not here distinguished from the main body of the species in variety *sempervirens.*

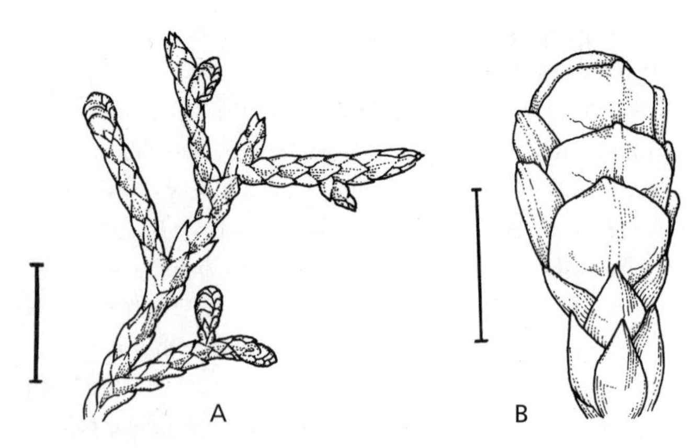

Twig with nearly mature pollen cones (A) and detail of single cone (B) of Mediterranean cypress (*Cupressus sempervirens* var. *sempervirens*) shortly before pollen release; scale, 1 cm (A), 2 mm (B).

Cupressus sempervirens var. *atlantica* (H. Gaussen) Silba

ATLAS CYPRESS

Branching frequently from just two rows of leaves. Scale leaves blue-green with light wax and resin from active glands. Seed cones 1.8–2.2 cm long, with three to five pairs of scales. High Atlas Mountains of southern Morocco. Synonyms: *Cupressus atlantica* H. Gaussen, *C. dupreziana* var. *atlantica* (H. Gaussen) Silba.

Cupressus sempervirens var. *dupreziana* (A. Camus) Silba

SAHARAN CYPRESS, ALGERIAN CYPRESS

Branching frequently from just two rows of leaves. Scale leaves dark green, not waxy, with evident but generally inactive glands. Seed cones 1.8–2.4 cm long, with (five) or six pairs of scales. Tamrit Plateau in the Sahara in southeastern Algeria. Synonym: *Cupressus dupreziana* A. Camus.

Cupressus sempervirens Linnaeus var. *sempervirens*

MEDITERRANEAN CYPRESS

Branching predominantly from all four rows of leaves. Scale leaves dark green, not waxy, with relatively inconspicuous, inactive glands. Seed cones (1.5–)2–3(–4) cm long, with

Flattened branchlet spray of Saharan cypress (*Cupressus sempervirens* var. *dupreziana*) beginning a new episode of growth with paler foliage at its tips, ×1.

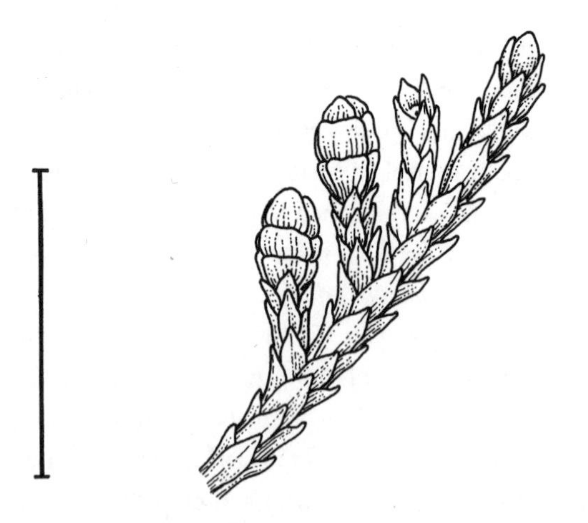

Twig with young seed cones of Mediterranean cypress (*Cupressus sempervirens* var. *sempervirens*) at about the time of pollination; scale, 1 cm.

(three or) four to six (or seven) pairs of scales. Throughout the range of the species west to Libya (or perhaps Tunisia) in North Africa. Synonym: *Cupressus sempervirens* var. *indica* Royle ex Carrière.

Cupressus torulosa D. Don in A. Lambert

HIMALAYAN CYPRESS, GALLA, SURAI (INDIA)

Tree to 25(–50) m tall, with trunk to 1.5(–3.5) m in diameter. Bark grayish brown, peeling in long, vertical strips. Crown broadly conical, rounded above, of short horizontal branches, sometimes drooping at the tips. Branchlets roundly four-sided, 1–1.5(–2) mm in diameter, branching from all four rows of leaves or irregularly only from two. Scale leaves on branchlets 1–2 mm long, dark green to grayish green with light waxiness, the edges minutely toothed, the back without a conspicuous gland. Pollen cones (3–)5–6(–8) mm long, 1–1.5 mm wide, with (6–)8–10(–12) pairs of pollen scales, each with three or four pollen sacs. Seed cones oblong or spherical, 1.2–2(–2.5) cm long, reddish brown at maturity, not waxy, with (three or) four to seven pairs of seed scales, each usually with a strong conical point on the face, the surface smooth or wrinkled. Seeds six to eight per scale, 4–5 mm long, the wings to 1.5 mm wide, light brown, not waxy, warty with resin pockets. Cotyledons (two or) three to five. Western Himalaya, from Chamba, Himachal Pradesh (India), to western Nepal. Forming groves or scattered in mixed conifer forests, particularly on limestone soils; 1,500–3,300 m. Zone 9. Synonyms: *Cupressus karnaliensis* Silba, *C. karnaliensis* var. *mustangensis* Silba.

Like other species of *Cupressus,* the wood of Himalayan cypress is fragrant, durable, and highly prized for a variety of indoor and outdoor uses. The range of Himalayan cypress is often described as extending east to China, but records from the eastern Himalaya all appear to belong to other species. Nonetheless, the Sino-Himalayan *Cupressus* species are in need of careful revision based on natural populations, a task not made easier by their frequent use in horticulture and forestry throughout the region. The species name (Latin for "with small swellings") refers to the somewhat beadlike swelling and thinning of the branchlets as the leaf pairs alternate along them, not to any twisting of the foliage.

Cupressus vietnamensis (Farjon & T. H. Nguyên) Silba

VIETNAMESE YELLOW CEDAR, VIETNAMESE GOLDEN CYPRESS, BÁCH VÀNG (VIETNAMESE)

Tree to 15 m tall, with trunk to 0.5 m in diameter. Bark reddish brown to purplish when fresh, weathering brown to grayish brown, becoming shallowly furrowed and peeling in long, narrow strips. Crown conical at first, becoming dome-shaped and irregular with age, with long, slender, upwardly angled to horizontal branches. Branches with juvenile and transitional foliage present on adult trees. Juvenile leaves 1.5–2 cm long, sticking straight out from the branchlets in alternating whorls of four separated by 4–5 mm. Branchlets bearing adult foliage flattened, 0.8–2(–3) mm wide, mostly arranged in flattened, frondlike sprays by branching from lateral leaf pairs, horizontal or drooping with upturned tips. Adult scale leaves on branchlets 1.5–3 mm long (to 5 mm on main shoots), light green at first, darkening to brownish green with age, occasionally with conspicuous, whitish, small stomatal patches, strongly differentiated into lateral and facial pairs, both types keeled, obscurely glandular. Tips of scale leaves elongate, narrowly and sharply triangular, loosely pointed forward or somewhat spreading, those of the facial leaves overlapping the base of the pair above them. Pollen cones 2.5–3.5 mm long, 2–2.5 mm wide, with five or six pairs of pollen scales, each with two (or three) pollen sacs. Seed cones nearly spherical, 9–12 mm in diameter, greenish brown at maturity, shiny or somewhat dulled by wax, with two (or three) pairs of seed scales, each with a prominent, pale, triangular point on the face. Seeds one to three per scale, 4.5–6 mm long, light brown or reddish brown, the wings about as broad as the body. Cotyledons two? Known only from the Bat Dai Son (mountains) in northern Ha Giang province, Vietnam. Scattered in mixed cloud forests on steep, rocky slopes and ridge tops with thin soils developed from limestone; 1,050–1,200 m. Zone 9? Synonyms: *Callitropsis vietnamensis* (Farjon & T. H. Nguyên) D. Little, *Xanthocyparis vietnamensis* Farjon & T. H. Nguyên.

Vietnamese yellow cedar caused a considerable stir when it was formally described in 2002 because it was assigned to a new genus. Unlike the two famous conifer genera first described in the middle and late 20th century, *Metasequoia* from China and *Wollemia* from Australia, the proposed *Xanthocyparis* was founded not just on a newly described species but also on another one that has been known for almost 200 years, Alaska yellow cedar (*Cupressus nootkatensis*). Unfortunately, because Alaska yellow cedar is the type and only species of a much earlier described genus, *Callitropsis* Oersted, *Xanthocyparis* is nomenclaturally superfluous and cannot be used for these two species if they are placed in a genus separate from *Cupressus.* The relationships of Vietnamese and Alaska yellow cedars are somewhat uncertain. While they are placed in *Cupressus* here, in line with several aspects of their morphology and biochemistry, their ability to cross with other species of *Cupressus,* and some DNA studies, other work suggests a more complicated picture in which each could represent a separate genus. Alternatively, since DNA studies show that they are most closely linked to the New World cypresses, if the latter were treated as a separate genus from the Old World cypresses, Alaska and Vietnamese yellow cedars would be placed in the New World genus even though they have two cotyledons like the Old World species. Even if each species were placed in its own genus, the name *Xanthocyparis* is not technically available, and Vietnamese yellow cedar would either need a new generic name or *Xanthocyparis* would have to be protected from the rules of nomenclature by formally conserving it nomenclaturally. The species itself requires urgent conservation action in the ordinary sense because it has been locally overexploited for its very fine, fragrant timber. The few hundred known remaining trees are generally in sites where it is too hard to cut them and remove the timber. Although searches in the vicinity of the known stands since 1999 had not revealed additional populations by the time the species was described in 2002, the Bat Dai Son lies within less than 15 km of the Chinese border and so Vietnamese yellow cedar might be expected in adjacent southeastern Yunnan, which has also suffered excessive deforestation.

Dacrycarpus (Endlicher) de Laubenfels

KAHIKATEA, DACRYCARP

Podocarpaceae

Evergreen trees and shrubs, sometimes of giant stature. Usually with a single straight trunk that may be cylindrical, with little taper, or fluted and buttressed. Bole free from branches for half or more of its height but, in shrubby individuals of exposed places, dividing near the ground into a few divergent limbs. Bark basically smooth but often with abundant small warts on young trees, remaining thin and breaking up into checkers or shedding irregular flakes and scales with age. Crown conical at first, becoming cylindrical with maturity and finally open, spreading and flat-topped or shallowly dome-shaped. Crown framework with a few major limbs and more numerous slender, rising branches. Lesser limbs breaking up into many branchlets densely clothed with compact foliage. Branchlets green for at least the first year, grooved between the attached leaf bases, of two types. Short shoots limited to a single increment of growth and bearing elongate foliage leaves, confined mostly to juveniles, where they make up a large percentage of the annual growth, but also borne sporadically and in varying proportions on adult trees. Long shoots with continuing growth bearing photosynthetic scale leaves and making up the bulk of adult foliage. Resting buds relatively inconspicuous and unspecialized, consisting of undeveloped foliage leaves. Leaves spirally arranged, closely and evenly spaced along and radiating all around the twigs, of two main types. Elongate leaves of short shoots flattened into two rows by twisting of the twigs between successive leaves and flexure of the lower portion of the leaf. These leaves flattened side to side rather than top to bottom, slightly sickle-shaped. Midrib raised and flanked by lines of stomates on both sides. Scale leaves of long shoots maintaining their radiating orientation, flattened top to bottom, triangular to sword-shaped, and variably curled forward with their tips rather spreading to tightly pressed to the twigs.

Plants dioecious. Pollen cones spherical to cylindrical, single at the tips of short, scaly, axillary stalks or of short shoots. Base of cone with a few bracts transitional between the scale leaves of the stalk and the numerous, spirally arranged pollen scales, each bearing two pollen sacs. Pollen grains medium to large (body 30–60 µm long, 50–85 µm overall), with three rounded air bladders evenly spaced around the body and nearly touching at their points of attachment. Each bladder about the same size as to substantially smaller than the body, with coarse, squiggly sculpturing on the bladders and minutely bumpy over the body. Cap thicker and more coarsely dotted than the germination face, sometimes with a three-pointed scar where the grain was in contact with its three sibling grains during development. Seed cones held upright at maturity, single at the tip of short leafy stalks in the axils of foliage leaves or at the ends of short shoots, maturing and falling in a single season. Cones highly modified and reduced, with a circle of variably developed, radiating free bracts at the base. Reproductive part with two to five bracts, these joining with the cone axis and each other, swelling, and becoming red to purple or black and juicy at maturity to form a warty podocarpium with leafy free bract tips. Upper one to three (or four) bracts fertile but only one usually maturing a single, plump, unwinged seed. Seed embedded in the leathery black or brown seed scale (the epimatium), covered with a thin, waxy film, the opening pointing down into the cone axis. Abortive seeds often enlarging considerably before their arrest and persisting as conical, black protuberances at the tip of the mature seed cone. Seeds roughly spherical, with a single or paired, generally low crest over the top formed by fusion of the fertile bract associated with the seed scale. Cotyledons two, each with two veins. Chromosome base number $x = 10$.

Wood hard, dense, on the heavy side for a softwood, not fragrant, the wide, white to creamy white sapwood slowly developing into pale to bright yellow or light brown heartwood. Grain fine and even, with growth increments delimited by narrow bands of inconspicuously darker latewood. Resin canals absent.

Stomates of scale leaves in patches confined to or more abundant on the upper side. Each stomate sunken beneath and partially hidden by the four to six subsidiary cells, which are sometimes shared between adjoining stomates in a line or flanked by a partial outer row of extra cells, or topped by a well-defined Florin ring. Midrib raised beneath and flat above, with one resin canal underneath and little development of sclerenchyma. Cross section of needle leaves with one resin canal beneath the midvein (but seeming to be on the trailing or outside edge of the leaf) and a wedge of sclerenchyma above (but seeming to be on the leading or inside edge of the leaf). Photosynthetic tissue with a thin palisade layer lining both the right and left faces.

Nine species in the southwestern Pacific region from northern Myanmar and southernmost China south through Indo-China and all of Malesia to New Caledonia and Fiji, and in New Zealand. References: Allan 1961, Buchholz and Gray 1948a, de Laubenfels 1969, 1972, 1978c, 1988, Fu et al. 1999f, T. H. Nguyên and Vidal 1996, Hill and Brodribb 1999, Hill and Carpenter 1991, Hill and Whang 2000, Pocknall 1981a, Pole 1997a, Salmon 1980, Smith 1979, Wells and Hill 1989a, b. Synonyms: *Bracteocarpus* A. V. Bobrov & Melikian, *Laubenfelsia* A. V. Bobrov & Melikian, *Podocarpus* sect. *Dacrycarpus* Endlicher.

Two species in the genus, common dacrycarp (*Dacrycarpus imbricatus*) and kahikatea (*D. dacrydioides*), are among the largest trees in the family Podocarpaceae, forest giants rivaled only by broadleaf miro (*Prumnopitys amara*) and rimu (*Dacrydium cupressinum*), which also grow in the same region. They are also common enough to be important timber trees. The species are all so similar in appearance and overlapping in habitat and habit (most can be either shrubs or trees depending upon ecological conditions) that they are typically not distinguished by local people when two or more are found in the same region. For instance, in New Guinea, with five species of dacrycarps, the greatest concentration of any region, the name *pau* has been recorded for four of the species among Enga speakers, while *umba* is used for at least three by Chimbu speakers. Species of *Dacrycarpus* have basically

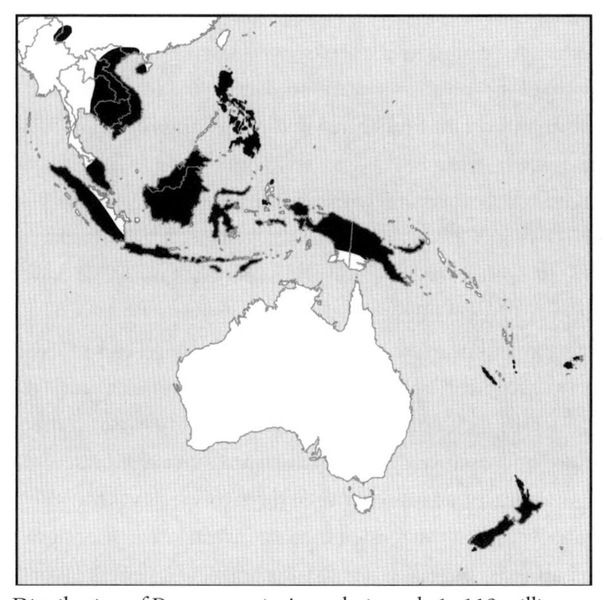

Distribution of *Dacrycarpus* in Australasia, scale 1 : 119 million

not entered into horticulture beyond the limits of a few botanical gardens, and there has been no cultivar selection.

The species of *Dacrycarpus* were long included as one of the more distinctive sections within *Podocarpus*. At the same time, many authors recognized similarities of these species to other genera with respect to some features that are not found in true *Podocarpus* species. Pollen grains normally with three air bladders, for instance, are elsewhere found only in *Microcachrys* and *Microstrobos* among extant Podocarpaceae. The warty podocarpium with several fertile bracts and the juvenile leaves flattened side to side are both features found also in *Acmopyle*. The adult foliage, on the other hand, is reminiscent of species included in *Dacrydium*. This similarity is reflected in the name of the genus (Greek for "teardrop fruit"), proposed as a sectional name in *Podocarpus* 40 years after the publication of the generic name *Dacrydium*. In fact, DNA studies show that *Dacrycarpus* is closely related to *Dacrydium* and *Falcatifolium* and only more distantly to *Acmopyle*, *Microstrobos*, and *Microcachrys*, and to *Podocarpus*, from which it was segregated in 1969. *Dacrycarpus* differs from all other genera of the Podocarpaceae in the fusion of the fertile bract with the seed scale to form a crest along the side of the epimatium that ends in a sideways beak.

Although kahikatea and the New Caledonian *Dacrycarpus vieillardii* were each segregated into a genus separate from the other *Dacrycarpus* species (and *D. dacrydioides* was even placed in a separate family) in the extremely subdivided revision of the Podocarpaceae and Taxaceae by morphologists Melikian and Bobrov (2000), their proposals are unlikely to be widely accepted. Their proposed classification is based almost entirely on narrowly defined details of the structure of the seed coats and contradicts an enormous amount of traditional morphological evidence as well as all published DNA studies based on analysis of several portions of both the chloroplast and nuclear genomes. Hence their names are all treated as synonyms here. Nonetheless, the sampling of species used in DNA studies will need to be greatly increased before relationships among the genera and species of Podocarpaceae are fully resolved.

While *Dacrycarpus* species are overwhelmingly tropical today, with kahikatea of New Zealand as the only species in a (very mild) temperate region, the fossil record of the genus is found almost exclusively in presently temperate regions. The fossil record of its foliage is one of the richest among the podocarps of the southwestern Pacific region. The oldest confirmed fossils are found in Eocene sediments 40 million years old or more in southern South America and in southeastern Australia (including Tasmania), both regions where the genus is extinct today. Somewhat older fossils assigned to extinct genera may also be related to *Dacrycarpus* but do not match any living species in the structure of their epidermis, which provides most of the characters used for assigning fossil foliage to this genus. The record in Australia is particularly rich, and the progressive reduction in leaf size and in the amount of the leaf surface occupied by stomates after the early Eocene has been interpreted as reflecting the increasing aridity of that continent. The genus seemingly became extinct in mainland Australia during the Miocene, about 20 million years ago, but it persisted in Tasmania much later and may not have succumbed until the Pleistocene ice ages of the last 2 million years, although there is some doubt about the identification of these late fossils. *Dacrycarpus* seems to be a relative latecomer to New Zealand, where the earliest confirmed foliage specimens date from the Miocene, and the fossil record there is surprisingly sparse considering the historical abundance of kahikatea in swampy ground on both islands.

Identification Guide to *Dacrycarpus*

The nine species of *Dacrycarpus* are difficult to identify just using their foliage, and their pollen cones greatly overlap in size, but if seed cones are present they may be distinguished by the cross-sectional shape of adult leaves, the width of most adult leaves, the length of the green bracts at the base of the podocarpium, and the seed length:

Dacrycarpus dacrydioides, leaf cross section triangular, width up to 0.6 mm, green bract less than 2.5 mm, seed up to 5 mm

D. imbricatus, leaf cross section triangular, width 0.6 mm or more or less, green bract 2.5–5 mm, seed 5–6 mm

D. vieillardii, leaf cross section diamond-shaped, width up to 0.6 mm, green bract less than 2.5 mm, seed 5–6 mm

D. steupii, leaf cross section diamond-shaped, width up to 0.6 mm, green bract 2.5–5 mm, seed 5–6 mm

D. cumingii, leaf cross section flattened side to side, width up to 0.6 mm, green bract more than 5 mm, seed 5–6 mm

D. kinabaluensis, leaf cross section flattened side to side, width at least 0.6 mm, green bract more than 5 mm, seed more than 6 mm

D. cinctus, leaf cross section flattened top to bottom, width up to 0.6 mm, green bract more than 5 mm, seed at least 6 mm

D. expansus, leaf cross section flattened top to bottom, width at least 0.6 mm, green bract 2.5–5 mm, seed 5–6 mm

D. compactus, leaf cross section flattened top to bottom, width at least 0.6 mm, green bract 2.5–5 mm, seed more than 6 mm

Dacrycarpus cinctus (Pilger) de Laubenfels

JUMBIRI DACRYCARP, SAREH (MALAY); GUGRAGOIN, JUMBIRI, KUBUK-KAJBEK, PAU, UMBA (PAPUAN LANGUAGES)

Shrub, or tree to 25(–33) m tall, with trunk to 0.9 m in diameter, often buttressed, dwarfed and shrubby in high montane shrublands. Bark dark brown, rough, shedding irregular large scales to reveal reddish brown patches. Crown open, shallowly dome-shaped to flat with slender, upwardly angled branches bearing numerous branchlets very densely clothed with foliage. Short shoots with spreading distichous leaves (in two flat rows) mostly confined to juvenile plants. Leaves of short shoots flattened side to side, up to 12 mm long and 0.8 mm wide at the middle of the shoots. Leaves of adult shoots sticking out from the twig at a forward angle and straight or curving gently forward, especially at the tip. Each leaf sword-shaped, slightly flattened top to bottom,1.5–4(–5) mm long, 0.4–0.6 mm wide, often with a thin layer of wax. Pollen cones 4–10 mm long, 2–3 mm thick, borne at the tip of a very short branchlet 2–3 mm long. Seed cones at the tip of a short branchlet 5–15 mm long, clasped beneath a circle of leafy, free bracts up to 6–10 mm long. Podocarpium warty, red, lightly covered with wax, 3–4 mm long. Combined seed coat and epimatium spherical, 6–7 mm in diameter, red at maturity, with a weak crest ending in a sharp beak. Eastern Malesia, from central Sulawesi (Indonesia) to the eastern tip of New Guinea (Papua New Guinea). Forming pure stands or codominant with other conifers and evergreen hardwoods in or above the canopy of montane forests, mossy forests, and sometimes swampy forests; (900–)1,800–3,000 (–3,600) m. Zones 9–10? Synonyms: *Bracteocarpus cinctus* (Pilger) A. V. Bobrov & Melikian, *B. dacrydiifolius* (Wasscher) A. V. Bobrov & Melikian, *Podocarpus cinctus* Pilger, *P. dacrydiifolius* Wasscher.

Although jumbiri dacrycarp is not as large a tree as common dacrycarp (*Dacrycarpus imbricatus*) or kahikatea (*D. dacrydioides*), it is still an important timber tree in the highlands of Papua New Guinea where it reaches its greatest abundance. It is generally found on well-drained sites with ample moisture. While it may be found in swampy forests, it is generally replaced by tjimba dacrycarp (*D. steupii*) in sodden soils of impeded drainage at lower elevations in New Guinea. At the other extreme of its ecological range, in and near subalpine shrublands at high elevations, it is generally replaced by the shorter leaved umbwa dacrycarp (*D. compactus*). In the zone of elevational overlap between jumbiri dacrycarp and umbwa dacrycarp, they are especially difficult to distinguish. Intergradation between the two species through hybridization has been suggested but not investigated. The species name, Latin for "belted," refers to the conspicuous circle of long, leafy bracts enveloping the seed cone like a grass skirt.

Dacrycarpus compactus (Wasscher) de Laubenfels
UMBWA DACRYCARP, KADZINAM, KAIPIK, UMBWA, UMBA-NIFIOGO (PAPUAN LANGUAGES)
Shrub, or tree to 15(–20) m tall, with trunk to 0.3(–0.6) m in diameter, usually not buttressed. Bark dark gray, warty, shedding scales to reveal pale reddish brown patches. Crown conical at first, becoming open and irregular, with slender horizontal or upwardly angled, crooked branches topped by dense, roundly flattened clusters of branchlets very densely clothed with foliage. Short shoots found in both juvenile and adult plants. Leaves of short shoots radiating all around the twigs rather than twisted into two rows, flattened side to side with a strongly raised midrib on both sides, 2–2.5 mm long and about 0.6 mm wide. Leaves of adult shoots sticking out from the twig and then curving strongly forward so as to clothe (and often tight against) the twigs. Each leaf narrowly triangular, flattened top to bottom but with a strongly raised midrib beneath, (1–)2–3(–5) mm long, (0.5–)0.6–1.0 (–1.5) mm wide, with a prickly tip. Pollen cones 7–8 mm long, 2–3 mm thick, borne at the tip of a very short branchlet 3–4 mm long. Seed cones at the tip of a short branchlet 6–17 mm long, clasped at the base by a circle of leafy, free bracts 4–5 mm long. Podocarpium warty, purple to black, 3–4 mm long. Combined seed coat and epimatium roughly spherical, 7–8.5 mm in diameter, purplish brown at maturity with a waxy coating, with a prominent crest ending in a small beak. Along the mountainous spine of New Guinea, from central Irian Jaya (Indonesia) to the Owen Stanley Range near Port Moresby, Papua New Guinea. Forming pure stands or mixed with other conifers and evergreen hardwoods in the canopy of subalpine forests or scattered as an emergent above subalpine heaths and grasslands; (2,800–)3,400–4,000(–4,300) m. Zone 9. Synonyms: *Bracteocarpus compactus* (Wasscher) A. V. Bobrov & Melikian, *Podocarpus compactus* Wasscher.

Umbwa dacrycarp is a common tree near the alpine tree line of New Guinea's high peaks, essentially replacing jumbiri dacrycarp (*Dacrycarpus cinctus*) in the upper montane and subalpine zone above 3,400 m, and sometimes the tallest tree in these forests. The two species have a broad but sparse elevational zone of overlap where they may hybridize with one another, producing intermediate trees that are difficult to assign to either species. Although umbwa dacrycarp occurs at higher elevations than jumbiri dacrycarp and is a smaller tree, the scientific name does not reflect its stature. Instead, it refers to the short, closely spaced leaves and the very dense branching in the crown, producing compact masses of short branchlets at the ends of the limbs that are tipped by pollen or seed cones in mature specimens. It has the largest seeds of any species of *Dacrycarpus*, and the mature seeds and podocarpium are a darker purplish red rather than the bright red found in some other species, so studies of the dispersal biology of umbwa dacrycarp would be interesting.

Dacrycarpus cumingii (Parlatore) de Laubenfels
PHILIPPINE DACRYCARP, SANGU (MALAY), IGEM (PILIPINO)
Tree to 20(–25) m tall, with trunk to 0.5(–0.75) m in diameter, usually not buttressed. Bark dark grayish brown, shedding flakes to reveal lighter patches. Crown shallowly dome-shaped, fairly dense, with upwardly angled branches bearing numerous compact branchlets densely clothed with foliage. Short shoots mostly confined to juvenile plants. Leaves of short shoots mostly in two flat rows, flattened side to side, to 12 mm long and about 1.2 mm wide in the middle of the shoots. Leaves of adult foliage shoots sticking out from the twig and curving gently forward but not tight to the twig. Each leaf slightly sickle-shaped, flattened side to side with just a hint of a raised midrib, (2–)3–6 mm long, 0.6–0.8 mm wide, with a prickly tip. Pollen cones 8–10(–40) mm long, 2–3.5 mm thick, borne at the tip of a very short branchlet 2–5 mm long. Seed cones at the tip of a short branchlet 6–10(–20) mm long, clasped to the tip by a circle of leafy, free bracts 7–13 mm long. Podocarpium warty, purplish red, 2–3 mm long. Combined seed coat and epimatium somewhat egg-shaped, 5–6 mm long, (3.5–)4–5(–5.5) mm thick, brown at maturity, with a prominent crest ending in a beak 1–1.5 mm long. Mountains of the main islands of the Philippines, with outliers in Borneo and Sumatra (Indonesia). Clumped or scattered in the canopy of primary high montane mossy forests; (1,000–)1,850–2,650(–3,300) m. Zone 9? Synonyms: *Bracteocarpus cumingii* (Parlatore) A. V. Bobrov & Melikian, *Podocarpus cumingii* Parlatore.

Philippine dacrycarp is most common in those islands but also reaches some prominence in the high mountains of the northern tip of Sumatra. In both regions it is generally found above common dacrycarp (*Dacrycarpus imbricatus*), a species reaching its greatest abundance and stature in lowland and lower montane rain forests. Because of its smaller stature and prominence on high peaks in rugged country, Philippine dacrycarp is much less well known than its lower elevation, commercially important neighbor. It has been treated as a variety of common dacrycarp but is not particularly closely related to that species having its closest affinity instead with the very local Kinabalu dacrycarp (*D. kinabaluensis*), an even higher-elevation species known only from Mount Kinabalu in Sabah (Borneo, Malaysia). The species name honors Hugh Cuming (1791–1865), a British adventurer and professional natural history collector who collected the type specimen during a voyage to the Philippines in 1836–1840 aboard his yacht *Discovery*, which he built himself.

Dacrycarpus dacrydioides (A. Richard) de Laubenfels

KAHIKATEA (MAORI AND ENGLISH), WHITE PINE

Tree to 45(–60) m tall, often free of branches for half or more of its height, sometimes rising 30 m before the first branch, with trunk to 1.3(–1.8) m in diameter, cylindrical to fluted and buttressed at the base. Bark reddish brown, smooth, horizontally banded and bumpy when young, weathering light gray and becoming roughened by squarish flakes on large trees. Crown often remaining dense throughout life, conical in youth, becoming variously egg-shaped with maturity, with a few, heavy, horizontal to upwardly angled branches drooping near their ends and bearing numerous stiff or drooping branchlets densely clothed with foliage. Short shoots bearing leaves flattened into two rows predominant on juvenile plants up to 2 m tall, becoming progressively less frequent with maturity. Leaves of short shoots more or less in two flat rows, flattened side to side, (2–)4–7 mm long and 0.5–1 mm wide, longest in the middle of the shoot. Transitional leaves similar but shorter, up to 4 mm long. Leaves of adult foliage shoots placed all around the twig and curved strongly forward to clasp the twig tightly or loosely. Each leaf triangular to slightly sickle-shaped, half-moon-shaped in cross section, with just a hint of a prominently raised midrib beneath, 1–2 mm long, 0.6–0.8 mm wide, often with a slightly prolonged tip. Pollen cones (4–)8–15(–20) mm long, 2.5–3.5 mm thick, borne at the tip of a very short branchlet 1–7 mm long. Seed cones at the tip of a short branchlet 5–10(–25) mm long, scarcely clasped at the base by a sparse circle of scalelike, free bracts 1.5–2 mm long. Podocarpium warty, passing from green through yellow and orange to red with maturity, 3–5 mm long. Combined seed coat and epimatium spherical to a little elongate, 3–5 mm in diameter, 4–5 mm long, black at maturity, shiny beneath a thin dusting of wax, with an inconspicuous crest barely protruding at the end as a beak. Abortive ovules often conspicuous, 0.5–1.0 mm long. All three main islands of New Zealand but most common on the western side of South Island and rare on Stewart Island. Often forming pure, dense stands, especially in swamps, but reaching its greatest size

Swamp-side natural stand of young mature kahikatea (*Dacrycarpus dacrydioides*).

on wet but not waterlogged soils, more scattered among other species in and above the canopy on drier ground; 0–500(–700) m. Zone 9. Synonym: *Podocarpus dacrydioides* A. Richard.

Kahikatea is one of the forest giants of New Zealand. It is the tallest tree in New Zealand, though far exceeded in girth and volume by kauri (*Agathis australis*). Because of the long, clean bole and substantial girth, it was an important timber tree before the rise of plantation culture of exotic conifers, especially Monterey pine (*Pinus radiata*), as native species became depleted. The odorless white or light-colored wood was used in a large range of applications, both by Maori people and by settlers of European origin. Originally used for carving, canoes, and for general construction, especially interior work, during the 1920s to 1940s the major consumer was the dairy industry with wooden butter boxes made of Kahikatea being standard both domestically and in the export market until they were replaced by cardboard. The timber decays rapidly in contact with either soil or water and is susceptible to powder post beetles when used in tight, protected spaces. The susceptibility to decay in water exists despite the generally wet habitats favored by the species.

Twig of kahikatea (*Dacrycarpus dacrydioides*) with permanent framework shoot supporting short-lived, flattened (distichous), determinate photosynthetic branchlets, ×1.

Old mature kahikatea (*Dacrycarpus dacrydioides*) in nature.

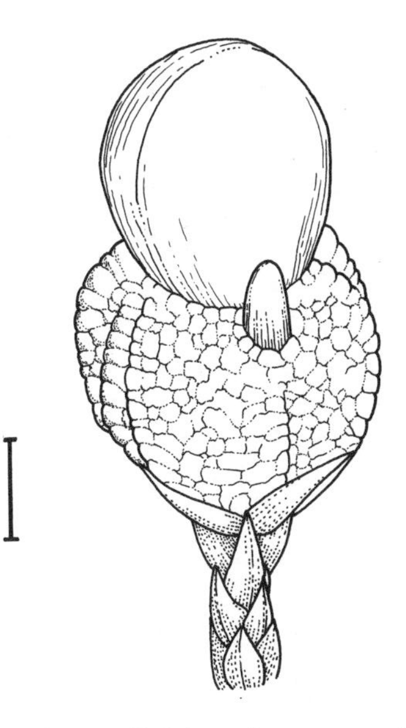

Fleshy mature seed cone of kahikatea (*Dacrycarpus dacrydioides*) at time of dispersal with retained aborted ovule in addition to the single viable seed; scale, 2 mm.

This predilection for damp soils is reflected in the drastic reduction in viability of the seeds accompanying any loss of moisture below 80%. Experiments suggest that, for this species at least, the podocarpium can transfer moisture to the seed after it is shed and prolong its period of germinability. At the same time, the podocarpium also has a role in dispersal by birds. Female trees heavily laden with seed cones are very conspicuous and may be visited by large flocks of kereru (*Hemiphaga novaeseelandiae novaeseelandiae*), a fruit-eating pigeon that is the only species capable of dispersing the seeds of many large-seeded New Zealand tree species, including miro (*Prumnopitys ferrugineus*) and several hardwoods.

Once germinated, the transition between the feathery foliage of seedlings and the scaly foliage of adults is especially dramatic. The scientific name, Greek for "like *Dacrydium*," refers to the adult foliage, much more like that of rimu (*Dacrydium cupressinum*) than that of species of *Podocarpus* to which kahikatea was assigned when it was first described. As a common, often dominant low-elevation species, kahikatea has been studied for many aspects of its biology, including pollination and the subsequent development of the seed cones, embryology, chemistry of the oils and pigments, and the conspicuous and numerous root nodules, which may help with nutrient uptake in waterlogged soils. For instance, an examination of the flavonoid pigments of the foliage found nine new compounds that had never before been isolated from any plant, including other New Zealand species of Podocarpaceae, out of the fifteen flavonoids recovered from extracts. Just as distinctively, and uniquely among 41 species of New Zealand conifers with a podocarpium and flowering plants with berries, the seeds of kahikatea, with their waxy coating, but not the podocarpium, reflect ultraviolet light, which is visible to many insects but not to vertebrates. The wax initially repels water when the seeds fall into a swamp or stream, increasing the time before they drown, during which they might reach a shallow site and germinate. As the species became depleted with overexploitation, large, dense stands became progressively more restricted to swampy forests of the Westland district of South Island. Unfortunately, riverside stands in this region are increasingly threatened by deepening water levels resulting from damming of western rivers. If the trees can survive the many hazards they face, they can live for as long as 700 years, although an age of 400–500 is much more typical of veteran giants.

Dacrycarpus expansus de Laubenfels

PAU DACRYCARP, P'AU (ENGA)

Tree to 25(–30) m tall, with trunk to 0.5(–0.6) m in diameter. Bark gray, flaking in scales. Crown cylindrical to shallowly dome-shaped, with rising branches bearing numerous branchlets densely clothed with foliage. Short shoots mostly confined to juvenile plants. Leaves of short shoots in two flat rows, flattened side to side, fairly straight to sickle-shaped, prickly, to 12 mm long and about 1.5 mm wide in the middle of the shoots. Leaves of adult foliage shoots sticking out from the twig and curving gently forward at a distance from it. Each leaf slightly sickle-shaped, flattened top to bottom, with a prominently raised midrib both above and beneath, of even size along the branchlet, 1.5–3(–4) mm long, (0.4–)0.6–0.8(–1.0) mm wide, with a pointed but not prickly tip. Pollen cones about 6 mm long, 3 mm thick, borne at the sides of very short branchlets 1–2 mm long. Seed cones at the tip of a very short branchlet 4–5 mm long, clasped over the podocarpium by a circle of leafy free bracts 3–4 mm long. Podocarpium warty, red, 2–3 mm long. Combined seed coat and epimatium somewhat egg-shaped, 5–6 mm long, 3–4 mm thick, brown at maturity, with a weak crest ending in a small beak about 1 mm long. Confined to the central highlands of Papua New Guinea in Enga province and adjacent highland portions of neighboring provinces. Forming dense pure stands or codominant with other conifers and broad-leaved evergreens in the canopy of or emergent above forests and the borders of savannas; (1,300–)2,000–2,750 m. Zone 9? Synonym: *Bracteocarpus expansus* (de Laubenfels) A. V. Bobrov & Melikian.

The five species of dacrycarp growing in New Guinea differ in ecological preferences so that they grow intermixed with one another to only a limited extent. While three of the species are restricted to or most prominent in undisturbed, primary forests and woodlands, pau dacrycarp (*Dacrycarpus expansus*) and tjimba dacrycarp (*D. steupii*) thrive in the vigorous young forests that follow natural and human disturbances. The more common and widespread species, tjimba dacrycarp, is found primarily on very wet sites, often with waterlogged soils. Pau dacrycarp replaces it on less sodden substrates. The two species are very similar in appearance and in most details but differ in the shape of their adult leaves in cross section and in the ring of free bracts encircling the base of the seed cones. The diamond-shaped leaves of tjimba dacrycarp contrast with the flattened ones of pau dacrycarp, and the bracts stick out beneath the podocarpium rather than clasping it.

Dacrycarpus imbricatus (Blume) de Laubenfels

COMMON DACRYCARP, MALAYAN YELLOWWOOD, JI MAO SONG (CHINESE), MAY HÜÖNG (THAI), SRÔ:L SÂ:R (CAMBODIAN), LÔ:NG LENG (LAOTIAN), THÔNG NÀNG (VIETNAMESE), JAMUJU, SAMPINUR, BUNGA, TJEMARA, RU BUKIT (MALAY), TUPI (PILIPINO), APÉ, JAMARI, ILJO, PAU, PAUPEEPEEN, UMBA (PAPUAN LANGUAGES), AMUNU, KAU TAMBUA (FIJIAN)

Shrub, or tree to 35(–50) m tall, shorter when open-grown, with trunk to 1.2(–2) m in diameter. Bark thin, fairly smooth, mottled gray, black, dark brown, and reddish brown through flaking in irregular patches and scales. Crown densely conical at first becoming variously cylindrical or egg- to dome-shaped with age and ultimately ragged and open, with a few major downswept, horizontal, and upright limbs breaking up into branches bearing many upright to drooping branchlets densely clothed with foliage. Short shoots predominant in juvenile plants and progressively reduced with age but still consistently, if sporadically, produced on mature trees, especially in shaded portions of the crown. Leaves of short shoots in two flat rows, flattened side to side, with a gently, continuous, forward S curve, longest near the middle or closer to the base of the shoot, (6–)10–17 mm long and 1–2(–2.5) mm wide, ending with a blunt point. Leaves of adult foliage shoots arranged all around the twig and sticking out from or tight against it, curved forward. Individual leaves variable among the four botanical varieties, triangular to sword-shaped, half-moon-shaped to triangular in cross section with a prominent midrib beneath and sometimes above, (0.1–)0.8–2(–5) mm long, 0.4–1.0 mm wide, not prickly. Pollen cones (5–)6–12 mm long, 2–2.5 mm thick, borne at the tip or sides of a very short branchlet 1–3 mm long. Seed cones at the tip of a short branchlet 3–10(–20) mm long, surrounded at the base by a circle of leafy, free bracts 1–3.5(–5) mm long. Podocarpium warty, red, 3–4 mm long, with one (or two) fertile bracts. Combined seed coat and epimatium roughly spherical, 5–6(–7) mm long, 4–6 mm in diameter, red to reddish brown at maturity, with a low crest whose free tip is hardly noticeable as a dimpled beak less than 1 mm long. Discontinuous throughout southeastern Asia, Malesia, and Melanesia, from northern Sumatra (Indonesia), northern Myanmar, southern China (southern Yunnan, Guangxi, and Hainan), and Luzon (Philippines) to Vanuatu and Fiji. Forming pure stands or mixed with other conifers or various evergreen hardwoods in or towering above the canopy of an enormous range of primary and secondary forest types, most frequently in montane rain forests of middle elevation but also occurring in wetter mossy forests (as in New Guinea) on the one hand or contrastingly in seasonally dry forests (as in Fiji) or at the margins of natural and postlogging savannas (as in Timor), sometimes on steep, slide-prone slopes (as in northern Sumatra) or in ravine forests (China and Indochina); (100–)700–2,700(–3,400) m, varying considerably throughout its range. Zone 10. Synonyms: *Bracteocarpus imbricatus* (Blume) A. V. Bobrov & Melikian, *Podocarpus imbricatus* Blume. Four varieties.

Common dacrycarp is, by far, the most widely distributed species of *Dacrycarpus,* occurring throughout the range of the genus except in New Caledonia, which it approaches to within 250 km, and in New Zealand, where the only temperate species, kahikatea (*D. dacrydioides*), occurs. Wherever it occurs, it is an important timber tree, and this is reflected in the great number of common names applied to it in the many languages found within its range, only a sampling of which is presented here. As with the other two conifers sharing a similar distribution, oleander podocarp (*Podocarpus neriifolius*) and dammar nagi (*Nageia wallichiana*), individuals of common dacrycarp vary a great deal from place to place, and some authors divide it into more than one species. With an even greater ecological range than either of the other two species, it is even more variable than they are, and this increases the difficulty of distinguishing it from the five other species of dacrycarp with which it occurs. Among these species, it has the shortest free bracts at the base of the podocarpium, and the others all have adult leaves with a different shape in cross section, as outlined in the identification guide. It may hybridize with at least one of these species, tjimba dacrycarp (*D. steupii*), in Borneo where intermediate specimens have been collected. Most of these species share the distinctively flattened short shoots of the juvenile plants (the Chinese common name, "chicken feather pine," emphasizes this juvenile stage) and are almost impossible to separate from common dacrycarp before maturity. A sixth species Kinabalu dacrycarp (*D. kinabaluensis*), occurs within the range of common dacrycarp on Mount Kinabalu in Borneo, but there appear to be no places where the two species occur together, Kinabalu dacrycarp always occurring at higher elevations.

Common dacrycarp itself varies in two features of the adult foliage that together serve to distinguish other species: the width of the leaves and whether they are held tight against the twig (are imbricate, hence the scientific name, from Latin *imbrex,* the half-cylindrical terracotta roofing tile so characteristic of Italy and Spain). On this basis, four botanical varieties of very unequal geographic distribution have been described, representing all four possible combinations. Geographically, they associate in two pairs. The two varieties at the southern fringe of the range (varieties *imbricatus* and *curvulus*), centered in Java, have adult leaves tight to the twigs while they are loose to spreading in the other two varieties (varieties *patulus* and *robustus*), partially overlapping with each other throughout the rest of the range. In each pair, one has coarser leaves, more than 0.6 mm wide, and the other has finer leaves, less than 0.6 mm wide. The two northern varieties appear to intergrade in Mindanao (Philippines). In general, the status and relationships of all the varieties should be further evaluated with fieldwork concentrated in their areas of overlap. Establishment of plantation culture involving this species in some areas could lead to further confusion surrounding its taxonomic structure if distant seed sources are used. Its conversion to use as a plantation tree is enhanced by its natural facility for germination and establishment in secondary and disturbed forests, predisposing it for success in the managed conditions of plantations.

Dacrycarpus imbricatus* var. *curvulus (Miquel) de Laubenfels
Tree to 8 m tall or a shrub, sometimes creeping on rock faces. Branchlets drooping. Adult leaves tightly clasping the twigs, 0.8–1 mm wide. Free bracts at base of seed cone clasping, 2.5–4.5 mm long. Java and northern tip of Sumatra. Exposed ridges and slopes; (1,350–)2,000–3,000(–3,400) m. Synonym: *Podocarpus imbricatus* var. *curvulus* (Miquel) Wasscher.

Dacrycarpus imbricatus (Blume) de Laubenfels **var. *imbricatus***
Tree to 50 m tall. Branchlets straight or drooping initially. Adult leaves tightly clasping the twigs, 0.4–0.6 mm wide. Free bracts at base of seed cone spreading at maturity, 2.5–4(–5) mm long. Java, Lesser Sunda Islands, and southwestern Sulawesi. Primary and secondary rain forests; (200–)700–2,700(–3,000) m.

Dacrycarpus imbricatus* var. *patulus de Laubenfels
Tree to 40 m tall. Branchlets straight. Adult leaves loose and spreading, 0.4–0.6 mm wide. Free bracts at base of seed cone spreading, 2–3 mm long. Throughout the range of the species except, apparently, in Java and the Lesser Sunda Islands, where it is replaced by varieties *imbricatus* and *curvulus,* and also absent from all of New Guinea (except the northern coast), where it is replaced by variety *robustus.* Primary and secondary rain forests; (0–)700–2,500(–3,000) m. Synonyms: *Bracteocarpus kawaii* (Hayata) A. V. Bobrov & Melikian, *Podocarpus kawaii* Hayata.

Dacrycarpus imbricatus* var. *robustus de Laubenfels
Tree to 45 m tall. Branchlets straight. Adult leaves loose and spreading, 0.6–0.8 mm wide. Free bracts at base of seed cone spreading, 2–3 mm long, or clasping and up to 5 mm. Northeastern part of the range of the species, from Luzon and northern Borneo to the eastern tip of New Guinea. Primary and secondary rain forests and mossy forests; (240–)700–2,700 (–3,300) m. Synonyms: *Bracteocarpus papuanus* (H. Ridley) A. V. Bobrov & Melikian, *Podocarpus papuanus* H. Ridley.

Dacrycarpus kinabaluensis (Wasscher) de Laubenfels

KINABALU DACRYCARP

Shrub, or tree to 13 m tall, with trunk to 0.3 m in diameter, often gnarled and contorted. Bark smooth, gray to black, flaky. Crown dense, irregular, conical to dome-shaped, with many short, crooked branches bearing numerous, tightly packed branchlets densely clothed with foliage. Short shoots nearly confined to juvenile plants. Leaves of short shoots in two flat rows, flattened side to side, (5–)10–15 mm long and 1–1.2 mm wide, largest in the middle of the shoot. Leaves of adult foliage shoots standing out away from and all around the twigs or lined up in five rows. Each leaf sickle-shaped, also predominantly flattened side to side, the midrib prominently raised on both sides, (1–)2–4(–6) mm long, (0.5–)0.8–1.0 mm wide, the tip curved forward and strongly prickly. Pollen cones about 8 mm long, 3 mm thick, borne at the sides of a very short branchlet about 3 mm long. Seed cones at the tip of a short branchlet 5–16 mm long, clasped to beyond the middle by a basal circle of free, leaflike bracts 5–8 mm long. Podocarpium warty, turning blue or purple at maturity, 2–4 mm long. Combined seed coat and epimatium nearly spherical to a little elongate, 5–7 mm long, 5–6 mm in diameter, reddish brown at maturity, with a low, inconspicuous crest whose free tip forms a prominent sideways beak about 1 mm long. Known only from Mount Kinabalu in Sabah (Malaysia), northern Borneo (hence the name). Forming pure stands or codominant with other trees and shrubs in upper montane forests and subalpine shrublands, right up to the alpine tree line, often on serpentine-derived soils; (2,100–)2,700–3,700(–4,000) m. Zone 9? Synonyms: *Bracteocarpus kinabaluensis* (Wasscher) A. V. Bobrov & Melikian, *Podocarpus imbricatus* var. *kinabaluensis* Wasscher.

The high-mountain Kinabalu dacrycarp has the most restricted distribution of any species of *Dacrycarpus,* seemingly entirely confined to the upper reaches of Mount Kinabalu. Common dacrycarp (*D. imbricatus*) is also found on the slopes of Mount Kinabalu but always at lower elevations, with a gap of 1,000 m on the Gurulau spur, for instance. The variety of common dacrycarp found on Mount Kinabalu, variety *patulus,* is known to occur as high as 3,000 m elsewhere but rarely exceeds 2,500 m and does not grow above 2,400 m on Mount Kinabalu. Hybridization between the two species has been suggested but, with the elevational gap manifested at most localities, seems unlikely without further documentation. Kinabalu dacrycarp is most closely related to Philippine dacrycarp (*D. cumingii*), a species of mossy forests in high mountains of the Philippines (primarily) but generally at lower elevations than Kinabalu dacrycarp. Besides reaching the highest elevation of any species of *Dacrycarpus,* if only by 50 m over umbwa dacrycarp (*D. compactus*) of New Guinea, only Kinabalu dacrycarp and New Caledonian dacrycarp (*D. vieillardii*), a species of lowland forests, are reported to grow regularly on serpentine-derived soils, but these two species do not appear to be closely related. While Kinabalu dacrycarp is a small tree in the high montane forests and mossy forests of the mountain, on exposed ridges at the same elevations as these forests and, especially, in the subsummit dwarf forest and scrublands above them, it becomes dwarfed and tightly compacted, no more than 2 m tall as the major component of an impenetrable, dwarfed canopy.

Dacrycarpus steupii (Wasscher) de Laubenfels

TJIMBA DACRYCARP, TJIMBA-TJIMBA (MALAY), NAK, MIEJOOP, APÉ, PAU (PAPUAN LANGUAGES)

Tree to 20(–36) m tall, with trunk to 0.5(–1) m in diameter. Bark smooth, dark brown, weathering gray, peeling in elongate flakes to reveal reddish or pink patches. Crown conical at first, becoming cylindrical to dome-shaped with age, the branches progressing from downwardly angled to horizontal to upwardly angled from bottom to top, with numerous stiff, upwardly arching branchlets densely clothed with foliage. Short shoots mostly limited to juvenile plants. Leaves of short shoots in two flat rows, flattened side to side, 4–8 mm long and 0.6–1 mm wide, longest near the middle of the shoot. Leaves of adult foliage shoots sticking out all around the twig and curving gently forward near the prickly tip. Each leaf narrowly triangular, diamond-shaped in cross section, the corners extended as ribs, 2–3(–4) mm long, 0.4–0.6 mm wide. Pollen cones 8–12 mm long, about 2 mm thick, borne at the tip or sides of a very short branchlet 2–5 mm long. Seed cones at the tip of a very short branchlet 3–5(–10) mm long, surrounded at the base by a circle of spreading, free, leafy bracts 3–4(–5) mm long. Podocarpium warty, red, 2–3 mm long. Combined seed coat and epimatium nearly spherical to a little elongate, 5–6 mm long, 4–6 mm in diameter, brown at maturity, with a low, inconspicuous crest ending in a short, narrow beak less than 1 mm long. Along the whole length of New Guinea from the Vogelkop to the eastern tip, with outliers on either side of the Makassar Straight in eastern Kalimantan and western Sulawesi (Indonesia). Forming pure stands with a closed canopy or scattered among grasses and sedges on waterlogged soils and in mountain swamps, or in especially wet mossy forests at high elevations, often following disturbance; (860–)1,500–2,500(–3,420) m. Zone 9? Synonyms: *Bracteocarpus steupii* (Wasscher) A. V. Bobrov & Melikian, *Podocarpus steupii* Wasscher.

Tjimba dacrycarp has had a somewhat checkered taxonomic history. It was first described from a single specimen in Sulawesi, where it has a very limited range. Later, when the species was transferred from *Podocarpus* to the new genus *Dacrycarpus,* many collections from New Guinea and the Philippines were assigned to it. The Philippine specimens were predominantly juvenile and were subsequently reassigned, in part, to Philippine dacrycarp (*D. cumingii*), a rather similar-looking species that is most readily distinguished by its longer, clasping, free bracts surrounding the seed cone from its base and by the side-to-side flattening of the adult foliage leaves. Tjimba dacrycarp is also similar to a geographically closer neighbor, pau dacrycarp (*D. expansus*), which is restricted to the eastern half of New Guinea. Both species are common and often dominant in secondary forests but pau dacrycarp regenerates in less sodden sites than those favored by tjimba dacrycarp. Furthermore, pau dacrycarp has adult foliage leaves flattened top to bottom and the free bracts clasp the base of the seed cone, although they are no longer than those of tjimba dacrycarp. The species has also been confused with the form of common dacrycarp usually found in New Guinea (*D. imbricatus* var. *robustus*), trees that are more frequently found in primary forest at generally lower elevation and never in as wet sites as tjimba dacrycarp. Furthermore, the wider, half-moon-shaped adult foliage leaves and the longer podocarpium separate mature trees of common dacrycarp from tjimba dacrycarp. The species name honors Ferdinand Steup (1898–1971), who collected plants in the region (1927–1942) and who first described this tree but mistakenly identified it as common dacrycarp (under the synonymous name *P. papuanus*).

Dacrycarpus vieillardii (Parlatore) de Laubenfels

NEW CALEDONIAN DACRYCARP

Tree to 15(–25) m tall, with trunk to 0.3 m in diameter. Bark thin, smooth, brown, weathering light gray and becoming mottled also with dark gray and reddish brown through the flaking away of small scales. Crown conical, dense, and symmetrical at first, becoming more or less cylindrical and very open with age, with upright to horizontal branches bearing numerous stiff, upright branchlets densely clothed with foliage. Short shoots mostly confined to juvenile plants, though also borne sporadically on shaded portions of the crown of adults. Leaves of short shoots in two flat rows, flattened side to side, (1.5–)3–10 mm long and 0.5–1.0 mm wide, longest in or a little before the middle of the shoot. Leaves of adult foliage shoots standing out all around the twig and curved gently forward without clasping the twig. Each leaf sword- to slightly sickle-shaped, diamond-shaped in cross section, with a prominently raised midrib above, 2–4 mm long and 0.4–0.6 mm wide, longest in the middle of the growth increment, with a tiny prickle at the tip. Pollen cones 7–13 mm long, 1–2 mm thick, borne at the tip or sides of a very short branchlet 1–4 mm long. Seed cones at the tip of a short branchlet 6–12(–20) mm long, ringed at the base by a circle of spreading, free bracts 1–2 mm long. Podocarpium warty, red, 2–3 mm long, with a waxy film. Combined seed coat and epimatium nearly spherical to egg-shaped, 5–6 mm long, 4–5 mm in diameter, purple at maturity, with a low, double crest barely protruding at the tip as a tiny beak. Southern three-quarters of New Caledonia but most common in the southernmost quarter. Often forming pure stands or codominant with other species in swamps and along riversides, commonly over serpentine-derived substrates; (0–)150–750(–900) m. Zone 10. Synonyms: *Laubenfelsia vieillardii* (Parlatore) A. V. Bobrov & Melikian, *Podocarpus vieillardii* Parlatore.

New Caledonian dacrycarp shares its preference for swampy ground with its close relatives kahikatea (*Dacrycarpus dacrydioides*) of New Zealand and tjimba dacrycarp (*D. steupii*) of New Guinea, but the latter typically occurs at much higher elevations. It has longer adult foliage leaves than does kahikatea, with a smaller podocarpium and larger seeds. The species that approaches most closely to New Caledonian dacrycarp geographically is common dacrycarp (*D. imbricatus*), which in the form of variety *patulus* occurs within about 400 km of New Caledonia on Erromango and Anatum Islands of southern Vanuatu, while the nearest populations of kahikatea in New Zealand are more than 1,500 km away and those of tjimba dacrycarp (*D. steupii*) in New Guinea more than 2,000 km. Considering that there is a gap of more than 2,000 km between the stands in Vanuatu and Fiji and the closest populations of common dacrycarp in New Guinea and New Britain, and that the gap between Vanuatu and Fiji itself is about 800 km, the fact that common dacrycarp has not colonized New Caledonia is another testimony to the isolated nature of the flora of this island. Of course, the causes could be predominantly ecological, and the bird-dispersed seeds of common dacrycarp might well arrive in New Caledonia upon occasion but fail to become established. Even though New Caledonia is rich in conifers (it has one of richest conifer assemblages in the world, perhaps the richest for its size), they are mostly restricted to serpentine-derived soils, on which common dacrycarp is not known to grow. Although New Caledonian dacrycarp grows, in part, in similar habitats to New Caledonian corkwood (*Retrophyllum minor*) and cattail rimu (*Dacrydium guillauminii*), it is much more common and widespread than these two species and does not immediately face the same conservation threats being experienced by them. The species name honors Eugène Vieillard (1819–1896), who collected the type specimen during his long sojourn in New Caledonia between 1857 and 1869 or so.

Dacrydium Solander ex J. G. Forster

RIMU, RU, DACRYDIUM

Podocarpaceae

Evergreen trees and shrubs of varying habit. Trunk varying from short and contorted to tall and columnar, sometimes fluted and buttressed. Bark fibrous, thin and hard, peeling in long strips or irregular to elongate flakes and becoming pockmarked or shallowly and narrowly ribbed and furrowed. Crown conical to dome-shaped, often with extra branches between the main tiers of whorled branches. Branches and branchlets strongly rising to weeping. Branchlets all elongate, without distinction into short and long shoots, usually hairless, remaining green for at least the first year, evidently grooved between the elongate, attached leaf bases. Resting buds not well developed, consisting of a cluster of arrested ordinary foliage leaves or slightly differentiated scale leaves. Leaves spirally arranged and radiating all around the twigs, fairly uniform in length along a growth increment. Leaves strongly differentiated in size and often in form between juvenile and adult foliage, the transition either abrupt or via a gradual progression through intermediate leaf forms. Juvenile leaves needlelike, thin and narrow, tapering very gradually but continuously from the base to the blunt or sharp tip, straight or a little curved forward. Adult leaves scale- to needlelike, straight or variously curved forward, often circular, semicircular, or diamond-shaped in cross section. Leaf tip blunt or pointed but rarely prickly, the base usually running onto the twig without a definite petiole.

Plants dioecious. Pollen cones egg-shaped to cylindrical, single at the tips of branchlets or in the axils of ordinary foliage leaves, or a few loosely clustered. Each pollen cone with a few bracts at the base and 40 or more densely spirally arranged pollen scales, each scale bearing two pollen sacs and with a scarcely to enormously elongated tip. Pollen grains medium (body 30–50 µm long, 35–70 µm overall), with two small, round air bladders, or with a slightly inflated fringe extending all around the body, leaving just a small area of the furrow uncovered. Surface fairly smooth in the furrow region, minutely bumpy or pitted over the thick cap, and coarsely sculptured with thick wrinkles on the

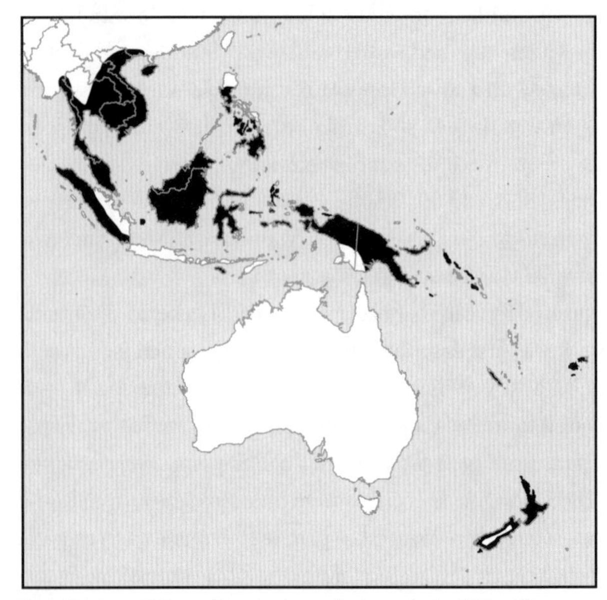

Distribution of *Dacrydium* in Australasia, scale 1 : 119 million

bladders or fringe and internal sculpturing completely filling the space. Seed cones single at the tips of ordinary branchlets in the axils of foliage leaves, highly modified and reduced. Each cone with a few small free bracts at the base and with about 12 bracts in the reproductive part that become a podocarpium. These bracts fused with the cone axis and with each other, becoming swollen, red, and juicy only just as the seed matures, even then retaining conspicuous free tips. Fertile bracts one to three, each with a single plump, unwinged seed encircled at the base (at maturity) by a thin, green, cuplike seed scale (the epimatium). Opening of the ovule pointing down into the cone axis at the time of pollination, when fully enveloped by the epimatium, but gradually straightening and outgrowing the epimatium so that at maturity, in the second year, the seed points away at an angle to the cone axis. Seeds more or less egg-shaped, conspicuously flattened so that they have a ridge going up each side, the opening of the micropyle at the seed tip short, straight, and blunt. Cotyledons two, each with two veins. Chromosome base number $x = 10$.

Wood moderately hard and heavy, fragrant, often with a strong distinction between sap- and heartwood but often also with a gradual transition. Sapwood narrow or broad, white to light brown. Heartwood yellow to reddish brown or bright red, often streaked. Grain very fine and even, lacking obvious growth rings in most species, although narrow annual bands of scarcely differentiated latewood are often present and annual growth rings may be well marked in *Dacrydium cupressinum.* Resin canals absent but with abundant individual resin parenchyma cells.

Lines of stomates on both inner (upper) and outer (lower) surfaces or (less often) only on the inner, usually arranged in more or less well defined stomatal bands (each containing two to six lines), those of the outer surface often short and confined to the region near the base. Each stomate sunken beneath and partly hidden by the one or two circles of four or five (or six) subsidiary cells, plugged by wax and surrounded by a Florin ring. Outer subsidiary cells and ordinary epidermal cells often with highly wrinkled cells walls. Leaf cross section with a single, prominent midvein located centrally, accompanied beneath by one resin canal and flanked by transfusion tissue at the sides and (often) above. Photosynthetic tissue radiating all around the midvein.

Twenty-one species in southeastern Asia, Malesia, New Caledonia, Fiji, and New Zealand. References: Allan 1961, de Laubenfels 1969, 1972, 1978c, 1988, Fu et al. 1999f, Hill and Brodribb 1999, Hill and Christophel 2001, Quinn 1970, 1982, Salmon 1980, Smith 1979, Tengnér 1965, 1967, Wells and Hill 1989a, b. Synonyms: *Corneria* A. V. Bobrov & Melikian, *Gaussenia* A. V. Bobrov & Melikian, *Metadacrydium* Baumann-Bodenheim ex Melikian & A. V. Bobrov.

The overall modern distribution of *Dacrydium* is similar to that of its closest relatives, *Dacrycarpus* and *Falcatifolium,* although the latter does not extend as far in the northwest (only in southern Malaya in mainland Asia) and in the southeast (it is not in Fiji or New Zealand). A few of the species, including *Dacrydium cupressinum* and *D. elatum,* are important timber trees producing some of the finest softwood lumber in their native regions. The scientific name of the genus, Greek for "little tear," refers to droplets of resin exuded from the bark. Rimu (*D. cupressinum*) is the only species in the genus with a (modest) history in horticulture, and there has been no cultivar selection in this essentially tropical genus.

Dacrydium is the second largest genus in the family Podocarpaceae and, like the large *Podocarpus,* it was long used in a broad sense that included many species now placed in separate genera, most of which are thought to be only distantly related to the species still retained in *Dacrydium.* The genus *Dacrydium* was then used as a catchall for podocarp species with scale leaves and upright seeds with a reduced epimatium, in contrast to the broad leaves, inverted seeds, and full-blown epimatium of *Podocarpus* (in the broad sense). Even before the initiation of dismemberment of the genus in 1969, it was recognized as heterogeneous, with at least three groups of species differing in pollen types, chromosome base number, and in details of foliage and of seed cones. With the removal of *Falcatifolium* in 1969, followed by *Halocarpus, Lagarostrobos,* and *Lepidothamnus* in 1982, what remains in *Dacrydium* is much more coherent.

Despite this coherence, there has been a proposal (Melikian and Bobrov 2000) to further subdivide *Dacrydium* into four genera so that rimu (*D. cupressinum*) would be the sole remaining species in *Dacrydium,* but this proposal has not been accepted by other conifer taxonomists. Basically, these segregate genera would all be each other's closest relatives (although the proposing authors think that *D. cupressinum* stands somewhat apart from the others) and the groupings conflict with some of the other available evidence. That said, there are precious few data known that could be used to investigate the relationships among species of *Dacrydium* so a convincing phylogenetic arrangement of species with the genus is not yet possible.

Like the related *Dacrycarpus* and *Falcatifolium, Dacrydium* has a fossil record of leafy shoots in Australia, from which the genus is now absent. These fossils belong to several species that extend from the mid-Eocene (about 45 million years ago) to the Miocene (about 20 million years ago), after which the genus disappeared from Australia, presumably because of the increasing aridity of the continent (living species of *Dacrydium* all have a high moisture requirement). These Australian records are the only known fossilized shoots of *Dacrydium,* but pollen grains assigned to the genus have a slightly wider range in time and space. Pollen records of *Dacrydium* extend from the Paleocene (more than 56 million years ago) to the Pliocene (less than 5 million years ago) in Australia and are also found in Pliocene deposits in New Guinea.

Identification Guide to *Dacrydium*

The 21 species of *Dacrydium* can only be identified securely when mature seed cones and pollen cones are present. For convenience of identification, since cones are not always present, they may be divided into three groups based on the length of fully adult leaves:

Group A, leaves less than 2 mm

Group B, leaves 2–6 mm

Group C, leaves more than 6 mm

Group A. These four species may be distinguished by the width of their adult leaves, the length of the free tips of the uppermost podocarpial bracts beneath the mature seed, and the length and the width of most mature pollen cones:

Dacrydium leptophyllum, leaves less than 0.4 mm wide, bract tips unknown, pollen cones unknown

D. nausoriense, leaves 0.4–0.6 mm wide, bract tips about 2 mm, pollen cones less than 3 mm by 1.5–2 mm

D. elatum, leaves 0.4–0.6 mm wide, bract tips about 2 mm, pollen cones 3–6 mm by up to 1.5 mm

D. novoguineense, leaves at least 0.6 mm wide, bract tips about 3 mm, pollen cones 6–10 mm by up to 1.5 mm

Group B. These 11 species may be distinguished by the width of their adult leaves, the length of the free tips of the uppermost podocarpial bracts beneath the mature seed, and the length of most mature pollen cones:

Dacrydium gracile, leaves up to 0.4 mm wide, bract tips about 3 mm, pollen cones less than 9 mm

D. magnum, leaves up to 0.4 mm wide, bract tips about 5 mm, pollen cones more than 9 mm

D. medium, leaves 0.4–0.6 mm wide, bract tips about 3 mm, pollen cones up to 9 mm

D. nidulum, leaves 0.4–0.6 mm wide, bract tips about 4 mm, pollen cones at least 9 mm

D. pectinatum, leaves 0.4–1 mm wide, bract tips about 3 mm, pollen cones at least 9 mm

D. cupressinum, leaves 0.6–1 mm wide, bract tips about 1.5 mm, pollen cones at least 9 mm

D. lycopodioides, leaves 0.6–1 mm wide, bract tips about 2.5 mm, pollen cones less than 9 mm

D. spathoides, leaves 0.6–1 mm wide, bract tips about 3 mm, pollen cones unknown

D. cornwallianum, leaves 0.6–1 mm wide, bract tips about 4 mm, pollen cones at least 9 mm

D. balansae, leaves at least 1 mm wide, bract tips about 4 mm, pollen cones at least 9 mm

D. araucarioides, leaves at least 1 mm wide, bract tips about 5 mm, pollen cones at least 9 mm

Group C. These six species may be distinguished by the length and the width of their adult leaves, the length of the free tips of the uppermost podocarpial bracts beneath the mature seed, and the length and the width of most mature pollen cones:

Dacrydium beccarii, leaves 6–10 mm by up to 0.4 mm, bract tips about 2 mm, pollen cones 6–10 mm by 2.5–3 mm

D. ericoides, leaves 6–10 mm by at least 0.6 mm, bract tips about 4 mm, pollen cones 6–10 mm by 2–2.5 mm

D. xanthandrum, leaves 6–10 mm by at least 0.6 mm, bract tips 3–4 mm, pollen cones 10–15 mm by 2–3 mm

D. gibbsiae, leaves 6–10 mm by at least 0.6 mm, bract tips 5 mm or more, pollen cones more than 15 mm by at least 5 mm

D. comosum, leaves more than 10 mm by at least 0.6 mm, bract tips about 2 mm, pollen cones 6–10 mm by 2.5–3 mm

D. guillauminii, leaves more than 10 mm by at least 0.6 mm, bract tips about 10 mm, pollen cones 10–15 mm by 3–5 mm

Dacrydium araucarioides Ad. Brongniart & Gris

CANDELABRA RIMU

Plate 26

Tree to 6 m tall, with trunk to 0.2 m in diameter. Bark thin, brown, weathering gray and flaking in rough scales. Crown thin, broadly dome-shaped to flat-topped, with circles of branches extending out nearly horizontally, turning up at the ends, and bearing numerous upturned branchlets densely clothed with foliage, the effect like a candelabra made of green snakes since the twigs are completely hidden by the scaly leaves. Transition between juvenile and adult foliage gradual. Juvenile leaves to 12 mm long. Adult leaves scalelike, densely overlapping, sticking out from the twig at a forward angle and then gradually curling back in toward the twig at the tip, 3–5 mm long, 1–1.4 mm wide, wider than thick, half-moon-shaped in cross section. Midrib with a shallow but sharp keel along the outer face, prominent and flanked by broad grooves containing lines of stomates on the inner face. Pollen cones appearing thinner than their stalks and accompanying leafy bracts, 1–2 cm long, (1.5–)2.5–3 mm in diameter. Pollen scales triangular, curled inward at their tips like the leaves, without a prolonged tip. Seed cones on a branchlet with leaves about 3 mm long, shorter than average for adult foliage leaves. Free tips of the bracts making up the podocarpium like the foliage leaves but straighter, conspicuously longer than the leaves of the supporting branchlet and progressively longer toward the tip of the cone, the uppermost to 5 mm long and loosely wrapping the seeds. Mature podocarpium bright red, with one or two (or three) fertile bracts. Epimatium cuplike, about 1 mm deep. Seed roughly egg-shaped, 4–4.5 mm long, 2.5–3 mm thick, with a small, rounded beak 0.1–0.2 mm long. Southern half of New Caledonia, from just north of Houaïlou to the Plaine des Lacs near the southern tip. Common but scattered as an emergent in maquis minière, dense shrublands growing on serpentine-derived soils; 200–1,000 m. Zone 9. Synonym: *Metadacrydium araucarioides* (Ad. Brongniart & Gris) Baumann-Bodenheim ex Melikian & A. V. Bobrov.

Candelabra rimu really lives up to its scientific name (Latin for "like *Araucaria*"), the trees looking vegetatively very much like just another of the many species of *Araucaria* in New Caledonia, albeit a dwarf one. Of course, the massive seed cones or large pollen cones of an *Araucaria* contrast strongly with the tiny, red, berrylike seed cones or relatively inconspicuous small pollen cones of *Dacrydium araucarioides.* Candelabra rimu is one of the five species of conifers restricted to the maquis and one of the characteristic species in this vegetational formation. Some species of *Araucaria* can also be found in the maquis (although they are more common in rain forests), but when they do they are usually larger trees than are individuals of candelabra rimu. Few rain-forest species enter the maquis because land supporting this type of vegetation is dry as well as laced with toxic metals.

Dacrydium balansae Ad. Brongniart & Gris

BALANSA RIMU

Tree to 12(–22) m tall, with trunk to 0.3 m in diameter. Bark brown and smooth with many small warty lenticels initially, weathering gray and breaking up into thick blocks. Crown conical, thick, with upwardly angled branches bearing many upright, snaky branchlets densely clothed with foliage. Twigs partially hidden by the somewhat loose, scaly leaves, appearing a little rough. Transition between juvenile and adult foliage gradual. Juvenile leaves to 13 mm long. Adult leaves often a little waxy, scalelike, overlapping loosely, sticking out from the twig at a forward angle and then abruptly curling back in toward the twig at the tip, 3–4.5 mm long, 1–2 mm wide, about as thick as wide or a little wider, half-moon-shaped in cross section. Midrib with a shallow but sharp keel along the outer face, prominent and flanked by narrow grooves containing lines of stomates on the inner face. Pollen cones thinner than to about as wide as their stalks and accompanying leafy bracts, 1–1.5 cm long, 1.5–2.5 mm in diameter. Pollen scales triangular, curled inward at their tips like the leaves, without a prolonged tip. Seed cones on a branchlet with leaves 3.5–4 mm long, a little shorter than average for adult foliage leaves. Free tips of the bracts making up the podocarpium like the foliage leaves, a little longer than the leaves of the supporting branchlet, 3.5–4 mm long, the uppermost loosely curled over the seed. Mature podocarpium red, with one (or two) fertile bracts. Epimatium cuplike, slightly fleshy, very shallow, about 1 mm deep. Seed roughly egg-shaped, 4–5 mm long, 3–3.5 mm thick, with a small, rounded beak 0.2–0.3 mm long. Throughout most of New Caledonia, from the Massif d'Ouazangou-Taom southward. Found occasionally in most vegetation types on serpentine-derived soils, from maquis minière scrublands to rain forests, most commonly in dry forests but attaining its greatest stature with increased moisture; 150–900 m. Zone 9. Synonym: *Metadacrydium balansae* (Ad. Brongniart & Gris) Baumann-Bodenheim ex Melikian & A. V. Bobrov.

Balansa rimu is more widespread than its close relative candelabra rimu (*Dacrydium araucarioides*) and usually grows under moister conditions, hence its generally larger stature. The two species are very similar to one another, and identification may well be complicated by occasional hybridization between them. The species name honors Benedict Balansa (1825–1892), who gathered the type material during a collecting trip in New Caledonia in 1868–1872, one of many he undertook, beginning in his 20s, to places as disparate as Algeria, Asia Minor, Paraguay, and Vietnam, where he died.

Dacrydium beccarii Parlatore

ELFIN RU, KAYU EMBUN, EKOR KUDA, SAMPINUR TALI (MALAY), NETUKURIA, MEJOOP (PAPUAN LANGUAGES)

Shrub, or tree to 20(–35) m tall, with trunk, when present, to 0.3 m in diameter. Bark thin, smooth, brown, weathering gray, flaking in scales, and becoming fissured with age. Crown shallowly dome-shaped, irregular, compact, with numerous upwardly angled branches bearing many upright, stiff branchlets densely clothed with and hidden by foliage. Transition between juvenile and adult foliage gradual. Juvenile leaves to 17(–20) mm long. Adult leaves needlelike, sword-shaped, rather fine in texture, densely overlapping, sticking out from the twig at a forward angle, 5–8(–10) mm long, 0.3–0.4 mm wide, a little wider than thick, triangular in cross section, with a straight, prickly tip. Midrib prominently raised beneath. Pollen cones appearing thinner than their stalks and accompanying leafy bracts, 7–10 mm long, 2.5–3 mm in diameter. Pollen scales narrowly triangular, with a prolonged tip about 1 mm long. Seed cones on a very short branchlet with reduced leaves about 1 mm long. Free tips of the bracts making up the podocarpium resembling the foliage leaves but shorter, a little longer than the leaves of the supporting branchlet, 1–2 mm long, surrounding the base of the epimatium. Mature podocarpium red, with one or two (or three) fertile bracts. Epimatium cuplike, about 1 mm deep. Seed dark brown, roughly egg-shaped, about 4 mm long by 2–3 mm thick, with a tiny beak. Across the central band of Malesia, from Sumatra and southern Malaya east through Borneo, southern Philippines, Sulawesi, the Moluccas, and New Guinea to Guadalcanal (Solomon Islands), more scattered eastward. Often a dominant in shrublands or low forests of wet, exposed mountain ridges on a variety of soil types, including volcanic rocks, sandy peats, and serpentine-derived soils; (600–) 1,000–2,000(–2,500) m. Zone 10?

In line with its generally short stature, elfin ru has about the shortest, finest leaves among the *Dacrydium* species with needle- rather than scalelike leaves. The result is a branchlet that appears to be covered with green fur. It has the widest geographic distribution of any species of *Dacrydium,* rivaled only by the closely related kerapui ru (*D. xanthandrum*), a very similar but generally larger species with longer, flatter needles in a looser arrangement. The two species have similar ranges and occupy similar habitats so that a search for hybrids between them might be instructive. The species name honors Odoardo Beccari (1843–1920), an Italian botanist who collected extensively in Malesia between 1865 and 1878, including the type specimen of this species, collected in Sarawak.

Dacrydium comosum Corner

WOOLLY RU

Shrub to 4 m or tree to 12 m tall, with trunk to 0.1(–0.2) m in diameter. Bark thin, smooth, flaking and weathering gray. Crown very dense and compact, shallowly dome-shaped to flat, with upright branches bearing numerous erect branchlets densely clothed with foliage so as to appear woolly (hence the scientific name, Latin for "hairy"). Transition between juvenile and adult foliage gradual and primarily by reduction in length. Juvenile leaves prickly, to 33 mm long. Adult leaves needlelike, narrowly sword-shaped, loosely overlapping, sticking out widely all around the twig, fairly straight or a little curved forward near the prickly tip, 12–20 mm long, 0.6–1 mm wide, much wider than thick, flat. Midrib weakly raised. Pollen cones appearing much thinner than the branchlets with their spreading leaves and a little thinner than their stalks and accompanying leafy bracts, 8–10 mm long, about 3 mm in diameter. Pollen scales triangular with a narrowly prolonged, erect tip 1.5–2 mm long. Seed cones on a branchlet with leaves about 4 mm long. Free tips of the bracts making up the podocarpium the foliage leaves but shorter, shorter than those of the supporting branchlet, about 2 mm long, surrounding the base of the epimatium. Mature podocarpium red, with one or two fertile bracts. Epimatium cuplike, about 1 mm deep. Seed light brown, egg-shaped, 4–5 mm long, 2.5–4 mm thick, with a tiny beak. Known only from the mountains near Kuala Lumpur and on Gunong Tahan in Pahang and Selangor States, central Malaya (West Malaysia). Scattered to locally dominant in low, mossy forests or shrublands on exposed, windswept ridges; 1,400–2,200 m. Zone 9? Synonym: *Corneria comosa* (Corner) A. V. Bobrov & Melikian.

Woolly ru is a rare species overall, but when present it can make up a significant portion of the canopy. It has the longest needles among the Malesian species of *Dacrydium* with narrow, needlelike adult leaves: elfin ru (*D. beccarii*), Kinabalu ru (*D. gibbsiae*), kerapui ru (*D. xanthandrum*), and others. It is also one of the smallest of these species, though all are short trees or shrubs of wet, mossy forests on exposed mountain ridges, usually above 1,500 m. It is ironic, perhaps, that the large forest giants in the genus, by contrast, all have tiny, scalelike leaves making up their adult foliage.

Dacrydium cornwallianum de Laubenfels

SWAMP RU

Tree to 30 m tall, with trunk to 0.4 m in diameter. Bark reddish brown, thin, smooth, flaking in small scales, and weathering grayish brown. Crown narrowly cylindrical, with many short horizontal branches bearing numerous, upwardly angled branchlets densely clothed with and hidden by foliage. Transition between juvenile and adult foliage gradual. Juvenile leaves prickly, to 12 mm long. Adult leaves scalelike, densely overlapping, sticking out from the twig at a forward angle and curving continuously forward and then inward at the prickly tip, 2–5 mm long, 0.6–0.8 mm wide, twice as wide as thick. Cross section half-moon-shaped, with a sharp ridge along the outer face and a shallowly raised midrib flanked by shallow grooves containing lines of stomates on the inner face. Pollen cones appearing conspicuously thinner than their stalks and accompanying leafy bracts, about 12 mm long, 1.8 mm in diameter. Pollen scales triangular, without a prolonged tip. Seed cones on a short branchlet with reduced leaves at least 1.5 mm long. Free tips of the bracts making up the podocarpium resembling the foliage leaves, generally longer than the leaves of the supporting branchlet, progressively longer toward the tip and surrounding the epimatium and base of the seed, up to 4 mm long. Mature podocarpium red, with one or two fertile bracts. Epimatium cuplike, about 1 mm deep. Seed brown, roughly egg-shaped, about 5 mm long, 3 mm thick, with a tiny beak. Scattered across the highlands of the western two-thirds of New Guinea, from near Mendi (Southern Highlands province, Papua New Guinea) westward. Forming pure stands or dominating the canopy of swamp forests; (750–)1,450–2,300(–2,770) m. Zone 9? Synonyms: *Corneria cornwalliana* (de Laubenfels) A. V. Bobrov & Melikian, *Dacrydium nidulum* var. *araucarioides* de Laubenfels.

Swamp ru is closely related to the more widespread samiampi ru (*Dacrydium nidulum*) and was originally described as a variety of that species. Samiampi ru has looser, more widely spreading leaves than does swamp ru and, in New Guinea, grows at lower elevations in primary rain forests. A few other conifers may be found with swamp ru on the same waterlogged, peaty soils, including tjimba dacrycarp (*Dacrycarpus steupii*), kebu podocarp (*Podocarpus pseudobracteatus*), and Papuan incense cedar (*Papuacedrus papuana*). The species name honors author Zoe Cornwall (b. ca. 1940), who was de Laubenfels's assistant in New Guinea when he studied it and decided that it should be separated taxonomically from *Dacrydium nidulum*.

Dacrydium cupressinum Solander ex J. G. Forster

RIMU (MAORI AND ENGLISH), [NEW ZEALAND] RED PINE

Tree to 35(–60) m tall, with trunk to 1.5(–2.5) m in diameter, sometimes free of branches for 20 m or more. Bark smooth, brown, warty with rows of tiny lenticels, weathering gray and flaking off in irregular, generally elongate scales. Crown narrowly conical and weeping in youth, becoming egg-shaped to spherical with age, rather shallow in dense stands, with horizontal to upwardly angled branches bearing numerous drooping branchlets densely clothed with and largely hidden by foliage. Transition between juvenile and adult foliage gradual. Juvenile leaves on long, sparsely branched, hanging branchlets, 4–7(–9) mm long. Adult leaves scalelike, narrowly triangular, loosely overlapping, sticking out from the twig at a forward angle, slightly incurved at the prickly tip, 2–3(–4) mm long, 0.5–0.8(–1) mm wide, a little wider than thick, triangular in cross section. Midrib with a sharp keel beneath and scarcely raised above. Cones often borne on young trees with foliage transitional between juvenile and adult leaves as well as on mature trees with adult foliage. Pollen cones appearing thicker than their stalks and accompanying leafy bracts, (5–)8–15 mm long, 4–6 mm in diameter. Pollen scales triangular, with a conspicuous, prolonged, needlelike green tip 0.5–1.5 mm long. Seed cones on a very short branchlet with reduced leaves 1–1.5 mm long. Free tips of the bracts making up the podocarpium resembling the foliage leaves but shorter, about as long as the leaves of the supporting branchlets, 1–1.5 mm long, barely overlapping the base of the epimatium. Mature podocarpium red, with one fertile bract. Epimatium cuplike, 3–4 mm deep. Seed deep bluish black, egg-shaped, (4–)6–8 mm long, 3–4 mm thick, with a broad, dimpled beak less than 1 mm long. Widely distributed throughout the three main islands of New Zealand, mostly within 50 km of the coast on South Island but well distributed within the interior of North Island. Forming small pure stands or scattered as a dominant or codominant in or above the canopy of temperate lowland and montane rain forests and reaching the subalpine zone on Stewart Island; 0–750(–950) m. Zone 9.

Along with kauri (*Agathis australis*) and totara (*Podocarpus totara*), rimu is one of the premier native timber trees of New Zealand. The wood, with contrasting brown sapwood and red heartwood is beautiful, especially when marbled with yellow and orange color variations. Although it is very resinous, it is not

Giant ancient rimu (*Dacrydium cupressinum*) in nature.

durable in outdoor use. Nonetheless, the tree was so abundant originally that the wood was used for such applications as fence posts and railway bridges, even though they had to be replaced within a decade. Interior uses in framing, finishing and furniture are much more common, and the applications that take advantage of its attractive grain are now mostly fulfilled in the form of veneer. These forest giants can be very long-lived, with a maximum age approaching 1,200 years, although half of that is a more typical longevity. The virgin stands of great trees were quickly depleted during the 19th and early 20th centuries, and large, ancient trees were still being cut in the 1960s. Rimu has always accounted for approximately half of all timber produced from native trees in New Zealand, but nowadays native trees make up less than 10% of the annual cut, having been replaced by plantations of exotic conifers, like Monterey pine (*Pinus radiata*). Although rimu is still an abundant tree in New Zealand, the forests are now dominated by younger, second- or third-growth trees, many still in the much admired youth phase, with its neat, conical shape and graceful, weeping branches. The scaly leaves of the adult foliage is somewhat reminiscent of that of northern hemisphere junipers and cypresses (hence the scientific name, Latin for "like *Cupressus*").

Old mature rimu (*Dacrydium cupressinum*) in nature.

Rimu is the only species of *Dacrydium* from a temperate region and is also the species whose biology has been most thoroughly investigated. The other six New Zealand species once included in *Dacrydium* are all now assigned to other genera (*Halocarpus, Lepidothamnus,* and *Manoao*), each differing from rimu in many aspects of their biology, including such hidden features as embryonic development, chromosomal arrangements, oil chemistry of the heartwood, and pigments in the leaves. It is the only species retained in *Dacrydium* by some authors, but most conifer taxonomists keep it with its tropical relatives.

As a common, often dominant large tree, rimu plays a major role in many New Zealand forest communities. For example, its intermittent reproduction tends to be synchronized locally. Several species of parrots, pigeons, and other birds eat and disperse the seeds (which are at the lower end of the size range among New Zealand podocarps) along with the podocarpium that they

Twig from young rimu (*Dacrydium cupressinum*) with transitional foliage, ×1.

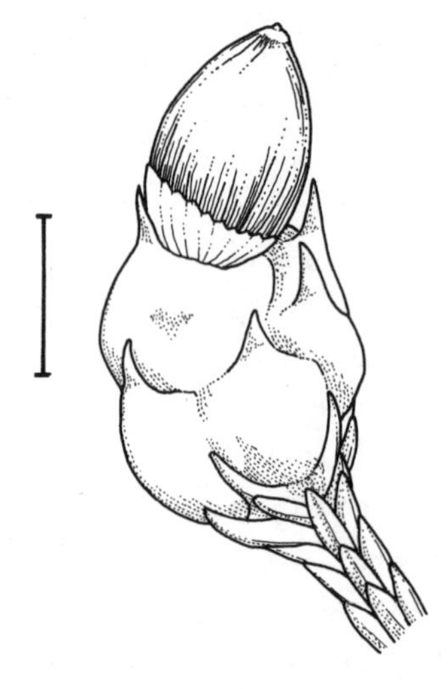

Seed cone of rimu (*Dacrydium cupressinum*) at tip of branchlet with ordinary adult foliage; scale, 2 mm.

pick out of the foliage in the canopy. On Stewart Island at least, it appears that one species of bird, the highly endangered kakapo (*Strigops habroptilus*), a ground-dwelling parrot, even times its breeding cycle to link with that of rimu. During mast years, at irregular intervals of 1–5 years, the ground beneath rimu stands can be littered with fleshy fallen seed cones, while in poor years such few cones as fall may not even have good development of the podocarpium. Pollen cones also show fluctuations in abundance, more regular than the seed cones perhaps, with a cycle of 5–6 years. The tree is also a host for a variety of generalist and specialist herbivorous insects, including wood-borers and foliage and seed eaters. The young trees are even browsed by introduced herbivores like deer, goats, and Australian opossums.

Surprisingly for such a common and widespread tree, spanning a considerable environmental range in New Zealand, relatively little genetic variation has been detected compared to most northern hemisphere conifers studied. The repeated cycles of glaciation that periodically reduced the favorable environment for forest growth in New Zealand during the last million years may well have produced population bottlenecks for rimu that reduced its variability. Comparable processes have been proposed for some northern hemisphere species with limited variation, like red pine (*Pinus resinosa*) in northeastern North America and Torrey pine (*P. torreyana*) in southern California.

Dacrydium elatum (Roxburgh) N. Wallich ex W. J. Hooker

MOUNTAIN RU, BẠCH ĐÀN (VIETNAMESE), SRÔ:L KRÂHÂ:M (CAMBODIAN), RU BUKIT, SEMPILOR, SAMPINUR TALI (MALAY)

Tree to 30(–40) m tall, with cylindrical, unbuttressed trunk to 1 m in diameter. Bark thin, soft, reddish brown to brown, weathering grayish brown, becoming shallowly fissured, and shedding elongate scales to reveal pink, dark red, and warm brown patches. Crown conical, regular, and dense in youth, becoming narrowly and deeply dome-shaped with maturity, and finally ragged and open, with strong, upwardly angled branches (often drooping in the lower crown) bearing many stiff, upright, threadlike branchlets densely clothed with and hidden by grayish green foliage. Transition between juvenile and adult foliage sometimes gradual over time but appearing abrupt because of the continued occurrence of branchlets bearing juvenile foliage on adult trees, especially in the lower crown. Juvenile leaves (6–)8–12(–18) mm long, diamond-shaped in cross section. Adult leaves scalelike, triangular, densely overlapping, tightly pressed forward against the twig, pointed but not prickly, 1–1.5 mm long, 0.4–0.6 mm wide, wider than thick, triangular in cross section. Midrib raised as a sharp keel beneath. Pollen cones appearing thinner than the supporting branchlet and accompanying leafy bracts, 4–6(–10) mm long, 1–1.2 mm in diameter. Pollen scales triangular, the tip pointed but not prolonged. Seed cones on a short branchlet with slightly reduced leaves about 1 mm long. Free tips of the bracts making up the podocarpium resembling the foliage leaves but slightly longer, longer than the leaves of the supporting branchlet, 1.5–2 mm long, scarcely surrounding the base of the epimatium. Mature podocarpium red, slightly swollen and juicy, with one fertile bract. Epimatium cup-like, asymmetrical, about 1.5 mm deep. Seed ripening from reddish to black, egg-shaped, 4–5 mm long, 2.5–3 mm thick, with a tiny, short, broad, dimpled beak less than 0.5 mm long. Discontinuous in Southeast Asia and western Malesia, from northern Thailand and northern Vietnam through Indochina and the southern Malay Peninsula to the central western coast of Sumatra and

Mature mountain ru (*Dacrydium elatum*) in nature.

northern Borneo. Forming pure stands or scattered as a dominant or codominant in open, secondary rain forests on many substrates including sandy high ground in peat swamps, often mixed with other conifers, such as *Dacrycarpus imbricatus*, or with evergreen hardwoods; (0–)500–2,000(–3,000) m. Zone 10. Synonyms: *Corneria elata* (Roxburgh) A. V. Bobrov & Melikian, *C. pierrei* (P. Hickel) A. V. Bobrov & Melikian, *Dacrydium beccarii* var. *subelatum* Corner, *D. junghuhnii* Miquel, *D. pierrei* P. Hickel.

Mountain ru is the most abundant species of *Dacrydium* in Malaya and is also an important timber tree in Vietnam and, to a lesser extent, in Sarawak. The wood has many uses, including cabinetry and the making of masts. The high resin content inspires its use as torches and incense sticks, and the oil itself is extracted and used in traditional medicine. The composition of the distilled oil closely resembles that of the commercial cedarwood oil extracted from the unrelated junipers in North America, with a high content of cedrol and cedrene, sesquiterpenes (oils with 15 carbon atoms) otherwise found primarily in most members of the Cupressaceae (including *Juniperus* species) and in the true cedars (*Cedrus*) of the Pinaceae, after which these chemical compounds are named. Some of the common names, like *ru* and *sempilor*, make reference to the general resemblance of the threadlike adult twigs of this species to those of species of *Casuarina* and related genera, very common flowering plants of seashores in the region that are so similar in appearance to conifers that they bear the common name Australian pines. Belying its scientific name (Latin for "lofty"), which is often justified by well-grown trees on favorable sites, mountain ru is also a common dominant, along with common dacrycarp (*Dacrycarpus imbricatus*) and oleander podocarp (*Podocarpus neriifolius*), of dwarfed rain forests with a canopy height of only about 10 m on infertile sites that are either unusually dry or occupy sodden depressions. Because of its success in disturbed forests, mountain ru has proven amenable to cultivation and is used as a plantation species to a limited extent. It has been confused over the years with several other species of *Dacrydium* whose adult foliage is much looser, including elfin ru (*D. beccarii*), samiampi ru (*D. nidulum*), and melur ru (*D. pectinatum*). Thus older literature referring to its occurrence in China, the Philippines, and Sulawesi east to Fiji is actually based on these species.

Dacrydium ericoides de Laubenfels

HEATH RU, SEMPILOR (MALAY)

Tree to 17 m tall, with trunk to 0.3 m in diameter. Bark thin, grayish brown, scaly. Crown dome-shaped, with short, rising branches bearing many drooping branchlets densely clothed with foliage. Transition between juvenile and adult foliage gradual and scarcely noticeable. Juvenile leaves to 12 mm long. Adult leaves needlelike, straight, sword-shaped, well separated along and sticking straight out from the twig, with a prickly tip, 5–10 mm long, 0.7–1 mm wide, much wider than thick, very broadly triangular in cross section. Midrib sharply raised beneath and flanked by shallow grooves with lines of stomates above. Pollen cones appearing thinner than their stalks and accompanying leafy bracts, 7–10 mm long, 2–2.5 mm in diameter. Pollen scales triangular, with a triangular, prolonged tip about 1 mm long. Seed cones on a very short branchlet with reduced leaves 2–3 mm long. Free tips of the bracts making up the podocarpium resembling the foliage leaves but shorter, longer than the leaves of the supporting branchlets, 3–4 mm long, encircling the epimatium and much of the seed. Mature podocarpium red, with one (or two) fertile bracts. Epimatium cuplike, about 1 mm deep. Seed dark, roughly egg-shaped, about 4 mm long, 3 mm thick, with a tiny beak. Known only from a few localities on Bukit Dulit and the Merurong Plateau in northern Sarawak (Malaysian Borneo). Scattered as a codominant in the canopy of low, mossy forests on exposed mountain ridges; 1,000–1,500 m. Zone 10. Synonym: *Corneria ericoidea* (de Laubenfels) A. V. Bobrov & Melikian.

Like its close relatives elfin ru (*Dacrydium beccarii*), Idenburg ru (*D. spathoides*), and other *Dacrydium* species with thin, narrow, needlelike leaves, heath ru is an inhabitant of moss-draped rain forests on mountain ridges exposed to strong winds but with a perpetually saturated atmosphere. Several of these species, as presently understood, have restricted geographic distributions, and their interrelationships with one another merit further attention. The leaf form of these species, so strongly contrasting with that of the scale-leaved species in the genus, can be reminiscent of that of northern hemisphere heaths and heathers (hence the scientific name, Latin for "like *Erica*").

Dacrydium gibbsiae Stapf

KINABALU RU

Bushy tree to 12(–20) m tall, with short trunk to 0.3 m in diameter. Bark thin, gray, and scaly. Crown compact, with short branches bearing numerous stiff, short branchlets densely clothed with and hidden by the foliage. Twigs coarse, drooping in youth to upright with maturity. Transition between juvenile and adult foliage gradual. Juvenile leaves to 16(–20) mm long. Adult leaves needlelike, spearhead-shaped, stiff, densely overlapping, sticking out from the twig at a forward angle, 5–8 mm long, 0.8–1.0(–1.3) mm wide, three to four times as wide as thick, shallowly triangular in cross section. Leaf tip prickly, bent forward parallel with the twig or tucked in toward it. Midrib sharply keeled beneath and slightly raised above with shallow grooves bearing lines of stomates on either side. Pollen cones appearing slightly thinner than their stalks and accompanying leafy bracts, 2–2.5 cm long, (4–)5–7 mm in diameter. Pollen scales with a narrowly triangular, prolonged tip 5–6 mm long. Seed cones on a very short branchlet with slightly reduced leaves. Free tips of the bracts making up the podocarpium resemblig the foliage leaves but narrower, a little longer than the leaves of the supporting branchlet and progressively longer upward, 5–7 mm long, enclosing the epimatium and much of the seed. Mature podocarpium reddish, slightly swollen, with one or two (or three) fertile bracts. Epimatium cuplike, 1–1.5 mm deep. Seed shiny brown, egg-shaped to spherical, 4–5 mm long, 3–4 mm thick, with a wide, ringlike beak. Known only from Gunung (Mount) Kinabalu in Sabah (Malaysia), northern Borneo. Scattered to locally codominant in the canopy of the low, wet, mossy forest phase of upper montane rain forests on exposed ridges or rare in the lower montane zone, usually on serpentine-derived soils; (1,400–)2,000–3,500(–3,700) m. Zone 9. Synonym: *Corneria gibbsiae* (Stapf) A. V. Bobrov & Melikian.

With several peaks over 4,000 m in the summit plateau region, Mount Kinabalu is by far the tallest mountain in the Malesian region outside of New Guinea. As one of the few regional mountains extending into the upper montane and subalpine zones, Mount Kinabalu houses many endemic species, including four or five species of conifer found nowhere else. These species, including Kinabalu ru, are mostly found in the upper montane zone that is absent from nearby peaks. Of the five species of *Dacrydium* growing on Mount Kinabalu, only kerapui ru (*D. xanthandrum*) joins Kinabalu ru in the upper montane zone but does not accompany it even higher into the subalpine zone. These two species contrast completely in their foliage, with the coarse twigs of Kinabalu ru appearing reptilian and tightly clothed by the stiff, overlapping leaves, while the thinner twigs of kerapui ru appear furry with the needlelike leaves extending straight out. The species name honors Lilian Gibbs (1870–1925), who collected the type specimen in 1910, along with about a thousand other specimens, and wrote the first paper on the vegetation of Mount Kinabalu in 1914.

Dacrydium gracile de Laubenfels

SLENDER RU

Tree to 30 m tall, with trunk to 0.4 m in diameter. Bark thin, gray, and scaly. Crown narrow and irregular, with horizontal to upwardly angled branches bearing numerous upturned branchlets openly clothed with foliage. Transition between juvenile and adult foliage gradual. Juvenile leaves to 18 mm long. Adult leaves needlelike, sword-shaped, somewhat openly spaced along the twigs, sticking out at a forward angle and then curving gently forward, 3–6(–9) mm long, about 0.4 mm wide, twice as wide as thick, triangular in cross section, with a prickly tip. Midrib prominently raised beneath. Pollen cones appearing much thinner than their stalks and accompanying leafy bracts, 6–7 mm long, about 2 mm in diameter, the bracts 3–5 mm long. Pollen scales triangular, with small, prolonged tip to 1 mm long. Seed cones on a very short branchlet with reduced leaves about 1 mm long. Free tips of the bracts making up the podocarpium resembling the foliage leaves but shorter, conspicuously longer than the leaves of the supporting branchlet, 2–3 mm long, surrounding the epimatium but not the seed. Mature podocarpium red, with one (or two) fertile bracts. Epimatium cuplike, about 1 mm deep. Seed dark brown, roughly egg-shaped, about 4 mm long, with a tiny beak. Rare and scattered in northern Borneo, from Gunung Kinabalu (Sabah) south to Bukit Balu (Sarawak), both East Malaysia. Scattered in the canopy of lower montane rain forests or, occasionally, a constituent of lower-stature heath forests on sandstone substrates; (950–)1,200–1,600(–1,800) m. Zone 10. Synonym: *Corneria gracile* (de Laubenfels) A. V. Bobrov & Melikian.

The branchlets and foliage of slender ru are less robust looking than those of some of its relatives (hence the scientific name, Latin for "slender"), especially Kinabalu ru (*Dacrydium gibbsiae*) from higher up on Mount Kinabalu and the more closely related islet ru (*D. magnum*) from farther away in the Moluccas and Solomon Islands. Although generally a rare species and not recognized and described as a new species until 1988, slender ru is moderately common on the lower slopes of Mount Kinabalu, and the type was collected in the vicinity of the much-visited park headquarters of this long-established national park. It had been collected much earlier but was misidentified as one of the other species on the mountain and not recognized as distinct.

Dacrydium guillauminii J. Buchholz

CATTAIL RIMU, SWAMP DACRYDIUM, QUEUES DE CHAT (FRENCH)

Upright shrub 1–2 m tall or rarely a tree to 6 m, usually with several trunks to 0.1(–0.15) m in diameter. Bark brown and smooth with many small, warty lenticels, weathering gray and breaking up into thin blocks divided by shallow fissures. Crown conical, with upright branches bearing scattered branchlets densely clothed with hairlike foliage. Transition between juvenile and adult foliage gradual. Juvenile leaves to 17 mm long. Adult leaves needlelike, densely inserted, sticking out from the twig at a forward angle and then continuing fairly straight, 8–20 mm long, 0.8–1 mm wide, much wider than thick, shallowly half-moon-shaped in cross section. Midrib narrowly raised on both the inner and outer faces, the lines of stomates on either side of the midvein on the inner face not in grooves. Pollen cones appearing thinner than but not very distinct from their stalks and accompanying leafy bracts, 1–1.5 cm long, 2.5–5 mm thick, tapering from near the base toward the tip. Pollen scales narrowly triangular, with a very prolonged, leaflike tip 2–6 mm long, progressively shorter from the base to the tip of the cone. Seed cones on a branchlet with leaves on the shorter side of average for adult foliage leaves. Free tips of the bracts making up the podocarpium similar to the leaves of the supporting branchlet or somewhat shorter, the uppermost extending beyond the seeds and nestling them. Mature podocarpium red, slightly swollen, with one to five fertile bracts. Epimatium cuplike, thin, 1–1.5 mm deep, enclosing about a third of the seed. Seed roughly egg-shaped, 4–5 mm long, 1.5–2 mm thick, with a small, pointed beak 0.2–0.3 mm long. Known only from the Plaine des Lacs in southernmost New Caledonia along a few km of the Rivière des Lacs downstream from the Chutes de la Madeleine as well as along the shores of the main lakes drained by this river: Lac en Huit, Lac Intermédiaire, and Grand Lac. Growing in small populations in swampy ground over serpentine substrates at the waters edge or a little away from it; 200–250 m. Zone 10. Synonym: *Gaussenia guillauminii* (J. Buchholz) A. V. Bobrov & Melikian.

Cattail rimu is one of the rarest conifers in the world. Like its slightly more common neighbor New Caledonian corkwood (*Retrophyllum minor*) it is a rheophyte, growing with its feet in the water at the margin of water bodies. Bourgeoning tourism and changing water levels are both threats to this highly restricted species, another example of the risks associated with extremely narrow habitat specialization. The much more common candelabra rimu (*Dacrydium araucarioides*) grows nearby in the maquis minière scrublands on dry ground surrounding the lakes and rivers. These two species have completely different adult foliage, with candelabra rimu having the typical scale leaves of many species in the genus while cattail rimu has adult foliage that appears to be a retention of the juvenile foliage form found in all species of the genus (it is thus considered pedomorphic). Despite this seemingly major developmental difference between the two species, very rare hybrids between them have been reported (four hybrids all told, in 1994 and 1996), but these need to be investigated experimentally. The species name honors André Guillaumin (1885–1974), one of the pioneers in the study of the flora of New Caledonia, who collected it (although not the type specimen) during one of the many stays that contributed to his long string of publications on the flora of the island from 1911 to 1973.

Dacrydium leptophyllum (Wasscher) de Laubenfels ex Silba

MOUNT GOLIATH RU

Possibly a shrub or small tree with a compact, much-branched, dense crown. Twigs very slender and flexible, upright, straight or gently curved inward. Transition between juvenile and adult foliage gradual. Juvenile leaves to 3 mm or more. Adult leaves scale-like, densely and tightly overlapping, sticking out from the twig and then bending strongly forward, 1–2 mm long, 0.2–0.3(–0.4) mm wide, two to three times as wide as thick, broadly triangular in cross section, with a prickly tip pointing forward or curved back into the twig. Midrib sharply keeled beneath and nearly flat or with shallow grooves containing lines of stomates on either side above. Pollen cones and seed cones not seen or described elsewhere. Known only from Mount Goliath, in east-central Irian Jaya, western New Guinea. Presumably in subalpine forest; 3,000–3,600 m. Zone 9? Synonyms: *Bracteocarpus leptophyllus* (Wasscher) A. V. Bobrov & Melikian, *Corneria leptophylla* (Wasscher) A. V. Bobrov & Melikian, *Podocarpus leptophyllus* Wasscher.

As far as is known, and unless more localities are discovered, this may be one of the rarest conifers in the world, being found at the top of a single mountain peak. Very few specimens have been gathered, and even a basic morphological understanding of the species is among the most incomplete of any living conifer. Even the correct genus for the species remains uncertain in the absence of collections including reproductive structures. The species name, Greek for "slender leaf," reflects the unusually small size of the leaves in comparison to those of species of *Podocarpus,* where it was first placed, even if they are not that unusual in *Dacrydium.*

Dacrydium lycopodioides Ad. Brongniart & Gris

CLUB MOSS RIMU

Tree to 25(–30) m tall, with trunk to 0.6 m in diameter. Bark brown, smooth at first, with numerous horizontally arranged small, warty lenticels, darkening with age except where mottled with paler brown by the flaking off of scales. Crown conical at first, becoming more irregular with age, with upwardly angled branches bearing numerous upright, somewhat sinuous branchlets densely clothed with foliage. Twigs partially hidden by the slightly overlapping leaves. Transition between juvenile and adult foliage gradual. Juvenile leaves (6–)9–12 mm long. Adult leaves needlelike and sword-shaped, sticking out from the twig at a wide forward angle, with a continuous gentle forward curve, 2.5–5 mm long, 0.7–0.8 mm wide, much wider than thick, thin and slightly curved in cross section with a prickly tip. Midrib prominent on the inner face, flanked by lines of stomates in shallow grooves. Pollen cones appearing thinner than their stalks and accompanying leafy bracts, 4–7 mm long, 1–1.2 mm in diameter. Pollen scales egg-shaped, with a narrowly triangular tip, not curled in at the tip. Seed cones on a branchlet with leaves 1–2 mm long, shorter than most adult foliage leaves. Free tips of bracts making up the podocarpium like the foliage leaves, progressively longer upward, the uppermost 2.5 mm long and loosely wrapping the epimatium and the base of the seed. Mature podocarpium red, with one (or two) fertile bracts. Epimatium unevenly cuplike, 1–1.5 mm deep. Seed roughly egg-shaped, shiny brown, 3–4 mm long, 2–2.7 mm thick, with a pointed beak 0.3–0.6 mm long, Discontinuous on individual mountains in the southern third of New Caledonia from Mount Nakada near Canala to Mount Mou and Mount Ouin near Nouméa. Scattered and sometimes abundant in the canopy of moist montane forests of upper slopes; 900–1,400 m. Zone 9. Synonym: *Gaussenia lycopodioides* (Ad. Brongniart & Gris) A. V. Bobrov & Melikian.

Unlike the other three species of *Dacrydium* in New Caledonia, club moss rimu is not always associated with serpentine-derived soils. Although it is not as widely distributed as candelabra rimu (*D. araucarioides*), it can be a common canopy member on those mountain peaks where it occurs. This is the case at the summit of Mount Mou, where the type specimen was collected and where it has been collected many times since. The thin but tough leaves give the branchlets an appearance much like that of many temperate and tropical species of club moss (hence the scientific name, Latin for "like *Lycopodium,*" itself Greek for "wolf foot"). These leaves are much less drought resistant than the closely overlapping scale leaves of candelabra rimu from drought prone maquis minière scrublands but are still much more tolerant than those of species, like cattail rimu (*D. guillauminii*), that never face drought.

Dacrydium magnum de Laubenfels

ISLET RU

Tree to 30 m tall, with trunk to 0.6 m in diameter. Bark thin, gray, and scaly. Crown shallowly dome-shaped, with horizontal branches turning up at the ends and bearing numerous crowded branchlets densely clothed with and hidden by foliage. Transition between juvenile and adult foliage gradual. Juvenile leaves to 17(–20) mm long. Adult leaves needlelike, sword-shaped, densely overlapping, of uniform length along the branchlet, sticking out at a forward angle and then curving forward parallel and close to the twig, 3–6 mm long, 0.3–0.4 mm wide, a little wider than thick, half-moon-shaped in cross section, the tip prickly, straight or curved slightly inward. Midrib slightly raised as a sharp keel beneath. Pollen cones appearing thinner than their stalks and accompanying leafy bracts, 10–16 mm long, 2–2.5 mm in diameter. Pollen scales triangular with a narrow, straight, prolonged tip 1.5–2 mm long. Seed cones on a branchlet with slightly reduced leaves 2–4 mm long. Free tips of the bracts making up the podocarpium resembling the foliage leaves but shorter, longer than the leaves of the supporting branchlet, 3–5 mm long, surrounding the epimatium and the base of the seed. Mature podocarpium red to reddish brown, with one or two fertile bracts. Epimatium cuplike, 1–1.5 mm deep. Seed brown, roughly egg-shaped, 4.5–5 mm long, with a tiny beak. Known only from a few small islands east and west of New Guinea: Obi (Indonesia: Moluccas south of Halmahera), Sudest Island (Papua New Guinea: southern Louisiade Archipelago), and Choiseul, Santa Isabel, and Guadalcanal (Solomon Islands). Scattered to gregarious in the canopy of lowland and lower montane moist forests, shorter along with the canopy when growing on exposed ridges; 60–1,200 m. Zone 10. Synonyms: *Corneria magna* (de Laubenfels) A. V. Bobrov & Melikian, *Dacrydium beccarii* var. *rudens* de Laubenfels.

Islet ru appears to be closely related to elfin ru (*Dacrydium beccarii*), Kinabalu ru (*D. gibbsiae*), slender ru (*D. gracile*), and scrub ru (*D. medium*) but differs from all of them in having nearly full-sized leaves on reproductive branchlets (hence the scientific name, Latin for "big"). It also has longer pollen cones than most of its relatives. Its absence from the lowlands and lower mountains of New Guinea in between its widely separated eastern and western areas of distribution is puzzling. Perhaps it has been overlooked or confused with some phases of one of the seven other species of *Dacrydium* recorded for New Guinea.

Dacrydium medium de Laubenfels

SCRUB RU, SANGU (MALAY)

Low shrub to tree to 10(–20) m tall, with trunk, often distorted, to 0.2 m in diameter. Bark thin, gray, and scaly. Crown egg- to dome-shaped, compact, dense, with short, crooked branches bearing numerous upright branchlets densely clothed with and partly hidden by foliage. Transition between juvenile and adult foliage gradual. Juvenile leaves to 20 mm long. Adult leaves needlelike, sword-shaped, loosely overlapping, uniform in size along and sticking out from the twig at a forward angle and then curving strongly forward (less commonly straight), 3–6(–8) mm long, 0.5–0.6 mm wide, about twice as wide as thick, broadly triangular in cross section, with a straight or tucked-in prickly tip. Midrib raised beneath. Pollen cones appearing thinner than their stalks and accompanying leafy bracts, 7–9 mm long, 2–2.5 mm in diameter. Pollen scales triangular with a relatively wide, prolonged tip 1.5–2 mm long but tucked in at the end. Seed cones on a short branchlet with reduced leaves about 2 mm long. Free tips of the bracts making up the podocarpium resembling the foliage leaves but shorter, a little longer than the leaves of the supporting branchlet, 2–3 mm long, progressively longer upward, surrounding the epimatium and base of the seed. Mature podocarpium dark red, with one (or two) fertile bracts. Epimatium cuplike, about 1.5 mm deep. Seeds brown, egg-shaped, about 5 mm long, with a tiny beak. Mountains of central Malaya (West Malaysia) and of the northern fifth of Sumatra (Indonesia). Scattered and often dominant as an emergent over dry montane scrubland on impoverished, rocky or sandy soils; (950–)1,400–2,200(–2,600) m. Zone 10? Synonym: *Corneria media* (de Laubenfels) A. V. Bobrov & Melikian.

The shrubby, gnarled trees of scrub ru were originally identified as elfin ru (*Dacrydium beccarii*), which also occurs in the Malayan mountains but on wetter sites so that it also contrasts in its longer, outstretched needles. Because of its longer ordinary foliage leaves, elfin ru also has a more dramatic contrast between them and the shorter needles of the branchlets supporting the pollen and seed cones than is found in scrub ru. Outside of its stunted growth habit, scrub ru is a rather ordinary species within *Dacrydium* (hence the scientific name). Eastward, in Borneo and the Philippines, it is replaced by melur ru (*D. pectinatum*), a species with wider ecological and morphological variation than scrub ru but rather similar when growing on the drier kinds of sites characteristic for scrub ru. Melur ru is generally found at somewhat lower elevations, typically has somewhat more spreading, shorter leaves, a shorter tip to the pollen scales, and slightly smaller seeds. Further assessment of the relationship of scrub ru to these two species and others in the genus is warranted.

Dacrydium nausoriense de Laubenfels

NAUSORI RU, YAKA, TANGITANGI (FIJIAN)

Tree to 25 m tall, with trunk to 0.9(–1.2) m in diameter, cylindrical, without fluting or buttresses. Bark thin, weathering gray and flaking in thick scales to reveal brown patches. Crown open, cylindrical, with short, horizontal to upwardly angled branches turned up at the ends and bearing tufts of upright branchlets densely clothed with and nearly hidden by foliage. Transition between juvenile and adult foliage abrupt. Juvenile leaves to 9 mm long. Adult leaves between needle- and scalelike, sword-shaped, loosely overlapping, sticking out at a forward angle and always a little spreading from the twig, (0.5–)0.7–1(–1.4) mm long, 0.4–0.5 mm wide, a little wider than thick, triangular in cross section, the blunt tip curled forward and then inward. Midrib sharply raised beneath and flat or slightly raised between shallow stomatal grooves above. Pollen cones appearing about as thick as their stalks and accompanying leafy bracts, tiny, 2–2.5 mm long, 1.5–2 mm in diameter. Pollen scales broadly triangular, without a conspicuously prolonged tip. Seed cones on a short branchlet with unreduced leaves about 1 mm long. Free tips of the bracts making up the podocarpium resembling the foliage leaves but longer, progressively longer upward, to 2 mm long, tightly enclosing the epimatium. Mature podocarpium not greatly swollen, passing through red and purple to black, with one (or two) fertile bracts. Epimatium cuplike, 0.5–1 mm deep. Seed dark reddish brown, egg-shaped, 3–4 mm long, 2–3 mm thick, with a short, broad, dimpled beak to 0.5 mm long. Locally endemic in Fiji, restricted to the Nausori Highlands of western Viti Levu (hence the scientific name), where abundant, and possibly in the Sarava region of Vanua Levu. Scattered as a dominant in dense, seasonally dry forest; (180–)500–800 m. Zone 10. Synonym: *Corneria nausoriense* (de Laubenfels) A. V. Bobrov & Melikian.

Doubt has been expressed about the separateness of Nausori ru from its neighbor in Viti Levu, samiampi ru (*Dacrydium nidulum*), a more widespread species. Fijians make no distinction between the two in their common names, and it has been suggested that *D. nausoriense* is simply a drought-tolerant ecotype of *D. nidulum*. However, each taxon is fairly uniform within its respective portion of the island, and there is no intergradation across the (now denuded) Sigatoka River valley, which cleanly separates their areas of distribution. The larger pollen cones, gradual transition between juvenile and adult foliage, and larger, straight leaves and podocarpial bracts of samiampi ru show no overlap with the conditions found in Nausori ru. *Dacrydium nausoriense* is at least as closely related to melur ru (*D. pectinatum*) as to samiampi ru. Occurrence of the species in Vanua Levu was proposed when the species was first described in 1969, vigorously denied in Smith's flora of Fiji in 1979, and reasserted by M. F. Doyle in 1998. Relatively few specimens of *Dacrydium* have been collected on Vanua Levu, and additional collections would be desirable, focusing on highlands at the drier, western end of the island. Because of the seasonality of rainfall in the Nausori Highlands, Nausori ru has annual growth rings, albeit not very distinct ones. Nonetheless, it is possible to read these rings, as confirmed by radiocarbon dating, and large trees appear to reach an age of about 300–400 years, growing at a steady rate of about 2.5 mm increase in diameter each year, although mortality increases greatly beyond about 200 years. The resulting wood is valuable, and Nausori ru is cut extensively for its timber, used primarily in furniture making. Given the limited range of distribution of the species, an area of only a little more than 1,000 square kilometers, it is likely that all large trees will be cut, but regeneration seems adequate and the species is not in immediate danger of extinction as long as the land is not converted from forest to grassland.

Dacrydium nidulum de Laubenfels

SAMIAMPI RU, SAMIAMPI, KASUARI, TJIKWAL, JAMMARI, BINBAN, UIER (PAPUAN LANGUAGES), YAKA, TANGITANGI, LEWENINI (FIJIAN)

Tree to 30(–35) m tall, with trunk to 0.9(–2) m in diameter. Bark thin, smooth, brown, dotted with warty lenticels, weathering grayish brown and shedding thin scales or plates. Crown cylindrical, dense, with numerous short or upwardly angled branches bearing clustered tufts of flexible, upright branchlets somewhat openly clothed with and scarcely hidden by foliage. Transition between juvenile and adult foliage gradual. Juvenile leaves to 12(–20) mm long. Adult leaves needlelike, narrowly triangular, hardly overlapping, sticking out from the twig at a forward angle and continuing out straight to the blunt or minutely prickly tip or curved slightly inward or outward, (1–)2–5(–7) mm long, 0.3–0.7 mm wide, about twice as wide as thick, triangular in cross section. Midrib prominently and sharply raised beneath. Pollen cones appearing thinner than their stalks and accompanying leafy bracts, 8–12(–18) mm long, 1–2 mm in diameter. Pollen scales triangular, with a prolonged tip about 1 mm long. Seed cones on a short branchlet with reduced leaves 1.5–3 mm long. Free tips of the bracts making up the podocarpium resembling the foliage leaves, conspicuously longer than the leaves of the supporting branchlet and progressively longer upward, 2.5–4 mm long, enclosing the epimatium and nestling the seed (hence the scientific name, Latin for "little nest"). Mature podocarpium red, with one or two fertile bracts. Epimatium cuplike, 1–1.5 mm deep. Seed dark brown, shiny, egg-shaped, (2–)3–4 mm long, 1.5–2.5 mm thick, with a tiny, dimpled beak. Across eastern Malesia from Sumba (Lesser Sunda Islands), central Sulawesi, and Halmahera (Moluccas) to the eastern tip of New Guinea, and also in Fiji (on Vanua Levu, Viti Levu, Ovalau, and Kadavu), but not in the intervening Solomon Islands or Vanuatu. Scattered or gregarious in the canopy of primary and secondary lowland or lower montane rain forests on many substrates, from waterlogged to seasonally dry; 0–600(–1,200) m. Zone 10. Synonym: *Corneria nidula* (de Laubenfels) A. V. Bobrov & Melikian.

Samiampi ru occupies the eastern segment of a string of species that stretches across most of the range of the genus *Dacrydium.* The most widespread of these, besides samiampi ru, are mountain ru (*D. elatum*) in the west and melur ru (*D. pectinatum*) in between. Differences among these species are minor, and they might be better treated as geographically replacing varieties or subspecies of a single species. However, such a change would require a more detailed examination of a broader segment of the genus. Samiampi ru is rather scattered within its range but may be locally common, as it is in Fiji and in western New Guinea (Irian Jaya, Indonesia). In Fiji it is an important timber tree, found throughout the main islands except in the Nausori Highlands of Viti Levu, where it is replaced by the locally endemic Nausori ru (*D. nausoriense*). Both have attractive figure (that is partly due to fungal staining) and are highly prized for making furniture. Oil extracted from the wood contains diterpenes (compounds with 20 carbon atoms) that are found in other species of *Dacrydium* as well as in species belonging to other podocarp genera, emphasizing the lack of unusual characteristics in this rather typical species for its genus.

Dacrydium novoguineense L. Gibbs

MUNUMP RU, MUNUMP, KAOWIÉ, ARU (PAPUAN LANGUAGES)

Tree to 30 m tall but commonly much smaller, sometimes as short as 1.5 m, with trunk to 0.5 m in diameter. Bark thin, weathering gray and shedding irregular patches. Crown dome-shaped, dense, with upwardly angled branches bearing dense tufts of upright, cordlike branchlets densely clothed with and completely hidden by foliage. Transition between juvenile and adult foliage abrupt and early so that modest-sized trees may lack juvenile foliage entirely. Juvenile leaves to 10(–16) mm long. Adult leaves scalelike, the exposed portion diamond-shaped, densely overlapping, closely pressed against the twig, (0.8–)1.0–1.5(–2) mm long, 0.4–0.7(–1.0) mm wide, wider than thick, half-moon-shaped in cross section, the minutely prickly tip curved inward. Midrib sharply keeled near the base beneath. Pollen cones appearing just a little thinner than the supporting branchlet and accompanying leafy bracts, 5–10 mm long, about 1.5 mm in diameter. Pollen scales triangular, without a prolonged tip. Seed cones on a short branchlet with ordinary foliage leaves. Free tips of the bracts making up the podocarpium more needlelike than the foliage leaves, spreading, longer upward, the upper ones to 3 mm long, surrounding the epimatium and base of the seed. Mature podocarpium red, quite swollen and juicy, with one fertile bract. Epimatium cuplike, about 1.5 mm deep. Seed dark brown, egg-shaped, about 5 mm long, 3 mm thick, with a broad, low, dimpled beak. Eastern Malesia, from Sulawesi and the Moluccas, through Irian Jaya (all Indonesia) to the vicinity of Mount Wilhelm, Bismarck Range, Papua New Guinea. Scattered as an emergent above forests and scrublands of low stature, including mossy forests on ridges and open, regenerating sites in the wake of fires; (700–)1,500–2,200(–3,000) m. Zone 9. Synonym: *Corneria novoguineense* (L. Gibbs) A. V. Bobrov & Melikian.

Munump ru is very similar to mountain ru (*Dacrydium elatum*) in its tight, scalelike, adult foliage following an abrupt transition from the needlelike leaves of the juvenile phase. These species differ most obviously in the larger seed of munump ru surrounded by longer podocarpial bracts. Mountain ru, despite its common name, is also typically found at lower elevations than munump ru. Both are more tolerant of disturbance than most species of *Dacrydium* and are found frequently in secondary forests. Munump ru is much more abundant in New Guinea (from which it was first described, hence the scientific name) than in the more isolated stations in the western portion of its range.

Dacrydium pectinatum de Laubenfels

MELUR RU, MELUR, TJEMANTAN, SEMPILOR (MALAY), LU JUN SONG (CHINESE)

Tree of varying stature, to 30(–40) m tall, with trunk to 1.5(–3) m in diameter. Bark thin, reddish brown, weathering gray, mottled with reddish brown and yellowish brown by shedding of thin, irregular flakes and eventually shallowly furrowed. Crown cylindrical to dome-shaped, with upwardly angled branches, in whorls when young, much more irregular with age, bearing open clusters of gently drooping to upright branchlets somewhat openly clothed with and slightly hidden by foliage. Transition between juvenile and adult foliage gradual. Juvenile leaves 15–20 mm long. Adult leaves needlelike, narrowly triangular, slightly overlapping, gently S-curved, first outward and then forward at the rounded tip topped by a tiny, incurved prickle, (2–)3–5 mm long, 0.4–0.6 (–0.8) mm wide, about as thick, diamond-shaped in cross section. Midrib sharply keeled both above and beneath. Pollen cones appearing thinner than their stalks and accompanying leafy bracts (6–)8–12 mm long, 1.5–2 mm in diameter. Pollen scales broadly triangular, without a narrow, prolonged tip. Seed cones on a very short branchlet with reduced leaves to 2 mm long. Free tips of the bracts making up the podocarpium resembling the foliage leaves but shorter, a little longer than the leaves of the supporting branchlets, to 3 mm long, clasping the epimatium and the base of the seed. Mature podocarpium red, swollen and juicy, with one or two fertile bracts. Epimatium cuplike, about 1.5 mm deep. Seed shiny dark brown, egg-shaped to nearly spherical, 4–5 mm long, 3–4 mm thick, with a short, wide, dimpled beak. North-central Malesia, from southern Hainan (China) and the Sierra Madre of central Luzon (Aurora province) through the Philippines and Borneo (Malaysia, Brunei, and Indonesia) to Belitung, between Borneo and Sumatra. Scattered in and above the canopy of lowland and lower montane rain forests or gregarious or even forming pure stands of shorter to stunted trees on especially wet or dry sites over peat, sand, or serpentine-derived soils, in most cases on flatlands or gentle slopes; 0–1,600(–2,100) m. Zone 9. Synonyms: *Corneria pectinata* (de Laubenfels) A. V. Bobrov & Melikian, *Dacrydium pectinatum* var. *robustum* de Laubenfels.

Like mountain ru (*Dacrydium elatum*) to the west and samiampi ru (*D. nidulum*) to the east, melur ru is a large forest tree of modest elevations that can become ecologically more prominent as the trees themselves attain more modest dimensions on sites less favorable overall to exuberant forest growth. Where it is abundant it can be an important timber tree, and the attractive, strong wood is used in construction, furniture making, and boatbuilding as well as for veneers incorporated into plywood. The tree has been so overexploited in the mountains of Hainan that the formerly extensive forests there have almost been eliminated, and the species is officially considered vulnerable to extinction in China. It is also fast growing and benefits from full sunlight when young so that plantation culture has been suggested to relieve human pressure on the natural forests. The species name, Latin for "comblike," reflects the toothlike regularity of the spreading leaves, in contrast to the tightly pressed scale leaves of mountain ru, with which this species was originally confused.

Dacrydium spathoides de Laubenfels

IDENBURG RU

Tree to 34 m tall, with trunk to 0.5 m in diameter. Bark thin, dark gray, shedding in thin plates and exuding red resin. Transition between juvenile and adult foliage gradual. Juvenile leaves to 7 mm or more long. Adult leaves needlelike, sword-shaped, densely overlapping, sticking out from the twig at a forward angle and then straight or slightly curved to the minutely prickly tip, 2–4 mm long, 0.8–0.9 mm wide, four to five times as wide as thick, crescent-shaped in cross section. Midrib raised beneath and flanked by grooves containing lines of stomates above. Seed cones on a short branchlet with reduced leaves less than 2 mm long. Free tips of the bracts making up the podocarpium resembling the foliage leaves but shorter, longer than the leaves of the supporting branchlet, longer upward, 2–3 mm long, enclosing the epimatium and about half of the seed (hence the scientific name, Latin for "spathelike," referring to an enlarged bract enclosing a flower cluster). Mature podocarpium red, somewhat swollen, with one or two fertile bracts. Epimatium cuplike, about 1.5 mm deep. Seed brown, egg-shaped, 4–5 mm long by 2–2.5 mm thick, with a sharp beak. Known only from the highlands of New Guinea south of the Taritatu (Idenburg) River, eastern Irian Jaya (Indonesia). Scattered in the canopy of mossy forests; 2,100–2,200. Zone 10. Synonym: *Corneria spathoidea* (de Laubenfels) A. V. Bobrov & Melikian.

The very local Idenburg ru is known only from a handful of specimens collected in the vicinity of Barnhard Camp during the

Archbold expeditions to New Guinea. It differs from kerapui ru (*Dacrydium xanthandrum*) and heath ru (*D. ericoides*) primarily in its shorter foliage leaves and upright podocarpial bracts. Its relationship to these species would be worth investigating if it were ever rediscovered and new collections were made.

Dacrydium xanthandrum Pilger

KERAPUI RU, KERAPUI, SERINGOUN (MALAY)

Tree to 30(–36) m tall, reaching maturity at as little as 2 m when stunted on exposed ridges, with trunk to 0.7 m in diameter. Bark thin, reddish brown, weathering grayish brown, streaked by lines of warty lenticels, becoming mottled by the shedding of thick flakes and scales. Crown cylindrical to dome-shaped, with upwardly angled branches bearing tufts of flexible, upright branchlets somewhat openly clothed with and not hidden by foliage. Twigs whiplike, continuing unbranched for several growth increments or gathered in clumps through profuse branching. Transition between juvenile and adult foliage gradual and due primarily to progressive, modest shortening. Juvenile leaves to 20 mm long. Adult leaves needlelike, sword-shaped, hardly overlapping, sticking almost straight out from the twig at a slightly forward angle, 6–10 mm long, (0.5–)0.6–0.8 mm wide, three to four times as wide as thick, flat to crescent-shaped in cross section, with a prickly tip continuing straight or turned a little outward. Midrib deeply raised beneath and flanked by shallow grooves containing lines of stomates above. Pollen cones appearing a little thinner than their stalks and accompanying short, leafy bracts, 5–13 mm long, 2–3 mm in diameter. Pollen scales triangular with a narrowed, prolonged tip 0.5–1.5 mm long. Seed cones on a very short branchlet with reduced leaves about 4 mm long. Free tips of the bracts making up the podocarpium resembling the foliage leaves but shorter, even shorter than the leaves of the supporting branchlet, 2–4 mm long, spreading outward beneath the epimatium. Mature podocarpium red, slightly swollen and juicy, with one or two fertile bracts. Epimatium asymmetrically cup-shaped, 1.5–2 mm deep. Seed shiny dark brown, egg-shaped, 3–4.5 mm long by 2–2.5 mm thick, with a low, broad, dimpled beak. Scattered across the northern and central tiers of Malesia, from northern Sumatra (Indonesia) and southern Malaya (Malaysia) east to Bougainville (Solomon Islands) and north to central Luzon (Philippines). Most commonly a dominant or codominant tree or shrub in the canopy of low, mossy forests on wet ridge tops with accumulating peat over varied substrates, but also scattered in the canopy of adjacent primary montane rain forests; (500–)1,000–2,700(–3,000) m. Zone 9. Synonym: *Corneria xanthandra* (Pilger) A. V. Bobrov & Melikian.

Kerapui ru and elfin ru (*Dacrydium beccarii*) are the most widely distributed of the long-needled *Dacrydium* species and have very similar geographic ranges and habitat preferences, though kerapui ru reaches higher elevations, and elfin ru more commonly grows in montane scrublands. The length of their needles is also basically the same, but those of elfin ru are only about half as wide as those of kerapui ru, and they are more densely arranged, projecting forward to much more closely clothe the twigs rather than standing out openly. Thus the twigs of elfin ru are reminiscent of a cat's tail while those of kerapui ru more closely resemble a bottle brush. The codistribution of such similar species is rather curious and merits a closer scrutiny of their ecological requirements and preferences. The scientific name, Greek for "yellow man," refers to the color of the conspicuous pollen cones, hardly unusual in *Dacrydium* or related podocarp genera.

Diselma J. Hooker

DISELMA

Cupressaceae

Evergreen shrubs or small trees with fibrous bark peeling in strips on multiple main branches from the base. Branchlets alternate, radiating in three dimensions, four-angled. Scaly winter buds absent. Leaves in alternating, overlapping pairs, scalelike, keeled, completely clothing the branchlets, rarely with a resin gland on the outer face. Free tips of leaves shorter than to about as long as the attached leaf bases, tightly pressed against the twig.

Plants dioecious. Pollen cones single at the ends of branchlets, with three or four alternating pairs of pollen scales, each scale with two pollen sacs. Pollen grains small (25–30 μm in diameter), spherical, minutely bumpy, sometimes also with a short germinal papilla. Seed cones numerous, single at the tips of ordinary branchlets, maturing in a single season, roughly spherical with two alternating pairs of seed scales and a central column. Cone scales formed by intimate fusion of the bract and seed scale, woody but thin, radiating from the base and touching but not overlapping when closed. The upper pair of scales each two-seeded, larger than the lower, sterile pair (hence the scientific name, Greek for "two seats"). Seeds oval with a prominent pollination tube at the tip, two equal wings extending the length of and as wide as the body, and a third smaller or rudimentary wing. Wings formed as outgrowths of the seed coat. Cotyledons two, each with one vein. Chromosome base number $x = 11$.

Wood light, moderately hard, with light-colored sapwood sharply contrasting with purplish brown heartwood. Grain very fine and even, with distinct, very narrow growth rings marked by a fairly abrupt transition to a few rows of latewood tracheids with hardly any central lumen (opening). Resin canals absent but with numerous individual resin parenchyma cells scattered through the growth increment or concentrated into narrow, open bands.

Stomates forming two broad, triangular bands on the hidden, inner (upper) face of the leaves and patches at the base of the outer (lower) face where they are covered by the tips of the previous leaf pair on the twig. Each stomate sunken in a narrow pit between and virtually hidden by the four to six surrounding subsidiary cells, which are topped by a steep, complete Florin ring. Leaf cross section with a single-stranded midvein directly above a single large resin canal (that may not extend far into the free leaf tip) and flanked by small wedges of transfusion tissue. Photosynthetic tissue forming a thin palisade layer beneath the epidermis and accompanying thin hypodermis only in the free tip, the rest made up of spongy mesophyll.

One species in Tasmania. References: Clifford and Constantine 1980, W. M. Curtis 1956, de Laubenfels 1965, Farjon 2005b.

As long believed, *Diselma* has been shown using DNA sequences to be closely related to South American *Fitzroya* but is closest to African *Widdringtonia.* It is somewhat like a miniaturized version of *Widdringtonia,* in much the same way as *Microbiota* resembles a smaller version of *Platycladus. Diselma archeri* is too small to yield lumber and has no other economic uses, despite its ecological importance in subalpine Tasmania, it is not in general cultivation and there has been no cultivar selection. Unlike many southern hemisphere members of the Cupressaceae, there is no known fossil record of *Diselma,* so the taxonomic and geographic history of its divergence from a common ancestor with *Fitzroya* and *Widdringtonia* is entirely speculative.

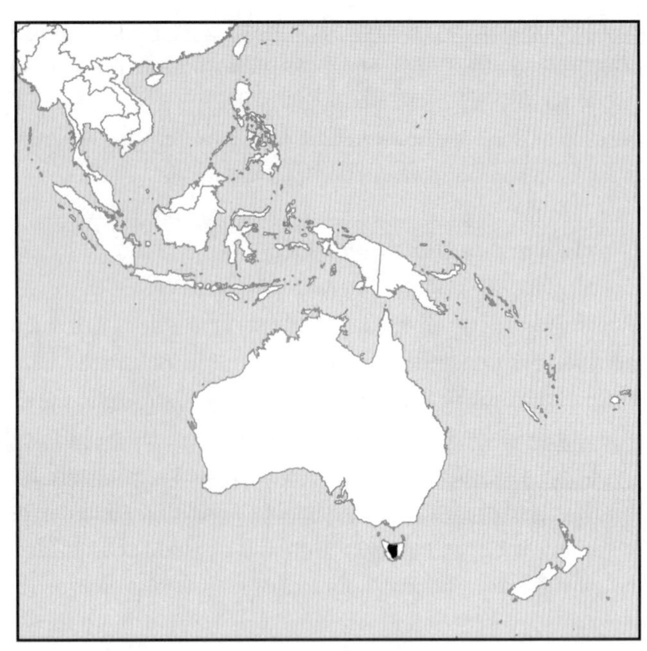

Distribution of *Diselma archeri* in Tasmania, scale 1 : 119 million

Diselma archeri J. Hooker

CHESHUNT PINE, CHESTNUT PINE

Erect shrub to 2.5 m, or sometimes a tree to 6 m tall, with trunk to 30 cm in diameter. Bark reddish brown when fresh, darkening and then weathering grayish brown, rough with persistent scale leaf bases and then accumulating flakes and scales. Crown irregular, broadly dome-shaped in overall outline but made up of individual cones, each with numerous rising branches bearing dense, rounded tufts of branchlets at their tips. Branchlets short, turning brown by the second or third year (or later), completely hidden by the leaves. Adult leaves blunt, 1–2 mm long, dark green. Pollen cones 2–3 mm long. Seed cones 4–5 mm across, with a club-shaped central column 4–5 mm long, about as long as the scales and seeds. Seeds longer than the scales, about 5 mm

Weather-beaten old shrub of Cheshunt pine (*Diselma archeri*) near timberline in nature.

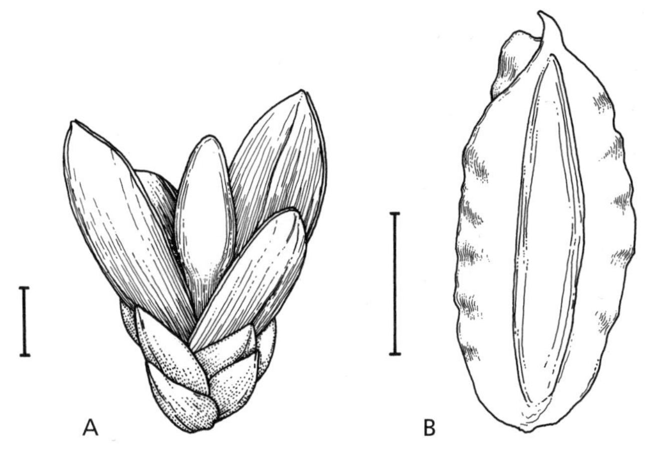

Open mature seed cone after seed dispersal (A) and face view of dispersed seed (B) of Cheshunt pine (*Diselma archeri*); scale, 1 mm.

long. Central and western Tasmania (Australia). Mostly alpine and subalpine shrublands but also extending down into high-elevation temperate rain forest; (800–)1,000–1,400 m. Zone 8?

Cheshunt pine often grows with other alpine conifers in Tasmania, Tasmanian dwarf pine (*Microstrobos niphophilus*), strawberry pine (*Microcachrys tetragona*), and mountain plum pine (*Podocarpus lawrencei*), and is rather similar to the first two in general appearance, although they are members of the Podocarpaceae rather than the Cupressaceae. Though small in stature, the oldest known individual is about 550 years old. It usually does not regenerate well after rare fires in these cold heathlands. In forests, it grows in the understory of *Athrotaxis* and *Nothofagus* species. In cultivation, it is most successful in cool, wet climates. Cheshunt pine has entered cultivation under the garbled name chestnut pine.

Falcatifolium de Laubenfels

SICKLE PINE

Podocarpaceae

Evergreen trees and shrubs, of varying habit from prostrate to upright, occasionally reaching the canopy, often with multiple stems. Bark fibrous, smooth, flaking in small patches. Crown conical to narrowly to very broadly and shallowly dome-shaped. Vegetative branchlets all elongate, without distinction into long and short shoots, hairless, remaining green for a least the first year, grooved between the elongate, attached leaf bases. Resting buds inconspicuous and unspecialized, consisting of loosely arranged scale leaves that persist at the base of new growth. Leaves spirally attached to the twigs, of three main forms with some transitional leaves. Seedling leaves radiating all around the twigs, flattened top to bottom, narrowly sword-shaped, with stomatal bands on either

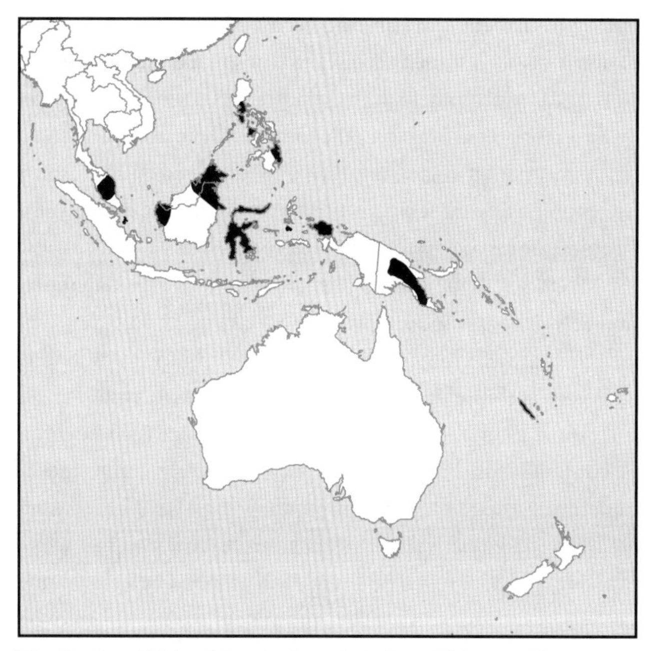

Distribution of *Falcatifolium* in Australasia from Malaya and Luzon to New Caledonia, scale 1 : 119 million

side of the prominent midrib beneath. Scale leaves of adults radiating all around the twigs, confined primarily to reproductive shoots and to the bases of growth increments. Juvenile and adult needlelike foliage leaves flattened into two rows by the flexure of the blade near its base. Each leaf flattened side to side (and thus with a left and right side rather than a top and bottom side), somewhat sickle-shaped (hence the scientific name, Latin for "sickle leaf"), the base curved down toward the base of the twig and the tip curved forward toward its tip.

Plants dioecious. Pollen cones cylindrical, single or in clusters of two to four at the ends of short, scaly stalks in leaf axils or (less frequently) at the ends of shoots. Without a conspicuous circle of distinctive bracts at the base and with numerous, densely spirally arranged pollen scales, each bearing two pollen sacs. Pollen grains medium (body 30–35 μm long, 60–70 μm overall), with two round air bladders that extend along the sides as well as out from the ends of the body so that, from above, the body appears to be surrounded by a single, large, elliptical sac. Surface coarsely sculptured with squiggly lines except over the top of the body, which is finely dotted. Seed cones single at the ends of short scaly stalks in the axils of foliage leaves or (less frequently) at the ends of shoots, these stalks curling back at maturity, maturing and falling in a single season. Cones highly modified and reduced, the reproductive part with 8–12 bracts that unite with the cone axis at maturity to form a red, swollen, juicy podocarpium bristling with the conspicuous free tips of the bracts. Seeds single, egg-shaped but slightly flattened and unwinged, embedded at the base in each of one (or two) humped seed scales (the epimatium). Hump becoming more prominent during maturation as the opening of the ovule is reoriented away from the base and the mature seed finally points away at an angle from the cone axis, Seeds with a pair of ridges uniting over the tip. Cotyledons two, each with two veins. Chromosome base number $x = 10$.

Wood soft, light, fragrant, light brown, with little differentiation between sapwood and heartwood. Grain fine and even, with indistinct growth rings. Resin canals absent but with individual resin cells and abundant nonresinous wood parenchyma.

Stomates in numerous lines of varying lengths and degrees of continuity on both surfaces. Each stomate partially tucked underneath the four to six surrounding subsidiary cells, which are often shared between adjacent stomates in a row, and which may or may not be (even on a single leaf) topped by a Florin ring. Leaf cross section with a single, prominent or obscure midvein narrowly raised as a midrib on both sides, accompanied beneath (seemingly the outer side of the midvein) by a singe large resin canal, flanked by wedges of transfusion tissue, and with scattered accessory transfusion tissue extending throughout the center of the leaf. Photosynthetic tissue with a palisade layer covering both sides of the leaf inside the epidermis and its associated incomplete hypodermis, leaving little room for spongy mesophyll.

Five species in northern Malesia, from central Malaya (Malaysia) and southern Luzon (Philippines) to eastern New Guinea (Papua New Guinea) and in New Caledonia. References: de Ferré et al. 1977, de Laubenfels 1969, 1972, 1978c, 1988, Hill and Scriven 1999, Lee 1952, Stockey and Ko 1988, Stockey et al. 1992.

Although it was the first of the several new genera whose separation from *Dacrydium* since 1969 has generally been accepted by most taxonomists, *Falcatifolium* is actually much more closely related to the species of *Dacrydium* in the restricted sense now generally adopted than are the genera segregated more recently. While they were separated from *Dacrydium* more than 10 years after the separation of *Falcatifolium*, DNA studies show that *Halocarpus, Lagarostrobos, Lepidothamnus,* and *Manoao* are only distantly related to *Dacrydium* while *Falcatifolium* is its closest relative. *Falcatifolium* was long recognized as a distinctive group within *Dacrydium* before its generic segregations, even though it was never given formal botanical status as a section or subgenus. It is distinguished from *Dacrydium* by the position of the cones, the form of the epimatium, and the distinctive form of the leaves.

The more or less sickle-shaped leaves flattened side to side rather than top to bottom are highly unusual in conifers, and the only other extant genera with full-blown development of this leaf form are also podocarps, *Acmopyle* and juvenile foliage of *Dacrycarpus* (some species of *Picea* are somewhat diamond-shaped in cross section but have little or no vertical blade development). Because of the particular evolutionary relationships of these three genera to each other, to *Dacrydium,* and to *Podocarpus* and its closest relatives, it is not clear whether this leaf form was inherited from a common ancestor or evolved independently more than once.

The living species of *Falcatifolium* are very closely related and homogeneous, differing primarily in the size, proportion, and detailed shape of the foliage leaves. The genus is not in general cultivation, and there has been no cultivar selection. The fossil record is very sparse, but there are confirmed occurrences in the mid-Eocene of southeastern Australia, some 45 million years ago, so the separation from the other genera must be older than that. The same sediments, and slightly more recent Oligocene sediments in Tasmania, contain shoots of an extinct genus, *Sigmaphyllum,* with leaves of the same form, but a distinct epidermal structure from that of *Falcatifolium* species.

Identification Guide to *Falcatifolium*

Please note that the pollen or seed cones, or both, are only necessary for distinguishing the first two species (which have completely distinct natural geographic ranges), but the cone measurements, if one has them, may help to confirm identifications of the other three species. The five species of *Falcatifolium* may be distinguished by the length and width of most adult leaves, the length of most mature pollen cones, and the length of the mature podocarpium:

Falcatifolium papuanum, leaves up to 2 cm by 2.5–4 mm, pollen cones less than 1.5 cm, podocarpium more than 5.5 mm

F. taxoides, leaves up to 2 cm by 2.5–4 mm, pollen cones 1.5–2.5 cm, podocarpium 3.5–5.5 mm

F. angustum, leaves 2–4 cm by up to 2.5 mm, pollen cones less than 1.5 cm, podocarpium unknown

F. gruezoi, leaves 2–4 cm by 4–6 mm, pollen cones 2.5 cm or more, podocarpium less than 3.5 mm

F. falciforme, leaves at least 4 cm by at least 6 mm, pollen cones 2.5 cm or more, podocarpium 3.5–5.5 mm

Falcatifolium angustum de Laubenfels

NARROW-LEAF SICKLE PINE

Tree to 20 m tall, with trunk to 0.3 m in diameter. Bark purplish brown, weathering gray and flaking in irregular scales. Crown narrow, the drooping branchlets densely clothed with foliage. Leaves crowded on the twigs and radiating around them rather than flattened in a plane as they are in the other species of the genus, 1.5–3.5

cm long (to 7 cm in juveniles), 1–2.5 mm wide (hence the scientific name, Latin for "narrow"). Individual needles nearly parallel-sided and straight for most of their length, tapering abruptly to the prickly tip and more gradually to the wedge-shaped base that runs rather smoothly onto the twig. Midrib prominently raised on both the right and left sides. Pollen cones about 1 cm long and 2 mm across. Seed cones not seen or previously described. Coast of western Borneo in Sarawak (Malaysia). Scattered among hardwoods in low-growing keranga forests of dry coastal flats; 90–240 m. Zone 10?

This extremely rare species is poorly known, with only a handful of specimens collected from just two widely separated localities very near the coast of Sarawak. There are national parks in both areas, so the species may not be in immediate danger of extinction despite its limited range. Nonetheless, the conservation and taxonomic statuses of narrow-leaf sickle pine warrant further study. It has the narrowest adult leaves in the genus and, with their orientation all around the twigs, it approaches long-leaved species of the related genus *Dacrydium*, such as cattail rimu (*D. guillauminii*), in general appearance, if not in detail.

Falcatifolium falciforme (Parlatore) de Laubenfels

MALAYSIAN SICKLE PINE, MALAYAN YEW, KAYU CHINA, IGUH GAWAH (MALAY)

Shrub, or tree to 15 (–36) m tall, with trunk to 0.5 m in diameter. Bark smooth, bright purplish brown, weathering purplish gray and flaking to reveal dark red inner bark. Crown conical to cylindrical, with horizontal branches bearing alternate branchlets densely clothed with foliage. Leaves just overlapping along the twigs, in two loose rows, leathery, (2–)4–7 cm long (to 12 cm in juveniles), 5–9 mm wide (to 12 mm in juveniles). Individual needles widest before or near the middle and smoothly and gently S-curved throughout their length (hence the scientific name, Latin for "sickle-shaped"), tapering very gradually to the forwardly curved triangular tip and more abruptly to the asymmetrical roundly wedge-shaped base with just a hint of a petiole. Midrib prominently raised on both the right and left sides. Pollen cones 2–4 cm long and 2.5–3.5 mm across. Seed cones with a podocarpium 4–5 mm long, the seed 6–7 mm long, black at maturity. Southern Malay Peninsula to Borneo (mostly in the northwestern half), with a few small islands in between. Scattered in the understory of lowland and montane rain forests, mostly on poor substrates including serpentine-derived soils, or reaching the canopy of short forests on exposed ridges and, more rarely, of tall forests on deep, rich soils; (400–)800–2,100 m. Zone 10. Synonym: *Dacrydium falciforme* (Parlatore) Pilger.

Malaysian sickle pine is the westernmost species of the genus. At one time all the sickle pines throughout Malesia were included within this species, but in 1969, trees and shrubs growing east of Borneo were segregated as two additional species. These species are all closely related, and their relationships merit further study. Malaysian sickle pine is moderately common in the understory of Malayan and Bornean montane forests (for example, in the lower montane forest of Mount Kinabalu, Sabah), often accompanying its larger and more widespread relatives mountain ru (*Dacrydium elatum*) and common dacrycarp (*Dacrycarpus imbricatus*). However, it is much less well known than either of these two important timber trees with respect to its ecological preferences, reproductive biology, regeneration, and other aspects of its biology. Presumably associated with its ecological tolerances, it has the largest leaves among the *Falcatifolium* species, especially when it grows as a small tree of the shady, moist understory. The large leaves, for a species related to *Dacrydium* and *Dacrycarpus*, is probably why it was originally described (in 1868) as a species of the more distantly related *Podocarpus* and then assigned to *Nageia* before being transferred to *Dacrydium* in 1903, where taxonomists were content to keep if for more than 60 years until it was made the type of the newly recognized genus *Falcatifolium* in 1969. The young leaves, like those of other species in the genus, are bright red when they first expand and remain so for a few months. The red color is produced by two anthocyanins (the common red and blue pigments of flowers, including morning glories, garden geraniums, delphiniums, and peonies), which also color the podocarpium of *Falcatifolium* species and other podocarps. One of the anthocyanins, podocarpin A, is named after the family and apparently restricted to it (or nearly so) while the other, cyandidin-3-glycoside, is also found in many flowering plants as well as in red tissues of other conifers. The wood chemistry is also typical for the family, and the oil that has been extracted from the wood contains some diterpenes (compounds with twenty carbon atoms arranged in several ring structures) that are widespread in conifer wood oils and others that are special to the family, like podocarpic acid. The individual compounds found in Malaysian sickle pine and in New Caledonian sickle pine (*F. taxoides*) are shared with other species in the closely related genera *Dacrydium* and *Dacrycarpus* but also with some podocarps that are only distantly related, like *Parasitaxus* and *Prumnopitys*.

Falcatifolium gruezoi de Laubenfels

PHILIPPINE SICKLE PINE

Low shrub, to tree to 12(–24) m tall, with trunk to 0.2 m in diameter. Bark thin, smooth, reddish brown to purplish brown, weathering gray and flaking to reveal patches of bright color. Crown conical, with slender, spreading branches bearing numerous branchlets densely clothed with foliage. Leaves just overlapping along the twigs, in two loose rows, leathery and with a thin dusting of wax, (0.6–)1.3–3.5 cm long (to 7.5 cm in juveniles), (2.5–)4–7 mm wide. Individual needles widest before or near the middle and smoothly and gently S-curved throughout their length, tapering very gradually to the forwardly curved, sharply pointed tip and more abruptly to the asymmetrically roundly wedge-shaped base on a very short but distinct petiole. Midrib prominently raised on both the right and left sides. Pollen cones (1.7–)2.5–6 cm long and (1.5–)2–3 mm across. Seed cones with a podocarpium 2–3.5 mm long, the seed 6–7 mm long, black at maturity. Philippines (from southern Luzon southward) to central Sulawesi and the Moluccas (Obi) with an outlier in Sarawak. Scattered among broad-leaved shrubs in the canopy of montane heathlands on exposed ridges and at the edge of clearings; (700–) 1,200–2,200 m. Zone 9? Synonyms: *Falcatifolium falciforme* var. *usan-apuensis* de Laubenfels & Silba ex Farjon, *F. usan-apuensis* (de Laubenfels & Silba ex Farjon) de Laubenfels & Silba.

Philippine sickle pine is found in scattered localities on widely separated mountains. It is closely related to Malaysian sickle pine (*Falcatifolium falciforme*), which usually grows in forests rather than shrublands, but differs from that species most obviously in its smaller, waxy leaves and podocarpium. One specimen from Sarawak, within the range of Malaysian sickle pine, has been treated as a small-leaved variant of that species or as a separate species by some authors but is here included in Philippine sickle pine because of its greater resemblance to this species. The taxonomic, ecological, and geographic limits between Philippine and Malaysian sickle pines require further study in the field. The species name honors William Gruezo (b. 1951), a mycologist who collected the type specimen in one of the best-known localities for the species, Mount Halcon on Mindoro in the Philippines, where botanical specimens of it had first been collected in 1895.

Falcatifolium papuanum de Laubenfels

NEW GUINEA SICKLE PINE, MUNGAG, TUGL (PAPUAN LANGUAGES)

Tree to 22 m tall, with trunk to 0.4 m in diameter, rarely dwarfed and shrubby in exposed locations. Bark reddish brown, weathering gray to dark brown and flaking in distinct scales. Crown broadly dome-shaped or flattened, with slender, rising branches bearing numerous branchlets densely clothed with foliage. Leaves slightly overlapping to densely, crowded along the twigs, in two loose rows, sometimes thinly waxy on the side facing down, (0.6–)1–2 cm long, (1.8–)2–4 mm wide. Individual needles usually inconspicuously widest before (or near) the middle, curved away from the twig at the base but straight from there and usually not curving forward at the tip, tapering very gradually and then more abruptly to the roundly triangular tip with a distinct prickle and more abruptly to the asymmetrically roundly wedge-shaped base on a very short but well-defined petiole. Midrib more prominently raised on the side facing up. Pollen cones 0.5–1.5 cm long and 2–2.5 mm across. Seed cones with a podocarpium 6–7 mm long, the seed 6–7 mm long, black at maturity. Endemic to New Guinea (hence the scientific name), mostly along the spine of Papua New Guinea but with an outlier in the Vogelkop (Doberai Peninsula), Irian Jaya. Scattered among hardwoods and other conifers in the understory of cool, wet montane forests and rarely in shrubby heathlands on wet, exposed ridges; (1,500–)1,800–2,400 m. Zone 9? Synonym: *Falcatifolium sleumeri* de Laubenfels & Silba.

Like other species of *Falcatifolium*, New Guinea sickle pine is typically an understory tree, often common but overlooked among a myriad of larger, more conspicuous or economically more important trees. It is most similar to the geographically more widely separated New Caledonian sickle pine (*F. taxoides*) rather than to its nearer neighbor, Philippine sickle pine (*F. gruezoi*). It is difficult to separate from New Caledonian sickle pine because they both have straight leaves of similar size, but New Guinea sickle pine has prickly leaves and a larger podocarpium in the seed cones. The single dwarf specimen from the Doberai Peninsula of western Irian Jaya remains enigmatic in its wide geographic separation from the rest of the known range of the species and in being at the lower end of the range of the species in its overall stature and in the size of its leaves. Finding other such individuals in similar exposed and stressful localities elsewhere would help to clarify its status.

Falcatifolium taxoides (Ad. Brongniart & Gris) de Laubenfels

NEW CALEDONIAN SICKLE PINE

Shrub, or tree to 15 m tall, with trunk to 0.2 m in diameter. Bark thin, smooth, light reddish brown, weathering light gray and flaking sparingly. Crown conical to dome-shaped, open, with slender horizontal to gently rising branches bearing alternating branchlets densely clothed with foliage. Leaves not touching or slightly overlapping to crowded along the twigs, strictly in two rows or more loosely so or even radiating all around the twigs, shiny bright green on the side facing up and white with wax on the side facing down, (0.3–)1–2(–3) cm long, (2–)2.5–4(–6) mm wide. Individual needles fairly parallel-sided or inconspicuously widest near or before the middle, curved away from the twig at the base but straight from there and usually not curving forward at the tip, tapering very gradually and then more abruptly to the rounded, blunt tip and more abruptly to the asymmetrically wedge-shaped base on a short but distinct petiole. Midrib scarcely noticeable.

Pollen cones 1.5–2.5 cm long and 1.5–2 mm across. Seed cones with a podocarpium 4–5 mm long, the seed 6–7 mm long, black at maturity. Endemic to and found along the whole length of New Caledonia except the northernmost tip. Scattered among hardwoods and other conifers in the understory of montane rain forests on soils derived both from serpentine and from granitic substrates; (100–)800–1,200(–1,400) m. Zone 10. Synonym: *Dacrydium taxoides* Ad. Brongniart & Gris.

New Caledonian sickle pine is common but not abundant throughout the forests of the main island that are wet enough to experience relatively infrequent fires, but not generally in the dampest mossy forests. In many localities throughout the island, it grows with or near New Caledonian acmopyle (*Acmopyle pancheri*). This species has similar leaves similarly flattened side to side and in two rows, but the leaves differ because the lines of stomates are only on the side that faces down, and they are confined to short shoots, with scale leaves on the long shoots. The superficially yewlike appearance of the foliage is responsible for the scientific name, but leaves of yew (*Taxus*), like those of most of the many conifers with yewlike foliage, are flattened top to bottom rather than side to side. Perhaps the most notable feature of New Caledonian sickle pine is that it is the sole host of coral pine (*Parasitaxus ustus*), the only known parasitic conifer, whose lurid purple stems can be found sprouting up from the ground around trunks of its host. What affect this parasitism may have on the growth and reproduction of infected trees is unknown. Chemical constituents known from New Caledonian sickle pine do not explain the specificity of the parasitic relationship with coral pine because they are fairly typical for podocarps, and the wood chemistry, especially, is quite similar to that of Malaysian sickle pine (*F. falciforme*). There are no known occurrences of *Falcatifolium* on any of the islands in between New Caledonia and the western end of New Guinea, where its closest relative, New Guinea sickle pine (*F. papuanum*), occurs. New Caledonian sickle pine is thus the most geographically isolated species in the genus.

Mature seed cone of New Caledonian sickle pine (*Falcatifolium taxoides*); scale, 5 mm.

Fitzroya J. Hooker

ALERCE, PATAGONIAN CEDAR

Cupressaceae

Evergreen trees with a single massive trunk clothed with fibrous, deeply furrowed bark peeling in vertical strips. Crown conical to dome-shaped, sparsely branched with slender horizontal or slightly drooping branches, some of which thicken as major limbs. Branchlets alternate, radiating in three dimensions. Winter buds of unexpanded leaves. Leaves in alternating trios, scalelike to clawlike, standing out from the twigs with a long free tip and without glands.

Plants usually dioecious. Pollen cones single in the leaf axils near the ends of the branchlets, with five to eight alternating trios of pollen scales. Each scale with two to six pollen sacs on the inner face of a triangular blade at the tip of a short, slender stalk. Pollen grains small (30–40 µm in diameter), spherical, minutely bumpy. Seed cones well separated, single at the tips of ordinary branchlets, maturing in a single season, roughly spherical with three or four alternating trios of seed scales, the inner trio forming a sterile central column, the fertile middle trio(s) and the sterile outer trio radiating from the base of the cone. Fertile scales woody, touching but not overlapping when closed, consisting of closely united bract and seed scales, with a prominent bract point below the tip. Seeds oval, one or two per fertile scale, with two or three equal wings (or one smaller or rudimentary) extending the length of and as wide as the body. Seed wings formed as outgrowths from the seed coat. Cotyledons two, each with one vein. Chromosome base number $x = 22$.

Wood unscented, light, soft, and highly decay resistant. Sapwood narrow, yellowish white, sharply contrasting with the pale to dark reddish brown to reddish orange heartwood often contrastingly streaked with lighter and darker hues, but losing its brightness with age. Grain fairly even and moderately fine, with evident growth rings marked by a somewhat abrupt transition to a narrow band of much smaller and thicker-walled latewood tracheids. Resin canals absent but with scattered individual resin parenchyma cells more frequent in the outer part of the growth increment.

Stomates in two prominent white stomatal bands both above and beneath. Each stomate sunken beneath and almost hidden by the four to six subsidiary cells that are covered by a very thick cuticle that does not rise further to form a Florin ring. Leaf cross section with a single-stranded midvein above a single large resin canal and flanked by wedges of transfusion tissue. Photosynthetic tissue with a weakly developed palisade layer beneath the epidermis and adjoining thin, nearly continuous (except under the stomatal bands) hypodermis.

One species in the southern Andes and coast ranges of South America. References: de Laubenfels 1965, Farjon 2005b, Hill and Whang 1996, Lara and Villalba 1993, Rodríguez R. et al. 1983.

The name of the genus honors British Vice-Admiral Robert Fitz Roy (1805–1865), who was captain of the *Beagle* during its second voyage around the world (1831–1836), with a chief goal of to mapping the coasts and waters of southern America, where the tree was an important and conspicuous forest component. This expedition, of course, is much better known for the inspiration it gave to the ship's naturalist, Charles Darwin, then in his 20s. Although the genus is in cultivation to a limited extent in botanical and estate gardens in moderate temperate regions of high precipitation, there has been no cultivar selection.

Fitzroya differs from Australian *Callitris* and *Actinostrobus*, the other southern hemisphere Cupressaceae with leaves in whorls of three, in its expanded leaf tips with prominent stomatal bands, and its rounded seed cone scales. DNA studies agree with the long-held speculation based on morphology that it is closest to the Tasmanian *Diselma* (and to African *Widdringtonia*), despite the fact that these genera have a paired leaf arrangement. The only known fossil *Fitzroya* has been described from the Oligocene of Tasmania, where it grew with other plants now also found growing with *Fitzroya* in South America.

Distribution of *Fitzroya cupressoides* in Chile and adjacent Argentina, scale 1 : 35 million

Fitzroya cupressoides (G. Molina) I. Johnston

CHILEAN ALERCE, PATAGONIAN CEDAR, FITZROYA, ALERCE, LAHUÉN (SPANISH)

Plate 27

Tree to 50(–70) m tall, with trunk to 3.5(–5) m in diameter, clear of branches for up to 25 m. Bark brownish red, to 5 cm thick. Crown narrowly cylindrical to conical, very sparse in extreme age. Leaves 2.5–4 mm long, dark green outside the stomatal bands. Pollen cones 7–8 mm long, straw colored. Seed cones 6–8 mm long. Seeds 2.5–4 mm long. Southern Andes and Coast Ranges of Chile and adjacent Argentina. Forming solid stands on moist to waterlogged soils or mixed with other temperate conifers and southern beech (*Nothofagus*); 100–1,500 m. Zone 8.

Chilean alerce is the largest tree in temperate South America and the second oldest known in the world after the bristle cones pines. One individual examined in a study of climate change in the Andes exceeded 3,600 years. The trees have a high moisture requirement, growing in a region with much rainfall and sodden soils, but they are less fussy in cultivation as long as there is no prolonged freezing and dry periods are not too hot. The wood is excellent and has been exploited commercially since 1599. The trees, once abundant in their very limited range, have been officially protected since 1976 but are still threatened by poaching

Natural stand of all ages of Chilean alerce (*Fitzroya cupressoides*).

and deliberate burning. There is also unexplained natural decline in many stands, with numerous persistent snags and reduced regeneration. This decline is reminiscent of that experienced by the ecologically somewhat equivalent Alaska yellow cedar (*Cupressus nootkatensis*) in the North Pacific and does not seem to be due to recent climate change. Populations from Argentina are genetically differentiated in DNA and enzymes from the more numerous ones in Chile, pointing to separate refuges east and west of the Andes during the most recent glaciation. There is also a lesser north-to-south gradient within Chile. Leaf oils are also variable within the species, but not with any geographic pattern. In contrast to the substantial natural genetic variation in Chilean alerce that is fairly typical for conifers, trees in cultivation appear to have a very narrow genetic base and little variation, pointing to a history of vegetative propagation.

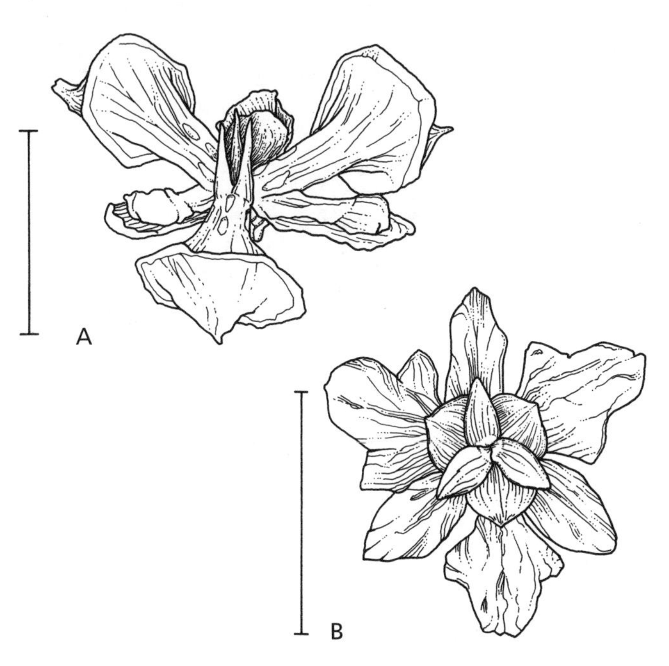

Side (A) and bottom (B) views of open mature seed cone of Chilean alerce (*Fitzroya cupressoides*) with three-pronged columella after seed dispersal; scale, 5 mm.

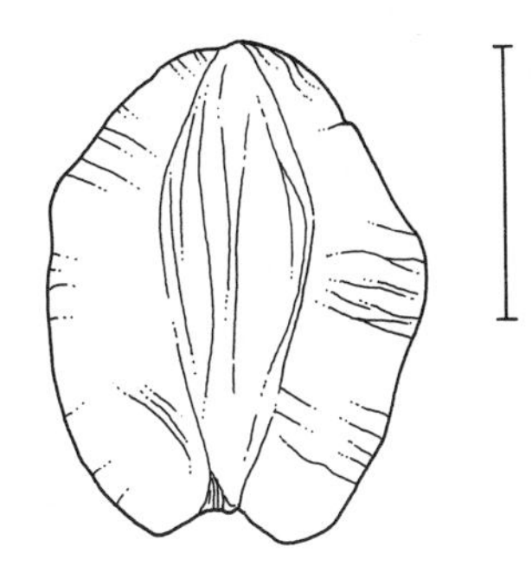

Face view of dispersed seed of Chilean alerce (*Fitzroya cupressoides*); scale, 2 mm.

Fokienia A. Henry & H. H. Thomas

FUJIAN CEDAR

Cupressaceae

Evergreen trees with a straight single main trunk clothed with smooth to furrowed, fibrous bark peeling in vertical patches or strips. Crown cylindrical to dome-shaped, densely branched with horizontal or gently drooping branches. Branchlets in flattened fernlike sprays with alternate branching. Scaly winter buds lacking. Seedling leaves in alternating quartets, needlelike, standing out from and well spaced on the stem. Seedling phase short-lived with even the first lateral branches bearing juvenile rather than seedling leaves. Juvenile and adult leaves in alternating pairs but appearing nearly whorled, scalelike, the facial and lateral pairs dissimilar, clothing the branchlets, without resin glands. Facial leaves wedge-shaped, the broadly pointed tip overlapping the bases of the next facial leaf and lateral pair. Lateral leaves completely separated by the facial leaves, broadly and prominently keeled, their tips rounded in adults and sharply pointed in juveniles.

Plants monoecious. Pollen cones numerous, single (or up to three together) at the ends of short branchlets, spherical, with (three to) five or six alternating pairs of pollen scales, each with three pollen sacs. Pollen grains small (30–35 µm in diameter), nearly spherical, with a short germinal papilla, otherwise almost featureless. Seed cones well separated, single at the ends of slender short branchlets, maturing in two seasons, nearly spherical but slightly longer than wide, with six to eight alternating pairs of peltate, woody seed scales. Each cone scale formed by complete fusion of the seed scale and subtending bract, with a hexagonal or pentagonal face crossed by a horizontal furrow with a central point representing the free bract tip. Middle three to five pairs of scales each with two seeds. Seeds oval with two very unequal wings derived from the seed coat in the upper half, the outer wing (facing away from the adjacent seed) up to the same size as the seed body, the inner wing much smaller. Cotyledons two, each with one vein. Chromosome base number $x = 11$.

Wood of medium weight and strength, with a narrow band of light reddish brown sapwood sharply contrasting with the dark yellowish brown heartwood. Grain fairly even and very to moderately fine, with well-defined growth rings marked by an abrupt transition to a narrow band of much smaller but not much thicker walled latewood tracheids. Resin canals absent but with numerous individual resin parenchyma cells scattered through the growth increment or occasionally concentrated in loose bands.

Stomates in very prominent white stomatal zones on the undersides of the branchlets. Each stomate set beneath and partially hidden by the five to seven surrounding subsidiary cells, which

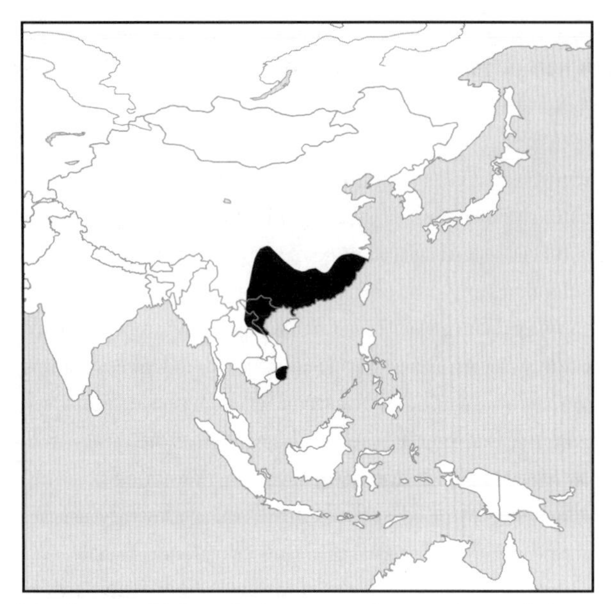

Distribution of *Fokienia hodginsii* in southeastern Asia, scale 1 : 120 million

are often shared between neighboring stomates, are topped by a steep, often interrupted Florin ring, and also bear other prominent individual papillae. Leaf cross section with a single-stranded midvein above a single resin canal and flanked by short wedges of transfusion tissue. Photosynthetic tissue forming a thin palisade layer beneath the epidermis and adjacent thin hypodermis on the face of the leaves occupying the upper side of the branchlets with spongy mesophyll extending down to the stomatal zones.

One species in eastern Asia. References: Farjon 2005b, Fu 1992, Fu et al. 1999g, McIver 1992, McIver and Basinger 1990, Yu and Zeng 1992.

The foliage of *Fokienia* (the name honors coastal Fujian province of China, from which it was originally described) somewhat resembles that of *Thujopsis,* though of thinner texture, while the seed cones are similar to those of *Chamaecyparis,* to which it is closely related according to DNA sequences. With its requirement for a warm, very humid climate, *F. hodginsii* is grown to a very limited extent in horticulture, and there has been no cultivar selection.

Fossil specimens of *Fokienia* with foliage and seed and pollen cones are found in the Paleocene of western Canada, while similar foliage is widespread, if uncommon, at the same time elsewhere in western North America and has also been found in northwestern China and much later, during the Miocene, in the northwestern United States. These fossils differ from the living plant in having the branchlets with opposite branching and the seed cones in pairs, replacing branchlets. The seed wings are also equal and extend all the way around the seed. There are no known fossils with intermediate features.

Mature Fujian cedar (*Fokienia hodginsii*) in nature.

Fokienia hodginsii (S. Dunn) A. Henry & H. H. Thomas

FUJIAN CEDAR, FUJIAN BAI (CHINESE)

Tree to 30 m tall, with trunk to 1 m in diameter. Bark purplish brown, weathering grayish brown, sparsely scaly above, shallowly furrowed lower down. Crown conical at first, cylindrical with age. Branchlet sprays soft but stiff, of varied orientation, with prominent waxy white stomatal patches occupying most of the lower surface. Scale leaves bluish green above and on the nonstomatal regions beneath. Lateral leaves of juveniles with spreading tips, 5–10 mm long, longer than the facial ones, at 4–7 mm. Facial and lateral leaves of adults about equal, 2–7 mm long, without glands. Leaf tips turned inward, those of facial leaves overlapping the base of the next pair. Pollen cones 4–5 mm long, yellow. Seed cones 1.5–2.5 cm long, 1.2–2.2 cm across. Seed body 4–5 mm long, 3–4 wide. China south of the Changjiang (Yangtze River) and adjacent northern Laos and Vietnam, reappearing in southern Vietnam south to Lâm Dông province. Subtropical montane forests with other conifers and broad-leaved evergreens; (100–) 350–1,800(–2,000) m. Zone 9.

Fujian cedar has a valuable wood with the typical "cedar" qualities of the cypress family, aromatic and with fine, straight grain. Although widespread across southern China, it is now rare

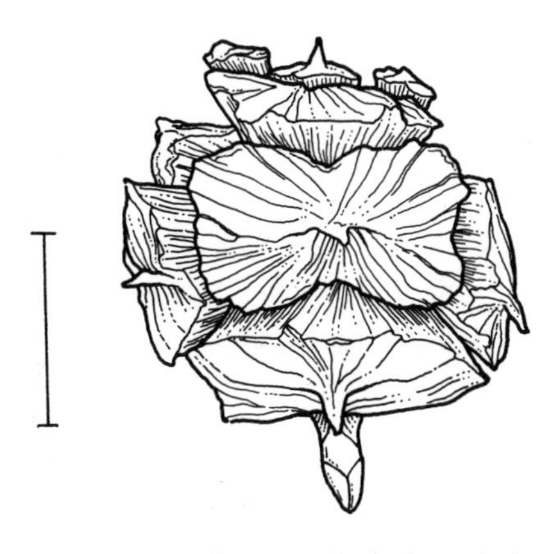

Open mature seed cone of Fujian cedar (*Fokienia hodginsii*) after seed dispersal; scale, 1 cm.

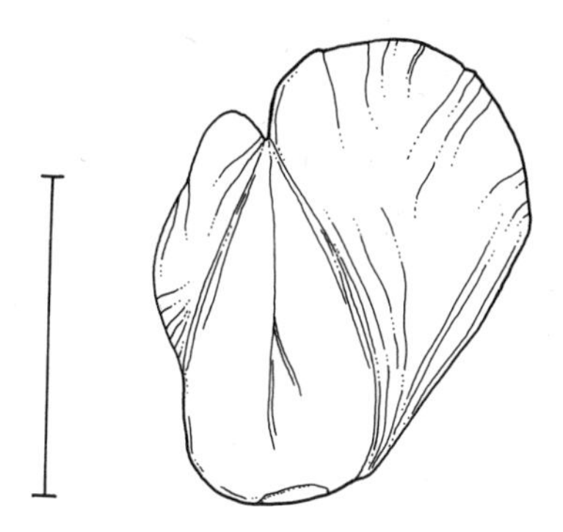

Face view of dispersed seed of Fujian cedar (*Fokienia hodginsii*); scale, 5 mm.

due to habitat loss and overexploitation. It has been suggested as a candidate for afforestation throughout its native region. In cultivation, it thrives only in humid, warm climates since its broad foliage is very susceptible to frost and desiccation. With its conspicuous, waxy white stomatal patches, the foliage is reminiscent of that of hiba arborvitae (*Thujopsis dolabrata*) and Korean arborvitae (*Thuja koraiensis*), while the shape of the leaves and their apparently whorled arrangement is like that of Chinese incense cedar (*Calocedrus macrolepis*), which it also resembles in its unequal seed wings. It differs from all of these in its seed cone shape and the details of the seed wings, which point to a closer relationship to *Chamaecyparis*, a relationship supported also by DNA studies. The striking differences between juvenile and adult leaves of Fujian cedar led early authors to describe them as different species.

Glyptostrobus Endlicher

CHINESE SWAMP CYPRESS

Cupressaceae

Irregularly deciduous trees with a single, straight trunk, clothed with thin, fibrous bark peeling in vertical strips. Crown conical to dome-shaped, to open, intermittently branched from the base of the trunk with upwardly angled branches. Shoot system differentiated into annually deciduous short shoots and persistent long shoots, these remaining green 2–3 years. Leaves spirally arranged, those of seedlings and young trees flat, linear, and needlelike, radiating on long shoots and flattened into two rows on short shoots. Those of adult trees of two intergrading types: scalelike and densely overlapping on long shoots and reproductive shoots, clawlike and twisted into two rows on short shoots.

Plants monoecious. Pollen cones in small groups, each single at the end of a short, sparsely branched reproductive branch near the young seed cones, spherical, with 15–20 spirally arranged pollen scales, of which the lower three or four are sterile. Fertile pollen scales with a thin stalk, leaflike blade, and 5–7(–10) spherical pollen sacs at their junction. Pollen grains small (25–40 µm diameter), spherical, with a short, curved germination papilla, but otherwise almost featureless. Seed cones single at the ends of short reproductive branchlets more or less pear-shaped, woody, maturing in a single season and remaining intact as their scales spread to release the seeds. Cone scales 20–22(–40), overlapping, spirally arranged, elongate, spoon-shaped (with just the middle four to six scales fully fertile). Each scale with the fertile and bract portions intimately fused, the bract smaller and represented by a triangular protrusion on the external face, the fertile portion ending in about five to eight laterally fused teeth (hence the scientific name, Greek for "carved cone") and bearing two seeds in notches on the inner face. Seeds oval, with a single wing derived from the seed coat extending downward from one side at the end of the seed body. Cotyledons (four or) five, each with one vein. Chromosome base number $x = 11$.

Wood moderately light and soft, very decay resistant, with pale whitish sapwood sharply contrasting with reddish brown heartwood. Grain very even and moderately fine, with clearly evident growth rings marked by a somewhat gradual transition to a variable but often narrow band of much smaller, thicker-walled latewood tracheids. Resin canals absent but with numerous individual resin cells scattered through the growth increment.

Stomates arranged in broad bands on either side of the midrib both above and beneath. Each stomate sunken beneath (but not hidden by) the narrow inner circle of four to six subsidiary cells, which have a smooth surface, without a Florin ring. Leaf cross section commonly somewhat diamond-shaped with a single stranded midvein above a single large resin canal and flanked by wedges of transfusion tissue and sometimes with an extra, smaller resin canal out near the leaf edge on each side. Photosynthetic tissue relatively homogeneous inside the epidermis and adjacent partial (primarily at leaf edges), thin

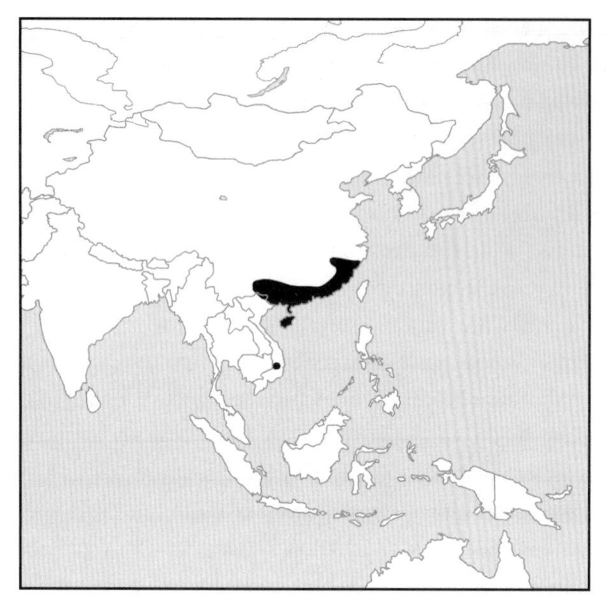

Distribution of *Glyptostrobus pensilis* in southeastern Asia, scale 1 : 120 million

hypodermis, without a well-defined palisade layer, so wholly composed of loosely packed spongy mesophyll without any particular orientation.

One species in southern China. References: R. W. Brown 1936, Brunsfeld et al. 1994, Eckenwalder 1976, Farjon 2005b, L. K. Fu 1992, Fu et al. 1999h, Henry and McIntyre 1926, Price and Lowenstein 1989, Takaso and Tomlinson 1990, Wittlake 1975, Ying, et al. 1993.

Glyptostrobus is generally considered most closely related to *Taxodium* but also shows many similarities to *Cryptomeria,* and one could postulate a common ancestor of the three that most resembled the latter genus, with *Taxodium* arising last from a *Glyptostrobus*-like ancestor through increasing specialization for swamp life. Like its relatives, it used to be assigned to the family Taxodiaceae, but there is ample reason to merge these genera into a larger family Cupressaceae. *Glyptostrobus pensilis* is scarcely frost tolerant and is not particularly handsome, so it is rarely cultivated outside of botanical gardens and there has been no cultivar selection.

The earliest known fossils definitely assignable to *Glyptostrobus* based on the characteristic seed cones are found in Paleocene sediments of coastal Alaska and the Rocky Mountain Fort Union Group, dating to some 60 million years ago. Like *Taxodium* and *Metasequoia,* with which it often grew in swamp forests, *Glyptostrobus* was widespread during the Tertiary in Eurasia and North America. While it disappeared from North America after the Miocene, it remained in Europe and Japan into the Pliocene.

Glyptostrobus pensilis (Staunton) K. Koch

CHINESE SWAMP CYPRESS, GLYPTOSTROBUS, SHUI SONG (CHINESE)

Tree to 25 m tall, with trunk to 1.2 m in diameter, only slightly flaring at the base. Roots bearing low, rounded aerial outgrowths ("knees") when growing by or in water. Bark light reddish brown, weathering very pale tan or gray, shallowly ridged and furrowed. Crown rather irregular, even in youth, sparse and open, with numerous but thin and airy tufts of branchlets at the tips of the thin branches. Branchlets completely hidden by the attached

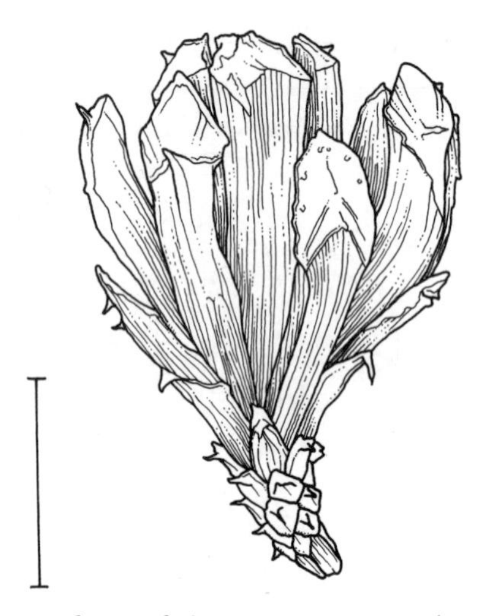

Open mature seed cone of Chinese swamp cypress (*Glyptostrobus pensilis*) after seed dispersal; scale, 1 cm.

Natural stand of mature Chinese swamp cypress (*Glyptostrobus pensilis*) during the growing season.

bases of elongate leaves or by the overlapping blades of scale leaves. Needle- and clawlike leaves 9–20 mm long, scale leaves 2–3 mm long, those of long shoots persisting 2–3 years. Pollen cones about 4 mm long. Seed cones 1.5–2.5 cm long, 1.3–1.5 cm across. Seeds about 0.7 mm long with the wing extending about the same length beyond the base. Primarily in southern China (from southeastern Yunnan to Fujian, most abundant in coastal lowlands of Fujian and Guangdong provinces) with a few localities in Vietnam. Swamps and other wet places, such as margins of rice paddies: 0–700(–1,000) m. Zone 9. Synonyms: *Glyptostrobus heterophyllus* (Brongniart) Endlicher, *G. lineatus* (Poiret) Druce.

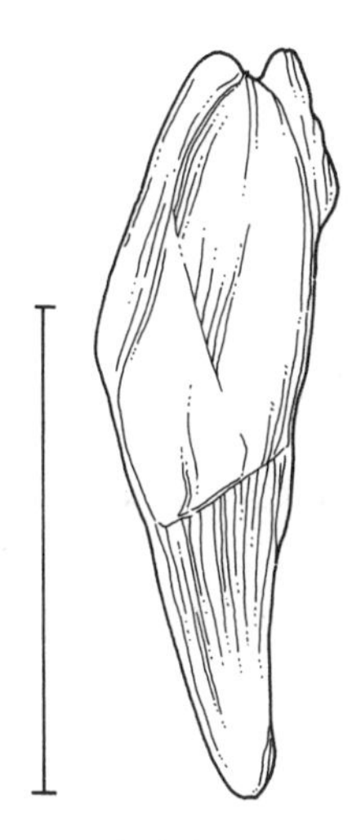

Face view of dispersed seed of Chinese swamp cypress (*Glyptostrobus pensilis*); scale, 1 cm.

Natural stand of mature Chinese swamp cypress (*Glyptostrobus pensilis*) in winter.

Chinese swamp cypress is rare in the wild today and is rather uncommon in cultivation outside China, where centuries of cultivation beside rice paddies obscures the original native range of the tree. Historically, many trees cultivated in Europe as Chinese swamp cypress were actually the North American pond cypress (*Taxodium distichum* var. *imbricarium*). The aerating, root-borne knees of Chinese swamp cypress are less spectacular than those of its American cousin, bald cypress (*T. distichum*), being more reminiscent of camels' humps than conical spires.

Halocarpus Quinn

PINK PINE

Podocarpaceae

Evergreen trees and shrubs. Trunk(s) single or multiple, short or extended, straight, bowed, or crooked, cylindrical or asymmetrical. Bark obscurely fibrous, smooth or with horizontal ridges at first, later shedding in irregular flakes and scales. Crown dome-shaped or almost flat-topped, often extending to the ground and spreading much wider than high, dense in youth, remaining so or opening up with age, with heavy, horizontal to upwardly angled branches. Branchlets all elongate, without distinction into short and long shoots, hairless, remaining green for at least the first year, shallowly grooved between the attached leaf bases in juvenile foliage, completely hidden by the scale leaves in adult foliage. Resting buds not differentiated, consisting solely of the new, as yet unexpanded ordinary foliage leaves. Leaves spirally attached. Juvenile foliage sharply distinct from and giving way abruptly without extended transitional forms to adult foliage. Commonly with equally abrupt reverse transitions and both foliage types usually found on mature trees. Juvenile leaves needlelike, sword-shaped, flattened top to bottom, standing out sharply from the twigs and usually gently curved downward along their length. Adult leaves scalelike, tightly pressed against the twigs and overlapping, the exposed portions more or less diamond-shaped, keeled or rounded, with a thin, papery fringe at the tip.

Plants dioecious. Pollen cones one (to three) at the tips of otherwise ordinary foliage shoots. Each pollen cone cylindrical, with a few bracts at the base and with 10–20 spirally arranged, roundly triangular pollen scales, each bearing two pollen sacs. Pollen grains small to medium (body 20–45 µm long, 35–75 µm overall), with two round, internally strongly striated air bladders. Bladders partially tucked underneath the conspicuously larger, minutely bumpy, oval body around the germination furrow, but also extending well beyond its ends. Seed cones single at the tips of otherwise ordinary foliage shoots, highly modified and reduced, maturing in one or two seasons. Cone with about 5–15 closely spaced, spirally arranged bracts, which may become slightly fleshy but do not unite into a podocarpium. Middle (one or) two to five bracts fertile, from which one to three (to five) seeds mature. Each fertile bract bearing a single seed that has its opening pointed back down into the cone axis and is completely covered by but not united with the seed scale (the epimatium). Seed and epimatium hard, flattened, ribbed, and with a thick crest extending around the sides and over the top, nestled at the base in a thick, fleshy, cuplike aril (hence the scientific name, Greek for "halo fruit"). Cotyledons two, each with two veins. Chromosome base number $x = 9$, 11, or 12.

Wood fragrant, dense and hard, strong or brittle, and durable, the creamy white sapwood sharply or gradually contrasting with the light brown to pinkish brown heartwood. Grain very fine and even, with clear growth rings marked by variable widths of darker latewood. Resin canals and individual resin cells both absent.

Stomates in irregular patches on both sides that are not arranged in lines or in any one direction. Each stomate sunken beneath and largely hidden by the four to six inner subsidiary cells, which may be surrounded by a less organized circle of another four to six, and topped by a sunken, interrupted Florin ring. Cuticle in some species thicker than the epidermal cells that support it. Midvein single, inconspicuously buried in the leaf tissue of the adult leaves, with one resin canal immediately beneath it. Photosynthetic tissue on both sides of the leaf, not surrounded by hypodermis inside the epidermis.

Three species in New Zealand. References: Allan 1961, Quinn 1982, Salmon 1980.

The species in this small, geographically restricted genus of small trees and shrubs are all morphologically very similar to one another. Each species, however, has a genetic complement with a different number of chromosome pairs. These differences are actually underlain by a basic similarity since each complement is based on sixteen chromosome arms that are grouped together differently to make chromosomes in the three species. Similar differences in chromosome number in most other podocarp species are based on a complement of 20 chromosome arms. Nonetheless, DNA studies show that *Halocarpus* is reasonably closely related to *Lagarostrobos* of Tasmania, *Manoao* of New Zealand, and *Parasitaxus* of New Caledonia, all of which have the 20 chromosome arms standard in the family.

The loss of chromosome arms compared to the standard 20 found throughout most of the family might be a point in favor of a result found in only some DNA studies that suggests that *Halocarpus* is especially closely related to the bizarre celery pines (*Phyllocladus*), which have an intermediate number of arms (18). Both of these genera also have the mature seeds seated in an aril, an uncommon structure in the family, but their arils might not be directly comparable structures. While both develop late in the maturation of the seeds, the aril of *Halocarpus* is an outgrowth of the epimatium while that of *Phyllocladus* arises from the base of the seed itself. Whether there is an underlying common evolutionary origin for the arils of these two genera is worthy of further investigation and might shed some light on whether they are mutually closest relatives within the family as well. If so, features in *Halocarpus* might help in interpreting the highly unusual morphology of *Phyllocladus*. The species of *Halocarpus* are not in general cultivation, and no cultivar selection has taken place. There is no known fossil record of *Halocarpus* that could contribute to understanding these problems or reveal whether the genus was once more widely distributed.

Identification Guide to *Halocarpus*

The three species of *Halocarpus* may be distinguished by the length of the longer juvenile needle leaves (which are present on mature plants as well), confirmed by whether the branchlets are angular or rounded when rolled between the fingers (reflecting whether the adult scale leaves are prominently keeled or not), and by either the length of most pollen cones or by the color of the aril in the mature seed):

Halocarpus bidwillii, longer juvenile needles up to 1 cm, branchlets round, pollen cones up to 5 mm, aril white or pale yellow

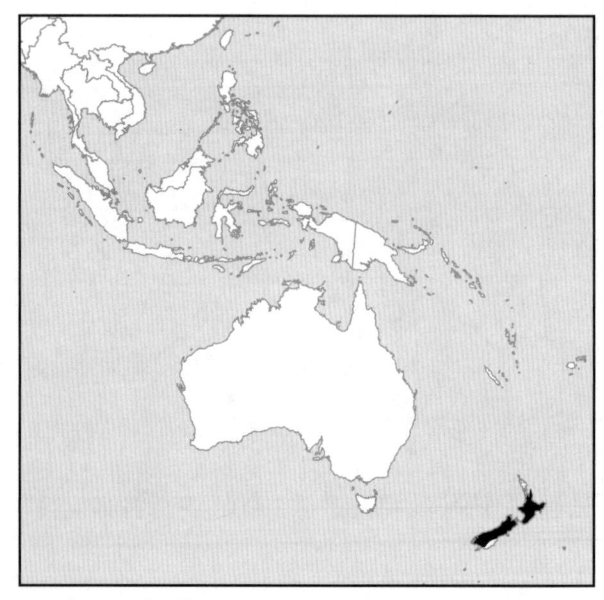

Distribution of *Halocarpus* in New Zealand, scale 1 : 119 million

H. biformis, longer juvenile needles 1–2 cm, branchlets round, pollen cones up to 5 mm, aril orange

H. kirkii, longer juvenile needles more than 2 cm, branchlets angular, pollen cones more than 5 mm, aril orange or dark yellow

Halocarpus bidwillii (J. Hooker ex T. Kirk) Quinn

BOG PINE, NEW ZEALAND MOUNTAIN PINE, TARWOOD

Shrub (or occasionally tree) to 4 m tall, with trunks to 0.4 m in diameter. Bark thin, smooth and reddish brown at first, weathering gray and flaking profusely in small, irregular scales. Crown dense, conical at first and in closed canopy vegetation, becoming broadly dome-shaped in open sites, with widely spreading branches. Lower branches rooting and turning up at the ends to become new trunks and, in turn, sending out horizontal, rooting branches, thus forming roughly circular patches up to 10 m in diameter, sometimes with a hollow center where the original trunks have died and rotted away. Adult branchlets upright, profusely alternately branched, round or slightly squared off to the touch, 1–1.5 mm in diameter.

Large, old natural clone of bog pine (*Halocarpus bidwillii*) near timberline.

Dense natural stand of bog pine (*Halocarpus bidwillii*).

Adjacent young and mature seed cones of bog pine (*Halocarpus bidwillii*), the latter with a prominent puffy aril; scale, 1 cm.

Juvenile needles 0.5–1 cm long, 1–1.5 mm wide. Adult scale leaves various shades of yellowish green, 1–2 mm long (to 3 mm on main shoots), rounded to weakly keeled, lasting 3–5 years or more. Pollen cones dull red to reddish brown, (2–)3–4(–5) mm long, 1.5–2.5 mm in diameter. Seed cones bronzy, with one to three fertile bracts, each seed with a brilliant white or slightly yellowish, puffy, cuplike aril 2–4 mm wide and 1–1.5 mm deep. Seed and epimatium greenish red to purplish black, prominently ribbed, 2–3(–4) mm long. Scattered in the mountains of New Zealand from Cape Colville (at 36°30′S) on North Island southward and in the lowlands of southwestern South Island and Stewart Island. Locally common in subalpine and montane open or closed shrublands on both sodden and rocky sites and in lowland low bog forests and thickets; (0–)250–1,500 m. Zone 7. Synonym: *Dacrydium bidwillii* J. Hooker ex T. Kirk.

Bog pine is the shortest of the *Halocarpus* species and has never had much economic importance, except as firewood, for which it was much more suitable than nearby yellow silver pines (*Lepidothamnus intermedius*), with their sparking, resinous wood. The growth form varies considerably depending on the density of the vegetation in which the shrubs grow, much more symmetrical and spreading when they are in the open and not hemmed in by competitors. Some of the variation in habit might also be due to hybridization with pink pine (*H. biformis*), which has been suggested but not studied experimentally. Some additional variation in the species is probably genetic since there is substantial geographic variation in the oils found in the foliage. Bog pine was long confused with pink pine and with other scaly-leaved podocarp species in the genera *Lepidothamnus* and *Manoao* and was not formally described and named until almost 40 years after the first specimens were collected. The species name honors John Bidwill (1815–1853), an English-born botanist who collected the type specimen and wrote an account of his *Rambles in New Zealand* (1841) before becoming director of the Botanical Gardens, Sydney, Australia.

Halocarpus biformis (W. J. Hooker) Quinn

PINK PINE, YELLOW PINE

Tree to 10 m tall, with a short, crooked trunk to 0.6(–1) m in diameter, or a shrub in exposed locations, sometimes not exceeding 1 m. Bark thin, smooth, and reddish brown at first, weathering gray and flaking away in small, thick scales. Crown conical at first, becoming compactly to openly and irregularly dome-shaped with age, with thick, upwardly angled branches forking repeatedly and bearing alternately arranged, short to long branchlets. Adult branchlets upright, crowded or more openly branched, notably squared off to the touch, (2–)3–4 mm in diameter. Juvenile needles 1–2 cm long, 1.5–3 mm wide. Adult scale leaves yellowish green to dark green, 1–2 mm long (to 3 mm on main shoots), strongly keeled, lasting 3–4 years or more. Pollen cones red, fading to brown, 2.5–5 mm long, 2–4 mm in diameter. Seed cones tinged red, like many branchlet tips, with one to three fertile bracts, each seed with an orange or dark yellow, puffy, donut-shaped aril 1.5–2 mm wide and 0.5–1.5 mm deep. Seed and epimatium black, weakly ribbed, 2–3 mm long. Discontinuous through the mountains of New Zealand from about 37°S on North Island south to Stewart Island, more common and at lower elevations southward. Scattered in montane and subalpine forests and shrublands on rich soils and in lowland forests in southwestern South Island and Stewart Island; 0–1,400 m. Zone 8. Synonym: *Dacrydium biforme* (W. J. Hooker) Pilger.

As might be expected, pink pine has handsome pink heartwood when fresh. As one of the most, if not the most, durable softwood timbers in New Zealand, it was much appreciated for use as fence posts and in other applications in contact with soil, but the tree is not large enough to have ever been of commercial significance. Because it is very slow growing and can live 400 years or more, pink pine has been investigated for its potential in tree ring studies. Not surprisingly, given the wet climates in which it grows, its ring width depends more on temperature during the growing season than on precipitation, and the broader band of earlywood is more responsive than the narrow latewood within each annual ring. Thus an observed correlation of growth with the Southern Oscillation index, which measures the strength of the El Niño–Southern Oscillation oceanic and climate phenomenon, is probably mediated by greater sunlight and warmer temperatures during these dry years in the otherwise wet and often overcast and cool sites where pink pine grows. In wetter sites, trees of pink pine may surround the bogs where bog pine (*Halocarpus bidwillii*) grows, and the two species may hybridize. Pink pine was the first species of *Halocarpus* to be formally described and named, and the name, Latin for "two forms," reflects the extreme difference in form of the juvenile and adult foliage and the abrupt transition between them. Rather unusually among conifers, the juvenile and adult foliage also differs in many chemical constituents. This is true of the other two *Halocarpus* species as well. Based on pollen deposited at the time, one or more of these *Halocarpus* species joined mountain toatoa (*Phyllocladus alpinus*) as the first woody plants to invade grasslands in northern South Island as glaciers retreated 12,000 years ago. The type specimen of pink pine was collected in 1791 at the opposite end of South Island in Fiordland National Park in the southwest, where the species comes down to sea level. These specimens were collected by the well-traveled Archibald Menzies, who also collected the type specimen of Douglas fir (*Pseudotsuga menziesii*) in the fjord lands of northwestern North America.

Halocarpus kirkii (F. J. Mueller ex Parlatore) Quinn

MONOAO (MAORI AND ENGLISH)

Tree to 25(–30) m tall, with a columnar trunk to 1 m in diameter, rapidly dividing into major limbs at about half its height. Bark smooth with irregularly spaced narrow ridges at first, light to dark brown, weathering grayish brown and shedding in large, warty scales. Crown dense and conical at first, becoming dome-shaped and more open with age, with strongly upwardly angled branches forking extensively and bearing alternately arranged branchlets. Adult branchlets whiplike, upright, relatively sparsely branched, rounded or slightly squared off to the touch, 1–2 mm in diameter, sometimes reverting to juvenile foliage at their tips. Juvenile needles (1.5–)2.5–4 cm long, 1–3 mm wide. Adult scale leaves rich dark green, 2–3 mm long (to 5 mm on main shoots), rounded to weakly keeled, lasting 2–3 years or more. Pollen cones dark brown to black, 5–10 mm long, 4–5 mm in diameter. Seed cones red, with one to five fertile bracts, each seed with an orange or dark yellow, puffy, shallow aril 2.5–5 mm wide and 1–1.5 mm deep. Seed and epimatium black, strongly ribbed, (3–)5–7(–9) mm long. Rare and local in lowlands and mountains of northwestern North Island, New Zealand, north of about 37°20′S. Scattered or rarely gregarious in the canopy of lowland and lower montane mixed forests; 0–700 m. Zone 9. Synonym: *Dacrydium kirkii* F. J. Mueller ex Parlatore.

Monoao is one of the rarest conifers in New Zealand. It has about the same overall distribution as kauri (*Agathis australis*) but is much more local and rarely, if ever, dominates forest stands. Nonetheless, because the light brown heartwood was one of the finest softwoods in New Zealand, large trees were essentially eliminated in areas of easy access for forestry. Young trees are very handsome, thus monoao has had some horticultural use. The transition between juvenile and adult foliage is particularly dramatic because monoao has the longest juvenile leaves in the genus. It is also the only *Halocarpus* species that commonly shows a reversion to juvenile foliage in branchlets with adult foliage. In bog pine (*H. bidwillii*) and pink pine (*H. biformis*), branchlets with juvenile foliage, when present on mature trees, usually emerge directly from older twigs and lower branches inside the crown and not as a continuation of growth of branchlets with adult foliage. Monoao does not usually grow with either bog pine or pink pine, so no hybrids with these species have been reported or suspected. The species name honors Thomas Kirk (1828–1898), English-born New Zealand botanist and author of *The Forest Flora of New Zealand* (1889), who collected the type specimen on Great Barrier Island.

Juniperus Linnaeus

JUNIPER, CEDAR

Cupressaceae

Evergreen trees or tall to dwarf shrubs with a single main trunk or multistemmed from the base. Bark fibrous, usually furrowed and peeling in vertical strips but sometimes flaking and forming rectangular blocks. Shoot system weakly differentiated into main axes and side shoots (branchlets), usually branching in three dimensions. Leaves in alternating pairs or trios, of two main types. All juvenile junipers and adults of all species of section *Juniperus* and of a few species of section *Sabina* bearing flattened, linear, needlelike leaves with (one or) two prominent stomatal bands above. Needle leaves of section *Juniperus* jointed to the twig, while those of section *Sabina* run down onto it without a joint. Scalelike leaves of most adults of section *Sabina* closely enclosing the twig, the tips either pressed against it or pointing outward, sometimes with a prominent round or elongate resin gland in the middle or near the tip or base of the outer face. Successive pairs or trios of leaves on main shoots (whip leaves) well separated from one another with the conspicuous, long leaf bases clothing the stem. Leaves on branchlets crowded, their tips hiding the short bases of the next pair or trio along the twig.

Plants usually dioecious, but some species monoecious or with various proportions of plants with either pollen or seed cones or both together. Pollen and seed cones single (pollen cones clustered in *Juniperus drupacea*), attached directly in the axils of needlelike leaves in section *Juniperus* and at the ends of otherwise ordinary branchlets in section *Sabina*. Pollen cones spherical to oblong, with three to seven alternating pairs or trios of pollen scales. Each scale with two to eight (often four) pollen sacs on the inner side of the base of the roughly heart-shaped blade at the tip of a thin stalk. Pollen grains small (20–40 μm in diameter), spherical, sometimes with an obscure, smooth germination pore, otherwise almost featureless or with minute bumps. Seed cones spherical or approximately so, berrylike (a galbulus), composed of one to three alternating pairs or trios of fused seed scales of varying textures from fleshy to woody, from pink or red to blue or black at maturity, often with a thin waxy coating, maturing in one or two seasons and not opening at maturity. Seeds up to three per scale but only one per cone in many species, ovoid to angular, without wings, often with a prominent attachment scar. Cotyledons two, four, or six, each with one vein. Chromosome base number $x = 11$.

Wood fragrant, variously soft to moderately hard and medium to heavy, with white to light brown or yellowish brown sapwood sharply contrasting with the rich brown, reddish brown, or purplish brown heartwood, often streaked with brighter or darker hues and with an irregular boundary. Grain typically very fine to fairly fine and very even to somewhat wavy and irregular, sometimes with weak or false rings but usually with well-defined growth rings marked by an abrupt (or sometimes gradual) transition to much smaller but often not much thicker walled latewood tracheids. Resin canals absent but with abundant to sparse individual resin parenchyma cells scattered through the growth increment or more often concentrated in narrow open bands near their middle or in the outer half.

Stomates of scale leaves forming a large patch on each side of the midrib on the upper (inner) face and smaller patches on the hidden portions of the lower face. Each stomate sunken beneath and largely hidden by the four to six surrounding subsidiary cells, which are topped by a continuous but uneven, low Florin ring that may be surrounded by a shallow moatlike furrow. Leaf cross section with a single-stranded midvein above a single large resin canal (that may scarcely extend into the free leaf tip) and flanked by wedges of transfusion tissue. Photosynthetic tissue forming a thin, ill-defined palisade layer beneath the epidermis and adjacent single-layered, continuous (except beneath the stomates), hypodermis on the lower (outer) leaf face (or sometimes on both faces). Palisade layer much looser than usual and giving way smoothly to the spongy mesophyll that fills the bulk of the leaf volume.

Fifty-four species throughout most of the northern hemisphere and in East Africa. References: Adams 1993, 1994, 1995, Farjon 1992, Farjon 2005b, Fu et al. 1999g, Zanoni and Adams 1979. Synonyms: *Arceuthos* F. Antoine & Kotschy, *Sabina* P. Miller.

Juniperus has by far the most species and the broadest ecological and geographic range of any genus in the Cupressaceae. The relative drought tolerance and fleshy, bird-dispersed cones are often cited as factors responsible for this diversity, but the related genera *Cupressus* and *Callitris* are equally drought tolerant, and *Taxus* is similarly bird-dispersed, yet each of these has far fewer species than *Juniperus*. The name *Juniperus* is the classical Latin common name for a species in the genus, probably either common juniper (*J. communis*) or prickly juniper (*J. oxycedrus*), both of which may be found around Rome.

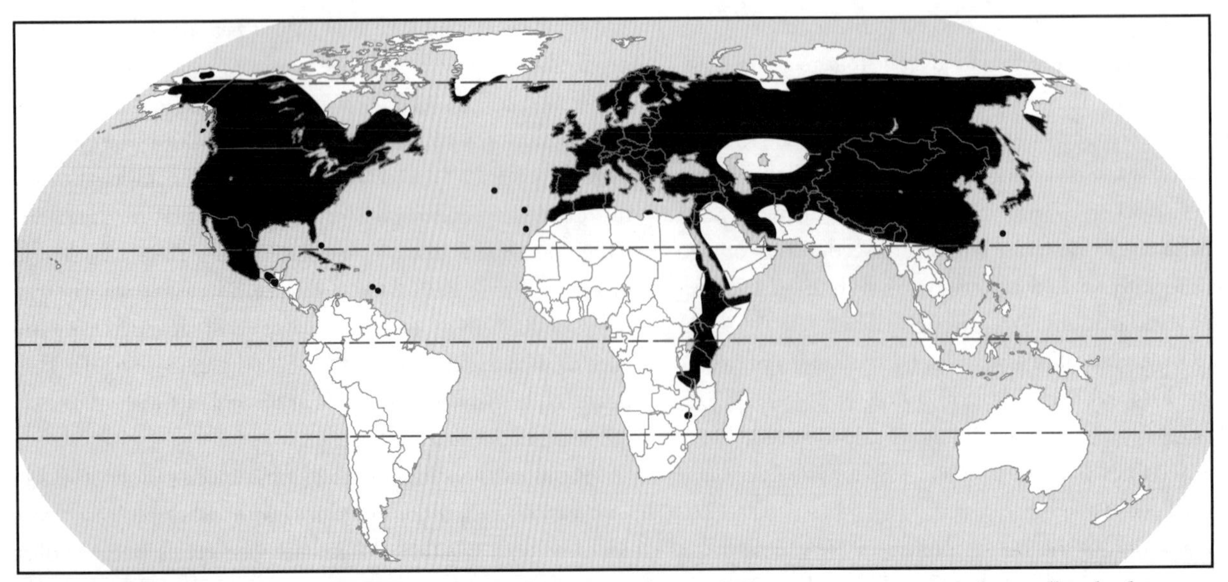

Distribution of *Juniperus* across the northern continents and in eastern Africa, including some species restricted to smaller islands, scale 1 : 200 million

The natural variability of the genus is augmented in cultivation, where about a dozen species gave rise to most of the numerous cultivars available. Cultivar selection has been very wide ranging, with variations on growth habit from the narrowest, pencil-like spires to cascading creepers closely hugging the ground, including upright weepers, teardrops, and broadly cylindrical spreaders, of every size from tight, dwarf buns to significant (if rarely stately) trees. There are even contortionists, like the well-known Hollywood juniper (*Juniperus chinensis* 'Kaizuka'), so prominent in the urban landscape of California. Foliage variants include juvenile and changeable forms (in section *Sabina*), variations in needle length (section *Juniperus*), and a wide spectrum of colors from pale to dark greens and light grayish greens to intense blues on the one hand or yellowish greens to deep golden sunshines on the other, not to mention variegation or transient blushes of white to silver or gold. No genus of conifers provides more variety of landscape forms and uses in its cultivars than *Juniperus.*

Some of the cultivars are selections from natural hybrid swarms, and natural hybridization between species is common in the genus in places where the ranges of related species overlap. Variability, hybridization, and inconvenient political boundaries all led to continuing taxonomic confusion in *Juniperus,* although analyses of DNA and volatile oils help clarify many difficulties.

DNA studies also show that the genus arose from a *Cupressus*-like ancestor and that it may be more closely related to either the New World cypresses or to the Old World ones (the results are ambiguous) than these two groups of cypresses are to each other, the latter result being perhaps the most unexpected one found to date by molecular studies of conifers.

Fragments of juniper twigs and seed cones found in ice-age pack-rat middens in southwestern North America are very helpful in reconstructing changes in plant communities and vegetation during glacial and interglacial cycles. Older fossils, however, have been elusive. Most fossils reported as Tertiary records of *Juniperus* subsequently prove to belong to other conifers, like *Glyptostrobus.* The wide distribution and large number of species in the genus could thus be more recent geologically than the diversification of many other conifers. A record from the late Oligocene or early Miocene, about 23 million years ago, is the oldest reasonably secure fossil, considerably later than a much less convincing Paleocene record more than 55 million years old.

Geographic Guide to *Juniperus*

Europe, North Africa, and western Asia
- *Juniperus communis*
- *J. drupacea*
- *J. excelsa*
- *J. foetidissima*
- *J. oblonga*
- *J. oxycedrus*
- *J. phoenicea*
- *J. procera*
- *J. sabina*
- *J. thurifera*

Azores Islands
- *Juniperus brevifolia*

Canary Islands
- *Juniperus cedrus*
- *J. phoenicea*

East Africa
- *Juniperus procera*

Siberia, Mongolia, northern China, and Korea
- *Juniperus chinensis*
- *J. communis*
- *J. davurica*
- *J. rigida*
- *J. sabina*

Japan, Sakhalin, and Kurile Islands
- *Juniperus chinensis*
- *J. communis*
- *J. conferta*
- *J. rigida*
- *J. taxifolia*
- *J. thunbergii*

Himalaya and central Asia
- *Juniperus excelsa*
- *J. indica*
- *J. pingii*
- *J. pseudosabina*
- *J. recurva*
- *J. semiglobosa*
- *J. squamata*

Central and southern China and Taiwan
- *Juniperus chinensis*
- *J. communis*
- *J. convallium*
- *J. formosana*
- *J. komarovii*
- *J. recurva*
- *J. saltuaria*
- *J. squamata*
- *J. tibetica*

Eastern North America (east of the Rocky Mountains)

Juniperus ashei
J. communis
J. horizontalis
J. scopulorum
J. virginiana

Western North America, including Baja California

Juniperus californica
J. coahuilensis
J. communis
J. deppeana
J. flaccida
J. horizontalis
J. monosperma
J. occidentalis
J. osteosperma
J. pinchotii
J. scopulorum

Mexico and Guatemala

Juniperus angosturana
J. ashei
J. blancoi
J. coahuilensis
J. comitana
J. deppeana
J. durangensis
J. flaccida
J. gamboana
J. jaliscana
J. monticola
J. pinchotii
J. saltillensis
J. standleyi

West Indies

Juniperus barbadensis
J. bermudiana
J. gracilior
J. saxicola

Taxonomic Guide to *Juniperus*

Despite *Juniperus* being the third most species-rich genus of extant conifers, little effort over the years has been put into devising anything but the most rudimentary formal classification for it. Traditionally, junipers were distributed among three sections, subgenera, or even genera. One of these traditional groups, containing only *J. drupacea* and thus monotypic, is clearly just an extreme within the variation found in one of the other two groups and hence is not generally accepted as distinct today. Thus the species are assigned to one of two groups, distinguished by leaf type and apparent cone position, as noted in the accompanying identification guide. These two sections are very unequal in numbers of species: 11 in section *Juniperus* versus 43 in section *Sabina.* Although the distinctiveness even of these groups has been questioned, unless and until DNA evidence shows conclusively that the smaller section *Juniperus* is embedded within section *Sabina* phylogenetically, maintaining these two sections as independent lineages seems most in accord with available evidence. There have been a few proposals for recognition of subsections within the sections (or sections if the two primary subdivisions are considered subgenera). All such proposals to date seem artificial and do little to advance understanding of taxonomic structure within the genus. Nonetheless, there surely is taxonomic structure within *Juniperus,* so one of the major challenges in conifer taxonomy is to devise a meaningful taxonomic subdivision of section *Sabina* (especially) that makes sense of its confusing morphological patterns.

Juniperus sect. *Juniperus* (synonyms: *Arceuthos* F. Antoine, sect. *Caryocedrus* Endlicher, sect. *Communes* Antoine, sect. *Oxycedrus* Spach, subg. *Caryocedrus* (Endlicher) H. Gaussen, subg. *Oxycedrus* (Spach) H. Gaussen)

Juniperus brevifolia
J. cedrus
J. communis
J. conferta
J. drupacea
J. formosana
J. oblonga
J. oxycedrus
J. rigida
J. taxifolia
J. thunbergii

Juniperus sect. *Sabina* Spach (synonyms: *Sabina* (Spach) F. Antoine, subg. *Sabina* (Spach) H. Gaussen)

Juniperus angosturana
J. ashei
J. barbadensis
J. bermudiana
J. blancoi
J. californica
J. chinensis
J. coahuilensis
J. comitana
J. convallium

J. davurica
J. deppeana
J. durangensis
J. excelsa
J. flaccida
J. foetidissima
J. gamboana
J. gracilior
J. horizontalis
J. indica
J. jaliscana
J. komarovii
J. monosperma
J. monticola
J. occidentalis
J. osteosperma
J. phoenicea
J. pinchotii
J. pingii
J. procera
J. pseudosabina
J. recurva
J. sabina
J. saltillensis
J. saltuaria
J. saxicola
J. scopulorum
J. semiglobosa
J. squamata
J. standleyi
J. thurifera
J. tibetica
J. virginiana

Identification Guide to *Juniperus*

The 54 species of *Juniperus* may be assigned to one of two sections as a convenient first step in identification. The sections may be distinguished by the form of the adult leaves, by their type of attachment, and by the apparent position of the pollen and seed cones:

section *Juniperus,* leaves needlelike, jointed, pollen and seed cones in the axils of leaves (but at the end of a short, inconspicuous branchlet)

section *Sabina,* leaves scalelike (sometimes needlelike), running down without a joint, pollen and seed cones at the tips of conspicuous branchlets

Section *Juniperus.* These 11 species may be distinguished by the length of most leaves, whether they are straight or curved forward, and by the color and typical length of the ripe seed cones:

Juniperus brevifolia, leaves less than 12 mm, curved, seed cones reddish brown, 9–12 mm

J. taxifolia, leaves up to 12 mm, straight, seed cones purple, 9 mm or less

J. thunbergii, leaves up to12 mm, straight, seed cones purple, 9–12 mm

J. cedrus, leaves up to 12 mm, straight, seed cones reddish brown, 9–12 mm

J. oxycedrus, leaves up to 12 mm, straight, seed cones reddish brown, 12 mm or more

J. oblonga, leaves more than 12 mm, straight, seed cones brown, less than 9 mm

J. rigida, leaves 12 mm or more, straight, seed cones purple, less than 9 mm

J. communis, leaves 12 mm or more, straight or curved, seed cones blue, up to 9 mm

J. formosana, leaves 12 mm or more, straight, seed cones reddish brown, 9–12 mm

J. conferta, leaves 12 mm or more, curved, seed cones purple, 9–12 mm

J. drupacea, leaves more than 12 mm, straight, seed cones brown, more than 12 mm

Section *Sabina.* For convenience of identification, these 43 species may be divided into five groups based on whether the adult leaves are needle- or scalelike and minutely toothed or not, and whether the seed cones typically have one seed or more:

Group A, leaves needlelike, untoothed, one or more seeds

Group B, leaves scalelike, untoothed, usually one seed

Group C, leaves scalelike, untoothed, usually two or more seeds

Group D, leaves scalelike, toothed, usually one seed

Group E, leaves scalelike, toothed, usually two or more seeds

Group A. These six species may be distinguished by the length of the adult leaves, the extent of furrowing beneath the leaves, and the color of the ripe seed cones (with *Juniperus saxicola* the only New World species in this group, the others all Asian):

Juniperus saxicola, leaves less than 6 mm, not furrowed, seed cones blue

J. pingii, leaves less than 6 mm, not furrowed, seed cones purple-black

J. davurica, leaves 6–8 mm, not furrowed, seed cones brown

J. chinensis, leaves 6–8 mm, not furrowed, seed cones blue-black

J. squamata, leaves 6–8 mm, furrowed the whole length, seed cones blue-black

J. recurva, leaves 8–13 mm, furrowed for half their length, seed cones purple-brown

Group B. These three species may be distinguished by the length of most scale leaves and by the diameter of the ripe seed cones:

Juniperus saltuaria, scale leaves less than 2 mm, seed cones less than 7 mm

J. komarovii, scale leaves 2 mm or less, seed cones more than 7 mm

J. tibetica, scale leaves 2 mm or more, seed cones more than 7 mm

Group C. These 15 species may be further distributed between two subgroups based on whether the tips of the scale leaves are tight to the branchlets or loose to spreading:

Group C1, scale leaves tight

Group C2, scale leaves loose to spreading

Group C1. These eight species may be distinguished by the maximum width of the first-year branchlets, the maximum length of the scale leaves on branchlets, and the maximum length of the seed cones:

Juniperus barbadensis, widest branchlets less than 1.1 mm wide, longest scale leaves less than 1.1 mm, largest seed cones less than 6 mm

J. sabina, widest branchlets less than 1.1 mm wide, longest scale leaves 1.1–1.5 mm, largest seed cones less than 6 mm

J. chinensis, widest branchlets less than 1.1 mm wide, longest scale leaves 1.1–1.5 mm, largest seed cones 6–9 mm

J. blancoi, widest branchlets less than 1.1 mm wide, longest scale leaves more than 1.5 mm, largest seed cones less than 6 mm

J. bermudiana, widest branchlets 1.1–1.5 mm wide, longest scale leaves 1.1–1.5 mm, largest seed cones less than 6 mm

J. scopulorum, widest branchlets 1.1.–1.5 mm wide, longest scale leaves more than 1.5 mm, largest seed cones 6–9 mm

J. semiglobosa, widest branchlets more than 1.5 mm wide, longest scale leaves more than 1.5 mm, largest seed cones 6–9 mm

J. foetidissima, widest branchlets more than 1.5 mm wide, longest scale leaves more than 1.5 mm, largest seed cones more than 9 mm

Group C2. These nine species may be distinguished by the maximum width of the first-year branchlets, the maximum length of the scale leaves on branchlets, and the maximum length of the seed cones, with one pair (*Juniperus scopulorum* and *J. horizontalis*) further distinguished by whether the branchlets carrying the seed cones are straight or curved and by whether the plant is an upright tree or a spreading shrub:

Juniperus procera, widest branchlets less than 1.1 mm wide, longest scale leaves less than 1.1 mm, largest seed cones 6–9 mm

J. gracilior, widest branchlets less than 1.1 mm wide, longest scale leaves 1.1–1.5 mm, largest seed cones 6–9 mm

J. davurica, widest branchlets less than 1.1 mm wide, longest scale leaves more than 1.5 mm, largest seed cones less than 6 mm

J. thurifera, widest branchlets less than 1.1 mm wide, longest scale leaves more than 1.5 mm, largest seed cones 6–9 mm

J. excelsa, widest branchlets 1.1–1.5 mm wide, longest scale leaves 1.1–1.5 mm, largest seed cones more than 9 mm

J. virginiana, widest branchlets 1.1–1.5 mm wide, longest scale leaves more than 1.5 mm, largest seed cones less than 6 mm

J. scopulorum, widest branchlets 1.1–1.5 mm wide, longest scale leaves more than 1.5 mm, largest seed cones 6–9 mm, on straight stalks, plant upright

J. horizontalis, widest branchlets 1.1–1.5 mm wide, longest scale leaves more than 1.5 mm, seed cones 6–9 mm, on curved stalks, plant spreading

J. foetidissima, widest branchlets more than 1.5 mm wide, longest scale leaves more than 1.5 mm, largest seed cones more than 9 mm

Group D. These 13 species may be further distributed between two subgroups based on whether the tips of the scale leaves are tight to the branchlets or spreading away from them:

Group D1, scale leaf tips tight

Group D2, scale leaf tips spreading

Group D1. These eight species may be distinguished by the maximum width of the first year branchlets, the maximum length of the scale leaves on the branchlets, and the color of the ripe seed cones:

Juniperus comitana, widest branchlets less than 1.1 mm wide, longest scale leaves less than 1.5 mm, seed cones dark violet

J. convallium, widest branchlets less than 1.1 mm wide, longest scale leaves less than 1.5 mm, seed cones brown

J. gamboana, widest branchlets less than 1.1 mm wide, longest scale leaves 1.5–2 mm, seed cones red

J. indica, widest branchlets 1.1–1.5 mm wide, longest scale leaves 1.5–2 mm, seed cones blue

J. angosturana, widest branchlets 1.1–1.5 mm wide, longest scale leaves more than 2 mm, seed cones brown

J. californica, widest branchlets more than 1.5 mm wide, longest scale leaves 1.5–2 mm, seed cones blue

J. saltillensis, widest branchlets more than 1.5 mm wide, longest scale leaves 1.5–2 mm, seed cones purple

J. osteosperma, widest branchlets more than 1.5 mm wide, longest scale leaves more than 2 mm, seed cones brown

Group D2. These five species may be distinguished by the maximum width of the first year branchlets, the maximum length of the scale leaves on the branchlets, and the color of the ripe seed cones:

Juniperus ashei, widest branchlets 1.1–1.5 mm wide, longest scale leaves 1.5–2 mm, seed cones blue

J. pinchotii, widest branchlets 1.1–1.5 mm wide, longest scale leaves more than 2 mm, seed cones red

J. pseudosabina, widest branchlets more than 1.5 mm wide, longest scale leaves 1.5–2 mm, seed cones brown

J. coahuilensis, widest branchlets more than 1.5 mm wide, longest scale leaves more than 2 mm, seed cones red

J. monosperma, widest branchlets more than 1.5 mm wide, longest scale leaves more than 2 mm, seed cones brown

Group E. These nine species may be distinguished by the maximum width of the first-year branchlets, the maximum length of the scale leaves on the branchlets, and the color of the ripe seed cones, with one pair (*Juniperus flaccida* from southwestern North America and *J. phoenicea* from the Mediterranean region) further distinguished by whether the tips of the scale leaves on the branchlets spread away from the twig or are tight against it:

Juniperus thurifera, branchlets less than 1.1 mm wide, longest scale leaves 1.5 –2 mm, seed cones purple

J. flaccida, widest branchlets less than 1.1 mm wide, longest scale leaves 1.5–2 mm, seed cones reddish brown, scale leaf tips loose

J. phoenicea, widest branchlets less than 1.1 mm wide, longest scale leaves 1.5–2 mm, seed cones reddish brown, scale leaf tips tight

J. jaliscana, widest branchlets 1.1–1.5 mm wide, longest scale leaves less than 1.5 mm, seed cones reddish brown

J. monticola, widest branchlets 1.1–1.5 mm wide, longest scale leaves less than 1.5 mm, seed cones bluish black

J. deppeana, widest branchlets 1.1–1.5 mm wide, longest scale leaves 1.5–2 mm, seed cones reddish brown

J. standleyi, widest branchlets 1.1–1.5 mm wide, longest scale leaves 1.5–2 mm, seed cones blue

J. durangensis, widest branchlets more than 1.5 mm wide, longest scale leaves 1.5–2 mm, seed cones reddish brown

J. occidentalis, widest branchlets more than 1.5 mm wide, longest scale leaves more than 2 mm, seed cones bluish black

Juniperus angosturana R. P. Adams

ANGOSTURA JUNIPER, SLENDER ONE-SEED JUNIPER

Juniperus sect. *Sabina*

Shrub, or tree to 10 m tall, usually multitrunked from near the base, with single trunk to 45 cm in diameter. Bark longitudinally furrowed or checkered, to 1 cm thick. Branches spreading crookedly, arising haphazardly to form an irregular, rounded crown. Branchlets rounded, short (6–12 mm long) and slender (1–1.3 mm thick). Adult leaves in alternating pairs (or trios), scalelike, 1–2.5 mm long, dark green, with or without resin glands, the edges minutely toothed, the tip rounded to pointed and with a tiny prickle, pressed forward. Pollen and seed cones on different plants. Pollen cones oblong, 2.5–3 mm long, with five or six alternating pairs of pollen scales. Seed cones oblong or egg-shaped, 4–4.5 mm long, bluish brown with a thin waxy coating, maturing in 1 year. Seed one, about 4 mm long, orangish brown, the pale attachment scar not extending beyond the base. Northeastern Mexico, in the Sierra Madre Oriental from northwestern Nuevo León to northern Hidalgo. Dry grasslands, brushlands, and woodlands; (1,050–)2,200–2,800 m. Zone 9? Synonym: *J. monosperma* var. *gracilis* M. Martínez.

Aging mature Ashe juniper (*Juniperus ashei*) in nature.

Angostura juniper belongs to the one-seeded group that includes roseberry juniper (*Juniperus coahuilensis*), one-seed juniper (*J. monosperma*), and Pinchot juniper (*J. pinchotii*). It hybridizes with roseberry juniper where the ranges of the two species overlap in Nuevo León and Coahuila, Mexico. Although it was originally described as a variety of one-seed juniper, it is widely separated from that species and differs in its much finer foliage and smaller seed cones as well as in leaf oils and DNA. The species name derives from the type locality, Angostura in the state of San Luis Potosí.

Juniperus ashei J. Buchholz

ASHE JUNIPER, MOUNTAIN CEDAR, ROCK CEDAR

Juniperus sect. *Sabina*

Tree to 15(–18) m tall, with trunk to 1.1 m in diameter, usually single-stemmed for the first 1–3 m. Bark reddish brown, weathering pinkish gray, thin, shallowly furrowed. Crown irregular, thin, and rounded, composed of horizontal to ascending branches bearing erect, four-angled branchlets. Branchlets slender, 0.7–1.4 mm thick. Leaves in alternating pairs, dark green, scalelike in adults, those of long shoots 3–6 mm long. Scale leaves of branchlets 1–2 mm long with a minutely toothed edge, a blunt or pointed, slightly spreading tip, and a prominent, round, closed oil gland on the central keel near the base. Pollen and seed cones on separate plants. Pollen cones single at the tips of branchlets, oblong, 2–3 mm long, with three to five alternating pairs of pollen scales. Seed cones nearly spherical, 6–9 mm long, dark blue with a waxy coating, single at the tips of straight branchlets, maturing in a single season. Seed usually one, 4–6 mm long, glossy dark brown, with a prominent attachment scar about half as long as the seed. Southern Missouri to northern Coahuila (Mexico). Open uplands and canyons on rocky, limestone soils; 150–600(–1,800) m. Zone 6?

Ashe juniper is the common juniper of central Texas and the Ozarks in the south-central United States, where, as mountain or rock cedar, it is cut for fence posts and distilled to yield cedarwood oil. Adams (1977) showed that, contrary to early reports, Ashe juniper does not commonly hybridize with its eastern and western neighbors, eastern red cedar (*Juniperus virginiana*) and Pinchot juniper (*J. pinchotii*). It is most closely related to Saltillo juniper (*J. saltillensis*) of adjacent northern Mexico. The largest known individual in the United States in combined dimensions when measured in 1999 was 17.4 m tall and 1.1 m in diameter, with a spread of 14.6 m. The species name honors William Ashe (1872–1932), the American forest botanist who collected the type specimen but is better known, perhaps, for the numerous synonymous hawthorn (*Crataegus*) species he described.

Juniperus barbadensis Linnaeus

CARIBBEAN JUNIPER

Juniperus sect. *Sabina*

Tree to 15 m tall, with single trunk to 60 cm in diameter. Bark pale brown, weathering light gray, thin, fibrous, peeling in long, narrow strips. Branchlets four-angled, very slender (less than 1 mm in diameter). Leaves in alternate pairs, light green, scalelike in adults, those of branchlets about 1 mm long, rounded and inconspicuously glandular on the back, with a smooth edge and a blunt or pointed tip tightly pressed against the branchlet. Pollen and seed cones on the same plant. Pollen cones oblong, (2.5–)4–5 mm long, with six or seven alternating pairs of pollen scales. Seed cones flattened spherical, 4–5 mm long, 5–8 mm wide, reddish blue to dark blue with a waxy coating, maturing in a single season. Seeds two to four, 2–3 mm long, light yellowish brown, the paler attachment scar extending no more than about a fourth the way up the sides. Bahamas to Jamaica and St. Lucia. Dry open woodlands on exposed limestone; 0–1,600 m. Zone 9? Two varieties.

Caribbean juniper is the most widely distributed of the four West Indian juniper species but, like the other three, is a rare species of considerable conservation concern in most of its range. It had already been extirpated from its type locality in Barbados in the 18th century, and it is far from secure even in its areas of greatest abundance in Cuba. Its relationships to the other West Indian species, Bermuda redcedar (*Juniperus bermudiana*), Hispaniolan juniper (*J. gracilior*), and prickly redcedar (*J. saxicola*), and to the mainland eastern redcedar (*J. virginiana*) are not fully resolved and require more expansive phylogenetic research to clarify. It may be conspecific with Hispaniolan juniper as part of a geographic replacement series among varieties, but Hispaniolan juniper has long scale leaves with spreading tips and larger seed cones, so keeping them separate poses few problems in identification.

Juniperus barbadensis Linnaeus **var.** ***barbadensis***

BARBADOS CEDAR, LESSER ANTILLES PENCIL CEDAR

Old leaves of long shoots with a conspicuous, sunken gland extending almost to the tip. Formerly uncommon on Barbados and St. Lucia, this species has been eliminated by sugarcane cultivation from all of its range except a small area at the summit of Petit Piton in St. Lucia.

Juniperus barbadensis **var.** ***lucayana*** (N. Britton) R. P. Adams

ANTILLEAN REDCEDAR, GREATER ANTILLES JUNIPER

Old leaves of long shoots with inconspicuous glands not extending to near the tip. Northern Bahamas, Cuba, and Jamaica. Synonyms: *Juniperus barbadensis* var. *jamaicensis* Silba, *J. lucayana* N. Britton.

Juniperus bermudiana Linnaeus

BERMUDA REDCEDAR

Juniperus sect. *Sabina*

Tree to 16 m tall, with trunk to 60 cm in diameter. Bark dark red, weathering grayish brown, thin, shallowly furrowed between peeling strips. Branches spreading. Branchlets four-angled, relatively coarse, 1.3–1.6 mm thick. Adult leaves of branchlets scalelike, in alternating pairs, gray-green to blue-green, about 1(–2) mm long, rounded or grooved externally, the edges smooth, the tips blunt or pointed, pressed against the twig, the glands inconspicuous (sunken and reaching almost to the tip in leaves of long shoots). Pollen and seed cones on separate plants. Pollen cones oblong, 4–6 mm long, with six to eight alternating pairs of pollen scales. Seed cones almost spherical to somewhat flattened, 4–5 mm long, dark blue with a waxy coating, maturing in a single season. Seeds one or two (or three), 2–3 mm long, shiny brown, the pale attachment scar extending about a third of the way up the sides. Bermuda. Formerly on limestone slopes and in swamps; 0–50 m. Zone 9.

Formerly important in shipbuilding and furniture making, most Bermuda redcedar trees were killed by scale insects introduced from the mainland United States. Other stands were eliminated by habitat conversion, and the state of the few remnant old trees in out of the way locations is precarious.

Juniperus blancoi M. Martínez

BLANCO JUNIPER, TÁSCATE (SPANISH)

Juniperus sect. *Sabina*

Shrub, or tree to 15 m tall, with trunk to 45 cm in diameter, usually branching well above the base. Bark brown, weathering grayish brown, thin, longitudinally peeling. Crown conical with horizontal or upturned branches. Branchlets somewhat four-angled, very fine, about 1 mm thick. Adult leaves scalelike, in alternating pairs, 1.5–2.0 mm long, green or yellowish green, the tip blunt or pointed, pressed against the stem, with a smooth margin, rounded on the back and with an elongate resin gland occupying about one-quarter of the center of the exposed portion. Pollen and seed cones on separate plants. Pollen cones nearly spherical, 2–3 mm long, with three or four alternating pairs of pollen scales.

Old mature tree form of Azores juniper (*Juniperus brevifolia*) under more favorable conditions.

Shrubby natural stand of Azores juniper (*Juniperus brevifolia*) under harsh conditions.

Seed cones wider than long, often bilobed to accommodate the two divergent seeds, about 3–6 mm long and 5–9 mm wide, dark bluish black with a thin waxy coating and a straight stalk. Seeds (1–)2(–5), 2–5 mm long, the prominent attachment about half as long as the seed. Western Mexico, from southern Sonora and Chihuahua to the state of Mexico. Moist woodlands of ravines; (1,200–)1,500–2,500(–2,900) m. Zone 8? Synonyms: *J. blancoi* var. *mucronata* (R. P. Adams) Farjon, *J. mucronata* R. P. Adams.

Blanco juniper is a very local species found in widely scattered sites along the Sierra Madre Occidental. It is the only purely Mexican species with smooth leaf margins. It is most closely related to Rocky Mountain juniper (*Juniperus scopulorum*), which extends into northern Mexico, and could be treated as a distinctive southern variety of that species. The species name honors Cenobio Blanco, the Mexican forest engineer who collected the type specimen near El Salto, Durango, in 1945 and further documented the trees with photographs.

Juniperus brevifolia (Seubert) F. Antoine

AZORES JUNIPER, CEDRO (PORTUGUESE)

Juniperus sect. *Juniperus*

Tree or shrub, to 6 m tall, with trunk to 50 cm in diameter. Bark reddish brown, thin, peeling in vertical strips. Crown broad, with a dense branching habit. Adult leaves needlelike, closely spaced, jointed to the twig, in alternating trios, 3–10 mm long, curved forward and overlapping, dark green, with two broad white stomatal bands above (on the inner face), the tip blunt. Pollen and seed cones on separate plants. Pollen cones slightly oblong, 3–5 mm long, with three to five alternating trios of pollen scales. Seed cones almost spherical but a little longer than broad, 8–10 mm long, dark reddish brown, maturing in 2 years. Seeds three, 5–6 mm long, three-angled and furrowed, light brown, the paler attachment scar just at the base. Azores. Open woodlands of mountain slopes; (240–)300–800(–1,525) m. Zone 9. Synonym: *J. oxycedrus* var. *brevifolia* Seubert.

Buried logs of Azores juniper far larger than any living trees suggest that the species was overexploited in the past and is capable of attaining larger dimensions than are seen today. Plantation culture of faster-growing exotic conifers and *Eucalyptus* species has also reduced the available habitat for Azores juniper. It is closely related to the other Macaronesian juniper, Canary Island juniper (*Juniperus cedrus*), and to prickly juniper (*J. oxycedrus*). It is also the juniper species found farthest out into the ocean, being about 1,000 km from the nearest stations of Canary Island juniper in Madeira and 1,750 km from mainland stands of prickly juniper on the Atlantic coast of Portugal. It was first described as a short-leaved variety of prickly juniper, hence the scientific name, Latin for "short leaf."

Juniperus californica Carrière

CALIFORNIA JUNIPER

Juniperus sect. *Sabina*

Shrub averaging 3.5 m, or tree to 8(–10) m tall, usually with several trunks from near the base to 40(–50) cm in diameter. Bark thin, pinkish brown, weathering gray, peeling in vertical strips. Crown dense, irregular, more or less spherical, with spreading or ascending branches. Branchlets cylindrical, relatively coarse, about 2 mm thick. Adult leaves of branchlets scalelike, in alternating trios, light green, 1–2(–3) mm long with a conspicuous oblong gland, the edges minutely toothed, the tips blunt or pointed, pressed against the twigs. Pollen and seed cones usually on separate plants. Pollen cones plumply egg-shaped, 4–6 mm long, with 9–10 alternating pairs of pollen scales. Seed cones spherical, usually 8–12(–18) mm long, bluish brown with a waxy coating, maturing in 1 year. Seeds one (or two) 5–7 mm long, shiny light brown, with a paler attachment scar two-thirds the length of the seed. Central Baja California, Mexico (including Cedros and Guadalupe Islands), to northern California and western Arizona. Open woodlands on dry, rocky hillsides; (250–)500–1,300(–1,600) m. Zone 9.

California juniper is the only juniper species found in the dry inner Coast Ranges of California and Baja California. Unlike other western junipers, it usually grows below the elevation band of pinyon pines (*Pinus* subsection *Cembroides*) and hence is not typically a species of pinyon-juniper woodland but rather of the drier desert shrublands downslope from them. Since it is usually of small stature, it has little economic importance. In fact, the largest known individual in combined dimensions when measured in 1976 was 10.1 m tall, 0.8 m in diameter, with a spread of 12.2 m.

Juniperus cedrus P. Webb & Berthelot

CANARY ISLAND JUNIPER, CEDRO CANARIO (SPANISH)

Juniperus sect. *Juniperus*

Tree to 15(–30) m tall, with trunk to 50(–100) cm in diameter, usually much smaller today in nature. Bark pinkish to purplish brown, weathering grayish brown, thin, shallowly ridged and furrowed, peeling in narrow strips. Crown narrowly conical, with slender, graceful branches from near the base bearing numerous drooping branchlets. Branchlets short, about 1 mm thick, six-angled, with a waxy coating. Adult leaves needlelike, standing out from and jointed to the twig, in alternating trios, 4–12(–20) mm long, with two stomatal bands above, dark green with a waxy coating at first beneath, the tip pointed but not prickly. Pollen and seed cones on separate plants. Pollen cones spherical, 3.5–5 mm long, with three to five alternating trios of pollen scales. Seed cones spherical, 8–12 mm long, reddish brown, with a bluish waxy coating, maturing in 1 year. Seeds one (to three), 5–10 mm long, light brown, the paler attachment scar extending halfway or more up the seed. Canary Islands (La Palma and Tenerife, formerly also on Gomera and Gran Canaria) and Madeira. Subalpine woodlands and upper ends of valleys; (450–)2,000–3,000 m. Zone 9. Synonyms: *J. cedrus* subsp. *maderensis* (C. Menezes) Rivas Martínez, Wildpret & P. Pérez, *J. oxycedrus* subsp. *maderensis* C. Menezes.

Canary Island juniper is a handsome tree of rapid growth that has largely been eliminated from its native home by overexploitation for its valuable wood. It could disappear from the wild but retains a minor role in horticulture, particularly in Mediterranean climate regions. It deserves to be planted more often. Furthermore, with its excellent quality wood and the fastest growth rate among the junipers, it could also serve as a candidate for timber plantations as a complement to warm climate pines.

Juniperus chinensis Linnaeus

CHINESE JUNIPER, YUAN BAI (CHINESE), BYAKUSHIN (JAPANESE), HYANG NAMU (KOREAN)

Plate 28

Juniperus sect. *Sabin*

Shrub, or tree to 20 m tall, often multistemmed, with trunk to 1.5 m in diameter. Bark reddish brown, longitudinally furrowed and flaking. Crown quite variable in form from spirelike to broad and spreading or even creeping, with spreading or ascending branches. Branchlets slender, to 1 mm thick, four- or six-angled. Adult leaves in alternating pairs, scalelike, about 1.5 mm long, bright green to bluish green with a pale smooth edge and an indented gland on the back, the tip blunt and pressed against the twig. Juvenile leaves usually also present and sometimes predominant on adult plants, in alternating trios (or pairs), needlelike, standing out from, and without a distinct attachment joint to, the twig, 6–12 mm long, with two bluish white stomatal bands above, ending in a sharp tip. Pollen and seed cones usually on separate plants. Pollen cones oblong, 3–4 mm long, with five to seven alternating pairs of pollen scales. Seed cones spherical, 6–9 mm in diameter, blue-black, covered with mealy whitish wax. Seeds (one or) two or three (to five), 3–5 mm long, grooved and pitted, brown, with a small basal attachment scar. Southeastern Russia (Sakhalin Island) through Japan and Korea to Mongolia and eastern and central China and Taiwan; (0–)300–1,600(–2,300) m. Zone 3. Synonyms: *J. gaussenii* W. C. Cheng, *Sabina chinensis* (Linnaeus) F. Antoine, *S. gaussenii* (W. C. Cheng) W. C. Cheng & W. T. Wang. Three varieties.

Chinese juniper is rather similar to the North American eastern redcedar (*Juniperus virginiana*) and Rocky Mountain juniper

(*J. scopulorum*) in general appearance, and it is sometimes difficult to assign cultivars to one or another of these species. Chinese juniper itself has produced numerous cultivars, both on its own and in the form of Pfitzer juniper (*J.* ×*pfitzeriana*), its hybrid with savin (*J. sabina*). While many of these cultivars have the normal two sets of chromosomes, others have three or four, and they vary in hardiness, habit, and proportion of juvenile and adult leaves. The varieties (especially variety *procumbens*) are often recognized as separate species, and the relationships among these taxa are in need of further study at the population level. Unfortunately, the taxonomic picture is somewhat confounded by the long history of cultivation in its homeland in eastern Asia so that truly wild populations are not always easy to recognize.

Juniperus chinensis Linnaeus **var. *chinensis***
YUAN BAI (CHINESE), IBUKI (JAPANESE)
Generally of tree habit and with adult leaves predominant. Japan (central and southern Honshū on the Pacific Ocean side, and Shikoku and Kyūshū), Korea, northeastern China, and Mongolia. Generally at low to moderate elevations. Widely cultivated in China beyond its natural range and used there for interior work and small items on account of its fragrant, reddish wood.

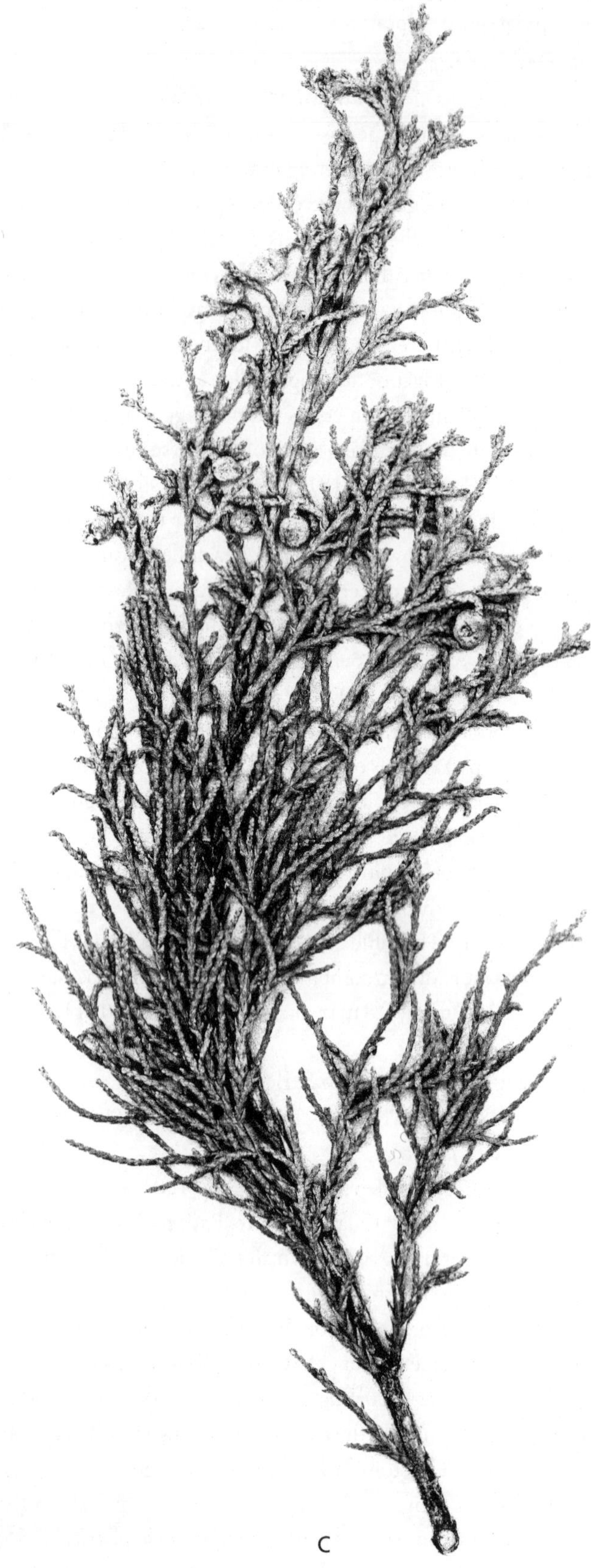

Densely branched twigs with juvenile (A), transitional (B), and adult (C) foliage of Chinese juniper (*Juniperus chinensis*), the latter bearing seed cones near the junction between the present and previous years' growth increments, ×1.

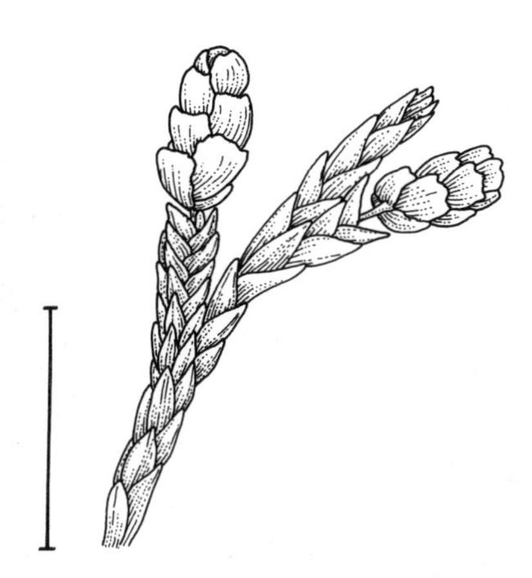

Twig with mature pollen cones of Chinese juniper (*Juniperus chinensis*); scale, 5 mm.

Juniperus chinensis* var. *procumbens (P. Siebold & Zuccarini) Endlicher

SONARE (JAPANESE), SOM HYANG NAMU (KOREAN)

Prostrate shrub, spreading to 2 m and up to 75 cm tall, with juvenile leaves predominant. Japan (northwestern Kyūshū) and southern Korea (Taehuksando Island). Seashores. Synonyms: *Juniperus procumbens* P. Siebold & Zuccarini, *Sabina procumbens* (P. Siebold & Zuccarini) J. Iwata & Kusaka.

Juniperus chinensis* var. *sargentii A. Henry

YAN BAI (CHINESE), SHINPAKU (JAPANESE), NUN HYANG NAMU (KOREAN)

Creeping shrub, spreading to 3 m and up to 80 cm tall, the branches sometimes turned up at the tips, with adult foliage predominant. Southeastern Russia (Sakhalin), Japan, Korea, and Taiwan. Open vegetation on rocky substrates from near sea level to subalpine fell fields at 2,300 m. Synonyms: *Juniperus chinensis* var. *taiwanensis* R. P. Adams & C. F. Hsieh, *J. chinensis* var. *tsukusiensis* Masamune, *J. sargentii* (A. Henry) G. Koidzumi, *Sabina sargentii* (A. Henry) K. Miyabe & Tatewaki.

Juniperus coahuilensis (M. Martínez) H. Gaussen ex R. P. Adams

ROSEBERRY JUNIPER, TÁSCATE (SPANISH)

Juniperus sect. *Sabina*

Shrubs dividing at the base or small trees to 8(–12) m tall, with trunk to 1 m high and 20(–100) cm in diameter. Bark thin (about 5 mm thick), light gray with a reddish cast, peeling in vertical strips. Crown broad and irregular, composed of ascending or spreading branches. Branchlets in dense masses, four- or six-angled, 1 cm or less long, 0.7–2 mm thick. Adult leaves in alternating pairs or trios, scalelike, bright green to yellowish green, those of permanent branchlets 4–6(–9) mm long. Scale leaves of branchlets 1–2(–3) mm long, the edge minutely toothed, glands oblong, the tips usually short-pointed and spreading away from the twig. Pollen and seed cones on separate plants. Pollen cones almost spherical or slightly elongated, 2–4 mm long, with six alternating pairs of pollen scales. Seed cones spherical to slightly pointed, 6–7 mm long, orange to red, appearing rose-pink because of the thin waxy coating, maturing in 1 year. Seeds one (or two), 4–5 mm long, brown with a paler attachment scar about half the length of the seed. Southwestern United States to northern Mexico. Dry grasslands and open oak juniper woodlands; (880–)1,200–2,000(–2,200) m. Zone 7? Synonyms: *J. erythrocarpa* Cory (misapplied), *J. erythrocarpa* var. *coahuilensis* M. Martínez. Two varieties.

Roseberry juniper is very closely related to Pinchot juniper (*Juniperus pinchotii*) and hybridizes with it where their ranges overlap, as in Big Bend National Park in Texas. Many authors do not distinguish this species from Pinchot juniper, but when it has been recognized, it has usually been called *J. erythrocarpa* Cory. The type specimen of that name, however, belongs in Pinchot juniper, so *J. erythrocarpa* is a synonym of *J. pinchotii* and the less familiar name, *J. coahuilensis,* must be used. Likewise, *J. pinchotii* var. *erythrocarpa,* applied to this juniper by Silba (1986), is also formally a synonym of Pinchot juniper, even though Silba was describing roseberry juniper. Ashe juniper (*J. ashei*) also occurs with roseberry juniper and Pinchot juniper in Big Bend National Park but does not hybridize with either of them. The northernmost populations of roseberry juniper differ from the main body of the species in volatile oils, DNA, and statistics of morphology and are recognized as a distinct variety. The largest known individual in the United States in combined dimensions when measured in 2001 was 8.5 m tall, 1.0 m in diameter, with a spread of 10.1 m.

Juniperus coahuilensis* var. *arizonica R. P. Adams

Glands of leaves on permanent long shoots about half as long as the attached leaf base. Southeastern Arizona and southwestern New Mexico.

Juniperus coahuilensis (M. Martínez) H. Gaussen ex R. P. Adams **var. *coahuilensis***

Glands of leaves on permanent long shoots about two-thirds as long as the attached leaf base. Southwestern United States (western Texas) and northern Mexico (Sonora and Coahuila through Chihuahua and Nuevo León to southern Durango and northeastern Zacatecas).

Juniperus comitana M. Martínez

MAYAN JUNIPER, SICOP (MAYAN)

Juniperus sect. *Sabina*

Tree to 12 m tall, with trunk to 80 cm in diameter. Bark grayish brown, peeling in longitudinal strips and only 5 mm thick. Crown irregular, rounded, and conical, with irregularly placed, outspread or rising branches. Branchlets slightly four-angled, very slender, up to 1 mm thick, 1.5–2 cm long. Adult leaves mostly in alternating pairs, scalelike, 1–1.5 mm long, bright green, with a depression on the back that may or may not contain a small gland, the edges paler and minutely toothed, the tip pointed and pressed against the twigs. Pollen and seed cones on separate plants. Pollen cones oblong, 4–6 mm long, with six alternating pairs of pollen scales. Seed cones oblong to spherical, 5–8(–10) mm long, dark violet with a waxy coating, maturing in 1 year. Seeds one (or two), 4–6 mm long, yellowish brown, the paler attachment scar small. Southeastern Mexico (Chiapas) and Guatemala. Dry forests and woodlands on rocky, limestone substrates; 1,200–2,300 m. Zone 10?

Mayan juniper is the southernmost of the New World one-seeded junipers, a group of six species that extend north to Montana. It is uncommon and local throughout its limited range and appears to have little economic importance. Despite the lack of significant direct exploitation, the species is threatened by general logging and habitat conversion accompanying human population growth.

Juniperus communis Linnaeus

COMMON JUNIPER, ALMINDELEG ENE (DANISH), KOTIKATAJA (FINNISH), MOZHZHEVEL'NIK OBYKNOVENNYI (RUSSIAN), GENÉVRIER COMMUN (FRENCH), ENEBRO (SPANISH), IPAR IPURUA (BASQUE), GEMEINER WACHOLDER (GERMAN), PAPRASTASIS KADAGYS (LITHUANIAN), KÖZÖNSÉGES BORÓKA (HUNGARIAN), JAŁOWIEC POSPOLITY (POLISH), ADI ARDIC (TURKISH), AKHSALI (OSSETIAN), XIAN BEI CI BAI (CHINESE), RISHIRI-BYAKUSHIN (JAPANESE)

Juniperus sect. *Juniperus*

Of quite variable habit, low or tall shrub to 4–5 m high or tree to 15 m tall, usually multistemmed, with stems to 20(–30) cm in diameter. Bark reddish brown, peeling in longitudinal strips or flakes. Crown ranging from wide and flattened to columnar, with spreading branches. Branchlets with three ridges. Adult leaves in alternating trios, needlelike, (4–)10–20(–25) mm long, 1–2 mm wide, dark green, standing out from and with a jointed attachment to the twig, the edges smooth, the upper (forward) side flat, with a single broad or narrow white-waxy stomatal band, the lower side with an elongate gland, the tip spinelike. Pollen and seed cones on separate plants. Pollen cones oblong, 4–5 mm long, with five or six alternating trios of pollen scales. Seed cones spherical when fully mature, otherwise slightly elongated, 6–9(–13) mm long, bluish black, maturing in 2–3 years. Seeds two to four, 4–5 mm long, with resinous pits, dark brown, the paler attachment scar

Branching twigs of common juniper (*Juniperus communis*) with newly emerging branchlets (A), pollen cones releasing pollen (B), and three generations of seed cones (C), the latter currently receptive, immature at 1 year old, and mature at 2 years old, ×0.5.

Mature seed cone of common juniper (*Juniperus communis*) in a position corresponding to and replacing an axillary shoot; scale, 1 cm.

extending more than halfway up the seed. Widespread across the northern hemisphere at middle to high latitudes. Varied open, dryish habitats in boreal and montane regions; 0–2,800(–4,000) m. Zone 2. Four varieties.

Common juniper is the most widely distributed conifer in the world and one of the most widely distributed woody plants in the northern hemisphere, with a tremendous longitudinal, latitudinal, and elevational range. It is the most northerly juniper and one of the handful of most northerly conifers in the world. The extensive ecogeographic amplitude and the accompanying morphological variation have provided ample opportunity for selection and breeding of cultivars. The enormous range also brings common juniper into contact with an vast number of language communities. As a result, it probably has the largest number of common names of any conifer species, only a small sample of which are listed above, chosen for variation, the names in many languages being fairly similar to one another.

Common juniper is distinguished from the other species in section *Juniperus* by the undivided (except sometimes at the base) stomatal band on the flat or shallowly grooved upper side of the leaves. Although Franco (1962) assigned all of the arctic alpine occurrences around the world, with their ground-hugging habit and short needles, to a single subspecies, they may have been derived independently in each region. This might be resolved with molecular biological techniques. Such studies might also help with other taxonomic uncertainties in this group. Acceptance of an inclusive species is countered by at least as many authors who divide it into two, three, or more species, only partially corresponding to the subspecies recognized here. Clearly more work is needed on this widespread complex, which, if it is indeed a single species, is unique among conifers for its presence in both the Old and New Worlds.

The "berries" of common juniper are the main flavoring of gin and also have medicinal applications, but the plant is rarely large enough to be a significant source of wood products, except for small turned objects and a varnish derived from the wood oils. The largest known individual in the United States in combined dimensions when measured in 1993 was 14.0 m tall, 0.3 m in diameter, with a spread of 8.5 m.

Juniperus communis subsp. *alpina* (Suter) L. J. Čelakovský

Ground-hugging or low, spreading shrub to 1(–2) m high with upturned branchlets. Needles closely spaced, curved forward, 5–15 mm long, 1–2 mm wide, the stomatal band at least twice as wide as each green margin. High mountains of southern Eurasia and western North America (British Columbia, Canada, to central California) and low elevations of cool maritime and arctic regions in Eurasia, Iceland, and southern Greenland. Synonyms: *Juniperus communis* var. *hondoensis* Satake, *J. communis* var. *jackii* Rehder, *J. communis* var. *montana* W. Aiton, *J. communis* var. *nana* (Willdenow) Boissier, *J. communis* var. *nipponica* (Maximowicz) E. H. Wilson, *J. communis* var. *saxatilis* P. Pallas ex Willdenow, *J. rigida* subsp. *nipponica* (Maximowicz) Franco, *J. sibirica* Burgsdorff.

Juniperus communis Linnaeus subsp. *communis*

Upright shrub, or tree to 15 m tall, with upturned branchlets. Needles widely spaced, usually straight, 6–20(–25) mm long, 1–1.5 mm wide, the stomatal band about twice as wide as each green margin. Throughout most of Europe and Siberia east to the Lena River. At higher elevations in the south than in the north, but lower than the other two Eurasian subspecies.

Juniperus communis subsp. *depressa* (Pursh) Franco

Ground-hugging shrub to small tree (rarely) to 10 m tall, the branchlets turned up. Needles closely to widely spaced, curved gently forward, up to 15 mm long, 1.6 mm wide, the stomatal band narrower than, to as wide as, each green margin. North America, from Alaska to Newfoundland, south in the Rocky Mountain region to Arizona and New Mexico, absent from the high arctic, the Great Plains, the Pacific Coast region, and the southeastern coastal plain of the United States. Synonym: *Juniperus communis* var. *megistocarpa* Fernald & H. St. John (individuals from the Gulf of St. Lawrence region in eastern Canada with very large seed cones to 12 mm in diameter).

Juniperus communis* subsp. *hemisphaerica (J. & C. Presl) C. Nyman

Dense, rounded shrub to 2.5 m high with upturned branchlets. Needles closely spaced, straight, 4–12(–20) mm long, 1.3–2 mm wide, the stomatal band at least twice as wide as each green margin. Thinly scattered in mountains around the Mediterranean from Spain (where most abundant) and Morocco east to the Crimea (Ukraine) and Turkey.

Juniperus conferta Parlatore

JAPANESE BEACH JUNIPER, HAI-NEZU (JAPANESE)

Juniperus sect. *Juniperus*

Shrub to 0.5 m high, spreading to several meters. Branches dense, mat forming. Branchlets turned up at the tips. Adult leaves closely spaced, in alternating trios, needlelike, 10–17 mm long, 1–1.5 mm wide, curled forward along the twigs and attached to them by a jointed base, dull green, with a deep groove above containing a narrow, waxy, white, stomatal band, generally triangular in cross section, the tip prickly. Pollen and seed cones on separate plants. Pollen cones slightly longer than spherical, 3–5 mm long, with three or four alternating trios of pollen scales. Seed cones spherical, purplish black with a thin waxy coating, fleshy, about 1 cm across, maturing in 18 months. Seeds three, 5–7 mm long, three-angled and pitted, brown with a lighter attachment scar. Along the coast, central Sakhalin Island (Russia) and Japan (south through western Hokkaidō to south-central Honshū). Stabilized sand dunes by the seashore; 0–50(–100) m. Zone 6. Synonyms: *J. litoralis* Maximowicz, *J. rigida* subsp. *conferta* (Parlatore) Kitamura.

Twig of Japanese beach juniper (*Juniperus conferta*) with 2 years of growth, ×1.

Juniperus convallium Rehder & E. H. Wilson

TWIGGY JUNIPER, MEKONG JUNIPER, MI ZHI YUAN BAI (CHINESE)

Juniperus sect. *Sabina*

Tree to 12(–20) m tall, with trunk to 50 cm in diameter. Bark light gray, peeling in vertical strips. Crown dense, narrow, with upwardly arching branches. Branchlets closely spaced, not angled, slender, ca. 1 mm thick, erect to drooping, sometimes waxy. Adult leaves in alternating pairs or trios, scalelike, 1–1.5 (–2) mm long, dull green, often very waxy, with a depressed oil gland on the back, the edges pale and hardened, the tip blunt to pointed, pressed against the twig. Pollen and seed cones on the same or separate plants. Pollen cones nearly spherical, 2–3 mm long, with three or four pairs or pollen scales. Seed cones on a curved stalk, slightly elongated, (4–)6–8(–10) mm long, chestnut brown to almost black, usually without a waxy coating, maturing in 1 year. Seed one, 4–7 mm long, nearly spherical, pitted or furrowed, brown, with a paler attachment scar just at the base. Southwestern China: southern Gansu to northwestern Yunnan and Xizang (Tibet). Dry mountain slopes; (2,200–) 2,500–4,000(–4,430) m. Zone 6. Synonyms: *J. convallium* var. *microsperma* (W. C. Cheng & L. K. Fu) Silba, *J. mekongensis* V. Komarov, *J. microsperma* (W. C. Cheng & L. K. Fu) R. P. Adams, *J. ramulosa* Florin, *Sabina convallium* (Rehder & E. H. Wilson) W. C. Cheng & L. K. Fu, *S. microsperma* (W. C. Cheng & L. K. Fu) W. C. Cheng & L. K. Fu.

Twiggy juniper is an uncommon and somewhat obscure but distinctive juniper species of the steep-walled arid valleys of the headwaters of the principal Southeast Asian rivers, from which it takes its scientific name, Latin for "of closed valleys." It has little economic importance and is not in general cultivation.

Juniperus davurica P. Pallas

DAHURIAN JUNIPER, VERESK KAMENNYI (RUSSIAN), XING AN YUAN BAI (CHINESE)

Juniperus sect. *Sabina*

Spreading shrub to 0.5(–1) m from a short trunk. Bark purplish brown, weathering gray, thin, flaking. Crown flat to the ground, with branches gently rising at the tips. Branchlets four-sided, slender, about 1 mm thick. Adult leaves in alternating pairs, scalelike near the tips of the twigs, otherwise mostly needlelike but without a distinct joint where they attach to the twig. Needlelike leaves 3–9 mm long, with two broad white stomatal bands above scarcely separated from one another. Scalelike leaves 1–3 mm long, with a prominent resin gland on the rounded face. Pollen and seed cones on separate plants. Pollen cones quite oblong, 4–5 mm long, with six to nine alternating pairs of pollen scales. Seed cones on a curved branchlet, spherical, 5–6 mm through, dark brown with a bluish waxy coating, maturing in 1 year, Seeds usually two to four, about 5 mm long, shallowly grooved and pitted, light brown, with a paler small attachment scar. Far eastern Russian east of Lake Baikal (known as Dahuria, hence the scientific and common names), northern Mongolia, northeastern China. Rocky slopes and streams of mountains, 400–1,400 m. Zone 6. Synonyms: *J. sabina* var. *davurica* (P. Pallas) Farjon, *Sabina davurica* (P. Pallas) F. Antoine.

Dahurian juniper resembles savin (*Juniperus sabina*) and is sometimes united with it, but the juvenile foliage persists for much longer, with scalelike leaves appearing only at the tips of some branchlets, especially those bearing pollen or seed cones. Its relationship to savin should be explored with detailed population studies.

Juniperus deppeana Steudel

ALLIGATOR JUNIPER, TÁSCATE, CEDRO CHINO (SPANISH)

Juniperus sect. *Sabina*

Tree to 8(–30) m tall, with single or multiple trunks to 0.5(–2) m in diameter. Bark reddish brown, weathering dark gray, to 20 cm thick, usually strongly and regularly blocky and checkered (giving rise to the English common name). Crown broad, irregular, and open, with spreading or rising branches. Branchlets four- (or six-) sided, slender, 1–1.5 mm thick. Adult leaves in alternating pairs (or trios), scalelike, 1–2 mm long, often bluish green with a thin waxy coating, the edges minutely toothed, the tip pointed or with a short bristle pressed against the twig. Pollen and seed cones on separate plants. Pollen cones quite oblong (2.5–)3–6 mm long, with four to eight alternating pairs of scales. Seed cones on a straight or curved branchlet, spherical, 8–15(–20) mm through, reddish brown with a thin or thick waxy coating, maturing in 2 years. Seeds (one to) three to five (or six), 6–10 mm long, dark brown, with a pale attachment scar extending more than halfway up from the base. Southwestern United States (Arizona, New Mexico, and trans-Pecos Texas) south in Mexico through both the Sierra Madre Occidental and the Sierra Madre Oriental to the state of Puebla. Rocky slopes of foothills and mountains in various oak-pinyon-juniper woodlands and savannas; (750–)1,500–2,700(–3,200) m. Zone 8. Four varieties.

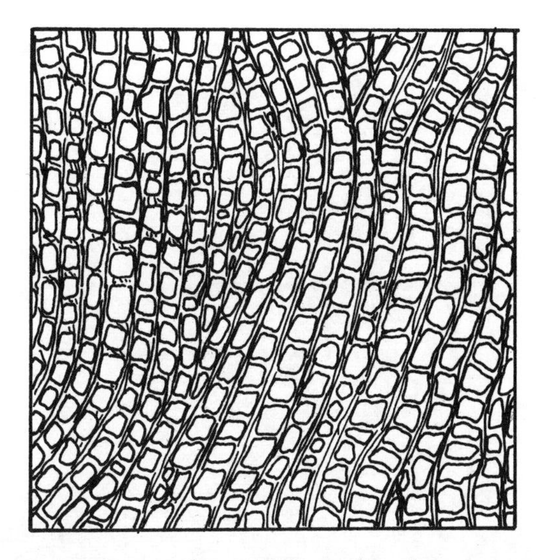

Typical checkered bark pattern of alligator juniper (*Juniperus deppeana* var. *deppeana*).

A number of geographic races of this most variable New World species have been described, differing primarily in stature and habit, in seed cone size and amount of waxy coating, and in how much of the trunk has checkered bark. Scattered individuals with furrowed bark may be derived by hybridization, but there is no independent evidence of this origin. The largest known individual in the United States in combined dimensions when measured in 1995 was 14.0 m tall, 2.7 m in diameter, with a spread of 14.9 m.

Juniperus deppeana Steudel **var. *deppeana***

Tall shrub, or tree to 10 m tall, with checkered bark throughout. Scale leaves with an oval oil gland that may rupture. Seed cones 8–15 mm in diameter, with a thin coating of wax. Southwestern United States and eastern Mexico (Coahuila to Puebla). Synonym: *Juniperus deppeana* var. *pachyphlaea* (J. Torrey) M. Martínez.

Juniperus deppeana **var. *robusta*** M. Martínez

Tree to 10–25(–30) m tall, usually with checkered bark throughout but rarely (in Durango) with longitudinally furrowed bark except at the very base of the trunk. Scale

Large old alligator juniper (*Juniperus deppeana* var. *robusta*) in nature.

leaves occasionally with a small, round oil gland. Seed cones 8–15 mm in diameter, with a thin to moderate coating of wax. Western Mexico (Sonora to Durango). Synonyms: *Juniperus deppeana* var. *pattoniana* (M. Martínez) Zanoni, *J. pattoniana* M. Martínez (based on individuals with furrowed bark).

Juniperus deppeana* var. *sperryi D. Correll

Tree to about 15 m tall, with longitudinally furrowed bark. Branchlets with the drooping posture of *Juniperus flaccida* (all other variants have upright branchlets) but apparently not derived by hybridization with that species, based on comparisons of oil content in foliage. Scale leaves with an oval oil gland that may rupture. Seed cones about 11 mm in diameter, with a thin coating of wax. Very rare in western Texas and in northeastern Sonora (Mexico). Synonym: *Juniperus deppeana* f. *sperryi* (D. Correll) R. P. Adams.

Juniperus deppeana* var. *zacatecensis M. Martínez

Shrub to 3–5(–8) m, with checkered bark throughout. Scale leaves usually with an elongate, closed gland near their base. Seed cones 13–20 mm in diameter, with a thick, white coating of wax. Eastern Durango and western Zacatecas, Mexico.

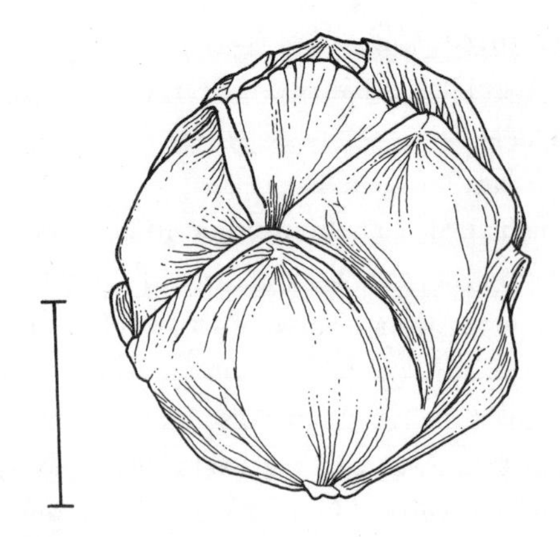

Mature seed cone of Syrian juniper (*Juniperus drupacea*) with prominent pattern of seed scales; scale, 1 cm.

Juniperus drupacea Labillardière

SYRIAN JUNIPER, DOUFRANE (ARABIC), ANDIZ (TURKISH)

Juniperus sect. *Juniperus*

Tree to 15(–40) m tall, with trunk to 1.2(–2) m in diameter. Bark brownish gray, peeling in narrow vertical strips. Crown narrowly conical when young, broadening greatly with age, with gently rising branches. Branchlets with three ribs and three grooves. Adult leaves in alternating trios, needlelike, 1.5–2.5 cm long, 1.5–4 mm wide, bright green, standing out from and with a jointed attachment to the twig, ridged beneath, with two prominent white stomatal bands above, the tip prickly. Pollen and seed cones on separate plants. Pollen cones in tight clusters of three to six in the axils of leaves, nearly spherical, 4–7 mm long, with four to six alternating trios of pollen scales. Seed cones single in the axils of leaves, nearly spherical to slightly elongate, 2.0–2.5 cm in diameter, brown or bluish brown with a heavy waxy coating, maturing in 2 years. Seeds three, united as a central pit, thus resembling a cherry with its pit (or drupe, hence *drupacea*), about 1.5 cm long. Greece (Peloponnesus), southern Turkey, western Syria, and Lebanon. Rocky slopes in forest or brushland; 500–1,900 m. Zone 8. Synonym: *Arceuthos drupacea* (Labillardière) F. Antoine & Kotschy.

The uncommon Syrian juniper is one of the most distinctive needle-leaved junipers. It has the longest needles and largest seed cones in the genus, and the clustered pollen cones are unique among junipers while the united seeds are unique among living conifers. The pulp surrounding the seeds is used to produce the traditional Turkish fruit concentrate called *pekmez*. The outlines of the individual scales that make up the seed cone are clearly marked on its surface. These indications of structure are nearly obliterated during cone development in most other juniper species. The wood of Syrian juniper is of high quality, with the decay resistance so common in the family, but the trees generally do not grow in exploitable stands and large trees are rare, so there is little commercial use of the species. Trees in cultivation typically retain the narrow growth habit of youth.

Juniperus durangensis M. Martínez

DURANGO JUNIPER, CEDRO, TÁSCATE (SPANISH)

Juniperus sect. *Sabina*

Shrubby tree to 7 m tall, usually dividing near the base, with short trunk to 50 cm in diameter. Bark yellowish brown, weathering grayish brown, thin, peeling in vertical strips. Crown irregular and open, with spreading or rising, twisted branches. Branchlets congested, usually four- or six-angled, 1.5 mm thick or more. Adult leaves in alternating pairs (or trios), scalelike, dark gray-green, 1–2 mm long, strongly rounded, the edge minutely toothed, the tip usually blunt. Pollen and seed cones on separate plants. Pollen cones plumply egg-shaped, 2–3 mm long, with five or six alternating pairs of pollen scales. Seed cones on a short, straight branchlet, round or with swellings over the seeds, 4–7 mm through, reddish brown with a thin, bluish, waxy coating. Seeds one or two (to four), 3–4 mm long, reddish brown, the pale attachment scar extending up to halfway up the side. Sierra Madre Occidental of western Mexico from northeastern Sonora to northern Jalisco, southern Zacatecas, and Aguascalientes. Open mixed woodlands on dry, rocky sites; 1,600–2,900 m. Zone 9.

The relationships of Durango juniper, like those of many Mexican juniper species, are unclear. It has short, closely spaced branchlets and shorter leaves than are found in many of the other species. The scale leaves of the side branchlets are only about 1 mm long and wider than long. These features are hardly exclusive to Durango juniper and do not point to any particular closest relatives. Although moderately widespread in northwestern Mexico, especially in the state of Durango from which it gets its name, it is not common and has little economic significance in a land of abundant pine (*Pinus*) timber.

Juniperus excelsa M. Bieberstein

TETHYAN JUNIPER

Juniperus sect. *Sabina*

Tree to 25 m tall, with trunk to 2.5 m in diameter, or dwarfed at the alpine timberline. Bark reddish brown, weathering dark grayish brown, longitudinally furrowed. Crown broadly conical to irregular, with branches initially outstretched and rising but eventually crooked and drooping with age. Branchlets cylindrical or four-angled, 0.7–1.3 mm in diameter. Adult leaves in alternating pairs, mostly scalelike, 0.6–1.6 mm long, light green to yellowish green, with a prominent, often resinous, round or oblong oil gland, edges smooth, the tip curved in toward the twig or turned outward. Pollen and seed cones on the same or separate plants. Pollen cones solitary at the tips of short branchlets, shortly oblong, 3–4 mm long, with four or five alternating pairs of pollen scales. Seed cones solitary in leaf axils near the tips of branchlets, spherical, 6–14 mm through, purplish brown, often with a waxy coating, maturing in 2 years. Seeds two to eight, 4–6 mm long, reddish brown with a pale attachment scar scarcely extending beyond the base. Tethyan region from the Balkan Peninsula and the Crimea (Ukraine) to Oman, Pakistan, and Uzbekistan. Dry rocky slopes of hills and mountains in juniper woodlands and conifer and mixed forests; 100–3,900 m. Zone 7. Two subspecies.

Despite its scientific name (Latin for "tall"), Tethyan juniper is not among the largest species in the genus, but it is far taller than most of the other eastern Mediterranean species. It is closely related to the even taller East African juniper (*Juniperus procera*), and the two were sometimes united in a single species. Conversely, the two subspecies were often treated as separate species, and additional segregate species were often recognized in the Russian taxonomic tradition of narrow species concepts. In a detailed study of the sort too rarely applied to Asian junipers, Farjon (1993) closely examined Tethyan juniper from throughout its range, assessing proposed distinguishing characters of the segregates, and showed how the two subspecies intergrade in their region of overlap around the Caspian Sea.

Juniperus excelsa M. Bieberstein **subsp. *excelsa***

GRECIAN JUNIPER, CRIMEAN JUNIPER, BOYLU ARDIÇ (TURKISH), DEDALI-GVIA (GEORGIAN), BURS (TADZHIK)

Branchlets less angular, 0.7–1 mm thick, often in flattened arrays. Scale leaves of branchlets 0.6–1.1 mm long, the tip pressed against the twig. Seed cones 6–11 mm through, with (two or) three to six (to eight) seeds. Western part of the range of the species east to Turkmenistan. In moister habitats (500–1,000 mm annual rainfall), where it overlaps with subspecies *polycarpos*; 100–2,300(–2,700) m. Synonym: *Juniperus isophyllos* K. Koch.

Juniperus excelsa **subsp. *polycarpos*** (K. Koch) Takhtajan

PERSIAN JUNIPER, TURKESTAN JUNIPER, KARA-ARCHA (KIRGHIZ)

Branchlets notably angular, 1–1.3 mm thick, branching in three dimensions. Scale leaves of branchlets 1.2–1.6 mm long, the tip bending away from the twig. Seed cones 6–14 mm through, with (two or) three or four (to six) seeds. Eastern part of the range of the species west to the Caucasus. In drier habitats (around 500 mm annual rainfall), where it overlaps with subspecies *excelsa*; 500–3,000(–3,900) m. Synonyms:

Juniperus macropoda Boissier; *J. polycarpos* K. Koch; *J. seravschanica* V. Komarov, *J. turcomanica* B. Fedtschenko.

Juniperus ×fassettii B. Boivin

Low shrubs often with the foliage and cone features of Rocky Mountain juniper (*Juniperus scopulorum*) but the habit of creeping juniper (*J. horizontalis*) or influenced by creeping juniper and not reaching tree form. This hybrid occurs naturally in the northern Rocky Mountains of Canada and the United States where the parent species overlap locally. The resulting hybrid swarms contain individuals with varied combinations of characters of the parent species. A number of cultivars were selected from these wild populations. Zone 3. Synonym: *J. scopulorum* var. *patens* Fassett.

Juniperus flaccida D. F. L. Schlechtendal

DROOPING JUNIPER, CEDRO LISO (SPANISH)

Juniperus sect. *Sabina*

Tree to 12(–20) m tall, unbranched for 1–2(–5) m, with trunk to 50 cm in diameter. Bark dark reddish brown, to about 1 cm thick, with flaky, interlacing ridges between longitudinal furrows. Crown broad and rounded, with spreading branches. Branchlets about 0.8 mm in diameter, limp and drooping (hence both the common and scientific names), often branching to form flattened arrays. Adult leaves in alternating pairs, scalelike, bright yellowish green to ashy gray, 1.5–2 mm long, often with a prominent round or elongate oil gland that may have a crust of dried resin, edges with very tiny teeth visible only at 40× magnification, the tip extended and pointed, standing out from the twig or bent forward along it. Pollen and seed cones on the same plant. Pollen cones oblong, about 2.5 mm long, with six alternating pairs of pollen scales. Seed cones on straight branchlets, spherical, 8–15(–20) mm in diameter, dark reddish brown with a thin bluish or purplish waxy coating, maturing in 2 years. Seeds 1–13, 5–6 mm long, reddish brown with a paler attachment scar just at the base. Big Bend National Park, western Texas, and northeastern Sonora (Mexico), south in Mexico through the Sierra Madre Oriental and Central Plateau, and scattered in the Sierra Madre Occidental to western Oaxaca. Dry, sunny, open juniper and mixed woodlands and grasslands, often on calcareous substrates; (900–)1,200–2,500(–2,900) m. Zone 9. Three varieties.

Drooping branchlets and large cones make this one of the more distinctive New World junipers. These and other distinctive features of drooping juniper have made it hard to discern its relationships to other species, although they probably lie within the region. It is also one of the most common and widespread junipers in eastern Mexico but is more local in the west. The three recognized varieties differ in their leaves and seed cones. The largest known individual in the United States in combined dimensions when measured in 1982 was 16.8 m tall, 0.8 m in diameter, with a spread of 10.7 m.

Juniperus flaccida D. F. L. Schlechtendal **var. *flaccida***

Scale leaves of branchlets with the tips pressed against the twig. Seed cones typically 10–15 mm in diameter, smooth, without obvious fissures between the fused scales, mostly with six to eight seeds. This is the common variety found throughout most of the range of the species except for portions of the Central Plateau and Guerrero in Mexico.

Juniperus flaccida **var. *martinezii*** (Pérez de la Rosa) Silba

Scale leaves of branchlets with slightly spreading tips. Seed cones 5–9 mm in diameter, smooth, without obvious fissures between the fused scales, with one or two (or three) seeds. Central part of the range of the species on the Central Plateau of Mexico from Aguascalientes and northeastern Jalisco to southern San Luis Potosí and northern Puebla. Synonym: *J. martinezii* Pérez de la Rosa. The proper placement of this variety is ambiguous. It resembles the one-seeded roseberry juniper (*Juniperus coahuilensis*) in some ways and the drooping *J. flaccida* var. *flaccida* in others. Analyses of foliage oils and DNA favor association with drooping juniper.

Juniperus flaccida var. poblana M. Martínez

Scale leaves of branchlets with the tips spreading outward. Seed cones up to 20 mm in diameter, with conspicuous fissures between the cone scales, often with 8–10 seeds. Southern part of the range of the species in Mexico, from central Jalisco and Morelos to western Oaxaca. An uncommon variant of uncertain status, usually found with or near variety *flaccida*.

Juniperus foetidissima Willdenow

STINKING JUNIPER, KOKULU ARDIÇ (TURKISH), TVIA (GEORGIAN) TSRTNENI (ARMENIAN), ERKE ARDYSH (TATAR)

Juniperus sect. *Sabina*

Tree to 15(–20) m tall (or a creeping shrub near timberline), with single trunk to 2(–3.5) m in diameter. Bark gray, furrowed and peeling in long vertical strips. Crown dense and irregular, with spreading branches. Branchlets dense, four-sided, coarse, 1.2–2 mm thick, bad smelling when crushed (hence both the common and scientific names). Adult leaves in alternating pairs, scalelike, those of branchlets 2–3 mm long, usually without conspicuous glands, the edges smooth, the tip bluntly pointed, pressed against the twig or slightly spreading. Pollen and seed cones on the same or different plants. Pollen cones nearly spherical, 2–3 mm long, with four to six alternating pairs of pollen scales. Seed cones on a very short straight branchlet, spherical, 5–13 mm through, dark blue or blackish with a waxy coating, maturing in 2 years. Seeds one or two (or three), stuck together, the combined mass 5–7 mm thick, light brown, with an inconspicuous basal attachment scar. Balkan Peninsula, Turkey, Cyprus, northern side of the Black Sea, and the Caucasus. In open places of mixed woodlands and forests, usually in the mountains. Near sea level (in the Crimea) to almost 2,000 m (in Cyprus). Zone 9. Synonyms: *J. foetidissima* var. *pindicola* Formánek, *J. sabinoides* Grisebach.

Stinking juniper has much the same range and habitat as the related Grecian juniper (*Juniperus excelsa* subsp. *excelsa*) and often grows with it, though it may be found on drier sites. It differs in its coarser, fetid foliage with thicker branchlets and longer scale leaves, and in having fewer seeds packed together in a coherent mass. Are there any ecological consequences, especially in palatability to herbivores, of the chemical differences in foliage between stinking juniper and Grecian juniper (and other co-occurring species) that manifest themselves in the fetidness of damaged tissue in the former?

Juniperus formosana Hayata

TAIWAN JUNIPER, CI BAI (CHINESE)

Juniperus sect. *Juniperus*

Shrubby, usually multistemmed tree to 5(–16) m tall, with trunk to 40 cm in diameter. Bark gray-brown, peeling in longitudinal strips. Crown narrow, with spreading or rising branches. Branchlets triangular, drooping. Adult leaves in alternating trios, needlelike, jointed at their attachment to the stem, (4–)12–25 mm long, 1–2 mm wide, with two prominent white stomatal bands above, green or waxy blue-green and with a central ridge beneath, sharp-pointed. Pollen and seed cones on separate plants. Pollen cones single in the leaf axils, egg-shaped, 1.5–2 mm long, with five or six alternating trios of pollen scales. Seed cones single on a short stalk in the leaf axils, spherical or somewhat elongate, (6–)8–10(–12) mm long, orange to reddish brown, without a waxy coating, maturing in 2 years, the top with three radiating furrows between the seed scales. Seeds three, 3–4 mm long, light brown, with three or four resin pockets in the basal attachment scar. Widespread in China, from eastern Xizang (Tibet) and Gansu to Jiangsu and Guangdong, central ranges of Taiwan. Understory tree in conifer and mixed forests of mountains, primarily *Chamaecyparis* forests in Taiwan; (400–)2,300–3,000(–4,000) m. Zone 9. Synonym: *J. formosana* var. *concolor* Hayata.

Taiwan juniper is a highly ornamental species, commonly planted in China but little known elsewhere. Specimens with the smallest leaves may be found in some dwarfed individuals at the alpine timberline. Taiwan juniper is closely related to other members of the complex of needle-leaved species in section *Juniperus* occupying the maritime regions of eastern Asia, such as temple juniper (*J. rigida*) and Ogasawara juniper (*J. taxifolia*), and the relationships among and classification of these taxa is in need of further study. Taiwan was known as Formosa (Latin for "beautiful") during the Japanese occupation, when the species was described.

Juniperus gamboana M. Martínez

GAMBOA JUNIPER, NUHKUPAT (TZELTAL MAYAN)

Juniperus sect. *Sabina*

Tree to 12 m tall, unbranched for 1–2 m, with trunk to 50 cm in diameter. Bark ashy gray, breaking up into blocks up to 15 mm thick. Crown broadly egg-shaped, with spreading or ascending, irregular branches. Branchlets cylindrical, slender, about 1 mm thick. Adult leaves mostly in alternating pairs, scalelike, 1.5–2 mm long, yellowish green, with a round resin gland in the center, the edges minutely toothed, the tip pointed, pressed against the twig. Pollen and seed cones on separate plants. Pollen cones single at the tips of branchlets, oblong, about 6 mm long, with six to eight alternating pairs of pollen scales. Seed cones single at the tips of short, straight, lateral branchlets, spherical, 5–8(–9) mm in diameter, yellowish red with a thin waxy coating at maturity, maturing in 1 year. Seeds one (or two), 4–6 mm long, with two or three longitudinal furrows, chestnut brown with a paler attachment scar up to two-thirds the length of the seed. Southeastern Mexico (Chiapas) and western Guatemala (Huehuetenango). Pine-oak-juniper woodlands of the mountains on limestone; 1,600–2,300 m. Zone 9?

The uncommon Gamboa juniper combines the alligator-skin bark of alligator juniper (*Juniperus deppeana*) with the one-seeded red seed cones of roseberry juniper (*J. coahuilensis*). Although the morphology and leaf oils of Gamboa juniper somewhat resemble those of the nearby one-seeded, violet-coned Mayan juniper (*J. comitana*), DNA analyses point toward a close relationship to alligator juniper. It may have the same kind of relationship to alligator juniper as *J. flaccida* var. *martinezii* has to the other varieties of drooping juniper (*J. flaccida*).

Juniperus gracilior Pilger

HISPANIOLAN JUNIPER, SABINA (SPANISH), CÈDRE (FRENCH)

Juniperus sect. *Sabina*

Tree to 15 m tall, or a creeping shrub at high elevations, with trunk to 2 m in diameter. Bark brown, weathering grayish brown, peeling in flakes to leave irregular plates. Crown rounded, with spreading or rising branches. Branchlets four-sided, slender, 0.7–1.0 mm thick, spreading or drooping. Adult leaves in alternating pairs (or trios), scalelike, 1–1.5 mm long, with or without conspicuous resin glands, the edges smooth, the tip abruptly long- or short-pointed. Pollen and seed cones on separate plants. Pollen cones 1–3 mm long, nearly spherical, with three to five alternating pairs of pollen scales. Seed cones single on a straight branchlet 1–2 mm long, spherical or bilobed, 5–7 mm long, reddish brown with a bluish waxy coating, maturing in 1 year. Seeds one or two, 3–4 mm long, light brown, the basal attachment scar rough and pitted. Hispaniola. Seasonally deciduous dry forest to subalpine scrub; 1,000–2,550 m. Zone 9? Three varieties.

Natural stand of creeping juniper (*Juniperus horizontalis*) hugging a rock pavement.

Old creeping juniper (*Juniperus horizontalis*) in nature with snakelike trunk and many mature seed cones.

Besides the three varieties of this species, Antillean redcedar (*Juniperus barbadensis* var. *lucayana*) is the only other juniper found on Hispaniola (in northern Haiti). It may now be extirpated in Haiti but, if present, can be distinguished by the shorter, rounder leaf tips and more numerous seeds (two to four) in the cones. All known populations of this species are threatened by uncontrolled fires and habitat conversion accompanying population growth. The species name (Latin for "thinner") reflects the finer texture of the branchlets compared to Bermuda redcedar (*J. bermudiana*) or eastern redcedar (*J. virginiana*), though they are no more slender than those of the geographically closer Antillean redcedar.

Juniperus gracilior Pilger **var. *gracilior***
Tree to 10 m tall with drooping branchlets. Glands on long shoot leaves inconspicuous. Western Dominican Republic; 1,000–1,700 m.

Juniperus gracilior **var. *ekmanii*** (Florin) R. P. Adams
Tree to 15 m tall with stiff branchlets. Glands on long shoot leaves conspicuous, oblong. Massif de la Selle, Haiti; 1,700–2,100 m. This variety has been driven to the verge of extinction by logging and human-set fires.

Juniperus gracilior **var. *urbaniana*** (Pilger & E. Ekman) R. P. Adams
Prostrate shrub. Glands on long shoot leaves conspicuous, elongate. Pic la Selle, Haiti. Found only on a limestone patch near the summit of the mountain; 2,300–2,550 m. Synonym: *Juniperus barbadensis* var. *urbaniana* (Pilger) Silba. This variety, although not exploited locally, has also been decimated by human-set fires.

Juniperus horizontalis Moench

CREEPING JUNIPER, SAVINIER (FRENCH)
Plate 29
Juniperus sect. *Sabina*
Low shrub to 0.3(–1) m, spreading along the ground for 1 m or more. Bark on older branches peeling in longitudinal strips. Crown flat, with numerous branches spreading or turned up at the tips. Branchlets dense, three- or four-angled, erect, 1–1.5 mm thick. Adult leaves in alternating pairs or trios, scalelike, 1.5–2 mm long, bluish green, with a conspicuous, dry, oblong glandular pit, the edges smooth, the tip rounded to pointed, spreading out from the twig. Pollen and seed cones on separate plants. Pollen cones single at the tips of short branchlets, oblong, 2.5–3.5 mm long, with four alternating pairs of pollen scales. Seed cones single at the tips of short curved branchlets, spherical or oblong, 5–8

Densely branching shoot of creeping juniper (*Juniperus horizontalis*) with 4–5 years of growth, ×0.5.

(–10) mm long, bluish black with a bluish waxy coating, maturing in 2 years. Seeds one to three (to six), 3.5–5 mm long, with several resin pockets, dark brown, the paler attachment scar extending up the sides up to a third their length. Northern North America from south-central Alaska to Newfoundland and Nova Scotia and south discontinuously to southeastern Wyoming, northeastern Illinois, and northeastern Massachusetts. Many cold, open habitats, from dry to wet and calcareous to acidic, generally rocky, with little organic matter; 0–1,000 m. Zone 3.

This widespread boreal juniper is the North American equivalent of the Eurasian savin (*Juniperus sabina*) and at one time was included in that species. It has generated numerous cultivars, partly through local hybridization with the closely related Rocky Mountain juniper (*J. scopulorum*) and eastern redcedar (*J. virginiana*). A number of these cultivars bear persistently juvenile foliage of needlelike leaves 2–6 mm long.

Juniperus indica A. Bertoloni

BLACK JUNIPER, BHIL (HINDI)

Juniperus sect. *Sabina*

Tree to 20 m tall or shrub to 1 m at high elevations and dry sites, with trunk to 1.2 m in diameter. Bark glossy dark reddish brown, dulling with exposure, deeply furrowed, peeling in long strips. Crown narrowly conical, with short, gently ascending branches. Branchlets stiff, irregularly branched, four-sided, 1–1.5 mm thick. Juvenile leaves in alternating trios, needlelike, 3–6 mm long, with two white stomatal bands above, grooved beneath, rather frequent in rapidly growing portions of mature trees. Adult leaves in alternating pairs, scalelike, bright green, about 1.5 mm long, with an inconspicuous elongate gland, the edges white and minutely toothed, the tip pointed and curved forward against the twig. Pollen and seed cones on separate plants. Pollen cones single at the tips of branchlets, nearly spherical, 2–3 mm long, with three or four alternating pairs of pollen scales. Seed cones single at the tips of straight or curved, short branchlets, slightly oblong, 8–10 (–13) mm long, shiny dark blue to black, without wax, maturing in 2 years. Seed one, 5.5–9 mm long, with two or three resin pits, light brown, the attachment scar basal. Himalayan ranges from the Indus River (India) to northwestern Yunnan (China). In mixed forests with birches and rhododendrons; 2,600–4,800(–5,050) m. Zone 9. Synonyms: *J. indica* var. *caespitosa* Farjon, *J. wallichiana* J. Hooker, *Sabina indica* (A. Bertoloni) L. K. Fu & Y. F. Yu, *S. wallichiana* (J. Hooker) W. C. Cheng & L. K. Fu.

Black juniper is a variable species that is often a tree in the eastern Himalaya but usually shrubby in the west. The shrubby forms were confused with the central Asian Turkestan juniper (*Juniperus pseudosabina*), which they closely resemble. Turkestan juniper has blunter leaves and slightly longer, narrower, black seed cones, often with two seeds. The lowest shrubby forms, at high elevations in the central Himalaya were distinguished as a variety, but most of their features seem to be habitat related and their characteristics are those expected in a subalpine ecotype.

Juniperus jaliscana M. Martínez

JALISCO JUNIPER, ENEBRO (SPANISH)

Juniperus sect. *Sabina*

Tree to 10 m tall, with erect trunk to 50 cm in diameter. Bark cinnamon brown, 1–2 cm thick, corky, fibrous and furrowed. Crown dense, narrow, rounded or irregular, with numerous rising branches. Branchlets cylindrical, 1–1.3 mm thick, sparsely branched. Adult leaves mostly in alternating pairs, scalelike, green, 0.7–1.5(–2) mm long, with a narrow gland more than half as long as the leaf, the edges minutely toothed, the tip blunt and curved forward along the twig. Pollen and seed cones presumably on separate plants. Pollen cones unknown. Seed cones single at the tips of short, straight branchlets, irregularly lumpy, 7–8 mm long, reddish brown with a thin waxy coating, maturing in 2 years. Seeds (2–)5–7(–11), 3–4 mm long, the pale attachment scar half as long as the seed. Western Mexico, known only from two localities, one in western Durango, the other in western Jalisco. Pine-oak-juniper woodlands on calcareous substrates; (430–)1,335–1,600(–2,670) m. Zone 9?

Jalisco juniper is one of the rarest and least known species in the genus. It may be closest to the more southerly Mayan juniper (*Juniperus comitana*), to the more northerly Rocky Mountain juniper (*J. scopulorum*), or to the geographically nearby Blanco juniper (*J. blancoi*) or Durango juniper (*J. durangensis*), but its relationships may remain obscure until new populations are discovered and sampled.

Juniperus komarovii Florin

KOMAROV JUNIPER, TA ZHI YUAN BAI (CHINESE)

Juniperus sect. *Sabina*

Tree to 10(–20) m tall, with trunk to 1.2 m in diameter. Bark warm brown, weathering grayish brown to gray, peeling in narrow strips. Crown dense, with outstretched, rising or hanging branches. Branchlets cylindrical to four-sided, upright, 1.2–1.5 mm in diameter. Adult leaves of branchlets in alternating pairs (or trios), scalelike, grayish green, 1–1.5 mm long (those of long shoots to 3.5 mm), with a conspicuous, round gland, the edges smooth, the tip pointed, pressed against the twig. Pollen and seed cones on the same plant. Pollen cones plumply egg-shaped, 2–2.5 mm long, with five alternating pairs of pollen scales. Seed cones egg-shaped, 9–10 mm long, bluish black with a pale blue waxy coating, maturing in 1 year. Seed one, 6–7.5 mm long, light brown, with a pitted attachment scar at the base. West-central China, in Gansu, Qinghai, and Sichuan. Montane forests; 3,000–4,000 m. Zone 5. Synonyms: *J. glaucescens* Florin, *Sabina komarovii* (Florin) W. C. Cheng & W. T. Wang.

The Swedish botanist Rudolf Florin (1894–1965), who described this species, and the Russian botanist Vladimir Komarov (1869–1945), for whom he named it, both described many Chinese juniper species, most of which are now considered synonyms of earlier described species or even of other species they themselves had described. In fact, Komarov juniper may prove to be a weeping variety of Tibetan juniper (*Juniperus tibetica*), described by Komarov just 3 years before this one. There is not yet enough material of Komarov juniper available to clarify its relationship with Tibetan juniper, but the drooping branches make its identification fairly easy.

Juniperus monosperma (Engelmann) C. Sargent

ONE-SEED JUNIPER

Juniperus sect. *Sabina*

Shrubby tree to 6(–15) m tall, with short trunk to 1.4 m in diameter. Bark brown to gray, thin, peeling in long, narrow strips. Crown open, broad and rounded, with upwardly angled branches. Branchlets concentrated at the tips of the branches, four- (or six-) angled, 1.3–2 mm thick. Adult leaves in alternating pairs or trios, scalelike, 1–3 mm long, yellowish green to dark green, with a prominent sunken gland near the base, the edge minutely toothed, the tip pointed and spreading away from the twig, sometimes overlapping the gland on the next leaf along. Pollen and seed cones on separate plants. Pollen cones single at the tips of branchlets, slightly oblong, 3–4 mm long, with four to six alternating pairs of pollen scales. Seed cones single at the tips of straight branchlets, nearly spherical, (4–)6–8 mm long, dark brownish blue with a waxy coating, maturing in 1 year. Seeds one (to three), 4–5 mm long, rich brown, the paler attachment scar about a third the length of the seed. Southwestern United States, from southeastern Utah and southern Colorado to northeastern Arizona and western Texas. Pinyon pine–juniper woodlands on dry, rocky slopes; 900–2,300 m. Zone 7.

One-seed juniper and Utah juniper (*Juniperus osteosperma*) are the common junipers of the Colorado Plateau region that includes Grand Canyon National Park. Together with Colorado pinyon (*Pinus edulis*), they cover vast tracts of land on plateaus and slopes between the deserts and montane forests. Although both junipers are often or usually one-seeded, Utah juniper differs from one-seed juniper in having inconspicuous leaf glands,

Mature one-seed juniper (*Juniperus monosperma*) in nature.

pollen and seed cones on the same tree, and paler brown seed cones at least some of which mature in 2 years. The largest known individual in combined dimensions when measured in 1981 was 8.8 m tall, 1.4 m in diameter, with a spread of 8.5 m.

Juniperus monticola M. Martínez

MOUNTAIN JUNIPER, MEXICAN JUNIPER, TLÁSCAL (SPANISH)

Juniperus sect. *Sabina*

Shrub, or shrubby tree to 6 m tall, or a dwarfed shrub at and above timberline, usually branched from the base, with trunk to 0.6 m in diameter. Bark reddish brown, weathering grayish brown, thin, 5–8 mm thick, fibrous, peeling in long, narrow strips. Crown irregular, with spreading to rising, often crooked branches. Branchlets sparse and elongate to very dense and short, four-angled, 1.5 mm thick, sometimes arranged in two-dimensional arrays. Adult leaves in alternating pairs, scalelike, 1–1.5 mm long, grayish green, often with a waxy coating, sometimes with an obscure gland near the base, the edge minutely toothed, the tip blunt and pressed against the twig. Pollen and seed cones on separate plants. Pollen cones single at the tips of branchlets, oblong, about 4 mm long, with six alternating pairs of pollen scales. Seed cones single at the tips of curved branchlets, spherical or irregularly lumpy, 5–10 mm long, dark bluish black or brownish blue, with a thin waxy coating, maturing in 1 year. Seeds (two or) three or four (to eight), 4–5 mm long, shiny dark brown, the paler attachment scar about half the length of the seed. Mountains of central Mexico (hence the scientific name, Latin for "mountain inhabiting"), from southern Coahuila and western Jalisco to Cofre de Perote, Veracruz, with an outlier in western Guatemala (Huehuetenango). Alpine grasslands and understory of subalpine forests on rocky substrates; (2,400–) 3,000–4,500 m. Zone 7? Synonyms: *J. monticola* f. *compacta* M. Martínez, *J. monticola* f. *orizabensis* M. Martínez, *J. sabinoides* (Kunth) C. Nees, not Grisebach.

Mountain juniper is a common species near timberline on the high mountains of central Mexico. Often, shrubs of strikingly different appearance due to the compactness of their branchlets grow near each other on the same mountainside. Some of these were described as botanical formae, but the shrubs are otherwise similar in morphology and chemistry, and there seems to be little need for formal recognition. The natural variability, however, might form a basis for cultivar selection. It was one of the first juniper species in Mexico to receive a botanical description, but the name only dates to 1946 because all prior choices turned out to be names that had already been used for other species or were otherwise technically unavailable for mountain juniper.

Juniperus oblonga M. Bieberstein

CAUCASIAN JUNIPER

Juniperus sect. *Juniperus*

Low shrub, or shrubby tree to 1.2(–5) m tall, with trunks to 10(–20) cm. Bark reddish brown, weathering gray, thin, fibrous, peeling in scales and then narrow strips. Crown broadly rounded, flat, or somewhat drooping, with spreading branches turned up at the tips. Branchlets triangular, sometimes drooping. Adult leaves in alternating trios, needlelike, angled backward from and jointed at their attachment to the twig, 16–20(–25) mm long, 1–1.5 mm wide, bright green, with a broad white waxy stomatal band above surrounding a distinct midrib and a prominent ridge along the whole length beneath. Pollen and seed cones on separate plants. Pollen cones single in the leaf axils, oblong, 4–5 mm long, with five or six alternating trios of pollen scales. Seed cones single in the leaf axils, spherical to oblong (hence the scientific name) and slightly crested at the tip, to 7 mm long, blackish brown, sometimes with a thin waxy coating, maturing in 2 years. Seeds (one to) three, 3.5–4 mm long, dark brown. Caucasus from northeastern Turkey and Georgia to Azerbaijan and northwestern Iran. Understory of varied woodlands and forests without apparent substrate preferences; foothills to timberline; (200–)500–3,000 m. Zone 8? Synonym: *J. communis* var. *oblonga* (M. Bieberstein) J. Medwedew.

The taxonomic status of Caucasian juniper is rather uncertain. Many authors treat it as a separate variety within common juniper (*Juniperus communis*) or submerge it into the widespread high montane *J. communis* subsp. *alpina.* This last is probably not a single phylogenetic entity anyway, so it would seem preferable to assess the relationships of its geographically widespread components before adding even more elements to it. Caucasian juniper is clearly close to common juniper and may well be part of the variation in that species, but there are too many unresolved questions regarding the classification of this whole group to feel comfortable submerging Caucasian juniper without further evidence.

Juniperus occidentalis J. Hooker

WESTERN JUNIPER

Juniperus sect. *Sabina*

Tree to 20(–30) m tall, unbranched for up to 1.5 m, with trunk to 1.5(–3.9) m in diameter. Bark brown or reddish brown, to 25 mm thick, fibrous, furrowed, peeling in longitudinal strips. Crown irregular and open, with spreading or rising branches. Branchlets three- or four-angled, stiff, 1–2 mm thick. Adult leaves in alternating pairs or trios, scalelike, 1–3 mm long, pale green, with a prominent resin gland often bearing dried white resin, the edges minutely toothed, the tip triangular, bent forward tightly against the stem. Pollen and seed cones on the same or separate plants. Pollen cones single at the tips of branchlets, plumply egg-shaped, 3–4 mm long, with four to six trios of pollen scales. Seed cones solitary at the tips of straight branchlets, slightly elongate, (5–)6–9(–10) mm long, bluish black with a thick waxy coating, maturing in 2 years. Seeds (one or) two or three, (2–)4–6.5 mm long, brown, the paler attachment scar extending up to a fourth of the length. Western United States (hence the scientific name, Latin for "western") from southeastern Washington and southwestern Idaho to southern California. Open woodlands of dry foothills and mountain slopes; (ca. 200–)1,000–3,000 m. Zone 7. Two subspecies.

This species overlaps with two others, California juniper (*Juniperus californica*) in California and Utah juniper (*J. osteosperma*) near the California-Nevada border but is easily distinguished from both of these, which have brown cones maturing in just 1 (to 2) years. It is known to hybridize with Utah juniper in northwestern Nevada. There are two geographic subspecies. The largest known individual in combined dimensions when measured in 2003 was 23.8 m tall, 3.9 m in diameter, with a spread of 17.1 m.

Young mature western juniper (*Juniperus occidentalis*) in nature.

Juniperus occidentalis* subsp. *australis Vasek
SIERRA JUNIPER
Pollen and seed cones equally divided between the same and different plants. Seed cones averaging 7.5 mm long. Southern portion of range north to southern Lassen County, California. Massive, picturesque old individuals are one of the characteristic landscape features in rocky openings of the upper montane forest in the high Sierra Nevada of California.

Juniperus occidentalis J. Hooker **subsp. *occidentalis***
NORTHWESTERN JUNIPER
Pollen and seed cones on different plants. Seed cones averaging 8.5 mm long. Northern portion of range south to central Lassen County, California. Overgrazing of arid grasslands in southeastern Oregon during the late 19th century led to increased survivorship of juniper seedlings, which have subsequently drastically reduced water availability and the productivity of accompanying grasses. Ancient trees exceeding 1,000 years old are confined to ungrazed slopes surrounding these young woodlands.

Juniperus osteosperma (J. Torrey) E. Little
UTAH JUNIPER, INTERMOUNTAIN JUNIPER
Juniperus sect. *Sabina*
Shrubby tree to 3–6(–12) m tall, often multitrunked, the main trunk, when present, to 0.8(–2.2) m in diameter. Bark gray-brown to ashy white, thin, shredding in long strips. Crown open, broad and rounded, with spreading or rising branches. Branchlets three- or four-angled, stiff, up to 2 mm thick. Adult leaves usually in alternating pairs, scalelike, 1–3 mm long, yellowish green, with an inconspicuous, embedded resin gland, the edges minutely toothed, the tip triangular and bent forward tightly against the twig. Pollen and seed cones on the same plant. Pollen cones single at the tips of branchlets, slightly oblong, 2.5–4.5 mm long, with 5–9(–12) alternating pairs of pollen scales. Seed cones single at the tips of straight branchlets, nearly spherical, (6–)8–10(–18) mm in diameter, tan to brown with a conspicuous waxy coating, maturing in 1–2 years. Seeds one (or two), 4–6(–8) mm long, brown, the paler attachment scar extending halfway up the seed. Western United States from northeastern California, eastern Idaho, and south-central Montana to southern California, central Arizona, and northwestern New Mexico. Juniper and pinyon pine–juniper woodlands on dry, rocky slopes and plateaus; (450–)1,300–2,700 m. Zone 6.

Utah juniper is the most common tree in the Great Basin of Utah and Nevada, joining Colorado pinyon (*Pinus edulis*) and single-leaf pinyon (*P. monophylla*) to form vast tracts of pinyon pine–juniper woodlands at middle elevations on the numerous ranges and also occurring below them on the slopes. Its distribution overlaps those of one-seed juniper (*Juniperus monosperma*) and western juniper (*J. occidentalis*), and it hybridizes with the latter in northwestern Nevada. The scientific name ("bone seed") is derived from the (not unusually) hard seeds. The largest known individual in combined dimensions when measured in 2002 was 12.2 m tall, 2.1 m in diameter, with a spread of 17.1 m.

Juniperus oxycedrus Linnaeus

PRICKLY JUNIPER, ZIMBRO-BRAVO, PIORRO (PORTUGUESE), CADA (SPANISH), CÀDEC (CATALAN), HEGO-ORREA (BASQUE), SMRIK (CROATIAN), SRVENA FEN'A (SERBIAN), MOZHZHEVEL'NIK KRASNYI (RUSSIAN), KATRAN ARDICI (TURKISH), GIKHI (ARMENIAN), DYSHI-ARDYSH (TATAR)

Juniperus sect. *Juniperus*

Shrub, or tree to 10(–15) m tall, with trunk to 30(–120) cm in diameter. Bark light gray, thin, peeling in longitudinal strips. Crown bushy, irregular, with spreading to stiffly upright branches, the tips sometimes drooping. Branchlets roundly three-sided, upright. Adult leaves in widely spaced, alternating trios, needlelike, standing out from and jointed at their attachment to the twig, (4–)8–20 (–25) mm long, 1–2.5 mm wide, dark green, with two waxy, white, shallow stomatal furrows above and a ridge below, the tip spiny. Pollen and seed cones on separate plants. Pollen cones single on a very short stalk in the leaf axils, egg-shaped, 1–2 mm long, with four to eight alternating trios of pollen scales. Seed cones single on short stalks in the leaf axils, approximately spherical, 6–15(–20) mm long, red to reddish brown, glossy or occasionally with a thin waxy coating, maturing in 2 years. Seeds (one or) two or three (or four), 5–12 mm long, brown. Mediterranean region and adjacent lands from Portugal east to the western Caucasus. Dry open scrublands, woodlands, and clearings; sea level to 1,000(–1,900) m. Zone 9.

Prickly juniper is the warm temperate replacement for common juniper (*Juniperus communis*). It differs in the paired rather than single stomatal bands and reddish rather than bluish mature seed cones. There are several subspecies that differ in habit, leaf size, and leaf tip shape. These may be treated as separate species by some authors. All are viciously prickly (hence the common and scientific names, Latin for "sharp cedar").

Juniperus oxycedrus* subsp. *badia (F. Gay) Debeaux

Conical tree to 15 m tall with upright branches dangling at the tips. Leaves (8–)12–20 mm long, 1.2–2 mm wide, the tip triangular. Seed cones 10–13 mm in diameter, with a thin waxy coating when young. Western end of the range of the species, replacing subspecies *oxycedrus* in woodlands of the interior of the Iberian Peninsula in Spain and Portugal.

Juniperus oxycedrus* subsp. *macrocarpa (J. E. Smith) J. Ball

Upright or trailing shrub to 2 m. Leaves 8–20(–25) mm long, 1.5–2.5 mm wide, the tip spiny. Seed cones (10–)12–15 (–20) mm in diameter, with a persistent, thin, waxy coating. In the vicinity of the seashore throughout the range of species, except in southwestern Portugal.

Juniperus oxycedrus Linnaeus **subsp. *oxycedrus***

Upright shrub to 4 m with upright or dangling branches. Leaves 8–15(–25) mm long, 1–1.5(–2) mm wide, the tip spiny. Seed cone (6–)8–10 mm in diameter, without a waxy coating. Dry, rocky woodland and scrub throughout the range of the species except in central Spain and Portugal. Synonyms: *Juniperus deltoides* R. P. Adams, *J. oxycedrus* var. *brachyphylla* H. Loret.

Juniperus oxycedrus* subsp. *transtagana Franco

Upright shrub to 2 m with dense branches. Leaves 4–12 mm long, 1–1.5 mm wide, the tip triangular. Seed cone 7–10 mm in diameter, without a waxy coating. On dunes and in scrub along and beside the seashore, replacing subspecies *macrocarpa* in southwestern Portugal. Synonym: *Juniperus navicularis* Gandoger.

Young mature prickly junipers (*Juniperus oxycedrus*) in nature.

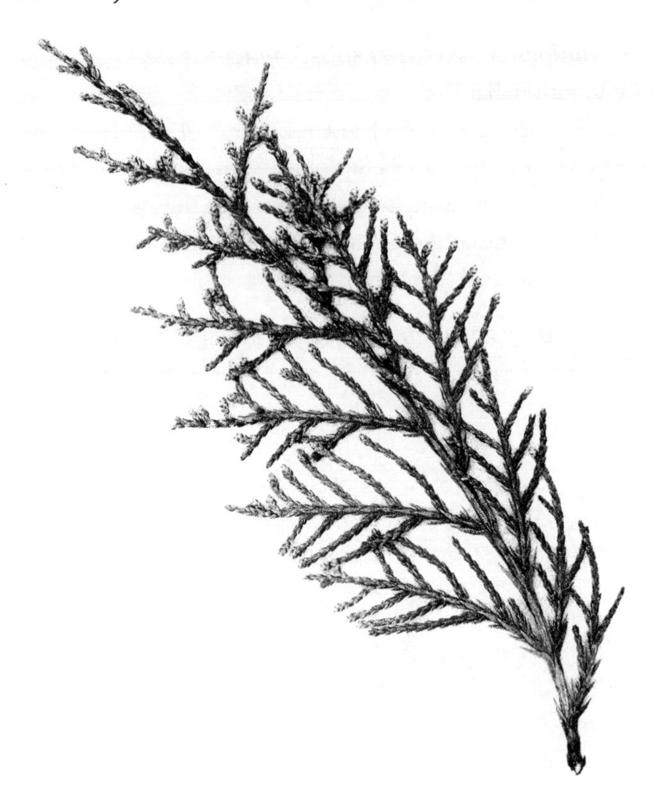

Twig of Pfitzer juniper (*Juniperus ×pfitzeriana*) with very regular flattened branchlet sprays bearing pollen cones, ×1.

Juvenile foliage of Pfitzer juniper (*Juniperus ×pfitzeriana*), ×1.

Juniperus ×pfitzeriana (F. Späth) P. A. Smith

PFITZER JUNIPER

Upright, spreading shrub (*J. chinensis* × *J. sabina*) to 3 m. Branches rising and arching to form a broad, flat crown to 5 m wide. Branchlets coarser than in savin (*J. sabina*), nodding. Adult leaves in alternating pairs or trios, scale- or needlelike but then running smoothly onto the twig without a joint, the scalelike leaves generally predominating, blue-green, green, or golden green. Pollen and seed cones on separate plants, those with pollen cones predominating in cultivation. Zone 3. Synonym: *J. ×media* Melle.

The original Pfitzer juniper was collected in the wild in Mongolia and was later described from the Pfitzer nursery in Germany as a variety of Chinese juniper (*Juniperus chinensis*) before being merged with three other cultivar groups under *J. ×media* by van Melle. This treatment is widely used in horticulture, but unfortunately, it cannot be maintained for two main reasons. First, another *J. ×media*, a hybrid between Turkestan juniper (*J. pseudosabina*) and central Asian pencil juniper (*J. semiglobosa*), had already been described and thus takes precedence over van Melle's name. Second, most of the cultivars other than the Pfitzer junipers assigned by van Melle to his *J. ×media* are probably not of hybrid origin and are actually cultivars of undiluted Chinese juniper. Using the name *J. ×pfitzeriana* and applying it only to the Pfitzer junipers solves these difficulties.

Juniperus phoenicea Linnaeus

PHOENICIAN JUNIPER, ZIMBRO-DAS-AREIAS (PORTUGUESE), SABINA SUAVE (SPANISH), SAVINA (CATALAN), MITER FENIZIARRA (BASQUE), ARÂR (ARABIC), FINIKE ARDICI (TURKISH)

Juniperus sect. *Sabina*

Shrub, or tree to 12 m tall, with trunk to 1 m in diameter. Bark dark grayish brown, fibrous, peeling in narrow strips. Crown dense, conical at first, broadening in age, with numerous, rising branches. Branchlets cylindrical, 1–1.5 mm thick. Juvenile needlelike leaves also common in adult plants. Adult leaves in alternating pairs or trios, scalelike, 1–2 mm long, green to blue-green, with a shallow, oblong, variably active glandular pit, the edges with a narrow, dry, minutely toothed margin, the tip broadly triangular and pressed forward against the twig. Pollen and seed cones usually on the same plant. Pollen cones single at the tips of branchlets, oblong, 4–6 mm long, with five to seven alternating pairs of pollen scales. Seed cones single at the tips of short, straight, branchlets, spherical to egg-shaped, 8–15 mm long, shiny dark brownish red or sometimes with a thin waxy coating, maturing in 2 years. Seeds three to nine, 3.5–7 mm long. Mediterranean region from Portugal, the Canary Islands, and Morocco to westernmost Asiatic Turkey, Cyprus, and northeastern Saudi Arabia. Dry rocky woodlands, scrub, and dunes; 0–1,000(–2,400) m. Zone 8. Two subspecies.

Juniperus phoenicea Linnaeus **subsp. *phoenicea***

Tips of scale leaves broadly or narrowly triangular. Seed cones nearly spherical, 8–12 mm long. Throughout the range of the species.

Juniperus phoenicea* subsp. *turbinata (Gussone) C. Nyman

Tips of scale leaves narrowly triangular. Seed cones egg-shaped, 12–15 mm long. Coastal sand dunes in the western portion of the range of the species east to Corsica and Sardinia. Synonym: *Juniperus phoenicea* subsp. *mediterranea* P. Lebreton & Thivend.

Juniperus pinchotii Sudworth

PINCHOT JUNIPER, REDBERRY JUNIPER

Juniperus sect. *Sabina*

Shrubby tree to 6 m tall, divided from near the base, with trunks to 30 cm thick. Bark light gray or brown, thin, peeling in narrow strips. Crown broad, irregular, with upright to spreading branches. Branchlets four- (or three-) angled, 0.7–1.5 mm thick. Adult leaves in alternating pairs or trios, scalelike, 1–2.5 mm long, yellow-green, with an active oblong gland, the edge minutely toothed, the tip narrowly triangular and spreading from the twig. Pollen and seed cones on separate plants. Pollen cones single at the tips of branchlets, oblong, 3–4 mm long, with four or five alternating pairs of pollen scales. Seed cones single at the tips of short, straight branchlets, spherical, 6–8(–10) mm long, bright or metallic red, without a waxy coating, maturing in 1 year. Seeds one (or two), 4–5 mm long, rich brown, the paler attachment scar about half the length of the seed. South-central United States, in central and western Texas, and adjacent Oklahoma, New Mexico, and Coahuila (Mexico). Juniper-pine-oak woodlands on dry calcareous soils; 300–1,000(–2,100) m. Zone 9. Synonym: *J. erythrocarpa* Cory.

Pinchot juniper overlaps with and is replaced by one-seed juniper (*Juniperus monosperma*) toward the northwest and Ashe juniper (*J. ashei*) toward the southeast. It has been reported to hybridize with those species, but no evidence of hybridization can be found in the leaf oils. In contrast, these chemicals suggest hybridization with roseberry juniper (*J. coahuilensis*), which overlaps with and replaces Pinchot juniper toward the southwest. The type specimen of *J. erythrocarpa* is actually a tree of Pinchot juniper even though the name *J. erythrocarpa* was incorrectly used for roseberry juniper. The largest known individual in the United States in combined dimensions when measured in 2003 was 8.5 m tall, 1.1 m in diameter, with a spread of 10.1 m.

Juniperus pingii W. C. Cheng in de Ferré

PING JUNIPER, CHUI ZHI XIANG BAI (CHINESE)

Juniperus sect. *Sabina*

Shrub, or shrubby tree to 4–9(–30?) m tall, with trunk to 30(–50) cm. Bark brown, weathering grayish brown, thin, peeling in flakes and later in narrow strips. Crown conical to broadly and shallowly dome-shaped, with numerous horizontal or upwardly angled branchlets, sometimes drooping at the tips. Branchlets rounded to four- or six-angled, (1–)2–4 mm thick. Adult leaves in closely spaced, alternating trios (or pairs), needle- to scalelike, standing out from the twig or curled forward along it, (1–)3–5(–7) mm long, (0.5–)1–1.5 mm wide, grayish green with wax on the outer face and with two white stomatal bands separated by a narrow green midrib beneath on the inner face, without glands, the edges smooth, the tip pointed, sometimes prickly. Pollen and seed cones on the same plant. Pollen cones single at the tips of short branchlets in the leaf axils, spherical, 3–4 mm long, with two or three alternating trios of pollen scales. Seed cones single at the tips of short branchlets in the leaf axils, spherical to egg-shaped, 5–9(–11) mm long, shiny purplish black, maturing in 2 years. Seed one, 5–7(–9) mm long, brown, the paler attachment scar very large and pitted. Mountains and high plateaus of western China from southern Gansu to northwestern Yunnan and south-central Xizang (Tibet). Upper montane and subalpine forests and alpine shrublands; (2,600–)3,000–4,500(–4,900) m. Zone 6? Synonyms: *J. carinata* (Y. F. Yu & L. K. Fu) R. P. Adams, *J. chengii* L. K. Fu & Y. F. Yu, *J. pingii* var. *carinata* Y. F. Yu & L. K. Fu, *J. pingii* var. *chengii* (L. K. Fu & Y. F. Yu) Farjon, *J. pingii* var. *miehei* Farjon, *J. pingii* var. *wilsonii* (Rehder) Silba, *Sabina pingii* (W. C. Cheng in de Ferré) W. C. Cheng & W. T. Wang, *S. wilsonii* (Rehder) W. C. Cheng & L. K. Fu.

Ping juniper is very closely related to Himalayan drooping juniper (*Juniperus recurva*) and to scaly juniper (*J. squamata*), two other high-elevation one-seeded junipers of the Sino-Himalayan region. This group encompasses a full range of growth habits from low, spreading shrubs to good-sized trees, with branching habits ranging from stiffly upright to weeping. Although the extremes of these species are fairly readily identifiable, the wide range of growth forms, needle sizes and forms, and seed cone sizes within each blur the distinctions among them. In general, Ping juniper has shorter, broader, more scalelike needles that lack the grooves found on the outer faces of needles of Himalayan drooping juniper and scaly juniper. Several varieties were described for Ping juniper, but given the variability characterizing all members of the complex, and the uncertainties of their separation from one another, it seems premature to carve up the variation among varieties, some of which are very narrowly circumscribed and others much more broadly. Ping juniper has little economic significance within its native range and has contributed little to horticulture, although there are a few cultivars in the trade. Distinguishing these from cultivars of scaly juniper is difficult.

Juniperus procera C. F. Hochstetter ex Endlicher

EAST AFRICAN JUNIPER, OL DARAKWA (MASAI, ARUSHA), NDERAKWA (CHAGGA), [M]SELEMUKA (KINGA, NYAKYUSA, WANGI)

Juniperus sect. *Sabina*

Tree to 40 m tall, usually with a single trunk to 1.5 m in diameter. Bark grayish brown, fibrous, deeply furrowed. Crown broad, irregular, with crooked branches, rising at first, drooping with age. Branchlets four-angled, 0.6–1 mm thick, partly arranged in flattened sprays. Adult leaves in alternating pairs, scalelike, 0.5–1 mm long, light green of yellowish green, with an active, narrowly elongate gland, the edges smooth, the tip narrowly triangular and separated from the twig. Pollen and seed cones on separate plants. Pollen cones single at the tips of branchlets, nearly spherical, 3–5 mm long, with five or six alternating pairs of pollen scales. Seed cones single at the tips of short, straight branchlets, spherical, (3–) 5–7(–8) mm long, brown to purplish black with a thin, bluish, waxy coating, maturing in 1 year. Seeds one to four, 4–5 mm long, yellowish brown, the paler attachment scar just at the base. Southwestern Saudi Arabia through the eastern African mountains and highlands to central Zimbabwe. Open, seasonally dry evergreen forest and savanna; 1,000–2,700(–3,600) m. Zone 9.

East African juniper is the only species of juniper to reach the southern hemisphere. It can be abundant and form pure stands in regions of moderate rainfall (1,000–1,200 mm per year) but decreases in frequency with greater or lesser rainfall. The wood quality is excellent, and the tree has been heavily exploited where it is abundant. As is typical for junipers, regeneration is excellent in the face of human disturbances, but old-growth stands remain only in protected areas and are not secure even there.

Juniperus pseudosabina F. E. L. Fischer & C. Meyer

TURKESTAN JUNIPER, URYUK ARCHA (TURKIC), ARSA (ALTAIC)

Juniperus sect. *Sabina*

Trailing shrub to small tree to 10(–18) m tall, with trunk to 0.5(–1) m in diameter. Bark dark to light reddish brown, weathering grayish brown, fibrous, flaking and then partially peeling in short strips. Crown dense, with spreading or ascending branches. Branchlets four-angled, 1.5–2 mm thick. Juvenile needlelike leaves sometimes persisting on adult plants. Adult leaves in alternating pairs, scalelike, 1.5–2(–3) mm long, gray-green to yellowish green, usually with a prominent oblong gland, the edges minutely toothed, the tip broadly triangular, separated from the twig. Pollen and seed cones on the same plant. Pollen cones single at the tips of branchlets, slightly oblong, 2–3 mm long, with three or four alternating pairs of pollen scales. Seed cones single at the tips of short branchlets, slightly elongate, (5–)8–15 mm long, brownish black to black, glossy in the absence of a waxy coating, maturing in 1 year. Seeds one (or two), 5–10 mm long, brown, the light brown attachment scar very large. Mountains of central Asia from the Pamir Alai and Tian Shan Ranges of Tadzhikistan and Kyrgyzstan to the Yablonony Range east of Lake Baikal in the Buryatia region of Russia and the Hentiyn Range in adjacent Mongolia. Forming juniper woodlands and scrublands on rocky soils in the alpine and subalpine zones; (900–)2,500–3,200(–4,100) m. Zone 6. Synonyms: *J. centrasiatica* V. Komarov, *J. pseudosabina* var. *turkestanica* (V. Komarov) Silba, *J. turkestanica* V. Komarov, *Sabina centrasiatica* (V. Komarov) W. C. Cheng & L. K. Fu, *S. pseudosabina* (F. E. L. Fischer & C. Meyer) W. C. Cheng & W. T. Wang, *S. pseudosabina* var. *turkestanica* (V. Komarov) Chun Y. Yang.

Low, shrubby forms of Turkestan juniper predominate in the eastern portion of its range and at higher elevations, while larger tree forms are found only in the western ranges, where they are exploited for timber. The fleshy seed cones are sweet, hence the Turkic common name ("peach juniper"). The seeds, once stripped of flesh, are strung as beads. The species name, Greek for "false sabina," reflects its deceptive resemblance to savin (*Juniperus sabina*), another ground-cover species with a predominance of scalelike leaves.

Juniperus recurva Buchanan-Hamilton ex D. Don

HIMALAYAN DROOPING JUNIPER, BETAR (WESTERN HIMALAYA), CUI ZHI BAI (CHINESE)

Juniperus sect. *Sabina*

Low spreading shrub, or tree to 25(–45) m tall, with a straight trunk to 2(–3.5) m in diameter. Bark reddish brown, weathering grayish brown, thin to thick, fibrous, peeling in narrow, vertical strips. Crown dense, broadly conical, with spreading to down-curved branches. Branchlets round, hanging. Adult leaves in dense, alternating trios, directed forward, needlelike, (1.5–)3–8 mm long, gray-green to dark green, with two narrow, waxy white or obscure stomatal bands beneath, running down onto the twig without a joint, the edges dry and white, the tip sharp. Pollen and seed cones on the same or separate plants. Pollen cones single at the tips of branchlets or in leaf axils, oblong, 4–6 mm long, with five to eight alternating pairs of pollen scales. Seed cones single at the tips of short branchlets in the leaf axils, oblong, 8–10(–13) mm long, purple-brown to black, glossy without a waxy coating, maturing in 2 years. Seed one, 6–9 mm long, with large resin pits, light brown, the paler attachment scar just at the base. Himalaya from western Nepal to Sichuan and Yunnan provinces, China. Various cool, moist habitats, from subalpine scrub to mixed temperate evergreen forest; (2,000–)2,500–4,600 m. Zone 8. Synonym: *Sabina recurva* (Buchanan-Hamilton ex D. Don) F. Antoine. Two varieties.

Himalayan drooping juniper is quite variable in stature across its range, and this has given rise to several cultivars. In contrast to Turkestan juniper (*Juniperus pseudosabina*) of the drier ranges north of the Himalaya, Himalayan drooping juniper attains its greatest stature at the eastern end of its range, and these trees are distinguished at the varietal level from the rest of the species. It is closely related to Ping juniper (*J. pingii*) and scaly juniper (*J. squamata*), and the great variability found within each of them sometimes makes it hard to assign particular cultivars or wild specimens to species with any confidence. The largest individuals of Himalayan drooping juniper are larger than any individual trees of the other two species. It is an important temple ornament within its native range and is a modest contributor to general horticulture, though nowhere near as common in gardens as scaly juniper. It is not immediately obvious what aspect of this tree merits the name "recurved," but perhaps it refers to the drooping rather than upright branchlets.

Juniperus recurva* var. *coxii (A. B. Jackson) Melville
COFFIN JUNIPER
Tree to 30 m tall, with a single trunk. Leaves 6–8 mm long, the stomatal bands prominent, the tip with an elongate point. Northern Myanmar and western Yunnan (China) to eastern Nepal. Because of its size and the fragrant, dense, uniform wood, this variety has been much (over-) exploited as a coffin wood. Few if any trees of the maximum dimensions cited are still extant. Synonyms: *Juniperus coxii* A. B. Jackson, *Sabina recurva* var. *coxii* (A. B. Jackson) W. C. Cheng & L. K. Fu.

Juniperus recurva Buchanan-Hamilton ex D. Don **var. *recurva***
Shrub, or tree to 10 m tall, often with multiple trunks. Leaves 3–6 mm long, the stomatal bands inconspicuous, the tip with a short point. Throughout the range of the species, intergrading with and replaced by variety *coxii* at the easternmost end.

Juniperus rigida P. Siebold & Zuccarini

TEMPLE JUNIPER, DU SONG (CHINESE), NUGANJU NAMU (KOREAN), NEZU (JAPANESE)
Juniperus sect. *Juniperus*
Tree to 10(–15) m tall, with a single trunk to 0.5 m in diameter. Bark dull gray, fibrous, shallowly fissured. Crown conical, with graceful arching branches. Branchlets dangling on older trees, three-angled, reddish brown and hairless between the well-spaced leaves. Adult leaves in alternating trios, needlelike, 1–2(–3) cm long, standing straight out from and jointed at their attachment to the twig, deep green to gray-green with a narrow, deep, white stomatal furrow above and a broad, flat ridge below, the edges smooth, the tip sharply pointed. Pollen and seed cones usually on separate plants. Pollen cones single in leaf axils of the preceding year, oblong, 3–5 mm long, with three to five alternating trios of pollen scales. Seed cones single at the tips of very short branchlets in the leaf axils, spherical, 6–8(–10) mm long, purplish black with a thin, pale blue waxy coating, maturing in 2 years. Seeds (two or) three, 5–7 mm long, with large resin pits, brown, the paler attachment scar extending a little up the sides. Japan (from northern Honshū to central Kyūshū), Korea, northeastern China, and extreme southeastern Russia (southern Ussuri district), Among grasses and shrubs on dry, rocky (often calcareous) mountain slopes and hillsides; (0–)100–1,000(–2,200) m. Zone 6.

Twig of temple juniper (*Juniperus rigida*) with 2 years of growth and a seed cone, ×1.

Temple juniper, as its common name suggests, is widely cultivated in temple and other gardens in Japan for its bright foliage and graceful weeping habit. A number of cultivars have arisen from wild selections and in cultivation. The species name refers to the stiff leaves, giving the prickles at their tips the backbone they need to make grasping the foliage uncomfortable.

Juniperus sabina Linnaeus

SAVIN OR SAVIN JUNIPER, MITER ARRUNTA (BASQUE), MOZHZHEVEL'NIK KAZATSKII (RUSSIAN), GYUVA (GEORGIAN), KARA ARCHA (KAZAKH), KONIN-ARTSA (MONGOLIAN), CHA ZI YUAN BAI (CHINESE)
Juniperus sect. *Sabina*
Spreading or upright shrub, rarely treelike to 5(–12) m tall, with trunk to 0.3(–1) m. Bark reddish brown, fibrous, peeling in strips. Crown broad and flat, with spreading branches, upturned at the tips. Branchlets upright, four-angled, 0.6–1 mm thick with an unpleasant, pungent odor when bruised. Needlelike juvenile leaves sometimes appearing on adult plants. Adult leaves in alternating pairs, scalelike, 1–1.3 mm long, dark to light green, with a prominent oblong resin gland, the edges smooth, the triangular

tip pressed against the twig. Pollen and seed cones on separate plants. Pollen cones single at the tips of branchlets, much elongated, 3–5 mm long, with about five pairs or trios of pollen scales. Seed cones single at the tips of short, curved branchlets, spherical or slightly broader than long, 4–6(–8) mm long, bluish black with a thin waxy coating, maturing in 1–2 years. Seeds (one or) two (to four), 3–5 mm long, yellowish brown, the paler attachment seen just at the base. Widely but discontinuously distributed across south-central Eurasia from Spain and Algeria to northwestern China. Dry rocky slopes of mountains and steppes; (700–)1,100–3,350 m. Zone 3. Synonyms: *J. chinensis* var. *arenaria* E. H. Wilson, *J. erectopatens* (W. C. Cheng & L. K. Fu) R. P. Adams, *J. ×kanitzii* Csató, *J. sabina* var. *arenaria* (E. H. Wilson) Farjon, *J. sabina* var. *erectopatens* (W. C. Cheng & L. K. Fu) Y. F. Yu & L. K. Fu, *J. sabina* var. *yulinensis* (T. C. Chang & Chun G. Chen) Y. F. Yu & L. K. Fu, *Sabina vulgaris* F. Antoine.

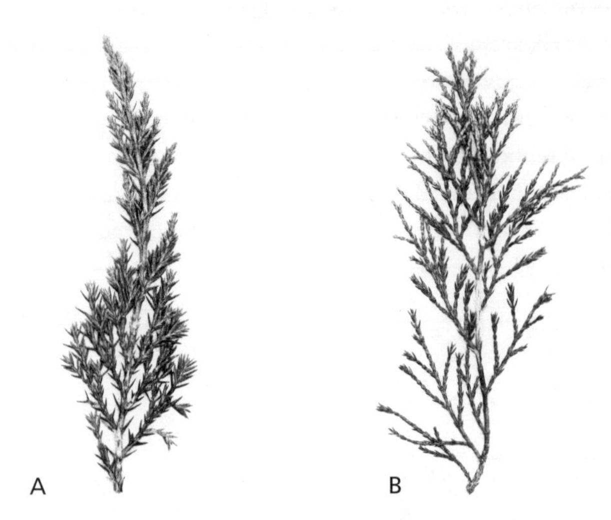

Branching twigs of savin (*Juniperus sabina*) with juvenile (A) and predominantly adult (B) foliage, ×0.5.

Natural stand of savin (*Juniperus sabina*) clothing the tops of rock outcrops.

Also a shrub of rocky barrens, savin is very similar to the North American creeping juniper (*Juniperus horizontalis*), which was originally described as a variety. Along with Chinese juniper (*J. chinensis*) it is one of the parents of Pfitzer juniper (*J. ×pfitzeriana*) and probably hybridizes with other species as well, giving rise to some of the variability that has been exploited in cultivar selection. The foliage has been used medicinally but is toxic in quantity. In addition to the common names given above, the numerous names present in western European languages are almost all variants on the theme of the English and Latin names. The technical synonymy of savin is also very large, and the attribution of some of its proposed variants here, as opposed to attaching to a related species, like Chinese juniper, is far from clear. As new evidence comes into play in the future, realignments of many of the confusing central Asian junipers are to be expected.

Juniperus saltillensis M. Hall

SALTILLO JUNIPER

Juniperus sect. *Sabina*

Shrubby tree to 7 m tall, divided from or near the base, with trunks to 30 cm in diameter. Bark whitish gray, thick and fibrous, peeling in long, soft strips. Crown broad and irregular, with rising branches. Branchlets numerous, spreading, sometimes drooping at the tips, cylindrical, 1–2 mm thick. Adult leaves in alternating trios (or pairs), scalelike, 1–2 mm long, light grayish green, with a prominent, flat or raised, round resin gland, the edges minutely toothed, the tip triangular and tightly pressed forward against the twig. Pollen and seed cones on separate plants. Pollen cones single at the tips of branchlets, oblong, 2–3.5 mm long, with four to six alternating pairs of pollen scales. Seed cones solitary at the tips of short branchlets, nearly spherical, 5–7 mm long, purple beneath a thick, pale blue, waxy coating, maturing in 1 year. Seeds one (to three), 4–5 mm long, dark brown, the pale attachment scar half the length of the seed. North-central Mexico, from northeastern Chihuahua and Zacatecas to south-central Nuevo León, including the type locality near Saltillo (hence the scientific name). Pinyon-juniper woodlands and wooded grasslands on calcareous substrates; 1,550–2,900 m. Zone 9. Synonym: *J. ashei* var. *saltillensis* (M. Hall) Silba.

Saltillo juniper is closely related to Ashe juniper (*Juniperus ashei*) and may be ancestral to it. The compounds in the leaf oils of Ashe juniper are a subset of those found in the Mexican species, reflecting a pattern common in progenitor/derived species pairs. Some DNA data, however, cast doubt on this and show no especially close relationship between Ashe juniper and Saltillo juniper.

Juniperus saltuaria Rehder & E. H. Wilson

SICHUAN JUNIPER, FANG ZHI BAI (CHINESE)

Juniperus sect. *Sabina*

Tree to 15(–20) m tall, with single trunk to 30 cm in diameter. Bark grayish brown, fibrous, peeling in narrow, vertical strips. Crown dense and conical, opening and broadening with age, with straight, upright or rising branches. Branchlets four-angled, short, curved, about 1 mm thick. Adult leaves in alternating pairs (or trios), scalelike, 1–2(–4) mm long, deep green, with an inconspicuous resin gland, the edges smooth, the tip triangular, curved forward, and loosely pressed against the twig. Pollen and seed cones on the same plant. Pollen cones single at the tips of branchlets, nearly spherical, 2–3 mm long, with four or five alternating pairs of pollen scales. Seed cones single at the tips of short, straight, upright branchlets, egg-shaped, 5–6 mm long, shiny black, without a waxy coating, maturing in 1 year. Seed one, 3–5 mm long. Central China, from southern Gansu to eastern Xizang (Tibet). Forming pure stands or mixed with other trees of the montane temperate evergreen forest; (2,100–)3,000–4,000(–4,600) m. Zone 6. Synonym: *Sabina saltuaria* (Rehder & E. H. Wilson) W. C. Cheng & W. T. Wang.

Sichuan juniper was locally common enough to have supplied construction timber in parts of its range and is one of the species recommended for reforestation in China. It is related to Turkestan juniper (*Juniperus pseudosabina*) and black juniper (*J. indica*) and replaces them at the eastern ends of their ranges.

Juniperus saxicola N. Britton & P. Wilson

PRICKLY REDCEDAR

Juniperus sect. *Sabina*

Tree to 8 m tall, often multistemmed, with trunks to 20 cm in diameter. Bark reddish brown, thin, peeling in scales and then in narrow strips. Crown dense, rounded to irregular, with numerous spreading to rising branches. Branchlets hidden by the elongate attached leaf bases, slender, long. Adult leaves needlelike, in alternating pairs, running down onto the stem without a joint, angled forward away from the twig, (4–)5–7(–9) mm long, without a gland, the edges smooth, the tip sharp. Pollen and seed cones probably on separate plants. Pollen cones unknown. Seed cones single at the tips of short branchlets, spherical or broader, 4–5 mm long, dark blue with a waxy coating, maturing in 1 year. Seeds two, about 4 mm long, light brown. Sierra Maestra, southeastern Cuba. Rocky openings in cloud forest on ridge tops; 1,200–1,800 m. Zone 9. Synonyms: *J. barbadensis* subsp. *saxicola* (N. Britton & P. Wilson) Borhidi, *J. barbadensis* var. *saxicola* (N. Britton & P. Wilson) Silba.

Prickly redcedar is the only New World species of section *Sabina* that does not develop scalelike adult leaves. Retention of juvenile foliage throughout the life of the plant may be related to the prevalence of fires in its restricted range. In both volatile leaf oil composition and some DNA data, prickly redcedar appears to fall in between the other strictly West Indian junipers, Caribbean juniper (*Juniperus barbadensis*), Bermuda redcedar (*J. bermudiana*), and Hispaniolan juniper (*J. gracilior*) on the one hand and the continental eastern redcedar (*J. virginiana*) on the other. All these species have scalelike adult leaves, but they also display varying proportions of shoots with needlelike leaves mixed in with fully adult foliage, especially prominent in some individuals of eastern redcedar and its cultivars.

Juniperus scopulorum C. Sargent

ROCKY MOUNTAIN JUNIPER

Juniperus sect. *Sabina*

Tree to 13(–20) m tall, with trunk to 2 m in diameter. Bark warm brown to dark brown, weathering dark grayish brown, thin, furrowed, peeling in thin rectangular plates or long, narrow strips. Crown conical or rounded, with spreading to rising branches. Branchlets stiff or drooping, four- or six-sided, 0.7–1.5 mm thick, often branching partially in a flattened sprays. Adult leaves in alternating pairs or trios, scalelike, 1–3(–4) mm long, green or blue-green, with a prominent but dry oblong resin gland, the edges smooth, the tip triangular, pressed forward against the twig or spreading from it. Pollen and seed cones on separate plants. Pollen cones single at the tips of branchlets, oblong, 1.5–3.5 mm long, with three or four alternating pairs of pollen scales. Seed cones single at the tips of short, straight branchlets, spherical or kidney-shaped, (4–)6–8(–9) mm in diameter, blue-black beneath the thick, pale blue, waxy coating, maturing in 2 years. Seeds one or two (to four), 4–5 mm long, light brown, the darker attachment scar a third to half the length of the seed. Rocky mountains (hence the scientific name, Latin for "of the rocks or crags") and outlying uplands of western North America from central British Columbia and western North Dakota to central Sonora, Chihuahua, and northern Coahuila (Mexico). Open juniper-pine-oak woodlands on rocky slopes; (0–)1,200–2,700 m. Zone 6.

Rocky Mountain juniper overlaps with and is replaced by creeping juniper (*Juniperus horizontalis*) northward and by eastern redcedar (*J. virginiana*) eastward. It hybridizes with both of them, and individuals from hybrid swarms are one of the main sources of its many cultivars. Other than its extensive use in cultivation, its role as an alternate host of cedar-apple rust (*Gymnosporangium juniperi-virginianae*), and its ability to invade overgrazed, marginal rangelands, the very common Rocky Mountain juniper has little present economic significance. The largest known individual in the United States in combined dimensions when measured in 1989 was 12.2 m tall, 2.0 m in diameter, with a spread of 6.4 m.

Well branched twig of Rocky Mountain juniper (*Juniperus scopulorum*) with about four poorly delimited years of growth bearing nearly mature pollen cones, ×1.

Juniperus semiglobosa E. Regel

HEMISPHERIC REDCEDAR, CENTRAL ASIAN PENCIL JUNIPER, SAUR-ARCHA (ALTAIC), KUN LUN DUO ZI BAI (CHINESE)

Juniperus sect. *Sabina*

Tree to 20 m tall, or a low shrub at the alpine timberline, usually with single trunk to 1(–2) m in diameter. Bark reddish brown to gray-brown, fibrous, furrowed, peeling in narrow vertical strips. Crown conical, broadening and opening with age, with outstretched, rising to spreading branches. Branchlets widely spaced, cylindrical to weakly four-angled, 1–2 mm thick, Adult leaves in alternating trios or pairs, scalelike, 1–2 mm long, light green to yellowish green, with an active, oblong resin gland, the edges smooth, the tip triangular, curved in and pressed against the twig. Pollen and seed cones on separate plants. Pollen cones single at the tips of branchlets, slightly oblong, 3–5 mm long, with four or five alternating pairs of pollen scales. Seed cones single at the tips of straight branchlets, lumpy with seeds, 4–8 mm across, bluish black, often with a thin waxy coating, maturing in 2 years. Seeds (one or) two or three (or four), 3–6 mm long, brown, the attachment scar extending partway up the side. High mountains of central Asia from Uzbekistan to northern Uttar Pradesh, India, and westernmost China. Open juniper woodlands of rocky soils in high, dry valleys; (1,500–)1,600–4,300 m. Zone 6. Synonym: *Sabina semiglobosa* (E. Regel) L. K. Fu & Y. F. Yu.

Hemispheric redcedar has a red heartwood (hence the Altaic common name, "red juniper") that has been used for making pencils. The Latin name reflects the overall semispherical shape of the seed cones. It often grows with Persian juniper (*Juniperus excelsa* subsp. *polycarpos*) and Turkestan juniper (*J. pseudosabina*), with which it is sometimes confused, just as several species of juniper may grow together in southwestern North America.

Juniperus squamata Buchanan-Hamilton ex D. Don

SCALY JUNIPER, FLAKY JUNIPER, PADMA CHUNDER (HINDI), GAO SHAN BAI (CHINESE)

Juniperus sect. *Sabina*

A shrub 15 cm to 1.5 m high or sometimes a tree to 12(–28?) m tall, with several trunks to 0.3(–0.5) m in diameter. Bark reddish brown, weathering grayish brown, thin, flaking and then peeling in narrow longitudinal strips. Crown broad and low, with branches usually trailing or rising in tree forms. Branchlets upright or hanging, cylindrical, slender. Adult leaves in alternating trios, closely spaced, needlelike, curved forward and overlapping (hence "scaly" and *squamata*), 3–5(–7) mm long, grayish green or blue-green with a broad, shallow, white, waxy stomatal groove above divided by a thin midrib, without a gland, the edges smooth, the tip prolonged into a point. Pollen and seed cones on separate plants. Pollen cones single at the tips of branchlets, spherical, 2.5–4 mm long, with three or four alternating trios of pollen scales. Seed cones single at the tips of straight branchlets, egg-shaped or spherical, 5–8 mm long, dark bluish black, maturing in 2 years. Seed one, 4–6 mm long, light brown, with darker pits and a narrow, lighter attachment scar. Widespread in the Himalaya from Afghanistan eastward and across China to Taiwan. Forming dense thickets in alpine scrub and groves of trees in subalpine woodlands; (1,300–)1,600–4,600(–4,850) m. Zone 4. Synonyms: *J. baimashanensis* Y. F. Yu & L. K. Fu, *J. kansuensis* V. Komarov, *J. morrisonicola* Hayata, *J. squamata* var. *fargesii* Rehder & E. H. Wilson, *J. squamata* var. *hongxiensis* Y. F. Yu & L. K. Fu, *J. squamata* var. *parviflora* Y. F. Yu & L. K. Fu, *Sabina squamata* (Buchanan-Hamilton ex D. Don) F. Antoine, *S. squamata* var. *fargesii* (Rehder & E. H. Wilson) L. K. Fu & Y. F. Yu.

Natural stand of hemispheric redcedar (*Juniperus semiglobosa*).

Alpine natural stand of shrubby scaly juniper (*Juniperus squamata*).

Branched twig of scaly juniper (*Juniperus squamata*) with about 5 years of growth marked by variations in needle length and color, ×0.5.

Scaly juniper commonly forms dense thickets mixed with common juniper (*Juniperus communis*) in the alpine zone of the Himalaya and the mountains of China. It is among the highest-elevation woody plants in this part of Asia. The dwarf habit has resulted in numerous cultivars. The resinous wood, like that of several other junipers, is burned as incense. Scaly juniper is very closely related to Ping juniper (*J. pingii*) and Himalayan drooping juniper (*J. recurva*) and intergrades with these species to varying extents so that identification may be difficult at times.

Juniperus standleyi Steyermark

STANDLEY JUNIPER, HUITÓ (SPANISH)

Juniperus sect. *Sabina*

Shrub, or tree to 15 m tall, or prostrate above the alpine timberline, with trunk, when present, often contorted, to 0.5 m in diameter. Bark reddish brown to dark brown, fibrous, furrowed, peeling in strips. Crown broad, flattened or irregular, with spreading or rising branches. Branchlets numerous, partly in flattened sprays, slightly four-angled, 1–1.5 mm thick. Adult leaves in opposite pairs, scale-like, 1.5–2 mm long, dark green to yellowish green with a waxy coating, with a round or oblong glandular pit, the edges minutely toothed, the tip blunt, pressed forward against the twig. Pollen and seed cones on separate plants. Pollen cones single at the tips of short branchlets, egg-shaped, 2–3.5 mm long, with three to five (to seven) alternating pairs of pollen scales. Seed cones single at the tips of short, curved branchlets, spherical or lumpy, 7–9 mm long, dark blue beneath the thin, pale blue waxy coating, maturing in 1 year. Seeds three or four (to six), 4–5 mm long, brown, the paler attachment scar about two-thirds the length of the seed. High mountains of western Guatemala and adjacent Chiapas, Mexico. Scattered as a dwarf shrub above timberline but forming dense groves of trees near the lower limit of its range; 3,000–4,250(–4,600) m. Zone 8?

Standley juniper appears to be most closely related to the other alpine junipers of Mexico, Durango juniper (*Juniperus durangensis*) in the north and mountain juniper (*J. monticola*) in the central portion of the country. It differs from them in the tendency for the branchlets to be arranged in flattened sprays. Durango juniper differs also in its brown cones bearing only one or two seeds, and mountain juniper differs in having flatter leaves that do not bulge out from the twig.

Juniperus taxifolia W. J. Hooker & G. Arnott

OGASAWARA JUNIPER, SHIMA-MURO (JAPANESE)

Juniperus sect. *Juniperus*

Shrub to 3 m, or shrubby tree to 13 m tall, with trunks to 15(–30) cm in diameter. Bark brown, weathering grayish brown, fibrous, peeling in thin scales. Crown low, irregular, with contorted, spreading or rising branches. Branchlets three-angled, slender, straight and rising. Adult leaves in alternating trios, needlelike, 7–14 mm long, light green, sticking out from and jointed at their attachment to the twig, with two separate, white, waxy stomatal bands above, slightly ridged below, the edges smooth, the tip broadly triangular and soft. Pollen and seed cones on separate plants. Pollen cones single in the axils of leaves, egg-shaped, 3–4 mm long, with three or four alternating trios of pollen scales. Seed cones single at the tips of short branchlets in the axils of leaves, spherical, 8–9 mm in diameter, purplish brown, without a waxy coating, maturing in 3 years. Seeds three, 4–5 mm long, light brown, the paler attachment scar extending up to halfway up the side. Ogasawara Islands, southern Japan. Open, stony soils; 0–300 m. Zone 9.

Ogasawara juniper is closely related to Ryukyu juniper (*Juniperus thunbergii*) and is often united with it. It may also be confused with other shoreline junipers in Japan, such as Japanese beach juniper (*J. conferta*). The genetic relationships and phylogeography of the various Japanese members of section *Juniperus* merit further study using numerous populations. The species name, Latin for "yew leaf," reflects the much flatter, broader needles, with two well-separated stomatal bands, compared to many other members of section *Juniperus*.

Juniperus thunbergii W. J. Hooker & G. Arnott

RYUKYU JUNIPER, OSHIMA-HAINEZU (JAPANESE)

Juniperus sect. *Juniperus*

Low shrub creeping to 10 m across. Bark reddish brown, fibrous. Crown flat, with crooked, trailing branches. Branchlets three-sided, upright. Adult leaves in alternating trios, needlelike, 6–12 mm long, light green, sometimes with a thin waxy coating, sticking out from and jointed at their attachment to the twig, with two separate, white, waxy stomatal bands above, slightly ridged below, the edges smooth, the tip narrowly triangular and not prickly. Pollen and seed cones on separate plants. Pollen cones single in the axils of leaves, oblong, 4–5 mm long, with four or five alternating trios of scales. Seed cones single at the tips of very short branchlets in the leaf axils, spherical, about 10 mm long, purplish brown, maturing in 3 years. Seeds three, 5–6 mm long, light brown, with a paler attachment scar. Southern Japan, Pacific Ocean shores of southern Honshū and islands from Tanega-shima to the Okinawa Islands. Seashores; 0–50(–100) m. Zone 9. Synonym: *J. taxifolia* var. *luchuensis* (G. Koidzumi) Satake.

Ryukyu juniper is so closely related to Ogasawara juniper (*Juniperus taxifolia*) that the two are often treated as varieties of a single species or merged entirely. Neither species has a large range, but Ryukyu juniper is more widely distributed than Ogasawara juniper yet has less variation in its growth habit and habitat. Additional genetic evidence might help in trying to resolve their taxonomic status. The species name honors Carl Thunberg (1743–1825), author of the first botanical flora of Japan, who collected the type specimen.

Juniperus thurifera Linnaeus

SPANISH JUNIPER, TRABINA, SABINA ALBAR (SPANISH), INTZENTSU-MITERRA (BASQUE), GENÉVRIER THURIFÈRE (FRENCH), TAOUALT (BERBER)

Juniperus sect. *Sabina*

Tree to 20 m tall, with trunk to 1(–2) m in diameter. Bark dark brown, weathering grayish brown, fibrous, furrowed, peeling in narrow strips. Crown conical, broadening and flattening greatly with age, with spreading or rising branches. Branchlets partly in flattened sprays, four-angled, about 1 mm thick, aromatic when crushed (hence the scientific name, Latin for "incense bearing"). Juvenile leaves common on adult plants, in alternating pairs, needlelike. Adult leaves in alternating pairs, scalelike, 1.5–2.5 mm long, bluish green, with an oblong glandular furrow, the edges smooth or minutely toothed, the narrowly pointed tip loose. Pollen and seed cones on separate plants. Pollen cones single at the tips of branchlets, spherical, 2–3 mm long, with five or six alternating pairs of pollen scales. Seed cones single at the tips of curved

Young mature Spanish juniper (*Juniperus thurifera*) in nature.

Large Spanish juniper (*Juniperus thurifera*) in nature.

branchlets, spherical, (6–)8–12 mm long, dark purple with a thin waxy coating, maturing in 2 years. Seeds (one or) two to four, (5–)6–10 mm long, shiny brown, the paler attachment scar extending to about a third of the length. Western Mediterranean in northern Morocco, northern Algeria, Spain, southern France, and Corsica. Open woodlands and scrublands of dry slopes in the mountains; (200–)500–2,000(–3,000) m. Zone 9. Synonyms: *J. thurifera* var. *africana* J. Maire, *J. thurifera* var. *gallica* Coincy.

Spanish juniper is fairly common in Spain but much rarer in North Africa, where it has suffered from direct exploitation for wood and from herbivory by the large populations of goats, and in the Alpes Maritimes in southeastern Franc, where succession to pine and oak has been underway. Sex ratios in these places are slightly male biased in young stands and slightly female biased in older ones, but not enough to make much difference in regeneration potential.

Juniperus tibetica V. Komarov

TIBETAN JUNIPER, DA GUO YUAN BAI (CHINESE)

Juniperus sect. *Sabina*

Tree to 15–30(?) m tall or a low shrub at the alpine timberline, with trunk to 1(–2) m in diameter. Bark bright reddish brown, weathering grayish brown, fibrous, peeling in narrow strips. Crown dense, with numerous erect, spreading or drooping branches. Branchlets numerous, four-angled, 1 mm thick. Juve-

Large old Tibetan juniper (*Juniperus tibetica*) in nature.

nile needlelike leaves often present along with adult foliage. Adult leaves in alternating pairs, scalelike, (1.5–)2–3 m long, gray-green to dark green, sometimes with a conspicuous, elongate, resin gland, the edges smooth, the tip triangular, pressed against the branchlet. Pollen and seed cones on the same plant. Pollen cones single at the tips of branchlets, nearly spherical, 1.5–2 mm long, with three or four alternating pairs of pollen scales. Seed cones single at the tips of branchlets, spherical, (5–)8–15(–18) mm long, reddish brown, glossy, without a waxy coating, maturing in 2 years. Seed one, 7–11 mm long, brown, with a paler attachment scar reaching close to the tip. Western China from Sichuan and nearby Xizang (Tibet), to Qinghai and Gansu. Mountain slopes; (2,200–)2,800–4,500(–4,800) m. Zone 5. Synonyms: *J. distans* Florin, *J. potaninii* V. Komarov, *J. przewalskii* V. Komarov, *J. zaidamensis* V. Komarov, *Sabina przewalskii* (V. Komarov) W. C. Cheng & L. K. Fu, *S. tibetica* (V. Komarov) W. C. Cheng & L. K. Fu.

Populations of Tibetan juniper from the northwestern portion of its range centered on Lake Qing-hai in Qinghai and Gansu provinces are often recognized as a separate species under the name *Juniperus przewalskii*. While they appear to be too weakly differentiated to maintain at the species level, they might merit recognition as a variety within Tibetan juniper.

Openly branched twig of eastern redcedar (*Juniperus virginiana*) with juvenile foliage, ×0.5.

Juniperus virginiana Linnaeus

EASTERN REDCEDAR

Juniperus sect. *Sabina*

Tree to 30 m tall, with single trunk to 1.7 m in diameter, fluted with age. Bark bright reddish brown, only slowly weathering grayish brown, thin, fibrous, furrowed, peeling in long, narrow, vertical strips. Crown dense, conical, usually narrow at first and broadening with age, with rising, spreading, or hanging branches. Branchlets stiff or drooping, cylindrical or slightly four-angled, 0.7–1.3 mm thick. Needlelike juvenile leaves commonly present on adult trees. Adult leaves in opposite pairs, scalelike, 1–3 mm long, dark green, with an inactive, oblong resin gland, the edges smooth, the tip blunt or triangular, curved away from the twig. Pollen and seed cones on separate plants. Pollen cones single at the tips of branchlets, egg-shaped, 3–5 mm long, with five to eight alternating pairs of pollen scales. Seed cones single at the tips of straight (or curved) branchlets, spherical, 3–6(–10) mm long, blue black with a prominent bluish white waxy coating, maturing in 1 year. Seeds one or two (or three), 1.5–4 mm long, light brown, the paler attachment scar extending up to halfway along the side. Eastern North America from southwestern North Dakota and southeastern Texas through southernmost Canada (southern Ontario and southwestern Quebec) to southwestern Maine and Florida. Usually in relatively open habitats on dry to wet soils; 0–100(–1,400) m. Zone 5. Two varieties.

Well branched twigs of eastern redcedar (*Juniperus virginiana*) with pollen (A) and seed (B) cones, ×0.5.

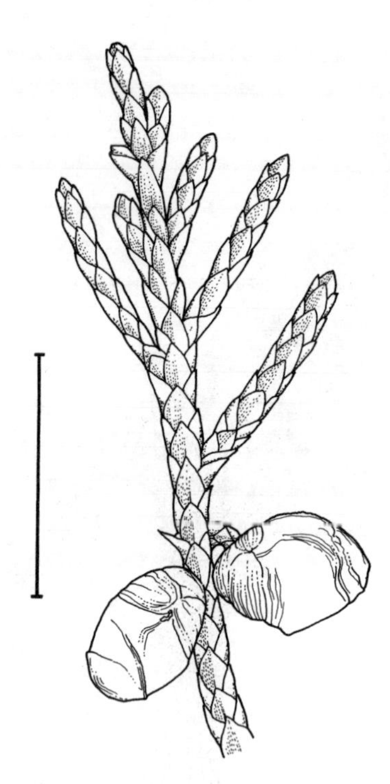

Mature seed cones of eastern redcedar (*Juniperus virginiana*); scale, 1 cm.

Eastern red cedar commonly invades abandoned agricultural land and pastures ("old fields"), often in a pattern that reflects dispersal by birds eating the berrylike cones. It overlaps with and is replaced by its close relatives creeping juniper (*Juniperus horizontalis*) northward and Rocky Mountain juniper (*J. scopulorum*) westward and hybridizes with both of them. Some of the numerous cultivars are derived from these hybrids. Hybridization with Ashe juniper (*J. ashei*) toward the southwest, which had been postulated on the basis of leaf and cone traits, has not been supported by variation in leaf oils. There are two geographic varieties, with variety *silicicola* forming a transition to the West Indian junipers of the Bahamas and Cuba. The largest known individual in combined dimensions when measured in 1997 was 17.4 m tall, 2.0 m in diameter, with a spread of 21.0 m. Another individual, almost as large, was 27.5 m tall and 1.5 m in diameter, with a crown spread of 13.7 m when measured in 1993.

Juniperus virginiana* var. *silicicola (J. K. Small) E. Murray
SOUTHERN REDCEDAR
Tips of scalelike leaves broadly to narrowly triangular. Pollen cones 4–5 mm long. Seed cones 3–4 mm long. Seeds 1.5–3 mm long. Immediate coastal region of the southeastern United States from North Carolina to western Florida and inland in the latter state. Synonym: *Juniperus silicicola* (J. K. Small) L. H. Bailey.

Juniperus virginiana Linnaeus **var. *virginiana***
Tips of scalelike leaves narrowly triangular. Pollen cones 3–4 mm long. Seed cones 4–6(–10) mm long. Seeds 2–4 mm long. Throughout the range of the species except at the southern fringe occupied by variety *silicicola*. Synonym: *Juniperus virginiana* var. *crebra* Fernald & Griscom.

Keteleeria Carrière
KETELEERIA
Pinaceae

Evergreen trees with a straight, single trunk bearing irregular tiers of rising to horizontal branches to form a conical crown when young, generally becoming broad and domelike or flat with maturity. Bark not fibrous, grayish brown and smooth at first, darkening and becoming flaky with age and then ridged and furrowed at the base of mature trunks. Branchlets all elongate, without distinction into long and short shoots, hairy or not, grooved between raised leaf bases bearing a slightly raised leaf attachment. Winter buds well developed, scaly, the scales remaining on the twigs for several years. Leaves spirally arranged, sometimes radiating all around the twigs but more commonly sticking straight out to the sides and sometimes above as well. Each leaf needlelike, flat, generally sword-shaped, usually straight or slightly curved, long pointed in juvenile foliage and on sprouts, bluntly pointed, rounded, or shallowly notched in adult foliage, tapering to a short petiole.

Plants monoecious. Pollen cones in rings of 4–10 from single buds at or near the tips of shoots, each on a short, slender stalk and with numerous, spirally arranged pollen scales, each bearing two pollen sacs. Pollen grains large to very large (body 55–125 µm long, some the largest among all living conifers), with two round air bladders that have a relatively small area of attachment to the large, oval, minutely bumpy body, the bladders with more wrinkled sculpturing. Seed cones cylindrical or slightly egg-shaped, single at the tips of ordinary, stout branchlets, upright both at pollination and at maturity, maturing in a single season and then releasing the seeds, but remaining on the tree and gradually breaking up in place or breaking off above the bottom scales, with numerous, densely spirally arranged, woody, thin seed scales. Seed scales of various shapes from circular to triangular but always with a thin stalk flanked by two small, downwardly directed lobes. The bract roughly half as long as but often not visible between the scales in the intact mature cone, with a three-lobed blade at the end of a longer (or nearly equal), stalk united with the stalk of the seed scale. Seeds two per scale, wedge-shaped, the asymmetrical wing derived from the seed scale and as long as it, cupping the seed body loosely on one side

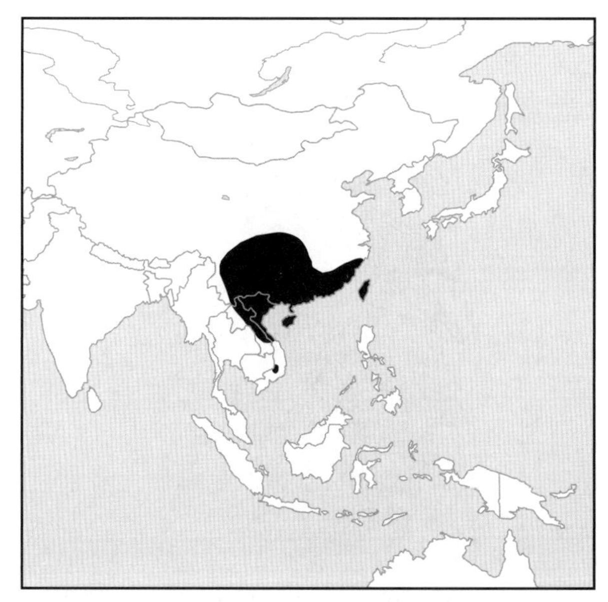

Distribution of *Keteleeria* in southeastern Asia, scale 1 : 120 million

and wrapping around to cover about one quarter to one third of the other side. Free part of wing twice as wide as the body and about two to three times as long. Cotyledons two to four, each with one vein, remaining within the seed coat underground. Chromosome base number $x = 12$.

Wood soft, slightly fragrant, light brown with a distinct reddish brown heartwood. Annual rings well developed and with abundant latewood. Vertical resin canals often present, especially in response to injury, but resin ducts absent from the rays.

Stomates arranged in two waxy or dark green bands beneath and sometimes also with a few to many incomplete lines above. Each stomate sunken deep in a pit beneath the four (or five) neatly arranged subsidiary cells that are covered with a very thick cuticle but are not circled by a further thickened Florin ring. Leaf cross section with a single-stranded midvein above a strand of transfusion tissue and with a resin canal on either side inside the lower epidermis near the needle edge. Photosynthetic tissue with a well-defined single palisade layer beneath the upper epidermis and adjacent, nearly continuous (except beneath lines of stomates, when present) hypodermis, the remainder consisting of spongy mesophyll connecting the palisade to the stomatal bands beneath.

Two species in China and Indochina. References: Farjon 1989, 1990, Fu et al. 1999c, Li and Keng 1994.

Trees of *Keteleeria* occur at generally lower elevations and in drier habitats than the true firs (*Abies*) and spruces (*Picea*) occupying the same general region. The name honors Jean-Baptiste Keteleer (1813–1903), a Belgian horticulturist and nurseryman. Both species remain relatively uncommon in cultivation although *K. davidiana,* more widespread in nature, is also more frequently encountered in collections. Consequently, there has been no cultivar selection.

Although most closely related to *Abies,* as evidenced by general morphology, immunological reactions, and DNA studies, *Keteleeria* has more in common with some pine (*Pinus*) species ecologically and in its wide-crowned growth habit. The species were originally confused with *Abies* but were recognized as a distinct genus within 20 years of their introduction to Europe. They differ from *Abies* most obviously in their clustered pollen cones and the fact that their seed cones do not break up with maturity.

Like Douglas firs (*Pseudotsuga*), the two species of *Keteleeria* vary greatly in seed cone characteristics: the length and width of the whole cone, the shape of the seed scales and attached bracts, and the corresponding shape of the seed wing. The needles also vary with age of the tree and with climatic conditions. As a result, some 16 species and a few additional varieties were named, several in the period 1975–1983. The treatment in the *Flora of China* (Fu et al. 1999c) accepted five species and four additional varieties, while Farjon (1989), in a detailed revision of the genus, accepted the three species commonly recognized before Flous's (1936) revision. Just two species are recognized here, one with two weakly differentiated varieties, and the additional proposed species are considered extreme variants within these two species. Little is known about the range of variation in natural populations of *Keteleeria.* Progress in understanding the taxonomic structure of the genus, however, will depend on detailed population studies rather than the limited available herbarium material. Unfortunately, many populations were decimated by forest clearance if not by specific exploitation.

Fossils of *Keteleeria* are known from Oligocene deposits more than 23 million years old in both Europe and western North America. The genus disappeared from North America before the close of the Miocene (more than 5 million years ago) but lasted into the succeeding Pliocene in Europe, by which time there are records of the genus in Asia as well.

Identification Guide to *Keteleeria*

The two species of *Keteleeria* are difficult to separate vegetatively but may be distinguished by the seed cones, particularly whether the seed scales are broadest above or below the middle and whether the seed wing is broadest above or below the middle:

Keteleeria davidiana, seed scales and seed wing broadest below the middle

K. fortunei, seed scales and seed wing broadest above the middle

Keteleeria davidiana (M. Bertrand) Beissner
COMMON KETELEERIA, TIE JIAN SHAN (CHINESE)
Tree to 25(–45) m tall, with trunk to 1.5(–2.5) m in diameter. Bark dark grayish brown, scaly, furrowed. Crown regular in youth, breaking up and broadening with age, with heavy, rising or horizontal branches bearing horizontal, or rarely dangling, side branches. Young shoots yellowish gray to reddish brown, often orange-brown, rarely greenish brown, often hairy at first, sometimes densely and persistently rusty-haired. Buds 3–6 mm long, plump, not resinous. Needles dark green or sometimes a little paler with wax, (1.5–)3–7 cm long (to 14 cm on juvenile foliage). Each needle with 12–16 lines of stomates in each light green to waxy greenish white stomatal band beneath and 0–10 lines on each side of the midrib above, often just near the tip. Leaf edges flat or a little curled under. Tip of adult leaves notched, rounded or short-pointed. Pollen cones in clusters of three to eight, 10–15 mm long, yellowish brown. Seed cones (5–)10–20(–25) cm long, (3–)4–6(–7) cm across, green or waxy bluish green before maturity, ripening reddish brown. Seed scales quite variable in shape and toothing around the edge but usually distinctly narrowed toward the tip and broadest below the middle, often curling back at maturity. Seed body wedge-shaped, 9–16 mm long, the wing 10–15 mm longer, broadest below the middle. Central and south-central China, Taiwan, Laos, and Vietnam. Usually mixed with broad-leaved evergreens in open forests and woodlands on slopes; (200–)700–1,500(–2,900) m. Zone 7. Two subspecies.

This most widespread species of *Keteleeria* varies a great deal in leaf, twig, and cone characters from place to place, and some of the variants were given taxonomic recognition, but there seems to be little geographic or taxonomic coherence to most of them. Trees in the southwestern portion of the range, however, seem to differ fairly consistently in the length of their needles and shape of their seed scales from trees to the east and north. These regional variants are here recognized as subspecies, with subspecies *evelyniana* being one of the most southerly Old World members of the Pinaceae. Some of the other proposed varieties or species may represent intergrades or hybrids at the border between the two subspecies in western Guizhou and Guangxi. Despite the extensive variation in the species, little of this variation has entered general cultivation, even though *K. davidiana* is more widely cultivated than *K. fortunei.* The species name honors Armand David (1826–1900), a French missionary and naturalist in China who collected the type specimen but who is most famous, perhaps, for bringing Père David's deer (*Elaphurus davidianus*) to the attention of western zoologists. Because of its rarity in cultivation, there is no established English common name for this species. The Chinese name *shan,* "hard iron" (the general term for conifers with needle leaves attached singly rather than clustered, as well as referring specifically to the unrelated China fir, *Cunninghamia lanceolata*), refers to the relatively hard, dense wood, which is the strongest large softwood in much of southern China and is employed for a great range of construction and carpentry uses. This exploitation has not overly depleted the species because it is one of the few conifers that will sprout new trunks after harvest.

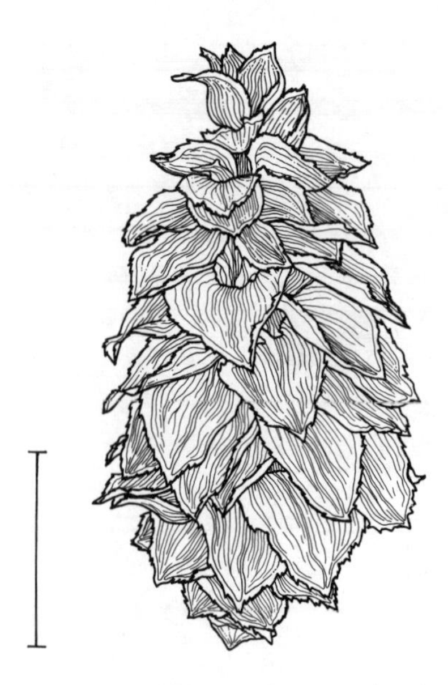

Open mature seed cone of common keteleeria (*Keteleeria davidiana* subsp. *davidiana*) after seed dispersal; scale, 5 cm.

Keteleeria davidiana (M. Bertrand) Beissner **subsp. *davidiana***
COMMON KETELEERIA, TA JIAN SHAN (CHINESE)
Adult needles 1.5–5 cm long, rounded or notched at the tip. Seed scales with straight or outwardly rounded edges above the widest point. Northern Taiwan and central China, from southeastern Gansu and southern Shaanxi to northern Guangxi. Synonyms: *Keteleeria calcarea* W. C. Cheng & L. K. Fu, *K. chien-peii* Flous, *K. davidiana* var. *calcarea* (W. C. Cheng & L. K. Fu) Silba, *K. davidiana* var. *formosana* (Hayata) Hayata, *K. davidiana* var. *pubescens* (W. C. Cheng & L. K. Fu) Silba, *K. formosana* Hayata, *K. pubescens* W. C. Cheng & L. K. Fu.

Keteleeria davidiana* subsp. *evelyniana
(M. T. Masters) Eckenwalder
LONGLEAF KETELEERIA, YUN NAN YOU SHAN (CHINESE)
Adult needles (2–)4–8 cm long, the tip with a short point. Seed scales with the edges curved inward above the widest point. Laos, Vietnam, and southwestern China, in Hainan, Yunnan, and adjacent Sichuan and Guizhou. Synonyms: *Keteleeria esquirolii* H. Léveillé, *K. evelyniana* M. T. Masters,

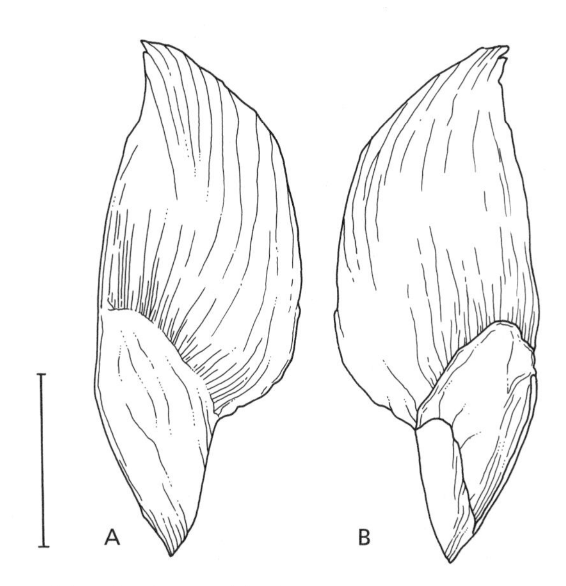

Upper (A) and undersides (B) of a dispersed seed of common keteleeria (*Keteleeria davidiana* subsp. *davidiana*); scale, 1 cm.

K. evelyniana var. *hainanensis* (W. Y. Chun & Tsiang) Silba *K. evelyniana* var. *roulletii* (A. Chevalier) Silba, *K. evelyniana* var. *xerophila* (Ji R. Hsüeh & S. H. Hao) Silba, *K. hainanensis* W. Y. Chun & Tsiang, *K. roulletii* (A. Chevalier) Flous, *K. xerophila* Ji R. Hsüeh & S. H. Hao.

Remnant mature longleaf keteleeria (*Keteleeria davidiana* subsp. *evelyniana*) on rocky outcrops in an otherwise fully agricultural landscape.

Keteleeria fortunei (A. Murray bis) Carrière

FOOTHILL KETELEERIA, YOU SHAN (CHINESE)

Tree to 30 m tall, with trunk to 1(–1.5) m in diameter. Bark yellowish brown to dark grayish brown, rough, furrowed, peeling in flakes. Crown conical and regular in youth, becoming broadly egg-shaped with age, with heavy, horizontal to steeply rising branches bearing horizontal side branches. Young shoots yellowish brown to reddish orange, variably and transiently hairy or, more commonly, hairless. Buds 3–5 mm long, egg-shaped, not resinous. Needles dark green above, 1–3(–4) cm long. Each needle with 12–20 lines of stomates in each whitish green stomatal band beneath and (none or) one or two (to five) lines on each side of the midrib above, especially near the tip. Leaf edges generally flat. Tip of adult leaves rounded. Pollen cones in clusters of three to eight, 10–15 mm long, yellowish brown. Seed cones (5–)10–20 cm long, (3.5–)5–6.5 cm in diameter, green or purplish green before maturity, ripening chocolate brown. Seed scales variable in shape and toothing around the edge, often roughly circular and broadest at or above the middle, sometimes curling back at maturity. Seed body wedge-shaped, 10–13 mm long, the wing 10–20 mm longer, broadest above the middle. Southern China, from southern Zhejiang and Fujian west to eastern Yunnan. Scattered with a few other conifers among broad-leaved evergreens in mixed evergreen forests on slopes of hills and low mountains; 200–800(–1,400) m. Zone 7. Synonyms: *Keteleeria cyclolepis* Flous, *K. fortunei* var. *cyclolepis* (Flous) Silba, *K. fortunei* var. *oblonga* (W. C. Cheng & L. K. Fu) L. K. Fu & N. Li.

This rare species is also uncommon in cultivation although more common than *Keteleeria davidiana* in Australia. It is also a staple of temple grounds in its native region, and such a tree, at Fuzhou in Fujian province, was one of the first seen by Robert Fortune (1812–1880), the British plant collector for whom this species is named. Temple trees are the best-protected populations and could be a seed source for reforestation with this commercially valuable species. The high oil content of the wood, which gives the tree its Chinese name, "oil" *shan*, makes it waterproof, and thus among other uses the timber is important in shipbuilding. The varieties (or species) recognized by some authors completely intergrade and overlap and so are not accepted here. Perhaps some of the variability in both this species and in *K. davidiana* is due to hybridization between them in their zone of overlap in Yunnan, Guizhou, and Guangxi. This possibility is worth investigating both experimentally and in the field.

Lagarostrobos Quinn

HUON PINE

Podocarpaceae

Evergreen trees and shrubs. Trunk cylindrical or fluted, dividing near the ground or free of major limbs for a third or more of its height. Bark obscurely fibrous, thin, smooth at first, later flaking in overlapping scales. Crown fairly open to dense, conical to irregularly and narrowly dome-shaped, with nearly horizontal to strongly upwardly angled branches. Branchlets all elongate, without distinction into long and short shoots, hairless, grooved between the attached leaf bases and remaining green for at least the first year (juvenile foliage) or completely hidden (adult foliage). Resting buds indistinct, consisting solely of as yet unexpanded ordinary foliage leaves. Leaves spirally attached. Juvenile foliage only moderately distinct from and giving way gradually but fairly quickly to adult foliage. Juvenile leaves scalelike, roundly triangular, a little flattened top to bottom, standing out from the twigs at a forward angle. Adult leaves scalelike, tightly pressed against and completely covering the twigs, overlapping, the exposed portion unevenly diamond-shaped, roundly to sharply keeled, with a toothed marginal frill.

Plants dioecious (a few individuals monoecious). Pollen cones single at the tips of otherwise ordinary short foliage shoots. Each pollen cone cylindrical, with a few bracts at the base and with 10–16 spirally arranged, triangular pollen scales, each with two pollen sacs. Pollen grains small to medium (body 20–40 µm long, overall 45–60 µm), with two relatively inconspicuously internally wrinkled air bladders that are much smaller than and mostly tucked under the very slightly elongate, minutely bumpy body, and accompanied by a prominent, small boss jutting out at the junction of the bladders with the cap of the body. Seed cones single and bent downward at the tip of otherwise ordinary foliage shoots, highly modified and reduced, with about 6–14 loosely spirally arranged bracts that become somewhat fleshy but do not unite into a podocarpium, maturing in a single season. Middle 4–8(–10) bracts fertile, from which one to five (to seven) seeds mature. Each fertile bract bears one seed that is nestled in the thin, papery, asymmetrically cup-shaped seed scale (the epimatium) and that has its opening pointed forward along the cone axis. Seed without an aril, hard, egg-shaped and somewhat flattened, wrinkled, ending in a broad, dimpled, straight beak. Cotyledons two, each with two veins. Chromosome base number $x = 15$.

Wood moderately soft and light, extremely durable, sweetly fragrant, the very pale sapwood sharply contrasting with the light brown to yellowish brown heartwood that is sometimes dotted with black. Grain very fine and even, with clear growth rings set off by a fairly broad band of slightly darker latewood. Resin canals and individual resin parenchyma cells both absent.

Stomates in patches on both sides but not arranged in lines or in any one direction. Each stomate sunken, plugged with wax that is not walled in by a Florin ring. Midvein single, inconspicuously buried in the leaf tissue of the adult leaves, with one resin canal immediately beneath it. Photosynthetic tissue forming a palisade layer on both sides of the leaf, without an intervening hypodermis inside the epidermis.

One species in Tasmania, Australia. References: W. M. Curtis 1956, K. Hill 1998: R. Hill and Brodribb 1999, Molloy 1995, Playford and Dettmann 1978.

Huon pine (*Lagarostrobos franklinii*), the sole living species of *Lagarostrobos* (the name, Greek for "slack cone," reflects the droopy seed cone and its relatively widely spaced bracts), was once included in *Dacrydium* but is only distantly related to the species in that genus. Instead, *Lagarostrobos* is closely related to *Manoao* of New Zealand and to *Parasitaxus* of New Caledonia but differs enough to justify the maintenance of a separate genus. Although *L. franklinii* is cultivated to a limited extent in botanical gardens and arboreta, it is not in general cultivation, and there has been little cultivar selection, with the exception of one particularly weeping form.

Acceptance of *Lagarostrobos* as a segregate genus seems especially apt since it has been recognizable since the mid-Cretaceous, some 90 million years ago. These early records are pollen grains that closely resemble those of Huon pine, which is the only living conifer with a tubercle sticking out at the base of each air bladder. These pollen grains are from sediments widely distributed across the southern

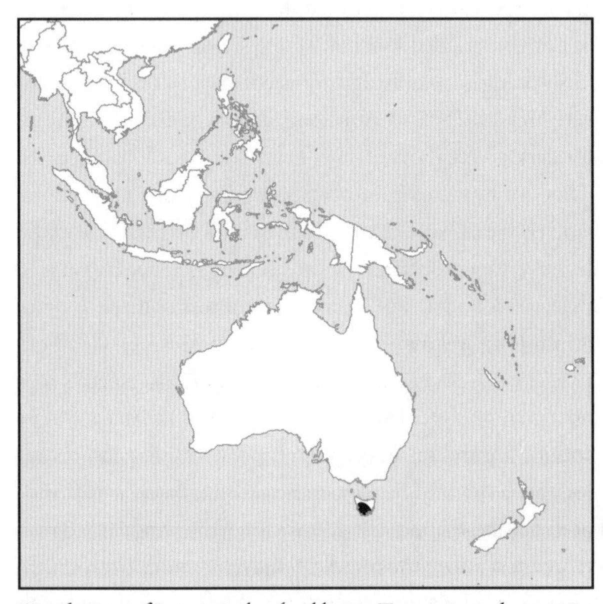

Distribution of *Lagarostrobos franklinii* in Tasmania, scale 1 : 119 million

hemisphere, including southern Australia, New Zealand, southern South America, and parts of Antarctica. Everywhere they occur, the sediments and associated species indicate a wet habitat like that inhabited today by Huon pine. This pollen became progressively less common in mainland Australia during the early Tertiary as the continent dried, and it disappeared completely there after the Miocene, some 5 million years ago or more. Two other similar types of tubercle-bearing pollen grains are known from late Cretaceous and lower Tertiary sediments of the southern hemisphere, and these may well represent now extinct lineages of *Lagarostrobos.*

In contrast to the widespread and long-lasting distribution of pollen grains probably belonging to *Lagarostrobos,* the only known or recognized macrofossils attributed to *Lagarostrobos* are found in its present home, Tasmania. The oldest shoots are found in Oligocene sediments (about 30 million years old), while recognizable macrofossils of the living Huon pine, are first known from the early Pleistocene (less than 2 million years ago), a striking contrast to the long presumed pollen record of this species (or a very close relative).

Lagarostrobos franklinii (J. Hooker) Quinn

HUON PINE

Tree to 25(–38) m tall, much shorter or even shrubby on dry hillsides or in subalpine shrublands. Trunk to 1(–1.8) m in diameter, round or somewhat fluted, sometimes with a corkscrew spiral. Bark light brown weathering light gray, smooth or variously warty at first, becoming densely scaly and shedding in thick flakes, sometimes with vertical fissures. Crown conical and gracefully drooping in youth, becoming deep and narrowly to broadly dome-shaped with age, with heavy, crooked, forking, steeply to shallowly upwardly angled branches arching at the end and bearing numerous, repeatedly alternately branched, upright to drooping branchlets densely clothed with and hidden by foliage. Juvenile leaves triangular, 1.5–5 mm long, spreading from the twig at a forward angle. Adult leaves yellowish green to dark green and dotted with white stomates, 1–2 mm long, sharply or more roundly keeled, lasting 4–5 years or more. Pollen cones dull red, 3–6.5 mm long and 1–1.5 mm in diameter. Seed cones dull white with a yellowish or greenish tinge, droopy, a little fleshy, 4–8 mm long, each seed

Scattered Huon pines (*Lagarostrobos franklinii*) in mixed montane forest, with three of many marked by arrowheads.

cupped for a quarter to a third of its length in a one-sided, jaggedly toothed, papery pink epimatium less than 1 mm deep. Seeds red with patches of a thin, waxy film, smooth to wrinkled (on drying), 2–2.5 mm long. Southwestern quarter of Tasmania, Australia. Most abundant and of largest size in wet soils along streamsides and lakeshores in mixed temperate rain forests but surviving also on drier sites as a stunted tree or shrub; 0–750(–1,030) m. Zone 8. Synonym: *Dacrydium franklinii* J. Hooker.

Huon pine was the source of one of the most highly prized softwood timbers in Tasmania. The wood shrinks very little in drying and is very resistant to decay in contact with water and soil and so was much used in boatbuilding. Its easy working qualities and ability to take a good finish also made it popular for interior work, and rare logs with attractive figure are coveted for carving, turning, and other decorative uses. Growth is slow so large trees were essentially eliminated from accessible stands. Many of these trees were of great age. Individuals can easily live 1,000 years, with the maximum documented age reaching nearly 2,300 years. In addition, lower branches in contact with the soil can root, turn up, and form new trunks. This process can produce large clonal groves, particularly in subalpine shrublands. One of these clones is estimated to be more than 10,000 years old.

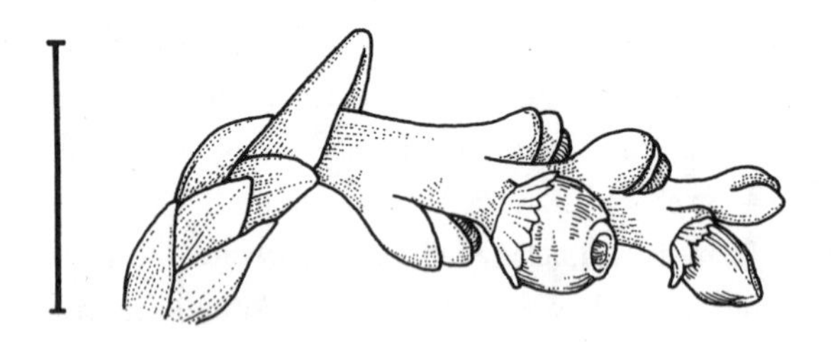

Seed cone of Huon pine (*Lagarostrobos franklinii*) with individual seeds in varying states of maturity; scale, 5 mm.

Tattered crown of old Huon pines (*Lagarostrobos franklinii*) in nature.

With such a long potential life span, Huon pine needs few successful episodes of reproduction to maintain populations. In fact, it is a masting species with seed production only at intervals of 5–7 years. The seed, when produced, is largely dispersed by water in which it can float for up to 2 months before losing viability. Upland stands away from water may have originally become established by relatively rare instances of dispersal by wind or by fruit-eating birds, such as parrots. Although Huon pine is predominantly dioecious, about 4% of individuals bear both pollen and seed cones during mast years, so isolated trees still have a small probability of setting seed. Even without monoecy and selfing, inbreeding through mating of close relatives appears to be common, and the species has a relatively low level of genetic variation compared to most conifers. This would be consistent with periods of extreme range constriction during the glaciations that covered much of Tasmania during the Pleistocene ice ages.

Young trees are very handsome in cultivation, with a graceful, weeping habit, the slender outer branches hanging for as much as 2 m. Even the main leader of the tree, like that of hemlocks (*Tsuga*) in the northern hemisphere, bends over in a graceful ark. One particularly narrow clone has been selected as 'Pendulum.' The species name honors John Franklin (1786–1847), English naval captain, arctic explorer, and governor of Tasmania (1836–1843), who died with all of his men while exploring the Northwest Passage through the Canadian arctic.

Larix P. Miller

LARCH

Pinaceae

Deciduous trees with a straight, single trunk bearing spirally arranged or regular tiers of three to five slender, short to long, horizontal branches, forming a conical to broadly conical crown. Bark nonfibrous, smooth at first, becoming scaly and sometimes shallowly ridged and furrowed at the base of large trees. Branchlets hairy or not, of two types: long shoots, which extend the growth of the tree and have well-separated needles, and short shoots, long-lived, highly condensed spurs arranged spirally around the long shoots and bearing a new tuft of (10–)20–50(–60) needles each year. Winter buds well developed, usually small, scaly, resinous or not. Leaves spirally arranged, densely so on short shoots and much more sparsely on long shoots, and sticking straight out all around the twigs. Each leaf needlelike, soft and flexible, flatly or a little plumply triangular or horizontally diamond-shaped in cross section, straight or slightly curved, the tip pointed but innocuous, the base abruptly narrowed to a very short petiole.

Plants monoecious. Pollen cones numerous, single at the tips of short shoots, upright, with numerous densely spirally arrange pollen scales, each bearing two pollen sacs. Pollen grains large to very large (55–115 µm in diameter), almost spherical, practically featureless except for a slightly thickened ring around one of the hemispheres. Seed cones spherical or egg-shaped to oblong, single at the tips of short shoots and with a narrower stalk, upright at pollination and at maturity in a single season, remaining intact, the 20–120 densely spirally arranged seed scales then spreading to release the seeds. Seed scales rounded-rectangular or rounded-triangular to heart-shaped, often slightly indented at the tip, the tongue-shaped bracts with a central bristle (continuing the midvein), attached only to the base of the scales and hidden by to strongly protruding between them. Seeds two per scale, oval, tilted with respect to the much larger asymmetric wing derived from the seed scale that cups the seed body on one side and overlaps a little around the edges on the other side. Cotyledons four to eight, each with one vein. Chromosome base number $x = 12$.

Wood hard, not fragrant, waxy or greasy to the touch, the yellowish brown to reddish brown or even quite red heartwood distinct from the light brown sapwood. Grain uneven, with a sharp transition between earlywood and latewood. Vertical resin canals few, small, and irregularly distributed, with horizontal water-conducting cells (tracheids) in the rays.

With tightly spaced lines of stomates confined to two stomatal bands on the lower side of the needle or with a few, less prominent lines of stomates above. Each stomate sunken beneath the four (to six) elongate surrounding subsidiary cells, which are covered by a thin cuticle and lack a Florin ring. Leaf cross section with a (sometimes interrupted) single resin canal following the edge of the needle on either side of the single-stranded but split midvein. Photosynthetic tissue mostly of spongy mesophyll but sometimes with one palisade layer in the center of the upper face. Hypodermis in one layer above and below the midrib and in one to three layers inside the corners at the outer edge of the leaf.

Ten species across the northern hemisphere in the boreal zone and in mountains south to the northwestern and northeastern United States, the Alps, and the Himalaya. References: Farjon 1990, Franco 1993b, Fu et al. 1999c, Gower and Richards 1990, Komarov 1986, Lee 1979, LePage and Basinger 1995a, Mill 1999b, Ostenfeld and Larsen 1930, W. H. Parker 1993, Schmidt 1995, Schorn 1994, Yamazaki 1995.

Larix is the classical Latin name for the European larch (*L. decidua*), which the Romans encountered in the Alps. Several larch species and hybrids have long been grown for forestry and horticulture, and a modest number of cultivars have been selected. These include blue foliage, dwarfs (most of which are derived from witches'-brooms), and weeping forms.

Larches are by far the most widely distributed, species-rich, and abundant of the five living genera of deciduous conifers today. The other genera (*Glyptostrobus, Metasequoia, Pseudolarix,*

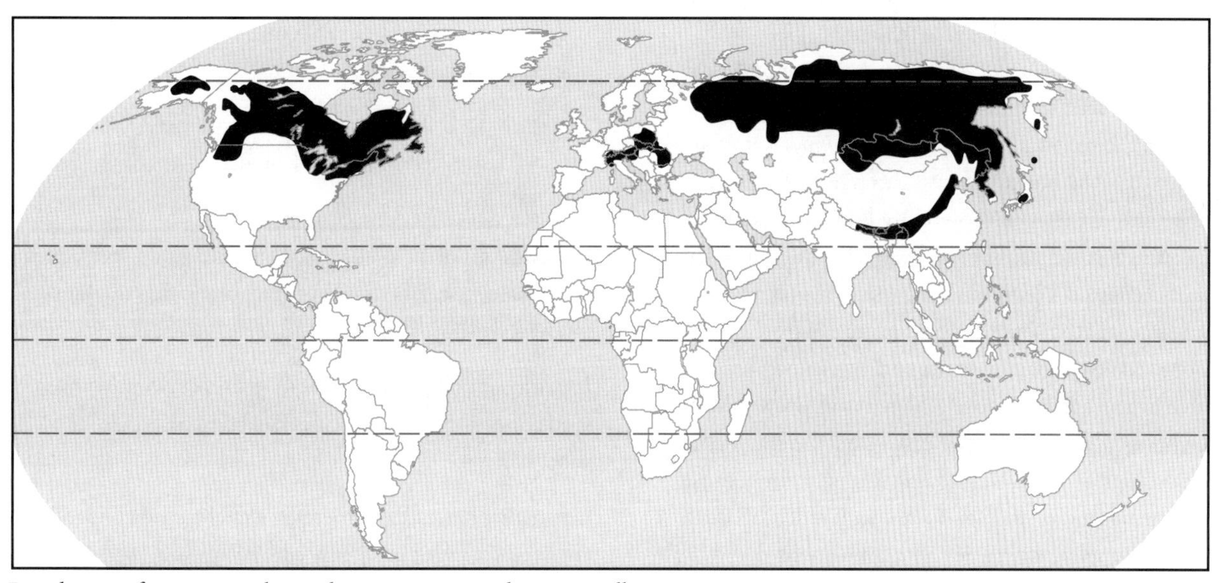

Distribution of *Larix* across the northern continents, scale 1 : 200 million

and *Taxodium*) were more widely distributed in the geologic past but are each now restricted to single (or two) species and a local region in warm temperate portions of China or the southeastern United States. In contrast, *Larix* is well established across the northern hemisphere boreal forests, and each of the three northern species has an extensive distribution. The seven more southerly mountain species, however, have restricted ranges like those of the species in the other deciduous genera. These deciduous genera are all unrelated to one another (except *Glyptostrobus* and *Taxodium*), and each is most closely related to, and phylogenetically nested among, evergreen genera, thus indicating that deciduousness arose repeatedly among conifers.

Each of the deciduous genera lived in the high arctic in the past, which may be where they evolved deciduousness to cope with the darkness of an arctic winter not yet frigid in the early Tertiary. *Larix* appears to be younger than the other five genera, with earliest known fossils dating only from the Eocene, less than 50 million years ago, while the others date back at least to the Cretaceous, 70 million years ago and more. The earliest larch fossils are in the high arctic, well north of their present northern limit. Almost everywhere that larches grow today in the north they are accompanied by evergreen conifers, mostly true firs (*Abies*), spruces (*Picea*), and pines (*Pinus*). Thus deciduousness in larches may be a holdover from an arctic past rather than a response to current conditions.

Many of the more southerly larches grow near species of Douglas fir (*Pseudotsuga*), their evergreen closest relatives. This relationship has long been known because of similarities in many aspects of structure and function, including wood, seed cones, and pollen and pollination mechanisms. DNA studies and other molecular investigations also point to the close relationship of these two genera and to their relative distance from the other pinoid genera, *Pinus, Picea,* and *Cathaya* (although they are closest to the latter).

The closeness of *Larix* and *Pseudotsuga* led taxonomists to believe that the southerly larches with three-pointed bracts sticking out between the seed scales (like those of Douglas firs) are more primitive than the boreal species with hidden bracts, and most classifications of the genus emphasized the distinction between long- and short-bracted species by placing them in separate botanical sections. DNA studies suggest, however, that the long-bracted larches of the New World are more closely related to their short-bracted neighbor than to the Old World long-bracted species, which in turn are more closely related to their own short-bracted neighbors. This uncertainty cast upon the traditional view of relationships by DNA studies is itself uncertain because the DNA segments studied to date do not vary much between species, because sampling of species and populations could be improved, and because larches can hybridize naturally and thus confuse molecular results as well as morphological ones. Clearly, more needs to be done on clarifying relationships among the larch species and sorting out their evolutionary history.

Identification Guide to *Larix*

The larches are very difficult to identify using only features of the twigs and needles, so this identification guide also employs characteristics of the mature seed cones. Because of variability within species and the extensive similarities between species, identifications using this guide may seem uncertain, so it is important to read the more detailed descriptions to confirm a tentative identification. The 10 species of *Larix* may be distinguished by whether the bracts peek conspicuously between the seed scales in the upper half of the seed cone or are hidden and, if they peek out, in what pattern; the length of longer ordinary seed cones; whether the seed scales curl outward or inward at the tip or remain relatively flat when they spread at maturity; the color intensity of the first year branchlets; and whether the branchlets are hairy or not:

Larix laricina, bracts hidden, seed cones less than 2.5 cm, seed scales flat, branchlets medium, not hairy

L. gmelinii, bracts hidden, seed cones 2.5–4 cm, seed scales flat or slightly curled inward, branchlets pale, not to densely short-hairy

L. kaempferi, bracts hidden, seed cones 2.5–4 cm, seed scales curled outward, branchlets usually pale, not or slightly short-hairy

L. sibirica, bracts hidden, seed cones 4–5 cm, seed scales curled inward, branchlets pale, not or slightly short-hairy

L. decidua, bracts hidden, seed cones 4–5 cm, seed scales flat, branchlets light, not hairy

L. lyallii, bract bristle tip only peeking out, seed cones 4–5 cm, seed scales curled outward, branchlets medium, densely woolly

L. occidentalis, bract bristle tip only peeking out, seed cones 4–5 cm, seed scales curled outward, branchlets medium, moderately short-hairy

L. mastersiana, bract blade emerging and curled downward, seed cones 4–5 cm, seed scales flat, branchlets pale, densely short-hairy

L. potaninii, bract blade emerging straight, seed cones 5–6 cm, seed scales flat or slightly curled outward, branchlets dark, not hairy

L. griffithii, bract blade emerging and curled downward, seed cones more than 6 cm, seed scales curled outward slightly, branchlets usually medium or sometimes dark, usually not hairy

Larix ×*czekanowskii* Szafer

Tree to 25(–30) m tall, with trunk to 1 m in diameter. Natural hybrids between Siberian larch (*Larix sibirica*) and Dahurian larch (*L. gmelinii*) occurring throughout their region of overlap between the mouth of the Yenisei River and Lake Baikal in Siberia (Russia). The characteristics generally overlap with those of the two parent species, which are closely related. The seed cones are about 3 cm long, small for Siberian larch and large for Dahurian larch as it grows in the zone of overlap. The seed scales differ from those of both parents, bending outward at maturity, somewhat like those of Japanese larch (*L. kaempferi*), rather than being flat or bending inward slightly. As well, the seed scales are a little reddish hairy on the outer face, falling in between the densely hairy scales of Siberian larch and the generally hairless (in this region) ones of Dahurian larch. The tree does not seem to show the hybrid vigor displayed by the Dunkeld larch (*L.* ×*marschlinsii*).

Larix decidua P. Miller

EUROPEAN LARCH, COMMON LARCH, MÉLÈZE CUMMUN (FRENCH), LÄRCHE (GERMAN), LARICE (ITALIAN), MODRZEW (POLISH), MODRIN (CZECH)

Plate 30

Tree to 35(–46) m tall, with trunk to 1(–1.8) m in diameter. Bark reddish brown to light gray, becoming furrowed, between irregular, interlacing, scaly, flat-topped ridges. Crown narrowly conical at first, broadening with age and becoming flat-topped and wide spreading in open grown trees, with numerous long, horizontal branches upswept at the tips. New branchlets light yellowish or pinkish brown, hairless and without wax. Buds small, about 3 mm long, dark reddish brown, hairless, not resinous. Needles of spur shoots straight, (12–)20–40(–50) on each spur, soft, (1.5–)2–3(–4) cm long and 0.5–1(–1.5) mm wide, light green, turning orange-yellow in autumn before falling. Midrib raised beneath and with a green stomatal band consisting of one to three lines of stomates on either side. Pollen cones almost spherical, 4–6 mm in diameter, elongating with age, light reddish brown. Seed cones half-again as long as wide, (1–)2–4(–5) cm long, with (25–)35–50(–70) seed scales, green with reddish edges before maturity, ripening brown and weathering gray, on a stout, curved stalk 5–10 mm long. Seed scales roundly diamond-shaped, the tip rounded, not curling outward, finely hairy on both sides. Bracts about 5–6 mm long, shorter than and mostly hidden by the seed scales, rounded and continuing into a bristle tip another 2–3 mm long. Seed body 3–4 mm long, without resin pockets, the firmly clasping wing another 5–10 mm longer. Discontinuously distributed in mountains of central and eastern Europe, from southeastern France and southwestern Italy to eastern Poland and central Romania, naturalized elsewhere in Europe and in North America. Usually mixed with other subalpine conifers in the southern mountains, also occurring at low elevations, especially northward; (200–)400–1,800(–2,500) m. Zone 2. Synonym: *Larix europaea* Lamarck & A. P. de Candolle. Two varieties.

European larch, with its rapid growth, good form in forests, and hard, durable wood, is an important timber tree. It was widely planted for this purpose in the past although it has now been largely superseded by *Larix* ×*marschlinsii,* its hybrid with Japanese larch (*L. kaempferi*), which has a faster growth rate. Other successful hybrids were produced with Siberian larch (*L. sibirica*), Dahurian larch (*L. gmelinii*), and western larch (*L. occidentalis*). Interest in the parent species has, perhaps, been revived some-

Old European larch (*Larix decidua*) in nature.

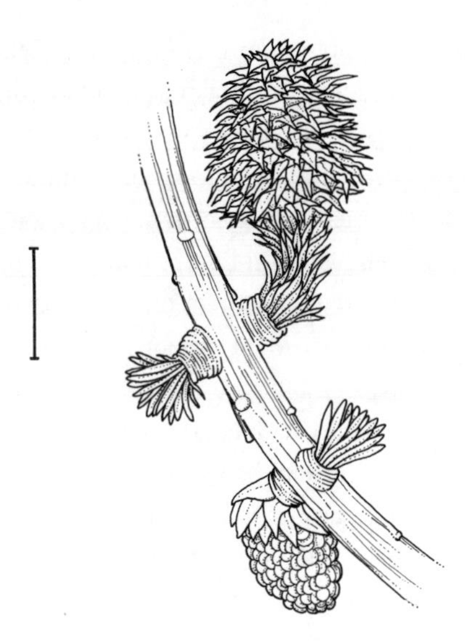

Mature pollen cone and receptive seed cone of European larch (*Larix decidua*) before the seed scales emerge from and overtop the bracts; scale, 5 mm.

Twig of European larch (*Larix decidua*) with pollen and seed cones at time of pollination and many dead spur shoots, ×0.5.

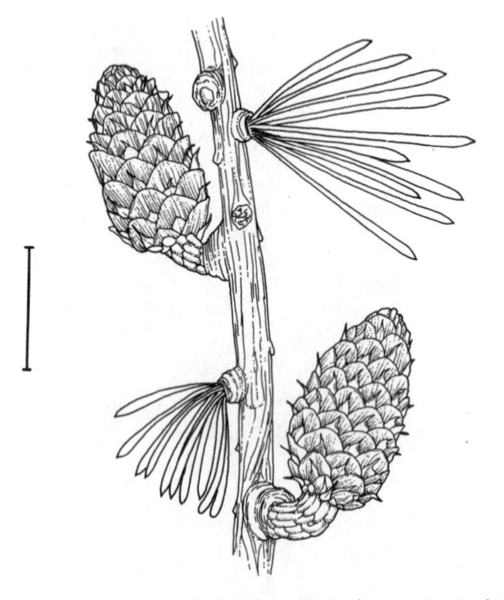

Green mature seed cones of European larch (*Larix decidua*) before seed dispersal; scale, 2 cm.

what by genetic engineering with the introduction of genes for herbicide and insect resistance. In contrast to its timber potential at moderate elevations, trees found near the alpine timberline are often slow growing and misshapen. Such trees can easily live to be 1,000 years old and more, paralleling the great longevity of other timberline species. This makes them suitable for studies of long-term climatic change.

At the other end of the age spectrum, larch plantlets were grown using tissue culture in numerous experiments. Even when the starting material is female gametophytes (the nutritive tissue of conifer seeds), which have a single set of chromosomes, the plantlets that result have the normal two sets of chromosomes. Thus the chromosome number spontaneously doubles during the formation of plantlets from the female gametophyte tissue, mimicking in part the process of self-fertilization. Although self-fertilization is possible in natural populations of European larch, by the time stands are mature, essentially the only trees remaining are those that arose through outcrossing between different individuals.

The closest relative of European larch is its nearest neighbor, Siberian larch (*Larix sibirica*), which approaches European larch without overlapping in northeastern Europe. It has often been suggested that the eastern variety of European larch (variety *polonica*) arose through hybridization between trees resembling the alpine variety (variety *decidua*) and Siberian larch, largely because of the intermediate size of their seed cones. However, studies of enzyme variants show that European larch and Siberian larch are sharply distinct from one another and that variety *polonica* shows no signs of the distinctive characteristics of Siberian larch. As might be expected of a tree that spans environments ranging from the Polish plains to the tree line of the Alps, there is a great deal of variation with European larch, some of which is reflected in the available selection of cultivars. The eastern European larches in particular, with their isolation in geographically separate and ecologically varied environments, stimulated the description of numerous botanical varieties. Nonetheless, Ostenfeld and Larsen (1930), the last monographers of *Larix* before Farjon (1990), made a good case for recognizing only two weakly distinguishable varieties of European larch, the alpine variety *decidua*, and *polonica* to the north and east. This treatment is followed here with different geographic limits for the two varieties.

Larix decidua P. Miller **var. decidua**
Twigs yellowish to pinkish brown. Needles with a pronounced ridge over the midrib beneath. Seed cones mostly 2.5 cm long or more, the seed scales roundly triangular, variably hairy on the outer face. Western, continuous portion of the range of the species east to Austria and northern Slovenia.

Larix decidua var. polonica (Raciborski ex Wóycicki) Ostenfeld and Syrach-Larsen
Twigs very pale yellowish brown. Needles with a less raised midrib beneath. Seed cones less than 2.5 cm long, the seed scales almost circular, definitely hairy on the outer face. Eastern, fragmented portion of the range of the species, in the Czech Republic, Slovakia, Poland, Romania, and barely into adjacent Ukraine. Synonym: *Larix decidua* var. *carpatica* Domin.

Larix gmelinii (F. Ruprecht) Kuzeneva

DAHURIAN LARCH, LISTVENITSA DAURSKAYA (RUSSIAN), SSISSI (TUNGUSIC), TIT (YAKUT), LUO YE SONG (CHINESE), IP'KAL NAMU (KOREAN), KUI (AINU), GUI-MATSU (JAPANESE)

Tree to 30(–36) m tall, with trunk to 1(–1.5) m in diameter. Bark grayish brown to gray over reddish highlights, smooth at first, becoming progressively more scaly and finally breaking up into flat-topped ridges divided by shallow furrows at the base. Crown conical at first, open, broadening and becoming flat-topped with age, with long, slender horizontal branches turned up at the ends. New branchlets pale yellowish brown to reddish brown often waxy, hairless to densely covered with conspicuous reddish brown hairs, prominently grooved. Buds small, about 2–3 mm long, dark brown, variably resinous. Needles of spur shoots straight, 20–35 on each spur, soft, (1–)1.5–3(–4) cm long and 0.5–0.8(–1.0) mm wide, rich, bright green, turning yellow in autumn before falling. Midrib raised beneath and with narrow, white stomatal bands on either side and no or one or two inconspicuous, interrupted lines of stomates above. Pollen cones a little oblong, 5–7 mm long, yellowish brown. Seed cones oblong to spherical, (0.8–)1.5–3.5(–4.5) cm long, with 15–45(–60) seed scales, dark purplish green with red edges before maturity, ripening reddish brown to purplish brown, on a slender curved stalk exceptionally to 2 cm long. Seed scales roundly five-sided, the tip rounded or slightly notched, opening rather flat, hairless to densely reddish woolly on the outer face. Bracts about 3–4 mm long, shorter than and hidden by the seed scales, squared off or three-pronged at the end with a bristle tip up to 1(–2) mm long. Seed body 2–4 mm long, without resin pockets, the firmly clasping wing another 5–7 mm longer. Northeastern Eurasia, eastward across eastern Siberia from its line of contact with *Larix sibirica* to the Russian Far East, south to Kamchatka, Sakhalin, and the southern Kurile Islands, the Korean Peninsula and vicinity, and near Lake Baikal, with an outlier in Hebei, Shanxi, and Henan provinces (China). Forming pure stands or mixed with other boreal forest conifers in the arctic lowlands to slopes of mountains southward on a wide variety of sites, including bogs and peat lands; 300–1,800(–2,800) m. Zone 2. Two varieties.

The Dahurian larch is a widespread and important tree of the eastern Taiga. It displays a great deal of variation in seed cone size and in the coarseness, color, and hairiness of the twigs across its extensive range. On this account, a number of different varieties were recognized, usually three to four, and these are sometimes treated as separate species. Careful consideration of the variation suggests that only two varieties should be given formal recognition. The plants in Korea and on the Kurile Islands, often recognized as separate varieties, seem to be simply an extreme expression of the hairiness on the twigs that is found to varying degrees throughout the range of variety *gmelinii*. The rusty hairiness becomes more prevalent and denser southward, with Korea and Japan representing the southern geographic limits and the hairiest twigs. The other variety, variety *principis-rupprechtii*, is separated from the main range of the species in east-central China. It has coarser twigs and larger seed cones that exceed the largest cones in the considerable range found in variety *gmelinii*. Dwarf plants of the alpine tree line were also given formal botanical recognition by some authors, but they are simply a response to the harsh alpine environment and their characteristics are not genetically fixed. Dahurian larch is closely related to Siberian larch (*Larix sibirica*), and the two species hybridize naturally all along their 2,500-km-long line of contact to form *L.* ×*czekanowskii*. It has also been crossed artificially with *L. decidua*, but the resulting hybrids are less vigorous than the trees of *L.* ×*marschlinsii* (*L. decidua* × *L. kaempferi*), and they are not often encountered. When they were first described, these hybrids were given a botanical name but it was not validly published according to the botanical *Code*, so the name has no standing. The species name honors Samuel G. Gmelin (1743–1774), a Russian naturalist who collected plants in Siberia, including the southern dry steppe region near Mongolia formerly known as Dahuria (hence the common name).

Larix gmelinii (F. Ruprecht) Kuzeneva **var. *gmelinii***

First-year branchlets relatively slender, those of long shoots up to 1.3 mm thick and those of spur shoots 2–3 mm thick. Seed cones (0.8–)1.5–2.5 cm long. Throughout the range of the species except in east-central China. Synonyms: *Larix cajanderi* H. Mayr, *L. dahurica* Turczaninow ex Trautvetter, *L. gmelinii* var. *japonica* (Maximowicz ex E. Regel) Pilger, *L. gmelinii* var. *olgensis* (A. Henry) Ostenfeld and Syrach-Larsen, *L. kamtschatica* (F. Ruprecht) Carrière, *L. kurilensis* H. Mayr, *L. olgensis* A. Henry.

Larix gmelinii* var. *principis-rupprechtii (H. Mayr) Pilger

First-year branchlets relatively stout, those of long shoots 1.4–2.5 mm thick and those of spur shoots 3–4 mm thick. Seed cones (2–)2.5–4.5) cm long. East-central China southwest of Beijing. Synonym: *Larix principis-rupprechtii* H. Mayr.

Larix griffithii J. Hooker

SIKKIM LARCH, HIMALAYAN LARCH, BINYA (NEPALESE), SAH (LEPCHA, FROM SIKKIM), ZANG HONG SHAN (CHINESE)

Tree to 20(–25) m tall, with trunk to 0.5(–0.8) m in diameter. Bark gray or grayish brown, smooth at first, becoming flaky, and finally breaking up into broad, scaly plates, separated by wide, shallow furrows. Crown conical at first, broadening and becoming egg-shaped with age, rather sparse, with gently rising branches bearing long, hanging twigs. New branchlets quite variable in coloration, from yellowish brown to purplish brown, sometimes waxy, variably hairy, especially in the obvious grooves between the raised leaf bases. Buds small, about 2–3 mm long, reddish brown to purplish brown, smooth or hairy, somewhat resinous. Needles of spur shoots straight or slightly curved, (10–)25–40(–50) on each spur, soft, (1–)2–3.5(–5.5) cm long and (0.6–)1–2 mm wide, bright green, turning bright golden brown in autumn before falling. Midrib prominently raised beneath and slightly so near the base above, with none to two incomplete and inconspicuous lines of stomates on each side above and 3–5 in a narrow white stomatal band on each side beneath. Pollen cones oblong (6–)10–20 mm long, yellowish brown. Seed cones oblong, widest near the middle, (4.5–)5–9(–11) cm long, with 60–100 seed scales, purplish green with bright reddish purple bracts before maturity, ripening dark to light brown with purplish brown bracts, on a short, curved stalk to 5(–7) mm long. Seed scales heart-shaped to elongately heart-shaped with a shallow notch or squared off, curling back a little at maturity, minutely yellowish hairy on the outer face at first. Bracts 15–20 mm long, much longer than the seed scales and curling down conspicuously to cover them, more or less spearhead-shaped, the long bristle tip to 5 mm long and either continuing down with the bract or curled back up. Seed body 4–5 mm long, spotted, the firmly clasping wing another 6–12 mm longer. In a narrow band through the eastern Himalaya, from central Nepal to the mountains of the borderlands of southwestern China in southeastern Xizang (Tibet) and northwestern Yunnan. Forming pure stands near the alpine tree line or mixed with other conifers and hardwoods through the subalpine cloud forest zone on screes and other rocky soils; (2,400–)3,000–4,000(–4,100) m. Zone 7. Synonyms: *Larix griffithiana* Carrière, *L. griffithii* var. *speciosa* (W. C. Cheng & Y. W. Law) Farjon, *L. kongboensis* R. Mill, *L. speciosa* W. C. Cheng & Y. W. Law.

Sikkim larch was the first of the Sino-Himalayan large-bracted larches to be given a botanical name. Unfortunately, two slightly different names have seen almost equal use, both commemorating William Griffith (1810–1845), the English physician and botanist who collected the tree in Bhutan in 1838, during 6 years of collecting throughout the Himalaya, and described it briefly in his journals, which were published posthumously in 1847. The *International Code of Botanical Nomenclature* provides for two different forms of commemorative names, as either an adjective or a possessive noun. In the case of Sikkim larch, those authors who favor *Larix griffithiana* (the adjectival form) argue that it was published a month before *L. griffithii* (the possessive noun) in 1855. However, as others showed, Joseph Hooker actually published the name *L. griffithii*, with a botanically acceptable description of the tree, in the journal of his travels in the Himalaya, published in January 1854, a full year and a half before he published it again in his more formal botanical account of Himalayan plants. Thus the rival name should once and for all be put to rest.

There is also some taxonomic confusion surrounding this species. Two segregate species or botanical varieties were named from the eastern portion of the range of the species in southwestern China, both of which are included here within *Larix griffithii* without any formal status. The distinguishing features of the trees from northwestern Yunnan and southeastern Xizang (Tibet) recognized by the authors of the *Flora of China* as *L. speciosa* are so slight (stouter spur shoots with more prominent remnant bud scales and narrower bracts) that recognition even as a variety (*L. griffithii* var. *speciosa*) seems unwarranted. The other proposed segregate, named *L. kongboensis* for the Gong-po district of southeastern Xizang from which the type material was collected, seems more distinct. It generally has shorter needles, pollen cones, and seed cones than is typical for *L. griffithii*, and the number of needles on the spur shoots is greater. Its measurements, however, do not completely fall outside of the range known elsewhere in *L. griffithii*. Furthermore, the proposed species is known only from three collections, each of which shows different features. Thus the short pollen cones come from a single collection while the short needles and seed cones are found on another. In the absence of additional material it seems prudent to regard these specimens as somewhat aberrant examples in an environment differing from that found in the main geographic range of the species farther west.

Sikkim larch is most closely related to the other two Sino-Himalayan large-bracted larch species, Chinese larch (*Larix potaninii*) and especially Sichuan larch (*L. mastersiana*). These three species do not appear to be closely related to the North American large-bracted species, subalpine larch (*L. lyallii*) and western larch (*L. occidentalis*). Like all of the montane larches, Sikkim larch has a much more restricted geographic distribution than the species of the northern boreal forests, Dahurian larch (*L. gmelinii*), Siberian larch (*L. sibirica*), and tamarack (*L. laricina*).

Larix kaempferi (A. Lambert) Carrière

JAPANESE LARCH, KARAMATSU (JAPANESE)

Tree to 25(–33) m tall, with trunk to 1(–1.3) m in diameter. Bark gray with red highlights, smooth at first, becoming flaky and finally deeply furrowed between flat-topped ridges at the base of large trunks. Crown narrowly conical at first, broadening and flattening greatly with age in open-grown trees, with numerous slender, long, horizontal branches. New branchlets yellowish brown to reddish brown, often waxy, slightly hairy or hairless, obviously grooved between the attached leaf bases. Buds conical to almost spherical, small, about 3–5 mm long, dark reddish brown, a little resinous. Needles of spur shorts straight, 20–35 on each spur, soft, (1–)1.5–2.5(–3.5) cm long, and 0.7–1 mm wide, bluish green, turning bright yellow in autumn before falling. Midrib raised beneath, with a few obscure lines of stomates above with about five lines of stomates in a white stomatal band on either side beneath. Pollen cones oblong, 5–6(–10) mm long, yellowish brown. Seed cones broadly egg-shaped to almost spherical, 1.5–3(–3.5) cm long, with (20–)30–40(–50) seed scales, bluish green with red edges before maturity, ripening brown, on a stout, curved stalk about 5 mm long. Seed scales round, the tip flat or slightly notched, curled back conspicuously at maturity. Bracts 5–7 mm long, half to two-thirds as long as the seed scales and hidden by them, except at the base of the cones, elongately triangular, the tip rounded, with a short bristle tip 1–2 mm long. Seed body 3–4 mm long, without resin pockets, the firmly clasping wing another 7–11 mm longer. Mountains of central Honshū (Japan), from Gifu to Fukushima prefectures. Forming pure stands or mixed with other conifers mostly on volcanic soils in montane and subalpine forests, and even forming an alpine scrub on Mount Fuji; (500–)1,200–2,700(–2,900) m. Zone 6. Synonym: *Larix leptolepis* (P. Siebold and Zuccarini) G. Gordon.

Twig of Japanese larch (*Larix kaempferi*) with new long shoot growth at its tip and numerous spur shoots, ×0.5.

Natural stand of Japanese larch (*Larix kaempferi*) in winter.

Japanese larch is abundant within its limited natural range on all of the mountain ranges in central Honshū near Tokyo. It yields an excellent, durable, hard timber and so is planted north of its natural range in Japan. Its rapid growth has also made it an important plantation tree elsewhere in eastern Asia and in Europe and North America. In this capacity, it has largely been superseded by the Dunkeld larch (*Larix ×marschlinsii*), its hybrid with European larch (*L. decidua*). It has also been crossed successfully with its close relatives, the other two northern Eurasian larch species, Siberian larch (*L. sibirica*) and Dahurian larch (*L. gmelinii*).

Because of its importance in forestry, many aspects of the reproductive biology and genetic variation of Japanese larch have been studied, including the timing of artificial pollination to optimize seed set. Provenance studies in several countries, including Japan, reveal little genetic differentiation among populations, as might be expected from the limited geographic range. Cold hardiness, which increases along with increasing latitude of origin, is one of the few obvious differences between populations, and there are no consistent geographic variants in morphology. The species does have a reasonable store of genetic variation within populations, however, and many cultivars have been selected in Japan and elsewhere. The needles yield a number of new terpenoid chemicals in their oils not found in other larch species. The species name honors Engelbert Kaempfer (1651–1716), one of the first European botanists to visit Japan.

Larix laricina (Du Roi) K. Koch

TAMARACK, AMERICAN LARCH, EASTERN LARCH, MÉLÈZE LARICIN (FRENCH)

Tree to 25(–35) m tall, with trunk to 0.6(–1.1) m in diameter, often much smaller at the northern limit of its range. Bark remaining scaly through life, weathering gray over reddish brown. Crown narrowly conical at first, becoming very narrowly egg-shaped above a long, bare, cylindrical trunk, with numerous slender, short, horizontal branches. New branchlets orange-brown, slightly waxy, hairless. Buds small, about 3 mm long, dark reddish brown, smooth, somewhat resinous, those of the spur shoots surrounded by a ring of hairs. Needles of spur shoots straight, (10–)15–40(–60) on each spur, soft, (1–)2–3(–5) cm long and 0.5–0.8 mm wide, light green, turning bright yellow in autumn before falling. Midrib flat and without stomates above, raised beneath and with one (to three) inconspicuous lines of stomates on either side. Pollen cones spherical, 3–4 mm in diameter, yellowish brown. Seed cones nearly spherical or a little elongate, 1–2 cm long, with 9–15(–20) seed scales, green with a reddish tinge before maturity, ripening light brown with paler edges, on a slender curved stalk 4–5 mm long. Seed scales circular to roundly fan-shaped, the tip rounded, not curled outward. Bracts hairless, 2.5–3.5 mm long, much shorter than and hidden by the seed scales, squared off or rounded with a bristle tip to 1 mm long. Seed body (1–)2–3 mm long, without resin pockets, the firmly clasping wing another (3–)4–6 mm longer. Across northern North America south of the arctic tundra, generally east of the Rocky Mountains, from central Alaska to Newfoundland, south to south-central Alberta, southeastern Manitoba, central Minnesota, northern Indiana, Ohio, Pennsylvania, and New Jersey, with outliers in southwestern Pennsylvania and adjacent Maryland and West Virginia. Growing in sparse to dense pure stands or mixed with other boreal conifers and hardwoods, primarily on boggy sites; 0–600(–1,200) m. Zone 1.

Tamarack (from a French Canadian rendering of an Algonquin name for the tree) is one of the most tolerant conifers for waterlogged soils in cold regions. Although it is most abundant on such sites in the North America boreal forest, it actually grows far better on well-drained upland sites. Its general absence from upland boreal forests today may be due to attacks by the larch sawfly, which has frequently reached epidemic proportions since the late 1800s. Originally thought to have been introduced to Canada

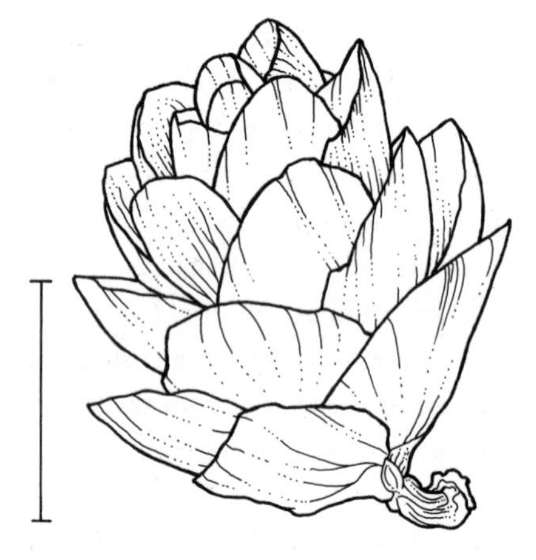

Open mature seed cone of tamarack (*Larix laricina*) after seed dispersal; scale, 1 cm.

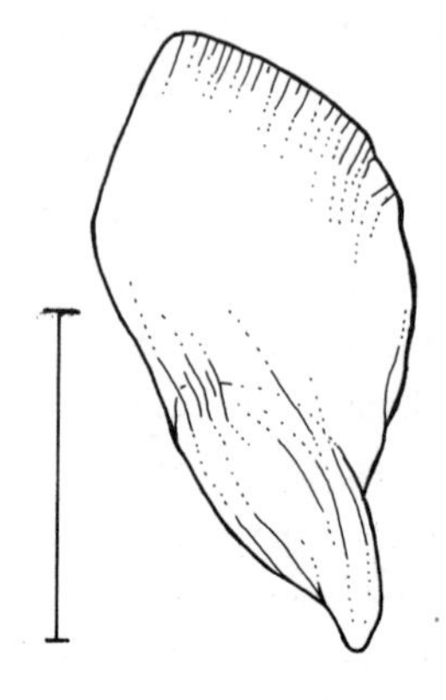

Upper side of dispersed seed of tamarack (*Larix laricina*); scale, 5 mm.

from Europe at about that time, careful study of tamarack growth ring series extending back 300 years suggest that the sawfly was already present at low levels in the 18th century. Ironically, the increase in severity of attacks at the end of the 19th century appears to have coincided with a warming climate at the end of the Little Ice Age that otherwise led to improved growth rates of tamarack.

Despite the enormous mortality of trees caused by severe sawfly outbreaks, tamarack populations, like those of other conifers, maintain a great deal of genetic variation. This is partly due to strong inbreeding depression that ensures that most successfully established seedlings are the result of outcrossing between unrelated individuals. Heavy seed crops typically occur at intervals of 3–5 years. Even though outcrossing is the norm for the species, there is still strong local differentiation for growth characteristics related to frost hardiness, a critically important adaptation in boreal trees. Most taxonomic features, however, show either large scale clines of continuous variation or little geographic patterning. There is thus little basis for recognition of varieties, even for the geographically separated portion of the range of the species in Alaska. There has been a modest amount of selection of cultivars with dwarf growth habits or variation in foliage color.

The seed cones most closely resemble those of the northeastern Asian *Larix gmelinii,* but DNA studies suggest that tamarack is more closely related to the other two (larger-coned) North American species, *L. lyallii* and *L. occidentalis.* A peculiarity of the seed cones is that many individuals produce occasional proliferous cones, ones with a leafy shoot growing out of the tip. This reversion to vegetative growth is brought to an end when the seed cone matures and dries up, as does the leafy shoot at its tip. Tamarack is usually a relatively small tree, but the largest known individual in the United States in combined dimensions when measured in 2003 was 18.9 m tall, 1.2 m in diameter, with a spread of 21.0 m. The greatest known longevity is 335 years, but the usual maximum age is 150–180 years. The wood is often highly resinous, imparting a measure of resistance to decay, so the primary use is for posts, poles, and railroad ties in contact with the soil. The scientific name means "larchlike" in Latin, referring to the resemblance to European larch, which Linnaeus had called *Pinus larix.*

Larix lyallii Parlatore

SUBALPINE LARCH, ALPINE LARCH

Tree to 15(–30) m tall, with trunk to 0.6(–2) m in diameter. Bark smooth and gray at first, becoming gray or brown and scaly and finally breaking up into small, irregularly square, scaly plates separated by shallow furrows. Crown narrowly conical at first, becoming more cylindrical with age, the trunk often forked near the alpine tree line, with numerous short, horizontal branches. New branchlets orange brown, densely woolly for the first year or two. Buds small, about 3 mm long, not resinous, the scales densely fringed with hairs. Needles of spur shoots straight, (25–)30–40 on each spur, soft, (1.5–)2–3.5 cm long and 0.6–0.8(–1) mm wide, bluish green, turning golden yellow in autumn before falling. Midrib projecting both above and beneath, with a few lines of stomates on either side. Pollen cones cylindrical, yellowish brown, 10–15 mm long. Seed cones oblong, (2.5–)3.5–5 cm long, with (40–)45–55(–60) seed scales, reddish green with reddish purple bracts before maturity, ripening dark purplish brown, on a short, thick, curved stalk about 5 mm long. Seed scales a little longer than wide, shallowly notched at the tip, curling outward at maturity, densely woolly on the outside. Bracts 12–18 mm long, slightly longer than the seed scales, emerging conspicuously beyond the scales with a curved-back bristle tip 1–5 mm long. Seed body 3–4.5 mm long, the strongly clasping wing another 6–9 mm longer. Confined to the northern Cascade Range and central Rocky Mountains of western North America in British Columbia and Alberta and Washington, Idaho, and Montana. Forming open pure stands or in groves mixed in with other conifers in the subalpine zone, especially on rocky slopes; (1,500–)2,000–2,800(–3,000) m. Zone 3.

Subalpine larch is the most geographically restricted of the western North American subalpine conifers. Its sites are usually in roadless areas and require long hikes to visit, but they are, nonetheless, popular destinations near some resort areas because of the trees' brilliant fall color. The trees are leafless 8 months of the year but still begin their seasonal activity earlier than other subalpine species, often well before the snow pack is fully melted. The subalpine larch has no commercial importance but plays a major role in stabilization of steep, rocky slopes, both in summer and during avalanche season.

Like many other subalpine conifers, subalpine larch is long-lived, most commonly living 400–500 years but often exceeding this age. The oldest trees are estimated to reach 1,000 years, although exact counts are impossible because of heart rot in larger trees. Since growth rates of subalpine larch are more sensitive to annual weather variations than those of its neighbors, this species has contributed to the record of climatic fluctuations over the past 650 years. The largest known individual in the United States in combined dimensions when measured in 1993 was 28.7 m tall, 1.9 m in diameter, with a spread of 17.1 m.

Subalpine larch is closely related to western larch (*Larix occidentalis*), which grows at lower elevations in the same region. The two species can hybridize naturally when they are growing near each other, and this cross has also been achieved artificially, producing hybrids with a higher growth rate than either parent. Although they both seem to resemble the long-bracted larch species of Asia, DNA studies suggest that they are much more closely related to the other North American species, tamarack (*L. laricina*). The species name honors David Lyall (1817–1895), the Scottish surgeon and naturalist who collected the first botanical specimens in 1858, as a member of the British Columbia Boundary Commission.

Larix* ×*marschlinsii Coaz

DUNKELD LARCH

Tree to 38 m or more tall, with trunk to 1 m or more in diameter. This hybrid occurs naturally wherever the two parents, Japanese larch (*Larix kaempferi*) and European larch (*L. decidua*), are grown together and is also made artificially on a commercial scale because it is one of the best-known examples of heterosis or hybrid vigor among trees. It is a fine timber tree, growing faster than either parent in both height and diameter, especially when young. The wood loses nothing of the strength and durability of those of its parents as a result of its rapid growth. The hybrids were first grown from seed of Japanese larch growing at Dunkeld in Scotland in 1900 and are still produced there, where many fine examples are to be found.

In botanical characteristics these trees generally resemble their parents or show intermediacy. For example, the first-year branchlets are pale like those of European larch but have a little bit of white hairiness inherited from Japanese larch, while their waxy coating is thinner than in that species. The needles more closely resemble those of Japanese larch, bluish green and longer, on average, than those of European larch, but the lines of stomates are largely confined to the lower surface, like those of European larch. The seed cones are almost as large as those of European larch, generally about 2.5–4 cm long, but the seed scales curl a little outward at the tips, like those of Japanese larch. The parent species are similar enough, however, that it is often difficult to distinguish young hybrids from either parent, and thus various features of the oil chemistry or DNA fragments have been proposed as useful adjuncts for identification.

Despite the ease with which the cross using Japanese larch as a seed parent can be made, the reciprocal cross with European larch as seed parents has a much lower yield of seed, and embryological studies show that a high proportion of those embryos abort, leading to failure of seed production. It is possible to grow embryos on tissue culture media, where both parents and hybrids produce similar results. The compatibility of the two parent species in producing Dunkeld larches is partly due to the fact that their chromosomes are very similarly arranged, so the hybrids show normal chromosome pairing during pollen or seed production. Thus the differences between the parent species are primarily differences in the forms (alleles) of particular genes present in each species. The hybrids get a different allele of these genes from each parent and thus have a higher level of heterozygosity than either parent. This is hypothesized to be a cause for the heterosis of the hybrids.

Dunkeld larch is usually called *Larix* ×*eurolepis*, a name combining elements from names of the parent species that are themselves synonyms, *L. europaea* for *L. decidua* and *L. leptolepis* for *L. kaempferi*. *Larix* ×*marschlinsii* is an older name than *L.* ×*eurolepis*, but it has been thought to refer to the hybrid between Japanese larch and Siberian larch (*L. sibirica*). The planting in Switzerland from which *L.* ×*marschlinsii* was described had both Siberian and European larches growing near the Japanese larch that was the seed parent of the hybrid. Many later authors assumed that Siberian larch was the pollen parent, but careful assessment of the characteristics of the hybrid convince most contemporary conifer taxonomists that it was actually European larch. *L.* ×*marschlinsii* thus becomes the correct name for Dunkeld larch, which arose spontaneously nearly simultaneously in Scotland (1900) and Switzerland (1901) and was described independently based on these two origins. Synonym: *Larix* ×*eurolepis* A. Henry.

Larix mastersiana Rehder & E. H. Wilson

SICHUAN LARCH, MASTERS LARCH, CHUANG HONG SHAN (CHINESE)

Tree to 20(–25) m tall, with trunk to 0.8 m in diameter. Bark grayish brown over dark brown, smooth at first, becoming flaky and finally breaking up into irregular, flat-topped plates. Crown conical at first, broadening and flattening with age, with numerous long, thin horizontal to gently upswept branches bearing short to long, hanging branchlets. New branchlets yellowish brown to light brown, not waxy, covered with short yellowish brown hairs at first, particularly in the shallow grooves, the spur shoots remaining densely yellow hairy at the tips. Buds small, about 2–3 mm long, light reddish brown, somewhat resinous. Needles of spur shoots straight or a little curved, (20–)25–40(–50) on each spur, soft, 1–2.5(–3.5) cm long and 1–1.2 mm wide, bright green, turning clear yellow in autumn before falling. Midrib raised both above and beneath, with three to five lines of stomates forming a greenish white stomatal band on either side beneath and 0–2 faint and incomplete lines of stomates on either side above. Pollen cones oblong, 10–15 mm long, yellowish brown. Seed cones oblong, (2.5–)3–4(–4.5) cm long, with (30–)40–60(–70) seed scales, pale purplish brown before maturity, ripening light brown with dark brown bracts, on a short, slender, curved stalk to 5 mm long. Seed scales heart-shaped with a shallow notch, not curling outward or inward, densely covered with brown hairs on the outer face at first. Bracts spearhead-shaped, 10–15 mm long, longer than the seed scales, emerging between them and curling down to hide them, with a short bristle tip to 2 mm long. Seed body 2–3 mm long, solid in color, the firmly clasping wing another 5–9 mm longer. Western edge of the Sichuan Pendi (basin) in central Sichuan (China). Scattered among other conifers along streamsides in the mountains west of the basin; (2,000–)2,300–3,500 m. Zone 7.

There were once extensive pure stands of this very local species, but the wood is highly prized for homebuilding and general construction, and these stands were within easy reach of Chengdu, the capital and second largest city in Sichuan. The accessible stands were heavily exploited during the first decades of the 1900s, and now the species is endangered in nature, largely restricted to scattered trees in remote locations. Some trees are protected in the Wolong Nature Reserve that was set up to preserve the giant panda. However, because of its fine timber, Sichuan larch is also used to a certain extent in plantation forestry, so it is not in immediate danger of extinction whatever the fate of the wild populations.

Sichuan larch is closely related to Sikkim larch (*Larix griffithii*) of the Himalayan region, from which it differs in its shorter seed cones and hairy spur shoots. It is hardier than Sikkim larch but is rarely found in cultivation in either Europe or North America. The species was named, about 10 years after his death, for Maxwell T. Masters (1833–1907), a British botanist who did extensive morphological and taxonomic work with conifers, including naming several Chinese and Japanese species, and combining the names of others into the forms used today.

Larix occidentalis T. Nuttall

WESTERN LARCH

Tree to 50(–70) m tall, with trunk to 1.5(–2.3) m in diameter. Bark reddish brown and smooth at first, becoming scaly and finally breaking up into irregular, flat-topped gray ridges separated by reddish furrows. Crown narrowly cylindrical, with numerous short horizontal branches above a straight, clear trunk. New branchlets orange-brown, hairy at first, becoming hairless during the first summer. Buds small, about 3 mm long, dark brown, resinous, those of spur shoots fuzzy. Needles of spur shoots straight, (12–)25–35(–45) on each spur, soft, (1.5–)2.5–4(–5) cm long and 0.5–0.8(–1) mm wide, pale green, turning brilliant orange-yellow in autumn before falling. Midrib raised beneath and with a few lines of stomates in a narrow band on either side. Pollen cones cylindrical, yellowish brown, about 8–15 mm long. Seed cones oblong, (2–)2.5–4.5(–6) cm long, with (30–)40–60 seed scales, green with a reddish tinge before maturity, ripening reddish brown, on a gently curved, coarse stalk (2.5–)4–12 mm long. Seed scales roundly triangular, the tip slightly notched, straight or curled back at maturity, fuzzy on the inner face. Bracts often curled back, 1–2 cm long, about as long as the seed scales, elongately triangular and tipped by a bristle that extends about 1 cm beyond the seed scale, pale against the darker seed scales. Seed body 3–4 mm long, without resin pockets, the firmly clasping wing another 6–10 mm longer. Pacific Northwest of North America east of the Cascade Range, in southeastern British Columbia and southwestern Alberta and in eastern Washington, northeastern Oregon, northern Idaho, and northwestern Montana. Usually mixed with other conifers though often making up the majority of a stand on deep soils of gentle mountain slopes, flats, and valleys; (400–)600–1,600(–2,100) m. Zone 5.

Western larch, like several tree species in the Pacific Northwest, is the largest in its genus. The largest known individual in the United States in combined dimensions when measured in 1999 was 49.4 m tall, 2.2 m in diameter, with a spread of 10.4 m. This larch produces good cone crops every 1–10 years and generally regenerates after fires since it is relatively shade intolerant. It is the fastest growing of the conifers in its region during its first century of growth and so tends to dominate the stands for 200–300 years but eventually is replaced by more shade-tolerant species. It typically lives to 400 years, with the oldest recorded age about 900 years. Stands containing these old trees are particularly important nesting sites for hole-nesting birds, about a quarter of the bird species in this region.

Western larch generally grows at lower elevations and on more favorable sites than its close relative, subalpine larch (*Larix lyallii*). There are a few places where the two species come close together, and natural hybrids are documented from such sites as well as being produced by deliberate crosses. Hybridization has not contributed much to variation in western larch, and it generally has slightly less genetic variation than the other conifers of the region. In comparison to these evergreen species, western larch produces more photosynthetic leaf surface area each year for a tree of a given diameter. However, lodgepole pine (*Pinus contorta*) and Douglas fir (*Pseudotsuga menziesii*), its principal associates, have more total leaf surface area per trunk diameter, although much of this area resides in less photosynthetically efficient older leaves. On the other hand, the soft, deciduous leaves of western larch are more prone to killing fungal diseases than the tougher leaves of the evergreen conifers, although the fastest-growing individuals appear better able to avoid the effects of the fungi. Both the common and scientific names (Latin for "of the west") reflect the discovery of this species in western North America long before the discovery of subalpine larch.

Larix* ×*pendula (Solander) R. A. Salisbury

Tree to 30 m tall, with trunk to 0.6 m in diameter. Crown of short horizontal branches bearing somewhat hanging branchlets. This hybrid between European larch (*Larix decidua*) and tamarack (*L. laricina*) appears to have arisen first in the 18th century after the American tamarack was introduced into Europe. It also appears to have occurred spontaneously in the vicinity of Lake Ontario (United States and Canada) with the introduction and planting of numerous European larches. It does not share the hybrid vigor of the Dunkeld larch (*L.* ×*marschlinsii*) and so is not used in forestry, although there are a few cultivars available in horticulture.

The tree generally resembles European larch but differs in its weeping branchlets that are orange-brown, like those of tamarack. The needles are a little broader than those of tamarack at about 1 mm across and a little blunter than those of European larch. The seed cones are the most distinctive feature, being intermediate in length (about 2–3 cm versus 1–2 cm in tamarack and up to 4–5 cm in European larch) and in number of seed scales (20–30 versus 10–20 in tamarack and 40–50 in European larch). The cones resemble those of tamarack in shape, since the lower scales, like those of tamarack and unlike those of European larch are conspicuously larger than the middle scales. Like the European larch, however, some of the lower bracts may just peak out between the scales.

Larix potaninii Batalin

CHINESE LARCH, HONG SHAN (CHINESE)

Tree to 35(–50) m tall, with trunk to 0.8(–1.5) m in diameter. Bark smooth at first, becoming scaly, and finally breaking up into narrow, flat-topped gray plates separated by deep reddish brown furrows. Crown broadly egg-shaped, fairly open, with long, slender, gently upswept branches. New branchlets pale yellowish brown to (more commonly) dark reddish brown, turning gray in the second and third year, not waxy, hairless or with a few hairs in the shallow grooves. Buds small, about 3 mm long, plumply egg-shaped, dark reddish brown, somewhat resinous. Needles of spur shoots straight, (20–)30–40(–50) on each spur, soft, (1–)1.5–3(–3.5) cm long and about 1–1.5 mm wide, bright green, turning brilliant golden yellow in autumn before falling. Midrib prominently raised beneath and not or weakly so above, with one or two obscure lines of stomates on either side above and a prominent white stomatal band on each side beneath. Pollen cones about 10 mm long, pale yellowish brown. Seed cones fairly cylindrical although a little narrower above, (2–)3–5(–7.5) cm long, with 35–65(–90) seed scales, purple with red bracts before maturity, ripening dark brown with purplish brown bracts, on a stout, curved stalk about 5 mm long. Seed scales roundly rectangular, about as long as wide to half again longer, the tip squared off, opening widely and flat or a little curled back, variously hairy on the outer face. Bracts 1–2 cm long, slightly longer than and obviously sticking up straight between the seed scales, tongue- or bladelike with a sharp bristle tip 1–3 mm long. Seed body 3–4 mm long, without resin pockets, the firmly clasping wing another 5–8 mm longer. From Mount Everest (Nepal and China) eastward discontinuously through the eastern Himalaya and in the mountains of western China from southeastern Xizang (Tibet) and northwestern Yunnan north through western Sichuan to southern Gansu and southwestern Shaanxi. Forming pure stands in the subalpine zone but more commonly mixed with other montane conifers below this belt and reaching its best development along streamsides at middle elevations; (2,500–)3,000–4,000(–4,800) m. Zone 5. Two varieties.

Chinese larch is distinguished from the other long-bracted larches of the Sino-Himalayan region by having the bracts stick out straight next to the seed scales rather than bending back away from the scales. It is an important timber tree in the core of its range in Sichuan province. There is modest variation in twig color, seed cone length, hairiness of the seed scales, and bract shape among the discontinuous portions of its distribution on isolated mountain ranges. As many as four species or varieties have been recognized, basically replacing each other geographically. Much more needs to be known about variation throughout the range of the species before there can be much confidence in any division of the species into varieties. For the time being only two varieties are accepted here, each of which is itself variable, the northern variety *potaninii* and southern variety *australis*. The species name honors Grigorii N. Potanin (1835–1920), a Russian botanist and explorer of eastern Siberia, Mongolia, and China who in 1893 collected the type specimen near Kangding (formerly Tatsienlu) in western Sichuan province, the region from which most later material has been collected.

Larix potaninii* var. *australis A. Henry ex Handel-Mazzetti

Spur shoots (2–)4–8 mm in diameter, hairless or sparsely hairy at the tips. Seed cones (2–)4–6.5(–7.5) cm long and 2.5–3.5 cm in diameter, scales with dense, soft, upright hairs on the outer face. Southern portion of the range of the species, from Mount Everest to northwestern Yunnan (China). Synonyms: *Larix himalaica* W. C. Cheng & L. K. Fu, *L. potaninii* var. *himalaica* (W. C. Cheng & L. K. Fu) Farjon & Silba, *L. potaninii* var. *macrocarpa* Y. W. Law.

Larix potaninii Batalin **var. *potaninii***

Spur shoots 3–4(–5) mm in diameter, densely hairy at the tips. Seed cones (2–)2.5–5 cm long and 1.5–2.8 cm in diameter, scales with soft, upright hairs or stiff flattened hairs on the outer face. Northern portion of the range of the species from southern Gansu and southwestern Shaanxi through Sichuan to northern Yunnan. Synonym: *Larix potaninii* var. *chinensis* L. K. Fu and Nan Li.

Larix sibirica Ledebour

SIBERIAN LARCH, LISTVENITSA (RUSSIAN), KARAGAI (TATAR)

Tree to 35(–45) m tall, with trunk to 1(–1.5) m in diameter, often tapering rapidly in the lower bole. Bark bright reddish brown, scaly at first, becoming deeply furrowed between narrow, flat-topped ridges. Crown narrowly conical at first, becoming flat-topped and broad with age, especially in open stands, with numerous long, slender horizontal branches. New branchlets pale yellowish brown to grayish yellow, distinctly grooved between the attached leaf bases, noticeably hairy at first. Buds small, about 3 mm long, dark brown darkening toward the base, not resinous. Needles of spur shoots straight, (20–)30–40(–50) on each spur, soft, (2–)2.5–3.5(–5) cm long and (0.1–)0.5–1 mm wide, bright green, turning bright yellow in autumn before falling. Midrib slightly raised beneath and with one or two lines of stomates on either side above and a few lines of stomates in a pale green stomatal band on either side beneath. Pollen cones almost spherical or a little longer, 5–7(–10) mm long, yellow. Seed cones egg-shaped, (2–)2.5–4.5(–5.5) cm long, with (20–)25–40(–50) seed scales, reddish green before maturity, ripening light brown, on a thick, curved stalk 5–10 mm long. Seed scales roundly triangular to circular, the tip rounded to slightly notched, curled a little inward at maturity, conspicuously reddish hairy on the outer face. Bracts 3–5 mm long, much shorter than and hidden by the seed scales, more or less triangular with a bristle tip 1–2 mm long. Seed body 4–5 mm long, without resin pockets, the firmly clasping wing another (5–)8–12(–17) mm longer. Widespread in the taiga zone of north-central Eurasia from northwestern Russia across western Siberia (hence the scientific and common names) to the Yenisei River in the north and near Lake Baikal in the south, with outliers in mountains of Mongolia and northwestern China. Forming pure, open stands or mixed with other boreal conifers and hardwoods on a great variety of sites and substrates but reaching its best development on warm, sandy sites, generally at higher elevations southward; 0–2,400 m. Zone 2. Synonyms: *Larix russica* (Endlicher) J. Sabine ex Trautvetter, *L. sukaczewii* Dylis.

Siberian larch falls geographically and genetically between European larch (*Larix decidua*) and Dahurian larch (*L. gmelinii*) and crosses readily with these species as well as with Japanese larch (*L. kaempferi*). The chemistry of its turpentine is most similar to that of Dahurian larch, and both have much higher levels of Δ^3-carene and β-pinene than does European larch, which has a turpentine dominated by α-pinene. The whole eastern boundary of the range of Siberian larch contains natural hybrids with Dahurian larch, and these are designated *L.* ×*czekanowskii*. Although the eastern variety of European larch (variety *polonica*) has similarly been considered of hybrid origin between European larch and Siberian larch, genetic evidence contradicts this idea. Apparently then, Siberian larch does not naturally overlap or hybridize with European larch. However, when the two species are grown together at a 50:50 ratio in plantations for hybrid seed production, almost all of the resulting seed, on both Siberian larches and European larches, have Siberian larch as the pollen parent.

There is extensive natural variability in Siberian larch across its range, and a number of varieties were described, but there appears to be little justification for formal recognition of variants within this species. There is even less cause to recognize the western and eastern portions of the range as separate species since all suggested differences between the larches in these regions greatly overlap. The southern portion of the range, at least, has somewhat less

Mature Siberian larch (*Larix sibirica*) in nature in winter with numerous persistent seed cones.

genetic diversity, as measured by enzyme variants, than has been found in other larches and in conifers in general. The tendency for early flushing at lower latitudes makes this species susceptible to frost injury in western Europe, but it is an important plantation tree in Russia and Scandinavia and is grown in amenity plantings on the Canadian Prairies. A few cultivars have been selected.

Lepidothamnus R. Philippi

SOUTHERN PYGMY PINE

Podocarpaceae

Evergreen small trees and dwarf shrubs. Multistemmed or with a single, straight, often short, cylindrical to fluted trunk. Bark obscurely fibrous, thin, peeling in small irregular flakes. Crown open, upright or spreading with creeping to strongly upwardly angled branches. Branchlets all elongate, without distinction into short and long shoots, hairless, remaining green for several years, with conspicuous short grooves between the attached leaf bases, where visible. Resting buds poorly developed, consisting only of undeveloped ordinary foliage leaves. Leaves spirally attached, round or variably roundly triangular in cross section, needlelike and well separated in juveniles, scalelike and closely spaced in adults (hence the scientific name, Greek for "scaly shrub"), with a gradual transition between juvenile and adult foliage.

Plants monoecious or dioecious. Cones borne on shoots with adult, juvenile, or intermediate foliage. Pollen cones single at the tips of otherwise ordinary foliage shoots, cylindrical, without a distinct ring of bracts at the base and with 15–45 spirally arranged, roundly triangular pollen scales, each bearing two round pollen sacs. Pollen grains small to large (body 25–55 μm long, 50–100 μm overall), with two round, internally wrinkled air bladders. Bladders almost as large as or noticeably smaller than and partially tucked under (and around the longitudinal germination furrow of) the minutely bumpy, round to slightly oblong body. Seed cones single at the tips of otherwise ordinary foliage shoots, highly modified and reduced, with three to six closely spaced, spirally arranged bracts which become variably fleshy and may coalesce with each other and the axis into a berrylike podocarpium at maturity. Upper one (or two) bracts fertile, each bearing one upright seed surrounded at the base by a thin, cuplike seed scale and with a curled-over beak. Seeds maturing in two seasons. Cotyledons two, each with two veins. Chromosome base numbers $x = 14, 15$.

Wood, when present, hard and heavy, strong and durable, very resinous, with pale sapwood contrasting with the reddish yellow heartwood. Grain very fine and even, with evident growth rings marked by gradual narrowing of the latewood cells. Resin canals and individual longitudinal resin parenchyma cells both absent.

Stomates arranged in patches rather than distinct lines on both leaf surfaces. Each stomate deeply sunken beneath and largely hidden by the four to six inner subsidiary cells, which are surrounded by an outer circle of a comparable number of cells and are topped by a complete, roughly circular Florin ring. Midvein single, without a conspicuous surface midrib, with or without one resin canal directly beneath it, and with compact bands of transfusion tissue on either side. Photosynthetic tissue on both surfaces.

Three species in New Zealand and in southern South America. References: Allan 1961, Covas 1995, Hoffmann J. 1994, Markham et al. 1989, Pole 1992a, 1997a, Quinn 1982, Rodríguez and Quezada 1995, Salmon 1980.

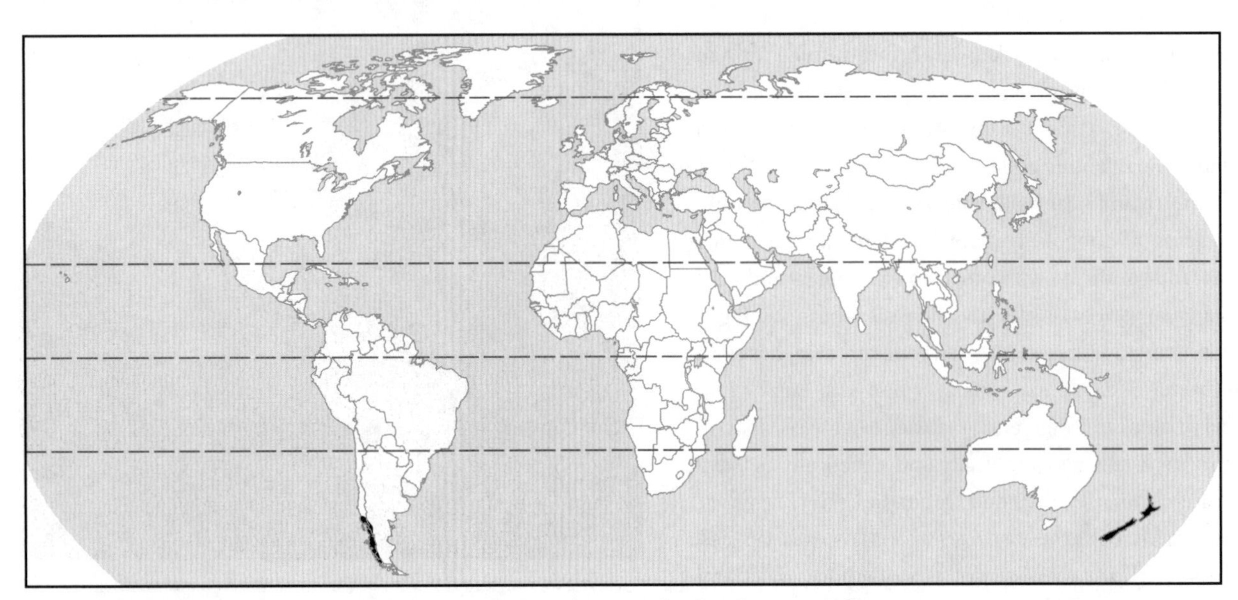

Distribution of *Lepidothamnus* in Chile, adjacent Argentina, and New Zealand, scale 1 : 200 million

This small genus contains two of the smallest living conifer species in the world, and the third is no forest giant. At least two also have among the smallest genomes (quantities of DNA in their chromosomes) known for conifers. The two smaller species, as two of the naturally most dwarf extant conifers, make interesting subjects for bog gardens in wet, mild temperate regions, but they are scarcely in cultivation and there has been no cultivar selection in the genus.

The species of *Lepidothamnus* are generally similar in appearance to species of several other small podocarp genera of New Zealand and Australia: *Halocarpus, Lagarostrobos,* and *Manoao.* DNA studies yield varied results but generally show that it is loosely associated with these genera. Although it is not as closely related to these genera as they are to each other, it is still much more closely related to them than it is to the species of *Dacrydium,* the genus in which all of these species were generally included before 1982. In that year, Quinn segregated these small genera from *Dacrydium* (or reinstated in the case of *Lepidothamnus,* which had had been formally described and named in 1861 but which had quickly fallen out of use). They differ from *Dacrydium,* and from each other, in many features of seed cones, epidermal structure, embryo development, wood anatomy, chromosome arrangements, and chemistry of their oils and pigments. All these features had been described previously but had only been considered indications of variation within *Dacrydium.* As it turned out, however, these differences reflect a much deeper division among all the species once included in *Dacrydium,* separating *Lepidothamnus* and its relatives into a different fundamental grouping of genera within the family than the one including *Dacrydium* as presently understood.

One of the distinctive characteristics of *Lepidothamnus* is the hooked beak of the seed (the micropyle), through which pollen enters the immature ovule during pollination and fertilization. With this structure, the ovules of *Lepidothamnus* preserve the hanging pollination droplet typical of the podocarps, even though they are upright rather than being pointed down into the cone axis as they are in the majority of podocarp species. Thus the pollen grains, with their air bladders, can still float up into the micropyle even though the ovules are reoriented compared to other podocarps.

The known fossil record of *Lepidothamnus* is very sparse but dates back to the late Eocene in Tasmania, more than 35 million years ago. These fossils are most similar to the largest extant species, yellow silver pine (*L. intermedius*) of New Zealand, and the genus is no longer present in Australia today.

Identification Guide to *Lepidothamnus*

The three species of *Lepidothamnus* may be distinguished by their growth habit, how tightly the adult scale leaves are pressed against the twigs, and whether the seed cones have a swollen podocarpium:

Lepidothamnus fonkii, dwarf shrub, adult scale leaves tight, podocarpium absent

L. laxifolius, dwarf shrub, adult scale leaves spreading away, podocarpium present

L. intermedius, tree, adult scale leaves loose, podocarpium absent

Lepidothamnus fonkii R. Philippi

CHILEAN PYGMY CEDAR, CHILEAN RIMU, CIPRÉS ENANO (SPANISH)

Dwarf shrub to 0.3(–0.5) m, without a significant main trunk. Bark thin, smooth, and brown. Crown spreading, with creeping branches bearing upright, much branched shoots densely clothed with foliage. Juvenile leaves 4–5 mm long. Adult scale leaves 1.5–3 mm long, tightly pressed against and completely clothing and hiding the twigs. Plants dioecious. Pollen cones 5–6 mm long, with 10–15 pollen scales. Bracts of seed cones remaining separate and dry or scarcely fleshy at maturity. Seeds shiny yellowish brown, 3–4 mm long, surrounded at the base by a thin, tough, cuplike aril less than 1 mm deep, with a smooth or slightly toothed rim. Southern Andes and coastal range of Chile and adjacent Argentina from 40°S to 55°S. Forming mixed cushions with other dwarf shrubs in low heathlands on wet, peaty soils and in subantarctic tundra; (0–)500–600 m. Zone 7. Synonym: *Dacrydium fonkii* (R. Philippi) J. Ball.

Like its close relative pygmy pine (*Lepidothamnus laxifolius*) of New Zealand, Chilean pygmy cedar is one of the smallest living conifers. It is the least known of the distinctive assemblage of

Dense clone of Chilean pygmy cedar (*Lepidothamnus fonkii*) in nature.

southern Andean conifers. While it appears to be quite rare, this might be due in part to its being overlooked in its purely vegetative stage, when without cones it might not be noticed among other heathlike shrubs. It differs from its relatives in New Zealand in a chromosomal rearrangement that means it has 14 pairs of chromosomes rather than the 15 found in pygmy pine and yellow silver pine (*L. intermedius*). Both chromosome base numbers are based on the same number of chromosome arms (20) found in the majority of podocarps, just attached to one another to make different numbers of whole chromosomes. Another feature in which Chilean pygmy cedar both resembles and differs from other podocarps is in the presence of small, woody nodules underground. These structures are found in most podocarps and some other conifers but their function and origin are uncertain. The nodules in *L. fonkii* are unusual because they occur on stems as well as on the roots, to which they are restricted in other conifers. This is probably because Chilean pygmy cedar has horizontal branches that creep beneath the surface of sphagnum bogs. Although these stem nodules might be consistent with a possible function as mycorrhizal structures, they cast some doubt on a popular idea of their origin as stunted branch roots, although they do not rule out that possibility. The species name honors Francisco Fonk, a physician who collected many plant specimens in Chile between 1852 and 1858, including the type specimen of this species.

Lepidothamnus intermedius (T. Kirk) Quinn

YELLOW SILVER PINE, [NEW ZEALAND] YELLOW PINE

Tree to 10(–15) m tall, with trunk to 0.6(–1) m in diameter and often forking repeatedly near the base. Bark thin, tight, dark brown, weathering gray and also persistently mottled with reddish brown irregular patches after flaking. Crown conical at first, becoming spreading, broadly and somewhat conically dome-shaped, with thin, strongly upwardly angled branches bearing numerous slender, upright or drooping branchlets densely clothed with foliage. Juvenile leaves standing straight out from the twigs or strongly curved forward, 9–15 mm long. Adult scale leaves keeled, yellowish green to brownish green, 1.5–3(–6) mm long, loosely held against the twig or somewhat spreading. Plants predominantly dioecious. Pollen cones 5–6 mm long, with 20–40 pollen scales. Bracts of seed cones dry or somewhat fleshy at maturity, yellow or pinkish orange, together about 1.5–4 mm long. Seeds shiny black to violet-black, 3–5 mm long, surrounded at the base by a thin, yellow, cuplike aril about 1 mm deep with an irregular rim. Sparsely distributed throughout almost the whole length of New Zealand, more common southward and on South Island mostly toward the west. Scattered to dominant in swamp forests and other low forests on wet or boggy sites with high rainfall from the lowlands to the subalpine zone; 0–900(–1,250) m. Zone 8. Synonym: *Dacrydium intermedium* T. Kirk.

Although yellow silver pine is a much larger plant than the other two species of *Lepidothamnus,* it shares their preference for areas with very wet climates. Perhaps because of the resulting reduction in sunlight and sodden soils, it is a slow growing tree. As a consequence, it produces a dense wood that is tough and durable in outdoor use. The wood had many applications until the species was depleted and, along with the other fine New Zealand native softwoods, replaced by plantations of Monterey pine (*Pinus radiata*) and other introduced species. Even during its period of unfettered exploitation it was not recommended for firewood since the extremely high resin content causes it to spark dangerously. Where it grows near pygmy pine, the two species can cross with one another, producing sterile hybrids. The scientific name reflects the general intermediacy of yellow silver pine in habit and foliage between pygmy pine and taller-growing species like silver pine (*Manoao colensoi*) and rimu (*Dacrydium cupressinum*), all of which were considered members of a single genus at the time the species was formally described and named. The name is equally reflective of the fact that individuals with foliage intermediate between extreme adult and juvenile forms are often fully mature and bear either pollen or seed cones.

Lepidothamnus laxifolius (J. Hooker) Quinn

PYGMY PINE, MOUNTAIN RIMU

Dwarf shrub to 0.3(–1) m, sometimes with cones on shoots as little as 5 cm high, without a single main trunk. Bark smooth, dark brown to gray, becoming a little scaly. Crown low, open, and irregular, with ground-hugging or clambering slender, flexible branches creeping to 1 m or more, rooting into the soil, and bearing upright, profusely branched branchlets at their tips and along the way. Juvenile leaves standing straight out from the twigs or curved gently forward, yellowish green to bluish green, (3–)5–8(–12) mm long. Adult scale leaves variously 1–4 mm long, usually standing straight out or slightly forward from the twigs rather than hugging them (hence the scientific name, Latin for "loose leaf"), sometimes tightly pressed against them, especially on twigs ending in pollen or seed cones. Plants monoecious or dioecious. Pollen cones 4–6(–9) mm long, with 15–20 pollen scales. Bracts of seed cones usually uniting at maturity into a red, swollen, juicy podocarpium 3–5 mm long but sometimes remaining separate and dry. Seeds shiny brown, 3–5 mm long, surrounded at the base by a thin, red, cuplike aril about 1–1.5 mm deep with a toothed rim. Mountains of New Zealand from the southern half of North Island south to Stewart Island, where it occurs in lowland sites. Mixed with other shrubs and sometimes clambering upon them in montane and

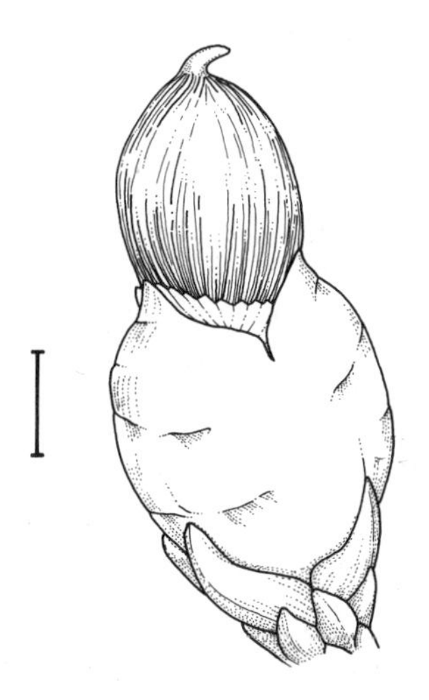

Mature fleshy seed cones of pygmy pine (*Lepidothamnus laxifolius*) with the aril just appearing between the podocarpium and the epimatium; scale, 1 mm.

subalpine scrublands on various nutrient-poor sites, both sodden peat lands and dry, fixed dunes; (10–)800–1,300 m. Zone 7. Synonym: *Dacrydium laxifolium* J. Hooker.

Pygmy pine lives up to this name since it has the smallest minimum mature size among all known living conifers. When John Bidwill, who collected the type specimen in 1839, first saw this species, he thought it was a moss. In some localities, there are taller shrubs that do not quite match pygmy pine. These were long thought to be hybrids with New Zealand mountain pine (*Halocarpus bidwillii*) during the time when both species were placed (mistakenly) in the genus *Dacrydium*. These plants have, indeed, been shown to be hybrids of pygmy pine but not with mountain pine. Instead, they are hybrids with the much more closely related small tree, yellow silver pine (*Lepidothamnus intermedius*), the only other New Zealand species in this genus. The chromosome structure of these hybrids has been studied, and it has been shown that, while the two parent species have the same chromosome number (15 pairs of chromosomes in most cells, a unique number among New Zealand podocarps), two of these chromosomes show rearrangements that distinguish the species. As a result of these differences, reproductive cells of the hybrids cannot undergo normal meiosis and the hybrids are sterile, eliminating the possibility of gene exchange between the parent species via the hybrids.

Libocedrus Endlicher

INCENSE CEDAR

Cupressaceae

Evergreen trees and shrubs with a single, straight trunk clothed in fibrous, furrowed bark peeling in long vertical strips. Densely branched from the base with short, horizontal or upsweeping branches forming a conical crown. Branchlets usually in flattened, fernlike sprays with opposite or alternate branching only from the lateral leaf pairs, but sometimes in more three-dimensional arrangements of branchlets that are squarish in cross section. Without definite winter buds. Seedling leaves in alternating quartets, needlelike, standing out from and well spaced on the stem, seedling phase short-lived, juvenile side branchlets appearing by the second year. Juvenile and adult leaves in alternating pairs, scalelike, dense, the bases running down a little onto the branchlets, the alternate pairs either all alike or differentiated into lateral and facial pairs. Lateral leaves of juveniles usually larger and more spreading than those of adults, generally keeled and folded around the branchlet, at least at the base. Facial leaves more or less oval, the successive pairs just overlapping or well separated by the lateral pairs. Leaves without glands.

Plants monoecious or dioecious. Pollen cones numerous but single at the tips of branchlets, oblong, with 5–12 alternating pairs of pollen scales, each scale with a heart-shaped blade bearing three to six pollen sacs at the base, around the short stalk. Pollen grains small (20–40 μm in diameter), nearly spherical, minutely bumpy and with an ill-defined, minute germination pore or slight projection. Seed cones typically in two to four pairs at the base of vigorous lateral shoots, each single at the end of a short branchlet, oblong, maturing in a single season. Cone with two obvious alternating pairs of thin woody scales and five smaller pairs grading into the leaves below, only the upper pair with two seeds each and about twice as long as the second pair. Each cone scale apparently with the bract completely fused to the seed-bearing portion and ending in a long, upcurved spine halfway up the scale or a little higher. Seeds oval, with two very unequal wings derived from the seed coat at the upper end, the outer wing a mere fringe, the inner expanding over the scale and overlapping with the wing of its neighbor. Cotyledons two, each with one vein. Chromosome base number $x = 11$.

Wood fragrant, light, soft, and moderately to extremely resistant to decay, with yellowish to light pinkish brown sapwood sharply contrasting with the dull red or dark yellow to dark brown heartwood. Grain fine to very fine and even, with well-defined or somewhat cryptic growth rings marked by a fairly gradual transition to a narrow band of much smaller and somewhat thicker walled latewood tracheids. Resin canals absent but with sparse to abundant individual resin parenchyma cells scattered through the growth increments or somewhat concentrated in open bands.

Stomates grouped in (often waxy) patches or bands primarily on protected surfaces of leaves, either on the inner face, hidden in grooves, or on the side facing down in horizontal sprays. Each stomate rectangular or a little rounded in outline, sunken beneath and largely hidden by the four to six surrounding subsidiary cells, which are often shared within and between rows and usually (except in *Libocedrus uvifera*) circled by a tall, steep, nearly continuous Florin ring. Subsidiary cells and other epidermal cells in stomatal zones also sometimes carrying additional rounded papillae. Midvein single-stranded above a single large resin canal and flanked in part by wedges of transfusion tissue. Photosynthetic tissue forming a palisade layer beneath the epidermis and accompanying thin hypodermis all around the external face or only on the up-facing side of flattened sprays, the remaining leaf volume occupied by looser spongy mesophyll, especially near the stomatal bands and patches.

Six species in Chile, New Caledonia, and New Zealand. References: Biloni 1990, de Laubenfels 1972, Enright and Hill 1995, Farjon 2005b, Hill and Carpenter 1989, McIver and Aulenback 1994, Rodríguez R. et al. 1983, Salmon 1980, Tomlinson et al. 1993. Synonyms: *Pilgerodendron* Florin, *Stegocedrus* Doweld. See also *Austrocedrus, Calocedrus,* and *Papuacedrus.*

Until the 1950s, most botanists included all of the incense cedars in both the northern and southern hemisphere, with their fragrantly resinous wood, in a single genus, *Libocedrus* (from the Greek "teardrop cedar" for the resinous exudate), the only conifer thought to have such a bihemispheric distribution. Then Li (1953) and Florin and Boutelje (1954) presented convincing arguments for breaking up this genus, and most subsequent authors followed their lead (de Laubenfels 1988 is one exception). DNA data generally support most of the distinctions. The northern hemisphere species, placed in the genus *Calocedrus,* are really more closely related to other northern genera, like *Platycladus,* than they are to any of the southern incense cedars, just as Li suggested. While the southern species are all related to one another, there is ample justification for separating the South American *Austrocedrus* and the New Guinean *Papuacedrus.* Since the New Zealand *L. plumosa* is the type species of the genus, the New Zealand and New Caledonian species retain the name *Libocedrus.* Some species of *Libocedrus* are cultivated to a limited extent outside of their native ranges in botanical collections indoors or outdoors in suitable mild, wet places, but no cultivar selection has taken place.

Both *Austrocedrus* and *Papuacedrus* have short projections of the bracts on their seed scales, at the top and bottom in the two genera, respectively, while the species of *Libocedrus* have a much longer prickle near or above the middle. Although the bract and seed scale in *Libocedrus* and its relatives are here referred to conventionally as fused, this is a theoretical interpretation of the evolutionary origin of these structures. The actual development of the seed cones presents no clear fusion of two separate structures but rather differential growth of the free and united portions of the bract and seed scale.

Ironically, the segregate that superficially appears most distinct and the one first generally accepted as such, *Pilgerodendron* (synonym, *Libocedrus uvifera*), is actually firmly nested among the five New Caledonian and New Zealand species in the DNA studies and so is included in *Libocedrus* here. Although its seed cones are similar to those of the other species, it has square branchlets that lack the wide lateral leaves often considered typical of incense cedars. However, *L. bidwillii* of New Zealand has similar adult foliage, and *L. chevalieri* of New Caledonia also has weak differentiation of lateral and facial leaf pairs. These three species occur at the ecological extremes of the range of the genus. *Libocedrus uvifera* and *L. bidwillii* occur in the coldest climates, and *L. chevalieri* in the driest. Expanded lateral leaves, like those of the other three species of *Libocedrus* (and the related *Papuacedrus*) are found in areas with a warm, moist climate.

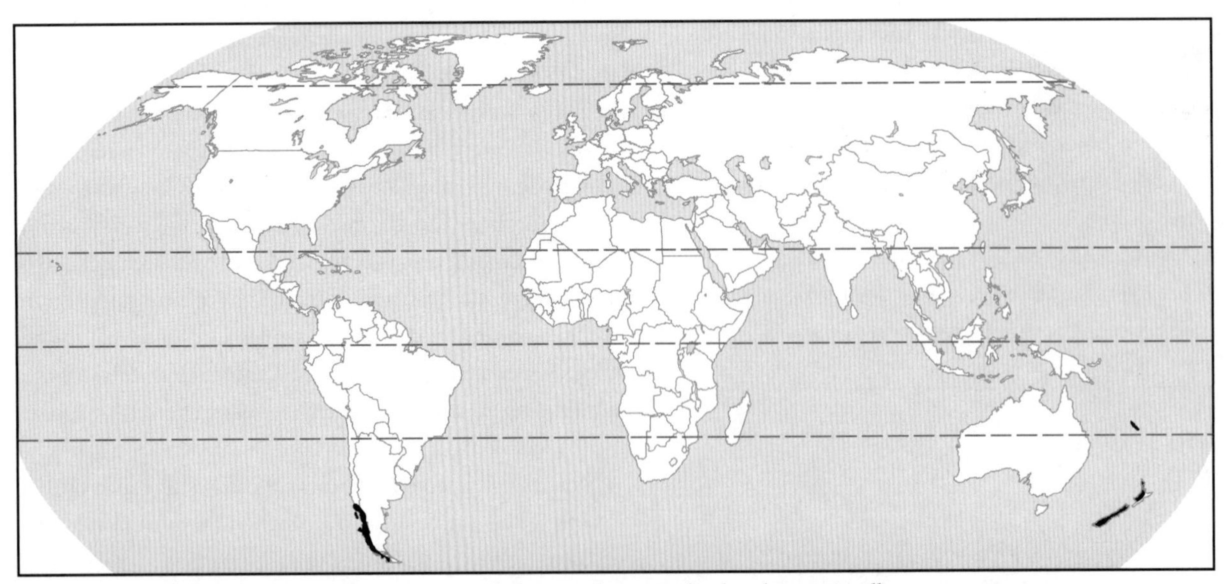

Distribution of *Libocedrus* in Chile, Argentina, New Caledonia, and New Zealand, scale 1 : 200 million

Fossils assigned to *Libocedrus* have been found in Oligocene sediments of Tasmania and later Mio-Pliocene deposits in New Zealand. Since the earlier fossils are contemporaneous with Tasmanian fossils assigned to *Austrocedrus* and *Papuacedrus,* they also support the separation of these genera.

Identification Guide to *Libocedrus*

The six species of *Libocedrus* may be distinguished by whether the facial and lateral leaves of fully adult foliage are similar or dissimilar, whether the juvenile leaves show the same distinction, whether the tip of each facial leaf touches the base of the next one, the length of the pollen cones, and the length of the seed cones:

Libocedrus uvifera, facial leaves similar to laterals in adults and juveniles, facial leaf tip touching, pollen cones less than 6 mm, seed cones up to 10 mm

L. bidwillii, facial leaves similar to laterals in adults, dissimilar in juveniles, facial leaf tip touching, pollen cones at least 6 mm, seed cones less than 10 mm

L. chevalieri, facial leaves similar to laterals in adults, dissimilar in juveniles, facial leaf tip touching, pollen cones more than 6 mm, seed cones at least 10 mm

L. yateensis, facial leaves dissimilar to laterals in adults and juveniles, facial leaf tip touching, pollen cones at least 6 mm, seed cones up to 10 mm

L. austrocaledonica, facial leaves dissimilar to laterals in adults and juveniles, facial leaf tip not touching, pollen cones more than 6 mm, seed cones at least 10 mm

L. plumosa, facial leaves dissimilar to laterals in adults and juveniles, facial leaf tip not touching, pollen cones up to 6 mm, seed cones at least 10 mm

Libocedrus austrocaledonica Ad. Brongniart & Gris

DWARF INCENSE CEDAR

Shrub to 2(–6) m, with several stems generally less than 15 cm in diameter, branching from near the base to form an open crown. Bark brown, peeling in thin flakes. Shoots arranged in fernlike sprays with a definite top and bottom side, composed of flattened branchlets arising in opposite pairs from the axils of all pairs of lateral leaves on either side of a specialized side shoot. Facial and lateral leaves very different from each other in both juvenile and adult foliage. Lateral leaves much larger than the facial ones, bowed upward, dark green above and with a broad, pale stomatal zone beneath. Lateral leaves of juveniles 4–6 mm long, pointed and spreading like saw teeth, those of adults 3–4 mm long, rounded and pressed against the back edge of the next leaf. Tips of facial leaves mostly falling far short of touching the bases of the next ones along the branchlet. Pollen cones (5–)8–12 mm long, with 8–12 pairs of pollen scales. Seed cones 10–12 mm long, the spiny free tips of the bracts 7–10 mm long. Seeds about 6 mm long, the large wing about twice as long and 1.5 times as wide as the body. Southern and central New Caledonia. Understory of montane rain forests on peaty soils; (700–)750–1,300(–1,400) m. Zone 9. Synonym: *Stegocedrus austrocaledonica* (Ad. Brongniart & Gris) Doweld.

Dwarf incense cedar is quite similar to kawaka (*Libocedrus plumosa*) of New Zealand and, like it, grows only in very humid habitats. It differs in being a shrub rather than a tree and in having larger pollen cones. The shrubs are too small to be of any economic significance in New Caledonia. Although they are potentially handsome ornamentals, they are so sensitive to strong sunlight and drought that they are unlikely to succeed outdoors anywhere but in the most sheltered and humid tropical locations. They might succeed as greenhouse subjects.

Libocedrus bidwillii J. Hooker

PAHAUTEA (MAORI AND ENGLISH), MOUNTAIN CEDAR, KAIKAWAKA (MAORI)

Tree to 20(–28) m tall, with trunk to 1(–1.5) m in diameter. Bark reddish brown to gray, peeling in long vertical strips. Crown dense, shallow, conical, with horizontal or gently ascending branches. Shoots arranged in fernlike sprays composed of flattened (juvenile) or squarish (adult) branchlets arising alternately or in opposite pairs from the axils of lateral leaves on either side of a specialized side shoot. Facial and lateral leaves grayish green, very different in juvenile foliage, basically similar in adult foliage. Lateral leaves of juveniles much larger than the facial ones, 3–4 mm long, bluntly pointed and bent forward against the back edge of the next leaf, those of adults 2–3 mm long, rounded, and pressed against the twig. Tips of the facial leaves rounded, overlapping the bases of the next ones along the branchlet. Pollen cones four-angled, 6–10 mm long, with three to seven alternating pairs of pollen scales. Seed cones 6–8 mm long, the spiny free tips of the bracts 4–5 mm long. Seeds about 3 mm long, the expanded portion of the larger wing protruding from the upper half of and a little longer than the body. South Island and western North Island of New Zealand. Canopy member or emergent in montane and subalpine mixed forests; 250–1,200 m. Zone 8.

Pahautea, which closely resembles Mount Humboldt incense cedar (*Libocedrus chevalieri*) of New Caledonia, grows at lower elevations on the wetter western side of the South Island than elsewhere in its range. Like *L. chevalieri* and Fuegian incense cedar

Natural stand of mature pahautea (*Libocedrus bidwillii*).

(*L. uvifera*) of South America, it shows that those species of southern incense cedar growing away from warm, humid environments, either colder or drier, lack the broad lateral leaves found in the species living in more favorable habitats. In peat at high elevations near the top of the subalpine forest, pahautea becomes shrubby, but still taller than many of its hardwood associates. Despite its significant stature, the wood of pahautea is of much lower quality than that of several of its podocarp neighbors. It was used for applications, like fence posts, where durability in the ground rather than good mechanical qualities were required. Its light weight encouraged some use in packaging for hand transport. What little exploitation there was ceased when the finest remaining stands were incorporated into nature reserves.

Libocedrus chevalieri J. Buchholz

MOUNT HUMBOLDT INCENSE CEDAR

Large shrub, or tree to 5 m tall, with multiple trunks to 10 cm in diameter. Bark tan to dark brown, peeling in elongate strips. Crown compact, with numerous branches densely clothed with foliage. Shoots arranged in fernlike sprays without a definite top and bottom side, composed of flattened (juvenile) to four-angled (adult) branchlets arising alternately or in opposite pairs from the axils of lateral leaves. Facial and lateral leaves very different in juvenile foliage and similar in adult foliage. Lateral leaves of juveniles much larger than the facial ones, 2.5–5 mm long, pointed and spreading like slightly forward-directed saw teeth, those of adults 2–3 mm long with rounded tips touching the trailing edge of the next leaf. Facial leaves keeled, each with a pointed tip that overlaps the base of the next one along the branchlet. Pollen cones 8–10 mm long, with 8–12 pairs of pollen scales. Seed cones 10–12 mm long, the spiny free tips of the bracts 6–7 mm long. Seeds about 6 mm long, the large wing about twice as long and wide as the body (or more). Northern and southern New Caledonia. Part of the canopy of dense, scrubby maquis on peaty soils on Mount Humboldt and a very few other peaks, (650–)1,450–1,620 m. Zone 9. Synonym: *Stegocedrus chevalieri* (J. Buchholz) Doweld.

Mount Humboldt incense cedar is similar to pahautea (*Libocedrus bidwillii*) of New Zealand in its loss of differentiation between lateral and facial leaves in the adult phase. Loss of expanded facial leaves may be attributed in this case to a drier as well as colder (as for *L. bidwillii*) environment than the warm humid habitats required by those species of *Libocedrus* with broad lateral leaves. Although its mountain summit habitats receive abundant precipitation, the steep rocky slopes on which it grows are not water retentive. Mount Humboldt incense cedar differs from *L. bidwillii* in having larger seed cones. The species is too small and too rare to have ever attracted commercial or even local exploitation.

Libocedrus plumosa (D. Don) C. Sargent

KAWAKA (MAORI AND ENGLISH), NEW ZEALAND CEDAR

Tree to 25 m tall, with trunk to 1.2 m in diameter. Bark reddish brown, peeling in long vertical strips. Crown dense, conical, with numerous horizontal or gently drooping branches densely clothed with tufted foliage near their ends. Shoots arranged in fernlike sprays with a definite top and bottom side, composed of flattened branchlets arising in opposite pairs from the axils of all pairs of lateral leaves on either side of a specialized side shoot. Facial and lateral leaves very different in both juvenile and adult foliage, the lateral leaves much larger than the facial ones. Lateral leaves of juveniles 4–6 mm long, pointed and pointing forward, those of adults 2–3 mm long, rounded and pressed against the back edge of the next leaf. Tips of facial leaves falling far short of touching the bases of the next ones along the branchlet. Pollen cones four-angled, 4–6 mm long, with four to seven alternating pairs of pollen scales. Seed cones 10–15 mm long, the spiny free tips of the bracts 4–6 mm long. Seeds about 5 mm long, the large

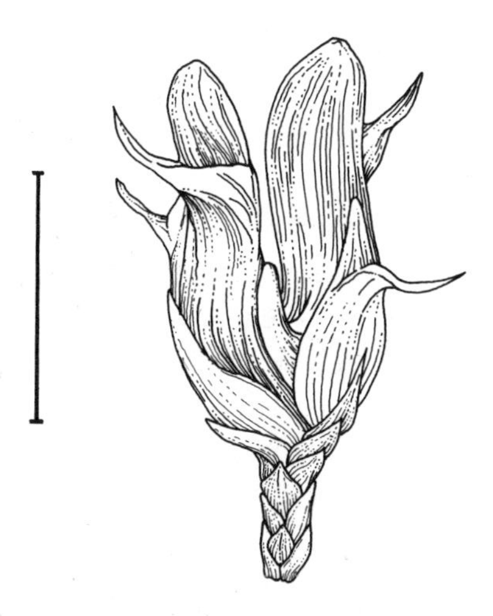

Open mature seed cone of kawaka (*Libocedrus plumosa*) after seed dispersal; scale, 1 cm.

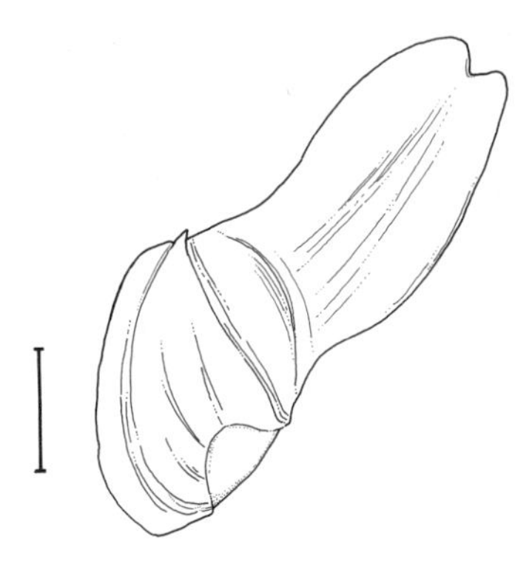

Dispersed seed of kawaka (*Libocedrus plumosa*); scale, 2 mm.

wing about twice as long and wide as the body (or more). Northwestern portion of both North Island and South Island in New Zealand. Local canopy emergent in lowland and montane mixed forest, often with *Agathis australis;* 0–600 m. Zone 8.

Kawaka is much more restricted in distribution and more tropical than the other New Zealand incense cedar, pahautea (*Libocedrus bidwillii*). The adult foliage of *L. plumosa* is similar to the juvenile foliage of *L. bidwillii.* Although the adult (and juvenile) foliage of the two species looks quite different close up, the overall appearance of the two trees is nonetheless quite similar. The foliage of kawaka is even more similar to that of dwarf incense cedar (*L. austrocaledonica*) from New Caledonia, but the latter is an understory shrub that never reaches much less overtops the canopy like the New Zealand species. The wood of kawaka, like that of pahautea, had little commercial importance, even when most trees were not yet confined to nature reserves and the lowland tropical rain forests were being felled and cleared more or less indiscriminately, because it is much inferior to that of kauri (*Agathis australis*) and many accompanying podocarps.

Young mature Fuegian incense cedar (*Libocedrus uvifera*) in nature.

Libocedrus uvifera (D. Don) Pilger

FUEGIAN INCENSE CEDAR, CIPRÉS DE LAS GUAITECAS, TEN (SPANISH)

Tree to 20(–40?) m tall, with trunk to 1(–1.5) m in diameter, or shrubby in its most exposed and harshest sites. Bark peeling in narrow strips, reddish chestnut. Crown open, with thin, ascending branches later flattening or drooping. Branchlets four-angled, upwardly directed and without distinct upper and lower sides, emerging singly from all four sides of parent twigs, or partially confined to just two sides to form frondlike sprays. All leaves similar in both juveniles and adults, without distinction into facial and lateral pairs, 2–3(–6) mm long, loosely overlapping or slightly spreading, pointed, keeled, yellow-green to bluish green. Pollen cones slightly four-angled, about 4–6 mm long, with 6–10 alternating pairs of pollen scales. Seed cones 8–12 mm long, the spiny free tips of the bracts 3–4 mm long. Seeds about 3 mm long, the larger wing about as wide and about 1.3 times as long as the body. Southern Chile and southern Argentina from about 40°S to 55°S on Tierra del Fuego and Isla del Estado. Forming groves

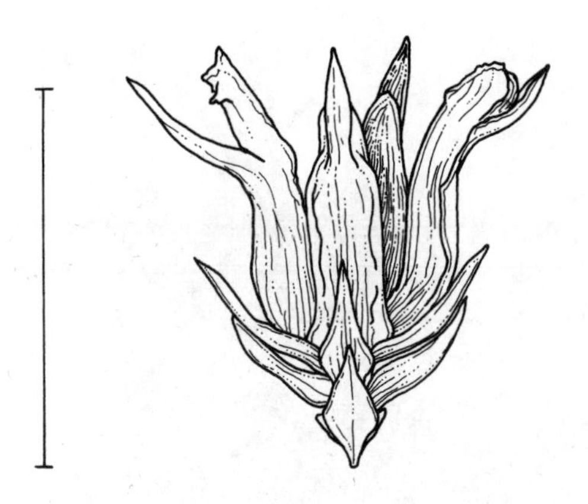

Open mature seed cone of Fuegian incense cedar (*Libocedrus uvifera*) after seed dispersal; scale, 1 cm.

in temperate rain forests and bogs with *Fitzroya* in the north and *Nothofagus* species in the south; 0–600(–750) m. Zone 8. Synonym: *Pilgerodendron uviferum* (D. Don) Florin.

Fuegian incense cedar is the most southerly conifer in the world, the only one to reach Tierra del Fuego. However, it reaches its greatest abundance, forming dense *cipresales,* far north of there, in the drenched coastal forests of the Chonos Archipelago and the adjacent mainland of Chile from about 42°S to 44°S. This was the first southern incense cedar to be separated from *Libocedrus* because it generally lacks the flattened sprays of branchlets found in other species, even in the juvenile stage, and most authorities still place it in its own genus, *Pilgerodendron.* However, this difference is rather superficial and its pollen and seed cones are typical of *Libocedrus* and different from those of *Austrocedrus* or *Papuacedrus.* Furthermore, its DNA structure places it squarely among the New Caledonian and New Zealand species. Several genera of conifers and hardwoods (including the southern beeches, *Nothofagus*) have similar or even more extensive distributions on both sides of the South Pacific, and even more did so in the geologic past. Like that of other incense cedars, the wood of *L. uvifera* is fragrant and highly resistant to decay.

Libocedrus yateensis Guillaumin

RIVIÈRE BLEUE INCENSE CEDAR

Shrub, or tree to 12 m tall, with one or more stems to 30 cm in diameter. Bark tan to bright reddish brown, peeling in thin vertical strips. Crown dense, conical to dome-shaped, with numerous crooked, ascending branches densely clothed with foliage. Shoots arranged in fernlike sprays with a top and bottom side, composed of flattened branchlets arising in opposite pairs from the axils of all pairs of lateral leaves on either side of a specialized side shoot. Facial and lateral leaves very different in juvenile foliage, less so in adult foliage, the lateral leaves larger than the facial ones. Lateral leaves of juveniles 4–6 mm long, pointed and curved forward, those of adults 2–3 mm long, rounded and pressed against the back edge of the next leaf. Tips of the keeled facial leaves just touching the bases of the next ones along the branchlet in adults, but not in juveniles. Pollen cones 6–10 mm long, with 8–12 alternating pairs of pollen scales. Seed cones 9–10 mm long, the spiny free tips of the bracts 10–12 mm long. Seeds about 6 mm long, the large wing two-thirds or as long and about as wide as or a little wider than the body. Southern and eastern New Caledonia. Floodplain rain forest; 150–250(–650) m. Zone 9. Synonym: *Stegocedrus yateensis* (Guillaumin) Doweld.

As an occasional occupant of the rain-forest canopy, Rivière Bleue incense cedar has smaller adult facial leaves than the understory dwarf incense cedar (*Libocedrus austrocaledonica*). It is the only southern incense cedar found in lowland tropical rain forest along streamsides. The alluvial soils it grows on are reminiscent of the floodplain habitat of the similar-appearing but unrelated Atlantic white cedar (*Chamaecyparis thyoides*) or western red cedar (*Thuja plicata*) of the northern hemisphere. Although the trees are largely confined to protected areas, they are susceptible to the frequent fires that ravage many parts of New Caledonia, fires that have increased in frequency and intensity, with population growth and the development of the mining industry on the island.

Manoao Molloy

SILVER PINE

Podocarpaceae

Evergreen trees and shrubs. Trunk cylindrical or somewhat fluted, dividing near the ground or free of major limbs for a third or more of its height. Bark obscurely fibrous, fairly thick, rough from an early age, shedding in thick flakes and larger scales. Crown dense, more or less cylindrical, with numerous strongly upwardly angled branches. Branchlets all elongate, without distinction into long and short shoots, hairless, grooved between the attached leaf bases and remaining green for at least the first year (juvenile foliage) or completely hidden (adult foliage). Resting buds indistinct, consisting solely of as yet unexpanded ordinary foliage leaves. Leaves spirally attached. Juvenile foliage clearly distinct from and giving way more or less gradually via transitional forms to adult foliage. Juvenile leaves needlelike, sword-shaped, flattened top to bottom, standing out straight from the twigs. Adult leaves scalelike, tightly pressed against and completely covering the twigs, overlapping, the exposed portions diamond-shaped to six-sided, rounded to variably keeled, with a thin, papery margin.

Plants dioecious. Pollen cones single at the tips of otherwise ordinary foliage shoots. Each pollen cone cylindrical, with a few

Distribution of *Manoao colensoi* in northern and central New Zealand, scale 1 : 119 million

bracts at the base and with (6–)10–16 spirally arranged, roundly triangular pollen scales, each with two pollen sacs. Pollen grains small to large (body 25–45 µm long, 50–70 µm overall), with two somewhat kidney-shaped, internally strongly wrinkly air bladders, extending beyond and partially tucked under the conspicuously larger, minutely bumpy, oval body. Seed cones one (or two) at the tips of otherwise ordinary foliage shoots, highly modified and reduced, with about three to eight somewhat loosely spirally arranged bracts, which do not swell and unite into a podocarpium. Middle two to five (or six) bracts fertile, from which one or two (to five) seeds mature. Each fertile bract bears one seed that is nestled in the somewhat fleshy, asymmetrically cup-shaped seed scale (the epimatium) and that has its opening pointed up and away at an angle to the cone axis. Seed without an aril, hard, not noticeably flattened, finely ribbed, ending in a broad, straight beak, maturing in two seasons. Cotyledons two, each with two veins. Chromosome base number $x = 10$.

Wood dense, hard, heavy, brittle, and durable, not fragrant but greasy and covered with crystals, the narrow white sapwood sharply contrasting with the yellow heartwood, which is sometimes mottled with pinkish brown or gray. Grain even, moderately coarse, usually with evident growth rings marked by a narrow band of slightly darker latewood. Resin canals and individual resin parenchyma cells both absent.

Stomates in patches on both sides, each with a random orientation and not arranged in lines. Each stomate sunken beneath the four to six subsidiary cells, surrounded by a nearly continuous Florin ring, and plugged with wax or leaf surface fungi. Midvein single, inconspicuously buried in the leaf tissue of the adult leaves, with one resin canal immediately beneath it. Photosynthetic tissue on both sides of the leaf, without a layer of hypodermis inside the epidermis.

One species in New Zealand. References: Allan 1961, Molloy 1995, Quinn 1982, Salmon 1980.

Manoao is a Maori name, reflecting the adoption of Maori names as the common names of many trees and other plants and animals by all New Zealanders. Unfortunately, this name is likely to be confused with monoao, the usual common name for the related and rather similar but much more restricted and rarer *Halocarpus kirkii.* When silver pine (*Manoao colensoi*), the sole species in this genus, was removed from *Dacrydium* by Quinn in 1982, it was placed in *Lagarostrobos* along with Huon pine (*L. franklinii*) of Tasmania. Silver pine was then separated from *Lagarostrobos* by Molloy in 1995, based on a wide range of differentiating features. Nonetheless, these two genera are very closely related, as DNA studies show, but DNA data also show that the New Caledonian *Parasitaxus ustus* is just as closely related to each of them as they are to each other. Since few taxonomists would presently be willing to combine the parasitic *Parasitaxus* with the other two species, maintenance of silver pine in its own genus seems advisable. If accumulating evidence favored combining all three species into a single genus, it would have to be called *Parasitaxus,* the oldest available name but not one terribly descriptive of silver pine and Huon pine. (The rules of botanical nomenclature explicitly state that names need not be appropriate, however.) Silver pine is scarcely in cultivation, and no cultivar selection has taken place. No fossils of *Manoao* have been recognized, although some 30-million-year-old (early Oligocene) fossil shoots from northwestern Tasmania were originally described as being intermediate between silver pine and Huon pine and might belong to either *Lagarostrobos* or *Manoao.*

Manoao colensoi (W. J. Hooker) Molloy

SILVER PINE, WESTLAND PINE

Tree to 12(–20) m tall, with trunk to 0.6(–1) m in diameter, but sometimes stunted and shrubby, especially on poor sites. Bark brown, with many horizontally elongate warty lenticels and shallow longitudinal furrows when young, weathering silvery gray (hence the common name) and flaking in irregular, warty scales to reveal reddish brown patches. Crown conical in youth, becoming more open, irregular, and egg-shaped in age, with each major upwardly angled limb forking repeatedly and bearing tufts of foliage. Adult branchlets upright, repeatedly and somewhat openly alternately branched, round or somewhat angular to the touch, 1–1.5(–2) mm in diameter. Juvenile foliage long persistent and, along with intermediate foliage, sometimes bearing pollen or seed cones. Juvenile leaves 5–12(–15) mm long, 0.7–1.5 mm wide. Adult scale leaves yellowish green or bright green to bluish green, 1–2.5 mm long, rounded to roundly keeled, lasting 4–6 years or more. Pollen cones with pinkish red to red pollen sacs on green scales, browning after release of pollen, 3–6 mm long, 1.5–2.5 mm in diameter. Seed cones green, to yellowish green, each seed cupped for about half its length on one side in a green to yellowish green epimatium about 4–5 mm long. Seeds purplish black with a thin waxy film, finely ribbed, (2–)3–5 mm long. New Zealand from about 35°S to 44°S, widely and sparsely scattered in the mountains of North Island and fairly continuously distributed along the western side of South Island. Scattered or gregarious in the canopy of low, mixed lowland and montane forests, particularly in boggy sites; 0–500(–950) m. Zone 8. Synonyms: *Dacrydium colensoi* W. J. Hooker, *Lagarostrobos colensoi* (W. J. Hooker) Quinn.

The wood of silver pine is, together with that of totara (*Podocarpus totara*), the most durable softwood in New Zealand, so its main use has been as long-lasting fence posts, particularly following the depletion of totara. This use and its employment as firewood both fail to take full advantage of a very handsome wood, but one not easily worked before mid–20th century power tools. With restricted supplies from extant stands, it now sees limited employment for carving, turnery, and furniture. The slow and climatically responsive growth of the tree, coupled with a substantial longevity that often exceeds 400 years and can reach nearly 600, has made it a candidate for dendroclimatological studies.

The foliage, in addition to containing general types of chemical compounds common across the Podocarpaceae, has yielded a group of nitrogen-containing toxins (homoerythrina alkaloids) not known in other podocarps but present in a very few other conifers and scattered unrelated plants. Another peculiarity of silver pines is their ability, extremely rare in conifers, to form clonal groves by sending up suckers from the long roots or buried branches that spread widely near the surface of bogs. This habit facilitates the colonization of these substrates, where seedling establishment is difficult. The species name honors William Colenso (1811–1899), British-born minister and printer who also collected many plants in New Zealand, including the type of this species. Due in part to the confusion of common names, but as much to the considerable similarity of silver pine to species of *Halocarpus* and *Lepidothamnus* growing with it in wet forests and bogs, the epithet *colensoi* was long misapplied in the 19th and early 20th centuries to pink pine (*H. biformis*), and silver pine went under another name.

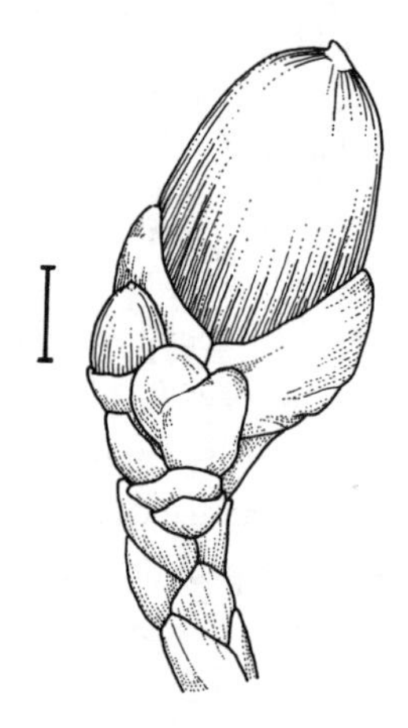

Mature seed cone of silver pine (*Manoao colensoi*) with large, puffy aril-like epimatium; scale, 1 mm.

Metasequoia Miki ex H. H. Hu & W. C. Cheng

DAWN REDWOOD

Cupressaceae

Deciduous trees with smoothish, thin, fibrous bark peeling in long strips on a strongly tapering trunk with a flaring, deeply fluted base. Branches numerous, angled upward to form a conical crown, broadening and flattening with age. Shoot system differentiated into permanent long shoots and deciduous short shoots that are shed together with their leaves in the autumn, both types produced in opposite pairs. Leaves in alternating pairs, initially at right angles to one another but coming to lie in a single plane on either side of the twig by twisting of the petioles and of the shoot axis alternately to the left and the right. Leaf blade needlelike, commonly linear, flat, soft in texture, abruptly narrowed to the short petiole, the tip rounded, bright green above.

Plants monoecious. Pollen cones single and well spaced in the axils of leaves on branched, dangling, specialized reproductive shoots. Each cone with about seven alternating pairs of bud scales at the base and about 20 irregularly arranged, elliptical pollen scales on thin stalks, each with three pollen sacs. Pollen grains small (20–35 µm in diameter), flattened spherical, with a straight or hooked germination papilla, the surface covered with minute bumps. Seed cones squarish spherical to top-shaped, woody, single at the ends of specialized reproductive short shoots bearing five or six pairs of short leaves that are shed before maturity of the cone in its first season. Cones emerging directly from buds on the long shoots that bear seven or eight alternating pairs of bud scales. Each cone with 8–14 alternating pairs of shield-shaped cone scales (of which the lower and upper two or three pairs are small or sterile, or both). Each cone scale with a transversely diamond-shaped, pentagonal, or hexagonal external facet, the fertile and bract portions about equal and intimately fused, with a single row of (five or) six to eight (or nine) seeds. Seeds elliptical, flattened, with a pair of flat wings derived from the seed coat and slightly longer and about two to four times wider than the seed body. Cotyledons two, each with one vein. Chromosome base number $x = 11$.

Wood not notably fragrant, light and soft, with pale whitish brown sapwood sharply contrasting with reddish brown heartwood of varying shades. Grain even and fairly coarse, with well-defined growth rings marked by a gradual transition to a fairly wide band of much smaller and thicker-walled latewood tracheids. Resin canals absent but with sparse to abundant individual resin parenchyma cells scattered though some growth increments and concentrated in open bands in others, generally more abundant in younger stems.

Stomates longitudinally to irregularly oriented, confined to two paler bands beneath, each band with 4–7(–10) lines. Individual stomates sunken beneath the (four or) five or six (to eight) surrounding subsidiary cells, which are topped by a thick Florin ring and often carry other papillae as well. Leaf cross section with a single-stranded midvein above a smaller resin canal, flanked by wings of transfusion tissue, and with an additional resin canal lying on either side out at the leaf margin. Photosynthetic tissue of fairly homogeneous spongy mesophyll without a distinct palisade layer of columnar cells beneath the upper epidermis and accompanying, almost continuous hypodermis.

One species in central China. References: Bartholomew et al. 1983, Basinger 1984, Farjon 2005b, Florin 1952, Fu et al. 1999h, Fuller 1976.

After *Metasequoia glyptostroboides* was discovered in the 1930s, it was introduced into cultivation on a massive scale, at least in botanical gardens initially. Perhaps, in part, because it proved to be slow to reach full reproductive maturity (seed cones are formed many years before pollen cones, so seed produced until the 1970s was generally nonviable), there has been little cultivar selection. The few available cultivars emphasize ease of vegetative propagation and narrower crowns.

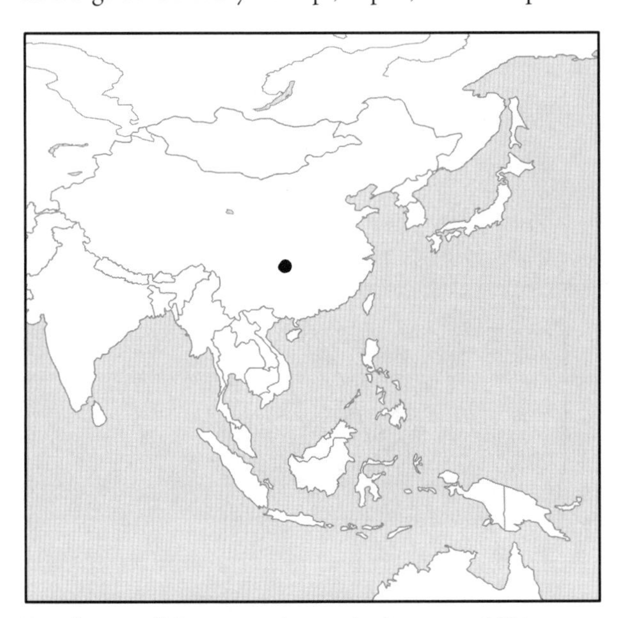

Distribution of *Metasequoia glyptostroboides* in central China, scale 1 : 120 million

Metasequoia is one important link in the unity of the family Cupressaceae because it joins the decussate leaf arrangement (alternating pairs) found in cypresses and junipers with the typical needlelike leaf form found in many redwoods. Although the genus shares annual shedding of its foliage with *Taxodium* (bald cypresses), it is actually much more closely related to *Sequoia* (redwoods) and *Sequoiadendron* (giant sequoias). In the past, some suggested that *Metasequoia* might have been partly ancestral to *Sequoia* through hybridization followed by polyploidy, but this view is not now generally accepted. Nonetheless, the evident similarity to *Sequoia* explains the scientific name, using Greek for "near."

It was a dominant tree in many parts of the northern hemisphere throughout the late Cretaceous and Tertiary (but not, apparently, in western Europe). The fossil species were also deciduous, and this allowed them to grow in the arctic far north of related evergreen trees like the redwood (to almost 80°N on Axel Heiberg Island and northernmost Greenland). The earliest recognizable species, from the mid-Cretaceous of Alaska and Siberia, had not yet fully established the opposite leaf arrangement that characterizes all later species. Several of these were described, based on variations in shape and arrangement of various organs and on different kinds of characters available for comparison because of different modes of preservation. Since there is just a single living species, however, and since this species consists of just a few thousand individuals in a small region of China, compared to the millions of hectares inhabited by Tertiary *Metasequoia*, nobody knows how much variation to expect within a species of this genus. Thus there is no certainty that the number of described fossil species correspond to the actual number of Tertiary species represented by these fossils. After the northern hemisphere cooled following the Eocene (from about 35 million years ago) and the arctic became too cold to support forests, the range of the genus became progressively more restricted. *Metasequoia* finally disappeared from North America and most of Eurasia after the Miocene, persisting longest (outside of its west-central Chinese last refuge) in Japan, where it lasted into the Pliocene.

Metasequoia glyptostroboides H. H. Hu & W. C. Cheng

DAWN REDWOOD, SHUI SHAN (CHINESE)

Plates 31–33

Tree to 35(–60) m tall, with a straight, fluted trunk to 2 m in diameter. Bark reddish brown to gray, peeling in thin strips from

Upper side of twig of dawn redwood (*Metasequoia glyptostroboides*) with current year's deciduous branchlets, some with immature pollen cones in the leaf axils, on a shoot of the previous year, ×0.5.

Upper side of twig of dawn redwood (*Metasequoia glyptostroboides*) with slightly immature seed cones replacing photosynthetic lateral shoots, ×0.5.

Twig of dawn redwood (*Metasequoia glyptostroboides*) with current season's growth, including immature seed cones on a shoot of the previous year; scale, 5 cm.

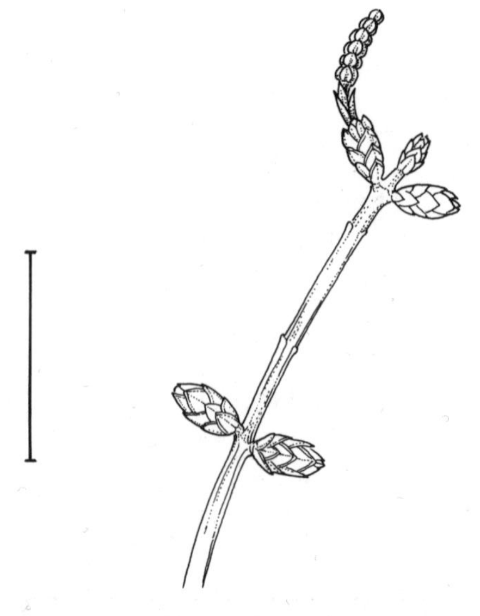

Dormant winter twig of dawn redwood (*Metasequoia glyptostroboides*) with a newly emerging seed cone at the time of pollination; scale, 1 cm.

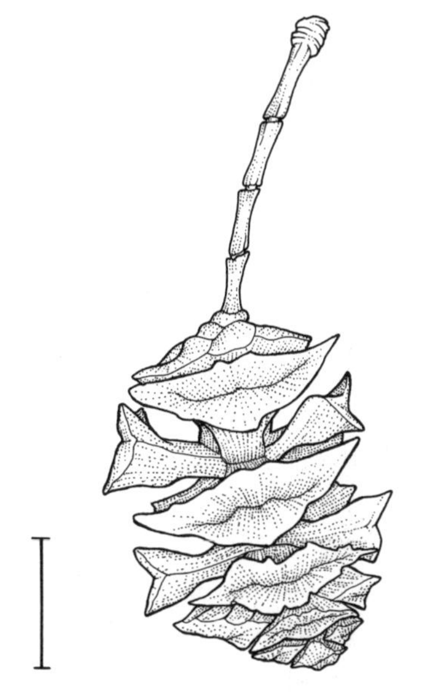

Open mature seed cone of dawn redwood (*Metasequoia glyptostroboides*) after seed dispersal; scale, 1 cm.

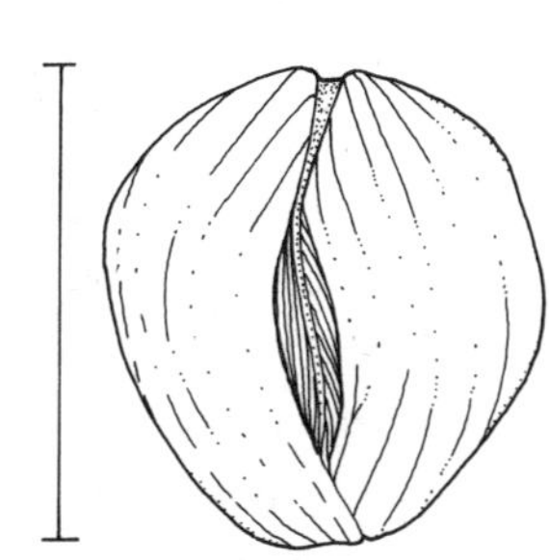

Face view of dispersed seed of dawn redwood (*Metasequoia glyptostroboides*); scale, 5 mm.

The "discovery tree" of dawn redwood (*Metasequoia glyptostroboides*) in Modaoxi, Hubei province, from which the type specimen was collected.

narrow ridges between shallow furrows. Crown fairly dense, conical at first and for many years, with numerous slender horizontal branches, a few of which become major upswept limbs of an ultimately broadly domed or flat crown. Deciduous short shoots mostly 10–15 cm long, with 25–35 pairs of leaves spreading distichously out to the sides, emerging from winter buds whose scales remain at the base of the shoots until they fall. Leaves longest through the middle two-thirds of the short shoot, markedly shorter at the base and tip, light green when new, turning reddish brown before falling in autumn. Leaves of middle portion of short shoot straight or slightly curved back, substantially parallel-sided for much of their length, tapering abruptly to the broadly pointed tip and the rounded base, 8–15 mm long (to 30 mm in juveniles), 1.2–2(–2.5) mm wide. Pollen cones about 5 mm long, in opposite pairs, with 7–10 pairs of pollen scales that become displaced from their opposite arrangement with growth. Seed cones with 8–12(–14) pairs of cone scales, (1.5–)2–3 cm long, 1–2.5 cm across, on stalks 1–7 cm long. Face of cone scales with a distinct notch all the way across between the bract portion and the seed scale portion, mostly 8–15 mm wide, 2–4 mm high. Seeds about 5 mm long, the two slightly longer wings each about twice as wide as the seed body. Central China, mostly in southwestern Hubei province. Streamsides at (750–)800–1,300(–1,500) m. Zone 5.

The discovery of a living species of *Metasequoia* in a remote region of China at the end of the Second World War, just a few years after its recognition in the fossil record, initiated a flurry of botanical and popular attention that has not since been matched, even by conifer finds that are scientifically as interesting. (Media coverage of the discovery of Wollemi pine, *Wollemia nobilis,* in 1994 was just as intense but did not last anywhere near as long before it faded from the headlines.) There was an atmosphere of truly international cooperation surrounding the discovery, even at this very difficult time, leading to a very openhanded distribution of seeds to botanical gardens in North American, Europe, and Asia. As a result, dawn redwood, which is, in addition, a widely adaptable, fast-growing, and attractive ornamental, became botanically well known in a very short time, with numerous studies of many aspects of its biology and cultivation. Hundreds of technical and popular articles have been dedicated to it and more appear each year.

The forester who first collected leafless specimens thought they might belong to the related genus *Glyptostrobus,* hence the species name. Originally discovered as a single, large, protected tree in a village south of the Changjiang (Yangtze River) this proved to be one of only about a dozen outlying trees outside the confines of what is usually referred to as the Shui Shan Valley (now the Xiaohe Commune), a closed valley about 25 km long in which about 2,000 mature individuals occur. Here, in its only native region, natural regeneration has essentially ceased since the discovery, with increasing human population leading to more intensive cultivation and deforestation. All of the trees are protected, and there is a permanent staff of conservation foresters in the valley to monitor the health of the population, but the accompanying vegetation that once sheltered young seedlings has disappeared. The few scattered giants that dot the valley floor here and in the outlying localities are estimated to be about 400–450 years old. They are still vigorous and generally succumb to accident rather than age.

Dawn redwoods in cultivation, like related giant sequoias (*Sequoiadendron giganteum*), are very fast growing, and some exceed 35 m in height, even though the oldest were planted out in 1949 and 1950. They start to produce their attractive seed cones at the age of about 10–15 years, but their seeds are generally sterile until the trees are old enough to also produce pollen cones, about 25–30 years. The trunks taper quickly above the buttressed base. One of the characteristic attractive features of the young adult trees is the way the narrow buttresses (or flutes) come together to form an arch over the indented insertions of the slender side branches. Some cultivars were selected for short branches and narrow growth habits in an attempt to allow homeowners to plant this attractive tree, which is otherwise too large for an ordinary house garden.

Microbiota V. Komarov

MICROBIOTA, SUBALPINE ARBORVITAE

Cupressaceae

Evergreen shrubs with smooth bark, densely branched with shallowly arching branches. Branchlets alternate, partially arranged in flattened, fernlike sprays. Without definite winter buds. Juvenile and adult leaves in alternating, overlapping pairs. Juvenile leaves needlelike, sometimes found on adult plants beneath the seed cones. Adult leaves scalelike, the facial and lateral pairs similar, with an oblong gland but without prominent stomata on the outer face, the pointed tip pressed against the twig or much elongated and standing out on shaded branchlets. Free leaf tips with a multicellular frill along the outer edges, which helps enclose the stomatal zone against the twig.

Plants monoecious or seemingly dioecious because of irregular production of cones of one or other sex. Pollen cones single at the tips of branchlets, with three or four crisscross pairs of bracts at the base and five or six crisscross pairs of functional pollen scales. Each heart-shaped scale bearing two large pollen sacs at the base, beneath the short stalk. Pollen grains small (30–35 μm in diameter), spherical, minutely bumpy but otherwise nearly featureless. Seed cones single at the tips of short branchlets, spherical, with one or two alternating pairs of very thin woody scales, maturing in a single season. Each cone scale with a prickle near the tip representing the free tip of the otherwise completely fused bract, spreading widely at maturity. Only one scale fertile and bearing a single nearly spherical to oval, wingless seed with a conspicuous short, straight beak at the tip. Cotyledons two, each with one vein. Chromosome base number $x = 11$.

Wood light and soft, with a narrow band of whitish sapwood sharply contrasting with the reddish brown heartwood. Grain fine and even to somewhat wavy, with well-defined growth rings marked by an abrupt transition to a narrow band of smaller and thicker-walled latewood tracheids. Resin canals absent but with sparse individual resin parenchyma cells scattered through the growth increment singly or in small clusters.

Stomates confined to broad bands covering most of the inner (upper) face of both kinds of leaves, the whole face strongly papillate. Each stomate with guard cells tucked beneath the (four to) six to eight surrounding subsidiary cells and topped by a high, steep, often at least partially interrupted Florin ring. Leaf cross section with a single midvein above a single large resin canal (neither of which may enter far into the free tip, if at all) and flanked by wedges of transfusion tissue. Photosynthetic tissue with a palisade layer around the outer (lower) face beneath the epidermis and accompanying, continuous hypodermis (which becomes progressively thicker up into the free leaf tip), the remainder occupied by spongy mesophyll extending to the stomatal zones. Palisade tissue thicker and more prominent in the lateral leaves and in the facial leaves on the upward-facing side of the branchlets than in the downward-facing facial leaves, where it may be weakly and discontinuously developed, especially in the portion attached to the twig.

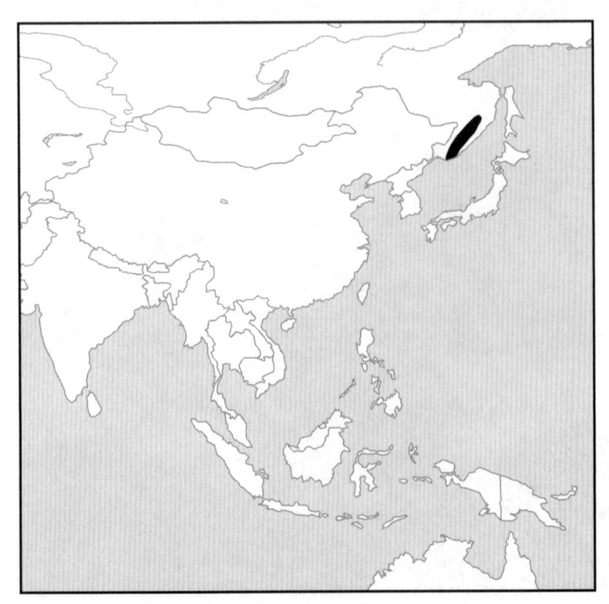

Distribution of *Microbiota decussata* in the Russian Far East, scale 1 : 120 million

One species in extreme southeastern Russia. References: Farjon 2005b, Komarov 1986.

Although some liken *Microbiota* to a *Juniperus* with seed cones that open, it is really much closer in morphology and DNA structure to *Platycladus.* As reflected in its name (Greek for "little" *Biota,* the latter a synonym of *Platycladus*), *Microbiota* is much like a miniaturized version of *Platycladus* with its shrubby habit and tiny cones. *Microbiota decussata* grows as a neat, circular, mound and is increasingly chosen in place of some spreading junipers. Despite this, there has been almost no cultivar selection, most of the plants on the market being a single clone. The foliage is not particularly distinctive and the seed cones are not very durable, so it is not surprising that the genus has no known fossil record.

Microbiota decussata V. Komarov

SUBALPINE ARBORVITAE, MICROBIOTA, DWARF ARBORVITAE

Spreading shrub to 0.5(–1) m tall, 1.5 m or more across, the original trunk very short. Bark remaining smooth by flaking, bright reddish brown at first, darkening through purplish brown to dark grayish brown. Crown with numerous, gently arching, slender branches densely clothed with branchlets, forming a shallow dome. Branchlets cylindrical, attached alternately out to either side of their parent twigs to form largely two-dimensional sprays. Leaves about 2 mm long, yellowish green, turning purplish brown in winter, with a prominent pointed tip spreading conspicuously away from the twig or pressed forward against it and overlapping the base of the next leaf along it. Pollen cones oblong, 2–3 mm long, with five or six pairs of pollen scales, straw-colored. Seed cones 4–6 mm long, with two pairs of very thin cone scales surrounding the single seed. Seeds 2–3 mm long, egg-shaped, smooth, and wingless but with an evident, tiny (micropylar) beak. Sikhote-Alin district, southeastern Russian Far East. Alpine and subalpine shrublands and forests; (340–)1,000–1,500(–3,000) m. Zone 2.

Although highly restricted in the wild, being known on only a few "bald" mountains above the timberline or within the adjacent subalpine forest, this species is quite adaptable in cultivation and has become an established alternative to creeping junipers, like *Juniperus horizontalis,* which it closely resembles in general appearance. The striking winter bronzing adds a further decorative touch to a well-proportioned ground cover. It is not tolerant

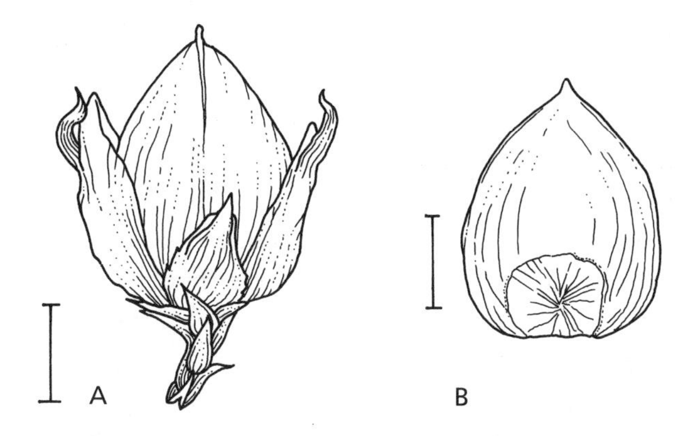

Thin-textured open mature seed cone with single seed (A) and dispersed seed (B) of subalpine arborvitae (*Microbiota decussata*); scale, 1 mm.

Upper side of flattened branch spray of subalpine arborvitae (*Microbiota decussata*) with 2 years of growth and both alternate and opposite branching, ×0.5.

Spreading mature shrub of subalpine arborvitae (*Microbiota decussata*) in nature.

of drought and so must be kept well watered. Individuals with spreading leaf tips have a much rougher (and prickly!) appearance than those in which they are tight to the twigs. Pollen and seed cones are scarcely seen in cultivation.

Microcachrys J. Hooker

MICROCACHRYS

Podocarpaceae

Evergreen shrubs without any upright main trunks. Bark fibrous, flaking in scales and then peeling in long strips and becoming shallowly furrowed with age. Juvenile and adult foliage similar. Branchlets weakly differentiated into long shoots of main axes and primarily photosynthetic short shoots, hairless, completely hidden by the leaves until thickening growth brings them into view. Resting buds not well developed, consisting solely of young, unexpanded foliage leaves. Leaves attached in pairs, each pair at right angles to the preceding one, scalelike, overlapping, closely hugging the twigs, edged by a thin, toothed, papery fringe.

Plants dioecious (rarely monoecious?). Pollen cones single at the tips of otherwise ordinary foliage shoots, egg-shaped to cylindrical. Each pollen cone, with a few bracts at the base and with (12–)16–24(–28) seemingly spirally arranged (by crowding of alternating whorls of 4) pollen scales. Each scale roundly and broadly triangular, with two pollen sacs. Pollen grains small (body 25–40 µm long, 30–45 µm overall), with (two or) three (to six) round, nearly smooth air bladders, extending a little beyond and partially tucked under the conspicuously larger, almost featurelessly smooth, squashed-ball-shaped body. Seed cones single at the tips of short, otherwise ordinarily foliage branchlets that may alternate in close proximity to one another along a slightly older shoot, somewhat fleshy but recognizably conelike in appearance. Cones with (12–)16–24(–28) seemingly spirally arranged (by crowding of alternating whorls of four) juicy fertile bracts, which do not unite into a podocarpium. Each fertile bract bears one seed near its tip that is pointed back down into the cone axis and is nestled in an asymmetrical, cup-shaped, fleshy seed scale (epimatium). Seed without an extra aril, hard, ending in a tiny, dimpled beak, maturing in two seasons. Cotyledons two, each with two veins. Chromosome base number $x = 15$.

Wood moderately dense and hard, slightly fragrant, waxy, the off-white sapwood weakly contrasting with the pale brown heartwood. Grain very fine and even, with well-defined, narrow growth rings marked by a gradual transition to a very narrow band of slightly darker latewood. Resin canals absent but sometimes with scattered individual resin cells.

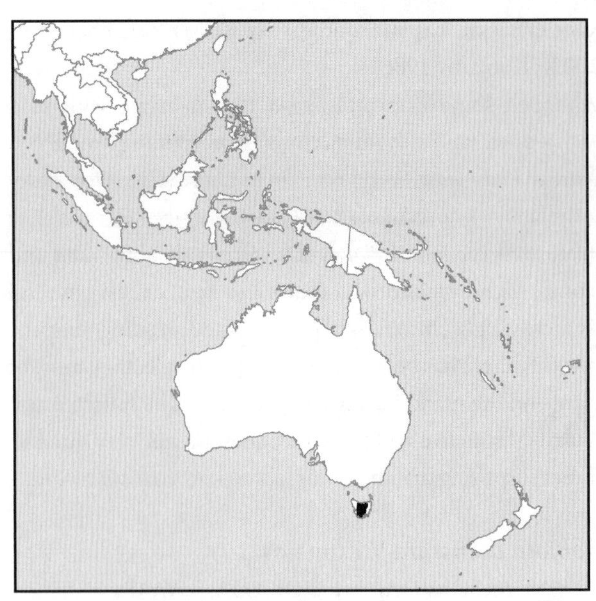

Distribution of *Microcachrys tetragona* in Tasmania, scale 1 : 119 million

Without stomates on the outer face but with densely packed parallel lines of stomates on the inner face except over the midvein. Each stomate sunken below the level of but visible between the four (to six) surrounding subsidiary cells, which are topped by a prominent, continuous, sharply raised Florin ring. Midvein single, slightly grooved above, not evident beneath, with one large resin canal immediately below it flanked by small wedges of transfusion tissue. Photosynthetic tissue with a palisade layer on the lower (outer) side adjacent to a layer of hypodermis just inside the epidermis.

One species in Tasmania, Australia. References: Cookson and Pike 1954, Curtis 1956, K. Hill 1998, R. Hill and Brodribb 1999. Synonym: *Pherosphaera* W. Archer.

Although with its dense, dark green foliage and bright red seed cones, potentially a handsome subject for rock gardens and as a substitute for creeping junipers in moderate climates, strawberry pine (*Microcachrys tetragona*) is little cultivated, and there has been no cultivar selection. The genus name, Greek for "little catkin," refers to the pollen and seed cones, small in comparison to those of the unrelated Tasmanian cedars of the genus *Athrotaxis* (Cupressaceae), in which strawberry pine was placed when in was first described. Based on both morphological features and DNA studies, *Microcachrys* is clearly related to *Microstrobos*, another Australian genus, but the relationships of these two genera to other podocarps are less clear. In phylogenetic trees of the family, they branch near the base of the group of genera including both *Dacrydium* and *Podocarpus*. Sharing this near basal position with them are *Saxegothaea* of Chile and *Acmopyle* of New Caledonia and Fiji, two genera with little superficial resemblance to *Microcachrys* and *Microstrobos*. The relative positions of these three phylogenetic branches (considering *Microcachrys* and *Microstrobos* as a single branch) in DNA studies vary with the exact genes sampled. Because the arrangement of branches near the base of a phylogenetic tree is important in assessing the pattern of evolution of characters (like scale leaves versus needle leaves), further study of this aspect of podocarp relationships is warranted.

Unfortunately, the fossil record of *Microcachrys* shoots, which might shed some light on character evolution, is very sparse. Outside of relatively recent specimens from Tasmanian ice-age deposits, less than 2 million years old, there is just a single shoot from Miocene coals (10–20 million years old) near Melbourne, Victoria, Australia. Sparse as they may be, these fossils are nonetheless interesting because some come from lowland vegetation while the sole living species, strawberry pine (*M. tetragona*), presently inhabits the subalpine zone. In contrast to this meager macrofossil record, pollen with mostly three air bladders closely resembling the distinctive pollen of the living strawberry pine is known from sediments as old as the Jurassic, more than 150 million years old, and from localities all across southern Australia, in New Zealand, Kerguelen, and possibly in India, disappearing from New Zealand and most of mainland Australia after the Oligocene, some 23 million years ago. This extended fossil record is in keeping with the relatively basal position of strawberry pine in phylogenies of the Podocarpaceae, implying that its ancestors diverged from other podocarps fairly early in the history of the family.

Microcachrys tetragona J. Hooker

STRAWBERRY PINE, CREEPING PINE

Low shrub, spreading to 1(–2) m, though usually smaller. Bark yellowish brown to purplish brown, weathering brownish gray to dark brown. Crown flat, with crooked creeping branches hugging the ground and bearing numerous branchlets completely clothed with foliage. Branchlets horizontal to upright, partially arranged in comblike sprays, squared off or more rounded in cross section. Leaves bright green throughout or with a yellowish green base and a waxy white, rounded keel, 0.8–1.5 mm long (to 3 mm on main shoots), lasting 3 years or more. Pollen cones with yellow pollen sacs on yellowish brown scales, (2–)3–4.5 mm long, 1.5–2.5 mm in diameter. Seed cones bright red, nearly spherical or a little elongate, (2.5–)4–6(–8) mm long and 2.5–4(–6) mm in diameter, each seed enclosed for about a quarter of its length or more in the thin, red epimatium. Seeds black, somewhat triangular in cross section, smooth, about 1–2 mm long. Widespread in the western two-thirds of Tasmania, Australia. Often dominant in subalpine shrublands on substrates ranging from dry rocky fell fields to sodden peat lands; 1,000–1,350(–1,500) m. Zone 8. Synonym: *Pherosphaera hookeriana* W. Archer.

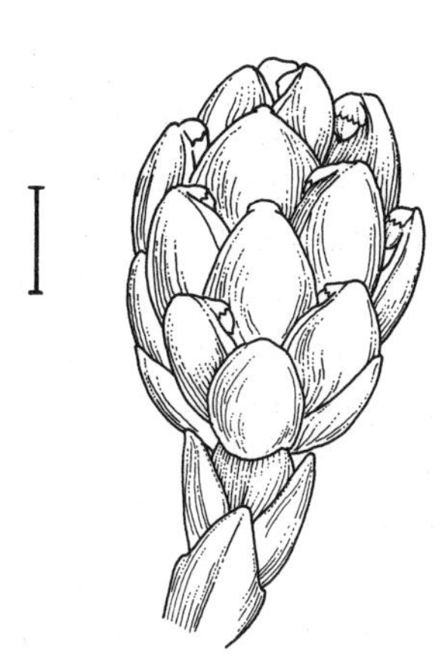

Fleshy mature seed cone of strawberry pine (*Microcachrys tetragona*) with many bracts and seeds; scale, 1 mm.

Strawberry pine is unique among living scale-leaved podocarps in having leaves arranged in alternating pairs lined up in four straight rows (hence the scientific name, Greek for "four-angled"), a contrast to the spiral leaf arrangement of species of *Athrotaxis*, the genus to which strawberry pine was assigned when it was first described. In this it resembles its neighbor Cheshunt pine (*Diselma archeri*) of the Cupressaceae more than it does another neighbor and close relative, the spirally scale-leaved Tasmanian dwarf pine (*Microstrobos niphophilus*). All three are characteristic species of subalpine heathlands and were much confused in early botanical literature. Strawberry pine is the most common species among them and the only one that is a spreader, the other two being upright shrubs. Although the branches are almost always ground hugging in nature, plants in cultivation are sometimes trained upright, and such branches reach a length of 3 m, much longer than their usual natural spread. The seed cones are very handsome when fresh, reminiscent of strawberries as the common name suggests, and a heavily coning shrub is a striking garden subject. Unfortunately, strawberry pine only thrives in a moderate, temperate climate with year round moisture.

Microstrobos J. Garden & L. A. Johnson

MICROSTROBOS

Podocarpaceae

Evergreen shrubs of varying habit and crown form, without a single main trunk. Bark obscurely fibrous, scaly, finally becoming somewhat furrowed. Branchlets all elongate, without distinction into long and short shoots, hairless, grooved between the attached leaf bases and remaining green for the first year but generally hidden by the foliage. Resting buds indistinct, consisting solely of as yet unexpanded ordinary foliage leaves. Leaves spirally attached. Juvenile and adult foliage similar. Leaves of both types scalelike or needlelike, a little flattened top to bottom, rounded to roundly angled on the outer face, standing out from or pressed against the twigs, with a very narrow, thin, papery margin.

Plants dioecious. Pollen cones single at the tips of otherwise ordinary foliage shoots. Each pollen cone spherical to egg-shaped, without distinctive bracts at the base and with about 8–25 spirally arranged, rounded pollen scales, each scale with two pollen sacs. Pollen grains small (body 25–40 µm long, 30–45 µm overall), with three round, smooth and inconspicuously internally sculptured air bladders, which extend a little beyond and are tucked under and touch each other beneath the conspicuously larger, essentially smooth, somewhat squashed spherical body. Seed cones single at the tips of very short, initially conspicuously down-curved shoots bearing ordinary foliage leaves, which later straighten and right the cones. Each seed cone somewhat reduced and modified but still conelike, with about 6–16 fairly tightly spirally arranged bracts that do not swell and unite to form a podocarpium. Middle two to eight bracts fertile, each bearing one upright seed that is apparently inserted directly on the bract without any obvious seed scale (epimatium). Seed hard, a little flattened, with three ribs, ending in a broad, cup-shaped, straight beak, maturing in a single season. Cotyledons two, each with two veins. Chromosome base number $x = 13$.

Wood dense, the narrow, light brown sapwood sharply contrasting with the darker heartwood. Grain very fine and even, with evident growth rings marked by a fairly abrupt transition to a narrow band of darker latewood. Resin canals absent but with a few individual resin cells.

Inner (upper) face of leaves with many lines of stomates making up a single, broad stomatal band. Each stomate sunken beneath and largely hidden by the four (to six) subsidiary cells, which are topped by a prominently raised, continuous or interrupted Florin ring and plugged with wax. Midvein single, inconspicuously buried in the leaf tissue of the scale leaves or raised as a midrib in the needlelike leaves, with one resin canal immediately beneath it and small wedges of transfusion tissue at the sides. Photosynthetic tissue with a palisade layer on the outer (lower) face of the leaf beneath the epidermis and accompanying hypodermis, the remaining tissue consisting of spongy mesophyll reaching the stomates.

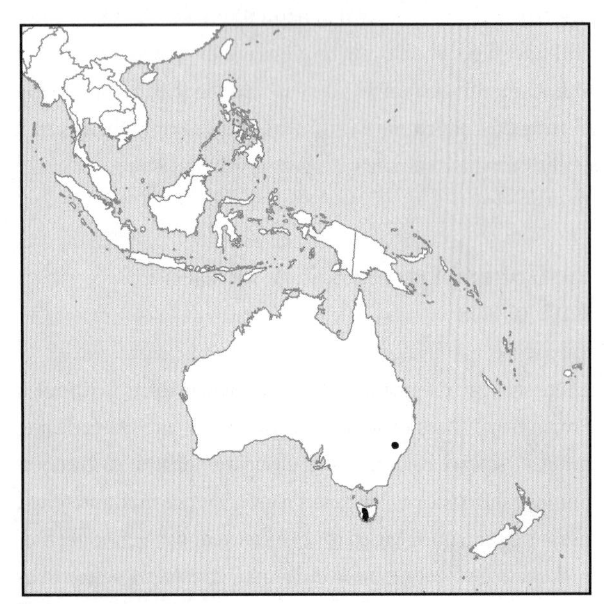
Distribution of *Microstrobos* in southeastern Australia, scale 1 : 119 million

Two species in southeastern Australia. References: Curtis 1956, Harden and Thompson, 1990, K. Hill 1998, R. Hill and Brodribb 1999. Synonym: *Pherosphaera* W. Archer, misapplied.

The three scale-leaved, shrubby subalpine conifer species of Tasmania, belonging to three different genera in two families (Podocarpaceae and Cupressaceae), are so similar in general appearance that it is hardly surprising that they were confused with one another when they were first described botanically in the mid-19th century based on just a handful of specimens. For a century more, the two species of *Microstrobos* were mistakenly referred to as *Pherosphaera,* mistakenly because the type specimens of the latter name are actually samples of strawberry pine (*Microcachrys tetragona*), another of the confusing Tasmanian species. Thus, although the name *Pherosphaera* was universally used for the plants now referred to *Microstrobos* and was never applied to strawberry pine, it is formally a synonym of *Microcachrys* rather than of *Microstrobos.* This interpretation of the names, first advanced in 1951, was challenged in 2005 by a careful but not unequivocal analysis of wording in the original literature. Whether we return to use of the name *Pherosphaera* for this genus must await decisions by the nomenclatural section of an International Botanical Congress. Like the closely related *Microcachrys, Microstrobos* bears fairly conelike seed cones, unlike the highly modified and reduced, variously fleshy, bird-dispersed ones characteristic of most podocarps. Nonetheless, the seed cones are rather small compared to those of many conifers in other families, especially Araucariaceae in the southern hemisphere or Pinaceae in the northern (hence the scientific name, Greek for "small cone"). The two extant species of *Microstrobos* are sparingly cultivated, primarily in botanical gardens, and there has been no cultivar selection.

Since pollen grains of *Microstrobos* can be difficult to distinguish from those of *Microcachrys,* the fossil pollen record of the genus is somewhat ambiguous. Fossil twigs of the two genera, in contrast, are readily distinguishable, and a somewhat meager fossil record of shoots of *Microstrobos* has been discovered in Tasmanian sediments dating back to the Eocene, some 50 million years ago. Some of these fossils occur with assemblages of plants that imply much warmer conditions than any now occupied by the extant Tasman dwarf pine (*Microstrobos niphophilus*).

Identification Guide to *Microstrobos*

The two species of *Microstrobos* are readily distinguished by the length of most adult leaves and whether they are closely pressed against the twigs or spread away from them:

Microstrobos niphophilus, leaves less than 2 mm, tight against twigs

M. fitzgeraldii, leaves more than 2 mm, spreading from twigs

Microstrobos fitzgeraldii (F. J. Mueller) J. Garden & L. A. Johnson

CASCADE PINE

Low shrub to 1 m high with no main trunk, spreading to 2 m, creeping and falling over rocks or with upturned tips or sometimes with erect branches. Bark reddish brown, weathering gray. Crown dense, with profusely branched, stems bearing numerous, spirally attached, flexible, drooping branchlets densely clothed with foliage. Leaves needlelike, spreading away from the twig at a forward angle and then bending in near the tip, dark green and with an angular keel on the outer (lower) face and waxy greenish white on the stomatal band of the inner (upper) face, 2.5–3.5 mm long, lasting 3–6 years or more. Pollen cones red, held upright, spherical to egg-shaped, 5–7 mm long, 3–5 mm in diameter, with 10–15(–25) pollen scales. Seed cones pinkish red, spherical to egg-shaped, (2–)3–4.5 mm long, (1.5–)2–2.5 mm thick, with four to eight fertile scales. Seeds dark brown, 1–2 mm long. With a very limited range in the Blue Mountains about 100 km west of Sydney, New South Wales, Australia. Largely confined to sandstone cliffs and caves within the mist zone of waterfalls; 700–900(–1,000) m. Zone 8. Synonym: *Pherosphaera fitzgeraldii* (F. J. Mueller) F. J. Mueller.

Mature shrub of cascade pine (*Microstrobos fitzgeraldii*) in nature on its preferred substrate of wet rocks.

Cascade pine is a very rare species from a habitat rarely favored by conifers. It is restricted to the concentration of waterfalls in the vicinity of Katoomba, where it dangles over rock ledges on south-facing cliffs. With this kind of very narrow ecological niche, it is hard to imagine how the species can disperse to new sites in the semiarid world of present day Australia, and its rarity is not surprising. Despite the extreme rarity of cascade pine, the pioneers of chemotaxonomy in Australia (and the world), Baker and Smith, during their work for *A Research on the Pines of Australia* (1910), managed to collect almost 70 kg of branchlets in their efforts to extract and identify the oils present in the leaves. That represents a lot of shorn plants, which may or may not have recovered from the stress. Luckily, there have been considerable improvements in both analytical techniques and conservation sensibilities since then. The species name honors Robert Fitzgerald (1830–1892), Irish-born Australian surveyor (rising to the post of Deputy Surveyor-General for New South Wales) who collected the type specimen while searching for orchids, his real passion, about which he corresponded with Darwin and wrote a two-volume set, *Australian Orchids.*

Microstrobos niphophilus J. Garden & L. A. Johnson

TASMANIAN DWARF PINE, DROOPING PINE, MOUNT MAWSON PINE

Dense shrub to 2.5 m tall and spreading to 5 m, without a main trunk. Bark reddish brown, weathering gray. Crown broadly dome-shaped from the ground, with many, densely branched upright, slender branches to about 8 cm in diameter bearing numerous short, spirally attached stiff branchlets densely clothed with foliage and forming clumps at the periphery of the bush. Leaves scalelike, tight to the twig, dark green or with a coppery flush, prominently keeled, (0.5–)1–1.5 mm long (to 2.5 mm on main shoots), lasting 3–5 years or more. Pollen cones brown with a red or purple blush, held upright, spherical to egg-shaped, 1.5–2 mm long, about 1.5 mm in diameter, with about 8–12 pollen scales. Seed cones remaining turned down after pollination or turning upright with maturity, dull brownish red, spherical to egg-shaped, 2–4 mm long, 2–3 mm thick, with (two or) three to five (to eight) fertile scales. Seeds shiny dark brown, 1–1.5 mm long. Mountains of central, southern, and western Tasmania, Australia. Scattered to gregarious in wet soils of subalpine habitats (hence the scientific name, Greek for "snow loving") of varying shrub density; 1,000–1,350(–1,500) m. Zone 8. Synonym: *Pherosphaera hookeriana* W. Archer, misapplied.

Tasmanian dwarf pine is rather infrequently encountered, but where it does occur, it can dominate or share dominance with other subalpine conifers (*Athrotaxis, Diselma* and *Microcachrys*). Although branches are slender, growth is so slow that a branch 8 cm in diameter may be 150 years old or more, and even the fastest-growing ones are likely more than 100 years. Tasmanian dwarf pine differs vegetatively from both of its very similar neighbors, Cheshunt pine (*Diselma archeri*) and strawberry pine (*Microcachrys tetragona*), in its spiral leaf arrangement, while their leaves are in opposite pairs. Pencil pine (*Athrotaxis cupressoides*), which also has spirally arranged scale leaves, has much coarser twigs than the 1–1.5 mm diameter twigs of Tasmanian dwarf pine. Of course, the seed cones of all of these vegetatively similar shrubs are quite distinct from one another. Still, Tasmanian dwarf pine generally looks more like these neighbors than it does like its closest relative, the only other species of *Microstrobos,* cascade pine (*M. fitzgeraldii*) from New South Wales.

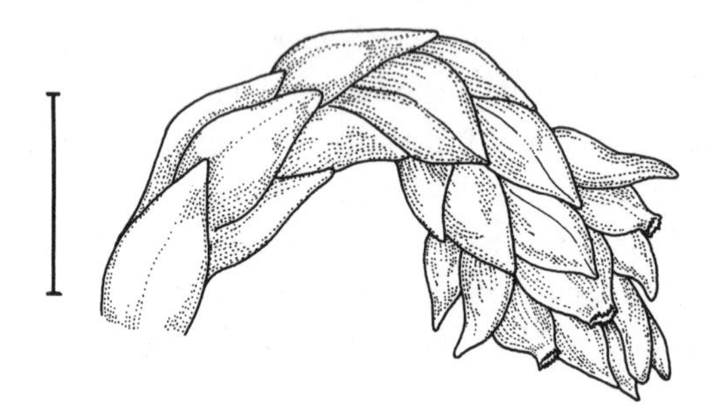

Nearly mature seed cone of Tasmanian dwarf pine (*Microstrobos niphophilus*); scale, 1 mm.

Large subalpine patch of Tasmanian dwarf pine (*Microstrobos niphophilus*) in boggy soil.

Nageia J. Gaertner

NAGI

Podocarpaceae

Small to large evergreen trees, with trunk single and straight or variously forked and crooked in unfavorable habitats. Bark fibrous, peeling in flakes to present a mottled, smooth surface. Crown narrow, forming an irregular ball or cylinder constructed initially of horizontal tiers of branches densely or more openly clothed with foliage. Branchlets all elongate and bearing foliage leaves, without distinction into long and short shoots, hairless, remaining green for at least the first year, more or less grooved between the very elongate attached leaf bases. Resting buds well developed, tightly wrapped by elongate green bud scales. Leaves evenly spaced along the twigs, in alternating pairs but aligned in two rows within a plane by twisting of the petioles and of the branchlets between the leaves, all leaves on one side of the twig facing up while those on the other face down as the twig lies flat, the members of a pair not always lined up perfectly across the twig. Individual leaves broad and multiveined, variably egg-shaped in outline, rather thick and leathery.

Plants monoecious. Pollen cones cylindrical, single or in clusters of up to seven at and near the end of a short, leafless stalk in the axil of a foliage leaf. Each cone with a few alternating pairs of sterile scales at the base and with numerous, densely crowded pollen scales. Each scale bearing two pollen sacs that are large compared to the somewhat heart-shaped blade. Pollen grains medium to large (body 40–45 µm long, 65–75 µm overall), with two kidney-shaped air bladders that are about the same size as or a little smaller than the oval, minutely bumpy body, the bladders with coarser squiggly ridges. Seed cones single or in clusters of two to five on short, branched or unbranched scaly stalks in the axils of foliage leaves, highly modified and reduced. Each cone with several alternating pairs of bracts on the cone axis, these either scaly and unmodified or (in two species) becoming united, fleshy, and swollen (the podocarpium). With a single, plump, unwinged seed embedded in an inverted position in each of one or two (or three) fleshy, green to black seed scales (the epimatium) in the axils of the uppermost bracts, maturing and falling in a single season. Seeds nearly spherical, without a crest or beak. Cotyledons two, each with two veins. Chromosome base numbers $x = 10$ and 13.

Wood moderately hard and heavy, fragrant, decay resistant in soil and water, yellowish brown to gray, with a core of darker heartwood. Grain very fine and even, often with little distinction between early and latewood but sometimes with well-defined growth rings delimited by a narrow band of darker latewood. Resin canals absent.

Stomates numerous, forming many incomplete parallel lines either on both surfaces or just on the lower side. Each stomate sunken beneath and largely hidden by the four (to six) surrounding subsidiary cells, which support a well-developed, complete, nearly circular Florin ring that appears higher because it is surrounded by a depressed moat. Leaf cross section with one resin canal beneath each of the parallel veins, which are sometimes flanked by wedges of weakly developed transfusion tissue. Photosynthetic tissue either with a well-developed palisade layer on one side beneath the epidermis and adjacent hypodermis or with a more poorly developed palisade on both faces.

Five species in southern and eastern Asia and throughout Malesia, from southern and northeastern India (Kerala and Assam), the southern half of China, and southern Japan to the eastern tip of New Guinea. References: de Laubenfels 1969, 1987, 1988, D. Z. Fu 1992, L. K. Fu 1992, Fu et al. 1999f, Hill and Pole 1992, Li and Keng 1994, Mill 1999a, 2001, T. H. Nguyên and Vidal 1996, Page 1988, K. C. Sahni 1990, Yamazaki 1995. Synonyms: *Decussocarpus* sect. *Dammaroides* (G. Bennett ex Horsfield) de Laubenfels, *Podocarpus* sect. *Dammaroides* G. Bennett ex Horsfield, *Podocarpus* sect. *Nageia* (J. Gaertner) Endlicher.

Nageia (the name references the Japanese name for one of its species, nagi, *N. nagi*) is the only genus of Podocarpaceae with multiveined leaves, which resemble those of the unrelated kauris (*Agathis*, Araucariaceae) closely enough that the two may be confused, especially since some species of the two genera have overlapping distributions. Although the leaves of the two genera are both in pairs and are often similar in shape and texture, the veins in leaves of *Nageia* converge at the tip, while those of *Agathis* continue nearly parallel to the tip. The resting buds of leafy specimens also help to distinguish the two genera, since those of *Nageia* are pointy with narrow, pointed bud scales while those of *Agathis* are rounded with broad, rounded bud scales. Of course, if specimens are fertile, there is no possibility of confusion since the pollen cones of *Agathis* are much larger than those of *Nageia* while the seed cones of the kauris are typical conifer cones with numerous flattened seed scales bearing winged seeds. Since *N. nagi* has a long history of cultivation, selection for various morphological traits (including leaf shape and growth habit) has taken place, though few such cultivars are available outside of eastern Asia.

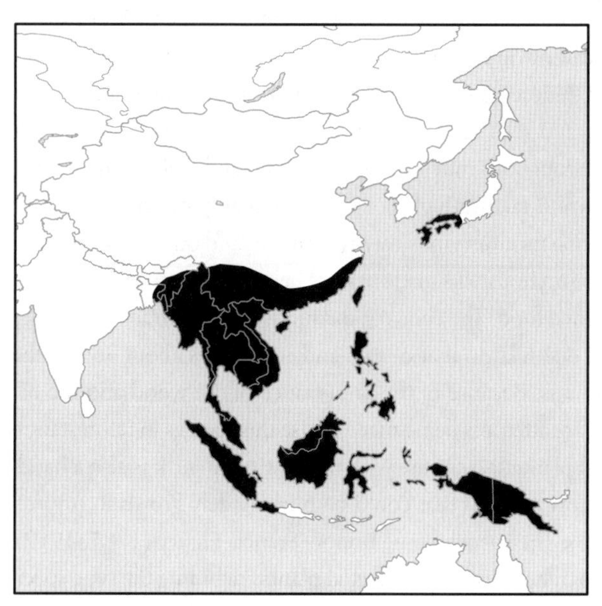

Distribution of *Nageia* in southeastern Asia and Malesia, scale 1 : 120 million

Nageia is closely related to *Afrocarpus* and *Retrophyllum*, as discussed under the latter, and beyond these two genera to *Podocarpus*, in which the *Nageia* species were long included. The closeness of these relationships, based on traditional characters as well as on DNA studies, makes it untenable to separate *Nageia* from these genera into a family of its own as has been proposed by some recent authors.

Other authors recognize more species in the genus than are accepted here, subdividing *Nageia nagi* into two to three species or more. The variability expressed by this species in cultivation makes it seem unwise to subdivide it without further study. There would, in fact, be something to be said for the opposite viewpoint of reducing the number of species accepted, with *N. fleuryi* and *N. nagi* on the one hand and *N. motleyi* and *N. wallichiana* on the other, differing in little but size of various parts. In the absence of additional evidence, the five species accepted here seem reasonably defensible. Likewise, attempts to apportion these few species between two botanical sections, based on the presence or absence of a swollen podocarpium, break down because of the intergradation of other characters, with *N. maxima* making a bridge between the two groups.

Unlike many other genera of podocarps (and conifers in general), there is no recognized fossil record of *Nageia*, so nothing is known of its Tertiary distribution and history. Some described broad-leaved podocarp fossils might belong to this genus but they are not well enough preserved for a definitive assessment of their affinities.

Identification Guide to *Nageia*

The five species of *Nageia* may be distinguished by the length of most leaves on adult trees and by whether lines of stomates are confined to the lower surface or are found on both sides, further confirmed by the presence or absence of a fleshy, swollen podocarpium and by the length and width of most mature pollen cones:

> *Nageia nagi,* leaves up to 8.5 cm, stomatal lines on lower side, podocarpium absent, pollen cones up to 2 cm by 3.5–5 mm
>
> *N. motleyi,* leaves less than 8.5 cm, stomatal lines on both sides, podocarpium present, pollen cones up to 2 cm by at least 5 mm
>
> *N. fleuryi,* leaves 8.5–17 cm, stomatal lines on lower side, podocarpium absent, pollen cones more than 2 cm by 3.5–5 mm
>
> *N. wallichiana,* leaves 8.5–17 cm, stomatal lines on both sides, podocarpium present, pollen cones up to 2 cm by 3.5–5 mm
>
> *N. maxima,* leaves at least 17 cm, stomatal lines on both sides, podocarpium absent, pollen cones up to 2 cm by less than 3.5 mm

Young mature Vietnamese nagi (*Nageia fleuryi*) in nature.

Nageia fleuryi (P. Hickel) de Laubenfels

VIETNAMESE NAGI, KIM GIAO NÚI (VIETNAMESE), CHANG YE ZHU BAI (CHINESE)

Tree to 25(–30) m tall, with trunk to 0.7 m in diameter. Bark remaining smooth, thin, flaking to leave purplish brown patches on a grayish background. Crown thin and open, with many, slender, crooked branches somewhat openly clothed with foliage. Leaves shiny dark green above with lines of stomates only on the lower side, (8–)10–18 cm long, (2–)3–5 cm wide. Blade widest near or a little before the middle, clearly widest there or fairly parallel-sided for up to a third of their length, tapering fairly gradually to a narrowly triangular, long drawn-out tip and more abruptly to the broadly wedge-shaped base narrowing to a petiole (2–)5–7 (–10) mm long. Pollen cones in clusters of (one to) three to six on a very short stalk 2–5 mm long. Each cone (1.5–)2–5(–6.5) cm long by about 4 mm in diameter. Seed cones usually single on a stalk 1.5–2.5(–3) cm long, without a fleshy podocarpium. Combined seed coat and epimatium blue-black at maturity with a thin waxy film, 15–18(–20) mm in diameter. Scattered in southernmost China (southeastern Yunnan, Guangxi, Guangdong, and Hainan), northern Taiwan, and through Vietnam south to Lâm Dông province. Scattered among hardwoods, forming small groves, or rarely dominating stands in the canopy or subcanopy of tropical broad-leaved evergreen forests and montane rain forests, with best growth on deep, rich, well-drained soils; (200–)500–1,000(–1,500) m. Zone 9. Synonyms: *Decussocarpus fleuryi* (P. Hickel) de Laubenfels, *Podocarpus fleuryi* P. Hickel.

Vietnamese nagi is closely related to nagi (*Nageia nagi*) and may be thought of as its tropical counterpart. The two species differ most obviously in the much larger leaves of Vietnamese nagi (the Chinese name means long-leaf nagi), but the pollen cones, seeds, and seed cone stalk are all also longer in this species. It is rare throughout much of its range but can be abundant locally. The wood takes a high polish and is highly prized for interior trim, fine furniture, musical instruments, and other craft items. In Vietnam, chopsticks made of it are believed by some to neutralize poisoned food. The culinary and industrial oil extracted from the seeds is similar to that of nagi and constitutes about 36% of the seed mass. Because of the importance of the wood and the oil, there are attempts being made to use this species in reforestation and in plantation culture within its native range. It is also a handsome ornamental that could be more widely grown in tropical regions. The species name honors Francis Fleury (1887–1919), a French botanist who collected plants, including the type specimen, in Vietnam in the early 20th century.

Nageia maxima (de Laubenfels) de Laubenfels

GIANT-LEAF NAGI, LANDIN PAYA (MALAY)

Shrub, or tree to 10 m tall, with trunk to 0.2 m in diameter. Bark pale gray, smooth, flaky. Crown narrowly dome-shaped, with short, slender horizontal branches densely clothed with foliage. Leaves widely spaced on the twigs, shiny dark green above with many lines of stomates on the lower surface and only a few above, (8–)18–34 cm long, (3–)6–9(–10) cm wide. Blade widest near or a little below the middle, tapering gradually to the narrowly triangular, prolonged tip and more abruptly to the broadly and roundly wedge-shaped base on a short petiole to 10 mm long. Pollen cones in clusters of up to nine on a short stalk to 10 mm long. Each cone 1–2 cm long by 2.5–3 mm in diameter. Seed cones in groups of one to five on a stalk 1–1.5 cm long, without a fleshy podocarpium. Combined seed coat and epimatium blue-black at maturity with a thin waxy film, 16–18 mm in diameter. Known only from Sarawak in Borneo. Clumped in the understory of peaty swamp forests and wet rain forests; 0–120 m. Zone 10. Synonym: *Decussocarpus maximus* de Laubenfels.

Giant-leaf nagi is the rarest species in the genus and has one of the more restricted ranges among the conifers. Nonetheless, because of its small stature and the lack of sustained exploitation of its preferred swamp habitats, it is not particularly threatened by human activities. The species stands out in having the largest leaves among all of the conifers (hence the species name). This is somewhat surprising in relation to its close relative and near neighbor bawa nagi (*Nageia motleyi*), which sometimes grows in similar boggy habitats but has the smallest leaves in the genus, only a fifth as long as those of giant-leaf nagi. Although *N. maxima* is also closely related to and very similar to dammar nagi (*N. wallichiana*) in most respects, it can be separated readily by its larger leaves with fewer lines of stomates above, by its more slender pollen cones, and by the lack of a fleshy podocarpium.

Nageia motleyi (Parlatore) de Laubenfels

BAWA NAGI, KAYU BAWA, KAYU SERIBU, PODO KEBAL MUSANG, KEBAL AGAM, MEDANG BALOH (MALAY AND OTHER LANGUAGES OF WESTERN MALESIA)

Tree to 40(–54) m tall, with trunk to 1 m or more in diameter. Bark smooth, flaking to reveal brown patches on a gray background. Crown cylindrical, with short, slender, horizontal branches densely clothed with foliage. Leaves closely spaced on the twigs, bright green above at first, with many lines of stomates both above and beneath, 3–5(–7.5) cm long, 1.5–2.5(–3) cm wide. Blade widest near the middle, tapering gradually and then more abruptly to the triangular apex, sometimes with a short prolongation of the tip, and more abruptly to the roundly wedge-shaped base on a very short petiole to 3 mm long. Pollen cones single and nearly stalkless, 1.5–2 cm long by 5–6 mm in diameter. Seed cones single on a very short stalk to 5 mm long, usually with a conspicuously swollen, juicy podocarpium 8–12 mm long made up of five to nine united bracts. Combined seed coat and epimatium blue-black at maturity with a thin waxy film, 13–16 mm in diameter. Southern half of Malay Peninsula, adjacent Sumatra, and southern Borneo, mostly near the coasts. Scattered in the canopy of primary and secondary rain forests on a variety of substrates and situations, from well-drained, even dryish slopes to waterlogged peat swamps; 15–500(–1,000) m. Zone 10. Synonyms: *Decussocarpus motleyi* (Parlatore) de Laubenfels, *Podocarpus motleyi* (Parlatore) Dummer.

Bawa nagi is closely related to and similar in most respects to dammar nagi (*Nageia wallichiana*) but has much smaller leaves on average with a less prolonged tip as well as a greater number of stomates on the upper leaf surface, solitary rather than clustered pollen cones, and a much shorter stalk beneath the seed cones. It is common in some places, but the relative accessibility of its stands at low elevations coupled with its excellent quality as a timber tree makes it of real conservation concern. The species name honors James Motley (ca. 1821–1859), a British mining engineer and mine manager in Borneo who collected plants, including the types of bawa nagi, and coauthored a volume of *Contributions to the Natural History of Labuan* before he was murdered.

Nageia nagi (Thunberg) O. Kuntze

NAGI (JAPANESE AND ENGLISH), COMMON NAGI, ZHU BAI (CHINESE)

Tree or tall shrub to 20(–25) m tall, sometimes multitrunked but usually with a single main trunk to 0.5(–1) m in diameter. Bark thin, smooth, purplish red or reddish brown, weathering to gray and flaking to produce a mottled appearance. Crown dense, narrow, conical to cylindrical, with numerous short, slender, rising branches densely clothed with foliage. Leaves well spread out along the twigs, shiny bright or dark green to grayish green above, paler beneath, with lines of stomates only on the lower side, (2–)4–8(–9) cm long, (0.7–)1.5–2.5(–3) cm wide. Blade widest near or a little below the middle, tapering rather quickly to the briefly and broadly prolonged triangular tip and to the roundly wedge-shaped base on a short petiole 3–8 mm long. Pollen cones single or in clusters of 2–6(–10) on a short stalk to 10 mm long. Each cone (0.5–)1–2(–2.5) cm long by about 4 mm in diameter. Seed cones one (or two) on a stalk 0.5–1.5 cm long, without a fleshy podocarpium. Combined seed coat and epimatium blue-black at maturity but sometimes appearing bluish gray because of a heavy waxy coating, usually spherical but sometimes pear-shaped, 10–14(–16) mm in diameter. Southern Japan from Wakayama prefecture in southern Honshū southward, Taiwan, and southeastern China, from Zhejiang west to southeastern Sichuan and south to Hainan. Scattered among hardwoods in primary and secondary warm temperate to subtropical broad-leaved evergreen forests on hillsides and in valleys; 0–1,200(–1,600) m. Zone 8. Synonyms: *Decussocarpus nagi* (Thunberg) de Laubenfels, *D. nagi* var. *formosensis* (Dummer) Silba, *Nageia formosensis* (Dummer) C. Page, *N. nagi* var. *formosensis* (Dummer) Silba, *N. nankoensis* (Hayata) R. Mill, *Podocarpus formosensis* Dummer, *P. nagi* (Thunberg) Pilger, *P. nagi* var. *angustifolius* Maximowicz, *P. nagi* var. *nankoensis* (Hayata) Masamune ex Y. Kudô & Masamune, *P. nagi* var. *rotundifolius* Maximowicz, *P. nankoensis* Hayata.

Nagi has a long history of cultivation in Japan and China, where it is planted in gardens, cemeteries, and temple grounds, and selection has broadened the range of leaf and seed sizes and shapes compared to other species in the genus. This long interaction with people makes interpretation of natural variation more difficult and casts doubt on the recognition of botanical varieties within the species or their segregation into separate species. For example, plants with narrower than average leaves from the north of Taiwan and those with broader leaves from the south are both segregated as varieties or species by some authors, but neither variant seems outside of the natural variation found elsewhere within the range of the species. Nagi is most closely related to Vietnamese nagi (*Nageia fleuryi*), which differs in having larger, more pointed leaves, plumper seeds, and longer pollen cones.

The soft wood, pale yellowish brown with tight grain, while not as highly prized as that of other species in the genus, is very durable in water and is widely used locally in construction, furniture making, and craftwork. Like *Nageia fleuryi*, the seeds contain a high concentration of an edible and industrial oil. Furthermore, the root bark has been found to produce a rich array of norditerpene dilactones, compounds with 20 carbon atoms in a complicated structure of four rings. These compounds have a wide variety of biological activities, including effectiveness as anticancer agents, insecticides, deterrents to grazing animals, and plant growth regulators. The Chinese name, "bamboo cypress," aptly describes this broad-leaved conifer with parallel veins, while the Japanese name, which means lull or calmness, might refer to medicinal qualities of the plant.

Nageia wallichiana (C. Presl) O. Kuntze

DAMMAR NAGI, NIRAMBALI (TAMIL), SOPLONG (KHASI), THITMIN MA (BURMESE), ROU TUO ZHU BAI (CHINESE), KIM GIAO (VIETNAMESE) CHHAMCHHA: SÂ:R (CAMBODIAN), MALA ALMACIGA (PHILIPPINE SPANISH), DAMAR LAKI LAKI, KEBAL MUSANG, MENGILAN, SIBULU SOMAK (MALAY), AUGOMA, BARARANG, DURWE, MEJERKA, MEWONGO, NIWOB, ORSONKOBU, WARAMIRA, WASWAYANGUMI (PAPUAN LANGUAGES)

Tree to 35(–54) m tall, with trunk to 0.7(–1.2) m in diameter, free from branches for half or more of its height and sometimes buttressed. Bark smooth, light brown, weathering gray, flaking in irregular patches to produce brown and white mottling, and becoming irregularly and shallowly furrowed at the base of large trunks. Crown dense to open, narrow, conical to cylindrical, with numerous short, horizontal to gently rising branches bearing straight or droopy branchlets openly clothed with foliage. Leaves separated from one another along the slender twigs, shiny yellowish green or bright green above at first, darkening and dulling with age, with more numerous lines of stomates beneath than above, (6–)9–16(–23) cm long, (1.3–)3–5(–9) cm wide. Blade widest near or a little before the middle, tapering steadily to the narrowly triangular, prolonged tip of variable length and abruptly to the broadly and roundly wedge-shaped base on a short petiole (2–)5–10 mm long. Pollen cones in clusters of (one to) three to five (to seven) on a short stalk to 10 mm long. Each cone (0.5–)1–2 cm long by 3–4.5 mm in diameter. Seed cones one (or two) on a stalk (0.5–)1–2 cm long, with a conspicuously swollen, juicy, reddish black podocarpium (7–)10–12(–18) mm long made up of five to nine united bracts. Combined seed coat and epimatium dark reddish purple to blue-black at maturity with a thin waxy covering, 15–20 mm in diameter. Southern Asia and Malesia, from southwestern and northeastern India (Kerala and Assam), through southeastern

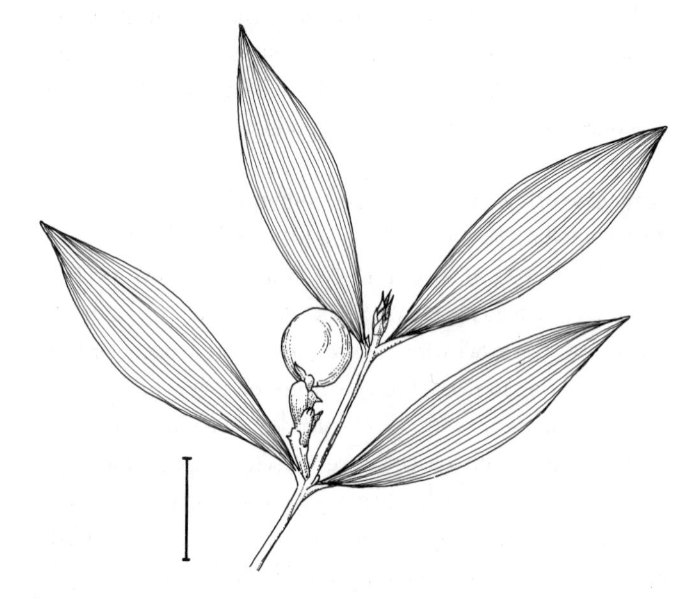

Twig with mature seed cone, including a small podocarpium, of dammar nagi (*Nageia wallichiana*); scale, 3 cm.

Asia (including Yunnan, southwestern China) and all of Malesia (including the Philippines) to the eastern tip of New Guinea (Normanby Island, Milne Bay province, Papua New Guinea). Scattered thinly to frequently in the canopy of various lowland and montane rain forests and subtropical evergreen broad-leaved forests on a wide range of substrates, from dry, rocky or sandy soils to peats; (5–)200–1,500(–2,100) m. Zone 10. Synonyms: *Decussocarpus wallichianus* (C. Presl) de Laubenfels, *Nageia blumei* (Endlicher) G. Gordon, *Podocarpus blumei* Endlicher, *P. wallichianus* C. Presl.

Dammar nagi is the most widespread species in the genus, with the ranges of bawa nagi (*Nageia motleyi*) and giant-leaf nagi (*N. maxima*) embedded entirely within its range and only that of nagi (*N. nagi*) lying entirely beyond its borders. As a conspicuous and important timber tree throughout this linguistically fragmented region, it has dozens of recorded common names, of which only a small selection is listed here. The fragrant, gray timber is used in carpentry, especially for domestic interior finishing, as well as in general construction and in the making of dugout canoes. An extract from the bark is used locally as a tonic. Similar extracts from the wood contain a wide range of complex terpenoid oils including the biologically active norditerpene dilactones better documented from *N. nagi*.

Historically, dammar nagi was divided between two species on the basis of the elongation of the leaf tip and the stiffness of the leaf, but these features respond readily to position in the crown and canopy and show no consistent relationship to any other characters. With its fleshy podocarpium, dammar nagi helps to link *Nageia, Retrophyllum,* and *Afrocarpus,* generally lacking this structure, with their closest relative *Podocarpus,* which always has it. The species name honors Nathaniel Wallich (1786–1854), who described this species based on his collections from northern India under a name that had previously been used for another species and was thus technically unavailable, necessitating a later renaming of the species.

Neocallitropsis R. Florin

CHANDELIER CYPRESS, NEOCALLITROPSIS

Cupressaceae

Evergreen trees or shrubs, with a single, straight trunk (or contorted and divided where damaged by fire) clothed in fibrous, furrowed bark. Crown open, composed of well-spaced whorls of gently rising branches with upturned ends. Branchlets cylindrical, densely clothed with leaves, all elongate, without differentiation into long and short shoots. Resting buds unspecialized, consisting solely of immature and embryonic foliage leaves. Leaves needle-like, in alternating quartets, keeled. Juvenile leaves long-triangular, standing out from the stem. Adult leaves shorter, arching forward and loosely overlapping the succeeding quartets.

Plants monoecious. Pollen cones sparse, single at the tips of otherwise ordinary branchlets, oblong, with alternating quartets of pollen scales. Each scale appearing similar to the foliage leaves from the outside, but a little shorter, and with 5–11 pollen sacs attached to the short stalk near the base of the triangular blade. Pollen grains small (30–40 µm in diameter), spherical, irregularly minutely bumpy and with a small, round germination pore. Seed cones sparse, single at the tips of short side branchlets, maturing in a single season. Each cone oblong, with foliage leaves grading into two alternating quartets of seed scales around a short central bump, only the inner quartet fertile. Each scale narrow, apparently intimately fused to the bract, which sticks out as a prickle just below the tip of the combined cone scale. Fertile scales each with two seeds, most of which often abort. Seeds oblong, with two equal, narrow wings derived from the seed coat running the length of the body or beyond. Cotyledons two, each with one vein. Chromosome base number x probably 11.

Wood fragrant and oily, light and soft, with grayish yellow sapwood weakly differentiated from the darker heartwood. Grain fairly even and fine, with inconsistent and often ill-defined growth rings set off by an almost imperceptible transition to smaller but scarcely thicker walled latewood tracheids. Resin canals absent but with few to many scattered individual resin parenchyma cells.

Stomates in several interrupted lines on both the inner (upper) and outer (lower) faces (or only the outer face of juvenile leaves), but not forming conspicuous bands. Each stomate deeply sunken and almost hidden beneath the four (to six) surrounding subsidiary

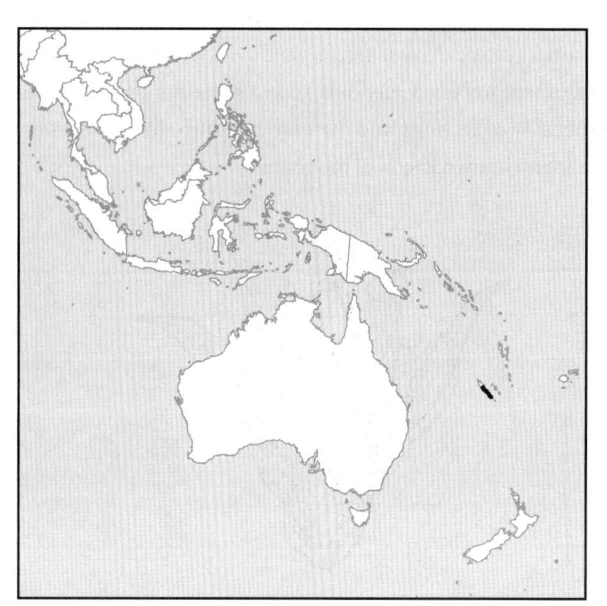

Distribution of *Neocallitropsis pancheri* in central and southern New Caledonia, scale 1 : 119 million

cells, which have a very thick cuticle surmounted by a low, circular or oval, nearly complete Florin ring. Leaf cross section approximately triangular, with a single-stranded midvein near the center above a larger resin canal and with up to five additional resin canals in an arc across the leaf at the outer face of the extensive wings of transfusion tissue. Photosynthetic tissue consisting almost entirely of a thick palisade layer on both faces (or only the upper face of juvenile leaves) beneath the epidermis and adjacent, nearly continuous hypodermis, which runs right to the edges of the stomates.

One species in New Caledonia. References: de Laubenfels 1972, Enright and Hill 1995, Farjon 2005b. Synonym: *Callitropsis* R. Compton.

Morphologically and in DNA sequences, *Neocallitropsis* is closely related to *Callitris,* which also occurs in New Caledonia as well as in Australia, and to *Actinostrobus,* as reflected in is name. When first described, it was called *Callitropsis* (Greek for "looks like *Callitris*"), but Florin added the *Neo-* ("new") when he realized that *Callitropsis* had still earlier been used to name a genus for Alaska yellow cedar (*Cupressus nootkatensis*), thus preempting its use for the New Caledonian plant. It differs from *Actinostrobus* and *Callitris* in its persistently juvenile morphology of elongate rather than scalelike leaves and from all species of both genera except *Callitris macleayana* in having its leaves and cone scales in quartets rather than trios (although occasional branches may have leaves or cone scales, or both, in trios).

The sole living species, *Neocallitropsis pancheri,* has a candelabra-like growth form similar to that of several other co-occurring trees emergent from the fire-prone, metal-laden maquis minière shrublands of serpentine soils in New Caledonia, including species of *Callitris, Agathis,* and *Araucaria* as well as the nonconiferous Casuarinaceae. *Neocallitropsis pancheri* is not in general cultivation, except in New Caledonia and in some botanical gardens and specialty collections, and there has been no cultivar selection. There are no known fossils of the genus.

Open mature seed cone of chandelier cypress (*Neocallitropsis pancheri*) after seed dispersal; scale, 1 cm.

Neocallitropsis pancheri (Carrière) de Laubenfels
CHANDELIER CYPRESS
Plates 34 and 35

Tree to 10 m tall, with trunk to 0.6 m in diameter. Bark light to dark brown, flaking after shedding the leaf bases, graying with age and peeling in thin vertical strips. Crown open, broadly triangular in youth, becoming flat-topped with age. Leaves of juveniles 8–14(–25) mm long, those of adults 4–6(–12) mm long, lasting 3–4 years. Pollen cones 8–10 mm long, with three to six (to eight) quartets of pollen scales. Seed cones (6–)10–15 mm long, the bract spines 2–3(–5) mm long. Seeds sharp-pointed, about 6 mm long, the wings about 0.6 mm wide. Southern fifth of New Caledonia and Mount Paéoua in the center. Rain forest and scrublands in river valleys and on serpentine soils; (30–) 800–950(–1,150) m. Zone 9. Synonyms: *Callitropsis araucarioides* R. Compton, *Neocallitropsis araucarioides* (R. Compton) R. Florin.

Trees of chandelier cypress resemble in general habit and foliage species of the unrelated genus *Araucaria,* the largest genus of conifers in New Caledonia This habit is common to many small trees of the fire-prone maquis minière, although few others resemble the araucarias as closely. Human activities, including dam building, eliminated or greatly reduced *Neocallitropsis pancheri* from its optimum habitat in river valleys so that few individuals today exceed 8 m in height. It also suffered during the first half of the 20th century from exploitation for the essential oil, used as an ingredient in perfumes. Trees in the most fire-prone habitats sustain considerable damage, and many become shrubby and irregular, with multiple stems replacing the original trunk. The handsome appearance of well-grown trees is appreciated in cultivation in New Caledonia and would be an asset in gardens in other wet tropical regions.

Nothotsuga H. H. Hu ex C. Page
BRACT HEMLOCK
Pinaceae

Evergreen, single-trunked or multistemmed trees. Bark scaly, rough, becoming ridged and furrowed with age. Crown egg-shaped and open, with slender horizontal to gently rising branches concentrated near the top of the trunk's annual growth increments and with a drooping leader. Lateral branching dense, with a tendency toward weakly developed persistent short shoots borne on the long shoots. Winter buds well developed, scaly. Leaves spirally arranged, sometimes partly two-ranked but mostly radiating around the twigs, widely spaced on long shoots. Each leaf needlelike, linear, flattened to shallowly triangular in cross section, abruptly narrowed to a short petiole that attaches to a small, weak, decurrent woody peg on the twig.

Plants monoecious. Pollen cones in umbrella-like clusters emerging from single terminal buds at the tips of twigs. Each pollen cone slender-stalked, oblong with 25–40 pollen scales. Each scale heart-shaped with two very large pollen sacs. Pollen grains with two bladders. Seed cones erect, single in the axils of leaves or at the tips of short axillary twigs, maturing in a single season but persisting several years after releasing the seeds, ellipsoid-cylindrical before opening. Bracts about half as long as the roundly diamond-shaped thin seed scales, just visible beyond the ones that overlap them. Seeds in pairs, each with a slightly longer asymmetric wing derived from the scale. Cotyledons not reported. Chromosome base number $x = 12$.

Wood with well-defined growth rings marked by a sharp transition between early- and latewood. Vertical resin canals in the latewood becoming more frequent with age.

With a few lines of stomates near the midrib above and more numerous ones filling most of the surface beneath. Structure of the stomates not reported. Leaf cross section with a single resin canal beneath the single-stranded midvein. Photosynthetic tissue with a partial palisade layer beneath the upper epidermis and adjoining layer of hypodermis pierced by openings connecting the stomates to the spongy mesophyll that fills the bulk of the leaf.

One species in south-central China. References: Farjon 1990, Frankis 1989, Fu et al. 1999c, Page 1988.

The sole species of *Nothotsuga* was originally described as a *Tsuga* and is often retained in that genus, with which it has many points of similarity. It differs from all hemlocks, however, in the position and arrangement of the pollen and seed cones, which more closely resemble those of *Keteleeria* and *Pseudolarix*. Wood and bark structures also show strong similarities to *Keteleeria* and *Tsuga* but support the separation of *Nothotsuga* from both genera. *Nothotsuga longibracteata* is not in general cultivation, and there has been no cultivar selection. The known fossil record is meager, but extends back to the Pliocene of Japan and Russia, about 3–5 million years ago.

Distribution of *Nothotsuga longibracteata* in southeastern China, scale 1 : 120 million

Nothotsuga longibracteata (W. C. Cheng) H. H. Hu ex C. Page

BRACT HEMLOCK, CHANG BAO TIE SHAN (CHINESE)

Erect or shrubby tree to 8–30 m tall, with trunk to 0.3–1.2 m in diameter. Bark brownish gray, rough, scaly, becoming furrowed with age. Crown conical at first, becoming flat-topped with age. Twigs yellowish brown or reddish brown, usually hairless but sometimes thinly hairy. Winter buds pointed, 2–4 mm long. Needles fairly uniform in length, 1.1–2.4 cm long, parallel-sided, the edge smooth, the tip bluntly pointed, with 5–12 stomatal lines above and two broad greenish white stomatal bands beneath. Pollen cones 5–10 mm long, yellow. Seed cones 2–5(–6) cm long on stalks 5–10 mm long, opening to 1.5–2.5(–3) cm wide, reddish purple before maturity, ripening dark brown. Seed scales 10–15(–22) mm long, broader than long. Bracts spoon-shaped, 7–18 mm long, visible among the seed scales but not protruding far. Seeds 4–8 mm long, the firmly attached wing another 6–12 mm longer. South-central China, from Guangxi and northeastern Guizhou to southern Fujian. Scattered in groups on moist mountains in broad-leaved evergreen or mixed forest; (300–)1,000–

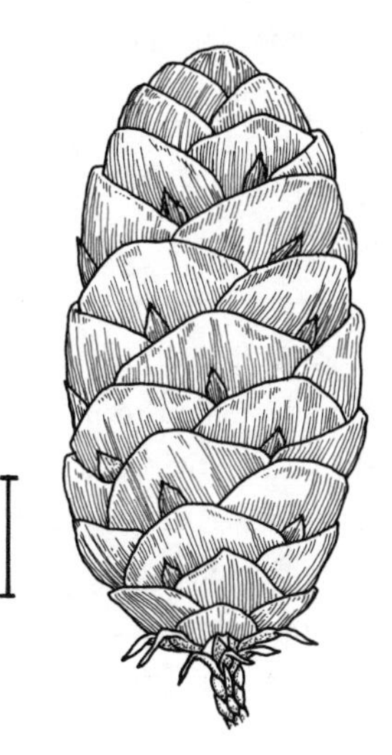

Mature seed cone of bract hemlock (*Nothotsuga longibracteata*) at the time of seed dispersal; scale, 1 cm.

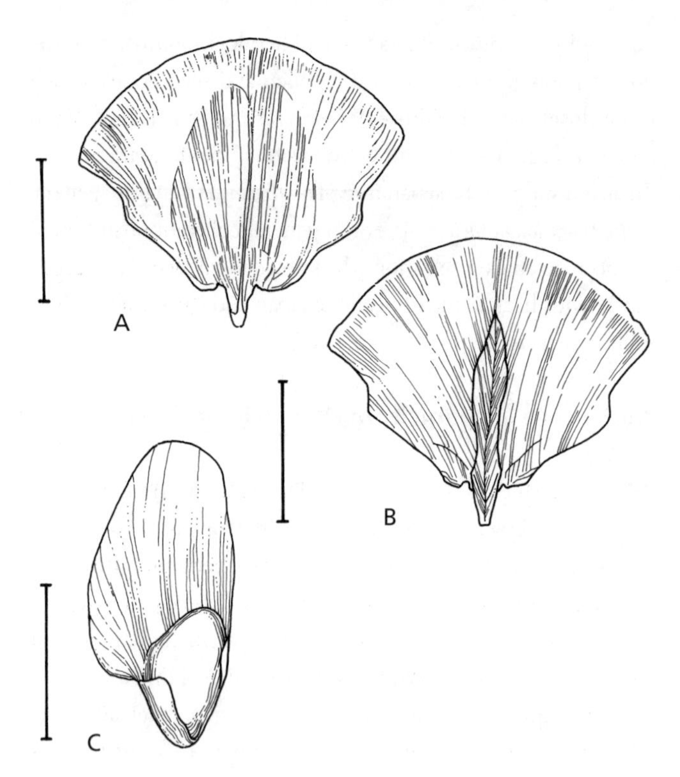

Upper (A) and undersides (B) of seed scale and underside of seed (C) of bract hemlock (*Nothotsuga longibracteata*); scale, 5 mm.

1,900(–2,300) m. Zone 6. Synonym: *Tsuga longibracteata* W. C. Cheng.

Bract hemlock is a rare species poorly known outside of its native distribution. The wood is appreciated for construction and carpentry. The tree was severely depleted by logging in northeastern Guizhou, the one area where it formed dense stands along with *Tsuga chinensis.* Attempts at recovery are hampered by difficulties in regeneration, including a high light requirement for seedlings, infrequent years of good seed production, and low viability of seeds.

Papuacedrus H. L. Li

PAPUAN INCENSE CEDAR

Cupressaceae

Evergreen trees with fibrous, furrowed bark peeling in vertical strips. Crown open, broadly conical to flat-topped, with long, slender, gently upwardly arching branches well dispersed along the trunk. Branchlets in fernlike, flattened sprays, the predominantly paired branchlets arising only in the axils of lateral leaves. Resting buds unspecialized, consisting solely of embryonic and immature ordinary foliage leaves. Seedling leaves in alternating quartets, needlelike, flat, standing out from and somewhat crowded on the stem. Seedling phase short-lived, confined to the main axis during the first year of growth, with lateral branchlets bearing juvenile leaves appearing by the second year. Juvenile and adult leaves in alternating pairs, scalelike, dense, flat or bowed upward on the branchlets, the bases of lateral leaves running down onto the branchlets and touching side to side except at the tip. Lateral and facial leaf pairs strongly distinct in juvenile foliage, becoming progressively less so with tree age, although still mostly quite distinct in maturity. Lateral leaves flattened side to side, more juvenile ones with large triangular tips standing out from the branchlet, later leaves with tips progressively reduced (except on main shoots) until the adult foliage with minute free tips. Facial leaves diamond-shaped and pressed against the twig, successive pairs completely separated by the bases of the lateral leaves, slightly to much smaller than the lateral leaves in juvenile foliage but less distinct from the smaller adult lateral leaves. Exposed lateral branchlets in the crown of mature canopy trees may be squarish rather than flattened in cross section and without obvious distinction between the facial and lateral leaf pairs.

Plants monoecious. Pollen cones crowded, but each single at the tip of a short branchlet, oblong, squarish in cross section. Each cone with 8–10 alternating pairs of pollen scales often arranged like four or five aligned quartets or so crowded as to appear irregular, each scale with two to four pollen sacs. Pollen grains small (25–30 µm in diameter), spherical, minutely bumpy and sometimes with an ill-defined germination pore. Seed cones crowded but each single at the tip of a short branchlet, maturing in a single season. Each cone oblong, with two alternating pairs of thin, woody cone scales. Each scale with the bract portion fused to the seed-bearing portion and ending in a blunt tip near the bottom of the scale, the upper pair of scales fertile, about twice as long as the sterile lower pair. Each fertile scale with two seeds. Seeds oblong, with two very unequal wings developed from the seed coat in the upper half, the outer wing a mere fringe, the inner expanded over the far half of the seed scale. Cotyledons two, each with one vein. Chromosome base number not reported but probably $x = 11$.

Wood fragrant, light and soft, with whitish brown sapwood somewhat contrasting with the darker heartwood. Grain very even and moderately coarse, essentially without growth rings or these weakly marked by a few, irregular, slightly smaller latewood tracheids. Resin canals absent but with a few individual resin parenchyma cells scattered through the wood.

Stomates in lines and patches of varying density and extent on all leaf faces, the patches on the lower sides of the branchlets often whitened with wax. Each stomate sunken well beneath the four to six (or seven) surrounding subsidiary cells, which are commonly shared between adjacent stomates and topped by a nearly continuous Florin ring but do not generally bear additional papillae. Leaf cross section with a single-stranded midvein inside (morphologically above) and pressed against a large resin canal in the lower part of the leaf, both embedded in a cylinder of transfusion tissue. With up to eight additional resin canals (or none) spaced evenly around the whole periphery. Photosynthetic tissue forming a prominent palisade layer beneath the epidermis and adjacent thick but incomplete hypodermis over the whole leaf surface only on the upper face of the branchlet. Only one of each pair of facial leaves with a palisade layer while the other has spongy mesophyll extending all the way to the surface. Lateral leaves with palisade tissue on either the right or left side (the opposite ones for the two members of a pair) of these laterally flattened leaves rather than the evolutionarily upper (inner) side, which remains attached to the branchlet.

One species in New Guinea and the Moluccas. References: Boutelje 1955, de Laubenfels 1988, Enright and Hill 1995, Farjon 2005b, Florin and Boutelje 1954, Hill and Carpenter 1989, Li 1953, McIver and Aulenback 1994, Offler 1984.

Papuacedrus is the most northerly of the southern incense cedars (which also include *Austrocedrus* and *Libocedrus*), barely crossing the equator. It differs from the related genera in a number of small characteristics but most obviously in the very large juvenile lateral leaves, the distinctive shape of the adult lateral leaves, and in the position of the stout bract tip in the lower half of the seed scale rather than near the middle (as in *Libocedrus*) or near the end (as in *Austrocedrus*). Although these differences are small, information from DNA structures agrees with the separation of *Papuacedrus* from similar genera. As with many other tropical montane conifer genera, *P. papuana* is not in cultivation, except in some botanical collections, and no cultivar selection has taken place.

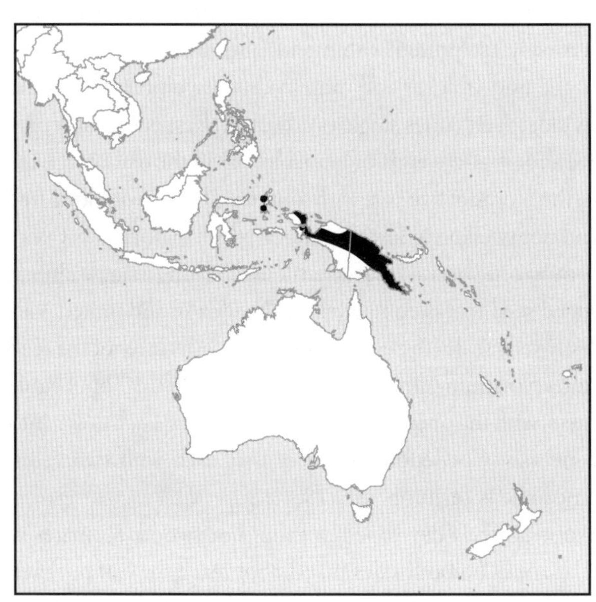

Distribution of *Papuacedrus papuana* in New Guinea and nearby Halmahera, scale 1 : 119 million

Recognition of three distinct genera is also supported by the presence of fossils of all three in the Tertiary of Tasmania. Fossilized shoots of *Papuacedrus* have been recorded from Oligocene sediments (about 30 million years old) at two sites in Tasmania. These are the only known fossils of the genus.

Papuacedrus papuana (F. J. Mueller) H. L. Li

PAPUAN INCENSE CEDAR, KAIPIL, WONGA (PAPUAN LANGUAGES)

Tree to 35(–50) m tall, with trunk to 1 m in diameter, or less than 10 m tall in subalpine scrub. Bark reddish brown and scaly at first, darkening with age. Crown pyramidal in youth but spreading, dome-shaped, and with drooping branches in age. Shoots arranged in flattened, fernlike sprays with branchlets arising alternately or in opposite pairs more or less evenly on the front and rear of the parent side twigs. Lateral scale leaves up to 10(–16) mm long in juveniles with the tip spreading to 3(–6) mm, progressively smaller and tighter to the branchlet in more mature

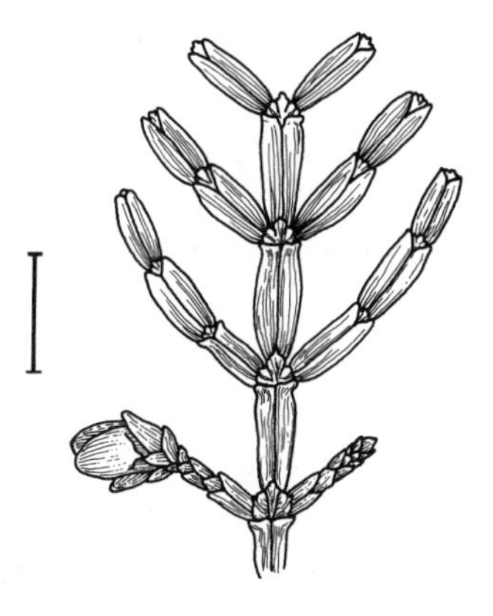

Face view of foliage spray of Papuan incense cedar (*Papuacedrus papuana*) with exclusively opposite branching and a mature seed cone on a modified branchlet; scale, 1 cm.

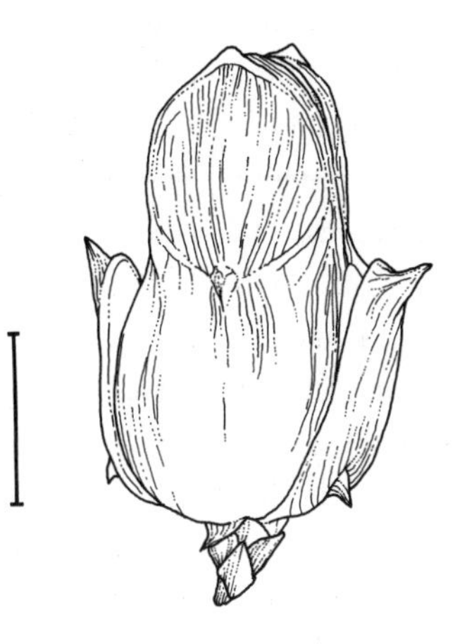

Face view of mature seed cone of Papuan incense cedar (*Papuacedrus papuana*) before seed dispersal; scale, 5 mm.

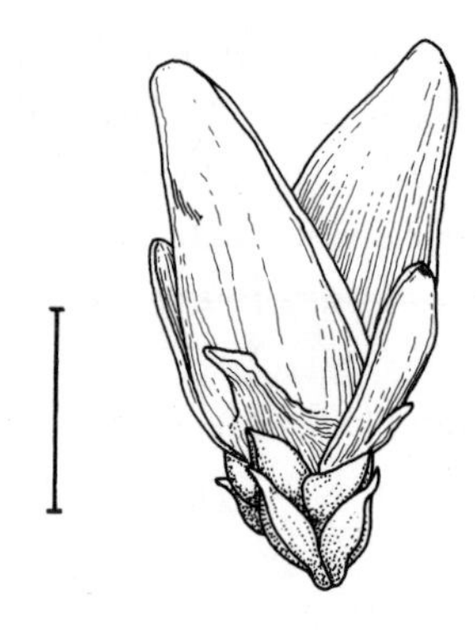

Oblique view of mature seed cone of Papuan incense cedar (*Papuacedrus papuana*) after seed dispersal; scale, 2 mm.

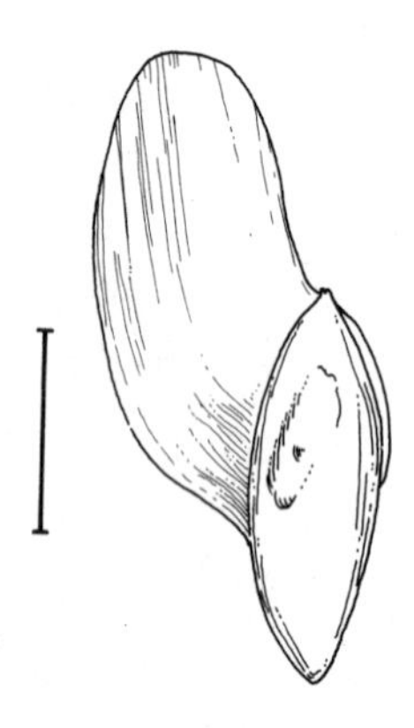

Dispersed seed of Papuan incense cedar (*Papuacedrus papuana*); scale, 1 mm.

trees, down to 1 mm long in the most exposed adult foliage Facial leaves much smaller and not changing as much in size with maturity, 1–3.5 mm long. Pollen cones 4–25 mm long, with 4–10 pairs or quartets of pollen scales. Seed cones 8–15 mm long, the seed scales often with fine radiating ridges extending from the bract tip. Triangular free tip of bract 1–2 mm long. Seeds 2–3 mm long, the larger wing extending straight out from the egg-shaped body by more than twice its width. Central mountain ranges along the whole length of New Guinea, also in the Moluccas. Canopy or subcanopy tree in montane rain forests and emerging through the canopy in subalpine scrub, forming pure stands or mixed with other conifers and hardwoods, (600–)1,500–3,300(–3,900) m. Zone 9. Synonyms: *Libocedrus arfakensis* L. Gibbs, *L. papuana* F. J. Mueller, *L. torricellensis* Lauterbach, *Papuacedrus arfakensis* (L. Gibbs) H. L. Li, *P. papuana* var. *arfakensis* (L. Gibbs) R. J. Johns, *P. torricellensis* (Lauterbach) H. L. Li.

The foliage of the juvenile phase varies across New Guinea. The widely spreading lateral leaves are quickly supplanted by leaves with their tips tucked in on trees growing in the western quarter of the island. This led to recognition of separate species or varieties for the eastern and western trees, but this separation does not seem warranted since the adults are not distinctive. The species is quite adaptable in New Guinea, growing on a variety of substrates at a broad range of elevations. Although it is capable of regenerating in closed forests, like those of *Nothofagus*, the "southern beech," it can make pure stands in response to fires. Its adaptability in cultivation is essentially untested.

Parasitaxus de Laubenfels

CORAL PINE

Podocarpaceae

Evergreen, red or purple, coral-like, parasitic shrub with numerous relatively short-lived stems emerging from the roots or lower stems of other conifers. Bark thin, smooth or scaly. Crown conical to cylindrical, with a few major branches that are themselves highly and repeatedly branched. Branchlets uniform, not evidently differentiated into long and short shoots, all with little elongation between successive leaves. Resting buds poorly defined, consisting of temporarily arrested foliage leaves. Leaves spirally inserted, all scalelike, fairly uniform in size, the base attached for a short distance along the twig, generally hidden by the blades of lower leaves, the blades flattened top to bottom, with a keel along the outer (lower) side.

Plants monoecious. Pollen cones single at the ends of branchlets or in the axils of foliage leaves. Each cone oval, without a set of markedly distinct basal bracts, and with just a few spirally arranged pollen scales, each scale bearing two pollen sacs. Pollen grain medium (body 45–50 µm long), with two hemispherical air bladders that are noticeably smaller than the ovoid body and diverge at about 120° away from the germination furrow. Bladders coarsely sculptured within and fairly smooth outside, the cap of the body thicker and more coarsely sculptured than the furrow side. Seed cones single on short, scaly, specialized reproductive shoots at the ends of branchlets or in the axils of foliage leaves, the reproductive part with four to six bracts with long free tips and not swelling to form a podocarpium, maturing in 1 year. Uppermost one (or two) bracts fertile and bearing a single plump, unwinged seed that is completely embedded in and united with the fleshy seed scale (the epimatium). The fleshy layer of the seed coat and epimatium surrounding a hard inner shell of about equal thickness, with the opening of the ovule pointing back down into the cone axis. Cotyledons probably two, each with two veins. Chromosome base number $x = 18$.

Wood scarcely developed, without heartwood. Stomates in irregular zones on both sides that do not reach the leaf tip. Each stomate sunken beneath the (three or) four or five (or six) surrounding subsidiary cells and surrounded by a Florin ring. Mid-

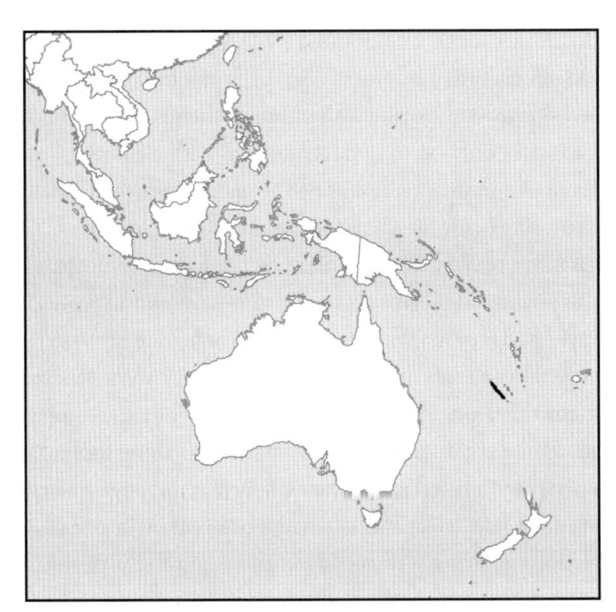

Distribution of *Parasitaxus ustus* in New Caledonia, scale 1 : 119 million

vein single-stranded, extending only a little way into the blade accompanied beneath by a large resin canal and sometimes with two others off to the sides at the edge of the small wings of transfusion tissue. Functional photosynthetic tissue lacking, the bulk of the leaf occupied by more or less homogenous spongy mesophyll.

One species in New Caledonia. References: de Laubenfels 1972, Gray 1960. Synonym: *Podocarpus* sect. *Microcarpus* Pilger.

Parasitaxus is the only known parasitic conifer (hence the scientific name, Latin for "parasitic yew," although it is not related to yews and does not even resemble them). It parasitizes New Caledonian sickle pine (*Falcatifolium taxoides*), another member of the Podocarpaceae. It produces roots under the bark of the host tree, usually growing on the roots but sometimes emerging out of the bottom of the trunk. The connection between the host and parasite is accompanied by a mycorrhizal fungus that may be involved in the transfer of nutrients between the two plants, as occurs with some hemiparasitic flowering plants. Like many flowering plant parasites, the leaves and branchlets are not green, but those of *Parasitaxus* do contain chlorophyll, which is not the case with the most extreme flowering plant parasites, like dodders (*Cuscuta*) in the morning glory family (Convolvulaceae) or Indian pipes (*Monotropa uniflora*) in the heath family (Ericaceae).

Although *Parasitaxus* was long treated as a botanical section (section *Microcarpus*) of *Podocarpus*, DNA studies suggest that it is more closely related to *Manoao* (silver pine) and *Lagarostrobos* (Huon pine), genera once included in *Dacrydium* and only distantly related to *Podocarpus*. It lacks the podocarpium found in *Podocarpus* and its close relatives, and its pollen cones and pollen scales, as well as its exclusively scalelike leaves, also differ greatly from those of *Podocarpus* to such an extent that it is not obvious why it was ever included in that genus. The anatomy of the stems and the cellular patterns of the epidermis are consistent with a relatively close relationship to *Manoao* and *Lagarostrobos*, but the pollen and seed cones are not all that similar. Furthermore, the arrangement of the chromosomes is more similar to that of *Podocarpus* and its relatives than it is to *Dacrydium* and the small genera formerly included within it. Clearly, more research is needed, not only on the relationships of *Parasitaxus* but also on the functioning and origin of its parasitic habit. Such work is not helped by the fact that the sole living species has not proven to be amenable to cultivation, a difficulty shared with some other parasitic plants. *Parasitaxus ustus* is thus not in general cultivation and no cultivar selection has taken place. There is no known fossil record of *Parasitaxus* that might help in clarifying its relationships.

Parasitaxus ustus (Vieillard) de Laubenfels

CORAL PINE, CEDRE RABOUGRI (FRENCH)

Shrub to 1.5(–3) m tall, with stems to about 3 cm in diameter. Bark reddish brown to grayish brown. Crown with upwardly directed branches, branching alternately all around and bearing many branchlets completely clothed with foliage. Twigs reddish purple, rather fleshy and brittle in the first year. Resting buds not differentiated from the ends of the twigs, about 1–1.5 mm long. Leaves dense, reddish purple (hence the scientific name, Latin for "burnt"), lasting 3–5 years or more. Free, triangular tips of scale leaves 1–2 mm long, hiding the attached bases, tightly held against the twigs or a little spreading, particularly at branch points. Pollen cones 3–4 mm long and 1.5–2 mm wide, on a short, leafy stalk to 6 mm long. Pollen scales with the upturned blade almost semicircular, 1.5–2.5 mm long, minutely toothed and with a keel on the upper half ending in a short, blunt point. Seed cones on a short, leafy stalk to 5 mm long. Combined seed coat and epimatium fleshy, becoming wrinkled when dry, pale bluish white with a thick coating of wax over a reddish purple skin, nearly spherical, 2.5–4 mm in diameter, with a tiny beak at the tip that becomes more pronounced with drying. Along the whole length of New Caledonia and on the nearby Île des Pines to the southeast. Forming patches in the understory of rain forests in association with its host; (150–)400–900(–1,250) m. Zone 9. Synonym: *Podocarpus ustus* (Vieillard) Ad. Brongniart & Gris.

Coral pine is a rare species that occurs throughout the range of its host, New Caledonian sickle pine (*Falcatifolium taxoides*), one of the commoner rain-forest conifers of New Caledonia, but in only a fraction of the stands of this species and only patchily in those stands in which it occurs. This rarity, in combination with

Twig with mature seed cone of coral pine (*Parasitaxus ustus*); scale, 2 mm.

the unusual coppery coloration of the foliage and the stunted growth habit of the plants (the French common name means "stunted cedar"), are, perhaps, some of the reasons why the indigenous Kanak people have treated it as a sacred plant. Biochemical studies have not been conducted that might reveal additional, more practical reasons. The fleshy twigs become so brittle on drying that they often detach from their parent shoots, and herbarium specimens are, consequently, often rather fragmentary, with the taxonomically important pollen and seed cones falling away and becoming lost. In the field, the plants produce abundant seed cones, so the rarity of the plant is not due to limited seed production. It is unusual among members of the Podocarpaceae in producing pollen and seed cones on the same individual, although this is the more common condition in conifers as a whole.

Phyllocladus L. Richard & A. Richard

CELERY PINE, TOATOA

Podocarpaceae

Evergreen trees and shrubs, with multiple trunks or a single, upright, cylindrical to slightly fluted main trunk. Bark obscurely fibrous, thin, smooth and with numerous warty lenticels or scaly and sometimes becoming shallowly and irregularly ridged and furrowed. Crown dense, broadly conical to dome-shaped, initially with regularly spaced whorls of upwardly angled branches. Branchlets dramatically differentiated into long and short shoots. Long shoots forming the trunk and branching framework of the plant, of ordinary twiggy appearance, hairless, grooved between the attached bases of the scale leaves, remaining green for at least the first year. Shoot shoots in the form of phylloclades: flattened, leaflike, and photosynthetic twigs (hence the scientific name, Greek for "leaf stem"). Phylloclades lobed or toothed and, like true foliage leaves, either simple (a single segment attached directly to a long shoot) or compound (with few to many segments attached pinnately to a common axis). Each phylloclade segment flattened side to side, usually with photosynthetic tissue on both faces, stomates only on the underside, and with a more or less well developed midrib, representing the main axis of the branchlet, and bearing many branch veins extending out to minute scale leaves at the periphery. Both simple and compound phylloclades, as well as phylloclade segments, may resume growth and produce new phylloclades or whole long shoots and hence are not, strictly speaking, short shoots, but often grow as if they were. Resting buds at the tip of long shoots, well developed, tightly enclosed by specialized bud scales. Leaves spirally attached on long shoots, in a flattened plane around phylloclades, scalelike and often quickly shed in adults, needlelike and flattened top to bottom in juveniles. Juvenile foliage giving way quickly in the second year or much later to scale leaves and phylloclades of adult foliage.

Plants monoecious or dioecious or sometimes even with a few cones of mixed sex. Pollen cones in spirally arranged clusters of 1–25 on short stalks or directly in the axils of bracts near the end of a short branch. Each pollen cone cylindrical, without a ring of bracts at its base and with numerous (25–75), spirally arranged, roundly triangular pollen scales, each bearing two pollen sacs. Pollen grains small (body 15–35 μm long, 20–40 μm overall), with two small, smooth or wrinkled air bladders tucked underneath around the germination furrow and barely protruding beyond the ends of the minutely bumpy, oval body. Seed cones single or in tight clusters of two to four on short stalks or directly attached in various positions on the phylloclades, in notches at the tips or sides of simple phylloclades or of segments of compound phylloclades, or replacing whole lower phylloclade segments. Each seed cone compact, with 3–40 spirally arranged bracts, of which the middle 1–10(–20) are fertile and bear one ovule apiece. Bracts often uniting and becoming fleshy with maturity, what could be the equivalent of seed scales (or epimatium) developing late in the form of an aril. Seeds without wings, egg-shaped, slightly flattened, the opening pointed upward rather than down into the cone axis, surrounded at the base by a thin, leathery or fleshy cuplike aril, maturing in one (or two) season(s). Cotyledons two, each with two veins. Chromosome base number $x = 9$.

Wood heavy, strong, hard, and fragrant, the white to very light brown sapwood strongly contrasting with the small core of yellowish brown to orange-brown heartwood. Grain fine and even, with well-developed growth rings marked by obvious bands of darker latewood. Resin canals and individual resin parenchyma cells both absent.

Midvein of juvenile leaves single, prominently raised beneath, with one or two resin canals immediately beneath it. Juvenile

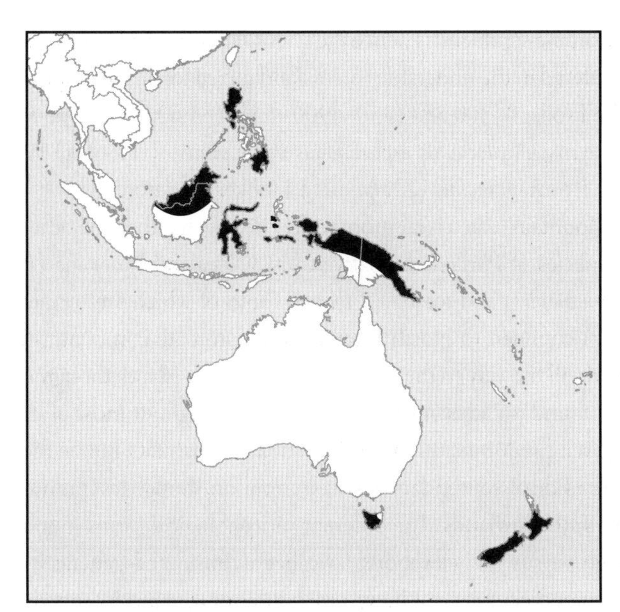

Distribution of *Phyllocladus* in Australasia, scale 1 : 119 million

leaves with photosynthetic tissue along the upper side and lines of stomates in stomatal bands on either side beneath, each stomate surrounded by a sunken Florin ring.

Five species in Malesia, New Zealand, and Tasmania. References: Allan 1961, Cookson and Pike 1954, de Laubenfels 1969, 1988, K. Hill 1998, R. Hill 1989, R. Hill and Brodribb 1999, Keng 1978, Molloy 1996.

The common name celery pine refers to the appearance of compound phylloclades, which are somewhat reminiscent of the leaves of celery above the stalk. Despite their distinctive and attractive appearance, species of *Phyllocladus* are cultivated to a very limited extent outside New Zealand and Australia. The few cultivars selected have blue-green phylloclades because of extra wax. Natural hybridization occurs between some species and may become a source of new cultivars in the future.

The extraordinary phylloclades of the celery pines are unique among the conifers making this group one of the most distinctive conifer genera. To this can be added other features unusual among the Podocarpaceae, including the aril of the seeds and the reduced air bladders of the pollen grains with a corresponding shift in mode of pollination. Small wonder, then, that many authors separate *Phyllocladus* into its own family, called Phyllocladaceae, while still acknowledging its closeness to Podocarpaceae. A few workers went even further and essentially denied any more particular connection of *Phyllocladus* to the Podocarpaceae than it has to some other groups. For instance, Keng (1974), emphasizing the phylloclade, suggested that *Phyllocladus* represents a direct link to the Paleozoic progymnosperms, a seedless group generally viewed as containing the ancestors of all seed plants. Similarly, Melikian and Bobrov (2000), emphasizing the aril and other seed features, placed *Phyllocladus* close to the yews (Taxaceae). When all features are taken together, however, including wood anatomy, chemistry of the wood and foliage, structure of the pollen and seed cones, embryo development, cotyledon structure, and organization of the chromosomes, the inclusion of *Phyllocladus* within the family Podocarpaceae seems inescapable. DNA studies generally confirm this assessment but suggest conflicting placements within the family. Depending on the genes examined, it has been seen either as sister group to all of the remaining genera or as a member of one of the two main groups of genera within the family, the one containing *Prumnopitys* and the scale-leaved genera like *Halocarpus* and *Parasitaxus*. In either case, it would not be closely related to the other genera with condensed, multiscaled cones, *Microcachrys* and *Saxegothaea*, which belong to the other grouping in the family, the one containing *Podocarpus* itself.

The pollination mechanism in *Phyllocladus*, with pollen captured directly in a small, upright pollination droplet, is different from that in the vast majority of Podocarpaceae, which produce a large, hanging drop in which pollen grains that have landed on the dry seed cone are later swept up and float into the opening (micropyle) of the ovule. The method of pollen capture in *Phyllocladus* is more similar to that found in many unrelated conifers in the families Cupressaceae and Taxaceae but is also not all that dissimilar from the one found in *Acmopyle*, another genus whose ancestors may have had a fairly early divergence within the Podocarpaceae. After successful pollination and fertilization, embryo development follows a pattern found in all other genera of and known only from Podocarpaceae in which one group of cells (the E tier) within the proembryo all have two nuclei rather than the single one found in other conifers. Thus the preponderance of evidence favors inclusion of celery pines within Podocarpaceae but in a relatively basal position.

The distinctive phylloclades of the genus are found as fossils in sediments dating back to the late Eocene (about 40 million years ago) in southeastern Australia, both in Tasmania, where the genus persists to this day, and on the mainland, where it became extinct, presumably due to increasing aridity since all of the living species live in wet climates. The earliest records from New Zealand, where three of the five extant species grow, only date from the Miocene, about 20 million years ago, based on both phylloclades and wood. Pollen grains apparently referable to *Phyllocladus* are more widespread and extend further back in time. Even the records from the Jurassic, 150 million years ago, could really belong to *Phyllocladus* and not to some extinct genus if the celery pines are the sister group to all other extant Podocarpaceae, as some DNA studies suggest.

Identification Guide to *Phyllocladus*

The five species of *Phyllocladus* may be distinguished by whether the phylloclades are simple (consisting of a single segment, which may be lobed) or compound (with five or more pinnately arranged segments), the length of most whole phylloclades (not just the simple segments of compound ones), and the length of the scale leaves at the base of young phylloclades (they may be shed as the phylloclades mature):

Phyllocladus alpinus, phylloclades simple, less than 9 cm, scale leaves at the base less than 1.5 mm

P. aspleniifolius, phylloclades simple, less than 9 cm, scale leaves at the base 1.5–3.5 mm

P. trichomanoides, phylloclades compound, less than 9 cm, scale leaves at the base 1.5–3.5 mm

P. hypophyllus, phylloclades compound, 9–15 cm, scale leaves at the base 1.5–3.5 mm

P. toatoa, phylloclades compound, at least 15 cm, scale leaves at the base more than 3.5 mm

Phyllocladus alpinus J. Hooker

MOUNTAIN TOATOA

Shrub, or tree to 8(–9) m tall, with a short trunk to 0.4 m in diameter. Bark dark brown and smooth at first, weathering gray and becoming roughened by scaly patches. Crown irregularly conical, with numerous, upwardly angled, slender to stout branches. Phylloclades attached singly or in rings of three to five on the long shoots, simple, (0.4–)1–3(–6) cm long, often waxy bluish green when young, weathering yellowish green to dark green. Individual phylloclades highly varied in shape, from egg- through diamond-shaped to almost sword-shaped, with slightly toothed to deeply lobed edges. Scale leaves of adult long shoots and buds to 1 mm long. Plants monoecious. Pollen cones in clusters of (one or) two to five (to seven) at the tip of a short branchlet (which may subsequently grow out beyond the cones). Each cone 3–6 mm long (to 12 mm after releasing pollen), 1.5–2 mm thick, bright red before pollen release, on a very short stalk 1–2 mm long. Seed cones single or in clusters of two to four or perhaps more attached to the sides of phylloclades near their base or even wholly replacing the expanded portion. Each seed cone roughly spherical, about 6–7 mm in diameter, with 6–12 bracts, of which one to six are fertile and mature one to three (to five) seeds. Bracts swollen, juicy, red to brown at maturity. Seeds hard, shiny black, 2.5–4 mm long, enclosed for up to two-thirds of their length by a white aril. Mountains of both main islands of New Zealand and adjacent lowlands of southwestern South Island. Exclusive or prominent in the canopy of subalpine scrublands (hence the scientific name) and low forest or scattered in the understory of lowland mixed forests on impoverished soils; 0–300 and (450–)900–1,500(–1,600) m. Zone 8. Synonyms: *Phyllocladus aspleniifolius* var. *alpinus* (J. Hooker) H. Keng, *P. trichomanoides* var. *alpinus* (J. Hooker) Parlatore.

Mountain toatoa is far more widely distributed than the other two species of *Phyllocladus* native to New Zealand, tanekaha (*P. trichomanoides*) and toatoa (*P. toatoa*), both of which are (or are nearly) confined to North Island. Mountain toatoa, in contrast, extends all the way to the southern tip of South Island, though it is not found on adjacent Stewart Island. Throughout most of its range, it is predominantly a shrub of subalpine habitats, but on the western side of South Island it is also common, though less prominent, in the lowlands. The different morphological forms found at high versus low elevations have been considered genetically determined ecotypes, although they intermingle and intergrade in intermediate or mixed habitats. Further study of the pattern of variation in the species and its genetic basis is warranted and could lead to formal recognition of varieties or subspecies while, in practice, some of this variation has been exploited through selection of cultivars. Mountain toatoa is the most frost hardy species of *Phyllocladus* in cultivation, and selections from its highest stands in southern South Island might further extend its hardiness rating.

It is possible that some of the variation in *Phyllocladus alpinus* is due to hybridization with the other two New Zealand species, but this concept has not been supported by limited genetic studies to date. Occasional natural hybrids among these three species in their regions of overlap have been reported but still require confirmation. Mountain toatoa is much more closely related to the Tasmanian celery-top pine (*P. aspleniifolius*) than it is to the other New Zealand species and is sometimes considered a variety of the same species. Since no other conifer species are common to the two regions, these two celery pines are kept separate here, but a convincing estimate of their time of physical and genetic separation might help settle this taxonomic question. (While the landmasses of Australia and New Zealand separated from one another in the late Cretaceous, some 75 million years ago, there is suggestive evidence for long-distance dispersal between them after that.) There is no apparent justification for combining *P. alpinus* with tanekaha, which has compound phylloclades. Both species display rather modest genetic variation and differ from one another in most of the enzymes that are variable.

Although mountain toatoa is a shrub, it can live to be 350 years old, the shortest known maximum longevity in the genus. The actual life span may be much greater, however, because individuals are clonal, with lower branches rooting in where they contact the soil. Through this kind of spread, mountain toatoa comes to dominate subalpine scrubland with time, even if few seedlings become established following a major disturbance, like a fire. The thin branches are very flexible, in part because of elasticity of the wood, and thus shed snow well as well as resisting wind damage, leading to a major role in vegetating avalanche tracts, which few other New Zealand shrubs can tolerate. Although the seed cones are berrylike and attractive to birds, like those of most Podocarpaceae, the seeds also pop out of them as they dry if they have not been removed by a potential disperser. The visual appearance and contrast of the black seed, white aril, and red seed cone at maturity is very striking and presumably enhances the chances of effective bird dispersal. It is also rather ornamental in cultivated plants, particularly in selections with persistently glaucous, bluish green phylloclades.

Phyllocladus aspleniifolius (Labillardière) J. Hooker

CELERY-TOP PINE

Tree to 20(–30) m tall, with a cylindrical, straight trunk to 1 m in diameter. Bark thin, marked by numerous warty lenticels, reddish brown to dark brown and smooth at first, thickening, weathering gray to very dark gray, breaking up into rectangular scales, and finally becoming deeply furrowed with age. Crown dense and conical in youth, becoming more open and irregularly cylindrical to dome-shaped, with short, slender, branches passing from upwardly angled to downswept. Phylloclades attached singly or in rings of two to five on the long shoots, simple, (1.5–)2.5–5(–8) cm long, shiny dark green. Individual phylloclades generally diamond-shaped overall, mostly shallowly toothed in the upper half (coarsely toothed in juveniles) but varying to deeply lobed. Juvenile leaves (1.5–)7–10 mm long. Scale leaves of adult long shoots and buds 2–3 mm long. Plants monoecious or sometimes dioecious. Pollen cones single or in clusters of two or three (to five) at the tip of a short branchlet. Each cone 3–5 mm long, 1–2 mm thick, green to pink or yellow just before shedding pollen, on a very short stalk 1–2 mm long. Seed cones single or, more commonly, in clusters of (two or) three or four attached to the sides of phylloclades or wholly replacing the expanded portion. Each seed cone roughly spherical, 3–5 mm in diameter, with (one to) three to eight bracts of which one to five are fertile and mature one to three seeds. Bracts swollen, juicy, pink to red at maturity. Seeds hard, greenish black, 2.5–3.5 mm long, enclosed for about three-quarters of their length by a white aril. Throughout Tasmania (Australia) and a few offshore islands but rare in the northeast and most common in the mountains of the west. Scattered or clumped in the canopy or understory of various mixed cool temperate rain forests, wet sclerophyll forests dominated by *Eucalyptus,* and low wet forests; 0–750(–1,200) m. Zone 9. Synonyms: *Phyllocladus glaucus* Carrière, *P. trichomanoides* var. *glaucus* (Carrière) Parlatore.

Celery-top pine was the first species of *Phyllocladus* to be given a formal botanical description and name (as a species of *Podocarpus,* hence the scientific name, Latin for "*Asplenium* leaf" because the "leaves," actually the phylloclades, looked more like the pinnae of the fern wall rue, *A. ruta-muraria,* than they did like those of the species of *Podocarpus* to which it was compared). It is the most widely distributed conifer species in Tasmania, covering the widest ecological range as well. The ecological diversity may be enhanced by gene flow between populations while the persistence of populations within various vegetation types is promoted by relatively continuous regeneration on an unusually wide variety of microsites raised above the forest floor. Unusually among conifers, then, regeneration does not necessarily require episodic disturbances, although such disturbances promote more extensive pulses of regeneration.

Pollen records show that celery-top pine attained ecological prominence soon after deglaciation of Tasmania 13,000 years ago and has, despite fluctuations in abundance, retained that prominence ever since. It has been an important timber tree, the easily worked, durable, dimensionally stable wood being used in many outdoor applications, including boatbuilding and for railway sleepers and flooring. The growth rings are very evident, and this, coupled with their responsiveness to climatic factors and the wide geographic and ecological range of the trees and their substantial longevity (a maximum age of about 900 years, the oldest known in *Phyllocladus*), has made celery-top pine a prime candidate for development of long-term dendroclimatological chronologies for Tasmania. It is closely related to mountain toatoa (*P. alpinus*) of New Zealand, and the two are sometimes treated as varieties of a single species.

Phyllocladus hypophyllus J. Hooker

MALESIAN CELERY PINE

Tree to 30 m tall though much shorter on exposed ridge tops and shrubby near the subalpine tree line, with a short cylindrical to fluted trunk to 1 m in diameter. Bark thin, marked by numerous large, warty lenticels, reddish brown to dark brown, breaking up into irregularly rectangular scales and revealing tan inner bark. Crown conical at first, becoming narrowly to broadly dome-shaped with age, with whorls of slender, spreading or contorted, upwardly angled branches. Phylloclades attached singly or in rings of two to five on the long shoots, compound, 10–15 cm long overall, with 5–10 alternately pinnate segments. Each segment (1.5–)3–6(–8) cm long (to 10 cm in juveniles), egg-shaped to diamond-shaped, variously roundly to sharply toothed or deeply lobed, dark green (or sometimes thinly waxy at first) above and paler green to conspicuously waxy greenish white beneath (hence the scientific name, Latin for "under leaf"). Juvenile leaves 5–8 mm long. Scale leaves of adult long shoots and buds 2–3 mm long. Plants usually dioecious. Pollen cones single or, more commonly, in clusters of 2–8(–15) at the tip of a short branchlet or at the base of a growth increment. Each cone 10–15 mm long, 2.5–4 mm thick, yellowish brown just before shedding pollen, on a long stalk 5–25 mm long. Seed cones one (to three) in a notch near the tip of a phylloclade segment or on a highly reduced phylloclade, even to the extent of completely lacking an expanded portion. Each seed cone egg-shaped, 5–10 mm long, 4–6 mm in diameter, with 5–15 bracts of which (one or) two or three (to five) are fertile and mature one to three seeds. Bracts swollen, juicy, bright red fading to tan at maturity. Seeds hard, shiny brown to tan, (3–)5–8 mm long, enclosed for about half their length by the white aril. Common through the mountains of northern and central Malesia from the northern tip of Luzon (Philippines) to the western tip of Borneo (Indonesia) to the eastern tip of New Guinea (Papua New Guinea). Scattered or somewhat gregarious in a wide range of generally wet, mixed forests and scrublands, from lower montane forests to mossy forests and subalpine dwarf forests and scrublands on exposed ridges, with a roughly inverse relationship between stature and relative prominence in the community; (900–)1,500–3,400(–4,000) m. Zone 10. Synonyms: *Phyllocladus hypophyllus* var. *protractus* O. Warburg, *P. major* Pilger, *P. protractus* (O. Warburg) Pilger.

With such a wide ecological range, Malesian celery pine displays considerable variation, particularly in the shape of the phylloclades and the amount of waxiness on their lower surface. Because of these and other variations, the trees from the three main distributional areas (Borneo, New Guinea, and the Philippines) are sometimes assigned to three separate species. However, there seems to be so much variability within these regions and overlap between them that assignment to a single species, as is done by most workers, appears fully justified. Whether a workable scheme of varieties could be devised within *Phyllocladus hypophyllus* is worthy of further study, including genetic studies across the ecological as well as geographic range of the species. The ecological component of variation in Malesian celery pine might well be studied on Mount Kinabalu in Borneo (Sabah, Malaysia), from which it was first described and where it is common and found over almost the whole elevational gradient for the species.

Phyllocladus toatoa Molloy

TOATOA (MAORI AND ENGLISH)

Tree to 12(–25) m tall, with a cylindrical, rapidly tapering trunk to 0.4(–0.9) m in diameter. Bark thin, smooth at first with a scattering of warty lenticels and irregular, horizontal, low ridges, dark brown weathering grayish brown to silvery brown, flaking in rectangular scales to reveal dark orange patches and finally becoming deeply furrowed. Crown conical at first, becoming irregularly cylindrical to dome-shaped with age, with whorls of stout, generally strongly upwardly angled branches and many scattered, late developing, extra branches between the regular flushes. Phylloclades attached singly or in rings of (2–)5–7(–10) at the tips of growth increments, compound, (5–)12–20(–30) cm long overall (down to 1.5 cm in juveniles), with 5–10(–14), predominantly alternately pinnate segments. Each segment 2–5(–8) cm long, roundly diamond-shaped to fan-shaped, roundly and shallowly toothed (sharply toothed in juveniles) to deeply lobed, more or less yellowish green above and thinly to prominently waxy bluish green beneath. Juvenile leaves (5–)10–15 mm long. Scale leaves of adult long shoots and buds (0.5–)10–15(–20) mm long. Plants predominantly dioecious but individuals of one sex often with a few cones of the other. Pollen cones single or, more frequently, in clusters of (5–)10–20 at the tip of a short branchlet or at the base of a growth increment. Each pollen cone (1–)1.5–2.5(–3) cm long, 4–7(–10) mm thick, yellowish brown to pink or brownish red just before shedding pollen, on a long, heavy stalk 5–10(–15) mm long. Seed cones in open groups of (three or) four to seven (or eight), each individually replacing a lower phylloclade segment. Each seed cone spherical to a little elongate, (5–)7–13 mm long, 5–9 mm in diameter, with 10–20(–25) bracts of which 8–20 are fertile and mature (1–)3–15(–20) seeds. Bracts swol-

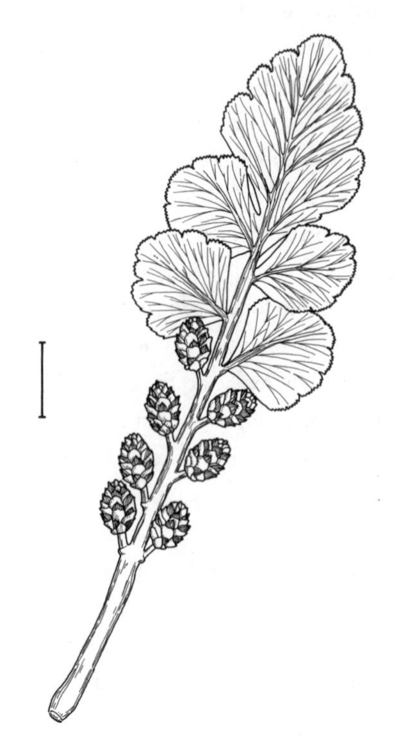

Reproductive phylloclade with mature seed cones of toatoa (*Phyllocladus toatoa*); scale, 2 cm.

len, juicy, pinkish red to brownish red at maturity. Seeds hard, shiny dark brown to black, (2–)3–4 mm long, enclosed for about half their length by the thick white aril. Northern two-thirds of North Island and some adjacent offshore islands, New Zealand. Scattered or, less commonly, somewhat gregarious in the understory of mixed lowland and montane forests and in the canopy of stunted forests and scrublands on exposed ridges, usually on damp to wet substrates; 0–600(–1,000) m. Zone 10. Synonym: *Phyllocladus glaucus* Carrière (misapplied).

Toatoa is the most ornamental of the *Phyllocladus* species in cultivation, although it is very tender, even with selections from its harshest localities. It is the most narrowly distributed celery pine in New Zealand but is common in much of its range and, with many occurrences in protected areas as well as abundant regeneration in disturbed areas, is essentially not of conservation concern. Although the tree is moderately common and its white wood is tough and strong, it is generally too small and not concentrated enough to have ever been of commercial importance as a timber tree, unlike its frequent neighbor, kauri (*Agathis australis*). It is much less long lived than kauri, reaching a maximum age of about 500 years and usually dying long before that.

Throughout its range, toatoa overlaps with the other two species of *Phyllocladus* in New Zealand, alpine toatoa (*P. alpinus*) and tanekaha (*P. trichomanoides*), essentially falling between them in elevation of greatest abundance. Hybrids with both species have been reported from disturbed sites in the presence of each pair of parents but have been little studied and not formally described. For more than 125 years and following the lead of New Zealand forest botanist Thomas Kirk, this species was mistakenly called *P. glaucus*. The name *P. toatoa* was proposed in 1996 for this long-known species because the type specimen of *P. glaucus* actually came from a cultivated Tasmanian celery-top pine (*P. aspleniifolius*), the only species of *Phyllocladus* that had been introduced to Europe by 1855.

Phyllocladus trichomanoides D. Don

TANEKAHA (MAORI AND ENGLISH), CELERY PINE

Plate 36

Tree to 20(–25) m tall, with a straight cylindrical trunk to 0.8(–1) m in diameter and free of branches for half its height. Bark thin, smooth or with a roughened surface, greenish brown to dark gray or black, thickening, flaking, and weathering lighter with age and sometimes becoming furrowed. Crown conical at first, becoming narrowly and deeply cylindrical with age, with whorls of slender, horizontal to gently upwardly or downwardly angled branches. Phylloclades attached singly or, more frequently, in rings of (two to) four to eight at the tips of growth increments or of other phylloclades, compound, (2.5–)3–8(–16) cm long overall, with (6–)9–12(–15) alternately pinnate segments. Each segment (1–)1.5–2.5(–4) cm long, roundly diamond- to egg-shaped, minutely angularly toothed to deeply lobed (especially sharply cut and toothed in juveniles), bluish green with a variable film of wax above and beneath. Juvenile leaves 8–10(–20) mm long. Scale leaves of adult long shoots and buds 2–6(–10) mm long. Plants predominantly monoecious. Pollen cones single or in clusters of (2–)5–10(–15) at the tip of a growth increment or of a phylloclade. Each cone (5–)7–11 mm long by 3.5–4.5 mm thick, reddish purple to red just before shedding pollen, on a short stalk 2–7(–10) mm long. Seed cones in open groups of (three to) five to eight, each attached singly or two or three together at the sides of (or totally replacing) the reduced segments of a specialized reproductive phylloclade. Fertile phylloclades 3–5 cm long, arranged in a whorl replacing a whorl of ordinary photosynthetic phylloclades at the tip of a growth increment. Each seed cone roughly spherical, 2–5 mm in diameter, with two to five (to eight) bracts of which one to five are fertile and mature one to three seeds. Bracts, swollen, juicy, reddish purple to bluish purple at maturity. Seeds hard, shiny bluish black, 2–3 mm long, enclosed for about half their length by the thin, white aril. Discontinuous in New Zealand throughout

Vigorous mature tanekaha (*Phyllocladus trichomanoides*) in nature.

northern and central North Island and the northern edge of South Island. Lowland mixed forests; 0–800 m. Zone 10.

Tanekaha, as the most frequent *Phyllocladus* in lowland forests, is also normally the largest celery pine in New Zealand. It is a handsome, fairly slow growing tree with a very regular branching pattern when young made up mostly of glossy phylloclades. The frondlike phylloclades are the source of the scientific name, Latin for "like *Trichomanes,*" a genus of ferns with leaflets that may be similar in shape, if not texture, to the individual segments of tanekaha phylloclades. Since tanekaha is tolerant of both sun and shade, it is highly suited to cultivation in wet, temperate but frost-free climates. While it is slow growing, it can live to about 500 years (although 250 years is more typical), so large trees were formerly available for commercial exploitation. The wood is not only very fine-grained and strong but also the most flexible among native New Zealand softwoods, a property appreciated in the construction of boats, fishing rods, and spears. The wood was also put to the usual range of uses for softwoods. Annual variation in the clearly defined growth rings in response to varying climate, coupled with moderate longevity and reasonable durability of the timber, including soundness of heartwood in living trees, give tanekaha a modest potential for dendrochronological studies, in kauri (*Agathis australis*) forests, for instance.

Besides the wood, the bark and phylloclades have been exploited as sources of tannins and of dye. These products did not require large, mature trees so harvesting could take advantage of the extensive regeneration of this species in disturbed forests resulting from dispersal of the abundant seeds, smallest amongst all the New Zealand podocarp species. Tannins from the phylloclades have a rather complex composition, with some compounds found in all investigated species of *Phyllocladus* and other related compounds found only in tanekaha. Following up on Maori traditional practice, some of these compounds are being investigated for medicinal uses so that, even though natural tannins are no longer used in tanning leather, tanekaha phylloclades may once again become an economic product.

Twig of tanekaha (*Phyllocladus trichomanoides*) with 2 years of growth, including immature seed cones, ×1.

Picea A. Dietrich

SPRUCE

Pinaceae

Evergreen trees with a straight, single trunk bearing regular annual tiers of three to five horizontal branches from the base to form a symmetrical, spirelike to broad conical crown which usually flattens somewhat at the top in extreme age. Bark generally smooth at first, nonfibrous and flaking in small, irregular patches, usually reddish in overall hue. Branchlets all elongate, without distinction into long and short shoots, hairy or not, prominently grooved between raised leaf bases. Each base bearing a distinct woody peg (sterigma) to which the needle is attached. Winter buds well developed, scaly. Leaves spirally arranged and radiating all around the twigs, though often somewhat parted on the lower side, needlelike, often square in cross section but sometimes flat or rounded, straight or curved, the tip pointed, sometimes painfully so, the base narrowed to a short petiole and jointed at the raised peg.

Plants monoecious. Pollen cones single, hanging down along the twigs from scaly buds in the axils of needles of the previous year. Each cone with numerous, densely spirally arranged pollen scales, each scale bearing two pollen sacs. Pollen grains large to very large (body 50–120 µm long, 65–130 µm overall), larger than those of *Pinus* and a little smaller on average than those of *Abies,* with two round air bladders a little smaller than the oblong body and diverging at a right angle or a little less. Surface of the body minutely warty and the bladders with more wrinkled sculpturing. Seed cones cylindrical or oblong, single on short stalks in the axils of leaves of the previous year among the upper branches of the crown, upright at pollination but dangling at maturity. Maturing and usually falling in a single season, the numerous, densely spirally arranged seed scales then spreading to release the seeds. Seed scales round to somewhat diamond-shaped, the tiny, hidden bract attached only at the base. Seeds two per scale, wedge-shaped, the asymmetrical wing derived from the seed scale, cupping the seed body below and much longer and a little wider than it. Cotyledons 5–15, each with one vein. Chromosome base number $x = 12$.

Wood odorless to slightly fragrant, light but unusually hard and strong, not very decay resistant, with white to yellowish sapwood either indistinguishable from or gradually blending into the nearly white through pinkish yellow to light brown heartwood.

Grain very even and fine to fairly coarse, usually with well-defined growth rings marked by a very gradual transition to a variable-width band of much smaller and thicker-walled, spirally thickened, latewood tracheids, which like those of the earlywood, are exceptionally long. Resin canals lined with thick-walled cells, not especially abundant but scattered through the growth rings or more frequent in the latewood, sometimes arranged in open bands. Individual resin parenchyma cells normally absent.

Usually with lines of stomates in a central, nonwaxy band on each of the four sides, but with just two bands in flattened needles. Each stomate deeply sunken beneath and almost hidden by the four to six surrounding subsidiary cells, which are covered by a very thick cuticle without a Florin ring. Leaf cross section with a (sometimes interrupted) single pair of resin canals, usually near the side corners of the square on either side of the apparently single-stranded midvein embedded in a cylinder of transfusion tissue. Photosynthetic tissue not (or only weakly) organized into a definite palisade layer but all mesophyll cells appearing to radiate from the midvein out to the epidermis and underlying hypodermis, which consists of a single layer overall that builds up to two or three layers at the corners.

Twenty-nine species across the northern hemisphere, throughout the boreal forest and in mountains southward. References: Farjon 1990, Franco 1993a, Fu et al. 1999a, Komarov 1986, Lee 1979, Martínez 1963, Mikkola 1969, K. C. Sahni 1990, Schmidt 1989, Taylor 1993b, von Schantz and Juvonen 1966, J. W. Wright 1955, Yamazaki 1995.

Spruces are the most prominent trees of the vast boreal forests of the northern hemisphere, extending southward in the mountains or in cooler than normal habitats. They have an overall more northerly range than the similarly distributed silver firs (*Abies*) and a wider range than the larches (*Larix*), extending much farther south. Most spruce species experience snowy winters, and their conical growth habit minimizes snow accumulation and persistence. This same conical form and the density of their evergreen foliage makes them important horticultural subjects in cool climates. They are, in fact, among the most commonly cultivated conifers in these regions, and numerous cultivars have been selected, mostly from five species (*Picea abies, P. glauca, P. omorika, P. orientalis,* and *P. pungens*) but with a smattering among a dozen others. Cultivar selection has been highly, varied embracing extreme variations in habit (such as narrowly upright, cascading weeping, or sparsely branched), color variations (from golden to cold pale blue), reproductive variation (early and abundant seed cone production or altered cone form), and a myriad of dwarfs varying in shape, density, branching form, and needle length and color.

Spruces are also important in forestry because their exceptionally long wood fibers, requiring little destructive bleaching, are good for paper pulp. The wood is also used for general construction, and some species are cultivated in plantations for this purpose, like *Picea sitchensis* in Scotland. Spruces are also among the trees most strongly affected by forest dieback in central Europe (*P. abies*) and eastern North America (*P. rubens*), which is generally attributed to industrial and automotive pollution.

Although there are some uncertainties in spruce classification, recognition of species has been far more stable than it has in the associated silver firs. The relationships among the species, however, are not settled, and the traditional classification into three botanical sections based on the shape of the needles and of

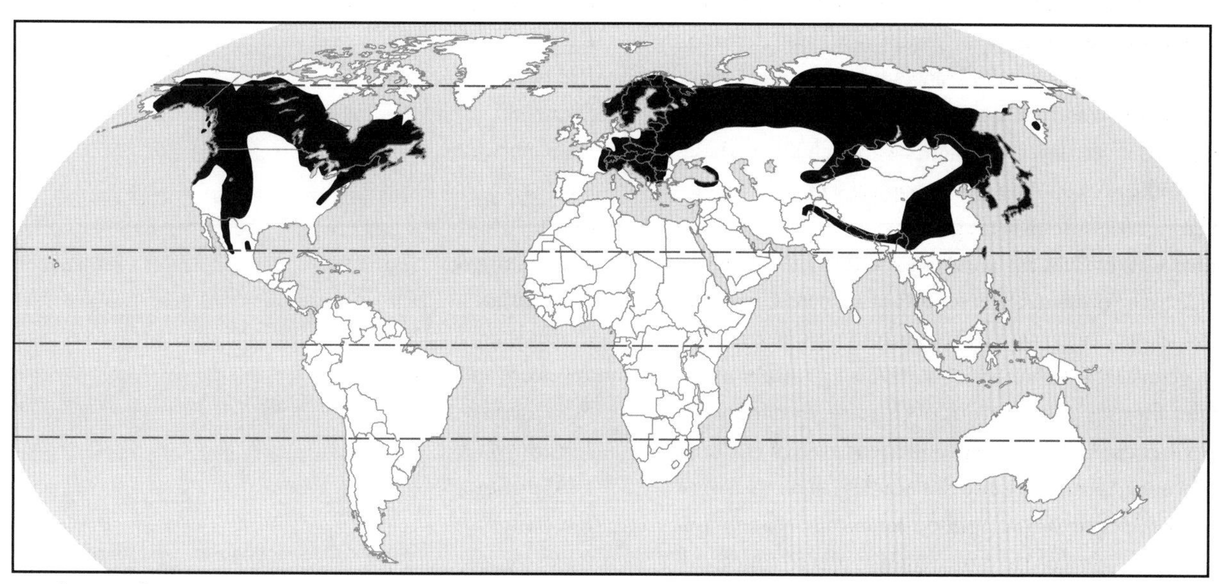

Distribution of *Picea* across the northern continents, scale 1 : 200 million

the scales of the seed cones is not consistent with the whole range of available evidence.

Crossing behavior among the species is particularly perplexing. A great deal of information on crossability has accumulated through efforts at spruce improvement for forestry. These tree breeding programs have not actually had much practical impact on forestry practice, despite the apparent superiority of certain species hybrids. Crossability among spruce species ranges from fully interfertile to completely incompatible in all attempts. The degree of crossability seems to have only a small correlation with the degree of relatedness of the species. Sometimes closely related species are readily intercrossable (like *Picea glauca, P. engelmannii,* and *P. sitchensis*), and distantly related species may be reproductively isolated from one another (like *P. chihuahuana* and *P. abies*). Just as frequently, however, it appears that closely related species are largely reproductively isolated from one another (like *P. engelmannii* and *P. pungens,* or *P. mariana* and *P. rubens*), while relatively unrelated species may still be crossable. A remarkable example of the latter is the distinctive and rare Serbian spruce (*P. omorika*), which has the widest range of crossability of any spruce, crossing freely or with varying degrees of difficulty with many species pairs that cannot cross with each other. The reported crossability of different spruce species is somewhat inflated by the tendency to produce normal seed cones and heavy seeds even when there are no viable embryos. Acceptance of percentage germination (and later hybrid verification) as the appropriate measure of crossability in *Picea* drastically lowers the overall rate of crossability reported but still leaves a very high proportion of species with at least some crossability in artificial pollinations. Despite the frequency of successful species combinations in breeding trials, hybrids are less common in nature than those of silver firs. They occur most frequently between related species with abutting (parapatric) ranges, like *P. glauca* and *P. engelmannii.* Related species with broadly overlapping (sympatric) ranges, like *P. engelmannii* and *P. pungens,* or *P. mariana* and *P. rubens,* typically evolved reproductive barriers and exhibit very low crossabilities with each other. These evolutionary influences on crossability drastically limit its usefulness for inferring relationships in the genus.

Definite fossils of *Picea* are found in early Tertiary sediments and continue to the present. Older reports of the genus from the Cretaceous lack its diagnostic features and are now assigned to the extinct genus *Pityostrobus,* which contains (probably unrelated) species showing combinations of characters not found in any of the living genera. The earliest confirmed fossils of *Picea* are from late Eocene deposits. Because the needles fall off the twigs as they dry (a problem in preparing good botanical specimens today), most of the fossil record of spruce consists of seed cones and individual needles. Occasionally, however, twigs enter sediments before the needles drop, and some of these intact branchlets from the Miocene of Oregon bear needles up to 6 cm long, longer than those of any living species. These have been interpreted as coming from a warmer climate than any supporting spruce species today, so the ecological diversity of the genus may have been broader in the Tertiary than it is today. Another indication of this is that spruces were present (as a minor component) in the high arctic *Metasequoia* forests of the early Tertiary, far north of their present northern limit, under climatic conditions that are not duplicated anywhere on Earth today. One fossil species, *P. critchfieldii* Jackson & Weng, is the only tree species known to have become extinct at the end of the last glaciation in North America. It had been widespread and dominant in forests south of the ice in the eastern United States at the height of glaciation just 18,000 years ago before suddenly vanishing as the climate warmed.

Geographic Guide to *Picea*

Europe and western Asia
- *Picea abies*
- *P. omorika*
- *P. orientalis*

Siberia, Russian Far East, northern China, and Korea
- *Picea abies*
- *P. asperata*
- *P. jezoensis*
- *P. koyamai*
- *P. meyeri*
- *P. schrenkiana*
- *P. wilsonii*

Japan, Sakhalin, and Kurile Islands
- *Picea alcoquiana*
- *P. glehnii*
- *P. jezoensis*
- *P. koyamai*
- *P. maximowiczii*
- *P. polita*

Himalaya
- *Picea brachytyla*
- *P. smithiana*
- *P. spinulosa*

Southwestern and central China
- *Picea asperata*
- *P. brachytyla*
- *P. likiangensis*
- *P. neoveitchii*
- *P. purpurea*
- *P. wilsonii*

Taiwan
Picea morrisonicola
Eastern and boreal North America (east of the Rocky Mountains)
Picea glauca
P. mariana
P. rubens
Western North America
Picea breweriana
P. engelmannii
P. pungens
P. sitchensis
Mexico
Picea chihuahuana
P. engelmannii
P. mexicana

Identification Guide to *Picea*

Confident identification of spruce trees to species is usually very difficult using only leaf and shoot characteristics, so features of mature seed cones are also employed in this identification guide. Fortunately, shed cones are found beneath mature trees more often than not, even if they cannot be found still attached to the shoots. For convenience of identification, the 29 species of *Picea* may be divided into six artificial groups based upon the length of the longest ordinary needles and whether the needles have a distinct top and bottom side (bifacial) or all four faces are similar (square):

Group A, longest needles no more than 1.6 cm, bifacial

Group B, longest needles no more than 1.6 cm, square

Group C, longest needles 1.7–2.5 cm, bifacial

Group D, longest needles 1.7–2.5 cm, square

Group E, longest needles more than 2.5 cm, bifacial

Group F, longest needles more than 2.5 cm, square

Group A. These three species may be distinguished by the length of the central bud and the length of a middling seed cone:

Picea purpurea, central bud less than 6 mm, seed cone less than 5 cm

P. likiangensis, central bud up to 6 mm, seed cone more than 5 cm

P. wilsonii, central bud more than 6 mm, seed cone more than 5 cm

Group B. These six species may be distinguished by the length of the central bud and the length of a middling seed cone:

Picea mariana, central bud up to 5 mm, seed cone less than 3 cm

P. maximowiczii, central bud up to 5 mm, seed cone 5–7 cm

P. orientalis, central bud up to 5 mm, seed cone more than 7 cm

P. glehnii, central bud 5–6 mm, seed cone 3–5 cm

P. alcoquiana, central bud 5–6 mm, seed cone more than 7 cm

P. rubens, central bud at least 6 mm, seed cone 3–5 cm

Group C. These five species may be distinguished by the length of the central bud, whether the new branchlets are hairy or not, and the length, width, and color of a ripe, open middling seed cone:

Picea morrisonicola, central bud up to 5 mm, branchlets hairless, open seed cone less that 7 cm by up to 3 cm, dark brown

P. omorika, central bud up to 5 mm, branchlets hairy, open seed cone less than 7 cm by less than 3 cm, dark brown

P. jezoensis, central bud more than 5 mm, branchlets hairless, open seed cone less than 7 cm by up to 3 cm, pale brown

P. sitchensis, central bud more than 5 mm, branchlets hairless, open seed cone at least 7 cm by up to 3 cm, pale brown

P. brachytyla, central bud more than 5 mm, branchlets hairless or slightly hairy, open seed cone more than 7 cm by at least 3 cm, dark brown

Group D. These seven species may be distinguished by the length of the central bud and the length and width of a ripe, open, middling seed cone:

Picea glauca, central bud less than 7 mm, open seed cone less than 8 cm by up to 2.5 cm

P. abies, central bud less than 7 mm, open seed cone less than 8 cm to more than 12 cm by 2.5–4 cm

P. neoveitchii, central bud less than 7 mm, open seed cone 8–12 cm by more than 4 cm

P. koyamai, central bud at least 7 mm, open seed cone less than 8 cm by 2.5–4 cm

P. asperata, central bud more than 7 mm, open seed cone 8–12 cm by 2.5–4 cm

P. polita, central bud more than 7 mm, open seed cone 8–12 cm by more than 4 cm

P. chihuahuana, central bud more than 7 mm, open seed cone at least 12 cm by more than 4 cm

Group E. These three species may be distinguished by the color of the branchlets and whether the tips of the seed scales are rounded or have a narrowed projection:

Picea breweriana, branchlets reddish brown, seed scale tips rounded

P. schrenkiana, branchlets yellowish gray, seed scale tips rounded

P. spinulosa, branchlets yellowish gray, seed scale tips projecting

Group F. These five species may be distinguished by the length of the central bud, whether the twigs are hairy or not, the length of a middling seed cone, and whether the seed scales are rounded or have a narrowed projection:

Picea mexicana, central bud less than 6 mm, twigs hairless, seed cone less than 6 cm, seed scales projecting

P. engelmannii, central bud up to 6 mm, twigs hairy, seed cone up to 6 cm, seed scales projecting

P. pungens, central bud more than 6 mm, twigs hairless, seed cone 6–10 cm, seed scales projecting

P. meyeri, central bud more than 6 mm, twigs hairless to hairy, seed cone 6–10 cm, seed scales rounded

P. smithiana, central bud more than 6 mm, twigs hairless, seed cone more than 10 cm, seed scales rounded

Picea abies (Linnaeus) H. Karsten

NORWAY SPRUCE, FICHTE (GERMAN), EPICÉA COMMUN (FRENCH), SMEREK (CZECH), JODLA (POLISH), GRAN (NORWEGIAN), KUUSI (FINNISH), EL' EVROPEISKAYA (RUSSIAN)

Plate 37

Tree to 50(–60) m tall, with trunk to 1.5(–2) m in diameter. Bark breaking up somewhat into gray plates at the base of old trees. Crowns quite variable, narrowly or broadly conical, with upswept or stiffly outstretched branches bearing horizontal or dangling side branches. New branchlets orange-brown, hairless to densely hairy. Buds 4–7 mm long, slightly resinous or not. Needles dark to bright green, 1–2.5(–3) cm long, curved forward, square, with two to four lines of stomates on each side, not prickly. Pollen cones (8–)12–25 mm long, purplish red. Seed cones 5–16(–20) cm long, green before maturity, ripening medium brown. Seed scales roughly egg- or diamond-shaped, woody and stiff. Seed body 3–5 mm long, the wing 10–15 mm longer. Northern Eurasia, from the Maritime Alps and northern Balkan Peninsula north to Scandinavia and east across European Russia and Siberia to the Lena River, south to the Altai and Amur regions. In pure stands or mixed with other boreal and montane conifers and hardwoods on moist soils; 0–3,000 m. Zone 5. Three subspecies.

Norway spruce has the broadest natural range in the genus and is also the most widely cultivated species of *Picea* in both forestry and horticulture and has the greatest number of cultivars. It is one of the important timber species of the world, exploited for construction wood, pulp, firewood, and other products. There is a great deal of variation in morphological and ecological characteristics across the range, but much of this variation may be found within populations as well as from place to place. Trees of the western and eastern parts of the range of the species are distinct enough to be considered different subspecies here and are often treated as separate species. They intergrade in northeastern Europe, and the intermediates are treated as a third, hybrid (designated notho-) subspecies. *Picea abies* is one of the tallest tree species in Europe, giving it one of its abandoned synonyms, *excelsa,* lofty. Individuals may live to be 300 years old or more, but premature death became all too prevalent in the spruce forests of central Europe during the period of forest decline in the second half of the 20th century. The species name reflects early confusion between the spruces (*Picea*) and silver firs (*Abies*) before the generic concepts were stabilized in their present form.

Picea abies (Linnaeus) H. Karsten **subsp. *abies***

NORWAY SPRUCE, ETC.

Twigs hairless or slightly hairy. Needles dark green. Seed cones 10–20 cm long. Seed scale margins cutoff triangular. Western portion of the range of the species in Europe. Synonyms: *Picea abies* subsp. *alpestris* (Brügger) Domin, *P. abies* var. *acuminata* (G. Beck) Dallimore & A. B. Jackson, *P. excelsa* J. Link, *P. montana* Schur.

Picea abies* nothosubsp. *fennica (E. Regel) Parfenov

Trees intermediate between subspecies *abies* and *obovata* from their zone of contact in northeastern Europe. Synonym: *P.* [×]*fennica* E. Regel

Picea abies* subsp. *obovata (Ledebour) Hultén

SIBERIAN SPRUCE, EL' SIBIRSKAYA (RUSSIAN), KARA-SHERSE (TATAR), XIAN BEI YUN SHAN (CHINESE)

Twigs densely fuzzy. Needles bright green. Seed cones 5–8 cm long. Seed scale margins round or even notched. Eastern portion of the range of the species west to northern Scandinavia. Synonym: *Picea obovata* Ledebour.

Upper side of twig of Norway spruce (*Picea abies* subsp. *abies*) with 2 years of growth, ×1.

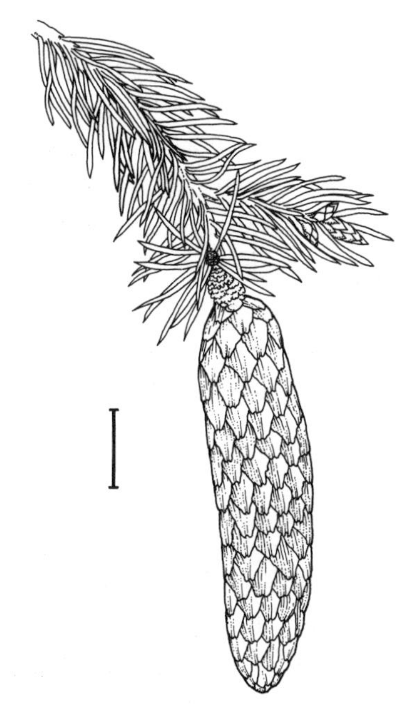

Twig of Norway spruce (*Picea abies* subsp. *abies*) with unopened mature seed cone; scale, 2 cm.

Open natural stand of mature Siberian spruce (*Picea abies* subsp. *obovata*).

Picea* ×*albertiana S. Brown

ALBERTA WHITE SPRUCE, INTERIOR SPRUCE

Tree to 30 m tall, with characteristics combining those of the parent species or intermediate between them. It combines the hairy twigs of Engelmann spruce (*Picea engelmannii*) with needles that are generally more like those of white spruce (*P. glauca*) though a little longer, on average, at 2–2.5 cm. The seed cones are shorter than those of *P. engelmannii* and a little broader than those of *P. glauca,* and the scales have a somewhat angular margin compared to rounded in *P. glauca* and cutoff triangular in *P. engelmannii.* This hybrid occurs in the zone of contact between its closely related, but ecologically distinct, parents along the Rocky Mountain axis in Montana, Alberta, and British Columbia. Hybrid traits may be found some distance away from this overlap region but not full-blown hybrids. Much of the economically important spruce resource of interior British Columbia consists of stands of hybrids or introgressants of white spruce with Engelmann spruce traits. The reproductive cycle of Alberta white spruce in this region has been studied carefully with an eye to enhancing seed orchard cone production. Synonyms: *P. glauca* var. *albertiana* (S. Brown) C. Sargent, *P. glauca* var. *porsildii* Raup.

Picea alcoquiana (J. G. Veitch ex Lindley) Carrière

ALCOCK SPRUCE, IRAMOMI, MATSUHADA (JAPANESE)

Tree to 30(–35) m tall, with trunk to 1 m in diameter. Bark pale gray to grayish brown, breaking up into vertical plates on old trees. Crown broadly conical to egg-shaped, with long, thin, horizontal branches, rising when young, sweeping gently down with age, although still turning up at the tips. New branchlets yellowish brown to reddish brown, generally hairless but sometimes thinly hairy on vigorous shoots or those bearing seed cones, especially in the grooves between the leaf bases. Buds 5–6 mm long, slightly resinous. Needles dark bluish green, sometimes with a waxy film, 1–1.5(–2) cm long, curved slightly forward, square, with one to three lines of stomates on the sides facing the twig and three to six lines on the other two sides, prickly on young trees but just pointed on older ones. Pollen cones 10–15 mm long, red. Seed cones (4–)7–10(–12) cm long, reddish purple before maturity, ripening cinnamon brown. Seed scales broadly spoon- to diamond-shaped, thinly woody and pliable, straight or curled back at the tip. Seed body 3–4 mm long, the wing 8–12 mm longer. Mountains of central Honshū (Japan). Scattered among *Abies homolepis, A. veitchii, Picea jezoensis,* and other conifers and hardwoods in subalpine forest; (700–)1,000–2,000(–2,200) m. Zone 6. Synonyms: *P. alcoquiana* var. *reflexa* (Shirasawa & M. Koyama) Fitschen, *P. bicolor* (Maximowicz) H. Mayr, *P. bicolor* var. *reflexa* Shirasawa & M. Koyama.

This rare subalpine spruce co-occurs with the much more common Yezo spruce (*Picea jezoensis*), leading to persistent confusion in the identity and correct name of *P. alcoquiana* because John Gould Veitch (1839–1870) sent back specimens and seeds of both species to England from Japan under the same name in 1860. John Lindley (1799–1865) described the species in 1861 under the name that Veitch had chosen to honor Sir Rutherford Alcock (1809–1897) but drew his description and illustration from both species. Because of this, many authors use the next oldest available name for the species, *P. bicolor,* but *P. alcoquiana* remains the correct name because the original material representing Alcock spruce has been chosen as the type of the name in accordance with rules of botanical nomenclature. Trees from the western slopes of the Akaishi Mountain Range in central Honshū having seed cones with the scale tips curled back have been called variety *reflexa,* but it seems unnecessary to give them formal recognition.

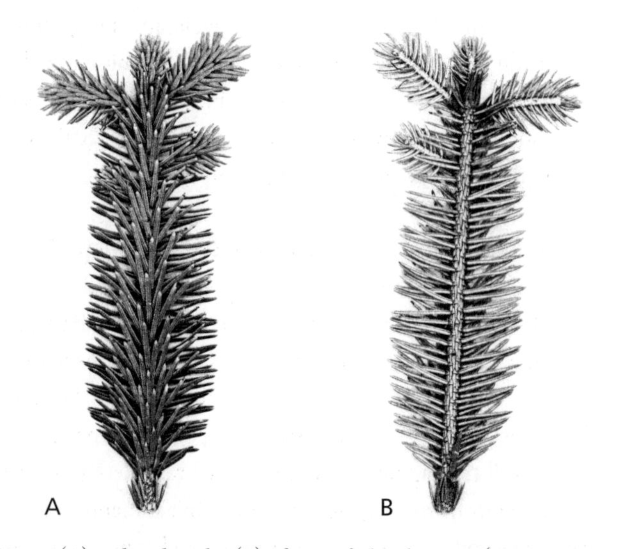

Upper (A) and undersides (B) of twig of Alcock spruce (*Picea alcoquiana*) with 2 years of growth, ×0.5.

Picea asperata M. T. Masters

DRAGON SPRUCE, YUN SHAN (CHINESE)

Tree to 45 m tall, with trunk to 1(–1.5) m in diameter. Bark grayish brown, furrowed between irregular plates on old trees. Crown conical to spirelike, with horizontal, gently rising, or gently downswept branches. New branchlets yellowish brown to reddish brown, thinly waxy or not, hairless or thinly hairy. Buds (4–)6–12 mm long, resinous. Needles dull green or a little bluish with wax, (0.5–)1–2(–2.5) cm long, curved slightly forward, square, with three to eight lines of stomates on each side, pointed to prickly. Pollen cones 10–20 mm long, red. Seed cones (5–)8–12(–16) cm long, green or reddish purple before maturity, ripening reddish brown. Seed scales rounded diamond-shaped, thick and stiff, the tips rounded, minutely toothed, or notched. Seed body (2–)3–4 mm long, the wing 8–14 mm longer. Mountains of west-central China, from southern Qinghai to southwestern Shaanxi, south through Sichuan. In pure stands or mixed with other upper montane conifers and hardwoods; (1,500–)2,400–3,600(–4,000) m. Zone 6. Three varieties.

Dragon spruce is a common species of central China that replaces the closely related *Picea abies* there and shows some of the same range of variability in cone scales and other features. In Chinese, it is the typical spruce, with most other species named "something or other" *yun shan.* The scientific name, Latin for "rough," refers to the prickly leaves. It is an important commercial species, not only for the many uses of its wood but also for resins from the trunk and fragrant oils distilled from the leaves and twigs.

Picea asperata M. T. Masters **var. asperata**

DRAGON SPRUCE, YUN SHAN

Branchlets hairy or not, yellowish or reddish brown, not waxy. Needles not waxy, pointed. Edges of seed scales smooth or minutely toothed. Throughout the range of the species. Synonyms: *Picea asperata* var. *ponderosa* Rehder & E. H. Wilson, *P. gemmata* Rehder & E. H. Wilson, *P. retroflexa* M. T. Masters.

Picea asperata var. aurantiaca (M. T. Masters) B. K. Boom

ORANGE-TWIG DRAGON SPRUCE, BAI PI YUN SHAN (CHINESE)

Branchlets not hairy, orange, waxy. Needles waxy, prickly. Edges of seed scales smooth or minutely toothed. Southwestern Sichuan. Synonym: *Picea aurantiaca* M. T. Masters.

Picea asperata var. heterolepis (Rehder & E. H. Wilson) L. K. Fu & N. Li

NOTCH-SCALE DRAGON SPRUCE, LIE LIN YUN SHAN (CHINESE)

Branchlets not hairy, orange-brown, not waxy. Needles a little waxy. Edges of seed scales notched. Western Sichuan.

Synonyms: *Picea asperata* var. *notabilis* Rehder & E. H. Wilson, *P. heterolepis* Rehder & E. H. Wilson.

Picea brachytyla (Franchet) E. Pritzel
SARGENT SPRUCE, MAI DIAO SHAN (CHINESE)
Tree to 35(–40) m tall, with trunk to 1(–1.2) m in diameter. Bark in old trees remaining flaky or breaking up into dark gray blocks divided by furrows. Crown broadly conical, opening up with age, with thin, horizontal or slightly downswept branches bearing dangling side branches. New branchlets very pale yellowish brown, hairless or thinly hairy. Buds 4–8 mm long, slightly resinous. Needles dark green to slightly bluish green with wax above, white below, (0.8–)1–2.2(–2.5) cm long, curved forward, flat, with five or six lines of stomates in each of two stomatal bands on the side facing the twig and without stomates on the outer side, pointed to a little prickly. Pollen cones 10–25 mm long, red. Seed cones (4–)6–10(–15) cm long, green or with a purplish blush before maturity, ripening brown or purplish brown. Seed scales angularly egg-shaped, thin but stiff, the tips often curled back a little. Seed body 3–4 mm long, the wing about 7–9 mm longer. Central China, from western Henan to southeastern Xizang (Tibet) and northwestern Yunnan, and adjacent Myanmar and India. Forming pure groves among other conifers and hardwoods in montane forests; (1,300–)2,300–3,000(–3,800) m. Zone 6. Three varieties.

Sargent spruce produces excellent timber used for a variety of purposes, including making aircraft and musical instruments. Much of the need is supplied by plantation culture as the tree is rare in nature. It has a close association with *Sinarundinaria nitida,* the favored bamboo food of the giant panda, so the Wolong Nature Reserve also protects this attractive spruce. Like red spruce (*Picea rubens*) in North America, *P. brachytyla* regenerates within forests and does not need the large disturbances (fires, windthrow) required by many other spruces. The scientific name means "short knob."

Picea brachytyla (Franchet) E. Pritzel **var. brachytyla**
SARGENT SPRUCE, MAI DIAO SHAN, YUAN BIAN ZHONG (CHINESE)
Mature bark grayish brown, furrowed between square blocks. Needles mostly 1–2 cm long, up to 2 mm wide, thin. Seed cones up to 12 cm long, green before maturity. Eastern part of the range of the species from western Henan to Sichuan. Synonyms: *Picea brachytyla* var. *pachyclada* (Patschke) Silba, *P. brachytyla* var. *rhombisquamea* Stapf.

Picea brachytyla var. complanata (M. T. Masters) W. C. Cheng ex Rehder
YOU MAI DIAO SHAN (CHINESE)
Mature bark pale gray or gray, usually remaining flaky. Needles mostly 1–2 cm long, up to 2 mm wide, thick. Seed cones up to 15 cm long, purplish green before maturity. Western part of the range of the species in Xizang (Tibet), Yunnan, and adjacent India.

Picea brachytyla var. farreri (C. Page & K. Rushforth) Eckenwalder
FARRER SPRUCE, MIAN DIAN YUN SHAN (CHINESE)
Mature bark grayish brown, breaking up into irregular patches. Needles mostly 2–2.5 cm long, about 1 mm wide, thick. Seed cones up to 14 cm long, green or purplish green before maturity. Northern Myanmar and adjacent western Yunnan (China). Synonym: *Picea farreri* C. Page & K. Rushforth.

Picea breweriana S. Watson
BREWER SPRUCE
Tree to 40(–55) m tall, with trunk to 1(–1.5) m in diameter. Bark breaking up with age into reddish scaly plates separated by furrows. Crown narrowly cylindrical, with horizontal branches bearing very long dangling side branches hanging down as much as 2.5 m. New branchlets reddish brown, minutely fuzzy. Buds 5–7(–8) mm long, resinous. Needles dark green above, grayish green with wax beneath, (1.5–)2–3(–3.5) cm long, straight or with a gentle backward curve, flattish or oval, with four to six lines of stomates in the two stomatal bands on the side facing the twig and without stomates on the outer side, bluntly pointed. Pollen cones 20–25 mm long, light brown. Seed cones (6.5–)8–12 cm long, purplish green before maturity, ripening reddish brown.

Mature Brewer spruce (*Picea breweriana*) in nature.

Twigs of Brewer spruce (*Picea breweriana*) with 2 years of growth and some leaf undersides showing, ×0.5.

Open mature cone of Brewer spruce (*Picea breweriana*) after seed dispersal; scale, 5 cm.

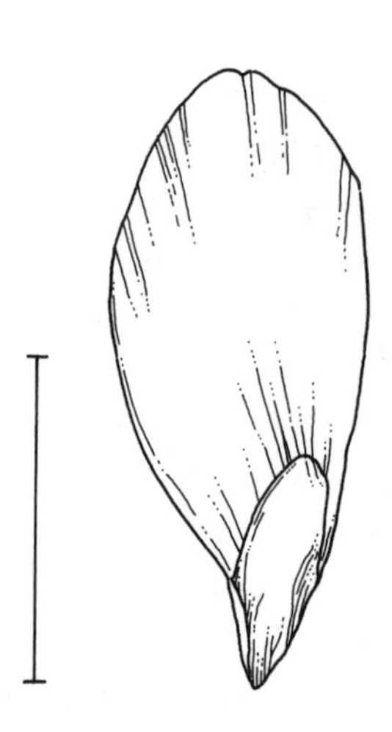

Underside of dispersed seed of Brewer spruce (*Picea breweriana*); scale, 5 mm.

Seed scales broadly egg-shaped, thin and stiff, those near the base of the cone often curled back. Seed body 3–4 mm long, the wing 7–9(–12) mm longer. Klamath Mountain region of southwestern Oregon and northwestern California. Forming pure groves or more commonly mixed with any of 15 other conifers and as many broad-leaved evergreen trees and shrubs in montane forests on rocky ridges and slopes; (550–)1,500–2,300 m. Zone 6.

Brewer spruce is fairly common within its very local range, a region of high snowfall and deep snow accumulation but little summer rainfall. At its western limit, it reaches within 22 km of the Pacific Ocean but does not overlap with the coastal rain-forest species, Sitka spruce (*Picea sitchensis*). It appears to be excluded from the coastal forests by the much greater growth rates of the coastal conifers like redwood (*Sequoia sempervirens*), Douglas fir (*Pseudotsuga menziesii*), and Sitka spruce. DNA studies suggest that *P. sitchensis* is its closest relative, and they can cross successfully in plantations, but there are no known natural hybrids. DNA data also suggest that *P. breweriana* is not closely related to similar-looking weeping spruces from the Old World, like Serbian spruce (*P. omorika*), Sargent spruce (*P. brachytyla*), and East Himalayan spruce (*P. spinulosa*), all treated by some as species of a flat-needled section *Omorika* within the genus. Instead, it appears that the weeping habit, and associated flat needles with stomates only on the side toward the ground, has arisen repeatedly in midlatitude mountainous regions with high snowfall.

Deep snowpacks, in combination with growth on slopes, give most trees of Brewer spruce a curved trunk as they recover from being pushed over by the snow as saplings. Snow accumulation falls off dramatically at the inland limit of Brewer spruce, where it rarely grows near Engelmann spruce (*Picea engelmannii*) and then without forming hybrids. Deliberate attempts to cross these two species were unsuccessful. A garden hybrid with Caucasian spruce (*P. orientalis*) has been reported and given a name with no botanical standing but has not been verified. Brewer spruces may live 900 years and can regenerate within mature forests, so as long as exploitation in its rugged homeland remains light, the future of this species is assured despite its narrow range. The two largest known individuals in combined dimensions when measured in 1993 and 1999 respectively were 53.7 m tall, 1.6 m in diameter, with a spread of 16.8 m, and 41.8 m tall, 2.2 m in diameter, with a spread of 13.4 m. The species name honors William H. Brewer (1828–1910), who was botanist for the California Geological Survey when he collected the first botanical specimen of it in 1863, on the flanks of Mount Shasta.

Picea chihuahuana M. Martínez

CHIHUAHUA SPRUCE, PINABETE (SPANISH)

Tree to 30(–45) m tall, with trunk to 0.7(–1.2) m in diameter. Bark gray, flaky, with small checkers separated by shallow, narrow grooves. Crown broadly conical, with gracefully arched branches of very dif-

Natural stand of Chihuahua spruce (*Picea chihuahuana*).

Upper (A) and undersides (B) of twigs of Chihuahua spruce (*Picea chihuahuana*) with 2 years of growth, ×0.5.

ferent lengths at a given level, bearing horizontal or slightly drooping side branches. New branchlets pale yellowish brown, hairless. Buds (4–)7–8(–10) mm long, resinous. Needles grayish green with wax, (1.2–)1.5–2(–2.5) cm long, straight or gently curved forward, diamond-shaped, with three to six lines of stomates in a band along the center of each of the four faces, very prickly. Pollen cones (1–)2–3 cm long, reddish. Seed cones (8.5–)10–14(–16) cm long, green before maturity, ripening chestnut brown. Seed scales egg-shaped, thin and flexible. Seed body (3–)4–6(–8) mm long, the wing 7–13 mm longer. Sierra Madre Occidental in southwestern Chihuahua and western Durango and Sierra Madre Oriental in Nuevo León (Mexico). Usually mixed with other conifers and a few hardwoods along streams in montane forest; (2,150–) 2,300–2,700(–3,200) m. Zone 7. Two varieties.

Chihuahua spruce is one of the rarest and most vulnerable of the world's spruce species. There are fewer than 40 known stands, and the largest of these has fewer than 2,500 trees, while smaller stands may have only a few dozen. Regeneration is generally poor due to deliberate fires, which seedlings cannot tolerate, and livestock grazing. The smaller populations are highly inbred and have very few viable seeds in their cones. Compared to other conifers there is considerable differentiation between stands in enzyme variants. This implies restricted or no genetic contact among the isolated populations. Spruce pollen in Pleistocene sediments of lakes in central Mexico suggests that Chihuahua spruce may have had a much wider distribution in Mexico during the coolest periods of glacial expansion that only subsequently became restricted.

Its relationships to other spruce species are uncertain. It is clearly not closely related to other North American spruces, and opinion generally seeks an alliance among the Asian species, either with dragon spruce (*Picea asperata*) and its relatives or with Sargent spruce (*P. brachytyla*) and its relatives. DNA studies support the latter relationship, but Chihuahua spruce has not been crossed successfully with any of these species that have been tried, although crosses with Sargent spruce itself have not been attempted. In fact, Chihuahua spruce has only been crossed successfully with one other species, the promiscuous Serbian spruce (*P. omorika*), which is not considered a close relative.

Picea chihuahuana M. Martínez **var. *chihuahuana***

Needles in cross section a little higher than wide, stiff and fiercely prickly. Seed cones usually less than 11 cm long with smooth-edged scales less than 22 mm long. Seed body 3–5 mm long. Sierra Madre Occidental.

Picea chihuahuana* var. *martinezii (T. F. Patterson) Eckenwalder

Needles in cross section markedly wider than high, flexible and less prickly. Seed cones usually more than 11 cm long with minutely toothed scales more than 26 mm long. Seed body 5–8 mm long. Sierra Madre Oriental (with only two known stands). Synonym: *Picea martinezii* T. F. Patterson.

Picea engelmannii C. Parry ex Engelmann

ENGELMANN SPRUCE

Plates 6 and 38

Tree to 50(–60) m tall, or a spreading shrub above the alpine timberline, with trunk to 2(–2.5) m in diameter. Bark remaining flaky with age to uncover reddish brown patches beneath the generally gray exterior. Crown conical, flattening at the top with age, with densely spaced tiers of gently rising branches bearing horizontal side branches. New branchlets pale yellowish brown, usually finely hairy. Buds 3–6 mm long, a little resinous. Needles yellowish green to bluish green with wax, 1.5–3(–5) cm long, straight or curved gently forward, square, with (one or) two or three stomatal lines on each side, pointed but not prickly. Pollen cones 10–15 mm long, red. Seed cones (2.5–)3–6(–8) cm long, reddish green before maturity, ripening greenish brown. Seed scales round or cutoff diamond-shaped, thin and flexible. Seed body 2–3 mm long, the wing about 7–10 mm longer. Central British Columbia south in the Cascade Ranges to northwestern California and in the Rocky Mountains to southeastern Arizona and southern New Mexico. Forming pure stands or mixed with subalpine fir (*Abies lasiocarpa*) and less commonly with other conifers in subalpine forest and extending above timberline as a low shrub; (600–)1,000–3,300(–3,700) m. Zone 3.

Engelmann spruce is an important timber tree throughout the Rocky Mountains, growing with subalpine fir above other montane conifers like lodgepole pine (*Pinus contorta*) to the upper limits of tree growth and beyond. It is well adapted to its harsh subalpine habitat. The oldest known individual, more than 850 years, lives at 3,500 m, only 10–30 m below the alpine timberline in central Colorado. The largest known individual in combined dimensions when measured in 1998 was 47.6 m tall, 2.1 m in diameter, with a spread of 8.2 m. The species name honors George Engelmann (1809–1884), a physician and botanist in St. Louis, Missouri, who undertook pioneering studies in North American cacti and yuccas as well as of conifers.

Traditionally, Engelmann spruce has been considered closely related to other western North American spruces, including white spruce (*Picea glauca*), Mexican spruce (*P. mexicana*), blue spruce (*P. pungens*), and Sitka spruce (*P. sitchensis*). In nature, it only overlaps with two of these, white and blue spruce, and its interactions with these are very different, even though it overlaps with both of them from above. It is highly interfertile with white spruce, and there are numerous natural hybrids over a large area in western Montana, Alberta and British Columbia, enough to have been given taxonomic recognition as *P.* ×*albertiana.* Artificial crosses of this parentage often display hybrid vigor and are being tried for timber and pulp production in eastern Canada. In contrast, natural hybrids with blue spruce are extremely rare, even where the species grow in intimate contact throughout a huge range of overlapping distribution in the central Rocky Mountains. Furthermore, artificial crosses almost always fail due to strong genetic barriers leading to embryo abortion. These same kinds of barriers are found in attempts to cross Engelmann spruce with Mexican spruce, which is so similar morphologically that it is often considered a variety of *P. engelmannii,* but they are effectively reproductively isolated.

DNA studies shed some surprising light on the relationships of Engelmann spruce. This evidence confirms that Mexican spruce is the closest relative of Engelmann spruce and that white spruce is also a near relative, but these three species do not seem particularly close to the other species traditionally placed in section *Casicta,* including Yezo spruce (*Picea jezoensis*) in Asia as well as blue spruce and Sitka spruce. They might even be more closely related to red spruce (*P. rubens*) and black spruce (*P. mariana*) than to those with which they are traditionally allied. Clearly, more work on spruce phylogeny is required, even though the relationships of these species have been studied intensively since the 1960s. Some of the discrepancy between the different lines of evidence used so far to study the affinities of Engelmann spruce may be due to the fact that population-level studies involving numerous individuals from different sources have only been applied to its supposed North American relatives. These studies need to be expanded on a worldwide basis.

Picea glauca (Moench) A. Voss

WHITE SPRUCE, SKUNK SPRUCE, ÉPINETTE BLANCHE (FRENCH)

Plates 39 and 40

Tree to 30(–40) m tall, with trunk to 1 m in diameter. Bark becoming gray and more flaky with age, revealing reddish patches of new bark. Crown narrowly conical, sometimes extremely so, with gently rising branches, passing through horizontal and becoming gently depressed with age and bearing horizontal or gently drooping side branches. New branchlets yellowish to slightly reddish brown, usually hairless. Buds (3–)4–6 mm long, not resinous. Needles variably waxy (hence the scientific name) to give colors from bright green to intensely blue but usually dull gray-

Underside of twig of white spruce (*Picea glauca*) with 2 years of growth, ×0.5.

Twig of white spruce (*Picea glauca*) with immature seed cones after pollination; scale, 2 cm.

ish green, (1–)1.5–2(–2.5) cm long, straight or curved gently forward, square, with two to four lines of stomates on each side, sharp to blunt. Pollen cones 10–20 mm long, red. Seed cones (2.5–)3–6(–8) cm long, green before maturity, ripening medium brown. Seed scales round, thin but stiff and brittle. Seed body 2–4 mm long, the wing 5–8 mm longer. All across northern North America from Alaska to Newfoundland south to northwestern Montana, the upper Great Lakes, the Adirondack Mountains of northern New York, and northern New England, with outliers in central Wyoming and the Black Hills of South Dakota. Often forming pure stands but also mixed with other conifers or poplars (*Populus balsamifera* and *P. tremuloides*) and paper birch (*Betula papyrifera*) in the boreal forest; 0–1,500 m. Zone 1. Synonym: *Picea glauca* var. *densata* L. H. Bailey.

White spruce is the premier timber tree of the North American boreal forest, growing on a wider variety of sites and to a larger size than any other tree in this region. It is so abundant that its most important pests are finely attuned to its annual cycle. For instance, each life stage of the spruce budworm matures in time to feed on a particular growth phase of the spruce. Despite its enormous geographic range there is not a corresponding increase in regional variation compared to other conifers. One striking feature that varies geographically is the shape of the crown. The narrowest spires are found farthest north where the winters are snowiest and harshest. There is also much variation in foliage color that has served well in cultivar selection, but this variation is not geographic. White spruce is the second smallest North American spruce species after black spruce (*Picea mariana*). The largest known individual in combined dimensions when measured in 2001 was 39.6 m tall, 1.0 m in diameter, with a spread of 9.2 m.

Individual trees with the greatest amount of heterozygosity (possession of two different forms of a single gene) are generally the fastest growing, and artificial hybrids with Engelmann spruce (*Picea engelmannii*), which are heterozygous for many genes, may show considerable hybrid vigor. White spruce can be crossed fairly readily with two other western spruces as well, blue spruce (*P. pungens*) and Sitka spruce (*P. sitchensis*). Its range overlaps with those of Engelmann spruce and Sitka spruce, and it hybridizes naturally with both of these to form *P.* ×*albertiana* and *P.* ×*lutzii*, respectively. It does not overlap with or hybridize with blue spruce in nature. Natural hybridization with black spruce, with which it shares the full expanse of the boreal forest, is extremely rare, and such hybrids are not particularly vigorous, showing that hybrid vigor depends on favorable combinations of genes, not just heterozygosity.

Picea glehnii (Friedr. Schmidt) M. T. Masters

SAKHALIN SPRUCE, AKA-EZOMATSU (JAPANESE), CHIKAN SHUNKU (AINU), EL' GLENA (RUSSIAN)

Tree to 30(–40) m tall, with trunk to 1(–1.6) m in diameter. Bark reddish brown overlain with gray and breaking up into irregular small plates at the base of old trees. Crown cylindrical and open, with slender, short horizontal branches turned up at the tip and bearing dangling, short side branches. New branchlets reddish brown, densely hairy. Buds 4–6 mm long, resinous. Needles dark green on the side away from the branchlet, a little waxy on the inner side, (0.3–)0.6–1.2(–1.8) cm long, curved forward, square, with one or two lines of stomates on the two outer faces and two to four lines on the two inner faces, pointed but not prickly. Pollen cones

Upper (A) and undersides (B) of twigs of Sakhalin spruce (*Picea glehnii*) with 2 years of growth, ×0.5.

Underside of twig of Yezo spruce (*Picea jezoensis*) with the most recent year of growth, ×0.5.

7–14(–20) mm long, red. Seed cones 3–5.5(–8.5) cm long, purple (rarely green) before maturity, ripening grayish brown. Seed scales round, thin but stiff, the outer edge smooth or irregular. Seed body 2–3.5 mm long, the wing 5–7 mm longer. Hokkaidō and nearby Honshū, and southern Kurile Islands (Japan) and southern Sakhalin (Russia). Forming pure stands or growing with Yezo spruce (*Picea jezoensis*), Sakhalin fir (*Abies sachalinensis*), and a few other conifers and hardwoods in northern mixed forest and boreal forest, especially on wet soils; 0–1,600 m. Zone 6.

Sakhalin spruce produces a white, satiny timber that is highly prized for construction, much more so than its more common associate, Yezo spruce (*Picea jezoensis*). Its Japanese name, "red Yezo spruce," distinguishes it from the latter by contrasting its reddish bark and branchlets to the grayer Yezo spruce. It hybridizes with Yezo spruce to produce *P.* ×*notha,* but how common this hybrid is in nature remains to be investigated. It is not closely related to Yezo spruce, and DNA studies show that it is one of the close Asian allies of Norway spruce (*P. abies*). The species name honors Peter von Glehn (1835–1876), who accompanied the describer, Friedrich Schmidt, on the Russian Geographical Society expedition to eastern Asia in 1861, during which he collected the type specimen on Sakhalin.

Picea jezoensis (P. Siebold & Zuccarini) Carrière

YEZO SPRUCE, HOKKAIDŌ SPRUCE, EZO-MATSU (JAPANESE), SHUNG (AINU), EL' AYANSKAYA (RUSSIAN), KAMUNBI NAMU (KOREAN), YU LIN YUN SHAN (CHINESE)

Tree to 40(–50) m tall, with trunk to 1(–1.5) m in diameter. Bark breaking up on large trees into interrupted, narrow, dark grayish brown ridges between deep furrows. Crown broadly conical or egg-shaped, with outstretched, long, horizontal branches bearing short dangling side branches. New branchlets pale yellowish brown, hairless and shiny. Buds 5–8 mm, slightly resinous or not. Needles dark green on the outer side, greenish white to snowy white on the inner side, (0.8–)1.5–2.5(–3) cm long, gently curved forward, flat, with six or seven lines of stomates in each of two stomatal bands on the inner side, pointed or slightly prickly. Pollen cones 15–20 mm long, purplish red. Seed cones (3–)4–7.5(–9) cm long, reddish purple or green before maturity, ripening light brown. Seed scales cutoff diamond-shaped, minutely toothed at the tip, thin and flexible. Seed body 2–3 mm long, the wing 5–6 mm longer. Northeastern Asia, from southern Kamchatka and the western shores of the Sea of Okhotsk (Russia) south to central Honshū (Japan), Korea, and Nei Mongol (China). Forming pure stands or more commonly mixed with other conifers and some hardwoods in montane and subalpine forest; (50–)300–2,500 m. Zone 5. Three varieties.

This handsome species is the most widely distributed of the purely Asian spruces. It is an important timber and pulp species, usually being more common than the other spruces with which it is associated. It has often been viewed as the closest Asian relative of the western North American spruces related to Sitka spruce (*Picea sitchensis*) and Engelmann spruce (*P. engelmannii*), resembling the former particularly closely in its needles and seed cones.

Mature Yezo spruce (*Picea jezoensis*) in nature.

DNA studies suggest a particular affinity with blue spruce (*P. pungens*) and, to a lesser extent, Sitka spruce, but only a distant relationship to Engelmann spruce and white spruce (*P. glauca*). It can be crossed most easily with white spruce, so each of these lines of evidence (morphology, DNA studies, and crossability) implies different relationships among its North American allies. It is not closely related to its nearest neighbor in Hokkaidō and vicinity, Sakhalin spruce (*P. glehnii*), which is close to Norway spruce (*P. abies*), but nonetheless, these species probably hybridize in nature to form *P. ×notha*. There is a fair amount of variation throughout the range of *P. jezoensis*, so a number of varieties and even separate species were described from it, although field study of the species throughout its range in five different countries will be required to achieve a convincing classification of varieties. The species name is derived from another name for the island of Hokkaidō, Yezo.

Picea jezoensis* var. *hondoensis (H. Mayr) Rehder
TOHI (JAPANESE)
Second year branchlets reddish brown. Needles 0.8–1.5 cm long with snowy white stomatal bands. Seed cones 4–7 cm long, the scales 8–10 mm wide. Honshū (Japan).

Picea jezoensis (P. Siebold & Zuccarini) Carrière **var. *jezoensis***
Second-year branchlets orange-brown. Needles 1.5–2.5 cm long, with greenish white stomatal bands. Seed cones 4–9 cm long, the scales 6–8 mm wide. Most of the range of the species except where the other varieties occur. Synonyms: *Picea ajanensis* F. E. L. Fischer, *P. jezoensis* var. *microsperma* (Lindley) W. C. Cheng & L. K. Fu, *P. kamchatkensis* Lacassagne.

Picea jezoensis* var. *komarovii (V. N. Vassiljev) W. C. Cheng & L. K. Fu
CHANG BAI YU LIN YUN SHAN (CHINESE)
Second-year branchlets yellowish brown. Needles 1–2 cm long, with greenish white stomatal bands. Seed cones 3–4 cm long, the scales 7–8 mm wide. Southern and eastern Jilin (China) and adjacent North Korea and Russia. Synonym: *Picea jezoensis* var. *koreana* Uyeki.

Picea koyamai Shirasawa
KOYAMA SPRUCE
Tree to 20(–30) m tall, with trunk to 0.6(–0.8) m in diameter. Bark dark grayish brown, remaining flaky but also becoming shallowly furrowed at the base of large trees. Crown conical, fairly open and thinning further with age, with horizontal or gently rising branches bearing short horizontal or slightly hanging side branches. New branchlets yellowish brown to reddish brown, usually hairless, a little hairy in the grooves of weak side branchlets. Buds 6–8(–13) mm long, resinous. Needles green to a little bluish green with wax, (0.6–)1–2(–2.5) mm long, straight or curved, diamond-shaped, with two to five lines of stomates on each side, blunt to pointed. Pollen cones 10–25 mm long, red. Seed cones 4–8(–9.5) cm long, green or with a reddish tinge before maturity, ripening yellowish brown. Seed scales egg-shaped to nearly circular, thinly woody and stiff, seed body (2–)3–4 mm long, the wing 8–12 mm longer. Central Honshū (Japan), North Korea, and nearby China and Russia. Forming pure stands or mixed with other spruces and other montane conifers; (400–)1,000–2,000 m. Zone 6. Two varieties

Based on the major lines of evidence available today (morphology, crossing behavior, and DNA studies), Koyama spruce is closely related to dragon spruce (*Picea asperata*) and thus to Norway spruce (*P. abies*). It is not especially close to Alcock spruce (*P. alcoquiana*), with which it hybridizes naturally in central Honshū to form *P. ×shirasawae*. It is nowhere common within its range but is especially rare in Japan, where just a few groves occur in the Yatsugatake Mountains, accompanied by Alcock spruce and their hybrids. On the mainland it is exploited for a variety of forest products but is not singled out among associated spruces. The species name honors Mitsuo Koyama (1885–1935), the Japanese botanist who discovered it in 1911, the last Japanese spruce species to be recognized, while the mainland stands were assigned incorrectly to various other previously described species. The relationships among the island and mainland stands are controversial. Many authors

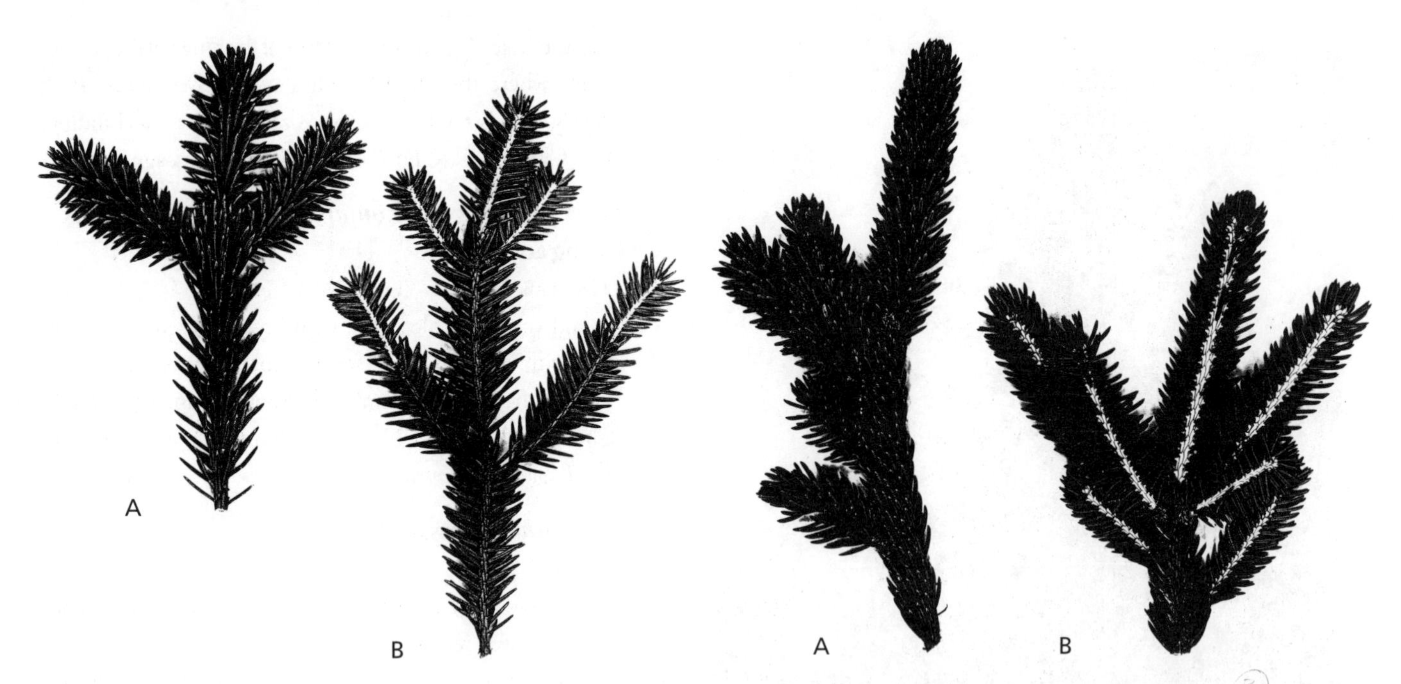

Upper (A) and undersides (B) of twigs of Korean spruce (*Picea koyamai* var. *koraiensis*) with 2 and 3 years of growth, respectively, ×0.5.

Upper (A) and undersides (B) of twigs of Koyama spruce (*Picea koyamai* var. *koyamai*) with 2 years of growth, ×0.5.

separate the mainland trees as one or more distinct species, but the differences are small and the obvious close relationship favors retention as weakly distinct varieties of a single species.

Picea koyamai* var. *koraiensis (T. Nakai) T. N. Liou & W. T. Wang

KOREAN SPRUCE, CHONGBI NAMU (KOREAN), HONG PI YUN SHAN (CHINESE), EL' KOREISKAYA (RUSSIAN)

Needles spreading relatively evenly all around the horizontal twigs. Northeastern China (Heilongjiang, Jilin, and Liaoning), North Korea, and adjacent far southeastern Russia. Synonyms: *Picea koraiensis* T. Nakai, *P. koraiensis* var. *pungsanensis* (Uyeki) Farjon, *P. pungsanensis* Uyeki.

Picea koyamai Shirasawa **var. *koyamai***

KOYAMA SPRUCE, YATSUGATAKE-TŌHI (JAPANESE)

Needles concentrated to the sides of and above the horizontal twigs. Central Honshū (Japan).

Picea likiangensis (Franchet) E. Pritzel

LIJIANG SPRUCE, LI JIANG YUN SHAN (CHINESE)

Tree to 50 m tall, with trunk to 2.5 m in diameter. Bark gray with orange-brown patches soon breaking up into irregular blocks separated by deep vertical furrows. Crown conical, with long, thin, upswept branches bearing short side branches all around. New branchlets pale brownish yellow, variably hairy or not. Buds 4–6 mm long, resinous. Needles dark green on sides without stomatal lines, bluish green with wax when stomatal lines are present, 0.6–1.5 cm long, curved forward, square or somewhat flattened, with four to seven lines of stomates on the two inner faces and none to four lines on the outer ones, sharp to blunt. Pollen cones 20–25 mm long, red. Seed cones 4–10(–12) cm long, red, yellow, or green before maturity, ripening medium brown to reddish brown. Seed scales variably shovel-shaped, with a toothed tip, thin and flexible. Seed body 2–4 mm long, the wing 5–10 mm longer. Southwestern China, from southern Qinghai to northwestern Yunnan. Forming pure stands or mixed with other conifers in subalpine forest; 2,500–4,100 m. Zone 6. Two varieties.

Lijiang spruce is a major subalpine tree of the disconnected steep mountains of southwestern China, including the Lijiang ("beautiful river") range, which gives it its name. It is one of the most variable spruce species from place to place and even at a single locality. Variation is particularly marked in hairiness of twigs, shape of needles, and cone size. At least some of the variation is probably due to hybridization with dragon spruce (*Picea asperata*) and Sargent spruce (*P. brachytyla*), which co-occur with it. It can be hybridized successfully with both of these species under controlled conditions. Some of the described varieties appear to be of hybrid origin, but details of parentage are not clear and the possible hybrids are included within variety *rubescens* for now, pending further study.

Picea likiangensis (Franchet) E. Pritzel **var. *likiangensis***

LI JIANG YUN SHAN, YUAN BIAN ZHONG (CHINESE)

New shoots slender, sparsely hairy. Needles slightly flattened top to bottom, with two to four lines of stomates on each

Natural stand of Lijiang spruce (*Picea likiangensis*).

outer face. Seed cones 7–12 cm long, dark purple before ripening. Southern part of the range of the species.

Picea likiangensis var. *rubescens* Rehder & E. H. Wilson

CHUAN XI YUN SHAN (CHINESE)

New shoots stout, densely hairy, the hairs sometimes glandular. Needles square to slightly flattened side to side or top to bottom, with none to four lines of stomates on each outer face. Seed cones 4–9 cm long, yellowish green or reddish green to dark purple before ripening. Throughout the range of the species. Synonyms: *Picea balfouriana* Rehder & E. H. Wilson, *P. hirtella* Rehder & E. H. Wilson, *P. likiangensis* var. *balfouriana* (Rehder & E. H. Wilson) H. Hillier, *P. likiangensis* var. *hirtella* (Rehder & E. H. Wilson) W. C. Cheng, *P. likiangensis* var. *linzhiensis* W. C. Cheng & L. K. Fu, *P. likiangensis* var. *montigena* (M. T. Masters) Tang S. Liu, *P. montigena* M. T. Masters, *P. purpurea* var. *balfouriana* (Rehder & E. H. Wilson) Silba, *P. purpurea* var. *hirtella* (Rehder & E. H. Wilson) Silba.

Picea ×*lutzii* E. Little

LUTZ SPRUCE

Tree to 30 m tall, resembling its parents (*Picea glauca* × *P. sitchensis*) in the characteristics they share and combining or falling between those by which they differ. The needles are sharp-pointed but not as prickly as those of Sitka spruce (*P. sitchensis*) because they are less stiff. They are a little flattened top to bottom and have fewer lines of stomates on the two outer faces, one or two (to four), than on the inner ones, (three to) five or six (to eight). The seed cones resemble those of Sitka spruce, with pale brown, thin, toothed seed scales, but these are rounded as in white spruce (*P. glauca*), and the cones are shorter, only 3–7 cm long. The hybrid occurs naturally in the relatively narrow region where the ecologically highly differentiated parents overlap, inland from the panhandle of southeastern Alaska and northern coastal British Columbia, for example, along the Skeena River. It has also been produced in experimental crossing programs. It is named for Harold J. Lutz, the Yale University forestry professor who collected the type material in Alaska in 1950 and 1951.

Picea mariana (P. Miller) N. Britton, Sterns & Poggenburg

BLACK SPRUCE, ÉPINETTE NOIRE (FRENCH)

Plate 41

Tree to 30 m tall but often dwarfed and less than 10 m on extreme sites, with trunk to 0.3(–0.6) m in diameter. Bark grayish brown, remaining thin and scaly with age. Crown very narrowly cylindrical, often open below and with a denser topknot at the summit, with short, strongly downswept branches turning up at the tips and bearing short side branches in all directions, the lowest branches often rooting into wet mossy substrates to form clonal clumps of trees. New branchlets yellowish brown to orange-brown, with many short, sometimes glandular hairs, especially in the grooves between the leaf bases. Buds 3–5(–6) mm long, not resinous, the outermost scales long and needlelike. Needles initially bluish green with wax (0.6–)0.8–1.5(–2) cm long, curved gently forward, square or slightly flattened, with one or two lines of stomates on the two outer faces and three or four lines on the inner ones, usually blunt. Pollen cones 10–15 mm long, reddish brown. Seed cones (1.5–)2–3(–4) cm long, dark purple before maturity, ripening reddish brown. Seed scales egg-shaped, the edge round and minutely toothed, thin but woody and brittle. Seed body 2–3 mm long, the wing 2–5 mm longer. All across northern North America, from Alaska to Newfoundland south to central British Columbia across to southern Manitoba and the Great Lakes states across to the mid-Atlantic US states. Growing in pure stands or mixed with tamarack (*Larix laricina*) in bogs and with any of the other North American boreal conifers and hardwoods on other site types throughout the boreal forest; 0–800(–1,500) m. Zone 1.

Given the enormous distributional range of this species, it exhibits remarkably little morphological variation. There may, however, be modest differentiation of enzymes between upland and lowland stands, although different studies come to contrasting conclusions about this. Black spruce is intolerant of shading, and regeneration often depends on the frequent fires of the boreal forest. Like many other fire-dependent conifers, the seed cones of black spruce are serotinous ("late"), remaining closed on the tree for up to 20 years until the remaining seeds are released in the aftermath of a fire. Over most of its range, black spruce grows with or near white spruce (*Picea glauca*), but natural hybridization between them is extremely rare. It also contacts two other relatives of white spruce, Engelmann spruce (*P. engelmannii*) and Sitka spruce (*P. sitchensis*), apparently without hybridizing with either, although artificial hybrids between them can be made with great difficulty. The fourth species with which it overlaps is its close relative red spruce (*P. rubens*), similar in morphology and DNA sequences but ecologically distinct. They hybridize consistently but sparsely from Ontario to Nova Scotia, especially after fires open upland red spruce stands adjacent to lowland black spruce.

Upper side of twig of black spruce (*Picea mariana*) from a bog with 3 years of growth and retained unopened mature seed cones, ×0.5.

Underside of twig of black spruce (*Picea mariana*) from an upland site with 2 years of growth, ×0.5.

The hybrids are usually competitively inferior to both parents and disappear from the maturing stands with time. Artificial hybrids between the two species, however, occasionally show some hybrid vigor and may have a place in plantation culture.

Black spruce is one of the most important sources of wood pulp for paper in North America. The species name means "of Maryland," even though *Picea mariana* does not occur in the modern state, because 18th century botanists used the name for a much broader area. The largest known individual in the United States in combined dimensions when measured in 1989 was 23.8 m tall, 0.5 m in diameter, with a spread of 6.4 m.

Picea ×*mariorika* B. K. Boom

Tree to 20 m tall. A garden hybrid easily made between its geographically widely separated but phylogenetically closely related parents (*Picea mariana* × *P. omorika*). Its twigs have a combination of the short hairs of black spruce (*P. mariana*) with the glandular hairs of Serbian spruce (*P. omorika*). The needles are flatter than those of black spruce but usually have the one or two lines of stomates on the two outer faces found in black spruce, while Serbian spruce has stomates only on the inner faces. The seed cones are 3–5 cm long, intermediate between the shorter cones of black spruce (1.5–4 cm) and the longer ones of Serbian spruce (4–6.5) cm.

Picea maximowiczii E. Regel ex M. T. Masters

MAXIMOWICZ SPRUCE, HIME-BARAMOMI (JAPANESE)

Tree to 25(–40) m tall, with trunk to 0.8(–1.6) m in diameter. Bark grayish brown, becoming broadly and irregularly ridged between deep furrows on large trunks. Crown conical, with densely spaced rising branches bearing side branches all around. New branchlets pale grayish brown to yellowish brown, usually hairless. Buds 2.5–5 mm long, resinous. Needles dark green, (0.6–)1–1.5(–2) cm long, curved forward, square, with two to four lines of stomates on each side, sharp, especially on shaded branches. Pollen cones 10–15 mm long, pale brown. Seed cones (2.5–)4–7 cm long, pale green before maturity, ripening shiny brown. Seed scales broadly egg-shaped, the exposed edge rounded-off triangular, thin but woody and stiff. Seed body (2.5–)3–5 mm long, the wing 5–6 mm longer. Extremely rare on a few mountain ranges in central Honshū (Japan). Mixed with other conifers and hardwoods in montane forest and grassy openings on rocky slopes; 1,100–2,000 m. Zone 7. Synonym: *Picea maximowiczii* var. *senanensis* Yasaka Hayashi.

Although Maximowicz spruce is a rare species, it varies enough that individuals with shorter needles and seed cones growing among those with longer needles and cones were distinguished as a separate variety, with little apparent justification. Despite its limited range in central Honshū, Maximowicz spruce overlaps with four of the other five species of Japanese spruce (all except Sakhalin spruce, *Picea glehnii*) but is not known to hybridize naturally with any of them. DNA studies suggest that it is most closely related to Alcock spruce (*P. alcoquiana*), one of the species that it grows with in the Yatsugatake Mountains, and that it also has affinities with Sargent spruce (*P. brachytyla*) in China and Chihuahua spruce (*P. chihuahuana*) in Mexico. Because of its rarity and small stature, it has little current economic importance. The species name honors Carl Maximowicz (1827–1891) of the botanical garden in St. Petersburg, Russia, who received herbarium specimens and seeds of it sent from Japan in 1865 by Chonosuke (Tschonoski) Sukawa (1841–1925), who in turn had several plants named after him by Maximowicz.

Picea mexicana M. Martínez

MEXICAN SPRUCE, CIPRÉS (SPANISH)

Tree to 25(–30) m tall, with trunk to 0.6(–1) m in diameter. Bark pale grayish brown, remaining scaly with age. Crown narrowly conical, with horizontal to upraised branches bearing short side branches all around. New branchlets grayish yellow, hairless. Buds 4–5 mm long, resinous. Needles bluish green with wax, (1.5–)3.5–4 cm long, square, with three or four lines of stomates on each side, prickly. Pollen cones 10–15 mm long, red. Seed cones 5–6 cm long, green before maturity, ripening yellowish brown. Seed scales roughly diamond-shaped, minutely toothed at the tip, thin and flexible. Seed body 3–4 mm long, the wing 6–8 mm longer. Mexico, very local in southwestern Chihuahua and west-central Nuevo León. Mixed with other conifers in subalpine forest; 2,500–3,500 m. Zone 7. Synonyms: *Picea engelmannii* subsp. *mexicana* (M. Martínez) P. A. Schmidt, *P. engelmannii* var. *mexicana* (M. Martínez) R. J. Taylor & T. F. Patterson.

Mature Mexican spruce (*Picea mexicana*) in nature.

Mexican spruce is very closely related to Engelmann spruce (*Picea engelmannii*), as evidenced by both morphological similarity and close similarity in DNA studies. These similarities led many to treat Mexican spruce as a variety or subspecies of *P. engelmannii*. However, these spruces cross only with some difficulty, at best producing less than 5% as many germinable seeds as crosses within Mexican spruce and often failing entirely. Crossability is higher (5–9%) with the related white spruce (*P. glauca*), Sitka spruce (*P. sitchensis*), and even the more distant red spruce (*P. rubens*), and is highest (at least 10%) with the distantly related Chinese Lijiang spruce (*P. likiangensis*). These crossability relationships argue in favor of keeping Mexican spruce separate from Engelmann spruce, although they appear to be mutually closest relatives.

Mexican spruce is one of the most localized spruce species worldwide and correspondingly vulnerable. Although its subalpine habitat puts it under relatively light pressure from exploitation, the highly restricted populations are subject to extreme risk from grazing and accidental fires.

Picea meyeri Rehder & E. H. Wilson

MEYER SPRUCE, BAI QIAN (CHINESE)

Tree to 30 m tall, with trunk to 0.6(–1) m in diameter. Bark grayish brown, becoming more scaly with age. Crown conical, becoming predominantly cylindrical, with long, thin horizontal to gently depressed branches bearing short, droopy side branches. New branchlets pale yellowish brown, variably hairy or not. Buds 6–10 mm long, resinous and slightly hairy. Needles grayish green with wax, (0.8–)1.5–3 cm long, curved forward, square, with four to eight lines of stomates on each side, pointed but not prickly. Pollen cones 20–25 mm long, red. Seed cones 6–9(–12) cm long, green or red before maturity ripening yellowish brown to reddish brown. Seed scales fan-shaped, smooth-edged, narrowly ribbed, woody and stiff. Seed body 3–4 mm long, the wing 7–10 mm longer. North-central China, in Hebei, Shanxi, and nearby Nei Mongol. Forming pure stands or mixed with other conifers in high montane and subalpine forests on mountain slopes; 1,600–2,700 m. Zone 6.

Upper (A) and undersides (B) of twigs of Meyer spruce (*Picea meyeri*) with 2 years of growth, ×0.5.

Lichen-draped mature Taiwan spruce (*Picea morrisonicola*) in nature.

Meyer spruce falls geographically between Koyama spruce (*Picea koyamai*) and dragon spruce (*P. asperata*) in the chain of species related to Norway spruce (*P. abies*) that is found all across northern Eurasia. It is also intermediate in elevational range between these two neighbors, falling between the modest elevations of *P. koyamai* var. *koraiensis* to the northeast and the subalpine stands favored by the three varieties of *P. asperata* in the high mountains of southwestern China. Like other members of the group, Meyer spruce produces an excellent timber, which is harvested both from natural stands and from plantations. The species name honors Frank N. Meyer (1875–1918), a collector for the US Department of Agriculture who collected the type specimen in 1908 and later died during a collecting trip in China.

Picea morrisonicola Hayata

TAIWAN SPRUCE, TAI WAN YUN SHAN (CHINESE)

Tree to 40(–50) m tall, with trunk to 1(–1.5) m in diameter. Bark grayish brown, remaining scaly with age. Crown roughly cylindrical, with long, thin horizontal branches bearing short, horizontal side branches. New branchlets pale yellowish brown, hairless. Buds (2–)3–5 mm long, slightly resinous. Needles dark green, 0.8–1.8 cm long, curved forward, diamond-shaped and slightly flattened top to bottom, with two or three lines of stomates on the two outer faces and four or five lines on the two inner ones, blunt or pointed but not prickly. Pollen cones 10–15 mm long, red. Seed cones (4–)5–7.5 cm long, purplish green before maturity, ripening brown. Seed scales egg-shaped, with a smooth edge, woody and stiff. Seed body 3–4 mm long, the wing 6–8 mm longer. High peaks of the central range of Taiwan, including Yushan (called Mount Morrison in English, hence the scientific name). Forming pure stands or more commonly mixed with other montane and subalpine conifers on slopes and in ravines; (2,000–)2,400–2,900(–3,000) m. Zone 7.

Taiwan spruce is restricted to Taiwan and is the only spruce species found on that island. It is an uncommon tree, but the timber is of good quality and is used for a variety of purposes. In keeping with its geographic isolation, it is not especially closely related to any other species, but morphological similarities and DNA studies suggest a loose affinity with Sargent spruce (*Picea brachytyla*) and its allies, including another southerly outlier of the genus, Chihuahua spruce (*P. chihuahuana*).

Picea* ×*moseri M. T. Masters

Tree to 20 m tall. A garden hybrid (*Picea jezoensis* × *P. mariana*) with the short needles of black spruce (*P. mariana*) but somewhat flattened and with silvery white stomatal bands on the inner faces. The foliage is very dense and borne on sparsely hairy branchlets with little tendency to dangle despite the usual association of prominent stomatal bands on two-sided leaves with a weeping habit in spruces, including Yezo spruce (*P. jezoensis*). The two parents are presumably not closely related, so this is an example of a successful wide cross.

Picea neoveitchii M. T. Masters

VEITCH SPRUCE, DA GUO QING QIAN (CHINESE)

Tree to 15(–20) m tall, with trunk to 0.5 m in diameter. Bark rough gray, scaly. Crown conical, with long, slender branches that pass from angled upward at first, through horizontal, to gently downswept, and bear short side branches to the sides and above. New branchlets pale yellowish brown, usually hairless. Buds 5–6 mm long, slightly resinous. Needles dark green, 1.5–2.5 cm long, curved forward, square, with four to seven lines of stomates on each face, stiff and pointed to a little prickly. Pollen cones 15–20 mm long, red. Seed cones 8–14 cm long, green before maturity, ripening pale brown. Seed scales broadly diamond-shaped, woody and stiff. Seed body 5–6(–7) mm long, the wing another 10–12 mm longer. Central China, from southwestern Henan and western Hubei to southern Gansu. Generally mixed with other conifers in montane forest on slopes and in ravines; 1,300–2,200 m. Zone 7.

Veitch spruce is a rare species of the Qinling Range and associated mountains where, unlike many other spruces, it usually occurs as scattered individuals within mixed forests. It is under threat due to overgrazing, fires, and overexploitation for its fine wood. There is some compensation by cultivation in plantations, but the natural stands are dwindling. The species name honors John Gould Veitch (1839–1870), who collected in China and Japan in 1860. It is called "new" (*neo-*) because Veitch fir (*Abies veitchii*) had been transferred to the genus *Picea* earlier, during the confusion of names of these two genera, thus preempting the simple form of the name. DNA studies suggest that Veitch spruce is closest to Wilson spruce (*P. wilsonii*), among the species related to Sargent spruce (*P. brachytyla*) and tigertail spruce (*P. polita*). This whole group of East Asian spruces is in need of taxonomic revision.

Picea* ×*notha Rehder

Tree to 30 m tall. Presumably a natural hybrid (*notha* is a Greek word for "hybrid") but first discovered among a batch of seedlings grown from seed of Sakhalin spruce (*Picea glehnii*) sent to the Arnold Arboretum in Massachusetts from Japan. The two parent species (*P. glehnii* × *P. jezoensis*) often grow together in Hokkaidō. The twigs are thinly shaggy, intermediate between the densely hairy Sakhalin spruce and the hairless Yezo spruce (*P. jezoensis*). The needles are flattened top to bottom, like those of Yezo spruce, but they have a few lines of stomates on the outer faces rather than none. The seed scales are intermediate in width between those of the two parents and are thinner and more flexible that those of Sakhalin spruce but less wavy than those of Yezo spruce. Zone 6.

Dense natural stand of young mature Serbian spruce (*Picea omorika*).

Picea omorika (Pančić) Purkyně

SERBIAN SPRUCE, PANČIĆ OMORIKA (SERBIAN)

Tree to 30(–50) m tall, with trunk to 1 m in diameter. Bark reddish brown, remaining thin and scaly with age. Crown narrowly conical, often spirelike, with numerous short, downswept branches turning up at the ends and bearing dangling side branches. New branchlets grayish brown to orange brown, densely hairy at first. Buds 3–4(–8) mm long, mostly not resinous. Needles shiny dark green above (on the outer face), waxy grayish white beneath (on the inner face), 1–2 cm long, straight and angled forward or curved forward, flat, with two bands of four to six lines of stomates on the inner face and without them on the outer one, blunt or slightly pointed. Pollen cones 20–25 mm long, pale reddish purple. Seed cones (2–)4–6.5 cm long, deep blackish violet before maturity, ripening chocolate brown. Seed scales fan-shaped to circular, thin but woody and stiff. Seed body 2–4 mm long, the wing 5–8 mm longer. Tara and Javor Mountains of western Serbia (Yugoslavia) and adjacent Bosnia and other ranges near the Drina River. Forming pure stands or mixed with Norway spruce (*Picea abies*) and other conifers in montane forest on steep limestone and serpentine slopes; (300–)800–1,600(–1,700) m. Zone 5.

Serbian spruce has one of the most limited natural ranges among the spruce species. This range has been further depleted by fires and overexploitation since its discovery by Josif Pančić in 1877, after first learning of its existence in 1855. The species name is the Serbian name for "spruce" in general, encompassing both Serbian spruce and the more common Norway spruce (*Picea abies*). Before the ice ages, Serbian spruce was widespread across Europe. Its progressive restriction to its present scattered stands, many with few individuals, led to a reduction in genetic variation within populations and a higher differentiation between populations compared to the norm in conifers. Although small population sizes often lead to inbreeding and homozygosity, this is not happening in Serbian spruce because outbred individuals with greater heterozygosity are more vigorous than their inbred siblings and gradually replace them as a cohort of seedling ages. This species is now widely cultivated in temperate regions for its elegant spirelike form and, despite its restricted natural range, has spawned a number of distinctive cultivars.

Underside of twig of Serbian spruce (*Picea omorika*) with 2 years of growth, ×0.5.

Although the flattened needles with conspicuous stomatal bands confined to the inner face suggest a relationship to the weeping spruces, like Brewer spruce (*Picea breweriana*), Sargent spruce (*P. brachytyla*), and Sikkim spruce (*P. spinulosa*), DNA studies link it with the North American black spruce (*P. mariana*) and red spruce (*P. rubens*). These two species are very unlike Serbian spruce in their needles but rather similar in their seed cones. Furthermore, they fairly freely intercross with Serbian spruce, which also produces vigorous hybrids with white spruce (*P. glauca*) and Sitka spruce (*P. sitchensis*). Despite these successes in crosses with geographically distant species, Serbian spruce is essentially reproductively isolated from its nearest neighbor, Norway spruce (*P. abies*), and no natural hybrids between these species are known.

Picea orientalis (Linnaeus) J. Link

CAUCASIAN SPRUCE, ORIENTAL SPRUCE, EL' VOSTOCHNAYA (RUSSIAN), NADZVI (GEORGIAN), MAKHRI (ARMENIAN), DOĞU LADINI (TURKISH)

Two mature trees and background forest of Caucasian spruce (*Picea orientalis*) in nature.

Tree to 35(–50) m tall, with trunk to 2 m in diameter. Bark grayish brown, flaking, becoming rough with small, scaly plates. Crown densely conical, with numerous gently downswept branches turning up at the ends and bearing short, dangling side branches. New branchlets pale yellowish brown, covered with brownish hairs. Buds 3–5 mm long, not resinous. Needles shiny dark green, 0.6–1 cm long, angled or curved forward, square, with one or two lines of stomates on the two outer faces and two to five lines on the inner ones, blunt. Pollen cones 10–20 mm long, pink. Seed cones (4.5–)6–10 cm long, purple before maturity, ripening reddish brown. Seed scales broadly egg-shaped to circular, thin but woody and stiff. Seed body 3–4.5 mm long, the wing 6–9 mm longer. Caucasus and Pontic Mountains around the eastern end of the Black Sea in Russia, Georgia, Armenia, and Turkey (the old Orient, "east," of the species name). Forming pure stands or mixed with other conifers and hardwoods, often as a dominant on moist, shaded slopes or in ravines; (50–)1,000–2,100 m. Zone 5.

The three spruce species of western Eurasia, Caucasian spruce, Norway spruce (*Picea abies*), and Serbian spruce (*P. omorika*) are all rather distinct from one another. In fact, each is more closely related to species of eastern Asia than to either of the other western species. DNA studies suggest that Caucasian spruce is closest to tigertail spruce (*P. polita*) and its relatives. The needles are the shortest in the genus and clothe the branchlets densely, giving the tree an elegant appearance and hence making it a popular ornamental. It is a tree of mature forests rather than a pioneer species and can live to be 400(–500) years old. Although the wood is soft, the white color and large size of the tree make it an important commercial species.

Underside of twig of Caucasian spruce (*Picea orientalis*) with 2 years of growth, ×1.

Picea polita (P. Siebold & Zuccarini) Carrière

TIGERTAIL SPRUCE, HARI-MOMI, BARA-MOMI (JAPANESE)

Tree to 30(–40) m tall, with trunk to 1(–1.3) m in diameter. Bark grayish brown, becoming shallowly furrowed between narrow, irregular, scaly ridges. Crown broadly conical, with long, horizontal branches bearing short, horizontal to upright side branches, becoming dangling with extreme age. New branchlets yellowish brown, hairless and shiny (hence the species name, Latin for "polished"). Buds 8–12 mm long, a little resinous or not. Needles dark green, mostly 1.5–2 cm long, curved forward at the tips, sticking out to the sides and twisted upward above the twigs, square, with four to six lines of stomates on each face, stiff and fiercely prickly. Pollen cones 30–35 mm long, reddish purple. Seed cones (5–)7–10(–12) cm long, green before maturity, ripening yellowish brown to reddish brown. Seed scales fan-shaped with a rounded or rounded-triangular, minutely toothed or smooth edge, thin but woody and stiff. Seed body 5–7 mm long, the wing 10–13 mm longer. Mountains on Pacific Ocean side of Japan in central Honshū (south to Kii Peninsula), Shikoku, and Kyūshū. Occasionally forming pure stands but more commonly mixed with other conifers and hardwoods on volcanic soils; (400–)600–1,700(–1,850) m. Zone 6. Synonym: *Picea torano* (P. Siebold ex K. Koch) Koehne.

Tigertail spruce is the most bristly species of an often prickly genus. It is also the most southerly of the Japanese spruces but does not compensate for this with an increase in elevation. Despite its distinctive appearance, DNA comparisons suggest that it is closely related to purple-cone spruce (*Picea purpurea*) from China and Caucasian spruce (*P. orientalis*) from Turkey. Its wood is rather typical spruce wood, and this, in combination with its rarity and modest dimensions, means that it is not sought out as a timber species. Many authors use the second oldest name for this species, *P. torano,* citing a nomenclatural difficulty with *P. polita.* The point is open to interpretation, however, and the technical legitimacy of *P. polita* is accepted here.

Picea pungens Engelmann

BLUE SPRUCE, COLORADO BLUE SPRUCE

Tree to 40(–50) m tall, with trunk to 1.5 m in diameter. Bark grayish brown to dark gray, becoming fissured between irregular, small, scaly plates. Crown conical, dense, with numerous generally horizontal branches and intermingled extra short branches on the trunk, the main branches bearing short, stiff side branches all

Underside of twig of blue spruce (*Picea pungens*) with 2 years of growth, ×0.5.

Dried twig of blue spruce (*Picea pungens*) with most leaves shed from three growth increments, ×0.5.

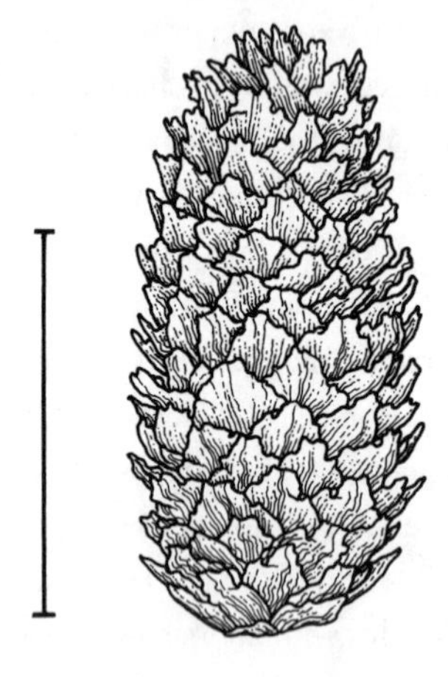

Open mature cone of blue spruce (*Picea pungens*) after seed dispersal; scale, 5 cm.

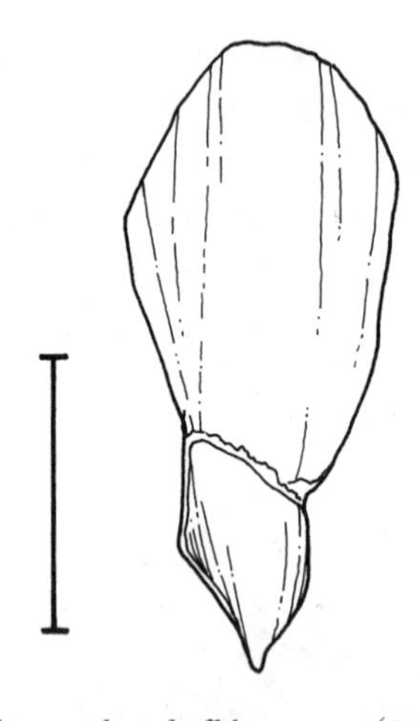

Underside of dispersed seed of blue spruce (*Picea pungens*); scale, 5 mm.

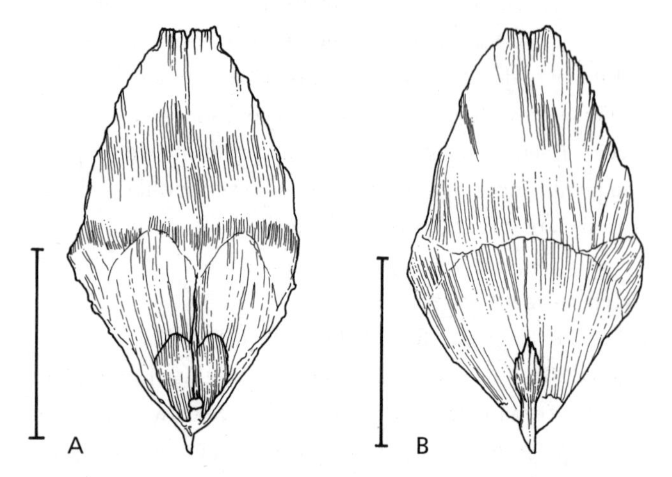

Upper (A) and undersides (B) of removed seed cone scales of blue spruce (*Picea pungens*); scale, 1 cm.

around. New branchlets yellowish brown, not hairy. Buds (5–)6–12 mm long, not resinous. Needles variably bluish green to grayish green with wax, 1.5–3 cm long, angled a little forward, square in cross section, with three to six lines of stomates on each face, stiff and prickly (hence the species name, Latin for "piercing"). Pollen cones 20–30 mm, red. Seed cones (5–)7–10(–12) cm long, green before maturity, ripening light brown. Seed scales cutoff diamond-shaped, the narrow tip minutely and irregularly toothed, thin and flexible. Seed body 3–5 mm long, the wing 5–7 mm longer. Rocky Mountains from northwestern Wyoming and northeastern Idaho (and possibly north-central Montana) to southeastern Arizona and southern New Mexico. Growing sparsely with other conifers on moist slopes or in pure stands along streams in the montane forest zone; 1,800–3,000(–3,350) m. Zone 3.

Blue spruce joins white spruce (*Picea glauca*) and Norway spruce (*P. abies*) as one of the three most widely and frequently cultivated spruces in northeastern North America. It has proven surprisingly adaptable, considering its dryish, montane natural habitat, and numerous cultivars have been selected, most emphasizing the intensely blue-gray new needles that give the species its common name. Its range is included within that of Engelmann spruce (*P. engelmannii*), and it has often been considered a lower elevation derivative of that species. DNA studies suggest that it is much more closely related to such Asian species as Lijiang spruce (*P. likiangensis*) and Yezo spruce (*P. jezoensis*) than to any North American species, but far more evidence is needed to settle the relationships of blue spruce. Natural hybridization with Engelmann spruce is extremely rare, and this combination is very difficult to make in controlled crosses. Because of its general uncommonness, it is not an important commercial species. The largest known individual in combined dimensions when measured in 2001 was 38.7 m tall, 1.5 m in diameter, with a spread of 13.1 m.

Picea purpurea M. T. Masters

PURPLE-CONE SPRUCE, ZI GUO YUN SHAN (CHINESE)

Tree to 40(–50) m tall, with trunk to 1(–2) m in diameter. Bark dark gray, becoming rough but remaining scaly with age. Crown conical, with fairly short horizontal to downswept branches turning up at the tips and bearing side branches all around. New branchlets pale yellowish brown, densely hairy. Buds 3–5 mm long, resinous. Needles shiny bright or dark green above, white beneath, 0.7–1.2(–1.4) cm long, angled forward and tightly pressed against the twigs and covering them, flattened top to bottom, without or with one or two incomplete lines of stomates on the two outer faces and four to six lines in the stomatal bands on the inner faces, blunt. Pollen cones 15–25 mm long, red. Seed cones 2.5–4(–7) cm long, dark reddish purple before maturity, ripening purplish brown. Seed scales diamond-shaped with a projecting tip, thin and flexible. Seed body 2.5–4 mm long, the wing 3–5 mm longer. West-central China in southeastern Qinghai, southern Gansu, and northwestern Sichuan. In pure stands or mixed with other conifers and a few hardwoods in subalpine and upper montane forest on north-facing slopes; 2,600–3,800 m. Zone 6. Synonym: *Picea likiangensis* var. *purpurea* (M. T. Masters) Dallimore & A. B. Jackson.

Although often associated with Lijiang spruce (*Picea likiangensis*), DNA studies suggest that purple-cone spruce belongs to the group of species related to Sargent spruce (*P. brachytyla*). Within this group it appears closest to tigertail spruce (*P. polita*) and Caucasian spruce (*P. orientalis*). Characteristics of the needles and seed cones, however, more closely resemble those of relatives of Lijiang spruce, especially blue spruce (*P. pungens*) and Yezo spruce (*P. jezoensis*). The high-quality timber is also like that of Lijiang spruce, so the relationships of this species are in need of further study.

Picea rubens C. Sargent

RED SPRUCE, ÉPINETTE ROUGE (FRENCH)

Tree to 40 m tall, with trunk to 1(–1.5) m in diameter. Bark reddish brown to grayish brown, becoming strongly but irregularly ridged and furrowed on large trees. Crown shallow, broadly conical, with widely spaced, long, horizontal branches turning up at the tips and bearing horizontal or slightly drooping side branches. New branchlets yellowish brown, hairy, especially in the grooves between the leaf bases. Buds 5–8 mm long, with a few bristlelike scales outermost, not conspicuously resinous. Needles yellowish green to dark green, (0.8–)1–1.6(–3) cm long, curved a little forward and upward, square, with two to four lines of stomates on each side, pointed. Pollen cones 15–25 mm long, red. Seed cones 2.3–4.5(–6) cm long, green or purplish green before maturity, ripening chocolate brown. Seed scales broadly egg-shaped, thin but woody, stiff, and brittle. Seed body 2–3.5 mm long, the wing 4–8 mm longer. Eastern North America, from Cape Breton Island, Nova Scotia, west to south-central Ontario, south to the southern Appalachians in western North Carolina and eastern Tennessee. Forming pure stands or mixed with Fraser fir (*Abies fraseri*) in high montane forest in the south but usually mixed with hardwoods as single trees or small groves at moderate elevations in the north; 0–1,500(–2,000) m. Zone 3.

Red spruce differs profoundly in ecology from its close relative and neighbor, black spruce (*Picea mariana*). While black spruce is a pioneer species with a high light requirement and considerable tolerance for a dry atmosphere, red spruce in the north is a tree of mature forests, with gap regeneration and a requirement for high atmospheric humidity. Their seeding strategies are also different, with red spruce releasing all of its seeds each year, while black spruce retains some seeds within unopened cones for many years on the tree. Unfortunately, these characteristics, in combination with lack of a soil seed bank, led to the elimination of red spruce from many stands after thoughtless clear-cut harvesting for the excellent timber. Sometimes, if there are some residual red spruce trees left after such harvesting, they may cross with nearby black spruces and produce a transient generation of natural hybrids. There is controversy over whether these hybrids go on to backcross with the parent species, leading to introgression, but the evidence for this is rather unsatisfactory, and introgression is, at best, much rarer than some enthusiastic authors thought. Red and black spruces are also closely related to Serbian

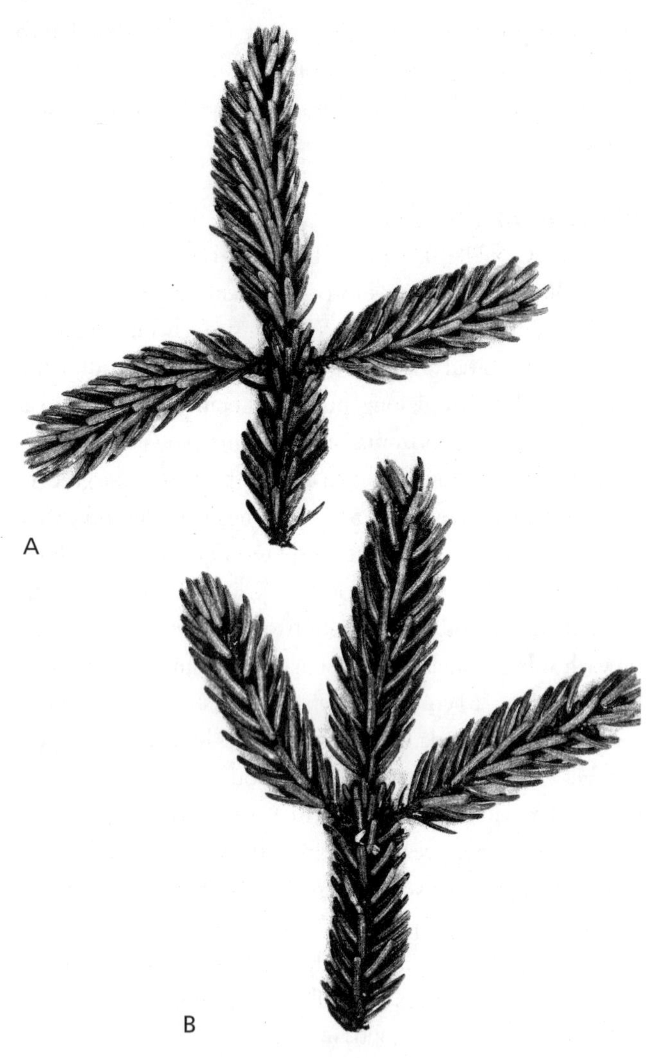

Upper (A) and undersides (B) of twigs of red spruce (*Picea rubens*) with 2 years of growth, ×1.

Twig tip of red spruce (*Picea rubens*) with open mature seed cones after shedding seeds; scale, 1 cm.

spruce (*P. omorika*), and both species can hybridize successfully with it. Based on DNA studies, these three species are not closely related to other spruces and form a group of their own that has not been given formal taxonomic recognition. The largest known individual in combined dimensions when measured in 1986 was 37.5 m tall, 1.37 m in diameter, with a spread of 11.9 m. The common and scientific names both refer to the reddish cast of the bark in comparison to white and black spruces.

Picea ×saaghyi Gáyer

Tree to 30 m tall. A garden hybrid combining the characteristics of the parent species. The twigs are hairless like both parents but intermediate in color, darker than in Yezo spruce (*Picea jezoensis*) but more yellowish than is usual in white spruce (*P. glauca*). The needles are sharp, as they often are in both species, and flattened top to bottom, dark above and whitened beneath like those of Yezo spruce, but with a few incomplete lines of stomates on the outer faces and stiff like those of white spruce.

Picea schrenkiana F. E. L. Fischer & C. Meyer

SCHRENK SPRUCE, EL' SCHRENKA (RUSSIAN), XUE LING YUN SHAN (CHINESE)

Tree to 40(–60) m tall, with trunk to 1(–1.5) m in diameter. Bark reddish brown to dull, dark brown, breaking up into small, scaly plates with age. Crown dense, conical to narrowly cylindrical, with upswept, horizontal, or downswept branches bearing numerous short side branches all around, opening up somewhat with age. New branchlets greenish yellow, yellow, or grayish yellow, hairy or not. Buds 5–10(–12) mm long, slightly resinous or not. Needles bright green, (1–)1.5–3(–3.5) cm long, curving or angled slightly forward and sometimes upward, flattened diamond-shaped, with four to six lines of stomates on each of the two outer faces and five to eight lines on the inner faces, blunt to pointed. Pollen cones 10–25 mm long, yellowish pink to yellowish red. Seed cones (5–)7–10(–11.5) cm long, green or purple before maturity, ripening dull dark brown, sometimes with a purplish tinge. Seed scales broadly egg-shaped, smooth-edged, woody and stiff. Seed body 3–5 mm long, the wing 8–10 mm longer. Mountains of central Asia, from Ningxia and western Nei Mongol (China) to southeastern Kazakhstan and Kyrgyzstan. Forming pure forests or mixed with a few other conifers and hardwoods on steep montane to subalpine slopes; (1,000–)1,500–3,500(–3,800) m. Zone 6. Two varieties.

Schrenk spruce is one of the most important trees of the forest zones of the arid central Asian mountain ranges, from the Helan Shan and Lupan Shan to the Dzungarian Ala Tau and the Tian Shan. It is an important and valuable timber tree in these regions of limited wood resources. As well as being geographically isolated from other spruce species (except for some overlap with *Picea asperata*), DNA studies suggest that Schrenk spruce has no close relatives, although it may be loosely associated with the Sargent spruce (*P. brachytyla*) group or perhaps the Lijiang spruce (*P. likiangensis*) group. As would be expected of an inhabitant of a series of isolated mountain ranges separated by large sweeps of unfavorable habitat, Schrenk spruce varies from range to range, and a number of segregates were named, but only two varieties within the species are accepted here. The species name honors Alexander G. von Schrenk (1816–1876), the Russian botanist who collected the type material.

Picea schrenkiana* var. *crassifolia (V. Komarov) V. Komarov

QING HAI YUN SHAN (CHINESE)

Tree to 25 m tall. Branchlets greenish yellow, not dangling. Basal bud scales curled back. Needles often more than 2 cm long. Eastern part of range of species west to northeastern Qinghai (China). Synonym: *Picea crassifolia* V. Komarov.

Picea schrenkiana F. E. L. Fischer & C. Meyer **var. *schrenkiana***

QUE LING YUN SHAN (CHINESE)

Tree to 40(–60) m tall. Branchlets yellow or grayish yellow, often dangling. Basal bud scales tight to the bud. Needles usually less than 2 cm long. Western part of the range of the species east to Xinjiang (China). Synonyms: *Picea robertii* Vipper, *P. schrenkiana* subsp. *tianschanica* (F. Ruprecht) Bykov, *P. schrenkiana* var. *tianschanica* (F. Ruprecht) W. C. Cheng & S. H. Fu, *P. tianschanica* F. Ruprecht.

Picea* ×*shirasawae Yasaka Hayashi

HIME-MATSUHADA (JAPANESE)

Tree to 30 m tall. A natural hybrid between the extremely localized Koyama spruce (*Picea koyamai*) and the somewhat more widespread Alcock spruce (*P. alcoquiana*). It generally resembles the latter more closely, but has the slightly shorter needles and rounded seed scales of Koyama spruce. It differs from both parents in having some hairs in the grooves between the leaf bases on the twigs and in having longer seed cones, up to 12 cm long. It is confined to the Yatsugatake Mountains, the only place where both parents grow together, and to the nearby Akaishi Mountains where Koyama spruce no longer occurs. The parent species are not particularly closely related, and the rarity of *P. koyamai* may have promoted hybridization with Koyama spruce as the seed parent. Synonyms: *P. alcoquiana* var. *acicularis* (Shirasawa & M. Koyama) Fitschen, *P. bicolor* var. *acicularis* Shirasawa & M. Koyama.

Picea sitchensis (Bongard) Carrière

SITKA SPRUCE, TIDELAND SPRUCE

Tree to 70(–95) m tall or dwarfed and shrublike on high barren outcrops, with trunk to 5(–6.5) m in diameter, often further flared and buttressed at the base. Bark gray, remaining relatively smooth and scaly even on the largest trees. Crown narrowly cylindrical, with relatively slender, horizontal branches drooping at the tips and bearing stiffly horizontal to hanging side branches, the lowest branches sometimes rooting to start new trees. New branchlets very pale pinkish brown, hairless. Buds (4–)5–10 mm long, generally not resinous. Needles dark green to yellowish green above, bluish green beneath with wax, (1–)1.5–2.5(–3) cm long, curved slightly forward, flattened diamond-shaped, without or with one to three incomplete lines of stomates on the two outer faces and with three to five lines in the stomatal bands on the two inner faces, stiff and prickly. Pollen cones 20–35 mm long, red. Seed cones (4–)5–9(–10) cm long, green before maturity, ripening light brown, seed scales roughly diamond-shaped, minutely and irregularly toothed, thin and flexible to a little stiff. Seed body 2–4 mm long, the wing 5–8 mm longer. Pacific Coast fog belt of North America within 200 km of the ocean or less, from Kodiak Island and the adjacent Alaska Peninsula, Alaska, through British Columbia, to central Mendocino County, California. Forming pure stands or more commonly mixed with several other conifers

Middle-aged mature Sitka spruce (*Picea sitchensis*) in nature.

Underside of twig of Sitka spruce (*Picea sitchensis*) with 2 years of growth, ×0.5.

and a few hardwoods on flats and coastward slopes, rarely forming shrubby thickets, as on rocky outcrops above the Juneau Ice Field in Alaska; 0–500(–1,200) m. Zone 7.

Sitka spruce, named after Sitka Island in Alaska, from which the type specimen came, is one of the forest giants of the rain-soaked and fog shrouded forests along the Pacific Coast of North America. It is the largest spruce species and one of the largest conifers in the world, only redwood (*Sequoia sempervirens*), giant sequoia (*Sequoiadendron giganteum*), Douglas fir (*Pseudotsuga menziesii*), Montezuma bald cypress (*Taxodium mucronatum*), and possibly New Zealand kauri (*Agathis australis*) exceeding it in height or diameter. The two largest known individuals in the United States in combined dimensions when both were measured in 1998 were 62.2 m tall, 5.1 m in diameter, with a spread of 28.4 m, and 58.4 m tall, 5.4 m in diameter, with a spread of 29.3 m. The tallest Sitka spruce in Canada, the Carmanah Giant on Vancouver Island, measured in 1988 at 95.7 m tall but only 3.05 m in diameter, may be the tallest tree standing in Canada, now that taller Douglas firs have fallen or been felled. Because of the large dimensions of the clear lumber that it produces, Sitka spruce is an important timber tree. It is also one of the most valuable plantation trees in Great Britain, especially in Scotland. It joins British individuals of its fellow North American Pacific conifers, grand fir (*Abies grandis*) and Douglas fir, as the tallest trees in Europe at 60 m and more, exceeding the tallest native species.

Long thought to be closely related to the Asian Yezo spruce (*Picea jezoensis*), because of its flat needles and similar seed cones, and to the North American Engelmann spruce (*P. engelmannii*), blue spruce (*P. pungens*), and white spruce (*P. glauca*), DNA studies suggest that the closest relative of Sitka spruce is in another western North American species, the extremely localized Brewer spruce (*P. breweriana*), and that it is rather remote from white spruce and Engelmann spruce, though not, perhaps, from blue spruce and Yezo spruce. However, Sitka spruce crosses readily with all of these in experimental pollinations as well as with additional species not considered closely related, like Serbian spruce (*P. omorika*). In nature, Sitka spruce approaches or overlaps locally with white spruce, Engelmann spruce, and Brewer spruce. It hybridizes freely with white spruce on the Kenai Peninsula (Alaska) and in the Skeena River valley (British Columbia) to form *P.* ×*lutzii.* It may also hybridize with Engelmann spruce in British Columbia, but this is less clear. Although their ranges come within 25 km of each other, Sitka spruce never quite grows together with Brewer spruce and no wild hybrids between these two species are known.

Picea smithiana (N. Wallich) Boissier

MORINDA SPRUCE, WEST HIMALAYAN SPRUCE, HIMALAYAN WEEPING SPRUCE, MORINDA (HINDI), CHANG YE YUN SHAN (CHINESE)

Tree to 50(–70) m tall, with trunk to 1.5(–2.3) m in diameter. Bark light grayish brown, breaking up into irregular, scaly plates separated by shallow furrows on large trees. Crown conical to cylindrical with numerous thin, horizontal to drooping branches bearing hanging side branches. New branchlets shining tan or grayish tan, hairless. Buds 6–12 mm long, slightly resinous. Needles dark green, (2.5–)3.5–4.5(–5.5) cm long, angled and curved slightly forward, diamond-shaped and slightly flattened side to side, with two to five lines of stomates on each face, pointed but not prickly. Pollen cones 20–30 mm long, reddish yellow. Seed cones (8.5–)10–18 cm long, green before maturity, ripening shiny rich brown. Seed scales broadly egg-shaped, smooth-edged, woody and stiff. Seed body 5–6(–7) mm long, the wing 10–14 mm longer. Western Himalaya, from eastern Afghanistan to central Nepal and adjacent China in Xizang (Tibet). Forming pure stands or mixed with other upper montane and subalpine conifers and hardwoods; (2,150–)2,400–3,400(–3,700) m. Zone 8.

Underside of twig of Morinda spruce (*Picea smithiana*) with 2 years of growth.

Morinda spruce is a handsome tree with weeping branches and branchlets bearing the longest needles among the extant spruce species. Where abundant, as in Kashmir (India), it is heavily exploited for a variety of uses, including paper pulp, railway ties, floor and ceiling joists, and wooden aircraft. Although it is rare in Xizang (Tibet), its high value may lead to plantation culture there. It is not especially closely related to any other spruce species, but DNA studies suggest a possible linkage with the similarly weeping Serbian spruce (*Picea omorika*) and its nonweeping North American relatives, black spruce (*P. mariana*) and red spruce (*P. rubens*). There are two plausible contemporary Smiths for whom Wallich might have named this species: James Edward Smith (1759–1828), founding president of the Linnean Society of London, who died 4 years before publication of the species, and Mr. Smith (or Smythe), gardener at Hopetoun in Scotland, who raised the first trees from seed sent there in 1818.

Picea spinulosa (W. Griffith) A. Henry

SIKKIM SPRUCE, EAST HIMALAYAN SPRUCE, EHSING (DZONGKHA), XU MI YUN SHAN (CHINESE)

Tree to 60(–70) m tall, with trunk to 2.5 m in diameter. Bark gray or grayish brown, becoming narrowly ridged and furrowed at the base of large trunks. Crown cylindrical, with long, slender, downswept branches bearing dangling side branches. New branchlets tan to yellowish gray, hairless. Buds 5–8(–12) mm long, slightly resinous or not. Needles dark green above, waxy white beneath, (1.5–)2–3(–3.5) cm long, slightly curved forward, diamond-shaped and a little flattened top to bottom, without or with one to three incomplete lines of stomates on the two outer faces and with four to seven lines in the stomatal bands on the two inner faces, prickly (hence the species name, Latin for "with small spines"). Pollen cones 15–25 mm long, pink. Seed cones (4–)5–10(–12) cm long, green with purple trim before maturity, ripening shiny reddish brown. Seed scales angular egg-shaped with a squared off projecting tip, woody and stiff. Seed body 3–5 mm long, the wing 6–10 mm longer. Eastern Himalaya, from Sikkim (India) and adjacent China in Xizang (Tibet) through Bhutan to Arunachal Pradesh (India). Forming pure stands or, more commonly, mixed with other high montane and subalpine conifers and hardwoods; 2,400–3,700 m. Zone 8.

Although somewhat similar to the nearby Morinda spruce (*Picea smithiana*) and Sargent spruce (*P. brachytyla*) in appearance, Sikkim spruce is not closely related to either of these weeping species. Evidence from DNA is somewhat ambiguous, supporting either an isolated position for the species or a relationship to Lijiang spruce (*P. likiangensis*), Yezo spruce (*P. jezoensis*), and blue spruce (*P. pungens*). Its affinities warrant further study. Like other large spruces, its wood finds general use in construction, and the trees are cultivated to a limited extent in forest plantations.

Picea wilsonii M. T. Masters

WILSON SPRUCE, QING QIAN (CHINESE)

Tree to 50 m tall, with trunk to 1.5 m in diameter. Bark dark gray, roughening but remaining scaly with age. Crown conical to cylindrical, with horizontal, spreading branches bearing horizontal or drooping side branches. New branchlets pale yellowish gray, hairless or with a few tiny hairs at first. Buds 6–8 mm long, not resinous. Needles dark green, (0.5–)1–1.5(–2) cm long, curved and angled forward, diamond-shaped and somewhat flattened top to bottom, with two to four lines of stomates on the two outer faces and four or five lines on the inner faces, blunt to prickly. Pollen cones 20–30 mm long, pink. Seed cones 4–8 cm long, green before maturity, ripening light yellowish brown. Seed scales egg-shaped, smooth-edged, thin but woody and a little stiff. Seed body 3–4.5 mm long, the wing 9–11 mm longer. North-central China from Qinghai and Sichuan to Nei Mongol, Shanxi, and Hubei. Scattered among other spruces and hardwoods in upper montane

Sparse natural stand of Wilson spruce (*Picea wilsonii*).

forest; 1,400–3,000 m. Zone 6. Synonyms: *Picea wilsonii* var. *shansiensis* Silba, *P. wilsonii* var. *watsoniana* (M. T. Masters) Silba.

Wilson spruce shows considerable variation in stature and needle length and thickness across its broad range. Much of this variation can be attributed to climatic differences from site to site, particularly in moisture availability during the growing season, and so does not merit formal taxonomic recognition. It is harvested along with its more common associated spruce species and used indifferently for the same purposes: pulp and general carpentry. DNA studies imply a relationship to Sargent spruce (*Picea brachytyla*) and its allies, particularly the rare Veitch spruce (*P. neoveitchii*), which shares part of its range. The species name honors Ernest H. Wilson (1876–1930), who collected seeds of many trees and shrubs in central China, first for the Veitch nursery firm in England and later for the Arnold Arboretum of Harvard University.

Pinus Linnaeus

PINE

Pinaceae

Evergreen single or multistemmed trees or shrubs. Bark smooth at first, nonfibrous and flaking in small irregular patches, highly varied among the species at maturity, in some remaining smooth, in others becoming variously ridged and furrowed, and in others divided into regular or irregular plates, sometimes massive, shieldlike ones. Saplings and young trees initially bearing regular annual or twice yearly tiers of about three to five rising or horizontal branches from the base to form a symmetrical, conical or cylindrical crown that may broaden and flatten with age, becoming irregular and sometimes even tattered. Branchlets strongly differentiated into two kinds, short shoots and long shoots. Short shoots are the most extreme ones among the conifers, reduced to a bundle (fascicle) of (one or) two to eight needles on an abortive growth tip and surrounded at the base by a persistent or shed sheath of papery scale leaves. Long shoots are ordinary woody shoots extending the growth of the tree but without green needles (except in seedlings and trunk sprouts), instead clothed with papery scale leaves, smooth or furrowed between the attached bases of these scales. Most scale leaves with short shoots in their axils, the rest with either pollen or seed cones, or nothing at all. Winter buds well developed, scaly, usually resinous. Foliage leaves in spirally arranged bundles radiating all around the long shoots, needlelike, each bundle circular in cross section and the individual needles then shaped like pie pieces in cross section. Needles usually longer than those of any other Pinaceae, straight or twisted, uniform in thickness for most of their length, abruptly narrowed to the pointed tip and the slightly narrowed base.

Plants monoecious. Pollen cones numerous and crowded all around the base of new long shoots as they emerge from the terminal buds, single in the axils of scale leaves and surrounded at the base by other scale leaves. Each cone cylindrical, with numerous spirally arranged pollen scales, each scale bearing two pollen sacs. Pollen grains small to large (body 30–75 µm long, 40–80 µm overall), with two round air bladders diverging from the larger spherical to oval body at 90° (a right angle) to each other or more. Pollen body minutely warty, the bladders with a more wrinkled sculpture and often set off from the body by a distinct ridge. Seed cones quite varied in size and shape, from almost spherical to nearly cylindrical, often strongly asymmetrical, with or without a scaly stalk, often in a circle of two to five around the branchlet near the tip of a year's growth. Each cone single in the axil of a scale leaf, standing straight out from the twigs at pollination, stiffly outstretched, angled forward or backward, or hanging at maturity. Maturing in two (or three) seasons and falling after releasing the seeds or remaining intact many years until opened by fire. Seed scales numerous, densely spirally arranged, with a tiny bract attached to the base and never protruding between the scales, woody, either thin and relatively uniform (often in subgenus *Strobus*) or with a diamond-shaped, shieldlike thickening (umbo) on the variably thickened exposed face near the tip of the scale (subgenus *Pinus* and some subgenus *Strobus*). The umbo, when pres-

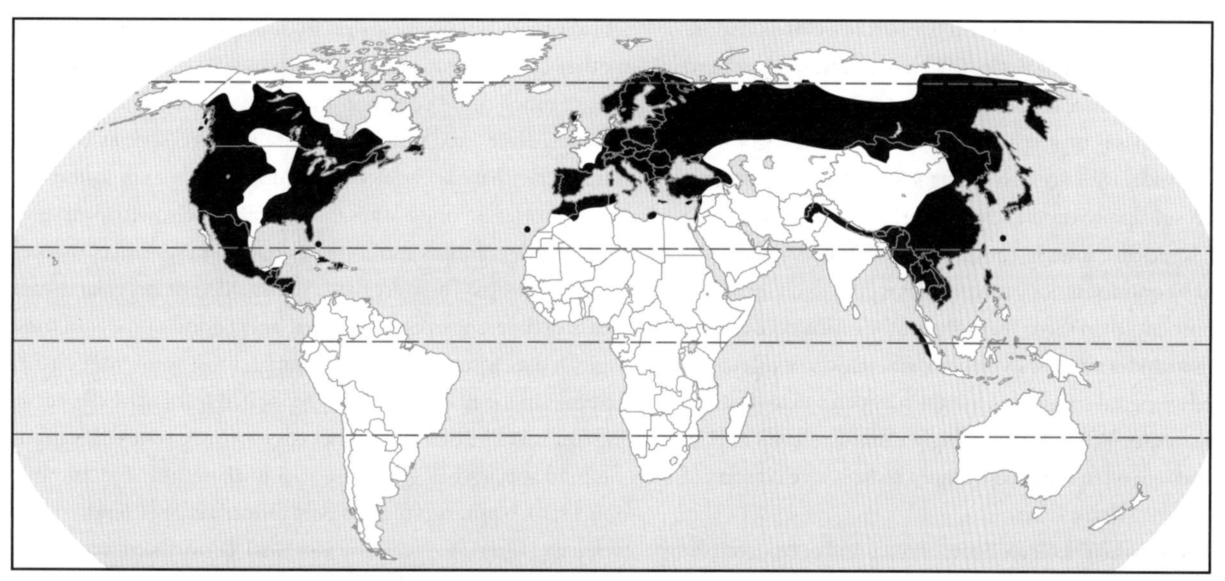

Distribution of *Pinus* across the northern continents and south of the equator in Sumatra, with restricted species on some oceanic islands, scale 1 : 200 million

ent, usually ending in a varied point, from short, to pricklelike, to massive claws, but slender points may be shed or break off, and in some species they never form and the umbo has a flat top. Seeds two per scale, oblong, the asymmetrical wing derived from the seed scale and clasping the seed body loosely or tightly via two thin arms running the length of the body, sometimes reduced to just these two arms and hence the seeds seemingly wingless. Cotyledons 2–15(–18), each with one vein. Chromosome base number $x = 12$.

Wood soft (subgenus *Strobus*) to hard (subgenus *Pinus*), with a more or less pronounced resinous odor and reddish brown to brown heartwood well differentiated from the light brown or yellowish brown sapwood. Grain even to uneven, with a gradual (subgenus *Strobus*) to abrupt (subgenus *Pinus*) transition between earlywood and latewood. Vertical resin canals numerous and large, with horizontal water-conducting cells (tracheids) in the rays.

Lines of stomates confined to the inner faces adjoining the other needles of a bundle (some subgenus *Strobus*) or also present on the outer face (subgenus *Pinus* and some subgenus *Strobus*). Each stomate sunken deeply beneath and sometimes almost hidden by the 4–11 (depending on the species) surrounding subsidiary cells, which are often topped by a variously formed (again depending on the species) Florin ring whose height may be enhanced by an encircling furrow. Leaf cross section with a variable number of resin canals in various positions with respect to the epidermis and a single- (subgenus *Strobus*) or double-stranded (subgenus *Pinus*) midvein embedded in a cylinder of transfusion tissue. Photosynthetic tissue without a well-defined palisade layer, uniform all around the needle between the large central vascular cylinder and the epidermis and adjacent layers of hypodermis that are continuous except beneath the stomatal lines.

Ninety-seven species, throughout much of the northern hemisphere from the boreal zone to the tropics, extending south to Nicaragua and Cuba in the New World and Mediterranean North Africa, the Himalaya, and Malesia in the Old World. References: Carvajal and McVaugh 1992, Coode and Cullen 1965, Critchfield and Little 1966, de Laubenfels 1988, Farjon 1990, 2005a, Farjon and Styles 1997, Fu et al. 1999c, Gaussen et al. 1993, Harlow 1931, Komarov 1986, Kral 1993, Lee 1979, Li and Keng 1994, Little and Critchfield 1969, Martínez 1963, Meikle 1977, Miller 1976, Mirov 1953, 1967, Perry 1991, Reidl 1966, Richardson 1998, K. C. Sahni 1990, Shaw 1909, 1914, Yamazaki 1995, Zohary 1966.

Pinus is the largest and most widespread genus of conifers in the northern hemisphere and one of the two largest in the world, with (like the southern hemisphere *Podocarpus*) about one hundred species, almost twice as many as the next largest genus (*Juniperus*). Its ecological range is astonishingly broad, from boreal forests and alpine shrubberies to lowland tropical savannas, and from swamp margins to desert slopes. It is the most ecologically diverse genus, not only among conifers but perhaps among all woody plants. There are few areas of the northern hemisphere out of sight of some species of pine. Nevertheless, there are some common elements in the ecology of the species, one of the most important being that the majority of the species grow in areas subject to periodic fires. They are typically pioneer species with a high light requirement for establishment. The disturbance that provides these open conditions more often than not is fire. Once established, a number of species can live for hundreds of years and become part of a closed, mature forest. Even these long-lived species will eventually disappear from the canopy in the absence of fire. The longest-lived individual trees of all, the Great Basin bristlecone pines (*Pinus longaeva*) of California and Nevada, with ages exceeding 4,000 years, are denizens of very open subalpine woodlands of desert mountain ranges and thus never experience intense competition for light. Another ecological feature common among pine species is growth on soils even poorer in nutrients than those that support the majority of conifer species. This includes sandy soils within a wide variety of vegetation types as well as special conditions, like the hardpan soils of Californian pygmy forests. When there are higher nutrient levels in cool climates, pines are often replaced by more competitive conifers of other genera, like *Abies* or *Picea.*

In addition to their ecological importance and considered on a worldwide basis, pines have the greatest economic value and cultural significance among all the conifers. They are sources of many forest products, including construction timber, plywood, pulp, resins, and edible seeds. The latter gives the genus its name, *Pinus* being classical Latin for the Mediterranean stone pine (*P. pinea*), exploited throughout the Mediterranean region for food. Resins or naval stores, once so important for caulking wooden ships, are obtained both as by-products of pulp production and lumber production and by directly tapping the trees, taking advantage of the resin canals which run both vertically and horizontally throughout pine trunks. Pines have become the most important timber-producing conifers far beyond their natural distribution, having been planted throughout the temperate southern hemisphere and even in the tropics, where few other conifers are amenable to plantation culture. They are so successful in some places, as in South Africa and New Zealand, that some species, Monterey pine (*P. radiata*) in particular, can become invasive pests in natural vegetation. Scots pine (*P. sylvestris*), which sees use primarily as a Christmas tree in Canada and the United States, is likewise widely naturalized in northeastern North America. The economic and cultural importance of pines also makes them popular subjects of scientific and technological study, and there is about as much technical literature on pines as on all of the other conifer genera combined.

Not surprisingly, given their cultural significance, ornamental long needles compared to other conifers, and tolerance of neglect, many pine species have also entered into general cultivation for horticulture. Some, such as Austrian pine (*Pinus nigra*), Scots pine (*P. sylvestris*), and mugo pine (*P. mugo*), are among the most commonly cultivated ornamental conifers in temperate North America and Europe. Since these and others have been cultivated for centuries here and in eastern Asia and elsewhere, there has been very extensive cultivar selection, including cultivars of hybrid origin. Pine cultivars pretty much cover the range found in other conifers, but selections with modified habit, shape, and size are much more common than shifts in needle color. There are blues, silvers, and variegated golds in several species, but even more species have seen selections with twisted or contorted needles or shorter or longer ones than the norm. Growth form variations include columns and cones, weepers and spreaders, umbrellas and tabletops. There are many rock garden or larger landscape dwarfs of different sizes and shapes, and a good number of these originated as witches'-brooms although most were from chance seedlings in the nursery bed or propagated from naturally occurring runts. While there are hundreds of named pine cultivars, no one pine species has given rise to anywhere near the numbers found in such other conifers as Port Orford cedar (*Chamaecyparis lawsoniana*), sugi (*Cryptomeria japonica*), or northern white cedar (*Thuja occidentalis*), so there is plenty of room for further experimentation and selection in this genus.

It is clear that *Pinus* is closest to *Picea, Cathaya, Pseudotsuga,* and *Larix,* and some authors place all these genera together in a subfamily, Pinoideae, but the relationship is not close, and other authors restrict subfamily Pinoideae to *Pinus* alone while some even make it the sole genus of a restricted Pinaceae, with all other genera in a family Abietaceae. Few go so far, but all recognize the comparatively isolated position of this genus, which has almost as many species as the other ten genera combined and occupies more of the Earth's surface.

Because there are so many species of pines (for a conifer anyway, nothing like 500 species of oaks, 750 species of figs, or 1,700 species of the potato genus), there has also been much effort expended on elaborate classifications of the genus involving botanical subgenera, sections, subsections, and sometimes series. In broad outline, the classification presented by Little and Critchfield (1969) has been widely followed and is used here, with modifications and simplifications demanded by more recent studies. The subsection names given after each species name below show which species are related closely to each other. Generally speaking, facility of species hybridization follows these subsections and few species will hybridize outside of their own subsection, so the great majority of reported natural and synthetic pine hybrids are between species within a subsection. There are commonly crossing barriers between species within subsections as well, so members of a subsection are often not freely intercrossable. DNA studies and other biochemical and molecular investigations shed new light on relationships among pine species, confirming many traditional groups, including the two subgenera of hard or yellow (subgenus *Pinus*) and soft or white (subgenus *Strobus*) pines, but also suggesting revisions. Perhaps the most dramatic of these changes concerns the uniquely flat-needled *P. krempfii* of Vietnam, so unusual that some authors placed it in a separate genus, *Ducampopinus,* while Little and Critchfield used the same name for a subgenus containing only this species. DNA studies confirm a minority opinion based on traditional structures, including the undivided midvein, that this two-needled pine belongs among the largely five-needled species of subgenus *Strobus.*

DNA studies have not yet made much of an impact on the most vexed problems in pine taxonomy, those involving the recognition of botanical species, subspecies, and varieties. Even in the well-studied pine floras of North America and Europe, there are extensive disagreements about whether to recognize certain kinds of pines as species, varieties or subspecies, or even as variants not worthy of formal recognition. The situation is even worse in the rich pine floras of China and Mexico, which need extensive new investigations at the population level involving the full array of modern botanical techniques. Floristic studies of both of these regions (Fu et al. 1999c for China, and Perry 1991 and Farjon and Styles 1997 for Mexico), while presenting important advances over previous work, still leave many unanswered questions. At least some of the taxonomic complexity in Mexican pines is probably due to natural hybridization between species, but this has scarcely been investigated and even then in only a very few species. So, despite the fact that pines are the most studied of conifers and the most widely cultivated on both a small scale and in massive plantations, there is still much to learn about the species and their relationships.

Discerning these relationships is not made any easier by the fact that *Pinus* is the oldest modern genus of Pinaceae, as far as known, with many fossils from throughout the northern hemisphere, beginning in the early Cretaceous, over 100 million years ago. By the Eocene, some 50 million years ago, many modern sections and subsections were in evidence, so there has been a great deal of time for evolution and confounding extinction within the genus. Its age also obscures its relationships to the other genera of Pinaceae, and many extinct Cretaceous species (placed in the fossil genus *Pityostrobus*) combine characters of *Pinus* and various other genera.

Geographic Guide to *Pinus*

Europe, North Africa, and western Asia
- *Pinus cembra*
- *P. halepensis*
- *P. heldreichii*
- *P. mugo*
- *P. nigra*
- *P. peuce*
- *P. pinaster*
- *P. pinea*
- *P. sylvestris*

Canary Islands
- *Pinus canariensis*

Siberia, Russian Far East, Mongolia, northeastern China, Korea, and Japan
- *Pinus bungeana*
- *P. densiflora*
- *P. koraiensis*
- *P. luchuensis*
- *P. parviflora*
- *P. pumila*
- *P. sibirica*
- *P. sylvestris*
- *P. thunbergii*

Himalaya
- *Pinus bhutanica*
- *P. gerardiana*
- *P. roxburghii*
- *P. wallichiana*

Southwestern and central China, and Taiwan
- *Pinus armandii*
- *P. densata*
- *P. massoniana*
- *P. parviflora*
- *P. squamata*
- *P. tabuliformis*
- *P. taiwanensis*
- *P. yunnanensis*

Southeast Asia and Malesia
- *Pinus dalatensis*
- *P. kesiya*
- *P. krempfii*
- *P. merkusii*
- *P. parviflora*

Eastern and boreal North America
- *Pinus banksiana*
- *P. echinata*
- *P. elliottii*
- *P. glabra*
- *P. palustris*
- *P. pungens*
- *P. resinosa*
- *P. rigida*
- *P. strobus*
- *P. taeda*

P. virginiana

Western North America and Baja California

Pinus albicaulis
P. aristata
P. arizonica
P. attenuata
P. balfouriana
P. cembroides
P. contorta
P. coulteri
P. culminicola
P. edulis
P. engelmannii
P. flexilis
P. jeffreyi
P. lagunae
P. lambertiana
P. leiophylla
P. longaeva
P. monophylla
P. monticola
P. muricata
P. ponderosa
P. quadrifolia
P. radiata
P. sabiniana
P. strobiformis
P. torreyana

Mexico (except Baja California) and Guatemala

Pinus arizonica
P. ayacahuite
P. caribaea
P. cembroides
P. culminicola
P. devoniana
P. douglasiana
P. durangensis
P. engelmannii
P. greggii
P. hartwegii
P. herrerae
P. jaliscana
P. lawsonii
P. leiophylla
P. lumholtzii
P. maximartinezii
P. maximinoi
P. montezumae
P. nelsonii
P. oocarpa
P. patula
P. pinceana
P. ponderosa
P. praetermissa
P. pringlei
P. pseudostrobus
P. rzedowskii
P. strobiformis
P. strobus
P. tecunumanii
P. teocote

West Indies

Pinus caribaea
P. cubensis
P. occidentalis
P. tropicalis

Taxonomic Guide to *Pinus*

The classification here was proposed by Gernandt et al. (2005) on the basis of phylogenetic relationships as revealed by DNA studies. It is consistent with traditional morphological characters but differs considerably from the generally used previous standard classification by Little and Critchfield (1969) and somewhat less from an earlier attempt at a classification based on molecular data done by Price et al. (1998). Compared to these two classifications, and virtually every other prior classification, this one recognizes far fewer lowest-level groupings of species (subsections), and the species composition and affinities of these groupings to one another are moderately to substantially different from earlier treatments. The traditional subsections abandoned in this classification lack phylogenetic coherence themselves or cause phylogenetic incoherence (paraphyly) in other subsections. They are listed here as synonyms of the accepted subsections. The subsections help in predicting character distributions and crossing behavior and may be useful when making species identifications.

Pinus subg. *Pinus* (synonym: subg. *Diploxylon* Rehder)
sect. *Pinus*
subsect. *Pinaster* J. C. Loudon (synonyms: subsect. *Canarienses* J. C. Loudon, subsect. *Halepenses* J. Burgh, subsect. *Pineae* E. Little & W. Critchfield)
Pinus canariensis
P. halepensis
P. heldreichii

P. pinaster
P. pinea
P. roxburghii
subsect. *Pinus* (synonym: subsect. *Sylvestres* J. C. Loudon)
Pinus densata
P. densiflora
P. kesiya
P. luchuensis
P. massoniana
P. merkusii
P. mugo
P. nigra
P. resinosa
P. sylvestris
P. tabuliformis
P. taiwanensis
P. thunbergii
P. tropicalis
P. yunnanensis
sect. *Trifoliae* Duhamel
subsect. *Contortae* E. Little & W. Critchfield
Pinus banksiana
P. contorta
P. virginiana
subsect. *Australes* J. C. Loudon (synonyms: subsect. *Attenuatae* J. Burgh, subsect. *Leiophyllae* J. C. Loudon, subsect. *Oocarpae* E. Little & W. Critchfield)
Pinus attenuata
P. caribaea
P. cubensis
P. echinata
P. elliottii
P. glabra
P. greggii
P. herrerae
P. jaliscana
P. lawsonii
P. leiophylla
P. lumholtzii
P. muricata
P. occidentalis
P. oocarpa
P. palustris
P. patula
P. praetermissa
P. pringlei
P. pungens
P. radiata
P. rigida
P. taeda
P. tecunumanii
P. teocote
subsect. *Ponderosae* J. C. Loudon (synonyms: subsect. *Pseudostrobi* J. Burgh, subsect. *Sabinianae* J. C. Loudon)
Pinus arizonica
P. coulteri
P. devoniana
P. douglasiana
P. durangensis
P. engelmannii
P. hartwegii
P. jeffreyi
P. maximinoi
P. montezumae
P. ponderosa
P. pseudostrobus
P. sabiniana
P. torreyana
Pinus subg. *Strobus* J. Lemmon (synonym: subg. *Haploxylon* Rehder)
sect. *Parrya* H. Mayr
subsect. *Balfourianae* Engelmann
Pinus aristata
P. balfouriana
P. longaeva
subsect. *Nelsoniae* J. Burgh
Pinus nelsonii
subsect. *Cembroides* Engelmann
Pinus cembroides
P. culminicola
P. edulis
P. lagunae
P. maximartinezii
P. monophylla
P. pinceana
P. quadrifolia
P. rzedowskii
sect. *Quinquefoliae* Duhamel (synonym: sect. *Strobus* R. Sweet)
subsect. *Gerardianae* J. C. Loudon
Pinus bungeana
P. gerardiana
P. squamata
subsect. *Krempfianae* E. Little & W. Critchfield
Pinus krempfii

subsect. Strobus J. C. Loudon (synonym: subsect. Cembrae J. C. Loudon)

Pinus albicaulis
P. armandii
P. ayacahuite
P. bhutanica
P. cembra
P. dalatensis
P. flexilis
P. koraiensis
P. lambertiana
P. monticola
P. parviflora
P. peuce
P. pumila
P. sibirica
P. strobiformis
P. strobus
P. wallichiana

Identification Guide to *Pinus*

Because there are so many species of *Pinus* with overlapping vegetative characteristics, it is also necessary to use characteristics of pollen and seed cones for identification. Even so, attempting to identify a pine can be a frustrating experience, especially when there are no geographic restrictions. This guide should allow you to boil an identification down to two or three species, even if you are not confident in a definitive identification. Then you can carefully read the descriptions of these two or three species for additional decisive differences. At this point, geographic distribution may prove helpful for trees of known origin, or hardiness zones may prove conclusive. Most of the measurements in the identification guide refer to average or predominant counts or sizes rather than to the full possible range, so choose representative leaves or cones rather than extremes. Even so, some species are so variable that they are included in more that one identification group.

The 97 species of *Pinus* may first be assigned to the two subgenera of hard pines and soft pines by the number of strands (vascular bundles) in the needle midvein and whether the sheaths surrounding the bases of the needle bundles curl back and are shed during the first year or whether they persist until the needles are shed:

Pinus subg. *Strobus* (soft pines), one strand in the needle midvein, needle bundle sheaths shed

Pinus subg. *Pinus* (hard pines), two strands in the needle midvein, needle bundle sheaths persistent

One species of subgenus *Strobus, Pinus nelsonii,* has persistent sheaths but is unique among three-needled pines in that the needles in a bundle stick together along their whole length long after they have reached full size. In contrast, one species of subgenus *Pinus, P. leiophylla,* has sheaths that are shed at maturity. It differs from soft pines (subgenus *Strobus*) not only in having two vascular bundles in the midveins of each of the three or five needles in a bundles but also in its typical hard pine (subgenus *Pinus*) seed cone, with the exposed (epiphyses) faces of the seed scales diamond-shaped with a central umbo.

Subgenus *Strobus* (soft pines). Among the 33 additional species in subgenus *Strobus* (besides *Pinus nelsonii*), *P. monophylla* is unique in having just a single needle in most bundles. *Pinus krempfii* stands out with its flattened pair of needles, each typically 1.5–4 mm wide. The remaining 31 species may be divided into five groups for ease of identification based on the number of needles in most bundles and the length of larger, but still typical needles:

Group A, needles fewer than five in a bundle, up to 6 cm

Group B, needles fewer than five in a bundle, 6 cm or more

Group C, needles five in a bundle, up to 6 cm

Group D, needles five in a bundle, 6–11 cm

Group E, needles five in a bundle, at least 11 cm

Group A. Among these five species, only *Pinus culminicola* has lines of stomates confined to the inner faces of the needles. The remaining species, all of which have lines of stomates on the outer faces as well as the inner faces (though often fewer), may be distinguished by the predominant needle number and the length of the pollen cones:

Pinus edulis, needles two in a bundle, pollen cones more than 6 mm

P. lagunae, needles three in a bundle, pollen cones up to 6 mm

P. cembroides, needles three in a bundle, pollen cones more than 6 mm

P. quadrifolia, needles four in a bundle, pollen cones more than 6 mm

Group B. These three species may be distinguished by the length of most seed cones and whether the umbo ends in a weak or strong prickle:

Pinus bungeana, seed cones up to 8 cm, prickle strong

P. pinceana, seed cones up to 10 cm, prickle weak

P. gerardiana, seed cones more than 12 cm, prickle strong

Group C. Among these nine species, two species stand out because they have seed cones that remain closed after maturity while in all the rest they open soon after maturity. *Pinus pumila* has seed cones less than 5 cm long and lines of stomates confined to the inner faces of the needles while *P. albicaulis* has seed cones at least 5 cm long and lines of stomates on both the inner and outer faces. The remaining seven species may be distinguished by the length of the longest needles, whether there are lines of stomates on both the inner and outer faces or only on the inner faces, the length of the seed cones, and for the first three species, the length of the prickle on the umbo:

Pinus balfouriana, needles up to 4 cm, stomatal lines on inner faces only, seed cones 6 cm or more, prickle 1 mm or less

P. longaeva, needles up to 4 cm, stomatal lines on inner faces only, seed cones 6 cm or more, prickle 1–6 mm

P. aristata, needles up to 4 cm, stomatal lines on inner faces only, seed cones 6 cm or more, prickle 6 mm or more

P. culminicola, needles more than 4 cm, stomatal lines on inner faces only, seed cones less than 6 cm

P. parviflora, needles more than 4 cm, stomatal lines on inner faces only, seed cones 6 cm or more

P. quadrifolia, needles more than 4 cm, stomatal lines on inner and outer faces, seed cones mostly less than 6 cm

P. flexilis, needles more than 4 cm, stomatal lines on inner and outer faces, seed cones more than 6 cm

Group D. These 12 species may be assigned to three subgroups based on whether the seed cones remain closed after maturity or open up and whether the lines of stomates are confined to the inner faces of the needles or are found on both the inner and outer faces:

Group D1, seed cones closed, stomatal lines on inner faces only

Group D2, seed cones open, stomatal lines on inner faces only

Group D3, seed cones open, stomatal lines on inner and outer faces

Group D1. These three species may be distinguished by the length of the longer seed cones and the predominant needle length:

Pinus cembra, seed cones up to 8 cm, needles less than 9 cm

P. sibirica, seed cones 8–13 cm, needles less than 9 cm

P. koraiensis, seed cones 8–14 cm, needles 9 cm or more

Group D2. These six species may be distinguished by the length of the ordinary seed cones, whether the umbo is on the face or at the tip of the seed scale, and by the usual length of the seed body:

Pinus dalatensis, seed cones up to 8 cm, umbo at tip of scale, seed body less than 9 mm

P. armandii, seed cones mostly 8–15 cm, umbo at tip of scale, seed body 9–15 mm

P. rzedowskii, seed cones 8–15 cm, umbo on face of scale, seed body up to 9 mm

P. strobus, seed cones mostly 8–15 cm, umbo at tip of scale, seed body up to 9 mm

P. monticola, seed cones mostly 15–30 cm, umbo at tip of scale, seed body less than 9 mm

P. maximartinezii, seed cones 15–30 cm, umbo on face of scale, seed body more than 18 mm

Group D3. These three species may be distinguished by the length of the seed cones:

Pinus peuce, seed cones 8–15 cm

P. strobiformis, seed cones 15–30 cm

P. lambertiana, seed cones 30 cm or more

Group E. Among these five species, *Pinus squamata* stands out because it has lines of stomates on both the inner and outer faces of the needles, rather than just the inner, and because the umbo is prominent and protrudes from the face of the seed scale rather than being a thin continuation of the tip. *Pinus armandii* is also distinctive within the group for its seed cones that are robustly egg-shaped rather than more cylindrical, and its large seeds (the body more than 1 cm long), with a wing much shorter than the length of the body or even absent, while the others have seeds with the body less than 1 cm, bearing wings much longer than the body. The other three species may be distinguished by the usual length of the needles, the usual length of the seed cones, and whether the scales at the base of the seed cone point forward toward the tip or curl back around the stalk:

Pinus ayacahuite, needles up to 18 cm, seed cones more than 20 cm, scales at base of seed cone curled back

P. wallichiana, needles up to 18 cm, seed cones more than 20 cm, scales at base of seed cone pointed forward

P. bhutanica, needles up to 24 cm, seed cones less than 20 cm, scales at base of seed cone pointed forward

Subgenus *Pinus* (hard pines). Among the 62 species of subgenus *Pinus* (besides *P. leiophylla*), *P. durangensis* stands out because the bundles predominantly have six needles, while all other species have a majority of bundles with five or fewer needles. For convenience in identification, the 61 remaining species may be divided into three groups by the predominant number of needles in each bundle:

Group F, needles two in a bundle

Group G, needles three in a bundle

Group H, needles four or five in a bundle

Group F. These 26 species may have three needles (or even more) in some bundles, but the most common number is two. These species may be assigned to four subgroups based on the predominant length of the needles:

Group F1, needles up to 8 cm

Group F2, needles 8–11 cm

Group F3, needles 11–15 cm

Group F4, needles 16 cm or more

Group F1. Among these eight species, *Pinus banksiana* stands out with the shortest needles, rarely more than 4 cm long, while the others usually exceed this length. The remaining seven species may be distinguished by the predominant length of the needles, the length of mature pollen cones, and the length of most seed cones:

Pinus sylvestris, needles up to 6 cm, pollen cones 3–6 mm, seed cones up to 6 cm

P. mugo, needles up to 6 cm, pollen cones 6–10 mm, seed cones less than 6 cm

P. contorta, needles up to 6 cm, pollen cones 10–15 mm, seed cones less than 6 cm

P. glabra, needles at least 6 cm, pollen cones 10–15 mm, seed cones up to 6.5 cm

P. heldreichii, needles at least 6 cm, pollen cones 10–15 mm, seed cones at least 6.5 cm

P. virginiana, needles at least 6 cm, pollen cones 15–20 mm, seed cones less than 6.5 cm

P. pungens, needles at least 6 cm, pollen cones 15–20 mm, seed cones more than 6.5 cm

Group F2. These six species may be distinguished by the length of the mature pollen cones, whether the needles are flexible (up to 1 mm in diameter) or stiff (1–2 mm in diameter), the position of the resin canals in the needles, whether midway between the needle surface and the midvein (medial) or touching the outer surface (marginal), and whether the seed cones fall soon within a year after maturity or persist for two or more seasons:

Pinus densiflora: pollen cones 5–10 mm, needles flexible, resin canals marginal, seed cones falling

P. densata: pollen cones 10–20 mm, needles stiff, resin canals marginal, seed cones persisting

P. taiwanensis: pollen cones 10–20 mm, needles stiff, resin canals medial, seed cones persisting

P. thunbergii: pollen cones 10–20 mm, needles stiff, resin canals medial, seed cones falling

P. nigra: pollen cones 20–30 mm, needles stiff, resin canals medial, seed cones falling

P. echinata: pollen cones 20–30 mm, needles flexible, resin canals medial, seed cones persisting

Group F3. These nine species may be distinguished by the length of mature pollen cones, the length of average seed cones, whether the cones fall (if they do) with the stalk or fall off the stalk leaving a few basal seed scales behind, and whether the resin canals in the needle are next to the needle surface (marginal), next to the midvein (internal), or in between (medial):

Pinus tabuliformis, pollen cones 5–10 mm, seed cones 3–10 cm, falling with stalk, resin canals marginal

P. halepensis, pollen cones 5–10 mm, seed cones 6–10 cm, falling off stalk, resin canals marginal

P. luchuensis, pollen cones 10–20 mm, seed cones 3–4 cm, falling with stalk, resin canals marginal and medial

P. cubensis, pollen cones 10–20 mm, seed cones 4–6 cm, falling with stalk, resin canals internal

P. densata, pollen cones 10–20 mm, seed cones 4–6 cm, falling with stalk, resin canals marginal

P. resinosa, pollen cones 10–20 mm, seed cones 4–6 cm, falling off stalk, resin canals marginal

P. muricata, pollen cones 10–20 mm, seed cones 6–10 cm, never falling, resin canals medial

P. pinaster, pollen cones 10–20 mm, seed cones 10–20 cm, falling with stalk, resin canals medial

P. pinea, pollen cones 10–20 mm, seed cones 10–20 cm, falling off stalk, resin canals marginal

Group F4. These four species may be distinguished by the predominant length of the seed cones, the length of mature pollen cones, and whether the resin canals in the needle are at the surface (marginal), midway between the surface and the midvein (medial), or stretching all the way between (septal):

Pinus massoniana, seed cones less than 10 cm, pollen cones 15–20 mm, resin canals marginal

P. merkusii, seed cones up to 10 cm, pollen cones 20–30 mm or more, resin canals medial

P. tropicalis, seed cones less than 10 cm, pollen cones 20–30 mm, resin canals septal

P. elliottii, seed cones 10 cm or more, pollen cones 30–40 mm, resin canals medial

Group G. Among these 21 species, *Pinus lumholtzii* stands out as the only one whose needles hang straight down from the twigs rather than sticking out around the twigs stiffly or with a gentle droop. The remaining 20 species may be assigned to three subgroups based upon their average needle length:

Group G1, needles less than 16

Group G2, needles 16–24 cm

Group G3, needles more than 24 cm

Group G1. These 10 species may be distinguished by the length of fully elongated pollen cones, the length of average seed cones, whether these open within a couple of years of maturity or remain closed many years, whether those that fall, fall with the stalk or off the stalk (leaving a few basal seed scales behind), whether the prickle on the umbo persists or breaks off (is deciduous), and whether the resin canals in the needle are next to the midvein (internal), deep within the leaf tissue (medial) or at the surface (marginal):

Pinus greggii, pollen cones 1–2 cm, seed cones less than 7 cm, closed, not falling, prickle deciduous, resin canals medial

P. herrerae, pollen cones 1–2 cm, seed cones less than 7 cm, open, falling with stalk, prickle deciduous, resin canals internal

P. teocote, pollen cones 1–2 cm, seed cones less than 7 cm, open, falling with stalk, prickle deciduous, resin canals medial

P. rigida, pollen cones 1–2 cm, seed cones up to 7 cm, open or closed, falling with stalk, prickle persistent, resin canals medial

P. radiata, pollen cones 1–2 cm, seed cones 7 cm or more, closed, not falling, prickle deciduous, resin canals medial

P. attenuata, pollen cones 1–2 cm, seed cones 7 cm or more, closed, not falling, prickle persistent and stout, resin canals medial

P. yunnanensis, pollen cones 2–3 cm, seed cones about 7 cm, open, falling with stalk, prickle deciduous, resin canals medial and marginal

P. kesiya, pollen cones 2–3 cm, seed cones up to 7 cm, open, falling with stalk, prickle deciduous, resin canals marginal

P. taeda, pollen cones 3–4 cm, seed cones 7 cm or more, open, falling with stalk, prickle persistent, resin canals medial

P. ponderosa, pollen cones 3–4 cm, seed cones 7 cm or more, open, falling off stalk, prickle persistent and stout, resin canals medial

Group G2. These eight species may be distinguished by the length of the fully extended pollen cones, the length of average seed cones, whether these fall along with the stalk or off of it leaving a few basal scales behind, whether the prickles are slender, stout, or massive, whether they are persistent or soon shed (deciduous), and whether the resin canals in the needles are touching the midvein (internal) or the needle surface (marginal) or in between (medial)

Pinus sabiniana, pollen cones up to 2 cm, seed cones more than 15 cm (and massive), falling off stalk, prickles massive, persistent, resin canals medial

P. pringlei, pollen cones 2–3 cm, seed cones less than 10 cm, falling off stalk, prickles slender, deciduous, resin canals internal

P. kesiya, pollen cones 2–3 cm, seed cones less than 10 cm, falling with stalk, prickles slender, deciduous, resin canals marginal

P. yunnanensis, pollen cones 2–3 cm, seed cones less than 10 cm, falling with stalk, prickles slender, deciduous, resin canals marginal and medial

P. caribaea, pollen cones 3 cm or more, seed cones up to 10 cm, falling off stalk, prickles slender, persistent, resin canals internal

P. ponderosa, pollen cones 3 cm or more, seed cones up to 15 cm, falling off stalk, prickles stout, persistent, resin canals medial

P. elliottii, pollen cones 3 cm or more, seed cones 10–15 cm, falling with stalk, prickles stout, persistent, resin canals medial

P. jeffreyi, pollen cones 3 cm or more, seed cones 15 cm or more, falling off stalk, prickles stout, persistent, resin canals medial

Group G3. These five species may be distinguished by the length of the fully extended pollen cones, the length of average seed cones, whether the resin canals in the needles are next to the midvein (internal), the outer surface of the needle (marginal), or in between (medial), and by the length of the fresh sheath around the needle bundle before it weathers:

Pinus roxburghii, pollen cones up to 20 mm, seed cones up to 20 cm, resin canals marginal, fresh sheath 2–3 cm

P. canariensis, pollen cones 15–25 mm, seed cones less than 20 cm, resin canals marginal, fresh sheath 1–2 cm

P. coulteri, pollen cones 20–25 mm, seed cones more than 20 cm and massive, resin canals medial, fresh sheath 2–4 cm

P. engelmannii, pollen cones 25 mm or more, seed cones less than 20 cm, resin canals medial, fresh sheath 3–5 cm

P. palustris, pollen cones 25 mm or more, seed cones 15–25 cm, resin canals internal, fresh sheath 2–3 cm

Group H. For convenience of identification, these 16 species may be divided into three subgroups by the predominant needle length:

Group H1, needles up to 15 cm

Group H2, needles 15–20 cm

Group H3, needles more than 20 cm

Group H1. These five species may be distinguished by the length of average seed cones and by the average needle length:

Pinus herrerae, seed cones less than 5 cm, needles more than 12 cm

P. praetermissa, seed cones 5–7 cm, needles about 12 cm

P. arizonica, seed cones 7–10 cm, needles up to 12 cm

P. jaliscana, seed cones 7–10 cm, needles more than 12 cm

P. hartwegii, seed cones more than 10 cm, needles more than 12 cm

Group H2. These six species may be distinguished by the average length of the seed cones and whether the resin canals in the needles are next to the midvein (internal), midway between the midvein and the needle surface (medial), or extend all the way in between (septal):

Pinus herrerae, seed cones less than 5 cm, resin canals internal

P. jaliscana, seed cones 5–8 cm, resin canals septal

P. occidentalis, seed cones 5–8 cm, resin canals internal

P. lawsonii, seed cones 5–8 cm, resin canals internal and medial

P. tecunumanii, seed cones 5–8 cm, resin canals medial

P. patula, seed cones more than 8 cm, resin canals medial

Group H3. These eight species may be distinguished by the average length of the seed cones and the length of the fully extended pollen cones:

P. oocarpa, seed cones up to 7 cm, pollen cones 1–2 cm

P. patula, seed cones 7–10 cm, pollen cones 1–2 cm

P. douglasiana, seed cones 7–10 cm, pollen cones 2–3 cm

P. maximinoi, seed cones 7–10 cm, pollen cones 3–4 cm

P. pseudostrobus, seed cones 10–20 cm, pollen cones 2–3 cm

P. montezumae, seed cones 10–20 cm, pollen cones 3–4 cm

P. torreyana, seed cones 10–20 cm, pollen cones 4–6 cm

P. devoniana, seed cones 20–30 cm, pollen cones 3–4 cm

Pinus albicaulis Engelmann

WHITEBARK PINE

Pinus subg. *Strobus* sect. *Quinquefoliae* subsect. *Strobus*

Tree to 20 m tall at lower elevations, progressively shortened upward and forming a spreading shrub at the alpine tree line. Multistemmed or with a single trunk to 1.5(–2.5) m in diameter. Bark pale grayish white at first (hence the scientific and common names), breaking up into small gray or light tan blocks with age. Crown oval or rounded, with numerous horizontal or upswept, slender branches. Twigs reddish brown, minutely hairy at first, becoming gray and bald during the second season. Buds loose, 5–10 mm long, not resinous. Needles in bundles of five, (3–)4–7(–9) cm long, stiff, lasting 4–8 years, dark yellowish green. Individual needles with lines of stomates on all three faces, an undivided midvein, and two to four resin canals touching the epidermis. Sheath 8–12 mm long, soon shed. Pollen cones 10–15 mm long, red. Seed cones (4–)5–8 cm long, egg-shaped to almost spherical, with (7–)30–60 scales, purplish green before maturity, ripening purplish brown to almost black, remaining closed and breaking up on the tree, often picked apart by seed-eating birds and mammals, unstalked or very short stalked. Seed scales wedge-shaped, thick, with a prominent triangular umbo at the tip. Seed body 7–11 mm long, egg-shaped, unwinged. Mountains of western North America from central British Columbia south to east-central California, northeastern Nevada, and northwestern Wyoming. Forming sparse, pure stands or mixed with other subalpine and high montane conifers on rocky soils; (1,100–)1,500–3,500(–3,700) m. Zone 4.

Mature whitebark pine (*Pinus albicaulis*) in nature.

This timberline pine is the only North American member of the subalpine stone pines within the larger subsection *Strobus,* the other four all being Eurasian. It is the most genetically, as well geographically, isolated species in the subsection and does not cross in hybridization experiments with the other species. There is, however, very strong evidence from DNA sequences that it has hybridized with neighboring sugar pine (*Pinus lambertiana*) populations. Despite whitebark pine's geographically recent arrival in North America, its chloroplast genome, shared with its Asian relatives, has permeated sugar pine throughout the northern portion of the larger species' range and has replaced the chloroplast genome sugar pine inherited from its North American ancestors and still carries in the southern half of its range. Trees (or timberline shrubs) of whitebark pine often grow in clumps widely separated from other clumps. Genetic analyses show that these clumps frequently consist of related individuals, possibly deriving from singe cones cached by the seed-eating Clark's nutcrackers that are also their major seed predator. They have no other effective dispersal mechanism. Whitebark pine is the most susceptible of all the white pine species to white pine blister rust, and many stands have been decimated by this fungus. The largest known individual in the United States in combined dimensions when measured in 1980 was 21 m tall, 2.7 m in diameter, with a spread of 14.3 m.

Pinus aristata Engelmann

COLORADO BRISTLECONE PINE

Pinus subg. *Strobus* sect. *Parrya* subsect. *Balfourianae*

Tree to 20(–25) m tall, with trunk to 1.2 m in diameter. Bark grayish brown to gray, becoming shallowly fissured between irregular, flat-topped, vertical ridges. Crown cylindrical or rounded, with numerous, stiffly upraised branches densely clothed with foliage. Twigs light reddish brown and minutely hairy initially, becoming gray and bald with age. Buds about 10 mm long, resinous. Needles in bundles of five, each needle (2–)3–4 cm long, stiff, lasting

Twig of Colorado bristlecone pine (*Pinus aristata*) with 4 years of growth, ×0.5.

10–15(–20) years, dark bluish green, speckled with dried resin drops. Individual needles with lines of stomates only on the inner faces, an undivided midvein, a single small resin canal just beneath a central groove running down the outer face, and none to three shorter resin canals nearby. Sheath (5–)10–15 mm long, curling back and soon shed. Pollen cones about 10 mm long, bright yellow. Seed cones 6–11 cm long, egg-shaped and flat at the base, with 100 scales or more, purple before maturity, ripening tan to grayish brown, opening widely to release the seeds and then falling, unstalked or very short stalked. Seed scales paddle-shaped, thin but thicker at the exposed tip, with a diamond-shaped umbo on the exposed face bearing a prickle (*arista* in Latin, hence the scientific and common names) 4–10 mm long. Seed body 5–6 mm long, the easily detachable wing 7–15 mm long. Rocky Mountains of central Colorado to northern New Mexico and the San Francisco Peaks of Arizona. Forming pure stands or mixed with other subalpine and upper montane conifers; 2,500–3,500 m. Zone 5.

Pinus aristata is a characteristic species of the timberline forest of the central and southern Rocky Mountains. It is immediately distinguishable from the two more westerly members of the subalpine subsection *Balfourianae* by the single resin duct below a groove on the outer face of the needle that often ruptures to exude spots of resin. The chemical composition of this resin and that of the wood is sharply distinct from the corresponding resins of the related foxtail pine (*P. balfouriana*) and Great Basin bristlecone pine (*P. longaeva*). It grows in less arid habitats than *P. longaeva,* which was not recognized as a separate species until 1970 and, while long-lived, does not attain the enormous age of that species, with maximum recorded ages of about 2,000 years. The largest known individual in combined dimensions when measured in 1985 was 23.2 m tall, 1.1 m in diameter, with a spread of 11.9 m.

Pinus arizonica Engelmann

ARIZONA PINE, PINO AMARILLO (SPANISH)

Pinus subg. *Pinus* sect. *Trifoliae* subsect. *Ponderosae*

Tree to 30(–36) m tall, with trunk to 0.8(–1.2) m in diameter. Bark thick, reddish brown to dark brown, broken up into a checkerboard of small, irregular, scaly plates separated by narrow, dark furrows. Crown tapering gradually, fairly open, with slender, downswept branches densely clothed with foliage at the tips. Twigs purplish brown to reddish brown, usually with a thin waxy coating, rough with the bases of scale leaves, hairless. Buds 12–25 mm long, scarcely resinous. Needles in bundles of (three or) four or five, mostly five, each needle (5–)8–13(–17) cm long, stiff or slightly droopy, lasting 2–4 years, yellowish green or bluish green. Individual needles with variably conspicuous lines of stomates on all three faces, and three to five (to seven) resin canals midway between the leaf surface and the two-stranded midvein at the three corners and below the outer face. Sheath (8–)12–25 mm long, the lower (5–)8–10 mm persisting and falling with the bundle. Pollen cones 15–25 mm long, yellow. Seed cones 5–10 cm long, egg-shaped, often slightly asymmetric, with 75–150 seed scales, green before maturity, ripening shiny reddish brown, opening widely to release the seeds and then falling off the short, thick stalk, leaving a few basal scales behind. Seed scales paddle-shaped, the exposed face horizontally diamond-shaped, low or domed, crossed by a ridge and topped by a flat umbo with a short, stout prickle. Seed body 3–7 mm long, the easily detachable wing 12–20 mm longer. Southeastern Arizona (hence the scientific and common names) and southwestern New Mexico, south in the Sierra Madre Occidental to Durango (Mexico). Forming pure stands or mixed with other pines and other trees on dry slopes; 2,000–3,000 m. Zone 7. Synonyms: *P. arizonica* var. *cooperi* (C. Blanco) Farjon, *P. cooperi* C. Blanco, P. *ponderosa* var. *arizonica* (Engelmann) G. R. Shaw.

The taxonomic status and identity of Arizona pine is the most confused among the close relatives of ponderosa pine (*Pinus ponderosa*) in subsection *Ponderosae.* Authors are about equally divided in using the name for a predominantly five-needled pine, as done here, or for a three- or four-needled pine, which is treated here as *Pinus ponderosa* var. *stormiae.* Those who apply the name to the five-needled pine, in accordance with the type specimen, call it either an independent species or a variety of *P. ponderosa.* Although closely related to ponderosa pine, there is ample evidence for keeping Arizona pine separate. It is a rare species in the southwestern United States and hybridizes there with *P. ponderosa* var. *scopulorum* and with Apache pine (*P. engelmannii*). It is much more common in western Mexico, especially in the state of Durango, where it was described as a separate species and then as a separate variety, but there do not seem to be enough differences to warrant a separate designation for these southern trees. Throughout the Sierra Madre Occidental where it occurs, there is confusion and possibly hybridization with other members of subsection *Ponderosae.* The whole group of Ponderosa pine relatives needs further careful study. The largest known individual in the United States in combined dimensions when measured in 1998 was 38.7 m tall, 1.2 m in diameter, with a spread of 17.3 m.

Pinus armandii Franchet

CHINESE WHITE PINE, ARMAND PINE, YUAN BIAN ZHONG (CHINESE)

Pinus subg. *Strobus* sect. *Quinquefoliae* subsect. *Strobus*

Tree to 25(–50) m tall, with trunk to 1(–2) m in diameter. Bark dark grayish brown, flaking and ultimately breaking up into small, square blocks at the base of large trees. Crown cylindrical, rather open, with widely spaced tiers of long, thin, horizontal branches bearing foliage only near their ends. Twigs grayish green to light grayish brown, hairless to transiently minutely hairy. Buds about 7–10 mm long, slightly resinous. Needles in bundles of five (to seven), each needle (3–)6–15(–18) cm long, thin, flexible, and often drooping, with a kink near the base, lasting 2–3 years, dark green. Individual needles without stomates on the outer face, whitish green with wax over the stomates on the inner faces, with an undivided midvein, two small resin canals beneath the outer face either touching the epidermis or completely within the leaf tissue, usually with a third resin canal inside the leaf tissue near the corner between the two inner faces, and rarely with up to four additional resin canals scattered within the leaf tissue beneath any or all of the faces. Sheath 2–7 mm long, soon shed. Pollen cones 15–30 mm long, pale green with a reddish blush. Seed cones (5–)6–16(–20) cm long, egg- to elongate egg-shaped, with 70–100 scales, green or purple and sometimes a little waxy before maturity, ripening yellowish brown or reddish brown, gaping and releasing the seeds at maturity, on a stalk (1–)2–4 cm long. Seed scales angularly egg-shaped, thick and a little fleshy, particularly around the seeds, the diamond-shaped or triangular tip tapering from the thickest point to a diamond-shaped or triangular umbo that is straight to curled back. Seed body 8–15 mm long, plump, wingless or with a short, firmly attached wing 1–4(–7) mm long. Southern Japan, central and southern China. Usually mixed with other conifers and hardwoods on rocky slopes; (100–)1,000–2,500(–3,000) m. Zone 7. Five varieties.

This is the most widely distributed white pine (subgenus *Strobus*) in China and has a corresponding broad elevational and climatic range, with associated morphological variability. This has led to recognition of up to five species, particularly for geographically isolated local variants, but these are treated here as varieties of one species. Despite their geographic and climatic differences, they grow in similar stands on similar sites and have substantial points of agreement in their morphology. In particular, they share rather similar seed cones with distinctive seed scales that gape at maturity to release the wingless or ineffectively short-winged seeds. This combination of characteristics links the subalpine stone pines with other species of subsection *Strobus,* and consequently *Pinus armandii* was often included in either subsection *Strobus* or subsection *Cembrae* by different authors (when the latter was considered distinct from the former) as well as in an intermediate subsection that is also no longer recognized by botanists. Limber pine (*P. flexilis*) in North America is the most similar appearing species but is not closely related. The only species with which *P. armandii* has been crossed successfully is the very dissimilar North American sugar pine (*P. lambertiana*). Both crossing programs and DNA studies involving *P. armandii* should be expanded to include all of the varieties in order to examine their relationships to one another and to other species of white pine. The lumber serves a variety of general carpentry uses as well as providing pulp and turpentine. Turpentine is also derived from the edible seeds. The species name honors Père Armand David (1826–1900), a French missionary and naturalist in southwestern China who collected the type specimen in Yunnan but is most famous, perhaps, for Père David's deer.

Pinus armandii var. *amamiana* (G. Koidzumi) Hatusima

AMAMI-GOYŌ (JAPANESE)

Twigs grayish brown, transiently hairy. Needles 3–8 cm long, with three resin canals, the two near the outer face touching the epidermis. Seed cones 5–7 cm long, reddish brown, the seed scales slightly curled under at the wide umbo. Seeds black, thick-shelled, unwinged or rarely with an inconspicuous, ridgelike wing. Yaku Shima and Tanegashima south of Kyūshū (Japan). Synonym: *Pinus amamiana* G. Koidzumi.

Pinus armandii Franchet var. *armandii*

HUA SHAN SONG (CHINESE)

Twigs grayish green, hairless. Needles 8–15 cm long, with three (to seven) resin canals, those near the outer face usually deep inside the leaf tissue. Seed cones 10–20 cm long, yellowish brown, the seed scales straight of slightly curled under at the narrow umbo. Seeds pale brown to black, thick-shelled, unwinged, ridged, or rarely with a narrow wing to

2 mm. Southwestern Xizang (Tibet) to southern Gansu to southern Shanxi south to Yunnan, central Guizhou, and western Hubei (China).

Pinus armandii var. dabeshanensis (W. C. Cheng & Y. W. Law) Silba

da bie shan wu zheng song (Chinese)

Twigs yellowish brown, hairless. Needles 5–12(–14) cm long, with two or three resin canals, the two near the outer face touching the epidermis. Seed cones 11–14 cm long, yellowish brown, the seed scales prominently curled under below the wide umbo. Seeds pale brown, thin-shelled, the wing 1–2 mm long. Dabie Shan (mountains) of southwestern Anhui and eastern Hubei (China). Synonyms: *Pinus dabeshanensis* W. C. Cheng & Y. W. Law, *P. fenzeliana* var. *dabeshanensis* (W. C. Cheng & Y. W. Law) L. K. Fu & N. Li.

Pinus armandii var. fenzeliana (Handel-Mazzetti) Eckenwalder

HAI NAN WU ZHEN SONG (CHINESE)

Twigs pale reddish brown, hairless. Needles 10–18 cm long, with three resin canals, the two near the outer face touching the epidermis. Seed cones 6–9 cm long, yellowish brown, the seed scales prominently curled under well back from the narrow umbo. Seeds reddish brown, thin-shelled, the wing 2–4(–7) mm long. Southeastern Sichuan through Guizhou and Guangxi to Hainan (China). Synonym: *Pinus fenzeliana* Handel-Mazzetti.

Pinus armandii var. mastersiana (Hayata) Hayata

TAI WAN GUO SONG (CHINESE)

Twigs grayish brown, hairless. Needles 8–15 cm long, with (two or) three resin canals, the two near the outer face touching the epidermis. Seed cones 10–20 cm long, reddish brown, the seed scales curled under below the narrow umbo. Seeds dark brown, thick-shelled, with an inconspicuous, ridgelike wing. Taiwan. Synonym: *Pinus mastersiana* Hayata

Pinus attenuata J. Lemmon

KNOBCONE PINE

Pinus subg. *Pinus* sect. *Trifoliae* subsect. *Australes*

Tree to 25(–35) m tall, with trunk to 0.8(–1.1) m in diameter. Bark dark grayish brown, smooth and scaly above, divided below into irregular blocks by shallow, narrow furrows. Crown narrow, thin, and irregular, with branches horizontal to angled steeply upward and thinly to moderately thickly clothed with foliage at the tips. Twigs reddish brown, hairless. Buds 15–20 mm long, resinous. Needles in bundles of three, each needle (7–)9–18(–20) cm long, stiff and sometimes slightly twisted, lasting 4–5 years, yellowish green. Individual needles with lines of stomates on all three faces, a two-stranded midvein, and two to five resin canals deeply embedded in the leaf tissue at the outer corners and elsewhere. Sheath 10–20 mm long, persisting and falling with the bundle. Pollen cones 10–15 mm long, light reddish brown. Seed cones (8–)10–15 cm long, held flush to the stem, narrowly and pointedly egg-shaped, strongly asymmetric, with 75–150 or more seed scales, green before maturity, ripening light yellowish brown, remaining attached and closed for up to 20 years or more, opening after fires to release the seeds, on a stalk up to 1 cm long. Seed scales paddle-shaped, the exposed portion low and rounded on the side facing the shoot and near the tip, progressively larger and more pyramid-shaped toward the base on the outer side (hence the common name), with a thick, strong, clawlike umbo. Seed body 5–7 mm long, almost black, the easily detachable wing 12–20 mm long. Coast Ranges and Sierra Nevada of Pacific North America from southwestern Oregon, through California to northwestern Baja California (Mexico). In fire-prone shrublands (chaparral) on dry foothill slopes; 250–1,200(–1,800) m. Zone 7.

Knobcone pine is the most interior of the three California closed-cone pine species and grows on the driest, harshest sites. It is a characteristic fire-succession species of California chaparral. The seed cones may remain closed for so long that they become embedded in the growing branches that bear them. Once the bundles of needles have fallen, the stacked rings of two to four cones around the branches, each matured in a different year, become very evident. The passage of a brush fire kills the tree and opens all of the accumulated cones, releasing the whole crop of seeds onto the newly cleared seed bed. Thus the stands of this species are generally uniform in age, although overlapping fires may produce a mosaic of patches of different ages. Knobcone pine in closely related to the other two California closed-cone pines, bishop pine (*Pinus muricata*) and Monterey pine (*P. radiata*), and has been hybridized successfully with both of them. The hybrids with Monterey pine have been given the name *P.* ×*attenuradiata*. Although knobcone pine barely overlaps with Monterey pine and not at all with bishop pine in nature today, DNA studies strongly suggest that there has been natural hybridization with both species in the past. This has resulted in the transfer of cellular organelles from knobcone pine to Monterey pine at the Cambria population and to southern California populations of bishop pine, all of which are otherwise typical of their species. The scientific name (Latin for "tapering") was chosen to describe both the seed cones and the shape of the crown. The largest known individual in combined dimensions when measured in 1976 was 35.7 m tall, 1.1 m in diameter, with a spread of 20.1 m.

Pinus ×attenuradiata Stockwell & Righter

Tree to 25 m tall or more, with trunk to 0.7 m in diameter or more. Deliberate hybrid between knobcone pine (*Pinus attenuata*) and Monterey pine (*P. radiata*), also occurring naturally at Año Nuevo Point in California. It is one of the fastest growing of all pines, resembling Monterey pine in this and in its tall, straight trunk but with a spindly branching habit more like that of knobcone pine. The delayed onset of rough bark, essentially confined to the lower trunk, is also derived from the knobcone pine parent. The twigs are reddish brown and hairless like those of both parents, while the buds are less resinous than those of either parent. Needles of the two parent species are very similar, both usually in bundles of three, mostly 9–15 cm long, yellowish green, with lines of stomates on all three faces. Needles of the hybrids are similar but a little more flexible and droopy because they are thinner than those of either parent, typically 1.1–1.2 mm thick rather than 1.3–1.8 m, and are darker, dusty green. Pollen cones resemble those of both parents, 10–15 mm long and pinkish brown. The asymmetrical seed cones are more similar to those of knobcone pine than to those of Monterey pine. Like those of both parents, they have a modal length around 11 cm, but they are proportionately more slender than those of Monterey pine and the lower seed scales on the outer side have the more pointy, conical projections of knobcone pine rather than the more rounded, domelike ones of Monterey pine. The seed cones are straighter than those of knobcone pine, however, and also remain closed many years on the tree.

First developed in 1927, this was the first artificial hybrid produced at the US Forest Service's Institute of Forest Genetics at Placerville, California. It is also one of the best in terms of forestry potential because it combines the rapid growth and good form of the coastal Monterey pine with the greater cold and drought tolerance of the interior knobcone pine. Unlike many other hybrids, this one is reasonably intermediate in many of the traits that distinguish its parents. This suggests that the differences between knobcone and Monterey pines are largely quantitative and determined by the action of numerous genes of small effect rather than major mutations. Thus, because of what it says about the relationships of its parent species, this hybrid has considerable potential scientific as well as economic value. In addition to its natural occurrence in native stands in California, it has also occurred spontaneously in New Zealand (and perhaps in other southern hemisphere countries), where both parent species have been introduced as plantation trees.

Pinus ayacahuite C. G. Ehrenberg ex D. F. L. Schlechtendal

MEXICAN WHITE PINE, ACALOCOTE (SPANISH)

Pinus subg. *Strobus* sect. *Quinquefoliae* subsect. *Strobus*

Tree to 40(–50) m tall, with trunk to 1.5(–2) m in diameter. Bark grayish green and smooth at first, flaking and then ultimately becoming grayish brown and breaking up into small, scaly blocks divided by shallow cracks at the base of large trees. Crown conical at first, broadening and becoming flat-topped and irregular with age, fairly open, even in youth, with widely spaced horizontal to gently drooping branches well clothed with foliage. Twigs light grayish brown, hairless or transiently minutely hairy. Buds 6–12(–15) mm long, slightly resinous or not. Needles in bundles of five (or six), each needle (8–)10–18(–20) cm long, very soft and flexible, straight or slightly drooping, lasting 2–3 years, bluish green to grayish green with wax. Individual needles with lines of stomates only on the inner faces, an undivided midvein, two small resin canals touching the epidermis of the outer face, and sometimes with up to four additional resin canals around the periphery, starting with either of the inner faces near where they come together. Sheath 1.5–2(–3) cm long, soon shed. Pollen cones 7–15 mm long, yellowish brown. Seed cones (10–)25–45(–50) cm long, taperingly cylindrical and nearly straight to obviously curved, with 100–150 seed scales, green before maturity, ripening light brown, opening widely to release the seeds and then falling, on a stout stalk to 2.5 cm long. Seed scales narrowly diamond-shaped, the hidden part thin, the exposed part thicker and triangular to elongated, slightly to sharply and strongly curved back, with a narrow, diamond-shaped umbo at the tip. Seed body 8–15 mm long, with a firmly attached wing 10–30 mm long. Mountains of southern Mexico and northern Central America, from western Michoacán, southern Hidalgo, and northwestern Veracruz (Mexico) to southern Guatemala, with outliers in northwestern El Salvador and western Honduras. Sometimes in groves but usually mixed with and often towering above other pines and conifers in montane forest on deep moist soils; (1,500–)2,000–3,200(–3,600) m. Zone 7. Two varieties.

The relationship between Mexican white pine and the more northerly southwestern white pine (*Pinus strobiformis*) is unclear, and some taxonomists consider them part of a single species. Mexican white pine has much longer seed wings than southwestern white pine and lacks the strongly curled seed scales of the latter in most of its range. In the northern variety *veitchii,* however, the seed wings are a little shorter than those of the more common variety *ayacahuite,* and the seed scales are prominently bent back, even more so, perhaps, than in *P. strobiformis.* Trees of variety *veitchii* differ from those of variety *ayacahuite* only in their seed cones and seeds, so far as known. They may be of hybrid origin between Mexican and southwestern white pines, but this has not been verified by any detailed studies, and the species do not grown together today. Except for variety *veitchii,* the two species are amply distinct, and there seems inadequate justification for combining them given present evidence. They have not been crossed successfully, although *P. ayacahuite* has been crossed with the related limber pine (*P. flexilis*) as well as with eastern white pine (*P. strobus*) and Himalayan white pine (*P. wallichiana*). Mexican white pine is a large tree providing excellent timber and so has been subject to heavy and threatening exploitation. The scientific name is derived from an Aztec name for the tree that honors its stature and habitat, *ayauhquahuitl*, "cloud tree." The 16th century monk Bernardino de Sahagún, who described it in his *History of the Things of New Spain*, wrote (in Aztec) that "it stands towering, highest of all."

Pinus ayacahuite C. G. Ehrenberg ex D. F. L. Schlechtendal **var. *ayacahuite***

Seed cones up to 40 cm long with up to about 120 seed scales, which have a triangular tip only gently curled back. Seed body 8–10 mm long, the wing 20–35 mm long. Throughout the range of the species. Synonym: *Pinus ayacahuite* var. *oaxacana* Silba.

Pinus ayacahuite* var. *veitchii (Roezl) G. R. Shaw

Seed cones up to 50 cm long, with up to about 150 seed scales, which have an elongate tip sharply and strongly angled back. Seed body 10–12 mm long, the wing 10–20 mm long. Northern part of the range of the species south to Morelos and northwestern Puebla (Mexico). Synonyms: *Pinus ayacahuite* var. *loudoniana* (G. Gordon) Silba, *P. veitchii* Roezl.

Pinus balfouriana Greville & J. Balfour

FOXTAIL PINE

Pinus subg. *Strobus* sect. *Parrya* subsect. *Balfourianae*

Tree to 20(–25) m tall, with trunk to 1.5(–2.6) m in diameter. Bark gray or bright reddish brown, becoming narrowly and irregularly ridged and furrowed on breaking up into broader plates. Crown narrowly to broadly conical, becoming irregular with age, with numerous short, upwardly arching or long, gently downswept branches densely clothed with foliage (hence the common name, since these branches look, fancifully, like green foxtails). Twigs reddish brown, hairless or minutely hairy at first, becoming yellowish gray and bald with age. Buds 8–10 mm long, resinous. Needles in bundles of five, each needle 1.5–4 cm long, stiff, straight, and remaining tight together, lasting 7–30 years, dark bluish to yellowish green, without resin flakes. Individual needles with lines of stomates only on the inner faces, an undivided midvein, and two medium-sized resin canals touching the ungrooved epidermis of the outer face. Sheath scales 5–10 mm long, curling back and soon shed. Pollen cones 6–10 mm long, red. Seed cones 6–9(–11) cm long, egg-shaped and sharply tapered at the base, with 70–90 scales, purple before maturity, ripening brown to reddish brown, opening widely to release the seeds and then falling intact with the 7- to 18-mm-long stalk. Seed scales paddle-shaped, thin but thicker at the exposed tip, with diamond-shaped umbo on the exposed face bearing a minute, weak prickle to 1 mm long. Seed body about 7–10 mm long, pale with dark speckles, the easily detachable wing 10–20 mm long. Klamath Mountains and southern Sierra Nevada of California. Often forming pure subalpine stands in the south or mixed with other montane conifers in the north, usually on upper slopes or exposed ridges; (1,500–)2,000–3,500(–4,000) m. Zone 6. Two subspecies.

Foxtail pine is found in two small, ecologically and geographically distinct areas separated by some 500 km. The trees in the two areas differ enough that they are treated as separate subspecies or varieties by many authors, as they are here. The southern populations in the Sierra Nevada are just across the Owens Valley from the related ancient Great Basin bristlecone pines (*Pinus longaeva*) of the Schulman Grove in the White Mountains. Their foliage is very similar to that of their neighbors but has small resin canals in the needles and conspicuous waxy bands over the stomatal lines not found in *P. longaeva.* The longevity of the needles, like that of the trees (at up to about 2,000 years), is less than in Great Basin bristlecone pine although still quite respectable. The stalked seed cones of *P. balfouriana* have fewer seed scales than those of *P. longaeva* and the Rocky Mountain bristlecone pine (*P. aristata*). The southern and northern populations of foxtail pine are freely intercrossable with one another, and the southern populations (subspecies *austrina*) are also freely intercrossable with *P. longaeva* while the northern populations (subspecies *balfouriana*) have reduced fertility in this combination. All cross with *P. aristata* only with great difficulty. The largest known individual in combined dimensions when measured in 1982 was 23.2 m tall, 2.6 m in diameter, with a spread of 10.4 m. Few other pines reach this diameter despite this being a relatively short, slow-growing species. The species name honors John H. Balfour (1808–1884) director of the Royal Botanic Garden, Edinburgh, who described it using the name provided by the collector, John Jeffrey.

Pinus balfouriana* subsp. *austrina R. Mastrogiuseppe & J. Mastrogiuseppe

Bark bright reddish brown. Branches upwardly arching. Needles yellowish green, the resin canals often conspicuously unequal in diameter. Seed cones reddish brown. Seeds generally smaller, commonly 7–8 mm. Southern portion of the range of the species in Fresno, Inyo, and Tulare Counties, California.

Pinus balfouriana Greville & J. Balfour **subsp. *balfouriana***

Bark gray with pink highlights. Branches gently downswept. Needles bluish green, the resin canals usually nearly equal in diameter. Seed cones brown. Seeds generally larger, commonly 9–10 mm. Northern portion of the range of the species in Siskiyou, Trinity, and Tehama Counties, California.

Pinus banksiana A. Lambert

JACK PINE, PIN GRIS (FRENCH)

Pinus subg. *Pinus* sect. *Trifoliae* subsect. *Contortae*

Tree to 20(–27) m tall, with trunk to 0.5(–1) m in diameter, but often short and spindly in dense stands on nutrient poor soils. Bark bright orange-brown and scaly when young, becoming dark gray and broken up into wavy, vertically elongate blocks separated by shallow furrows. Crown narrowly conical, with numerous short, horizontal or gently upswept branches. Thinly clothed with foliage, often retaining dead branches many years in crowded stands. Twigs yellowish brown initially but soon becoming grayish brown, hairless, roughened with scale leaves, often flushing twice in a single season. Buds 5–15 mm long, resinous. Needles in bundles of two, each needle 2–4(–5) cm long, stiff, sometimes slightly twisted, lasting 2–3 years, yellowish green. Individual needles with lines of stomates on both faces, a two-stranded midvein, and usually just two resin canals embedded in the green leaf tissue near the outer corners. Sheath 3–6 mm long, only the short outer scales remaining after the first season and persisting and falling with the bundle. Pollen cones 10–15 mm long, yellowish brown. Seed cones 3–7 cm long, egg-shaped, slightly asymmetrical, and often curved forward toward the twig, with 50–80 seed scales, green before maturity, ripening orange-brown but then often remaining unopened for years and turning gray, then only opening widely to release the seeds after a fire, almost stalkless. Seed scales wedge-shaped, woody and stiff, the exposed portion mostly low, but sometimes domed on lower scales of the outer side of the seed cone, with a small, central, diamond-shaped umbo sometimes bearing a tiny prickle. Seed body 3–5 mm long, the easily detachable wing 10–12 mm long. Northern North America, from the McKenzie River near Great Bear Lake and central Alberta east to Nova Scotia and south to southern Wisconsin and Michigan. Usually forming dense, even-aged pure stands but also mixed with other trees on sandy, infertile sites in the boreal forest; 0–600(–800) m. Zone 1.

Jack pine is the most northerly pine in North America and the sole pine species of most of the North American boreal forest. It has some ecological correspondence to the unrelated Scots pine (*Pinus sylvestris*) of northern Eurasia, but forests of the latter often have a richer herbaceous understory. Like other boreal forest conifers, it is highly dependent on fires for regeneration. It retains the last 10–20 years worth of cones unopened so that fires release an enormous number of seeds. One of North America's rarest birds, Kirtland's Warbler, breeds only in young stands of jack pines established after large fires. The seeds are an important food for another pair of bird species, the red and white-winged crossbills, whose crossed bill tips are adapted for extracting the seeds from between the tight seed scales of the closed cones. The bill size and structure of crossbills vary with the shape and strength of the seed cones of jack pine and other conifers that they encounter habitually in a given area. Jack pine is closely related to lodgepole pine (*P. contorta*), which replaces it in the montane forests of western North America. The two species overlap and hybridize in Alberta and the southwestern Northwest Territories and have also been deliberately crossed with success. Attempts to cross jack pine with the other species in subsection *Contortae,* the southern scrub pine (*P. virginiana*), have not been successful although genetic evidence confirms that they are closely related. Jack pine is generally a small tree, and the largest known individual in the United States in combined dimensions when measured in 1995 was 17 m tall, 0.9 m in diameter, with a spread of 18.6 m. The species name honors Joseph Banks (1743–1820), who participated in an important scientific expedition to Newfoundland and Labrador, which jack pine does not quite reach.

Pinus bhutanica A. Grierson, D. Long & C. Page

BHUTAN PINE

Pinus subg. *Strobus* sect. *Quinquefoliae* subsect. *Strobus*

Tree to 25 m tall or more, with trunk to 0.5 m or more in diameter. Bark light grayish brown, flaky, shallowly furrowed at the base of large trees. Crown open, narrowly conical, with slender, wavy, horizontal to drooping branches sparsely clothed with foliage near the tips. Twigs brown, heavily waxy, covered with short, sticky hairs. Buds 6–8 mm long, somewhat resinous. Needles in bundles of five, each needle (12–)15–24(–28) cm long, strongly hanging, lasting 2 years, bright light green on the outer face, strongly whitened with wax over the stomatal bands on the inner faces. Individual needles with an undivided midvein, two resin canals next to the outer face, either touching the epidermis or a little separated from it, and with one (or two) resin canals touching the epidermis of one or both inner faces near the angle where they come together. Sheath (1.8–)2–3 cm long, bright golden brown, soon shed. Pollen cones probably 10–20 mm long at maturity, yellowish brown. Seed cones 12–20 cm long, taperingly cylindrical and a little curved, with 60–80 seed scales, maturing light brown, opening widely to release the seeds and then falling, on a stalk 4.5–6 cm long. Seed scales paddle-shaped, thin and flexible, the exposed portion straight and a little cupped, ending in a narrow diamond-shaped umbo that bends inward a little. Seed body 6–8 mm long, the firmly attached wing about 20 mm long. Western and eastern Bhutan (hence the scientific and common names) and adjacent Arunachal Pradesh (India) and Xizang (Tibet) (China). Usually mixed with evergreen hardwoods in moist montane forests, though attaining its best development in slightly drier sites; 1,750–2,400 m. Zone 8?

This close relative of Himalayan white pine (*Pinus wallichiana*) is embedded within its range and replaces it to a large extent in eastern Bhutan, though typically growing at slightly lower elevations and in moister forest. Bhutan pine has a more weeping habit than *P. wallichiana,* with longer, decisively dangling needles that have a slightly different arrangement of resin canals, but the cones and seeds of the two are very similar. The relationships and status of *P. bhutanica* warrant further study. Despite its rarity, it does not appear to be particularly threatened by exploitation because of its scattered occurrence and the readier availability of *P. wallichiana* and chir pine (*P. roxburghii*).

Pinus bungeana Zuccarini ex Endlicher

LACEBARK PINE, BAI PI SONG (CHINESE)

Pinus subg. *Strobus* sect. *Quinquefoliae* subsect. *Gerardianae*

Tree to 20(–30) m tall, with trunk to 1.5(–3) m in diameter, often dividing near the base. Bark dark and flaking in multicolored, irregular patches when young (hence the common name), becoming progressively whiter with age until the predominant impression on large, old trees may be of a smooth, bone white trunk (the Chinese name is "white bark pine"). Crown dense, conical when young, becoming broad, flat-topped, and irregular with age, with numerous long, slender, upwardly angled, sinuous branches moderately clothed with foliage near the tips. Twigs grayish green, hairless and shiny. Buds about 6–10 mm long, slightly resinous or not. Needles in bundles of three, each needle 5–10 cm long, stiff and straight, lasting 4–5 years, bright dark green to yellowish green. Individual needles with stomates on all three faces, an undivided midvein, and three to seven large resin canals underneath all faces next to the epidermis or occasionally one or two of them away from it. Sheath 1–2 cm long, soon shed. Pollen cones 6–15 mm long, yellowish brown. Seed cones (3–)4.5–6(–8) cm long, egg-shaped with a flat base, with 30–50 seed scales, light green before maturity, ripening yellowish brown, opening widely to release the seeds, remaining attached for a year or two before falling, almost stalkless or with a short, slender to stout stalk up to about 1 cm long. Seed scales angularly egg-shaped, thin but woody and inflexible, the exposed portion thicker, with a ridge across the face separating it into a flat, crescent-shaped rim and a trapezoidal lower portion, the diamond-shaped umbo at the center of the ridge and sticking straight out in a flattened, stout prickle. Seed body 7–12 mm long, egg-shaped, the readily detached wing 2–5 mm long. Scattered in mountains of central China, mostly around the basin of the Huang He (Yellow River), from central Shanxi west to southern Gansu south to northern Sichuan and western Hubei. Mixed with other conifers and hardwoods in open woodlands on moist, well-drained, rocky soils; (500–)1,000–1,800 (–2,800?) m. Zone 7.

Although fairly widespread in central China, the native distribution of lacebark pine is highly fragmented. The tree is rare overall and nowhere common in nature. However, it is an important cultivated ornamental in northern China inside and beyond its native range and has been so for centuries. The multiple white trunks of ancient trees are highly prized in temple and palace grounds as well as in public spaces. There are numerous individuals 200–300 years old and some reputed to be as old as 900 years. Many fine old trees have individual names, like the great ginkgos in Japan. The white bark for which it is cultivated becomes increasingly prominent after the tree is about 50 years old, so the

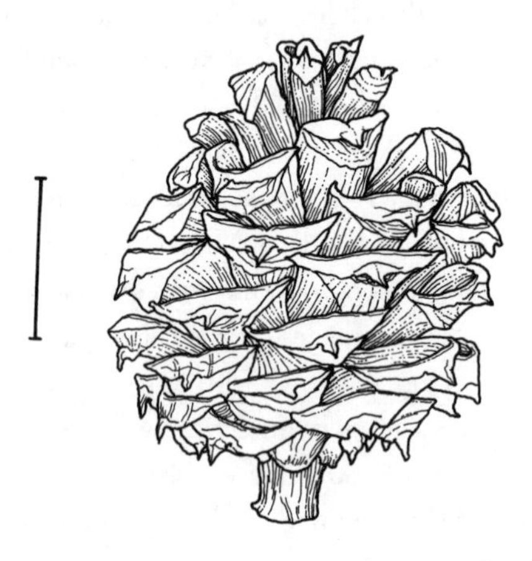

Open mature seed cone of lacebark pine (*Pinus bungeana*) after shedding seeds; scale, 2 cm.

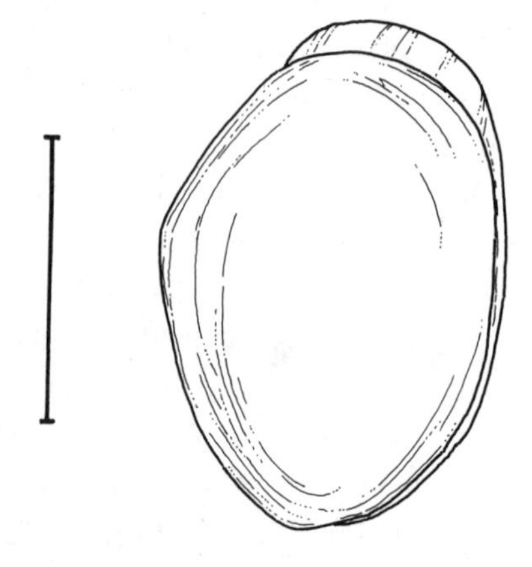

Seed of lacebark pine (*Pinus bungeana*) with rudimentary wing; scale, 5 mm.

mostly younger trees in the West retain mottled bark, highly reminiscent of that of plane trees (*Platanus*) and very ornamental in its own right. The multitrunked (shrublike) habit is due to the slow growth of seedlings and weak apical dominance. The leader may grow only a few centimeters in its first year and is subsequently overtaken by several lateral branches that grow upright and become trunks. Given its rarity in nature, the appreciation of its ornamental qualities, its slow growth, and its brittle wood, lacebark pine is not a commercial or local timber tree. The seeds are edible, although smaller than those of its close relative of the Himalaya, chilgoza pine (*Pinus gerardiana*), and are gathered as pine nuts. These two species are often considered intermediate between white (or soft) pines (subgenus *Strobus*) and yellow (or hard) pines (subgenus *Pinus*), but DNA studies show that the species of subsection *Gerardianae* are embedded firmly within subgenus *Strobus* at the base of the clade containing the true white pines and the montane stone pines (subsection *Strobus*). The species name of lacebark pine honors Alexander A. von Bunge (1803–1890), who collected the type specimen in 1831 while serving as Russian envoy to the imperial court of Beijing and who later wrote about the collections he made in northern China and Mongolia.

Pinus canariensis R. Sweet ex K. Sprengel

CANARY ISLAND PINE

Pinus subg. *Pinus* sect. *Pinus* subsect. *Pinaster*

Tree to 30(–40) m tall, with trunk to 2 m in diameter. Bark rich reddish brown weathering gray, thick, separated into small, flaky plates by shallow furrows. Crown conical at first, becoming flat-topped and open with age, with slender branches, originally angled upward but becoming downswept with age and densely clothed with foliage. Twigs yellowish brown, rough with bases of scale leaves, hairless. Buds 15–20 mm long, not resinous. Needles in bundles of three, each needle (15–)20–30 cm long, flexible and arched, lasting 2–3 years, shiny, bright, dark green or a little bluish. Individual needles with narrow lines of stomates on all three faces, two to four resin canals at the outer corners and inside the outer face, and a two-stranded midvein. Sheath 10–20 mm long, persisting and falling with the bundle. Even mature trees with slender, juvenile shoots emerging from trunks and large branches bearing single, blue-gray leaves 3–8 cm long. Pollen cones numerous, (10–)15–25(–30) mm long, purplish yellow. Seed cones (7–)10–17(–20) cm long, taperingly cylindrical, with 75–120 seed scales, grayish green before maturity, ripening shiny light yellowish brown, opening widely to release the seeds and then falling, leaving behind the relatively slender stalk 5–20 mm long. Seed scales elongate diamond-shaped, the exposed face projecting as a straight, diamond-shaped pyramid topped by an umbo sometimes with a small, transient prickle. Seed body (7–)9–12(–14) mm long, the firmly attached wing 12–25 mm longer. Western Canary Islands, from Gran Canaria to La Palma and Hierro. Forming pure stands of varying density on the dry, rocky slopes of the volcanic mountains above the belt of laurel forest; (500–)1,000–2,000(–2,200) m. Zone 9.

Normal twig of Canary Island pine (*Pinus canariensis*) with adult foliage on 2 years of growth, ×0.5.

Trunk-borne twig of Canary Island pine (*Pinus canariensis*) with juvenile foliage as single needles, ×0.5.

The range of Canary Island pine is the westernmost extension of the Mediterranean pines into the Atlantic. This pine has traditionally been considered to be closely related to the Himalayan *Pinus roxburghii*, but present evidence points to just as close a relationship to maritime pine (*P. pinaster*) and Mediterranean stone pine (*P. pinea*), both of which occur in nearby Morocco. The appearance of juvenile twigs emerging from the trunk is a distinctive feature of *P. canariensis* shared by few other pines, and it is the only three-needled pine in the western part of the Old World. Although as the premier softwood of the Canary Islands it has been heavily exploited for timber production and few large, old trees remain, this species is still common and many stands have been replanted.

Pinus caribaea P. Morelet
CARIBBEAN PINE, PINO COSTEÑO, PINO MACHO (SPANISH), HUHUB (MAYAN)
Pinus subg. *Pinus* sect. *Trifoliae* subsect. *Australes*
Tree to 30(–45) m tall, with trunk to 0.7(–1) m in diameter. Bark bright reddish brown, thick, with small, flaky plates separated by deep, narrow, dark furrows. Crown conical when young, becoming broadly round-topped and shallow with age, with slender, rising to horizontal branches thinly clothed with foliage. Twigs grayish to reddish brown, thick, strongly roughened by the bases of the scale leaves, hairless but initially a little mealy. Buds 15–25 mm long, slightly resinous. Needles in bundles of (two or) three (to five), each needle (5–)15–25(–30) cm long, stiff, lasting 2–3 years, variously yellowish green to dark green or bluish green. Individual needles with numerous, conspicuous lines of stomates on both the inner and outer faces, and two to eight resin canals touching the two-stranded midvein at its corners and outer face. Sheath 15–20 mm long at first, often weathering to 10–12 mm, persisting and falling with the bundle. Pollen cones 20–40 mm long, yellowish brown with a red blush. Seed cones (3–)6–12(–14) cm long, egg-shaped, with 90–180 seed scales, green before maturity, ripening shiny dark reddish brown, opening widely to release the seeds and then falling off the slender, 1–2.5 cm stalks. Seed scales nearly rectangular, the exposed face horizontally diamond-shaped, almost flat, the umbo ending in a small prickle. Seed body 4–7 mm long, the firmly clasping or easily detachable wing another (10–)15–20 mm longer. Turks and Caicos Islands and northern Bahamas to northern Central America. Typically forming pure or mixed open pine stands on fire-prone but seasonally flooded lowland flats; 0–600(–900) m. Zone 9. Three varieties.

Caribbean pine has the most southerly extension among the New World pines, reaching to about 12°N along the Caribbean coast of Nicaragua. It is an important timber species throughout its range but is not threatened because frequent fires help promote regeneration. It has also been introduced to plantation culture in such other tropical regions as East Africa, northern Australia, and southern China. There is modest variation throughout the range of the species, and three weakly distinguished varieties are recognized among the three regions in which it occurs. Possibly the increased needle number of variety *hondurensis* is due to hybridization with *Pinus oocarpa,* with which it commonly grows naturally. Until about 1950, slash pine (*P. elliottii*) of the southern United States was usually included as part of *P. caribaea,* and the two species are closely related, although distinct.

Pinus caribaea var. bahamensis (Grisebach) W. Barrett & Golfari
BAHAMAS PINE
Juvenile leaves green and transient. Some bundles with two needles usually present. Sheath shortening to 10–12 mm. Seed cones to 14 cm long. Seed wings easily detached. Turks and Caicos Islands and northern Bahamas (Abaco, Andros, Gran Bahama, and New Providence); sea level to about 30 m.

Pinus caribaea P. Morelet **var. caribaea**
Juvenile leaves green and transient. Almost all bundles with three needles. Sheath remaining 15–20 mm long. Seed cones to 10 cm long. Seed wings firmly attached. Western Cuba (Isla de la Juventud and Pinar del Río); 50–350 m.

Pinus caribaea var. hondurensis (Sénéclauze) W. Barrett & Golfari
HONDURAS PINE
Juvenile leaves blue green and long persistent. Some bundles with four or five needles usually present. Sheath remaining 15–20 mm long. Seed cones to 14 cm long. Seed wings firmly attached to easily detached. Southern Quintana Roo (Mexico) and central Guatemala through Belize and Honduras to northern Nicaragua; 0–900 m.

Pinus cembra Linnaeus
AROLLA PINE, SWISS STONE PINE, AROLE (FRENCH), ARVE (GERMAN)
Pinus subg. *Strobus* sect. *Quinquefoliae* subsect. *Strobus*
Tree to 20(–25) m tall, though often dwarfed at the alpine timberline. Trunk to 1(–1.5) in diameter. Bark gray and smooth at first, becoming grayish brown and scaly, flaking to reveal reddish brown patches. Crown densely conical when young, becoming cylindrical and finally very open, with slender rising branches becoming horizontal and turned up at the ends, densely clothed with foliage only near the tips. Twigs yellowish brown at first, becoming blackish brown, densely woolly with orange-brown hairs. Buds 6–10 mm long, resinous. Needles in bundles of five, each needle 5–8(–12) cm long, stiff, straight, and loose, lasting 3–6 years, bright to dark green on outer face. Individual needles with lines of stomates only on the waxy grayish green inner faces, and two small resin canals between the outer epidermis of the outer face and the undivided midvein without touching either. Sheath 20–30 mm long, soon shed. Pollen cones 10–20 mm long, red. Seed cones (5–)6–8 cm long, egg-shaped, with about 50 seed scales, violet before maturity, ripening warm reddish brown, remaining closed and disintegrating with release of the seeds during the spring following maturation, short-stalked. Seed scales

Twig of arolla pine (*Pinus cembra*) with many years of growth, ×0.5.

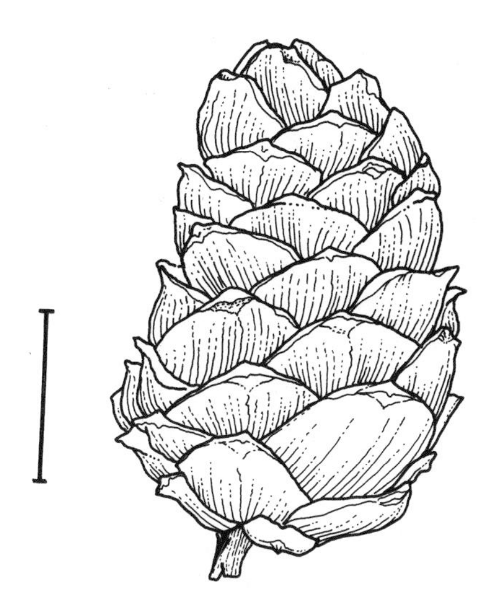

Closed mature seed cone of arolla pine (*Pinus cembra*); scale, 2 cm.

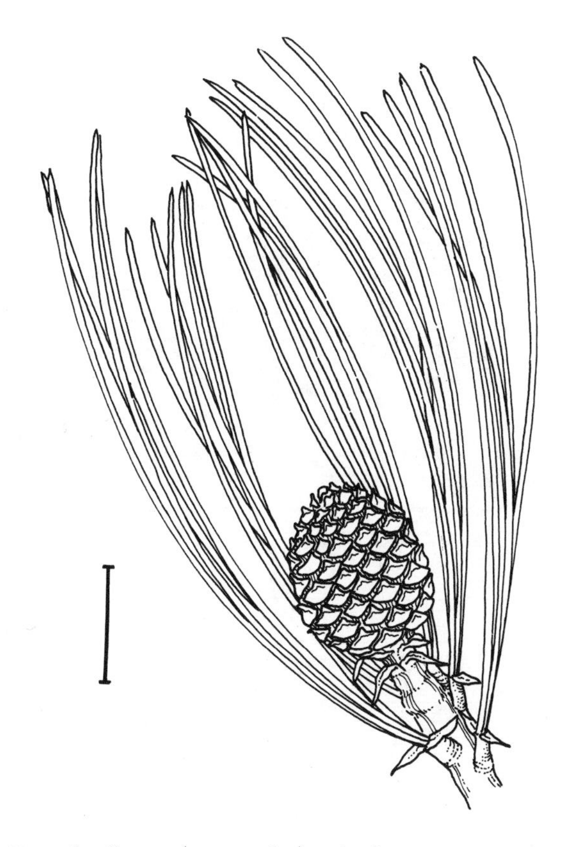

Twig of arolla pine (*Pinus cembra*) with a first-year immature seed cone; scale, 2 cm.

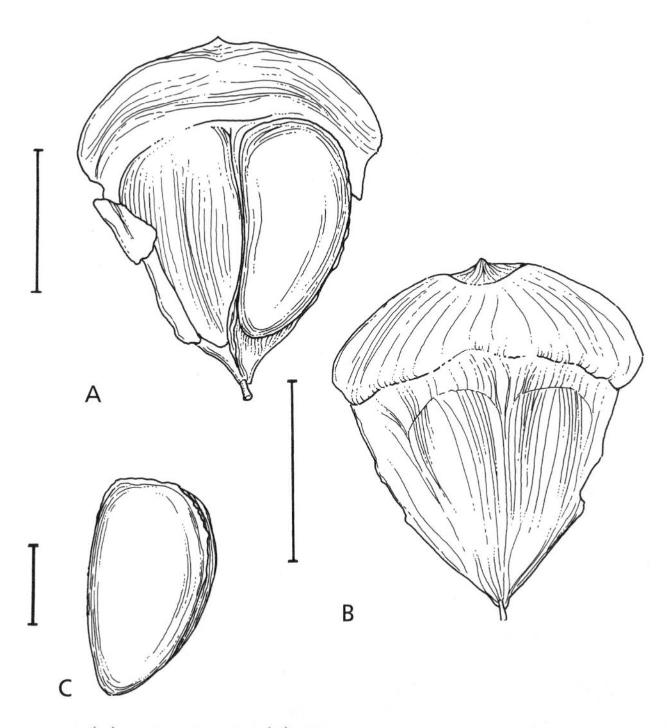

Upper (A) and undersides (B) of detached seed scale and separate seed (C) of arolla pine (*Pinus cembra*); scale, 1 cm.

fan-shaped, somewhat fleshy, especially at the exposed, triangular tip, with a triangular umbo right at the tip. Seed body 12(–14) mm long, unwinged or with a very narrow, easily detached wing. Alps and Carpathians of south-central Europe, from southeastern France to central Romania. Forming pure stands or mixed with other conifers in closed to sparse subalpine forests, woodlands, and scrublands; (1,300–)1,500–2,000(–2,750) m. Zone 5.

Arolla pine is the westernmost Eurasian member of the subalpine, edible-seeded stone pines. It is closest to Siberian stone pine (*Pinus sibirica*), with which it can hybridize, and which is also geographically closest, reaching northeastern Europe. It is not closely related to the only other European white pine, the nearby Macedonian pine (*P. peuce*), and will not cross with that species. Like other tree-line pines, *P. cembra* is slow growing and long-lived, not first maturing cones in its native stands until it is at least 50 years old and reaching an extreme age of 500 years or more. As in other members of the group, the seeds are dispersed by nutcrackers, squirrels, and other seed eaters that cache seeds, some of which are never retrieved.

Pinus cembroides Zuccarini

MEXICAN PINYON, PINO PRIETO, PINO PIÑONERO (SPANISH)

Pinus subg. *Strobus* sect. *Parrya* subsect. *Cembroides*

Tree to 15(–20) m tall, with trunk to 0.5(–1) m in diameter, often forking repeatedly. Bark dark brown to blackish brown, broken into irregular, low scaly plates by shallow furrows. Crown dense and egg-shaped when young, opening and becoming very wide spreading with age, with numerous upwardly angled branches sparsely clothed with foliage. Twigs reddish brown, hairless, graying with age. Buds 5–8(–12) mm long, variably resinous. Needles in bundles of two to four (or five), sticking together at first, each needle (2–)3–5(–6) cm long, gently curved and stiff or a little flexible, lasting 3–5(–7) years, dull dark green or yellowish green to grayish green. Individual needles with lines of stomates on both the inner and outer faces, an undivided midvein, and (one or) two resin canals visible beneath the epidermis of the outer face. Sheath 4–7 mm long curling back and persisting for a time before falling. Pollen cones 5–10 mm long, straw colored. Seed cones (2–)3–5 cm long, nearly spherical to broadly cone-shaped, with 20–40(–50) seed scales, yellowish green to purplish before maturity, ripening yellowish brown to light reddish brown, opening widely to release the seeds and then falling with the 2–5(–8) mm long stalk. Seed scales paddle-shaped, with deep pockets for the seeds extending almost the whole length, the exposed face diamond-shaped and rising in a shallow pyramid to a diamond-shaped, prickly umbo. Seed body (10–)12–14 mm long, the shell 0.5–1.1 mm thick, the rudimentary wing remaining attached to the seed scale. On either side of the Meseta Central of Mexico, from trans-Pecos Texas and northeastern Sonora south to southeastern Puebla (Mexico). Forming pure stands or mixed with other small trees in an open pinyon-juniper-oak woodland on dry slopes with thin, rocky soils between desert scrub and montane pine-oak forest; (700–)1,700–2,700(–3,000) m. Zone 7. Two subspecies.

This is the most common and widespread pinyon pine of Mexico and reaches the most southerly locations in the group but barely enters the United States in the Chisos and Davis Mountains of Texas. It is found throughout the foothills of both the Sierra Madre Occidental and the Sierra Madre Oriental and overlaps with several other pinyon species. Historically, the name has often been used to include all of the small-coned pinyons of the United

Mature Mexican pinyon (*Pinus cembroides*) in nature.

Open mature seed cone of Mexican pinyon (*Pinus cembroides*) after seed dispersal; scale, 3 cm.

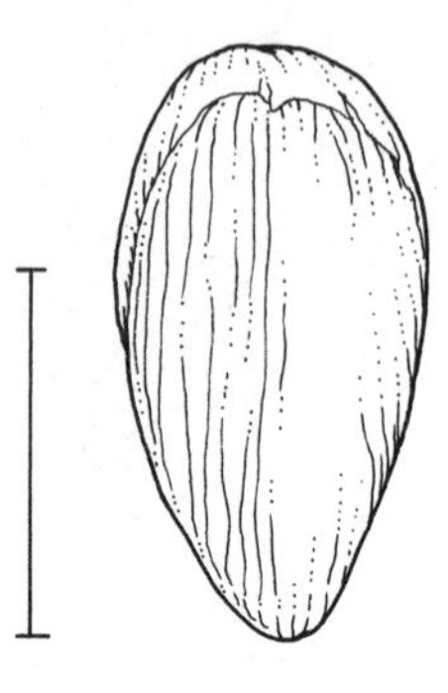

Seed of Mexican pinyon (*Pinus cembroides*); scale, 5 mm.

States and Mexico. Even when the species of the United States and northern Mexico with needles in bundles of one, two, or four or five were separated at the species level rather than treated as varieties, a broad concept of *Pinus cembroides* still embraced all of the predominantly three-needled trees until about 1980. From that point, it became increasingly clear that trees of different species were being indiscriminately included within the broad concept of *P. cembroides.* Some of these segregates were recognized as distinct before 1970, but were treated as varieties of Mexican pinyon. Although there has been a fair amount of agreement on which pinyons to separate from *P. cembroides,* it is much less clear how to relate them to each other and to other described pinyons. Based on morphological, biochemical, and DNA studies, most are here treated as varieties of *P. culminicola,* a species once considered to include only five-needled subalpine shrubs. With the trees of the southern tip of Baja California treated as separate species, Mexican pinyon is left with two subspecies distinguished by minor differences in morphology and wood oils. It is not known to hybridize with any of the varieties of *P. culminicola* with which it overlaps. The species name *cembroides* (Latin for "like cembra") links this pinyon with the European arolla pine, which it resembles only in having edible seeds and belonging to the white pines (subgenus *Strobus*). The largest known individual in the United States in combined dimensions when measured in 1982 was 20.1 m tall, 0.9 m in diameter, with a spread of 13.4 m.

Pinus cembroides Zuccarini **subsp. *cembroides***
Bark with horizontal furrows interrupting the vertical ridges. Needles in bundles of two to four, yellowish green to dull green. Northern portion of the range of the species, south to the states of Hidalgo and Mexico.

Pinus cembroides* subsp. *orizabensis D. Bailey
Bark with continuous, if irregular, vertical ridges not conspicuously interrupted by horizontal furrows. Needles in bundles of (two or) three or four (or five), bluish green. Southern portion of the range of the species, only in the states of Puebla, Tlaxcala, and Veracruz, Mexico. Synonym: *Pinus orizabensis* (D. Bailey) D. Bailey & F. Hawksworth.

Pinus contorta D. Douglas ex J. C. Loudon
LODGEPOLE PINE
Pinus subg. *Pinus* sect. *Trifoliae* subsect. *Contortae*
Tree to 35(–50) m tall, with trunk to 1(–2) m in diameter, or a miniature tree just 1–2 dm tall on soilless coastal hard pans. Bark variously pale reddish brown, gray, or blackish brown, flaking in small scales, remaining flaky or breaking up into irregular ridges separated by shallow furrows at the base of large trees. Crown narrowly conical to cylindrical, often spirelike in dense stands or multitrunked, open, and irregular on exposed headlands, with numerous thin, horizontal or upswept branches densely clothed with foliage only at the tips. Twigs yellowish brown to reddish brown, hairless, roughened by scale leaves and their bases. Buds 12–15 mm long, slightly resinous. Needles in bundles of two, each needle (2–)3–7(–8) cm long, stiff, often twisted, lasting 3–5(–8) years, dark green to yellowish green. Individual needles with lines of stomates on both faces, a two-stranded midvein, and usually two resin canals deep within the leaf tissue near the outer corners. Sheath 0.3–0.6(–1) cm long, only the shorter, outer scales persisting after the first year and falling with the bundles. Pollen cones 5–15 mm long, yellowish brown with a reddish blush. Seed cones (2–)3–5(–6) cm long, egg-shaped, often slightly asymmetrical, especially at the base, standing out from the twig or curved backward, with 75–90(–120) seed scales, green before maturity, ripening orange-brown, opening widely to release the seeds or, more commonly, remaining closed on the branches many years until opening after a fire, unstalked or with a short stalk to 1 cm. Seed scales paddle-shaped, woody and stiff, the exposed portion diamond-shaped, low or somewhat pyramidal near the base of the cone on the outer side, the diamond-shaped umbo with a slender, rigid prickle to 6 mm long. Seed body 3–5 mm long, mottled with black to all black, the easily

Natural stand of lodgepole pine (*Pinus contorta*).

detachable wing 10–12 mm long. Western North America, from southwestern Yukon and southeastern Alaska south to southern Colorado and northern Baja California (Mexico). Commonly forming pure stands but also mixed with other trees in many fire-prone habitats, from coastal flats and bogs to subalpine forests; 0–3,500 m. Zone 7. Three varieties.

Lodgepole pine occupies one of the widest ecological ranges, in both elevation and environment, of any of the pines, and has correspondingly broad morphological variation, particularly in growth habit. The scrubby twisted trees of the seashore that provided the scientific name of the species contrast with the tall, slender spires of the subalpine forests that were formerly used as poles for tepees and other skin lodges, the origin for the common name. Trees of the Rocky Mountains are tallest while those of the Sierra Nevada (California) have the largest diameters and are largest overall. The largest known individual in the United States in combined dimensions when measured in 2001 was 35.4 m tall, 2.0 m in diameter, with a spread of 15.2 m. Although the regional populations share many traits, including DNA patterns, that distinguish them from jack pine (*Pinus banksiana*), their closest relative, they have been distinguished as three or four varieties, subspecies, or species. The species is divided into three varieties here, treating the miniature trees of the pygmy forests of northern California as a southern extension of the shore pines rather than as a separate variety. Although all of the varieties are subject to periodic fires, the populations of the Sierra Nevada of California (variety *murrayana*) and the coast (variety *contorta*) do not share the persistently closed cones of the majority of the range (variety *latifolia*) and of jack pine. Lodgepole pines hybridize naturally with jack pines where they overlap in the foothills east of the Rocky Mountains in western Canada. Evidence from morphological characteristics, resins, enzymes, and DNA studies all confirm this hybridization, but they suggest different degrees of pervasiveness of hybrid characters within each species. Similar hybrids have been produced by controlled pollination, and these have been named *P. ×murraybanksiana.*

Pinus contorta D. Douglas ex J. C. Loudon **var. *contorta***

SHORE PINE

Shrubby, twisted, sometimes dwarf, trees exceptionally to 30 m. Bark ridged and furrowed. Needles 2–6(–7) cm long, dark green. Seed cones asymmetric, curved backward on the twig, usually opening after maturity but remaining attached for several years. Coastal forests up to 600 m from southern Alaska to northern California. Synonym: *Pinus contorta* var. *bolanderi* (Parlatore) J. Lemmon.

Pinus contorta* var. *latifolia Engelmann

LODGEPOLE PINE

Tall, spirelike trees, exceptionally to 50 m. Bark remaining predominantly scaly, even on large trunks. Needles (4–)5–8 cm long, yellowish green. Seed cones asymmetric, curved backward on the twig, remaining closed many years after maturity. Mountains in western North America up to 3,500 m, from the central Yukon and southern Northwest Territories (Canada), south in the Cascade Range to Washington and in the Rocky Mountains to central Utah, southern Colorado, and southwestern South Dakota.

Pinus contorta* var. *murrayana (Greville & J. Balfour) Engelmann

TAMARACK PINE, SIERRA LODGEPOLE PINE

Straight, often slender trees, exceptionally to 40 m. Bark remaining predominantly scaly, even on large trunks. Needles 5–8 cm long, yellowish green. Seed cones fairly symmetrical, standing out from the twig, often opening at maturity and soon falling. Pacific Mountains up to 3,500 m, from the Cascade Range of Oregon south through the Sierra Nevada of California to the Sierra San Pedro Martír of northern Baja California (Mexico).

Pinus coulteri D. Don

COULTER PINE

Pinus subg. *Pinus* sect. *Trifoliae* subsect. *Ponderosae*

Tree to 20(–25) m tall, with trunk to 0.8(–1) m in diameter. Bark dark brown to dark grayish brown, with interlacing vertical ridges divided by deep, darker furrows. Crown broadly conical, with numerous long, slender, gently upwardly arching branches, turning up and densely clothed with foliage at the tips. Twigs stout, reddish brown beneath a bluish white coating of wax, roughened by the bases of scale leaves. Buds 1.5–3(–4) cm long, resinous. Needles in bundles of three, each needle (15–)20–30 cm long, stiff and straight, lasting 3–4 years, light grayish green. Individual needles with obvious white lines of stomates on all three faces, a two-stranded midvein, and 2–10 resin canals around the midvein deep within the leaf tissue. Sheath 2–4 cm long, persisting and falling with the bundle. Pollen cones 20–25 mm long, dark yellowish brown. Seed cones 20–30(–35) cm long, egg-shaped, massive, with 120–240 seed scales, yellowish green before maturity, ripening yellowish brown to light brown, opening widely to release the seeds but then sometimes persisting several years before falling, leaving behind a few basal seed scales on the thick, persistent stalk to 3 cm long. Seed scales very thick and woody, diamond-shaped, the exposed portion forming a thick, protruding, triangular boss continuing into the long, heavy, curved, clawlike umbo. Seed body (10–)12–18(–22) mm

Natural stand of Coulter pine (*Pinus coulteri*).

long, uniformly dark brown, the easily detachable wing 18–30 mm long. Scattered from west-central California to northern Baja California. Most commonly mixed with lower montane and woodland trees and chaparral shrubs on dry sites in foothills and mountains; (300–)900–1,800(–2,150) m. Zone 8.

At about two kg, the seed cones of Coulter pine are the heaviest of any pine and among the heaviest of all conifers. The large seeds are edible and have been eaten by people as well as a range of seed-eating birds and mammals. The seed cones are similar to those of gray pine (*Pinus sabiniana*) of the foothills of the central valley of California, but the species are not intercrossable. Although usually a small tree of poor form, Coulter pine can grow well on good sites that also support Jeffrey pine (*P. jeffreyi*), a less similar species with which it has been crossed successfully. The largest known individual when measured in 2001 was 43.0 m tall, 1.2 m in diameter, with a spread of 22.0 m. Coulter pine is somewhat resistant to the fires that characterize many of its sites. With its large seed crops it can be an aggressive colonizer of sites cleared of understory by burning. However, saplings and adult trees do best in the absence of fires. The species name honors the Irish botanist and physician Thomas Coulter (1793–1843), who collected the type specimen in the Santa Lucia Mountains of California in 1831.

Pinus cubensis Grisebach

ORIENTE PINE

Pinus subg. *Pinus* sect. *Trifoliae* subsect. *Australes*

Tree to 30(–40) m tall, with trunk to 1 m in diameter. Bark grayish brown, thick, with rough, blocky ridges or small plates divided by deep furrows. Crown conical when young, becoming shallowly bowl-shaped and open with age, with slender horizontal branches sparsely clothed with foliage. Twigs reddish brown but whitened by a waxy coating, a little roughened with the bases of scale leaves, hairless. Buds 8–15 mm long, not resinous. Needles in bundles of two (or three), each needle (4–)8–15(–17) mm long, stiff and straight, lasting 2–3 years, dark green. Individual needles with numerous visible lines of stomates on both the inner and outer faces, and three to six resin canals touching the two-stranded midvein at the corners and inside both faces. Sheath 6–10 mm long, soon curling back but persisting and falling with the bundle. Pollen cones 15–20 mm long, yellow. Seed cones 4–7 cm long, broadly egg-shaped, with 50–90 seed scales, green before maturity, ripening dark yellowish brown, on slender stalks 1–2 cm long, opening widely to release the seeds and then falling. Seed scales paddle-shaped, the exposed face horizontally diamond-shaped, crossed by a ridge topped by a small diamond-shaped umbo with a short, easily broken prickle. Seed body 5–7 mm long, the firmly attached wing another 12–19 mm longer. Eastern Cuba, in Guantánamo, Holguín, and Santiago de Cuba provinces, all parts of the historic Oriente province. Forming pure, open pine forests in the mountains and formerly down to the coast, especially on the difficult serpentine soils that are so prominent in this part of Cuba; (100–)300–900(–1,200) m. Zone 9.

Classification of the pines of eastern Cuba has varied over the years. The trees of the eastern and western portions of the old Oriente province have minor and disputed differences from one another. Both are clearly related to the pines of Hispaniola and some early authors simply included them within *Pinus occidentalis* without distinction. Modern authors used almost every possible variation, combining or separating the three elements. The majority opinion is followed here, separating *P. cubensis* from Hispaniolan *P. occidentalis* and the pines of western Oriente (which are also included in *P. occidentalis*) on the basis of its smaller and stiffer needles in bundles of two and seed cones with small prickles. The problem deserves further careful comparison and assessment of many populations throughout eastern Cuba and Hispaniola. Although Oriente pine is heavily exploited for timber, it is holding its own because it experiences little competition on serpentine soils and is also an aggressive invader of degraded vegetation.

Pinus culminicola Andresen & J. Beaman

PLATEAU PINYON

Pinus subg. *Strobus* sect. *Parrya* subsect. *Cembroides*

Low, multitrunked shrub, or tree to 10 m tall, with trunk to 0.5 m in diameter. Bark dark grayish brown, scaly and broken up into irregular, small plates by shallow furrows. Crown spreading, rounded or flat-topped, often dense, with numerous slender, horizontal or upwardly angled branches densely clothed with foliage at the tips. Twigs light brown at first, becoming gray, hairless, generally stiff and roughened with scale leaves. Buds 5–10 mm long, not resinous. Needles in bundles of two to five (or six), sticking together during the first year, each needle (2–)3–5(–6) cm long, stiff and curved, lasting 2–5(–7) years, bluish green to grayish green. Individual needles with lines of stomates only on the inner faces, an undivided midvein, and two or three (to five) resin canals inside the epidermis of the outer face. Sheath 4–8 mm long, the scales separating or curling back and usually persisting the first year. Pollen cones 4–8 mm long, light purplish yellow. Seed cones (2–)2.5–4.5 cm long, spherical to somewhat flattened, with 20–30(–40) seed scales, purplish green before maturity, ripening yellowish brown to reddish brown, opening widely to release the seeds and then falling with the 3–8(–10) mm long, curved stalk. Seed scales paddle-shaped, the deep pockets for the seeds just reaching the diamond-shaped exposed portion that rises in a low pyramid on the face and ends in a large diamond-shaped umbo sometimes bearing a small, fragile prickle. Seed body 5–16 mm long, the shell (0.1–)0.3–1(–1.3) mm thick, the rudimentary wing remaining attached to the seed scale. Southeastern Arizona to the Edwards Plateau of Texas south through the Sierra Madre Occidental and Sierra Madre Oriental and interior ranges of Mexico to southern Zacatecas and southern San Luis Potosí. Woodlands and thickets, generally on dry soils over limestone; (450–)1,200–3,700 m. Zone 7? Four varieties.

Dense alpine scrub of Potosí pinyon (*Pinus culminicola* var. *culminicola*) on Cerro Potosí.

Mature shrub of Potosí pinyon (*Pinus culminicola* var. *culminicola*) in nature.

The four varieties of plateau pinyon vary in morphology, biochemistry, and ecology and are often treated as separate species, but DNA studies link them together and separate them clearly from Mexican pinyon (*Pinus cembroides*), with which most have been associated in the past. In part, they form an elevational-replacement series with taller stature, fewer needles, and larger seeds with thinner shells at lower elevations. Much more detailed studies need to be undertaken of all pinyons at the population level before there can be much confidence in the classification of these socially and economically important little trees of dryland North America. The most extensive studies so far have been on wood oils. Coordinated studies of morphology and DNA, however, are greatly needed. Plateau pinyon overlaps extensively with Mexican pinyon, but the two species do not appear to hybridize with one another, unlike several other combinations of overlapping pinyons. The scientific name, Latin for "peak dweller," refers particularly to variety *culminicola,* which reaches the highest elevations of any pinyon and experiences the most humid environment among them. The largest known individual when measured in 1999 was 9.8 m tall, 0.5 m in diameter, with a spread of 11.3 m.

Pinus culminicola **var.** ***bicolor*** (E. Little) Eckenwalder

BORDER PINYON

Tree to 10(–12) m tall. Needles in bundles of two to four (or five). Resin canals two. Seed body 10–12 mm long, the shell 0.7–1.2 mm thick. Northwestern portion of the range of the species in the Sierra Madre Occidental, from Arizona and New Mexico to northern Durango (Mexico); 1,500–2,600 m. Synonyms: *Pinus cembroides* var. *bicolor* E. Little, *P. discolor* D. Bailey & F. Hawksworth.

Pinus culminicola Andresen & J. Beaman var. *culminicola*

POTOSÍ PINYON

Shrub to 5 m. Needles in bundles of (four or) five (or six). Resin canals two. Seed body 4–7 mm long, the shell 0.8–1 mm thick. Top of high mountains of the borderland region between central Coahuila and central Nuevo León, Mexico; 3,000–3,700 m.

Pinus culminicola var. *johannis* (M. Robert) Silba

DWARF PINYON

Shrub to 4 m. Needles in bundles of two or three (or four). Resin canals two. Seed body 8–13 mm long, the shell 0.5–1(–1.3) mm thick. Southeastern portion of the range of the species, in the Sierra Madre Oriental and some interior ranges, from northwestern Coahuila south to southern Zacatecas and southern San Luís Potosí, Mexico; 2,300–3,000 m. Synonym: *Pinus johannis* M. Robert.

Pinus culminicola var. *remota* (E. Little) Eckenwalder

THIN-SHELL PINYON

Shrub, or tree to 9 m tall. Needles in bundles of two (or three). Resin canals two or three (to five). Seed body 12–16 mm long, the shell (0.1)0.3–0.6 mm thick. Northeastern portion of the range of the species, from trans-Pecos Texas and the Edwards Plateau south to southeastern Chihuahua and southeastern Coahuila (Mexico); (450–)1,200–1,800 m. Synonyms *Pinus catarinae* Robert-Passini, *P. cembroides* var. *remota* E. Little, *P. remota* (E. Little) D. Bailey & F. Hawksworth.

Pinus dalatensis de Ferré

VIETNAMESE WHITE PINE, THÔNG NĂM LÁ DA LAT (VIETNAMESE)

Pinus subg. *Strobus* sect. *Quinquefoliae* subsect. *Strobus*

Tree to 40 m tall, with trunk to 0.5 m in diameter. Bark light gray, flaking in reddish brown scales and becoming shallowly furrowed between small blocks. Crown irregularly cylindrical and open, with horizontal branches sparsely clothed with foliage near the tips. Twigs reddish brown, hairless to densely hairy, thinly coated with wax. Buds 6–12 mm long, not or sparsely resinous. Needles in bundles of five, each needle (3–)4–10(–14) cm long, thin and flexible, lasting 2–3 years, bluish green. Individual needles with lines of stomates only on the inner faces, an undivided midvein, two resin canals touching the epidermis of the outer face, and sometimes with a third resin canal touching the epidermis of one or the other inner face near their juncture. Sheath 1–1.5 cm long, soon shed. Pollen cones about 20 mm long, purple. Seed cones (5–)10.5–16(–23) cm long, taperingly cylindrical, with about 75 seed scales, green before maturity, ripening light brown, opening widely to release the seeds and then falling with the stout stalk, up to 5 cm long. Seed scales broadly paddle-shaped, thin and flexible, the exposed part a little thicker, straight, somewhat cupped, ending in a narrow, diamond-shaped umbo. Seed body 6–8(–10) mm long, the wing 15–20 mm longer. Highlands of central Vietnam. Growing in pure stands or, more commonly, mixed with other conifers and evergreen hardwoods on shallow, infertile soils; 1,500–2,500 m. Zone 9? Synonyms: *P. dalatensis* subsp. *procera* Businský, *P. dalatensis* var. *bidupensis* Businský.

Vietnamese white pine is one of the rarest and most narrowly localized conifer species. It was first described from a locality in the central highlands near the resort town of Dalat (hence the scientific name) and has since been collected elsewhere in the highlands. It is not known to be in cultivation outside Vietnam and so is very poorly known. There have been no experimental studies of its status and relationships, but no other nearby Asian species would seem to accommodate it comfortably as a variety.

Pinus densata M. T. Masters

CHINESE MOUNTAIN PINE, GAO SHAN SONG (CHINESE)

Pinus subg. *Pinus* sect. *Pinus* subsect. *Pinus*

Tree to 30 m tall, with trunk to 1(–1.3) m in diameter. Bark reddish brown and flaking at first, finally becoming grayish brown and breaking up into small, polygonal plates separated by narrow furrows. Crown conical at first, broadening and flattening with age, with numerous slender to heavy, horizontal branches well clothed with foliage near the ends. Twigs, coarse, shiny yellowish brown to reddish brown, hairless, rough with the bases of scale leaves. Buds 12–16 mm long, slightly resinous. Needles in bundles of two (or three), each needle (6–)8–14(–15) cm long, thick and stiff, slightly twisted, lasting 3(–4) years, bright green to dark green. Individual needles with lines of stomates on both the inner and outer faces, a two-stranded midvein, and three or four (to seven) resin canals touching the outer face near the corners and in between. Sheath (5–)8–12 mm long, weathering to 3–6 mm and persisting and falling with the bundle. Pollen cones 10–20 mm long, yellowish brown. Seed cones 4–6 cm long, slightly asymmetrical, egg-shaped to broadly egg-shaped, with 60–80 seed scales, green before maturity, ripening shiny rich brown, opening widely to release the seeds and then persisting a while before falling with the very short stalk to 1 cm long. Seed scales spoon-shaped, the exposed face diamond-shaped, protruding, especially at the base on the side away from the twig, crossed by a ridge topped by a diamond-shaped umbo bearing a short, persistent prickle. Seed body 4–6 mm long, the easily detachable wing another 15–20 mm longer. Mountains of southwestern China, from southern Qinghai through western Sichuan and eastern Xizang (Tibet) to northern Yunnan. Forming pure stands in the subalpine zone or mixed with other pines in upper montane forests (1,100–)2,600–3,500(–4,200) m. Zone 6. Synonym: *P. tabuliformis* var. *densata* (M. T. Masters) Rehder.

Chinese mountain pine has the highest elevational range among the Chinese hard pines (the Chinese name means "high mountain pine"). As is typical for a subalpine pine, the foliage densely clothes the twigs and the annual growth increments are relatively short, giving a condensed appearance to the tree (hence the scientific name). Because of the slow growth, the wood has narrow annual rings and it, too, is dense.

The status of Chinese mountain pine has been controversial over the years. Most authors treat it as a high-elevation variety of Chinese red pine (*Pinus tabuliformis*), which extends at lower elevations northeastward from the distribution area of *P. densata.* The other common opinion holds that it is a hybrid between *P. tabuliformis* and Yunnan pine (*P. yunnanensis*), but some authors who thought that still treated it formally as a variety of the Chinese red pine. In its simplest form, this hypothesis of hybridity is unlikely because the geographic range of *P. densata* is largely distinct from that of the other two species, it grows at much higher elevations than either of them, and it breeds true and does not segregate characters more typical of the other two species in its offspring. Genetic studies suggest that *P. densata* is an independent species that originated following hybridization between *P. tabuliformis* and *P. yunnanensis* rather than consisting directly of hybrids. While the frequency of variant enzymes is most similar to that in Chinese mountain pine, a few show frequencies clearly intermediate between those of the other two species. The distribution of chloroplast DNA variants is even more compelling. Even though they are only inherited from the pollen parent in conifers (unlike in flowering plants, where they are mostly maternally inherited), different trees of *P. densata* have the distinctive chloroplast types found in each of the other two species (and not shared with the third major Chinese hard pine that reaches the region, Masson pine, *P. massoniana*). In addition, some trees of Chinese mountain pine have a third kind of chloroplast that combines distinctive features of the others. When the original hybridization might have taken place and how species formation then occurred are speculative and require more detailed study.

Pinus densiflora P. Siebold & Zuccarini

JAPANESE RED PINE, AKA-MATSU (JAPANESE), SHONAMU (KOREAN), CHI SONG (CHINESE)

Pinus subg. *Pinus* sect. *Pinus* subsect. *Pinus*

Tree to 30(–35) m tall, with trunk to 1.5(–1.8) m in diameter. Bark bright reddish brown, smooth, and flaky at first, breaking up into scaly gray plates separated by deep reddish furrows on large trees. Crown broadly conical, becoming progressively more flat-topped with age, with tiers of upwardly angled, then horizontal, and finally gently drooping branches sparsely to densely clothed with foliage at the tips. Twigs yellowish brown, often with a thin dusting of wax, scarcely roughened by the well-spaced bases of scale leaves, hairless. Buds about 8–12 mm long, slightly resinous. Needles in bundles of two, each needle (5–)7–12(–15) cm long, stiff and straight or sometimes twisted, lasting 2(–3) years, dark green to bright bluish green. Individual needles with a few lines of stomates on both faces, and (2–)3–9(–12) resin canals surrounding the two-stranded midvein and touching the outer surface. Sheath 15 mm at first, becoming a little reduced but persisting and falling with the bundle. Pollen cones densely crowded at the beginning of each year's growth (hence the scientific name), 5–10 mm long, pale yellow. Seed cones 3–5.5 cm long, almost spherical, with (50–)70–90(–100) seed scales, green before maturity, ripening yellowish brown, opening widely to release the seeds and then falling with the short, straight or curved stalk. Seed scales oval, the exposed face horizontally diamond-shaped, flat or sometimes conically enlarged at the base of the seed cone, the umbo with a short, blunt tip. Seed body 3–7 mm long, the firmly attached wing another 10–15 mm longer. Throughout Japan (except Hokkaidō) and Korea, extending south near the coast in China to northeastern Jiangsu and northward to the Ussuri region of far eastern Russia. Forming pure stands or mixed with other conifers and hardwoods on rocky slopes and shores; (0–)100–900(–1,400) m. Zone 7. Synonyms: *P. densiflora* var. *ussuriensis* T. N. Liou & Z. Wang, *P. densiflora* var.

Mature Japanese red pine (*Pinus densiflora*) in nature.

zhangwuensis S. J. Zhang, C. X. Li & X. Y. Yuan, *P. funebris* V. Komarov, *P. sylvestris* var. *sylvestriformis* (Takenouchi) W. C. Cheng & C. D. Chu, *P. takahasii* T. Nakai.

Japanese red pine is one of the most important timber trees of Korea, making up some 40% of the forests there. Cool, wet conditions when the new needles first flush are the most influential factors in producing optimal growth. Perhaps responding to the maritime climate, it is one of the fastest-growing hard pines in the northeastern United States, especially near the coast. It is very closely related to Scots pine (*Pinus sylvestris*), as evidenced by both DNA studies and its morphological features, differing most obviously in its longer needles. It crosses artificially with *P. sylvestris* to yield vigorous, early coning hybrids, and also crosses successfully with several other geographically separated pines in subsection *Pinus,* including Austrian pine (*P. nigra*), Masson pine (*P. massoniana*), and Taiwan pine (*P. taiwanensis*). It also hybridizes naturally with Japanese black pine (*P. thunbergii*), which grows with it along the coast of Japan and Korea, to form *P.* ×*densithunbergii.* Although geographically variable, there seems little justification for giving formal taxonomic recognition to even the most distinctive variants.

Pinus ×*densithunbergii* Uyeki

Natural hybrids between Japanese red pine (*Pinus densiflora*) and Japanese black pine (*P. thunbergii*), growing with the parents where they occur together along the coasts of Korea and Japan. These trees are intermediate between the parent species in many features in which they overlap, such as in the paired needles about 9–10 cm long, pollen cones about 10 mm long, and seed cones about 5 cm long. The most distinctive feature is the position of the resin canals in the needles, near to but not touching the epidermis, while they do touch in *P. densiflora* and are deeply embedded in the leaf tissue of *P. thunbergii.*

Pinus devoniana Lindley

MICHOACÁN PINE, PINO LACIO, OCOTE MACHO (SPANISH)

Pinus subg. *Pinus* sect. *Trifoliae* subsect. *Ponderosae*

Tree to 30 m tall, with trunk to 1 m in diameter. Bark dark reddish to grayish brown, broken up into vertical rows of small, scaly, rectangular plates by deep, blackish furrows. Crown conical at first, broadening and rounding with age to become shallowly to deeply domed, with numerous horizontal to upwardly angled branches, densely clothed with foliage at the tips. Twigs dark brown, very coarse, shaggy with rings of persistent scale leaves and rough with the bases of scale leaves or these peeling away during the second year, hairless. Buds 15–40 mm, not conspicuously resinous. Needles in bundles of (four or) five (or six), each needle (20–)25–35(–45) cm long, stiff or slightly flexible, straight or slightly drooping, lasting 2–3 years, shiny green. Individual needles with evident lines of stomates on all three faces, and three or four (to six) resin canals deep within the leaf tissue at the corners and below the outer face or one or two touching the two-stranded midvein. Sheath 25–40(–45) mm long, weathering to 20–35 mm, dark brown to almost black, sticky with resin. Pollen cones densely crowded, 20–40 mm long, tan tinged with pale purple. Seed cones (15–)20–30(–37) cm long, elongately egg-shaped, slightly asymmetrical, often curved, with 150–250 seed scales, green before maturity, ripening light brown, opening widely to release the seeds and then falling, leaving behind a few basal scales attached to the stout, short stalk to 1.5(–2) cm long. Seed scales roughly rectangular with a shallowly triangular end, the exposed face diamond-shaped, slightly raised and crossed by a sharp ridge topped by a strong, variously raised umbo bearing a small, fragile prickle. Seed body (5–)6–8(–10) mm long, the clasping wing another 18–35 mm longer. Mountains of southern Guatemala and southern Mexico from central Chiapas north to southern San Luis Potosí and northern Nayarit. Usually mixed with other species in open pine or pine-oak forests on dry sites susceptible to fires; (700–)1,200–2,500(–3,000) m. Zone 9. Synonyms: *P. michoacana* M. Martínez, *P. montezumae* var. *macrophylla* (Lindley) Parlatore.

Michoacán pine, like a few other unrelated New World hard pines in seasonally dry, open, pine forests, has a persistent seedling "grass" stage that is highly resistant to fires until the tree can grow above the level of grass fires. The species is quite variable in its needles and seed cones, and many varieties and segregated species were named, but there are no breaks in the variation that warrant formal taxonomic subdivision. Some of the variability may be due to hybridization with Montezuma pine (*Pinus montezumae*), a very close relative with a broadly overlapping distribution. These two species intergrade so extensively that they might better be considered varieties of a single species, in which case the correct name for Michoacán pine would be *P. montezumae* var. *macrophylla.* The problem deserves careful study at the population level. With its grass stage, Michoacán pine appears to be more tolerant of disturbance than is Montezuma pine and so it freely invades logged or burned hillsides while Montezuma pine stays a species of more mature forests.

Pinus douglasiana M. Martínez

DOUGLAS PINE

Pinus subg. *Pinus* sect. *Trifoliae* subsect. *Ponderosae*

Tree to 30(–45) m tall, with trunk to 0.75(–1) m in diameter. Bark reddish brown, darkening with age and breaking up into large, elongate, scaly plates separated by deep, dark furrows. Crown conical at first, becoming shallowly dome-shaped with age, with numerous gently rising to horizontal branches densely clothed with foliage at the tips. Twigs dark brown, initially rough with the bases of scale leaves but these beginning to flake off at the end of the first year, hairless. Buds 12–25 mm long, not resinous. Needles in bundles of (four or) five (or six), each needle (18–)22–32(–37) cm long, stiff but gently drooping, lasting 2–3 years, yellowish green to light green, often bluish green with a thin, waxy coating on the inner faces. Individual needles with just a few lines of stomates on all three faces, and (two or) three resin canals at the corners midway between the outer leaf surface and the two-stranded midvein and separated by wedges of dense, hard sclereids. Sheath (10–)20–30(–35) mm long, persisting and falling with the bundle. Pollen cones numerous, crowded, 20–30 mm long, light brown with a reddish blush. Seed cones 7–10 cm long, egg-shaped, with 75–120 seed scales, green before maturity, ripening light brown to dark brown, opening widely to release the seeds and falling with the 10–15(–20) mm long stalk. Seed scales broadly wedge-shaped, the exposed face diamond-shaped, shallowly raised, crossed by a horizontal ridge that is topped by a strong, pointed umbo. Seed body 4–6 mm long, the clasping seed wing another 20–25 mm longer. Mountains of the Pacific slope of Mexico, from Sinaloa to Oaxaca and extending eastward in the Transverse Volcanic Ranges to the state of Mexico. Forming pine or pine-oak forests in combination with other species on well-drained slopes; (1,100–)1,700–2,500(–2,700) m. Zone 7.

Douglas pine is closely related to Maximino's pine (*Pinus maximinoi*) and false white pine (*P. pseudostrobus*) and is only distinguished from them with difficulty. Although these three species favor different modal elevations, their elevational ranges overlap and they may be found growing together. *Pinus pseudostrobus* lacks wedges of sclereids protruding into the leaf tissue, has the bases of the scale leaves persisting several years, and leaves the stalk behind on the twig with a few basal scales when the seed cones are shed. *Pinus maximinoi* has more flexible, thinner needles, 0.6–0.9 mm thick, versus (0.7–)0.9–1.2 m thick in *P. douglasiana*. The species name honors Margaret Douglas (fl. 1943), an enthusiast for the Mexican flora who supported Martínez, the author of the species, in his work on Mexican pines.

Pinus durangensis M. Martínez

DURANGO PINE, PINO BLANCO, PINO REAL (SPANISH)

Pinus subg. *Pinus* sect. *Trifoliae* subsect. *Ponderosae*

Tree to 30(–40) m tall, with trunk to 0.7(–1.0) m in diameter. Bark grayish brown, with elongate, scaly plates separated by shallow, dark furrows. Crown roughly spherical, with horizontal to strongly downwardly angled branches densely clothed with foliage at the tips. Twigs orangish brown to reddish brown with a thin dusting of wax, rough with the bases of scale leaves, becoming smooth by the third year, hairless. Buds 12–20(–25) mm long, scarcely resinous. Needles in bundles of five to seven (or eight), most frequently six, each needle (10–)15–24(–28) cm long, stiff and fairly straight, lasting 2(–3) years, shiny light green to waxy bluish green. Individual needles with a few lines of stomates on all three faces, and (one to) three (to five) resin canals midway between the outer surface and the two-stranded midvein at the corners and next to the outer face. Sheath (18–)25–30 mm at first, weathering to (10–)15–20 mm and persisting and falling with the bundle. Pollen cones 15–30 mm long, brownish yellow. Seed cones (5–)6–10(–13) cm long, egg-shaped, with (60–)75–120 seed scales, green before maturity, ripening straw-colored and then progressively darkening, opening widely to release the seeds but not falling for 1–2 years, at which time it leaves a few basal scales behind on the short (3–12 mm) stalk or falls with the stalk. Seed scales almost rectangular, the exposed face horizontally diamond-shaped, crossed by a sharp ridge that ends in a stout, prickly umbo. Seed body (3–)5–7 mm long, the easily detachable wing another 12–17 mm longer. Scattered in the Sierra Madre Occidental of Mexico, from northeastern Sonora to central Michoacán, but most abundant in Durango (hence the scientific name), with an outlier in the Sierra de Manantlán, Jalisco. Forming pure stands or mixed with other pines, oaks, and even other conifers at its upper and lower elevational ranges; (1,400–)1,800–2,500(–3,000) m. Zone 8. Synonym: *P. martinezii* E. Larsen.

With a most common frequency of six needles and occasionally up to eight needles, Durango pine has the most needles per bundle of any pine on average. Some of the variation in needle number is geographic, following the common pattern among pines with variable needle numbers of fewer needles northward. This is one reason why the relationship of this species to Arizona pine (*Pinus arizonica*), a somewhat more northerly species with four or five needles in a bundle, deserves closer study. The affinities of Durango pine are in doubt, with some authors recognizing a close relationship with ponderosa pine (*P. ponderosa*) and its close relatives in subsection *Ponderosae* (like *Pinus arizonica*), while others favor an alliance with the completely unrelated Mexican closed-cone pines in subsection *Australes*. It is treated as a member of subsection *Ponderosae* here, both because of the structure of the bark, foliage, and seed cones and because of similarities in DNA. Because of its excellent form, high-quality wood, and originally dense, pure stands, this species is heavily exploited for timber and the best stands are severely depleted.

Pinus echinata P. Miller
SHORTLEAF PINE
Pinus subg. *Pinus* sect. *Trifoliae* subsect. *Australes*
Tree to 35(–42) m tall, with trunk to 0.9(–1.2) m in diameter. Bark reddish brown to grayish brown with rusty highlights, broken up into flat, rectangular, scaly plates by broad, dark furrows. Crown egg- to dome-shaped, deep, with numerous slender, horizontal branches densely clothed with foliage at the tips. Twigs slender, reddish brown, often thinly waxy, hairless, rough with the bases of scale leaves. Buds 0.5–1 cm long, resinous. Needles in bundles of two (or three), each needle (4–)7–11(–13) cm long, flexible but straight or gently twisted, lasting 3–5 years, dark yellowish green to bluish green. Individual needles with thin, scattered lines of stomates on both the inner and outer faces, and one to four resin canals at the corners and in between, midway between the needle surface and the two-stranded midvein. Sheath 0.7–1.5 cm long, weathering to 0.5–1 cm long and persisting and falling with the bundle. Pollen cones 15–30 mm long, yellowish green, sometimes with a blush of purple. Seed cones 4–7 cm long, egg-shaped, with 75–100(–120) seed scales, green before maturity, ripening reddish brown, opening widely to release the seeds and then persisting several years before falling with the short (to 1 cm) stalk. Seed scales paddle-shaped, the exposed face horizontally diamond-shaped, low, crossed by a ridge topped by a large umbo bearing a down-curved, sharp spine (hence the scientific name, Latin for "prickly"). Seed body 6–7 mm long, the firmly attached wing another 12–18 mm longer. Southeastern United States, from southernmost New York, central Pennsylvania, and southern Ohio south to northern Florida and west to eastern Oklahoma and eastern Texas with a gap in the Mississippi River valley. Forming pure stands or mixed with other pines and hardwoods on dry, upland sites; 200–610 m. Zone 8.

Shortleaf pine has the broadest geographic range among the "southern pines," species of subsection *Australes* from the southeastern United States. It is an abundant tree, in part because of its ability to invade abandoned agricultural land (along with the loblolly pine, *Pinus taeda*) and survive the fires that frequently sweep through its well-drained habitats. It is ultimately replaced by oaks and other hardwoods if enough time passes after a severe fire. It produces an excellent timber but is greatly exceeded in growth rate by loblolly pine and slash pine (*P. elliottii*) and so is not a major plantation species. It has been crossed successfully with most of the other southern pines, including loblolly, slash, pitch (*P. rigida*), and table mountain (*P. pungens*) pines, and with longleaf pine (*P. palustris*) with considerable difficulty. Sometimes it hybridizes naturally with loblolly and pitch pines and perhaps with some of the others. The largest known individual in combined dimensions when measured in 1980 was 42 m tall, 1.1 m in diameter, with a spread of 23 m. When it died, the next largest tree was a full 10 m shorter.

Pinus edulis Engelmann
COLORADO PINYON, TWO-LEAF PINYON
Pinus subg. *Strobus* sect. *Parrya* subsect. *Cembroides*
Tree to 10(–20) m tall, with trunk to 0.6(–1.7) m in diameter, but often dividing into several trunks at or near the base. Bark light brown, becoming shallowly furrowed between irregularly shaped and uneven small plates. Crown egg-shaped when young, progressively broadening with age, with numerous, upwardly angled branches from the base moderately clothed with foliage. Twigs hairless, light reddish brown at first, becoming light gray. Buds 5–10 mm long, resinous. Needles in bundles of (one or) two (or three), each needle 2–4.5(–6) cm long, gently curved, stiff, and sticking together, lasting 4–9 years, bright light green to bluish green. Individual needles with lines of stomates on both the inner and outer faces, an undivided midvein, and two resin canals touching the epidermis of the outer face about midway between the corners and the midline. Sheath 5–7 mm long, curling back and soon shed. Pollen cones 7–10 mm long, yellowish brown. Seed cones (3.5–)4–5 cm long, broadly egg-shaped to nearly spherical, with 15–40 seed scales, green before maturity, ripening yellowish brown, opening widely to release the seeds and then falling, unstalked or very short stalked. Seed scales oblong, with a deep recess for the seeds and a thickened, pyramid-shaped exposed portion angling outward and tipped by a stiff diamond-shaped umbo. Seed body 10–15 mm long, the shell 0.5–0.6 mm thick, unwinged or with a short, undeveloped wing remaining attached to the seed scale. Southwestern United States, centered on the Four Corners area, from southwestern Wyoming, central Utah, and north-central Colorado to southeastern Arizona, southwestern New Mexico, and trans-Pecos Texas. Forming pure stands or mixed with junipers in pinyon-juniper woodland on dry slopes, hilltops, and tablelands; (900–)1,400–2,400(–3,200) m. Zone 7.

The edible pine nuts, or *piñones* (hence the scientific name), from this species have long been an important winter food for some peoples in the summer-rainfall deserts of the southwestern United States. Like the distantly related limber pine (*Pinus flexilis*) and southwestern white pine (*P. strobiformis*) of higher elevations in the same region, seeds of Colorado pinyon are apparently dispersed by Clark's nutcrackers and other seed-eating birds and mammals. Millions of these trees were cut for fuel and charcoal and to shore up the mine ceilings of late 19th century silver boom towns like Silver City, New Mexico. It is so similar to other pinyon pine species that the taxonomic assignment of many populations is in doubt, particularly in a band from eastern California to northwestern Chihuahua (Mexico). This uncertainty is enhanced by hybridization with singleleaf pinyon (*P. monophylla*) in its limited overlap in California and throughout northwestern and central Arizona. Reports of Colorado pinyon from northwestern Chihuahua are referable to border pinyon (*P. culminicola* var. *bicolor*), which sometimes has needles in bundles of two as well as three or four. The largest known individual in combined dimensions when measured in 1982 was 21 m tall, 1.7 m in diameter, with a spread of almost 16 m.

Pinus elliottii Engelmann

SLASH PINE

Pinus subg. *Pinus* sect. *Trifoliae* subsect. *Australes*

Tree to 35(–42) m tall, with trunk to 0.8(–1.1) m in diameter. Bark with small, scaly, reddish brown blocks separated by broad, shallow, dark gray furrows. Crown dome-shaped or tapering, deep, with numerous gently rising branches densely clothed with foliage only at the tips. Twigs orange-brown, coarse, hairless, rough with the bases of scale leaves. Buds 12–20 mm long, not resinous. Needles in bundles of two or three, each needle (10–)15–20(–30) cm long, stiff, slightly twisted, lasting 2(–3) years, yellowish green to bluish green. Individual needles with lines of stomates on both the inner and outer faces, and three (to nine) resin canals surrounding the two-stranded midvein midway between it and the needle surface. Sheath 12–20 cm long, weathering to 10–15 cm and persisting and falling with the bundle. Pollen cones 3–5(–8) cm long, purple. Seed cones (7–)9–15(–20) cm long, egg-shaped to cylindrical, with 100–130(–150) seed scales, green before maturity, ripening shiny, rich brown, opening widely to release the seeds and then falling the following year with the short stalk. Seed scales approximately square-sided, the exposed face narrowly horizontally diamond-shaped, fairly flat and crossed by a low ridge topped by a large, diamond-shaped umbo bearing a short, stout, straight prickle. Seed body 6–7 mm long, the firmly attached wing another 13–20 mm longer. Coastal plain of the southeastern United States from central South Carolina to southernmost Florida, southern Mississippi and easternmost Louisiana. Forming pure stands or mixed with other pines and hardwoods, particularly on seasonally flooded flats; 0–150 m. Zone 9. Two varieties.

Slash pine has the most restricted natural distribution among the four most economically important "southern" pines of subsection *Australes* native to the southeastern United States. Like shortleaf pine (*Pinus echinata*), longleaf pine (*P. palustris*), and loblolly pine (*P. taeda*), slash pine produces an excellent timber and abundant naval stores (oleoresin), but its most important use in its native range is for veneer for plywood. It is increasingly the most common of the four in international plantation culture in tropical to warm temperate lands because of its superior growth rate. It crosses fairly readily with each of the other three species, and the resulting hybrids are also of some importance in plantation culture. It crosses with difficulty with the co-occurring sand pine (*P. clausa*) and more northerly pitch pine (*P. rigida*). It is generally considered to be most closely related to Caribbean pine (*P. caribaea*) of the Bahamas, Cuba, and Central America, with which it is sometimes combined, and it hybridizes naturally with that species when the two are grown together. Trees of southern Florida, distinguished as a variety, are somewhat intermediate between the northern variety of slash pine and Caribbean pine in morphology and oleoresin chemistry, and resemble longleaf pine in having a fire-resistant seedling "grass" stage. The largest known individual in combined dimensions when measured in 2001, a northern tree, was 39.3 m tall, 1.2 m in diameter, with a spread of 18.0 m, while the largest known tree of the southern variety is only about half as tall. The species name honors Stephen Elliott (1771–1830), who first distinguished the tree as a variety of loblolly pine in his *Sketch of the Botany of South Carolina and Georgia.*

Pinus elliottii var. *densa* E. Little & K. Dorman

Seedlings with a "grass" stage lacking extension growth. Needles in bundles of two (or three), each needle with up to nine resin canals. Open seed cone with a rounded base. Southern half of Florida. Synonyms: *Pinus densa* (E. Little & K. Dorman) de Laubenfels & Silba, *P. densa* var. *austrokeysensis* Silba.

Pinus elliottii Engelmann var. *elliottii*

Seedlings with normal growth. Needles in bundles of (two or) three, each needle with up to five resin canals. Open seed cone with a flat base. Central Florida northward.

Pinus engelmannii Carrière

APACHE PINE, PINO REAL (SPANISH)

Pinus subg. *Pinus* sect. *Trifoliae* subsect. *Ponderosae*

Tree to 25(–35) m tall, with trunk to 0.8(–1) m in diameter. Bark brown, with narrow, elongate, scaly, flat plates divided by shallow furrows. Crown deeply domed, open, with fairly thick, horizontal branches densely clothed with foliage at the ends. Twigs grayish brown, very stout (1–2 cm through), very rough with the bases of scale leaves, hairless. Buds 15–25(–30) mm long, variably resinous. Needles in bundles of three (to five), each needle (18–)25–35(–45) cm long, stiff but drooping because of their length, lasting 2(–3) years, yellowish green to dull green or even slightly bluish green. Individual needles with numerous narrow but conspicuous lines of stomates on all three faces, and (2–)3–6(–14) resin canals at the corners and in between midway between the surface and the two-stranded midvein. Sheath 30–45 mm long at first, the lower 15–25 mm persisting and falling with the bundle. Pollen cones 20–40 mm long, yellowish brown. Seed cones (8–)10–15(–18) cm long, egg-shaped with a flat to conical base, a little asymmetrical, with 75–150 seed scales, green before maturity, ripening light brown to light yellowish brown, opening widely to release the seeds and then falling, leaving a few basal scales behind on the stout, usually extremely short (but up to 1.5 cm) stalk. Seed scales paddle-shaped, the exposed face projecting in a pyramid topped by an umbo with a sharp, stout spine. Seed body 5–8 mm long, the firmly attached or easily detachable wing another 18–25 mm longer. Southeastern Arizona and southwestern New Mexico through the Sierra Madre Occidental to southern Zacatecas, with outliers in the Sierra Madre Oriental in Coahuila and Tamaulipas (Mexico). Mixed with other species in open pine-oak woodlands on varied dry sites in the mountains; (1,200–)1,500–2,500(–3,000) m. Zone 8.

Apache pine is one of a handful of Mexican pines whose distributions include the mountain ranges of extreme southern Arizona and New Mexico. It is a close relative of ponderosa pine (*Pinus ponderosa*), with which it is fairly intercrossable. It crosses even more readily with Arizona pine (*P. arizonica*) and Montezuma pine (*P. montezumae*). It may be distinguished readily from ponderosa pine by its darker bark, much thicker twigs, and much longer needles. Both Arizona pine and Montezuma pine have five shorter needles in the bundles. As with most of the species making up the vast pine forests of the Sierra Madre ranges of Mexico, and in contrast to the species of the Caribbean lowlands, little work has been done on the ecology, genetics, and relationships of Apache pine. The scientific name honors George Engelmann (1809–1884), botanist at the Missouri Botanical Garden in St. Louis, who first described the species under a name that later proved to be unavailable. The largest known individual in the United States in combined dimensions when measured in 1998 was 32.9 m tall, 1.0 m in diameter, with a spread of 13.4 m.

Mature Apache pine (*Pinus engelmannii*) in nature.

Pinus flexilis E. James

LIMBER PINE

Pinus subg. *Strobus* sect. *Quinquefoliae* subsect. *Strobus*

Tree to 20(–26) m tall or dwarfed at the alpine tree line. Trunk often crooked, to 1.5(–2.2) m in diameter. Bark grayish white and smooth on young trees, becoming grayish brown and finally breaking up somewhat into shallow, interlaced ridges at the base of large trunks. Crown fairly open, even in youth, conical at first and then broadening and becoming rounded, with coarse, upwardly angled branches bearing dense tufts of foliage. Twigs flexible (hence the common and scientific names), pale reddish brown, usually transiently minutely hairy. Buds 8–10 mm long, resinous. Needles in bundles of five, each needle (3–)3.5–6 (–7.5) cm long, flexible but straight, lasting 5–6 years, dark green to bluish green with wax. Individual needles with fewer and less conspicuous lines of stomates on the outer face than on the inner faces, an undivided midvein, and two (or three) small resin canals touching the epidermis below the outer face. Sheath 10–15(–20) mm long, soon shed. Pollen cones about 15 mm long, pinkish yellow. Seed cones 7–11(–15) cm long, egg-shaped, with 40–60

Twig of limber pine (*Pinus flexilis*) with needles from 2 years separated by a bare region, which bore pollen cones, ×0.5.

seed scales, green before maturity, ripening pale yellowish brown, opening widely to release the seeds and then falling, unstalked or with a short stalk 3–6(–10) mm long. Seed scales pointedly egg-shaped, the exposed portion much thickened, straight, with a broad diamond-shaped umbo at the tip. Seed body 10–15(–18) mm long, plump, usually unwinged or with a short, ridgelike wing. Mountains of western North America, from southwestern Alberta to southern California, northern Arizona and northern New Mexico. Scattered singly or in groves among other subalpine conifers on rocky soils; (1,000–)1,500–3,650 m. Zone 2.

Although not as long-lived as its more famous neighbor, Great Basin bristlecone pine (*Pinus longaeva*), limber pine reaches a very respectable age, with trees known to be 1,700–2,000 years old in the Toiyabe Range in Nevada. Even with this longevity, the trees remain relatively small, and the largest known individual in the United States in combined dimensions when measured in 1988 was 17.7 m tall, 2.2 m in diameter, with a spread of 14.0 m. Despite its natural habitat in arid mountains, limber pine has had wide success in cultivation in moist, cool temperate lowlands and a number of cultivars have been developed. There is some confusion with cultivars derived from the closely related southwestern white pine (*P. strobiformis*) or from hybrids between the two species found where they overlap in northern New Mexico. This confusion extends to native stands as well, and *P. strobiformis* was often considered a variety of *P. flexilis*. They are readily distinguishable by foliage and seed cones, however, even though *P. flexilis* is most similar to *P. strobiformis* in the zone of overlap. Here, bluish green foliage like that of *P. strobiformis* is most common and most intense compared to plain green or even yellowish green needles farther north. Since bluer foliage is often favored in garden conifers, many of the cultivars come from in and near the overlap zone, leading to the confusion in their identity. In this region, *P. strobiformis* at its northern limit typically occurs at lower elevation than *P. flexilis* at its southern limit, so actual first-generation hybrids are rare. Limber pine was also crossed successfully with eastern white pine (*P. strobus*), western white pine (*P. monticola*), Himalayan white pine (*P. wallichiana*), possibly with Mexican white pine (*P. ayacahuite*), and with extremely limited success with whitebark pine (*P. albicaulis*). Some of these hybrids hold promise as ornamental and timber trees. Like those of the Chinese white pine (*P. armandii*), the seed cone and seed traits of limber pine fall between those of typical white pines, like *P. strobus, P. ayacahuite, P. monticola,* and *P. wallichiana,* and those of the alpine stone pines, like *P. cembra* (Swiss stone pine) and *P. albicaulis.* On this basis, *P. flexilis, P. strobiformis,* and *P. armandii* are sometimes placed in their own subsection rather than being grouped with the typical white pines in subsection *Strobus.* Most available evidence, including DNA studies and the pattern of crossability just mentioned, favors retention with the other white pines rather than a separate status, especially since limber pine is not crossable with *P. armandii.*

Pinus gerardiana N. Wallich ex D. Don

CHILGOZA PINE, CHILGOZA (HINDI)

Pinus subg. *Strobus* sect. *Quinquefoliae* subsect. *Gerardianae*

Tree to 18(–25) m tall, with trunk to 1 m in diameter. Bark with multicolored scales flaking in irregular patches and becoming whiter with age. Crown dense, conical in youth, broadening and becoming rounded with age, with numerous long, sinuous, upwardly arched branches openly clothed with foliage near the tips. Twigs grayish green to yellowish green, hairless. Buds about 6–10 mm long, slightly resinous. Needles in bundles of three, each needle (5–)6–10(–12) cm long, stiff and straight, lasting 2(–3) years, dark green. Individual needles with lines of stomates on all three faces, an undivided midvein, and four to seven large resin canals scattered around the periphery next to the epidermis. Sheath 1–2 cm long, not shed until the second year. Pollen cones 7.5–15 mm long, yellowish brown. Seed cones (9–)15–20(–23) cm long, elongately egg-shaped, with 75–90 seed scales, green before maturity, ripening light reddish brown, opening widely to release the seeds and then falling, short-stalked. Seed scales paddle-shaped, the exposed portion markedly thickened with a variable length, prominent hooked projection curving down from the outer face and ending in a prickly umbo. Seed body like a giant, black grain of long-grained rice, very thin-shelled, 20–25 mm long, the readily detachable wing 4–5 mm longer and often

Twig of chilgoza pine (*Pinus gerardiana*) with several years of growth, ×0.5.

remaining stuck to the seed scale. Western Himalaya, from southeastern Afghanistan to western Kashmir and eastern Himachal Pradesh (India). Mixed with other conifers and hardwoods to form open woodlands in dryish flat valleys within the mountains; (1,800–)2,000–3,000(–3,350) m. Zone 8.

Chilgoza pine closely resembles its relative, lacebark pine (*Pinus bungeana*), in bark and foliage but has larger, differently proportioned seed cones and seeds. The seeds are edible and highly prized, providing revenue to collectors who sell them in lowland markets. As the timber is not of very high quality, the trees are preserved for seed production. They are uncommon, however, being confined to semiarid interior valleys outside of the monsoon belt. Most seed harvesting is conducted in native stands with few plantations for seed production. Chilgoza pine is much less hardy than lacebark pine and is not much cultivated outside of its native region. The lack of hardiness in many temperate localities may also be due to inappropriate distribution of precipitation. Perhaps localities like Yosemite Valley in California, with deep snow cover in winter but dry summers, would show whether the tenderness of the species was really due to sensitivity to frost. The species name honors Captain Alexander Gerard (1792–1839), surveyor with the Bengal Native Infantry who encountered the tree during explorations in the Koonawur district in the Himalaya in 1821.

Pinus glabra T. Walter

SPRUCE PINE

Pinus subg. *Pinus* sect. *Trifoliae* subsect. *Australes*

Tree to 30(–40) m tall, with trunk to 0.8(–1) m in diameter. Bark gray, long remaining smooth and flaky but darkening and finally breaking up into blocky ridges divided by shallow furrows. Crown deep but sparse, gradually tapering to a rounded or flat top, with a few horizontal branches thinly clothed with foliage near the ends. Twigs purplish brown, sometimes a little waxy, hairless (hence the scientific name, Latin for "hairless," although this is true for most southern US pines), the bases of the scale leaves smooth. Buds 6–12 mm long, slightly resinous. Needles in bundles of two, each needle (4–)5–8(–10) cm long, flexible but not drooping, twisted, lasting 2–3 years, dark green. Individual needles with inconspicuous lines of stomates on both faces, and two (or three) resin canals at the corners (and in between) midway between the needle surface and the two-stranded midvein. Sheath 5–10 mm long, weathering to 3–7 mm and persisting and falling with the bundle. Pollen cones (5–)8–15 mm long, purplish tan. Seed cones (3–)4–7(–10) cm long, egg-shaped, with 60–90 seed scales, green before maturity, ripening reddish brown, opening widely to release the seeds and then persisting 3–4 years before falling with the short stalk (to 1 cm long). Seed scales broadly wedge-shaped, the exposed face horizontally diamond-shaped, fairly flat with a large, flat umbo tipped by a small, fragile prickle. Seed body 5–6 mm long, the firmly attached wing another 9–15 mm longer. Extreme southeastern United States, from southeastern South Carolina to northern Florida, west to southern Mississippi and eastern Louisiana. Scattered as single trees among hardwoods on moist, well-drained sites; 0–150 m. Zone 9.

The ecology of spruce pine is not like that of other "southern" pines, species of subsection *Australes* native to the southeastern United States. All the others are more or less tolerant of fires and depend on them for reproduction and the establishment of solid stands. In contrast, spruce pine, with its thin bark, inability to sprout after fires, and prompt seed release upon cone maturation, is eliminated from stands by fires. Instead, it is shade tolerant as a seedling and becomes a part of mature hardwood forests, most often on the upper terraces of floodplains that do not experience prolonged flooding. Its scattered occurrence, slow growth, and moderate wood quality keep it from being a major commercial species, unlike its closely related neighbors, shortleaf pine (*Pinus echinata*), longleaf pine (*P. palustris*), slash pine (*P. elliottii*), and loblolly pine (*P. taeda*). Nonetheless, it is under some threat through conversion of natural forests on many kinds of sites to plantations of these more commercially productive species. Because of its lack of commercial importance, there is relatively little study of this species compared to other southern pines. It crosses successfully with shortleaf pine, but no natural hybrids involving it are known. The largest known individual in combined dimensions when measured in 2003 was 47.6 m tall, 1.2 m in diameter, with a spread of 21.4.

Pinus greggii Engelmann ex Parlatore

GREGG PINE

Pinus subg. *Pinus* sect. *Trifoliae* subsect. *Australes*

Tree to 25 m tall, with trunk to 0.8 m in diameter. Bark grayish brown, long remaining smooth, becoming thick at the base of large trees and broadly ridged between deep furrows. Crown dome-shaped to cylindrical, open, with numerous horizontal branches well clothed with foliage. Twigs reddish brown to grayish brown, hairless, smooth, with shallow grooves between the bases of the scale leaves. Buds 8–15 mm long, not conspicuously resinous. Needles in bundles of three, each needle (7–)10–15 cm long, stiff and straight, lasting 3–4 years, shiny light green. Individual needles with several lines of stomates on all three faces, and two to four (to six) resin canals at the corners surrounding the two-stranded midvein midway to the needle surface. Sheath 8–10 mm long, weathering to (3–)5–8 mm or occasionally shed entirely, but usually persisting and falling with the bundle. Pollen cones 15–20 mm, yellowish brown. Seed cones in circles of three to six (to eight) around the twig, (6–)10–13(–15) cm long, egg-shaped to oblong, asymmetrical, especially at the base, with 60–120 seed scales, green before maturity, ripening shiny yellowish brown, remaining closed and attached to the branch by a very short, stout stalk for many years and gradually opening and releasing the seeds over time, beginning at the middle and often leaving the bottom scales closed like a handle. Seed scales broadly wedge-shaped, the exposed face diamond-shaped and flat with a small, flat umbo bearing a tiny, fragile prickle. Seed body 5–8 mm long, the clasping wing another 14–18(–20) mm longer. Sierra Madre Oriental of Mexico, from southeastern Coahuila and central Nuevo León to northern Puebla. Mixed with other trees in varied forest types; 1,300–2,700(–3,000) m. Zone 9.

Gregg pine is closely related to Mesoamerican closed-cone pine (*Pinus oocarpa*) and Mexican weeping pine (*P. patula*) and essentially replaces those species in northeastern Mexico. It is highly interfertile with *P. patula* in artificial crosses but crosses with *P. oocarpa* have not been attempted. It also shows some similarities to the Californian closed-cone pines, knobcone pine (*P. attenuata*) and Monterey pine (*P. radiata*), and early classifications placed all these species together in a single subsection of the genus. It has become clear that all these species belong in an expanded subsection *Australes* together with many open-coned pines. It is not clear, however, whether the various closed-cone species are most closely related to one another or whether they have been derived independently from open-coned ancestors. The species name honors Josiah Gregg (1806–1850), who collected the type specimen and died during a later collecting trip.

Pinus ×*hakkodensis* Makino

A natural hybrid from Japan between the shrubby, alpine Japanese mountain pine (*Pinus pumila*) and the subalpine tree Japanese white pine (*P. parviflora* var. *pentaphylla*). It shares numerous traits with both parents, shares others with one or the other parent, and is intermediate in a few others. The plants are generally shrubby like *P. pumila,* but the five needles are a little twisted, like those of *P. parviflora.* The resin canals resemble *P. pumila* in position, touching the epidermis of the outer face near the midvein rather than near the corners as in *P. parviflora,* but there are always two, rather than one or two as in *P. pumila,* or two or three as in *P. parviflora.* The seed cones are intermediate in length, at 4–6 cm, and the scales are thin and open at maturity in some cases, like those of *P. parviflora,* or thick and remain closed in others, like those of *P. pumila.* The seeds have short wings, shorter than the seed body, unlike either the wingless seeds of *P. pumila* or the seeds with wings longer than the seed body of *P. parviflora* var. *pentaphylla.* In this respect they resemble *P. parviflora* var. *parviflora,* which occurs way below them in the montane forest. DNA studies specifically addressing the situation suggest that the hybrids are produced when seed cones of *P. pumila* are pollinated by *P. parviflora* var. *pentaphylla,* which disperses its pollen a little earlier than *P. pumila* does, and so sometimes could reach early ovules of the alpine shrub.

Pinus halepensis P. Miller

ALEPPO PINE

Pinus subg. *Pinus* sect. *Pinus* subsect. *Pinaster*

Tree to 20(–40) m tall, often shrubby on harsh sites. Trunk to 1(–1.2) m in diameter, often dividing near the base. Bark reddish brown to silvery gray and long remaining smooth, finally becoming reddish brown, thick and deeply ridged and furrowed. Crown broadly conical to dome-shaped, flattening and opening up with age, with numerous upwardly angled to horizontal branches well clothed with foliage. Twigs reddish brown to grayish brown, hairless, smooth. Buds 7–12 mm long, not resinous. Needles in bundles of two (or three), each needle (3.5–)6–18(–23) cm long, slender and flexible to thick and stiff, lasting 2(–3) years, bright to dark green. Individual needles with inconspicuous, evenly spaced lines of stomates on both the inner and outer faces, and (1–)3–8(–11) large resin canals surrounding the two-stranded midvein but touching the outer surface of the needle. Sheath 7–13(–16) mm long, weathering to 3–9 mm and persisting and falling with the bundle. Pollen cones densely clustered, about 7–10 mm long, yellowish brown with a reddish flush. Seed cones (4.5–)6–10 (–12) cm long, conical to egg-shaped, with (60–)70–90 seed scales, green before maturity, ripening shiny reddish brown, opening only slowly when ripe and persisting several years before falling off the inconspicuous to pronounced thick stalk 3–22 mm long. Seed scales paddle-shaped, the exposed face diamond-shaped, crossed by a ridge, fairly flat to evidently bulging, the diamond-shaped umbo often a little indented, without a prickle. Seed body 5–9 mm long, the firmly attached wing another (12–) 15–24(–28) mm longer. Discontinuous throughout the Mediterranean and Black Sea basins and west to the western end of the Atlas Mountains in Morocco and east to Kurdistan in northern Iraq. Forming pure, open stands or mixed with other evergreens on dry slopes; 0–800(–1,150) m. Zone 8. Two subspecies.

Aleppo pine is the most widely distributed and abundant of the Mediterranean pines, making up as much as 50% of the forest stands in some regions. Because of the degree of variation that developed among the isolated populations, it is sometimes divided into as many as five separate species. There is a great deal of genetic variation across the range of the species in morphology, biochemical features, enzymes, and DNA, but this variation does not strongly support segregate species. Instead, most of the available evidence seems to favor recognizing just a single species with two subspecies that differ primarily in needle length and the way the seed cones are carried on the twigs. In places where the two subspecies grow together naturally or because one or both have been planted for forestry, they hybridize freely and completely intergrade. There are many studies of natural hybridization between the two subspecies, and hybrids with the related maritime pine (*Pinus pinaster*) have also been reported. Because of its drought tolerance, Aleppo pine is a major plantation species throughout its natural range, but especially in the eastern Mediterranean, which is drier than the west and has fewer additional pine species available. It has ornamental use but much less commercial value in other, moister Mediterranean climate regions like California and South Africa, where Monterey pine (*P. radiata*) and other species grow faster, larger, and with better wood quality. Both the species and common name honor the ancient northwestern Syrian city of Aleppo and surrounding province of Haleb, where forests of it once grew.

Pinus halepensis* subsp. *brutia (M. Tenore) Holmboe

CALABRIAN PINE

Smooth, scaly bark reddish brown. Twigs yellowish brown to reddish brown. Needles (6–)12–18(–23) cm long, dark green, generally thick and stiff. Sheath 9–16 mm long, weathering to 4–9 mm. Seed cones sticking out straight or forward on a stalk 3–11 mm long. Northeastern portion of the range of the species, from Crete, Cyprus, Lebanon, and northern Iraq, north to the northern shore of the Black Sea. Synonyms: *Pinus brutia* M. Tenore, *P. brutia* subsp. *eldarica* (J. Medwedew) Nahal, *P. brutia* var. *pendulifolia* Frankis, *P. brutia* subsp. *pityusa* (Steven) Nahal, *P. brutia* subsp. *stankewiczii* (Sukaczev) Nahal, *P. eldarica* J. Medwedew, *P. halepensis* var. *brutia* (M. Tenore) A. Henry, *P. halepensis* var. *eldarica* (J. Medwedew) Beissner, *P. halepensis* var. *pityusa* (Steven) G. Gordon, *P. pityusa* Steven, *P. pityusa* var. *stankewiczii* Sukaczev, *P. stankewiczii* (Sukaczev) Fomin.

Pinus halepensis P. Miller **subsp. *halepensis***

ALEPPO PINE

Smooth, scaly bark silvery gray. Twigs yellowish gray. Needles (3.5–)6–10(–15) cm long, light green, thin and flexible. Sheath 7–10 mm long, weathering to 3–7 mm. Seed cones angled downward on a stalk 10–22 mm long. Throughout the range of the species except the northeastern portion, adjoining or overlapping with subspecies *brutia* in western Greece and the northeastern end of the Mediterranean, from Lebanon to southeastern Turkey.

Pinus hartwegii Lindley

HARTWEG PINE, MESOAMERICAN SUBALPINE PINE

Pinus subg. *Pinus* sect. *Trifoliae* subsect. *Ponderosae*

Tree to 30 m tall, with trunk to 0.7(–1) m in diameter, shorter at higher elevations and shrubby at the alpine tree line. Bark grayish brown, thick, divided into irregular, scaly or rough plates of various sizes on different trees. Crown dense, conical at first, broadening and rounding with age, sometimes becoming almost spherical, with numerous gently rising to horizontal branches densely clothed with foliage at the tips. Twigs dark reddish brown with a thin coating of grayish wax at first, rough with the bases of scale leaves, hairless. Buds 2–3 cm long, not resinous. Needles in bundles of (three or) four or five (to seven), each needle (6–)10–17(–25) cm long, thick and stiff, microscopically toothed along the edges, lasting 2–3 years, light to dark green or sometimes a little waxy bluish green. Individual needles with numerous lines of stomates on all three faces, and (2–)4–8(–11) resin canals surrounding the two-stranded midvein midway between it and the leaf surface. Sheath 20–30(–40) mm long at first, weathering to 10–20 mm and persisting and falling with the bundle. Pollen cones 12–25 mm long, reddish brown. Seed cones (6–)8–12(–15) cm long, egg-shaped with a flat base, often slightly asymmetrical, with 130–220 seed scales, purplish green before maturity, ripening dark brown to black, opening widely to release the seeds and persisting a while before falling off the very short stalk to 10 mm and leaving a few basal scales behind. Seed scales square-sided with a bluntly triangular tip, the exposed face horizontally diamond-shaped, thin and low but crossed by a horizontal keel topped by a diamond-shaped umbo often bearing a stout prickle. Seed body 5–7 mm long, the clasping wing another (12–)15–20(–23) mm longer. Mountains of western Honduras, southern Guatemala, and northward discontinuously through southern and central Mexico to southern Chihuahua, southern Coahuila, and central Nuevo León. Forming pure open forests and thickets or mixed with other subalpine and upper montane species; (2,300–)2,700–4,000(–4,300) m. Zone 8. Synonyms: *P. donnell-smithii* M. T. Masters, *P. hartwegii* var. *rudis* (Endlicher) Silba, *P. montezumae* var. *hartwegii* (Lindley) Engelmann, *P. montezumae* var. *rudis* (Endlicher) G. R. Shaw, *P. rudis* Endlicher.

Hartweg pine is the highest growing of the Mesoamerican hard pines, clothing the upper reaches of the many volcanoes of the area with progressively shorter forests up to the alpine tree line. There is modest variation in the species in stature, needle length and number, and seed cone color. This provoked the recognition of three species by some authors instead of the one accepted here. A few careful studies show that this variation is so chaotic that there is little justification for subdividing the species even into varieties. The only consistent variation is the occurrence of predominantly three needles per bundle on the mountains surrounding the Valley of Mexico rather than the four or five needles that predominate elsewhere. These populations are otherwise typical of the species, and it seems unnecessary to give them formal taxonomic recognition. On the other hand, the obvious close relationship of *Pinus hartwegii* to Montezuma pine (*P. montezumae*) and Michoacán pine (*P. devoniana*) led other authors to consider it a variety (or two varieties) of *P. montezumae.* It is actually more distinct from that species and from *P. devoniana* than they are from each other. It is readily distinguished by its shorter needles and dark brown to black seed cones, even when it grows with its close relatives in the lower portion of its elevational range. It may, however, hybridize with either or both these species, a possibility not fully explored. The species name honors Karl Hartweg (1812–1871), the German botanist who collected the type specimen during his extensive travels in Mexico.

Pinus heldreichii H. Christ

BOSNIAN PINE, PINO LORICATO (ITALIAN)

Pinus subg. *Pinus* sect. *Pinus* subsect. *Pinaster*

Tree to 20(–30) m tall, with trunk to 1(–2) m in diameter. Bark long remaining light gray, smooth, and scaly, finally breaking up into small, scaly, angular plates. Crown conical, becoming more cylindrical with age, dense to open, with numerous slender, upwardly angled branches, densely clothed with foliage. Twigs gray, hairless, with obvious grooves between the bases of the scale leaves. Buds 12–15 mm long, not resinous. Needles in bundles of two, each needle 6–8.5(–10) cm long, thick and stiff, prickly at the tip, lasting 5–6 years, shiny dark green. Individual needles with inconspicuous lines of stomates on both faces, and 2–11 resin canals around the two-stranded midvein midway between it and the needle surface. Sheath 12–20 mm long, persisting and falling with the bundle. Pollen cones densely clustered, about 1 cm long, bright yellow. Seed cones (5–)7–8 cm long, egg-shaped, with (50–)80–100(–130) seed scales, purple or green before maturity, ripening light brown, opening widely to release the seeds and then falling with the extremely short stalk. Seed scales paddle-shaped, the exposed face diamond-shaped, slightly domed, with a horizontal ridge topped by a small umbo bearing a short, dull point or curved, sharp prickle. Seed body 6–7 mm long, the clasping wing another 15–25(–30) mm longer. Mountains of the northern Balkan Peninsula, particularly on the western side, and a few isolated localities in southern Italy. Forming pure stands or mixed with other trees in montane forests up to the alpine timberline, on various substrates, including serpentine; 1,300–2,300(–2,650) m.

Twig of Bosnian pine (*Pinus heldreichii*) with 2 years of growth, ×0.5.

Old mature Bosnian pine (*Pinus heldreichii*) in nature.

Zone 6. Synonyms: *P. heldreichii* var. *leucodermis* (F. Antoine) Markgraf ex Fitschen, *P. leucodermis* F. Antoine.

Although Bosnian pine is a rare and very local endemic in the Balkan region, its handsome bark and foliage have made it an important ornamental elsewhere in Europe. Available evidence concerning its relationships is contradictory. It is similar enough in morphology to its neighbor, Austrian pine (*Pinus nigra,* of subsection *Pinus*) to have once been considered a variety of that species, and the two can hybridize successfully. It has also been reported to have crossed successfully with red pine (*P. resinosa*), a North American member of subsection *Pinus.* DNA studies, however, strongly support a relationship with the other Mediterranean pines in subsection *Pinaster.* This problem merits further study, as does the appropriate taxonomic status of morphological variants within the species. Traditionally, Bosnian pine was subdivided into two varieties or even species, based on purple immature seed cones and umbos with curved prickles versus green immature seed cones and umbos with a straight point, the latter mostly confined to the vicinity of Mount Olympus in Greece. The forms intergrade, and in any event, the differences hardly seem to merit formal taxonomic recognition. Furthermore, studies of genetic variation in DNA reveal an absence of detectable variation, implying that the whole species was reduced to a single tiny population in the relatively recent past, much as postulated for red pine (*Pinus resinosa*) in North America. The species name honors Theodor von Heldreich (1822–1902), a German-born botanist who spent most of his life in Greece and was director of the botanical garden and natural history museum in Athens.

Pinus herrerae M. Martínez

HERRERA PINE, PINO CHINO (SPANISH)

Pinus subg. *Pinus* sect. *Trifoliae* subsect. *Australes*

Tree to 35 m tall, with trunk to 0.8(–1.0) m in diameter. Bark dark reddish brown, thick, very rough, with scaly but rough-topped, platelike ridges separated by shallow grooves. Crown approaching spherical, open with slender, horizontal branches well clothed with foliage. Twigs yellowish brown to reddish brown, sometimes with a light dusting of wax, the bases of the scale leaves flaking off after the first year, hairless. Buds 8–15 mm long, not resinous. Needles in bundles of (two or) three (to five), each needle (10–) 12–18(–20) cm long, soft and droopy, lasting (2–)3 years, light green to yellowish green. Individual needles with several lines of stomates on the outer face, a few on the inner faces, and (one or) two or three (or four) resin canals surrounding and right next to the two-stranded midvein. Sheath 14–20 mm long, weathering to 8–15 mm and persisting and falling with the bundle. Pollen cones densely clustered, 10–18 mm long, yellowish brown. Seed cones (2–)3–5 cm long, egg-shaped to almost spherical, with 35–85 seed scales, green before maturity, ripening dull light brown, opening widely to release the seeds and then falling with the 5–10(–15) mm long, backwardly curved stalk. Seed scales paddle-shaped, the exposed face irregularly crescent-shaped, low, crossed by a horizontal keel and topped by a diamond-shaped umbo bearing a slender, fragile prickle. Seed body (2.5–)3–4 (–5.5) mm long, the clasping wing another 5–8(–10) mm longer. Mountains of western Mexico from southern Guerrero and central Michoacán to western Chihuahua. Forming small groves among many other species in subtropical to warm temperate pine-oak forest; (1,000–)1,200–2,400(–2,600) m. Zone 9.

Although some authors suggest a relationship with lodgepole pine (*Pinus contorta*), the majority of authors and evidence link Herrera pine with other small-coned Mexican pines, such as Ocote pine (*P. teocote*) or Lawson pine (*P. lawsonii*), sometimes even treating it as a variety of the former. Like these relatives, it is a good source of resin and is sometimes heavily exploited for this and for its timber. The species name honors Mexican biologist Alfonso Herrera, a colleague of Martínez at the National Autonomous University in Mexico City.

Pinus ×holfordiana A. B. Jackson

A spontaneous garden hybrid that has also been made deliberately between two widely separated tropical mountain species of typical white pines (subsection *Strobus*), Mexican white pine (*Pinus ayacahuite*) and Himalayan white pine (*P. wallichiana*). The original trees were derived from seed of Mexican white pine. The reciprocal hybrid is very difficult to make in deliberate attempts to cross the species. It is a very fast growing tree, often extending 50–60 cm each year, with a more open habit than either parent. The general appearance is more similar to Himalayan white pine, but many of the detailed features are closer to those of Mexican white pine. It has the hairy twigs of *P. ayacahuite* combined with the five longer needles of *P. wallichiana.* The bark is orange-brown, much brighter than the dark bark of Mexican white pine. Seed cones vary among individual trees. Although they are usually broader than those of Himalayan white pine, they are sometimes as slender but then lack its "banana" shape, having a sharper tip. The seed parent of the original hybrids was a tree of *P. ayacahuite* var. *veitchii* and so has the seed scales strongly angled back. This trait was completely lacking in the hybrid offspring, which have straight scales. The seed body, 7–10 mm long, is at the upper size range of *P. wallichiana* and the lower range of *P. ayacahuite.* The seed wing is about the same length in both parents and hybrids, but those of Himalayan white pine and the hybrids are narrower than those of Mexican white pine and thus proportionately longer.

Pinus ×hunnewellii Alb. G. Johnson

A spontaneous garden hybrid also made deliberately between two naturally geographically widely separated typical white pines (subsection *Strobus*), *Pinus parviflora* × *P. strobus.* It is similar to both parents in their shared features and intermediate or generally favoring one or the other in the features by which they differ. It is faster growing than either parent and has a loose, open crown whose irregularity is accentuated by susceptibility to white pine weevil, which kills the leader. The irregularity also stems, in part, from the pronounced hybrid vigor of the trees, resulting in forking of lead shoots and even of seed cones. Furthermore, lower branches often turn up at the ends and form new leaders. The hybrids have reduced fertility compared to their parents, however, with 77% pollen fertility and only about 0.5% seed fertility. The bark remains scaly like that of Japanese white pine (*P. parviflora*) rather than ridged like that of eastern white pine (*P. strobus*). The twigs are minutely hairy, more persistently so than either parent. The five needles, 6.5–8.5 cm long, fall within the midrange of *P. strobus* and are longer than the normal range of *P. parviflora.* Like those of *P. parviflora,* however, they are twisted, curved, and more bluish green than those of *P. strobus.* Pollen cones are pink, like those of Japanese white pine, and are grouped in an elongate cluster, unlike the tight cluster of yellow pollen cones in eastern white pine. Seed cones are short-stalked with a flat base as in *P. parviflora* rather than long-stalked and tapered as in *P. strobus* but are the same length as those of eastern white pine and have more widely spreading seed scales than either parent. The seed body is intermediate in length, at about 7–9 mm, smaller than that of *P. parviflora* and longer than that of *P. strobus,* and has a long wing like that of the latter species, about twice as long as the seed body rather than just as long or less as in Japanese white pine. The first trees found had *P. strobus* as the seed parent. The reciprocal hybrid is rather difficult to make in deliberate attempts to cross the parent species, perhaps because of the difference in seed size.

Vigorous *Pinus ×hunnewellii,* the garden hybrid between Japanese white pine (*P. parviflora*) and eastern white pine (*P. strobus*).

Pinus jaliscana Pérez de la Rosa

JALISCO PINE

Pinus subg. *Pinus* sect. *Trifoliae* subsect. *Australes*

Tree to 30(–35) m tall, with trunk to 0.8(–1) m in diameter. Bark reddish brown to grayish brown, with long, scaly, flat plates separated by shallow dark furrows. Crown dome-shaped to spherical, open, with slender branches passing from rising through horizontal to gently downswept and sparsely clothed with foliage. Twigs reddish brown, the bases of the scale leaves shedding after the first year, hairless. Buds 7–15 mm long, not resinous. Needles in bundles of (three to) five, each needle 12–18(–22) cm long, fine but generally straight, lasting 2–3 years, light green to yellowish green. Individual needles with a few inconspicuous lines of stomates on all three faces, and (one or) two to four (or five) large resin canals usually spanning the whole space between the leaf surface and the two-stranded midvein. Sheath 10–15 mm long, wearing to 8–10 mm and persisting and falling with the bundle. Pollen cones 10–20 mm long, yellowish brown with a reddish blush. Seed cones (4–)6–8(–10) cm long, egg-shaped, sometimes slightly asymmetrical, with 75–100 seed scales, green before maturity, ripening shiny yellowish brown, opening widely to release the seeds, except for the basal scales, and then falling with the stout, curved stalk, (7–)10–15 mm long. Seed scales paddle-shaped, contracted beneath the exposed face, which is shallowly pyramidal, except for a marked protrusion on scales at the bottom on the side away from the twig, with a small, low umbo bearing a slender, fragile prickle. Seed body 3–6 mm long, the clasping wing another (10–)12–16(–20) mm longer. Mountains of western Jalisco, Mexico (hence the common and scientific names). Mixed with other pines and oaks on deep, well-drained soils; 800–1,650 m. Zone 9. Synonyms: *P. patula* var. *jaliscana* (Pérez de la Rosa) Silba, *P. macvaughii* Carvajal.

Jalisco pine is a local species that is presumed to be closely related to the more widespread Herrera pine (*Pinus herrerae*), of the mountains of western Mexico, and Mexican weeping pine (*P. patula*), of the mountains of eastern Mexico. It also displays similarities to a number of other Mexican pines in subsection *Australes,* so its affinities are in need of further study. It may be distinguished from neighboring *P. herrerae,* which has the resin canals in the needles touching the midvein (but not the leaf surface) and basal scales of the seed cones opening widely.

Pinus jeffreyi Greville & J. Balfour

JEFFREY PINE

Plate 42

Pinus subg. *Pinus* sect. *Trifoliae* subsect. *Ponderosae*

Tree to 40(–60) m tall, with trunk to 1.2(–2.5) m in diameter. Bark dark reddish brown with light brown highlights, divided into narrow vertical plates by slightly narrower furrows. Crown conical to cylindrical with a rounded top, with numerous short, horizontal to down-arched stout branches densely clothed with foliage near the tips. Twigs thick, purplish brown and with a thin waxy coating, rough with the bases of scale leaves, hairless. Buds 1.5–2.5(–3) cm long, not resinous. Needles in bundles of (two or) three, each needle (12–)15–25(–28) cm long, straight, slightly twisted, and a little stiff, lasting (2–)4–5(–7) years, grayish green to bluish green, sometimes a little yellowish. Individual needles with about twice as many lines of stomates on the outer face as on each of the inner faces, and two to five resin canals at the corners and scattered around the two-stranded midvein deep inside the leaf tissue. Sheath 20–25(–30) mm long at first, the lower 10–15 mm persisting and falling with the needles. Pollen cones 20–35 mm long, yellowish brown. Seed cones (10–)13–20(–25) cm long, elongated egg-shaped, with 135–150 seed scales, purplish green before maturity, ripening light reddish brown to tan, opening widely to release the seeds and then falling, leaving the bottom scales behind still attached to the stubby, thick stalk. Seed scales angularly paddle-shaped, thin but stiff, the exposed face diamond-shaped with a small, central umbo ending in a slender, downwardly incurved prickle. Seed body 9–12 mm long, the easily detachable wing 18–25 mm longer. Mountains from southwestern Oregon through California to northern Baja California (Mexico). Most commonly mixed with other montane conifers in open or dense stands; (1,100–)1,500–2,500(–3,000) m. Zone 6. Synonym: *P. jeffreyi* var. *baja-californica* Silba.

Jeffrey pine replaces its close relative ponderosa pine (*Pinus ponderosa*) at higher elevations in the Sierra Nevada and other mountains of California. The two species hybridize with difficulty, producing about 3% good seed when artificially pollinated, but rare natural hybrids have been found. It also crosses at about the same level with Coulter pine (*P. coulteri*), both naturally and artificially, and is the main link between the bigcone pines, like Coulter pine, and more typical members of subsection *Ponderosae.* Like the bigcone pines, Jeffrey pine has straight-chain hydrocarbons in its turpentine, particularly heptane, which give it a vanilla or fruity sweet odor. Ponderosa pine lacks these compounds and has higher levels of two others, Δ^3-carene and limonene, that give it a more resinous smell. A very few seeds have also been produced in attempted crosses between Jeffrey pine and three other typical members of subsection *Ponderosae.* The largest known individual in combined dimensions when measured in 1984 was 60 m tall, 2.5 m in diameter, with a spread of 27.4 m. The species name honors John Jeffrey (1826–1854), the Scottish gardener who collected the type specimen for the Oregon Botanical Association in Edinburgh before disappearing in Arizona.

Pinus kesiya Royle ex G. Gordon

KHASI PINE, BUENGET PINE, DING-SE (KHASI), TINYU (BURMESE), KA XI SONG (CHINESE), PÈ:K SA:M BAÏ (LAOTIAN), THÔNG BA LÁ (VIETNAMESE), SALENG (PILIPINO)

Pinus subg. *Pinus* sect. *Pinus* subsect. *Pinus*

Tree to 30(–45) m tall, with trunk to 0.6(–1) m in diameter. Bark thick, reddish brown to grayish brown with reddish highlights, breaking up into interrupted, platelike ridges separated by narrow, deep furrows. Crown open, narrowly dome-shaped, flattening with age, with numerous short, contorted, horizontal branches thinly clothed with foliage near the tips. Twigs yellowish brown to rich, bright brown, hairless, rough with the bases of scale leaves until these peel away in the second or third year. Buds 15–20 mm long, not resinous. Needles in bundles of (two or) three (or four), each needle (10–)12–22(–27) cm long, very slender and flexible, lasting 2 years, light green to grayish green. Individual needles with lines of stomates on both the inner and outer faces and three to six resin canals touching the needle surface all around the two-stranded midvein. Sheath 8–18(–24) mm long, weathering to 5–10 mm and persisting and falling with the bundle. Pollen cones densely crowded, (15–)25–30 mm long, yellowish brown. Seed cones (4–)5–7.5(–10) cm long, egg-shaped, symmetrical or slightly asymmetrical, with 55–110 seed scales, green before maturity, ripening medium brown, opening widely to release the seeds and then persisting several years before falling with the short stalk to 1 cm long. Seed scales paddle-shaped, the exposed face horizontally diamond-shaped, distinctly projecting, crossed by a strong ridge topped by a large, low umbo bearing a fragile, blunt prickle. Seed body 5–6(–8) mm long, the easily detachable wing another 12–16(–20) mm longer. Southeastern Asia, from northeastern India and southwestern China—southeastern Xizang (Tibet) and western Yunnan—through Myanmar, northern Thailand, and Laos to southern Vietnam (generally near Dalat) and also in northern Luzon Island (Philippines). Forming pure open stands on poor, fire-prone hillsides or mixed with hardwoods; (300–)700–2,000(–2,700) m. Zone 9. Synonyms: *P. insularis* Endlicher, *P. kesiya* var. *insularis* (Endlicher) H. Gaussen, *P. kesiya* var. *langbianensis* (A. Chevalier) W. C. Cheng & L. K. Fu, *P. khasya* Royle.

Khasi pine is closely related to Yunnan pine (*Pinus yunnanensis*), differing primarily in the finer needles and the seed scales with a definitely protruding and horizontally ridged exposed face. It has excellent, termite resistant wood and abundant resin production. These qualities led to experimental plantations in many tropical and warm temperate countries. With its very discontinuous distribution and climatically and ecologically varied occurrences, a great deal of genetic variation was discovered in traits important for forestry, like stem form and growth rate. There are also substantial differences in the proportions of different chemicals in the resins from different regions. The proportions found in the related Yunnan pine closely resemble those found in populations from southern Vietnam but not those of northern trees. The scientific and common names are derived from the Khasi Hills of Meghalaya state in India, at the northwestern limit of distribution of the species.

Pinus koraiensis P. Siebold & Zuccarini

KOREAN PINE, MANCHZHURSKII KEDR (RUSSIAN), HONG SONG (CHINESE), CHAS NAMU (KOREAN), CHOSEN-GOYO (JAPANESE)

Pinus subg. *Strobus* sect. *Quinquefoliae* subsect. *Strobus*

Tree to 30(–50) m tall, with trunk to 1(–1.5) m in diameter, often forking. Bark grayish brown and smooth at first, darkening, flaking, and becoming irregularly and shallowly furrowed with age. Crown deeply dome-shaped, with numerous thin branches angled upward and then gradually becoming horizontal, with foliage in tufts near the ends. Twigs rich reddish brown and densely minutely hairy at first, becoming gray and bald. Buds 10–18 mm long, sparsely to densely resinous. Needles in bundles of five, each needle (6–)8–12(–13) cm long, flexible, usually lasting 2–3 years, dark green on the outer face, whitish green with wax over the lines of stomates on the inner faces. Individual needles with an undivided midvein and three (to five) modest resin canals deep in the leaf tissue near the corners. Sheath 5–15 mm long, soon shed. Pollen cones 15–20 mm long, red. Seed cones 9–11(–14) cm long, cylindrical, with about 90 seed scales, green to purple before maturity, ripening reddish brown, the scales spreading enough to expose the seeds but not release them and falling the following year. Seed scales diamond-shaped, thick and somewhat fleshy where the seeds are nestled, with an elongate tip slightly curled back and ending in a small, triangular umbo. Seed body (12–)14–17 mm long, unwinged. Amur River region of far eastern Russian and northeastern China, south through Korea and in central Honshū (Japan). Usually mixed with other conifers in subalpine forests on rocky mountain slopes; (200–)600–2,000(–2,500) m. Zone 3.

This handsome tree has spawned a number of cultivars. It is a larger tree than most of the other mountain stone pines in subsection *Strobus* (except *Pinus sibirica*) and has considerable timber value as well as producing turpentine and edible and medicinal seeds. Although it is named for Korea, it was severely depleted there by excess harvesting early in the 20th century. The most extensive stands now lie in the main part of the range in Manchuria and nearby Russia. It is not crossable with other subalpine nut pines and has produced just a single hybrid with the western North American sugar pine (*P. lambertiana*), a less closely related member of subsection *Strobus*. It is possible that the five-needled stone pines are independently derived from within subsection *Strobus* by selection in subalpine habitats for animal-dispersed, unwinged seeds borne in soft, unopening seed cones.

Pinus krempfii P. Lecomte

FLAT-NEEDLE PINE, THÔNG-SRE (VIETNAMESE)

Pinus subg. *Strobus* sect. *Quinquefoliae* subsect. *Krempfianae*

Tree to 20(–30) m tall, with trunk to 0.5 m in diameter. Bark reddish brown, shallowly furrowed between low vertical ridges. Crown open and irregular, the branches densely to sparsely clothed with foliage near the tip of each year's increment and bare near its base. Twigs hairless. Buds about 5 mm long, not resinous. Needles in bundles of two, each needle 3–7(–14) cm long, 1.5–4(–7) mm wide, flat and straight to curved like a scythe, the pair separating like scissor blades, lasting 2–6 years or more, shiny dark green. Individual needles with lines of stomates only on the inner face, or sometimes with a few stomates near the tip of the outer face, an undivided midvein, 4–10 small resin canals along the whole width next to the epidermis of the inner face, and sometimes with two resin canals near the midvein next to the epidermis of the outer face. Sheath about 2–2.5 cm long, very soon shed. Pollen cones 5–9 mm long, light brown. Seed cones (5–)7–9 cm long, egg-shaped, with 40–60 seed scales, reddish brown, opening widely to release the seeds and then falling, on a slender stalk 2–3 cm long. Seed scales paddle-shaped, thickened at the exposed tip, with a diamond-shaped umbo on the outer face bearing a persistent, fragile prickle. Seed body 3–5 mm long, with an easily detached wing 6–7.5 mm long. Highlands of central Vietnam in Khanh Hoa and Lam Dong provinces. Scattered in the canopy of dense evergreen broad-leaved forests on deep, moist soils; 1,200–2,000 m. Zone 9.

This rare pine, so unusual in the form of its needles, conforms to the white or soft pines (subgenus *Strobus*) in all other respects, including the anatomy of the needles and the anatomy and chemistry of the wood. The needles vary greatly in width with age of the plant and position in the crown, the narrowest being scarcely wider than those of several other coarse-needled pine species. The anatomy of the needles varies with their width so that broader needles actually have fewer resin canals than narrow ones. The seed cones are particularly similar to those of the North American subalpine foxtail pine (*Pinus balfouriana*), and Pilger (1926) even included it in subsection *Balfourianae.* DNA studies clearly include it as basal to the typical white pines (subsection *Strobus*), refuting attempts by some botanists to place it in subgenus *Pinus* (hard or yellow pines), give it its own genus or subgenus, or derive it from hybridization between a pine and a member of some other genus, like *Keteleeria* or *Pseudolarix.* While it is not that distinct from other basal white pines, it is still often placed it in a subsection by itself (subsection *Krempfianae*) adjacent to subsection *Balfourianae.* This apparent intercontinental relationship is intriguing and mirrored by similarities between the Asiatic lacebark (*P. bungeana*) and chilgoza (*P. gerardiana*) pines (subsection *Gerardianae*) and North American pinyon pines (subsections *Cembroides* and *Nelsoniae*), especially certain Mexican species, like *P. nelsonii* and *P. pinceana.* These cross-continental connections were supported by early molecular data, but later DNA studies refuted them by showing that *P. krempfii* is phylogenetically closer to its continental neighbors the *Gerardianae* than it is to the American *Balfourianae* and that the latter are closer to their continental neighbors *P. nelsonii* and the *Cembroides.* Thus the intercontinental similarities are best interpreted as parallel independent evolutionary developments rather than evidence of repeated intercontinental migrations. The species name honors M. Krempf, who collected the type specimen in 1921.

Pinus lagunae (Robert-Passini) Passini

CAPE PINYON

Pinus subg. *Strobus* sect. *Parrya* subsect. *Cembroides*

Tree to 15(–21) m tall, with trunk to 0.8(–1.2) m in diameter, sometimes dividing near the base. Bark dark grayish brown, broken into thick, rectangular blocks divided by regular deep furrows at the base of mature trees. Crown rounded, thin and open, with widely spaced, gently rising thin branches densely clothed with foliage near the tips. Twigs reddish brown at first, hairless but roughed by tiny scale leaves. Buds 4–6 mm long, not resinous. Needles in bundles of two or three (or four), each needle (2.5–)4–8 cm long, flexible, lasting 3–4 years, grayish green. Individual needles with lines of stomates on both the inner and outer faces, an undivided midvein, and two resin canals beneath the epidermis of the outer face. Sheath 4–6 mm long, curling back and persisting only during the first year. Pollen cones 4–6 mm long, yellow. Seed cones 3–6 cm long, almost spherical or broader than long, with (15–)20–30 seed scales, green before maturity, ripening glossy light reddish brown, opening widely to release the seeds and then falling with the slender, 0.2–1.2 cm long stalk. Seed scales paddle-shaped, with deep seed cavities only inward from the diamond-shaped, thickened but only slightly raised exposed portion that ends in a diamond-shaped umbo bearing a small prickle on the face. Seed body (10–)11.5–15(–16) mm long, the shell (0.2–)0.4–0.7(–0.9) mm thick, the rudimentary wing remaining attached to the seed scale. Restricted to the Sierra de la Laguna (hence the scientific name) in the Cape region of Baja California Sur (Mexico). Forming pure stands or mixed with *Quercus devia* in pine-oak woodland on dry slopes and flats over limestone; (1,200–)1,500–2,000(–2,200) m. Zone 9? Synonyms: *P. cembroides* subsp. *lagunae* (Robert-Passini) D. Bailey, *P. cembroides* var. *lagunae* Robert-Passini.

This sole pinyon of southern Baja California has historically been included without distinction within *Pinus cembroides,* the commonest Mexican pinyon. It differs in a few morphological and chemical characteristics, including needle length and the main wood oil constituents. This led to its separation as a variety in 1981, a subspecies in 1983, and a species in its own right in 1987. Although the differences cited are slight, DNA studies support the separation of *P. lagunae* from *P. cembroides* and suggest that it may be more closely related to the other widespread Mexican pinyon, plateau pinyon (*P. culminicola*). Cape pinyon is the most local of the typical pinyons and is thus of conservation concern. While it is a small tree, it is taller than most pinyons and usually has a single, straight trunk. Preliminary experiments suggest that it is quicker growing than other pinyon species, even under conditions of low rainfall, so it has been recommended for plantation culture as a fuel, lumber, and food tree for arid regions.

Pinus lambertiana D. Douglas

SUGAR PINE

Pinus subg. *Strobus* sect. *Quinquefoliae* subsect. *Strobus*

Tree to 60(–85) m tall, often clear of branches for 30 m. Trunk to 2(–3.5) m in diameter. Bark dark grayish brown in the deep, interconnecting furrows between reddish brown, interrupted, narrow, flat-topped vertical ridges. Crown raggedly oval, open, with numerous, widely spaced, slender, unequally long, horizontal branches turning up at the tips when not weighted down by seed cones and bearing relatively sparse foliage. Twigs light reddish brown, smooth, minutely hairy. Buds 5–8 mm, resinous. Needles in bundles of five, each needle 5–8(–10) cm long, stiff, twisted, and slightly diverging, lasting 2–3(–4) years, bluish green with wax. Individual needles with lines of stomates on all three faces, an undivided midvein, two small resin canals touching the epidermis of the outer face, and sometimes with a third resin canal near the juncture of the two inner faces, either touching the epidermis of one of them or immersed in the leaf tissue between them. Sheath (10–)15–20 mm long, soon shed. Pollen cones 10–15 mm long, straw-colored. Seed cones very large, (25–)30–45(–60) cm long mostly straight and broadly cylindrical with a slight taper toward the tip, with 150–250 seed scales or more, green before maturity, ripening medium yellowish brown, opening widely to release the seeds and then falling, on stalks 6–15 cm long. Seed scales paddle-shaped, moderately thick and rigid, straight, the exposed portion a little cupped and with a sharp angle along the midline, with a narrow diamond-shaped umbo at the tip. Seed body 10–15(–20) mm long, the firmly attached wing 20–30 mm long. Mountains of Pacific North America, from central Oregon through California and westernmost Nevada to northern Baja California (Mexico). Almost always mixed with other conifers (and often towering over many of them) in montane forest; 350–3,200 m. Zone 7. Synonym: *P. lambertiana* var. *martirensis* Silba.

Sugar pine is the tallest pine species, standing up well among the other giant conifers of Pacific North America. The largest known individual in combined dimensions when measured in 1997 was 63.7 m tall, 3.5 m in diameter, with a spread of 18.0 m. David Douglas, who named the species in 1827 to honor Aylmer B. Lambert (1761–1842), author of the influential *A Description of the Genus Pinus,* reported a fallen tree with a diameter of 5.5 m. It takes 500 years or more to reach these enormous sizes, and individuals up to 760 years old have been reported. Such trees were heavily exploited during the early 20th century to construct such marvels as the largest pipe organ in the world, the Wanamaker Grand Court Organ in Philadelphia, Pennsylvania, built in 1904, which even boasts some pipes made of sugar pine logs large enough to hide a Shetland pony. The giants were mostly gone by about 1950, except in a very few protected groves in parks, victims of intense harvesting and of introduced pests and diseases, especially white pine blister rust, to which it is especially susceptible. The seed cones, longest among the pines, keep scale with the trees and bear edible seeds as large as those of many nut pines. The sugary sap, which gives the tree its common name, exudes from deep wounds that avoid contamination by the more shallowly seated resin. This sap has been used medicinally, as a laxative, as well as a sweet treat.

Although clearly a member of subsection *Strobus,* the relationships of sugar pine are not entirely clear. It has some similarity to its closest neighbor, western white pine (*Pinus monticola*), but will not cross with that species, nor with any of the other typical white pines. The seed cones are somewhat like a giant version of those of limber pine (*P. flexilis*), and it has similarities in enzyme systems with that species, but it will not cross with it either. Instead, a very few hybrid seeds were obtained in crosses with two Asiatic species, Chinese white pine (*P. armandii*) and Korean pine (*P. koraiensis*). It is clearly isolated within the subsection, and DNA studies do not fully resolve its affinities to other white pine species. However, they confirm the crossability of sugar pine with the mountain stone pines like *P. armandii* and *P. koraiensis* in a spectacular way. Sugar pine trees in the southern portion of the species' range have a different kind of chloroplast (with a different DNA genome sequence) than those living in the northern half. The southern chloroplasts are unique to this species and loosely related to those found in most other North American typical white pines. Those from the northern trees, in contrast, are virtually identical to those found in neighboring trees of whitebark pine (*P. albicaulis*), a mountain stone pine, and linked in turn to the Asian stone pines and not to any other North American species. Basically, the best interpretation of this result is that northern sugar pines acquired their chloroplasts through hybridization with whitebark pine, and the acquired chloroplasts completely replaced the original ones throughout the north, even though the two species do not generally grow together within their shared range. Since the chloroplasts of conifers are inherited paternally, unlike the maternally inherited ones in most flowering plants (or the mitochondria of people for that matter, recall "mitochondrial Eve"), we know that whitebark pine was the pollen parent in the interspecies crosses. Once the chloroplasts of Asian origin got into sugar pine trees, they spread, either for selective reasons, because the new chloroplasts conveyed some kind of advantage on either the pollen grains or the trees that contained them compared to the original chloroplasts, or for historical reasons, such as if the first trees of sugar pine to arrive in the area during their northward migration following deglaciation hybridized with whitebark pine and these hybrids served as the nucleus for further expansion. The true story of the events that led to this peculiar replacement of one chloroplast type by another may someday be revealed by an even more detailed analysis of a broader range of genetic markers.

Pinus lawsonii Roezl ex G. Gordon

LAWSON PINE

Pinus subg. *Pinus* sect. *Trifoliae* subsect. *Australes*

Tree to 25(–30) m tall, with trunk to 0.8 m in diameter. Bark dark brown, with flaky, interlacing, narrow, platelike ridges separated by paler, wide furrows. Crown conical and dense at first, becoming open and almost spherical with age, with stout, rising branches well clothed with foliage near the ends. Twigs reddish brown with a thin waxy coating, hairless, initially rough with the bases of scale leaves, but these flaking off after the first year. Buds 8–15 mm long, not resinous. Needles in bundles of (two or) three to five, each needle (12–)16–20(–25) cm long, very stiff and straight, lasting 2–3 years, grayish green, thinly to thickly coated with wax. Individual needles with numerous conspicuous lines of stomates on both the inner and outer faces, and (one to) three to six resin canals touching and surrounding the two-stranded midvein, or the two at the outer corners midway between the midvein and the needle surface. Sheath 15–25 mm long at first, weathering to 10–15 mm and persisting and falling with the bundle. Pollen cones 10–20 mm long, yellowish brown. Seed cones 5–8(–9) cm long, egg-shaped, often slightly asymmetrical, with 50–100(–150) seed scales, green before maturity, ripening dull yellowish brown, opening widely to release the seeds and then falling with the soft, backwardly curved, 6–12 mm long stalks. Seed scales roughly rectangular, the exposed face horizontally diamond-shaped, crossed by a horizontal ridge, mostly fairly flat but strongly protruding on the basal scales away from the twig, all with a large, outwardly hooked umbo. Seed body 4–5 mm long, the clasping wing another 10–16 mm longer. Mountains of southern Mexico, from central Oaxaca to southeastern Jalisco. Scattered trees or groves among other species in warm, moist pine-oak forest on lower and middle slopes; 1,300–2,500(–2,800) m. Zone 9. Synonym: *P. lawsonii* var. *gracilis* Debreczy & I. Rácz.

Lawson pine is similar to several of the species with which it sometimes grows, especially Herrera pine (*Pinus herrerae*) and Pringle pine (*P. pringlei*). Neither of these species has such conspicuously waxy needles, while the former differs in its slender, flexible needles and the latter has seed cones remaining on the tree for several years. The species name honors Charles Lawson (1794–1873), who owned a nursery in Edinburgh, Scotland, where he grew many western North American plants from seed, including Port Orford Cedar (*Chamaecyparis lawsoniana*).

Pinus leiophylla Schiede & Deppe ex D. F. L. Schlechtendal & Chamisso

CHIHUAHUA PINE, PINO CHINO (SPANISH)

Pinus subg. *Pinus* sect. *Trifoliae* subsect. *Australes*

Tree to 25(–35) m tall, with trunk to 0.8(–0.9) m in diameter. Bark dark grayish brown to almost black, with thick, scaly blocks or ridges separated by lighter, deep furrows. Crown conical when young, becoming openly and irregularly dome-shaped, with unevenly spaced, slender, upswept branches well clothed with spreadout foliage. Twigs reddish brown, often thinly waxy at first, slightly rough with the bases of scale leaves, hairless. Buds 6–15 mm long, slightly resinous. Needles in bundles of (two or) three to five (or six), each needle (4–)7–14(–17) cm long, stiff or slightly flexible, lasting 2–3 years, yellowish green or bluish green with a thin, waxy coating. Individual needles with narrow but conspicuous lines of stomates on both the inner and outer faces, and (one or) two to six (or seven) resin canals midway between the outer surface and the two-stranded midvein (or occasionally one or two right next to the midvein) at the corners and below the outer face. Sheath (10–)12–15(–20) mm long, peeling away completely and falling with the expansion of the needles. Pollen cones 10–20 mm long, yellowish brown. Seed cones (3.5–)5–7(–9) cm long, egg-shaped with a flat bottom, with (40–)50–80(–100) seed scales, green before maturity, ripening shiny light brown, opening widely to release the seeds and then persisting several years before falling with the slender, 1–2 cm long stalks. Seed scales pointedly rectangular, the exposed face diamond-shaped, low, crossed by a shallow ridge topped by a small diamond-shaped umbo bearing a small, sharp prickle. Seed body (2–)3–5 mm long, the clasping but easily detachable wing another (7–)10–15(–19) mm longer. Mountains of southern and western Mexico and the adjacent United States. Usually mixed with other pines and other trees in pine and pine-oak forests on rocky soils; (1,500–)1,600–2,800(–3,300) m. Zone 9. Two varieties.

Chihuahua pine has such distinctive features that it was traditionally placed in its own subsection along with Lumholtz weeping pine (*Pinus lumholtzii*). While it shares the shedding of the needle bundle sheaths with *P. lumholtzii* (a feature not found in any other New World hard pine), its ability to sprout from stumps (and from trunks and branches) and the prolonged development of the seed cones (which take 3 years to mature) are unique among New World pines. However, DNA studies suggest that western hemisphere hard or yellow pines (subgenus *Pinus*) all belong to two main groupings, and *P. leiophylla* has proven to be closely related to the pines of the southeastern United States and especially also to the various closed-cone pine groups that make up an expanded subsection *Australes*. This accords with its long retention of the open cones even after the seeds have been shed. *Pinus leiophylla* consists of two weakly differentiated and broadly overlapping varieties that also intergrade with one another, especially in the states of Durango and Zacatecas, Mexico. The scientific name (Greek for "smooth leaves") refers to the lack of the microscopic teeth along the edges of the leaves that are found in many pine species of western Mexico. The largest known individual in the United States in combined dimensions when measured in 1998 was 26.5 m tall, 1.0 m in diameter, with a spread of 10.4 m.

Pinus leiophylla* var. *chihuahuana (Engelmann) G. R. Shaw

Tree to 25 m tall. Needles most often three per bundle, with mostly five to eight lines of stomates on the outer face and usually four to six resin canals. Northern portion of the range of the species, from southeastern Arizona and southwestern New Mexico south through the Sierra Madre Occidental to southern Zacatecas (Mexico); mostly at 1,500–2,500 m. Synonym: *Pinus chihuahuana* Engelmann.

Pinus leiophylla Schiede & Deppe ex D. F. L. Schlechtendal & Chamisso **var. *leiophylla***
Tree to 35 m tall. Needles most often four per bundle, with mostly four to six lines of stomates on the outer face and usually two or three resin canals. Southern portion of the range of the species in the Sierra Madre del Sur, transverse ranges, and Sierra Madre Occidental north to east-central Sonora and west-central Chihuahua (Mexico); mostly at 2,000–3,000 m.

Pinus longaeva D. Bailey

GREAT BASIN BRISTLECONE PINE
Plate 43
Pinus subg. *Strobus* sect. *Parrya* subsect. *Balfourianae*
Tree to 16 m tall, with trunk to 2(–4) m in diameter. Bark bright reddish brown, breaking up with age into small blocks separated by shallow furrows. Crown conical to cylindrical when young, becoming irregular, twisted, and flat-topped, with numerous slender, twisted, drooping branches densely clothed with foliage. Twigs reddish brown and hairless or minutely hairy at first, becoming yellowish gray and bald with age. Buds 8–10 mm long, resinous. Needles in bundles of five, each needle 1.5–3.5 cm long, stiff, straight, and remaining tight together, lasting 10–35(–45) years, dark yellowish green, mostly without resin flakes. Individual needles with lines of stomates only on the inner faces, an undivided midvein, and two large, equal resin canals touching the ungrooved epidermis of the outer face and visible beneath it. Sheath scales 5–10 mm long, curling back and soon shed. Pollen cones 7–10 mm long, reddish purple or yellow. Seed cones 6–9.5(–11) cm long, egg-shaped and rounded at the base, often with over 100 scales, reddish purple or yellowish green before maturity, ripening reddish brown, opening widely to release the seeds and then falling intact, usually with a short stalk about 5–7 mm long. Seed scales paddle-shaped, thin but thicker at the exposed tip, with a diamond-shaped umbo bearing a slender prickle usually about 3 mm long. Seed body 5–8 mm long, pale with reddish speckles, the easily detachable wing 7–15 mm long. High mountains of the Great Basin in eastern California, Nevada, and Utah. Forming pure stands or mixed with other conifers in open subalpine forests and woodlands, mostly on limey soils; (1,700–)2,200–3,500 m. Zone 5. Synonym: *P. aristata* var. *longaeva* (D. Bailey) E. Little.

At almost 5,000 years old, the bristlecone pines of the desert White Mountains of California are the oldest known individual trees (hence the scientific name). The oldest individuals live under intensely arid conditions and have an extremely slow growth rate. Their needles are among the longest-lived of all leaves, exceeded only by those of *Welwitschia* in the Namib Desert of southwestern Africa, and perhaps by those of some palm trees. These needles, like all conifer needles lasting more than 1 year, add new phloem cells (the food conducting cells) each spring, but not water-conducting xylem. Under less harsh conditions, needles are retained for a shorter period, and the trees succumb much sooner to hazards of vigor, like heart rot. The tops of the old trees become ragged, and living bark is often confined to just a portion of the circumference, so they never become very tall. No matter how microscopically slow growth may be, however, 5,000 years is a long time and thus large individuals of *Pinus longaeva* are among the thickest pine trees. The largest known individual in combined dimensions when measured in 2003 was 15.9 m tall, 3.7 m in diameter, with a spread of 13.4 m. Great Basin bristlecone pine shares a growth peculiarity with the other species of subsection *Balfourianae* that is very rare in other pines. Almost half of the side branches come out of the center of a needle bundle, while in other species the growth tip between the needles dies early in the growth of the bundle and cannot give rise to a branch. This means that bristlecone pines have extra branches that did not arise from buds at the tips of the twigs, increasing the density of the living portions of the crown. Although Great Basin bristlecone pine was only separated taxonomically as a species from the similar Rocky Mountain bristlecone pine (*P. aristata*) in 1970, many features suggest that it is actually more closely related to the foxtail pine (*P. balfouriana*), with which it is fully interfertile, while it is essentially uncrossable with *P. aristata*. All three species are relatively distant from other white pines (subgenus *Strobus*) except, perhaps, Nelson pinyon (*P. nelsonii*), which has some similar morphological features but lacks the subalpine ecology of the *Balfourianae* and which has not yet been thoroughly studied.

Pinus luchuensis H. Mayr

LUCHU PINE, RYUKYU ISLANDS PINE, RYUKYU-MATSU (JAPANESE), MAACHI (OKINAWAN)
Pinus subg. *Pinus* sect. *Pinus* subsect. *Pinus*
Tree to 15(–25) m tall, with trunk to 1 m in diameter. Bark gray, long remaining smooth, eventually beginning to flake and then darkening to grayish black and breaking up into irregular plates. Crown flat-topped and open, with numerous long, slender, upwardly angled branches densely clothed with foliage at the tips. Twigs yellowish brown, with tufts of tiny hairs, rough with the bases of scale leaves. Buds 1–2 cm long, reddish brown, resinous. Needles in bundles of two, each needle 12–16(–20) cm long, slender but stiff and straight, lasting 2(–3) years, dark green. Individual needles with lines of stomates on both faces and two to six resin canals, the

two at the corners touching the needle surface and the others midway between the surface and the two-stranded midvein. Sheath 10–15 mm long, weathering to 5–10 mm and persisting and falling with the bundle. Pollen cones densely crowded, 12–20 mm long, reddish brown. Seed cones 3–4(–6) cm long, egg-shaped, with 60–80 seed scales, green before maturity, ripening dull yellowish brown, opening widely to release the seeds and then persisting a while before falling with the slender, short stalk to 1 cm long. Seed scales paddle-shaped, the exposed face irregularly fan-shaped, flat, crossed by a low ridge topped by a small, slightly protruding umbo. Seed body 3–4 mm long, the easily detachable wing another 9–12 mm longer. Ryukyu Islands of Japan, from Amami-Ō-shima in the north to Yonaguni Jima in the south. Forming pure, open stands from the coast to the mountain slopes; 0–300(–700) m. Zone 9.

Forests of Luchu pine once clothed much of the Ryukyu Islands (also called Luchu Islands, hence the scientific and common names), but as the primary natural source of timber on the islands they were severely depleted, particularly during the Second World War. The tree is still common, however, and is also grown in plantations. Away from the seashore it has a straight, upright trunk, contrasting strongly with its stunted habit in exposed, coastal prairies. It is ecologically similar to Japanese black pine (*Pinus thunbergii*), which replaces it northward. However, it is generally considered to be more closely related to Taiwan red pine (*P. taiwanensis*) of Taiwan and eastern China, which typically grows at much higher elevations. Little research has been conducted on the genetic variation and relationships of Luchu pine, so a study using contemporary techniques is warranted.

Pinus lumholtzii B. Robinson & Fernald

LUMHOLTZ WEEPING PINE, PINO TRISTE (SPANISH)

Pinus subg. *Pinus* sect. *Trifoliae* subsect. *Australes*

Tree to 20 m tall, with trunk to 0.6(–0.7) m in diameter. Bark dark grayish brown, very rough, with irregular, scaly ridges and deep furrows. Crown dome-shaped, open, often irregular, with numerous slender, horizontal or rising branches densely clothed at the tips with entirely hanging foliage. Twigs reddish brown with a distinct waxy layer, hairless, generally fairly smooth but shallowly grooved between the bases of the scale leaves, which begin to flake off after the first season. Buds 8–15 mm long, a little resinous. Needles in bundles of (two or) three (or four), each needle (14–)23–30(–40) cm long, flexible but hanging straight down, lasting 2 years, light green to yellowish green. Individual needles with numerous conspicuous lines of stomates on both the inner and outer faces, and 4–8(–10) resin canals surrounding the two-stranded midvein and touching it or more often midway between it and the needle surface. Sheath (20–)25–35 mm long initially, soon peeling back and shedding completely. Pollen cones 2–3 cm long, yellowish brown with a reddish flush. Seed cones (3–)4–6(–7) cm long, symmetrically egg-shaped, hanging like the needles, with (50–)60–80 (–100) seed scales, green before maturity, ripening grayish to reddish brown, opening widely to release the seeds and then falling together with the slender, curved, (5–)10–15 mm long stalk. Seed scales fairly rectangular, the exposed face spade-shaped, thickened at the tip, with a large umbo bearing a thin, fragile prickle. Seed body 3–5 mm long, the clasping wing another 8–14 mm longer. Mountains of western Mexico, from central Chihuahua to Jalisco and Guanajuato. Mixed with other species in dry to wet pine and pine-oak forests and woodlands; (1,500–)1,800–2,400(–2,900) m. Zone 9. Synonyms: *P. yecorensis* Debreczy & I. Rácz, *P. yecorensis* var. *sinaloensis* Debreczy & I. Rácz.

Lumholtz weeping pine appears similar to Mexican weeping pine (*Pinus patula*), a much larger tree that replaces it in the wetter mountains of eastern and southern Mexico. More detailed examination of such features as the falling sheath around the bundles of needles point to a closer relationship with the neighboring Chihuahua pine (*P. leiophylla*), and that is the closest relationship accepted by most authors. Studies show that these two species are not as distinct from other hard pines as once thought and that they are derived from the early diversification of the subsection *Australes*,

Characteristic mature Lumholtz weeping pine (*Pinus lumholtzii*) in nature.

one of the two major groupings of North American hard pines. The species name honors Carl Lumholtz (1851–1922), who led the scientific expedition on which the type was collected in 1893.

Pinus massoniana A. Lambert

MASSON PINE, MA WEI SONG (CHINESE), THÔNG DUÔI NGUA (VIETNAMESE)

Pinus subg. *Pinus* sect. *Pinus* subsect. *Pinus*

Tree to 30(–45) m tall, with trunk to 1(–1.5) m in diameter. Bark reddish brown, smooth, and peeling in strips above, graying and breaking up into vertically aligned, narrowly rectangular, flat-topped blocks flaking to reveal reddish brown patches and separated by deep furrows, sometimes persistently reddish brown and thinner. Crown dome-shaped, sometimes flattening with age, often open and irregular, with numerous upwardly angled to horizontal branches densely clothed with foliage. Twigs yellowish brown, sometimes a little waxy, hairless, rough with the bases of scale leaves. Buds 2–5 cm long, not resinous. Needles in bundles of two (or three), each needle (7–)14–20(–22) cm long, very slender and flexible, slightly twisted, lasting 2 years but produced twice each year, bright green. Individual needles with inconspicuous lines of stomates on both the inner and outer faces, a two-stranded midvein, and four to nine resin canals by the needle surface, mostly around the outer face. Sheath 10–20 mm long, weathering to 5–12 mm and persisting and falling with the bundle. Pollen cones densely crowded, 15–20 mm long, yellowish brown. Seed cones (2.5–)4–7(–9) cm long, egg-shaped to almost spherical, with (50–)70–100 scales, green before maturity, ripening rich brown, opening widely to release the seeds and then falling or persisting 2–3 years before falling with the very short stalks to 5(–10) mm long. Seed scales roughly rectangular, the exposed face horizontally diamond-shaped, flat to slightly raised, crossed by a weak ridge topped by a large, flat umbo bearing a fragile or persistent blunt or sharp prickle. Seed body 4–6(–9) mm long, the easily detachable wing another (12–)16–24(–30) mm longer. From Sichuan and eastern Yunnan eastward throughout southern China and on Taiwan and Hainan and in northern Vietnam (where perhaps not native). Forming pure forests and woodlands or mixed with other conifers, hardwoods, and bamboos; 0–1,500(–2,000) m. Zone 8. Synonyms: *P. massoniana* var. *hainanensis* W. C. Cheng & L. K. Fu, *P. massoniana* var. *shaxianensis* D. X. Zhou.

Mature Masson pines (*Pinus massoniana*) in nature.

Masson pine is the most important pine species (both ecologically and economically) of the warm temperate and subtropical regions of southern China. Exploitation of natural stands and plantation for timber occur primarily at the upper levels of the elevation range, while lower stands yield fuel wood and resins. The wood has various specialized uses, including as a substrate for growing edible mushrooms and as torches after girdling to allow accumulation of resin. The buds and needles are used in herbal medicine, and the pollen is sometimes used for decorating pastries. The species name honors Francis Masson (1741–1805), the first plant collector sent out by the Royal Botanic Gardens, Kew, who worked mostly in South Africa and never visited China but did send many plants to Aylmer Lambert.

Masson pine is most closely related to Japanese red pine (*Pinus densiflora*) and is not as close to the other major Chinese hard pines, Chinese red pine (*P. tabuliformis*), Chinese mountain pine (*P. densata*), and Yunnan pine (*P. yunnanensis*), which are themselves closer to Japanese black pine (*P. thunbergii*). Masson pine crosses successfully with both the Japanese red and black pines, but attempts to cross it with Chinese red pine, the North American red pine (*P. resinosa*), and the European black pine (*P. nigra*) and Scots pine (*P. sylvestris*) all failed. Despite the wide geographic range of Masson pine, there is relatively little variation from place to place. Trees from the island of Hainan tend to retain their reddish brown juvenile bark on larger trunks than those elsewhere, and some populations in Fujian province have larger seed cones and seeds than the norm, and umbos with a sturdy, sharp prickle. Both of these have been recognized as varieties but are so weakly distinguished that they hardly seem appropriate for formal recognition.

Pinus maximartinezii Rzedowski

KING PINYON, PIÑÓN REAL (SPANISH)

Pinus subg. *Strobus* sect. *Parrya* subsect. *Cembroides*

Tree to 10(–20) m tall, with trunk to 0.3(–0.6) m in diameter. Bark reddish brown, thin and smooth at first, becoming gray with age, thickening and cracking into tightly packed polygonal blocks separated by narrow furrows. Crown dome-shaped, with numerous horizontal to upright branches from the base bearing hanging side branches that are sparsely clothed with foliage just at the tips. Twigs reddish brown or grayish with wax, hairless. Buds 4–8 mm long, resinous. Needles in bundles of (three to) five, each needle 7–10(–13) cm long, straight but soft and flexible, lasting 1–2 years, bright green to bluish green on the outer face, the inner faces whitened with wax. Individual needles with lines of stomates on the inner faces and two large resin canals touching the epidermis of the outer face, one on either side of the undivided midvein. Sheath (6–)7–8 mm long, curling back at first and then shed. Pollen cones 8–10 mm long, pale yellowish brown. Seed cones 15–27 cm long, egg-shaped, very stout and heavy, with 60–110 seed scales, green before maturity, ripening light brown to reddish brown, opening widely but releasing the seeds only with difficulty, nearly stalkless. Seed scales diamond-shaped, the exposed portion angled out and back as a thick, broad pyramid ending in a large, bluntly triangular umbo up to 10 mm long. Seed body (20–)22–25(–28) mm long, mostly oblong, though a little wider at the outer end than at the inner, the seed coat very thick and hard, the vestigial wing remnant not separating from the seed scale. Cerro Piñones at the southern end of the Sierra de Morones near Juchipila, southern Zacatecas (Mexico). Mixed with other trees in open, pine-oak woodland on shallow, rocky soils; 1,600–2,250 m. Zone 8?

This rare nut pine is known from a single location with fewer than 2,500 adult trees. Virtually the whole seed crop is harvested annually for local consumption and for sale in the market town of Juchipila. The harvesting is also conducted destructively, by cutting off cone-bearing branches, so the trees themselves are further damaged each year. Any seeds that might escape and germinate are likely to be grazed off by livestock or burned off by fires during the long dry season. The future of this giant pinyon pine is thus in some doubt it its native home, although conservation stands are being established in other countries. Despite these limitations on natural regeneration, genetic studies show that the species contains a reasonable amount of genetic variation in a pattern that does not imply a great deal of inbreeding. One peculiarity, however, is that those enzymes that have more than one form in different individuals have only two variants, and these occur in about a 50:50 ratio. This is very

The only known natural stand of king pinyon (*Pinus maximartinezii*).

Mature king pinyons (*Pinus maximartinezii*), which bear the largest seed cones of any pinyon pine species.

unlike the situation in other pines, which often have several variants of variable enzymes, and implies that within the last few generations the species was drastically reduced in numbers, perhaps even to just a single individual, since such an individual can only have two different variants of any gene. If it did go through such a population bottleneck, the king pinyon has already demonstrated remarkable regenerative ability when left to its own devices, no doubt aided in its arid environment by the abundant stored reserves in the large seeds. These seeds house an embryo with the largest known number of cotyledons among the conifers, 19–24.

The species has other peculiarities as well, including its massive cones, but phylogenetic studies place it firmly with two other anomalous Mexican pinyons, *Pinus pinceana* and *P. rzedowskii*, outside the group of typical pinyon pines such as Colorado pinyon (*P. edulis*), Mexican pinyon (*P. cembroides*), and singleleaf pinyon (*P. monophylla*), within subsection *Cembroides*. There is a parallelism with, but no real affinity to, the Himalayan and Chinese species in subsection *Gerardianae*, lacebark pine (*P. bungeana*), southern lacebark pine (*P. squamata*), and chilgoza pine (*P. gerardiana*). Despite their morphological similarity to the anomalous pinyons, the Old World *Gerardianae* are actually more closely related to the typical white pines of subsection *Strobus* than to members of subsection *Cembroides*, and the pinyons are grouped instead with two other New World subsections, *Nelsoniae* and *Balfourianae*, of which the latter, at least, seems less similar overall. Whatever future research reveals about these relationships, DNA studies suggest that king pinyon is most closely related to the much smaller and smaller-coned weeping pinyon (*P. pinceana*), which lives in the Sierra Madre Oriental over 400 km to the north or east of Cerro Piñones. The species name honors Maximino Martínez (1888–1964), who greatly advanced the study of Mexican conifers, describing many new species and varieties, not all of which have stood the test of time.

Pinus maximinoi H. Moore

MAXIMINO PINE, PINO CANIS (SPANISH)

Pinus subg. *Pinus* sect. *Trifoliae* subsect. *Ponderosae*

Tree to 35(–50) m tall, with trunk to 0.9(–1) m in diameter. Bark grayish brown, with long, flat-topped ridges separated by deep furrows. Crown conical at first becoming cylindrical or dome-shaped, often remaining fairly dense, with numerous slender, horizontal branches well clothed with foliage at the ends. Twigs light brown, occasionally thinly waxy, hairless, the bases of the scale leaves fairly smooth, flaking away completely after the first year. Buds 12–20 mm long, not resinous. Needles in bundles of (four or) five (or six), each needle (15–)20–28(–35) cm long, thin, flexible, and drooping, lasting 2(–3) years, light yellowish green to bluish green. Individual needles with just a few lines of stomates on all three faces, and two or three (or four) resin canals at the corners of the needles midway between the needle surface and the two-stranded midvein and separated by bands of hard, dense, colorless sclereids. Sheath 11–19(–25) mm long, persisting and falling with the bundle. Pollen cones 3–4 cm, light reddish brown. Seed cones (4–)6–9(–12) cm long, egg- to broadly egg-shaped, slightly asymmetrical and curved, with 45–120(–160) seed scales, green before maturity, ripening light reddish brown, opening widely to release the seeds and then falling together with the stout, curved, 10–15 mm long stalks. Seed scales roughly rectangular, the exposed face diamond-shaped, scarcely raised, with a small umbo bearing a slender, fragile prickle. Seed body 4–7 mm long, the clasping wings another (13–)16–22 mm longer. Mountains of central and southern Mexico and northern Central America, from southern Sinaloa and Puebla (Mexico) south to Jinotega (Nicaragua). Mixed with other pines and evergreen or deciduous hardwoods in many kinds of forests, woodlands, and savannas; (450–)800–2,000(–2,800) m. Zone 9. Synonym: *P. pseudostrobus* var. *tenuifolia* (Bentham) G. R. Shaw.

Maximino pine is very closely related to Douglas pine (*Pinus douglasiana*) and false white pine (*P. pseudostrobus*) and has been

Mature Maximino pines (*Pinus maximinoi*) in nature.

confused or merged with each of these. It can be distinguished from both by its slender, droopy needles and further from the latter by its clear cells crossing the green of the needle cross sections and its shorter seed cones (these are up to 18 cm long in *P. pseudostrobus*). As a strongly Mesoamerican pine, it is one of only four species (with *P. caribaea, P. oocarpa,* and *P. tecunumanii*) to reach Nicaragua, at the southern limit of pine distribution in the New World. The species name honors Maximino Martínez (1888–1964), the Mexican botanist who did more than anyone else to document the conifer riches of his country.

Pinus merkusii Junghuhn & de Vriese

TENASSERIM PINE, SUMATRAN PINE, SHAJA (BURMESE), NAN YA SONG (CHINESE), PÈ:K SO:NG BAÏ (LAOTIAN), THÔNG HAI LÁ (VIETNAMESE), SRÂL (CAMBODIAN), DAMMAR BATU (MALAY), AGUU (PILIPINO)

Pinus subg. *Pinus* sect. *Pinus* subsect. *Pinus*

Tree to 30(–70?) m tall, with trunk to 1(–2) m in diameter. Bark thick, dark grayish brown, deeply ridged and furrowed, with the ridges breaking up into scaly, square blocks. Crown open, broadly conical to dome-shaped or flattened, with numerous spreading, horizontal to upwardly angled branches well clothed with foliage at the ends. Twigs dark brown, hairless, rough with the bases of scale leaves. Buds 1–2 cm long, not resinous. Needles in bundles of two, each needle (15–)17–25(–30) cm long, straight, thin and flexible or slightly thicker and a little stiff, falling during the second year, bright green. Individual needles with prominent lines of stomates on both faces, and two resin canals at the corners midway between the two-stranded midvein and the needle surface and sometimes with a third linking the inner face and midvein. Sheath (10–)12–18(–20) mm long, persisting and falling with the bundle. Pollen cones (2–)4–5 cm long, orange-yellow. Seed cones 5–10(–13) cm long, very narrowly egg-shaped when closed, approaching spherical when open, with 45–75 seed scales, green before maturity, ripening shiny reddish brown, opening widely to release the seeds and then falling with the slender stalk about 1 cm long. Seed scales wedge- to paddle-shaped, the exposed face equally diamond-shaped to barely five-sided, distinctly if modestly pyramidally projecting, with fine, radiating wrinkles, crossed by a sharp horizontal ridge topped by a large, usually indented, prickleless umbo. Seed body 5–8 mm long, the easily detachable wing another 17–25 mm longer. Scattered on hillsides and tablelands across Southeast Asia from southeastern Myanmar, northern Laos and Vietnam, and southern China (southern Guangxi, southwestern Guangdong, and Hainan), south through Thailand and Indochina and in northern Sumatra and the northern Philippines (Mindoro and extreme western Luzon). Forming pure, open pine savannas on poor soils where subjected to frequent fires or mixed with hardwoods that grow up around it in their absence; (50–)250–1,500(–2,000) m. Zone 9. Synonyms: *P. latteri* F. Mason, *P. merkusii* var. *latteri* (F. Mason) Silba, *P. merkusii* var. *tonkinensis* (A. Chevalier) A. Chevalier ex H. Gaussen.

The isolated stand of *Pinus merkusii* on Mount Kerintje in Sumatra, at about 2°S, is the only natural occurrence of *Pinus* in the southern hemisphere. With the exception of the Myanmar-Thailand border region, where narrowly continuous populations can occur, the range of Tenasserim pine consists largely of such geographically isolated occurrences. These populations differ from one another in growth form, bark thickness, needle length and thickness, distribution of stomates, and seed cone length. The differences are minor and do not justify proposed subdivision of the species into varieties, subspecies, or even species. The populations also differ in the frequencies of different enzyme variants, but these differences do not correspond to those found in the morphological features. On the whole, there is less genetic variation within populations of this species and more difference between regional populations than is typical among the pines or others conifers, suggesting that the different regions have been isolated from one another for some time. This contrasts with its geographically overlapping neighbor, Khasi pine (*P. kesiya*), which is more typical of pines in general. Although they are both members of subsection *Pinus,* Tenasserim pine is not particularly closely related to Khasi pine and, in fact, is the most genetically distinctive species in the whole subsection, while DNA studies link it to Masson pine (*P. massoniana*) of China and Cuban red pine (*P. tropicalis*). Where they occur near each other, *P. merkusii* generally grows at lower elevations than *P. kesiya.* Even when they overlap, they appear to be reproductively isolated from one another. As the main low-elevation pine of Southeast Asia and a species that responds well to a regime of fires, Tenasserim pine is an important commercial species as a timber producer and source of resins. It is widely planted in lowland Indonesia outside its natural range. The species name honors Merkus, the Dutch colonial governor of Indonesia at the time the species was described.

Pinus monophylla J. Torrey & Frémont

SINGLELEAF PINYON

Plates 44 and 45

Pinus subg. *Strobus* sect. *Parrya* subsect. *Cembroides*

Tree to 10(–20) m tall, with trunk to 0.5(–1.3) m in diameter, but often dividing into several trunks at or near the base. Bark grayish brown, dividing into wavy, narrow ridges separated by shallow furrows. Crown egg-shaped, becoming more dome-shaped and open with age, with numerous upwardly angled branches from near the base densely clothed with foliage near the ends. Twigs yellowish brown to reddish brown, hairless or slightly hairy at first, becoming gray and hairless with age. Buds 5–10(–15) mm long, resinous. Needles in bundles of one (or two), each needle (2–)4–6 cm long, very stiff and prickly, straight or slightly curved, lasting 4–8(–10) years, dull grayish green to dark green. Individual needles with lines of stomates all around, an undivided midvein, and (2–)4–12(–23) small to modest-sized resin canals. Sheath 5–11 mm long, curling back and soon shed. Pollen cones 7–10 mm long, reddish purple. Seed cones (4–)5–7(–8) cm long, egg-shaped to roughly spherical, with (15–)25–60 seed scales, green before maturity, ripening orange-brown, opening widely to release the seeds and then falling with the 0.5–1 cm long stalk. Seed scales paddle-shaped, with deep seed cavities, the diamond-shaped exposed portion rising pyramid-like on the face and ending in a large diamond-shaped umbo sometimes bearing a short, weak prickle. Seed body 13–18 mm long, the shell 0.3–0.5 mm thick, unwinged or with a rudimentary wing remaining attached to the scale. Basin and Range region of southwestern North America from south-central Idaho and east-central California to southwestern New Mexico and northern Baja California (Mexico). Forming pure stands or mixed with junipers and sometimes shrubby oaks in open pinyon-juniper woodlands in an elevational band overlapping with chaparral or sagebrush scrub below and montane pine forests above; (950–)1,200–1,600(–2,300) m. Zone 7. Three subspecies.

Although many pine species with needles in bundles of two or three may have individual bundles with solitary needles, whether through imperfect development or later loss, *Pinus monophylla* is the only pine characterized by development of a single needle in each bundle (hence the scientific and common names). Some authors treat the single-needled pinyons of Arizona and New Mexico as varieties of Colorado pinyon (*P. edulis*), but they fit more comfortably with the other varieties of *P. monophylla*. Nonetheless, *P. edulis* is fairly closely related to singleleaf pinyon and hybridizes with it wherever they overlap, from southeastern California to southwestern New Mexico, and this may have contributed to the characteristics of subspecies *fallax*.

The seeds of *Pinus monophylla* are larger than those of *P. edulis* and have been an important winter food for some peoples of Nevada and transmontane (eastern) California, where *P. monophylla* is the only pinyon pine species, although the seed fat content there is lower than that of *P. edulis*. Needles of singleleaf pinyon are common in ice-age nests of pack rats well south of the present distribution in Arizona, overlapping the present range of border pinyon (*P. culminicola* var. *discolor*) but showing no hint of hybridization with that species. Singleleaf pinyon also overlaps with its apparent closest relative in DNA studies, the four- or five-needled Parry pinyon (*P. quadrifolia*), in southern California and Baja California and hybridizes with it, yielding trees in which two- or three-needled bundles predominate. Such hybridization in the past may have contributed to the predominance of four-needled bundles in *P. quadrifolia*, but the subject deserves further study. The largest known individual in the United States in combined dimensions when measured in 1991 was 13.7 m tall, 1.3 m in diameter, with a spread of 12.2 m.

Pinus monophylla* subsp. *californiarum (D. Bailey) Zavarin

Needles 1.2–1.4 mm in diameter, with (9–)11–26(–35) lines of stomates and (5–)7–19(–24) resin canals. Sheath about 6 mm long, loosely curled back. Open seed cones about 4.5–6 cm in diameter. Seed body 13–16 mm long, the shell 0.3–0.4 mm thick. Southwestern portion of the range of the species in southern California and northern Baja California (Mexico). Synonym: *Pinus californiarum* D. Bailey.

Pinus monophylla* subsp. *fallax (E. Little) Zavarin

Needles 1.0–1.3 mm in diameter, with 14–18 lines of stomates and three to six resin canals. Sheath about 5–6 mm long, loosely curled back. Open seed cones about 4.5–7.5 cm in diameter. Seed body 13.5–16 mm long, the shell 0.4–0.5 mm thick. Southeastern portion of the range of the species from northwestern Arizona to southwestern New Mexico. Synonyms: *Pinus californiarum* subsp. *fallax* (E. Little) D. Bailey, *P. edulis* var. *fallax* E. Little.

Pinus monophylla J. Torrey & Frémont **subsp. *monophylla***

Needles 1.3–1.6 mm in diameter, with (12–)15–35(–40) lines of stomates and two to seven (to nine) resin canals. Sheath about 9 mm long, tightly curled back. Open seed cones about 6–7 cm in diameter. Seed body 16–18 mm long, the shell 0.25–0.35 mm thick. Northern portion of the range of the species, from southern Idaho and western Utah, through Nevada to south-central California.

Pinus montezumae A. Lambert

MONTEZUMA PINE, OCOTE HEMBRA (SPANISH)

Pinus subg. *Pinus* sect. *Trifoliae* subsect. *Ponderosae*

Tree to 30(–35) m tall, with trunk to 0.75(–1) m in diameter. Bark dark grayish brown, with blocky and interrupted, flat-topped but rough and scaly ridges separated by shallow to deep furrows. Crown dome-shaped to spherical, with heavy, upwardly angled to horizontal branches moderately clothed with foliage toward the tips. Twigs reddish brown, very rough with the bases of scale leaves, hairless. Buds 12–30 mm long, not conspicuously resinous. Needles in bundles of (three to) five (to seven), each needle (14–)20–30(–40) cm long, slender, flexible, and slightly drooping or a little thicker and stiff, lasting 2–3 years, dull light green to slightly bluish green with wax. Individual needles with several lines of stomates on all three faces, and (two or) three to six resin canals at the corners and surrounding the two-stranded midvein midway between it and the needle surface. Sheath 20–35 mm long at first, weathering to 15–25 mm long but persisting and falling with the bundle. Pollen cones 2–4 cm long, reddish brown. Seed cones (8–) 12–18(–23) cm long, egg-shaped to cylindrical, often slightly curved but otherwise nearly symmetrical, with 80–150(–250) seed scales, green before maturity, ripening shiny yellowish to medium brown, opening narrowly or widely to release the seeds and then falling, leaving behind a few basal scales on the persistent thick stalk that is 1–2 cm long or shorter. Seed scales wedge-shaped, the exposed face horizontally diamond-shaped and fairly flat, or conically projecting on the basal scales, with a small, low umbo bearing a tiny, fragile prickle. Seed body 5–7 mm long, the clasping wing another (12–)15–25 mm longer. Mountains of northeastern, central, and southern Mexico and southern Guatemala, from central Nuevo León and southern Durango (Mexico) to Baja Verapaz and Chimaltenango (Guatemala). Forming pure, open stands or commonly mixed with other conifers and hardwood in a wide range of forests and woodlands; (1,050–)2,000–3,000(–3,500) m. Zone 9. Synonyms: *P. gordoniana* K. Hartweg ex G. Gordon, *P. montezumae* var. *gordoniana* (K. Hartweg ex G. Gordon) Silba, *P. montezumae* var. *lindleyi* J. C. Loudon, *P. montezumae* var. *mezambranae* Carvajal, *P. oaxacana* var. *diversiformis* Debreczy & I. Rácz.

Montezuma pine is one of the more variable Mexican species, even when Hartweg pine (*Pinus hartwegii*) and Michoacán pine (*P. devoniana*), sometimes treated as varieties, are excluded. Many other varieties were named, but all occur scattered throughout the range with typical individuals, so there is little justification for recognizing them. Some of this variation, involving coarseness and stiffness of the needles, length of the seed cones, and thickness of the tips of the seed scales, could be due to hybridization with its closest relatives, *P. hartwegii* and *P. devoniana*. Hybridization with other co-occurring species of subsection *Ponderosae* is not out of the question, since *P. montezumae* crosses successfully with ponderosa pine (*P. ponderosa*) and Apache pine (*P. engelmannii*) in experimental trials. The possibilities of natural hybridization involving Montezuma pine certainly merit further investigation. The species produces an excellent, hard timber, and it is heavily exploited in the vicinity of Mexico City, where it reaches its greatest abundance. The species name honors Motecuhzoma Xocoyotzin ("Montezuma," 1368–1420), the Aztec emperor who first encountered the Spanish conquistadors.

Pinus monticola D. Douglas ex D. Don

WESTERN WHITE PINE

Pinus subg. *Strobus* sect. *Quinquefoliae* subsect. *Strobus*

Tree to 50(–75) m tall, with trunk to 1.5(–3.5) m in diameter. Bark grayish brown when young, becoming reddish brown or yellowish brown and broken up into a checkerboard pattern by regular cracks between four- (or more-) sided blocks. Crown narrowly conical at first, becoming oval to cylindrical, with numerous, regularly spaced, thin, sharply and later gently upswept branches, somewhat sparsely clothed with foliage. Twigs light reddish brown and minutely hairy at first, becoming purplish brown to grayish brown and hairless. Buds 4–10 mm long, slightly resinous. Needles in bundles of five, each needle (4–)5–10 cm long, twisted and straight but flexible, lasting 3–4 years, dark bluish green. Individual needles with stomatal lines confined to the inner faces, an undivided midvein, (one or) two resin canals touching the epidermis of the outer face, and occasionally a third resin canal in the angle between the inner faces. Sheath 10–15(–18) mm long, soon shed. Pollen cones 10–15 mm long, straw-colored. Seed cones 10–25(–35) cm long, taperingly cylindrical and slightly curved, with 90–160 seed scales, yellowish green before maturity, ripening pale, often yellowish, brown, opening widely to release the seeds and then falling, on a thick stalk to 2 cm long. Seed scales egg-shaped, the exposed part somewhat thickened but still a little flexible, with a small triangular umbo at the tip, the lowermost, narrow, sterile scales strongly curled back. Seed body 5–8 mm long, the firmly attached wing 18–26 mm long. Mountains of western North America (hence the scientific name, Latin for "mountain living"), Pacific mountains and lowlands from central British Columbia to central California and Rocky Mountains from central British Columbia to northeastern Oregon, northern Idaho, and northwestern Montana. Mostly mixed with other conifers on a variety of soils, but generally on gentle, moist slopes; 0–3,000(–3,300) m. Zone 7.

Western white pine is one of the typical white pines and has been crossed successfully with most of the other species of this subsection. Its closest neighbor and sometime associate, sugar pine (*Pinus lambertiana*), is a notable exception, and these two species are effectively incapable of exchanging genes. Like sugar pine, western white pine is highly susceptible to white pine blister rust, and one of the goals of crossing programs is to introduce resistance to this disease from related pines in the disease's homeland of Asia. The seed cones are like a larger, coarser version of those of the closely related eastern white pine (*P. strobus*), while their foliage is very similar. Hybrids between these two species are very vigorous and fertile, like many of the other combinations among typical white pines. Western white pine was and remains a particularly important timber tree in the Inland Empire of northern Idaho and adjacent Washington, Montana, and British Columbia. The largest known individual in the United States in combined dimensions when measured in 1991 was 46 m tall, 3.2 m in diameter, with a spread of 15.8 m, a stout but not particularly tall individual. Longevity is estimated to reach about 400 years.

Pinus mugo Turra

EUROPEAN MOUNTAIN PINE, SPIRKE (GERMAN)

Pinus subg. *Pinus* sect. *Pinus* subsect. *Pinus*

Compact spreading shrub to 3.5(–6) m tall, or a tree to 25 m. Trunk of tree forms to 0.6(–1.0) m in diameter. Bark reddish brown to dark grayish brown, thin, smooth, flaking in small scales, breaking up slightly into lines of irregular, small plates at the base of large trunks. Crown broadly conical and irregular in tree forms, compact and upright to creeping in shrubby forms. Twigs reddish brown to almost black, hairless, shallowly but sharply grooved between the bases of the scale leaves. Buds 6–15 mm long, very resinous. Needles in bundles of two (or three), each needle (2.3–)3–6(–7.5) cm long, thick, stiff, and twisted, often somewhat curved, lasting (2–)4–8(–10) years, dark green to grayish green or bright green. Individual needles with few to many conspicuous lines of stomates on both the inner and outer faces, a two-stranded midvein, and (1–)3–7(–16) small resin canals scattered all around the periphery of the needle. Sheath 12–20 mm long, persisting and falling with the bundle. Pollen cones 6–12 mm long, yellowish white to red. Seed cones (1.5–)2–6(–7) cm long, egg-shaped, with 60–90 scales, shiny yellowish brown to reddish brown or blackish brown, opening widely to release the seeds and then falling with the very short stalk to 7 mm long or persisting for up to 4 years. Seed scales paddle-shaped, the exposed face diamond-shaped, quite variable, from almost flat to greatly protruding in a downwardly curved pyramid, the large umbo also varying from indented to conical and tipped by a stout to slender, fragile prickle. Seed body 4–6 mm long, the easily detachable wing another 10–15 mm longer. Mountains of southern and central Europe, from northwestern Spain northeastward to southeastern Germany, south to central Italy and southeastward through the northern Balkan Peninsula to southwestern Bulgaria. Forming pure stands in montane bogs and shrubberies at the alpine tree line or mixed with other species on a variety of substrates, from calcareous to acidic and from sandy to sodden; (200–)1,000–2,300 m. Zone 3. Two subspecies.

The dwarf forms of mountain pine are far and away the most popular shrubby pines in cultivation. Fooled by the small selling size and slow growth of these mugo pines, homeowners and landscapers often plant them in unsuitable sites, which they outgrow all too soon. Numerous cultivars of varying shape and needle

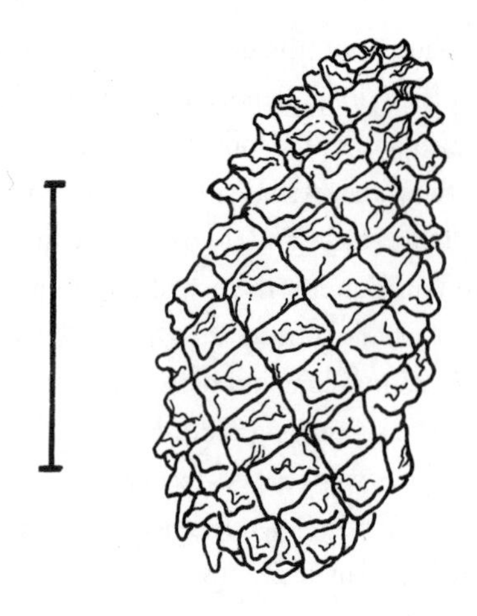

Unopened mature seed cone of European mountain pine (*Pinus mugo*); scale, 2 cm.

length and color have been selected. The tree forms are much less frequently encountered although they have escaped from cultivation in the lowlands of northern Europe, colonizing bogs and sand dunes. There is tremendous natural variation in growth and needle and seed cone characteristics in different parts of the range of mountain pine. Quite commonly, the tree forms of the western portion of the distribution are considered a separate species from the shrubby forms in the Alps and eastward, and additional species are sometimes recognized in the north. Since the tree forms freely hybridize with the shrubby forms in their broad region of overlap in central Europe and the two completely intergrade, showing almost every conceivable combination of the distinctive features, they are treated here as subspecies of a single species. Mountain pine also hybridizes extensively with Scots pine (*Pinus sylvestris*) where they grow together, forming *P.* ×*rhaetica,* but there is no free intergradation like that found between the two subspecies of *P. mugo.*

Pinus mugo Turra subsp. *mugo*

MUGO PINE, DWARF MOUNTAIN PINE

Generally shrubby or sometimes treelike in sheltered locations or when growing near subspecies *uncinata.* Seed cones symmetrical, 1.5–5 cm long, the exposed face of the seed scales flat or a little raised. Eastern portion of the range of the species, from the eastern border of France eastward. Synonyms: *Pinus montana* P. Miller, *P. mughus* Scopoli, *P. mugo* var. *mughus* (Scopoli) Zenari, *P. mugo* var. *pumilio* (Haenke) Zenari.

Pinus mugo subsp. *uncinata* (Ramond ex A. P. de Candolle) Domin

MOUNTAIN PINE, PINO NEGRO (SPANISH)

Generally a tree though sometimes shrubby on exposed sites or when growing near subspecies *mugo.* Seed cones asymmetrical, 2.5–7 cm long, the exposed face of the seed scales distinctly protruding, especially on the side of the cone away from the twig at the base. Western portion of the range of the species, from western Slovakia and Austria westward. Synonyms: *Pinus mugo* var. *rostrata* (F. Antoine) G. Gordon, *P. mugo* subsp. *rotundata* (J. Link) Janchen & H. Neumayer, *P. uncinata* Ramond ex A. P. de Candolle, *P. uncinata* var. *rotundata* (J. Link) F. Antoine.

Pinus muricata D. Don

BISHOP PINE

Pinus subg. *Pinus* sect. *Trifoliae* subsect. *Australes*

Tree to 20(–35) m tall, with trunk to 1(–1.4) m in diameter. Bark dark grayish brown, soon broken up into long vertical ridges divided by deep, dark furrows. Crown broadly dome-shaped, open, with horizontal to sharply rising branches bearing dense tufts of forwardly directed foliage only at the tips. Twigs brown, hairless, rough with scale leaves and their bases. Buds 1–1.5(–2.5) cm long, resinous. Needles in bundles of two (or three), each needle (7–)10–16 cm long, stiff and slightly twisted, lasting 2–3 years, bluish green or dark yellowish green. Individual needles with lines of stomates on both the inner and outer faces, a two-stranded midvein, and 2–4(–14) resin canals around the midvein deep within the leaf tissue. Sheath 1–1.5 cm long, becoming somewhat tattered but persisting and falling with the bundle. Pollen cones (5–)10–20 mm long, pale red. Seed cones 5–9 cm long, symmetrical or asymmetrical and egg-shaped, with 75–120 seed scales, green before maturity, ripening orange-brown, persisting and remaining closed many years after maturity and opening after fires, on short, curved stalks 0.5–1 cm long. Seed scales wedge-shaped, the exposed portion variably thin to strongly protruding as a long, conical projection, especially on the outer scales of asymmetrical cones near the base, the umbo forming a stiff prickle (hence the scientific name, Latin for "like the [spiky] *Murex* snail"). Seed body 5–7 mm long, very dark, the easily detachable wing 14–20 mm long. Discontinuous along the Pacific Coast of North America from northern California to northern Baja California (Mexico). Forming pure stands or mixed with other coastal conifers on various sites from dry ridges to bogs; 0–300 m. Zone 8. Synonyms: *P. muricata* var. *borealis* Axelrod, *P. remorata* H. Mason.

The largest known individual in combined dimensions when measured in 2001 was 21.4 m tall, 0.9 m in diameter, with a spread of 21.0. Although it is thus a smaller tree and lacks the worldwide economic importance of Monterey pine (*Pinus radiata*), its close relationship to that species, followed by the interest generated by early results, led to intensive studies of bishop pine. These studies, involving growth characteristics, crossing relationships, botanical features, wood and leaf oils, enzyme variants, seed proteins, and various kinds of DNA, present a complex and partly contradictory view of variation among populations within the species. Taken together, they define three regional races within its limited geographic range.

The northern race extends south to Sea Ranch in Sonoma County, California, where it meets the central race, which extends southward from there to include the small population in Monterey that is completely surrounded by Monterey pines. The southern race covers the remainder of the range, from San Luis Obispo (from which the species receives its common name since *obispo* is Spanish for "bishop") to the vicinity of San Vicente, Baja California, including stands on Santa Cruz and Santa Rosa Islands. None of these races is uniform, but the southern race is particularly variable in cone morphology and oil composition. The two island populations have a high frequency of trees with small, symmetrical, relatively smooth cones, and these have been treated as a separate species (*Pinus remorata*) or variety. They intergrade completely with trees bearing fiercely spiky cones, however, and there are no other features known to distinguish them, so this separation is not generally accepted by botanists. The sharpest morphological boundary separates the northern and central races at Sea Ranch. Within the space of about 3 km there is a complete turnover between trees with green needles belonging to the central race (which is indistinguishable from the southern race in its needles) and those of the northern race, whose needles are bluish green because the stomates have conspicuous plugs of wax. There is a changeover in oil composition here, too, and a small proportion of trees at the boundary are evidently hybrids, based on their intermediate oil composition. Despite this, breeding behavior points to the boundary between the central and southern races as being most important genetically. There is an almost complete barrier to crossability, stronger than that between most closely related pine species, between the northern and central races on the one hand and the southern race on the other. Furthermore, while the southern race can be crossed with both Monterey pine (*P. radiata*) and knobcone pine (*P. attenuata*), the northern races are isolated from these two species also.

The picture becomes even more complicated when the results of DNA studies are considered. They reveal somewhat different historical relationships among the races, depending upon what kind of DNA is examined. The DNA in the chloroplasts of trees belonging to the southern race is more closely related to that of Monterey pines than to that of trees belonging to the northern races. DNA from the mitochondria of southern trees is rather heterogeneous, with the island populations having their own unique variants, while mainland trees share variants with both Monterey pines and knobcone pines, but not trees of the northern races. Thus most of the genetic evidence favors separation of the two northern races from the southern one, but the central race is apparently morphologically indistinguishable from the southern one, while the northern race differs only trivially from these in needle color. Until someone can find convincing morphological differences between the three races it does not seem feasible to accept varieties within bishop pine because they would not be identifiable without genetic analyses. Work continues on this evolutionarily most interesting pine species and its closed-cone relatives.

Pinus ×murraybanksiana Righter & Stockwell

Slender tree to 25 m tall or more, including both artificial and natural hybrids between jack pine (*Pinus banksiana*) and lodgepole pine (*P. contorta*), with varying resemblances to the two parents. The parent species are rather similar to one another, so there are few characteristic features of the hybrids. They often grow more rapidly than either parent, although they are similar to jack pine in growth rate and in their gray bark. By contrast, the growth habit is more like that of lodgepole pine, with the branches stiff and upwardly angled rather than flexible and sometimes even drooping. The paired needles, 3–5 cm long, are a little longer on average than those of jack pine and a little shorter than those of lodgepole pine. The seed cones are variable with some of the variation due to variations in parentage. The natural hybrids involve the Rocky Mountains *P. contorta* var. *latifolia,* while artificial hybrids produced in California had local variety *murrayana* of the Sierra Nevada as seed parents. Because of differences in the seed cones of these two varieties, the natural and artificial hybrids differ in cone characteristics. Both have seed cones that remain on the trees for years after maturity, but in natural hybrids the cones are asymmetrical, usually curled, and remain closed at maturity. In contrast, the artificial hybrids have straight, nearly symmetrical seed cones that open at maturity but still persist after releasing their seeds. Both kinds of hybrids lack the prominent doming of the exposed portions of the lower seed scales on the side away from the twig that is usually found in lodgepole pine, and they have much smaller prickles, no larger than 1 mm long. Seeds resemble those of both parents and are generally viable so that advanced generation hybrids and back crosses may be formed.

The natural hybrids, which are restricted to the region of overlap of the parent species in the foothills of the Rocky Mountains in western Canada, were studied extensively with respect to morphology, wood oils, enzymes, and DNA variation. These hybrids are important in understanding the details of inheritance of DNA from chloroplasts and mitochondria in conifers, which differ from the patterns found in the more numerous flowering plants. They are also useful in unraveling the biosynthetic pathways and genetic control of terpenoids, the main chemicals of the wood oils. Little follow-up work was done on the artificial hybrids, and much still remains to be learned about the natural hybrids. For instance, hybrids from the northern area of range overlap in the southwestern Northwest Territories were not studied at all, yet they might constitute an interesting comparison to the much-studied southern hybrids of Alberta.

Pinus ×neilreichiana H. Reichardt

Natural and artificial hybrids between the Europe black pine (*Pinus nigra*) and Scots pine (*P. sylvestris*). The hybrids are intermediate between the parents in some features, such as needle length, at about 7–10 cm, but also bring together other characteristics of each parent. They may give an overall impression of being an odd specimen of either of the parent species. The growth rate in youth is faster than that of black pine and slower than that of Scots pine. The bark is variable but is typically redder than in black pine on the branches and grayer and darker than in Scots pine on the trunk. The buds, at about 8–10 mm long, are intermediate in length (as well as in diameter and shape) between those of the two parents. The paired needles are shorter and a little bluer than those of black pine and often are more obviously twisted. Correspondingly, they are longer and less blue than those of Scots pine and are noticeably stiffer. While black pine has resin canals midway between the needle surface and the two-stranded midvein, and Scots pine has them touching the needle surface, the hybrid has them in both positions. The needle sheaths and pollen cones are each intermediate between those of the parents at 10–15 mm long, versus smaller in Scots pine and larger in black pine. The size of seed cones in the parent species overlaps so strongly that the hybrids are not noticeably different from either, but they do not get as large as those of black pine, and the exposed face of the seed scale is more raised than that of Scots pine. The seed body is intermediate in length, at about 4.5–6 mm long compared to 3–5 mm in Scots pine and 5–7 mm in black pine. Hybrids occur spontaneously in plantations outside the natural range of one or both parent species, but they might occur naturally in the northwestern Balkan Peninsula or in Turkey, where black and Scots pine have their most extensive range overlaps. Different experimenters have greatly varied success in producing this hybrid artificially, and often the reciprocal crosses also have differing success rates. Some authors who attempted reciprocal crosses succeeded only with *P. nigra* as the seed parent, while others succeeded only with *P. sylvestris!*

Pinus nelsonii G. R. Shaw

NELSON PINYON

Pinus subg. *Strobus* sect. *Parrya* subsect. *Nelsoniae*

Tree to 10 m tall, with trunk to 0.3 m in diameter. Bark light gray and smooth at first, becoming scaly at the base and darker, with brown highlights. Crown very broad, with numerous long, slender branches that pass from upraised, through horizontal, to downswept with age, sparsely clothed with foliage at the tips. Twigs grayish white, hairless, sometimes thinly waxy. Buds 8–10 mm long, resinous. Needles in bundles of three (or four), each needle (4–)5–8(–10) cm long, stiff and straight to curved, remaining stuck together until just before falling after 2–3 years, dark green to grayish green with wax. Individual needles with lines of stomates on all three faces, an undivided midvein, and two large resin canals near the outer corners touching the epidermis of the outer face. Sheath 5–7(–9) mm long, persisting until the bundles are shed. Pollen cones 7–9 mm long, pinkish brown. Seed cones (5–)8–12(–14) cm long, cylindrical, with 50–100 seed scales, purple before maturity, ripening reddish brown, opening widely, but not enough for automatic release of the seeds, then falling off the stout stalk (2.5–)3.5–6.5 cm long. Seed scales egg-shaped, the exposed portion very thick, with a thick triangular projection extending down from the face and ending in a stout, triangular umbo. Seed body (10–)12–15 mm long, egg-shaped, the seed coat very thick and hard, the vestigial wing remnant not separating from the seed scale. Sierra Peña Nevada and other ranges of the Sierra Madre Oriental from southern Coahuila or southwestern Nuevo León to central San Luis Potosí (Mexico). Scattered in dry, open, scrubby pinyon-juniper-oak woodlands; (1,600–)1,800–2,400(–3,200) m. Zone 9.

Although rare, Nelson pinyon is less threatened than many other Mexican pines because as a small, sparsely seeding tree of remote localities it is not subject to the same intense exploitation as other species, most notably king pinyon (*Pinus maximartinezii*). The species name honors Edward W. Nelson (1855–1934), who procured the type specimen, among many other Mexican plants that he collected for the US Department of Agriculture.

Nelson pinyon was long included with other pinyon pine species in subsection *Cembroides*, having many morphological similarities to three anomalous, basal members of the group: king pinyon, weeping pinyon (*Pinus pinceana*), and Coalcomán pinyon (*P. rzedowskii*). DNA studies show fairly decisively, however, that it is actually more closely related to the bristlecone pines (subsection *Balfourianae*) than it is to the other pinyon pines, but distinct enough that it is placed in subsection *Nelsoniae* all by itself. These relationships effectively quash the idea that Nelson pinyon and the three other anomalous pinyon species form an intercontinental link to the Chinese and Himalayan edible-seed species in subsection *Gerardianae*. Instead, section *Parrya*, to which the pinyon pines and the *Balfourianae* belong, is entirely New World in distribution, while the other section within the white or soft pines of subgenus *Strobus*, section *Quinquefoliae*, is largely Old World with a few New World members. This exactly parallels the situation in the hard pines (subgenus *Pinus*) in which one section (*Trifoliae*) is entirely New World while the other (section *Pinus*) is largely Old World with just two New World species.

Mature Nelson pinyon (*Pinus nelsonii*) in nature.

Pinus nigra J. F. Arnold

BLACK PINE, AUSTRIAN PINE

Pinus subg. *Pinus* sect. *Pinus* subsect. *Pinus*

Tree to 30(–50) m tall, with trunk to 1.5(–2) m in diameter. Bark light to dark grayish brown or black (hence the scientific name, Latin for "black," in contrast to the bright orange bark of *P. sylvestris*) divided into narrow, interlaced ridges by deep furrows. Crown conically dome-shaped to almost spherical, often with a few heavy lower branches and numerous tiered, slender, upturned, horizontal upper branches densely clothed with foliage near the tips. Twigs orange-brown to dark brown or even black, hairless, rough with the bases of scale leaves. Buds (6–)10–18(–20) mm long, resinous. Needles in bundles of two, each needle (4–)7–14(–19) cm long, thick but slightly flexible, straight or curved, sometimes a little twisted, lasting (2–)3–5(–6) years, usually dark green. Individual needles with numerous, closely spaced, inconspicuous lines of stomates on both faces, and (2–)3–10(–17) resin canals all around the two-stranded midvein midway between it and the needle surface. Sheath 15–25 mm long, weathering to 4–6(–12) mm long and persisting and falling with the bundle. Pollen cones (15–)20–30(–35) mm long, yellowish brown. Seed cones sticking out from the twigs, (3–)5–8(–12) cm long, egg- to broadly egg-shaped with a flat base, with (60–)75–100 seed scales, green before maturity, ripening light brown to yellowish brown, shiny or

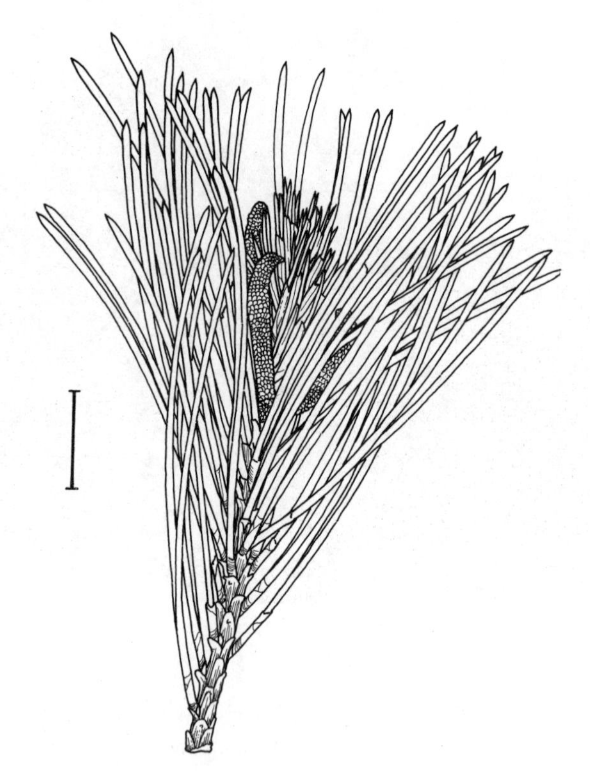

Twig of black pine (*Pinus nigra*) with pollen cones at time of pollination; scale, 2 cm.

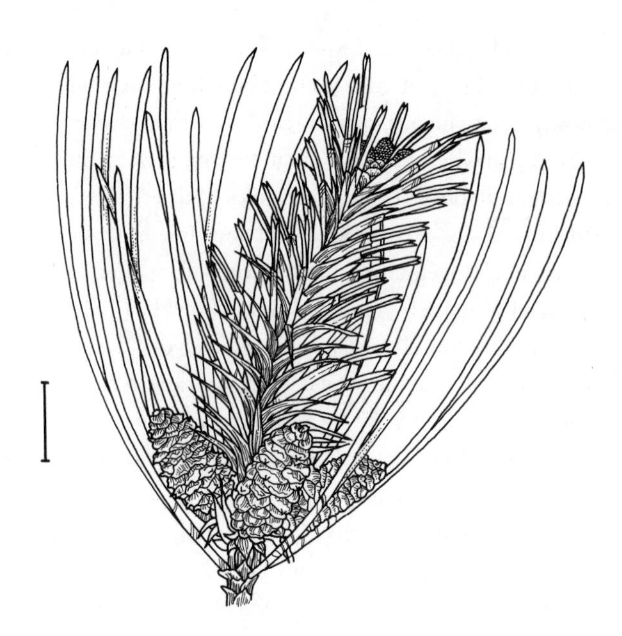

Twig of black pine (*Pinus nigra*) with newly pollinated and maturing seed cones; scale, 2 cm.

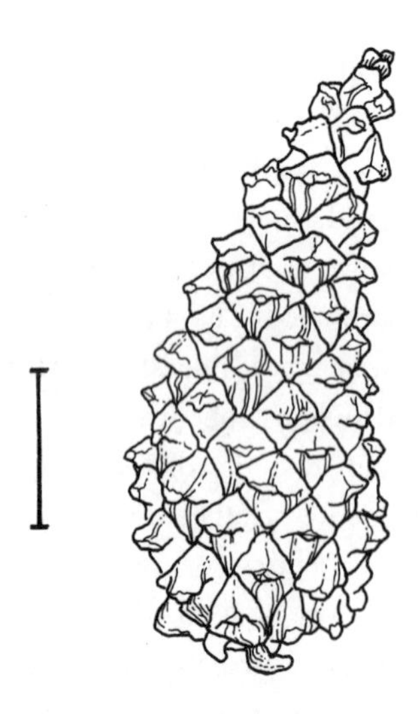

Unopened mature seed cone of black pine (*Pinus nigra*); scale, 2 cm.

not, opening widely to release the seeds and then falling with the inconspicuous stalk, 5–10 mm long. Seed scales spoon-shaped, the exposed face fan-shaped, crossed by a sharp horizontal ridge, flat or modestly protruding, with a small diamond-shaped umbo tipped by a tiny, blunt or sharp prickle. Seed body 5–7 mm long, the easily detachable wing another 12–22(–25) mm longer. Discontinuously distributed throughout southern Europe (north to eastern Austria) and Turkey, with outliers on the northern shore of the Black Sea and in northwestern Africa, in Morocco and Algeria. Forming pure stands or mixed with other conifers and hardwoods in a wide variety of environments, chiefly in the mountains; (200–)300–1,800(–2,200) m. Zone 4. Two subspecies.

Black pine is an important timber species that usually grows above the band of Aleppo pine (*Pinus halepensis*) in the numerous areas where the two species occur near each other. It is widely grown as an ornamental and to a certain extent in plantation forestry both within its home range and beyond in Europe and North America. Numerous cultivars were selected from the wild and from nursery beds. Furthermore, provenance tests seeking superior seed sources for plantation culture confirm the genetic basis of much of the morphological variation long recognized in the species. This variation, coupled with the fragmented geographic distribution, led to the description of dozens of natural variants of black pine as varieties, subspecies, or even species.

Nowadays, most often five or six subspecies are recognized, sometimes with included varieties, but these proposed subspecies intergrade with each other. Despite the distinctive growth characteristics of certain populations, like the superior qualities for forestry of Corsican black pines, there appears to be little justification for complex subdivision or the species when one considers it as a whole throughout its distribution. There does appear to be some justification for dividing black pine into just two subspecies, an eastern and a western one, rather than the five or six most commonly recognized. It might then be possible to recognize two to four varieties within each subspecies, corresponding to those more narrowly defined subspecies. However, much more thorough range-wide assessment is needed before a final decision on the utility of adopting narrowly defined taxonomic variants within *Pinus nigra*. A conservative subdivision into two subspecies is followed here, and even these two are only weakly differentiated. Perhaps the strongest argument for this treatment is that the narrowly drawn subspecies or varieties are still highly heterogeneous for morphological features not used in their definition, for turpentine and enzyme composition, and for DNA sequences. A recommendation, made in an account of variation in the species published in 1957, that distinctive seed sources be named simply by identifying their place of origin rather than giving them formal taxonomic recognition, still seems sensible today.

Black pine crosses successfully with several other species of subsection *Pinus* and even with some species of subsection *Pinaster*. The reported hybrids include crosses with the European mugo (*Pinus mugo*), Scots (*P. sylvestris*), Bosnian (*P. heldreichii*), and Aleppo (*P. halepensis*) pines, and the Eastern Asian Japanese red (*P. densiflora*), black (*P. thunbergii*), Taiwan (*P. taiwanensis*), and Chinese Mountain (*P. tabuliformis*) pines. Some of these occur spontaneously in nature or in plantations, especially the hybrids with Japanese red, Bosnian, and Scots pines. The latter is named *P. ×neilreichiana*. On the other hand, reported crosses with the North American red pine (*P. resinosa*), also a member of subsection *Pinus*, are extremely doubtful. Most attempts to cross these rather similar-looking species fail, but rare successes have been reported. Trees from one reportedly successful attempt described in detail as such hybrids were definitely shown not to be *P. nigra* × *P. resinosa*, using the enzyme variants they contained. Instead, these trees had an Asian pine as their pollen parent, very possibly Japanese red pine, which hybridizes easily with black pine in either direction. Pollen contamination is always of concern in artificial crossing programs.

Pinus nigra J. F. Arnold **subsp. *nigra***

Needles with the full range of length for the species, 4–19 cm, stiff because of two to five layers of thick hypodermis cells just beneath the surface. Seed cones with the full range of length for the species, 3–12 cm long. Eastern portion of the range of the species, from central and northeastern Italy eastward. Synonyms: *Pinus banatica* (Georgescu & Ionescu) Georgescu & Ionescu, *P. nigra* var. *austriaca* (Höss) J. C. Loudon, *P. nigra* var. *caramanica* (J. C. Loudon) Rehder, *P. nigra* subsp. *dalmatica* (Visiani) Franco, *P. nigra* subsp. *pallasiana* (A. Lambert) Holmboe.

Pinus nigra **subsp. *salzmannii*** (Dunal) Franco

Needles with a restricted range of length for the species, 8–16 cm long, flexible because the one or two layers of strengthening hypodermis cells beneath the surface are also relatively thin walled. Seed cones with a restricted range of length for the species, 4–8 cm long. Western portion of the range of the species, from Sicily, Calabria (the "toe" of Italy), and Corsica westward. Synonyms: *Pinus nigra* var. *calabrica* (J. C. Loudon) C. K. Schneider, *P. nigra* var. *cebennensis* (Godron) Rehder, *P. nigra* subsp. *laricio* (Poiret) J. Maire, *P. nigra* var. *maritima* (W. Aiton) Melville, *P. nigra* subsp. *mauretanica* (J. Maire & Peyerimhoff) Heywood.

Pinus occidentalis Swartz

HISPANIOLAN PINE

Pinus subg. *Pinus* sect. *Trifoliae* subsect. *Australes*

Tree to 30(–45) m tall, the long, straight trunk to 1(–1.5) m in diameter. Bark grayish brown, thick, deeply furrowed between narrow, scaly, flat-topped ridges. Crown oval to bowl-shaped, with slender, horizontal to upswept branches densely clothed with foliage at the tips. Twigs brown beneath a waxy blush, rough with the bases of scale leaves, hairless. Buds 8–15 mm long, slightly resinous. Needles in bundles of (two or) three to five (or six), each needle (10–)15–20 cm long, flexible and drooping, lasting 2–3 years. Individual needles with evident lines of stomates on both the inner and outer faces, and three to five resin canals touching the two-stranded midvein at its corners and sides. Sheath 8–15 mm long, persisting and falling with the bundle. Pollen cones 10–15 mm long, brownish yellow. Seed cones (4–)6–9(–11) cm long, egg-shaped, often a bit asymmetrical, with 50–100 seed scales, green before maturity, ripening shiny dark brown, opening widely to release the seeds and then persisting several years before falling with the slender, 1–2 cm long, curved to straight stalk. Seed scales paddle-shaped, the exposed face horizontally diamond-shaped or five-sided, crossed by a modest ridge topped by a small umbo with a stout, sharp prickle. Seed body 4–6 mm long, the easily detachable wing 9–16(–18) mm longer. Western Hispaniola (hence the scientific name, Latin for "western"), in Haiti and the western Dominican Republic, and eastern Cuba, in Granma and Santiago de Cuba provinces. Forming pure, open to dense stands or mixed with various hardwoods, primarily in the mountains today; (50–)900–2,700(–3,175) m. Zone 9. Two varieties.

This, the only pine found on the island of Hispaniola, was an important timber tree there and supported an export market into the second quarter of the 20th century. It is not faring so well today. Political anarchy in Haiti in the 1990s led to elimination of conservation measures and government regulation of forestry and then reckless depletion of the already much-reduced pine forests. Things are a little better in the Dominican Republic, but slash-and-burn subsistence agriculture by a rapidly growing population in both countries is eliminating forest land. Rampant wildfires, whether purposeful or caused by negligence, destroy many regenerating forests because young tees, lacking the thick, fire-resistant bark of mature specimens, are killed by them. Although not as fast growing as some other tropical pines, Hispaniolan pine is ripe for improvement programs because of its good form and excellent wood, one of the best among the tropical pines. The wood shows about three or four pairs of alternating dark and light bands each year, corresponding to wetter and drier growth periods. The presence of the species in Cuba is controversial, since some authors include the pines of the Sierra Maestra in the western part of the old Oriente province in *Pinus cubensis* (Oriente pine) and others consider them a separate species. They are considered here a weakly distinguished variety within *P. occidentalis.* They resemble Hispaniolan pine from Hispaniola and differ from Oriente pine in their longer, flexible needles, seed scale umbos with a stronger prickle, and seeds with an easily detachable wing. They differ from the pines of Hispaniola in having usually three needles per bundle rather than four or five, but Oriente pines have only two needles per bundle. The relationships among the pines of eastern Cuba and Hispaniola deserve much further study.

Pinus occidentalis* var. *maestrensis (Bisse) Silba
SIERRA MAESTRA PINE
Needles in bundles of (two or) three (or four), 15–20 cm long. Sierra Maestra of Granma and Santiago de Cuba provinces (Cuba); (300–)1,000–1,700(–2,000) m. Synonym: *Pinus maestrensis* Bisse.

Pinus occidentalis Swartz **var. *occidentalis***
HISPANIOLAN PINE
Needles in bundles of (three or) four or five (or six), 10–18 cm long. Western Hispaniola, from Azua and la Vega provinces in the central Dominican Republic westward; (50–)900–2,700(–3,175) m. Synonym: *Pinus occidentalis* var. *baorucoensis* Silba.

Pinus oocarpa Schiede ex D. F. L. Schlechtendal
MESOAMERICAN CLOSED-CONE PINE, PINO COLORADO (SPANISH)
Pinus subg. *Pinus* sect. *Trifoliae* subsect. *Australes*
Tree to 35(–55) m tall, with trunk to 0.7(–1.2) m in diameter. Bark dark grayish brown, thick, rough, with irregular, scaly ridges broken up into elongate plates by shallow furrows. Crown broadly dome-shaped, open, with irregularly forked, upraised branches densely clothed with foliage only at the tips. Twigs reddish brown, hairless, initially rough with the bases of scale leaves but these flaking off during or after the second season. Buds 12–25 mm long, not resinous. Needles in bundles of (three to) five, each needle (11–)17–25(–30) cm long, usually thick, stiff, and straight, lasting 2(–3) years, shiny light green to yellowish green. Individual needles with a few inconspicuous lines of stomates on all three faces, and four to eight resin canals surrounding the two-stranded midvein and usually touching both it and the outer needle surface. Sheath 18–27 mm long, persisting and falling with the bundle. Pollen cones 15–20 mm long, yellowish brown with a reddish blush. Seed cones 3–8(–10) cm long, very broadly egg-shaped to almost spherical when open, with 40–100(–130) seed scales, green before maturity, ripening light to dark yellowish brown, opening only gradually but ultimately very widely when ripe and remaining attached for several years before falling together with the long, slender, curved stalk (1–)1.5–3.5(–4.5) cm long. Seed scales very woody, irregularly rectangular, the exposed face roughly diamond-shaped to six-sided, fairly flat to strongly conically protruding and tipped by a modest umbo usually bearing a tiny, fragile prickle. Seed body 4–8 mm long, the clasping wing another (8–)12–18 mm longer, strongly thickened toward the seed body. Mountains of Mesoamerica and western Mexico, from Matagalpa (Nicaragua) and Olancho (Honduras) to northeastern Hidalgo and southeastern Sonora (Mexico). Forming pure stands or mixed with other species in open, fire-prone pine-oak woodlands; (200–)1,200–2,000(–2,700) m. Zone 9. Synonym: *P. oocarpa* var. *trifoliata* M. Martínez.

This fire-adapted tree has the widest geographic and elevational range of any Mesoamerican pine species but reaches its best development at middle elevations. It is an important timber tree in Central America and has been tried for plantation culture in other tropical mountain regions. Northward in Mexico, it becomes much more shrubby and less valuable, perhaps due in part to the kind of past depredations that have made the related Monterey pine (*Pinus radiata*) just a picturesque tree in its native stands but an important timber tree in plantations. *Pinus oocarpa* is often confused with its closest relatives, Mexican scrub pine (*P. praetermissa*) and Tecun Umán pine (*P. tecunumanii*), both in nature and in plantation trials. It is highly interfertile with another close relative, Mexican weeping pine (*P. patula*), in artificial crossing programs, although possible hybridization in nature has not been assessed. Perhaps the long seed cone stalks sometimes found in *P. patula* in the southern portion of its range are due to hybridization where the two species co-occur. Hybridization may also explain the trees of *P. oocarpa* with predominantly three needles per bundle instead of the usual five that are found widely scattered through the range of the species. There have also been reports of natural hybridization with the less closely related lowland Caribbean pine (*P. caribaea*), another member of subsection *Australes,* in their zone of elevational overlap in Honduras. Although some authors doubt whether this hybridization occurs in nature, such hybrids are made with difficulty artificially. The species name, Latin for "egg fruit," refers to the shape of the unopened seed cones and the fact that they long remain closed on the tree.

Pinus palustris P. Miller
LONGLEAF PINE
Plates 46 and 47
Pinus subg. *Pinus* sect. *Trifoliae* subsect. *Australes*
Tree to 40(–47) m tall, with trunk to 0.8(–1.2) m in diameter. Bark thick, orange-brown, broken up into small, scaly plates by broad, dark gray furrows. Crown deeply and openly dome-shaped, with gently rising, slender branches densely clothed with foliage only at the tips. Twigs very coarse, orange-brown, rough with the bases of scale leaves, hairless. Buds 3–4 cm long, silvery white, not conspicuously resinous. Needles in bundles of (two or) three, each needle (15–)20–30(–45) cm long, flexible and straight or drooping, slightly twisted, lasting about 2 years, shiny yellowish green. Individual needles with inconspicuous lines of stomates on both the inner and outer faces, and four to seven resin canals surrounding and touching the two-stranded midvein. Sheath 2–3 cm long, weathering to 1–2 cm and persisting and falling with the bundle. Pollen cones 2–6(–8) cm long, purple. Seed cones (12–)15–25 cm long, narrowly egg-shaped, with 75–150 seed scales, green before maturity, ripening dull brown, opening widely to release the seeds and then falling during the winter with the very short stalk. Seed scales broadly wedge-shaped, the exposed face horizontally diamond-shaped, flat, with a prominent umbo bearing a sharp, down-turned prickle. Seed body 9–12 mm long, the firmly attached wing another 25–30 mm long. Coastal plain of the southeastern United States, from southeastern Virginia to central Florida and southeastern Texas. Forming pure, open stands or mixed with other pines and oaks on fire-prone flatlands and gentle slopes; 0–700 m. Zone 8.

Longleaf pine is one of the four most economically important "southern" pines, the species of subsection *Australes* native to the southeastern United States. It is the most fire-tolerant among these species because of its thick bark and its seedling "grass" stage. This is a very condensed, slow-growing phase during which there is massive root growth but little top growth, with the buds protected from fire by a dense clothing of needles. This seedling phase greatly slows the growth of longleaf pine compared to the other three main species: shortleaf pine (*Pinus echinata*), slash pine (*P. elliottii*), and loblolly pine (*P. taeda*). Therefore, even though longleaf pine produces the finest timber among these species in addition to abundant oleoresin, vast areas of longleaf pine forest were harvested and converted to plantations of the other species, depending on the kind of site. Only about 3,900 hectares of old-growth longleaf pine remain from an original range of 37 million. This loss of longleaf pine forest has been slowed by government requirements to preserve habitat for an endangered bird species, the red-cockaded woodpecker, which breeds only in mature longleaf pine stands. Although longleaf pine depends on occasional fires to maintain its dominance by killing other, faster-growing species, it is not restricted to dry sites but also grows on seasonally flooded flats (hence the scientific name, Latin for "of marshes"). The largest known individual in combined dimensions when measured in 1999 was 36.6 m tall, 1.0 m in diameter, with a spread of 20.1 m. Longleaf pine crosses naturally with its neighbors, loblolly pine and slash pine, and was crossed artificially with shortleaf pine, although this cross was difficult to make. The natural hybrid with loblolly pine was one of the earlier interspecies pine hybrids to be recognized as such and was named *P. ×sondereggeri.* Studies of genetic variation in longleaf pine reveal greater genetic diversity in the western part of its range than in the east, suggesting that it expanded into its present range after the last glaciation from a single refuge in Texas.

Mature longleaf pine (*Pinus palustris*) in nature.

Pinus parviflora P. Siebold & Zuccarini
JAPANESE WHITE PINE, GOYO-MATSU (JAPANESE)
Pinus subg. *Strobus* sect. *Quinquefoliae* subsect. *Strobus*
Tree to 25(–30) m tall, with trunk to 1(–1.5) m in diameter. Bark dark gray to dark brown, flaking in scales and ultimately becoming narrowly ridged and furrowed at the base of large trunks. Young crown densely conical, becoming cylindrical and open with age, with slender, short, horizontal branches bearing foliage near their ends. Twigs yellowish brown to reddish brown, hairless to transiently densely hairy. Buds about 5 mm long, not resinous to slightly resinous. Needles in bundles of (two to) five, each needle (2.5–)4–6(–9) cm long, a little stiff and curved, lasting 3–4 years, dark green on the outer face, waxy whitish green or just light green on the inner faces. Individual needles with lines of stomates only on the inner faces, an undivided midvein, and two (or three) small resin canals, the two near the outer face touching the epidermis or not, the third, when present, deep within the leaf tissue at the corner joining the two inner faces or sometimes touching the epidermis on one side. Sheath 5–15 mm long, soon shed. Pollen cones 5–6 mm long, reddish brown. Seed cones (3–)6–11(–17) cm long, shortly cylindrical when closed but opening egg-shaped, with 30–80 seed scales, green before maturity, ripening yellowish brown, reddish brown, or dark brown, opening widely to release the seeds and persisting several years, almost unstalked or with a stalk to 2 cm long. Seed scales angularly egg-shaped, thin and flexible, straight or curled back below the small, triangular umbo. Seed body 8–10(–12) mm long, pale to dark brown, the firmly attached wing a little shorter than to about 1.5 times as long as the seed body, (3–)7–12(–16) mm long. Japan, Ullung Island (Korea), Taiwan, southern China, and northern Vietnam. Usually mixed with hardwoods on mountain slopes and hilltops; (60–)300–2,000(–2,500) m. Zone 5. Five varieties.

Japanese white pine has long been cultivated as an ornamental and as a favorite subject for bonsai in both Japan and China. There are many cultivars, including dwarf ones and several color variants. The species is named for the unusually small seed cones (Latin for "small flower") among the true white pines in subsection *Strobus.* Otherwise, *Pinus parviflora* is a fairly typical white pine and crosses successfully with several others, including Macedonian pine (*P. peuce*) from Europe, Himalayan white pine (*P. wallichiana*), and eastern (*P. strobus*) and western (*P. monticola*) white pines from North America. The species varies across its broad range, and the regional populations described as varieties here are considered separate species in many accounts of pines. More work needs to be done on the relationships of the varieties to each other and to other species. The three southern varieties are rare and have been severely depleted by general logging and habitat destruction. The wood thus obtained in mixture with other species has a variety of uses in general construction and carpentry. A single population of variety *kwangtungensis* on limestone hills in southwestern Guangxi is unusual in having most needles in bundles of two or three with obscure stomates and the smallest seed cones in the species at 3–4 cm long.

Pinus parviflora var. _kwangtungensis_ (W. Y. Chun ex Tsiang) Eckenwalder
HUA NAN WU ZHEN SONG (CHINESE),
THÔNG NĂM LÁ (VIETNAMESE)
Branchlets yellowish brown, hairless. Needles in bundles of (two, three, or) five, each needle 3.5–7 cm long, 1–1.5 mm wide, with two (or three) resin canals, the two near the outer face touching the epidermis. Seed cones (3–)5–9(–17) cm long on a stalk up to 2 cm long. Seed wing 12–18 mm long. Southeastern China, from Guizhou to Guangdong south to Hainan and in northern Vietnam, often on the summit of limestone hills. Synonyms: *Pinus kwangtungensis* W. Y. Chun ex Tsiang, *P. kwangtungensis* var. *variifolia* N. Li & Y. C. Zhong, *P. wangii* var. *kwangtungensis* (W. Y. Chun ex Tsiang) Silba.

Pinus parviflora var. _morrisonicola_ (Hayata) C. L. Wu
TAI WAN WU ZHEN SONG (CHINESE)
Branchlets reddish brown, transiently hairy. Needles in bundles of five, each needle (4–)6–9 cm long. 0.6–1 mm wide, with two resin canals touching the epidermis below the outer face. Seed cones 7–10(–11) cm long on a stalk 0.5–1 cm long. Seed wing 12–15(–20) mm long. Taiwan. Synonym: *Pinus morrisonicola* Hayata.

Pinus parviflora P. Siebold & Zuccarini **var. _parviflora_**
HIME-KO-MATSU (JAPANESE)
Branchlets yellowish brown, usually transiently hairy. Needles in bundles of five, each needle 3–6 cm long, 0.6–1 mm wide, with two (or three) resin canals, the two near the outer face touching the epidermis. Seed cones (4–)6–8 cm long on a stalk less than 0.5 cm long. Seed wing 3–7 mm long. Southern Japan from central Honshū to Shikoku and Kyūshū and also on Ullung Island (Korea). Mostly on the Pacific Ocean side of central Honshū, where it overlaps with variety *pentaphylla.* Synonym: *Pinus pentaphylla* var. *himekomatsu* (K. Miyabe & Y. Kudô) Makino.

Pinus parviflora var. _pentaphylla_ (H. Mayr) A. Henry
KITA-GOYO (JAPANESE)
Branchlets yellowish brown, transiently hairy. Needles in bundles of five, each needle 3–6 cm long, 0.6–1 mm wide,

with two (or three) resin canals, the two near the outer face touching the epidermis. Seed cones (4–)6–8 cm long on a stalk less than 0.5 cm long. Seed wing 10–12 mm long. Northern Japan from southern Hokkaidō to central Honshū, where mostly on the Sea of Japan side in the area of overlap with variety *parviflora*. Synonym: *Pinus pentaphylla* H. Mayr.

Pinus parviflora var. *wangii* (H. H. Hu & W. C. Cheng) Eckenwalder

MAO ZHI WU ZHEN SONG (CHINESE)

Branchlets reddish brown, transiently hairy. Needles in bundles of five, each needle 2.5–6 cm long, 1–1.5 mm wide, with three resin canals, the two near the outer face not touching the epidermis. Seed cones 4.5–9 cm long on a stalk 1.5–2 cm long. Seed wing 12–16 mm long. Southern Yunnan (China) and possibly central Vietnam, usually on limestone hills. Synonym: *Pinus wangii* H. H. Hu & W. C. Cheng.

Twig of Japanese white pine (*Pinus parviflora* var. *parviflora*) with 3 years of growth, ×0.5.

Opened seed cone of Japanese white pine (*Pinus parviflora* var. *parviflora*) after seed dispersal; scale, 1 cm.

Pinus patula Schiede & Deppe ex D. F. L. Schlechtendal & Chamisso

MEXICAN WEEPING PINE, PINO TRISTE (SPANISH)

Pinus subg. *Pinus* sect. *Trifoliae* subsect. *Australes*

Tree to 35(–40) m tall, with trunk to 1 m in diameter. Bark bright reddish brown and smooth above, browner and strongly ridged and furrowed only near the base of large trunks. Crown dome-shaped, very open, with few, slender, gently rising branches densely clothed with drooping foliage at the tips. Twigs drooping, orange-brown, rough with the bases of scale leaves, hairless. Buds 12–20 mm long, not resinous. Needles in bundles of three or four (or five), each needle (10–)15–25(–30) cm long, thin, flexible, and drooping, lasting 2–3 years, yellowish green to pale or dark green. Individual needles with several evenly spaced, conspicuous lines of stomates on all three faces, and (one to) three (or four) resin canals at the corners, usually midway between the needle surface and the two-stranded midvein. Sheath 20–30 mm long, weathering to 10–15 mm and persisting and falling with the bundle. Pollen cones 15–20 mm long, yellowish brown with a reddish blush. Seed cones (5–)7–10(–12) cm long, narrowly egg-shaped,

Mature Mexican weeping pine (*Pinus patula*) in nature.

slightly asymmetrical, with 75–120(–150) seed scales, light greenish brown before maturity, ripening shiny yellowish brown to brown, opening widely to release the seeds or more commonly remaining closed many years before opening and falling with the very short or sometimes up to 12(–20) mm long, curved stalk. Seed scales very woody, wedge-shaped, the exposed face flat or a little swollen on the bottom scales away from the twig, topped by a small umbo bearing a tiny, fragile prickle. Seed body 4–6 mm long, the clasping wing another 15–20 mm longer. Mountains of eastern Mexico, from western Tamaulipas and adjacent Nuevo León to Oaxaca and the state of Mexico. Forming pure stands or mixed with other pines and various hardwoods in moist pine-oak forests and cloud forests; (1,400–)1,800–2,800(–3,300) m. Zone 9. Synonym: *P. patula* var. *longipedunculata* Loock ex M. Martínez.

Although both *Pinus patula* and the relatively distantly related Lumholtz weeping pine (*P. lumholtzii*) of western Mexico are "weeping" pines, their weeping appearance is distinct. The needles of *P. lumholtzii* are thick and hang down stiffly in a rigid curtain on either side of the twig while the needles of *P. patula* are thinner and droop in a graceful, if steep, arch. *Pinus patula* is closely related to Gregg pine (*P. greggii*) and Mesoamerican closed-cone pine (*P. oocarpa*), two other closed-cone species, and is highly interfertile with these species in artificial crosses. Natural hybridization with *P. oocarpa* might be responsible for the sporadic occurrence of trees of *P. patula* with long seed cone stalks, but this possibility has not been seriously investigated. There is some evidence that *P. patula* might be moderately crossable with Ocote pine (*P. teocote*) and its relative Lawson pine (*P. lawsonii*), from an allied Mexican pine group within subsection *Australes*, but this has not been confirmed. It is barely crossable with Monterey pine (*P. radiata*) of the Californian closed-cone pine group of subsection *Australes*. There is some uncertainty about the relationship between *P. patula* and Tecun Umán pine (*P. tecunumanii*) of Chiapas (Mexico) and Central America. It is clear that they are each other's closest relatives within the Mesoamerican closed-cone pine group. Despite several studies involving comparisons of leaves, seed cones, and resins, however, it is still uncertain whether they are separate species or varieties of a single species. Both are proving to be important species in plantation culture in other tropical mountain regions. The scientific name, Latin for "spread out," may refer to the way the seed cones stand out from the twigs in contrast to other closed-cone pine species known at the time this one was first described.

Pinus peuce Grisebach

MACEDONIAN PINE

Pinus subg. *Strobus* sect. *Quinquefoliae* subsect. *Strobus*

Tree to 20(–30) m tall, with trunk to 1 m in diameter. Bark grayish brown, flaking in scales and becoming narrowly ridged and furrowed at the base of large trees. Crown densely cylindrical in young trees, opening up with age, with stiffly upswept short branches clothed with foliage toward the ends. Twigs stout, greenish brown at first, becoming yellowish gray, hairless. Buds about 10 mm long, resinous. Needles in bundles of five, each needle (6–)7–9(–12) cm long, stiff and straight, lasting 3–4 years, grayish green with wax. Individual needles with stomates on all three faces, an undivided midvein, and usually two resin canals of unequal diameter touching the epidermis below the outer face. Sheath about 18 mm long, soon shed. Pollen cones 10–15 mm long, yellowish green blushed red. Seed cones (7–)8–15(–20) cm long, cylindrical and curved, with 30–60 scales, green or purple before maturity, ripening light yellowish brown, opening widely to release the seeds and then falling, short-stalked. Seed scales egg-shaped, bowed outward, thin and flexible, with a straight, narrow, diamond-shaped umbo at the tip. Seed body 6–8 mm long, the firmly attached wing 12–16 mm long. Mountains of the southern Balkan Peninsula, mostly in southwestern Bulgaria and southwestern Yugoslavia (Montenegro and Kosovo) with outliers in adjacent Greece, Macedonia, and Albania. Forming pure stands or mixed with European silver fir (*Abies alba*) and Norway spruce (*Picea abies*) in dense to open mountain conifer forests on soils derived from granite and sandstone but not limestone; (600–)1,000–2,000(–2,200) m. Zone 5.

This handsome, hardy pine is another Balkan relict of a formerly broader distribution, like Serbian spruce (*Picea omorika*). It is well separated, geographically and taxonomically, from the other two European white pines, arolla pine (*Pinus cembra*) in the Alps and Siberian pine (*P. sibirica*) in northwestern Russia. Those are both stone pines in a different group within subsection (*Strobus*) from the typical white pines, and *P. peuce* will not cross with either of these. Instead, it forms hybrids readily with four other geographically much more distant species of typical white pines: eastern (*P. strobus*) and western (*P. monticola*) white pines of North America, and Japanese (*P. parviflora*) and Himalayan (*P. wallichiana*) white pines of Asia. The range of the species is divided into eastern and western segments with a gap of roughly 200 km on either side of the Vardar and Morava River valleys. Trees in these two regions differ in minor features, like needle length, but the differences are not enough to separate them as varieties, as some authors have done.

Pinus pinaster W. Aiton

MARITIME PINE, CLUSTER PINE, PINO RESINERO (SPANISH), PINHEIRO-BRAVO (PORTUGUESE)

Pinus subg. *Pinus* sect. *Pinus* subsect. *Pinaster*

Tree to 25(–40) m tall, with trunk to 1(–1.5) m. Bark bright reddish brown, thick, becoming deeply fissured between irregu-

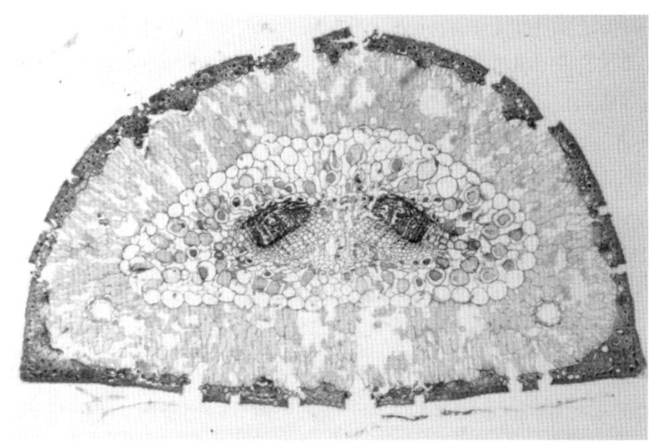

Cross section of preserved, stained, and sectioned needle of maritime pine (*Pinus pinaster*).

larly platelike, scaly ridges. Crown dome-shaped or flattened, becoming open with age, with upwardly angled branches densely clothed with foliage at the ends. Twigs reddish brown, hairless, somewhat rough with the bases of scale leaves. Buds 20–35 mm long, not resinous. Needles in bundles of two, 10–25 cm long, thick, stiff, and prickly, lasting 2(–3) years, shiny bright green. Individual needles with prominent lines of stomates on both faces, and 2–9(–15) resin canals at the corners and surrounding the two-stranded midvein midway between it and the needle surface. Sheath 20–30 mm long, persisting and falling with the bundle. Pollen cones about 1 cm long, pale yellowish brown. Seed cones clustered, (8–)9–18(–22) cm long, egg-shaped, slightly asymmetrical, with 120–150 scales, green before maturity, ripening glossy light brown, remaining closed for several years before opening and falling with the short stalk exceptionally to 15–20 mm long. Seed scales paddle-shaped, the exposed face diamond-shaped, with a pyramidal protrusion crossed by a ridge topped by a prominent umbo with a sharp, stout, short spine. Seed body 7–8 mm long, the clasping wing another 20–30 mm longer. Western Mediterranean in southwestern Europe and northwestern Africa, from northwestern Italy and Pantelleria Island (between Sicily and Tunisia) west to southwestern France, Portugal, and central Morocco. Forming pure, open stands or mixed with other trees on sand dunes and other substrates in coastal lowlands and at moderate elevations in the interior of Corsica, the Iberian Peninsula, and Morocco; 0–1,000(–2,000) m. Zone 8. Synonyms: *P. pinaster* subsp. *atlantica* E. Villar, *P. pinaster* subsp. *escarena* (Risso) K. Richter, *P. pinaster* subsp. *hamiltonii* (M. Tenore) E. Villar, *P. pinaster* var. *mesogeensis* (Fieschi & H. Gaussen) Silba, *P. pinaster* subsp. *renoui* (E. Villar) J. Maire.

Maritime pine has long been an economically important species for timber and for extraction of turpentine, both form natural populations and from plantations within its native range and in many other warm temperate, maritime regions. More recently, an extract of the bark containing polyphenols (which have antioxidant properties) has been promoted for medicinal purposes. There is considerable (if minor) variation in morphological features across the range of the species, and several subspecies were described. However, different kinds of data (morphology, oleoresin chemicals, and DNA sequences) suggest different subdivisions of the species. The differences between trees in different regions are probably due both to the pattern of recolonization of its range from two or more glacial refuges and to the influence of the varied environments characterizing the different regions. It would be surprising if there were not differences in appearance and growth between trees on sand dunes in southwestern France and those in the Atlas Mountains of Morocco, to give two examples of populations named as subspecies. Therefore, there seems to be little reason to give formal taxonomic recognition to the variation within maritime pine. It occasionally hybridizes naturally with a co-occurring related Mediterranean pine species, Aleppo pine (*Pinus halepensis*). The scientific name was used by Pliny in the first century AD to refer to a pine that grew naturally along the coast of Italy near Rome (Latin, "wild pine") that was distinct from the cultivated stone pine (*P. pinea*).

Pinus pinceana G. Gordon

WEEPING PINYON, PINO BLANCO (SPANISH)

Pinus subg. *Strobus* sect. *Parrya* subsect. *Cembroides*

Tree to 10(–12) m tall, with trunk to 0.3 m in diameter. Bark gray and smooth at first, becoming brownish gray, flaking, and finally breaking up into flat, rectangular plates separated by shallow, narrow, vertical and horizontal cracks. Crown broad and rounded, with numerous irregularly placed, arching branches dangling at the ends and clothed with downswept foliage to give a weeping appearance. Twigs light gray, hairless. Buds 3–6 mm long, not resinous. Needles in bundles of three (or four), each needle (5–)6–12(–14) cm long, straight and slender but stiff, lasting 2–3 years, grayish green. Individual needles with lines of stomates on the inner faces and sometimes with one or two inconspicuous lines of stomates on the outer face, an undivided midvein, and two large resin canals touching the epidermis of the outer face near the outer corners. Sheath about 10 mm long, curling back and soon shed. Pollen cones 8–10 mm long, purplish tan. Seed cones 5–10 cm long, variously egg-shaped, with 30–60 seed scales, purple before maturity, ripening shiny reddish brown, opening just enough to release the seeds and breaking off with part of the 1–2 cm long stalk. Seed scales diamond-shaped, weakly attached, with deep depressions for the seeds, the exposed face weakly or prominently thickened and angled outward, crossed by a sharp ridge peaking in a small diamond-shaped umbo. Seed body 10–14 mm long, wingless or with a

Mature weeping pinyon (*Pinus pinceana*) in nature.

rudimentary wing remaining attached to the seed scale. Discontinuous in the Sierra Madre Oriental of northeastern Mexico, from central Coahuila to northern Querétaro and Hidalgo. Scattered in dry, open woodland with numerous succulents on rocky slopes and in gulches; 1,400–2,300 m. Zone 8?

This small, rare pinyon pine occupies some of the most arid habitats in the group, often occurring well below the band of pinyon-juniper woodland dominated by Mexican pinyon (*Pinus cembroides*). It is the only member of the pinyons in subsection *Cembroides* with drooping twigs and needles, a weeping habit shared with one of its occasional associates, the drooping juniper (*Juniperus flaccida*). The seed cones are much more elongate than those of the typical pinyon pines, like *P. cembroides,* and resemble those of Nelson pinyon (*P. nelsonii*), another species of the Sierra Madre Oriental. Studies of wood oil composition and DNA suggest, however, that *P. pinceana* is not closely related to *P. nelsonii* but rather to king pinyon (*P. maximartinezii*), a species from the Sierra Madre Occidental with much coarser, more massive seed cones and needles in bundles of five rather than three. Like its western relative, weeping pinyon is threatened by livestock grazing and by exploitation, though for fuel rather than for its much smaller edible seeds. The species name honors Robert T. Pince (ca. 1804–1871), a nurseryman specializing in fuchsias in Devon, England, who has no apparent connection to the species, except for its being named after him.

Pinus pinea Linnaeus

MEDITERRANEAN STONE PINE, PINO PIÑONERO (SPANISH), PINHEIRO-MANSO (PORTUGUESE)

Pinus subg. *Pinus* sect. *Pinus* subsect. *Pinaster*

Tree to 25(–30) m tall, with trunk to 1.5 m in diameter. Bark with rich reddish brown, flat-topped, irregular plates separated by broad, dark brown furrows. Crown very wide spreading, broadly and shallowly domed, with numerous slender to thick, horizontal to upwardly angled, outstretched branches thickly clothed with foliage near the ends. Twigs yellowish brown to grayish brown, hairless, rough with the bases of scale leaves. Buds 6–12 mm long, not resinous. Needles in bundles of two, each needle (6–)8–15(–20) cm long, thick, stiff, and slightly twisted but not prickly, lasting (2–)3(–4) years, bluish green to bright green. Individual needles with numerous lines of stomates on both faces, a two-stranded midvein, and usually two (or three) interrupted resin canals along the midline of each face next to the surface. Sheath 10–15 mm long, weathering to 5–10 mm and persisting and falling with the bundle. Pollen cones 10–20 mm long, pale orange-brown. Seed cones (6–)8–12(–15) cm long, spherically egg-shaped, with 50–100 seed scales, green before maturity, ripening shiny, slightly reddish brown, opening widely to release the seeds and persisting 2–3 years before falling off the thick, short stalk often leaving a few basal scales behind. Seed scales broadly paddle-shaped, the exposed face five- or six-sided, prominently domed and crossed by ridges, the umbo large, flat, and surrounded by a channel. Seed body 15–22 mm long, plump, the easily detachable wing only up to another 10 mm longer. Scattered throughout the lands on the northern side of the Mediterranean Sea with outliers east to the Pontic Mountains southeast of the Black Sea in northeastern Turkey. Growing in pure open stands most often near the sea but also in the interior of Spain; 0–350(–1,000) m. Zone 8.

The dense, dark, umbrella-like crowns of this economically important, edible-seeded species are one of the characteristic sights of Mediterranean Europe, a striking contrast to the narrow spires of the Mediterranean cypress (*Cupressus sempervirens*). The large oily seeds are the pine nuts (*pinea* in Latin, hence the scientific name) of the Mediterranean, roasted and eaten as nuts after cracking and removing the thick shell or incorporated in cooking, as in the pesto sauce of Italy. Because of these nuts, the trees have been planted since ancient times, so the natural range of the species is not entirely clear. Although the wood is of good quality, the relatively slow growth of the trees compared to some other species of overlapping climatic tolerances, like Aleppo pine (*Pinus halepensis*) and Monterey pine (*P. radiata*), ensure that *P. pinea* is only a minor species in commercial timber plantations. Despite the long history of planting, there is still considerable genetic differentiation among different populations revealed by enzymes, more than usual with conifers. As well, there is less than usual variation within populations. Both of these results reflect the scattered distribution of small populations characteristic of this species. It is morphologically quite distinctive, and all attempts to cross it with other species have failed, so it has often been placed in a subsection all by itself. However, DNA studies link it convincingly with other species in the Mediterranean pine group that has begun to be generally recognized as subsection *Pinaster.* This grouping was also suggested in the past by botanists examining wood or seed cones.

Pinus ponderosa D. Douglas ex P. Lawson & C. Lawson
PONDEROSA PINE, WESTERN YELLOW PINE
Plate 48
Pinus subg. *Pinus* sect. *Trifoliae* subsect. *Ponderosae*
Tree to 50(–75) m tall, with trunk to 2(–2.8) m in diameter. Bark bright yellowish brown to reddish brown, thick, in mature trees divided into large, roughly vertically rectangular, flat-topped, scaly plates by deep, dark furrows. Crown deep, cylindrical or tapering, with numerous thick, horizontal branches densely clothed with foliage. Twigs orange-brown to reddish brown, coarse, rough with bases of scale leaves, hairless. Buds 12–25 mm long, very resinous. Needles in bundles of two to four (or five), each needle (7–)10–25(–30) cm long, stiff and straight to slightly twisted, lasting (2–)4–6(–8) years, yellowish green to dark green, sometimes a little waxy. Individual needles with inconspicuous lines of stomates on both the inner and outer faces, and 2(–10) resin canals midway between the outer surface and the two-stranded midvein at the corners and under both faces. Sheath 15–30 mm long at first, the lower 10–15 mm persisting and falling with the bundle. Pollen cones 15–35 mm long, yellowish brown or red. Seed cones 5–15 cm long, egg-shaped to almost spherical, sometimes slightly asymmetrical, with 75–120 seed scales, green before maturity, ripening yellowish brown, opening widely to release the seeds and soon falling off the short, thick stalk, leaving a few basal scales behind. Seed scales paddle-shaped to fairly rectangular, the exposed face horizontally diamond-shaped, crossed by a ridge topped by a prominent pyramidal umbo ending in a strong, sharp prickle. Seed body (3–)5–7(–9) mm long, the easily detachable wing 12–20(–25) mm longer. Widely but discontinuously distributed across western North America from north-central Nebraska to southern British Columbia south to southern California, northern Durango and western Tamaulipas (Mexico). Forming pure, open forests on dry soils or mixed with hardwoods and other pines and conifers; 0–3,000 m. Zone 6. Three varieties.

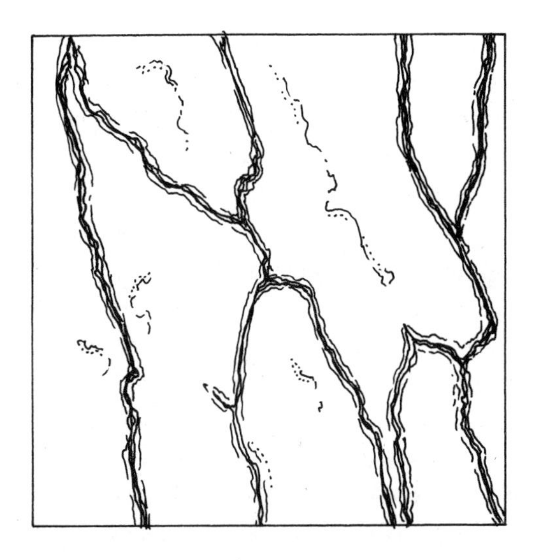

Bark plates of ponderosa pine (*Pinus ponderosa*) whose dimensions vary with age and environmental conditions.

Ponderosa pine is the second most widely distributed pine species in western North America after lodgepole pine (*Pinus contorta*), which is more northerly on average. As the most economically valuable hard pine in its region, ponderosa pine is one of the handful of most thoroughly studied pine species in the world. This is true even though it does not have the role in international plantation culture of other well-studied species, like Monterey pine (*P. radiata*) or loblolly pine (*P. taeda*). Despite a fairly clear picture of taxonomic structure in the northern and central portions of its range, with several well-defined geographic races assigned to two varieties, there is still much confusion in the south, especially with regard to the most closely related members of subsection *Ponderosae* in Mexico. A third three-needled variety of *Pinus ponderosa* in northern Mexico and the adjacent southwestern United States is recognized here, but the five-needled Arizona pine (*P. arizonica*), often considered a ponderosa pine variety, is accepted as a distinct species. The two northern varieties, presumably separated from one another during glacial periods, now meet and hybridize in the northern and central Rocky Mountains. Although the described geographic races generally occupy discrete, if abutting, ranges, the North Plateau race of the Pacific Northwest interior extends southward at high elevations into the range of the California race (both races belong to variety *ponderosa*) as far south as Lake Tahoe in California and Nevada. Since it is sharply distinct from the surrounding ponderosa pines of lower elevations, it is often recognized as a separate species, Washoe pine (*P. washoensis*), but genetic studies show that it fits squarely within the North Plateau race as a subalpine ecotype and it is included here within variety *ponderosa.* The races of *P. ponderosa* display a complicated pattern of crossability, with moderate crossing barriers in some combinations. There is also considerable crossability with the most similar other species of subsection *Ponderosae,* including Arizona pine (*Pinus arizonica*), Apache pine (*P. engelmannii*), and Montezuma pine (*P. montezumae*), and somewhat less with Durango pine (*P. durangensis*) and Jeffrey pine (*P. jeffreyi*).

Large trees of the California race, while not as large as the biggest sugar pines (*Pinus lambertiana*), are still giants, hence the scientific name "weighty," from Spanish. The largest known individual in combined dimensions when measured in 1997 was 69.2 m tall, 2.3 m in diameter, with a spread of 20.7 m. Trees of the other races are much smaller but can still reach heights of 45–50 m and diameters exceeding 1.5 m. The widely spaced, columnlike trunks of ponderosa pine, with their broad plates of bright bark, are a characteristic feature of lower and middle slopes of major mountain ranges throughout western North America. Isolated small groves are also found at lower elevations in seemingly unexpected places, like the wet Willamette Valley of western Oregon, or the Pine Ridge rising from the prairies of northwestern Nebraska. On the other hand, the species is absent from the greater part of the arid Intermountain region, in southeastern Oregon, southern Idaho, western Utah, and most of Nevada.

Pinus ponderosa D. Douglas ex P. Lawson & C. Lawson **var. *ponderosa***

PACIFIC PONDEROSA PINE

Tree to 75 m tall, with trunk to 2.8 m in diameter. Needles usually three per bundle, each needle 15–30 cm long. Pollen cones red. Seed cones 8–15 cm long. Western portion of the range of the species, from southern British Columbia and western Montana to southern California; 0–2,300 m. Synonym: *Pinus washoensis* H. Mason and Stockwell.

Pinus ponderosa* var. *scopulorum Engelmann

ROCKY MOUNTAIN PONDEROSA PINE

Tree to 45 m tall, with trunk to 1.5 m in diameter. Needles two or three per bundle, each needle (5–)10–17 cm long. Pollen cones yellowish brown. Seed cones 5–10 cm long. Eastern portion of the range of the species, from eastern Montana, western North Dakota, and north-central Nebraska south to eastern Nevada, Arizona, New Mexico, and western Texas; 1,000–3,000 m.

Pinus ponderosa* var. *stormiae (M. Martínez) Silba

SOUTHWESTERN PONDEROSA PINE, PINO REAL (SPANISH)

Tree to 35 m tall, with trunk to 1.2 m in diameter. Needles three or four (or five) per bundle, each needle (8–)10–25 cm long. Pollen cones yellowish brown. Seed cones 5–10(–15) cm long. Southern portion of the range of the species, from southern Arizona and New Mexico and western Texas south in the Sierra Madre Occidental to Durango and in the Sierra Madre Oriental to western Tamaulipas (Mexico); 1,300–3,000 m. Synonym: *Pinus arizonica* var. *stormiae* M. Martínez.

Pinus praetermissa Styles & McVaugh

MEXICAN SCRUB PINE, PINO PRIETO (SPANISH)

Pinus subg. *Pinus* sect. *Trifoliae* subsect. *Australes*

Bushy tree to 15(–20) m tall, with trunk to 0.3 m in diameter. Bark grayish brown, thin, breaking up into irregular, scaly ridges, separated by shallow furrows. Crown rounded, open, with scattered, slender branches thinly clothed with foliage. Twigs reddish brown, hairless, the scale leaf bases persistent but smooth. Buds 8–15 mm long, not resinous. Needles in bundles of (four or) five, each needle 8–16 cm long, thin and flexible, lasting 2–3 years, light green. Individual needles with a few, inconspicuous lines of stomates on all three faces, and one or two (to four) small resin canals touching the two-stranded midvein. Sheath 11–14 mm long, weathering to 7–12 mm and persisting and falling with the bundle. Pollen cones 10–15 mm long, yellowish brown with a reddish blush. Seed cones (3.5–)5–7 cm long, broadly egg-shaped to almost spherical, with 50–80(–120) seed scales, light brown before maturity, ripening light brown, opening widely to release the seeds and then falling, leaving a cluster of basal scales behind on the long, slender, curved stalk 3–3.5 cm long, or these bottom scales simply shed. Seed scales wedge-shaped, the exposed face five- or six-sided and scarcely raised, the small umbo often depressed and tipped by a tiny, fragile prickle. Seed body 5–6(–8) mm long, the clasping wing another (12–)15–20 mm longer. Mountains of the Pacific slope of central Mexico, from Sinaloa to Jalisco. Usually mixed with oaks or seasonally deciduous hardwoods and sometimes with other pines in open woodlands; 900–1,500(–1,900) m. Zone 9. Synonym: *P. oocarpa* var. *microphylla* G. R. Shaw.

This very local species of dry lower slopes was included with Mesoamerican closed-cone pine (*Pinus oocarpa*) until 1990 (hence the scientific name, Latin for "overlooked"). However, although it is closely related to *P. oocarpa,* the only real similarity is a close resemblance of their unopened seed cones. But cones of *P. oocarpa* long remain unopened on the tree while those of *P. praetermissa* are shed soon after opening at maturity. The foliage of the two is quite different, the needles of *P. praetermissa* being much shorter and softer than those of *P. oocarpa,* with fewer lines of stomates and fewer resin canals located differently in the needle. Because it is essentially a shrubby tree that usually does not grow in solid pine stands, there is little commercial exploitation, although no doubt there is some local usage. It has the most restricted distribution among the Mexican and central American hard pines (subgenus *Pinus*) though not as restricted as two pinyon pines (subgenus *Strobus*), king pinyon (*P. maximartinezii*) and Coalcomán pinyon (*P. rzedowskii*), of the same general region.

Pinus pringlei G. R. Shaw

PRINGLE PINE, PINO ROJO (SPANISH)

Pinus subg. *Pinus* sect. *Trifoliae* subsect. *Australes*

Tree to 25 m tall, with trunk to 0.9(–1) m in diameter. Bark reddish brown to grayish brown, divided into small, scaly plates by deep, reddish brown furrows. Crown irregularly dome-shaped, open, with few, thick, horizontal to gently rising branches sparsely clothed with foliage at the tips. Twigs stout and stiff, reddish brown with a thin waxy bloom, hairless, rough with the bases of scale leaves at first, but these flaking off after the first year. Buds 8–15 mm long, not resinous. Needles in bundles of (two or) three (or four), each needle (15–)18–25(–30) cm long, thick, stiff, and straight, lasting 2–3 years, bright green. Individual needles with obvious lines of stomates all the way around both the inner and outer faces, and (three or) four to seven (to nine) resin canals usually touching the two-stranded midvein at the corners and in between. Sheath 12–20 mm long, weathering to 10–15 mm and then persisting and falling with the bundle. Pollen cones 15–25 mm long, yellowish brown. Seed cones 5–8(–10) cm long, slightly asymmetrical, narrowly egg-shaped, with 50–100 seed scales, green before maturity, ripening shiny yellowish brown, opening from the tip a few scales at a time but eventually falling and leaving behind a few basal scales on the short, stout, curved stalk 5–10 mm long. Seed scales broadly wedge-shaped, the exposed face horizontally diamond-shaped, low but crossed by a ridge tipped by a small, low umbo bearing a tiny, fragile prickle. Seed body 4–6 mm long, the clasping wing thickened at the seed body, another 12–18 mm longer. Mountains of southwestern Mexico, from Michoacán to Oaxaca. Forming pure stands or mixed with other species in moist pine and pine-oak forests; (1,500–)1,750–2,200(–2,800) m. Zone 9.

Although fairly local in distribution, the fine, light-colored timber and abundant resin of Pringle pine make it an important commercial species. It shares these characteristics with its close relatives Ocote pine (*Pinus teocote*) and Herrera pine (*P. herrerae*), which also have the relatively low number (for Mexican pines) of three needles in a bundle. Pringle pine has thicker, stiffer needles than either of these other species. The species name honors Cyrus G. Pringle (1838–1911), who collected the type species as well as many other species during his extensive plant collecting trips to Mexico.

Pinus pseudostrobus Lindley

FALSE WHITE PINE, PINO LACIO, PINO LISO (SPANISH)

Pinus subg. *Pinus* sect. *Trifoliae* subsect. *Ponderosae*

Tree to 40(–50) m tall, with trunk to 1(–1.5) m in diameter. Bark grayish brown or dark reddish brown, long remaining smooth but finally breaking up into rough scaly ridges or broader, flat-topped plates separated by deep furrows. Crown cylindrical to dome-shaped or even spherical, dense to open, with numerous thin, gently rising branches clothed with drooping foliage at the tips. Twigs coarse, reddish brown, conspicuously bluish waxy at first, smooth or rough with the bases of scale leaves, hairless. Buds 12–20 mm long, not resinous. Needles in bundles of (four or) five (to eight), each needle (15–)20–30(–40) cm long, thin, flexible, and variously drooping, lasting 2–3 years, dark yellowish green to bright green or even a little waxy bluish green. Individual needles with several conspicuous lines of stomates on all three faces, and (two or) three or four (to six) resin canals at the corners and around the two-stranded midvein usually midway between it and the outer needle surface. Sheath 20–30(–35) mm long, weathering to 15–25 mm long and persisting and falling with the bundle. Pollen cones densely clustered, 2–3(–3.5) cm long, light reddish brown. Seed cones (6–)10–15(–18) cm long, broadly to narrowly egg-shaped or cylindrical, usually symmetrical and slightly curved, with 100–200 seed scales, green before maturity, ripening shiny to dull yellowish brown to reddish brown, opening widely to release the seeds and then falling, leaving behind a few basal scales on the short, stout stalks to 1(–2) cm long. Seed scales roughly rectangular to wedge-shaped, the exposed portion horizontally diamond-shaped, almost flat to greatly protruding, especially on the lower scales away from the twig, the umbo more prominent on more protruding scales, sometimes tipped with a small, fragile prickle. Seed body 5–7 (–10) mm long, the clasping seed wing another 18–25 mm long. Mountains of Central America and Mexico, from La Paz (Honduras) north to central Nuevo León and southern Sinaloa (Mexico). Commonly mixed with other species in moist or dry pine forests; (800–)1,800–2,800(–3,250) m. Zone 9. Two varieties.

Pinus pseudostrobus owes its scientific name (Latin for "false strobus") to its slender, flexible needles in bundles of five and its long seed cones with thin, flexible scales, both reminiscent of those of the eastern white pine (*P. strobus*), which belongs to the white pines (subgenus *Strobus*) rather than the hard pines (subgenus *Pinus*), to which *P. pseudostrobus* belongs. The length of its cones is exceptionally variable for a hard pine, as is the amount of protrusion of the seed scale tips. Among the Mesoamerican hard pines, only Montezuma pine (*P. montezumae*) approaches it in variability, and like this species, some authors recognize many varieties or even distinct species within false white pine. Most of this variation is fairly continuous, and some of it may be due to hybridization with the related Douglas pine (*P. douglasiana*) and Maximino pine (*P. maximinoi*) or other co-occurring species in subsection *Ponderosae.* Trees with highly protruding seed cone scale tips are concentrated in the southern portion of the range of the species and also have broader seed cones than the majority of populations. These two forms are treated as separate varieties, but their relationship merits further study. Northward in the Sierra Madre Oriental in eastern Mexico the needles gradually become shorter and stiffer and the seed cones longer, but there seems to be little value in giving these continuous changes formal taxonomic recognition as some authors have done. False white pine produces ample resin as well as fine timber and so has been heavily exploited in the areas near Mexico City, where it is most abundant. It is also used extensively in forestry trials in several tropical and subtropical countries.

Pinus pseudostrobus var. _apulcensis_ (Lindley) G. R. Shaw

Exposed portion of the seed scale and the umbo protruding prominently, especially near the base of the seed cone on the side away from the twig. Southern portion of the range of the species, from Santa Ana (El Salvador) to Guerrero and Puebla (Mexico). Synonyms: *Pinus oaxacana* Mirov, *P. pseudostrobus* var. *oaxacana* (Mirov) S. Harrison.

Pinus pseudostrobus Lindley **var. _pseudostrobus_**

Exposed portion of the seed scale and the umbo flat or a little protruding on the basal scales of the seed cone on the side away from the twig. Throughout the range of the species. Synonyms: *Pinus estevezii* (M. Martínez) J. P. Perry, *P. nubicola* J. P. Perry, *P. pseudostrobus* var. *coatepecensis* M. Martínez, *P. pseudostrobus* var. *estevezii* M. Martínez, *P. pseudostrobus* var. *laubenfelsii* Silba.

Pinus pumila (P. Pallas) E. Regel

DWARF SIBERIAN PINE, JAPANESE MOUNTAIN PINE, HAI-MATSU (JAPANESE), NUN CHAS NAMU (KOREAN), YAN SONG (CHINESE)

Pinus subg. *Strobus* sect. *Quinquefoliae* subsect. *Strobus*

Shrub to 2(–6) m tall, but spreading to 10 m, with creeping or upright trunks to 15 cm in diameter and covered with flaky, grayish or blackish brown bark. Crown round or flattened, with numerous stiffly upright branches bearing foliage toward the ends. Twigs brown and densely hairy at first, later reddening and then graying. Buds about 10 mm long, heavily resinous. Needles in bundles of five, each needle 3–7(–10) cm long, stiff, usually lasting 2–4 years, dark green and without stomates on the outer face, whitish green with wax on the inner faces. Individual needles with lines of stomates on the inner faces, an undivided midvein, and (one or) two small resin canals touching the epidermis of the outer face near the corners. Sheath 10–15 mm long, soon shed. Pollen cones 8–10 mm long, dark red. Seed cones 3–4.5(–5) cm long, egg-shaped, with about 30 seed scales, violet and then green before maturity, ripening light brown, remaining closed or gaping slightly and falling the next year, short-stalked. Seed scales fan-shaped, somewhat fleshy where the seeds are borne, the thin, woody tip short, sometimes curled a little back, and ending in a small, triangular umbo. Seed body 6–10 mm long, plumply egg-shaped, unwinged. Northeastern Asia, primarily throughout Russia east of the Lena River, and then scattered southward in high mountains in northern Mongolia, northeastern China, Korea, and northern and central Japan. Forming dense alpine and subalpine thickets or as an understory in open subalpine forests; (0–)500–2,500(–3,200) m. Zone 5.

Along with the unrelated European mugo pine (*Pinus mugo* subsp. *mugo*) and Mexican border pine (*P. culminicola*), this is one of the few truly shrubby pines (hence the scientific name, "dwarf" in Latin). It is not a timber species but does yield edible seeds and distilled oils as well as contributing to wildlife and soil conservation. Although often considered closely related to *P. sibirica,* another mountain stone pine in subsection *Strobus,* with which its range overlaps in the west, it has also been suggested that *P. pumila* is closer to the Japanese white pine (*P. parviflora*), a typical white pine. It has not been crossed successfully with *P. sibirica* or any other mountain stone pine, but there are reported natural hybrids with *P. parviflora,* known as *P.* ×*hakkodensis,* where the alpine range of *P. pumila* overlaps with the subalpine *P. parviflora* var. *pentaphylla.* DNA studies are consistent with this hybridization, which again calls into question the unity of the mountain stone pines, formerly included in their own subsection *Cembrae,* suggesting that the common characteristics of the species are due to convergent evolution. Except for hybridization with *P. parviflora* there is little genetic differentiation of shrubs of this species from place to place, so seed dispersal by nutcrackers and other animals appears to be effective in spreading migrants between populations.

Pinus pungens A. Lambert

TABLE MOUNTAIN PINE

Pinus subg. *Pinus* sect. *Trifoliae* subsect. *Australes*

Tree to 20(–30) m tall, with trunk to 0.4(–0.8) m in diameter, often crooked and irregular in cross section. Bark thin, reddish brown to dark gray, breaking up into flaky, elongate plates between shallow, dark furrows. Crown flat-topped, irregular, and open, with a few, thick, upwardly angled branches sparsely clothed with foliage. Twigs orange-brown, tough but snapping at their breaking point, rough with the bases of scale leaves, hairless. Buds 6–9 mm long, resinous. Needles in bundles of two (or three), each needle (3–)4–9(–10) cm long, twisted, stiff, and prickly (hence the scientific name, Latin for "piercing"), lasting (2–)3 years, dark green. Individual needles with conspicuous, narrow lines of stomates on both the inner and outer faces, and (0–)2–7(–11) resin canals at the corners and in between, midway between the needle surface and the two-stranded midvein. Sheath 5–10 mm long, weathering to 4–6 mm and persisting and falling with the bundle. Pollen cones 12–18 mm long, reddish purple. Seed cones (4–)6–8(–10) cm long, plumply egg-shaped, asymmetrical, with 90–140 seed scales, green before maturity, ripening shiny light yellowish brown, opening gradually from the middle over the course of 2–3(–5) years and then persisting another several years (up to about 30!) before falling with the very short stalks (to 1 cm long). Seed scales wedge-shaped, the exposed face horizontally diamond-shaped, protruding as a low pyramid crossed by a sharp ridge topped by a prominent umbo bearing a strong, sharp, upcurved spine (another justification for the scientific name). Seed body 4–6(–7) mm long, the firmly attached wing another (10–)15–22(–30) mm longer. Appalachian Mountains and adjoining Piedmont of the eastern United States, from central Pennsylvania to northeastern Georgia. As scattered small groves or single trees among oaks and other fire-tolerant hardwoods; (100–)350–1,100(–1,400) m. Zone 6.

This pine grows on dry, rocky slopes and ridge tops (hence the common name). It is the most montane of the pines of the southeastern United States. Like most of the other southern pines, table mountain pine depends on fires for reproduction. Although the trees can be killed by fire because of their thin bark, the thick-scaled seed cones and their slow release of seeds provide for regeneration after the fire. In fact, the parent trees have to be killed for optimal seedling and stand establishment. With its scattered occurrence in difficult terrain, small size, and indifferent wood quality, table mountain pine is definitely not a commercially important species. Nonetheless, it has been the subject of several studies of genetic variation and regeneration ecology. Despite its limited total population, it has the same high level of total genetic variation found in most other conifers. Its widely separated populations do have an effect, however, and there is much more difference between populations than conifers usually display. Furthermore, even different seed cones on a single tree frequently receive pollen from different trees. This pollen can come from another species, and natural hybrids with pitch pine (*Pinus rigida*), the only other southeastern pine that can grow as high as *P. pungens*, are known. In addition, table mountain pine crosses artificially with shortleaf pine (*P. echinata*), a coastal plain species. The largest known individual in combined dimensions when measured in 1984 was 28.7 m tall, 0.8 m in diameter, with a spread of 14.0 m.

Pinus quadrifolia Parlatore ex Sudworth

PARRY PINYON

Pinus subg. *Strobus* sect. *Parrya* subsect. *Cembroides*

Tree to 10(–16) m tall, with trunk to 0.5(–0.7) m in diameter. Bark reddish brown and smooth at first, flaking, becoming grayish brown, and finally breaking up into roughly rectangular blocks separated by shallow furrows. Crown dense, conical when young, broadening, rounding and becoming more open with age, with numerous upwardly angled to horizontal branches sparsely clothed with foliage at the ends. Twigs yellowish brown, hairless to minutely hairy, graying and balding with age. Buds 4–8(–14) mm long, slightly resinous. Needles in bundles of (one to) three to five (or six), sticking together during the first year, each needle (1.5–)3–5(–6) cm long, green to bluish green. Individual needles with stomatal bands usually confined to the inner faces, an undivided midvein, and (one or) two (or three) large resin canals visible beneath the epidermis of the outer face. Sheath 5–9 mm long, curling back and soon shed. Pollen cones 7–10 mm long, reddish purple. Seed cones (3–)4–6(–7) cm long, broadly egg-shaped to almost spherical, with 25–35(–50) seed scales, green before maturity, ripening yellowish brown, opening widely to release the seeds and then falling along with the short, slender stalk. Seed scales paddle-shaped, with deep seed cavities, the exposed portion diamond-shaped and pyramidally thickened on the face, ending in a strong diamond-shaped umbo. Seed body (12–)14–17(–18) mm long, the shell 0.2–0.3 mm thick, the rudimentary wing remaining attached to the seed scale. Peninsular ranges of southern California to the Sierra San Pedro Martír of northern Baja California (Mexico). Sometimes forming pure stands but more commonly mixed with other small trees on dry, rocky slopes in a band of pinyon-juniper-oak woodland between the chaparral and coastal sage scrub below and the montane conifer forest above; (900–)1,100–2,500(–2,700) m. Zone 7. Synonym: *P. juarezensis* Lanner.

Parry pinyon is one of the most variable pinyon pines in the number of needles in each bundle among trees within individual populations. Four is the most common number (hence the scientific name, "four leaves" in Latin), but some trees are almost entirely five-needled, and three-needled bundles are common, while bundles of one and two needles may also be found. This variability provoked the suggestion that *Pinus quadrifolia* arose through hybridization between an almost extinct five-needled species and singleleaf pinyon (*P. monophylla*), which has a limited overlap with Parry pinyon in both California and Baja California. The evidence supporting this suggestion is weak, and there appears to be no reason to separate predominantly five-needled trees from the more numerous individuals with four and five needles together. Artificial crosses between Parry pinyon and singleleaf pinyon yield trees with needles in bundles of two or three, supporting the interpretation of such trees in mixed stands as hybrids. There is also good anatomical and biochemical evidence of limited hybridization between these two species. According to DNA studies, these two pinyons are not each other's closest relatives, so the hybrids cannot be interpreted alternatively as ancestral to both one- and four- or five-needled populations. None of the other pinyon pines have ranges overlapping that of *P. quadrifolia.* The largest known individual in the United States in combined dimensions when measured in 1976 was 16.6 m tall, 0.7 m in diameter, with a spread of 12.8 m.

Pinus radiata D. Don

MONTEREY PINE, RADIATA PINE, INSIGNIS PINE

Pinus subg. *Pinus* sect. *Trifoliae* subsect. *Australes*

Tree to 30(–40) m tall, with trunk to 1(–2) m in diameter. Bark with irregular, dark grayish brown ridges separated by shallow to deep reddish brown furrows. Crown a deep, cylindrical dome or open and irregular on exposed headlands, with numerous upwardly angled branches densely clothed with tufts of foliage only near the tips. Twigs reddish brown, hairless, sometimes a little waxy. Buds 8–15 mm long, variably resinous. Needles in bundles of two or three (to five), each needle 8–15(–20) cm long, stiff and a little twisted, lasting 3–4 years, bright green to dark yellowish green. Individual needles with lines of stomates on both the inner and outer faces, a two-stranded midvein, and one to five (to eight) discontinuous resin canals deep within the leaf tissue, or even one or two touching the endodermis surrounding the midvein. Sheath (0.5–)1–2 cm long, persisting and falling with the bundle. Pollen cones 10–20 mm long, yellowish brown. Seed cones (5–)7–15(–19) cm long, asymmetrically or symmetrically egg-shaped, with 90–180 seed scales, green before maturity, ripening pale reddish brown, usually retained and remaining closed after maturity for several to many years until after a fire, on a short stalk to 1(–1.5) cm long. Seed scales paddle-shaped, the exposed portion thin and low on small, symmetrical cones, forming thick domes on the outer side of the base of large, asymmetrical cones, the umbo diamond-shaped, a little indented with a central bump or short-lived prickle. Seed body 5–8 mm long, mottled, the easily detachable wing 15–30 mm long. Adjacent to the Pacific Ocean in five areas from Año Nuevo Point to Cambria, California, and on Guadalupe and Cedros Islands, Baja California (Mexico). Forming open or dense pure coastal closed-cone pine forests in the fog belt; 10–800(–1,160) m. Zone 8. Two varieties.

No other tree with world significance as a timber species has such a restricted natural distribution. The five natural stands total about 6,000 hectares while forest plantations of the species total more than 4 million hectares. The main growing areas are in New Zealand, Chile, and Australia, but there are also sizable regions in Spain and South Africa, all far from the native range and its pests and diseases. It has even become an invasive species in its adoptive lands, displacing native vegetation both by its own dispersal and during preparation for its plantations. Each of the native stands has distinctive characteristics, particularly in the size and shape of the seed cones, and all have been involved to varying extents in breeding superior genotypes for the southern hemisphere plantations. Monterey pine is closely related to the other California closed-cone pines and, like them, largely depends on fires for natural regeneration. It can be crossed with knobcone pine (*Pinus attenuata*) and the southern populations of bishop pine (*P. muricata*). Hybrids with knobcone pine occur naturally in California and also among the plantation trees of New Zealand. Past natural hybridization with bishop pine may help explain some of the peculiarities in the genetic structure of the latter species. Despite natural variability and the extensive breeding work for forestry, few ornamental cultivars have been developed. The scientific name reflects how the mature, closed seed cones form rings radiating around the branches. The largest known wild individual in combined dimensions when measured in 2003 was 61 m tall, 2.8 in diameter, with a spread of 26.8 m. If left to their own devices, selected trees in the southern hemisphere might well someday exceed these dimensions.

Pinus radiata* var. *binata (Engelmann) J. Lemmon

Needles predominantly in bundles of two. Seed cones averaging 6–9 cm long, symmetrical, the scale tips low and not greatly thickened. Guadalupe and Cedros Islands, Mexico; 275–1,160 m. Synonyms: *Pinus muricata* var. *cedrosensis* J. T. Howell, *P. radiata* var. *cedrosensis* (J. T. Howell) Silba.

Pinus radiata D. Don **var. *radiata***

Needles predominantly in bundles of three. Seed cones averaging 9–15 cm long, more or less asymmetrical, the outer, basal scale tips prominently swollen and protruding. California mainland; 10–300(–440) m.

Pinus resinosa W. Aiton

RED PINE, PIN ROUGE (FRENCH)

Plate 49

Pinus subg. *Pinus* sect. *Pinus* subsect. *Pinus*

Tree to 25(–38) m tall, with trunk to 0.8(–1) m in diameter. Bark with variably reddish brown, elongate flat plates separated by dark brown to blackish furrows. Crown conical at first, becoming cylindrical in dense stands, with relatively few, slender, horizontal branches fairly densely clothed with foliage near the ends. Twigs orangish to reddish brown, rough with bases of scale leaves, hairless. Buds 12–20 mm long, resinous. Needles in bundles of two, each needle (10–)12–16(–18) cm long, stiff and brittle, lasting 3–4 years, glossy dark green. Individual needles with narrow lines of stomates on both surfaces, and two resin canals touching the inner face away from the corners, and one to four smaller resin canals spaced inside the outer face around the two-stranded midvein. Sheath 20–25 mm long at first, the lower 10–15 mm persisting and falling with the bundles. Pollen cones 10–15 mm long, purple. Seed cones 3.5–7 cm long, broadly egg-shaped, with 50–80 seed scales, green before maturity, ripening pale reddish brown, opening widely to release the seeds and then falling, leaving a few basal scales attached to the twig by a very short stalk. Seed scales paddle-shaped, the exposed face a little raised with a bluntly triangular umbo. Seed body 5–6 mm long, the easily detachable wing 13–15 mm longer. Great Lakes and maritime regions of the middle portion of eastern North America, from Nova Scotia and Massachusetts west to southeastern Manitoba and northeastern Minnesota, with outliers in Newfoundland, West Virginia and Illinois. Forming pure stands or mixed with other pines, oaks, aspens, and other trees on dry, fire-prone rocky or sandy soils; (0–)200–800(–1,300) m. Zone 2.

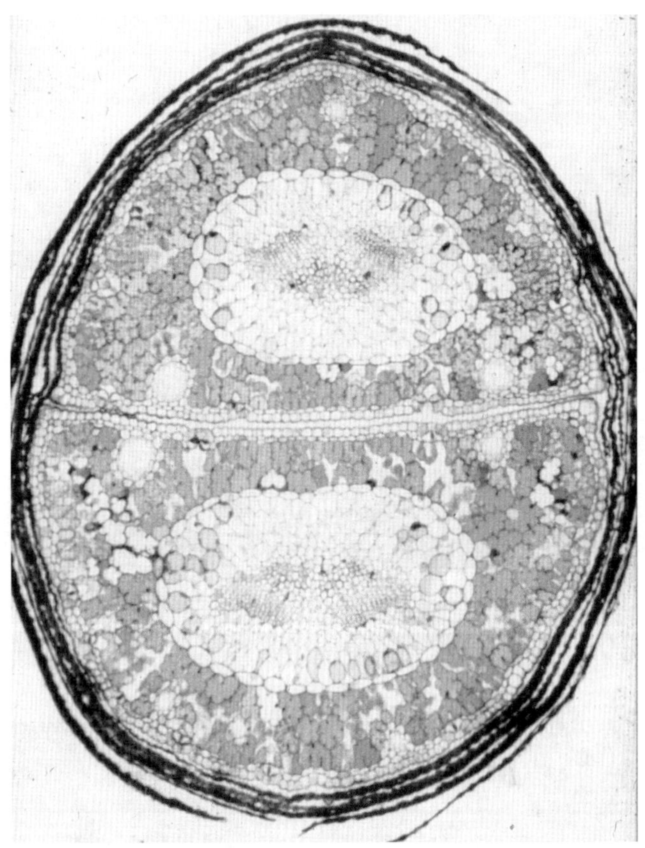

Cross section of preserved, stained, and sectioned fascicle of red pine (*Pinus resinosa*) showing the sheath of scale leaves surrounding the foliage needle pair.

Despite its rather typical appearance, *Pinus resinosa* is a very unusual pine in several respects. It and *P. tropicalis* of Cuba are the only New World members of subsection *Pinus* and, indeed, the only New World hard pines (subgenus *Pinus*) with close relatives in the Old World. Despite this, red pine is strongly reproductively isolated from other pines, including its Eurasian relatives. Out of attempts with many species, only a single cross with the similar and related Austrian pine (*P. nigra*) was ever reported as successful, and this has not been repeated. In fact, the supposed hybrids were later shown not to have had red pine as their pollen parent but rather an Asian pine, possibly Japanese red pine (*P. densiflora*), which crosses readily with Austrian pine. Most pines hybridize with varying degrees of success, but often relatively easily with other members of the same subsection.

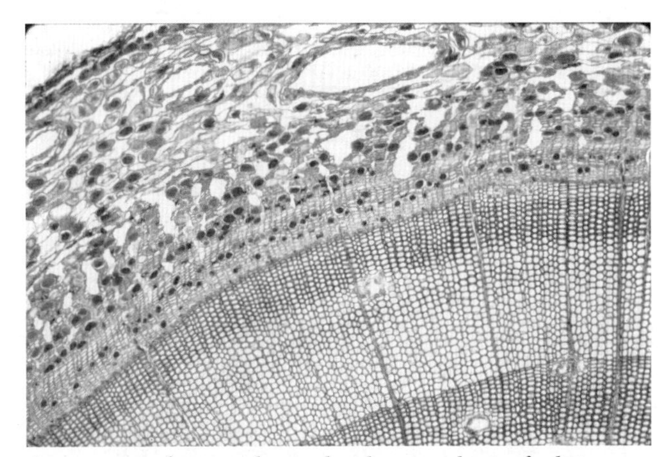

Cross section of preserved, stained, and sectioned twig of red pine (*Pinus resinosa*) showing both bark and wood tissues, the latter with parts of two growth rings containing resin canals.

On the other had, red pine is also distinctive in being capable of self-fertilization with little detrimental effect. Such natural selfing may reach 10% even in large populations. The lack of inbreeding depression is due, perhaps, to the very homozygous condition of its genes, without the deleterious recessive mutations found in other conifers. Thus red pine has extremely low genetic variation in morphological features, ecology, leaf oils, enzymes, and even in normally highly variable DNA regions. This is in contrast to most other pines and other conifers in general. In fact, it is the only widely distributed conifer to have such depleted genetic resources, although some rare, geographically restricted species like *Pinus torreyana* or *P. maximartinezii* are similarly homogeneous. This genetic uniformity implies a fairly recent (in evolutionary terms) reduction to a single small population, presumably in association with the last full glaciation.

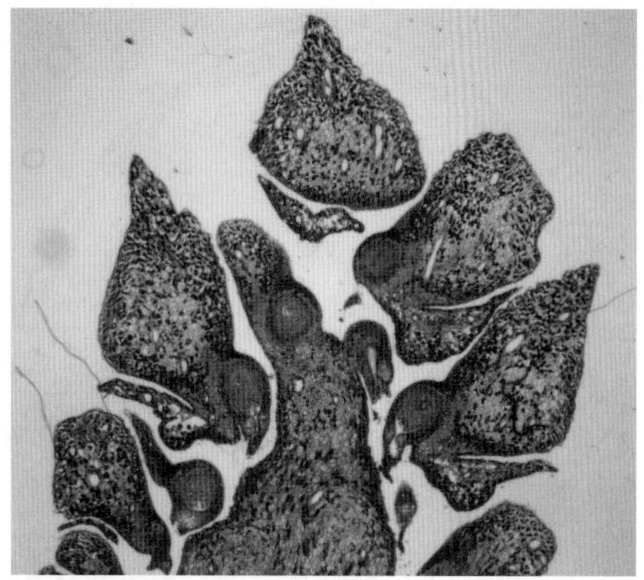

Cross section of preserved, stained, and sectioned young seed cone of red pine (*Pinus resinosa*) showing the thickened tips of the cone scales and the attached young ovules.

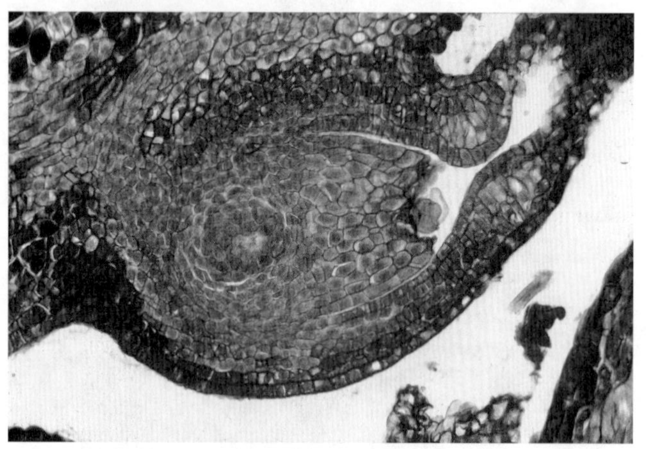

Cross section of preserved, stained, and sectioned ovule of red pine (*Pinus resinosa*) on its way to becoming a seed.

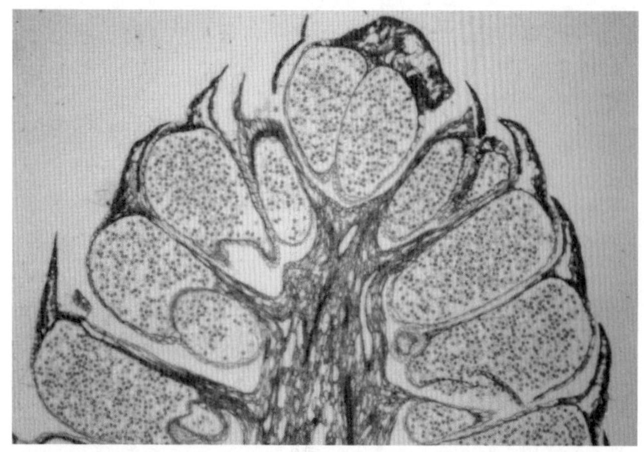

Cross section of preserved, stained, and sectioned pollen cone of red pine (*Pinus resinosa*) showing the pollen sacs stuffed with pollen grains.

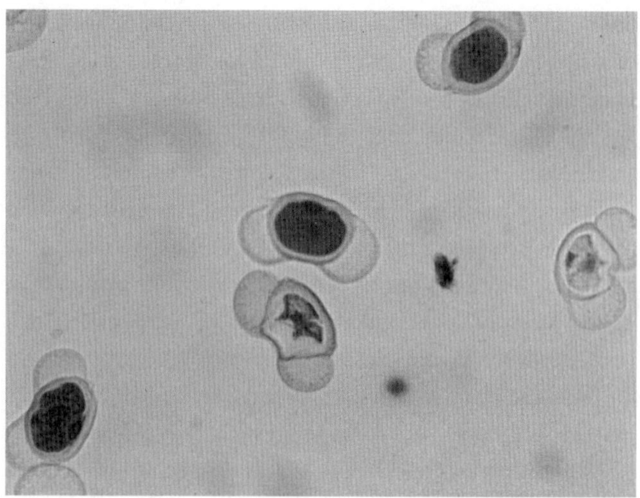

Stained pollen grains of red pine (*Pinus resinosa*) showing the air bladders (sacci).

Besides the ability to self-pollinate, which would allow a single seed to found a new colony, another consequence of genetic identity in the species might be its peculiar ability to root graft, also found in a few other, more variable conifers, like eastern hemlock (*Tsuga canadensis*). On average, every red pine in a dense stand may have its roots connected to those of three other trees. In a red pine plantation, and red pine is an important timber plantation species throughout its natural range, if the stand is thinned by girdling the saplings, the stumps can stay alive for decades as their root systems are fed by neighboring trees, which themselves show little reduction in growth as a result. This may also contribute to the fire resistance of red pine, which is a premier fire species, showing both resistance to, and most regeneration after, fires in a region where fires are mostly restricted to certain substrates. Yet another biological feature of red pine that may be influenced by its genetic uniformity is the fact that its needles (about 90% of them) are heavily infected inside with endophytic fungi, as many as seven different species at a time, a rate of infection far higher than in other pine species studied.

Lack of variation also negatively affects its resistance to pests. Unfortunately for its use as a plantation species and ornamental in the southern portion of its range and southward, it is there highly susceptible to the introduced European pine shoot moth (*Rhyacionia buoliana*), which disfigures it and reduces its commercial value. It was hoped that hybrids with Austrian pine (which is resistant to this insect) could be used as a means of introducing resistance to red pine, but the lack of confirmed crosses and inability to backcross preclude this breeding strategy with traditional techniques. The largest known individual red pine in combined dimensions when measured in 1993 was 37.8 m tall, 1.0 m in diameter, with a spread of 18.3 m.

Pinus ×*rhaetica* Brügger

Natural hybrids between European mountain pine (*Pinus mugo*) and Scots pine (*P. sylvestris*), occurring occasionally throughout the regions where the two species overlap, from the northern side of the Pyrenees in France, through the Alps to the Carpathians in Slovakia, usually at the junction between the subalpine habitat of *P. mugo* and the montane forests supporting *P. sylvestris*. The hybrids vary but generally fall between the two parents in their characteristics or else present combinations of characters not found in either parent. They are usually trees, though some are shrubby. The bark is often a darker red on the branches than in Scots pine, or even gray, but usually still smooth and scaly, while the trunk is grayer and more frequently divided into irregular plates. The paired needles run the gamut from the dark green most characteristic of *P. mugo* to the bluish green typical for *P. sylvestris*. If the trees more closely resemble mugo pine overall, then the needles are bluer, and if, on the other hand, they are closer to Scots pine, then the needles are brighter green. Again, if the trees are generally similar to mugo pine, then the first-year seed cones are curved back on their stalks rather than straight as they would be in *P. mugo*. And if they more closely resemble Scots pine, then the first-year seed cones are upright rather than recurved. Thus the hybrids typically resemble one or the other parent species but have characters that are discordant for the species they most resemble that, instead, are derived from the other parent. As a result, it is hard to distinguish the hybrids from extreme variants within the parent species unless they are more strongly intermediate. In any event, the hybrids appear to be rare, in part due to only slightly overlapping pollination periods for the parent species. Synonyms: *P.* ×*digenea* G. Beck, *P. sylvestris* var. *engadinensis* Heer.

Pinus rigida P. Miller

PITCH PINE, POND PINE, PINYON DES CORBEAUX (FRENCH)

Pinus subg. *Pinus* sect. *Trifoliae* subsect. *Australes*

Tree to 25(–30) m tall, with trunk to 0.8(–1.3) m in diameter, often crooked. Bark reddish brown to dark grayish brown, thin, with small, elongate, scaly plates separated by narrow, shallow furrows. Crown deeply dome-shaped, open and irregular, with slender horizontal to upwardly angled branches densely clothed with foliage just at the tips. Twigs coarse, orange-brown, sometimes with a thin coating of wax, hairless, a little rough with the bases of scale leaves. Buds 10–15(–20) mm long, resinous. Needles in bundles of (two or) three to four (or five), each needle (5–)7–16(–21) cm long, thick, stiff (hence the scientific name), and twisted, lasting 2–3(–4) years, yellowish green. Individual needles with inconspicuous lines of stomates on both the inner and outer faces, and 2–11 resin canals at the corners and in between midway between the needle surface and the two-stranded midvein or some touching the midvein. Sheath 9–20 mm, weathering to 5–12 mm and persisting and falling with the bundle. Pollen cones 18–30 mm long, pale yellowish brown. Seed cones (3–)5–8(–9) cm long, egg-shaped to broadly egg-shaped, with 90–140 seed scales, green before maturity, ripening orange brown, opening widely to release the seeds or remaining closed for many years and, in either case, remaining many years on the tree before falling with the very short, stout stalk occasionally up to 15 mm long. Seed scales paddle-shaped, the exposed face horizontally diamond-shaped, slightly pyramidal and crossed by a distinct ridge topped by a large umbo bearing a fragile to persistent, sharp prickle. Seed body 4–6 mm long, the firmly attached wing another 15–20 mm longer. Eastern United States, from central and western Florida north to western Kentucky, southern Ohio, southern Maine, northern Vermont, and northern New York and adjacent southern Ontario and southern Quebec, with a large gap through the piedmont region, from Georgia to Maryland. Forming pure stands on fire-prone sites that vary from rocky ridge tops to seasonally flooded, lowland ponds; 0–1,400 m. Zone 5. Two subspecies.

Pitch pine has the most northerly limit among the "southern" pines, species of subsection *Australes* native to the southeastern United States. As such it has had some role in plantation forestry, particularly in Korea, where its hardiness compensates for its poorer growth form, wood quality, and oleoresin production compared to the four species that are most economically important in the southeastern United States: shortleaf pine (*Pinus echinata*), longleaf pine (*P. palustris*), slash pine (*P. elliottii*), and loblolly pine (*P. taeda*). Artificial hybrids with the last species show increasing promise as a way of combining the hardiness of pitch pine with the faster growth and better timber qualities of loblolly pine. This hybrid also occurs naturally where the two species grow together, as do hybrids with table mountain pine (*P. pungens*) and shortleaf pine, and the latter hybrid has also been made artificially. Artificial hybrids with the Monterey pine (*P. radiata*) of California help confirm the placement of the California closed-cone pines in subsection *Australes* rather than in their own subsection, which was also suggested by some DNA studies.

The two subspecies of pitch pine are usually considered separate species since their geographic ranges are almost completely separate and they occupy different characteristic habitats. However, they are very similar morphologically, differing most obviously in needle length and whether the seed cones typically open at maturity, and they intergrade completely in their small region of overlap in southern New Jersey and the Delmarva Peninsula of Delaware and Maryland. Although the southerly pond pine (subspecies *serotina*) with persistently closed (serotinous) seed cones occurs almost exclusively in seasonally flooded sites, and pitch pine (subspecies *rigida*) usually grows on dry sites and usually has seed cones opening promptly at maturity, pitch pine is more variable and sometimes resembles pond pine in these characteristics. The proportion of serotinous cones in the northern pitch pine is strongly associated with the frequency of fires. Such cones are especially common in the frequently burned New Jersey Pine Barrens. Fires are generally more common in the range of pond pine than in that of pitch pine, and this by itself could explain the differences in frequency of serotinous seed cones in the two subspecies. Likewise, the habitat differences between them has much to do with the relative frequencies of these habitats within the geographic ranges they occupy. So despite their differences in appearance, habitat, and distribution, there is little justification for the traditional placement of these two pines in separate species. The differences, however, do seem to warrant treatment as subspecies rather than the rank of variety that is more commonly used for botanical variants of conifers.

A distinctive feature that unites the two subspecies is that their mature trunks bear weak shoots much younger than any mature branches at that level. These emerge from dormant buds that were originally on parts of the twigs not carrying bundles of needles. There are also many of these dormant buds buried in the base of the trunk at the ground level, but they do not arise from the roots. These buds allow the tree to sprout and regenerate after a crown fire that kills the aboveground portions of the trunk. Thus stands of pitch pines regenerate after fires both by seed and by vegetative sprouts. The largest known individual in combined dimensions when measured in 1999 was 30.2 m tall, 1.4 m in diameter, with a spread of 12.2 m. This tree belongs to the northern subspecies, while the largest pond pine, measured in 1992, was 27.1 m tall and 1.0 m in diameter, with a crown spread of 16.2 m.

Pinus rigida P. Miller **subsp. *rigida***

PITCH PINE

Needles (5–)7–12(–15) cm long with a sheath less than 1 cm long. Pollen cones about 2 cm long. Seed cones typically opening at maturity, about 3–4 cm wide at the base on a stalk less than 1 cm long and with the umbos bearing persistent prickles. Northern portion of the range of the species south to southern Delaware, northern West Virginia, northern and western Virginia, western North Carolina, and northeastern Georgia. Commonly on dry sites; 0–1,400 m.

Pinus rigida **subsp. *serotina*** (A. Michaux) R. T. Clausen

POND PINE

Needles (10–)13–16(–21) cm long with a sheath to 2 cm long. Pollen cones about 3 cm long. Seed cones typically remaining closed at maturity, about 3.5–5 cm wide on a stalk 1–1.5 cm long and with the umbos bearing fragile prickles. Southern portion of the range of the species north to southern New Jersey, eastern Virginia, eastern North Carolina, eastern South Carolina, southern Georgia, and central Mississippi. Commonly on seasonally flooded sites; 0–200 m. Synonyms: *Pinus rigida* var. *serotina* (A. Michaux) Hoopes, *P. serotina* A. Michaux.

Pinus roxburghii C. Sargent

CHIR PINE, HIMALAYAN LONGLEAF PINE, CHIR (HINDI)

Plate 50

Pinus subg. *Pinus* sect. *Pinus* subsect. *Pinaster*

Tree to 30(–55) m tall, with trunk to 1.2 m in diameter. Bark dark reddish brown to grayish brown, thick, breaking up into scaly, narrow, longitudinal plates separated by shallow, dark furrows. Crown shallowly dome-shaped to spherical, open, with relatively few, long, slender, horizontal or gently raised branches sparsely clothed with foliage at the tips. Twigs pale grayish brown, hairless, rough with the bases of scale leaves. Buds 1–2 cm long, not resinous. Needles in bundles of three, each needle (15–)20–30(–33) cm long, thick but flexible and drooping, lasting 1–3 years, light green. Individual needles with inconspicuous lines of stomates on all three faces, a two-stranded midvein, and two (to four) resin canals touching the needle surface at the corners and in between. Sheath 20–30 mm long, weathering to 12–20 mm and persisting and falling with the bundle. Pollen cones in dense clusters, 1–2 cm long, yellowish brown. Seed cones 10–20 cm long, egg-shaped, with 100–150 seed scales, green before maturity, ripening shiny brown to grayish brown, opening slowly after ripening and falling off the short, thick stalk to 15 mm long. Seed scales wedge-shaped to paddle-shaped, thick and woody, the exposed face horizontally diamond-shaped, protruding as a stout, hooked, pyramidal boss tipped by a small, triangular umbo without a prickle. Seed body 7–10(–13) mm long, the clasping wing another 20–25 mm longer. Across the Himalaya, especially the outer ranges, from the North-West Frontier province of Pakistan to the state of Arunachal Pradesh in northeastern India, more abundant and continuous in the western Himalaya. Forming pure, open forests in seasonally dry mountain valleys and slopes of the monsoon belt, or mixed with other species near the upper and lower limits of its distribution; (450–)600–1,500(–2,300) m. Zone 9.

In many winter-dry valleys at moderate elevations in the outer Himalaya, chir pine can be virtually the only tree and forms extensive forests. Because of its usually modest size and slow growth, it is not a particularly important timber species except where it is the only conifer. However, it is also the source of a turpentine that is used as a liniment for rheumatic pains. The annual increase in wood peaks at the age of about 40 years, while the size of the crown continues to increase for another 20 years before leveling off. Relatively little is known about genetic variation within chir pine but a recessive mutation that reduces chlorophyll production has been isolated from a few individuals planted in provenance trials established during the 1920s.

Likewise, the relationships of chir pine to other species of pine are somewhat uncertain. Traditionally, it was linked with Canary Island pine (*Pinus canariensis*), a species with which it supposedly was crossed successfully, and these two species were placed by themselves in their own subsection *Canarienses.* DNA studies support linkage with all of the Mediterranean pines in subsection *Pinaster.* Some earlier studies suggested that *Pinus roxburghii* might be the closest relative either of Old World hard pines (section *Pinus*) as a whole or of New World hard pines as a whole (section *Trifoliae*). Since these early placements were refuted by more extensive DNA studies and since there are clear morphological and crossing similarities with *P. canariensis,* chir pine is retained here in the expanded subsection *Pinaster.* Chir pine has not been crossed successfully with any other pine, including Canary Island pine, so crossability is of no help in settling its taxonomic position. The species name honors William Roxburgh (1751–1815), who wrote a flora of India in which he described chir pine under a scientific name that cannot be used under the rules of botanical nomenclature.

Pinus rzedowskii Madrigal & M. Caballero

COALCOMÁN PINYON, RZEDOWSKI PINYON

Pinus subg. *Strobus* sect. *Parrya* subsect. *Cembroides*

Tree to 25(–50) m tall, with trunk to 0.6(–1.2) m in diameter. Bark dark grayish brown, with elongate, rectangular, scaly plates separated by deep, narrow, dark furrows. Crown of mature trees open, irregularly egg-shaped, with thin, variably horizontal branches sparsely clothed with foliage near the tips. Twigs gray, roughened with short scale leaf bases, not hairy. Buds 8–10 mm long, not resinous. Needles in bundles of (three or) four or five, each needle 6–10 cm long, thin and flexible but straight, lasting 2–3 years, yellowish green or a little grayish with wax. Individual needles with lines of stomates only on the two inner faces, an undivided midvein, and two (or three) moderate resin canals touching the epidermis of the outer face beneath longitudinal grooves about halfway between the center and the edge and sometimes with one or two additional resin canals next to the inner faces. Sheath 7–9 mm long, curling back and persisting a while but finally shed before the bundle. Pollen cones 5–6 mm long, light purplish tan. Seed cones 10–15 cm long, egg-shaped, with 50–70 seed scales or more, green before maturity, ripening shiny yellowish brown, opening widely to release the seeds and then falling along with the 1.5–3 cm long, slender stalk. Seed scales paddle-shaped, woody and stiff, the diamond-shaped exposed portion thicker and raised to a large diamond-shaped umbo at the summit of a ridge crossing the face, the umbo sometimes bearing a small prickle. Seed body (6–)8–10 mm long, the clasping wing 20–35 mm long. Sierra Madre del Sur in the Coalcomán district of Michoacán (Mexico). Forming small pure stands in a general matrix of pine-oak forest on rocky, limestone slopes and hilltops; (1,700–)2,100–2,500 m. Zone 9?

Pinus rzedowskii differs from all other pinyon pines in its fully developed seed wings. This, together with the occasional presence of a prickle on the seed cone umbo and the common occurrence of needles in bundles of five, led to suggestions for its inclusion with the foxtail pines (subsection *Balfourianae*) or in a separate subsection by itself or together with the other long-coned Mexican pinyons, *P. maximartinezii, P. pinceana,* and *P. nelsonii.* These morphological features make it somewhat intermediate between the foxtail pines, like foxtail pine (*P. balfouriana*) and intermountain bristlecone pine (*P. longaeva*), and the typical pinyon pines, like Mexican pinyon (*P. cembroides*) and Colorado pinyon (*P. edulis*). Another suggestion was that the long-coned pinyons belong with the Asian lacebark pine (*P. bungeana*) and chilgoza pine (*P. gerardiana*) in subsection *Gerardianae.* DNA studies are consistent with the retention of *P. rzedowskii,* along with the other anomalous pinyons (except *P. nelsonii*), as basal branches to the typical pinyons within subsection *Cembroides.* The closest relatives of the pinyons appear to be the foxtail pines. Thus the similarities of these species to the Asian nut pines in subsection *Gerardianae* are due to convergent evolution.

There are only some 6,500 trees of this species known, scattered among 12 small populations in southern Michoacán ranging in size from 1 to 3,500 trees. The trees are threatened primarily by fires as the stands are remote and little exploited. Despite the rarity of the species, it has a great deal of genetic variation in its enzymes, variation more typical of widespread pine species and contrasting strongly with the extremely limited genetic variation in Torrey pine (*Pinus torreyana*), a rare and restricted hard pine from California with about the same total number of trees in two populations. Unlike the related king pinyon (*P. maximartinezii*), with its single known population, there is no evidence of genetic bottlenecks due to recent population crashes in *P. rzedowskii.* Instead, it displays random changes in enzymes from stand to stand due to the varying histories of each, largely isolated stand. The species name honors Jerzy Rzedowski (b. 1926), a Polish-born Mexican botanist who named the related *P. maximartinezii* among his numerous contributions to understanding the taxonomy and ecology of Mexican plants.

Pinus sabiniana D. Douglas ex D. Don
GRAY PINE
Pinus subg. *Pinus* sect. *Trifoliae* subsect. *Ponderosae*
Tree to 25(–50) m tall, with trunk to 1.2(–1.5) m in diameter, often forking near the base. Bark dark grayish brown, with irregular, scaly, vertical ridges, separated by shallow, reddish brown furrows. Crown open, conical at first, passing through cylindrical to vase-shaped, with seemingly irregular, upright to horizontal predominantly slender branches sparsely clothed with foliage. Twigs pale bluish tan with a thin, waxy coating, rough with bases of scale leaves, hairless. Buds 10–15 mm long, resinous. Needles in bundles of three, each needle (15–)20–25(–30) cm long, flexible and drooping, lasting 3–4 years, dusty grayish green. Individual needles with numerous prominent lines of stomates on all three faces, and 2–3(–10) resin canals at the corners and scattered around the two-stranded midvein deep inside the leaf tissue. Sheath 20–25 mm long, the lower portion persisting and falling with the bundles. Pollen cones 10–15 mm long, yellow. Seed cones 15–25 cm long, massively egg-shaped, with 90–120 scales, green before maturity, ripening tan to light brown, opening widely to release the seeds and then persisting 5 or more years, on stout stalks 2–5 cm long. Seed scales diamond-shaped, the exposed face projecting in a pyramid tipped by an umbo in the form of a thick, curved spine. Seed body 20–25 mm long, the easily detachable wing abut 10 mm longer. Foothills surrounding the Central Valley of northern and central California. Forming pure stands or mixed with oaks and other pines in grassy, open foothill woodlands; (30–)300–1,000(–2,000) m. Zone 8.

Characteristically forked gray pines (*Pinus sabiniana*) in nature.

The sparse narrow crown of long, grayish needles supported by a forked trunk that is characteristic of gray pine is also one of the most characteristic sights of the foothill savannas of California. Although rarely forming dense stands, the trees are common enough that heavy crops of the large seeds were an important food for native peoples in the region. The heavily spiky cones are smaller than those of its close relative, Coulter pine (*Pinus coulteri*), usually weighing 300–600 g but exceeding 1 kg upon occasion. They vary a great deal in size, asymmetry, and length and shape of the spikes, but this variation is not geographically coherent and no varieties are recognized. Gray pine is largely reproductively isolated, but a single hybrid has been produced by artificial crossing with *P. coulteri,* and a few more resulted from crosses with the third species in the bigcone group, Torrey pine (*P. torreyana*). Unlike *P. coulteri,* gray pine does not cross with Jeffrey pine (*P. jeffreyi*) or other more typical members of subsection *Ponderosae.* The scientific name honors Joseph Sabine (1770–1837), inspector-general of assessed taxes in Great Britain and also secretary of the Horticultural Society (forerunner of the Royal Horticultural Society) at the time that Douglas collected and described the species under their sponsorship. The largest known individual in combined dimensions when measured in 1986 was 49 m tall, 1.5 m in diameter, with a spread of 24 m. When it died, the next largest tree was a full 12 m shorter.

Pinus* ×*schwerinii Fitschen
A spontaneous garden hybrid that has also been made deliberately between two geographically and ecologically widely separated species of typical white pines (subsection *Strobus*), eastern white pine (*Pinus strobus*) of North America and Himalayan white pine (*P. wallichiana*). The first trees of this parentage found were seedlings of *P. strobus,* but the reciprocal cross is actually a little easier to produce experimentally. The hybrids have a broad, conical crown that resembles that of *P. wallichiana* because of the drooping needles. They exhibit hybrid vigor in both growth rate and in resistance to white pine blister rust. The bark remains smooth and light gray for an extended period. Twigs resemble those of Himalayan white pine in their green to greenish brown color and conspicuous waxy coating, and those of eastern white pine in being minutely hairy. The foliage is mostly concentrated at the end of each year's twigs. The five drooping needles, 8–12 cm long, fall between the normal ranges of the shorter needles of *P. strobus* and the longer ones of *P. wallichiana* and are bright green

to bluish green, resembling some forms of both parents in this respect. The number of lines of stomates on each inner face of the needles, three to six, is higher than in *P. strobus* (two to four) and lower than in *P. wallichiana* (five or six). The seed cones, 10–20 cm long and about 6 cm in diameter, with about 60 seed scales, fall within the extreme range of *P. strobus* but are larger than the norm for the species. They are decisively smaller than those of the Himalayan white pine but have a slender stalk 2–4 cm long, like that species and longer than found in eastern white pine. The seed scales open more widely than those of either parent species. The seeds, about 7–9 mm long with wings 20–25 mm long, fall near the upper end of the range of *P. strobus* and the lower end of *P. wallichiana,* which overlap considerably in seed size. This hybrid is a promising avenue for introducing rust resistance into *P. strobus* or as a plantation tree in its own right.

Pinus sibirica Du Tour

SIBERIAN STONE PINE, SIBIRSKII KEDR (RUSSIAN), XIBOLIYA WUZHENG SONG (CHINESE)

Pinus subg. *Strobus* sect. *Quinquefoliae* subsect. *Strobus*

Tree to 35(–40) m tall, or shrubby at the alpine timberline. Trunk to 1(–1.8) m in diameter. Bark light brown or grayish brown, flaking and becoming deeply and narrowly ridged and furrowed at the base of large trees. Crown dense, conical at first, becoming broadly and deeply domed with age, with numerous upwardly angled branches bearing dense foliage near the ends. Twigs yellowish brown, densely covered with pale rusty hairs. Buds 7–10 mm long, not resinous. Needles in bundles of five, each needle (5–)7–10(–13) cm long, stiff and somewhat twisted, usually lasting 2–5 years, dark green and without stomates on the outer face, grayish green with wax on the inner faces. Individual needles with lines of stomates on the inner faces, an undivided midvein, and three resin canals deep inside the leaf tissue at the corners of the needles. Sheath 15–20 mm long, soon shed. Pollen cones 10–20 mm long, red. Seed cones held upright, (5–)6–12(–13) cm long, elongate egg-shaped, with 50–60 seed scales, purple before maturity, ripening purplish brown, remaining closed or gaping slightly without releasing the seeds and shed the next year, short-stalked. Seed scales irregularly circular, somewhat fleshy at the base, a little thickened at the tip and ending in a small but distinct triangular umbo, densely covered with flattened-down hairs. Seed body 10–14 mm long, plumply egg-shaped, wingless. Primarily Siberian (hence the common and scientific names), from the Ural Mountains east to the southern Lena River (with isolated stands in the Kola Peninsula of northwestern Russia) south to the Altai Mountains of Russia, eastern Kazakhstan, and northern China (Xinjiang province), to central Mongolia, and to northeastern China (northwestern Heilongjiang and northeastern Nei Mongol). Growing alone or mixed with other trees in a wide range of habitats, from peat bogs and grassy plains to dry foothill slopes and subalpine screes; (100–)800–2,000(–2,400) m. Zone 3. Synonyms: *P. cembra* subsp. *sibirica* (Du Tour) P. Krylov, *P. cembra* var. *sibirica* (Du Tour) J. C. Loudon, *P. sibirica* var. *hingganensis* (H. J. Zhang) Silba.

Closely related to central European arolla pine (*Pinus cembra*), *P. sibirica* differs in its greater stature and in the thin, easily cracked seed coat that makes the seeds a much easier food source. It is widely separated from *P. cembra* geographically but crosses with that species and no other, as far as known, including the supposedly related dwarf Siberian pine (*P. pumila*), another mountain stone pine in subsection *Strobus,* with which its range overlaps from the Lake Baikal region eastward to the Lena River. Studies of enzymes reveal little genetic differentiation across the wide range of the species, with a few immigrants estimated to be brought into each population by nutcrackers and other seedeaters in each generation. People also gather the seeds and smash them to make a beverage, press them for edible oil, or eat them whole. The tree is an important source of timber throughout the Siberian lowlands but even more so in the highlands, where Scots pine (*P. sylvestris*) is less common. It is a handsome ornamental but has few cultivars and is not extensively planted for seed production since it does not begin to bear seeds in nature until it is 30–40 years old. It is long-lived, with a maximum age of about 500 years.

Pinus* ×*sondereggeri H. Chapman

Natural and artificial hybrids between longleaf pine (*Pinus palustris*) and loblolly pine (*P. taeda*), two of the most economically important "southern" pines, species of subsection *Australes* native to the southeastern United States. This is the only one of the numerous natural and artificial hybrid combinations among the southern pines that has been given a valid botanical name. The natural hybrids of this combination are scattered throughout the southeastern Coastal Plain wherever the two parents occur together but are most abundant, perhaps, in the portion of the range west of the Mississippi River in eastern Texas and western Louisiana. They also crop up in plantations of one of the species grown in the vicinity of the other. They combine some features of each of the parent species as well as having many intermediate characteristics. On the whole, they show more similarity to and more overlap in characteristics with loblolly pine than with longleaf pine. As seedlings, they are immediately distinguishable from the latter because they lack its condensed "grass" stage but instead begin ordinary height growth, even in their first year. Both parent species have needles in bundles of three, as do the hybrids, but the hybrids are intermediate in needle length, at 12–22 cm, between the very long needles of longleaf pine (mostly 20–30 cm, but up to 45 cm) and the shorter needles of loblolly pine (mostly 10–17 cm and up to 23 cm). These needles have their resin canals both touching the midvein and midway between it and the needle surface, while those of longleaf pine are mostly near the midvein and those of loblolly pine are mostly in the midway position. The seed cones, 10–20 cm long, are also intermediate in length between those of longleaf pine (mostly 15–25 cm long) and those of loblolly pine (mostly 6–12 cm long). These hybrids have no particular advantages in growth rate or wood qualities over the parent species, so they are not now important in plantation culture. However, the bypassing of the seedling grass stage may make crossing with these hybrids a potential means of speeding up the growth of longleaf pine, which is greatly slowed by its grass stage. Zone 9.

Pinus squamata X. W. Li & Ji R. Hsüeh

SOUTHERN LACEBARK PINE, QIAO JIA WU ZHEN SONG (CHINESE)

Pinus subg. *Strobus* sect. *Quinquefoliae* subsect. *Gerardianae*

Tree with scaly, grayish green bark darkening to deep brown and flaking to reveal paler inner bark. Twigs reddish brown, densely minutely hairy with pale hairs and with a waxy coating. Buds 6–10 mm long, resinous. Needles in bundles of (four or) five, each needle 9–17 mm long, slender but stiff. Individual needles with lines of stomates on all three faces, an undivided midvein, and three to five large resin canals scattered around the periphery next to the epidermis. Sheath soon shed. Seed cones about 9 cm long, pointy egg-shaped, with 100–120 seed scales, opening widely to release the seeds, on stalks 1.5–2 cm long. Seed scales tongue-shaped, the exposed portion thickened and crossed by a ridge bearing a blunt, triangular umbo. Seed body black, the easily detached wing about 1.6 cm long. Northeastern Yunnan (China); 2,200–2,250 m. Zone 9?

This may be the rarest pine species and one of the rarest of all conifers. It was described in 1992 from a single population of about 20 trees, and many of its features have not yet been described. Its flaky bark (hence the scientific name, Latin for "scaly"), needle structure, and general seed cone structure clearly ally it with lacebark pine (*Pinus bungeana*) and chilgoza pine (*P. gerardiana*), but it has darker bark, longer needles in bundles of five, like those of true white pines (subsection *Strobus*), and a long, functional seed wing, again like those of true white pines. This interesting combination of characteristics suggests that further knowledge of this rare species will have considerable value for studies of pine phylogeny and relationships.

Pinus strobiformis Engelmann

SOUTHWESTERN WHITE PINE, PINO ENANO (SPANISH)

Pinus subg. *Strobus* sect. *Quinquefoliae* subsect. *Strobus*

Tree to 25(–35) m tall, with trunk to 1(–1.5) m in diameter. Bark pale gray and smooth at first, becoming dark grayish brown and sinuously furrowed between uneven, flat-topped, narrow vertical ridges. Crown conical at first, rounding and becoming more irregular with age, with widely spaced, upwardly angled to horizontal slender branches loosely clothed with needles near their tips. Twigs light reddish brown and hairless to minutely hairy at first, graying and balding with age. Buds about 10 mm long, resinous. Needles in bundles of five, each needle (4–)5–9 cm long, flexible, slightly twisted, and spreading a little, lasting (2–)3–5 years, dark green to bluish green with wax over the stomates on all three faces. Individual needles with only none to two lines of stomates on the outer face, an undivided midvein, and two (to four) small resin canals touching the epidermis of the outer face. Sheath 15–20 mm, soon shed. Pollen cones 6–12 mm long, yellowish brown. Seed cones (9–)15–25(–30) cm long, roughly cylindrical and curved, with about 100 seed scales, green tinged with purple before maturity, ripening light brown, opening widely to release the seeds and then falling, on a stout stalk 1.5–2.5(–6) cm long. Seed scales diamond-shaped, the exposed tip thick, elongate, and usually curled or angled back to varying degrees, ending in a narrow, triangular umbo. Seed body 10–13 mm long, the wing reduced to a shallow ridge at the tip. Mountains of southwestern North America, from north-central Arizona and north-central New Mexico, or perhaps southwestern Colorado, south to southwestern Durango and western Tamaulipas, with outliers in central Jalisco and southwestern San Luis Potosí (Mexico). Forming small pure stands or more commonly mixed with other pines, other conifers, and various oaks and other hardwoods in montane to subalpine forests; 1,900–3,000(–3,500) m. Zone 7. Synonyms: *P. ayacahuite* var. *brachyptera* G. R. Shaw, *P. ayacahuite* var. *novogaliciana* Carvajal, *P. ayacahuite* var. *reflexa* (Engelmann) A. Voss, *P. ayacahuite* var. *strobiformis* (Engelmann) C. Sargent ex J. Lemmon, *P. flexilis* var. *macrocarpa* Engelmann, *P. flexilis* var. *reflexa* Engelmann, *P. reflexa* (Engelmann) Engelmann, *P. strobiformis* var. *carvajalii* Silba, *P. strobiformis* var. *potosiensis* Silba.

As evident from its synonyms, the status and identity of this species are often debated. It falls between limber pine (*Pinus flexilis*) and Mexican white pine (*P. ayacahuite*) both geographically and in its taxonomic features. It is quite variable, and there are disputes about whether certain populations, like those in Jalisco or on Cerro Potosí in Nuevo León, belong to this species, where they are sometimes treated as separate varieties, or to one or the other of its relatives. They are assigned to *P. strobiformis* here, but it is possible that some of the variability and resulting taxonomic confusion in this species and its relatives is due to past hybridization among them. Southwestern white pine crosses successfully with *P. flexilis* and with western white pine (*P. monticola*), but its crossability with *P. ayacahuite* is unknown. The scientific name refers to a general similarity to eastern white pine (*P. strobus*), a similarity shared with other members of subsection *Strobus,* including its closest relatives. Like *P. strobus,* the timber is of high quality (if much smaller) and much sought for general carpentry, but its general scarcity and frequent relative inaccessibility preclude it from being an important commercial species. The largest known individual in the United States in combined dimensions when measured in 1974 was 33.8 m tall, 1.5 m in diameter, with a spread of 18.9 m.

Pinus strobus Linnaeus

EASTERN WHITE PINE, WEYMOUTH PINE, PIN BLANC (FRENCH)

Plates 51 and 52

Pinus subg. *Strobus* sect. *Quinquefoliae* subsect. *Strobus*

Tree to 35(–60) m tall, with trunk to 1.5(–1.8) m in diameter. Bark light grayish brown, smooth at first, becoming broken into vertical, rectangular blocks by shallow to deep cracks. Crown conical at first, becoming flat-topped, cylindrical, and irregularly open with age, with wide-spreading, horizontal branches of variable length reasonably densely clothed with foliage near their ends. Twigs pale reddish brown, hairless or a little hairy at first. Buds 4–8(–15) mm long, slightly resinous or not. Needles in bundles of five, each needle (5–)6–10(–15) cm long, straight or arching, thin and flexible, lasting 2–3 years, dark green to bluish green. Individual needles with lines of stomates only on the inner faces, an undivided midvein, (one or) two resin canals touching the epidermis of the outer face, and sometimes with a third resin canal touching the epidermis of one or the other inner face near the angle where they come together. Sheath 10–15 mm long, soon shed. Pollen cones 5–15 mm long, straw-colored. Seed cones (6–) 8–20(–25) cm long, taperingly cylindrical and pointed at both ends, often curved, with 40–100 seed scales, yellowish green to bluish green or even purple-tinged before maturity, ripening light brown, opening widely to release the seeds, with a slender stalk (1–)2–4.5 cm long. Seed scales egg-shaped, thin and a little flexible, the exposed tips straight, the narrow, diamond-shaped umbo straight or slightly hooked inward. Seed body (5–)6–8(–9) mm long, egg-shaped, the firmly attached wing 18–25(–30) mm long. Eastern North America in southeastern Canada, northeastern and east-central United States, southern Mexico, and southwestern Guatemala. Forming pure stands, or more commonly mixed with red pine (*P. resinosa*) and other conifers and numerous hardwoods on a variety of soil types; 0–1,500(–2,200) m. Zone 3. Two varieties.

Twig of eastern white pine (*Pinus strobus*) with 4 years of growth and pollen cones, ×0.5.

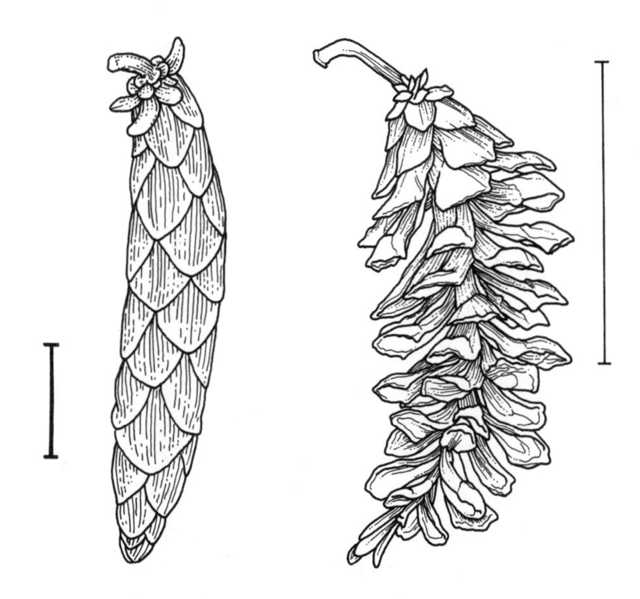

Unopened mature seed cone of eastern white pine (*Pinus strobus*); scale, 2 cm.

Open seed cone of eastern white pine (*Pinus strobus*) after seed dispersal; scale, 5 cm.

Eastern white pine is the tallest tree in eastern North America and often towers above associated hardwoods to form a super-canopy. This happens because it follows the common conifer pattern of establishing as a pioneer on disturbances but outliving the pioneer hardwoods that come in with it. When these associates die and their shade-tolerant successors replace them, the white pines keep on growing and thus persist as emergent dominants of a forest in which they cannot effectively regenerate. The giant trees whose felling provided much of the prosperity of the North American colonies in the 18th and early 19th centuries, as well as the masts of the British navy of the time, were 200- to 300-year-old pioneers on agricultural land abandoned by native slash and burn farmers after it had lost its fertility for growing corn. There are reports of felled trees from those days 80 m tall and 3.6 m in diameter, but the two largest, more recently known individuals in the United States in combined dimensions, when measured in 1984, were 61.3 m tall, 1.5 m in diameter, with a spread of 15.8 m, and 55.2 m tall, 1.6 m in diameter, with a spread of 19.5 m. Both have since died, and a new champion measured only 38 m tall in 2003. Eastern white pine is still an important timber tree today both from natural regeneration on agricultural land abandoned in the 19th and 20th centuries and from forest plantations. It is also grown extensively as a timber tree in central and southern Europe. It is susceptible to white pine blister rust, which kills the trees, and to white pine weevils, which deform them by killing the leader. That is one reason why so much research has been done on crossability among white pine species, a by-product of the search for rust resistant white pine hybrids suitable for eastern North America.

Eastern white pine crosses successfully with most of the other species in subsection *Strobus* that have been attempted, but not with sugar pine (*Pinus lambertiana*) or Chinese white pine (*P. armandii*). Although there is a great deal of variation in *P. strobus*, this is generally not geographic, and few attempts have been made to subdivide the species into varieties. One exception is the Central American population, separated by at least 2,000 km from the northern range of the species but growing with many other species, like sweet gum (*Liquidambar styraciflua*) and black cherry (*Prunus serotina*), with a similar disjunct distribution. These southern populations are weakly distinguishable from the northern ones and are usually recognized as a distinct variety, although some authors treat them as a separate species. This variety was depleted by general harvesting of accessible stands and further seriously endangered by local exploitation of more remote trees for their highly prized timber. The scientific name, *strobus*, is a classical Latin name for a conifer transferred to this species.

Pinus strobus var. *chiapensis* M. Martínez

PINABETE (SPANISH)

Needles 10 cm long on average with about 25 tiny teeth per centimeter of edge along the middle. Stalk of seed cone up to 4.5 cm long. Seed scales about 90 on average, the sterile ones at the base near the cone stalk not curled back. Western Guatemala and Chiapas to central Guerrero and northern Puebla (Mexico); at moderate elevations and often in cloud forest. Synonym: *Pinus chiapensis* (M. Martínez) Andresen.

Pinus strobus Linnaeus var. *strobus*

Needles 8 cm long on average with about 12 tiny teeth per centimeter of edge along the middle. Stalk of seed cone up to 3 cm long. Seed scales about 70 on average, the sterile ones at the base curled back toward the cone stalk. Newfoundland to southeastern Manitoba south to northern Georgia; from sea level to modest elevations in the southern mountains.

Pinus sylvestris Linnaeus

SCOTS PINE, EUROPEAN REDWOOD, PINO DE VALSAÍN (SPANISH), PINHEIRO-SILVESTRE (PORTUGUESE), PINO BRAVO (GALICIAN), LER GORRIA (BASQUE), PI ROIG (CATALÁN), PIN SILVESTRE (FRENCH), GEMEINE KIEFER (GERMAN), PREDE (LATVIAN), VANLIG FURU (NORWEGIAN), TALL (SWEDISH), MENTI (FINNISH), SOSNA LESNAYA (RUSSIAN), PITCHUI (GEORGIAN), TEGOSH (ARMENIAN), NARAT (TURKIC), BES (YAKUTIAN), OU ZHOU CHI SONG (CHINESE)

Plates 53 and 54

Pinus subg. *Pinus* sect. *Pinus* subsect. *Pinus*

Tree to 30(–40) m tall, shorter in arid regions or shrubby near the alpine tree line. Trunk to 1(–1.7) m tall, often forked or leaning. Bark bright reddish orange, smooth, and peeling in papery flakes on the upper trunk and branches, breaking up into gray, scaly, interrupted ridges separated by broad, reddish brown furrows on the lower trunk. Crown variably conical, through cylindrical, to nearly spherical, with contorted horizontal to upraised branches densely clothed with foliage only at the tips. Twigs greenish yellow to orange, hairless, slightly grooved between the bases of the scale leaves. Buds 5–8(–11) mm long, resinous. Needles in bundles of two (or three), each needle (0.5–)4–7(–12) cm long, somewhat flexible but straight, conspicuously twisted, very broad for their height (about two to three times as broad as high), prickly, lasting 2–3(–9) years, usually bluish green. Individual needles with numerous, conspicuous lines of stomates on both faces, a two-stranded midvein, and (2–)6–10(–22) resin canals touching the needle surface all around the periphery. Sheath 9–12 mm long, weathering to 3–6 mm and persisting and falling with the bundle. Pollen cones (2–)3–6(–8) mm long, pink or yellow. Seed cones (1.5–)3–5(–7) cm long, egg- to broadly egg-shaped, with 70–100 seed scales, green before maturity, ripening shiny or dull yellowish brown to reddish brown to grayish brown, opening widely to release the seeds and then falling with the slender, short, curved stalk

(2–)4–12(–15) mm long. Seed scales variably paddle-shaped, the exposed face unequally diamond-shaped, flat to conically protruding, especially near the base on the side away from the twig, crossed by a distinct ridge topped by a flat, small umbo bearing an inconspicuous, blunt prickle. Seed body (2–)3–5 mm long, the easily detachable wing another 10–15 mm longer. Discontinuously distributed all the way across Eurasia from near the Atlantic Ocean in Norway, Scotland, and northwestern Spain to the Pacific (Sea of Okhotsk) in Russia, with its broadest latitudinal distribution in western Eurasia from about 37°N in Spain and Turkey to above 70°N in Norway but mostly confined to a broken band between about 50°N and 65°N across central and eastern Eurasia to about 142°E. Forming immense, pure forests (hence the scientific name, Latin for "of forests"), mixed with birches and other boreal trees, or confined to isolated groves in the steppes south of the boreal forest in Siberia, on substrates ranging from dry sands to bogs; 0–2,100(–2,600) m, generally at lower elevations in the north and higher elevations in the south. Zone 2. Synonyms: *P. armena* K. Koch, *P. hamata* (Steven) Sosnowsky, *P. kochiana* Klotzsch ex K. Koch, *P. sosnowskyi* T. Nakai, *P. sylvestris* var. *aquitana* H. Schott, *P. sylvestris* var. *armena* (K. Koch) Fitschen, *P. sylvestris* var. *hamata* Steven, *P. sylvestris* var. *lapponica* E. M. Fries, *P. sylvestris* var. *mongolica* D. Litvinov, *P. sylvestris* var. *nevadensis* H. Christ, *P. sylvestris* var. *rigensis* (Desfontaines) P. Ascherson & K. Graebner, *P. sylvestris* var. *scotica* (Willdenow) H. Schott, *P. sylvestris* var. *sibirica* Ledebour.

Scots pine is far and away the most widely distributed pine species in the world and the second most widely distributed conifer after common juniper (*Juniperus communis*). In addition to its enormous geographic range of 150° of longitude and almost 35° of latitude, it has an elevational range from sea level to 2,600 m, and an ecological amplitude from the arctic to the Mediterranean mountains to the margins of the central Asian deserts on a wide variety of substrates supporting widely different vegetation and plant communities. It comes as no surprise, then, that Scots pine is highly variable in morphological characteristics, like needle length and color and bark color, thickness, and roughness, and in genecological traits, like growth rate, winter hardiness, and branching habit. These are the kinds of features that are assessed in provenance trials of a few dozen seed sources and that led foresters to identify many races of *Pinus sylvestris* that were then given botanical names. In fact,

Mature Scots pine (*Pinus sylvestris*) in nature.

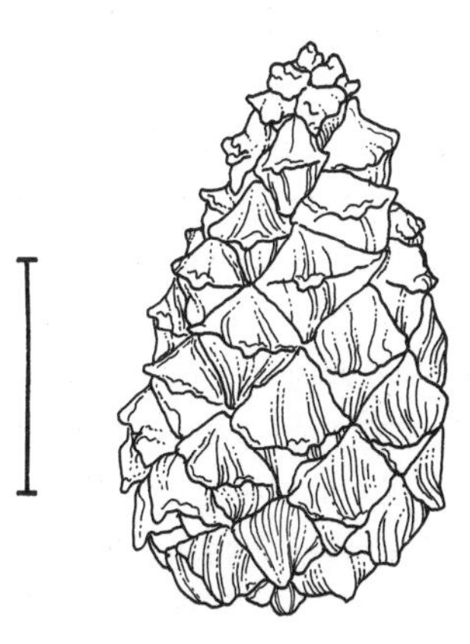

Unopened mature seed cone of Scots pine (*Pinus sylvestris*); scale, 2 cm.

Open seed cone after seed dispersal (A) and dispersed seed (B) of Scots pine (*Pinus sylvestris*); scale, 1 cm.

about 150 varieties have been described. This kind of narrow fragmentation serves no useful purpose in taxonomy. It is far better for practical purposes, like choosing the best seed sources for plantations in a given region, to simply identify desirable seed sources by their place of origin rather than giving them formal taxonomic recognition. With the present lack of complete range-wide studies involving taxonomically appropriate characters, there appears to be no immediate prospect of a consistent subdivision of the species as a whole into a few morphologically distinctive geographically coherent forms that could be widely accepted as varieties or subspecies. The best candidates, perhaps, are the populations from the Balkans through Turkey and the Crimea (Ukraine) to the Caucasus (Georgia and Russia), which have seed cones with a glossy, varnishlike surface, and those from northern China and vicinity with their unusually long needles. Even those, however, overlap with populations from the main range in these features and are scarcely distinguishable by any other characteristics. So until a convincing rationale for formal taxonomic subdivision is presented, the unifying features of the pure stands that separate it from its closest relatives are emphasized, and no varieties or subspecies are accepted here. The only synonyms listed here are those named varieties that are accepted by the major current regional floras.

All the natural variation provides a great opportunity for selection of cultivars, and many are available. The tree, whether in the form of cultivars, selected provenances, or relatively unselected forms, is very widely cultivated in temperate regions as an ornamental and for timber production. It escaped from cultivation and became naturalized outside of its natural range in Europe and in northeastern North America. Considering its ecological dominance and economic importance in northern and central Europe and across Siberia (Russia), a great deal is known about its patterns of growth, heartwood formation, biosynthesis and inheritance of leaf and wood oils, disease resistance and susceptibility, insect interactions, and fire history. Its economic importance over such a large range is also reflected in the large number of common names that it inspired across Eurasia, only a few of which are given here. The trees can live 400 years and often grow under marginal conditions, so they can be sensitive indicators of past climates and are used in dendroclimatology and dendrochronology. In the Alps and other central and southern European mountain ranges, Scots pine hybridizes with European mountain pine (*Pinus mugo*) to form *P.* ×*rhaetica*. It also crosses successfully in experimental trials with European black pine (*P. nigra*), Japanese black pine (*P. thunbergii*), and its close relative Japanese red pine (*P. densiflora*). All these are species of subsection *Pinus*, but many attempts with other species of the subsection fail, so Scots pine is relatively genetically isolated within the subsection.

Pinus tabuliformis Carrière

CHINESE RED PINE, CHINESE PINE, YOU SONG (CHINESE), MANCHU KOM SHOL (KOREAN)

Pinus subg. *Pinus* sect. *Pinus* subsect. *Pinus*

Tree to 25 m tall, with trunk to 0.8(–1.2) m in diameter. Bark dark grayish brown with reddish highlights, thin, passing from narrow ridges above to progressively larger, scaly plates lower down. Crown very broadly dome-shaped, or flat (hence the scientific name, Latin for "table-shaped"), with upwardly angled, long, heavy branches bearing thinner, downswept ones modestly clothed with foliage at the tips. Twigs coarse, yellowish brown to reddish brown, sometimes waxy, hairless, somewhat rough with the bases of scale leaves. Buds 12–18 mm long, slightly resinous. Needles in bundles of two (or three), each needle (6–)10–15 cm long, stiff or flexible, straight, lasting 2–3 years, bright green to dark green. Individual needles with inconspicuous lines of stomates on both the inner and outer faces, a two-stranded midvein, and (2–)5–9(–10) resin canals all around the periphery usually touching the surface. Sheath 10–20 mm long, weathering to 5–10 mm and persisting and falling with the bundle. Pollen cones in a short grouping, 5–10 mm long, yellowish brown. Seed cones 2.5–9 cm long, broadly egg-shaped, with 50–80 seed scales, green before maturity, ripening shiny yellowish brown to brown, opening widely to release the seeds and then persisting several years before falling with the short, slender stalk to 1 cm long. Seed scales paddle-shaped, the exposed face roughly horizontally diamond-shaped, a little to prominently protruding, crossed by a horizontal ridge topped by a large diamond-shaped umbo bearing a short, thick spine. Seed body 6–8 mm long, the easily detachable wing another 10–15 mm longer. Widely distributed across the hills and mountains of central China from central Qinghai and northern Sichuan east to Shandong and Jilin and adjacent Korea. Forming pure open or closed stands or mixed with other conifers and hardwoods in forests and woodlands on various substrates; (50–)1,000–2,500(–3,000) m. Zone 5. Two varieties.

Chinese red pine is the chief hard pine of northern China, and its considerable drought tolerance allows it to thrive all the way to the fringes of the arid zone. As well as producing timber, the species is exploited for its resin and for herbal medicine. Chinese red pine, especially in the most extreme flat-topped forms, is a treasured ornamental in China, and many fine trees are to be seen in and around Beijing. There is considerable variation across the range of the species, and several varieties have been described, but only two seem to merit formal recognition, while a third is better separated as an independent species. This latter, Chinese mountain pine (*Pinus densata*), is thought to have arisen after hybridization between *P. tabuliformis* and Yunnan pine (*P. yunnanensis*), but this cross is almost impossible to make under controlled conditions. Instead, Chinese red pine crosses readily with more closely related members of subsection *Pinus,* including Japanese black pine (*P. thunbergii*), Taiwan red pine (*P. taiwanensis*), and Khasi pine (*P. kesiya*). The affinities of these species, including *P. yunnanensis,* originally suggested by their close morphological similarity, is also supported by DNA studies. These studies reveal that the fourth major Chinese hard pine, Masson pine (*P. massoniana*), is more distant from this close-knit group of species.

Pinus tabuliformis* var. *henryi (M. T. Masters) C. T. Kuan
BA SHAN SONG (CHINESE)
Twigs reddish brown, usually whitened with wax. Needles 7–12 cm long, relatively slender, somewhat flexible. Seed cones 2.5–5 cm long. Exposed face of seed scales only slightly protruding. South-central portion of the range of the species, from southern Shaanxi through eastern Sichuan and western Hubei to Hunan. Synonym: *Pinus henryi* M. T. Masters.

Pinus tabuliformis Carrière **var. *tabuliformis***
YOU SONG (CHINESE)
Twigs yellowish brown, usually not waxy or only transiently so. Needles (6–)10–15 cm long, thick and stiff. Seed cones 4–9 cm long. Exposed face of the seed scales strongly projecting, especially near the base. Throughout the range of the species except where occupied by variety *henryi.* Synonyms: *Pinus tabuliformis* var. *mukdensis* (Uyeki ex T. Nakai) Uyeki, *P. tabuliformis* var. *umbraculifera* T. N. Liou & W. T. Wang.

Pinus taeda Linnaeus
LOBLOLLY PINE
Pinus subg. *Pinus* sect. *Trifoliae* subsect. *Australes*
Tree to 35(–45) m tall, with trunk to 1(–1.5) m in diameter. Bark dark brown, thick, scaly, breaking up into broad, interlacing ridges or irregular, small plates separated by narrow furrows. Crown dome-shaped, deep to shallow, generally dense, with numerous slender, upwardly angled branches densely clothed with foliage near the tips. Twigs yellowish brown to reddish brown, hairless, rough with the bases of scale leaves. Buds 10–15(–20) mm long, slightly resinous. Needles in bundles of (two or) three, each needle (8–)10–17(–23) cm long, thick but flexible, twisted, pointing a little forward, lasting (2–)3 years, dark yellowish green or sometimes grayish green. Individual needles with inconspicuous, evenly spaced lines of stomates on both the inner and outer faces, and two (to seven) resin canals at the corners and in between, mostly midway between the needle surface and the two-stranded midvein. Sheath 15–20(–25) mm long, weathering to 10–15 mm and persisting and falling with the bundle. Pollen cones (1.5–)2–4 cm long, yellowish brown, sometimes with a reddish blush. Seed cones (6–)7–10(–12) cm long, egg-shaped to cylindrical, with 120–160 seed scales, green before maturity, ripening dull yellowish brown, opening widely to release the seeds and then sometimes remaining for up to 3–4 years before falling with the extremely short stalk. Seed scales gently tapering, the exposed face horizontally diamond-shaped, from nearly flat on the upper scales to noticeably protruding at the base of the cone, crossed by a distinct ridge topped by an umbo bearing a strong, sharp, downwardly pointing spine. Seed body 5–7 mm, the firmly attached wing another 15–23 mm longer. Coastal Plain and Piedmont of the southeastern United States, from southwestern New Jersey to central Florida, west to eastern Texas and southeastern Arkansas, with a gap in the Mississippi River valley. Forming pure stands or mixed with other pines and hardwoods on a wide variety of sites from swamp margins to old fields; 0–700 m. Zone 7.

Grown for lumber, plywood, pulp, and oleoresin, loblolly pine is the most economically important plantation tree among the four major "southern" pines, species of subsection *Australes* native to the southeastern United States. As such, there are more studies of its growth patterns, geographic and genetic variation, reproductive biology, breeding strategies, disease and pest resistance, environmental tolerances, plantation management, and utilization than for any of the other three species: shortleaf pine (*Pinus echinata*), longleaf pine (*P. palustris*), and slash pine (*P. elliottii*). Loblolly pine was the first conifer in which maternal inheritance of mitochondria (the typical pattern in higher plants) was demonstrated and one of the few conifers for which there are extensive genetic maps covering all the chromosomes. The various studies suggest that loblolly pine survived the most recent glacial period in two separate refuges, one in southern Texas and adjacent Mexico and the other in southern Florida. These observations may explain the greater resistance to fusiform rust disease of populations west of the Mississippi River than of those eastward through the bulk of the range of the species.

Alternatively, this resistance has been explained by natural hybridization with the much more rust resistant shortleaf pine. Artificial hybridization with various species is used as a way of bringing the excellent forestry qualities of loblolly pine to a broader range of potential plantation sites. For example, in addition to rust resistance from shortleaf pine, breeders have attempted to introduce greater frost hardiness from the more northerly but commercially inferior pitch pine (*Pinus rigida*) for use in plantations in Korea and the northeastern United States, and greater tolerance for poor sites from slash pine for use in Zimbabwe. Natural hybrids with all three of these species are known, as well as with longleaf pine. The latter hybrid, one of the first natural pine hybrids to be described, is the only one to have been given a formal scientific name, *P.* ×*sondereggeri.*

The scientific name of the very resinous loblolly pine was simply borrowed by Linnaeus from the ancient Roman name for torches and the pitchy pines they were made from. The largest known individual in combined dimensions when measured in 2003 was 50.9 m tall, 1.4 m in diameter, with a spread of 21.7 m.

Pinus taiwanensis Hayata

TAIWAN RED PINE, TAI WAN SONG (CHINESE)

Pinus subg. *Pinus* sect. *Pinus* subsect. *Pinus*

Tree to 35(–50) m tall, with trunk to 0.8(–1) m in diameter. Bark dark grayish brown, broken up into small, irregular plates by narrow furrows. Crown cylindrical, flat-topped, open, with numerous slender, short, horizontal or gently rising branches sparsely clothed with foliage. Twigs yellowish brown to reddish brown, hairless, rough with the bases of scale leaves. Buds 10–15 mm long, sparsely resinous. Needles in bundles of two, each needle (4.5–)8–12(–17) cm long, slender but stiff, straight and scarcely twisted, lasting 2–3 years, glossy dark green. Individual needles with inconspicuous lines of stomates on both faces, and (two to) four to six (to eight) resin canals all around the two-stranded midvein, mostly midway to the needle surface. Sheath 10–15 mm long, weathering to (2–)5–10 mm and persisting and falling with the bundle. Pollen cones 1–2 cm long, yellowish brown to reddish brown. Seed cones 3–7 cm long, egg-shaped, with 40–100 seed scales, green before maturity, ripening orange-brown to rich brown, opening widely to release the seeds and then persisting a while before falling with the very short, stout stalk to about 1 cm long. Seed scales variably spoon-shaped, the exposed face horizontally diamond-shaped to five-sided, crossed by a ridge topped by a small, flat or slightly raised umbo bearing a tiny, fragile or persistent prickle. Seed body 5–6 mm long, the easily detachable wing another 10–15 mm longer. Abundant in the mountains of Taiwan (hence the scientific and common names) and scattered on the Chinese mainland west to Henan, Guizhou, and Guangxi. Forming pure, dense or open stands or mixed with other conifers and oaks and other hardwoods on mountain slopes; (600–)750–3,000(–3,400) m. Zone 9. Synonyms: *P. hwangshanensis* K. Hsia, *P. luchuensis* var. *hwangshanensis* (K. Hsia) M. H. Wu, *P. luchuensis* var. *taiwanensis* (Hayata) Silba, *P. taiwanensis* var. *damingshanensis* W. C. Cheng & L. K. Fu.

Taiwan red pine is the most widely distributed and abundant pine on that island and joins *Taiwania* and the two *Chamaecyparis* species among the most important native timber trees. It is much less common on the mainland, where it occurs as a secondary species within the range of the commercially much more important Masson pine (*Pinus massoniana*), which generally grows at lower elevations. The relative abundance on the mainland reverses the situation found on Taiwan, where Masson pine also co-occurs but is less frequently encountered. The mainland trees are sometimes distinguished as a separate variety or even species from those on Taiwan, but the proposed differences are few and have not been shown to be consistent. Taiwan red pine is less closely related to Masson pine than it is to more geographically distant species, like Yunnan pine (*P. yunnanensis*), Chinese red pine (*P. tabuliformis*), and Japanese black pine (*P. thunbergii*), as confirmed by DNA studies. It is generally considered closest, however, to Luchu pine (*P. luchuensis*) of the adjacent Ryukyu Islands, which grows under rather different ecological conditions. Taiwan red pine was sometimes considered a separate variety within the same species as Luchu pine, but their morphological and ecological differences seem sufficient to keep them apart unless more compelling evidence for merger is presented. The whole group of seven species closely related to Japanese black pine (including *P. taiwanensis*), which replace each other geographically across eastern and southeastern Asia, deserves closer study.

Pinus tecunumanii Eguiluz & J. P. Perry

TECUN UMÁN PINE

Pinus subg. *Pinus* sect. *Trifoliae* subsect. *Australes*

Tree to 50(–55) m tall, with trunk to 1.2(–1.4) m in diameter. Bark bright reddish brown, smooth and flaky above, becoming scaly and ultimately breaking up into grayish brown interlaced ridges divided by deep, reddish brown furrows at the base of large trunks. Crown narrow, deep, and dense, with numerous thick, horizontal branches densely clothed with foliage at the tips. Twigs reddish brown, sometimes a little waxy, hairless, rough with the bases of scale leaves. Buds 12–20 mm long, not resinous. Needles in bundles of (three or) four or five, each needle (14–)17–21(–25) cm long, soft, flexible, and gently drooping, lasting 2(–3) years, bright light green to yellowish green. Individual needles with evenly spaced lines of stomates covering all three faces, and (two or) three or four (or five) resin canals at the corners and in between, usually midway between the outer needle surface and the two-stranded midvein. Sheath 20–25 mm long at first, weathering to 12–18 mm and persisting and falling with the bundle. Pollen cones 15–20 mm long, yellowish brown. Seed cones 4–7(–8) cm long, egg-shaped, with 75–100(–140) seed scales, light brown before maturity, ripening shiny reddish to grayish yellow, opening gradually from the middle when ripe and falling after 2–3 years along with the slender, (5–)10–20(–25) mm long stalks. Seed scales broadly wedge-shaped, the exposed face horizontally diamond-shaped, generally low, with a shallow horizontal ridge topped by a prominent, flat, diamond-shaped umbo bearing a tiny, fragile prickle. Seed body 4–6(–7) mm long, the clasping wing another 9–14 mm longer. Mountains of Mesoamerica, from Chiapas (Mexico) to Cayo (Belize) and Matagalpa (Nicaragua). Forming pure stands or more often mixed with other pines and hardwoods in moist pine forests and woodlands and in cloud forests; (300–)500–2,600(–2,900) m. Zone 9. Synonyms: *P. oocarpa* var. *ochoterenae* M. Martínez, *P. patula* subsp. *tecunumanii* (Eguiluz & J. P. Perry) Styles.

Long simply subsumed within *Pinus oocarpa*, controversy has surrounded this pine since it was first recognized as a distinct species in 1953. At that time, the species was named in a manner that did not conform to the rules required for formal taxonomic recognition, and it was not until 1983 that the requirements were met and it received an acceptable botanical name as a species. However, as early as 1940, the populations in Chiapas at the northwestern end of the range were designated as a variety of *P. oocarpa*. In international plantation trials of *P. oocarpa* begun in the 1970s, certain sources consistently outperformed all others, and these proved to be *P. tecunumanii* rather than *P. oocarpa*. In fact, careful morphological and chemical study of these trials, as well as of wild trees in their native habitats, show that *P. tecunumanii* is much more closely related to *P. patula* than to *P. oocarpa*. Some authors would, therefore, unite Tecun Umán pine as a subspecies of *P. patula* or even submerge it entirely without nomenclatural recognition. Previous studies have not been comprehensive and their results are not completely unambiguous, so confusion still reigns with regard to the taxonomic limits, status, and geographic distribution of this pine. Acceptance of a separate species distributed from Chiapas to Nicaragua is similar to the treatment of Farjon and Styles (1997) except that the specimens from Oaxaca (Mexico) assigned by them to *P. tecunumanii* are here referred to *P. patula*. It is clear that more work is needed on the classification and relationships of an internationally increasingly important plantation tree, one of the finest timber trees among all of the tropical pines. The species name honors Tecun Umán, a 16th century Mayan leader in Guatemala, whence this pine was first described.

Pinus teocote Schiede & Deppe ex Chamisso & D. F. L. Schlechtendal

OCOTE PINE, OCOTE (SPANISH)

Pinus subg. *Pinus* sect. *Trifoliae* subsect. *Australes*

Tree to 20(–25) m tall, or shorter southward, with trunk to 0.7 (–0.8) m in diameter. Bark thick, dark grayish brown, broken up into broad, rough, interrupted ridges by deep, wide, reddish brown furrows. Crown dome-shaped to cylindrical, with numerous slender, horizontal to gently rising branches sparsely clothed with foliage at the tips. Twigs reddish brown with a dusting of wax, hairless, channeled between the long scale leaf bases which begin to peel off after the second year. Buds 8–15 mm long, not conspicuously resinous. Needles in bundles of (two or) three (or four), each needle (7–)10–15(–19) cm long, stiff and straight, lasting 2–3 years, bright light green. Individual needles with prominent lines of stomates all around both the inner and outer faces, and two to five (to seven) resin canals surrounding the two-stranded midvein midway to the needle surface. Sheath 12–20(–23) mm long at first, rupturing and weathering to (5–)10–15 mm and then persisting and falling with the bundle. Pollen cones 1–2 cm long, yellowish brown. Seed cones (2.5–)4–6(–7) cm long, egg-shaped to conical when fully open, sometimes slightly asymmetrical and curved, with 50–100 seed scales, green before maturity, ripening light brown, opening widely to release the seeds and then falling or persistent for a while before falling with the short, stout stalk 5–8(–12) mm long. Seed scales paddle-shaped, the exposed face horizontally diamond-shaped, only slightly raised, crossed by a ridge, the umbo large, flat to pointed and bearing a fragile prickle. Seed body 3–5 mm long, the clasping wing another (8–)10–15 mm longer. Mountains throughout Mexico, from western Chihuahua and northern Coahuila to southeastern Chiapas. Forming pure stands or mixed with other pines and oaks in open forests and woodlands; 1,000–3,000(–3,300) m. Zone 9.

As one of the principal pine species of the mountains surrounding the Valley of Mexico, *Pinus teocote* has been an important commercial species since Aztec times. The species name is the Spanish adaptation of its Aztec name, *teocotl*, "stone pine." It has the widest distribution of any strictly Mexican pine species and has a correspondingly wide ecological range. Across the varied conditions in which it grows, it displays considerable variation in stature as well as in the size and form of the needles and seed cones. The possibility of natural hybridization where it co-occurs with its close relatives Herrera pine (*P. herrerae*), Lawson pine (*P. lawsonii*), and Pringle pine (*P. pringlei*) requires investigation.

Pinus thunbergii Parlatore

JAPANESE BLACK PINE, KURO-MATSU (JAPANESE), KOM SHOL (KOREAN)

Plate 55

Pinus subg. *Pinus* sect. *Pinus* subsect. *Pinus*

Tree to 30(–35) m tall, with trunk to 1(–2) m in diameter. Bark dark gray with reddish gray flakes, divided into flat, irregularly polygonal plates by narrow, black furrows. Crown broadly conical at first, becoming flat-topped, irregular, and open with age, with slender horizontal to shallowly upwardly angled branches densely clothed with foliage near the ends. Twigs yellowish brown, hairless, rough with the bases of scale leaves. Buds white, (1–)1.5–2.5(–4) cm long, not resinous. Needles in bundles of two (or three), each needle (6–)9–12(–15) cm long, stiff and straight, lasting 3(–4) years, dark green. Individual needles with inconspicuous lines of stomates on both the inner and outer faces and (2–)4–10(–11) resin canals all around the two-stranded midvein midway out to the needle surface. Sheath 6–10 mm long, weathering to 4–6 mm long and persisting and falling with the bundle. Pollen cones densely crowded, 10–13 mm long, yellowish brown. Seed cones 4.5–6 cm long, egg-shaped to almost spherical, with (40–)50–60(–80) seed scales, grayish green before maturity, ripening yellowish brown, opening widely to release the seeds and then falling along with the short stalk to about 1 cm long. Seed scales variably spoon-shaped, the exposed face horizontally diamond-shaped to five-sided, almost flat, crossed by a ridge topped by a flat umbo bearing a tiny, fragile or persistent prickle. Seed body 3.5–6(–8) mm long, the easily detachable wing another 10–15 mm longer. Coastal region of southern Korea and of Japan from the northern tip of Honshū south to Takara Jima south of Kyūshū. Forming pure, open stands or mixed with other trees near the seashore on sandy or rocky soils; 0–300(–1,000) m. Zone 6. Synonym: *P. thunbergiana* Franco.

Japanese black pine is the quintessential shoreline tree in Japanese art, and its appearance is reproduced in Japanese gardens and bonsai. It is the most widely planted conifer in Japan after sugi (*Cryptomeria japonica*), and many cultivars have been selected over the centuries. It is also a valued ornamental throughout the temperate world, not requiring the coastal habitat of its native stands to grow successfully. In fact, it attains its best growth in sheltered sites away from the immediate influence of salt spray at the seashore. It is widely cultivated for timber in Japan on such sites. Perhaps because of its economic importance, it has been the subject of intensive genetic and molecular biological studies. It was the first conifer to have its chloroplast genome mapped in detail, and the DNA sequence data thereby derived are used to probe genetic variation in other pine species, and even in other members of the pine family. This is how some genetic variation was discovered in the North American red pine (*Pinus resinosa*) when all previous studies had found it to be genetically uniform.

In addition to its adaptability to cultivation as an ornamental and timber tree, *Pinus thunbergii* can be crossed successfully with many other species of subsection *Pinus*, including several Chinese species, but not with Scots pine (*P. sylvestris*). It crosses naturally with Japanese red pine (*P. densiflora*) where the two species grow together to form the hybrid *P.* ×*densithunbergii.* It also crosses naturally with the European black pine (*P. nigra*) when the two are grown together in plantations or ornamental plantings. Both these hybrid combinations grow faster than either parent, at least when they are young, but the hybrid with *P. densiflora* is easier to make artificially than the one with *P. nigra* and so is more promising for forestry use.

As well as crossing readily with other species, *Pinus thunbergii* begins reproduction at a very early age, seed cones sometimes appearing when the trees are only 3 years old. Heavy crops of seed cones are another of the features that have contributed to its popularity as an ornamental. The seed cones are interesting to morphologists and developmental biologists also, because they sometimes appear among or in place of pollen cones. The species name honors Pehr Thunberg (1743–1825), a Swedish botanist and student of Linnaeus who traveled in Japan and included the species in the first botanical flora of Japan, which he wrote, but referred it incorrectly to Scots pine.

Pinus torreyana C. Parry ex Carrière

TORREY PINE

Pinus subg. *Pinus* sect. *Trifoliae* subsect. *Ponderosae*

Tree to 25(–40) m tall, with trunk to 1(–2) m in diameter, often forked and deformed by wind and storms. Bark reddish brown to grayish brown, irregularly broken up into small, flat-topped, vertical blocks by shallow furrows. Crown conical at first, broadening, flattening, and opening with age, often becoming broader than high, with numerous, upwardly arched, thick branches sparsely clothed with foliage at the tips. Twigs grayish purple-brown with a thin waxy coating, rough with bases of scale leaves, hairless. Buds 15–25(–50) mm long, resinous. Needles in bundles of (three to) five, each needle (15–)18–30 cm long, stiffly outstretched and straight or slightly twisted, lasting 2–4 years, grayish yellow-green or bluish green. Individual needles covered with lines of stomates on all three faces, and with a two-stranded midvein and three to six resin canals at the corners and in between deep inside the leaf tissue. Sheath 2–3(–5) cm long, the lower 1–2 cm persisting and falling with the bundles. Pollen cones (2–)4–6 cm long, yellow. Seed cones 9–16 cm long, massively egg-shaped, with 60–80 seed scales, green before maturity, ripening dark reddish brown to chestnut brown, opening widely to release the seeds and persisting several years, with a few seeds remaining caught among the lower scales, on stout stalks 2–4 cm long. Seed scales angularly paddle-shaped, the exposed face projecting in a pyramid tipped by a stout, sharp, triangular umbo. Seed body 15–25 mm long, the easily detachable wing 10–15 mm longer. Santa Rosa Island and near Del Mar, coastal San Diego County, southern California. Forming pure, open stands among grass or shrubs; 50–150 m. Zone 9. Synonym: *P. torreyana* subsp. *insularis* J. Haller.

Torrey pine is one of the rarest pine species in the world, with fewer than 5,000 mature trees and as many saplings in one mainland and one island population, each divided between two main groves. The mainland and island populations have been distinguished as separate subspecies, but the morphological differences are small and inconsistent between trees in native stands and in cultivation. Trees in the island and mainland populations also have turpentines with almost the same chemicals, but two of these, β-phellandrene and lemon-scented limonene, differ modestly in quantity between them. A study of 59 enzyme genes in the two populations found only two that differed between them. More interestingly, this study found that all the trees are otherwise completely uniform in their enzyme composition, the lowest genetic variation ever seen in a conifer. This implies that the Torrey pine populations were reduced to very small numbers within the last 10,000 years and have not yet regained genetic variation via mutations. Against this background, there seems little reason to assign each population to its own subspecies. The small apparent differences are largely an artifact of the present lack of variability within each population and the reduction of the species as a whole to just two populations, without the intermediates that link the more strongly differentiated populations of the ecologically similar and unrelated bishop (*Pinus muricata*) and Monterey (*P. radiata*) pines, neither of which is usually subdivided taxonomically. Unlike these latter two species, there are no known fossil cones of Torrey pine in ice-age coastal sediments anywhere along the California coast.

Torrey pine is closely related to the other bigcone pines within subsection *Ponderosae*, Coulter pine (*Pinus coulteri*) and gray pine (*P. sabiniana*), and has a low crossability of about 2% with gray pine but is otherwise reproductively isolated from other pine species. The largest known individual in combined dimensions when measured in 1986 was 49 m tall, 1.5 m in diameter, with a spread of 24 m. This tree, a cultivated individual planted in 1890, is twice as tall as the tallest trees in the native groves, although only a little thicker. The species name honors John Torrey (1796–1873), the eastern American botanist to whom Charles Parry sent the original specimens and accompanying description for publication.

Pinus tropicalis P. Morelet

CUBAN RED PINE, TROPICAL PINE, PINO BLANCO, PINO HEMBRA (SPANISH)

Pinus subg. *Pinus* sect. *Pinus* subsect. *Pinus*

Tree to 30 m tall, with trunk to 1 m in diameter. Bark dark, reddish brown to grayish brown, with small, irregular, flat, scaly plates separated by deep, dark, narrow furrows. Crown conical at first, becoming broader, flatter, and more open with age, with gently rising to horizontal, thick branches sparsely clothed with foliage at the tips. Twigs coarse, shiny light reddish brown, rough with bases of scale leaves, hairless. Buds (10–)15–25(–30) mm long, sparsely resinous. Needles in bundles of two (or three), each needle (15–)20–30 cm long, stiff and straight, lasting 2(–3) years, bright pale green or yellowish green. Individual needles with inconspicuous lines of stomates on both the inner and outer faces, and 6–10 (mostly 8–9) large resin canals surrounding the two-stranded midvein between the vein boundary and the leaf surface. Sheath 15–20 mm long at first, the lower portion persisting and falling with the bundles. Pollen cones 20–30 mm long, yellow flushed with red. Seed cones 5–8 cm long, egg-shaped, with 50–80(–120) seed scales, green before maturity, ripening light reddish brown, opening widely to release the seeds and then persisting several years before falling with the short, stout stalks. Seed scales paddle-shaped, the exposed face horizontally diamond-shaped, a little raised, the umbo without a prickle. Seed body (2.5–)4–5 mm long, the clasping wing 12.5–16 mm longer. Western Cuba. Forming open, fire-prone savannas alone or mixed with *P. caribaea* on sandy or gravelly soils; 0–200(–300) m. Zone 10.

Cuban red pine and red pine (*Pinus resinosa*) are the only North American members of the otherwise Eurasian subsection *Pinus*. The history of this subsection in North America is unknown. Much more restricted in distribution than red pine, it is not known whether *P. tropicalis* also lacks genetic variation and inbreeding depression. *Pinus tropicalis* resembles some unrelated North American species, like longleaf pine (*P. palustris*), in its fire-resistant juvenile "grass" stage. This stage is not found in Caribbean pine (*P. caribaea*), with which it shares the characteristic pine savannas of western Cuba, including Pinar ["pine forest"] del Río province and Isla de la Juventud (formerly Isle of Pines). These savannas are wet during the summer and dry during the winter. Cuban red pine, although growing continuously, responds to these changes with light wood during the wet season and dark wood during the dry season, so broad annual rings are marked by paired light and dark bands. This pattern is often complicated by narrow false rings caused by heavy rains during the dry season or droughts during the wet season. Cuban red pine is also distinctive for its resin canals, the largest among the pines, at least relative to the appearance of the leaf cross section.

Pinus virginiana P. Miller

SCRUB PINE, POVERTY PINE

Pinus subg. *Pinus* sect. *Trifoliae* subsect. *Contortae*

Tree to 20(–35) m tall, with trunk to 0.5(–1) m. Bark orange-brown and flaky when young, becoming grayish brown and breaking up into scaly, irregular, narrow blocks separated by shallow furrows. Crown becoming rounded, spreading, and irregular with age, with numerous slender, outstretched branches turning up or down at the ends, sparsely clothed with foliage at the tips, and not dropping cleanly after death. Twigs reddish brown or purple-brown, sometimes with a pale waxy coating, hairless and smooth. Buds 0.6–1 mm long, resinous or not, the scales white-fringed. Needles in bundles of two, each needle (2–)4–9(–10) cm long, stiffly outstretched, somewhat twisted, lasting 2–4 years, dull dark green to yellowish green. Individual needles with lines of stomates on both the faces, a two-stranded midvein, and two (or three) resin canals usually buried deep in the leaf tissue near the outer corners. Sheath 3–10 mm, the long, inner scales tattering, the lower scales persisting and falling with the bundle. Pollen cones 10–20 mm long, yellowish brown. Seed cones 3–9 cm long, egg-shaped to conical, symmetrical, with 50–120 seed scales, green before maturity, ripening reddish brown, opening at maturity or remaining tightly closed for years, in either case persisting on the twigs at the end of short stalks to 1 cm long. Seed scales egg-shaped, stiff, with a distinct purple border around the far edge on the inner face, the exposed portion diamond-shaped, crossed side to side by a ridge bearing a diamond-shaped umbo ending in a sharp prickle. Seed body 4–5(–7) mm long, mottled or all dark, the easily detachable wing 15–20 mm long. Discontinuous in lowlands and foothills of the eastern United States from Long Island (New York) through southern Ohio to western Tennessee south to Florida. Generally in open, fire-prone sites from sandy flats and dunes and shaley barrens to dry hillsides and old fields; 0–750(–900) m. Zones 5–7. Two varieties.

This small, twiggy tree named for the state of Virginia, in the heart of its distribution, is almost useless for lumber because the persistent dead branches make the wood hopelessly knotty. However, it produces excellent pulp for paper on soils too poor to support reasonable growth of any of the other southern pines. Because of the density of the stands, they also produce acceptable yields in fuelwood and biomass plantations. Trees with dark green needles are also suitable as Christmas trees. So despite the poor sites and poor form of this species, it is produced in large numbers for plantation culture and has been subject to at least modest improvement efforts. The two varieties are usually treated as separate species, but they are very similar to each other morphologically, highly intercrossable, and as similar in their enzymes as are the varieties of lodgepole pine (*Pinus contorta*). Present evidence is a little ambiguous about the correct taxonomic placement of scrub pine. Morphologically, it is very similar to jack pine (*P. banksiana*) and lodgepole pine of subsection *Contortae*, but it is not crossable with either of these species. It is, however, slightly crossable with some of its neighboring species in subsection *Australes*. Despite these crossability results, DNA studies tilt the balance decisively in favor of placement in subsection *Contortae* rather than subsection *Australes*. The largest known individual in combined dimensions when measured in 1997 was 27.8 m tall, 0.8 m in diameter, with a spread of 12.8 m.

Pinus virginiana* var. *clausa (Engelmann) Eckenwalder

SAND PINE

Needles 5–9(–10) cm long, with 8–10 lines of stomates on the outer face, five to seven lines on the inner face, and little separation between the two stands of the midvein. Seed cones usually remaining closed after maturity until after a fire but opening promptly in some populations (known as Choctawhatchee sand pine) of western Florida. Umbo with a short, stout prickle. Southern portion of the range of the species in Florida and adjacent coastal Alabama. Confined to sand hills and dunes at up to 60 m, the highest elevation within its range. Synonyms: *Pinus clausa* (Engelmann) C. Sargent, *P. clausa* var. *immuginata* D. Ward.

Pinus virginiana P. Miller **var. *virginiana***

VIRGINIA PINE

Needles (2–)4–7 cm long, with 11–13 lines of stomates on the outer face, 8–11 lines on the inner face, and wide separation between the two strands of the midvein. Seed cones opening promptly after maturity. Umbo with a longer, more slender (but still broad-based) prickle. Northern portion of the range of the species south to northern South Carolina, northern Georgia, and central Alabama. In all habitats of the species over the full elevational range.

Pinus wallichiana A. B. Jackson

HIMALAYAN WHITE PINE, BLUE PINE, KAIL (HINDI, KASHMIRI), TONGSHI (BHUTANESE), GIAO SONG (CHINESE)

Pinus subg. *Strobus* sect. *Quinquefoliae* subsect. *Strobus*

Tree to 50(–70?) m tall, with trunk to 1(–1.5) m in diameter. Bark dark grayish brown, becoming flaky and finally furrowed between scaly, interrupted, narrow ridges. Crown dense, broadly conical, deeper or shallower depending on the density of the stands, with numerous upswept branches becoming horizontal with age and well clothed with foliage. Twigs yellowish green when fresh, hairless and slightly waxy, aging reddish brown and losing the wax. Buds 6–8 mm long, slightly resinous. Needles in bundles of five, each needle (6–)12–18(–23) cm long, flexible and gently drooping, lasting 3–4 years, bright green to dark green on the outer face, bluish green with a little wax on the inner faces. Individual needles with lines of stomates usually restricted to the inner faces, an undivided midvein, two resin canals touching the epidermis of the outer face, and often a third resin canal inside the leaf tissue near where the two inner faces come together. Sheath 1.4–2 cm long, soon shed. Pollen cones 10–20 mm long, yellowish brown. Seed cones (10–)20–30 cm long, taperingly cylindrical, noticeably curved, with 60–80 seed scales or more, bluish green before maturity, ripening light brown, opening widely to release the seeds, on a slender stalk (2–)3–4(–6) cm long. Seed scales angularly egg-shaped, thin, the exposed tip a little thicker, straight, slightly cupped, with a narrow, diamond-shaped umbo at the tip. Seed body (3–)6–8(–10) mm long, the firmly attached wing (10–) 15–25(–30) mm long. Himalaya, from eastern Afghanistan to southeastern Xizang (Tibet) and northwestern Yunnan (China). Often forming pure forests in broad dry valleys or mixed with other conifers and hardwoods in moister montane and subalpine forests; (1,600–)2,000–3,000(–3,900) m. Zone 7. Synonyms: *P. griffithii* McClelland, *P. wallichiana* var. *parva* K. Sahni.

Along with the unrelated, three-needled chir pine (*Pinus roxburghii*), a hard pine, this is one of the two most abundant and economically important Himalayan pines. It is used both for timber and as a source of turpentine. It grows mostly above the main belt of chir pine and, unlike that species, is not fire tolerant, although it seeds abundantly on newly burned lands. It is very similar in general features to the other typical white pines in subsection *Strobus*, and it crosses readily with most of them to form vigorous and fertile hybrids, including *P.* ×*holfordiana* (with *P. ayacahuite*) and *P.* ×*schwerinii* (with *P. strobus*). It has quite a wide range and spans a great deal of climatic variation, especially in amount and seasonal distribution of precipitation, but little is known of morphological and genetic variation across this range. It is commonly cultivated in moderate climates as a handsome and fast-growing ornamental, and a few cultivars have been developed. Its greatest importance in North America may be as a parent of hybrids resistant to white pine blister rust that can replace susceptible native species in timber plantations. The species name honors Nathaniel Wallich (1786–1854), superintendent of the Calcutta Botanic Gardens, who first gave the species a scientific name, but this was identical to a previously published name of another species (so it was a later homonym), which precludes its use for the Himalayan white pine.

Pinus yunnanensis Franchet

YUNNAN PINE, YUN NAN SONG (CHINESE)

Pinus subg. *Pinus* sect. *Pinus* subsect. *Pinus*

Tree to 30 m tall, or shrub not exceeding 2 m where subjected to frequent fires. Trunk to 1 m in diameter. Bark bright reddish brown, smooth, and peeling above, finally becoming dark gray and broken up into elongate, scaly plates by deep furrows. Crown dense, broadly conical, eventually flattening with age, with numerous short branches densely clothed with foliage near the ends. Twigs coarse, pinkish to reddish brown, hairless, rough with the bases of scale leaves. Buds 2–3 cm long, not resinous. Needles in bundles of (two or) three, each needle (7–)10–25(–30) cm long, slender, flexible or a little stiff, drooping, lasting 2(–3) years, light green. Individual needles with a few lines of stomates on both the inner and outer faces, and three to five resin canals all around the two-stranded midvein, the two at the outer corners usually midway to the needle surface, the others at the surface. Sheath 10–15 mm long, persisting and falling with the bundle. Pollen cones 15–30 mm long, yellowish brown. Seed cones (4–)5–9(–11) cm long, egg-shaped, with 75–100 seed scales, green before maturity, ripening glossy pale yellowish brown at first and progressively darkening to rich brown, opening widely to release the seeds or remaining closed for another year and persisting for up to several years before falling with the short, stout stalk to 1 cm long. Seed scales irregularly spoon-shaped, the exposed face horizontally diamond-shaped to six-sided, scarcely protruding, crossed by an inconspicuous ridge topped by a small, flat, diamond-shaped umbo bearing a tiny prickle. Seed body 4–5 mm long, the easily detachable wing another 12–16 mm long. Mountainous region of southwestern China, in Yunnan (hence the scientific and common names) and adjacent parts of Xizang (Tibet), Sichuan, Guizhou, and Guangxi. Forming pure, dense to open stands or mixed with other conifers and hardwoods on mountain slopes and in adjacent river valleys; (400–)600–2,600(–3,400) m. Zone 8. Two varieties.

Yunnan pine is very closely related to Khasi pine (*Pinus kesiya*), and the two species are unusual among Asian hard pines in subsection *Pinus* (and indeed among all the species of this subsection wherever they may occur in Eurasia and North America) in typically having three needles rather than two as the most frequent number in a bundle. They both have droopy needles, although those of *P. kesiya* are typically finer than those of Yunnan pine. The chemical composition of their oils is very similar, and both are exploited for their turpentines as well as for timber. Khasi pine generally has a more delicate appearance than Yunnan pine, with finer twigs and smaller cones, on average, in addition to the finer needles. The two species have slightly overlapping ranges, with *P. kesiya* extending southward and westward from the area of *P. yunnanensis.* Trees of Yunnan pine at lower elevations in the southeastern portion of its range have finer foliage than in other regions, and this may be due to some influence from *P. kesiya* via past hybridization.

At the other elevational and geographic extreme, on high, dry mountain slopes at the northern end of its range, Yunnan pine takes the form of a shrub with short, stiff needles and smaller seed cones that remain unopened after ripening. The same kinds of features can be found, for instance, in populations of the North American pitch pine (*Pinus rigida*) in the New Jersey Pine Barrens, where they are subjected to frequent, intense fires. In addition to the role of selection by fire, this variety might also be influenced by hybridization with the high-elevation Chinese mountain pine (*P. densata*), which itself might have originated following hybridization between Yunnan pine and the more northerly Chinese red pine (*P. tabuliformis*). Further studies are needed to understand the genetic basis of the elevational variants of Yunnan pine as well as to clarify its relationship to Khasi pine. More studies of its crossability with other species of subsection *Pinus* are also needed. It is known to be crossable with Japanese black pine (*P. thunbergiana*) as well as *P. tabuliformis*, but attempts to cross it with the European black pine (*P. nigra*) and Scots pine (*P. sylvestris*) failed. Few data are available on crossing relationships with the other Asian hard pines.

Pinus yunnanensis* var. *pygmaea (Ji R. Xue) Ji R. Xue
Shrub to 2 m tall. Needles 7–13 cm long, stiff and straight. Seed cones 4–5 cm long, remaining closed after ripening. Northern portion of the range of the species, in Sichuan and northern Yunnan; 2,200–3,100(–3,400) m.

Pinus yunnanensis Franchet **var. *yunnanensis***
Tree to 30 m tall. Needles 10–25(–30) cm long, somewhat flexible and drooping. Seed cones 5–9(–11) cm long, opening widely to release the seeds. Throughout the range of the species; (400–)600–2,600(–3,100) m. Synonym: *Pinus yunnanensis* var. *tenuifolia* W. C. Cheng & Y. W. Law.

Platycladus Spach

ORIENTAL ARBORVITAE

Cupressaceae

Evergreen trees and shrubs with one to many (clustered) trunks clothed with fibrous, shallowly furrowed bark peeling in papery vertical strips. At least when young, densely branched from the base with upwardly arcing branches that support a conical to egg-shaped crown. Branchlets alternate or a few opposite, in fernlike, flattened but three-dimensionally twisted sprays held vertically rather than horizontally (in contrast to *Thuja* or *Chamaecyparis*). Without definite winter buds, the shoot tips protected simply by the youngest ordinary foliage leaves. Seedling leaves in alternating quartets, needlelike, standing out from and well spaced on the stem. Seedling phase short-lived, with adult branchlets appearing by the second year. Adult leaves in alternating pairs, scalelike, dense, their bases running down onto the branchlets, the tips triangular and directed forward, those of the lateral pairs folded around the branchlet, the alternate pairs otherwise not much differentiated, the lateral leaves touching along their bases on main shoots but separated by the bases of the facial leaves on side branchlets. Each leaf with a glandular depression in the midline.

Plants monoecious. Pollen cones numerous, single at the ends of short branchlets, oblong, with four to six alternating pairs of pollen scales, each with three to six pollen sacs. Pollen grains small (25–40 μm in diameter), nearly spherical, minutely bumpy and sometimes with an ill-defined germination pore, otherwise nearly featureless. Seed cones moderately widely spaced, single at the tips of short branchlets, maturing in a single season, oblong, with three or four alternating pairs of overlapping, basally attached, oblong, woody (somewhat fleshy until maturity) seed scales. Each cone scale consisting of an intimately fused seed scale and bract, with a hooked triangular prickle below the tip (the free end of the bract). Outer two pairs of scales joined at the base, the other one or two pairs attached within, scales of the two middle pairs each with one or two seeds. Seeds oval, pointed, smooth, usually wingless or with the hint of a pair of narrow wings running the length of the body. Cotyledons two, each with one vein. Chromosome base number $x = 11$.

Wood fragrant, of medium weight and strength, with whitish sapwood sharply contrasting with the orange-brown to dark brown heartwood. Grain very fine and even, with well-defined growth rings marked by a gradual and then rapid transition to a very narrow band of smaller and slightly thicker walled latewood tracheids. Resin canals absent but with a moderate number of individual resin parenchyma cells scattered through the growth increment, especially near the latewood.

Stomates in two stomatal zones on either side that lack the waxy white patches found in some related genera. Each stomate sunken beneath and partially obscured by the four to six surrounding subsidiary cells, which are topped by a low Florin ring. Leaf cross sec-

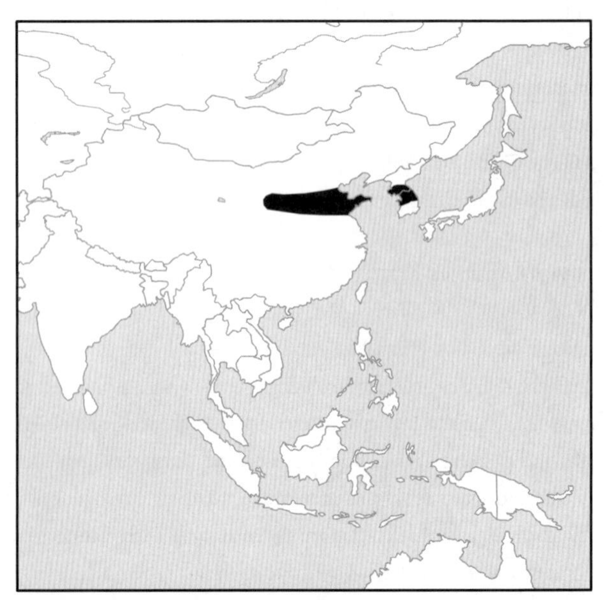

Distribution of *Platycladus orientalis* in northeastern China and Korea, scale 1 : 120 million

Upper side of flattened branch spray of Oriental arborvitae (*Platycladus orientalis*), ×0.5.

Branch spray of Oriental arborvitae (*Platycladus orientalis*) with unopened mature seed cones; scale, 5 cm.

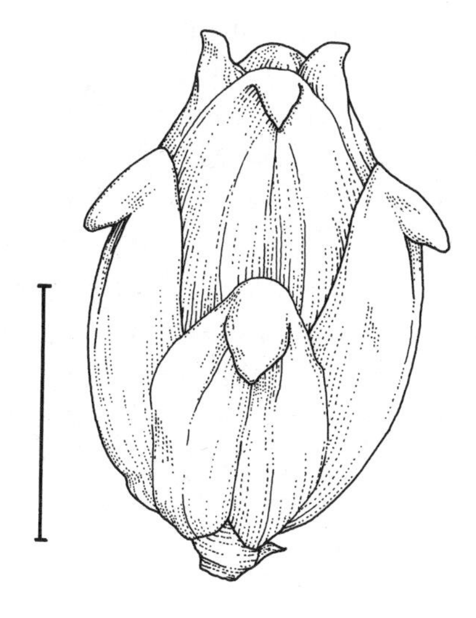

Unopened mature seed cone of Oriental arborvitae (*Platycladus orientalis*); scale, 1 cm.

tion with a single-stranded midvein above a large, saclike resin canal (visible at the leaf surface as the gland) and flanked by small wedges of transfusion tissue. Photosynthetic tissue with a variable palisade layer lining the exposed outer (lower) face beneath the epidermis and adjacent, thin layer of hypodermis that does not extend to the inner (upper) face on which the stomates are found.

One species in eastern Asia. References: Farjon 2005b, Fu et al. 1999g, Martin 1950, Xie et al. 1992. Synonym: *Biota* D. Don.

Oriental arborvitae has been cultivated for centuries, with many cultivars selected, though nowhere near as many as in *Thuja occidentalis* (northern white cedar). While *Platycladus orientalis* is generally a tree in nature, a large majority of the cultivars are dwarfs or shrubby. Cultivar selection has emphasized shape (mostly balls, cones, and narrow columns), juvenile and threadlike foliage, and foliage color variants (including blue, emerald and lime greens, yellows, golds and oranges, and bronze to purple winter foliage).

Often treated as a section of *Thuja,* this genus is more closely related to *Microbiota* and, perhaps, to *Tetraclinis.* DNA sequences suggest a rather distant relationship to *Thuja,* and there is no longer any

Open seed cone of Oriental arborvitae (*Platycladus orientalis*) after seed dispersal; scale, 1 cm.

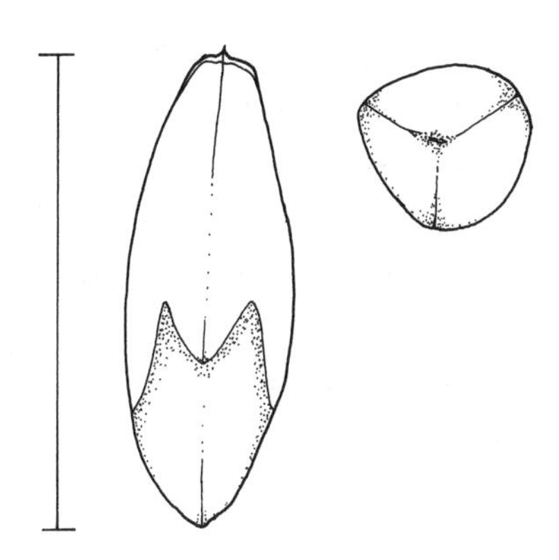

Dispersed, wingless seed of Oriental arborvitae (*Platycladus orientalis*) from side and tip view; scale, 5 mm.

justification for treating the sole extant species, *Platycladus orientalis*, as *Thuja orientalis*, as is still done in many references, particularly in the horticultural literature. The vertical arrangement of the foliage sprays, like pages in an open book stood on end, are diagnostic, as are the somewhat fleshy seed scales and wingless, plump seeds. Unlike many of its relatives, *Platycladus* has no known fossil record.

Platycladus orientalis (Linnaeus) Franco

Oriental arborvitae, ce bai (Chinese), ch'ukpaek namu (Korean)

Tree to 25 m tall, with trunk to 1 m in diameter or more, though often multistemmed or shrubby in cultivation. Crown dense and narrowly conical in youth, becoming open and irregular with age. Scale leaves generally 1–3 mm long (to 6 mm on main shoots), green to yellowish green, the stomates not evident without magnification. Pollen cones 2–3 mm long, yellowish. Seed cones 1–2 cm long, reddish brown at maturity, with a bluish white waxy coating on the outside of the scales, usually the two lower fertile scales with two seeds each and those of the upper pair each with a single seed. Seeds 5–7 mm long. China (Sichuan, Yunnan, and Hunan to Nei Mongol and Liaoning) and Korea. Many habitats, 50–3,300 m. Zone 6. Synonyms: *Biota orientalis* (Linnaeus) Endlicher, *Thuja orientalis* Linnaeus.

Oriental arborvitae has a wide but scattered distribution across China. Its occurrence in small, widely separated populations led to considerable differentiation across its range. This, in combination with its long history of centuries of cultivation in China, Korea, and Japan, and more recently in the West, led to production of numerous cultivars. In addition to the usual dwarf and color variants, many of these cultivars bear persistently juvenile foliage of needlelike leaves or elongated, threadlike branchlets reminiscent of the cultivars of *Chamaecyparis*. It may be the most adaptable of all conifers in ornamental cultivation, thriving in sites as disparate as the hot Pacific lowlands of Nicaragua and the seasonally cold landscape of Montreal, Canada.

Podocarpus L'Héritier ex Persoon

YELLOWWOOD, PODOCARP, TOTARA, PLUM PINE

Podocarpaceae

Evergreen trees and shrubs of varying habit. Trunk often forked and crooked or twisted but straight and single in some species. Bark fibrous and peeling in narrow strips. Crown typically broad and rounded, often with extra branches between the main tiers of whorled branches. Branchlets all elongate, without distinction into short and long shoots, usually hairless, remaining green at least the first year, obscurely to prominently grooved between the elongate, attached leaf bases. Resting buds well developed, the scales often green and leafy in texture. Leaves spirally arranged and radiating all around the twigs, mostly variably ribbon- and sword-shaped rather than needlelike, with a prominent midvein that is raised beneath and raised or channeled above. The leaf tip rounded or pointed, the base tapering abruptly to a short petiole.

Plants dioecious. Pollen cones cylindrical, single or in clusters of 2–8(–12) at the ends of (usually leafless) stalks or sessile in the axils of foliage leaves. Each cone with a few bud scales or bracts at the base and numerous, densely spirally arranged pollen scales, each scale bearing two pollen sacs. Pollen grains small to medium (body 20–50 μm long, 45–85 μm overall), with two round air bladders that may be smaller to larger than the slightly flattened pollen body and diverge at about 120° away from the germination furrow. Air bladders fairly smooth but with coarse internal sculpturing, the cap of the body thicker and more coarsely sculptured than the furrow side. Seed cones maturing and falling in a single season, single on leafless stalks in the axils of foliage leaves, highly modified and reduced. With or without two free bracteoles at the base, with two to five bracts united with the cone axis and becoming swollen, juicy, and berrylike (the podocarpium), and with a single plump, unwinged seed embedded in each of one to three fleshy, green to black seed scales (the epimatium). Seeds with or without a prominent crest, the opening of the ovule pointing down into the cone axis. Cotyledons two, each with two veins. Chromosome base numbers $x = 10, 11, 17, 18$, and 19.

Wood soft, light, usually not fragrant, yellowish brown to light brown, often with little distinction between sapwood and heartwood, but sometimes with a central core of reddish brown inner heartwood. Grain fine and even, often lacking obvious growth rings but the latewood sometimes faintly darker and denser. Resin canals absent but with individual resin cells.

Without stomates above and with several lines of stomates forming a broad stomatal band on either side beneath. Each stomate sunken beneath and largely hidden by the four (to six) surrounding subsidiary cells that may or may not have a discontinuous low Florin ring. Leaf cross section with one or three resin canals beneath and near the single-stranded midvein and sometimes with additional resin canals in other positions, with extensive transfusion tissue and accessory transfusion tissue. Photosynthetic tissue with a fairly thin to prominent palisade layer beneath the upper epidermis and its adjacent variable layer(s) of hypodermis and often strongly separated by the accessory transfusion tissue from the spongy mesophyll occupying the lower half above the stomatal bands.

Eighty-two species across the southern hemisphere and extending northward in the New World to the eastern coast of Mexico and throughout the West Indies, in Africa to just north of the equator in Cameroon and southernmost Sudan, and in Asia to northeastern India across to central Japan. References: Buchholz and Gray 1948a, c, de Laubenfels 1969, 1985, 1988, Fu et al. 1999f, Gray 1953b, 1955, 1956, 1958, R. Hill 1995, Markham et al. 1985b. Synonym: *Margbensonia* A. V. Bobrov & Melikian.

The name *Podocarpus,* Greek for "foot fruit," refers to the fleshy, swollen cone axis and bracts (together called a podocarpium) found in most species. Despite its large number of species, *Podocarpus* has contributed little to horticulture, especially in comparison to *Pinus, Juniperus, Picea,* and *Abies,* the large northern hemisphere genera. Only a few species are in general cultivation, and these have spawned few cultivars. Those that have been selected follow the usual patterns: variations in habit (dwarf and weeping), leaf length, and foliage color (blue and gold). None of the species is really hardy in cold temperate regions, but some of the hardier cultivars appear to be derived from hybrids between the New Zealand species growing on South Island. Podocarps are preeminently tropical, and the wide range of natural environments they inhabit, from the seashore to subalpine scrub, provides a wide-open field for bringing more than the existing handful of these handsome trees and shrubs into cultivation for many different garden conditions in the tropics.

Podocarpus and *Pinus,* the two largest genera of conifers, with 82 and 97 species respectively, present more contrasts than points of similarity. In addition to having the greatest number of species among the conifers of its hemisphere (*Podocarpus* in the south and *Pinus* in the north), each also has the widest range and overall greatest abundance. The resemblances essentially stop there, however. While species of *Pinus* are predominantly temperate, characteristically forming pure, open forests on dry soils, and often dependent upon fire for reproduction, podocarps are mostly found in tropical to warm temperate, moist or wet forests, where they constitute scattered individuals or small groves in the understory or canopy. Associated with these ecological differences, pines have desiccation-resistant, needlelike leaves while those of podocarps have broader photosynthetic areas at the expense of greater potential water loss through transpiration. The cones of pines are woody and somewhat fire resistant, with winged seeds that can be blown by the wind onto bare mineral earth cleared by fires. In contrast, the cones of podocarps are berrylike and attractive to bird and mammal dispersers, while the seeds have the food reserves needed for establishment within closed communities. Despite these striking differences in ecology and geography, the two genera evolved a similar number of species. Interestingly, the third largest genus of conifers, *Juniperus* (the junipers of the northern hemisphere), with about half as many species, are bird dispersed like the podocarps but have habitats more similar to those of the pines.

Species of *Podocarpus* can be thought of more as tropical and south temperate than as southern hemispherical. As well, they extend much farther into the northern hemisphere, at about 25°N in the New World, 5°N in Africa, and 35°N in Asia, than *Pinus* does into the southern hemisphere, with its minimal intrusion just southward over the equator into Sumatra. Nor was *Podocarpus* confined to the southern hemisphere in the geologic past, with credible records in eastern Asia and North America, although the vast majority of fossils that are definitely attributable to the genus are found in the far south. This reflects a distribution on the former continent of Gondwanaland, which was breaking up into the

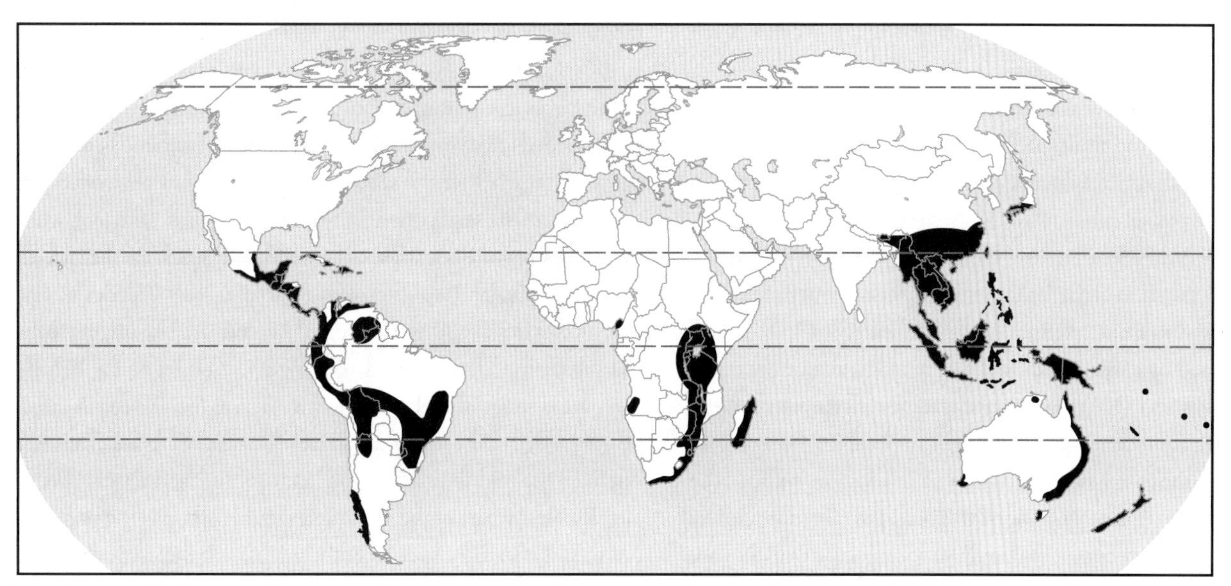

Distribution of *Podocarpus* across the southern continents and tropics, extending north to Japan and with some restricted species on oceanic islands, scale 1 : 200 million

modern southern continents during the Cretaceous, when the first known fossils of *Podocarpus* are recorded.

Although most of the living species of *Podocarpus* are rain-forest trees with large, sword-shaped leaves, some of the temperate species, like those found in New Zealand and Chile, have much smaller leaves that are very similar to those of yews (*Taxus*) or redwoods (*Sequoia*). One of the New Zealand species, the alpine totara (*P. nivalis*), is even a dwarf shrub growing above the alpine tree line. The tree species yield a fine timber, often sold as yellowwood because of its yellowish brown color (and not to be confused with many other trees known as yellowwoods, such as the North American *Cladrastris kentuckea*, a deciduous hardwood in the legume family). Before widespread introduction of Monterey pine (*Pinus radiata*) and other pine species as plantation timber trees in the southern hemisphere, yellowwoods were often the premier softwoods for construction, as they were in southern Africa. Pollination in the genus is more efficient than one might expect from seed cones with just a single ovule (or two) because pollen is scavenged from the flanks of the cone by a pollination drop that emerges from the micropyle (mouth) of the ovule.

About 30 species that were unquestioningly included within *Podocarpus* before 1969, some of which are still often so listed in horticulture, are here assigned to other genera. Buchholz and Gray, in their comprehensive revision of *Podocarpus*, beginning in 1948, followed modern tradition in treating these species as members of *Podocarpus*, although some had been separated in the 19th century. Within the genus in its broad sense, they recognized eight botanical sections. Only one of these, section *Podocarpus* (which they called *Eupodocarpus* under a different set of nomenclatural rules), is still included in *Podocarpus*, although it was by far the largest portion of the old genus in terms of number of species. The other seven sections are each treated as separate genera by some authors, or are grouped into as few as four genera by others. Although some contemporary taxonomists still adopt a broad generic concept, notably in South Africa, where *Afrocarpus* is generally not separated from *Podocarpus*, all subsequent evidence reinforces the heterogeneity of the old *Podocarpus* and the distinctiveness of its original elements. Separation of the 30 anomalous species from *Podocarpus*, first proposed in modern times by de Laubenfels (1969), is amply confirmed by careful studies of morphology, anatomy, pollen, biochemistry, and DNA. Even quite traditional characters of the shoots, leaves, and cones amply distinguish species of the segregate genera from those retained in *Podocarpus*. In addition, some of the segregate genera (formerly sections of *Podocarpus*) are more closely related to other members of the family Podocarpaceae than they are to the other species retained here in *Podocarpus* in the strict sense. *Parasitaxus ustus*, for instance, is so unlike any species of *Podocarpus* that it is hard to see how it was ever associated with this genus. The cross-references in the Index make it easy to find the correct genus for each species removed from *Podocarpus*. There is also further discussion of relationships under each of the segregate genera.

Classification within what is left of *Podocarpus* is also somewhat controversial. Buchholz and Gray initially distributed the species of *Podocarpus* in the strict sense (their section *Eupodocarpus*) among four subsections, and Gray later added two for four anomalous species. In a complete review of the genus, de Laubenfels (1985) drastically reorganized the species. He established two subgenera that seem very well founded. Subgenus *Foliolatus* is more homogeneous than subgenus *Podocarpus*. It contains only the species of Buchholz and Gray's subsection B (and Gray's small subsection F). Although this subgenus contains the majority of the species in the genus, all from Asia and the western Pacific, they are basically rather similar to one another. All have a pair of bracteoles at the base of the seed cone and a series of common leaf features, including a raised midrib (usually), resin canals confined to the vicinity of the midvein, and stomates without a Florin ring. Within this subgenus, de Laubenfels described nine botanical sections, but these are so similar that it is not yet clear whether they represent the most effective arrangement of the subgenus.

The species in subgenus *Podocarpus* are much less uniform. This subgenus includes subsections A, C, and D of Buchholz and Gray (and Gray's later-segregated small subsection E). All these species lack the bracteoles of subgenus *Foliolatus* and all have stomates with Florin rings, but some species have resin canals near the edge of the leaf and others do not, and many have a groove along the midrib above, while a few have a ridge. There are fewer species in this subgenus, and de Laubenfels also assigned them to nine botanical sections. The African species (formerly subsections A and E) differ from all others in having resin canals near the outer corners of the leaves in addition to those next to the midrib. These species were divided into two botanical sections by de Laubenfels, differing primarily in whether the seed and epimatium or the swollen axis and bracts of the seed cone are the main fleshy bird attractant, while the other structure is leathery. The temperate South Pacific species, from New Zealand, southeastern Australia, New Caledonia, and one species from southern Chile and Argentina, all belonging to the old subsection D, are treated as a single section by de Laubenfels. They are distinguished by having small leaves that lack resin canals near the edge of the leaf and also lack the support tissue extending out from the midrib that is found in most other species of *Podocarpus*. The fleshy part of the seed cones in this group is the podocarpium, as it is in most species of the genus, and the epimatium and seed coat are not fleshy. *Podocarpus smithii* from Queensland (Australia), newly described by de Laubenfels in 1985, was assigned to a section all by itself since it has larger leaves than the species in subsection D and a fleshy seed coat and epimatium. Except for *P. nubigenus* of subsection D, all New World species of *Podocarpus* were assigned by Buchholz and Gray to subsection C, exclusive to the Americas. These species were distributed among five sections by de Laubenfels, distinguished by characteristics of the buds and pollen cones. As with the sections proposed for subgenus *Foliolatus,* it is not clear yet whether these sections represent the most effective arrangement of the New World species. Thus, while the subgenera proposed by de Laubenfels are well supported, it is likely that new information will lead to changes in the assignments to sections used here, which are those proposed by de Laubenfels.

There is wide variation in the number of species recognized in the genus by different authors. Some of the difficulty in species delimitation may be due to hybridization. Natural hybridization has been well documented among the four New Zealand species, but there has not been a single published study looking for hybridization anywhere else within the range of the genus. This is an area that clearly needs more work and that could also contribute to improving the arrangement of species within the subgenera by helping clarify relationships. Another source of confusion in species boundaries is that all species go through substantial changes in leaf size and shape throughout their life as well as varying in different environments. This has often resulted in dividing parts of the morphological continuum for a single species into distinct species. Other such cases may well be lurking among the numerous species that are known from only one or a few specimens. The synonyms indexed are those that may be found in current literature. For additional synonyms used in older literature, please consult Farjon (1998), although several names accepted in that work are treated as synonyms here.

Potentially, any species in the genus might sometimes bear two seeds in a cone even when it normally has just one. The descriptions here only indicate whether such cones have been found or not. Thus, "with one fertile seed scale" means that a second seed has not been recorded, "with one (or two) fertile seed scales" means that it has, and "with one or two fertile seed scales" means that they are fairly common.

Geographic Guide to *Podocarpus*

Central America and the Andean region of South America

Podocarpus celatus
P. costaricensis
P. glomeratus
P. guatemalensis
P. matudae
P. nubigenus
P. oleifolius
P. parlatorei
P. rusbyi
P. salignus
P. sprucei

Venezuela, Guianas, Brazil, and northeastern Argentina

Podocarpus acuminatus
P. brasiliensis
P. buchholzii
P. lambertii
P. magnifolius
P. pendulifolius
P. roraimae
P. salicifolius
P. sellowii
P. steyermarkii
P. tepuiensis

West Indies

Podocarpus angustifolius
P. coriaceus
P. purdieanus
P. trinitensis

Africa

Podocarpus elongatus
P. henkelii
P. latifolius

Madagascar

Podocarpus capuronii
P. humbertii
P. madagascariensis
P. rostratus

Australia

Podocarpus dispermus
P. drouynianus
P. elatus
P. grayae
P. lawrencei
P. smithii
P. spinulosus

New Zealand

Podocarpus acutifolius
P. cunninghamii
P. nivalis
P. totara

New Caledonia

Podocarpus decumbens
P. gnidioides

P. longifoliolatus
P. lucienii
P. novae-caledoniae
P. polyspermus
P. sylvestris

Mainland Asia, Malesia, Melanesia, and Polynesia

Podocarpus affinis
P. annamiensis
P. archboldii
P. atjehensis
P. borneensis
P. bracteatus
P. brassii
P. confertus
P. costalis
P. crassigemmis
P. deflexus
P. gibbsiae
P. globulus
P. insularis
P. laubenfelsii
P. ledermannii
P. levis
P. macrocarpus
P. macrophyllus
P. micropedunculatus
P. nakaii
P. neriifolius
P. pallidus
P. pilgeri
P. polystachyus
P. pseudobracteatus
P. ridleyi
P. rubens
P. rumphii
P. salomoniensis
P. teysmannii

Taxonomic Guide to *Podocarpus*

This classification was devised by de Laubenfels in 1985. It is the first classification of the genus (with its present narrow scope) to use formal taxonomic categories. It has not yet been tested in practice by other authors or by using molecular data. Such testing will probably reduce the number of sections recognized or further group them. Nonetheless, for now this classification helps give some taxonomic structure to the otherwise overwhelming mass of *Podocarpus* species and hence is adopted here.

Podocarpus subg. *Podocarpus*
sect. *Podocarpus*
Podocarpus elongatus
P. latifolius
sect. *Scytopodium* de Laubenfels
Podocarpus capuronii
P. henkelii
P. humbertii
P. madagascariensis
P. rostratus
sect. *Australis* de Laubenfels
Podocarpus acutifolius
P. cunninghamii
P. gnidioides
P. lawrencei
P. nivalis
P. nubigenus
P. totara
sect. *Crassiformis* de Laubenfels
Podocarpus smithii
sect. *Capitulatis* de Laubenfels
Podocarpus glomeratus
P. lambertii
P. parlatorei
P. salignus
P. sellowii
P. sprucei
sect. *Pratensis* de Laubenfels
Podocarpus oleifolius
P. pendulifolius
P. tepuiensis
sect. *Lanceolatis* de Laubenfels
Podocarpus coriaceus
P. costaricensis
P. matudae
P. rusbyi
P. salicifolius
P. steyermarkii
sect. *Pumilis* de Laubenfels
Podocarpus angustifolius
P. buchholzii
P. roraimae
sect. *Nemoralis* de Laubenfels
Podocarpus acuminatus?
P. brasiliensis
P. celatus
P. guatemalensis

P. magnifolius
P. purdieanus
P. trinitensis

Podocarpus subg. *Foliolatus* de Laubenfels (synonym: *Margbensonia* A. V. Bobrov & Melikian)

sect. *Foliolatus* de Laubenfels
Podocarpus archboldii
P. borneensis
P. deflexus
P. insularis
P. levis
P. neriifolius
P. novae-caledoniae
P. pallidus
P. rubens

sect. *Acuminatis* de Laubenfels
Podocarpus dispermus
P. ledermannii
P. micropedunculatus

sect. *Globulus* de Laubenfels
Podocarpus annamiensis
P. globulus
P. lucienii
P. nakaii
P. sylvestris
P. teysmannii

sect. *Longifoliolatus* de Laubenfels
Podocarpus atjehensis
P. bracteatus
P. confertus
P. decumbens
P. gibbsiae
P. longifoliolatus
P. polyspermus
P. pseudobracteatus
P. salomoniensis

sect. *Gracilis* de Laubenfels
Podocarpus affinis
P. pilgeri

sect. *Macrostachyus* de Laubenfels
Podocarpus brassii
P. costalis
P. crassigemmis

sect. *Rumphius* de Laubenfels
Podocarpus grayae
P. laubenfelsii
P. rumphii

sect. *Polystachyus* de Laubenfels
Podocarpus elatus
P. macrocarpus
P. macrophyllus
P. polystachyus
P. ridleyi

sect. *Spinulosus* de Laubenfels
Podocarpus drouynianus
P. spinulosus

Identification Guide to All Species of *Podocarpus*

Following this guide to all the species of the genus is a guide to the most commonly cultivated ones. Correct identification of specimens of all 82 species of *Podocarpus* on a worldwide basis is a daunting task. To make the task more manageable the guide is restricted to mature individuals because of the greater range in character traits among juveniles. Juvenile trees will generally be difficult, if not impossible, to identify! An attempt is made to minimize the use of reproductive features in this identification guide and to restrict the use of internal leaf structures, which are prominent in some technical botanical keys. In practice, the resin canals are fairly conspicuous in a clean cross section of a leaf cut with a razor blade when examined with the 10× enlargement found in magnifying glasses. When reproductive features are used here, they generally include both pollen and seed cones, essentially as alternatives because of the separate sexes in these trees. It may well be necessary to refer to the geographic listing preceding this identification guide to help reduce the number of possible species under consideration, and to the detailed descriptions of the species in the text for final confirmation of the identity of an unknown specimen. For convenience of identification, the 82 species of *Podocarpus* may be divided into nine informal groups on the basis of the length and the width of the majority of leaves on a mature tree or shrub (there may be some overlap in species among groups):

Group A, leaves less than 4 cm by less than 3 mm

Group B, leaves less than 4 cm by 3–5 mm

Group C, leaves less than 4 cm by more than 5 mm

Group D, leaves 4–8 cm by less than 6 mm

Group E, leaves 4–8 cm by 6–9 mm

Group F, leaves 4–8 cm by more than 9 mm

Group G, leaves more than 8 cm by less than 9 mm

Group H, leaves more than 8 cm by 9–13 mm

Group I, leaves more than 8 cm by more than 13 mm

Group A. These five species may be distinguished by the width of their broadest leaves (as opposed to the majority of leaves, used to get to this group), the form of the midrib on the upper side of the leaves, and the number of resin canals beneath the midvein (there may be more in other positions):

Podocarpus gnidioides, leaves less than 3 mm wide, midrib grooved, resin canal one

P. rostratus, leaves less than 3 mm wide, midrib flat, resin canals three

P. acutifolius, leaves less than 3 mm wide, midrib raised or flat, resin canal one

P. nivalis, leaves more than 3 mm wide, midrib strongly grooved, resin canal one

P. lawrencei, leaves more than 3 mm wide, midrib flat or a little grooved, resin canal one

Group B. Among these species, only *Podocarpus brassii* and *P. gibbsiae* have needlelike bracteoles at the base of the berrylike podocarpium, and besides the characters included here, *P. brassii* has a narrowly and prominently raised midrib while that of *P. gibbsiae* is broad and shallow. The 13 species in this group may be distinguished by the length of most leaves, the width of wider ordinary leaves, the form of the midrib on the upper side of the leaf, the length of most pollen cones, and the length of the mature seed with its enclosing epimatium:

Podocarpus nivalis, leaves up to 2.5 cm by up to 4 mm, midrib grooved, pollen cones up to 1 cm, seed up to 7 mm

P. totara, leaves up to 2.5 cm by up to 4 mm, midrib flat, pollen cones 1–2 cm, seed up to 7 mm

P. tepuiensis, leaves up to 2.5 cm by more than 4 mm, midrib grooved, pollen cones 1–2 cm, seed up to 7 mm

P. humbertii, leaves up to 2.5 cm by more than 4 mm, midrib flat to slightly raised, pollen cones 1–2 cm, seed more than 10 mm

P. gibbsiae, leaves up to 2.5 cm by more than 4 mm, midrib slightly raised, pollen cones 1–2.5 cm, seed 7–10 mm

P. brassii, leaves up to 2.5 cm by more than 4 mm, midrib prominently raised, pollen cones more than 2.5 cm, seed 7–10 mm

P. glomeratus, leaves at least 2.5 cm by up to 4 mm, midrib grooved, pollen cones up to 1 cm, seed up to 7 mm

P. cunninghamii, leaves at least 2.5 cm by up to 4 mm, midrib grooved or flat, pollen cones 1–2.5 cm, seed up to 7 mm

P. lambertii, leaves at least 2.5 cm by up to 4 mm, midrib grooved or flat, pollen cones 1–2 cm, seed 7–10 mm

P. nubigenus, leaves at least 2.5 cm by up to 4 mm, midrib prominently raised, pollen cones 1–2.5 cm, seed 7–10 mm

P. sprucei, leaves at least 2.5 cm by more than 4 mm, midrib grooved or slightly raised, pollen cones up to 1 cm, seed 7–10 mm

P. angustifolius, leaves at least 2.5 cm by more than 4 mm, midrib flat to slightly raised, pollen cones 1–2 cm, seed 7–10 mm

P. capuronii, leaves at least 2.5 cm by more than 4 mm, midrib flat to slightly raised, pollen cones more than 2 cm, seed more than 10 mm

Group C. These nine species may be distinguished by the length of the longest leaves, the form of the midrib on the upper side of the leaf, the position of the widest point on the leaf, and the number of resin canals beneath the midvein:

Podocarpus acuminatus, leaves up to 4 cm, midrib grooved, leaf widest before the middle, resin canal one

P. buchholzii, leaves up to 4 cm, midrib grooved, leaf widest near the middle, resin canal one

P. roraimae, leaves up to 4 cm, midrib raised, leaf widest near the middle, resin canal one

P. brassii, leaves up to 4 cm, midrib narrowly raised, leaf widest before the middle, resin canals three

P. gibbsiae, leaves up to 4 cm, midrib broadly raised, leaf widest near or beyond the middle, resin canals three

P. rusbyi, leaves more than 4 cm, midrib grooved, leaf widest near the middle, resin canal one

P. angustifolius, leaves more than 4 cm, midrib flat, leaf widest near the middle, resin canal one

P. affinis, leaves more than 4 cm, midrib narrowly raised, leaf widest before the middle, resin canals three

P. pilgeri, leaves more than 4 cm, midrib narrowly raised, leaf widest near or beyond the middle, resin canals three

Group D. All the species in this group except *Podocarpus novae-caledoniae,* which has three resin canals, have a single resin canal beneath the midvein. The seven species in this group may be distinguished by the width of the widest leaves, the form of the midrib on the upper side of the leaf, the length of most pollen cones, and the length of most mature seeds with the enclosing epimatium:

Podocarpus parlatorei, leaves less than 5 mm wide, midrib weakly grooved or flat, pollen cones up to 12 mm, seeds up to 7 mm

P. lambertii, leaves less than 5 mm wide, midrib deeply grooved or flat, pollen cones up to 12 mm, seeds 7–10 mm

P. novae-caledoniae, leaves less than 5 mm wide, midrib weakly grooved, pollen cones 12–20 mm, seeds 7–10 mm

P. spinulosus, leaves less than 5 mm wide, midrib narrowly raised, pollen cones up to 12 mm, seeds more than 10 mm

P. elongatus, leaves at least 5 mm wide, midrib flat or narrowly raised, pollen cones 12–20 mm, seeds 7–10 mm

P. salignus, leaves at least 5 mm wide, midrib narrowly raised, pollen cones more than 20 mm, seeds up to 7 mm

P. drouynianus, leaves at least 5 mm wide, midrib broadly raised, pollen cones up to 12 mm, seeds more than 10 mm

Group E. These 22 species may be divided into two subgroups based on the length of the longest ordinary leaves:

Group E1, leaves up to 8 cm

Group E2, leaves more than 8 cm

Group E1. These 12 species may be distinguished by the width of their widest ordinary leaves, the form of midrib on the upper side, the number of resin canals beneath the midvein, the length of most pollen cones, the number of pollen cones in a cluster from a leaf axil, and the length of the mature seed with its enclosing epimatium:

Podocarpus steyermarkii, leaves up to 9 mm wide, midrib grooved, resin canal one, pollen cones about 2.5 cm, one, seed more than 8 mm

P. affinis, leaves up to 9 mm wide, midrib narrowly raised, resin canals three, pollen cones more than 2.5 cm, one, seed up to 8 mm

P. decumbens, leaves up to 9 mm wide, midrib broadly raised, resin canals three or more, pollen cones up to 2.5 cm, one, seed up to 8 mm

P. longifoliolatus, leaves up to 9 mm wide, midrib broadly raised, resin canals three or more, pollen cones up to 2.5 cm, usually one, seed more than 8 mm

P. trinitensis, leaves more than 9 mm wide, midrib grooved, resin canal one, pollen cones up to 2.5 cm, one, seed up to 8 mm

P. oleifolius, leaves more than 9 mm wide, midrib grooved, resin canal one, pollen cones more than 2.5 cm, one, seed up to 8 mm

P. latifolius, leaves more than 9 mm wide, midrib flat to slightly grooved or raised, resin canals three or more, pollen cones up to 2.5 cm, usually one, seed more than 8 mm

P. pallidus, leaves more than 9 mm wide, midrib flat to narrowly raised, resin canals three, pollen cones up to 2.5 cm, one, seed more than 8 mm

P. rubens, leaves more than 9 mm wide, midrib flat to narrowly raised, resin canals three, pollen cones more than 2.5 cm, up to three, seed more than 8 mm

P. polystachyus, leaves more than 9 mm wide, midrib narrowly raised, resin canals three, pollen cones more than 2.5 cm, more than three, seed more than 8 mm

P. guatemalensis, leaves more than 9 mm wide, midrib broadly raised, resin canal one, pollen cones more than 2.5 cm, one, seed up to 8 mm

P. costalis, leaves more than 9 mm wide, midrib broadly raised, resin canals three, pollen cones more than 2.5 cm, one, seed more than 8 mm

Group E2. These 10 species may be distinguished by the width of their widest ordinary leaves, the form of the midrib on the upper side, the number of resin canals beneath the midvein, the length of most pollen cones, the number of pollen cones in a cluster from a leaf axil, and the length of the mature seed with its enclosing epimatium:

Podocarpus pendulifolius, leaves up to 9 mm wide, midrib grooved, resin canal one, pollen cones up to 3 cm, one, seed less than 1 cm

P. salignus, leaves up to 9 mm wide, midrib narrowly raised, resin canal one, pollen cones up to 3 cm, more than three, seed less than 1 cm

P. insularis, leaves up to 9 mm wide, midrib narrowly raised, resin canals three, pollen cones more than 3 cm, up to three, seed less than 1 cm

P. sellowii, leaves more than 9 mm wide, midrib flat to grooved, resin canal one, pollen cones up to 3 cm, up to two, seed less than 1 cm

P. annamiensis, leaves more than 9 mm wide, midrib flat to broadly raised, resin canals three, pollen cones more than 3 cm, usually one, seed less than 1 cm

P. crassigemmis, leaves more than 9 mm wide, midrib narrowly raised, resin canals three, pollen cones more than 3 cm, usually one, seed at least 1 cm

P. sylvestris, leaves more than 9 mm wide, midrib broadly raised, resin canals three, pollen cones up to 3 cm, up to three, seed more than 1 cm

P. macrophyllus, leaves more than 9 mm wide, midrib broadly raised, resin canals three, pollen cones about 3 cm, more than three, seed about 1 cm

P. confertus, leaves more than 9 mm wide, midrib broadly raised, resin canals three, pollen cones more than 3 cm, usually one, seed at least 1 cm

P. elatus, leaves more than 9 mm wide, midrib broadly raised, resin canals three or more, pollen cones up to 3 cm, more than three, seed more than 1 cm

Group F. These 13 species may be distinguished by the length of their longest ordinary leaves, the width of the widest normal leaves, the form of the midrib on the upper side, the number of resin canals in the midvein, the length of most pollen cones, the number of pollen cones in a cluster from a leaf axil, the length of the mature seed with its enclosing epimatium, and (for *Podocarpus brasiliensis* and *P. oleifolius*) the form and position of the tips of the bud scales:

Podocarpus brasiliensis, leaves up to 8 cm by up to 12 mm, midrib grooved, resin canal one, pollen cones less than 3.5 cm, one, seed up to 8 mm, bud scale tips sharp and free

P. oleifolius, leaves up to 8 cm by up to 12 mm, midrib grooved, resin canal one, pollen cones less than 3.5 cm, one, seed up to 8 mm, bud scale tips blunt and tight

P. purdieanus, leaves up to 8 cm by more than 12 mm, midrib grooved, resin canal one, pollen cones less than 3.5 cm, one, seed up to 8 mm

P. celatus, leaves up to 8 cm by more than 12 mm, midrib grooved, resin canal one, pollen cones about 3.5 cm, one, seed 8–12 mm

P. borneensis, leaves up to 8 cm by more than 12 mm, midrib broadly raised, resin canals two or three, pollen cones at least 3.5 cm, up to three, seed up to 8 mm

P. globulus, leaves up to 8 cm by more than 12 mm, midrib broadly raised, resin canals three, pollen cones about 3.5 cm, usually one, seed 8–12 mm

P. sellowii, leaves more than 8 cm by up to 12 mm, midrib flat to grooved, resin canal one, pollen cones about 3.5 cm, one or two, seed 8–12 mm

P. macrocarpus, leaves more than 8 cm by up to 12 mm, midrib narrowly raised, resin canals three, pollen cones less than 3.5 cm, up to four, seed more than 12 mm

P. nakaii, leaves more than 8 cm by up to 12 mm, midrib broadly raised, resin canals three, pollen cones about 3.5 cm, up to three, seed 8–12 mm

P. elatus, leaves more than 8 cm by up to 12 mm, midrib broadly raised, resin canals three or more, pollen cones less than 3.5 cm, more than three, seed more than 12 mm

P. magnifolius, leaves more than 8 cm by more than 12 mm, midrib grooved, resin canal one, pollen cones at least 3.5 cm, one, seed 8–12 mm

P. matudae, leaves more than 8 cm by more than 12 mm, midrib narrowly raised, resin canal one, pollen cones more than 3.5 cm, usually one, seed more than 12 mm

P. smithii, leaves more than 8 cm by more than 12 mm, midrib broadly raised, resin canal one, pollen cones at least 3.5 cm, up to three, seed more than 12 mm

Group G. These 10 species may be distinguished by the length and width of the largest ordinary leaves, the form of the midrib on the upper side, the number of resin canals beneath the midvein, the length of most mature seeds with the enclosing epimatium, and the number of pollen cones in most clusters from a leaf axil:

Podocarpus drouynianus, leaves up to 12 cm by up to 6 mm, midrib narrowly raised, resin canal one, seeds 8–15 mm, pollen cones more than three

P. pendulifolius, leaves up to 12 cm by 6–9 mm, midrib grooved, resin canal one, seeds less than 8 mm, pollen cone one

P. salignus, leaves up to 12 cm by 6–9 mm, midrib narrowly raised, resin canal one, seeds less than 8 mm, pollen cones more than three

P. atjehensis, leaves up to 12 cm by 6–9 mm, midrib narrowly raised, resin canals three, seeds 8–15 mm, pollen cone one

P. polyspermus, leaves up to 12 cm by 6–9 mm, midrib broadly raised, resin canals three plus four above the midvein, seeds 8–15 mm, pollen cone one

P. madagascariensis, leaves up to 12 cm by at least 9 mm, midrib narrowly raised, resin canals three plus two at the corners, seeds more than 15 mm, pollen cones up to three

P. salomoniensis, leaves more than 12 cm by 6–9 mm, midrib broadly raised, resin canals three to five, seeds 8–15 mm, pollen cone one

P. salicifolius, leaves more than 12 cm by at least 9 mm, midrib flat, resin canal one, seeds 8–15 mm, pollen cone one

P. henkelii, leaves more than 12 cm by at least 9 mm, midrib narrowly raised, resin canals three plus two at the corners, seeds at least 15 mm, pollen cones more than three

P. elatus, leaves more than 12 cm by at least 9 mm, midrib broadly raised, resin canals three to five, seeds at least 15 mm, pollen cones more than three

Group H. For convenience in identification, these 17 species may be divided into two subgroups based on the width of the widest ordinary leaves:

Group H1, leaves up to 12 mm wide

Group H2, leaves more than 12 mm wide

Group H1. These eight species may be distinguished by the length of the longest ordinary leaves, the form of the midrib on the upper side, the length of most resting buds on leading shoots, the length of most pollen cones, and the length of most mature seeds with the enclosing epimatium:

Podocarpus coriaceus, leaves up to 12 cm, midrib grooved or flat, resting buds at least 5 mm, pollen cones at least 3 cm, seeds 5–10 mm

P. macrocarpus, leaves up to 12 cm, midrib narrowly raised, resting buds 2.5–5 mm, pollen cones less than 3 cm, seeds at least 15 mm

P. ridleyi, leaves up to 12 cm, midrib narrowly raised, resting buds at least 5 mm, pollen cones less that 3 cm, seeds 5–10 mm

P. pseudobracteatus, leaves up to 12 cm, midrib narrowly raised, resting buds at least 5 mm, pollen cones at least 3 cm, seeds 10–15 mm

P. elatus, leaves up to 12 cm, midrib broadly raised, resting buds less than 2.5 mm, pollen cones up to 3 cm, seeds at least 15 mm

P. confertus, leaves up to 12 cm, midrib broadly raised, resting buds at least 5 mm, pollen cones at least 3 cm, seeds 10–15 mm

P. salicifolius, leaves more than 12 cm, midrib flat, resting buds at least 5 mm, pollen cones more than 3 cm, seeds 5–15 mm

P. deflexus, leaves more than 12 cm, midrib broadly raised, resting buds 2.5–5 mm, pollen cones more than 3 cm, seeds 10–15 mm

Group H2. These nine species may be distinguished by the length of the largest ordinary leaves, the form of the midrib on the upper side, the length of most resting buds on leading shoots, the number of pollen cones in most clusters from a leaf axil, and the length of most mature seeds with the enclosing epimatium:

Podocarpus matudae, leaves up to 12 cm, midrib narrowly raised, resting buds 2.5–5 mm, pollen cone usually one, seeds 10–15 mm

P. lucienii, leaves up to 12 cm, midrib broadly raised, resting buds less than 2.5 mm, pollen cones three, seeds 15–20 mm

P. archboldii, leaves up to 12 cm, midrib broadly raised, resting buds 2.5–5 mm, pollen cone one, seeds 5–10 mm

P. smithii, leaves up to 12 cm, midrib broadly raised, resting buds 2.5–5 mm, pollen cones up to three, seeds less than 5 mm

P. levis, leaves more than 12 cm, midrib flat to slightly grooved or raised, resting buds at least 5 mm, pollen cones up to three, seeds 10–15 mm

P. neriifolius, leaves more than 12 cm, midrib narrowly raised, resting buds 2.5–5 mm, pollen cones up to three or rarely up to 6, seeds 10–15 mm

P. bracteatus, leaves more than 12 cm, midrib narrowly raised, resting buds at least 5 mm, pollen cones usually one or two, seeds 10–15 mm

P. grayae, leaves more than 12 cm, midrib broadly raised, resting buds 2.5–5 mm, pollen cones usually three or four, seeds 10–15 mm

P. micropedunculatus, leaves more than 12 cm, midrib broadly raised, resting buds at least 5 mm, pollen cones usually and no more than three, seeds 5–10 mm

Group I. These seven species may be distinguished by the length and width of the largest leaves, the form of the midrib on the upper side, the length of most resting buds on leading shoots, the number of pollen cones in most clusters from a leaf axil, the length of most pollen cones, and the length of most mature seeds with the enclosing epimatium:

Podocarpus costaricensis, leaves up to 13 cm by up to 18 mm, midrib grooved, resting buds at least 5 mm, pollen cone one, up to 3.5 cm, seeds unknown but probably less than 11 mm

P. matudae, leaves up to 13 cm by up to 18 mm, midrib raised, resting buds up to 5 mm, pollen cone usually one, more than 3.5 cm, seeds 11–18 mm

P. teysmannii, leaves up to 13 cm by more than 18 mm, midrib raised, resting buds less than 5 mm, pollen cone usually one, up to 3.5 cm, seeds less than 11 mm

P. laubenfelsii, leaves more than 13 cm by up to 18 mm, midrib raised, resting buds less than 5 mm, pollen cones at least three, up to 3.5 cm, seeds less than 11 mm

P. rumphii, leaves more than 13 cm by more than 18 mm, midrib slightly raised, resting buds less than 5 mm, pollen cones at least three, at least 3.5 cm, seeds 11–18 mm

P. dispermus, leaves more than 13 cm by more than 18 mm, midrib raised, resting buds about 5 mm, pollen cones up to three, less than 3.5 cm, seeds more than 18 mm

P. ledermannii, leaves more than 13 cm by more than 18 mm, midrib raised, resting buds at least 5 mm, pollen cones usually three, more than 3.5 cm, seeds 11–18 mm

Identification Guide to Cultivated Species of *Podocarpus*

Only a few species of *Podocarpus* are likely to be encountered in cultivation outside botanical gardens in the temperate northern hemisphere. Shortest leaved at up to 4 cm are *Podocarpus lawrencei*, *P. nivalis*, *P. acutifolius*, *P. totara*, *P. cunninghamii*, and *P. nubigenus*, then at 4-6 cm are *P. spinulosus*, and *P. elongatus*, followed at 6-10 cm by *P. salignus*, *P. drouynianus*, *P. macrophyllus*, and *P. elatus*, with *P. henkelii* and *P. neriifolius* being the longest at more than 10 cm. Note that plants in the commercial trade sold as *P. elongatus* are likely *Afrocarpus falcatus* (distinguished by extensive lines of stomates both on the upper and undersides of the leaf) and those sold as *P. spinulosus* are likely *P. elatus*.

Podocarpus acuminatus de Laubenfels

NEBLINA PODOCARP

Podocarpus subg. *Podocarpus* sect. *Nemoralis?*

Tree to 5 m tall. Resting buds spherical, 4–6 mm long, loosely constructed of triangular bud scales with a stiff, narrow point. Leaves 3–4 cm long (to 10 cm in juveniles), about 8 mm wide (to 13 mm in juveniles), the edges somewhat curled under, widest near or below the middle, with a prolonged pointed tip (hence the scientific name, acuminate is the technical term). Seed cones on a leafless stalk, without basal bracteoles, the reproductive part with three or four very unequal bracts, these and the axis becoming somewhat swollen. Fertile seed scales one or two, the combined seed coat and epimatium with a fleshy layer over a hard inner shell and with a prominent, small crest at the tip. Tepuis (isolated, flat-topped sandstone mountains) of the Neblina Massif of the Amazonas states of Brazil and Venezuela and of the Chimantá Massif of southeastern Bolívar state of Venezuela. Scattered in cloud forests and on rocky, open sites; 1,900–2,400 m. Zone 9?

Neblina podocarp is the most recently described *Podocarpus* species of the tepuis of the Guiana Highlands of southern Venezuela, Brazil, and the Guianas. It is also the least well known, and so there are many details to fill out about its morphological features and ecology.

Podocarpus acutifolius T. Kirk

PRICKLY TOTARA, WESTLAND TOTARA, NEEDLE-LEAVED TOTARA

Podocarpus subg. *Podocarpus* sect. *Australis*

Usually a spreading or upright shrub, or sometimes a tree to 9(–10) m tall, with trunk, when present, to 40 cm in diameter. Bark thin, reddish brown to gray, peeling in flakes. Crown conical in treelike forms, broadly dome-shaped in shrubs, with the ascending branches correspondingly short and stout or long and thin, rooting on contact with the ground. Branchlets numerous, scattered or clustered, well clothed with foliage. Twigs pale bluish green to dark green, prominently grooved between the attached leaf bases. Resting buds slightly pear-shaped, about 3–4 mm long, tightly wrapped by elongate bud scales with loose, pointed tips. Leaves evenly spaced along the branchlets, stiff and tough, variably light green above, green between the strongly contrasting, greenish white, broad stomatal bands beneath, bronzing during the winter, (1–)1.5–2.5 cm long (to 4 cm in juveniles), (0.7–)1.5–2.5(–3.5) mm wide. Blades nearly parallel-sided for much of their length, straight, tapering abruptly to the prickly tips (hence the scientific name, Latin for "sharp leaf") and to the base with a very short petiole to about 1 mm long. Midrib fairly flat above and prominently raised beneath, with one small resin canal beneath the midvein and with relatively weak supporting tissue extending out from it, the edges of the leaves more strongly reinforced. Pollen cones (0.6–)1–2(–2.5) cm long and 2–3(–4) mm wide, one to three (to five) at the end of a short, leafless stalk (2–)5–10(–15) mm long, sometimes concentrated near the ends of the branchlets. Pollen scales with a rounded, upturned tip less than 1 mm long and wide. Seed cones on a very short, leafless stalk to 2 mm long, without basal bracteoles, the reproductive part with two (or three) unequal bracts, these and the axis becoming very swollen and fleshy, about 6 mm in diameter, bright red at maturity. Fertile seed scales one (or two), the combined seed coat and epimatium with thin flesh over a hard shell, dark greenish brown, 3–5 mm long and 2–3 mm wide, with a crest extending along one side and over the top to form a blunt beak. Mountains and adjacent lowlands of the northern and western sides of South Island (New Zealand) south to about 44°S. Scattered in forests and scrublands of valleys in the montane and lowland zones; (100–)300–1,000 m. Zone 7.

Prickly totara is the least common of the New Zealand *Podocarpus* species and the only one confined to South Island. Within its restricted range, it reaches its greatest abundance in the Nelson Lakes district, where the type material was collected in 1874. It is a taller shrub than its more subalpine neighbor, snow totara (*P. nivalis*), with which it hybridizes, and such hybrids are the source of the cultivar 'Pendulus' that was historically incorrectly associated with totara (*P. totara*). Prickly totara also commonly forms hybrids with this latter species, and these are the source of the cultivar 'Aureus,' so named because of the bronzing of the leaves inherited from *P. acutifolius*. There are also known hybrids with the fourth New Zealand *Podocarpus* species, Hall totara (*P. cunninghamii*). The two latter hybrids are small trees or larger shrubs than is typical for prickly totara, and the tree forms in this species may have arisen through hybridization. Nowhere else have hybrids among *Podocarpus* species been described, and it may be that some of the species and botanical varieties of this genus described elsewhere are really hybrids. Further work is clearly needed as even the New Zealand hybrids have not been as thoroughly documented as they might be. Prickly totara is quite unusual among *Podocarpus* species, even those of New Zealand, in having well-marked growth rings with conspicuous latewood. Unfortunately, there is some confusion surrounding the correct name for this species, and some authors mistakenly applied the name *P. lawrencei* to it rather than to the related mountain plum pine of southeastern Australia. See *P. lawrencei* for further discussion of this problem.

Mature shrub of prickly totara (*Podocarpus acutifolius*) in nature.

Podocarpus affinis B. Seemann

KUASI PODOCARP, FIJIAN TOTARA, KUASI (FIJIAN)

Podocarpus subg. *Foliolatus* sect. *Gracilis*

Tree to 9 m tall. Branches of the crown with numerous short branchlets densely clothed with foliage at the tips. Twigs green, prominently grooved between the attached leaf bases. Resting buds roughly spherical, 1–2.5 mm long, tightly clothed by rounded bud scales with pointed free tips. Leaves standing out and forward all around and somewhat concentrated toward the tips of the twigs, leathery, lasting 2–4 years, dark green above, waxy bluish green beneath at first, becoming rusty on drying, 3–5(–6.5) cm long, 6–9(–11) mm wide. Blades widest below the middle, tapering gradually and then abruptly to the rounded tip and abruptly to the wedge-shaped base on a short petiole about 3–5 mm long. Midrib narrowly raised above and broadly so beneath, with three resin canals beneath the midvein, well-developed wings of support tissue extending out to the sides, and a partial layer of large hypodermis cells beneath the upper epidermis. Pollen cones 2–3.5 cm long and 2–3 mm in diameter, one directly in the axils of the foliage leaves. Pollen scales with a triangular, upturned tip less than 0.5 mm long. Seed cones on a short, leafless stalk to about 1 cm long, with a pair of triangular basal bracteoles 2–5 mm long, the reproductive part with two or three roughly equal bracts, these and the axis becoming red, swollen, and juicy, 4–7 mm long by 2–4 mm thick. Fertile seed scales two or three, the combined seed coat and epimatium leathery over a hard inner shell, (3–)5–7(–9) mm long by (2–)3–4(–5) mm thick, with a low crest along one side ending in a short beak. Mountains of Viti Levu, Fiji. Scattered or forming pure stands in wet, dense, low forests of mountain summits and exposed ridges; 600–960 m. Zone 9.

The rare kuasi podocarp is not closely related to the other and more common Fijian *Podocarpus* species, the widespread oleander podocarp (*P. neriifolius*). Instead, its affinities lie with several species from the Philippines and beyond, including Pilger podocarp (*P. pilgeri*). The two Fijian species are generally not found together, with *P. affinis* occupying higher elevations than *P. neriifolius*. Kuasi podocarp is too small and too rare to be of any commercial significance but is endangered by general habitat destruction. The species name, Latin for "related to," was chosen, perhaps, because Seemann, the author of the species, originally confused it with plum pine (*P. elatus*) of Australia.

Podocarpus angustifolius Grisebach

CUBAN PODOCARP, SABINA CIMARRONA (SPANISH)

Podocarpus subg. *Podocarpus* sect. *Pumilis*

Shrub, or tree to 7(–10) m tall, with trunk to 1 m in diameter but often multitrunked or branching from the base. Crown irregular, with numerous short branches bearing pairs or whorls of short (mostly 2–8 or more cm long) branchlets densely clothed with foliage. Twigs green, variously slender and flexible to thick and stiff, prominently grooved between the attached leaf bases. Resting buds egg-shaped to nearly spherical, loosely formed of triangular to rounded, stiff, keeled bud scales, at least the outer of which have a point at the tip. Leaves densely and evenly arranged along (at intervals of 2–10 mm) and around the twigs, standing out from them or directed forward, lasting 1–3 years, stiffly leathery or a little flexible, shiny dark green above, yellowish green beneath (1.5–)2–5(–7) cm long (to 10 cm in juveniles), 3–7(–10) mm wide (to 13 mm in juveniles). Blades straight or slightly curved to one side, flat or a little bowed upward on either side of the midrib, sometimes wrinkling on drying, the edges curled under, sometimes nearly parallel-sided but more often clearly widest near the middle, tapering smoothly to the triangular tip that usually ends in a sharp prickle and to the roundly wedge-shaped base on a very short petiole 1–3 mm long. Midrib flat to slightly raised above, sharply raised (to slightly grooved) beneath, with one resin canal beneath the midvein, wings of support tissue extending out to the sides, and a continuous layer of hypodermis beneath the upper epidermis. Pollen cones (0.8–)1–1.5(–2) cm long and 3–4 mm wide, single directly in the axils of foliage leaves. Pollen scales with an irregularly minutely sharp-toothed, upturned, roundly triangular tip about 1 mm long. Seed cones on a very short, leafless stalk 2–7(–10) mm long, without basal bracteoles, the reproductive part with two distinctly unequal bracts, these and the axis becoming swollen and juicy, red, nearly spherical to longer than thick, (3–)5–10(–15) mm long. Fertile seed scale one, the combined seed coat and epimatium leathery over a hard inner shell, 6–10 mm long by 3.5–5 mm thick, with a pronounced conical beak 1–2 mm long. Mountains of eastern- and westernmost Cuba, Blue Mountains of eastern Jamaica, mountains of Haiti, and Central Cordillera of the Dominican Republic. Scattered in various humid forest types, including pine forests, broad-leaved montane forests, and elfin forests. Sometimes on limestone substrates; (500–)750–1,750(–2,250) m. Zone 9. Two varieties.

As might be expected of a montane plant with a wide geographic and ecological range in the Greater Antilles, Cuban podocarp displays a great deal of variation from place to place. As a result, some authors divided it into as many as six more local species, some known only from their type specimens. There do not appear to be any substantive morphological discontinuities among the various populations, however, and they are treated here as a single species consisting of two weakly distinguished botanical varieties, one in western Cuba and the other covering the eastern portion of the range. The gap between the two botanical varieties is almost 900 km, while the largest gaps in the east, between Cuba and the northern peninsula of Haiti and between Jamaica and the southern peninsula of Haiti, are only about 250 km. Many of the local populations are rare and are endangered by cutting for carpentry and fuel. Like many other podocarps belonging to various genera, Cuban podocarp contains the insect hormones called ecdysterols as well as other compounds with complicated structures, including some known only from this species. That apparent restriction may not mean much since only a handful of *Podocarpus* species have had thorough chemical investigations. Such studies, along with those of DNA, might help decide whether it is more closely related to the Guiana Highlands species Buchholz podocarp (*P. buchholzii*) and Roraima podocarp (*P. roraimae*) or to its neighbor, Purdie podocarp (*P. purdieanus*).

Podocarpus angustifolius Grisebach **var. *angustifolius***

Leaves fairly parallel-sided, 3–4.5 mm wide. Western portion of the range of the species, in western Cuba (Pinar del Río).

Podocarpus angustifolius* var. *aristulatus
(Parlatore) Staszkiewicz

Leaves clearly widest near the middle, (3–)5–7(–10) mm wide. Eastern portion of the range of the species, in eastern Cuba (Guantánamo), Jamaica, and Hispaniola. Synonyms:

Podocarpus angustifolius var. *wrightii* Pilger, *P. aristulatus* Parlatore, *P. buchii* I. Urban, *P. ekmanii* I. Urban, *P. leonii* Carabia, *P. urbanii* Pilger, *P. victorinianus* Carabia.

Podocarpus annamiensis N. Gray

INDOCHINESE PODOCARP, THÔNG TRE NAM (VIETNAMESE), HAI NAN LUO HAN SONG (CHINESE)

Podocarpus subg. *Foliolatus* sect. *Globulus*

Tree to 12(–16) m tall, with trunk to 1 m in diameter. Bark pale to dark grayish brown, flaking in scales. Crown narrow with short, upright branches bearing paired or clustered branchlets densely clothed with foliage at their tips. Twigs coarse, green, grooved between the attached leaf bases. Resting buds spherical, 2.5–4 mm in diameter, tightly wrapped by roundly triangular bud scales, the outer scales erect. Leaves standing out evenly around the twigs and crowded near their tips, leathery, thick and stiff, shiny bright green above, paler beneath, lasting 2–4 years, 4–10.5 cm long (to 18 cm in juveniles), 5–10 mm wide (to 20 mm in juveniles). Blades straight or occasionally curved to one side, the margins curled under, widest below the middle, tapering gradually and then abruptly to the broadly and roundly triangular tip and more abruptly to the roundly wedge-shaped base on a petiole 2–6 mm long. Midrib broadly raised to nearly flat above, variously raised to grooved beneath, with two large resin canals flanking a small one beneath the midvein, wings of support tissue extending out to the sides, and with a strongly interrupted layer of hypodermis beneath the upper epidermis. Pollen cones pale yellow, 3–5 cm long and 3–5 mm wide, one (to three) directly in the axils of the foliage leaves or nearly so. Pollen scales with a tiny, upturned, triangular tip less than 0.5 mm long. Seed cones on a short, leafless stalk 2–10 mm long, with a pair of basal bracteoles, the reproductive part with two similar bracts, these and the axis becoming reddish orange to reddish purple, swollen, and juicy, (4–)8–11 mm long by 7–9 mm thick. Fertile seed scale one, the combined seed coat and epimatium dark purple with a waxy coating, 8–10 mm long by about 6 mm thick, rounded or sometimes with a small beak. Southern Hainan (China), eastern Myanmar, and coastal mountain range of Vietnam. Scattered in montane rain forests and seasonally drier forests on slopes and ridges with leached red or yellow soils; 600–1,700 m. Zone 9.

This modest-sized tree is so highly prized for its fine, even-grained wood, used in making carvings, writing brushes, and musical instruments, that it has almost been exterminated on the island of Hainan, where in situ and ex situ conservation measures have been initiated. Its status in the main portion of its range in Vietnam is uncertain because it is synonymized there with the much more widely distributed oleander podocarp (*Podocarpus neriifolius*). It differs from that species in its smaller, stiffer leaves with a blunter tip and in its tightly wrapped resting buds. Its seeds and their enclosing epimatium average smaller than those of oleander podocarp, which can reach 16 mm long. The species name refers to Annam, a once independent kingdom that became the middle portion of Vietnam.

Podocarpus archboldii N. Gray

SOA PODOCARP, SOA, SARAU (PAPUAN LANGUAGES)

Podocarpus subg. *Foliolatus* sect. *Foliolatus*

Tree to 40 m tall, with trunk to 1 m in diameter, smooth or buttressed. Resting buds 2–4 mm long, 3–4 mm in diameter, loosely constructed of upright, triangular bud scales with slightly spreading tips. Leaves sticking out all around the twigs, leathery, thin, shiny dark green above, paler and duller beneath, 7–12 cm long (to 18 cm in juveniles), 10–14 mm wide (to 16 mm in juveniles). Blades straight, the margins flat, nearly parallel-sided or slightly widest below the middle, tapering gradually and then abruptly to a roundly triangular tip and abruptly to a roundly wedge-shaped base on a petiole to 5 mm long. Midrib broadly and prominently raised above, more broadly raised beneath, probably with three resin canals beneath the midvein, wings of support tissue extending out to the sides, and scattered cells of hypodermis beneath the upper epidermis. Pollen cones about 4 cm long and 4 mm in diameter, one directly in the axils of the foliage leaves or, more often, on a short, leafless stalk to 5 mm long. Pollen scales with a tiny, upturned, triangular tip about 0.5 mm long. Seed cones on a short, leafless stalk 5–11 mm long, with a pair of needlelike basal bracteoles about 2 mm long, the reproductive part with two or three very unequal bracts, these and the axis becoming red, swollen, and juicy, 6–10 mm long. Fertile seed scale one, the combined seed coat and epimatium leathery over a hard inner shell, 7–8 mm long, without an evident crest or beak. Scattered intermittently along the mountain backbone of New Guinea from the Vogelkop in western Irian Jaya (Indonesia) to the Huon Gulf in eastern Papua New Guinea. Scattered among hardwoods in the canopy of midmontane rain forests; 700–1,650(–2,200) m. Zone 9. Synonym: *Margbensonia archboldii* (N. Gray) A. V. Bobrov & Melikian.

Soa podocarp has been confused with baula podocarp (*Podocarpus crassigemmis*) and oleander podocarp (*P. neriifolius*), both of which grow near it in New Guinea. Baula podocarp grows at higher elevations and has tight, spherical resting buds closely wrapped by the bud scales. Oleander podocarp, which generally occurs at lower elevations, is more closely related and much more similar, with rather subtle, somewhat overlapping differences. While leaf size and shape of the two species are almost the same, oleander podocarp has a narrow prolongation of the tip not found in soa podocarp, and the sides of the raised midrib are nearly vertical rather than strongly sloped. Oleander podocarp usually has pollen cones in clusters of three, without a measurable stalk, and the seed cones are not known to add a third, reduced bract to the podocarpium. The species name honors Richard Archbold (1907–1976), the American explorer and biologist who financed seven and led three expeditions to New Guinea, including the one on which the type was collected.

Podocarpus atjehensis (Wasscher) de Laubenfels

ACEH PODOCARP

Podocarpus subg. *Foliolatus* sect. *Longifoliolatus*

Tree to 8(–15) m tall, with trunk to 0.2 m in diameter. Branches stout. Resting buds spherical to somewhat elongated, 6–14 mm long, loosely constructed of enclosing to upright, roundly triangular bud scales with prolonged tips, the outer scales often curled outward and up to 20 mm long. Leaves densely attached along and around the twigs, standing out at first but later angling down and back along them, leathery, thick, and stiff, emerging pink, maturing, shiny dark green above and paler beneath, 7–11 cm long (to 18 cm in juveniles), 5–8.5 mm wide. Blades generally straight, widest below the middle but often nearly parallel-sided, tapering very gradually to the narrowly triangular tip and a little more quickly to the narrowly wedge-shaped base on a short petiole 3–4 mm long. Midrib narrowly but prominently raised above, sometimes within flanking channels, more broadly raised beneath, with three resin canals beneath the midvein, the central one sometimes smaller than the flanking ones, and with wings of support tissue extending out to the sides and an incomplete layer of hypodermis beneath the upper epidermis. Pollen cones 2–3.5 cm long and 4–4.5 mm wide, single directly in the axils of the foliage leaves. Pollen scales with a tiny, upturned triangular tip only 0.2 mm long. Seed cones on a leafless stalk 8–16 mm long, with a pair of sword-shaped basal bracteoles (2–)3–6 mm long, the reproductive part with two unequal bracts, these and the axis becoming swollen, juicy, and red, 7–11 mm long by 3–6 mm thick. Fertile seed scale one, the combined seed coat and epimatium leathery over a hard inner shell, coated with a film of bluish white wax, 9–11 mm long by 7–8 mm thick, rounded at the tip without an evident crest or beak. Known only from mountains of Aceh (formerly spelled Atjeh) province, northern Sumatra (hence the scientific name) and the Wissel Lakes district of western Irian Jaya, New Guinea (Indonesia). Gregarious or scattered in mossy forests and scrublands on poor soils; 1,800–3,300 m. Zone 9? Synonyms: *Margbensonia atjehensis* (Wasscher) A. V. Bobrov & Melikian, *P. neriifolius* var. *atjehensis* Wasscher.

Aceh podocarp is a rare tree of uncertain taxonomic status. It was originally described as a botanical variety of the widespread oleander podocarp (*Podocarpus neriifolius*) from high elevations in northern Sumatra, but its large resting buds, narrow, backswept leaves without a prolonged tip, and fatter pollen cones are beyond the range of features for that variable species. The backswept leaves are reminiscent of the wider ones of Teku teak (*P. deflexus*), another rare and local species that also occurs in northern Sumatra but generally at lower elevations. The known distribution of Aceh podocarp in Sumatra and New Guinea, without any intervening localities, combined with substantial differences in elevation between stands in the two regions, warrants closer scrutiny. The tree yields a white resin that is not exploited commercially, unlike that of the dammars (*Agathis*), and with its small stature it is not under direct threat from forest clearance. However, with so few populations it is vulnerable to the increasingly frequent fires associated with human activity in the region.

Podocarpus borneensis de Laubenfels

BULOH PODOCARP, BULOH, MENTADEH (MALAY), BUBUNG (IBAN)

Podocarpus subg. *Foliolatus* sect. *Foliolatus*

Tree to 15(–25) m tall, with trunk to 0.2 m in diameter or more. Resting buds conical, 4–10 mm long by 2–3 mm in diameter, with narrowly triangular to sword-shaped, upright outer bud scales. Leaves standing out straight all around the twigs, leathery, thick, and very stiff, shiny dark green above, paler beneath, (2.5–)3.5–7(–9) cm long (to 16 cm in juveniles), 8–14 mm wide. Blades generally straight, widest near or below the middle, tapering gradually and then abruptly to a roundly triangular or rounded tip and more abruptly to a roundly wedge-shaped base on a petiole 3–5 mm long. Midrib broadly and prominently raised above, often shallowly grooved beneath, with two or three resin canals beneath the midvein, wings of support tissue extending out to the sides, and scattered large, cells of hypodermis beneath the upper epidermis, with other sclereids scattered within the photosynthetic tissue. Pollen cones 3–5 cm long and 2–3 mm wide, one to three directly in the axils of the foliage leaves or on a very short, leafless stalk to 4 mm long. Pollen scales with a broadly triangular, upturned tip less than 0.5 mm long. Seed cones on a very short, leafless stalk to 2 mm long, with a pair of needlelike basal bracteoles to 2 mm long, the reproductive part with two slightly unequal bracts, these and the axis becoming swollen, juicy, and red, 6 mm long. Fertile seed scale one, the combined seed coat and epimatium leathery over a hard inner shell, 6–8 mm long by 5–6 mm thick, tipped by a small beak. Mountains running along the whole axis of Borneo (hence the scientific name) parallel to the northwestern coast and continuing this line on the Karimata Islands off the southwestern coast. Scattered or gregarious in a variety of montane habitats, from wet, rocky ridges to nutrient poor sands and in the understory of tall montane rain forests; (360–)700–2,100 m. Zone 9? Synonym: *P. polystachyus* var. *rigidus* Wasscher.

Buloh podocarp stands out for its very thick, stiff leaves and the grooved midrib underneath. The stiff texture is due to the mass of large, hard sclereids found not only in a partial layer of hypodermis beneath the epidermis but also in the photosynthetic tissue itself. Likewise, the grooved midrib beneath often suppresses the central resin canal that is found in leaves of most other members of *Podocarpus* subg. *Foliolatus*. It was originally described as a botanical variety of sea teak (*P. polystachyus*), a species largely confined to seacoasts and modest elevations nearby in the interior, primarily in Borneo and Malaya.

Podocarpus bracteatus Blume

UNUNG PODOCARP, KAYA UNUNG, KI PANTJAR (MALAY)

Podocarpus subg. *Foliolatus* sect. *Longifoliolatus*

Tree to 30(–40) m tall, with trunk to 1 m in diameter. Resting buds narrowly conical, tightly wrapped by upright, sword-shaped bud scales with free tips, the outer ones longer than the inner, 5–12 mm long. Leaves standing out all around the twigs, leathery, lasting 2–3 years, shiny dark green above, paler beneath, (6–)10–14(–17) cm long (to 23 cm in juveniles), 9–14 mm wide (to 20 mm in juveniles). Blades straight or curved to one side, widest below the middle, tapering gradually to a triangular tip and a little more abruptly to a roundly wedge-shaped base on a petiole 2–4 mm long. Midrib narrowly raised above, sometimes bordered by grooves, broader and less conspicuous beneath, with three resin canals beneath the midvein, sometimes with another one on either side within the wings of support tissue extending out to the sides, and with an incomplete layer of hypodermis beneath the upper epidermis. Pollen cones (2.5–)3.5–6(–9) cm long and 2.5–4 mm in diameter, one or two (or three) directly in the axils of the foliage leaves but elongating between the bud scales at the base with maturity to produced a scaly stalk to 8 mm long (hence the scientific name, Latin for "bracted"). Pollen scales with a triangular, upturned tip to 1 mm long. Seed cones on a long, leafless stalk 10–20 mm long, with a pair of needlelike basal bracteoles 4–5 mm long, the reproductive part with two to four bracts, these and the axis becoming swollen, juicy, and red, 10–14 mm long. Fertile seed scales one or two, the combined seed coat and epimatium leathery over a hard inner shell, 10–14 mm long by 6–8 mm thick, tipped by a small beak. Discontinuous in western and central Indonesia, mostly in Java but with outliers to northern Sumatra, central Sulawesi and Flores. Scattered in the canopy of middle montane rain forests; (400–)1,000–2,600 m. Zone 9. Synonyms: *P. neriifolius* var. *bracteatus* (Blume) Wasscher, *P. neriifolius* var. *brevipes* (Blume) Pilger.

Unung podocarp replaces the more widespread oleander podocarp (*Podocarpus neriifolius*) throughout the mountain forests of Java, where it reaches its greatest abundance. Unung podocarp is generally similar to oleander podocarp and has often been treated as a synonym. It differs in lacking a narrowed point prolonging the leaf tip, in the elongation of the axis within the lower part of the pollen cones, producing the appearance of a scaly stalk, and in its somewhat larger seed cones, both the podocarpium and the seed being longer on average than those of oleander podocarp. Both species are fine timber trees, and neither the trees nor the wood are usually distinguished by foresters and lumbermen.

Podocarpus brasiliensis de Laubenfels

SERRA PODOCARP, PINHEIRINO DA SERRA (PORTUGUESE)

Podocarpus subg. *Podocarpus* sect. *Nemoralis*

Tree to 15 m tall. Twigs usually paired, green, weakly grooved between the attached leaf bases. Resting buds roughly spherical, 3 mm in diameter, loosely formed of triangular bud scales with pointed, free, upright tips. Leaves somewhat distantly and evenly distributed around and along the twigs, lasting 2–3 years, dark green above and paler beneath, 5–8 cm long, 9–12 mm wide. Blades straight, the margin curled under slightly, nearly parallel-sided, tapering fairly abruptly near the ends to the broadly triangular tip and the roundly wedge-shaped base with a petiole about 3–8 mm long. Midrib narrowly grooved above and broadly raised beneath, with one resin canal below the midvein and wings of support tissue extending out to either side. Pollen cones 2.5–4 cm long and 2.5–3.5 mm in diameter, single directly in the axils of foliage leaves. Pollen scales with a tiny, smooth-edged, triangular, upturned tip to 1 mm long. Seed cones on a long, leafless stalk 8–12(–20) mm long, without basal bracteoles, the reproductive part with two very unequal bracts, these and the axis becoming a little swollen and fleshy, red, 7–8 mm long by 2–3 mm thick. Fertile seed scale one, the combined seed coat and epimatium with a leathery purple skin over a hard inner shell, 7–8(–10) mm long by 5–8 mm thick, with a short, blunt beak. Discontinuous in low mountains of high plains fringing the Amazon River basin on the north and south from southern Venezuela to the Federal District of Brazil. Often growing in gallery forest along streams; 800–1,800 m. Zone 10.

Serra podocarp is closely related to Guatemalan podocarp (*Podocarpus guatemalensis*) and Trinidadian podocarp (*P. trinitensis*), which differ most obviously in having their midribs raised above. The species is apparently common in some regions but is often overlooked by botanists and is underrepresented in collections.

Podocarpus brassii Pilger

CHUGA PODOCARP, CHUGA, BACELA (CHIMBU), MAJA (MONDO)

Podocarpus subg. *Foliolatus* sect. *Macrostachyus*

Spreading shrub, or tree to 30 m tall, with trunk, when present, to 0.75 m in diameter. Crown rounded to flat, with many short branches bearing clustered branchlets densely clothed with foliage. Twigs coarse, rigid, green, deeply and prominently grooved between the attached leaf bases. Resting buds spherical, 4–5 mm in diameter, the erect bud scales longer, up to 5(–8) mm long with their projecting tip. Leaves densely and evenly standing out or forward along and around the twigs, thick, leathery and very stiff, shiny dark green above, paler and duller beneath, 1–2(–2.5) cm long (to 4 cm in juveniles), 3–7.5 mm wide. Blades straight, the margins turned down, widest below the middle, tapering fairly quickly to the triangular tip, which may have a short point, and abruptly to the roundly wedge-shaped base on a very short petiole to 2 mm long. Midrib narrowly and shallowly raised above, more broadly raised beneath, with (one or) three resin canals beneath the midvein, wings of support tissue extending out to the sides, and a scattering of hypodermis cells beneath the upper epidermis. Pollen cones 2.5–3 cm long and 3.5–7 mm in diameter, one directly in the axils of the foliage leaves. Pollen scales with an upturned triangular tip 1–4 mm long. Seed cones on a thick short, leafless stalk 1–9 mm long, with a pair of needlelike basal bracteoles 2–3 mm long, the reproductive part with two (or three) unequal bracts, these and the axis becoming swollen, juicy, and dark purple with bluish wax, 5–9 mm long by 3–7 mm thick. Fertile seed scales one (or two), the combined seed coat and epimatium leathery over a hard inner shell, purplish brown, 7–10(–13) mm long by (5–)6–9 mm thick, without a crest or beak. Along the mountainous spine of New Guinea from central Irian Jaya (Indonesia) through Papua New Guinea to Mount Suckling in the east. Occupying and sometimes dominating many open, high-elevation habitats, from the margins of subalpine forests to alpine grasslands, scrublands, and rock fields, occasionally in mossy forests; (2,000–)2,600–3,750 m. Zone 9. Two varieties.

Underside of twig of chuga podocarp (*Podocarpus brassii*) with two growth increments, ×0.5 .

Chuga podocarp is the most alpine of the *Podocarpus* species of New Guinea, reaching the highest elevations, living in the most exposed environments, and having the smallest, toughest leaves. It cannot be confused with any other species on the island. There are two botanical varieties that differ only in their growth habit and in their pollen cones, so that vegetative or seed-bearing specimens cannot be distinguished in the herbarium. They occupy much the same range, but variety *brassii* is more common and grows in a wider range of habitats than variety *humilis,* which is more or less confined to scrublands, sometimes on saturated soils. The species name honors Leonard Brass (1900–1971), the Australian botanist and explorer who collected the type specimen and many other plants during the Archbold expeditions in New Guinea.

Podocarpus brassii Pilger **var. *brassii***

Tree but sometimes mature at only 3 m. Pollen cones 5–7 mm in diameter. Tip of pollen scales narrowly triangular, 3–4 mm long.

Podocarpus brassii* var. *humilis de Laubenfels

Spreading shrub no more than 30 cm tall, to bushy tree to 5 m. Pollen cones 3.5–5 mm in diameter. Tip of pollen scales triangular, to 1 mm long.

Podocarpus buchholzii de Laubenfels

BUCHHOLZ PODOCARP

Podocarpus subg. *Podocarpus* sect. *Pumilis*

Tree to 7 m tall, often dwarfed. Twigs green, prominently grooved between the attached leaf bases. Resting buds roughly spherical, about 3 mm across and 2 mm long (to 6 mm when the bud scales are leaflike), loosely formed of triangular to leaflike bud scales with a small bristle tip. Leaves densely and evenly distributed along and around the twigs, shiny dark green above, paler beneath, 2–3 cm long, (4–)5–8 mm wide. Blades straight, the edges slightly turned down, widest near the middle and from there tapering quickly to the bluntly and roundly triangular tip and to the wedge-shaped base with a short petiole to 3 mm long. Midrib narrowly channeled above and broadly and shallowly channeled between two small ridges beneath, with one resin canal beneath the midvein and wings of support tissue extending out to the sides. Pollen cones 1–1.5 cm long and 2–3 mm wide, single directly in the axils of foliage leaves. Pollen scales with a smooth-edged triangular tip about 1 mm long. Seed cones on a short, leafless stalk 8–10 mm long, without basal bracteoles, the reproductive part with two unequal bracts, these and the axis becoming modestly swollen and fleshy, 7–8 mm long by 2.5–4 mm thick. Fertile seed scale one, the combined seed coat and epimatium with a leathery skin over a hard inner shell, 7–9 mm long by about 5 mm thick, with a ridge along one side ending in a narrow, pointed beak. Tepuis (isolated, flat-topped, sandstone mountains) of the Chimantá Massif in the Guiana Highlands, Bolívar state, Venezuela. Scattered along streamsides and in other wet places in elfin cloud forests; 2,100–2,500 m. Zone 9. Synonym: *Podocarpus buchholzii* var. *neblinensis* Silba.

Buchholz podocarp is closely related to Roraima podocarp (*Podocarpus roraimae*), differing most obviously in its grooved rather than raised midrib and slightly broader leaves on average. It occurs entirely within the range of Roraima podocarp, and both species are found on some tepuis of the Chimantá Massif. Here, Buchholz podocarp occurs at generally higher elevations and in more sheltered sites with saturated soils. Because of its small size and the low stature and remote location of the vegetation of which it is part, Buchholz podocarp is not threatened by direct exploitation. The species name honors John T. Buchholz (1888–1951), an American plant embryologist who, together with Netta Gray (honored by *P. grayae*), began a worldwide taxonomic revision of *Podocarpus* in the broad sense, including groups now treated as separate genera.

Podocarpus capuronii de Laubenfels

CAPURON YELLOWWOOD

Podocarpus subg. *Podocarpus* sect. *Scytopodium*

Tree to 20 m tall. Twigs strongly grooved between the elongate, attached leaf bases. Resting buds nearly spherical, with tightly overlapping bud scales. Leaves standing out evenly around and along the twigs, angled slightly forward, leathery, bright dark green above, paler beneath, lasting 2–3 years, 2.5–5 cm long, 2.5–5.5 mm wide. Blades straight or curved to one side, the edges thin and slightly drooping, mostly parallel-sided, tapering abruptly to the pointed tip and more gradually to the virtually stalkless base. Midrib scarcely protruding above and strongly so beneath, with three resin canals beneath the midrib and two at the outer corners of the leaf and very rarely with additional ones scattered around the periphery. Pollen cones 2–3 cm long and 3.5–4.5 mm wide, one to three at the tip of a short, smooth stalk to 1 cm long. Pollen scales with a sharp, triangular, upright tip about 1.5 mm long and 1 mm wide. Seed cones on a short, bare stalk to 5 mm long, without basal bracteoles, the reproductive part with two unequal bracts, these and the axis becoming slightly swollen and leathery, 7–10 mm long by 3–6 mm thick. Fertile seed scale one, the combined seed coat and epimatium initially coated by a whitish waxy film, very fleshy and 16–18 mm long by 10–11 mm thick, crested along one side. Northern and central Madagascar, on and near the Massif du Tsaratanana and the Massif de l'Itremo. Short, wet forest of mountain ridges; 1,500–2,800 m. Zone 8? Synonym: *P. woltzii* H. Gaussen.

The rare Capuron yellowwood is only known from a few localities in the high mountains that make up the spine of Madagascar. It has the largest leaves and seeds among the three *Podocarpus* species found in these high-elevation forests and scrublands. Usually, the leaves have the typical distribution of resin canals for African podocarps, with three resin canals beneath the midrib and two at the outer edges of the leaf. Rare specimens contain additional resin canals scattered about the periphery of the leaf. In this respect they resemble Laurent yellowwood (*P. rostratus*), which grows nearby in some localities. Likewise, some specimens of Laurent yellowwood lack the extra resin canals. Could these unusual specimens of each species be due to hybridization between the two? The species name honors René Capuron (1921–1971), a French botanist who collected many specimens in Madagascar, including the type specimen of this species.

Podocarpus celatus de Laubenfels

ULCUMANU PODOCARP, ULCUMANU (QUECHUA, SPANISH)

Podocarpus subg. *Podocarpus* sect. *Nemoralis*

Tree to 20(–25) m tall. Twigs green, prominently grooved between the attached leaf bases. Resting buds about 2 mm long and 3 mm in diameter, loosely enfolded by upright, triangular bud scales with short free tips. Leaves evenly and openly arranged along and around the twigs, glossy dark green above, paler beneath, 5–7 cm long (to 18 cm in juveniles), 10–14 mm wide (to 20 mm in juveniles). Blades straight, flat and of even thickness out to the edges which are not turned down, widest near the middle (well below the middle in juveniles), from there tapering fairly quickly to the bluntly and roundly triangular tip (very gradually to a narrowly triangular tip in juveniles) and to the broadly and roundly wedge-shaped base. Midrib narrowly and sharply grooved above, broadly raised beneath, with wings of support tissue extending out to the sides. Pollen cones probably about 3–4 cm long by about 3 mm wide and single directly in the axils of foliage leaves. Seed cones on a short, leafless stalk 4–6 mm long, without basal bracteoles, the reproductive part with two unequal bracts, these and the axis becoming somewhat swollen and fleshy, 5–6 mm long by about 3 mm thick. Fertile seed scale one, the combined seed coat and epimatium plump, with a leathery skin over a hard inner shell, 9–10 mm long by 7–8 mm thick, rounded at the tip without an apparent crest. Widely and discontinuously distributed in the mountains fringing the Amazon River basin on the north and west, from southeastern Venezuela through the eastern Andean cordillera to western Bolivia. Scattered in the subcanopy of humid forests on moist slopes; (0–)400–1,200(–1,500) m. Zone 10.

Ulcumanu podocarp is closely related to Trinidadian podocarp (*Podocarpus trinitensis*), broadleaf podocarp (*P. magnifolius*), and Guatemalan podocarp (*P. guatemalensis*), which may be distinguished from it, respectively, by smaller seeds, broader leaves, and midribs raised above. Specimens of Ulcumanu podocarp have been mistakenly identified as belonging to these species, hence the scientific name, Latin for "hidden." It might actually be best to treat it as a variant of one of these species and, in fact, the whole group, including the other two described species in botanical section *Nemoralis,* is in need of careful revision, but existing material is sparse for many parts of the range of the group.

Podocarpus confertus de Laubenfels

BORNEO SCRUBLAND PODOCARP

Podocarpus subg. *Foliolatus* sect. *Longifoliolatus*

Tree to 36 m tall, with a slender trunk. Twigs paired or clustered, green, not prominently grooved between the attached leaf bases. Resting buds 6–10 mm long by 2–3 mm wide, loosely constructed of sword-shaped, upright bud scales. Leaves evenly standing out and a little forward around the twigs, more concentrated near their tips, leathery, lasting 3(–4) years, shiny dark green above, paler, duller, and a little brownish beneath, 5–12 cm long (to 20 cm in juveniles), 7–10 mm wide (to 12 mm in juveniles). Blades straight to strongly curved to one side, sometimes just near the tip, the edges turned under, fairly parallel-sided but widest near or below the middle, tapering very gradually to the narrowly pointed tip and more abruptly to the roundly wedge-shaped base on a petiole 3–6 mm long. Midrib broadly and prominently raised above, flat or grooved beneath, with three resin canals beneath the midvein at least, wings of support tissue extending out to the sides, and an uninterrupted layer of hypodermis beneath the upper epidermis. Pollen cones 3–4.5 cm long and 2.5–3.5 mm in diameter, one (or two) directly in the axils of the foliage leaves. Pollen scales with a triangular, upturned tip to 0.5 mm long. Seed cones on a short, leafless stalk 5–13 mm long, with a pair of sword-shaped basal bracteoles 5–6 mm long, the reproductive part with two unequal bracts, these and the axis becoming swollen, juicy, and red, 8–12 mm long. Fertile seed scale one, the combined seed coat and epimatium leathery over a hard inner shell, 10–11 mm long by 6–7 mm thick, rounded at the tip without an evident crest or beak. Scattered along the northwestern side of Borneo in Sabah and Sarawak (Malaysia) and in western Kalimantan (Indonesia). Usually forming pure polelike stands or mixed with other trees in open, stunted forests on infertile and often serpentine-derived soils, occasionally becoming a large tree on better sites; (100–)600–1,200 m. Zone 9.

Borneo scrubland podocarp is very similar in most features to buloh podocarp (*Podocarpus borneensis*), which grows throughout much the same range in sites without a serpentine bedrock. Buloh podocarp has, on average, shorter, wider, thicker leaves, the pollen cones sometimes in clusters of three, and smaller seeds with a smaller podocarpium on a shorter stalk. Although they are placed in different sections, their relationship merits closer study since their differences may be due, in part, to the different environments they inhabit. The species name is Latin for "crowded," but there is nothing especially crowded about the leaves or other organs of these trees compared to many other related species of the genus.

Podocarpus coriaceus L. Richard & A. Richard

ANTILLEAN PODOCARP, YACCA PODOCARP, WEEDEE, CAOBILLA (SPANISH), RAISINIER MONTAGNE, LAURIER-ROSE (FRENCH)

Podocarpus subg. *Podocarpus* sect. *Lanceolatis*

Tree to 10(–20) m tall or sometimes shrubby, with trunk to 0.3(–0.5) m in diameter, often crooked. Bark grayish brown, fibrous, smooth at first, becoming flaky and finally ridged and furrowed and peeling in strips. Crown narrowly cylindrical or spreading, with slender branches generally bearing paired branchlets densely clothed with foliage. Twigs green, coarse, prominently grooved between the attached leaf bases and somewhat angled. Resting buds nearly spherical or somewhat elongated, 5–7 mm long and 3–7 mm wide, tightly or somewhat loosely wrapped by roundly triangular or sometimes somewhat leaflike bud scales, with free, pointed tips. Leaves densely and evenly attached along and around the twigs, lasting 2–3 years, leathery (hence the scientific name, Latin for "leathery," a rather common condition in *Podocarpus,* but this was one of the first species in the genus to be described), dark green and a little shiny above, duller and a little paler to yellowish green beneath, (6–)10–12(–15) cm long (to 22 cm in juveniles), 9–12 mm wide (to 17 mm in juveniles). Blades straight or a little curved to one side, the edges turned down, widest near or below the middle, tapering very gradually to the narrowly and roundly triangular drawn-out tip and more abruptly to the roundly wedge-shaped base on a very short petiole 2–4(–6) mm long. Midrib flat or slightly grooved above, broadly raised beneath, with one resin canal below the midvein, wings of support tissue extending out to the sides, and a continuous layer of hypodermis beneath the upper epidermis that may be thicker near the edges and above the midvein. Pollen cones 2.5–4 cm long and 5–7 mm in diameter, single directly in the axils of foliage leaves. Pollen scales with a sparsely jaggedly toothed, broadly triangular upturned tip about 1 mm long. Seed cones on a short, leafless stalk 5–10(–14) mm long, without basal bracteoles, the reproductive part with two or three very unequal bracts, these and the axis becoming somewhat swollen and fleshy, bright to dark red, 6–9 mm long by 5–8 mm thick. Fertile seed scales one (or two), the combined seed coat and epimatium leathery over a hard inner shell about 1 mm thick, brown, 7–9 mm long by 6–7 mm across, with a low ridge along one side ending in a short, broad beak. Puerto Rico through the Lesser Antilles to Trinidad and Tobago. Scattered in lower montane rain forests and extending into adjacent forest formations; (100–)500–1,200 m. Zone 9.

At one time, most of the lower montane podocarps from the whole Caribbean region were identified with this species, but over time the populations in different regions were distinguished as separate species. Some of these, such as Purdie podocarp (*Podocarpus purdieanus*), are not actually particularly closely related to Antillean podocarp, while others, such as willowleaf podocarp (*P. salicifolius*) of Venezuela, are much closer. Although Antillean podocarp can produce excellent and valuable wood, the tree is often stunted and distorted so that it is not of much economic value. It is the only native conifer throughout its range, except in Trinidad where Trinidadian podocarp (*P. trinitensis*) also grows. Their ecological separation there, if any, is yet to be determined because researchers and foresters typically do not distinguish them, assuming that any podocarp encountered is Antillean podocarp.

Podocarpus costalis C. Presl

SEABLUFF PODOCARP, LAN YU LUO HAN SONG (CHINESE)

Podocarpus subg. *Foliolatus* sect. *Macrostachyus*

Shrubby tree to 5 m tall. Bark smooth, greenish, peeling in thin flakes. Crown broad, with numerous horizontal, contorted branches bearing branchlets densely clothed with foliage. Twigs short, stout, light green, inconspicuously grooved between the attached leaf bases. Resting buds nearly spherical, 2–4 mm long, tightly wrapped by triangular bud scales with spreading tips. Leaves densely and evenly standing out from or forward around and along the twigs or more concentrated near the tips, leathery, lasting 2–4 years, dull bluish green above, yellowish green beneath, (2.5–)4–7 cm long (to 9 cm in juveniles), (5–)7–10 mm wide (to 13 mm in juveniles). Blades straight, the margins flat or slightly turned down, widest beyond the middle, tapering abruptly to a roundly triangular, round, or even slightly notched tip and gradually to a wedge-shaped base on a short petiole 2–3 mm long. Midrib fairly prominently raised above, more broadly and shallowly raised beneath, with three resin canals beneath the midvein, wings of support tissue extending out to the sides, and a partial hypodermis beneath the upper epidermis. Pollen cones 2.5–3.5 cm long and 6–8(–10) mm wide, one directly in the axils of the foliage leaves. Pollen scales with a narrow, tonguelike, upturned tip about 3 mm long. Seed cones on a short, leafless stalk (2–)4–6(–10) mm long, with a pair of 1.5 mm long needlelike bracteoles that are quickly shed, the reproductive part with two unequal bracts, these and the axis becoming swollen and juicy, bright red to purple, (7–)10–15 mm long by 8–9 mm thick. Fertile seed scale one, the combined seed coat and epimatium leathery over a hard inner shell, dark blue, (7–)9–10 mm long by 6–8 mm thick, with an inconspicuous beak. Restricted to the Babuyan, Batan, and Lanyu Islands between Luzon (Philippines) and Taiwan, the Polillo Islands off the eastern coast of Luzon, and the Bucas Islands off northern Mindanao (Philippines). Coastal thickets among rocks on bluffs at the shoreline of small islands (hence the scientific name, Latin for "coastal"); 0–300 m. Zone 9.

This geographically and ecologically restricted species is somewhat similar in appearance and sometimes in habitat to the more widespread sea teak (*Podocarpus polystachyus*), but the two are not known to grow together naturally anywhere. Both are widely cultivated in their native regions for their compact, evergreen habit. Sea teak differs in its often larger, usually slightly pointed leaves that stand out more from the twig and have flat margins and in its thinner, clustered pollen cones. The locality on Lanyu (Orchid Island) off the southern tip of Taiwan is famous and has been much visited by botanists.

Podocarpus costaricensis de Laubenfels ex Silba

COSTA RICAN PODOCARP

Podocarpus subg. *Podocarpus* sect. *Lanceolatis*

Tree to 30 m tall, with trunk to 0.35 m in diameter or more. Bark light reddish brown, breaking up into somewhat interlacing ridges. Twigs green, grooved between the attached leaf bases. Resting buds 5–15 mm long, loosely constructed of upright, narrowly triangular to somewhat leaflike bud scales. Leaves somewhat openly and fairly evenly distributed around and along the twigs, but slightly more concentrated near their ends, lasting 1–2 years, shiny dark green above, paler and duller beneath, 6–13 cm long, 12–17 mm wide. Blades straight or slightly curved near the tip, the edges turned down, widest near the middle or a little below, tapering very gradually to the prolonged, narrowly triangular tip and more abruptly to the roundly wedge-shaped base on a short petiole 2–5 mm long. Midrib deeply grooved above, broadly raised beneath, with one resin canal beneath the midvein. Pollen cones yellowish green, 2.5–4(–6) cm long and 3–4 mm wide, single in the axils of foliage leaves. Pollen scales with a triangular, upturned tip about 1 mm long. Discontinuous in Central America in southern San José province (Costa Rica) and in southwestern Darién province (Panama). Scattered in lower montane rain forests and in disturbed sites; (70–)1,000–1,700 m. Zone 10?

Costa Rican podocarp is closely related to Matuda podocarp (*Podocarpus matudae*), and these trees could be accommodated readily within that species were it not for the difference in the midribs. Matuda podocarp has the midribs raised above as prominently as Costa Rican podocarp has them grooved. There are some species of *Podocarpus* in which the midrib may be either raised or grooved, but in these cases each condition is weak, not like the unambiguous structures in these two Central American trees. Additional stands should be sought in the intervening regions of Honduras and Nicaragua, but the middle-elevation forests are one of the most degraded environments in these countries, so we may never know if there were once populations with intermediate characteristics.

Podocarpus crassigemmis de Laubenfels

BAULA PODOCARP, BAULA, SULA (CHIMBU), KABOR, KAIP (ENGA)

Podocarpus subg. *Foliolatus* sect. *Macrostachyus*

Tree to 25(–38) m tall or dwarfed on exposed ridges, with trunk to 0.7(–0.9) m in diameter, sometimes fluted. Bark thin, gray, flaking in scales and becoming ridged between shallow furrows. Crown conical at first, with very regular tiers of horizontal branches, becoming dome-shaped and irregular with spreading branches bearing clustered branchlets densely clothed with foliage. Twigs coarse, green, grooved between the attached leaf bases. Resting buds spherical, (3–)4–5 mm in diameter (hence the scientific name, Latin for "thick-budded"), but the basally tightly clasping scales much longer, to 8(–10) mm long with their blunt tips curled outward. Leaves standing out or forward densely and evenly all around and along the twigs, thick and leathery, shiny dark green above, paler beneath, 3–9(–11) cm long (to 20 cm in juveniles), 4.5–12 mm wide (to 14 mm in juveniles). Blades usually straight, the edges turned down, widest at or below the middle, tapering gradually and then more abruptly to a narrowly triangular tip and gradually to a narrowly wedge-shaped base on a short petiole to 3(–5) mm long. Midrib narrowly and prominently raised above, broader beneath, with three resin canals beneath the midvein, wings of support tissue extending out to the sides, and a discontinuous hypodermis beneath the upper epidermis. Pollen cones 3–5 cm long and (3–)5–6(–8) mm wide, one (or two) on a short, leafless stalk 1–4(–7) mm long. Pollen scales with a narrowly triangular, upturned tip 2–3 mm long. Seed cones on a long, leafless stalk (3–)5–14 mm long, with a pair of needlelike basal bracteoles 2–3 mm long that may soon be shed, the reproductive part with two unequal scales, these and the axis becoming swollen and juicy, passing through deep red to dark purplish black, (7–)10–15 mm long. Fertile seed scale one, the combined seed coat and epimatium leathery over a hard inner shell, 11–15 mm long by 9–13 mm thick, without an evident crest or beak. Mountainous spine of New Guinea from western Irian Jaya (Indonesia) through Papua New Guinea to Mount Suckling in the east. Scattered to gregarious in or as an emergent over the canopy of upper montane rain forests and mossy forests; (1,800–)2,000–3,000(–3,500) m. Zone 9.

Baula podocarp somewhat resembles the more widespread oleander podocarp (*Podocarpus neriifolius*), which occurs at generally lower elevations and has predominantly larger leaves with a prolonged tip and flat margins. It is actually more closely related to neighboring chuga podocarp (*P. brassii*), which has much smaller leaves and occurs in more open habitats with a higher maximum elevation. It was originally confused with soa podocarp (*P. archboldii*), which occupies the same range at lower elevation and differs primarily in reproductive features: pollen scales with a much smaller tip and the podocarpium often with a third, smaller sterile bract.

Podocarpus cunninghamii Colenso

HALL TOTARA, THIN-BARK TOTARA, TOTARA KIRI KOTUKUTUKU (MAORI)

Podocarpus subg. *Podocarpus* sect. *Australis*

Tree to 20 m tall, with trunk to 0.9(–1.25) m in diameter. Bark reddish brown, persistently thin, papery, flaking in narrow, short strips or in large sheets. Crown conical at first, becoming broadly dome-shaped with age, with numerous crooked branches bearing drooping to stiffly outstretched branchlets densely clothed with foliage. Twigs olive green, deeply grooved between the attached leaf bases. Resting buds small and stout, about 2 mm in diameter, covered by tightly overlapping, broad bud scales with pointy tips. Leaves evenly spaced along the twigs, often flattened to the sides in juvenile trees but spreading all around the branchlets in adults, stiff, brownish green, (1–)2–3(–4) cm long (to 6.5 cm in juveniles but usually much shorter), (1.5–)3–4.5 mm wide (to 6 mm in juveniles). Blades broadest near or below the middle, from there tapering gradually and then abruptly to the short, sharply pointed tips and more rapidly to the rounded, short-stalked base. Midrib flat or grooved above and slightly raised beneath, with one resin canal beneath the midvein and with lines of stomates in slightly waxy bands on the lower surface. Pollen cones (1–)1.5–2.5 cm long (to 3 cm after the pollen is shed), and 6–8 mm wide, one to three (to five) directly in the axils of the leaves or at the end of a short, leafless stalk about 10 mm long. Pollen scales with a short, minutely toothed, upturned tip about 1.5 mm long and about 1 mm wide. Seed cones on a very short, leafless stalk 1–5 mm long, without basal bracteoles, the reproductive part with two (to four) unequal bracts, these and the axis becoming very swollen, juicy, and bright red at maturity,

5–7 mm long and 3–5 mm wide. Fertile seed scales one or two, the combined seed coat and epimatium hard, smooth, dark purplish green, 5–7 mm long by 2–4 mm thick, with a shallow furrow along one side and a pointed beak at the tip. Throughout New Zealand from the northern end of North Island near 35°S though South Island to the South Cape of Stewart Island near 47°S. Scattered among other trees in all forest zones but most common in upper montane forests; 0–600 m. Zone 8. Synonym: *P. hallii* T. Kirk.

Hall totara was long confused with totara (*Podocarpus totara*), and even the Maori do not seem to have traditionally distinguished the two species with separate names, despite the cultural importance of totara wood. It was first described sketchily in 1884 and was not given a thorough botanical description until 1889 (when it was independently named with the synonym that gives rise to the common name). Most older works on New Zealand plants used the later name because the present one was so briefly described that its identity seemed uncertain. Present opinion has swung to the use of *P. cunninghamii* because the description, however brief, contains a few features, like the thin, papery bark, that clearly distinguish Hall totara from totara. The species name honors Allan Cunningham (1791–1839), an English-born botanist who worked in both Australia and New Zealand.

Separation of Hall totara from totara is not helped by the fact that the two species can hybridize with one another. The largest reported tree of Hall totara, the Mota totara in Dean Forest, 30 m tall and 2.65 m in diameter in 1984, might be such a hybrid. Hall totara is generally a smaller tree than totara and tends to replace it at the higher elevations of their mutual range. Its wood is also generally considered inferior to that of totara though still of considerable value. Hall totara also hybridizes with the two shrubby New Zealand *Podocarpus* species at the higher end of its range, snow totara (*P. nivalis*) and prickly totara (*P. acutifolius*). Hybridization among many New Zealand plants has been hypothesized to have had ecological consequences, and this might be the case with Hall totara. It has the widest ecological range among the New Zealand *Podocarpus* species, extending from coastal forests to the subalpine zone and from wet forests to the driest regions of New Zealand. In the mountains of central Otago in the dry interior portion of South Island, Hall totara occurs only in sheltered sites at an unusually low tree line. This seems to be due to the occurrence of intermittent fires, and there are stumps of large trees that show that the region was covered by closed-canopy forest before the 13th century. Thus the species is not limited by either drought or temperature here, explaining the lack of success in using it in dendroclimatological studies in the region. That is somewhat disappointing because the tree is known to live more than 550 years and typically lives 450 years.

Podocarpus decumbens N. Gray

NEW CALEDONIAN CREEPING PODOCARP

Podocarpus subg. *Foliolatus* sect. *Longifoliolatus*

Creeping and clambering shrub to 0.4 m, the central stems to 5 cm in diameter. Bark thin, yellowish brown to reddish brown, weathering grayish brown and peeling in strips. Crown spreading with horizontal branches bearing upturned branchlets densely clothed with foliage near the tips. Twigs grayish green with wax, prominently grooved between the attached leaf bases. Resting buds elongate, 6–10 mm long (the outer scales sometimes to 18 mm), loosely constructed of straight, triangular to leaflike bud scales, progressively shortened inward in the bud. Leaves densely and evenly distributed around and along the twigs or concentrated at their tips, extending stiffly outward or a little forward, leathery, shiny green above, waxy grayish green beneath, (3–)5–8 cm long, (4–)6–8 mm wide. Blades straight or slightly curved to one side, the edges turning down gently, widest beyond the middle, tapering fairly quickly to the roundly triangular tip and much more gradually to the narrowly wedge-shaped base on a petiole 3–6 mm long. Midrib broadly raised above and with a longitudinally wrinkled raised area beneath when dry, with three resin canals beneath the midvein, two to four above it, sometimes with two additional ones in the support tissue extending out to the sides, and with a thin, sparingly doubled, interrupted hypodermis beneath the upper epidermis. Pollen cones (1–)1.5–3 cm long and 2.5–4 mm wide, single in the axils of foliage leaves, the enclosing lower bracts curling back. Pollen scales with a broadly triangular, upturned, straight tip. Seed cones on a short, leafless stalk about 6 mm long, with two (or three) fingerlike basal bracteoles, the reproductive part with three or four slightly unequal bracts, these and the axis becoming swollen and juicy, red, 6–7 mm long by 3–4 mm thick. Fertile seed scales one (or two), the combined seed coat and epimatium leathery over a hard inner shell, 6–7 mm long by 4–5 mm thick, with a low beak. Montagne des Sources and Mont Kouakoue in southern New Caledonia near Nouméa. Forming spreading clonal patches in open forests and maquis brushlands on serpentine-derived soils, especially with a build-up of peat; 800–1,000 m. Zone 9.

Like the slightly taller and unrelated *Podocarpus gnidioides* with which it sometimes grows, *P. decumbens* is an inhabitant of harsh scrublands. This contrasts with the rain-forest habitat of its closest relatives, *P. longifoliolatus* and *P. polyspermus*. It has an even more limited geographic range than these species but is common in places on Montagne des Sources, making spectacular clonal patches up to 20 m in diameter through rooting of the creeping stems (hence the scientific name, Latin for "reclining"). Other than its reduced stature and creeping habit, *P. decumbens* resembles *P. longifoliolatus* in most taxonomic characters but differs in its stiffer leaves and predominance of a single seed in the seed cones. Unexpectedly, perhaps, *P. longifoliolatus* grows at higher elevations than *P. decumbens*. Since the two species are not known to grow on the same mountains, their reproductive behavior in proximity to one another cannot be assessed.

Podocarpus deflexus H. Ridley

TEKU TEAK

Podocarpus subg. *Foliolatus* sect. *Foliolatus*

Shrub, or tree to 8(–13) m tall, with trunk to 0.1 m in diameter or more. Bark grayish brown, smooth, flaking in scales and becoming shallowly furrowed. Crown irregularly and openly dome-shaped, with slender, crooked, spreading branches bearing clustered branchlets densely clothed with foliage. Twigs coarse, green, grooved between the attached leaf bases and bumpy with the scars of fallen leaves. Resting buds about 3 mm in diameter, with a spherical core of tightly wrapped, roundly triangular bud scales surrounded by sword-shaped outer scales up to 12 mm long and strongly curled outward. Leaves densely and evenly attached along and around the twigs and sharply angled back along them (hence the scientific name, Latin for "bent away"), leathery, thick, and stiff, lasting (1–)2–3(–4) years, shiny dark green above, paler beneath, 10–22 cm long (to 27 cm in juveniles), 7–12 mm wide (to 15 mm in juveniles). Blades generally straight, the edges flat, nearly parallel-sided or widest below the middle, tapering gradually and then more abruptly to the triangular tip and to the narrowly wedge-shaped base on a petiole 6–12 mm long. Midrib broadly raised above, flatter to deeply grooved beneath, with three resin canals beneath the midvein, often with another one on either side within the wings of support tissue extending out to the sides, and with a partial hypodermis layer of scattered groups of large cells beneath the upper epidermis. Pollen cones (one to) three directly in the axils of the foliage leaves or on a very short, leafless stalk to 2 mm long. Seed cones on a thick, leafless stalk 9–15 mm long, with a pair of needlelike basal bracteoles 1.5–2 mm long, the reproductive part with (two or) three (or four) unequal bracts, these and the axis becoming swollen, juicy, and red, 9–15 mm long by 4–8 mm thick. Fertile seed scales one or two, the combined seed coat and epimatium leathery over a hard inner shell, 11–12 mm long by 6–9 mm thick, rounded at the tip without an evident crest or beak. Isolated stands on single mountains in north-central Malaya (Malaysia) and northern Sumatra (Indonesia). Emergent above the canopy of moist dwarf montane scrub; 1,400–2,100 m. Zone 9.

Teku teak, named from the upper Teku Valley in Pahang state in Malaya, has one of the most restricted distributions among the species of the genus. With its small size and that of the scrub in which it grows, it is not threatened by direct exploitation. However, fires set by people, like those that blackened the skies of southeastern Asia and Malesia in 1997, are a threat for a species with so few known populations. Although the leaf shape and cone features are very much like those of the more widely distributed oleander podocarp (*Podocarpus neriifolius*) of nearby rain forests, the large, stiffly hanging leaves set it apart from that species and all others of the region except the equally rare Aceh podocarp (*P. atjehensis*), being somewhat reminiscent of the unrelated hayuco podocarp (*P. pendulifolius*) of Venezuela.

Podocarpus dispermus C. White

ATHERTON PLUM PINE

Podocarpus subg. *Foliolatus* sect. *Acuminatis*

Tree to 20 m tall, with trunk to 0.5 m in diameter. Bark grayish brown, thin, fibrous. Crown irregularly dome-shaped, with thin branches bearing whorled branchlets sparsely clothed with foliage. Twigs light green, weakly grooved between the attached leaf bases. Resting buds pointed, about 5 mm long, with very narrow bud scales ending in a loose point. Leaves evenly spaced along the branchlets or concentrated at intervals, shiny dark green above, paler beneath, (7–)10–20 cm long, 18–25(–30) mm wide. Blades fairly parallel-sided, tapering through about a fifth of the length to the pointed, triangular tip and a little less to the rounded base ending in a short petiole to about 7 mm long. Midrib prominently raised above and slightly grooved beneath, usually with three resin canals beneath the midvein but sometimes with just one or with two extra ones flanking it, the upper side of the leaf with a discontinuous layer of hypodermis beneath the epidermis. Pollen cones 1–3 cm long and 2–3 mm wide, one to three directly in the axils of the leaves. Pollen scales with a pointed, broadly triangular, upturned tip less than 1 mm long. Seed cones on a short, leafless stalk to 15 mm long, with a pair of basal bracteoles, the reproductive part with three or four roughly equal bracts, these and the axis becoming very swollen and juicy, bright red, nearly spherical, 5–6 mm in diameter. Fertile seed scales (one or) two, the combined seed coat and epimatium thinly fleshy over a hard inner shell, dark purple, 20–25 mm long by 15–20 mm thick, without an obvious crest or beak. Atherton Tableland and vicinity near Cairns, Queensland (Australia). Scattered in the canopy of coastal and montane rain forests; 0–700 m. Zone 10. Synonym: *Margbensonia disperma* (C. White) A. V. Bobrov & Melikian.

Atherton plum pine is a rare species from a restricted range in northeastern Queensland, where it grows near other rare podocarps, including doubleplum pine (*Podocarpus smithii*), brown pine (*P. grayae*), and *Prumnopitys ladei*. The more widely distributed plum pine (*Podocarpus elatus*) and broadleaf miro (*Prum-*

nopitys amara) are also found here. Atherton plum pine is on the IUCN (International Union for the Conservation of Nature) list of threatened and endangered conifers without a specific designated status because the information on its circumstances are so limited. It is the smallest tree among the podocarps in these forests, never emerging above the general level of the rain-forest canopy. It has the widest leaves, proportionately, among these species. Like them, it has just a single vein in the leaf, so the wings of water-conducting accessory transfusion tissue extending out from the midvein are important for both support and transport in the expanded leaf blade. Atherton plum pine is not closely related to the other Australian species but is most similar in general appearance to plum pine. It differs most obviously in its bright red podocarpium rather than the purple one that gives plum pine its common name. It also differs from the other podocarps with which it grows (and from most other species of *Podocarpus*) in having a predominance of cones with two seeds rather than just one (hence the scientific name, Latin for "two-seeded").

Podocarpus drouynianus F. J. Mueller

EMU GRASS, EMU BUSH

Podocarpus subg. *Foliolatus* sect. *Spinulosus*

Upright shrub to 2(–3) m, with numerous slender, sparingly branched, upright stems densely clothed with foliage and arising directly from a permanent, woody, underground stem and root mass (a lignotuber). Bark reddish brown, thin, stringy. Twigs light green, shallowly grooved between the attached leaf bases. Resting buds pointed, 2–4(–5) mm long and a little narrower, tightly wrapped by narrow bud scales with spreading, free tips. Leaves evenly spaced along and around the twig, projecting stiffly upward, light green above, waxy bluish green beneath (2–)5–12 cm long, 2–6 mm wide. Blades sword-shaped, often with a distinct curve to one side, widest near the middle but tapering only very gradually to the triangular tip usually with a nonprickly point and to the wedge-shaped base attached more or less directly to the twig without a petiole. Midrib only slightly raised above and much more prominently so beneath, with one resin canal beneath the midvein and an incomplete layer of hypodermis beneath the upper epidermis. Pollen cones 4–12 mm long and 3–5 mm wide, 1–6 on a branched or unbranched, scaly stalk 1–2.5 cm long. Pollen scales with a tiny, bluntly triangular, untoothed, upturned tip about 0.5 mm long and wide. Seed cones on a long, leafless stalk to 2 cm long, with a pair of basal bracteoles, these shed by maturity, the reproductive part with two or three unequal bracts, these and the axis becoming very swollen and juicy, purple with a waxy coating, nearly spherical, (12–)20–25) mm long. Fertile seed scales one (or two), the combined seed coat and epimatium thinly fleshy over a hard inner shell, purplish black with a thin waxy coating, 10–15(–19) mm long by 7–15 mm thick, without a crest or beak. Southwesternmost Western Australia, mostly southwest of a line between Bunbury and Albany. As scattered shrubs or forming thickets in the shrubby, often heathlike understory of open jarrah (*Eucalyptus marginata*) and karri (*E. diversicolor*) forests; 0–250(–300) m. Zone 9. Synonym: *Margbensonia drouyniana* (F. J. Mueller) A. V. Bobrov & Melikian.

This is the only podocarp in the Mediterranean climate region of southwestern Australia, or in all of Australia west of the Great Divide in the four eastern states for that matter, with the exception of a few isolated populations of Gray's plum pine (*Podocarpus grayae*) in Arnhem Land, Northern Territory. Like its close relative, creeping plum pine (*P. spinulosus*), of eastern Australia, emu grass tolerates the fires to which its habitat is prone. The underground lignotuber from which it sprouts new shoots after a fire is a feature found in many unrelated shrubs with which it shares these seasonally dry forests. Through this lignotuber, emu grass is known to live to 300 years even though individual trunks have a short life span. Emu grass and creeping plum pine are the only members of botanical section *Spinulosus.* They stand out from all other members of the subgenus *Foliolatus* in having only one resin canal beneath the midvein instead of the usual three found in all the other species. Their long, narrow, willow- or grasslike leaves (hence the common name) are unusual among Australian podocarps. The cut shoots of emu grass, usually about 80 cm long, with their attractive, stiff, light green leaves, can stay fresh in water for 2 weeks. As a result, emu grass has become an important element in the cut-flower trade for the export market. It is more useful for this than creeping plum pine because its branches average longer and its leaves are not prickly. The shoots are harvested from the wild, but there appears to be relatively little effect on the native populations because of the sprouting propensity of the shrubs. The species name honors Edouard Drouyn de Lhuys (fl. 1863–1870), Minister of Foreign Affairs for France, after he got Emperor Napoleon III to give the author of the species a knighthood in 1863, for his service to French science.

Podocarpus elatus R. Brown ex Endlicher

PLUM PINE, BROWN PINE, SHE PINE, ILLAWARRA PLUM

Podocarpus subg. *Foliolatus* sect. *Polystachyus*

Tree to 30(–40) m tall, with trunk to 0.7(–0.9) m in diameter. Bark dark grayish brown, fibrous, narrowly fissured between narrow flat-topped ridges or these becoming scaly. Crown deeply dome-shaped with thin, straight, upwardly angled branches bearing single or whorled branchlets moderately clothed with foliage near the tips. Twigs yellowish green, prominently grooved between the attached leaf bases. Resting buds pointed, 1.5–2.5 mm long, 0.7–1.5 mm thick, tightly wrapped by triangular bud scales with a projecting tip. Leaves well spaced along the twigs and partially rearranged out from the side, shiny dark green to yellowish green above, pale green beneath, (4–)5–12(–16) cm long (to 20 cm in juveniles), 6–12 mm wide (to 18 mm in juveniles). Blades flat or the margins slightly thickened, nearly parallel-sided or widest near the middle in short leaves, tapering abruptly to the hard-pointed tip and to the wedge-shaped base on a short petiole to 3 mm long. Midrib prominently raised above and slightly grooved to slightly raised beneath, usually with three resin canals beneath the midvein but sometimes with just one or with two extra ones embedded in the wings of water-conducting accessory transfusion tissue extending out to the sides, the upper leaf surface with a uniform layer of hypodermis above the photosynthetic tissue. Pollen cones (0.5–)2–3 cm long (elongating up to 5 cm after the pollen is shed) and (1–)2–3(–5) mm wide, (1–)2–4(–10) directly in the axils of foliage leaves. Pollen scales with a broadly triangular, upturned, pointed tip less than 1 mm long. Seed cones on a short, leafless stalk to 10 mm long, with a pair of basal bracteoles that may be shed before maturity, the reproductive part with two or three similar-sized bracts, these and the axis becoming very swollen and juicy, dark purple to black with a thin waxy coating, spherical or a little longer than wide, (10–)15–25(–30) mm in diameter. Fertile seed scales one (or two), the combined seed coat and epimatium purple with a waxy coating, with a thin, resinous flesh around a hard inner shell, (12–)15–20(–22) mm long by (8–)12–15 mm thick, slightly crested along one side. East coast of Australia, from Princess Charlotte Bay on the York Peninsula of Queensland to near Bateman's Bay in southeastern New South Wales. Scattered in the canopy of coastal and near-coastal rain forests, reaching its best development on deep soils near rivers; 0–1,000 m. Zone 9. Synonym: *Margbensonia elata* (R. Brown ex Endlicher) A. V. Bobrov & Melikian.

Plum pine is the commonest and most widespread tall, rain-forest podocarp in Australia (hence the scientific name, Latin for "lofty"). It is much more common and with a more continuous distribution in New South Wales than in Queensland, where, especially in the north, its occurrences are scattered and confined to somewhat drier sites than in the south. It has only moderate importance as a timber tree now that the primary stands are largely depleted. Its wood is employed in making furniture, turned objects, and dock piles, among other uses. The latter use arises from its resistance to wood-boring marine organisms, and the same chemicals deter termites and other land-based wood-borers. The leaves of the tree are also protected from insect feeding by producing ecdysterone, an insect hormone that prevents them from molting. On the other hand, the podocarpiums, as dispersal units that depend on animals for their distribution, lack such nasty chemicals and are even eaten by people as a traditional and contemporary food. Called Illawarra plums, they give rise to the common name plum pine and have a slightly resinous taste. The trees are also cultivated as handsome ornamentals in warm climates, with their large, bright green leaves and dense, dome-shaped crowns in open-grown trees of parks and yards. They germinate readily from fresh seeds, but these are not very tolerant of drying before and during germination.

Podocarpus elongatus (W. Aiton) L'Héritier ex Persoon

CAPE YELLOWWOOD, YELLOW PINE, FERN PINE, BREADE RIVER YELLOWWOOD, WESTELIKE GEELHOUT (AFRIKAANS)

Podocarpus subg. *Podocarpus* sect. *Podocarpus*

Shrub, or tree to 6(–24) m tall, usually multitrunked. Bark greenish gray to dark gray, generally smooth because of the small diameter of the trunks, becoming shallowly furrowed with age and sometimes peeling in long, thin strips. Crown broad and rounded, with numerous thin, ascending branches bearing slightly drooping branchlets. Twigs yellowish green, shallowly grooved between the attached leaf bases. Resting buds nearly spherical, about 2–3 mm in diameter, with tightly overlapping, narrowly triangular bud scales with spreading tips. Leaves sometimes nearly in pairs, more crowded near the ends of the trigs, stiff, grayish green with a thin waxy coating, (1.8–)3–6(–7) cm long (to 13 cm in juveniles), (3–)4–6(–9) mm wide (to 13 mm in juveniles). Blades straight sided in the middle third or widest near the middle, tapering gradually and then abruptly to the sharply triangular tip and to the very short petiole. Midrib slightly raised near the base above and very prominently so beneath, with one resin canal beneath the midvein and two at the center corners of the leaf, and, unlike all other species of *Podocarpus*, often with one to six evident (at 10× magnification) rows of stomates on the upper surface in addition to the stomatal bands beneath. Pollen cones (1–)1.5–2(–2.5) cm long (to 4 cm after the pollen is shed) and 3–5 mm wide, one to five directly in the axils of the leaves or at the tip of a short, leafless stalk up to 7 mm long. Pollen scales with a broadly and roundly triangular, minutely to coarsely toothed tip 0.5–0.6 mm long and 0.6–0.8 mm wide. Seed cones on short leafless stalk to 6(–13) mm long, without basal bracteoles, the reproductive part with two or three unequal scales, these and the axis becoming quite swollen and fleshy, turning bright red at maturity, 6–8 mm in diameter. Fertile seed scales one (or two), the combined seed coat and epimatium thin and leathery with resinous inclusions, dark green mottled with dark purple and covered with a thin layer of wax, 6–12(–14) mm long and nearly as wide, with a very weak crest along one side. Mountains of the western Cape region of South Africa. Scattered among other trees mostly on sandy soils, attaining best development along streams and rivers, stunted on drier slopes; 0–500 m. Zone 10.

Cape yellowwood served as the botanical type and only species of *Podocarpus* when the genus was first described in 1807. It was also the first podocarp species to be described botanically, when William Aiton included it as a species of yew (*Taxus*) in the first edition of *Hortus Kewensis* in 1789. The leaves were longer than those of any of the other yew species that he described, hence the scientific name. Three other podocarps were also described as species of yew before the genus *Podocarpus* was recognized as distinct. However, even though Charles L'Héritier included just Cape yellowwood in his new genus, it was so distinctive and so quickly accepted that only a single podocarp species after 1807 was described as a yew. This was in 1819, when only one other early species had yet been transferred to the new genus.

Oddly enough, Cape yellowwood is neither the most common nor most imposing yellowwood of South Africa even though it was the first one described. It is the smallest of the South African yellowwoods and very often remains shrubby throughout its life and hence is of little significance as a timber tree. Perhaps the shrubby habit made the heavy crops of seed cones, with their bright red, berrylike podocarpium beneath the seed, more conspicuous than those of its more widespread and lofty relatives, true yellowwood (*Podocarpus latifolius*) and common yellowwood (*Afrocarpus falcatus*). These fleshy cones make Cape yellowwood the only conifer listed as a host of the Mediterranean fruit fly, *Ceratitis capitata* (Wiedemann), in Florida. It is very similar to true yellowwood in most botanical features and is sometimes confused with it but has shorter leaves, and the frequent occurrence—in (0–)20–80% of the leaves of each tree—of one or more lines of stomates in deep grooves on the upper side of the leaves is unique with the genus. Internally, the other African species of *Podocarpus* usually have three resin canals below the midvein, while Cape yellowwood has just one.

Podocarpus gibbsiae N. Gray

GIBBS PODOCARP

Podocarpus subg. *Foliolatus* sect. *Longifoliolatus*

Tree to 20 m tall, with trunk to 0.2 m in diameter. Branches numerous, bearing mostly single or paired branchlets densely clothed with foliage. Twigs coarse, green, grooved between the attached leaf bases. Resting buds elongated, 4–6(–9) mm long by 1.5–2.5 mm in diameter, loosely constructed of narrowly triangular, upright bud scales, the outermost sometimes leaflike. Leaves somewhat openly standing straight out around the twigs, denser near their tips, thick and leathery, shiny dark green above, duller beneath, lasting (1–)2–3 years, 1–3(–5) cm long (to 9 cm in juveniles), 4–7 mm wide (to 9 mm in juveniles). Blades mostly straight, the edges slightly turned down, widest near or beyond the middle, tapering fairly abruptly to the rounded or roundly triangular tip and more gradually to the wedge-shaped base on a short petiole 1–3 mm long. Midrib broadly and shallowly raised above, especially near the base, almost flat beneath, with three resin canals beneath the midvein, thick wings of support tissue extending out to the sides, and a nearly continuous layer of hypodermis beneath the upper epidermis. Pollen cones 1–2.5 cm long, 2.5–4 mm in diameter, one (or two) directly in the axils of foliage leaves. Pollen scales with an upturned, triangular tip to near 1 mm long. Seed cones on a very short, leafless stalk to 3 mm or more long, with a pair of needlelike basal bracteoles 3–4 mm long, the reproductive part with two unequal bracts, these and the axis becoming red, swollen and juicy, 8–10 mm long. Fertile seed scale one, the combined seed coat and epimatium leathery over a hard inner shell, 8–10 mm long. Confined to northern Borneo on Mount Kinabalu in Sabah (Malaysia). Scattered in the understory of lower and upper montane rain forests and in the canopy of mossy forests on exposed ridges, usually on serpentine-derived soils; 1,200–2,400 m. Zone 9.

The very local Gibbs podocarp is not closely related to any of the other small-leaved species in the genus, most of which are members of subgenus *Podocarpus*. It is most similar, perhaps, to the neighboring but commoner and more widespread Pilger podocarp (*P. pilgeri*), which has much more crowded, somewhat thinner, more pointed leaves and somewhat shorter, tighter resting buds. The species name honors Lilian Gibbs (1870–1925), the English botanist who first collected herbarium specimens of this species on Mount Kinabalu in 1909 and who published a long account of the flora and vegetation of this mountain in 1914 as well as an extensive study of the development of seed cones in *Podocarpus* and allied genera (1912) based on material she collected there and in Fiji and New Zealand.

Podocarpus globulus de Laubenfels

SAPIRO PODOCARP, SAPIRO (MURUT)

Podocarpus subg. *Foliolatus* sect. *Globulus*

Tree to 27 m tall, with trunk to at least 0.2 m in diameter. Twigs scattered, slender, green, grooved between the attached leaf bases, densely clothed with foliage. Resting buds spherical 2–3 mm in diameter, tightly wrapped by the roundly triangular bud scales. Leaves densely and evenly standing out from and a little forward around and along the twigs, a little more crowded near the tips, lasting 2–3 years, leathery, shiny dark green above, duller and waxy beneath, 3.5–8(–10) cm long (to 16 cm in juveniles), 9–15 mm wide (to 24 mm in juveniles). Blades straight or very slightly curved to one side, bowing up on either side of the midrib, the edges curled down, fairly parallel-sided or inconspicuously widest near the middle, tapering fairly abruptly to a short, roundly triangular apex and to a roundly wedge-shaped base on a petiole 2–3 mm long. Midrib broadly and shallowly raised above, more prominently raised beneath, probably with three resin canals beneath the midvein and wings of support tissue extending out to the sides. Pollen cones 2.5–4.5 cm long and 3–4 mm wide, one (to three) directly in the axils of the foliage leaves or on a very short, leafless stalk to 1 mm long. Pollen scales with a tiny, upturned, triangular tip less than 0.5 mm long. Seed cones on a very short, leafless stalk 3–4 mm long, with a pair of needlelike basal bracteoles about 2 mm long that are soon shed, the reproductive part with two somewhat unequal bracts, these and the axis becoming red, swollen, and juicy, 8–9 mm long. Fertile seed scale one, the combined seed coat and epimatium leathery over a hard inner shell, 8–10 mm long by 5.5–6 mm thick, tipped by a small beak. Northern Borneo, in Sabah and Sarawak (Malaysia). Scattered in the understory of lower montane rain forests or occupying the canopy of mossy forests on ridges and peaks, at least sometimes on serpentine-derived soils; 300–1,500 m. Zone 9?

Sapiro podocarp was described, in part, from specimens that had been assigned to the closely related Teijsmann podocarp (*Podocarpus teysmannii*) or the more distant oleander podocarp (*P. neriifolius*). Like Teijsmann podocarp, it differs conspicuously from oleander podocarp in its spherical, tight, resting buds (hence the scientific name) rather than pointed ones with loose, upright scales. Teijsmann podocarp differs from sapiro podocarp in its larger leaves with a protracted point and in the round tip of the combined seed coat and epimatium, without any discernible crest or beak. While sapiro podocarp is never found with Teijsmann podocarp, which occurs farther west, it may grow with the more widely distributed oleander podocarp. In these cases, oleander podocarp is often a much larger tree and more frequently enters the canopy.

Podocarpus glomeratus D. Don

INTIMPA PODOCARP, INTIMPA, HUAMPO (QUECHUA AND SPANISH), PINO DE MONTE (SPANISH)

Podocarpus subg. *Podocarpus* sect. *Capitulatis*

Shrub, or tree to 12 m tall, with trunk to 0.8(–2.5) m in diameter. Bark dark brown to grayish brown, scaly, with shallow furrows. Crown conical at first, becoming irregularly dome-shaped, with numerous thin branches bearing clustered branchlets densely clothed with foliage. Twigs green, prominently grooved between the attached leaf bases. Resting buds nearly spherical, 5–7 mm in diameter, densely wrapped by broad, even leaflike bud scales with short free tips curling outward. Leaves held out stiffly all around and along the length of the branchlets, lasting about 2 years, dark grayish green and a little shiny above, whitish green with wax beneath, (1.5–)2–5(–6) cm long, (2–)3–4 mm wide. Blades straight or slightly curved to one side, nearly parallel-sided, the edges turned down, tapering fairly abruptly to the long-pointed, sharp tip and to the roundly wedge-shaped base with a very short petiole less than 3 mm long. Midrib grooved above, inconspicuously raised beneath, with one resin canal beneath the midvein, strongly developed support tissue extending out to the sides, and a continuous (or sometimes interrupted) layer of hypodermis beneath the upper epidermis. Pollen cones 5–10 mm long and 1.5–2 mm wide, three to six (to eight) at the end of a short, leafless stalk 8–12(–20) mm long (hence the scientific name, Latin for "clustered"). Pollen scales with a short, triangular, minutely jaggedly toothed, upturned tip less than 1 mm long. Seed cones on a short, leafless stalk 2–5 mm long, without basal bracteoles, the reproductive part with two to four bracts, these and the axis becoming swollen and juicy, bright red, nearly spherical, 5–6 mm in diameter. Fertile seed scales one (or two), the combined seed coat and epimatium thinly fleshy and resinous over a hard inner shell, dark green, nearly spherical, 5–6(–8) mm in diameter, with a very small beak and little sign of a crest along one side. Discontinuous in the Andes from central Ecuador through Peru to central Bolivia. Scattered as single trees or growing in small groves in cloud forests and other montane forests, including subalpine forests; 2,500–3,600(–4,000) m. Zone 9. Synonym: *P. cardenasii* J. Buchholz & N. Gray.

Intimpa podocarp is an uncommon species of the cloud forests of the inner Andean ranges that usually grows as a very densely branched small tree but that may become shrubby in exposed sites at higher elevations. Although too small to be harvested as a timber tree, it was depleted in many regions by cutting for firewood and is endangered throughout much of its range. Unfortunately, seed viability may be as little as 5%, so natural regeneration is sparse. It is very closely related to huapsay podocarp (*Podocarpus sprucei*), which is almost confined to Ecuador and usually occurs on the more coastal side of the Andes. In contrast to intimpa podocarp, huapsay podocarp typically has wider leaves that are pale green but not waxy beneath, pollen cones on a branched, scaly stalk, and seed cones on slightly longer stalks, with a slightly larger podocarpium and seed. These two podocarps could be treated as subspecies of a single species, and Pilger, in his 1903 monograph of Podocarpaceae, considered *P. sprucei* an outright synonym of *P. glomeratus* without any separate taxonomic status. Further work on their relationship is warranted.

Podocarpus gnidioides Carrière

NEW CALEDONIAN ALPINE PODOCARP

Podocarpus subg. *Podocarpus* sect. *Australis*

Spreading, heathlike shrub to 0.5(–2) m, without a single main trunk. Bark reddish brown, peeling in long, thin fibrous shreds to reveal yellowish brown underlying strips. Crown flat, with numerous thin upright branchlets from spreading, horizontal branches. Twigs yellowish green, soon turning brown, strongly marked by the short, attached leaf bases, densely clothed with stiff, upwardly pointing foliage. Resting buds nearly spherical, 1–2 mm in diameter, with one or two outer scales extending to 2 mm. Leaves lasting 6–7 years or more, densely spaced all along the twig, shiny dark green above, the whole lower surface waxy between the down-rolled margins, 0.8–2.2 cm long (to 2.5 cm in juveniles), 2–2.5 mm wide. Blades widest beyond the middle, tapering gradually and then very abruptly to the rounded tip and very gradually to the wedge-shaped, almost stalkless base, the lines of stomates confined to the lower side. Midrib grooved above and slightly raised beneath, with a single resin canal beneath the midvein and prominent support tissue extending out toward the sides. Pollen cones 0.7–1.5 cm long and 1.5–2.5 mm wide, single at the tip of a short, naked stalk to 4 mm long. Pollen scales with a smooth-edged, roundly triangular, upturned tip 0.6–0.7 mm long and about 1 mm wide. Seed cones on a very short, leafless stalk to about 1 mm long, without basal bracteoles, the reproductive part with two nearly equal bracts, these and the cone axis becoming very swollen, juicy, and bright red at maturity, 3–5(–7) mm long. Fertile seed scales one or two, the combined seed coat and epimatium 5–7 mm long, 3–5 mm in diameter, with a wrinkled surface and a distinct crest along one side. Southern New Caledonia south of the Baie de St. Vincent. Forming thickets in dense shrublands (maquis) on serpentine soils on mountain peaks and ridges; (600–)750–1,400(–1,600) m. Zone 9

This species is only distantly related to the other six *Podocarpus* species of New Caledonia, all of which are members of subgenus *Foliolatus*. Instead, it is closely related to snow totara (*P. nivalis*) of New Zealand and also to mountain plum pine (*P. lawrencei*) of southeastern Australia, of which it was first described as a botanical variety. Somewhat mysteriously, this first treatment described it as a tree when all modern studies of the flora of New Caledonia found it growing strictly as a shrub confined to the fire-prone maquis minière of droughty, poisonous, serpentine soils. Both snow totara and mountain plum pine occasionally reach tree size, so perhaps *P. gnidioides* rarely did so too in sheltered localities before the island was swept by the intense fires that accompanied the development of the mining industry there. The species name means "like *Gnidia,*" a genus of flowering plants, many species of which have a similar heathlike growth habit.

Podocarpus grayae de Laubenfels

BROWN PINE, GRAY'S PLUM PINE

Podocarpus subg. *Foliolatus* sect. *Rumphius*

Tree to 30 m tall, with trunk to 0.7 m in diameter. Bark thin, dark gray, fibrous, scarcely peeling in short strips. Crown dome-shaped with slender, rising branches densely clothed with foliage at the tips. Twigs green, prominently grooved between the attached leaf bases. Resting buds nearly spherical, 2–3 mm in diameter, tightly wrapped by numerous broad, keeled bud scales with short free tips. Leaves evenly spaced along and around the twigs, dark green above, paler beneath, 8–19 cm long (to 25 cm in juveniles), 7–16 mm wide (to 19 mm in juveniles). Blades roughly parallel-sided but widest near the middle and from there tapering very gradually and then more abruptly to the roundly triangular, pointed tip and to the wedge-shaped base with a distinct petiole 4–10 mm long. Midrib straight, slightly and broadly raised above, more sharply so beneath, with three resin canals beneath the midvein and a continuous layer of hypodermis beneath the upper epidermis. Pollen cones 2–3 cm long and 3–4 mm wide, (one to) three or four directly in the axils of foliage leaves or on a very short, leafless stalk to 1 mm long. Pollen scales with a tiny, upturned triangular tip to 0.2 mm long. Seed cones on a short, leafless stalk 6–9 mm long, with a pair of basal bracteoles just 1 mm long, the reproductive part with two to four unequal bracts, these and the axis becoming swollen and juicy, red, 9–12 mm long. Fertile seed scales one or two, the combined seed coat and epimatium with a thin fleshy layer over a hard inner shell, dark red, 10–15 mm long by 7–9 mm thick, without an obvious crest or beak. Northeastern coastal region of Queensland (Australia) northward from Townsend with outliers in Arnhem Land, Northern Territory. Scattered as canopy trees in rain forests back from the coast; 0–600 m. Zone 10.

Brown pine, although not as local as the other rain-forest podocarps of Queensland, is still not very well known. First described as a new species in 1985, it was long misidentified as oleander podocarp (*Podocarpus neriifolius*) but is actually more closely related to Rumpf podocarp (*P. rumphii*). The other three species of *Podocarpus* that grow in the same vicinity all have pointy resting buds in contrast to the spherical ones of brown pine. In addition, it differs from plum pine (*P. elatus*) in having a red rather than purple podocarpium, from doubleplum pine (*P. smithii*) in having three resin canals beneath the midvein rather than a single one, and from Atherton plum pine (*P. dispermus*) in having seeds only about half as long. The species name honors Netta Gray (1913–1970), who from 1948 to 1962, at first with J. Buchholz and continuing after his death, wrote a 13-part monograph of *Podocarpus* in the broad sense that was current until 1969, including species now put in seven additional genera.

The name is sometimes spelled *grayi* or *grayii,* but this masculine form, which was used in the original description, is treated as a mistake to be corrected without fuss under the rules of botanical nomenclature.

Podocarpus guatemalensis P. Standley

GUATEMALAN PODOCARP, ALCAPRO NEGRO, CUHAU, OCOTILLO, QUAHAU (SPANISH)

Podocarpus subg. *Podocarpus* sect. *Nemoralis*

Tree to 25(–36) m tall, with trunk to 1(–1.5) m in diameter. Bark reddish brown, smooth and scaly at first, becoming deeply ridged and furrowed on large trees. Crown broadly dome-shaped, with upwardly angled branches bearing numerous branchlets densely clothed with foliage. Twigs green, prominently grooved between the attached leaf bases. Resting buds sometimes clustered, roughly spherical or egg-shaped, (1–)2–3(–8) mm long and wide, tightly wrapped by broadly and roundly triangular bud scales sometimes with free bristle tips. Leaves evenly and openly arranged along and around the twigs, shiny dark green above, paler beneath, (3–)4–8(–10) cm long (to 14 cm in juveniles), 7–10 mm wide (to 16 mm in juveniles). Blades straight or slightly curved to one side, flat, the thin edges turned down, widest below the middle and tapering gradually to the triangular apex (narrowly triangular in juveniles) and more abruptly to the roundly wedge-shaped base with a short petiole 2–5 mm long. Midrib raised above and flat or faintly grooved beneath, with one resin canal beneath the midvein, wings of support tissue extending out to the sides, and an interrupted and irregularly multiple layer of hypodermis beneath the upper epidermis. Pollen cones 3–4 cm long and 3–4 mm wide, single directly in the axils of foliage leaves. Pollen scales with a minute, broadly triangular, upturned tip about 1 mm long or less. Seed cones on a short, leafless stalk 3–7 mm long, without basal bracteoles, the reproductive part with two unequal bracts, these and the axis becoming somewhat swollen and fleshy, red, 6–7(–11) mm long by 2–4.5(–8) mm thick. Fertile seed scale one, the combined seed coat and epimatium leathery over a hard inner shell, 6–9(–12) mm long by 4–6(–8) mm thick, with a low, conical beak. Discontinuous through the Caribbean lowlands of Central and South America, from southernmost Mexico to the vicinity of Caracas, Venezuela, and in the Pacific lowlands of Columbia. Scattered in gallery forests, swamps, rain forests, and cutover areas; 0–700 (–1,100) m. Zone 10. Synonym: *P. allenii* P. Standley.

There is generally a strong elevational separation between Guatemalan podocarp and olive-leaf podocarp (*Podocarpus oleifolius*), the two most widely distributed neotropical species of *Podocarpus,* which have similar overall distributions but are not closely related. The latter is usually an inhabitant of cloud forests, in contrast to the lowland habitat of Guatemalan podocarp. There are a few places where the two species can be found near each other, but they are readily distinguishable because olive-leaf podocarp has tighter resting buds and a groove along the midvein above. Guatemalan podocarp is more closely related to Trinidadian podocarp (*P. trinitensis*), which replaces it in similar habitats in Trinidad and differs most obviously in its grooved midvein.

Podocarpus henkelii Stapf ex Dallimore & A. B. Jackson

HENKEL YELLOWWOOD, BASTERGEELHOUT (AFRIKAANS), UM SONTI (XHOSA, ZULU), NINJULA (BANTU)

Podocarpus subg. *Podocarpus* sect. *Scytopodium*

Tree to 30(–35) m tall, with trunk to 1.4(–1.9) m in diameter. Bark shaggy, shallowly furrowed, greenish tan to gray. Crown conical when young, becoming roundly cylindrical with age, with numerous, gently ascending, slender branches upturned at the ends and bearing drooping foliage. Twigs pale green, shallowly grooved between the moderately elongated, attached leaf bases. Resting buds nearly spherical, with tightly overlapping bud scales. Leaves sometimes nearly in pairs, more crowded near the ends of the twigs, drooping, shiny dark green, (6–)9–15(–21) cm long, (3–)6–10(–13) mm wide. Blades distinctly widest near the middle or below, often slightly curved to one side, tapering gradually to the long-pointed tip and more abruptly to the very short petiole. Midrib slightly protruding above and strongly so beneath, with three resin canals beneath the midrib and two at the outer corners of the leaf. Pollen cones (1.2–)1.5–2.5 cm long (to 4.5 cm after the pollen is shed) and 4–6(–8) mm wide, one to five directly in the axils of the leaves or at the tip of a very short, leafless stalk less than 5 mm long. Pollen scales with a bluntly triangular, slightly toothed tip 0.5–1 mm long and about 1 mm wide. Seed cones on a short, leafless stalk to 6 mm long, without basal bracteoles, the reproductive part with two to four unequal bracts, these and the axis becoming very slightly swollen and leathery, 6–8 mm long by 4–6 mm thick. Fertile seed scales one (or two), the combined seed coat and epimatium swollen and resinous but rather tough and leathery, olive green with a thin waxy coating, (12–)15–25(–35) mm long by (10–)15–20(–28) mm thick, with a very weak crest along one side. Discontinuous in eastern and southern Africa, in northern Malawi and western Tanzania and in Natal and adjacent eastern Cape province of South Africa. Montane or coastal rain forests; (100–)1,400–2,300 m. Zone 10. Synonym: *P. ensiculus* Melville.

Like the closely related Madagascar yellowwood (*Podocarpus madagascariensis*), this species is readily distinguished from the other podocarps with which it grows by its long, drooping leaves.

Its occurrence in widely separated regions led to the description of trees from eastern Africa as a separate species from those in southern Africa, but trees from the two regions have only slight, inconsistent differences, and the segregate species is rarely recognized today. Henkel yellowwood has some similarities to common yellowwood (*Afrocarpus falcatus*), especially the leathery rather than fleshy texture of the bracts of the seed cones and the arrangement of at least some leaves nearly in pairs. On this account, many South African botanists and foresters were reluctant to accept the separation of *Afrocarpus* from *Podocarpus*, but even when *A. falcatus* was included in *Podocarpus* it was not considered closely related to *P. henkelii* and was placed in a separate subgenus. Close examination reveals that *P. henkelii* has far more in common with the species belonging to the genus *Podocarpus* in the restricted sense, particularly in the structures of the pollen and seed cones and of the needles. Thus the resemblance to *A. falcatus* does not appear to point to any particularly close relationship between the species. The species name honors John Henkel (1871–1962), a South African forest administrator and author of an influential book, *The Woody Plants of Natal and Zululand*, who recognized its distinctness. The tree bears heavy seed crops every year, but many of the seeds, especially those that mature earliest, are heavily infested with seed-eating insects. Viable seed is very susceptible to desiccation, but with adequate moisture it germinates readily without dormancy to quickly produce handsome, graceful, columnar young trees with needles even longer than those of adults. For this reason it is grown as an ornamental to a limited extent in its native range and beyond.

Podocarpus humbertii de Laubenfels

HUMBERT YELLOWWOOD

Podocarpus subg. *Podocarpus* sect. *Scytopodium*

Tree to 15 m tall. Twigs grooved between the slightly elongate attached leaf bases. Resting buds nearly spherical, about 2 mm long, with tightly overlapping bud scales spreading at their pointy tips. Leaves sticking out all around and fairly evenly spaced along the twigs, shiny dark green above, duller and paler beneath, 0.8–1.2(–2.5) cm long, 2.5–5.5(–8) mm wide. Blades widest at the middle and tapering in a gradual curve to the broadly triangular tip and narrowly wedge-shaped, short-stalked base, the edge slightly rolled under near the base. Midrib flat or slightly raised above and more strongly raised beneath, with three resin canals beneath the midvein and two at the outer corners of the leaf. Pollen cones 8–20 mm long and 2.5–3 mm wide, one or two at the tip of a short, smooth stalk to 6 mm long. Pollen scales with a wavy-edged, triangular, upright tip about 1 mm long and 0.8 mm wide. Seed cones on a very short, bare stalk to 3 mm long, without basal bracteoles, the reproductive part with two unequal bracts, these and the axis becoming slightly swollen and leathery, 3–5 mm long by 1.5–2 mm thick, the free tip of the larger bract leaflike. Fertile seed scale one, the combined seed coat and epimatium fleshly and about 15 mm long by 8–10 mm thick, crested along one side. Northern Madagascar on the Massif du Tsaratanana and nearby ranges. Short, wet forests of mountain ridges; 1,600–2,800 m. Zone 8?

The rare Humbert yellowwood is known from only a few localities in the high mountains of the northern tip of the island. Compared to its close neighbors and relatives, the leaves are much shorter than those of Capuron yellowwood (*Podocarpus capuronii*) and a little shorter on average than those of Laurent yellowwood (*P. rostratus*) but are much wider than those of the latter species. The leaves of saplings are about twice as large as those of adult plants, which have a rather heathlike, compact growth form similar to that of other subalpine species, like the moderately related alpine totara (*P. nivalis*) of New Zealand. The species name honors Jean-Henri Humbert (1887–1967), a French botanist who collected many plants in Madagascar, including the type specimen of this species.

Podocarpus insularis de Laubenfels

DALA PODOCARP, DALA, TUNUM, IDA-AYEBO (PAPUAN LANGUAGES)

Podocarpus subg. *Foliolatus* sect. *Foliolatus*

Tree to 30(–40) m tall, with trunk to 0.6 m in diameter. Crown narrowly dome-shaped. Twigs green, grooved between the attached leaf bases. Resting buds rounded, 2.5–3 mm long, a little wider, tightly constructed of narrowly triangular bud scales curling out at the tips. Leaves sticking out all around the twigs, shiny dark green above, duller, paler, and sometimes turning rusty beneath, 5.5–9 cm long (to 15 cm in juveniles), 7–9 mm wide (to 14 mm wide in juveniles). Blades generally straight, widest near the middle, tapering quickly to the roundly triangular tip and to the roundly wedge-shaped base on a short petiole 3–4 mm long. Midrib narrowly and sharply raised above, more broadly and shallowly raised beneath, with three resin canals beneath the midvein, wings of support tissue extending out to the sides, and an interrupted layer of hypodermis beneath the upper epidermis. Pollen cones 3–3.5 cm long and 3.5–4 mm wide, one to three directly in the axils of the foliage leaves or on a short stalk. Pollen scales with a broadly triangular, upturned tip less than 0.5 mm long. Seed cones on a short, leafless stalk 7–10 mm long, with two basal bracteoles 1.5–2 mm long, the reproductive part with two unequal bracts, these and the axis becoming swollen, juicy, and red, 8–10 mm long. Fertile seed scale one, the combined seed coat and epimatium leathery over a hard inner shell, 9–10 mm long by 6–7 mm thick, rounded, without an evident crest or beak. Scattered sparsely in the mountains through the center of eastern New Guinea from easternmost Irian Jaya (Indonesia) eastward through Papua New Guinea and on most of the larger islands east of New Guinea from New Britain and other islands of Papua New Guinea through the Solomon Islands to Vanuatu (hence the scientific name, Latin for "of the islands"). Scattered in the canopy of lower montane rain forests of slopes, ridges, and wet ground; 0–1,680 m. Zone 9.

Dala podocarp has leaves intermediate in size among those of its closest relatives and neighbors in Papua New Guinea. Leaves of the New Guinea endemic soa podocarp (*Podocarpus archboldii*), of higher elevations, and the much more widespread oleander podocarp (*P. neriifolius*), which extends even farther east out into the Pacific, to Fiji, are larger while those of blushing podocarp (*P. rubens*), also of higher elevations, are smaller. The species also differ among themselves in the shape of the leaf tip, the position of the widest point on the leaf, and the prominence and steepness of the midrib. The rusty color of the leaf undersurface as its dries in dala podocarp also sets it off from the other species whose waxy coatings remain a dull whitish green.

Podocarpus lambertii Klotzsch ex Endlicher

LAMBERT PODOCARP, PINHEIRINHO (PORTUGUESE), PIÑEIRIÑO (SPANISH)

Podocarpus subg. *Podocarpus* sect. *Capitulatis*

Tree to 20(–30) m tall, with trunk to 0.7(–1.2) m in diameter. Bark grayish brown, fibrous and furrowed, peeling in narrow strips. Crown dome-shaped, with numerous branches bearing many clustered branchlets densely clothed with foliage at the tips. Twigs green, prominently grooved between the attached leaf bases. Resting buds spherical, 2–2.5 mm in diameter, wrapped by tightly overlapping, broad, rounded bud scales, the outer ones sometimes with free, pointy tips. Leaves evenly distributed along and around the branchlets, sticking straight out or directed a little forward, glossy dark green above, paler and dull beneath, (2–)3–5(–6) cm long, 2–4(–6) mm wide. Blades straight or a little curved to one side, widest at or below the middle, tapering gradually and then more abruptly to a short, triangular, prickly (or rarely blunt) point and more abruptly to the roundly wedge-shaped base attached directly to the twig almost without a petiole, the leaf edges flat. Midrib grooved or flat above, broadly raised beneath, with one resin canal beneath the midvein, wings of support tissue extending out to the sides, and a continuous layer of hypodermis beneath the upper epidermis. Pollen cones (5–)8–15 mm long and 1.5–2.5 mm wide, (one to) three to six at the end of a short, leafless stalk (5–)10–15 mm long. Pollen scales with a broadly triangular, hardly toothed, upturned tip about 0.5 mm long or less. Seed cones on a short, leafless stalk less than 1 cm long, without basal bracteoles, the reproductive part with three (or four) bracts, these and the axis becoming swollen and juicy, reddish purple, roughly spherical, (4–)5–8 mm in diameter. Fertile seed scales one (or two), the combined seed coat and epimatium spherical, slightly fleshy over an inner shell, 6–10 mm long, reddish green, with a weak crest along one side and forming a low, rounded beak, the whole becoming wrinkled past maturity. Eastern uplands of southern Brazil and adjacent Misiones province in northeastern Argentina. Scattered as individual trees in closed mixed forests or open woodlands, sometimes with Paraná pine (*Araucaria angustifolia*); (200–)500–1,000(–2,000) m. Zone 9. Synonym: *P. transiens* (Pilger) de Laubenfels ex Silba.

Lambert podocarp has the most southerly distribution among the *Podocarpus* species of eastern South America. It is closely related to Bolivian cerro podocarp (*P. parlatorei*) of the eastern slopes and foothills of the Andes, to willow podocarp (*P. salignus*) of Chile, and to the other species in section *Capitulatis*, with their occurrence in more seasonal environments than the rain-forest habitats characteristic for the majority of New World *Podocarpus* species. The seasonal habitats have historically been more accessible than the rain forests, and so all the species in this section were described by early in the 20th century while new rain-forest species were still being described at its end. Lambert podocarp grows over a large area of southern Brazil and has a fair amount of variation. Trees with larger leaves than usual also lack the prickly leaf tips found in most members of the species and have been separated as a botanical variety or even an independent species. They grow within the range of the more common forms, however, and have no differences other than those associated with leaf size and so are given no formal taxonomic recognition here. The species name honors Aylmer B. Lambert (1761–1842), whose *Description of the Genus Pinus* of 1824 is one of the most comprehensive early treatises on the conifers.

Podocarpus latifolius (Thunberg) R. Brown ex Mirbel

TRUE YELLOWWOOD, REAL YELLOWWOOD, OPREGTE GEELHOUT (AFRIKAANS), UMKHOBA (XHOSA AND ZULU), MUSENGERA (KIKUYU)

Podocarpus subg. *Podocarpus* sect. *Podocarpus*

Tree to 25(–35) m tall (or a stunted shrub as little as 2 m on dry, exposed slopes), with trunk to 1.2 m in diameter. Bark brown to grayish brown, smooth at first and peeling in flakes, becoming shallowly longitudinally furrowed between narrow strips. Crown shallowly dome-shaped, relatively small in forest grown trees, with thin, gently ascending branches, densely clothed at the tips with stiff, upright foliage. Twigs green, round to slightly angled, shallowly grooved between the slightly elongated, attached leaf bases. Resting buds nearly spherical, 2–3.5 mm in diameter, with tightly overlapping bud scales or the upper scales curled outward. Leaves sometimes nearly in pairs, often crowded near the ends of the twigs, rigid, shiny dark green or a little waxy, rarely with a few individual stomates on the upper surface but these never arranged in lines, (2–)4–8(–10) cm long (to 20 cm long in juveniles), 5–10(–14) mm wide. Blades nearly parallel sided for most of their length, straight or sometimes slightly curved to one side, tapering abruptly to the short, broadly triangular tip and to the shortly stalked base. Midrib slightly raised to slightly grooved above and prominently raised beneath, with three (to five) resin canals beneath the midvein and two at the outer corners of the leaf, very rarely with up to 11 additional resin canals scattered in the leaf tissue. Pollen cones 1–2.5 cm long (to 3.5 cm after the pollen is shed) and 2.5–4.5(–6) mm wide, pinkish, one (or two) directly in the axils of foliage leaves or at the tip of a very short, leafless stalk less than 5 mm long. Pollen scales with a roundly triangular, jaggedly toothed upturned tip about 0.6 mm long and broad. Seed cones on a slender, leafless stalk (0.2–)0.5–1(–2) cm long, without basal bracteoles, the reproductive part with two or three unequal bracts, these and the axis becoming greatly swollen and fleshy in a nearly spherical pink to deep purple red podocarpium 8–11(–14) mm in diameter. Fertile seed scales one (or two), the combined seed coat and epimatium leathery surrounding a hard, resinous shell, dark purplish black or somewhat waxy, 7–12(–14) mm long by 6–8(–12) mm thick, with a prominent crest along one side. Discontinuous in the mountains of sub-Saharan Africa, from southwestern Nigeria and northern Cameroon in the northwest and southernmost Sudan in the northeast to the southern coast of South Africa and the Cape of Good Hope. Most prominently forming pure stands or mixed with other species in wet montane high forests but also found as scattered shrubby trees on dry slopes; (0–)900–2,200(–2,600) m. Zone 10. Synonyms: *P. latifolius* var. *latior* Pilger, *P. milanjianus* Rendle, *P. thunbergii* W. J. Hooker, *P. ulugurensis* Pilger.

Originally treated as separate species in different portions of its range, *Podocarpus latifolius* is the most widespread podocarp in Africa as well as in southern Africa, where it is the most abundant yellowwood species and reaches its best development, as in the tall rain forests of Cape province and KwaZulu-Natal (South Africa). Although there is considerable variation in leaf size and shape, this variation is more closely tied to environmental conditions than to geographic distribution, so a single species without botanical varieties is now recognized. The species has also been confused, both taxonomically and in name, with both Cape yellowwood (*P. elongatus*) and common yellowwood (*Afrocarpus falcatus*). It has been supposed to hybridize with the former, but there is no real evidence for this. Whatever their length, the leaves are generally proportionately broader than those of the other African podocarps at the same developmental stage, hence the scientific name, Latin for "broad leaf."

True yellowwood is an important timber tree but has been severely depleted throughout most of its range by overexploitation. Although the wood is very attractive, its spiral grain causes it to warp if not carefully dried, and it is difficult to work, so an enormous quantity was used to make railway ties. These had to be replaced frequently unless they were creosoted because the timber is not durable in contact with the soil. Because of true yellowwood's relatively slow growth, plantation culture was never effectively undertaken with this species, and softwood lumber in southern and eastern Africa is now largely supplied by plantations of exotic pine species, some of which have also become invasive pests. Heavy seed crops of true yellowwood are only borne every 2–3 years, and even then much of the seed is destroyed by insect predation. Viability is lost quickly as the podocarpium dries and shrivels, but germination is rapid and high in fresh seed that has not been burrowed into by seed predators. If the trees can survive the cumulative hazards of youth, they can live 500 years. True yellowwood is only cultivated in horticulture to a limited extent and then primarily in South Africa. Fossil pollen records show that true yellowwood was once more widely distributed in West Africa than it is today. During a particularly wet period from about 3,700 to 3,400 years ago, it was found on many mountains from which it is now absent. It became restricted to its present sites by about 2,000 years ago after a period of progressive drying lasting about 1,000 years.

Podocarpus laubenfelsii Tiong

DE LAUBENFELS PODOCARP

Podocarpus subg. *Foliolatus* sect. *Rumphius*

Tree to 20(–35) m tall, with trunk to 0.6 m in diameter. Bark light brown, smooth, thin. Crown dome-shaped, with upwardly angled branches bearing paired or clustered branchlets well clothed with foliage. Twigs slender, green, grooved between the attached leaf bases. Resting buds rounded but a little longer than broad, 2.5–4 mm long, tightly clothed by triangular bud scales with spreading tips. Leaves openly and evenly standing out along and around the twigs, leathery, dark green above and paler beneath, (7–)12–19 cm long (to 24 cm in juveniles), 10–18 mm wide (to 24 mm in juveniles). Blades straight or a little curved to one side, especially at the tip, the edges curled under, fairly parallel-sided but inconspicuously widest below the middle, tapering very gradually and then quite abruptly to the prolonged, narrowly triangular tip and rather abruptly to the rounded base on a petiole 6–14 mm long. Midrib broadly raised above and beneath, probably with three resin canals beneath the midvein and possibly with a continuous layer of hypodermis beneath the upper epidermis. Pollen cones 2–4 cm long and 2.5–3.5 mm wide, three to five directly in the axils of the foliage leaves or, more frequently, on a short, leafless stalk to 6 mm long. Pollen scales with a triangular, upturned tip about 0.2 mm long. Seed cones on a leafless stalk 10–16 mm long, with two needlelike basal bracteoles about 1.5 mm long, the reproductive part with two slightly unequal bracts, these and the axis becoming swollen, juicy, and red, about 9 mm long. Fertile seed scale one, the combined seed coat and epimatium leathery over a hard inner shell, 8 mm or more long. Mountains of northern Borneo in Sabah, northern Sarawak (Malaysia), and East Kalimantan (Indonesia). Scattered in the canopy or emergent above primary lower montane rain forests on a range of aspects and substrates, from sandy soils to peaty ones; (600–)700–1,600 m. Zone 9?

This rare and local species is closely related to Rumpf podocarp (*Podocarpus rumphii*), differing most obviously in the presence of a short stalk at the base of most pollen cone clusters. It also occurs at higher elevations than the usual range for *P. rumphii.* Despite the general rarity of this tree, it is often noticed by botanical visitors to Borneo because it grows near the headquarters of Mount Kinabalu National Park in Sabah. The species name honors David de Laubenfels (b. 1925), the American botanist who collected the type specimen at the Kinabalu headquarters and whose reevaluation of the classification of podocarps in Malesia launched the contemporary interpretation of genera in the family, including *Podocarpus,* of which he published a revision in 1985.

Mature shrub of mountain plum pine (*Podocarpus lawrencei*) in nature.

Podocarpus lawrencei J. Hooker

MOUNTAIN PLUM PINE, TASMANIAN PODOCARP, MOUNTAIN YEW

Podocarpus subg. *Podocarpus* sect. *Australis*

Spreading or upright shrub to 1(–4) m, without a single main trunk, or sometimes a tree to 6(–20) m tall, with a short, crooked trunk to 0.3(–0.7) m in diameter. Bark smooth to slightly furrowed, fibrous, peeling in thin strips. Crown variable, completely prostrate and hugging and spreading over rocks to dome-shaped, with slender, spreading to rising branches bearing somewhat irregularly distributed, stiffly spreading or upright branchlets densely clothed with foliage. Twigs yellowish green to waxy bluish green, visibly grooved between the attached leaf bases. Resting buds nearly spherical, 1–2 mm in diameter, with tightly overlapping scales with spreading tips. Leaves evenly distributed along and stiffly radiating around the twigs, lasting 3 or more years, shiny dark green or sometimes a bluish green from wax above, grayish green and with stomates confined to broad, waxy bands on either side of the midrib beneath, (0.4–)0.6–1.2(–1.6) cm long, (1–)2–3(–4) mm wide. Blades nearly parallel-sided in the middle portion, tapering abruptly to the hard, rounded tip and gradually to the wedge-shaped base, nearly stalkless, straight or slightly curved to one side, the margins usually flat. Midrib flat or slightly grooved above, broadly raised beneath, with a single resin canal beneath the midvein and with prominent support tissue extending to the sides into the leaf tissue. Pollen cones (3–)4–7(–12) mm long and 3–5 mm wide, (one to) three to six at the end of a short, leafless stalk to 6(–10) mm long, dull red. Pollen scales with an upturned, smooth-edged, broadly triangular tip about 1 mm long and wide. Seed cones on a short, leafless stalk to about 3 mm long, without basal bracteoles, the reproductive part with two nearly equal bracts, these and the axis becoming very swollen, juicy, and bright red at maturity, (3–)4–6(–7) mm long. Fertile seed scales one (or two), the combined seed coat and epimatium leathery and wrinkled, greenish black, 3–5 mm long, with a prominent crest along one side and over the top that gives it a beaked or pear-shaped look. Southeastern Australia, from southeastern New South Wales through eastern Victoria to Tasmania. Mostly sheltered by and smothering rocks in alpine boulder fields and on talus slopes but also found in wet sclerophyll forests at lower elevations; (500–)900–1,800(–2,000) m. Zone 7. Synonym: *P. alpinus* R. Brown ex J. Hooker.

Mountain plum pine is a characteristic species of the alpine zone of the Snowy Mountains of southeastern mainland Australia and of various Tasmanian mountain ranges, spreading to form broad, thin blankets that cascade over boulders like green sheets. Even when growing in more sheltered habitats at lower elevations, mountain plum pines are still usually dense, upright shrubs

1–2 m tall. Nonetheless, trees do occur and the largest ones, some 20 m tall and 70 cm in diameter are found in eastern Victoria near the border with New South Wales. Despite its small size, mountain plum pine can live for more than 450 years and often reaches 200 years. It is one of the hardiest of the podocarps and is very slow growing in cultivation, maintaining a shrubby habit for a century or more. A few cultivars have been selected.

The species was long known as *Podocarpus alpinus* and is usually cultivated under that name. This was the earliest name applied to the species (in 1825), but unfortunately, the name did not meet the botanical requirements for publication until after the species was redescribed as *P. lawrencei* in 1845. There has been some confusion surrounding the use of the latter name, and some authors applied it to the somewhat related New Zealand species usually (and here) known as *P. acutifolius.* The application to mountain plum pine seems well founded, however, so the species is properly known as *P. lawrencei.* This species name honors Robert Lawrence (1807–1833), Tasmania's earliest resident plant collector, who collected the type material in Tasmania in 1831, just 2 years before he died by misadventure at an early age, as did so many of the famous collectors of the time. The species is only distantly related to the six other Australian *Podocarpus* species but, instead, is closely related to New Caledonian alpine podocarp (*P. gnidioides*) and to snow totara (*P. nivalis*) of New Zealand, neither of which is quite as dwarf as mountain plum pine.

Podocarpus ledermannii Pilger

LEDERMANN PODOCARP, SUA, BABAKO, NELIL (PAPUAN LANGUAGES)

Podocarpus subg. *Foliolatus* sect. *Acuminatis*

Tree to 25(–35) m tall, with trunk to 0.6 m in diameter. Bark smooth, gray, flaking in scales. Crown irregularly and openly dome-shaped, with sparse, thick branches bearing pairs of clustered branchlets. Twigs slender, green, grooved between the attached leaf bases. Resting buds elongate, pointed, 4–8(–10) mm long, 3–4 mm in diameter, loosely constructed of upright to spreading, narrowly triangular bud scales. Leaves openly and evenly arranged around and along the twigs or more concentrated near their tips, leathery, thin but stiff, (6–)10–20(–22) cm long, (12–)16–24(–28) mm wide. Blades straight or curved to one side, indistinctly widest near the middle or below, tapering gradually and then abruptly to a roundly triangular apex which usually has a prolonged tip and to a roundly wedge-shaped base on a petiole 3–5(–10) mm long. Midrib very prominently raised above and more shallowly and broadly raised beneath, with three resin canals beneath the midvein and rarely with two more embedded in the wings of support tissue extending out to the sides, the layer of large hypodermis cells beneath the upper epidermis usually incomplete. Pollen cones 4–5 cm long and 3–4 mm wide, (one to) three on a short, leafless stalk 1–4(–12) mm long. Pollen scales with an upturned, triangular tip less than 1 mm long. Seed cones on a short, leafless stalk (4–)6–9(–15) mm long, with two needlelike basal bracteoles 2–2.5 mm long, the reproductive part with two unequal bracts, these and the axis becoming swollen, juicy, and red, 9–16 mm long. Fertile seed scale one, the combined seed coat and epimatium leathery over a hard inner shell, 11–13 mm long by 9–10 mm thick, rounded at the tip without a beak. Scattered throughout New Guinea and in western New Britain. Generally confined to the understory or lower canopy of lowland and lower montane rain forests; 0–1,800(–2,200) m. Zone 9. Synonym: *P. idenburgensis* N. Gray.

Like the other *Podocarpus* species restricted to New Guinea, Ledermann podocarp is rather poorly known and often confused with its neighbors. Like the more widely distributed oleander podocarp (*P. neriifolius*), it is a tree of primary rain forests at modest elevations. The species name honors Carl Ledermann (1875–1958), the German botanist who collected the type specimen in Papua New Guinea on the German Sepik expedition of 1912–1913 and whose original set of specimens, left over from those that he himself destroyed in 1930, are housed in the Berlin herbarium.

Podocarpus levis de Laubenfels

MARISA PODOCARP, MARISA, SANRU (MALAY), WASIWARARE (PAPUAN LANGUAGES)

Podocarpus subg. *Foliolatus* sect. *Foliolatus*

Tree to 25(–35) m tall, with trunk to 0.4 m in diameter. Bark smooth, brown, peeling in flakes. Twigs green, grooved between the attached leaf bases. Resting buds egg-shaped, 3–9 mm long by 3–4 mm across, tightly wrapped by triangular bud scales, the outer ones spreading at their tips. Leaves sticking out evenly around the twigs, leathery and stiff, shiny dark green above, paler beneath, 8–14 cm long (to 20 cm in juveniles), 10–14 mm wide (to 15 mm in juveniles). Blades straight or curved to one side, inconspicuously widest below the middle, tapering gradually and then abruptly to the pointed or rounded tip and more abruptly to the roundly wedge-shaped base on a petiole 4–9 mm long. Midrib flat to slightly raised or channeled above, more broadly and prominently raised beneath, with three resin canals beneath the midvein, often with another one within the wings of support tissue extending out on each side, and with a continuous layer of hypodermis beneath the upper epidermis. Pollen cones 3–6(–8) cm long and 3–4.5 mm wide, one to three directly in the axils of the foliage leaves. Pollen scales with a broadly triangular, upturned tip less than 0.5 mm long. Seed cones on a short, leafless stalk to 10 mm long, with two basal bracteoles 2–4 mm long, the reproductive part with two or three bracts, these and the axis becoming swollen, juicy, and red, 5–6 mm long. Fertile seed scales one or two, the combined seed coat and epimatium leathery over a hard inner shell, (8–)12–13 mm long by 7–9 mm thick, rounded at the tip, without an evident crest or beak. Scattered discontinuously through northeastern Indonesia from northeastern Kalimantan (Borneo) through Sulawesi and Ambon to northeastern Irian Jaya (New Guinea). Scattered singly or gregariously in the canopy or subcanopy of lowland and lower montane rain forests; 0–1,650 m. Zone 9?

The continuous hypodermis (the extra layer of compact sclereids beneath the epidermis) of marisa podocarp and a few other species, like Rumpf podocarp (*Podocarpus rumphii*), give the upper side of the leaf, including the midrib, a smooth surface (hence the scientific name, Latin for "smooth") in contrast to the wrinkled surface found in the dried leaves of the many related species with only a partial layer of hypodermis. The long pollen cones and the common occurrence of two seeds in the seed cones are also minority conditions among the *Podocarpus* species of the region, especially in combination with a rounded leaf tip. Nonetheless, it is not a particularly distinctive species and its status merits further consideration.

Podocarpus longifoliolatus Pilger

MOUNT MOU PODOCARP

Podocarpus subg. *Foliolatus* sect. *Longifoliolatus*

Tree to 15(–20) m tall, with trunk to 0.5 m in diameter. Bark longitudinally furrowed, reddish brown to dark brown, weathering grayish brown. Twigs green, coarse, prominently grooved between the attached leaf bases and with prominent bud scars at intervals. Resting buds elongate, 6–12 mm long, loosely constructed of leaflike bud scales, the outer ones longest and curled back away from the bud. Leaves densely and evenly arranged along and around the twigs, extending straight out or a bit forward, bright green and a little shiny above, duller beneath, leathery, lasting 2–3 or more years, (5–)6–8(–10.5) cm long (to 14 cm in juveniles), 6–8 mm wide (to 11 mm in juveniles). Blades straight or a little curved to one side, the edges turned down, fairly parallel-sided in the middle portion, tapering gradually and then more abruptly to the roundly triangular tip, sometimes with a small prickle, and also tapering very gradually to the narrowly wedge-shaped base on a short, round petiole less than 10 mm long. Midrib broadly raised above and raised within a rimmed channel beneath, with three resin canals beneath the midvein, four above it, well-developed masses of support tissue extending out to the sides, and an interrupted, thin layer of hypodermis beneath the upper epidermis. Pollen cones 1.5–2.5 cm long and 2.5–3.5 mm wide, one (to three) directly in the axils of foliage leaves or on a short, leafless stalk to 5 mm long, the basal bud scales leaflike and curled out. Pollen scales with a broadly triangular, upturned tip about 1 mm long and wide. Seed cones on an angled, leafless stalk 9–17 mm long, with a pair of needlelike, down-curved, basal bracteoles 3.5–5 mm long, the reproductive part with three or four nearly equal bracts, these and the axis becoming swollen and juicy, bright red, 6–8 mm long by 3.5–4.5 mm thick. Fertile seed scales (one or) two (or three), the combined seed coat and epimatium leathery over a hard inner shell, waxy bluish green, 8–9 mm long by 5–7 mm thick, with a low beak that becomes less prominent with maturity. Known from just a few widely scattered mountains in central and southern New Caledonia. Scattered in the canopy or subcanopy of upper montane rain forests of low stature on serpentine-derived soils; (1,000–)1,100–1,200(–1,450) m. Zone 9.

Mount Mou podocarp replaces its close relative Mé Maoya podocarp (*Podocarpus polyspermus*) at higher elevations. The two species are very similar, but *P. polyspermus* has grayish green, generally larger leaves, straight bud scales, shorter pollen cone bud scales, and pollen scales with a longer, narrower, incurved tip. *Podocarpus longifoliolatus* appears to be rare overall but can be common in some localities, as on the summit of Mount Mou near Nouméa, where it has been collected repeatedly since the mid-19th century. Although the species name is Latin for "long-leaved," its leaves are no longer than those of svelte river podocarp (*P. novae-caledoniae*), the species with which the type specimens were filed before they were described as a separate species. The extra resin canals above the midvein are unusual within *Podocarpus* (they are not found in *P. novae-caledoniae,* for instance) but are one of the features distinguishing section *Longifoliolatus* from other sections in subgenus *Foliolatus,* as first noticed by Netta Gray in 1955.

Podocarpus lucienii de Laubenfels

LUCIEN PODOCARP

Podocarpus subg. *Foliolatus* sect. *Globulus*

Tree to 15 m tall. Bark thickly ridged and furrowed, pale brown to yellowish brown, weathering gray. Twigs clustered, green, slender, prominently grooved between the elongate leaf bases. Resting buds nearly spherical or a little elongate, 1–2 mm in diameter, tightly enclosed by the rounded bud scales. Leaves evenly distributed along and around the twigs, lasting 1–2 years, leathery, grayish green above with a silvery cuticle, paler beneath, 6–11(–14) cm long (to 18 cm in juveniles), 10–14(–16) mm wide (to 20 mm in juveniles). Blades straight, the thin edges gently down-curved to a little curled under, fairly parallel-sided, tapering abruptly at the ends to a semicircular tip and a roundly wedge-shaped base on a short petiole 1.5–6 mm long. Midrib broadly and inconspicuously raised above and narrowly and shallowly grooved beneath, with three resin canals beneath the midvein and well-developed wings of support tissue extending out to the sides. Pollen cones 15–20 mm long and 3–4 mm wide, three at the end of a short, leafless stalk 1–2 mm long. Pollen scales with a bluntly triangular, upturned tip 0.3–0.5 mm long. Seed cones on a short, leafless stalk (6–)9–12(–17) mm long, with two triangular basal bracteoles 1–2 mm long, the reproductive part with two unequal bracts, these and the axis becoming red, somewhat swollen, and juicy, (6.5–)9–11 mm long by 3.5–6 mm thick. Fertile seed scale one, the combined seed coat and epimatium somewhat fleshy over a hard inner shell, 14–20 mm long by 9–12 mm thick, with an inconspicuous, rounded crest at the tip. Scattered very discontinuously along the length of New Caledonia. Locally common in the understory of rain forests; (50–)200–900(–1,100) m. Zone 9.

Lucien podocarp is very closely related to kauri podocarp (*Podocarpus sylvestris*), from which it differs in its strikingly broader, duller green leaves and its much larger seeds. It is found in much the same habitat as false kauri but is less common and does not reach the largest dimensions of that species, so it is not exploited as a timber tree. Its name honors René Lucien, who was the proprietor of a lumber operation in southern New Caledonia and guided de Laubenfels in the northern Rivière Bleue region when he collected the type specimen.

Podocarpus macrocarpus de Laubenfels

LUZON PODOCARP

Podocarpus subg. *Foliolatus* sect. *Polystachyus*

Tree to 10(–20) m tall, with trunk to 0.3 m in diameter. Twigs paired or clustered, slender, green, prominently grooved between the attached leaf bases. Resting buds spherical to cylindrical, 2–4 mm in diameter, tightly or more loosely wrapped by upright to spreading triangular bud scales. Leaves shiny dark green above, paler and dull with wax beneath, leathery, thick, lasting 1–2 years, 6–10 cm long (to 15 cm in juveniles), 8–12 mm wide (to 14 mm in juveniles). Blades straight or somewhat curved to one side, the edges turned down, widest near the middle, tapering very gradually to the roundly triangular tip and to the roundly wedge-shaped base on a petiole 2–4 mm long. Midrib slightly but distinctly raised above and more broadly raised beneath, sometimes with a central groove, with three resin canals beneath the midvein, wings of support tissue extending out to the sides, and a nearly complete layer of hypodermis beneath the upper epidermis. Pollen cones about 2.5 cm long, (one or) two to four directly in the axils of the foliage leaves. Pollen scales with a triangular, upturned tip much less than 1 mm long. Seed cones on a short, leafless stalk 3–12 mm long, with a pair of needlelike basal bracteoles 1.5–2 mm long, the reproductive part with two unequal bracts, these and the axis becoming swollen, juicy, and purple, 10–12 mm long by 5–6 mm thick. Fertile seed scale one, the combined seed coat and epimatium leathery over a hard inner shell, 15–17 mm long by 10–12 mm thick (hence the scientific name, Greek for "big fruit"), rounded at the tip, without an evident crest or beak. Confined to mountains of northern Luzon (Philippines). Scattered in the understory of cloud forests; (100?–)2,000–2,100 m. Zone 9?

Like many other local *Podocarpus* species, Luzon podocarp was long confused with the more widely distributed oleander podocarp (*P. neriifolius*), which also occurs in Luzon. Oleander podocarp often has an extended leaf tip not found in Luzon podocarp, it usually has only three pollen cones in each cluster, and it has smaller seeds, usually 8–12 mm long. Luzon podocarp can be common within its limited range, and this, combined with its small size, means that it is not of immediate conservation concern despite its restricted distribution. It does not appear to be cultivated, even within its natural range.

Podocarpus macrophyllus (Thunberg) R. Sweet

BUDDHIST PINE, BIGLEAF PODOCARP, SOUTHERN YEW, LUO HAN SONG (CHINESE), NA HAN SHONG (KOREAN), KUSAMAKI (JAPANESE)

Plates 56 and 57

Podocarpus subg. *Foliolatus* sect. *Polystachyus*

Tree to 20 m tall or long remaining shrubby, with trunk to 0.6 m in diameter. Bark smooth, grayish brown, flaking in scales and becoming shallowly furrowed with age. Crown conical at first, remaining narrow and cylindrical or, more frequently, becoming irregularly dome-shaped and open, with numerous slender, horizontal to upwardly angled branches bearing paired or clustered twigs densely to somewhat openly clothed with foliage. Twigs thin, green to brown, prominently grooved between the attached leaf bases. Resting buds egg-shaped, 2–3 mm long and 1.5–2 mm in diameter, loosely constructed of upright, triangular bud scales with narrowed tips. Leaves pointing out and a little forward all around the twigs, often denser near the tips, leathery, lasting 2–3 years, shiny dark green (or bluish green or yellowish green) above, paler grayish green and duller beneath, (0.8–)3–10(–12) cm long (to 18 cm in juveniles, hence the scientific name, Greek for "big leaf"), (1–)4–10(–12) mm wide (to 14 mm in juveniles). Blades straight or slightly curved to one side, the edges flat or thickened, fairly parallel-sided or distinctly widest near, above, or below the middle, tapering abruptly to a narrowly triangular (or broadly triangular or even rounded) tip and more gradually to a narrowly wedge-shaped base on a petiole 2–4 mm long. Midrib prominently raised above and beneath, with three resin canals beneath the midvein, wings of support tissue extending

Twig of Buddhist pine (*Podocarpus macrophyllus*) with three episodes of growth, ×0.5.

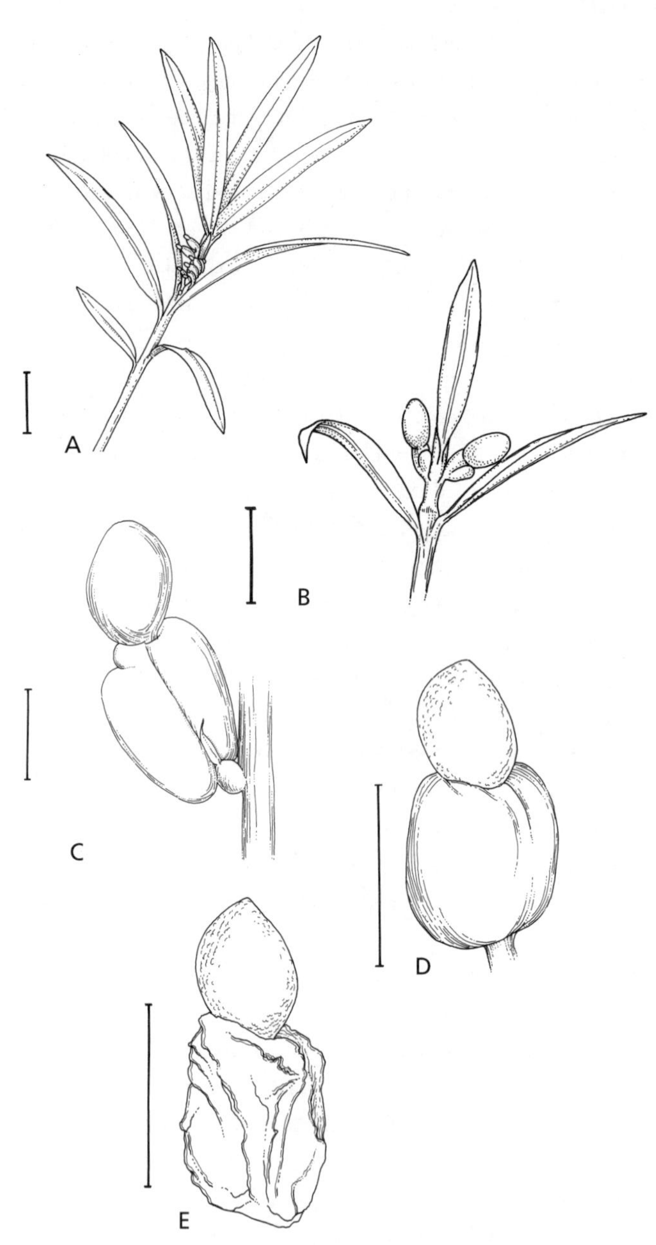

Developmental sequence of seed cones of Buddhist pine (*Podocarpus macrophyllus*) from shortly after pollination (A), through growth (B, C), to full ripeness (D), and then shriveling of the podocarpium (E); scale, 1 cm, except 5 mm in C.

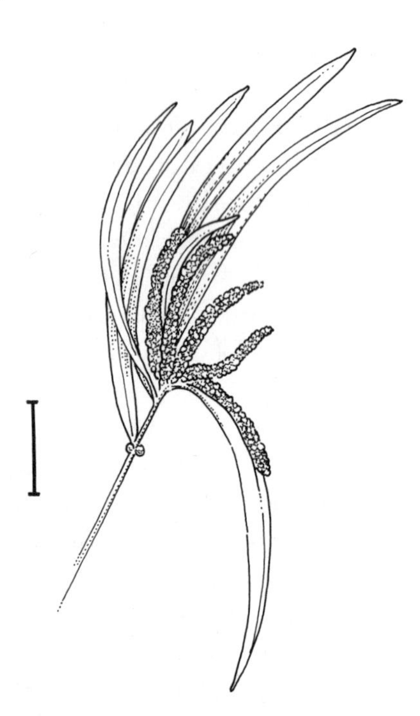

Twig of Buddhist pine (*Podocarpus macrophyllus*) with mature pollen cones at the time of pollination; scale, 1 cm.

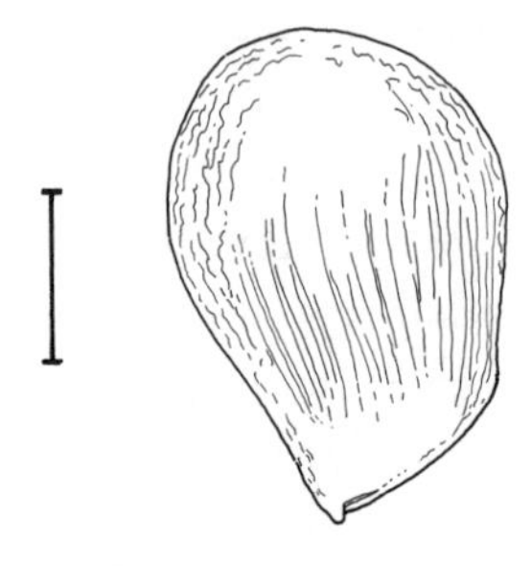

Seed of Buddhist pine (*Podocarpus macrophyllus*) detached from the podocarpium and without the epimatium; scale, 2 mm.

out to the sides, and an incomplete extra layer of small hypodermis cells between the upper epidermis and the palisade photosynthetic tissue, which consists of two or three layers of cells. Pollen cones 2–4(–5) cm long and 3–3.5 mm wide, (1–)3–5 on a very short, leafless stalk to 2 mm long. Pollen scales with a roundly triangular, upturned tip less than 1 mm long and ending in a point. Seed cones on a short, leafless stalk (2–)5–10(–17) mm long, with two or three needlelike basal bracteoles 3–4 mm long, the reproductive part with two or three bracts, these and the axis becoming swollen, juicy, and passing through red to dark purple, (6–)10–16(–20) mm long by (7–)10–15(–25) mm thick. Fertile seed scales one or two, the combined seed coat and epimatium leathery over a hard inner shell, greenish blue to purplish black with a thin, white waxy coating, 8–12 mm long by 6–9 mm thick, rounded at the tip without a noticeable crest or beak. Widespread in central and southern Japan (from central Honshū southward) and in central and southern China (from Jiangsu to Sichuan southward), very local in southernmost Korea, and perhaps in northern Myanmar. Scattered in the understory of warm temperate evergreen forests and in disturbed habitats like wooded roadsides and thickets; 0–500(–1,000) m. Zone 7. Four varieties.

Buddhist pine is the most widely cultivated species of *Podocarpus* in warm temperate climates, valued for its dense, dark green mass and tolerance of pollution and drought. Cultivation within and in the vicinity of its natural range obscured its exact boundaries and complicated its taxonomy. It even becomes naturalized far outside of its natural range, as in Florida. Over the centuries, there was interchange between cultivated selections and wild stands. Some taxonomists subdivide Buddhist pine into several species, which others consider botanical varieties. Some of these may actually be cultivars, and, indeed the species is in need of careful study to discern the range of natural variation and its association with geography before a truly consistent classification of botanical varieties within the species can be formulated. The botanical varieties provisionally accepted here vary in growth habit and in leaf size and shape. Besides these kinds of variations, cultivars have also been selected with variegated leaves and other modifications of leaf color. Although the trees are small, the wood has been used for a variety of purposes, including paper pulp, furniture, and various household utensils and tools. Because of its wide cultivation, it is the species of *Podocarpus* most often used by botanists in studies of such features as the development of leaves and seed cones.

Podocarpus macrophyllus* var. *chingii N. Gray

ZHU GUAN LUO HAN SONG (CHINESE)

Crown cylindrical, with upright branches. Twigs hairless. Leaves 0.8–3.5 cm long, 1–4 mm wide, the tip blunt. Jiangsu and Zhejiang (China). Synonyms: *Margbensonia chingiana* (S. Y. Hu) A. V. Bobrov & Melikian, *Podocarpus chingianus* S. Y. Hu.

Podocarpus macrophyllus (Thunberg) R. Sweet **var. *macrophyllus***

YUAN BIAN ZHONG (CHINESE), INU-MAKI (JAPANESE)

Crown rounded, with spreading or rising branches. Twigs hairless. Leaves 6–12 cm long, (5–)7–10(–12) mm wide, the tip pointed. Throughout the range of the species. Synonyms: *Margbensonia forrestii* (Craib & W. W. Smith) A. V. Bobrov & Melikian, *M. macrophylla* (Thunberg) A. V. Bobrov & Melikian, *M. sweetii* (C. Presl) A. V. Bobrov & Melikian, *Podocarpus forrestii* Craib & W. W. Smith.

Podocarpus macrophyllus* var. *maki P. Siebold & Zuccarini

DUAN YE LUO HAN SONG (CHINESE), RAKAN-MAKI (JAPANESE)

Crown rounded, with gently rising branches. Twigs hairless. Leaves (2.5–)4–8 cm long, 5–8 mm wide, the tip blunt or with a short point. Natural range uncertain. Synonyms: *Margbensonia chinense* (N. Wallich ex Jas. Forbes) A. V. Bobrov & Melikian, *M. maki* (P. Siebold & Zuccarini) A. V. Bobrov & Melikian, *Podocarpus chinensis* N. Wallich ex Jas. Forbes, *P. chinensis* var. *wardii* de Laubenfels & Silba, *P. macrophyllus* var. *angustifolius* Blume.

Podocarpus macrophyllus* var. *piliramulus Zhi X. Chen & Zhen Q. Li

MAO ZHI LUO HAN SONG (CHINESE)

Crown rounded, with spreading or rising branches. Twigs densely covered with short, dark hairs. Leaves 1.5–7 cm long, 2–4.5 mm wide, the narrowly rounded tip with a minute, blunt point. Hubei (China).

Podocarpus madagascariensis J. G. Baker

MADAGASCAR YELLOWWOOD, HETATRA (MALAGASY)

Podocarpus subg. *Podocarpus* sect. *Scytopodium*

Tree to 13 m tall, or sometimes much taller. Twigs shallowly grooved between the greatly elongated, attached leaf bases. Resting buds slightly squashed from spherical, with tightly overlapping, raised bud scales. Leaves spreading evenly around and along the twigs or concentrated near the tips, standing out stiffly or conspicuously drooping, lasting 1–2 years, leathery, dark green above, paler beneath, (4–)7–12(–13) cm long, (4–)6–10(–12) mm wide. Blades clearly widest at the middle and tapering gradually toward both the sharply to bluntly pointed tip and the very short stalked base, the edges a little rolled under. Midrib scarcely protruding above and strongly so beneath, with three resin canals beneath the midvein and one at the outer corner on each side, the interior tissue of the leaf largely taken up with strengthening sclereids and fibers. Pollen cones (1.2–)2–3.5(–5) cm long and (2.5–)3–4.5(–6) mm wide, one to three directly in the leaf axils or at the tip of a short, smooth stalk to 3 mm long. Pollen scales with a slightly toothed, upright tip in the shape of an equilateral triangle or a little narrower, the tip 0.5–1.5 mm long and 0.7–1 mm wide. Seed cones on a variable, bare stalk (6–)15–30(–40) mm long, without basal bracteoles, the reproductive part with three slightly unequal bracts, these and the axis becoming slightly swollen and leathery, 5–8 mm long by 5–8 mm thick. Fertile seed scale one, the combined seed coat and epimatium very fleshy, blackish, and 20–23(–28) mm long by 12–18 mm thick, weakly crested but enough so to make the seed appear slightly flattened. Throughout the mountainous spine of Madagascar (hence the scientific name), from the Massif du Tsaratanana in the north to the Massif de Beampingaratra in the south: in wet, mossy montane forests on various substrates of ridge tops and slopes; 800–2,000(–2,400) m. Zone 9? Synonyms: *P. madagascariensis* var. *procerus* de Laubenfels, *P. madagascariensis* var. *rotundus* L. Laurent.

The most common and widespread *Podocarpus* in the Malagasy Republic, Madagascar yellowwood differs from the other podocarps of the island in its longer, wider leaves and larger seeds. It is very closely related to Henkel yellowwood (*P. henkelii*) of eastern and southern Africa, which resembles it in these features and also in the characteristic drooping foliage. Variation in the size of the leaves and pollen cones in different habitats was used to divide the species into three botanical varieties. The biological status of these proposed varieties is unclear, and it seems preferable to withhold formal recognition until the situation can be clarified. Plants with shorter leaves occur on basaltic substrates, and ones with narrower leaves occur at lower elevations, but the consistency, stability, and genetic basis of these variations are unknown. If they are just extremes picked out of the ends of a continuum of variation, then there is little merit in distinguishing them under separate names.

Podocarpus magnifolius J. Buchholz & N. Gray

BROADLEAF PODOCARP, CINQUIMASE (SPANISH)

Podocarpus subg. *Podocarpus* sect. *Nemoralis*

Tree to 20(–30) m tall, with a straight main trunk. Crown narrowly cylindrical, with short, upwardly angled branches bearing relatively few branchlets densely clothed with foliage. Twigs coarse, green, prominently grooved between the attached leaf bases. Resting buds roughly spherical, about 5 mm long and 4 mm in diameter, loosely constructed of thick, narrowly triangular bud scales with minute, fingerlike free tips. Leaves densely and evenly distributed along and around the twigs, lasting 1–2 years, leathery, dark green above, paler beneath, (3–)5–9 cm long (to 21 cm in juveniles), 12–18(–24) mm wide (to 30 mm in juveniles). Blades straight, the edges curled under, widest near the middle in shorter leaves, nearly parallel-sided otherwise, tapering rapidly to the broadly and roundly triangular apex and stretched out into a narrowly triangular tip and to the roundly wedge-shaped base on a short, narrowly winged petiole 2–5 mm long (to 15 mm in juveniles). Midrib narrowly and deeply grooved above, broadly raised beneath, with one resin canal beneath the midvein, wings of accessory transfusion tissue extending out to the sides, many additional sclereids, and a continuous double layer of hypodermis beneath the upper epidermis. Pollen cones 4–5 cm long and 3–3.5 mm wide, single directly in the axils of foliage leaves. Pollen scales with an upturned, minutely toothed, triangular tip 1–2 mm long. Seed cones on a long, leafless stalk 1–2 cm long, without basal bracteoles, the reproductive part with two slightly unequal bracts, these and the axis becoming a little swollen and fleshy, bright red, 9–12 mm long by 3–5 mm thick. Fertile seed scale one, the combined seed coat and epimatium leathery over a hard inner shell, 10–11 mm long by 5–6 mm thick, tipped by a narrow, pointed beak. Widely and discontinuously distributed across the mountains and tablelands to the north and west of the Amazon and Orinoco River basins, from the coastal range near Caracas and the tepuis (isolated, flat-topped, sandstone mountains) of the Guiana Highlands in eastern Bolívar state in Venezuela to Panamá province, Panama, and to Lake Titicaca in western Bolivia. Scattered in the canopy or subcanopy of cloud forests or more open wet woodlands; 750–1,750(–2,000) m. Zone 9.

Broadleaf podocarp has the widest leaves, proportionately, among the large-leaved New World podocarps and, in fact, among all species in subgenus *Podocarpus.* The juvenile leaves are among the largest of all conifers, hence the scientific name, Latin for "large leaf." It is closely related to Ulcumanu podocarp (*P. celatus*), which has much the same overall distribution but generally at lower elevations. Besides having the leaves a little smaller on average, those of Ulcumanu podocarp differ in having flat rather than turned down edges and in lacking the extra projection on the leaf tip. Local regions supporting both species should be sought to gather data to help clarify their taxonomic and ecological relationships.

Podocarpus matudae C. Lundell

MATUDA PODOCARP, CIPRECILLO (SPANISH), CURUS-TÉ (MAYAN)

Podocarpus subg. *Podocarpus* sect. *Lanceolatis*

Tree to 20 m tall or more, with trunk to 1(–1.5) m in diameter. Bark furrowed, gray. Twigs green, prominently but narrowly grooved between the attached leaf bases. Resting buds elongated, the core about 3–5 mm long, the longer bud scales up to 15 mm long, loosely constructed of upright, narrowly triangular to somewhat leaflike bud scales. Leaves projecting around the twigs, widely spaced along them but more concentrated near the ends of growth, dark green above, a little paler beneath, leathery, flat, the edges slightly turned down, (4–)7–9(–12) cm long (to 20 cm in juveniles), 10–16 mm wide (to 19 mm in juveniles). Blades straight or somewhat curved to one side, widest near the middle or slightly below, tapering gradually to the narrowly prolonged, triangular tip and to the roundly wedge-shaped base with a short petiole 1–5 mm long. Midrib raised above, especially near the base, and weakly raised beneath, with one resin canal beneath the midvein, relatively little support tissue for the size of the leaf, and a continuous layer of hypodermis beneath the upper epidermis. Pollen cones (3–)4–6 cm long and 4–5 mm wide, one (or two) directly in the axils of foliage leaves or on a stalk to 2 mm long. Pollen scales with a minutely toothed, upturned, triangular tip 1–2 mm long. Seed cones on a leafless stalk (4–)8–18 mm long, without basal bracteoles, the reproductive part with two very unequal bracts, these and the axis becoming swollen and fleshy, reddish to dark brown, (4–)6–9 mm long by 4–6 mm thick. Fertile seed scale one, the combined seed coat and epimatium thinly fleshy over a hard inner shell, smooth or a bit wrinkled, (8–)12–15(–18) mm long by (7–)10–14 mm thick, with a low, blunt crest and a small beak. Mountains and nearby lowlands of northern Mesoamerica, from southern San Luis Potosí and Veracruz (Mexico) to Santa Ana (El Salvador) and Chiquimula (Guatemala) and possibly in adjacent Honduras. Scattered in the canopy of cloud forests and of mixed montane forests with pines, oaks, and other hardwoods; (200–)1,000–2,500(–3,500) m. Zone 9. Synonyms: *P. matudas* var. *jaliscanus* de Labenfels & Silba ex Silba, *P. reichei* J. Buchholz & N. Gray.

Young mature Matuda podocarp (*Podocarpus matudae*) in nature.

Matuda podocarp is the most northerly mainland podocarp in the New World, reaching near 22°N, a latitude matched by Cuban podocarp (*Podocarpus angustifolius*) in the West Indies. Compared to the other two *Podocarpus* species in northern Central America, it has a much more restricted distribution, since olive-leaf podocarp (*P. oleifolius*) and Guatemalan podocarp (*P. guatemalensis*) both extend from southern Mexico well into South America. Matuda podocarp differs from both of these species in its much longer, narrowly pointed bud scales, and olive-leaf podocarp further differs in having a groove above the midvein while Guatemalan podocarp usually has smaller pollen cones and seeds. Guatemalan podocarp is not particularly closely related to these neighbors but rather to the other long-budded species surrounding the Caribbean, Antillean podocarp (*P. coriaceus*) of the West Indies and Steyermark podocarp (*P. steyermarkii*) and willowleaf podocarp (*P. salicifolius*) of Venezuela. The species and common name honor Eizi Matuda (1894–1978), a Mexican botanist who made extensive collections in the southernmost state of Mexico, Chiapas, including the type specimen of this species.

Podocarpus micropedunculatus de Laubenfels

KAYU PODOCARP, KAYU TJINA (MALAY)

Podocarpus subg. *Foliolatus* sect. *Acuminatis*

Colonial shrub, or bushy tree to 7(–13) m tall, the main trunk to 0.2 m diameter when treelike, spreading by suckers from underground runners. Crown irregular. Resting buds elongate, 6–15 mm long, 1–3 mm diameter, loosely composed of upright, narrowly triangular bud scales. Leaves 8–17 cm long (to 18 cm in juveniles), 10–15 mm wide (to 21 mm in juveniles). Blades mostly straight, fairly parallel-sided, tapering abruptly to a triangular apex with a prolonged tip and to a roundly wedge-shaped base on a short petiole 3–5 mm long. Midrib prominently raised above, wider with a central groove beneath. Pollen cones 3.5–7.5 cm long and 3–4 mm wide, (one to) three directly in the axils of foliage leaves or on a very short, leafless stalk to 2 mm long. Pollen scales with an upturned, triangular tip less than 1 mm long. Seed cones on a very short, leafless stalk to 1 mm long (this small peduncle being the origin of the scientific name) with two needlelike basal bracteoles 3–4 mm long, the reproductive part with two unequal bracts, these and the axis becoming red, swollen, and juicy, 8–10 mm long. Fertile seed scale one, the combined seed coat and epimatium leathery over a hard shell, 8–10 mm long by 5–6 mm thick, tipped by a small beak. Confined to northern Borneo, in Sabah and Sarawak (Malaysia), scattered in the understory of mixed conifer and hardwood forests or forming thickets in disturbed sites on sandy soils and other substrates through clonal spread; 0–200(–500) m. Zone 9.

Among the podocarps, clonal spread is usually found in subalpine species rather than those of low elevations. Two Australian lowland species unrelated to kayu podocarp that share this characteristic are creeping plum pine (*Podocarpus spinulosus*) and emu grass (*P. drouynianus*), both of which also occur in open, fire-prone stands on sandy soils. This trait allows kayu podocarp to spread and increase its dominance in disturbed sites where it is already established in the understory prior to disturbance.

Podocarpus nakaii Hayata

Taiwan podocarp, tai wan luo han song (Chinese)

Podocarpus subg. *Foliolatus* sect. *Globulus*

Tree to 18 m tall, with trunk to 1.8 m in diameter. Bark pale gray, peeling in strips and becoming ridged and furrowed with age. Crown with slender branches bearing paired branchlets loosely clothed with foliage near the tips. Twigs green, shallowly grooved between the attached leaf bases. Resting buds spherical, 1.5–3 mm in diameter, tightly wrapped by triangular of rounded bud scales with spreading tips. Leaves rather openly and evenly standing out around the twigs and concentrated near their tips, leathery and flexible, reddish at first, becoming shiny bright green above, paler, duller, and slightly waxy beneath, 5–10(–11) cm long, 8–11 (–14) mm wide. Blades straight or curved to one side, flat or slightly thickened at the edges, widest below the middle, tapering very gradually and then abruptly to the bluntly or sharply triangular tip and more abruptly to the broadly and roundly wedge-shaped base on a short petiole to 5 mm long. Midrib broadly to inconspicuously raised above, more broadly raised beneath, with three sometimes inconspicuous resin canals beneath the midvein, wings of support tissue extending out to the sides, and just a scattering of sclereids in a partial hypodermis beneath the upper epidermis. Pollen cones 2–4(–5) cm long and 3–4 mm wide, one to three directly in the axils of foliage leaves. Pollen scales with a tongue-like, minutely toothed, upturned tip 1–2 mm long. Seed cones on a short, leafless stalk 2–7(–12) mm long, with a pair of persistent, thick, narrowly triangular basal bracteoles about 1.5 mm long, the reproductive part with two nearly equal bracts separated by inconspicuous grooves, these and the axis becoming reddish orange to red, swollen, and juicy, (4–)7–9(–12) mm long by (3–)5–8(–10) mm thick. Fertile seed scale one, the combined seed coat and epimatium leathery over a hard inner shell, green, 10–12(–15) mm long by 7–8(–10) mm thick, with a prominent crest along one side ending in a pointed beak. Central Taiwan. Scattered in lower montane, broad-leaved evergreen forest; 300–800 m. Zone 9. Synonym: *P. macrophyllus* var. *nakaii* (Hayata) H. L. Li & H. Keng.

Taiwan podocarp has often been confused with the more widespread species, oleander podocarp (*Podocarpus neriifolius*) and Buddhist pine (*P. macrophyllus*), both of which also occur in Taiwan, It differs from both in its spherical resting buds with tightly fitting bud scales and its more obviously triangular basal bracteoles beneath the podocarpium. Oleander podocarp has larger leaves on average, and those of Buddhist pine are usually narrower. The spherical resting buds, reflected in the name of the botanical section, *Globulus*, point to a relationship with Teijsmann podocarp (*P. teysmannii*) of western Malesia. The species name honors Takenoshin Nakai (1882–1952), a Japanese botanist who collected extensively in Taiwan during the Japanese occupation.

Podocarpus neriifolius D. Don

OLEANDER PODOCARP, MOUNTAIN TEAK, HALIS (HINDI), GUNSI (NEPALI), THITMIN (BURMESE), BAI RI CHING (CHINESE), SRÔ:L (CAMBODIAN), KA DO:NG (LAOTIAN), KIM GIAO (VIETNAMESE), S'A:NG KH'AM (THAI), JATI BUKIT (MALAY), KAYU TADJI (SUMATRAN), ANTOH (JAVANESE), ASIMBOLO (FIJIAN)

Podocarpus subg. *Foliolatus* sect. *Foliolatus*

Tree to 25(–45) m tall (or sometimes shrubby), with trunk to 0.8 (–1.8) m in diameter but often much more slender, sometimes buttressed. Bark reddish brown, fibrous, peeling in strips and weathering grayish brown and becoming ridged and furrowed with age. Crown deeply to shallowly dome-shaped, with numerous widely spreading or rising, often forked and crooked branches, often drooping at their ends and bearing clustered (or scattered) branchlets densely clothed with foliage. Twigs green, slender, narrowly and shallowly grooved between the attached leaf bases. Resting buds egg-shaped, pointed, 2–5(–9) mm long, 1.5–3 mm wide, tightly wrapped by narrowly triangular bud scales, the outer scales with long, narrow or leaflike tips extending beyond the bud and curled outward. Leaves sticking out all around the twigs and well separated along them, leathery, flexible, shiny dark bluish green to yellowish green above, duller and paler beneath (3–)7–16(–18) cm long (to 25 cm in juveniles), (5–)8–16(–20) mm wide (to 24 mm in juveniles).

Underside of twig of oleander podocarp (*Podocarpus neriifolius*) with four episodes of growth, ×0.5.

Blades straight to noticeably curved to one side, the thin edges flat or slightly turned down, nearly parallel-sided or widest near or below the middle, tapering abruptly or gradually and then more abruptly to a triangular or roundly triangular (or rarely rounded) tip, often with a narrowly triangular extension of variable length, and relatively more abruptly to the narrowly to broadly and roundly wedge-shaped base on a petiole 2–6(–9) mm long. Midrib sharply (or rarely broadly and shallowly) raised above and (0.3–)0.5–1 mm wide, broader and weak to prominent beneath, with three resin canals beneath the midvein, sometimes with another one on either side within the wings of support tissue extending out to the sides, rarely with another two to six above the midvein, and with an interrupted layer of small hypodermis cells beneath the upper epidermis. Pollen cones brown to yellowish green (1–)2.5–6 cm long and 2–3.5(–8?) mm wide, 1–3(–6) usually directly in the axils of the foliage leaves but sometimes on a very short, leafless stalk to 3 mm long or rarely elongating in the lower part among the bud scales to produce a short, scaly stalk to 10 mm long. Pollen scales with a triangular to tonguelike or toothed upturned tip (0.2–)0.5–0.8(–4?) mm long. Seed cones on a long, leafless stalk (3–)9–20(–25) mm long, with two needlelike basal bracteoles (1–)2–3(–6) mm long, the reproductive part with two (to four) bracts, these and the axis becoming swollen, juicy, and orange-red to bright red (or even deep purple), (5–)8–12(–18) mm long by (2–)5–8 mm thick. Fertile seed scales one (or two), the combined seed coat and epimatium bluish green to purplish red or bluish black and covered by a pale blue to gray dusting of wax, 8–14(–18) mm long by 5–8(–13) mm thick, rounded or with a small, blunt beak at the tip. Widely distributed and often common from Nepal, southern China, and Taiwan south and east through eastern India (including the Andaman Islands), Southeast Asia, the entire Malesian region (Malaysia, Singapore, Brunei, Indonesia, the Philippines, and Papua New Guinea), the Solomon Islands and Vanuatu to Fiji. Scattered to gregarious as an emergent or in the canopy or understory of mature rain forests dominated by hardwoods, often along streamsides or even at the edge of tidal swamps; 0–1,200(–2,100) m. Zone 10. Synonyms: *Margbensonia degeneri* (N. Gray) A. V. Bobrov & Melikian, *M. neriifolia* (D. Don) A. V. Bobrov & Melikian, *P. decipiens* N. Gray, *P. degeneri* (N. Gray) de Laubenfels, *P. epiphyticus* de Laubenfels & Silba, *P. fasciculus* de Laubenfels, *P. palawanensis* de Laubenfels & Silba, *P. spathoides* de Laubenfels, *P. subtropicalis* de Laubenfels.

Oleander podocarp is by far the most widely distributed species of *Podocarpus* in tropical Asia and Oceania. With the exception of the southern species in Australia, New Zealand, and New Caledonia, Tongan podocarp (*P. pallidus*) just to the east, and Buddhist pine (*P. macrophyllus*) to the north, the remaining *Podocarpus* species of the region, some 24 of them, all have distributions entirely embedded within the range of oleander podocarp. It is also far and away the most common *Podocarpus* species of the lowland rain forests here, the other species being found mostly at higher elevations or along seashores or in various specialized habitats, like swamps or the stunted forests found on serpentine-derived soils or sterile sands. Even so, oleander podocarp itself occupies a wide range of habitats and has even been found growing on other trees as an epiphyte. It is also commonly cultivated as an ornamental both within and outside of its native range, including treatment as a greenhouse subject in cooler climates.

Not unexpectedly for such a widespread species in a landscape consisting of thousands of islands of different sizes, there is a great deal of variation in morphological features, some of which occurs throughout the range and some much more restricted. For instance, the presence of extra resin canals in the leaf above the midvein is found only in Fiji, where some trees have them and others do not. Variation in this feature and in many others led to the description of numerous botanical varieties, some of which were treated as separate species by various authors. The problem, from the standpoint of a taxonomic treatment, is that most of these proposed segregants of oleander podocarp either lack any kind of geographic coherence or co-occur with and perhaps intergrade with trees having more common combinations of features. Thus recognition of botanical varieties does not seem warranted with the information on variation in the species currently available. A thorough study of this species throughout its range and in comparison with related species would be very welcome.

In many parts of its range it is an important timber tree, producing an excellent lumber that is used for all the usual purposes asked of softwoods, mostly in general carpentry, sculpture, and papermaking but also in some specialty applications such as musical instrument making and the construction of masts for sailing ships, still widely used in the region. A yellow resin sometimes exuded by the bark is also an occasional item of commerce, at least locally. Detailed anatomical study of the seed cones of this species supports the standard interpretation of the epimatium as a structure equivalent to the seed scales of members of the Pinaceae, an interpretation also supported by studies of cone development in other species of *Podocarpus.*

The scientific name, Latin for "oleander [*Nerium*] leaf," refers to the glossy, dark green, broad, flat, strap-shaped leaves, so reminiscent of those of its Mediterranean namesake, a completely unrelated flowering plant.

Podocarpus nivalis W. J. Hooker

ALPINE TOTARA, SNOW TOTARA

Podocarpus subg. *Podocarpus* sect. *Australis*

Low, spreading or somewhat upright shrub to 1(–3) m, with prostrate trunk about 15 cm in diameter. Bark pale reddish brown to grayish, furrowed and peeling in narrow strips. Crown outstretched to flatly dome-shaped, with spreading, contorted branches rooting in contact with the soil and with numerous stiffly upright branchlets densely and tightly clothed with foliage. Twigs yellowish green, prominently grooved between the attached leaf bases. Resting buds inconspicuous, nearly spherical, about 1 mm in diameter, tightly clothed by narrow, minutely and irregularly toothed bud scales. Leaves closely and evenly spaced all around the branchlets, stiffly angled forward, thick and hard with a thickened margin, glossy yellowish green, bright green, or bluish green above, and with narrow white stomatal bands on either side of the midrib beneath, largest near the middle of each growth increment, (0.5–)1–1.5 cm long (to 2.5 cm in juveniles), 2–4 mm wide. Blades straight or curved to one side, widest near the middle or beyond, tapering gradually or abruptly to a short, hard, pointed tip and to the nearly stalkless base. Midrib grooved above and prominently raised beneath, with one resin canal beneath the midvein. Pollen cones red, 4.5–10(–15) mm long (to 25 mm after the pollen is shed) and 1.5–3 mm wide, (one or) two to four on a very short, leafless stalk to 6 mm long. Pollen scales with a minute, untoothed, blunt tip less than 0.5 mm long and wide. Seed cones on a very short, leafless stalk to 4 mm long, without basal bracteoles, the reproductive part with two nearly equal bracts, these and the axis becoming bright red, nearly spherical, swollen and juicy, (3–)5–7 mm long by 5–6 mm thick. Fertile seed scales one (or two), the combined seed coat and epimatium smooth and hard, bronzy green, (3.5–)5–6 mm long by 2–3 mm thick, with a blunt crest along one side. Mountains of North Island and South Island (New Zealand) from near Auckland southward. Forming large, clonal patches in subalpine scrublands and talus slopes or as scattered individual shrubs in adjacent forest margins; (500–)650–1,800(–2,000) m. Zone 7.

This is one of the hardiest of the podocarps, and several cultivars have been selected that join other dwarf conifers in rock gardens of moderate temperate regions throughout the world. It was first described from the high subalpine zone near the permanent snows on Mount Tongariro (hence the scientific name, Latin for "of the snows") and is closely related to the other subalpine species in section *Australis:* New Caledonian alpine podocarp (*Podocarpus gnidioides*) and mountain plum pine (*P. lawrencei*) of Australia. It is also close to the other New Zealand *Podocarpus* species and hybridizes naturally with all three: the two tree species, totara (*P. totara*) and Hall totara (*P. cunninghamii*), and the bushy prickly totara (*P. acutifolius*). All are more upright than is typical for alpine totara, but you cannot just rely on growth habit to distinguish them as hybrids because alpine totara also naturally takes on a more upright growth habit in the shade of subalpine forest margins where the hybrids are most often found. Leaf length is helpful for distinguishing hybrids from *P. nivalis,* but the flavonoid pigments in the leaves are the surest way until appropriate DNA studies have been conducted. Hybrids with prickly totara and Hall totara are more common than those with totara, which is rare in the upper subalpine forests, where hybridization is most frequent. Some of the hybrids have been brought into cultivation, mistakenly treating them as cultivar selections of one of the parent species. Sometimes, as near Christchurch, hybrids are found in areas where one of the parents is absent. Such occurrences might be due to local extirpation of one of the parents, to long-distance pollination of these wind-pollinated shrubs, or to long-distance establishment of the bird-dispersed seed, thanks to the attractiveness of the berrylike podocarpium.

Podocarpus novae-caledoniae Vieillard

SVELTE RIVER PODOCARP

Podocarpus subg. *Foliolatus* sect. *Foliolatus*

Shrub, or tree to 3(–5) m tall, usually without a single trunk. Bark dark reddish brown, breaking up into elongate, thick scales. Crown dome-shaped, with numerous upright branches bearing scattered branchlets densely clothed with foliage. Twigs slender and often whiplike, green, weakly grooved between the attached leaf bases. Resting buds egg-shaped, 2–3 mm long, tightly clothed by the triangular bud scales with a prolonged tip. Leaves standing out forwardly and evenly around and along the twigs, stiff and leathery, lasting 2–3 years, waxy bluish green at first but weathering dark green above, (4–)6–8(–9) cm long, 3–5(–6.5) mm wide. Blades straight or often curved to one side, the thick edges turned down, fairly parallel-sided for most of their length, tapering abruptly to the roundly triangular tip, with or without a prickle, and more gradually to the narrowly wedge-shaped base on a short, sometimes indistinct petiole (1–)3–6(–9) mm long. Midrib shallowly grooved above and correspondingly raised beneath, with (1 or) three resin canals beneath the midvein, very strong wings of support tissue extending out to the sides, and a continuous (and partially doubled) layer of small hypodermis cells beneath the upper epidermis. Pollen cones 10–15(–20) mm long and 1.5–3 mm wide, (one or) two or three directly in the axils of foliage leaves. Pollen scales with a roundly triangular, upturned tip to 0.5 mm long. Seed cones on a slender leafless stalk 7–10 mm long, with a pair of needlelike basal bracteoles about 1.5 mm long, the reproductive part with two or three (or four) unequal bracts, these and the axis becoming swollen, juicy, and yellow through bright or dark red to purple, 7–9 mm long by 3–5 mm thick. Fertile seed scales one or two (or three), the combined seed coat and epimatium leathery over a hard inner shell, shiny, 6–8 mm long by 4–6 mm thick, with a well-developed crest along one side ending in a short, pointed beak. Southern quarter of New Caledonia (hence the scientific name) and on Île des Pins. Maquis shrublands, mostly on serpentine soils along creeks but sometimes away from streamsides; 50–500(–750) m. Zone 9. Synonym: *P. beecherae* de Laubenfels.

Although the long, narrow leaves of svelte river podocarp are unlike those of any other species of *Podocarpus* in New Caledonia, the bushes from Île des Pins have slightly wider leaves than those on the main island and were treated in the past as a separate botanical variety or even as belonging to kauri podocarp (*P. sylvestris*). Under exceptionally favorable conditions, this species may become a small tree, but even in its most common, moist, riparian habitat it is usually shrubby in habit. Despite being a common species in this habitat in southern New Caledonia, it is not found beyond this region, the richest in conifer species on the island. In addition to streamsides, individuals are found in full-blown maquis minière on rocky but well-watered uplands nearby. These upland individuals have traditionally been included within *P. novae-caledoniae* but were described as a separate species (*P. beecherae*) in 2003. They differ in several ways from the riparian plants, including subtle distinctions in leaf size and shape and in color, structure and shape of podocarpium and epimatium. Because of difficulties in distinguishing the two forms, particularly as dried herbarium specimens, and because of uncertainties surrounding their genetic basis, the upland plants are retained here in *P. novae-caledoniae* without taxonomic distinction until their status can be further clarified.

Podocarpus nubigenus Lindley

CLOUD PODOCARP, CHILEAN PODOCARP, MAÑIÚ MACHO (SPANISH), MANILIHUAN (ARAUCANIAN)

Podocarpus subg. *Podocarpus* sect. *Australis*

Tree to 25(–30) m tall, with trunk to 0.9(–2) m in diameter. Bark light reddish brown to gray, thin, peeling in scales or strips and becoming shallowly furrowed. Crown conical when young, becoming broadly conical or rounded and more irregular and open with age, with fairly heavy horizontal to rising branches bearing short branchlets attached in twos or threes and densely clothed with foliage. Twigs yellowish green, grooved between the attached leaf bases. Resting buds slightly elongated and pointed, with sharply pointed, somewhat loosely overlapping scales. Leaves evenly spaced along the length of the twigs and spreading out all around them, light green to yellowish green above, with stomates confined to waxy bluish gray bands on either side of the midrib beneath, lasting 3 years or more, (1.5–)2–4(–4.5) cm long, (2–)3–4(–5) mm wide. Blades nearly parallel-sided or widest near or below the middle, straight or slightly curved to one side, tapering abruptly to the sharply long-pointed tip and to the rounded or wedge-shaped base with a very short petiole. Midrib distinctly protruding above and hardly so beneath, with one resin canal beneath the midvein and well-developed support tissue extending out into the leaves and

Large mature cloud podocarp (*Podocarpus nubigenus*) in nature.

making them very hard and rigid. Pollen cones yellowish brown, (0.8–)1.5–2.5 cm long and 1.5–2.5(–3) mm wide, one to three (to six) at the end of a short, leafless stalk to 5 mm long, clustered at the tips of the branchlets. Pollen scales with a roundly triangular, minutely upturned tip about 1 mm long and wide, the edge stiffened and toothed. Seed cones on a very short, leafless stalk to 1 mm long, without basal bracteoles, the reproductive part with four or five very unequal bracts, these and the axis becoming fleshy and somewhat swollen, bright red at maturity, 5–7 mm long. Fertile seed scales one or two, the combined seed coat and epimatium thinly fleshy, glossy or dull dark brownish red, a little elongate, 8–10 mm long by 6–8 wide, distinctly wider than the podocarpium beneath, with an inconspicuous crest along one side and over the top. Southern Andes and southern coastal ranges of Chile from the Toltén River in Cantín province at about 39°S to Isla Madre de Dios in Ultima Esperanza below 50°S and barely into adjacent Patagonian Argentina between about 40°S and 44°S. Forming pure stands in bogs and scattered in the surrounding cool temperate rain forests in regions of very high rainfall; (0–)10–900(–1,250) m. Zone 7.

Cloud podocarp reaches the most southerly limit of all species in the Podocarpaceae, and only its neighbor Patagonian incense cedar (*Libocedrus uvifera*) in the family Cupressaceae, another frequenter of bogs, reaches farther south, gaining the title of the world's most southerly conifer. Its far southerly distribution makes it one of the hardiest of podocarps, and it is a handsome ornamental when cultivated within and near its native range. However, it has a very high moisture requirement and does very poorly in regions without constant atmospheric humidity derived from fog and heavy rainfall (hence the scientific name, "cloud bringer," from Latin and Greek). Its native stands are so waterlogged that it is scarcely threatened by habitat conversion. The wood is highly prized for furniture and other uses that show off its beauty, a basic yellowish tinge streaked with red, so large trees are now rare. Nonetheless, it is one of the most shade-tolerant Andean conifers, regenerates well in the small gaps left by removal of large trees, and has one of the fastest growth rates among the southern species, so it does not face the same threats as a species like alerce (*Fitzroya cupressoides*). Although it does not have the longevity of the latter species and of some of its other neighbors, cloud podocarp is still known to live 400 years and typically reaches an age of 200. It is the only New World member of section *Australis*, the most cool temperate group among the southern species of *Podocarpus*, although it is not especially closely related to its southwestern Pacific relatives in New Zealand, Australia, and New Caledonia, such as *P. gnidioides*, *P. lawrencei*, and *P. totara*. It is much closer to them, however, than to the numerous other American podocarps, most of which are species of the tropical mountains and lowlands farther north.

Podocarpus oleifolius D. Don

OLIVE-LEAF PODOCARP, PINO APARRADO, CHILCA REAL, DIABLO FUERTE, ESPINTANILLA, SAUCECILLO (SPANISH)

Podocarpus subg. *Podocarpus* sect. *Pratensis*

Tree to 20(–35) m tall, with trunk to 1(–1.5) m in diameter. Bark yellowish brown, fibrous, peeling in narrow strips and becoming gray and shallowly furrowed with age. Crown dome-shaped, with numerous contorted, short branches bearing clustered branchlets densely clothed with foliage. Twigs yellowish green to grayish green, strongly grooved between the attached leaf bases. Resting buds nearly spherical, 2.5–5 mm in diameter, tightly or a little loosely wrapped by broad, rounded bud scales with dry, papery edges. Leaves densely and evenly distributed along and around the twigs, stiff and leathery, shiny dark green to olive green above, paler green but not waxy beneath, (2.5–)5–8(–11) cm long (to 16 cm in juveniles), (5–)6–12(–14) mm wide (to 20 mm in juveniles). Blades straight or inconspicuously curved toward one side, the edges thin and turned down, widest near the middle and from there tapering gradually to the narrowly triangular tip and to the roundly wedge-shaped, nearly stalkless base. Midrib channeled above and broadly and shallowly raised beneath, with one resin canal beneath the midvein, a variety of different support tissues in the leaf, and a continuous layer of hypodermis beneath the upper epidermis. Pollen cones (1.4–)2.5–4 cm long and 2.5–4(–7.5) mm wide, single on a short, leafless stalk up to 7 mm long. Pollen scales with a broadly and roundly triangular, upturned tip about 1 mm long. Seed cones at the end of a leafless stalk 5–12(–20) mm long, without basal bracteoles, the reproductive part with two or three (or four) unequal seed scales, these and the axis becoming very swollen and juicy, bright reddish purple, 6–9(–12) mm long by 5–6 mm in diameter. Fertile seed scales one or two, the combined seed coat and epimatium thinly fleshy over a hard inner shell, purplish green, 6–8(–12) mm long by 4–6 mm thick, with a very weak crest along one side and forming a low beak. Mountains of Oaxaca and Chiapas (Mexico), through Central America and northern South America to western Santa Cruz (Bolivia) and northwestern Venezuela (on a line from Táchira state to the Federal District). As scattered trees or small groves in cloud forests and elfin forests (*ceja de monte*, eyebrow of the mountain), less commonly in lower montane evergreen dry forests; (800–)1,500–2,500(–3,500) m. Zone 9. Synonyms: *P. ingensis* de Laubenfels, *P. macrostachyus* Parlatore, *P. monteverdeensis* de Laubenfels, *P. oleifolius* var. *equdorensis* Silba, *P. oleifolius* var. *macrostachyus* (Parlatore) J. Buchholz & N. Gray.

This is the most widespread species of *Podocarpus* in the New World, being found in 12 countries along the cordilleras that form the mountain backbone of the two continents. The species is correspondingly variable, and several botanical varieties were described based on size and shapes of leaves, lengths of cone stalks, and size of seed cones. Most of these proposed variants have no geographic coherence, and some of the features commonly vary just as much within individual trees. Trees at the highest elevations, in the elfin forests of Peru and Bolivia, are most compact and have the shortest and proportionately broadest leaves, but there is little apparent discontinuity with the variation found in trees at lower elevations. This is the variant represented by the type specimen of the species, so some authors separate the more widespread variants as a botanical variety or even as a separate species (then called *P. macrostachyus*), but here they are treated as ecological variants without formal taxonomic recognition. One feature that varies quite a bit in olive-leaf podocarp is the amount of support tissue in the leaves. Those trees with more support tissue have leaves that dry with wrinkles crossing them, while those with less have leaves that dry flat. The species name (Latin for "leaf of olive") refers to a modest resemblance to leaves of the olive tree, a resemblance hardly unique for this species among podocarps. Olive-leaf podocarp and huapsay podocarp (*P. sprucei*) are protected in Podocarpus National Park in Ecuador, the only national park in the world set aside specifically for conservation of *Podocarpus* species.

Podocarpus pallidus N. Gray

TONGAN PODOCARP, UHIUHI (TONGAN)

Podocarpus subg. *Foliolatus* sect. *Foliolatus*

Tree to 20(–30) m tall or shrubby in disturbed habitats, with trunk to 0.2 m or more in diameter. Bark reddish brown, smooth, weathering gray. Crown conical at first, becoming dome-shaped with age, with regular circles of horizontal branches bearing clustered branchlets densely clothed with foliage. Twigs green, grooved between the attached leaf bases. Resting buds egg-shaped, 2–5(–7) mm long by 2 mm across, tightly wrapped by narrowly triangular bud scales with a central ridge continuing into a spreading or upright bristle tip. Leaves standing out all around the twigs and concentrated near their tips, lasting 3(–4) years, leathery, shiny pale green above (hence the scientific name, Latin for "pale"), duller with a dusting of wax beneath, 2.5–8(–11) cm long (–15.5 cm in juveniles), (6–)7–10(–14) mm wide (–17 mm in juveniles). Blades straight or slightly curved to one side, the edges turned down or not, nearly parallel-sided or widest before the middle, tapering abruptly to a variously prolonged, narrowly triangular tip, sometimes topped by a hardened point, and to a roundly wedge-shaped base on a thick petiole 1–4 mm long. Midrib narrowly and shallowly raised near the base above and passing to flat or grooved near the tip, more broadly raised beneath or collapsing to a shallow groove, with (one to) three resin canals beneath the midvein, sometimes with another one on each side embedded in the wings of support tissue extending out to the sides, and occasionally with two above the midvein and with an incomplete layer of hypodermis beneath the upper epidermis. Pollen cones yellowish white, 10–15 mm long and 3–4 mm wide, one directly in the axils of the foliage leaves. Pollen scales with a triangular, upturned tip less than 1 mm long. Seed cones on a short, leafless stalk 6–15 mm long, with two needlelike basal bracteoles 1–2 mm long, the reproductive part with two or three unequal bracts, these and the axis becoming swollen, juicy, and red, 7–10 mm long. Fertile seed scales one (or two), the combined seed coat and epimatium nearly spherical, fleshy over a hard inner shell, bluish or greenish gray, 10–13 mm long by 8–10 mm thick, rounded at the tip or with a low crest. Tonga, known from Vava'u and 'Eua, the only limestone islands in Tonga exceeding 100 m in elevation. Common but scattered in the canopy of the highest-elevation hardwood-dominated rain forests or on limestone cliffs overlooking the sea; (120–)200–250(–300) m. Zone 10.

Although it is the only conifer native to Tonga, there is so little natural forest left in these islands that Tongan podocarp is not a major timber species. Instead, the local requirement for softwood lumber is met by plantations of the exotic Caribbean pine (*Pinus caribaea*). This circumstance may not be much more favorable for the persistence of Tongan podocarp than excessive exploitation would be since plantations and volunteer colonies of invading pines may displace natural forest. The conservation status of the species may have to be carefully monitored. The Tongan podocarp was long misidentified as plum pine (*Podocarpus elatus*), a strictly Australian species that differs most obviously in its darker green leaves and paired pollen cones.

Podocarpus parlatorei Pilger

BOLIVIAN CERRO PODOCARP, PINO DEL CERRO (SPANISH)

Podocarpus subg. *Podocarpus* sect. *Capitulatis*

Shrubby tree to 10(–20) m tall, with trunk to 0.7(–1.5) m in diameter, often crooked and dividing near the base. Bark dark brown, deeply furrowed between narrow, shredding, flat-topped ridges. Crown conical when young, becoming irregularly dome-shaped, dense, with numerous slender branches bearing clusters of twigs openly clothed with foliage. Twigs green, prominently grooved between the attached leaf bases. Resting buds nearly spherical, about 2 mm in diameter, tightly enclosed by roundly triangular bud scales with a narrow, upright free tip. Leaves well spaced along and projecting all around the twigs, a little more crowded near the end of the growth increment, lasting 1–2 years, dark green above, waxy whitish green over the stomatal bands beneath (3–)5–8 (–12) cm long, 2–3(–4.5) mm wide. Blades straight or a little curved to one side, widest near the middle or below, tapering very gradually to the narrowly pointed, prickly tip and more abruptly to the wedge-shaped or round base with a very short petiole about 1–2 mm long. Midrib slightly grooved or flat above, narrowly raised beneath, with one resin canal below the midvein, well-developed support tissue extending out to the sides, and an uninterrupted layer of hypodermis beneath the upper epidermis. Pollen cones (6–)8–12(–15) mm long and 1.5–2.5 mm wide, (2–)3–5(–6), at the end of a short, leafless stalk 6–15 mm long. Pollen scales with a minutely jagged, upturned triangular tip less than 0.7 mm long. Seed cones on a short, leafless stalk 5–10(–20) mm long, without basal bracteoles, the reproductive part with (two or) three (or four) unequal bracts, these and the axis becoming somewhat swollen and juicy, reddish purple, (3–)4–6 mm long by 2–4 mm in diameter. Fertile seed scales one (or two), the combined seed coat and epimatium with a swollen fleshy layer over a hard inner shell, blackish green, 5–8 mm long by 4–6 mm wide, with an inconspicuous crest along one side. Southeastern Bolivia and northwestern Argentina. Forming groves in ravines and on slopes in cloud forests; (700–)1,200–2,500(–3,000) m. Zone 9?

Bolivian cerro podocarp, when large enough, yields a valuable timber with varied uses in carpentry and furniture making. Smaller trees are used for paper pulp. It is a handsome ornamental but is not much grown outside of its native range, where it is one of the characteristic species of the cloud forests of the lower eastern slopes of the Andes (a region called the cerro in Argentina and Bolivia, hence the common name). The scientific name honors Filippo Parlatore (1816–1877), the Italian botanist who first formally described the species as *Podocarpus angustifolia,* a name that, as it turned out, had already been used. It belongs to a group of species that are found mostly in seasonal habitats in South America, not the tropical lowland and montane rain forests in which most of the South American *Podocarpus* species are found. Related species include willow podocarp (*P. salignus*) in central Chile and Lambert podocarp (*P. lambertii*) in southeastern Brazil and northeastern Argentina.

Podocarpus pendulifolius J. Buchholz & N. Gray

HAYUCO PODOCARP, PINO HAYUCO (SPANISH)

Podocarpus subg. *Podocarpus* sect. *Pratensis*

Tree to 10(–25) m tall, with trunk to 0.5 m in diameter. Bark grayish brown, fibrous. Crown dome-shaped, with short, stout branches bearing clustered branchlets somewhat openly clothed with foliage. Twigs green, distinctly grooved between the attached leaf bases. Resting buds bluntly rounded, 2–3 mm long and 3–4 mm in diameter, tightly enclosed by broadly triangular, keeled bud scales with thin, dry edges. Leaves somewhat concentrated near the tips of the twigs, attached all around but hanging down stiffly on either side of the twig through bending of the 2–3 mm long petiole, shiny light green above, a little paler but not waxy beneath, (4–)7–9(–12) cm long, (5–)6–9 mm wide. Blades generally straight, widest near the middle, from there tapering very gradually to both the narrowly triangular tip and the narrowly wedge-shaped base. Midrib with a groove above bordered by low ridges and with a low, rounded ridge beneath, with one resin canal beneath the midvein and a partly doubled, continuous layer of hypodermis beneath the upper epidermis. Pollen cones (1.2–)1.5–2(–2.5) cm long and 4–6(–7) mm wide, one directly in the axils of the foliage leaves virtually without a stalk. Pollen scales with a bluntly triangular, minutely toothed, upturned tip 1–1.5 mm long. Seed cones on a leafless stalk 7–15(–18) mm long, without basal bracteoles, the reproductive part with two (or three) unequal bracts, these and the axis becoming very swollen and juicy, dark red to purple with a waxy coating, 7–9 mm long. Fertile seed scales one (or two), the combined seed coat and epimatium with a leathery skin over a hard inner shell, grayish green with a waxy coating, 7–9 mm long by 5–6 mm thick, with a low, rounded crest along one side and forming a low beak. Sierra Nevada and other ranges of the cordillera in western Venezuela in the states of Táchira, Mérida, and Trujillo. Scattered as single trees or small groves in rocky, exposed, open habitats and in cloud forests and in the evergreen dry lower montane forests below them; (1,400–)1,800–2,600(–3,200) m. Zone 10.

With its stiffly hanging leaves (hence the scientific name, Latin for "hanging leaf"), hayuco podocarp is one of the most distinctive species of *Podocarpus* in the New World. The only other tropical American podocarp with hanging leaves is another Venezuelan species, *P. salicifolius,* found nearby to the northeast in the Venezuelan coastal range, but the two are not closely related. Hayuco podocarp is, however, closely related to the widely distributed olive-leaf podocarp (*P. oleifolius*) but is found just in a very restricted part of the range of the latter, where it can be quite common. Where these two species grow together in the mountains of western Venezuela, hayuco podocarp often grows at somewhat lower elevations and in rockier, drier, more open sites. As a result, hayuco podocarp is usually a dwarf tree less than 10 m tall and only reaches its maximum height in very favorable sites, like those it shares with olive-leaf podocarp. Hybrids should be sought in these favorable sites and in other places where they grow together.

Podocarpus pilgeri Foxworthy

PILGER PODOCARP, XIAO YE LUO HAN SONG (CHINESE), JAMEGA, PULING, YAMGA (PAPUAN LANGUAGES)

Podocarpus subg. *Foliolatus* sect. *Gracilis*

Tree to 15(–25) m tall or shrubby at times, with trunk to 0.6 m in diameter. Bark smooth, gray. Crown spreading, with numerous slender branches bearing single, paired, or clustered branchlets densely clothed with foliage at the tips. Twigs pale brown, grooved between the attached leaf bases, sometimes minutely hairy. Resting buds pointed, (1–)2–3 mm long, tightly clothed by narrowly triangular bud scales with slightly spreading or tight tips. Leaves densely and evenly standing out and forward around and along the twigs or clustered at intervals or near the tips, thin to leathery, stiff, sometimes waxy-coated at first, shiny dark green above, duller beneath, (1–)2–5(–8) cm long, 3–8(–12) mm wide. Blades straight or curved to one side, the edges variably turned down, widest near the middle, tapering quickly to the triangular or rounded tip and to the triangular base on a short petiole 1–3 mm long. Midrib narrowly and sharply raised above, more broadly and shallowly raised beneath, with (one to) three resin canals beneath the midvein, wings of support tissue extending out to the sides, and an interrupted layer of large hypodermis cells beneath the upper epidermis. Pollen cones (1–)2–3.5 cm long (to 5 cm after the pollen is shed), (1.5–)3–4(–5.5) mm wide, one (or two) directly in the axils of the foliage leaves or becoming scaly-stalked (on a stalk usually 1–3 mm long but up to 10 mm) by elongation between the basal sterile scales. Pollen scales with a broadly triangular, upturned tip less than 1 mm long bearing a tiny projecting point. Seed cones on a short, leafless stalk 2–13 mm long, with a pair of needlelike bracteoles 1.5–2.5(–4.5) mm long, the reproductive part with two slightly unequal bracts, these and the axis becoming juicy and swollen, red to purple, (5–)6–9(–12) mm long by 3–8 mm thick. Fertile seed scale one, the combined seed coat and epimatium leathery, green to purple, 7–9(–12) mm long by 5.5–7 mm thick, rounded at the tip or with a small crest and beak. Widespread in mountains of southeastern Asia (southernmost China, Vietnam, Laos, Thailand, and Cambodia) and of eastern Malesia, from Luzon (Philippines) and eastern Borneo (Sabah and eastern Kalimantan) through Sulawesi, the Moluccas, and New Guinea to the Solomon Islands. Mostly in mossy subalpine forests but extending down into other shady, cool montane forests; (700–)1,000–3,000(–3,800) m. Zone 9. Synonyms: *P. brevifolius* (Stapf) Foxworthy, *P. glaucus* Foxworthy, *P. lophatus* de Laubenfels, *P. ramosii* R. Mill, *P. rotundus* de Laubenfels, *P. tixieri* H. Gaussen, *P. wangii* C. C. Chang.

As a broadly distributed inhabitant of widely separated subalpine and upper montane forests covering a broad elevational range as well, Pilger podocarp displays a great range of variation in leaf features and other characteristics. As a result, many of the populations were described as separate species. While it is true that many of these populations have distinctive combinations of characters, taken as a whole they mostly just recombine features also found in other populations. A range-wide revision of this species that could provide a consistent basis for recognizing botanical varieties, if possible, would be welcome. The species name honors Robert Pilger (1876–1953), a German botanist who completed the first modern revision of *Podocarpus* in 1903.

Podocarpus polyspermus de Laubenfels

MÉ MAOYA PODOCARP

Podocarpus subg. *Foliolatus* sect. *Longifoliolatus*

Tree to 15 m tall. Bark fibrous, shallowly furrowed, dark brown to yellowish brown at first, weathering grayish brown. Branching sparse, the branchlets densely clothed with foliage near their tips. Twigs coarse, green, prominently grooved and with prominent leaf scars. Resting buds elongate, 10–15 mm long, loosely constructed of upright bud scales, the outer longer and leaflike. Leaves extending out and a little forward all around the twigs, crowded at the tips, leathery, lasting 1–2 years, grayish green, (4–)8–12 cm long (to 21 cm in juveniles), (5–)6–8(–9) mm wide (to 13 mm in juveniles). Blades straight or a little curved to one side, the edges turned down, broadest beyond or near the middle, tapering gradually to a narrowly triangular, sharp point and even more gradually to the narrowly wedge-shaped base on a broad petiole to 1 cm long. Midrib broadly and prominently raised above and raised within a groove or simply and broadly grooved beneath, with three resin canals beneath the midvein and probably four above, abundant support tissue extending out to the sides, and an interrupted layer of hypodermis beneath the upper epidermis. Pollen cones 2–2.5 cm long and 3.5–5 mm wide, single directly in the axils of foliage leaves, the persistent bud scales at the base rounded and keeled, curved inward. Pollen scales with a narrowly triangular, inwardly curled, upturned tip about 1.5 mm long and 0.5 mm wide. Seed cones on a short, leafless stalk 8–12 mm long, with two needlelike basal bracteoles 3–5 mm long, the reproductive part with four (or five) nearly equal bracts with their free tips curled outward, these and the axis growing together, becoming dark red, swollen and juicy, 10–15 mm long by 6–12 mm thick. Fertile seed scales (one or) two (or three), the combined seed coat and epimatium of each leathery, 8–9(–14) mm long by 6–7 (–11) mm thick, the tip rounded, without a beak. Discontinuous in the mountains of northern and central New Caledonia. Scattered in the lower canopy and understory of montane rain forests, often on serpentine-derived soils; (50–)650–950 m. Zone 10.

Mé Maoya podocarp is very closely related to its neighbor, Mount Mou podocarp (*Podocarpus longifoliolatus*). They resemble each other in most respects, including the predominance of two seeds and occasional presence of three seeds per seed cone, which gives the species its scientific name, Greek for "many-seeded" (a slight exaggeration). The two can be distinguished from each other by the grayish green leaves of *P. polyspermus* (bright green in *P. longifoliolatus*) and by its upright bud scales (curled outward in *P. longifoliolatus*) as well as by the longer, narrower tips of the pollen scales and tighter buds of the pollen cones. *Podocarpus polyspermus* grows at lower elevations than *P. longifoliolatus*, and there appears to be a gap of at least 150 m between the upper limit of the former and the lower limit of the latter. While Mé Maoya podocarp is very discontinuous in its distribution, not being found in many seemingly suitable localities, it can be common in those localities where it is found.

Podocarpus polystachyus R. Brown ex Endlicher

SEA TEAK, SETADA, JATI BUKI, KAYU KARAMAT, KANDABANG (MALAY), ARBUDJIN (PAPUAN LANGUAGES)

Podocarpus subg. *Foliolatus* sect. *Polystachyus*

Tree to 6(–20) m tall, or shrubby on exposed headlands, with trunk to 0.5 m in diameter, sometimes buttressed. Bark brown to grayish brown, shedding in flakes and becoming shallowly furrowed with age. Crown broadly conical to dome-shaped, with slender horizontal branches upturned at the end and bearing paired or clustered branchlets densely clothed with foliage. Twigs slender, green, weakly grooved between the attached leaf bases. Resting buds egg-shaped to conical, 1.5–3 mm long and 1–2.5 mm in diameter, tightly wrapped by triangular bud scales with outwardly bent bristle tips or the outer scales somewhat needlelike, 2.5–4(–10) mm long, and upright. Leaves densely and evenly projecting along and around the twigs or more crowded near their tips, angled forward, lasting 2–3 years, leathery and stiff, shiny dark green above, pale whitish green and dull beneath, (3–)4–8(–10) cm long, (4–)6–10(–13) mm wide. Blades straight to strongly curved to one side, the edge flat, inconspicuously widest near the middle, tapering imperceptibly and then abruptly to the roundly triangular to rounded tip and more gradually to the wedge-shaped base on a petiole 1–3 mm long. Midrib sharply raised above, broader and lower or even grooved beneath, with three resin canals beneath the midvein, wings of support tissue extending out to the sides, and an incomplete layer of large hypodermis cells beneath the upper epidermis. Pollen cones 2–4(–4.5) cm long, 2.5–3.5 mm in diameter, (one to) three to five directly in the axils of the foliage leaves (hence the scientific name, Greek for "many wheat ears"). Pollen scales with a tiny, broadly triangular, upturned tip 1 mm or less long. Seed cones on a short, leafless stalk (1–)3–6 mm long, with a pair of needlelike basal bracteoles 1–2 mm long that are soon shed, the reproductive part with two (or three) similar or quite unequal bracts, these and the axis becoming swollen, juicy, and passing through red to purple, 7–10 mm long by 4–6 mm thick. Fertile seed scales one (or two), the combined seed coat and epimatium leathery over a hard inner shell, covered by a waxy film, 7–9 mm long by 5–7 mm thick, rounded at the end, without an evident crest or beak. Discontinuous from central peninsular Thailand and northern Luzon (Philippines) south to Bangka and Belitung Islands (between Sumatra and Borneo), southern Borneo, and the Vogelkop Peninsula of western Irian Jaya, New Guinea (all Indonesia). Primarily on sandy coastal dunes and sea bluffs but also on outcrops in mangroves and on low limestone hills away from the sea; 0–250(–1,000) m. Zone 9. Synonyms: *Margbensonia polystachya* (R. Brown ex Endlicher) A. V. Bobrov & Melikian, *M. thevetiaefolia* (A. Zippelius ex Blume) A. V. Bobrov & Melikian, *P. thevetiaefolius* A. Zippelius ex Blume.

This is the common seaside podocarp on the southern fringes of the South China Sea. It is often cultivated within its native range, making an upright small tree with dense, glossy foliage. It rather closely resembles the more widely cultivated Buddhist pine (*Podocarpus macrophyllus*) in general appearance and foliage, but the latter has the edges of the leaves conspicuously turned under. In reproductive condition, pollen cones of Buddhist pine are a little smaller than those of sea teak, while the seed cones are on a longer stalk and have longer basal bracteoles. Individuals of sea teak growing inland on limestone hills have especially crowded, stiff leaves at the shorter and wider extremes for the species. They were separated as a distinct botanical variety by some authors, but the same features can be found in some coastal plants, so the distinction is not maintained here. The species is not currently of conservation concern, although further development of beach resorts in the region could change its status.

Podocarpus pseudobracteatus de Laubenfels

KEBU PODOCARP, KEBU, KAIP (ENGA), PULING (CHIMBU)

Podocarpus subg. *Foliolatus* sect. *Longifoliolatus*

Tree to 15 m tall or a shrub mature at as little as 1 m, with trunk to 0.2 m in diameter. Twigs coarse. Resting buds elongate, 5–9(–14) mm long by 3–5 mm in diameter, loosely constructed of narrowly triangular to sword-shaped bud scales of variable length within a bud, the outer ones often spreading or curling back at their tips. Leaves extending out evenly around the twigs, leathery and thick, dark green above, paler beneath, 6–11(–15) cm long (to 22 cm in juveniles), 7–12 mm wide (to 17 mm in juveniles). Blades straight or curved to one side, the edges flat, inconspicuously broadest below or occasionally near the middle, tapering gradually to the triangular tip and more abruptly to the roundly wedge-shaped base on a petiole 2–4 mm long. Midrib narrowly and steeply raised above, almost flat or grooved beneath, with three resin canals beneath the midvein, wings of support tissue extending out to the sides, and an interrupted layer of hypodermis beneath the upper epidermis. Pollen cones 4–5 cm long, single directly in the axils of the foliage leaves or on a very short, leafless stalk to 2 mm long. Pollen scales with a triangular upturned tip less than 1 mm long. Seed cones on a short, leafless stalk (2–)5–10(–14) mm long, with a pair of needlelike basal bracteoles 2–3.5 mm long, the reproductive part with two unequal bracts, these and the axis becoming swollen, juicy, and darkening through orange and red to almost black, 6–11 mm long. Fertile seed scale one, the combined seed coat and epimatium leathery over a hard inner shell, 10–12 mm long by 7–9 mm thick, with a low, rounded beak. Discontinuous along the mountainous spine of New Guinea from the Vogelkop peninsula in Irian Jaya (Indonesia) to near Port Moresby (eastern Papua New Guinea). Scattered singly or gregariously in high mountain, broad-leaved mossy rain forests or in conifer-dominated swamps; (1,700–)1,850–2,850(–3,500) m. Zone 9? Synonym: *P. archboldii* var. *crassiramosus* N. Gray.

Kebu podocarp is one of five *Podocarpus* species naturally restricted to New Guinea. It was confused with two of these other endemic species, soa podocarp (*P. archboldii*) and baula podocarp (*P. crassigemmis*), as well as with the widespread oleander podocarp (*P. neriifolius*). It is generally similar to all of these but differs in its beaked seed. It is even more similar to unung podocarp (*P. bracteatus*) of Java, from which it takes its scientific name (Greek for "false *bracteatus*"). That species has wider leaves, longer basal bracteoles on the seed cones, and pollen cones elongating between their bud scales to produce a scaly stalk.

Podocarpus purdieanus W. J. Hooker

PURDIE PODOCARP

Podocarpus subg. *Podocarpus* sect. *Nemoralis*

Tree to 40 m tall or more, with trunk to 1 m in diameter. Crown with spreading branches bearing branchlets in pairs or trios densely clothed with foliage. Twigs green, grooved between the attached leaf bases. Resting buds approximately spherical, 2–3.5 mm long, fairly tightly wrapped by triangular to rounded, crisp bud scales with a rib ending in a free tip. Leaves evenly distributed around the twigs, dense at the tips and bases, more open in between, shiny dark green above, paler bright green to whitish green beneath, leathery, (3.5–)4–8(–9) cm long (to 14 cm in juveniles), (6–)8–14 mm wide (to 17 mm in juveniles). Blades straight or slightly curved to one side, flat or bowed up on either side of the midrib, wrinkling on drying, the edges often turned down, fairly parallel-sided though widest in the middle third, tapering very gradually and then abruptly to the rounded tip ending in a tiny point and a little more gradually to the roundly wedge-shaped base on a short petiole to 3 mm long. Midrib grooved above, with or without a rim on either side, and raised beneath mostly near the petiole, with one resin canal beneath the midvein, wings of support tissue extending out the sides, and a continuous and sometimes partly doubled layer of hypodermis beneath the upper epidermis. Pollen cones 12–18 mm long, single directly in the axils of foliage leaves. Seed cones on a short, leafless stalk 2–4(–7) mm long, without basal bracteoles, the reproductive part with two similar-sized to noticeably unequal bracts, these and the axis becoming swollen and juicy, red, 7–8 mm long. Fertile seed scale one, the combined seed coat and epimatium leathery over a hard inner shell, 7–8 mm long by 4–5 mm thick, with a broadly conical, low beak. Mountains of central and eastern Jamaica and the central mountain range of the Dominican Republic. Scattered among evergreen hardwoods in lower montane rain forests; (600–)800–1,100(–1,200) m. Zone 9. Synonym: *P. hispaniolensis* de Laubenfels.

Purdie podocarp is an uncommon tree that grows in a habitat, the lower montane rain forest, that is subject to overexploitation in most of its range because of ready accessibility. However, the Cockpit Country of central Jamaica, at the western edge of its range, is a rugged landscape with much natural vegetation and many endemic species, so it is probably relatively secure there. The taxonomic concept of Purdie podocarp has, for some authors, embraced all the podocarps in the Greater Antilles west of Puerto Rico. For others, it has been restricted just to the trees of Jamaica. Here, more recently discovered populations in the Dominican Republic are included, but the more widespread shrubs of higher elevations are excluded as Cuban podocarp (*Podocarpus angustifolius*). The islands of Jamaica and Hispaniola contain both species, but Cuba only has Cuban podocarp. The populations of Purdie podocarp on Jamaica and Hispaniola were described as separate species, but there is little difference between them. It is not clear how closely related Purdie podocarp and Cuban podocarp are to one another and whether they should really be placed in separate botanical sections. Some knowledge of the paleobotanical history of these species in the Greater Antilles as well as DNA studies might help clarify these issues. The species name honors William Purdie (ca. 1817–1857), superintendent of the botanical garden in Trinidad from 1846 until his death, who collected the type specimen during his extensive botanical excursions in tropical America.

Podocarpus ridleyi (Wasscher) N. Gray

RIDLEY PODOCARP

Podocarpus subg. *Foliolatus* sect. *Polystachyus*

Tree to 15(–24) m tall, with trunk to 0.3 m in diameter. Crown with stout branches bearing clustered twigs densely clothed with foliage near the ends. Twigs slender, grooved between the attached leaf bases. Resting buds egg-shaped in outline, surrounded by needle-like, upright bud scales to 8 mm long (to 13 mm when leaflike). Leaves densely and evenly standing out around the twigs, more concentrated near their tips, leathery, very thick, and stiff, shiny bright yellowish green above, duller beneath, (5–)8–12(–15) cm long (to 20 cm in juveniles), (6–)9–12(–14) mm wide (to 19 mm in juveniles). Blades straight or noticeably curved to one side, the edges turned down, widest below the middle, tapering very gradually to the sharply triangular tip and more quickly to the roundly wedge-shaped base on a very short petiole 2–3 mm long. Midrib weakly to moderately raised above, more broadly raised within a bounded channel beneath, with three resin canals beneath the midvein, another two within the wings of support tissue at the sides, and a continuous extra layer of hypodermis beneath the upper epidermis in alternating groups with thin and thick walls. Pollen cones 1.5–2.5 cm long and 2–3(–4) mm wide, one to three (or four) directly in the axils of the foliage leaves or on a very short, leafless stalk 2–4 mm long. Pollen scales with a tiny, upturned, triangular tip 0.1–0.2 mm long. Seed cones on a leafless stalk (1–)3–9(–12) mm long, with a pair of needlelike basal bracteoles 2–2.5 mm long, the reproductive part with two unequal bracts, these and the axis becoming swollen, juicy, and at least pink, (5–)8–9 mm long. Fertile seed scale one, the combined seed coat and epimatium leathery over a hard inner shell, about 7 mm long by 4 mm thick, without an evident crest or beak. Isolated mountains throughout Malaya (Malaysia). Scattered to gregarious and prominent in the canopy of short, lower montane rain forests on impoverished soils; (480–)650–1,000(–1,300) m. Zone 9? Synonyms: *Margbensonia ridleyi* (Wasscher) A. V. Bobrov & Melikian, *P. neriifolius* var. *ridleyi* Wasscher.

Among the other common *Podocarpus* species of the Malay peninsula, Teijsmann podocarp (*P. teysmannii*) differs in its tightly wrapped spherical resting buds while oleander podocarp (*P. neriifolius*) and sea teak (*P. polystachyus*) both usually have just three resin canals below the midvein, with the leaf tips being abruptly prolonged in the former and rounded in the latter. Historically, this was a common tree on some mountains, and its small stature as well as that of the forests in which it grows subjected it to little exploitation pressure. The species name honors Henry Ridley (1855–1956), long-time director of the Singapore Botanic Garden, who collected extensively throughout Malaya including this species, though not its type specimen, and wrote a five-volume flora of Malaya. He is best known to biologists and geographers, however, for his important 1930 treatise on phytogeography, *The Dispersal of Plants Throughout the World,* inspired in part by his experiences in the Malesian region.

Podocarpus roraimae Pilger

RORAIMA PODOCARP

Podocarpus subg. *Podocarpus* sect. *Pumilis*

Shrub, or tree to 10(–15) m tall. Crown with numerous contorted branches bearing many branchlets densely clothed with foliage. Twigs green, grooved between the attached leaf bases. Resting buds rounded, a little broader than long, about 2 mm long and 3 mm in diameter, loosely enclosed by green, triangular (or sometimes somewhat leaflike) bud scales with a free, pointy tip. Leaves closely and evenly spaced around and along the twigs, shiny dark green above, paler and duller beneath, (1.5–)2–3 cm long (to 6 cm in juveniles), 4–6(–7) mm wide (to 8 mm in juveniles). Blades straight, the surface flat, the edges turned down, widest near the middle and tapering rapidly to the rounded or broadly triangular tip and the roundly wedge-shaped base with a short petiole to 3 mm long. Midrib raised above (the ridge often dividing into two near the tip) and shallowly grooved beneath, with one resin canal beneath the midvein, thin wings of support tissues to the sides, a continuous layer of large hypodermis cells beneath the upper epidermis, and clumps of even larger sclereids near the lower epidermis. Pollen cones probably single and less than 2 cm long. Seed cones on a short, leafless stalk 4–5(–7) mm long, without basal bracteoles, the reproductive part with two unequal bracts, these and the axis becoming somewhat swollen and fleshy, 6–8 mm long by 3–4 mm thick. Fertile seed scale one, the combined seed coat and epimatium with a leathery skin over a hard inner shell, 6–8(–10) mm long by 4–5(–6) mm thick, with a crest along one side ending in a small beak at the tip. Tepuis (isolated, flat-topped sandstone mountains) of the Guiana Highlands of Venezuela from Cerro Duida in central Amazonas state to Auyan-tepui and Mount Roraima in southeastern Bolívar state and adjacent Guyana (and possibly Brazil). Scattered in the canopy of elfin forests, the shortest form of cloud forest on high, exposed ridges; 1,800–2,400 m. Zone 9.

Roraima podocarp is named for Mount Roraima, best known as the inspiration for Arthur Conan Doyle's 1912 novel, *The Lost World.* The species is actually more abundant in the region of the Chimantá Massif, which has a greater expanse of suitable elfin forest habitat than Mount Roraima. It may well be found also on the other high mountain masses of the region, Cerro Marahuaca, Meseta de Jána, Sarisarinama-jidi, and Aparaman-tepui, the only ones reaching the elevation zone where Roraima podocarp grows. It is closely related to Buchholz podocarp (*Podocarpus buchholzii*), which differs most obviously in having a grooved midrib above and grows at even higher elevations on the Chimantá Massif.

Podocarpus rostratus L. Laurent

LAURENT YELLOWWOOD

Podocarpus subg. *Podocarpus* sect. *Scytopodium*

Tree to 10(–30?) m tall. Bark fiery red, weathering gray. Twigs nearly smooth or slightly grooved between the elongate, attached leaf bases. Resting buds somewhat flattened, about 1 mm long, with tightly overlapping bud scales, the outer ones spreading slightly at the tips. Leaves angled forward evenly all around and along the twigs, lasting 2–4 years, dark green above, paler beneath, (0.7–)1–2.5 cm long (to 5 cm in juveniles), 0.8–1.5 mm wide (to 2 mm in juveniles). Blades mostly parallel-sided but reaching its greatest width near the middle, tapering abruptly to the short, pointed tip and to the virtually stalkless base. Midrib flat above and strongly protruding beneath, with three resin canals just beneath the midvein and 6–10 others scattered around the periphery of the leaf inside the epidermis. Pollen cones (1–)1.5–2 cm long and 2–3 mm wide, one to three at the tip of a short, smooth stalk to 1 cm long. Pollen scales with a rounded, narrowly triangular tip 0.8–1.1 mm long and 0.5–0.7 mm wide. Seed cones on a short, leafless stalk to 6 mm long, without basal bracteoles, the reproductive part with two unequal bracts, these and the axis becoming slightly swollen and leathery, 3.5–5 mm long by 3–3.5 mm thick. Fertile seed scale one, the combined seed coat and epimatium slightly fleshy and 12–14 mm long by 6–8 mm thick, crested along one side. Discontinuous through the central mountains of Madagascar, from the Massif du Tsaratanana to the Massif de l'Andringitra. In fire-prone subalpine heath scrub to montane forests; 1,200–2,500 m. Zone 8? Synonym: *P. perrieri* H. Gaussen & Woltz.

The rare Laurent yellowwood of the high mountains of Madagascar has the narrowest leaves among the Madagascan podocarps. The needles are very short, densely crowded, and heathlike in mature trees but much longer, threadlike, and widely separated in juveniles. The presence of extra resin canals around the periphery of the needle, in addition to the three canals typically found beneath the midvein in African podocarps, was once thought to be unique to this species but is now known to occur rarely in the related and sometimes co-occurring Capuron yellowwood (*Podocarpus capuronii*) as well. Because of these extra resin canals, N. Gray placed *P. rostratus* in a group by itself, but it is so closely related to the other Madagascan species that it is best accommodated with them in section *Scytopodium*. The species name refers to the beak of the pollen scales, shared with many other species.

Podocarpus rubens de Laubenfels

BLUSHING PODOCARP, KAIP (ENGA), UNGPOP, BIN, NELIT (PAPUAN LANGUAGES)

Podocarpus subg. *Foliolatus* sect. *Foliolatus*

Tree to 30 m tall, with trunk to 0.4 m in diameter. Crown dome-shaped with drooping branches bearing paired or clustered branchlets densely clothed with foliage. Twigs coarse, green, grooved between the attached leaf bases. Resting buds sharply conical, 2–3 mm long and 2–3(–5) mm wide, loosely constructed of upright, narrowly triangular bud scales up to 7 mm long with upright or spreading free tips. Leaves densely standing out or somewhat forward around the twigs, leathery, thick, and stiff, flushing bright red (hence the scientific and common name, Latin for "reddish"), becoming shiny dark green above, paler beneath, 3–6.5 cm long (to 8 cm in juveniles), 5–11 mm wide (to 14 mm in juveniles). Blades straight or slightly curved to one side, the edges thickened and curled under, somewhat parallel-sided or widest near the middle, tapering fairly quickly to the roundly triangular or rounded, sometimes slightly prolonged tip and more gradually to the roundly wedge-shaped base on a petiole 2–4 mm long. Midrib flat to squarely and narrowly raised above, more broadly raised beneath, with three resin canals beneath the midvein, wings of support tissue extending out to the sides, and a scattering (or very rarely a continuous layer) of hypodermis cells beneath the upper epidermis. Pollen cones 2–3.5 cm long and 2.5–4 mm wide, one to three directly in the axils of the foliage leaves or on a short, leafless stalk to 5 mm long. Pollen scales with a bluntly triangular, upturned tip to 0.5 mm long. Seed cones on a short, leafless stalk 4–9 mm long, with two needlelike basal bracteoles to 1.5 mm long, the reproductive part with two unequal bracts, these and the axis becoming swollen, juicy, and passing through red to purple, 6–8 mm long by 4–5 mm thick. Fertile seed scale one, the combined seed coat and epimatium leathery over a hard inner shell, 8–9 mm long by 5–6 mm thick, rounded or with an inconspicuous beak at the tip. Scattered across southern Indonesia and Papua New Guinea from the southern half of Sumatra, through Borneo, Sulawesi, the lesser Sunda Islands, and New Guinea to Normanby Island and western New Britain. Scattered to gregarious in the understory or canopy of lower montane rain forests, mossy forests, and swamp forests on slopes and ridges; (800–)1,500–3,000(–3,200) m. Zone 9? Synonyms: *P. indonesiensis* de Laubenfels & Silba, *P. neriifolius* var. *timorensis* Wasscher.

The red leaf flushes that give blushing podocarp its name are brighter and more conspicuous than those of the closely related and more widespread oleander podocarp (*Podocarpus neriifolius*), which grows generally at lower elevations. The leaves of oleander podocarp are larger than those of blushing podocarp and usually have a more broadly raised midrib on the upper side. The podocarpium of oleander podocarp remains red at maturity, and the pollen cones are slightly more slender. Blushing podocarp reaches its greatest prominence in forests of low stature on ridge tops where it can form a significant portion of the canopy.

Podocarpus rumphii Blume

RUMPF PODOCARP, MALAKA NAYAN (FILIPINO), KAYU CHINA (MALAY), WASABRARAN (BIAK OF NEW GUINEA)

Podocarpus subg. *Foliolatus* sect. *Rumphius*

Tree to 35(–45) m tall, with trunk to 0.8 m in diameter. Bark reddish brown, smooth, peeling in irregular flakes. Crown dome-shaped, with straight branches bearing clustered branchlets densely clothed with foliage at the tips. Twigs green, slender, prominently grooved between the attached leaf bases. Resting buds nearly spherical to roundly conical, 2.5–4 mm long, tightly wrapped by roundly triangular bud scales with a pointed tip that sometimes curls outward. Leaves openly distributed along and around the twigs, somewhat concentrated toward the tips, leathery, stiff, and tough, shiny dark green above, paler beneath, (6–)12–22 cm long (to 26 cm in juveniles), 10–23 mm wide (to 29 mm in juveniles). Blades straight or slightly curved to one side, nearly parallel-sided, tapering abruptly to the triangular tip and to the roundly wedge-shaped base on a petiole 4–10 mm long. Midrib broadly and shallowly raised above and less conspicuously so beneath, with three resin canals beneath the midvein, interrupted wings of support tissue extending out to the sides, and a nearly continuous layer of hypodermis beneath the upper epidermis. Pollen cones 3.5–4.5 cm long and 2–3 mm wide, (one to) three to five (to eight) directly in the axils of the foliage leaves or rarely on a very short stalk to 3 mm long. Pollen scales with a tiny, rounded, upturned tip less than 1 mm long. Seed cones on a short, leafless stalk (2–)7–10(–16) mm long, with two needlelike basal bracteoles about 1.5 mm long, the reproductive part with two or three very unequal bracts, these and the axis becoming swollen, juicy, and red, (6–)8–12 mm long by (3.5–)5–8 mm thick. Fertile seed scales one or two, the combined seed coat and epimatium waxy bluish green, leathery over a hard inner shell, (10–)12–15 mm long by 10–12 mm thick, without a crest or beak. Widely and discontinuously distributed on the Malesian islands and mainland, from Luzon (Philippines) to Malaya (Malaysia), Java (Indonesia), and Papua New Guinea. Scattered in the canopy or sometimes as an emergent in lowland or lower montane primary rain forests; 0–200(–1,550) m. Zone 9. Synonyms: *Margbensonia koordersii* (Pilger ex Koorders & Valeton) A. V. Bobrov & Melikian, *M. philippinensis* (Foxworthy) A. V. Bobrov & Melikian, *M. rumphii* (Blume) A. V. Bobrov & Melikian, *P. koordersii* Pilger ex Koorders & Valeton, *P. philippinensis* Foxworthy.

It is not clear whether Rumpf podocarp had a more continuous distribution in the recent past than it has today. It produces a fine timber and was heavily exploited throughout most of its range, so large trees are found only in protected areas. It is very closely related to De Laubenfels podocarp (*Podocarpus laubenfelsii*), which replaces it at higher elevations in Borneo, and to Gray's plum pine (*P. grayae*), which replaces it in Queensland. The species name honors Georg Rumphius (born Rumpf, 1628–1702), a Dutch naturalist with the Dutch East India Company on Ambon Island, whose *Herbarium Amboinense*, published posthumously in 1741, was the first botanical account of many Malesian plants.

Podocarpus rusbyi J. Buchholz & N. Gray

RUSBY PODOCARP

Podocarpus subg. *Podocarpus* sect. *Lanceolatis*

Shrub, or tree to 20 m tall. Branches with paired or clustered branchlets densely clothed with foliage. Twigs green, deeply grooved between the attached leaf bases. Resting buds elongated, often more than 10 mm long, tightly wrapped by narrowly triangular, somewhat leafy bud scales with long, sharp, free tips ending in a bristle. Leaves densely and evenly attached along and around the twigs, light green above at first and darkening with age, paler beneath, lasting 3–5 years, stiff and leathery, thinning toward the flat or turned down edges, 2–5 cm long, 4–7 mm wide. Blades straight, widest near the middle, tapering rapidly to the prolonged, narrowly triangular tip and to the wedge-shaped base on a very short petiole. Midrib grooved above and raised beneath, with one resin canal beneath the midvein and a continuous layer of hypodermis beneath the upper epidermis. Pollen cones about 1 cm long and 2.5 mm wide, one directly in the axils of foliage leaves. Pollen scales with an upturned triangular tip less than 1 mm long. Seed cones on a very short, leafless stalk to 4 mm long, without basal bracteoles, the reproductive part with two (or three) very unequal bracts, these and the axis becoming swollen and juicy, reddish purple, 6–7 mm long. Fertile seed scales one (or two), the combined seed coat and epimatium with a thin, leathery skin over a hard inner shell, green, 6–7 mm long by 5–6 mm thick, with a crest along one side ending in a short, blunt beak. Known only from central Bolivia, from La Paz to western Santa Cruz. Scattered in cloud forests; 2,800–3,300 m. Zone 9.

Rusby podocarp is a very local species, only collected a few times in Bolivian cloud forests, mostly in the vicinity of Mapiri. It is, perhaps, most closely related to *Podocarpus steyermarkii* of Venezuela, which is generally a little larger in most features of the resting buds, leaves, and cones. It is not, on the other hand, closely related to any other Bolivian species of *Podocarpus*. There are many examples of species or species pairs that jump the gap between some portion of the Andes and the mountains of the Guiana Highlands in and around southeastern Venezuela. The species name honors Henry Rusby (1855–1940), an American botanist, Dean of the College of Pharmacy of Columbia University and one of the founders of the New York Botanical Garden, who collected the type specimen during his explorations in South America, with an emphasis on medicinal plants.

Podocarpus salicifolius Klotzsch & H. Karsten ex Endlicher

WILLOWLEAF PODOCARP, PINABETE (SPANISH)

Podocarpus subg. *Podocarpus* sect. *Lanceolatis*

Tree to 10(–20) m tall, with trunk to 0.4 m in diameter. Twigs prominently grooved between the attached leaf bases. Resting buds elongated, 5–15 mm long (to 25 mm if the bud scales are leafy), often rather loosely constructed of narrowly triangular bud scales with spreading, prolonged tips. Leaves densely and evenly distributed along and around the twigs and often drooping or hanging from them, leathery, bright green above, paler beneath, (6–)8–13 cm long (to 18 cm in juveniles), 7–11 mm wide (to 16 mm in juveniles). Blades straight or gently curved to one side, smooth and not wrinkling when dry, fairly parallel sided though widest near the middle, tapering gradually to the sharply triangular tip and more abruptly to the wedge-shaped base and the short, winged petiole 2–3 mm long. Midrib nearly flat above and broadly raised beneath, with one resin canal beneath the midvein and with a continuous layer of hypodermis beneath the upper epidermis and an interrupted layer of larger hypodermis cells next to the lower epidermis. Pollen cones (4–)6–8 cm long and 3–4 mm wide, one directly in the axils of the leaves but the sterile scales shed by maturity to reveal a scarred stalk up to 8 mm long. Pollen scales with a tiny, upturned, sharply triangular tip 1–2 mm long. Seed cones on a long, slender, leafless stalk (1–)1.5–2.5(–3) cm long, without basal bracteoles, the reproductive part with two (or three) unequal bracts, these and the axis becoming somewhat swollen and fleshy, dark purple, (6–)8–10(–12) mm long by 5–7 mm wide. Fertile seed scales one (or two), the combined seed coat and epimatium with a leathery skin over a hard inner shell, 8–12 mm long by 7–9 mm thick with a very small crest. Found only in the coastal range of Venezuela, from eastern Yaracuy to western Miranda. Scattered or forming pure stands in disturbed sites in cloud forests; 1,400–2,140 m. Zone 9. Synonym: *P. pittieri* J. Buchholz & N. Gray.

With its drooping leaves, this species superficially resembles hayuco podocarp (*Podocarpus pendulifolius*) from the cordilleran ranges in western Venezuela, but the two are not particularly closely related. Hayuco podocarp may be distinguished easily by its smaller, tight, round, resting buds and the grooved midrib. While both grow in cloud forests, hayuco podocarp also reaches up another 800 m into the lower subalpine forests. They grow within about 200 km of each other at the closest approximation of their ranges. Since willowleaf podocarp grows near the most urbanized portions of Venezuela, it is threatened by general forest exploitation and habitat destruction even though it is not a commercial species. On the other hand, its predilection for disturbed sites gives it some advantage when exploitation is limited. The species name (Latin for "willow leaf") refers to the shape of the leaf, which is not all that unusual among *Podocarpus* species. It was long confused with the closely related *P. coriaceus,* which is found only in the West Indies and does not have drooping leaves, although the shape and texture are similar. Many plant species are found in both Venezuela and the West Indies, so the confusion was not unreasonable with dried herbarium specimens in which the drooping leaves were not obvious.

Podocarpus salignus D. Don

WILLOW PODOCARP, MAÑÍO, MAÑÍO DE LA FRONTERA (SPANISH)

Podocarpus subg. *Podocarpus* sect. *Capitulatis*

Tree to 20 m tall, with trunk frequently crooked, to 0.5(–1) m in diameter. Bark grayish brown, fibrous, shedding in elongate flakes. Crown conical at first, becoming dome-shaped with age, very dense, with numerous slender, flexible, and slightly drooping branches bearing scattered and clustered branchlets densely clothed with foliage. Twigs yellowish green, inconspicuously to prominently grooved between the attached leaf bases. Resting buds dome-shaped, 2–2.5 mm long and 3–4 mm wide, somewhat loosely wrapped by broadly triangular bud scales with long free tips. Leaves mostly evenly spaced along the twigs and projecting all around or twisted horizontally to the sides, gently drooping, glossy dark or bright green above, yellowish green beneath, (5–)6–10 (–12) cm long, (3–)5–7(–8) mm wide. Blades straight or curved to one side, nearly parallel-sided, tapering very gradually to the narrowly triangular tip and more abruptly to the roundly wedge-shaped base with a short petiole to about 4 mm long. Midrib narrowly and sharply raised above, scarcely so beneath, with one resin canal beneath the midvein, water-conducting and support cells extending out from the midvein in wings to the margin, and an interrupted layer of hypodermis cells beneath the upper epidermis. Pollen cones very thin and droopy, 2–3 cm long (to 3.5 cm after the pollen is shed) and just 1–1.5 mm wide, (one to) five or six directly in the axils of foliage leaves or on a very short, leafless stalk to 2 mm long. Pollen scales with a rounded, upturned tip about 0.5 mm long. Seed cones on a short, leafless stalk (8–)10–15(–20) mm long, without basal bracteoles, the reproductive part with two or three (or four) unequal bracts, these and the axis becoming very swollen, juicy, and red to purplish red at maturity, nearly spherical, 6–8(–10) mm in diameter. Fertile seed scales one or two, the combined seed coat and epimatium with a smooth, leathery skin over a hard inner shell, reddish, 6–8 mm long by 3–4 mm thick, with a rounded crest along one side ending in a beak. Central Chile from the Río Maule at about 35°25′S to the former Osorno province of south-central Chile at about 41°S. Growing as scattered trees or small groves in the understory of temperate rain forests dominated by *Nothofagus obliqua* and other hardwoods, mostly along streamsides or on gentle, wet slopes; (0–)100–1,000(–1,500) m. Zone 8. Synonym: *P. curvifolius* Carrière.

Unlike the other southern Andean conifers, willow podocarp is endemic to Chile, with no spillover into the Argentinean lake district. Although it has a very high moisture requirement in its native range, it grows well in cultivation in many drier warm temperate regions, like southeastern Australia, and even seems to prefer well-drained soils. The wood is of excellent quality and is used locally for a variety of purposes. The species is rarely large enough or common enough for commercial exploitation but is cut to a limited extent in the provinces of Cautín and Valdivia, where the trees reach their best development in size and abundance. It has, nonetheless, become rare because of this exploitation and because of general habitat destruction within its limited range. Remnant populations still contain a great deal of genetic variation, like many other conifers, and even cultivated stands in Britain have a broad genetic base. Use of cultivated trees to aid the conservation effort for this tree, however, is complicated by the fact that it hybridizes in cultivation with the New Zealand totaras, *Podocarpus totara* and *P. cunninghamii*. Such hybrids have been distributed inadvertently as willow podocarp.

With its dense, twiggy crown of drooping branches and narrow, drooping leaves, willow podocarp really does resemble its common namesake as well as earning its scientific name, Latin for "willowlike." It is not especially closely related to its nearest neighbor, cloud podocarp (*Podocarpus nubigenus*), but has its closest relationships, instead, with five species from seasonal habitats east of the Andes, including Lambert podocarp (*P. lambertii*) and Bolivian cerro podocarp (*P. parlatorei*), nearest by in Argentina.

Podocarpus salomoniensis Wasscher

SOLOMON ISLANDS PODOCARP

Podocarpus subg. *Foliolatus* sect. *Longifoliolatus*

Tree to 15(–20) m tall, usually with a single, straight trunk. Bark smooth, thin, yellowish brown, flaking in scales and becoming shallowly furrowed in large trees. Crown open, with spreading branches drooping at their tips and bearing few branchlets somewhat openly clothed with foliage. Twigs coarse, green, conspicuously but shallowly grooved between the attached leaf bases. Resting buds elongate, loosely constructed of upright scales, the inner needlelike, to 11 mm long, with incurved, tailed tips, the outer often leaflike, to 22 mm long. Leaves somewhat openly and evenly projecting around and along the twigs, the lower ones sometimes bent back away from the tip, lasting 2 years, shiny dark green above, paler and duller beneath, thick, stiff, and leathery, 12–18 cm long, 6–9 mm wide. Blades straight or curved to one side, the edges strongly turned down, appearing nearly parallel-sided but widest below the middle, tapering inconspicuously and then more abruptly to the narrowly triangular tip and gradually to the wedge-shaped base on a short petiole 2–3 mm long. Midrib prominently raised above, sometimes within a broader channel, more broadly raised at the base beneath and flattening out or even becoming grooved near the tip, with three resin canals beneath the midvein, another one on either side inside the wings of support tissue, and an incomplete layer of small hypodermis cells beneath the upper epidermis. Pollen cones 3–5 cm long and 3–4 mm in diameter, one directly in the axils of the foliage leaves. Pollen scales with a tiny, upturned, triangular tip less than 1 mm long. Seed cones on a long, leafless stalk 10–15 mm long, with a pair of needlelike basal bracteoles about 4 mm long, the reproductive part with four unequally developed bracts in two pairs of similar length, these and the axis becoming dark red, swollen and juicy, 8–9 mm long and 8–11 mm by 4–5 mm in cross section. Fertile seed scales two but often one-seeded by abortion, the combined seed coat and epimatium leathery over a hard inner shell, about 11 mm long by 8 mm thick, without any evident crest or beak. Apparently confined to San Cristóbal Island, eastern Solomon Islands. Scattered in the understory and canopy of lower montane rain forests; 400–900 m. Zone 9.

Solomon Islands podocarp is easily distinguished from the other *Podocarpus* species native to these islands, the widely distributed oleander podocarp (*P. neriifolius*), which has wider leaves usually lacking resin canals beside the midvein, pollen cones sometimes in threes, and seed cones with two (or rarely three) bracts making up the podocarpium. The limited natural distribution of this species makes it of conservation concern although its actual situation is little known.

Podocarpus sellowii Klotzsch ex Endlicher

SELLOW PODOCARP, PINHEIRINHO DA MATA (PORTUGUESE)

Podocarpus subg. *Podocarpus* sect. *Capitulatis*

Shrub, or tree to 8(–20) m tall, with trunk, when present, to 1 m in diameter. Bark reddish brown and fibrous. Crown narrowly dome-shaped, with numerous short branches bearing paired or clustered branchlets densely clothed with foliage. Twigs green, prominently grooved between the attached leaf bases. Resting buds elongated, conical, up to 2 cm long when some of the bud scales are leaflike, these otherwise narrowly elongated with a free tip. Leaves densely inserted all around the twigs, stiff and hard, lasting about 2 years, shiny dark green above, paler but not waxy beneath, (1.5–)5–10 cm long (to 20 cm in juveniles), (4.5–)7–12 mm wide (to 20 mm in juveniles). Blades straight, the edges turned downward, widest near the middle, tapering gradually and then more abruptly to the bluntly or sharply triangular tip and to the roundly wedge-shaped base attached to a very short petiole about 2–3 mm long. Midrib flat near the tip (or sometimes all the way along), grooved with raised edges closer to the base above and raised beneath, with one resin canal beneath the midvein, wings of support tissue out to the sides, extra individual sclereids and fibers above and below the midvein, and a continuous layer of hypodermis beneath the upper epidermis. Pollen cones (0.4–)1–3(–4) cm long and (1–)2–3(–4) mm wide, one or two directly in the axils of foliage leaves or three to six (to eight) at the end of a short, leafless stalk to 2(–15) mm long. Pollen scales with a broadly triangular, irregularly toothed, upturned tip about 2 mm long. Seed cones on a short, leafless stalk (2–)4–7(–12) mm long, without basal bracteoles, the reproductive part with two or three (or four) unequal bracts, these and the axis becoming swollen and juicy, bright red, (3–)5–8(–15) mm long and 1.5–5 mm wide. Fertile seed scales one (or two), the combined seed coat and epimatium thinly fleshy over a hard, inner shell, dark bluish green, somewhat elongated, (3–)7–10 mm long by (2.5–)5–7 mm wide, with an inconspicuous crest along one side ending in a low, blunt beak. Widespread in uplands of central and southern Brazil from Rondônia and Pará to northeastern Brazil south to Rio Grande do Sul. Scattered as individual trees or small groves in many humid forests, including wetter portions of the caatinga region, gallery forests, and in Atlantic rain forests; (200–)500–1,350 m. Zone 9. Synonyms: *P. aracensis* de Laubenfels & Silba, *P. barretoi* de Laubenfels & Silba ex Silba.

Sellow podocarp, an inhabitant of varied rain forests throughout much of Brazil south of the Amazon River basin, is correspondingly variable in its stature, leaves, and pollen and seed cones. It grows in more humid environments than the only other podocarp species in southern Brazil, Lambert podocarp (*P. lambertii*). Although these two are both placed in botanical section *Capitulatis,* they are not especially closely related, and Sellow podocarp is rather anomalous within the section. It has wider leaves and larger buds than the other species in the section and usually has most of its pollen cones single or in pairs rather than the clusters of three or more that are regular in the other species. It is endangered in some parts of its range, particularly in the Atlantic forest near Brazil's largest cities, a habitat that has suffered in general through its proximity to these population centers. In other areas, Sellow podocarp, although exploited as one of the few softwoods of the region, is under less threat. The species name honors Friedrich Sellow (1789–1831), the German botanist who collected the type specimen and died while collecting in southern Brazil.

Podocarpus smithii de Laubenfels

DOUBLEPLUM PINE, SMITH PINE

Podocarpus subg. *Podocarpus* sect. *Crassiformis*

Tree to 30 m tall, with a single, straight trunk. Bark reddish brown, thin, peeling in flakes or short strips. Resting buds pointedly egg-shaped, (2–)3.5–5 mm long and 2.5–4 mm in diameter, with broad, unequal, tightly overlapping, keeled bud scales ending in a short, straight point. Leaves arranged all around the twigs, dark green above and with a thin waxy coating at first, with broad gray stomatal bands on either side of the midrib beneath, 5–11 cm long (to 16 cm in juveniles), 8–14 mm wide (to 21 mm in juveniles). Blades widest near the base, tapering gradually and then more abruptly to the narrowly triangular tip and abruptly to the wedge-shaped base with a very short petiole. Midrib prominently raised above and beneath, with one resin canal below the midvein and water-conducting support tissue extending out to the sides. Pollen cones 3–4.5 cm long and 4–6 mm wide, one to three at the end of a short, leafless stalk to 9 mm long. Pollen scales with an upturned, elongate tip to 2 mm long. Seed cones on a short, leafless stalk to 9 mm long, without basal bracteoles, the reproductive part with two (or three) nearly equal bracts, these and the axis becoming bright red, swollen, and juicy, nearly spherical and 5–6 mm in diameter. Fertile seed scales one (or two), the combined seed coat and epimatium also fleshy over a hard inner shell, darker red than the podocarpium, 15–25 mm long by 10–15 mm thick, with a prominent crest along one side ending in a pointed beak. Confined to the Atherton Tableland near Cairns in Queensland (Australia). Scattered in montane rain forest on soils of granitic origin along creeks; 900–1,200 m. Zone 10?

The rare doubleplum pine is known from only a few collections from a restricted area of northeastern Queensland that is entirely frost-free and has the highest rainfall in all of Australia, about 300 cm per year, concentrated mostly in the summer. It is one of the most distinctive species in the whole genus *Podocarpus* since both the podocarpium and the combined seed coat and epimatium are juicy, red, and berrylike, with the seed portion the more prominent. In this latter respect there is some resemblance to section *Scytopodium* of Africa, but in those species the podocarpium is not berrylike and the leaves have three resin canals beneath the midvein and others at the corners. These differences are enough to suggest that the enlarged, fleshy seeds are independently derived between doubleplum pine and the African species. Nor is this species closely related to the only other Australian species in subgenus *Podocarpus,* the shrubby mountain plum pine (*P. lawrencei*) of southeastern Australia, which instead is related to the New Zealand totara species (*P. totara* and others) in section *Australis.* The other podocarp species that grow near it in the Atherton Tableland, *P. grayae, P. dispermus,* and *Prumnopitys amara* and *P. ladei,* are all much more distant evolutionarily. In fact, doubleplum pine is probably the most taxonomically isolated of all species still retained in the genus *Podocarpus,* the only known member of its section and entirely devoid of close relatives. The species name honors Lindsay Smith (1917–1970), who hosted de Laubenfels in 1964, when he collected the type specimen on Mount Lewis.

Podocarpus spinulosus (J. E. Smith) R. Brown ex Mirbel

CREEPING PLUM PINE, DAMSON PLUM PINE

Podocarpus subg. *Foliolatus* sect. *Spinulosus*

Upright shrub to 2(–3) m tall or rarely a small tree, lower and spreading when growing along exposed sea bluffs and dunes, with many separate stems or trunks from the base. Bark on larger stems reddish brown to grayish, thin, stringy. Crown from ground level, dome-shaped or flat with slender branches bearing a few widely spaced or clustered branchlets densely clothed with foliage. Twigs light green, shallowly grooved between the attached leaf bases. Resting buds pointed, 2–4(–6) mm long and a little narrowed, tightly wrapped by narrowly triangular bud scales with a long, free tip. Leaves evenly spaced along the twigs and standing stiffly upright all around them, shiny green above, paler but not waxy beneath, (2–)3–6.5(–8.5) cm long, (2–)3.5–4.5(–5) mm wide. Blades sword-shaped, straight or slightly curved to one side, widest below the middle, tapering very gradually and then abruptly to the triangular tip with a long, sharp point and fairly abruptly to the roundly wedge-shaped base attached to the twig by a very short petiole to 2 mm long. Midrib raised both above and below, with one resin canal beneath the midvein and a discontinuous layer of hypodermis beneath the upper epidermis, mostly above the midvein and at the leaf edge. Pollen cones 4–8(–10) mm long (to 15 mm after the pollen is shed), 1–3(–4) mm wide, (one or) two to four (or five) on a very short, leafless stalk to 3 mm long. Pollen scales with a tiny, upturned triangular tip less than 1 mm long. Seed cones on a short, leafless stalk to 10 mm long, with a pair of basal bracteoles, these falling at maturity, the reproductive part with (two or) three or four unequal bracts, these and the axis becoming swollen and juicy, dark purple to bluish black with a waxy coating, 6–10 mm long. Fertile seed scales one or two, the combined seed coat and epimatium with a thin fleshy layer over a hard inner shell, reddish purple, (8–)10–15(–20) mm long by 7–10 mm thick, with a crest along one side ending in a prominent beak. Discontinuously distributed in the coastal region of eastern Australia, from near Rockhampton in southeastern Queensland to the southeastern corner of New South Wales at its border with Victoria. Scattered among other heathlike shrubs on infertile, fire-prone sandy soils of coastal forests and shrublands; 0–250(–500) m. Zone 9. Synonym: *Margbensonia spinulosa* (J. E. Smith) A. V. Bobrov & Melikian.

Creeping plum pine is closely related to emu grass (*Podocarpus drouynianus*) of southwestern Australia, and the two species are the sole members of botanical section *Spinulosus*, distinguished from all other members of subgenus *Foliolatus* by their single resin canal and from most others by their grasslike appearance. Creeping plum pine differs from emu grass in its prickly leaves (hence the scientific name, Latin for "a little spiny") without a whitish waxy coating beneath and in the prominent beak on the mature seed. The podocarpium is edible (hence the common name) but is not of much importance today as a human food. The plant is, along with many others in the region, on the menu of *Syntherata janetta*, one of the handsome giant silkworm moths in the family Saturniidae. This suggests that creeping plum pine lacks the potent insect hormone ecdysterone found in plum pine (*P. elatus*) of the same area. The classic work of Baker and Smith at the beginning of the 20th century on the chemistry of Australian conifers also failed to isolate any essential oils from the leaves of this species, even though it has a resin canal running down the length of each leaf.

Podocarpus sprucei Parlatore

HUAPSAY PODOCARP, HUAPSAY (QUECHUA AND SPANISH)

Podocarpus subg. *Podocarpus* sect. *Capitulatis*

Shrub, or tree to 10 m tall, with trunk to 1 m in diameter. Bark reddish brown, shallowly furrowed, fibrous. Crown dome-shaped, with numerous crooked, forking branches bearing clustered branchlets densely clothed with foliage. Twigs green, prominently grooved between the attached leaf bases. Resting buds nearly spherical, 2–2.5 mm in diameter, the tightly enclosing bud scales irregularly toothed, with a keel ending in a free tip or not and grading from narrowly sword-shaped outer scales to broad inner ones. Leaves stiffly and evenly held out along and around the twigs, directed somewhat forward, lasting 2–3 years, shiny dark green above, pale bluish green but not waxy beneath, 2.5–4 cm long (to 7 cm in juveniles), 3–5.5 mm wide. Blades distinctly widest near the middle to nearly parallel-sided, tapering gradually and then abruptly to the short, pointed, sometimes prickly tip and to the wedge-shaped base attached directly to the twig or with a very short petiole to 3 mm long, the edges turned down. Midrib grooved or sometimes slightly raised above, prominently grooved beneath, with one resin canal beneath the midvein, wings of support tissue extending out to the sides, and a discontinuous layer of hypodermis beneath the upper epidermis. Pollen cones 6–10 mm long and 2–2.5 mm wide, occasionally one directly in the axil of a foliage leaf but more frequently (4–)6–10(–12) on a stalk 15–25 mm long that is branched and scaly at the end. Pollen scales with a crisp, bluntly triangular, upturned tip less than 1 mm long. Seed cones on a short, leafless stalk 3–7 mm long, without basal bracteoles, the reproductive part with three or four unequal bracts, these and the axis becoming swollen and juicy, red, 6–7 mm long and a little narrower. Fertile seed scales one (or two), the combined seed coat and epimatium thinly fleshy over a hard inner shell, nearly spherical, dark bluish green, 7–8 mm in diameter, with a low, narrow crest along one side but without a beak. Central and outer Andes from central Ecuador (Bolívar province) to adjacent northwestern Peru (Department of Piura). Scattered as individual trees or shrubs among more numerous hardwoods in montane forests and subalpine forests and shrublands; (2,000–) 2,500–4,000 m. Zone 9.

Huapsay podocarp is closely related to intimpa podocarp (*Podocarpus glomeratus*) and replaces it in the central and western Andean ranges. It has a much narrower range than intimpa podocarp, being almost confined to Ecuador, but has a more continuous distribution than intimpa podocarp and is probably more common within its range, in part because it is a smaller tree with a lower level of exploitation. Intimpa podocarp has, on average, narrower leaves than huapsay podocarp, most prickly, smaller clusters of pollen cones on an unbranched stalk, and smaller seed cones. The species name honors Richard Spruce (1817–1893), an English botanist who collected the type specimen during the 15 years that he spent collecting plants in South America, often while desperately ill and under the most trying conditions.

Podocarpus steyermarkii J. Buchholz & N. Gray

STEYERMARK PODOCARP

Podocarpus subg. *Podocarpus* sect. *Lanceolatis*

Tree to 25 m tall. Twigs coarse, prominently grooved between the attached leaf bases. Resting buds elongate, 5–16 mm long (to 25 mm when the bud scales are leaflike), loosely assembled from narrowly triangular to sword-shaped, green bud scales with forward pointed free tips. Leaves projecting out all around the twigs, distributed all along their length but more concentrated near the tips, shiny dark green above, paler green to silvery green beneath, leathery, wrinkled across the width when dry, 3.5–8 cm long (to 14 cm in juveniles), 6–9(–11) mm wide (to 14 mm in juveniles). Blades straight or slightly curved downward or to one side near the tip, the edges a little turned under, broadest near or below the middle, from there tapering very gradually to the prolonged, narrowly triangular tip and more abruptly to the wedge-shaped base with a very short petiole or attached almost directly to the twig. Midrib narrowly grooved above and broadly raised beneath, with one resin canal beneath the midvein, wings of support tissue extending out to either side, and a continuous layer of hypodermis beneath the upper epidermis. Pollen cones 2–3 cm long (to 5 cm after the pollen is shed) and 4–5 mm in diameter, single directly in the axils of foliage leaves. Pollen scales with a bluntly triangular, upturned tip less than 1 mm long. Seed cones on a long, leafless stalk 7–15 (–20) mm long, without basal bracteoles, the reproductive part with two or three unequal bracts, these and the axis becoming somewhat swollen and fleshy, bright red, (4–)6–9 mm long by 2–3 mm thick. Fertile seed scales one (or two), the combined seed coat and epimatium leathery over a hard inner shell, smooth or ridged, dark green covered by a silvery waxy layer, 7–10 mm long by 5–7 mm thick, with a short beak. Tepuis (isolated, flat-topped, sandstone mountains) of the Gran Sabana region and vicinity of southern Bolívar state, Venezuela. Scattered as canopy trees in cloud forests; 1,800–2,400(–2,600) m. Zone 9.

This is one of the more distinctive of the seven *Podocarpus* species found in the Guiana Highlands region. Its large, spearhead-shaped leaves with a groove down the middle and the elongate resting buds with numerous narrow, green bud scales readily separate it from its neighbors. It is not closely related to any of these species but rather to willowleaf podocarp (*P. salicifolius*) of the Venezuelan coastal range and to Antillean podocarp (*P. coriaceus*) of the West Indies, which differ in their less prolonged leaf tips and in having the midrib raised above, among other features. Steyermark podocarp appears to be a rare species, with the known collections widely scattered through the region. Ongoing intense collecting efforts in this rugged region may reveal a more continuous distribution. The species name honors Julian Steyermark (1909–1988), who collected the type specimen, one of the first botanists to make a concerted effort to collect on the tepuis, beginning with five expeditions in 1944 and 1953 and ultimately yielding almost 28,000 separate specimens from the region.

Podocarpus sylvestris J. Buchholz

KAURI PODOCARP, FALSE KAURI

Podocarpus subg. *Foliolatus* sect. *Globulus*

Tree to 18(–30) m tall but sometimes only to 4 m, with trunk to 2 m in diameter. Bark reddish brown, scaly or furrowed, weathering grayish brown. Crown densely branched. Twigs slender, clustered, green, prominently grooved between the elongate attached leaf bases. Resting buds nearly spherical, 1–2 mm long, tightly wrapped by rounded bud scales. Leaves standing out from and evenly distributed around the twigs, concentrated near the tips, shiny bright green to grayish green above, paler dull green beneath, lasting 1–2 years, 5–9 cm long (to 17 cm in juveniles), 7–10 mm wide (to 15 mm in juveniles). Blades straight or slightly and simply or intricately curved to one side, flat or the thin edges slightly drooping, fairly parallel-sided or widest near the middle, tapering gradually and then abruptly to the roundly triangular, pointed or blunt tip and to the broadly to narrowly wedge-shaped base on a short petiole 1.5–4.5 mm long. Midrib slightly and broadly raised above, slightly channeled with low raised borders beneath, with three resin canals beneath the midvein, wings of support tissue extending out to the sides, and an incomplete layer of small hypodermis cells beneath the upper epidermis. Pollen cones (8–)10–18(–22) mm long and 2–3 mm wide, (1–)3 on a very short, leafless stalk about 1 mm long. Pollen scales with a bluntly triangular, stiff, upturned tip 0.3–0.5 mm long. Seed cones on a short, leafless stalk 5–10 mm long, with two tiny, needlelike basal bracteoles 1–2 mm long, the reproductive part with two unequal bracts, these and the axis becoming red, very swollen, and juicy, 6–7(–9) mm long by 6–8(–12) mm thick. Fertile seed scale one, the combined seed coat and epimatium leathery over a thin, hard, inner shell, 10–13(–16) mm long by 7–9 mm thick, topped by a minute, weak beak. Discontinuous along the length of New Caledonia. Usually scattered and often common in the understory of dense rain forests (hence the scientific name, Latin for "of the forest") but sometimes reaching the canopy on fertile soils or, in contrast, taking on a dwarf shrubby habit on exposed ridges; (50–)150–1,000(–1,350) m. Zone 9.

Kauri podocarp is very closely related to the rare Lucien podocarp (*Podocarpus lucienii*), which differs in its broader leaves and larger seed cones. It is less closely related to two other New Caledonian *Podocarpus* species with which it has been confused, Mount Mou podocarp (*P. longifoliolatus*), which has leafy bud scales, and svelte river podocarp (*P. novae-caledoniae*), which has much narrower leaves. Kauri podocarp is the most widespread and common species of *Podocarpus* in New Caledonia and has been exploited for its excellent, reddish brown timber, which resembles that of kauris (*Agathis*).

Podocarpus tepuiensis J. Buchholz & N. Gray

TEPUI PODOCARP

Podocarpus subg. *Podocarpus* sect. *Pratensis*

Shrubby tree to 10(–15) m tall, with trunk to 0.3 m in diameter. Bark thin, grayish brown. Crown broadly and irregularly conical, with numerous contorted branches bearing clustered branchlets densely clothed with foliage. Twigs green, prominently grooved between the attached leaf bases. Resting buds a little flattened, 2–3 mm long and 3 mm in diameter, tightly wrapped by rounded to broadly triangular, keeled bud scales with a pointed tip. Leaves densely and evenly arranged along and around the twigs, shiny dark green above, paler beneath, 1.5–2.5(–3) cm long (to 5 cm in juveniles), (2–)4–5(–6) mm wide (to 7 mm in juveniles). Blades generally straight, thinning to the flat or turned-down edge, broadest near the middle and from there tapering gradually and then more abruptly to the roundly triangular blunt tip and to the wedge-shaped or rounded base attached directly to the twig or with a very short petiole less than 2 mm long. Midrib deeply and narrowly grooved above, broadly and prominently raised beneath, with one resin canal beneath the midvein and continuous layers of hypodermis beneath both the upper and lower epidermis. Pollen cones 0.8–1.5 cm long and about 1.5 mm wide, one directly in the axils of foliage leaves. Pollen scales with a tiny, triangular, upturned tip less than 1 mm long. Seed cones on a very short, leafless stalk to 2 mm long, without basal bracteoles, the reproductive part with two (or three) unequal bracts, these and the axis becoming swollen and juicy, reddish purple, 4–5 mm long. Fertile seed scales one (or two), the combined seed coat and epimatium with a thin leathery skin over a hard inner shell, dark bluish green, 5–7 mm long by 4–5 mm thick, with a well-defined crest ending in a blunt beak. Mountains of southern Venezuela (southern Bolívar and northeastern Amazonas) and adjacent west-central Guyana and apparently disjunct to the northern end of Cordillera del Cóndor, province of Morona-Santiago, Ecuador. Scattered as single trees and small groves in open forests on infertile, rocky soils, especially along streamsides; (100–)1,000–2,000(–2,500) m. Zone 10.

This is one of a handful of *Podocarpus* species found on the geologically ancient, flat-topped, isolated, sandstone mountains of Venezuela known as tepuis (hence the common and scientific name). Tepui podocarp has the smallest leaves among these species, although those of Buchholz podocarp (*P. buchholzii*) and Roraima podocarp (*P. roraimae*) are not much larger, all three being far below the norm for tropical American *Podocarpus* species. It is possible that tepui podocarp is misplaced in botanical section *Pratensis* and might better join the other small-leaved tepui species in section *Pumilis.* The widely separated populations from Ecuador on the eastern flank of the Andes in the Cordillera del Cóndor also occur on infertile soils derived from sandstone. Populations collected in the western Amazonas lowlands of Venezuela near the Colombian border are ecologically more anomalous. They grow on the margins of savannas that are flooded during the summer rainy season, almost 1,000 m lower than the next lowest populations.

Podocarpus teysmannii Miquel

TEIJSMANN PODOCARP, SIKOEJOE LAUT,
KALEK ROTAN (MALAY)

Podocarpus subg. *Foliolatus* sect. *Globulus*

Tree to 12 m tall, with trunk to 0.3 m in diameter. Twigs often paired, stout, green, prominently grooved between the attached leaf bases. Resting buds spherical, 1.5–3 mm in diameter, tightly wrapped by the roundly triangular, dry-margined bud scales. Leaves openly and evenly standing out around and along the twigs or more concentrated near their tips, thick and leathery, shiny dark green above, shiny and paler beneath, 8–13 cm long (to 16 cm in juveniles), 14–21 mm wide (to 27 mm in juveniles). Blades straight, the edges flat, fairly parallel-sided or widest below the middle, tapering abruptly to the roundly triangular tip with a short, extended point and a little more gradually to the roundly wedge-shaped base on a short petiole 4–7 mm long. Midrib broadly and prominently raised above, broad and highest near the base beneath and flattening or even becoming channeled near the tip, with three resin canals beneath the midvein, wings of support tissue extending out to the sides, and a very incomplete layer of hypodermis beneath the upper epidermis. Pollen cones about 2.5 cm long, one (or two) directly in the axils of foliage leaves. Pollen scales with a small, upturned triangular tip less than 0.5 mm long. Seed cones on a short, leafless stalk 6–11 mm long, with a pair of needlelike basal bracteoles about 1 mm long that are soon shed, the reproductive part with two unequal bracts, these and the axis becoming red, juicy, and swollen, 7–9 mm long. Fertile seed scale one, the combined seed coat and epimatium leathery over a hard inner shell, 8–10 mm long by 5.5–6 mm thick, the tip rounded, without a crest or beak. Southern end of the Malay Peninsula (Malaysia) and Sumatra and adjacent islands (Indonesia). Scattered in the understory of primary or second growth rain forests, often near the seashore; 0–800(–1,140) m. Zone 9. Synonym: *P. neriifolius* var. *teysmannii* (Miquel) Wasscher.

Teijsmann podocarp has been included as a botanical variety of oleander podocarp (*Podocarpus neriifolius*), but its tight, rounded resting buds are completely unlike those of that widespread and highly variable species. It is fairly rare and this, combined with its small stature, precludes any economic exploitation directed specifically to it. The species name honors Johannes Teijsmann (1809–1882), a Dutch botanist who collected the type specimen and many other plants in Sumatra and neighboring islands.

Podocarpus totara G. Bennett ex D. Don

TOTARA (MAORI AND ENGLISH), COMMON TOTARA, MAHOGANY PINE, TAITURA, TAITEA (MAORI)

Podocarpus subg. *Podocarpus* sect. *Australis*

Tree to 30(–40) m tall, with trunk to 2(–3.6) m in diameter, often fluted. Bark thick, reddish brown, fibrous, weathering gray, peeling in long, persistent strips, and finally becoming deeply furrowed. Crown conical at first, becoming deeply dome-shaped with age and ultimately opening out and becoming irregular, with massive horizontal to upwardly angled branches bearing numerous thin branchlets densely clothed with foliage. Twigs yellowish green, prominently grooved between the attached leaf bases. Resting buds pointed, 1.5–2 mm long, with tightly overlapping, narrow bud scales with elongate, pointed tips extending beyond the mass of the bud. Leaves densely and evenly spaced along the branchlets, standing out stiffly all around them and directed slightly forward or flattened more or less into a plane on either side of the twig by twisting of the leaf bases, somber brownish green to yellowish green or bluish green above, with broad, grayish green stomatal bands occupying most of the lower surface, (1–)1.5–2.5(–3) cm long, (2.5–)3–4 mm wide (but just 1–2 mm wide in juveniles). Blades generally straight, fairly parallel-sided, tapering abruptly to the short, sharply pointed tip and more gradually to the almost stalkless base. Midrib flat above and only slightly raised beneath, with one resin canal beneath the midvein. Pollen cones 1–2 cm long and 3–4(–6) mm wide, one to three (or four) on a very short, leafless stalk to 2 mm long. Pollen scales with a broad, rounded, minutely toothed, upturned tip about 0.5 mm long and about 1 mm wide. Seed cones on a very short, leafless stalk to 3 mm long, without basal bracteoles, the reproductive part with two or three (or four) bracts, these and the axis becoming very swollen and juicy at maturity, bright red, almost spherical, 4–6 mm in diameter. Fertile seed scales one or two, the combined seed coat and epimatium with a thin black skin over a hard inner shell, 3–5 mm long by 2–3 mm thick, with a low crest along one side and a low, blunt beak. Throughout North Island and South Island (New Zealand) but most abundant in the center of North Island. Sometimes forming pure stands but more commonly a dominant or subdominant in mixed, lowland, montane, and even subalpine podocarp and hardwood forests; 0–480(–600) m. Zone 9.

Like many other New Zealand trees, totara carries a Maori name as its English common name as well, but it is unusual in also

Large old totara (*Podocarpus totara*) in nature.

Twig of totara (*Podocarpus totara*) with mature and immature seed cones; scale, 4 cm.

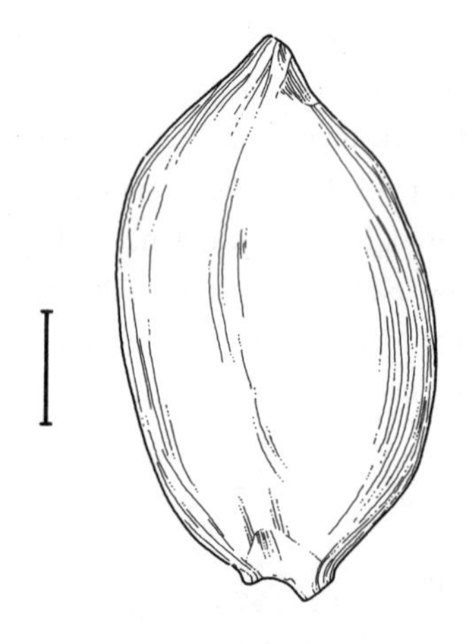

Mature seed of totara (*Podocarpus totara*) removed from the podocarpium; scale, 1 mm.

having had that name used as the scientific name. This reflects the cultural and commercial importance of this tree in New Zealand. While the qualities of its wood are generally a little inferior to those of kauri (*Agathis australis*) in most respects, except for durability in contact with the soil, the latter species has such a restricted range that totara was the premier timber tree of the islands. In contrast to podocarp species of Africa, generally known as yellowwoods for their heartwood color, the heartwood of the New Zealand totaras is reddish brown and excellent for all uses where attractively figured grain is not required. In traditional Maori societies it was the most important wood for making sea- and river-going canoes of all sizes, with their intricately carved bows. It was also used for the facades of public buildings and in other applications requiring carving. Milling of trees for timber through to the mid-20th century and conversion of forests to farmland demarcated by totara fence posts depleted this species. Now the forest giants that could be hollowed out into great, seaworthy canoes for a hundred rowers and passengers are all but gone except for parks and protected areas. Forestry in New Zealand is now largely built on plantations of introduced trees, especially Monterey pine (*Pinus radiata*) from California.

The Pouakani totara, on the western side of North Island, is the largest known living individual at about 40 m tall and 3.6 m in diameter and has been estimated to be 1,800 years old. The oldest documented age, however, is near 900 years, and the typical life span for an old tree is about 600 years. The tree grows best and lives longest on deep soils of river valleys, but it will even thrive in the subalpine forests, although here it is usually outnumbered by its close relative, Hall totara (*Podocarpus cunninghamii*), as also in the southernmost part of their mutual range. Natural regeneration is more common under the shade of hardwoods in the mixed forests in which they grow than under their own shade or that of the other podocarps in these forests.

Totara is commonly cultivated in New Zealand and in other moist, warm temperate regions as a shapely if dark tree or perhaps even more commonly as a hedge plant that closely resembles yews (*Taxus*) and that, like them, responds well to clipping. There is significant natural variation for frost tolerance in totara, correlated primarily with elevation, and it might be possible to find hardier strains than are currently in common cultivation, but not much gain can be expected in this direction. Several cultivars have been selected, some of them derived from the natural hybrids known to form with all three other New Zealand *Podocarpus* species. The most common hybrids are with its closest relative and generally nearest neighbor, Hall totara. Hybrids with the two shrubby species are found primarily at the upper elevational limit for totara. The existence of all of these hybrids has been documented using flavonoid pigments, but little work has been done on the pattern of natural hybridization in the field, an exception being a study of hybridization between totara and prickly totara (*P. acutifolius*) in Westland of South Island.

Podocarpus trinitensis J. Buchholz & N. Gray

TRINIDADIAN PODOCARP

Podocarpus subg. *Podocarpus* sect. *Nemoralis*

Tree to 20 m tall, with trunk to 0.8 m in diameter, becoming fluted with age. Bark smooth, dark grayish brown. Twigs green, prominently grooved between the attached leaf bases. Resting buds roughly spherical, about 3 mm long, loosely constructed of triangular bud scales with a fingerlike free tip. Leaves distributed around the twigs, slightly closer together near the ends of the growth increments, leathery, flat or a little wrinkled, shiny dark green above, paler beneath, 3–6 cm long, 7–10 mm wide. Blades straight or a little curved to the side, slightly turned down at the edges, widest near the middle and from there tapering fairly quickly to the roundly triangular tip and the similarly shaped base with a short, narrowly winged petiole 1–3 mm long. Midrib with an elevated groove above and broadly raised beneath, with one resin canal beneath the midvein, wings of support tissue extending out to the sides, and a continuous layer of hypodermis beneath the upper epidermis. Pollen cones 2 cm long or more. Seed cones on a short, leafless stalk 7–9 mm long, without basal bracteoles, the reproductive part with two unequal bracts, these and the axis becoming somewhat swollen and juicy, 6–8 mm long by 2.5–4 mm thick. Fertile seed scale one, the combined seed coat and epimatium leathery over a hard inner shell, 7–8 mm long by 5–6 mm thick, with or without a weak crest. Northern Range of Trinidad (hence the name). In gallery forests and on mountain slopes; 250–750 m. Zone 10.

Trinidadian podocarp is closely related to several species of *Podocarpus* growing in Venezuela, including Ulcumanu podocarp (*P. celatus*) and broadleaf podocarp (*P. magnifolius*). The former has larger seeds and a more pronounced groove down the midrib, while the latter has larger leaves with a tailed leaf tip. Many of the plants growing in Trinidad are closely related to or even identical with species growing in Venezuela. Of course, many species have an even wider distribution, but where there is a localized affinity for the island plants, it lies with Venezuela.

Prumnopitys R. Philippi

PLUM-SEEDED YEW, MIRO

Podocarpaceae

Evergreen trees of varying habit. Trunk cylindrical to irregular, columnar and limbless to a great height or breaking up into branches near the base. Bark obscurely fibrous, thin, scaly, shedding continuously in irregular patches ranging in size from flakes to small plates to produce a mottled appearance, sometimes becoming shallowly furrowed into narrow, vertical strips with age. Crown dense, spherical, egg-shaped, or cylindrical, compact to spreading, becoming irregular with age, with numerous slender, upwardly angled branches, often rapidly subdividing toward the periphery. Branchlets all elongate, without distinction into short and long shoots, hairless, remaining green for at least the first year, obscurely to prominently grooved between the elongate, attached leaf bases. Resting buds small, well developed, generally spherical and surrounded by specialized, green bud scales. Leaves spirally attached but often presented primarily in two flat rows by bending of the petioles, usually needlelike (broadly sword-shaped in *Prumnopitys amara*) and strongly resembling those of yews (*Taxus*), straight or slightly to moderately sickle-shaped, flattened top to bottom.

Plants dioecious (or rarely some individuals monoecious). Pollen cones single at the tip of or in a leaf axil on an ordinary foliage shoot or, more commonly, spirally arranged in groups of up to about 40 in the axils or more or less reduced leaflike bracts on specialized axillary reproductive shoots. Each pollen cone cylindrical, usually without a ring of bracts at the base and with numerous, spirally arranged, roundly triangular pollen scales, each bearing two pollen sacs. Pollen grains medium to very large (body 25–70 μm long, 45–115 μm overall), with two round, coarsely wrinkled air bladders distinctly smaller than the minutely bumpy body or rarely uniting to form a single bladder completely encircling the body around the equator. Seed cones generally single on a specialized axillary reproductive shoot clothed with scale leaves or variously reduced foliage leaves. Individual cones highly modified, rather open, with 1–12 widely spaced, spirally arranged fertile bracts, neither these nor the axis becoming fleshy (there is no podocarpium), each bract with a single seed completely embedded in the fleshy seed scale (the epimatium), the opening of the ovule pointing down into the cone axis. Combined seed coat and epimatium with a crest culminating in a pointed, single or double beak and with a thick, hard, fruit-stone-like layer inside the juicy, fleshy layer (hence the scientific name, Greek for "plum pine"). Seeds maturing and falling in one or two seasons. Cotyledons two (or three), each with two veins. Chromosome base numbers $x = 18$ and 19.

Wood relatively hard and heavy, sometimes with an unpleasant odor, the white to light brown sapwood sharply contrasting with the yellow to reddish brown to dark brown heartwood, which often has red or black streaks. Grain fine and even, usually with evident growth rings marked by narrow bands of darker latewood. Resin canals absent but often with scattered or clumped individual resin parenchyma cells.

Leaf surface with or without accompanying scattered stomates above and with several discontinuous lines of stomates forming a broad, pale, stomatal band on either side of the midrib beneath. Each stomate filled with a plug of wax and sunken deeply beneath and mostly hidden by the four (to six) surrounding subsidiary cells, which have a very thick cuticle topped by a low Florin ring surrounded by a sunken channel. Midvein single, prominent, weakly raised beneath and raised to grooved above, with one (to three) small resin canals immediately beneath it, small bands of transfusion tissue on either side (extending out to the edge of the leaf in *Prumnopitys amara*), and occasionally with scattered clumps of accessory transfusion tissue extending all the way out to the leaf edge. Photosynthetic tissue forming a well-developed palisade layer unaccompanied by a hypodermal layer beneath the upper epidermis and connecting extensively to the spongy mesophyll or separated by the accessory transfusion tissue.

Eight species in the western Pacific from Sumatra (Indonesia) and Luzon (Philippines) to New Zealand and in the New World cordilleran region from Costa Rica and Venezuela to Chile and Argentina. References: Allan 1961, Buchholz and Gray 1948a, b, Conran et al. 2000, Covas 1995, de Laubenfels 1972, 1978a, 1982, 1988, Gray and Buchholz 1951a, b, Hill and Brodribb 1999, Kelch 2002, Rodríguez and Quezada 1995, Salmon 1980, Sinclair et al. 2002, Torres-Romero 1988, Veillon 1962. Synonyms: *Podocarpus* sect. *Stachycarpus* Endlicher, *Podocarpus* sect. *Sundacarpus* J. Buchholz & N. Gray, *Prumnopitys* sect. *Sundacarpus* (J. Buchholz & N. Gray) de Laubenfels, *Stachycarpus* Tieghem, *Stachypitys* A. V. Bobrov & Melikian, *Sundacarpus* (J. Buchholz & N. Gray) C. Page, *Van-tieghemia* A. V. Bobrov & Melikian.

Only a few species of *Prumnopitys* are occasionally cultivated outside their homelands, including *P. andina* and *P. taxifolia.* Even within their native regions, they are rather infrequent, and no cultivar selection has taken place. With their handsome, dark green crowns, interesting bark, and conspicuous pollen cones or colorful, juicy seed cones, they are worthy of greater attention in the tropics and subtropics.

Several DNA studies show that there are two major groupings of podocarp genera, one containing *Podocarpus* and *Dacrydium* and related genera, embracing the majority of species in the family, while the other contains a smaller, less clearly defined grouping of genera, most of which have scaly leaves. Of the six currently recognized podocarp genera formerly included as botanical sections with *Podocarpus,* four are at least moderately closely related to this large genus and belong in the same grouping while the other two are not at all related to it and belong with the other group of genera. *Parasitaxus,* formerly treated as *Podocarpus* sect. *Microcarpus,* closely resembles the scaly-leaved genera, especially *Manoao* and *Lagarostrobos,* and so occasions little surprise in the reassessment of its taxonomic position. However, *Prumnopitys* superficially resembles several genera in the *Podocarpus* group, so much so that some authors were reluctant to accept its separation from *Podocarpus* despite well-known differences in biochemistry (such as flavonoid pigments) and morphology (such as complete lack of a podocarpium). An alliance with the scaly-leaved genera was never suspected before DNA studies revealed it.

DNA studies also show fairly decisively that broadleaf miro (*Prumnopitys amara*) is firmly embedded among the other *Prumnopitys* species and should not be set off from them in a separate botanical section or even a separate genus. This separate section or genus, named *Sundacarpus,* was proposed and then fairly widely accepted largely because broadleaf miro has much larger, more *Podocarpus*-like leaves than the other species, all of which have rather yewlike foliage. The seed cones and pollen cones of broadleaf miro are very similar to those of the other species of *Prumnopitys,* with DNA studies showing that the unusual leaves of *P. amara* arose within the genus rather than reflecting significant separation. However, even though the unity of *Prumnopitys* and its relationship to the scaly-leaved genera now seem well established, the relationships among the species within the genus is a completely open question and worthy of detailed study.

There is a modest fossil record of *Prumnopitys* known from the Australian region but not from South America. The oldest known fossils are found in Paleocene sediments (about 60 million years old) of New Zealand, where the genus still occurs. The only other Tertiary occurrences of *Prumnopitys* from New Zealand date from about 20 million years ago, in the Miocene, and these are considered so similar to one of the two extant species in New Zealand, matai (*P. taxifolia*), that they are assigned to that species. All other known fossils of *Prumnopitys* are found in southern Australia (Tasmania and Victoria), outside the present range

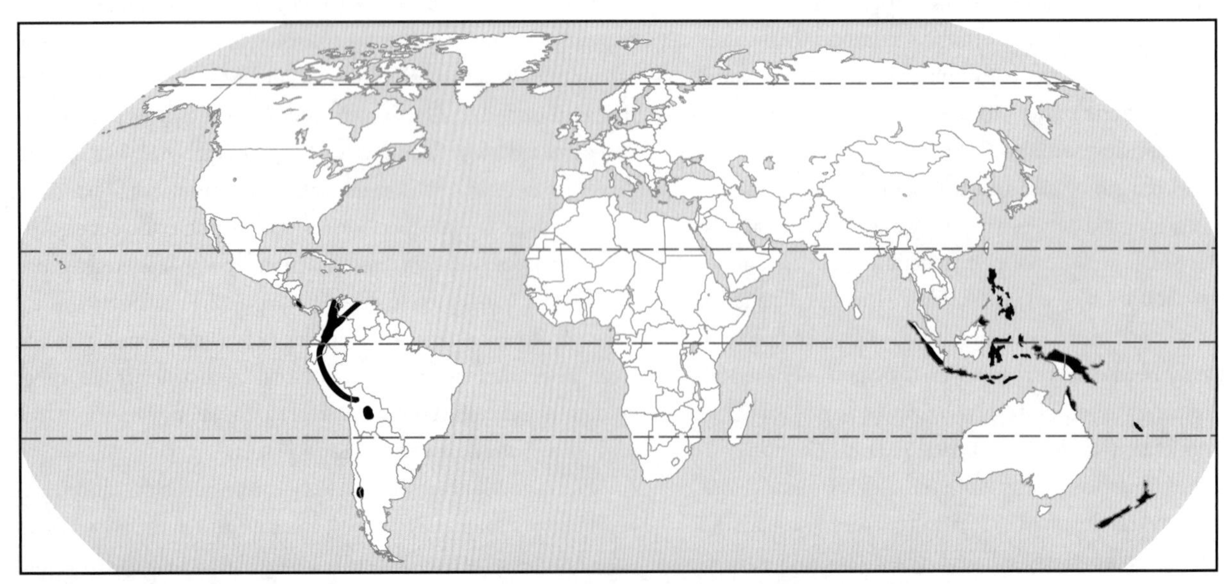

Distribution of *Prumnopitys* in Central and South America and Australasia, scale 1 : 200 million

of the genus, and date to the Eocene, roughly 50 million years ago. Like so many other podocarp fossils from the early Tertiary of southern Australia, these fossil leaves have numerous stomates on both leaf surfaces, indicating a much wetter climate in Australia's past than it presently experiences.

Identification Guide to *Prumnopitys*

It is easy to distinguish *Prumnopitys amara* from all others in the genus by its much larger leaves, rarely as small as 4 cm long and 6 mm wide, while none of the other species ever has leaves larger than 3 cm long and 5 mm wide (and usually much smaller than that). The other seven species are much more similar to each other but may be distinguished by the length and the width of the wider adult leaves, the form of the midrib and the number of stomates on the upper side of the leaves, and the length and color of most mature seeds with their epimatium:

Prumnopitys taxifolia, leaves up to 15 mm by up to 2 mm, midrib slightly raised, stomates on upper side none, seeds up to 11 mm, black

P. ferruginoides, leaves up to 15 mm by more than 2 mm, midrib essentially flat, stomates on upper side few, seeds 11–15 mm, red

P. ladei, leaves up to 15 mm by more than 2 mm, midrib flat to slightly raised, stomates on upper side many, seeds more than 20 mm, purplish black

P. montana, leaves 10–20 mm by more than 2 mm, midrib grooved, stomates on upper side none to few, seeds 11–15 mm, orange

P. harmsiana, leaves 10–20 mm by at least 2 mm, midrib prominently raised, stomates on upper side none, seeds up to 11 mm, yellow

P. andina, leaves more than 15 mm by up to 2 mm, midrib essentially flat, stomates on upper side none, seeds 15–20 mm, dark purple

P. ferruginea, leaves more than 15 mm by at least 2 mm, midrib slightly raised, stomates on upper side none, seeds 15–20 mm, red

Prumnopitys amara (Blume) de Laubenfels

BROADLEAF MIRO, QUEENSLAND BLACK PINE, SUNDA PODOCARP, KI MARAK (SUNDANESE)

Tree to 40(–60) m tall, with trunk to 1.4(–2) m in diameter, sometimes becoming deeply fluted and buttressed at the base of old individuals. Bark reddish brown, flaking in small, rectangular scales and finally weathering gray and breaking up into small, irregular checkers separated by shallow furrows. Crown conical at first, with symmetrical, whorled branches, becoming much more irregular, open, and dome-shaped with age, with few, large, upwardly directed to horizontal branches bearing whorled branchlets well clothed with foliage. Twigs green in the first year, turning brown, whorled on upright branches but often in pairs or trios on horizontal ones, weakly joined to the larger supporting branch. Resting buds nearly spherical, about 2 mm in diameter, the bud scales nearly equal in length. Leaves fairly evenly and openly spaced along the branchlets, persisting 1–2 years, shiny dark green above (when not covered by mosses and other epiphytes), the broad stomatal bands waxy, light bluish green beneath, (4–)8–12(–20) cm long (averaging shorter but sometimes reaching 25 cm in juveniles), (6–)8–14 mm wide (to 20 mm in juveniles). Blades fairly parallel-sided, straight or, more frequently, modestly curved to one side, tapering abruptly to a shorter or extended, tonguelike tip and to a short petiole to 6 mm long which detaches readily from the twig, leaving a flat, round scar, the margins a little curled under. Midrib grooved above, raised beneath, accompanied by a large resin canal, rarely with an extra resin canal beneath the main one or with an additional one on either side, the stomates scattered rather than being arranged in continuous lines. Pollen cones (0.6–)1.5–3(–3.5) cm long and (2–)2.5–3(–4) mm wide, 1–6(–10) on a short, leafless stalk to 7 mm long. Pollen scales with a sharply triangular, minutely toothed, upturned tip less than 1 mm long and wide. Seed cones (one or) two or three (to five) along a scaly stalk 0.5–3(–5) cm long, each in the axil of a deciduous scale, the bracts of the

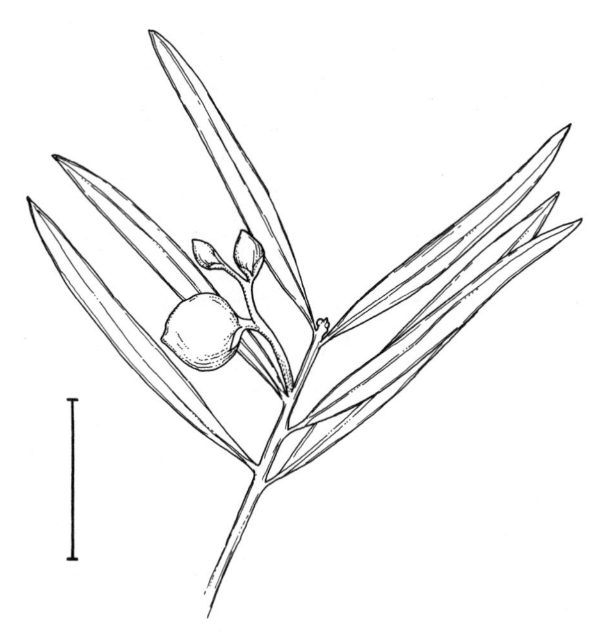

Twig of broadleaf miro (*Prumnopitys amara*) with seed cones in varying stages from newly pollinated to nearly mature; scale, 4 cm.

reproductive part enlarging only slightly at maturity and not merging with the axis. Fertile seed scale one, the combined seed coat and epimatium nearly spherical to noticeably elongate at maturity with a low crest along one side and a pronounced beak, smooth at first but wrinkling with drying, passing from red through dark purple to black, with a thin waxy coating, (12–)15–25(–30) mm long by 10–15(–25) mm thick, with a fleshy layer 3–4 mm thick over a hard, inner shell 1–2 mm thick. Fairly continuous throughout the Malay Archipelago from northern Sumatra and northern Luzon to southern New Ireland and northeastern Queensland but absent from Borneo (except in Sabah), northern Sulawesi, and many small islands. Scattered as canopy members or standing above the canopy in lowland and montane rain forests and other wet-forest types; (0–)600–2,000(–3,600) m. Zone 10. Synonyms: *Podocarpus amarus* Blume, *Sundacarpus amarus* (Blume) C. Page.

Broadleaf miro is one of the premier softwood timber trees throughout the Malesian region, even with the economic dominance of angiosperm dipterocarps. It is often common, even in secondary forests, so it is not immediately threatened by overexploitation. Despite the large and discontinuous range of this species, there is no apparent consistent geographic differentiation. That is not true of the common names, however, and there are dozens or even hundreds of different names in use on different islands and in different regions of the large islands. The species name (Latin for "bitter") refers to the taste of the fresh leaves, which are also described as bittersweet, or sweet! This suggests an interesting project for someone with a culinary bent.

Prumnopitys andina (Poeppig ex Endlicher) de Laubenfels

CHILEAN MIRO, PLUM-FRUITED YEW, LLEUQUE (SPANISH FROM ARAUCANIAN)

Bushy tree to 7(–15) m tall (to more than 21 m in cultivation), with a commonly forked, often crooked trunk to 0.5(–1) m in diameter. Bark thin, smooth, dark brown at first, weathering bright gray with reddish and bluish highlights. Crown dense, conical at first, becoming more rounded and irregular with age, with numerous short, horizontal to upwardly angled branches bearing richly branched, alternating, or slightly offset, predominantly horizontal branchlets densely clothed with foliage. Leaves lasting 2–3 years, extending out to the sides of the twigs in two even to ragged rows, angled slightly forward and usually curving gently outward, (10–)15–25 mm long, 1.5–2 mm wide. Blades widest near or below the middle, tapering very gradually to the roundly triangular, pointed but not prickly tip and to the rounded base on a very short, slender petiole. Midrib essentially flat on the rich green upper surface, which has no stomates, and raised beneath between waxy bluish green stomatal bands. Internal strengthening fibers and sclereids absent, the leaves thus drying flat. Pollen cones 5–8 mm long, 2.5–3 mm thick, in groups of 10–20, each attached singly along an axillary reproductive shoot 1–2.5 cm long, with leafy bracts 2–3.5 mm long. Seed cone axis 2–3 cm long, leafy at the base and with five to seven fertile bracts above, from which one to three (to five) seeds mature. The combined seed coat and epimatium passing through yellowish green and bluish to deep reddish purple at maturity, (1–)1.5–2 (–2.5) cm long, 10–15 mm thick with a flattened, conical beak 2–3 mm long. In a relatively small area of the southern Andes (hence the scientific name) and coastal range in central Chile (Maule region to northern Aisén) and just over the border in the Río Aluminé valley of the lake district of Argentina (Neuquén province). Growing sparsely or rarely in small groves in mixed stands with Chilean incense cedar (*Austrocedrus chilensis*) and various broad-leaved trees, including southern beeches (*Nothofagus*); 500–1,100 m. Zone 8. Synonyms: *Podocarpus andinus* Poeppig ex Endlicher, *Prumnopitys elegans* R. Philippi, *P. spicata* (Poeppig) Molloy & Muñoz-Schick, *Stachycarpus andinus* (Poeppig ex Endlicher) Tieghem.

Although Chilean miro is a rare tree in nature, it is the only species of *Prumnopitys* in general cultivation in Europe and North America, where it is also quite infrequent. The handsomely mottled, easily worked wood takes a fine finish and is highly prized for

Mature Chilean miro (*Prumnopitys andina*) in nature.

cabinetry and furniture. While it is not exploited commercially because of its rarity, even the local usage to which it is subjected poses some threat to the persistence of the species. The seeds, including the juicy epimatium, have a pleasant, slightly resinous flavor and are eaten locally under the name *uva de cordillera* (mountain grape). Vegetatively, Chilean miro is surprisingly similar to its more common neighbor, Prince Albert yew (*Saxegothaea conspicua*), and nonreproductive specimens of the latter are sometimes optimistically identified as the former so that Chilean miro is probably even rarer than commonly thought. There has been some dispute about the correct scientific name for Chilean miro. Setting aside the longevity in horticulture of the always incorrect name *P. elegans*, it has been argued that an alternative name for the New Zealand matai (*P. taxifolia*) is illegitimate, causing a cascade of name changes that would replace the name for Chilean miro. This argument seems to be incorrect, and there appears to be no real impediment to the continued use of *P. andina* for the Chilean species.

Prumnopitys ferruginea (G. Bennett ex D. Don) de Laubenfels

MIRO (MAORI AND ENGLISH), [NEW ZEALAND)]
BROWN PINE
Plate 58

Tree to 25(–30) m tall, with a smooth or slightly buttressed cylindrical trunk to 1 m in diameter, often unbranched for a third or more of its height. Bark thin, densely warty, dark brown at first, weathering grayish to almost black, shedding profusely in thick flakes and sometimes becoming deeply furrowed with age. Crown dense, conical at first, becoming deeply to shallowly cylindrical or even dome-shaped with age, with numerous short, upwardly angled branches bearing alternating or slightly offset horizontal branchlets densely clothed with foliage. Leaves lasting 2–3 years or more, extending out to the sides of the twigs in two quite even rows, angled forward, often arched along their length, straight to curved outward, (1.5–)2–2.5 cm long (to 3 cm in juveniles), 2–3 mm wide. Blades widest near the middle, tapering steadily to the rounded tip (sharply pointed in juveniles) and to the wedge-shaped base on a short, relatively thick petiole. Midrib flat to slightly raised on the yellowish green to bright green upper surface, which has no stomates (rarely a few right along the midvein), and raised beneath between paler, yellowish green but not waxy stomatal bands. Internal strengthening fibers and sclereids absent, the leaves thus drying flat. Pollen cones 0.5–1.5(–2) cm long, 2–3(–4) mm thick, attached singly at the tips of or in axils along otherwise unmodified shoots with ordinary foliage leaves. Seed cone axis 0.5–1 cm long, scaly and with one (or two) fertile bracts at the tip, from which one (or two) seeds mature, the scales progressively elongating upward near the fertile bract(s), to 3 mm long. The combined seed coat and epimatium bright red with a thin waxy film at maturity, (1–)1.5–2(–2.5) cm long, 6–12(–15) mm thick with a conical beak 1–2 mm long. Common throughout the three main islands of New Zealand but progressively more abundant southward. Forming pure stands or mixed with other conifers and evergreen hardwoods in the canopy of dense, shady, moist lowland forests: 0–700(–1,000) m. Zone 9. Synonyms: *Podocarpus ferrugineus* G. Bennett ex D. Don, *Stachypitys ferruginea* G. Bennett ex D. Don) A. V. Bobrov & Melikian.

Although it yields a hard, handsome wood, miro was less valued historically as a timber tree in New Zealand than rimu (*Dacrydium cupressinum*), with which it was frequently interchangeably harvested, or its closer relative matai (*Prumnopitys taxifolia*), because the wood can warp considerably during drying. It is the strongest native New Zealand softwood, however, and took on increasing prominence as other species were depleted, partially replacing matai in flooring, for instance. The seeds are the largest among all the New Zealand podocarps and also exceed the size of the podocarpium in those species that have one. They are a favorite food of New Zealand pigeons (*Hemiphaga novaeseelandiae*), which are the only native pigeons in the islands and which eat

Old mature miro (*Prumnopitys ferruginea*) in nature.

flowers, fruits, and foliage of a wide range of native and introduced trees and shrubs. The fleshy part of the seed coat and the epimatium are laced with resin-containing canals, which may be the main attractant for the pigeons. The red color of the seeds, however, derives from cyanidin, the most common red anthocyanin pigment throughout the Podocarpaceae. The chemistry of this species is, in general, fairly typical for the family in its flavonoids (including anthocyanins) and in the diterpenes (20-carbon oils) of the resins in the foliage and heartwood. Passage through the pigeon guts has essentially no effect on germination, which takes from 18 months to 4 years or more to complete. Fallen seeds have a good chance of survival to germination since they are not removed and eaten by introduced rodents to any significant extent.

As might be expected from a large-seeded species, miro is very shade tolerant. Its seedlings are quite unlike those of matai and of most other podocarp species in having a leader that droops from its inception and bears two-ranked leaves like those of lateral branchlets rather than having them radiate all around the shoot. In this respect it is similar to the unrelated hemlocks (*Tsuga*) in the northern hemisphere, also highly shade-tolerant trees with a long residence in the understory before reaching the canopy of dense mixed forests. At some sites on South Island, the upper elevational limit of miro (and of some other species) appears to have risen more than 60 m in the last 150 years or so as the climate of New Zealand has warmed by 0.5°C or more. This period is only about a quarter of the normal life span of these long-lived trees, which can even live to more than 750 years. The species name, Latin for "rusty," refers to the color of the foliage in dried herbarium specimens, contrasting strongly with the blackening foliage of dried matai, although it would seem to be at least as appropriate for the contrast of the seeds with those of matai.

Prumnopitys ferruginoides (R. Compton) de Laubenfels

NEW CALEDONIAN MIRO

Tree to 15(–20) m tall or shrubby at high elevations, with a short to relatively tall, upright trunk to 0.4 m in diameter. Bark thin, rough, yellowish brown to reddish brown when fresh, weathering dark brown or gray to almost black and shredding partially in small, irregular scales so that some flakes accumulate on top of the others. Crown densely conical at first, becoming more open, irregular, and cylindrical with age, with irregularly placed, short, slender upwardly angled branches bearing many alternating, horizontal to upright branchlets densely clothed with foliage. Leaves lasting 1–2 years, extending out to the sides of the twigs in two somewhat ragged rows or nearly radiating all around on more upright twigs, angled forward, straight, arched up to the midvein, 9–15(–18) mm long (to 30 mm in juveniles), 2–3.5 mm wide (to 5 mm in juveniles). Blades widest near or beyond the middle, tapering rapidly to the rounded tip with a blunt point (tapering gradually and then more abruptly to a sharp point in juveniles) and more gradually to the roundly wedge-shaped base on a short, slender petiole. Midrib essentially flat on the rich green upper surface with a scattering of stomates and raised beneath between slightly waxy, bluish green stomatal bands. Large internal strengthening fibers and sclereids extending out from the midvein, the leaves drying with numerous small wrinkles as a consequence. Pollen cones 3–4 mm long, about 1.5 mm thick, attached singly in the axils of ordinary foliage leaves on otherwise unmodified shoots. Seed cone axis 0.5–1 cm long, scaly and with one (or two) fertile bracts at the tip, from which one (or two) seeds mature, the scales progressively elongating upward near the fertile bract(s), to 3 mm long. The combined seed coat and epimatium red with a thin waxy film at maturity, 1–1.5 cm long, 6–10 mm thick, with a broadly conical beak about 1 mm long. Scattered on mountains in the southern three-quarters of New Caledonia, from Mount Panié southward and most common in the southern quarter. Scattered to gregarious in the canopy or subcanopy of rain forests and high mountain shrublands, primarily on serpentine-derived soils; (150–)400–1,400(–1,640) m. Zone 10. Synonyms: *Podocarpus distichus* J. Buchholz, *P. ferruginoides* R. Compton, *Stachypitys disticha* (J. Buchholz) A. V. Bobrov & Melikian, *S. ferruginoides* (R. Compton) A. V. Bobrov & Melikian.

New Caledonian miro generally resembles the New Zealand species for which it was named, miro (hence the scientific name, Latin for "like *ferruginea*"), but has notably shorter leaves, pollen cones, and seeds. The most distinctive feature, however, is the presence of large support fibers in the leaf tissue, causing a wrinkling on drying found elsewhere in the genus only in Queensland miro (*Prumnopitys ladei*). It is a common species throughout the rain forests of New Caledonia, with an elevational range that exceeds those of all other conifers on the island, with the possible exception of New Caledonian sickle pine (*Falcatifolium taxoides*). Like most of the other podocarps in New Caledonia, this species has little commercial value as a timber tree, and it does not seem to be threatened by such exploitation as it experiences.

Prumnopitys harmsiana (Pilger) de Laubenfels

YELLOW MIRO, ROMERILLO HEMBRA, GRANADILLO, PINO, CHAQUIRO (SPANISH)

Tree to 20(–35) m tall, with a (usually) slender, cylindrical trunk to 0.7(–1.7) m in diameter, Bark thin, smooth, bright reddish brown when fresh, weathering dark grayish brown and shedding in small scales. Crown dense, conical at first, becoming cylindrical with age, with many slender, upwardly angled to horizontal branches bearing alternating horizontal branchlets densely to more openly clothed with foliage. Leaves lasting 3–5 years, extending out to the sides of the twigs in two quite even flat rows, angled forward and then curved sharply outward near the base, continuing straight out or bent forward in a gentle S curve, flat, (8–)10–18(–22) mm long (to 27 mm in juveniles), (1.5–)2.5–4(–5) mm wide. Blades nearly parallel-sided in the middle three-fifths, sometimes widest before or beyond the middle, tapering abruptly to the roundly triangular tip with a sharp or blunt point and more gradually to the roundly wedge-shaped base on a short, slender petiole. Midrib roundly raised on the dark green upper surface, which has no stomates, and equally raised beneath between paler green, but not waxy, stomatal bands. Internal strengthening fibers and sclereids absent, the leaves thus drying flat. Pollen cones about 8–12 mm long, 2–2.5 mm thick, in groups of 10–15(–20), each attached singly along an axillary reproductive shoot 3–8 cm long, with scaly bracts about 2 mm long. Seed cone axis 1.5–3 cm long, leafy or scaly at the base and with four to six fertile bracts above, from which 1–3 seeds mature. The combined seed coat and epimatium greenish yellow at maturity, nearly spherical, 8–11 mm in diameter, with a flattened conical beak less than 1 mm long. Discontinuous through the Andean region of northern South America from northern Colombia and the vicinity of Caracas (Venezuela) to southern Peru and adjacent westernmost Bolivia. Scattered singly or in groves in primary and secondary montane forests; (1,000–)1,500–2,200(–2,800) m. Zone 10. Synonyms: *Podocarpus harmsianus* Pilger, *P. utilior* Pilger, *Prumnopitys utilior* (Pilger) Melikian & A. V. Bobrov.

Yellow miro generally grows at lower elevations in the Andes than its close relative northern miro (*Prumnopitys montana*), which occurs throughout the same region. Besides this ecological difference, the two species are readily distinguished by the midvein, grooved above in northern miro. There is a moderate amount of morphological variation in the different parts of its range, particularly in leaf size, which also varies considerably within individual trees. It has been locally valued as a timber tree at times, as on the eastern slope of the Andes in Junín province (Peru), but it is too uncommon to be a major commercial species. The species name honors Herman Harms (1870–1942), a colleague of Pilger at the Botanical Museum in Berlin who was a long-time editor of *Das Pflanzenreich,* the monograph series in which the species was published.

Prumnopitys ladei (F. M. Bailey) de Laubenfels

QUEENSLAND MIRO

Tree to 25 m tall, with a smooth or buttressed trunk to 1.7 m in diameter, Bark thin, smooth, reddish brown, weathering gray and flaking in thin scales. Crown dense, cylindrical, with upwardly angled branches bearing alternating horizontal branchlets densely clothed with foliage. Leaves lasting 3–4 years, extending out to the sides of the twigs in two even rows, just overlapping their neighbors or not, angled slightly forward, generally slightly curved outward, flat, (4–)10–16(–24) mm long, (1.5–)2–3.5(–5) mm wide. Blades widest near or beyond the middle, often nearly parallel sided in the middle half, tapering very gradually and then abruptly to the rounded tip with a blunt or prickly point, and abruptly to the roundly wedge-shaped base on a very short, broad petiole. Midrib essentially flat to slightly raised on the bright green upper surface, which has numerous scattered stomates, and raised beneath between paler, but not waxy stomatal bands. Large internal strengthening fibers and sclereids extend out from the midvein to the edge of the leaf leading to obvious wrinkling on drying. Pollen cones not seen or otherwise described. Seed cone axis about 1.5 cm long, scaly at the base and with one (or two) fertile bracts at the tip, from which one (or two) seeds mature, the scales progressively elongating upward near the fertile bract(s), to 7 mm long. The combined seed coat and epimatium dark purplish black with a thin waxy film at maturity, 1.5–2.5 cm long, 12–16 mm thick, with a broad, conical beak 1–2 mm long. Known only from Mount Lewis and Mount Spurgeon in the Atherton Tableland southwest of Cairns, Queensland (Australia). Scattered in the canopy of rain forests on granite-derived soils; 1,000–1,200 m. Zone 10. Synonyms: *Podocarpus ladei* F. M. Bailey, *Stachypitys ladei* (F. M. Bailey) A. V. Bobrov & Melikian.

Queensland miro is the rarest species of *Prumnopitys* and one of the rarer conifers overall. It has the largest trunk diameter among the small-leaved species of the genus, although it may be exceed by the large-leaved broadleaf miro (*P. amara*), which grows near it in the Atherton Tableland. While it would be interesting to know whether these two species interact in the Queensland rain forests, Queensland miro is much more similar to and closely related to the geographically more distant miro (*P. ferruginea*) of New Zealand and the New Caledonian miro (*P. ferruginoides*), especially to the latter. Queensland miro has a maximum known longevity of 600 years, which is exceeded by more than 150 years by its smaller-diameter and slower-growing close relative, *P. ferruginea.* The species name honors F. Lade, who collected the type specimen on Mount Spurgeon in 1902.

Prumnopitys montana (Humboldt & Bonpland ex Willdenow) de Laubenfels

NORTHERN MIRO, ROMERILLO, TRENSA, PINO COLORADO, PINO REJO, PINO LASO, CHAQUIRO (SPANISH)

Tree to 25–35 m tall, or dwarfed and shrubby at high elevations, with a cylindrical or somewhat fluted trunk to 0.6(–1) m in diameter. Bark thin, smooth, reddish brown when fresh, weathering gray to almost black and peeling in irregular scales or plates. Crown broadly dome-shaped, with slender horizontal branches bearing alternating or slightly offset horizontal branchlets densely to more openly clothed with foliage. Leaves lasting 3–5 years, touching side to side or more distantly spaced, extending out to the sides of the twigs in two even rows, angled slightly forward and then curving gently outward near the base, straight from there or curving slightly forward near the tip, (5–)10–18(–25) mm long, (1–)1.5–3.5(–4) mm wide. Blades widest beyond the middle, tapering abruptly to the rounded or roundly triangular tip with a sharp or blunt point and much more gradually and then abruptly to the wedge-shaped base on a very short, relatively broad petiole. Midrib grooved on the dark green upper surface, which has no or a scattering of a few stomates, and raised beneath between paler green or sometimes waxy stomatal bands. Internal strengthening fibers and sclereids absent, the leaves thus drying flat, though often bowed upward on either side of the midvein. Pollen cones (2.5–)6–16 mm long, (1.4–)1.8–2.5 mm thick, in groups of (6–)10–25(–35), each attached singly along an axillary reproductive shoot 2.5–5 cm long, with leafy bracts (1–)2–3(–4.5) mm long. Seed cone axis 1–3(–6) cm long, variably leafy at the base, the leaves progressively shorter upward, and with one (to three) fertile bracts above, from which one (to three) seeds mature. The combined seed coat and epimatium maturing yellowish orange with a thin, waxy film, (10–)12–16(–20) mm long, 6–12 mm thick, with a conical to flattened or double beak 0.5–2 mm long. Discontinuous in the mountains of Central and South America, in central Costa Rica and from northwestern Venezuela and northern Colombia south through the Andes to central Bolivia. Scattered individually or in groups in the canopy of primary and secondary wet upper montane forests and cloud forests, and of dwarfed forests and subalpine scrublands on wet, exposed ridges and at the margin of páramo; (1,400–)2,000–3,200(–3,600) m. Zone 9. Synonyms: *Podocarpus montanus* (Humboldt & Bonpland ex Willdenow) C. Loddiges ex N. Britton, *P. standleyi* J. Buchholz & N. Gray, *Prumnopitys exigua* de Laubenfels ex Silba, *P. standleyi* (J. Buchholz & N. Gray) de Laubenfels, *Van-tieghemia densifolia* (Kunth) A. V. Bobrov & Melikian, *V. meridense* (J. Buchholz & N. Gray) A. V. Bobrov & Melikian, *V. montana* (Humboldt & Bonpland ex Willdenow) A. V. Bobrov & Melikian.

Northern miro generally occurs at much higher elevations than its close relative yellow miro (*Prumnopitys harmsiana*), hence the scientific name, Latin for "of the mountains." Consequently, it frequently grows as a short, slender tree or bushy treelet in more exposed, harsher locations. When growing in more favorable sites at lower elevations, however, it becomes a substantial tree (though not as large as yellow miro) and is harvested for its fine-grained but hard and strong, handsome wood. It regenerates well after forest clearance and, in some regions, has extended its range down the mountains into secondary forests and disturbed land. It is grown to a limited extent as an ornamental in its native range. Following pruning under these circumstances, individuals may produce both pollen and seed cones, unlike the vast majority of podocarps, including other *Prumnopitys* species. Specimens may also be found with conelike galls that may be mistaken for reproductive structures during casual examination. The species varies considerably throughout its range, and populations at the northern and southern ends were often treated as separate species. Trees from Costa Rica (described as *P. standleyi*), for instance, have smaller than usual seeds and bracts in the lower part of the seed cone axis, and leaves with a few stomates above and waxy stomatal bands beneath. Trees from Bolivia (described as *P. exigua*) are even less distinct, sharing with Costa Rican plants the smaller seeds and bracts but lacking the leaf distinctions. After a thorough investigation of the species throughout its range, it might prove advisable to recognize some regional variants as varieties or subspecies, but these would not necessarily duplicate the segregate species that have been proposed to date.

Prumnopitys taxifolia (J. Banks & Solander ex D. Don) de Laubenfels

MATAI (MAORI AND ENGLISH), [NEW ZEALAND]
BLACK PINE

Tree to 25 m tall, with a short or extended, cylindrical or somewhat fluted trunk to 1.3(–2) m in diameter. Bark thin, smooth, bright red when fresh, weathering dark brown to black, peeling continuously in thick flakes of varying size to produce intricate, colorful mottling. Crown round, open, and irregular when young, becoming denser and dome-shaped to cylindrical with maturity, with many slender to thick, spreading to gently rising branches irregularly bearing elongate branchlets sparsely to densely clothed with foliage. Leaves lasting 2–3 years, extending out to the sides of the twigs in two even or, more frequently, ragged rows or nearly radiating all around on some twigs, angled slightly forward or standing out at nearly right angles, mostly straight, flat or bowed up along their length, 10–15(–20) mm long (5–10 mm in juveniles), 1–2 mm wide. Blades widest near or before the middle, nearly parallel-sided, tapering abruptly at the end to a broadly rounded tip with a short prickle and more gradually to the wedge-shaped base on a short, broad petiole. Midrib scarcely raised on the slightly bluish green upper surface without any stomates and raised beneath between strikingly waxy, pale bluish green stomatal bands. Without large internal strengthening fibers and sclereids, the leaves thus drying flat. Pollen cones 1–2 cm long, 2.5–4

Twig of matai (*Prumnopitys taxifolia*) with reproductive shoots bearing newly pollinated and mature seed cones; scale, 1 cm.

mm thick, in groups of (10–)20–30(–40) each attached singly along an axillary reproductive shoot 3–5 cm long with scalelike to leafy bracts 2–6 mm long. Seed cone axis 2.5–5 cm long, axillary, scaly at the base and with 3–12 scalelike fertile bracts above, from which 1–6(–12) seeds mature. The combined seed coat and epimatium deep purplish black with a thin waxy film at maturity, nearly spherical, 6–10 mm in diameter, with a short, broadly conical beak about 1 mm long, often depressed in a dimple. Found throughout New Zealand, most abundantly in the central part of North Island and the western side of South Island. Scattered or forming groves in the canopy of wet and dense or more open, drier mixed lowland forests of flats and hills; 0–550 m. Zone 9. Synonym: *Podocarpus spicatus* R. Brown ex Mirbel.

Matai, though never cut in quantities approaching those for some of the other native softwoods of New Zealand, like kauri (*Agathis australis*) or rimu (*Dacrydium cupressinum*), was nonetheless a highly respected timber tree, yielding valued lumber with a wide variety of uses. The dimensional stability of the wood with changing humidity, coupled with its durability, resilience, and ability to take a fine finish, made it especially popular for flooring in public buildings, a use for which it is still cut to a limited extent. Unlike some of the other species, the wood smells quite unpleasant when it is just sawn. It is a tree of generally more open forests than its close relative miro (*Prumnopitys ferruginea*). Its less shade-tolerant seedlings have a completely different growth habit: scraggly, with long, whiplike twigs clothed with distantly spaced scale leaves and giving rise to occasional short branchlets bearing photosynthetic needle leaves. These leaves and those of adults are somewhat similar to those of yews, hence the scientific name, Latin for "yew leaf." The juvenile phase is long-lived and gives way only gradually to adult trees. These are very slow growing, increasing by only 0.5–2 mm in diameter each year. Commensurately, they are very long-lived, with an average life span of 600 years and sometimes surpassing 1,000 years.

The seeds, although only half as long as those of miro, are still the next largest podocarp seeds in the New Zealand flora and are also an important food for fruit-eating birds, which disperse them. Up to a quarter of the seeds may be destroyed by a caterpillar that specializes in feeding on them. A smaller proportion can be destroyed by a gall-forming fly larva that replaces the embryo and alters the development of the surrounding seed coat as a larval feeding chamber. Perhaps as a counter to such attacks, the foliage, at least, has chemicals with insecticidal and insect hormonal activities as well as an ability to inhibit the growth of seedlings of other forest species. Oils from the leaves contain a rich mixture of potentially physiologically active compounds, including both sesquiterpenes and diterpenes (with 15 and 20 carbon atoms, respectively), and these compounds vary considerably in composition from place to place. The sap of the leaves also contains several antioxidant biflavones, made up of twinned flavonoid pigment molecules. Other flavonoid pigments are similar to but less varied than those found in miro, and both species have just a fraction of the range of flavonoids found in other New Zealand podocarp genera. However, they contain one group, those with two sugars attached together at one position, not found in any of the other species, reinforcing the separation of Matai and miro from *Podocarpus,* to which they were once assigned.

Pseudolarix G. Gordon

GOLDEN LARCH

Pinaceae

Deciduous tree with a single straight or forking trunk bearing regular tiers of widely spreading horizontal branches to form a broad, rounded crown. Bark smooth at first, nonfibrous, flaking and becoming ridged and furrowed with age. Branchlets strongly differentiated into persistent, spurlike short shoots and branch-building, stretched-out long shoots. Winter buds well developed, scaly. Leaves spirally arranged and radiating all around the twigs, those of the short shoots tightly crowded in pseudowhorls, those of the long shoots widely spaced. Individual leaves needlelike, soft, straight, flat, the base tapering to a slightly raised attachment scar on the twig.

Plants monoecious. Pollen cones in a cluster from a single bud at the tip of a short shoot, cylindrical and borne on a slender, leafless stalk. Each cone with numerous, densely spirally arranged pollen scales, each bearing and dominated by two large pollen sacs. Pollen grains large (body 45–80 μm long, 70–95 μm overall), with two round air bladders attached to a minutely bumpy oval body, the bladders with a rougher surface. Seed cones single at the tips of short shoots, proportionately short, upright, maturing in a single season and shedding the seeds and scales together. Individual cones with a relatively few spirally arranged triangular seed scales spreading open at an angle to the short cone axis even before maturity, the minute triangular bract (more than half as long as the seed scale in an extinct species) attached only at the base. Seeds two per scale, the body oval, with a much larger asymmetrical, triangular wing derived from the scale and completely covering it while tightly cupping the body on one side and overlapping a little on the other side. Cotyledons four to seven, each with one vein. Chromosome base number $x = 22$.

Wood soft, light, the yellowish brown heartwood little differentiated from the tan sapwood. Vertical resin canals lacking in normal wood and not produced in response to wounding. Some cells at the edges of the rays near ring boundaries are packed with small crystals.

Stomates arranged in about 12 lines confined to two broad bands covering most of the underside. Each stomate sunken beneath and partly hidden by the four (to six) surrounding subsidiary cells, which are often shared between adjacent stomates in the same line and which have a very thin cuticle (typical of deciduous conifers) showing no trace of a Florin ring. Leaf cross section with a single-stranded midvein flanked by small wedges of transfusion tissue and surrounded by a ring of large bundle sheath cells, and with one (to three) pairs of very small resin canals at the outer edges of the leaf. Photosynthetic tissue with a very thick palisade layer beneath the upper epidermis, which is partially lined by a discontinuous thin hypodermal layer, and with a spongy mesophyll consisting of very irregular cells continuing down to the stomatal bands.

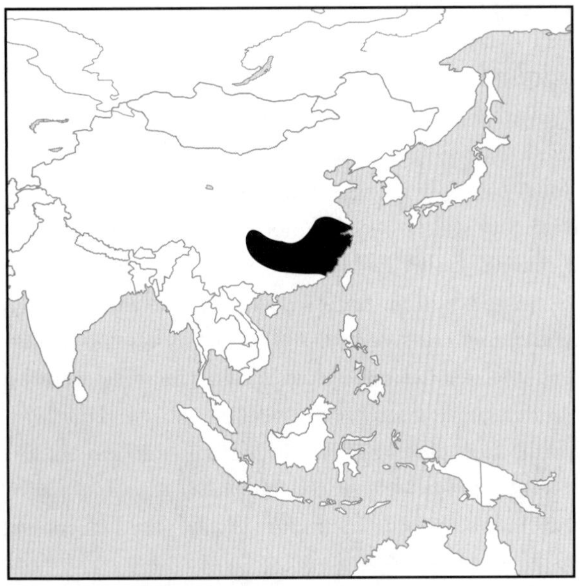
Distribution of *Pseudolarix amabilis* in China, scale 1 : 120 million

One species in China. References: Farjon 1990, Fu et al. 1999c, LePage and Basinger 1995b, T. S. Ying et al. 1993, Yu and Zeng 1992. Synonym: *Chrysolarix* H. Moore.

Published in 1858, the name *Pseudolarix* has enough technical nomenclatural difficulties that H. E. Moore (1965) proposed a substitute name, *Chrysolarix*. Not all taxonomists agreed with his arguments, and he, himself, later accepted the original name. The difficulties were formally set aside by nomenclaturally conserving the name with golden larch rather than Japanese larch (*Larix kaempferi*), as the type. *Pseudolarix amabilis* is well established in cultivation but not nearly frequently enough used as a handsome specimen for parks and larger gardens. The few cultivars that have been selected are dwarfs of varying habit.

Contrary to its name, *Pseudolarix* (Greek and Latin for "false larch") is not related to the larches (*Larix*), which it resembles only in having deciduous needles on long-lived spur shoots. All other morphological features, including the structure and arrangement of the pollen and seed cones and the anatomy of the needles, ally *Pseudolarix* with the abietoid genera of the Pinaceae: *Abies*, *Cedrus*, *Keteleeria*, *Nothotsuga*, and *Tsuga*. DNA studies and other molecular analyses point to a particularly close relationship to the hemlocks (*Tsuga* and *Nothotsuga*), which is somewhat ironic because *Larix* is closest to *Pseudotsuga* (the Douglas firs). Nonetheless, the chromosome base number $x = 22$ is unique in Pinaceae, which otherwise have a base number $x = 12$ (13 in *Pseudotsuga menziesii*). This suggests a polyploid origin for the golden larch, a phenomenon very rare among the conifers, the most spectacular example being the redwood (*Sequoia sempervirens*), a hexaploid (six sets of chromosomes from up to three different parent species).

The distinctiveness of *Pseudolarix* is reinforced by one of the oldest fossil records in the family Pinaceae, with convincing fossils from the early Cretaceous, while most other genera (except *Pinus*) are not recognizable until the Tertiary. Although the single living species is confined to China today, the genus was a little more diverse and much more widespread in the past, with many named species representing as few as two biological species. In the early Cretaceous the genus was confined to eastern Asia and western North America, spreading throughout boreal North America by the early Tertiary and reaching western Eurasia by the beginning of the Miocene. By the end of the Miocene, this broad distribution was fragmenting, and *Pseudolarix* disappeared from North America during the Miocene, from Japan at the end of the Pliocene, and from Europe soon thereafter.

Pseudolarix amabilis (J. Nelson) Rehder

GOLDEN LARCH, JIN QIAN SONG, JIN SONG (CHINESE)

Tree to 35(–45) m tall, with trunk to 2(–3) m in diameter. Bark reddish brown, grooved between narrow, scaly ridges. Crown broadly conical, with long, slender, horizontally spreading or rising branches bearing drooping side branches. Long shoots hairless, reddish brown at first, becoming dark gray with age. Short shoots with annual swellings bearing (10–)15–25(–30) needles (or their scars) separated by narrower rings of bud scars. Winter buds 2–3 mm long. Needles radiating straight out all around the short shoots, bright green, turning golden yellow in fall (hence the common names in both English and Chinese ("golden [coin] pine"), (2–)3–5.5(–7) cm long, (1.5–)2.5–3.5(–4) mm wide. Individual needles with five to seven lines of stomates in each pale green stomatal band beneath, the base gradually tapering, the tip bluntly pointed. Pollen cones 5–10 mm long on a stalk 5–10 mm long, yellowish green. Seed cones (4–)6–7.5(–8) cm long, 4–5(–5.5) cm across, lime green before maturity, ripening reddish brown. Seed body (5–)6–7 mm long, the wing 20–30 mm longer, as long as the seed scales. Eastern and central China, discontinuously distributed from eastern Sichuan east to southwestern Jiangsu, Zhejiang, and central Fujian. Mixed with other conifers or deciduous hardwoods in moist, mixed forests on acidic soils; 100–1,500(–2,300) m. Zone 6. Synonyms: *Chrysolarix amabilis* (J. Nelson) H. Moore, *Pseudolarix kaempferi* (A. Lambert) G. Gordon (misapplied).

Old mature golden larch (*Pseudolarix amabilis*) in nature.

Golden larch has been cultivated for centuries in China as a prized ornamental but was only introduced into world horticulture in 1853–1854. It has not fared so well in its few natural stands, where it has suffered overexploitation for its timber, being used in cabinetry and for boat and bridge building. Despite episodic seed crops that are heavy only every 3–5 years and low to moderate germination percentages, golden larch has a high regeneration capacity in forests of moderate density and is also being promoted for reforestation projects in China. The species name means "lovely," an accurate description of this handsome tree. The technical difficulties with the name of this plant arose because Gordon mistakenly used the species name of Japanese larch (*Larix kaempferi*) when describing golden larch as a new genus. The solution to this

Twig of golden larch (*Pseudolarix amabilis*) with leaves of the current year only, ×0.5.

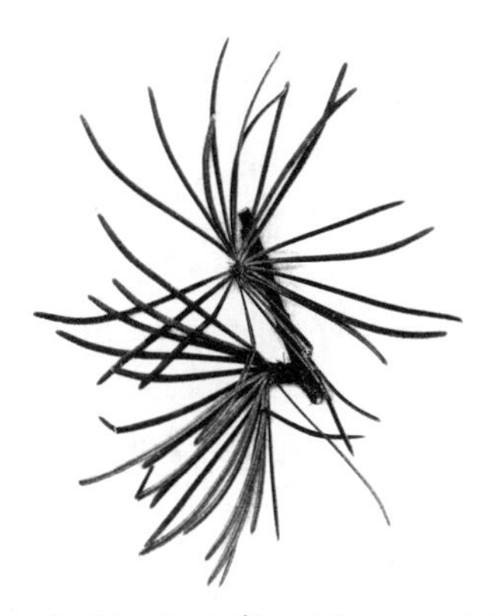

Single spur shoots of golden larch (*Pseudolarix amabilis*) with scars from 4 years of foliage pseudowhorls, ×0.5

Maturing seed cone at the tip of a spur shoot on a twig of golden larch (*Pseudolarix amabilis*); scale, 2 cm.

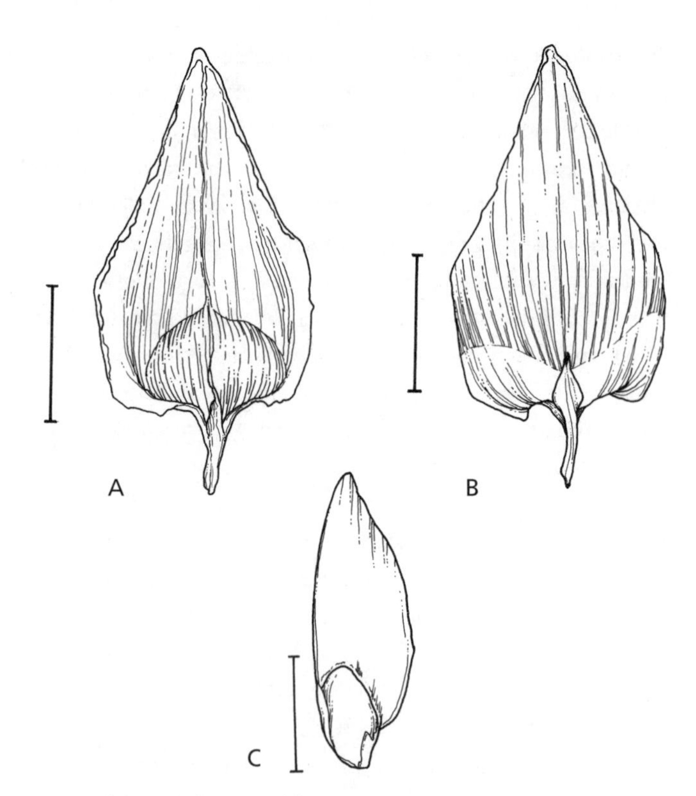

Upper (A) and undersides (B) of shed seed scale and underside of dispersed seed (C) of golden larch (*Pseudolarix amabilis*); scale, 1 cm.

Closed mature seed cone of golden larch (*Pseudolarix amabilis*) before shattering; scale, 1 cm.

problem was officially to assign the plant that Gordon described (golden larch) as the type of the genus rather than the plant whose species name he used (Japanese larch).

Pseudotaxus W. C. Cheng

WHITE-CUP YEW

Taxaceae

Evergreen shrubs. Trunks usually several from the base, slender, often contorted and branching repeatedly from near the base. Bark fibrous, peeling in thin, narrow strips. Crown dense, dome-shaped, with numerous thin branches bearing single, paired, or clustered branchlets. Branchlets all elongate, without distinction into long and short shoots, hairless, turning brown soon after expanding in the first year, grooved between and completely clothed by the elongate attached leaf bases. Resting buds well developed, small, with specialized hard, brown, triangular bud scales persisting at the base of the annual growth increment. Leaves spirally and evenly attached all along the twigs, secondarily rearranged into two uneven rows extending out on either side by twisting and bending of the petioles. Each leaf needlelike, sword-shaped, straight to strongly curved, flattened top to bottom.

Plants dioecious. Pollen cones in a double row all along the lower side of the current growth increment, each single in the axil of an ordinary foliage leaf. Each pollen cone cryptically compound, with a short axis bearing three or four crisscross pairs of bud scales at the base above a very short stalk and two additional bracts higher up among the highly reduced simple cones, which consist solely of a stalk with a whorl of four to six pollen sacs at the tip, each thus resembling a single pollen scale with radiating pollen sacs. Pollen grains small (20–35 µm in diameter), nearly spherical with a conspicuous germination papilla and a minutely and fairly evenly bumpy surface but otherwise almost featureless. Seed cones single in axils of and held beneath ordinary foliage leaves. Each seed cone without any trace of seed scales, consisting of six to eight crisscross pairs of outwardly keeled scalelike bracts at the base of a single seed surrounded by but free from a thickly fleshy, white, cup-shaped aril. Seeds plump, hard, slightly flattened, the

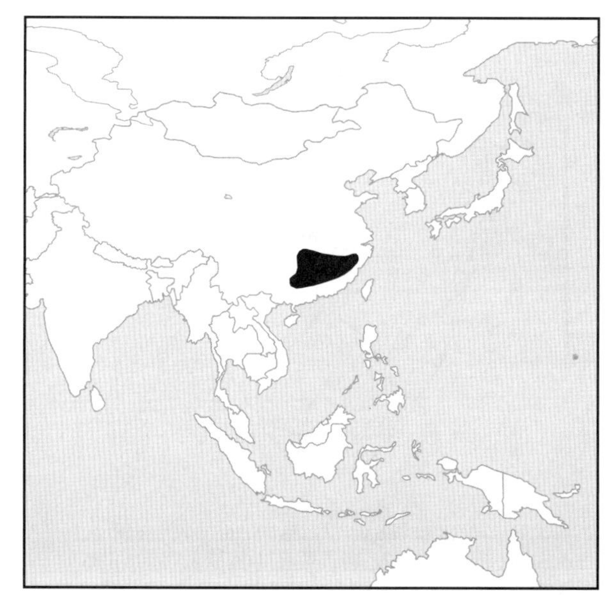
Distribution of *Pseudotaxus chienii* in southeastern China, scale 1 : 120 million

seed opening straight opposite and pointing away from the stalk, maturing and falling with the aril in a single season. Cotyledons two, each with one vein. Chromosome base number $x = 12$.

Wood of medium weight and strength, very light brown. Grain fine and even, with well-defined growth rings marked by a broad band of denser latewood. Resin canals and individual resin parenchyma cells both absent, the tracheids (wood cells) with evenly spaced spiral wall thickening, sometimes at a very shallow angle.

Without stomates above and with a broad, waxy white stomatal band on either side of, and clearly broader than, the green midrib region and flanked in turn by a narrower green marginal zone. Each stomatal band consisting of 15–20 closely spaced, straight to wavering, intermittently interrupted lines of stomates. Each stomate aligned lengthwise in its row, shallowly sunken beneath and partially hidden by the four to six surrounding subsidiary cells, of which the end ones are often shared between adjacent stomates in a line, without a Florin ring, the other epidermal cells (in contrast to *Taxus*) also generally lacking papillae, except right along the edge of the leaf. Midvein single, prominent, narrowly and sharply raised above, broadly and shallowly raised beneath, without resin canals, flanked by cylinders of transfusion tissue. Photosynthetic tissue forming a double palisade layer covering the upper side of the leaves beneath the upper epidermis, which is not accompanied by hypodermis, the palisade giving way to looser spongy mesophyll throughout the remainder of the leaf down to the stomatal region.

One species in eastern China south of the Changjiang (Yangtze River). References: Florin 1948b, L. K. Fu 1992, Fu et al. 1999e, T. S. Ying et al. 1993. Synonym: *Nothotaxus* Florin.

As might be expected from the name (Greek and Latin for "false yew"), *Pseudotaxus* is the closest relative to the true yews (*Taxus*) and, except for the white aril and without detailed examination, is superficially identical to them. The sole extant species of *Pseudotaxus,* in fact, was originally described as a species of *Taxus* in 1934, 13 years before it was nearly simultaneously independently transferred to a previously unrecognized new genus by both Cheng and Florin, who used different names, Cheng's being the earlier by less than 3 months. Despite their similarities, the generic separation of *Pseudotaxus* from *Taxus* is reinforced by the many small features that differ from those in all species of yews, including paired (rather than single) bracts at the base of the pollen and seed cones, the presence of two bracts among the pollen stalks, differences in pollen structure, quickly browning twigs, and waxy white (rather than pale yellowish green) stomatal bands without papillae on the cell surfaces (except at the leaf edge). DNA studies confirm both the close relationship to and the generic separation from *Taxus* species. These two genera are sister taxa, and their closest relative is the much more distinct New Caledonian yew (*Austrotaxus*). *Pseudotaxus chienii* is barely in cultivation, and there has been no cultivar selection. There is no known fossil record for *Pseudotaxus,* although it might be possible to recognize the relatively distinctive pollen grains of the genus if they were ever found in Tertiary deposits.

Pseudotaxus chienii (W. C. Cheng) W. C. Cheng

WHITE-CUP YEW, BAI DOU SHAN (CHINESE)

Shrub to 4 m tall, with multiple trunks to about 0.3 m in diameter. Bark grayish brown to dark brown. Leaves shiny dark to rich green above, (0.5–)1–2(–2.5) cm long, (2–)2.5–4(–4.5) mm wide, often somewhat sickle-shaped, usually widest near the base, tapering very gradually and then abruptly to the rounded tip topped by a prickle and abruptly to the rounded base on a very short petiole up to 1 mm long, the leaf edges slightly turned down. Pollen cones about 4–5 mm long, including the basal scales, and about 3–4 mm in diameter, with 6–12 pollen scales (actually extremely reduced simple cones). Seeds shiny chestnut brown, 5–8 mm long, 3–5 mm in diameter, seated in a thick, white aril a little longer and wider and more rounded than the seeds. Very discontinuous in southeastern China, from Hunan to Zhejiang, south to Guangxi and Guangdong. Scattered in the understory of humid montane hardwood forests with either an evergreen or deciduous canopy, sometimes in more exposed sites on rock outcrops or drier soils; 500–1,300 m. Zone 8. Synonyms: *Nothotaxus chienii* (W. C. Cheng) Florin, *Pseudotaxus liana* Silba (as "*liiana*").

White-cup yew is a rare species that is threatened more by general habitat destruction than by direct exploitation since the small-

Twig of white-cup yew (*Pseudotaxus chienii*) with mature seed cones; scale, 1 cm.

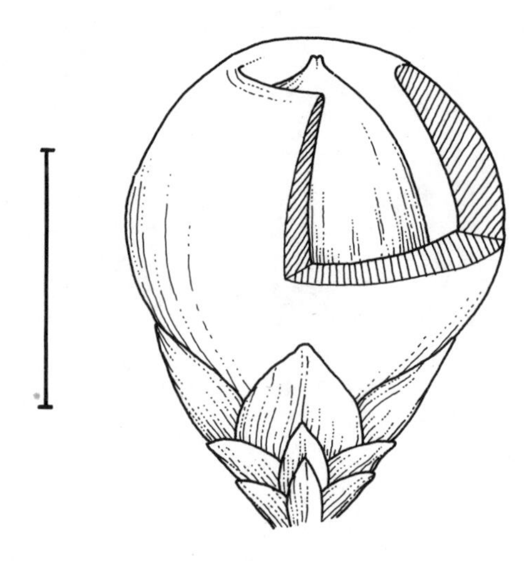

Mature seed cone of white-cup yew (*Pseudotaxus chienii*) with aril partially cut away to show tip of seed nestled within; scale, 5 mm.

Mature shrub of white-cup yew (*Pseudotaxus chienii*) in nature.

dimension wood is used to only a limited extent for carving diminutive objects. Only a very few stands are protected in nature reserves, and white-cup yew is scarcely cultivated in China and is not in general cultivation elsewhere, although its white arils and white stomatal bands give it considerable ornamental potential. The species name honors Sung Su Chien (b. 1885), the botanist who collected the extensive type material, including specimens with pollen and seed cones at the time of pollination and later the mature seeds, all from the northeastern edge of the range of the species in Zhejiang.

Pseudotsuga Carrière

DOUGLAS FIR

Pinaceae

Evergreen trees with a straight, single trunk bearing numerous, spirally arranged, horizontal branches not concentrated into regular tiers. Bark nonfibrous, smooth at first, flaking in scales and then quickly breaking up and finally becoming deeply furrowed between long, rectangular, blocky ridges. Crown deep, conical at first, becoming cylindrical with age. Branchlets all elongate, without distinction into long and short shoots, smooth or shallowly furrowed between the attached leaf bases, usually transiently hairy. Winter buds well developed, scaly, not conspicuously resinous. Leaves spirally arranged and radiating all around the twigs, though usually somewhat parted on the upper side. Individual leaves needlelike, flexible, flat or a little plump, usually straight, the tip notched, rounded, or pointed but not prickly, the base narrowed to a short, distinct petiole attached firmly to the twig on a low scar.

Plants monoecious. Pollen cones single, emerging from scaly buds in the axils of a few needles of the previous year that are back from the tip of the twig. Each cone with numerous, spirally arranged pollen scales, each bearing two pollen sacs. Pollen grains very large (75–115 μm in diameter), nearly spherical, quite large, almost featureless but with a broad, low, thin-walled germination dome, differing from other Pinaceae (except *Larix*) in lacking either a pair of air bladders or a puffy frill. Seed cones oblong, dangling singly at the tips of short side branches, maturing in a single season, the numerous, densely spirally arranged seed scales then spreading to release the seeds. Seed scales variable within species, fan-shaped to diamond-shaped or circular, the three-tipped, strap-shaped bracts attached only at the base of the scales and emerging prominently between them. Seeds two per scale, wedge-shaped, the asymmetric

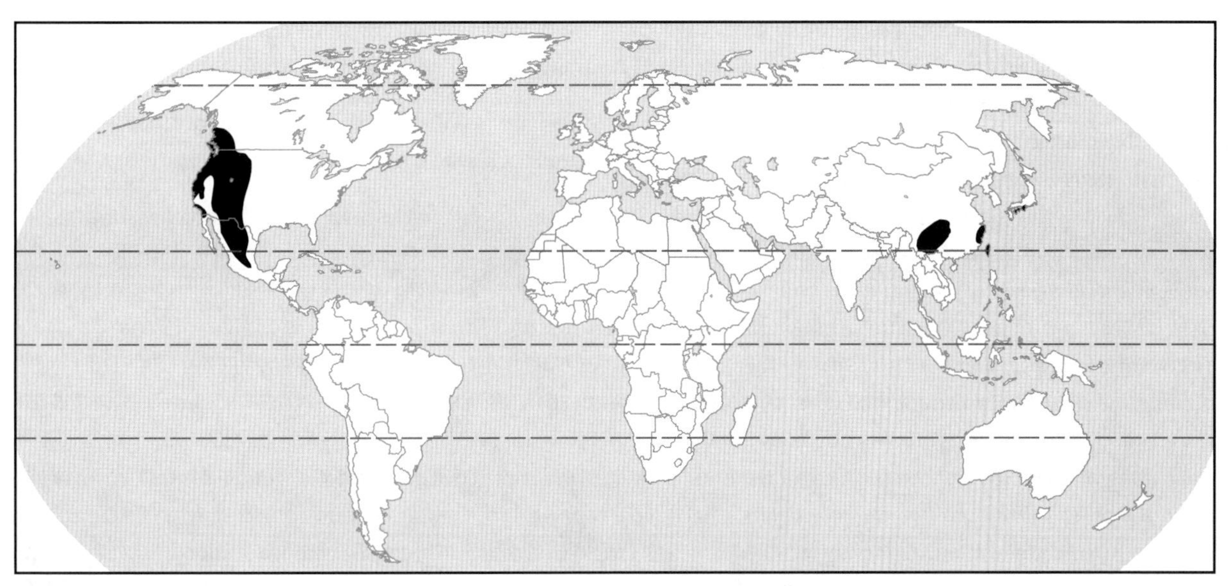

Distribution of *Pseudotsuga* in western North America and eastern Asia, scale 1 : 200 million

wing derived from the seed scale, a little larger than the seed body and cupping it on one side. Cotyledons (3–)6–9(–14). Chromosome base numbers $x = 12$ and 13.

Wood moderately hard and heavy, fragrant, the reddish brown heartwood contrasting with the light brown sapwood. Grain uneven, with a sharp transition between early- and latewood. Vertical resin canals small, few, and irregularly distributed, the vertical water-conducting cells (tracheids) with spiral thickening, and both horizontal tracheids and resin canals present in the rays.

Lines of stomates closely spaced and regular, confined to two stomatal bands on the lower side. Each stomate sunken beneath and largely hidden by the four (or five) surrounding subsidiary cells, which are often shared between stomates in the same and adjacent lines and which are covered with a very thick cuticle showing no trace of a Florin ring. Leaf cross section with a single resin canal on each side of the single-stranded midvein near the outer margins of the needle just inside the lower epidermis. Photosynthetic tissue with (one or) two layers of elongate palisade cells beneath the upper epidermis and (in some species) accompanying single hypodermal layer (double along the leaf midline), the remaining photosynthetic volume filled with looser spongy mesophyll radiating around the midvein and packed more irregularly between the palisade parenchyma and the stomatal bands.

Four species in western North America and eastern Asia. References: El-Kassaby et al. 1983, Farjon 1990, Fu et al. 1999c, Gernandt and Liston 1999, Li and Keng 1994, Lipscomb 1993, Martínez 1963, Strauss et al. 1990, Yamazaki 1995.

Although *Pseudotsuga* is a small genus, the Douglas firs have engendered a great deal of scientific interest, primarily because one species, *the* Douglas fir (*P. menziesii*), is the single most economically important tree species in the world. The other species are all rare or local tress with little commercial importance. In addition to its importance in forestry, Douglas fir is also the sole species of the genus in general use in horticulture. The moderate number of cultivars selected to date emphasize needle length and color (primarily ranging from green to blue) and growth habit, including dwarfs of various shapes and weeping forms.

Pseudotsuga is closely related to the larches (*Larix*), and these two genera share a pollination mechanism unusual in the Pinaceae, where pollen grains with two air bladders are the norm. Instead of landing on a pollination droplet and being drawn directly into the ovule with the help of the air bladders, pollen grains of *Pseudotsuga* and *Larix* germinate where they fall on the bracts of the receptive seed cones, and the pollen tubes have to grow from there into the ovules to effect fertilization. Élie Carrière named *Pseudotsuga* (Greek and Japanese "false hemlock") 12 years after *Tsuga* to emphasize the differences between these two genera, even though *Pseudotsuga* had pollen somewhat similar to that of hemlocks. The two genera are unrelated within Pinaceae, however, since *Tsuga* is an abietoid genus closely related to *Pseudolarix*, while *Pseudotsuga* is a pinoid closely related to *Larix*. Confusion in naming reigns! The confusion also extends to the common name. The name Douglas fir may also be seen hyphenated as Douglas-fir because these trees are not true firs (*Abies*). In fact, as just noted, they do not belong to the same subfamily of Pinaceae as the true firs (subfamily Abietoideae) but rather to the subfamily containing pines (*Pinus*), spruces (*Picea*), and larches (*Larix*), the subfamily Pinoideae. Thus an alternate common name, Oregon spruce, is slightly more accurate botanically, although these trees are not true spruces either.

Both *Pseudotsuga menziesii* and *P. sinensis* are quite variable across their respective ranges in western North America and China, and many varieties and even species have been proposed to accommodate this variation. However, there do not seem to be any strong breaks in the variation pattern of either species that would support recognition of additional species. Because of a lack of easily interpretable morphological features linking the species differentially, their relative relatedness is a little unclear. DNA studies seem to show that the North American species are each other's closest relatives, as are the two Asian species, so speciation has been independent on the two continents.

Douglas fir (*Pseudotsuga menziesii*) itself is also unusual among Pinaceae in its chromosome base number $2x = 13$ pairs rather than the 12 pairs common in the family, including all other examined species of *Pseudotsuga*. This unusual chromosome constitution arose by the splitting of one large, symmetrical pair of chromosomes found in the other species into two one-sided pairs, thus adding a single pair to the count without adding any new genes to the chromosomes. Douglas fir was also the tree in which the peculiar conifer trait (at least compared to most green land plants) of inheritance of chloroplasts via the pollen parent rather than the seed parent was first firmly established. This was also one of the two species (along with Monterey pine, *Pinus radiata*) in which another peculiarity of conifer chloroplasts was discovered, the loss of a piece of DNA (the "inverted repeat") found in all other green land plants except many members of the pea family, which have lost it independently. This species was also one of the first conifers in which inheritance of mitochondria was studied. It proved to be via the seed parent in Douglas fir, establishing a difference between members of the Pinaceae and the Cupressaceae, in which mitochondrial inheritance, like that of chloroplasts, is via pollen.

Unlike many other conifer genera, including a number of presently highly localized genera, like *Cathaya, Glyptostrobus, Keteleeria, Metasequoia, Pseudolarix,* and *Sequoia* that occurred widely across the northern hemisphere in the Tertiary, fossils of *Pseudotsuga* are confined to the region now occupied by the genus. They date back to the Oligocene, about 30 million years ago. This is relatively recent for a conifer genus, and if it really arose then, this might explain the closeness of the relationship among its species.

Identification Guide to *Pseudotsuga*

The four species of *Pseudotsuga* may be identified by the width of the needles, whether the tip of the needle is pointed, rounded, or notched, whether the bracts of the seed cones are straight or curl back, the length of most seed cones, and whether the twigs have tiny hairs or not:

Pseudotsuga menziesii, needles up to 1.5 mm wide, pointed or rounded, bracts straight or curved outward, seed cones 5–10 cm, twigs hairy or not

P. macrocarpa, needles up to 1.5 mm wide, pointed, bracts straight, seed cones at least 10 cm, twigs hairy or not

P. japonica, needles at least 1.5 mm wide, notched, bracts curled back, seed cones up to 5 cm, twigs hairless

P. sinensis, needles at least 1.5 mm wide, notched, bracts curled back, seed cones 5–8 cm, twigs usually hairy

Pseudotsuga japonica (Shirasawa) Beissner

JAPANESE DOUGLAS FIR, TOGA-SAWARA (JAPANESE)

Tree to 30 m tall, with trunk to 1(–1.6) m in diameter. Bark dull reddish brown to grayish brown, fairly thin, broken up into narrow, elongate plates arranged in interlacing, longitudinal rows. Crown broad, flat-topped, with few, long, slender, upwardly angled branches. Twigs yellowish brown, graying with age, hairless, with fine grooves between the leaf bases. Buds 4–6 mm long. Needles mostly sticking out to the sides, 1.5–2.5(–3) cm long, 1.5–2 mm wide, notched at the tip, shiny light to bright green above, the stomatal bands very white beneath. Pollen cones 10–15 mm long. Seed cones (3.5–)4–5(–6) cm long, 2–3(–4) cm across, purple before maturity, ripening light to dark brown, the seed scales rigid, the bracts curled back. Seeds 6–9 mm long, the wings 10–13 mm long. Southern Japan (Shikoku and nearby Kii Peninsula of Honshū). Scattered in mixed conifer forests; 500–1,100 m. Zone 6.

Japanese Douglas fir is a rare species in its native land, although where it occurs it may make up 10% or more of forests dominated by southern Japanese hemlock (*Tsuga sieboldii*) or other conifers. It is much slower growing than the North American *Pseudotsuga menziesii* or the Japanese sugi (*Cryptomeria japonica*), so it is not grown in timber plantations. In a country rich with conifers, this rare species is not a commercially important tree. Little is known of genetic variation within the species. It is most closely related to the Chinese Douglas fir (*P. sinensis*), and these two are rather distant from *P. menziesii* and the other North American species, bigcone Douglas fir (*P. macrocarpa*).

Pseudotsuga macrocarpa (G. Vasey) H. Mayr

BIGCONE DOUGLAS FIR

Plate 59

Tree to 30(–45) m tall, with trunk to 2(–2.3) m in diameter, the bark thick, dark gray with reddish brown highlights, with broad, flat-topped, interlacing ridges separated by deep furrows. Crown thin, very broadly conical, with widely spaced, widely spreading, long, thin branches in the lower crown and progressively shorter ones above. Twigs reddish brown, minutely hairy at first, inconspicuously ridged. Buds 7–8 mm long. Needles sticking out all around the twig but most densely to the sides, (2–)2.5–4.5 cm long, 1–1.5(–2) mm wide with a short point at the tip, dull bluish green above, the stomatal bands grayish green beneath. Pollen cones 15–25 mm long. Seed cones (9–)10–15(–20) cm long, 4–6(–7) cm across, yellowish green before maturity, ripening dark brown, the seed scales hard and rigid, the bracts straight. Seeds (9–)10–13 mm long, the wings 10–20 mm long. Mountain ranges of southern California, from central Santa Barbara County, east to southwestern San Bernardino County, and south to central San Diego County. Scattered as single trees or small groves emerging above shrubby chaparral on dry but sheltered slopes, or mixed with pines and other conifers in the lower portion of the montane yellow pine forest; (200–)1,000–2,000(–2,400) m. Zone 7.

Although a tree of scattered occurrence and limited total geographic range, bigcone Douglas fir is not a rare species, nor is it particularly threatened by the rampant residential development found in southern California because most of its existing stands are on public land. It has an important role in slope stabilization following fires in its native habitat, and because of its drought tolerance, it might prove useful in similar circumstances in other Mediterranean climate or dry, warm temperate areas. Unlike its relative to the north, Douglas fir (*Pseudotsuga menziesii*), bigcone Douglas fir has no commercial exploitation, and relatively little work has been done on any aspect of its biology. The oils in the leaves and living tissue beneath the bark were investigated to a limited extent but reveal little difference from place to place within the range and also limited (but consistent) differentiation from *P. menziesii.* The two species nowhere grow together although they approach within about 35 km of each other in Santa Barbara County. They have been crossed with difficulty and with a yield of just a few seeds. The chromosome complement of bigcone Douglas fir is very similar to that of the two Asian species of the genus and so differs from the unusual complement of Douglas fir. Despite this, DNA studies show that *P. macrocarpa* is actually more closely related to *P. menziesii* than it is to the Japanese *P. japonica* or the Chinese *P. sinensis.* The largest known individual in combined dimensions when measured in 1973 was 44.2 m tall, 2.1 m in diameter, with a spread of 25.9 m. The species name (Latin for "big fruit") comes from the much larger size of the seed cones in comparison to those of *P. menziesii* (and to the Asian species, but these were not described until 20–30 years after *P. macrocarpa*).

Mature bigcone Douglas fir (*Pseudotsuga macrocarpa*) in nature.

Open seed cone of bigcone Douglas fir (*Pseudotsuga macrocarpa*) after seed dispersal; scale, 5 cm.

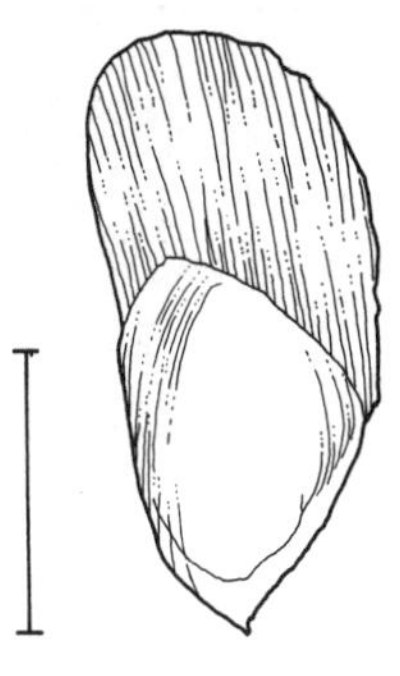

Underside of dispersed seed of bigcone Douglas fir (*Pseudotsuga macrocarpa*); scale, 1 cm.

Pseudotsuga menziesii (Mirbel) Franco

DOUGLAS FIR

Plate 60

Tree to 90(–100) m tall, with trunk to 3(–4.5) m in diameter. Bark reddish brown to grayish brown or even blackish, thick, breaking up with age into scaly, broad, interlacing ridges separated by deep furrows. Crown narrowly conical or narrowly egg-shaped to cylindrical with numerous short, slender, initially upwardly arched branches passing through horizontal to downswept with age. Twigs pale greenish yellow to reddish purple and minutely hairy, becoming gray and hairless with age, with inconspicuous, shallow grooves between the leaf bases. Buds 6–10 mm long, shiny reddish brown, not resinous. Needles sticking out to the sides and above, (1.5–) 2–3(–4) cm long, 1–1.5 mm wide, pointed or rounded at the tip,

Underside of twig of Douglas fir (*Pseudotsuga menziesii*) with 2 years of growth, ×0.5.

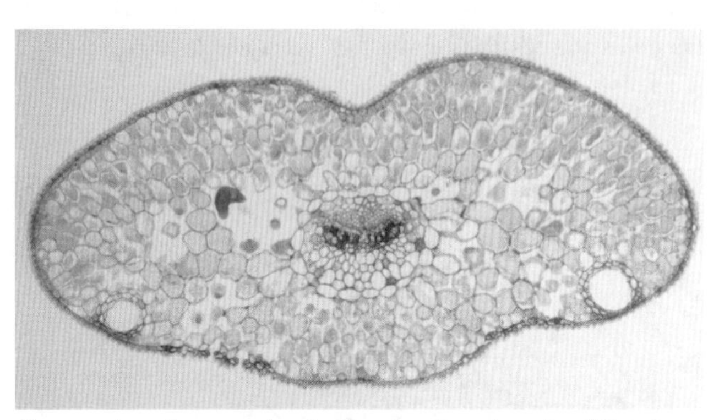

Cross section of preserved, stained, and sectioned needle of Douglas fir (*Pseudotsuga menziesii*), differing from eastern hemlock (*Tsuga canadensis*) in various ways, including having two resin canals.

Twig of Douglas fir (*Pseudotsuga menziesii*) with closed mature seed cone; scale, 2 cm.

Twig of Douglas fir (*Pseudotsuga menziesii*) with 3 years of growth and mature pollen cones at the time of pollination, ×0.5.

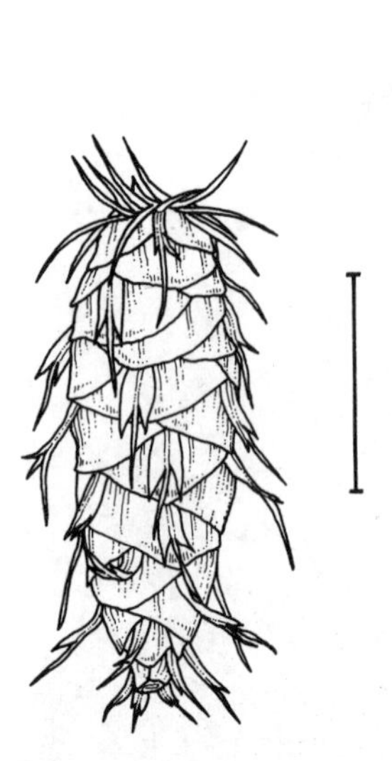

Closed mature seed cone of Douglas fir (*Pseudotsuga menziesii*); scale, 5 cm.

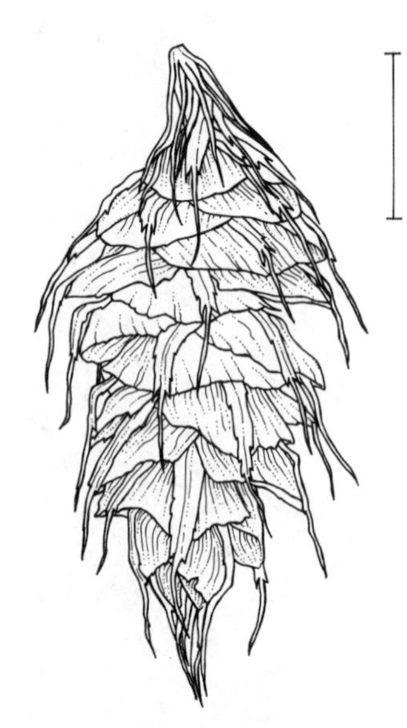

Open seed cone of Douglas fir (*Pseudotsuga menziesii*) after seed dispersal; scale, 2 cm.

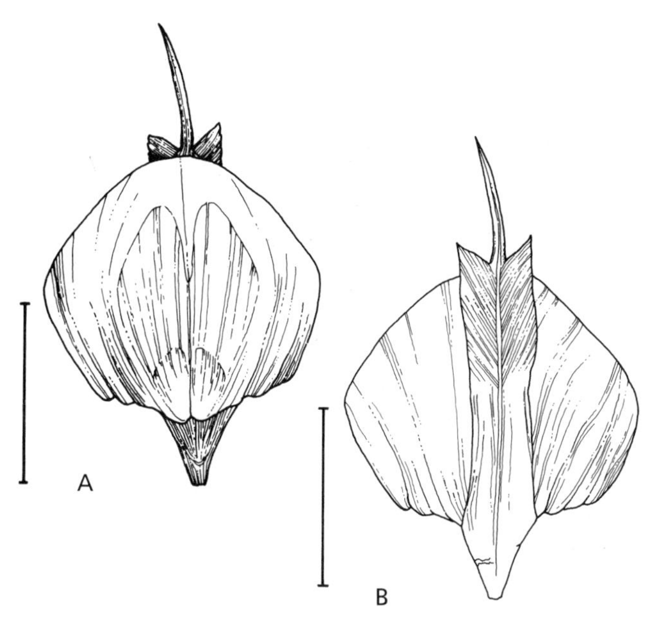

Upper (A) and undersides (B) of detached seed scales and bracts of Douglas fir (*Pseudotsuga menziesii*); scale, 1 cm.

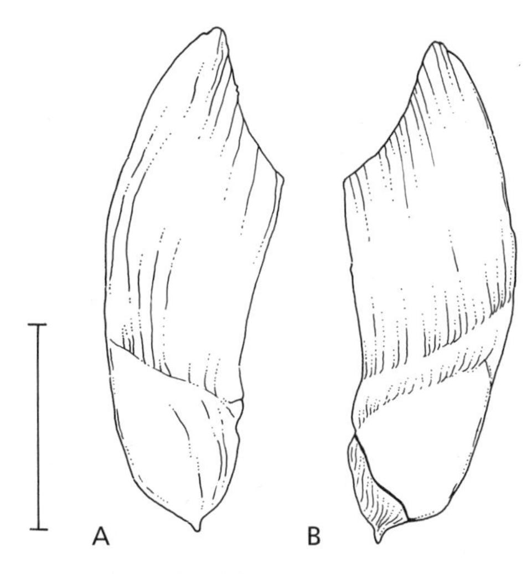

Upper (A) and undersides (B) of dispersed seeds of Douglas fir (*Pseudotsuga menziesii*); scale, 5 mm.

yellowish green or dark green to bluish green or grayish green, the stomatal bands pale whitish green. Pollen cones (10–)15–20 mm long, yellowish brown blushed with red. Seed cones 4–9(–10) cm long, 3–4 cm across, green before maturity, ripening yellowish brown to purplish brown, the seed scales flexible, the bracts pointed forward or sticking out. Seeds 5–7(–8) mm long, the wings 10–12(–14) mm long. Western North America from central British Columbia south through the Coast and Cascade Ranges and Sierra Nevada to southern California and through the Rocky Mountains and other cordilleran ranges to northern Oaxaca (Mexico). Forming enormous pure stands (especially after fires) or mixed with various other conifers and a few hardwoods in lowland and montane forests on a wide variety of substrates; 0–3,000 m. Zone 3 (variety *glauca*) or 7 (variety *menziesii*). Two varieties.

As the single most economically important tree species in the world, an enormous amount has been written about the growth, reproduction, natural enemies, and genetic variation of Douglas fir. It differs from the other three species of *Pseudotsuga*, all rather rare and localized, in being not only widespread and abundant but even broadly dominant. It is also unique in the whole family Pinaceae in having 13 pairs of chromosomes, which occurred by splitting one of the 12 pairs found in all the other species into two pairs. Whether this change has anything at all to do with its ecological success is uncertain.

What is clear is that Douglas fir contains a great deal of genetic variation in morphological, ecological, and biochemical characteristics, with very different variation patterns in different features. Variation in size and shape of seed cones, seed scales, and bracts has led, in the past, to recognition of numerous segregate species. Few authors accept these today, recognizing that these variants occur mixed throughout the range of the species in no discernible geographic pattern. An exception is the difference in average seed cone size and the position of the bracts that helps to distinguish the Pacific slope variety *menziesii* from the interior and cordilleran variety *glauca*. Much morphological variation in seed cones, twigs, and needles appears unrelated to geographic distribution or ecological conditions. Growth characteristics, however, such as dates of bud burst and bud formation and rate of growth, seem to be strongly selected under the local conditions of each population (or region). With all the available natural variation in needle and growth characteristics, it is not surprising that numerous cultivars have been selected.

Variation in biochemical characteristics presents a somewhat different picture. Analyses of leaf and wood oils document the existence of a series of intergrading regional chemical races within each of the two recognized botanical varieties. These studies further confirm the intergradation of the two varieties along their extensive band of contact throughout the interior of British Columbia. This is one of the reasons that the two varieties are not recognized as separate species. Studies of chloroplast DNA variation produced similar results, with intergrading regional races within each variety. In fact, Douglas fir was the first conifer in which the paternal inheritance of chloroplast DNA was demonstrated, in 1986. Only later was it shown that this pattern of inheritance is found in all conifers, in contrast to flowering plants in which it is almost always maternal. One interesting (and unexplained) inconsistency between the patterns of geographic variation revealed by the different kinds of features studied so far is that the populations of the Sierra Nevada of California clearly belong to the Pacific slope variety in their morphological features and DNA variants while their leaf and wood oils more closely resemble those of the interior variety.

Another kind of geographic pattern is revealed by studies of enzyme variants. The taxonomic unity of Douglas fir is strongly supported by these studies, which reveal only very modest separation among three enzyme races. The Pacific slope variety *menziesii* contains a single enzyme race, while the interior variety *glauca* has separate races north and south of about 44°N. Most interestingly, while the nearly continuous Pacific and northern interior races are rich in enzyme variants and show little differentiation between populations, the situation is quite different for the southern interior race. Here, the populations are smaller and more discontinuous than in the northern and western portion of the range of the species. In line with theoretical expectations, these populations have less genetic variation within populations and more differentiation between populations than in the other two races, although these effects are much less pronounced than they are in many other plants, like self-pollinating annuals.

With its difference in chromosome complement, Douglas fir is reproductively isolated from the other species of the genus. Attempted crosses with both the North American bigcone Douglas fir (*Pseudotsuga macrocarpa*) and the Chinese Douglas fir (*P. sinensis*) have been unsuccessful, while crosses between trees of varieties *menziesii* and *glauca* yield full seed set. Nonetheless, DNA studies show that *P. macrocarpa* and *P. menziesii* are sister species that differentiated after their common ancestor separated from the common ancestor of the two Asian species.

The largest known individual in the United States in combined dimensions when measured in 1998 was 91.8 m tall, 4.1 m in diameter, with a spread of 19.8 m. It, and every other giant Douglas fir (with claims of trees up to 126.5 m tall in the past), belongs to the Pacific slope variety *menziesii,* while trees of the interior variety *glauca* rarely exceed 40 m tall. The species name honors Archibald Menzies (1754–1842), the Scottish physician and naturalist who collected the first botanical specimen of the species in 1792 while ship's surgeon with the Vancouver expedition. The common name honors David Douglas (1798–1834), the Scottish plant collector whose seed lot introduced the species into cultivation in 1827 and who was killed by a boar when he fell into a game pit in Hawaii. From this small horticultural beginning, the species became one of the most important trees in plantation forestry throughout suitable portions of the temperate world.

***Pseudotsuga menziesii* var. *glauca* (Beissner) Franco**

ROCKY MOUNTAIN DOUGLAS FIR

Needles dark green in the north, progressively more bluish green or grayish green with wax southward. Seed cones 4–7 cm long, the bracts sticking out to the sides. Eastern portion of the range of the species, east of the Pacific crest from central British Columbia south through the interior mountains to northern Oaxaca (Mexico), very discontinuous throughout the south; 600–3,000 m. Synonyms: *Pseudotsuga flahaultii* Flous, *P. guinieri* Flous, *P. macrolepis* Flous, *P. menziesii* var. *caesia* (F. Schwerin) Franco, *P. menziesii* var. *oaxacana* Debreczy & I. Rácz, *P. rehderi* Flous.

Pseudotsuga menziesii* (Mirbel) Franco var. *menziesii

COAST DOUGLAS FIR

Needles yellowish green to light green without a waxy coating. Seed cones (5–)6–9(–10) cm long, the bracts pressed forward along the scales. Western portion of the range of the species, west of the Pacific crest from Douglas Channel, west-central British Columbia, south through the Pacific mountains to central California, with outliers in Santa Barbara and Fresno Counties; 0–1,800 m. Synonyms: *Pseudotsuga douglasii* (D. Don) Carrière, *P. taxifolia* (A. Lambert) N. Britton.

Pseudotsuga sinensis Dode

CHINESE DOUGLAS FIR, HUANG SHAN (CHINESE)

Tree to 40(–50) m tall, with trunk to 0.8(–1) m in diameter. Bark dark brown to dark grayish brown, thick, with scaly, flat-topped ridges separated by deep furrows. Crown conical when young, broadening and becoming deeply to shallowly dome-shaped with age, with numerous horizontal to upwardly angled branches. Twigs reddish brown to yellowish gray, sparsely to densely hairy at first, at least on side shoots, with fine grooves between the leaf bases. Buds 6–7 mm long. Needles mostly sticking out to the sides, or sometimes all around the twig, (0.7–)1–4(–5.5) cm long, 1.5–3.2 mm wide, notched at the tip, yellowish green to dark green or bluish green above, the stomatal bands white to dirty grayish green. Pollen cones 10–15 mm long. Seed cones 3.5–8 cm long, 2–5.5 cm across, purplish green before maturity, ripening dark reddish brown, the seed scales thin but rigid, the bracts curled back. Seeds 7–10 mm long, the wings 10–20 mm long. Discontinuously distributed across southern and central China (hence the scientific name, Latin for "of China") from northern Yunnan to Zhejiang and in Taiwan. Usually mixed with other conifers and deciduous or evergreen hardwoods on slopes of hills and mountains; (400–) 800–2,800(–3,300) m. Zone 8. Five varieties.

Like each of the species of *Pseudotsuga* other than *P. menziesii, P. sinensis* is a rare plant that is usually scattered among other trees rather than forming extensive stands. Despite this, the trees are still harvested indiscriminately where they are accessible and used for pulp as well as a source of timber for a variety of unspecialized carpentry and construction uses. It is listed as an endangered species in

Mature shortleaf Chinese Douglas firs (*Pseudotsuga sinensis* var. *brevifolia*) in nature.

Mature Taiwan Douglas fir (*Pseudotsuga sinensis* var. *wilsoniana*) in nature.

China, but few conservation measures for it have been put in place. The range is divided into five separated regions, each of which has trees that are distinguishable from those in the other regions. On this basis, many authors treated the populations of each region as constituting a separate species. The differences among them are so minor, however, comparable to those found among populations of *P. menziesii*, that they are best treated as varieties of a single species.

Pseudotsuga sinensis var. _brevifolia_ (W. C. Cheng & L. K. Fu) Farjon & Silba

DUAN YE HUANG SHAN (CHINESE)

Needles 0.7–1.5(–2) cm long and 2–3 mm wide. Seed scales roundly diamond-shaped. Guangxi and adjacent Guizhou; 400–1,300 m. Synonym: *Pseudotsuga brevifolia* W. C. Cheng & L. K. Fu.

Pseudotsuga sinensis var. _forrestii_ (Craib) Silba

LAN CANG HUANG SHAN (CHINESE)

Needles (2.5–)2.8–5.5 cm long and 1.3–2 mm wide. Seed scales nearly circular. Northwestern Yunnan and adjacent Sichuan; 2,400–3,000(–3,300) m. Synonym: *Pseudotsuga forrestii* Craib.

Pseudotsuga sinensis var. _gaussenii_ (Flous) Silba

HUA DONG HUANG SHAN (CHINESE)

Needles 1.5–3 cm long and 1.5–2 mm wide. Seed scales very broadly fan-shaped. Western Zhejiang and adjacent Anhui, Jiangxi, and Fujian; 600–1,500 m. Synonym: *Pseudotsuga gaussenii* Flous.

Pseudotsuga sinensis Dode **var. _sinensis_**

HUANG SHAN (CHINESE)

Needles (1.3–)2–3 cm long and 1.5–2 mm wide. Seed scales variably kidney-shaped to roundly fan-shaped. Northeastern Yunnan to southern Shaanxi and northwestern Hubei; (800–)1,000–2,800 m.

Pseudotsuga sinensis var. _wilsoniana_ (Hayata) L. K. Fu & N. Li

TAI WAN HUANG SHAN (CHINESE)

Needles 1.5–2.5(–4.5) cm long and 1–1.5(–2) mm wide. Seed scales kidney-shaped to roundly wedge-shaped. Taiwan; (800–)1,000–2,500(–2,700) m. Synonym: *Pseudotsuga wilsoniana* Hayata.

Retrophyllum C. Page

RETROPHYLLUM, FLIP-LEAF

Podocarpaceae

Evergreen trees and shrubs, sometimes with swollen, fluted, or buttressed trunks. Bark fibrous, peeling in short strips and often becoming strongly ridged and furrowed with age. Crown broadly dome-shaped and spreading, consisting initially of regularly alternating pairs of branches. Branchlets variably differentiated into long and short shoots, these generally similar or the long shoots with scale leaves, the short shoots frondlike, flattened, and shed intact together with their leaves. Both shoot types hairless, remaining green for at least the first year, prominently grooved between the attached leaf bases. Resting buds naked, not very evident. Foliage leaves of horizontal shoots in pairs, usually (except in *Retrophyllum minor*) brought into a single plane by twisting of the petioles and of the twigs between the leaves, those to the right of the twig all with their upper surface facing up while those on the left have their lower surface showing (hence the scientific name, Latin for "backward leaf"), more or less sword-shaped.

Plants dioecious. Pollen cones cylindrical, single or in clusters of two to five at the ends of short, axillary scaly stalks or of leafy short shoots. Each cone with a few alternating pairs of bracts at the base and numerous, densely crowded pollen scales, each bearing two pollen sacs. Pollen grains medium (body 35–40 µm long, 50–60 µm overall), with two round air bladders that are smaller than the body, body and bladders relatively smooth externally, the bladders also with coarse internal sculpturing. Seed cones single or in pairs on short, scaly stalks in the axils of foliage leaves. Individual cones highly modified and reduced, with a single, plump, unwinged seed embedded in an inverted position in each of one or two fleshy, green to red or blue-black seed scales (the epimatium) in the axils of somewhat leaflike bracts that may become slightly fleshy but do not enlarge into a conspicuous, colorful, berrylike podocarpium, maturing and falling in a single season. Seeds usually with a crest along one side and over the tip. Cotyledons two, each with two veins. Chromosome base number $x = 10$.

Wood soft, light, sweetly fragrant, light brown, with a core of yellowish brown to reddish brown heartwood. Grain fine and even, with growth rings delimited by narrow bands of denser latewood. Resin canals absent.

Both leaf faces with interrupted lines of stomates not organized into distinct stomatal bands. Each stomate sunken beneath and largely hidden by the four (to six) surrounding inner subsidiary cells, which rise around the opening in a prominent, continuous Florin ring that is surrounded by a sunken moat. Leaf cross section with a single midvein flanked by wings of transfusion and accessory transfusion tissue extending up to halfway to the margin and with one to three (to five) resin canals beneath the midvein and none to three additional ones on each side extending the length of the leaf halfway between the upper and lower surfaces. Photosynthetic tissue of adult foliage leaves forming a palisade layer one or more cells thick covering both surfaces beneath the epidermis and adjacent discontinuous patches (between the stomates) of hypodermal cells.

Four species in the southwestern Pacific from Fiji and New Caledonia to the Moluccas and in western South America. References: Conran et al. 2000, de Laubenfels 1969, 1972, 1982, 1987, Gray 1962, Gray and Buchholz 1948, Herbert et al. 2002, Hill and Brodribb 1999, Hill and Pole 1992, Page 1988, Sinclair

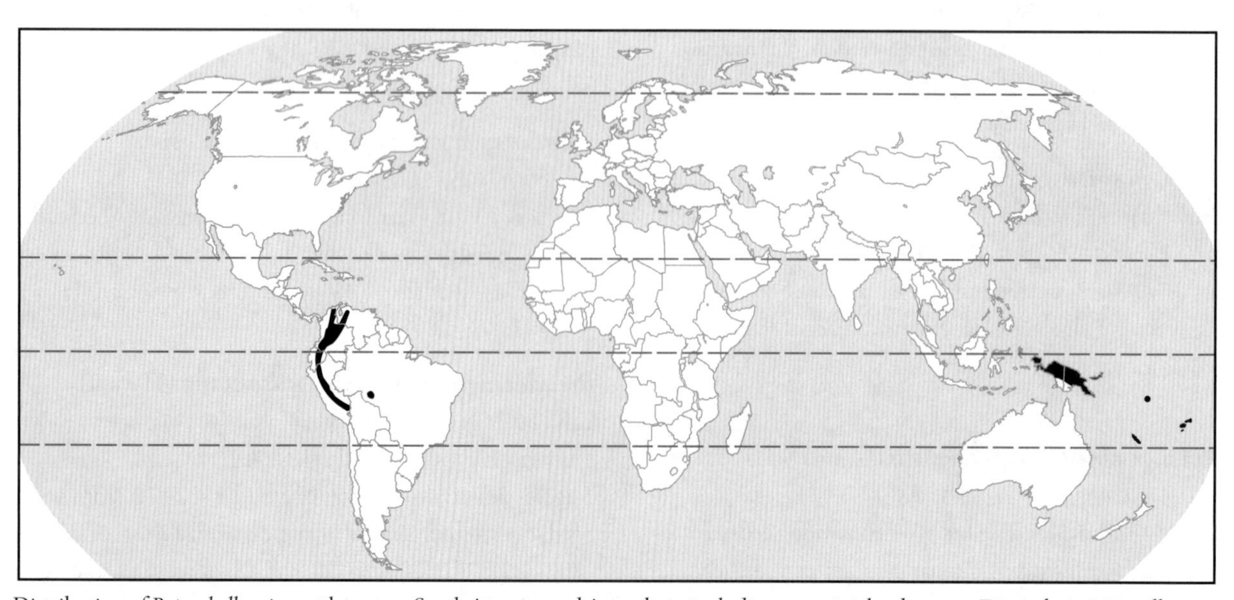

Distribution of *Retrophyllum* in northwestern South America and Australasia, including oceanic islands east to Fiji, scale 1 : 200 million

et al. 2002. Synonyms: *Decussocarpus* de Laubenfels, *Nageia* sect. *Polypodiopsis* (M. Bertrand) de Laubenfels, *Podocarpus* sect. *Polypodiopsis* M. Bertrand.

This small group of species was long included as a distinctive botanical section within the broadly defined *Podocarpus* that was current until 1969. In that year, de Laubenfels included them as a section within a separate genus (originally *Decussocarpus,* later replaced with *Nageia*) that also included sections *Nageia* and *Afrocarpus* (here all recognized as separate genera). All three sections usually lack the united, swollen bracts of the seed cone that make up the podocarpium that gives *Podocarpus* its name. They also share leaves attached in pairs on opposite sides of the twigs, in contrast to the individually spirally arranged leaves of *Podocarpus* species in the restricted sense generally accepted today. Finally, Page treated each of these sections as a separate genus, based on differences in leaf structure and base chromosome numbers. This separation is adopted here because of the coherence and ready recognizability of these three groups, but recognition as separate sections within a single genus would also be justifiable taxonomically. This is because the three groups are mutually each other's closest relatives, as shown by several DNA studies involving different genes, while the species within each group have likewise been shown to be closest to other members of the same group. Since each group is thus monophyletic, as are all three groups taken together, it is a matter of preference rather than scientific necessity whether they are recognized as separate genera or sections within a single genus. The various DNA studies also show that *Retrophyllum, Nageia,* and *Afrocarpus* are most closely related to *Podocarpus,* so their historical inclusion within that genus was not simply mistaken, even though relationships are better expressed by separating them.

The distinctive shoots of *Retrophyllum,* with the uniformly paired leaves each with a single midvein and all facing up on one side of the twig and down on the other, are unique among living conifers. Similar fossils can be traced as far back as the Eocene, more than 35 million years ago, in southern South America, New Zealand, and across southern Australia, but some of these may represent extinct genera since they differ in important details of the leaf and epidermal structure from any living species. Unlike some other southern conifer genera with species in both South America and Australia in the past and present, such as *Libocedrus* or *Prumnopitys,* no species of *Retrophyllum* are found today in New Zealand or eastern Australia. Species of *Retrophyllum* are not in general cultivation, and no cultivar selection has taken place.

Identification Guide to *Retrophyllum*

Adult specimens of the four species of *Retrophyllum* may be distinguished by the length of most foliage leaves, whether these stick out in two rows that lie in a single plane or stick out in four rows around the short shoots, whether the short shoots generally attach to leafy or to scaly shoots, the length of most pollen cones, and the length of the combined seed coat and epimatium for most seeds:

Retrophyllum comptonii, leaves less than 15 mm, in two rows, shoots scaly, pollen cones less than 8 mm, seed coat and epimatium 20–25 mm

R. rospigliosii, leaves more or less than 15 mm, in two rows, shoots leafy, pollen cones 8–12 mm, seed coat and epimatium at least 25 mm

R. minor, leaves more or less than 15 mm, in four rows, shoots scaly, pollen cones up to 8 mm, seed coat and epimatium at least 25 mm

R. vitiense, leaves more than 15 mm, in two rows, shoots scaly, pollen cones more than 12 mm, seed coat and epimatium less than 20 mm

Retrophyllum comptonii (J. Buchholz) C. Page

NEW CALEDONIAN FLIP-LEAF

Tree to 20(–30) m tall, with trunk to 0.8 m in diameter or more, unbranched for half or more of its height. Fresh bark tan to brown, becoming gray to dark gray and deeply furrowed with age. Crown spreading with numerous slender, rising branches bearing horizontally arranged short shoots. Resting buds protected by ordinary scale leaves. Leaves of long shoots and at the base of short shoots and reproductive shoots often scalelike, triangular, spreading, 2–4 mm long. Leaves of short shoots expanded and photosynthetic, 6–15 mm long (to 30 mm in juveniles), 2–5 mm wide (to 6 mm in juveniles). Blades of short shoot leaves widest near the middle, tapering gradually and then more abruptly to the broadly and roundly triangular tip and base, attached directly to the twig without a stalk. Midrib weak, narrow, and becoming inconspicuous among longitudinal wrinkles in dried leaves, with one resin canal beneath the midvein. Similar leaves also found in alternating pairs that are not flattened into a plane on some upright shoots of mature trees. Pollen cones 4–6 mm long (sometimes elongating to 12 mm after release of the pollen) and 2.5–3 mm wide. Seed cones on a stalk about 10 mm long or more bearing two to four alternating pairs of scaly or leafy sterile bracts before the fertile pair. Combined seed coat and epimatium pear-shaped, 20–25 mm long and (13–)15–20 mm in diameter, covered with a thin, waxy coating at first and maturing bright to dark red. Throughout the length of New Caledonia, from Mount Taom and Mount Ignambi in the northwest to Port Boisé in the southeast. Scattered in the canopy of montane or sometimes lowland rain forest, often on serpentine; (180–)750–1,400(–1,600) m. Zone 10. Synonyms: *Decussocarpus comptonii* (J. Buchholz) de Laubenfels, *Nageia comptonii* (J. Buchholz) de Laubenfels, *Podocarpus comptonii* J. Buchholz.

This is one of the most widely distributed and abundant of the many endemic conifer species of New Caledonia. It is the largest podocarp in New Caledonia and yields a fine timber. Despite this, it is not a specific target of exploitation because the trees do not form solid stands but occur scattered among other species in the rain forest. It is most closely related to the other *Retrophyllum* species of New Caledonia, the much more local, shrubby New Caledonian corkwood (*R. minor*) of the swampy Plaine des Lacs region in the south. Although it is common, there appears to be at least some isolation and genetic differentiation among local populations, a topic that merits further exploration. The species name honors Robert Compton (1886–1979), who collected specimens from large trees on Mount Mou in 1914 as *R. minor* and thus thought the latter name a misnomer.

Retrophyllum minor (Carrière) C. Page

NEW CALEDONIAN CORKWOOD, BOIS BOUCHON (FRENCH)

Dwarf tree or shrub to 3(–5) m tall (hence the scientific name), with trunk often conspicuously swollen, to 0.3 m in diameter. Bark deeply furrowed, dark brown or gray, often stained by mineral-laden waters. Crown spreading, dense or open, with slender branches rising from the summit of the swollen base and bearing tufts of upright or spreading branchlets. Resting buds with a few pairs of ordinary scale leaves. Leaves of long shoots and at the base of short shoots and reproductive shoots scalelike, roundly triangular, spreading, 1–4 mm long and often shed. Leaves of short shots expanded and photosynthetic, in alternating pairs, not twisted into a common plane as in other species of the genus, 7–20 mm long (to 40 mm in juveniles), 2.5–5.5 mm wide. Blades of short shoot leaves broadest near or before the middle, tapering gradually or abruptly to the triangular tip and more abruptly to the wedge-shaped base, attached directly to the twig without a stalk. Midrib region broad, often dividing up into three ridges on drying, with one resin canal beneath the midvein. Pollen cones (4–)5–8 mm long and 2–2.5(–3) mm wide. Seed cones on a short stalk about 4 mm long bearing two to four alternating pairs of scaly sterile bracts before the fertile pair. Combined seed coat and epimatium pear-shaped, (20–)25–28 mm long and (11–)15–17 mm in diameter, covered with a thin waxy coating at first and maturing brownish red. Confined to the Plaine des Lacs and nearby districts of southern New Caledonia. Growing in waterlogged soils or standing water at the edges of lakes and slow moving rivers over serpentine substrates; 0–250 m. Zone 10. Synonyms: *Decussocarpus minor* (Carrière) de Laubenfels, *Nageia minor* Carrière, *Podocarpus minor* (Carrière) Parlatore.

Twig of New Caledonian corkwood (*Retrophyllum minor*) with mature seed cone; scale, 2 cm.

The ecology of New Caledonian corkwood is very unlike that of its closest relative, New Caledonian flip-leaf (*Retrophyllum comptonii*), and the other two species of the genus, all of which are large rain-forest trees. The pronounced swelling of the trunk is like that of the unrelated pond cypress (*Taxodium distichum* var. *imbricarium*) of the Cupressaceae, which can grow along similar lake margin habitats in the southeastern United States. The swelling is achieved by an increase in corky and air-transmitting tissues, explaining the vernacular names. An additional adaptation for this habitat may be found in the seeds, which float with their corky stony layer after the epimatium and outer fleshy layer are shed or consumed by birds. Unfortunately, this rare species, with fewer than 2,500 living individuals, is threatened indirectly by changes in mining practices that alter the water levels so crucial in its narrow ecological niche.

Retrophyllum rospigliosii (Pilger) C. Page

AMERICAN FLIP-LEAF, ROMERÓN, ROMERILLO MACHO, PINO LASO, PINO SILVESTRE, ULCUMANU (SPANISH)

Tree to 35(–45) m tall, with trunk to 1.2(–2) m in diameter, often free of branches for half to two-thirds of its height. Bark reddish brown, smooth to shaggy, weathering dark gray and peeling in large scales or plates. Crown spreading, becoming rather open with age, with a few large horizontal or rising branches subdividing into many slender branches densely clothed with short shoots. Resting buds loosely constructed of scale leaves. Leaves of both long and short shoots expanded and photosynthetic, (7–)10–20(–23) mm long, (2–)3–5 mm wide (to 6 mm in juveniles), smallest on the most exposed branches of the crown.

Blades widest near or before the middle, tapering abruptly to the roundly triangular tip and rounded or roundly wedge-shaped base with just a hint of petiole. Midrib narrow and inconspicuous, with (one to) three (to five) resin canals beneath the midvein and one to three additional ones out to each side. Pollen cones (5–)8–12 mm long and 2.5–3(–3.5) mm wide. Seed cones on a stalk (5–)10–15(–20) mm long bearing two to five alternating pairs of leafy or scaly sterile bracts before the fertile pair, the combined seed coat and epimatium egg-shaped, (20–)25–30 mm long and (12–)15–20 mm in diameter, covered with a thin waxy coating at first and maturing dark red to blue-black. Discontinuous throughout the northern Andean region from western Venezuela (Trujillo, Mérida, and Táchira), through western Colombia, Ecuador, and Peru to southwestern Brazil (Rondônia). Scattered or forming small groves in primary or secondary rain forests; (250–)1,500–2,400(–3,750) m. Zone 9. Synonyms: *Decussocarpus piresii* Silba, *D. rospigliosii* (Pilger) de Laubenfels, *Nageia piresii* (Silba) de Laubenfels, *N. rospigliosii* (Pilger) de Laubenfels, *Podocarpus rospigliosii* Pilger, *Retrophyllum piresii* (Silba) C. Page.

American flip-leaf is the largest native conifer of the northern Andes, with a highly prized wood widely used in both carpentry and general construction. It does not appear to be particularly threatened by this level of exploitation, in part because it is a fairly common species, at least in some regions. More importantly, perhaps, it is one of the few native conifers of the region that seems to benefit from forest clearance. Although not drought tolerant, seedlings have a high light requirement and regenerate vigorously in clearings as long as there is sufficient soil moisture. Regeneration following clearance may be partly responsible for downward expansion of the elevational range of this species. It appears to have a very broad range of temperature tolerance since it has, by far, the widest elevational range of any Andean conifer, from tropical forests in the Amazonian lowlands of Ecuador and Brazil (as low as 250 m) to cold subalpine forests of high mountain passes in Columbia (as high as 3,750 m). Few conifers anywhere can match this kind of range. One specimen from low elevations in Brazil was separated as a distinct species, but it falls well within the range of variation of the species as a whole and is geographically close to some locations in Peru. At one time it was thought that the ecological versatility of this species was due, in part, to nitrogen-fixing root nodules, but its nodular roots harbor mycorrhizal fungi, not nitrogen-fixing bacteria like those in nodules of leguminous plants. American flip-leaf is less closely related to the three Old World species of *Retrophyllum* than these are to one another, but there is no doubt about the unity of the genus, including fossil representatives in both hemispheres. The species was discovered in 1918 during the very first expedition sent out by the fledgling Museum of Natural History at the Universidad Nacional Mayor de San Marcos in Lima under its inaugural director, physician Carlos Rospigliosi Virgil.

Retrophyllum vitiense (B. Seemann) C. Page

PACIFIC FLIP-LEAF, SALUSALU, NDAKUA SALUSALU (FIJIAN), MUGO, LEHIL (PAPUAN LANGUAGES),

Tree to 30(–43) m tall, with trunk to 1(–1.3) m in diameter, often unbranched for half or more of its height and sometimes buttressed. Bark reddish brown to brown, weathering light to dark gray but remaining relatively smooth through continued flaking. Crown spreading, composed of a few large major limbs, quickly ramifying and bearing numerous gently drooping branches densely clothed with branchlets. Resting buds tightly enclosed by several pairs of specialized bud scales 1.5–2 mm long. Leaves of long shoots and at the base of short shoots and reproductive shoots often scalelike, broadly and roundly triangular, generally tight to the twigs and not spreading, 1–2 mm long, frequently shed and leaving only the attached base. Leaves of short shoots expanded and photosynthetic, 15–30 mm long (to 40 mm in juveniles), 3–5.5 mm wide (to 8 mm in juveniles). Blades of short leaves widest below (or near) the middle, tapering gradually to the pointed tip and more abruptly to the rounded base with just a trace of a petiole. Midrib narrow and variously weak to prominent, with one resin canal beneath the midvein. Pollen cones (12–)15–20(–25) mm long and 2–2.5 mm wide. Seed cones on a stalk (2–)6–10 mm long bearing three or four alternating pairs of sterile bracts before the fertile pair, the combined seed coat and epimatium pear-shaped, 10–15(–20) mm long and 8–13 mm in diameter, covered with a thin, waxy coating at first and maturing dark red to purplish red. Discontinuously distributed from the Moluccas (Morotai), through New Guinea, New Britain, and the Solomon Islands (Santa Cruz Islands) to Fiji. Scattered in rain forests of varied composition but often associated with species of kauri (*Agathis*); (0–)500–1,800(–2,000) m. Zone 10. Synonyms: *Decussocarpus vitiensis* (B. Seemann) de Laubenfels, *Nageia vitiensis* (B. Seemann) O. Kuntze, *Podocarpus filicifolius* N. Gray, *P. vitiensis* B. Seemann.

Pacific flip-leaf is especially common in parts of New Guinea and in Fiji, with the species taking its name for the largest island in Fiji, Viti Levu. It is locally an important timber tree in Fiji, and the fragrant wood has a variety of applications in carpentry. The resin responsible for the fragrance of the foliage as well as that of the wood is also used to start fires, like that of Melanesian kauri (*Agathis macrophylla*). Pacific flip-leaf is most closely related to the two New Caledonian *Retrophyllum* species and much less so to the fourth living species in the genus, the South American *R. rospigliosii,* as evidenced both by the organization of the shoot systems and leaf types and by DNA studies.

Saxegothaea Lindley

SAXEGOTHAEA

Podocarpaceae

Evergreen trees with a dense crown composed of closely spaced branches extending almost to the ground, commonly hiding the single, straight trunk, which may become relatively massive for the height. Bark obscurely fibrous, peeling in irregular patches. Branchlets all elongate, without distinction into short and long shoots, hairless, turning brown after the first year, distinctly grooved between the attached leaf bases. Resting buds poorly differentiated, consisting solely of as yet unexpanded normal foliage leaves. Leaves spirally attached but partially rearranged into two ragged rows extending out to the sides of the twig by bending of the petioles, needlelike and resembling those of yews (*Taxus*), straight or slightly sickle-shaped, flattened top to bottom.

Plants monoecious. Pollen cones single or spirally arranged in groups of up to about 16 at the end of an otherwise unmodified shoot, each at the end of a short bare stalk and solitary or in pairs in the axil of an ordinary or slightly reduced foliage leaf. Individual pollen cones cylindrical, with a few scalelike bracts at the base and with numerous, spirally arranged triangular pollen scales, each bearing two pollen sacs. Pollen grains large (50–75 µm in diameter), roughly spherical, without air bladders, very smooth, and with a slightly protruding germination point. Seed cones single on a short scaly stalk at the tip of an otherwise ordinary branchlet, fully condensed, consisting of closely spirally arranged bracts intimately united with the somewhat fleshy seed scales, each scale with a single ovule embedded near the base with its opening pointing down into the cone axis. Seed scales becoming partially united with one another near the bases at maturity. Seeds skirted by a thin, fleshy aril, unwinged, maturing in a single season. Cotyledons two, each with two veins. Chromosome base number $x = 12$.

Wood soft and light, with little differentiation between sap- and heartwood, both light brown to yellowish brown, sometimes with darker streaks or even a reddish brown false heartwood due to bacterial or fungal staining. Grain moderately fine and even, with inconspicuous growth rings marked by a few rows of small but thin-walled latewood cells. Resin canals absent but with scattered individual resin cells.

Leaves without stomates above and with 8–12 somewhat irregular lines of stomates forming a broad, pale stomatal band on either side of the midrib beneath. Each stomate sunken beneath and largely hidden by the four to six surrounding subsidiary cells, which rise in a steep, narrow, lobed, and discontinuous Florin ring. Midvein single, prominently raised above and beneath, with one small resin canal immediately beneath it and with wings of transfusion tissue extending out to the sides. Photosynthetic tissue covering the upper side of the leaf.

Distribution of *Saxegothaea conspicua* in Chile and adjacent Argentina, scale 1 : 35 million

One species in the Andes of southern South America. References: Barrett 1998, Conran et al. 2000, Covas 1995, Hoffmann J. 1994, Kelch 2002, Rodríguez and Quezada 1995, Sinclair et al. 2002.

Many features of *Saxegothaea* (named for Queen Victoria's husband, Prince Albert of Saxe-Coburg-Gotha, who lived 1818–1861), including chromosome structures, wood anatomy, and embryo development as well as DNA studies, clearly ally it with the Podocarpaceae. Nonetheless, its compact, ordinary looking seed cones and pollen grains without even a hint of air bladders are very unlike the conditions found in *Podocarpus* and most other Podocarpaceae, so its exact relationships to other genera in the family are uncertain. John Lindley, who originally described and named the genus in 1851, thought it provided a link between *Podocarpus* and more typical conifers, in essence suggesting that it was what used to be referred to as primitive within the family. In fact, this idea is generally corroborated by DNA studies, which are consistent with a very early divergence of the ancestors of *Saxegothaea* within the lineage that led to the group of genera that includes *Podocarpus* and *Dacrydium*, one of two main lineages within the family. Some DNA studies even imply that its ancestors were the first to diverge from the ancestors of all the other genera in the family. There is no consensus on which of these patterns (if either) reflects the true relative position of *Saxegothaea* and its ancestors. In any event, the other genera with relatively compact, multiscaled seed cones, *Acmopyle, Microcachrys,* and *Phyllocladus,* also seem to have diverged relatively basally, so the highly reduced seed cones typical of *Podocarpus* and most other genera do not appear to be the ancestral condition within the family. Unlike for many other podocarp genera, there is no accepted fossil record for the genus that might help shed light on the timing of these events and hence on the likelihood of one or another evolutionary branching scheme. While *S. conspicua* has been cultivated for more than 150 years, it is uncommon in cultivation, and no cultivar selection has taken place.

Saxegothaea conspicua Lindley

PRINCE ALBERT YEW, SAXEGOTHAEA, MAÑIÓ HEMBRA, MANIÚ HEMBRA (SPANISH FROM ARAUCANIAN)

Tree to 20(–30) m tall but usually much smaller, with a cylindrical, usually straight and slender trunk to 1(–2) m in diameter. Bark thin, smooth, bright reddish brown when fresh, becoming scaly and weathering gray but remaining colorfully mottled through flaking. Crown conical at first, becoming irregular with age, with horizontal branches in whorls of three to five bearing numerous, closely spaced branchlets densely clothed with foliage. Leaves standing out from the twigs at the slightly forward angle, nearly overlapping or more widely spaced, lasting 4–6 years, dark green above, waxy bluish green beneath, (0.5–)1–2.5 cm long, 2–3(–3.5) mm wide. Blades nearly parallel-sided in the middle third or more, tapering abruptly to the roundly triangular tip with a blunt point and to the roundly wedge-shaped base on a slender petiole less than 1 mm long. Pollen cones (3–)4–6 mm long and 2.5–3 mm wide, on a stalk 0.5–2 mm long. Seed cones on a stalk about 1 cm long, broadly egg-shaped to nearly spherical, light bluish green with a thin waxy film at first, maturing yellowish brown, usually irregular, with 15–20 protruding, sharply triangular seed scales, (7–)10–15(–20) mm in diameter, usually with just one to four seeds maturing and being shed through the somewhat gaping scales. Seeds spherical to egg-shaped, slightly flattened, smooth, shiny brownish yellow to orange, 3–4 mm long. Occupying a limited range in the southern Andes and adjacent coastal range in Chile from just south of Concepción (36°S) to Chiloé Island (42°30′S), with an outlier in Aisén province (45°45′S) and also in a few locations in the adjacent lake district of Argentina in Chubut, Neuquén, and Río Negro provinces. Scattered singly or clumped in the canopy of humid mixed evergreen forests and swamp forests; (45–)250–1,000(–1,400) m. Zone 8.

Multitrunked mature Prince Albert yew (*Saxegothaea conspicua*) in nature.

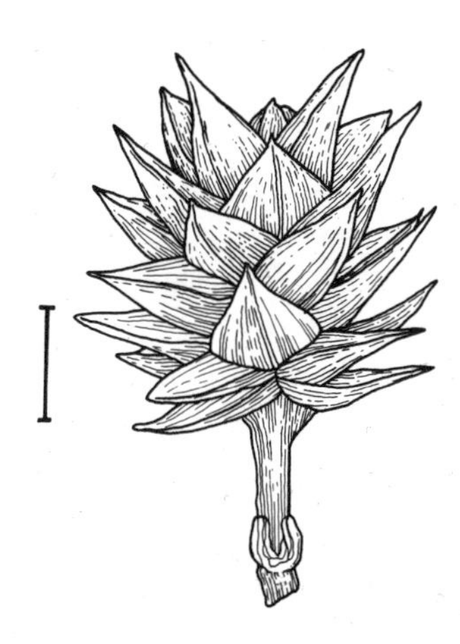

Closed mature seed cone of Prince Albert yew (*Saxegothaea conspicua*) before seed release; scale, 2 cm.

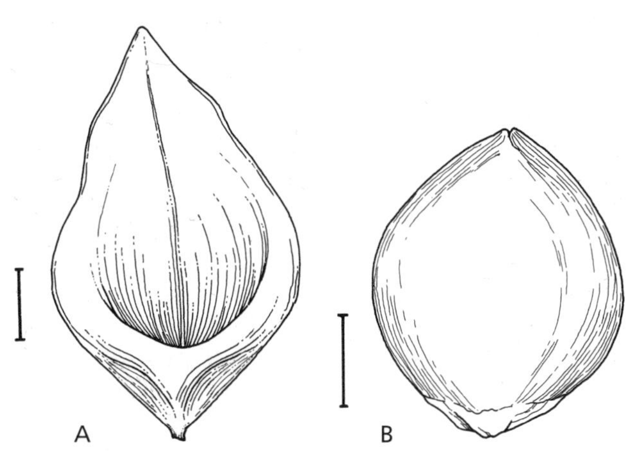

Upper side of seed cone scale (A) and removed seed (B) of Prince Albert yew (*Saxegothaea conspicua*); scale, 1 mm.

Prince Albert yew yields an excellent wood much appreciated locally in fine carpentry and for making barrels and veneers, but it is too rare in most regions to be an important commercial timber tree. When cut, the trees can regenerate from stump sprouts, so modest exploitation creates no special conservation concern for the species. It grows generally at a moderate rate of 2–3 mm in diameter per year and reaches a maximum known age of about 400 years. Its unusual seed cones and pollen grains compared to other podocarps are accompanied by an equally unusual type of pollination. In most podocarps, a very large pollination droplet extending from the ovule guides winged pollen grains into the micropyle by flotation. After landing on the seed scales, in contrast, the bladderless grains of *Saxegothaea* send out pollen tubes that cross the scale and enter the micropyle before penetrating the nucellus and egg cell (as in the other podocarps). Subsequent events of embryo development are rather similar to those of other podocarps, so it is largely the course of pollination that is unusual. The whole complex of unusual features in Prince Albert yew, however, which John Lindley thought linked different conifer groups together, led him to his scientific name for the species, Latin for "singular or remarkable."

Sciadopitys P. Siebold & Zuccarini

UMBRELLA PINE

Sciadopityaceae

Evergreen tree with fibrous, smoothish bark, peeling in narrow strips on the single trunk. Branches short, slender, densely arranged, sweeping gently upward, often initially in whorls of three to five above a whorl of needles. Branchlets with a swollen portion at the tip of each growth increment showing little internodal elongation and a variable-length, thinner portion beneath with extensive internodal elongation, each twig thus resembling both a short shoot and a long shoot in different portions, grooved between the attached leaf bases, hairless, green at first, becoming brown within the first year. Leaves spirally arranged, of two types: brown scale leaves scattered along the twigs and in two or three close, dense, pseudowhorls at the end of each growth increment; and radiating, linear, photosynthetic, double needles in the axils of the whorled scale leaves (hence the scientific name, Greek for "umbrella pine"). The needle leaves are conventionally interpreted as a pinelike dwarf shoot in which two needles have become fused side to side, the compound origin reflected in two widely separated, complete vascular bundles and a median groove both above and below.

Plants monoecious. Pollen cones in a dense, spherical cluster of 20–30 in the axils of scale leaves at the tip of a branchlet just above a whorl of double needles. Individual cones with numerous spirally arranged pollen scales, each scale with two pollen sacs. Pollen grains small to medium (30–40 µm in diameter), spherical, completely covered with rounded bumps, except at the thin, round, germination pore, this sometimes within a furrow. Seed cones woody, single at the tips of twigs, replacing an annual vegetative increment, maturing and opening in their second season. Individual cones with 50–85 overlapping, spirally arranged seed scales (of which the middle 30–60 are fertile). Each cone scale diamond-shaped externally, with the bract partially overlapping with and fused to the larger fertile portion for about half its length, the fertile portion bearing five to nine seeds in a single, angled row. Seeds elliptical, flattened, with lateral wings derived from the seed coat slightly narrower than the seed body but exceeding it in length. Cotyledons two, with one vein. Chromosome base number $x = 10$.

Wood sharply fragrant, light and soft, moderately decay resistant, the whitish sapwood only weakly distinguished from the light brown heartwood. Grain very even and fairly fine, with obvious growth rings set off by an abrupt transition to a variably narrow band of much smaller, thicker-walled latewood tracheids. Resin canals and individual resin parenchyma cells both absent.

Lower groove unique among living conifers in having a single stomatal band. Individual stomates only a little lowered beneath the level of the (four to) six to eight surrounding subsidiary cells, which bear prominent, rodlike papillae that may partially over-arch the opening. Midveins tilted away from each other, surrounded by transfusion tissue, but mostly extending in wedges out to the sides. Each needle with a pair of resin canals at the lower, outer corners and with none to eight additional canals spaced around the rest of the periphery. Photosynthetic

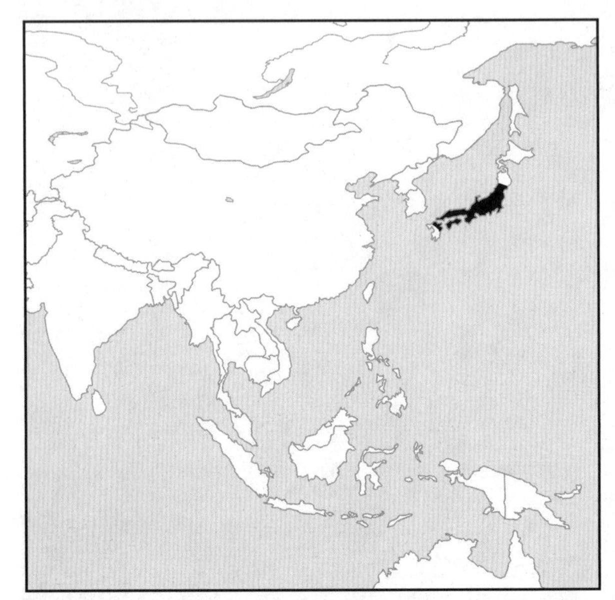

Distribution of *Sciadopitys verticillata* in central and southern Japan, scale 1 : 120 million

tissue with a single palisade layer all around the needle on both faces (but much thinner and interrupted beneath by the stomatal band) inside the epidermis and single extra layer of hypodermis.

One species in Japan. References: Farjon 2005b, Ohsawa 1994, Yamazaki 1995.

In modern times, *Sciadopitys* was typically aligned with the redwoods, which it broadly resembles in seed cone structure. However, even when included in the family Cupressaceae (or the former Taxodiaceae), because of the very distinctive leaves and growth pattern, it was often given a separate subfamily, equivalent to all other members of the family. Now, the weight of evidence from cone development, chromosomes, paleobotany, and DNA strongly supports assigning it to its own family. This evidence shows that it belongs to the group of families including Cupressaceae and Taxaceae but that it is sister to both and is no more closely related to one than it is to the other.

The nature of the double needles has also been controversial. Besides the conventional interpretation as a pair of needles fused side to side, they were considered to consist predominantly of highly modified stem tissue, which really does not explain their separate vascular bundles. Under either interpretation, they are technically referred to as cladodes, shoots that function like leaves. They have to be shoots because, like the seed scales of conifer cones, they are in the axils of leaves (scale leaves and bracts, respectively, for these two structures). Thus they are also technically short shoots, homologous to those of *Pinus,* but for ordinary communication it seems less confusing to refer to them as double needles, or even just needles, which they very closely resemble in structure, rather than as short shoots. Developmentally, they arise as a pair of primordia (embryonic leaves) on the bud apex in the axil of a scale leaf, followed by intercalary growth beneath those primordia, rather than by continued growth of the primordia themselves. Thus, even if they did evolve from a pair of leaves, they do not now grow by simple side-to-side fusion but by expansion of a region that is neither quite leaf nor stem in origin. This is a rather common theme in examining plant structures that are considered to represent evolutionary fusions of originally separate structures, like the bract and seed scale that make up many conifer cone scales, or the tubular corolla (petals) of a petunia flower.

Fossil leaves, cones, and pollen of *Sciadopitys* are found from the late Cretaceous through the Tertiary in the northern hemisphere. Similar leaves, with a single median band of stomata, are common in mid-Jurassic to early Cretaceous sediments of the arctic and were used to support an earlier occurrence of the family, because such leaves are not found in any other living conifer. Careful studies of their structure show that they are simple leaves, unlike the double needles of *Sciadopitys,* and they are now assigned to an extinct family, Miroviaceae, of uncertain relationship to other conifers. *Sciadopitys verticillata* has been grown in gardens for centuries in Japan, but few cultivars have been selected. Still, those few include dwarf globes, weepers, and narrow, upright cones and cylinders as well as yellow needles and ones that remain green in winter, when they typically bronze.

Sciadopitys verticillata (Thunberg) P. Siebold & Zuccarini

JAPANESE UMBRELLA PINE, KOYAMAKI (JAPANESE)

Tree to 30–45 m tall, with trunk to 2 m thick, sometimes forking near the base. Bark narrowly ridged and furrowed, reddish brown at first, soon weathering brownish gray to light gray. Crown narrowly conical, with numerous near whorls of slender branches turned up at the tips. Branchlets tan to yellowish brown, deeply grooved between the attached bases of the scale leaves. Scale leaves with the free tip about 1 mm long, the base running down the twig for 5 mm or more. Double needles persisting 2–3 years, in pseudowhorls of 15–35, (4–)8–14 cm long, 2–4(–7) mm wide, dark green above and below except for the 1 mm wide whitish stomatal groove running down the midline beneath, shallowly notched to rounded at the tip, without a distinctly narrowed petiole. Pollen cones spherical to oblong, 5–8 mm long. Seed cones oblong, (5–)8–12 cm long, 4–6 cm thick, the fan-shaped scales

Natural stand of young mature Japanese umbrella pine (*Sciadopitys verticillata*).

about 2.5 cm long and wide. Seeds 8–12 mm long and a little narrower including the wings. Japan, from central Honshū (where it is most abundant) to eastern Kyūshū. Forming pure stands or mixed with other conifers on rocky but moist mountain slopes at 600–1,200 m. Zone 6.

The English common name and the genus and species names all refer to the distinctive arrangement of the double needles, like the spokes on an umbrella, called verticillate in Latin-derived descriptive botanical terminology. *Sciadopitys* means "umbrella pine" in Greek. Although a well-respected tree in the wild in Japan

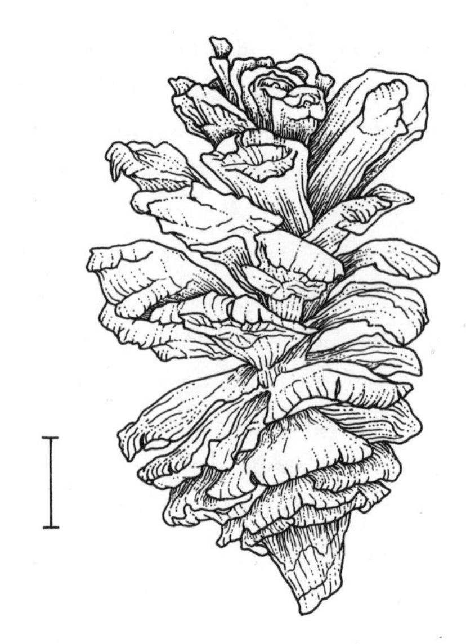

Open seed cone of Japanese umbrella pine (*Sciadopitys verticillata*) after seed release; scale, 1 cm.

Twig of Japanese umbrella pine (*Sciadopitys verticillata*) with 3 years of growth, ×0.5.

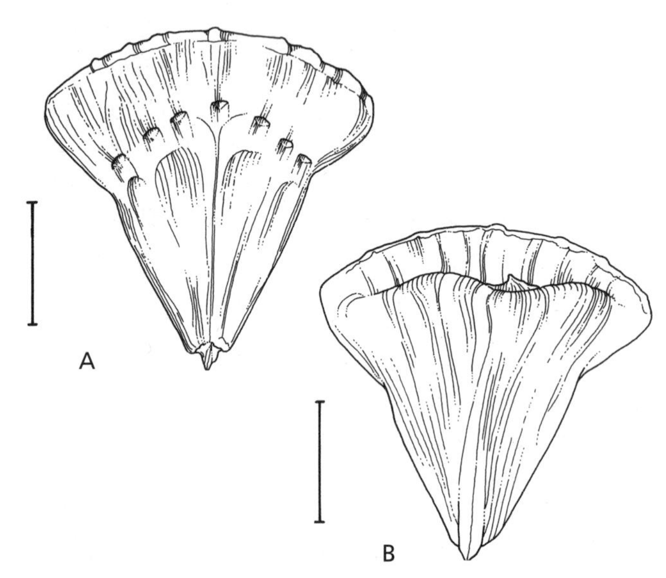

Upper (A) and undersides (B) of detached seed cone scale of Japanese umbrella pine (*Sciadopitys verticillata*); scale, 1 cm.

Tip of twig of Japanese umbrella pine (*Sciadopitys verticillata*) with cluster of mature pollen cones just before pollen release; scale, 3 cm.

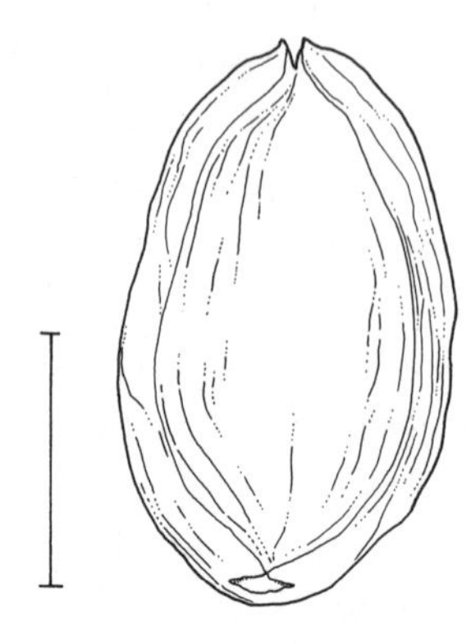

Dispersed seed of Japanese umbrella pine (*Sciadopitys verticillata*); scale, 5 mm.

(the Japanese common name refers to the richly templed Mount Koya, one of its sites of greatest natural abundance), it is not a particularly popular garden subject in its native land. In cultivation it is best grown there and elsewhere on sandy, fertile, acid soil in semishade. It also makes a striking subject for bonsai. As with *Cryptomeria, Thujopsis,* and many other Japanese conifers, there have been studies of geographic variation in Japanese umbrella pine. These reveal complex patterns of prehistoric migration in the species. The fragrant white wood resists water and so can be made into vessels.

Sequoia Endlicher

REDWOOD

Cupressaceae

Giant evergreen trees with a single, straight trunk bearing thick, deeply furrowed, fibrous bark. Branchlets weakly differentiated into long and short shoots, both elongate and clothed with dense or well-separated leaves, but short shoots having just a single year's growth without further extension in later years and in branched clusters falling intact after several years. Annual growth terminating in loose, scaly, winter buds. Leaves spirally arranged, the attached bases clothing the twigs, of two intergrading types in adult plants: flattened needlelike leaves and scale leaves. Scale leaves borne at both ends of each growth increment, on upright twigs, and on twigs ending in seed cones. Needle leaves longest in the middle portion of each twig, narrowing abruptly to a point at the tip, and having the petiole twisted to bring the needles into a flattened row on either side of the twig.

Plants monoecious. Pollen cones single at the tips of ordinary twigs and of very short reproductive shoots in leaf axils near the tips of the twigs. Individual cones with 6–15 spirally arranged pollen scales, each scale with two to six pollen sacs. Pollen grains small (30–40 μm in diameter), spherical, with a small germination papilla but otherwise almost featureless. Seed cones single at the tips of very short reproductive shoots at or near the ends of foliage twigs, maturing and opening in the first season, oblong or nearly spherical, with 15–30 spirally arranged, shieldlike seed scales. Each scale with (two to) five to seven seeds in one row (or with two additional seeds in a partial second row). Seeds flattened, the body oval and lens-shaped with two wings derived from the seed coat running along both sides of and narrower than the seed body. Cotyledons usually two, each with one vein. Chromosome base number $x = 33$ (in three sets of 11).

Wood essentially odorless, light, soft, and brittle, but extremely decay resistant, with a narrow band of white to light brown sapwood sharply contrasting with the red to purplish brown or dark brown heartwood. Grain very even to somewhat uneven and fine to moderately coarse, often attractively figured, with obvious growth rings of quite varied widths (including irregular and false rings), marked by a somewhat gradual to abrupt transition to a narrow band of much smaller, thicker-walled latewood tracheids. Resin canals absent but with numerous individual resin parenchyma cells scattered primarily through the outer half of the growth increment, including the latewood.

Stomates of needle leaves in a few irregular lines on both sides of the upper surface and in more lines collected in two bands beneath. Each stomate sunken beneath and partially hidden by the two (or three) circles of four to six surrounding subsidiary cells, which may rise in a low Florin ring around the opening. Leaf cross section with a single-stranded midvein flanked by small wedges of transfusion tissue above a large resin canal and with two additional resin canals above the lower epidermis out near the outer corners of the blade. Photosynthetic tissue forming a somewhat irregular palisade layer beneath the upper epidermis and accompanying incomplete layer of hypodermis and either grading into or giving way abruptly to the spongy mesophyll that extends down to the stomatal bands and lower epidermis, which lacks an adjacent hypodermal layer.

One species along the Pacific Coast of the southwestern United States. References: Chaney 1951, Eckenwalder 1976, Farjon 2005b, Watson 1993.

The name of the genus is generally thought to honor Sequoyah (c. 1765–1843), inventor of the Cherokee syllabary, who died just 4 years before its description, but Endlicher, who described the genus, did not explain his choice of name. *Sequoia*

Distribution of *Sequoia sempervirens* in coastal California and barely in adjacent Oregon, scale 1 : 60 million

sempervirens is fairly tender but is widely cultivated in mild, moist temperate regions. Few cultivars have been selected, but those include variations in habit (such as dwarf, compact, and weeping) and foliage form (persistently juvenile) and color (including blue and variegated).

Traditionally included in the family Taxodiaceae since 1926 and lending its common name to that family as "the redwood family," my suggestion in 1976 that this family be merged into an enlarged Cupressaceae has been supported by several subsequent studies, including comparison of DNA structures. *Sequoia, Sequoiadendron,* and *Metasequoia* are closely related, and a few authors treat all three living redwoods as species of *Sequoia.* The suggestion, made soon after the discovery of *Metasequoia,* that the dawn redwood is one of the ancestors of coast redwood, is not borne out by detailed analysis of the chromosomes nor by DNA studies.

Sequoia has rather generalized leaves and shoots that also resemble those of various Podocarpaceae and Taxaceae. As a result, many similar fossil shoots from the Triassic and Jurassic once assigned to *Sequoia* are now considered not to be referable to the redwood. The oldest fossils reasonably securely recognizable as redwoods date from the Cretaceous, like those of most related genera. The discovery of *Metasequoia* made paleobotanists realize that many of the Cretaceous and Tertiary fossils, especially those of northern lands, described as species of *Sequoia* really belong to *Metasequoia* or *Taxodium.* Nonetheless, the genus was widespread at midlatitudes in Eurasia and North America during the Tertiary and did not finally disappear from Europe and Japan until after the Pliocene. Specimens from the Pliocene of Japan were already hexaploids (with six sets of chromosomes), so we do not know when the increase in chromosome number from the typical diploid condition of other Cupressaceae occurred.

Sequoia sempervirens (D. Don) Endlicher

REDWOOD, COAST REDWOOD

Plate 61

Single-trunked or clustered tree to 112 m tall, with trunk to 9 m in diameter. Bark reddish brown at first, weathering grayish, to 35 cm thick, the ridges and furrows sometimes spiraling gently around the trunk. Crown more or less conical to cylindrical throughout the life span, composed of numerous horizontal to gently drooping branches. Needles straight or slightly curved outward, fairly parallel-sided for most of their length to recognizably widest near the middle, abruptly narrowed to the triangular tip topped by a weak prickle and to the rounded base, dark green (or sometimes with a waxy bloom) and with a few lines of stomates above, the stomatal bands beneath whitened, up to 30 mm long, lasting 3–5 years or more. Pollen cones 2–5 mm long. Seed cones 1.3–3.5 cm long, the external face of each cone scale (5–)8–10 mm wide and (3–)5–6 mm high. Seeds 3–6 mm long, about two-thirds as wide, including the wings. Seaward slopes of coastal mountains of southernmost Oregon and California south through Monterey County. Clothing slopes and valleys of the fog belt with solid stands or mixed with *Pseudotsuga menziesii* at sea level to 300(–1,000) m, reaching greatest size in deep alluvial soils of sheltered valleys. Zone 8.

Redwoods are the tallest living trees (with unconfirmed 19th century claims of taller eucalypts and Douglas firs) and can live for upward of 2,000 years. The bulkiest trees are not the tallest ones, and the largest known individual in combined dimensions when measured in 1997 was 97.9 m tall, 7.7 m in diameter, with a spread of 22.9 m. They are fire resistant and unusual among conifers in their ability to sprout from fire-killed or logged stumps, which may result in forming widening rings of genetically identical trees in successive generations of sprouting. The dormant buds that lead to this resprouting are also part of the massive burls found on the trunks of many redwood trees. Pieces of burls placed in water will sprout twigs, and such pieces are sold as houseplants even where the trees would not be hardy outdoors. Redwoods are also unique among conifers in their hexaploid (six-sets) chromosome constitution, which implies derivation by hybridization of at least two distinct ancestral species. They have a high moisture requirement and thrive today in summer-dry California only where coastal fogs maintain humidity and reduce summer temperatures. Studies in the redwood canopy by tree-climbing biologists disclosed a rich aerial ecosystem whose remnants rarely reach the forest floor in recognizable form and only began to reveal their complexities after the close of the 20th century. There is a nice popular account of their occurrence and history by J. J. Hewes (1981).

Preservation of old-growth redwood stands was long delayed because they are ecologically dominant and hence far from rare within their limited range and because of the economic value of the huge quantities of beautiful, decay-resistant wood produced by mature trees. Dedicated efforts by the Save the Redwoods League and other conservation organizations and individuals eventually turned public opinion overwhelmingly in favor of preservation, and many old-growth stands were incorporated into state and local parks, and eventually Redwood National Park. Even the second- (or third-) growth redwoods that constitute the vast majority of the Redwood Empire, however, are magnificent trees. Considering the absolute dependence of natural stands of redwood on the fog belt, what will happen to them if global warming drastically changes the fog regime?

Underside of twig of redwood (*Sequoia sempervirens*), ×0.5.

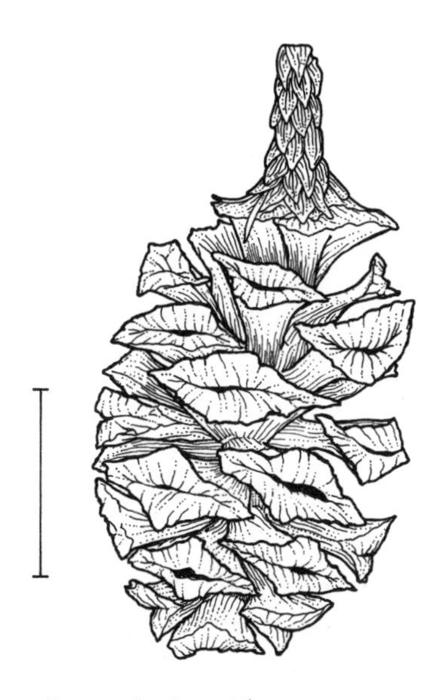

Open mature seed cone of redwood (*Sequoia sempervirens*) after seed dispersal; scale, 1 cm.

Natural stand of mature redwood (*Sequoia sempervirens*).

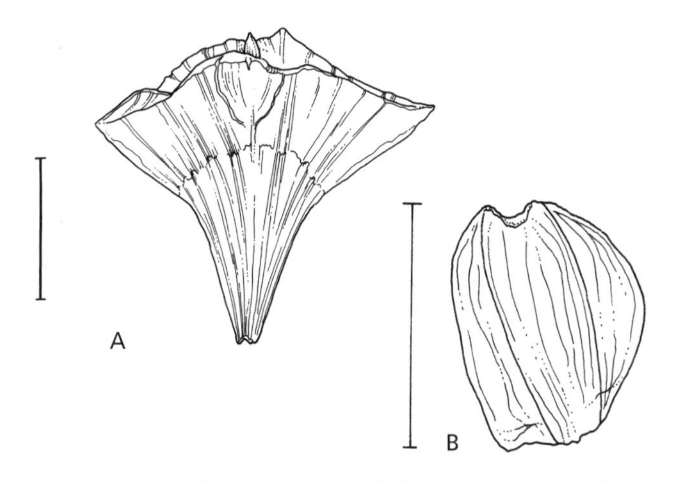

Upper side of detached seed cone scale (A) and dispersed seed (B) of redwood (*Sequoia sempervirens*); scale, 5 mm.

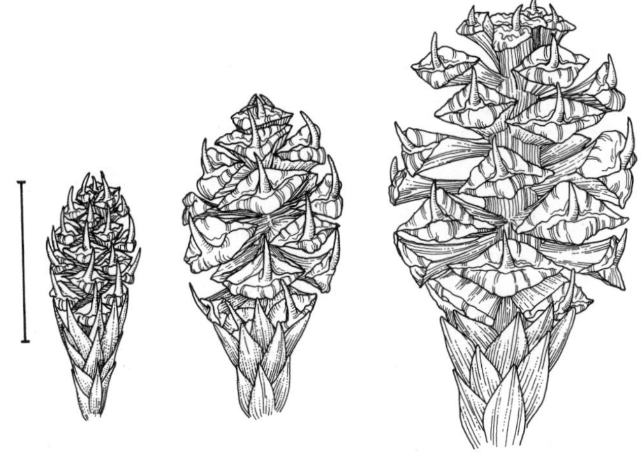

Developmental sequence of seed cones of redwood (*Sequoia sempervirens*) stopped at different times after pollination; scale, 1 cm.

Sequoiadendron J. Buchholz

GIANT SEQUOIA

Cupressaceae

Giant evergreen trees with thick, deeply furrowed, fibrous bark clothing the massive trunk. Crown cylindrical to narrowly dome-shaped, made up of relatively few massive limbs, typically turning up with elbowlike bends, the individual limbs as large as mature whole trees of many other conifers. Branchlets poorly differentiated into long and short shoots, densely clothed with leaves. Branched clusters of short shoots borne on a framework of permanent long shoots with slightly greater internodal elongation, falling intact after several years. Without definite winter buds. Leaves densely spirally arranged, awl-shaped to scalelike, the free tip spreading and longer than the attached base in young trees but gradually reduced and pressed forward with maturity.

Plants monoecious. Pollen cones unstalked, single at the ends of short shoots, spherical to oblong, with 12–20 spirally arranged pollen scales, each with two to five pollen sacs. Pollen grains small to medium (25–45 µm in diameter), flattened spherical with a pronounced germination papilla but otherwise almost featureless. Seed cones single at the ends of short shoots, maturing in their second year, remaining intact at maturity and persisting several years. Each cone oblong, woody, with 25–44 shieldlike seed scales. Each scale with the fertile portion and bract about equal and intimately fused to form a horizontally diamond-shaped external face bearing a triangular central protrusion from a horizontal groove. Seeds (3–)9–13 per seed scale in two rows, lens-shaped, with two equal wings derived from the seed coat along the whole length of and slightly wider than the body. Cotyledons usually three or four, each with one vein. Chromosome base number $x = 11$.

Wood odorless, light, weak, and brittle but very decay resistant, with a proportionately narrow band (of up to some 200 growth rings!) of almost white to very pale yellowish brown sapwood sharply contrasting with the red to purplish brown or dark brown, sometimes somewhat streaky heartwood. Grain very fine and even, with well-defined but somewhat wavy growth rings marked by an abrupt transition to a fairly narrow band of darker, smaller, thicker-walled latewood tracheids. Resin canals absent but with numerous individual resin parenchyma cells scattered primarily through the earlywood.

Stomates in irregular lines aggregated into two narrow bands on the inner face and bands or patches on the outer face, especially near the base. Each stomate sunken beneath and often almost hidden by the one to three concentric full or partial rings of four to six surrounding subsidiary cells that lack a Florin ring. Leaf cross section with a central, single-stranded midvein above a single small resin canal and flanked by wedges of transfusion tissue. Photosynthetic tissue with a poorly organized palisade layer (and scarcely identifiable as such) all the way around the leaf inside the epidermis and adjacent, very discontinuous, thin hypodermal layer reinforcing the corners.

One species in California. References: Buchholz 1939, Farjon 2005b, Watson 1993.

Buchholz used the name *Sequoiadendron* (hybrid Greek for "*Sequoia* tree") to preserve the traditional association of the giant sequoia with the redwood genus from which he had segregated it. The separation of *Sequoiadendron* from *Sequoia,* while almost universally accepted today, was quite controversial when first proposed in 1939 and was greeted with intense resistance. *Sequoiadendron* is more primitive than *Sequoia* in most respects and has additional similarities to many other genera of Cupressaceae. Nonetheless, DNA similarities and chromosome structures confirm the close relationship of *Sequoiadendron* to *Sequoia* and *Metasequoia,* and their morphology overlaps in a variety of details, though not enough to justify merging them all under *Sequoia,* as has been suggested.

Because of its rather generalized morphology, the fossil record of *Sequoiadendron* is not entirely clear but appears to extend back to the Cretaceous in Europe and North America, without representation in Asia. Despite the apparent absence from Asia, however, additional specimens were reported from late Cretaceous sediments in New Zealand, although there are no other southern hemisphere records of the genus from that time to the present. There are several other described genera of sequoioid cones from the Cretaceous of the northern hemisphere, and their relation-

Distribution of *Sequoiadendron giganteum* in the Sierra Nevada of California, scale 1 : 60 million

ships to each other and to the three extant genera of the group have yet to be worked out. Giant sequoia (*S. giganteum*), the sole living species, has been in cultivation since the late 19th century, but it has not been exceptionally variable in cultivation and few cultivars have been named. These include dwarf and weeping forms as well as those with variegated or unusually bluish foliage.

Sequoiadendron giganteum (Lindley) J. Buchholz

GIANT SEQUOIA, BIG TREE, WELLINGTONIA, SIERRAN REDWOOD

Plate 62

Tree to 90 m tall, with trunk to 11 m in diameter, flaring broadly and buttressed at the base at maturity, often fluted to the base of the crown. Bark bright to dark reddish brown, even in extreme old age, to 60 cm thick. Crown composed of a few major trunklike limbs and numerous horizontal branches, densely conical when young (for 100 years or more!), becoming oblong with age as the lower branches die. Twigs completely clothed by the attached leaf bases and partly hidden by the free tips. Leaves bluish green, with a stomatal band on each outer face and scattered stomates on the inner face, living 5–7 years or more, the free tip 3–8(–15) mm long. Pollen cones 4–8 mm long. Seed cones 4–9 cm long, remaining green for several years after ripening until the seeds are released. External face of each cone scale 15–30 mm wide and 5–10(–15) mm high. Seeds 3–6 mm long, about half as wide, including the wings. Restricted to California in the Sierra Nevada from Placer County to Tulare County (with the vast majority of the trees in Tulare County and adjacent Fresno County south

Middle-aged mature giant sequoia (*Sequoiadendron giganteum*) in nature.

Twig of giant sequoia (*Sequoiadendron giganteum*) with 2 years of growth, ×0.5.

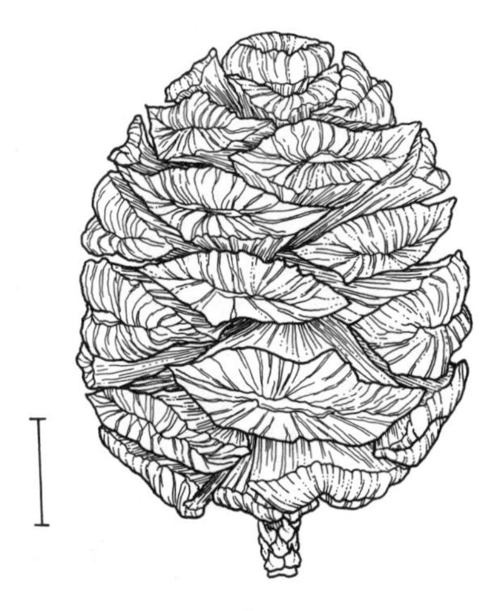

Open seed cone of giant sequoia (*Sequoiadendron giganteum*) after seed dispersal; scale, 1 cm.

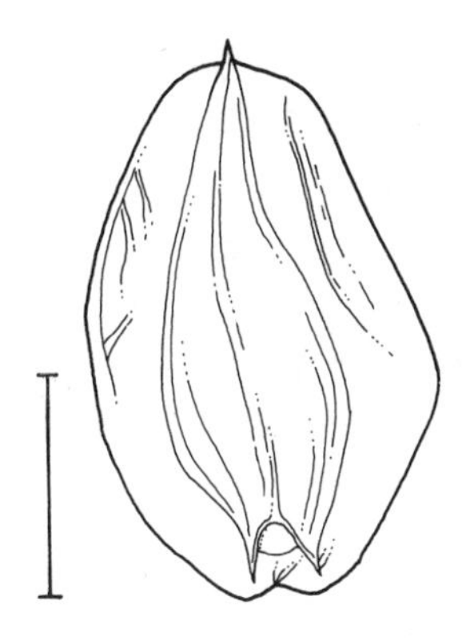

Underside of dispersed seed of giant sequoia (*Sequoiadendron giganteum*); scale, 2 mm.

of the Kings River). Forming groves on valley flats in montane mixed coniferous forest at (900–)1,800–2,300(–2,700) m. Zone 7. Synonym: *Sequoia gigantea* (Lindley) J. Decaisne.

Giant sequoias are the most voluminous living trees, combining great height (third or fourth tallest in the world) with great girth (third or fourth thickest). In the rating system of the American Forestry Association, combining girth, height, and crown spread, the US National Champion giant sequoia scores 1,300 points while no other trees but the redwoods reach even 1,000 points. The famous General Grant tree in King's Canyon National Park in California, the largest known individual in combined dimensions, measured 83.6 m tall, 8.3 m in diameter, with a spread of 32.6 m in 1999. Giant sequoias grow rapidly throughout their lives, and the largest trees may still add new wood annually at a rate equivalent to a telephone pole 45 cm in diameter and 8 m tall. Even in cultivation they display steady, rapid growth and may reach 25 m in 50 years and 50 m in 100 years, with a strongly tapered trunk. In nature, there are just 75 discrete groves. Their discovery, history, and beauty are well portrayed in a popular book by J. J. Hewes (1981).

Ironically, like many other large, long-lived, emergent conifers, the giant sequoia is a shade-intolerant successional species that depends on major disturbances (fire in this case) to eliminate competitors and allow successful reproduction. It can then live to be 3,500 years old, long outlasting any of the other conifers that become established in its shade. Because scars on the trunk caused by fires can be dated precisely, it has been possible to show that climatic changes over the last 1,000 years strongly influenced fire frequencies and intensities in the groves and thus the pattern of natural regeneration during this time.

Taiwania Hayata

TAIWANIA, TAIWAN REDWOOD

Cupressaceae

Evergreen trees with single, straight, fluted trunk clothed with thick, fibrous, red-brown bark peeling in strips. Crown shallow and open, composed of slender to moderately thick, gently upwardly angled branches. Branchlets very weakly differentiated into short-lived short shoots and long shoots that persist as the main branching framework. Short shoots shed intact, completely clothed by the persistent leaf bases, drooping gracefully from the horizontal branches, especially in young trees. Resting buds without specialized bud scales, formed solely by the normal spiral of variously immature ordinary foliage leaves surrounding the shoot apex. Leaves densely spirally arranged, of three types. Seedling leaves needlelike, flattened top to bottom with a prominent midrib beneath. Transitional juvenile leaves clawlike, flattened side to side and straight or slightly curved forward, progressively shorter with maturity of the tree. Adult leaves scalelike, diamond-shaped in cross section, curved forward and overlapping along the branch.

Plants monoecious. Pollen cones in radiating clusters of (one to) three to seven (to nine) from a common winter bud at the tips of branchlets. Pollen scales spirally arranged, 15–20, each with two or three pollen sacs. Pollen grains small (20–35 μm in diameter), spherical, with a small germination papilla but otherwise almost featureless. Seed cones single from winter buds at the tips of branchlets, more or less egg-shaped. Cone scales spirally arranged, 18–36, of which the middle 10–20 are fertile. Each scale thin and rounded, with an abruptly pointed tip, consisting mostly of the bract portion, with the fully fused seed-bearing portion reduced to a pad beneath the two seeds. Seeds lens-shaped, flattened, with an equal pair of wings derived from the seed coat, each wing longer than and about as wide as to a little wider than the body. Cotyledons two, each with one vein. Chromosome base number $x = 11$.

Wood odorless, light and soft but decay resistant, with nearly white sapwood sharply contrasting with the pale brown heartwood with purplish brown streaks. Grain very fine and even, with well-defined growth rings marked by a somewhat gradual and then abrupt transition to much smaller and noticeably thicker walled latewood tracheids. Resin canals absent but with sparse individual resin parenchyma cells scattered throughout the growth increment.

Stomates in waxy white bands on all four faces of both adult and juvenile leaves but not arranged in defined lines. Individual stomates sunken beneath and partially hidden by the innermost of one to three rings of four to six surrounding subsidiary cells. Leaf cross section with a nearly central single midvein right

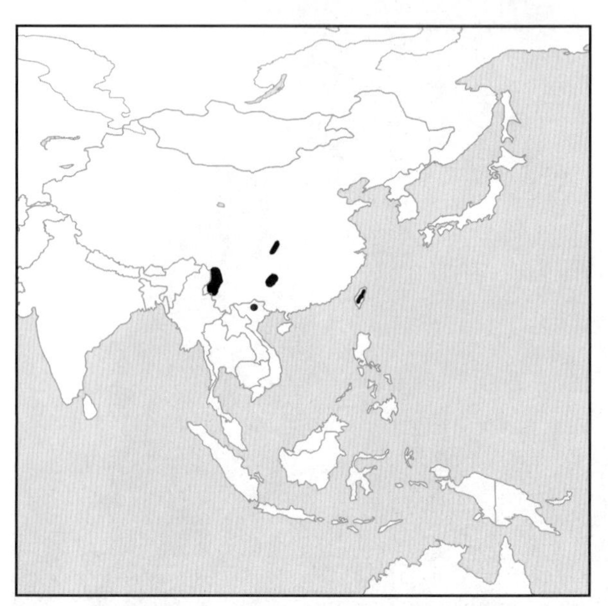

Distribution of *Taiwania cryptomerioides* in southeastern Asia, scale 1 : 120 million

above a single more or less equal-sized resin canal and flanked by prominent wings of transfusion tissue. Photosynthetic tissue very weakly organized into a thin, incomplete palisade layer all around the leaf periphery inside the epidermis and incomplete single hypodermal layer, with the rest consisting of spongy mesophyll surrounding the midrib and associated tissues.

One species in western China and Taiwan. References: Farjon 2005b, Fu et al. 1999h, Liu and Su 1983.

Taiwania is unusual among genera of the Cupressaceae in its superficially sprucelike cones, which differ, however, in having a very reduced seed-bearing portion. In this, it somewhat resembles *Cunninghamia* and, to a lesser extent, *Athrotaxis.* The foliage has some resemblance to that of *Cryptomeria* and somewhat less to that of *Sequoiadendron* but differs considerably in detail. While thus sharing features with a variety of genera of Cupressaceae, it does not seem to be particularly closely related to any one of them, and this lack of close relationships is also reflected in DNA studies (Brunsfeld et al. 1994). Although a single species is accepted here and in the *Flora of China,* many Chinese botanists, and a few elsewhere, have separated the mainland populations from those of Taiwan as distinct species. Although some material from continental Asia has larger cones and smaller leaves than the Taiwanese specimens, the difference appear minor, and there is too little material of the mainland trees to convincingly establish their full range of variation, so there appears to be little justification for distinguishing them. *Taiwania cryptomerioides* generally grows poorly in cultivation outside of its native range, and no cultivar selection has taken place.

Taiwania has a very sketchy fossil record with the oldest known specimens recorded in sediments of uncertain age (either late Paleocene or early Eocene), roughly 65 million years old, from Spitsbergen in the high arctic. From the Eocene, some 50 million years ago, into the Pliocene, some 4–5 million years ago, the genus was well established in Japan, but there is no fossil record from China.

Taiwania cryptomerioides Hayata

TAIWAN REDWOOD, TAIWANIA, TAIWAN SHAN, TU SHAN (CHINESE)

Tree to 60(–75) m tall, with a single, straight trunk to 3(–4) m in diameter. Bark reddish brown and smooth at first and flaking in small scales, remaining thin by peeling in narrow vertical strips, darkening with age and weathering gray in sheltered places. Crown narrowly conical to cylindrical at first, with numerous short horizontal branches, broadening, becoming shallower and dome-shaped with age. Twigs completely covered by and deeply grooved between the attached leaf bases and further partly to largely hidden by the free tips on adult shoots. Juvenile clawlike leaves stiff, straight or curved forward, tapering consistently from the base to the sharp, pointed tip, up to 2(–2.5) cm long, with a distinctly blue-green cast from wax over the stomatal bands, lasting many years (25–30 or more). Adult scale leaves 3–6 mm long, the free tips loose or tightly pressed against the twig. Pollen cones 4–7 mm long. Seed cones oblong, (0.8–)1.5–2.5 cm long, 5–10(–12) mm across. Seeds (3–)4–5(–7) mm long, a little more than half as wide as long, including the wing, which is continuous all around the body, except for the narrow notch at the micropyle. Taiwan, northern Vietnam, southwestern China, and adjacent Myanmar. Scattered in forests of other conifers or broad-leaved evergreens at (1,750–)1,800–2,600(–2,900) m. Zone 8. Synonym: *Taiwania flousiana* H. Gaussen.

Trees of Taiwan redwood from the Chinese mainland were distinguished as a separate species (*Taiwania flousiana*), but the differences between the various populations are minor. Although the wood is of high quality, the tree is nowhere so abundant as to make it a major timber species. Its primary historical use was as one of the premier coffin woods in a society that highly valued its pleasant fragrance and incorruptibility in soil as one honor accorded revered deceased ancestors. In part because of this, it was planted

Vigorous mature Taiwan redwood (*Taiwania cryptomerioides*) in nature.

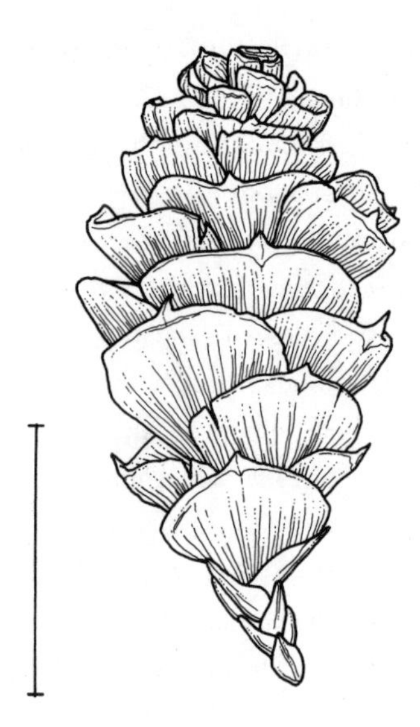

Open seed cone of Taiwan redwood (*Taiwania cryptomerioides*) after seed dispersal; scale, 1 cm.

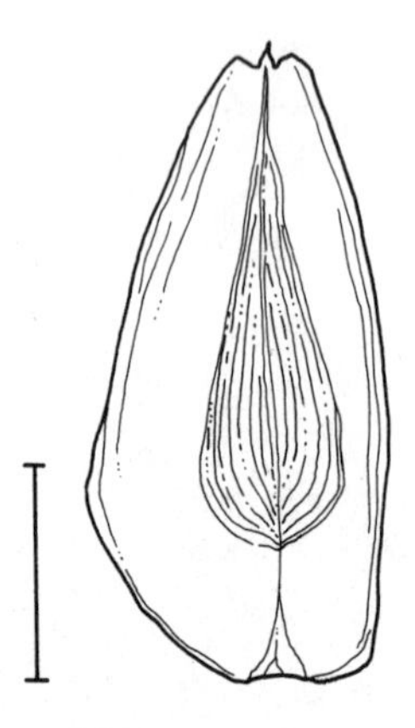

Underside of dispersed seed of Taiwan redwood (*Taiwania cryptomerioides*); scale, 2 mm.

in modest numbers far beyond its natural range, and preserved botanical specimens taken from such trees obscure the natural range of the tree, as they do with such other species as *Ginkgo biloba* and China fir (*Cunninghamia lanceolata*). It is only successfully cultivated in mild temperate climates with abundant moisture.

Taxodium L. Richard

BALD CYPRESS

Cupressaceae

Deciduous or subevergreen trees with a single, straight, buttressed trunk covered by fibrous, furrowed bark. Base of tree often surrounded by varying densities of cypress knees, which are more or less cone-shaped woody outgrowths from the roots, especially under swampy conditions. Densely branched from the base at first with upwardly arcing branches, the lower branches falling to leave a clear, straight bole. Shoot system moderately differentiated, with once-branched or unbranched deciduous (cladoptosic) determinate short shoots borne on permanent long shoots. Shoots emerging from definite winter buds. Leaves arranged spirally but often distichously spread out in one plane on either side of the twig by twisting of the petiole, standing out from the branchlets or tightly pressed forward along them, the attached leaf bases clothing the short shoots but less evident on long shoots. Leaves needle- to scalelike, predominantly of one form or the other in each species or variety, generally narrowly strap-shaped, except triangular when short, thin and soft, flattened. Autumn leaves turning from bright green to reddish brown, remaining attached to the short shoots and falling with them or shed individually on long shoots.

Plants monoecious. Pollen cones numerous, single and unstalked in the axils of scale leaves along dangling, branched, reproductive shoots. Individual pollen cones oblong, with 10–20 spirally arranged triangular pollen scales, each with 2–10 pollen sacs. Pollen grains small (25–35 µm in diameter), nearly spherical, with a small triangular germination papilla and minutely granular surface. Seed cones numerous, clustered but each single at the end of a short branchlet, maturing in a single season, remaining tightly closed until maturity and then disintegrating into individual scales and seeds, either in place or after being shed intact. Individual cones spherical to oblong, with 5–25 spirally arranged thin, peltate, resinous scales on thin stalks. Each cone scale with a roughly diamond-shaped bract tipped by a low, broadly conical boss on the face, fringed around the upper edge by rounded, toothlike lobes of the intimately fused seed scale and with (one or) two seeds. Seeds chunky, angular, wingless, resinous, readily floating in water. Cotyledons four to nine, each with one vein. Chromosome base number $x = 11$.

Wood commonly greasy and with a sour odor, occasionally dry and odorless, light to moderately heavy, soft, relatively strong and decay resistant, with pale yellowish white sapwood giving way gradually to the highly varied heartwood, ranging from yellowish or light brown through various reddish brown shades to almost black. Grain rather coarse and very even to somewhat uneven, with well-defined or somewhat irregular (because of partially missing or extra false) growth rings marked by a gradual or abrupt transition to a broad or narrow darker band of smaller, thicker-walled latewood tracheids. Resin canals absent but with numerous individual resin parenchyma cells scattered through the growth increment or loosely organized into open bands.

Stomates confined to two weakly distinguishable stomatal bands beneath or broadly distributed across both surfaces except over the midrib, only occasionally forming defined lines. Each irregularly oriented stomate sunken beneath and partially hidden by the four to six low surrounding subsidiary cells that entirely

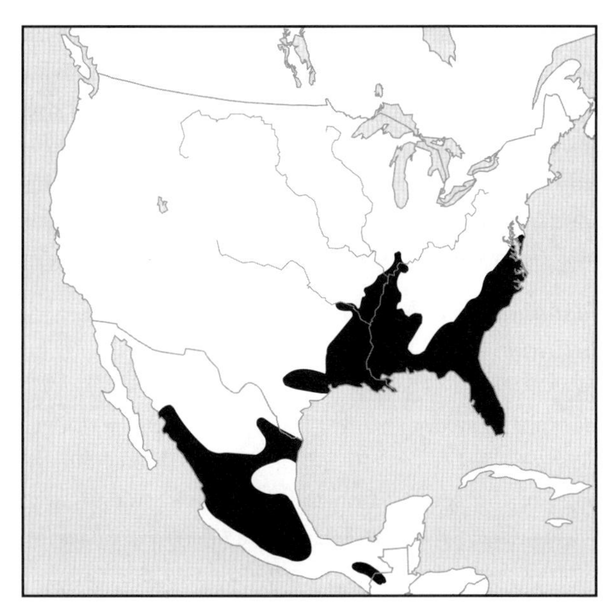

Distribution of *Taxodium* in southeastern North America south to Guatemala, scale 1 : 60 million

lack a Florin ring. Midvein single, often grooved above, and flat or slightly raised beneath, with or without a single resin canal immediately beneath it, and flanked by wedges of transfusion tissue. Photosynthetic tissue with a single palisade layer immediately beneath the upper epidermis without an intervening hypodermal layer, and with spongy mesophyll constituting the bulk of the tissue down to the lower epidermis.

Two species in southern North America. References: Farjon 2005b, Fu et al. 1999h, Martinez 1963, Watson 1985, 1993.

The generic name *Taxodium* (Greek for "yewlike") reflects the fact that this was the first conifer genus (among what later became a flood of genera) encountered by European botanists with foliage more or less reminiscent of that of common yew (*Taxus baccata*), which is rather distinctive when considering only European conifers. Despite having been introduced to cultivation more than 250 years ago, very few bald cypress cultivars have been selected, and these have almost all been based on variations in growth habit: columnar, broad, slightly drooping, or dwarf.

The distinctive cypress knees, woody conical projections rising upward from the root system, are found only in species of *Taxodium* and *Glyptostrobus* growing in swampy soils. They are generally thought to be involved in aeration of the root system, but this has yet to be demonstrated decisively. The Asian *Glyptostrobus* and *Cryptomeria* are the closest living relatives of this now exclusively North American genus. The nine phylogenetically most basal genera of the family Cupressaceae were traditionally recognized as a separate family Taxodiaceae, of which *Taxodium* is the type genus. In phylogenetic terms, however, *Taxodium, Glyptostrobus,* and *Cryptomeria* are actually more closely related to the larger crown group of Cupressaceae that includes the junipers (*Juniperus*), cypresses (*Cupressus*), incense cedars (*Libocedrus* and others), and 16 other genera in both the northern and southern hemispheres than they are to the other six genera (including the redwoods: *Sequoia, Sequoiadendron,* and *Metasequoia*) included with them in the segregate Taxodiaceae. This is why they are all included with Cupressaceae here, a rearrangement I suggested in 1976 that was strongly affirmed by DNA studies some 20 years later.

The oldest known *Taxodium* fossils are found in late Cretaceous deposits of North America, dating to more than 65 million years ago. It was much more widespread around the northern hemisphere during the Tertiary and persisted in Europe into the Pliocene, perhaps as recently as 2.5 million years ago. The trees were particularly abundant in the now arid Columbia Plateau region of Oregon and Washington during the late Tertiary, when ashfalls from the emerging volcanoes of the Cascade Range dammed streams, creating many swamps.

Identification Guide to *Taxodium*

The two species of *Taxodium* may be distinguished by the length of typical needles on flattened shoots and by when these shoots are shed:

Taxodium distichum, needles about 1.5 cm long, shed in autumn, followed by an extended "bald" period

T. mucronatum, needles about 1 cm long, shed with the emergence of the new shoots in early spring

Taxodium distichum (Linnaeus) L. Richard

BALD CYPRESS

Plate 63

Tree to 50 m tall, often with an abruptly swollen or buttressed trunk to 5 m in diameter. Bark reddish brown and smooth at first, remaining thin by peeling in narrow strips, turning brown, and then weathering brownish gray. Crown pyramidal when young, broadening and tattering with age into an irregular shallow dome or, more characteristically, a flat-topped inverted cone. Short shoots shed annually in the autumn, 5–10 cm long. Needles spreading distichously to either side of the shoots or radiating all around them, bright to somber green, turning reddish brown in the fall, 3–17 mm long. Pollen cones 2–3 mm long on reproductive shoots up to 25 cm long. Seed cones spherical, 1.5–4 cm in diameter with 5–10 seed scales. Seeds brown with white edges, 1–1.5 cm long. Southeastern United States coastal plain north to Delaware and west to east-central Texas and north in the Mississippi River valley to southern Indiana. Swamp forests on waterlogged soils, sea level to 160 m. Zone 5. Two varieties.

Their crowns draped with Spanish moss (a flowering plant, *Tillandsia usneoides*), their swollen trunks rising abruptly from the still waters of a deep swamp and surrounded by fields of knees, bald cypresses are among the most picturesque of conifers. The beauty and biological richness of these swamps is well portrayed in a book by John V. Dennis (1988). Aided by their aerating knees, bald cypress trees stand in the deepest waters endured by any of their regular associates, marching out furthest into the water at the edges of ponds, for instance. The trees are fairly fast growing, even under the relatively anaerobic conditions of a flooded swamp, and many of today's large trees sprang up after severe exploitation in the late 19th and early 20th centuries. The timber is exceptionally resistant to decay, and giant fallen logs are still dredged out of ancient river deposits in the Mississippi River valley as well as giving evidence in New England swamps of a wider prehistoric distribution. The largest known individual in combined dimensions when measured in 2001 was 25.3 m tall, 5.2 m in diameter, with a spread of 25.9 m.

Branched twig of bald cypress (*Taxodium distichum* var. *distichum*) with a single year of growth, ×0.5.

Reproductive twig of bald cypress (*Taxodium distichum* var. *distichum*) with pollen cones after pollen release, ×0.5.

Taxodium distichum (Linnaeus) L. Richard **var. *distichum***

SWAMP CYPRESS

Branchlets predominantly horizontal with spreading, distichous leaves (5–)10–17 mm long. Throughout the range of the species. Primarily in riverine swamps with flowing water. Synonym: *Taxodium distichum* var. *nutans* (W. Aiton) R. Sweet; until the type was examined, this name was considered the correct name for variety *imbricarium*.

Taxodium distichum* var. *imbricarium (T. Nuttall) H. Croom

POND CYPRESS

A smaller tree than variety *distichum*. Branchlets ascending, tightly clothed with spirally arranged, forwardly directed leaves 3–10 mm long. Confined to the more southerly and coastal part of the species range from North Carolina to Alabama. Primarily in relatively stagnant pond swamps. Less hardy than variety *distichum*. Zone 7? The name, variety *nutans* (W. Aiton) R. Sweet, was misapplied to the pond cypress

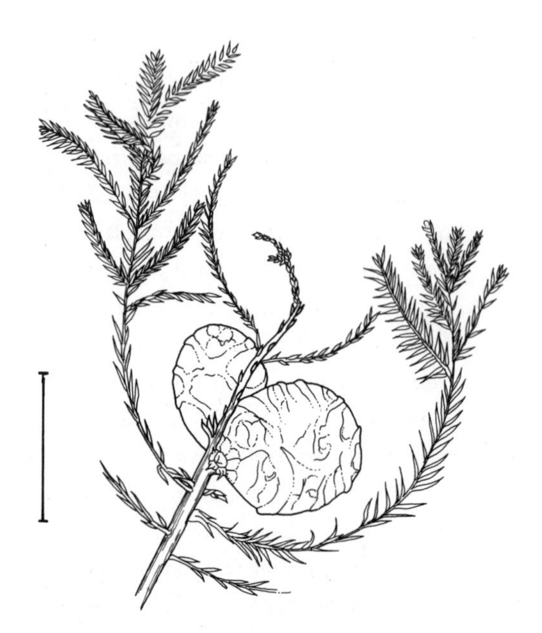

Twig of bald cypress (*Taxodium distichum* var. *distichum*) with closed mature seed cones before shattering; scale, 3 cm.

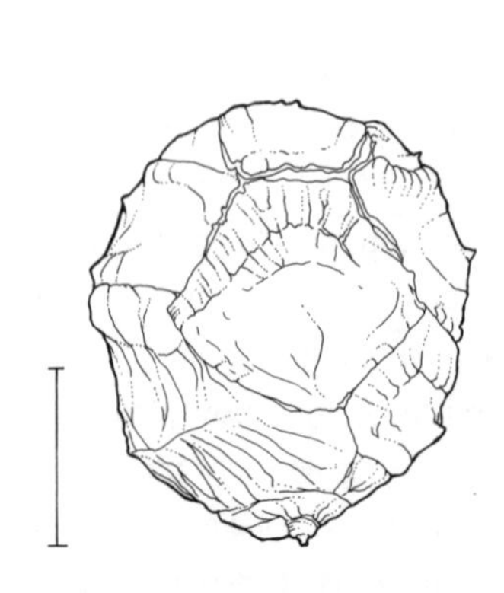

Closed mature seed cone of bald cypress (*Taxodium distichum* var. *distichum*) just at the point of shattering; scale, 1 cm.

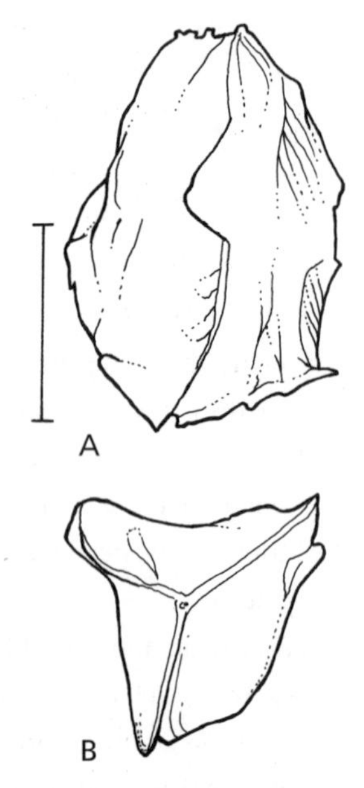

Side (A) and top (B) views of dispersed seed of bald cypress (*Taxodium distichum* var. *distichum*); scale, 5 mm.

but is technically a synonym of variety *distichum.* Synonym: *Taxodium ascendens* Ad. Brongniart.

Taxodium mucronatum M. Tenore

MONTEZUMA BALD CYPRESS, AHUEHUETE (SPANISH)

Tree to 35(–50) m tall, with trunk to 6(–13.6!) m in diameter, single or dividing near the base into two or three main trunks. Bark reddish brown, remaining so even in extreme age. Branchlets drooping to form a graceful, broad crown. Short shoots shed annually with or shortly before the emergence of the new shoots, 10–16(–22) cm long. Needles spreading distichously to either side of the twig, bright green, turning brown before falling, 6–12(–22) mm long. Pollen cones about 3 mm long, in drooping, branched clusters to 15 cm long. Seed cones oblong, 1.5–2(–3) cm long, with 20–25 seed scales. Seeds shiny brown, 5–10 mm long. Southern Texas through Mexico to Guatemala. Wet soils, primarily in the highlands, at (20–)300–2,500(–2,700) m. Zone 9. Synonym: *Taxodium distichum* var. *mexicanum* (Carrière) G. Gordon.

Montezuma bald cypress is the national tree of Mexico. It is long-lived, and many individual Montezuma bald cypresses have specific connections with the Spanish conquest of Mexico almost 500 years ago. The oldest accurately dated was 1,140 years old

Old mature Montezuma bald cypress (*Taxodium mucronatum*) in nature.

All-aged natural stand of pond cypress (*Taxodium distichum* var. *imbricarium*).

when assessed in 2003. The largest and most famous tree is the Arbol del Tule in southern Mexico, which has been measured several times over the past two centuries. In 2005 it was 35.4 m tall (previously recorded as 41.2), 11.5 m in average diameter (13.6 m one way and 9.3 m at right angles to that), with a spread of 43.9 m (previously 53.6 m). Its circumference of 46 m is the greatest girth of any single tree in the world (baobabs being its only rivals). Using the formula adopted by the American Forestry Association in assessing candidate champion trees for their Register of Big Trees of the United States, the Tule tree adds up to a whopping 1,577 points, more than the largest giant sequoia (*Sequoiadendron giganteum*). This is somewhat deceptive, however, and the Tule tree is not nearly as large as the biggest giant sequoias, such as the General Grant tree (the reigning champion) or the General Sherman tree, because its enormous girth is measured around the outsides of numerous, narrow buttresses or flutes, and its actual cross-sectional area is no more than about 80% of this perimeter and extends upward only 12–15 m before breaking up into separate, rapidly tapering trunks. In contrast, the great giant sequoias maintain their diameters for 100 m or more, so their volume is several times that of the Tule tree. That does not detract from its status as one of the world's great trees, of course. Because it quickly divides into several main trunks, some speculated over the years that the Tule tree actually consists of several trees grown together, but DNA fingerprints show that it is all one genetic individual (Dorado et al. 1996). Its age will probably never be known because it is hollow. Like those of their cousins, bald cypresses (*Taxodium distichum*), the branches of Montezuma bald cypresses are often draped with gray Spanish moss, and this, along with their size and age, may have contributed to the Aztec-derived name *ahuehuete*, which can be interpreted as "old man of the waters." In contrast to the Tule tree, some 1,000 km to the north near the Mexican border at the southern tip of Texas, the largest known individual in the United States in combined dimensions when measured in 2003 was just 20.7 m tall, 2.3 m in diameter, with a spread of 27.1 m. Montezuma bald cypress is cultivated in warm temperate climates, like southeastern Australia or California, and makes a handsome ornamental with a more drooping foliage habit than is found in bald cypress.

Taxus Linnaeus

YEW

Taxaceae

Evergreen trees and shrubs with single or multiple trunks. Bark smooth, thin, fibrous to corky, reddish brown, flaking and peeling in longitudinal strips to reveal purple-red patches. Crown densely branched in youth, the lower branches persistent in open-grown plants but large forest trees ultimately with a clear trunk. Branchlets all elongate, without distinction into long and short shoots, hairless, remaining green for the first year or two, clothed by and shallowly grooved between the elongate, attached leaf bases. Winter buds well developed, the thin, flexible scales often persistent at the base of the grown twigs. Leaves spirally attached but twisting at the base to lie flat (distichously) along either side of the twig (except forming a V in *Taxus cuspidata*), persisting several years. Each leaf needlelike, flattened top to bottom, sword-shaped, often slightly curved, leathery and flexible, dark green above, with two paler, often yellow-green or grayish green, stomatal bands along either side of the midrib beneath, abruptly narrowed to a distinct petiole and gradually or abruptly narrowed to the stiff, pointed tip. Midrib prominently raised above and prominently to inconspicuously raised beneath, the blade flat or bowed on either side and straight or slightly turned down or rolled at the edges.

Plants dioecious (or monoecious in *Taxus canadensis*). Pollen cones numerous along the branchlets, each single in the axil of a needle, short-stalked, emerging from a short-stalked, scaly bud. Reproductive portion of each cone spherical to oblong, with 4–16 densely clustered pollen scales. Each scale shieldlike, nearly radially symmetrical, with (two to) four to nine pollen sacs arranged all around a common stalk. Pollen grains small (20–40 µm in diameter), roughly spherical to somewhat flattened or irregular, without air bladders, with an ill-defined germination bump, and minutely bumpy over the whole surface. Seeds not in obvious cones, single at the ends of short stalks in the axils of the needles, maturing the first season, then reappearing in the same location for several years. Each smooth, round, hard, dark seed seated in an open, red, fleshy cup (aril) attached only to the seed stalk and with a few small scales at the base. Cotyledons usually two, each with one vein. Chromosome base number $x = 12$.

Wood heavy, hard, strong, and durable, with an abrupt transition between the creamy white to pale yellowish brown sapwood and the red to orange-brown hardwood. Grain fine and fairly even, with generally obvious growth rings (somewhat obscure in *Taxus baccata*) marked by a gradual to abrupt transition to a narrow to broad band of darker latewood. Resin canals and individual resin parenchyma cells both absent, the tracheids (wood cells) singly spirally thickened.

Stomates confined to the stomatal bands beneath in crowded or more open and irregular lines. Each stomate sunken beneath and largely hidden by the four to six surrounding subsidiary cells, which are topped by a prominent, steep, interrupted, knobby Florin ring and with most other epidermal cells in the stomatal bands covered with prominent, knoblike papillae. Midvein single, not reinforced by sclereids, without accompanying resin canals, and flanked by small bands of transfusion tissue. Photosynthetic tis-

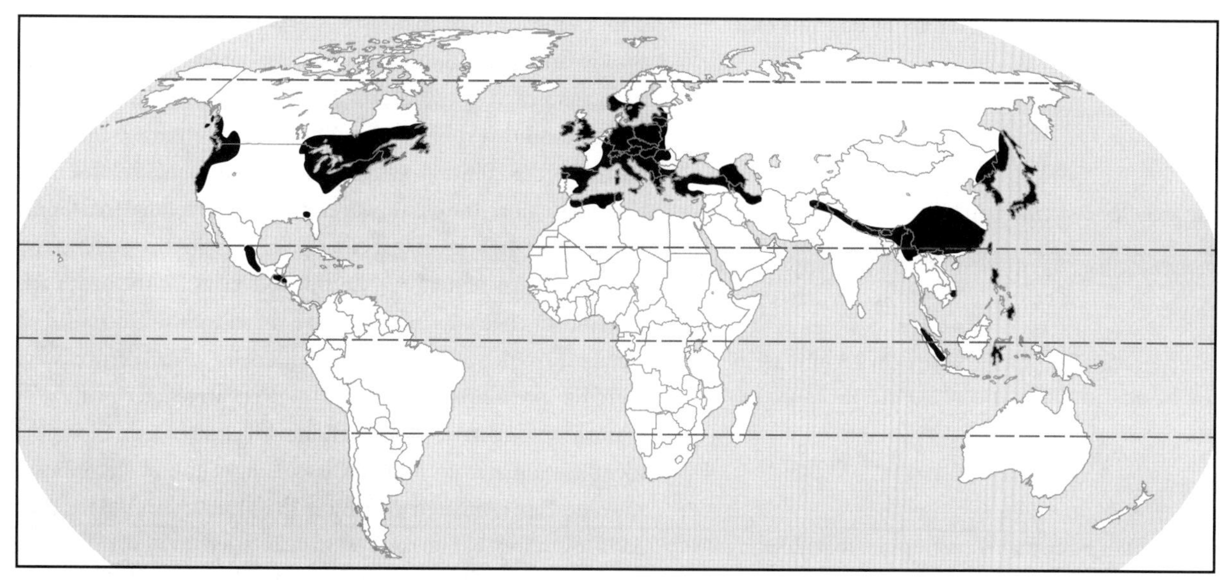

Distribution of *Taxus* in the northern continents and south of the equator in Indonesia, scale 1 : 200 million

sue with one to three layers of palisade cells lining the whole upper surface directly beneath the epidermis without an intervening hypodermal layer and without sclereids among the rounded or slightly horizontally elongated spongy mesophyll cells that extend from the palisade parenchyma to the lower epidermis.

Eight species in North America and Eurasia. References: Chadwick and Keen 1976, de Laubenfels 1988, Fu et al. 1999e, Hils 1993, Li et al. 2001a, K. C. Sahni 1990, Yamazaki 1995.

Yews are important in cultivation because of their tolerance for severe pruning, allowing them to be maintained at any size and to be shaped into dense hedges and topiaries. Their ability to sprout from dormant buds in bare wood after pruning is responsible for this horticultural prominence and is rather unusual among conifers. The more than 200 cultivars selected since the late 18th century cover the gamut of expected variations in habit (such as narrowly upright, like the well-known Irish yew, to spreading, drooping, and various shapes of dwarfs), in foliage form (shorter, more compact, and distorted) and color (mostly variations on yellow, gold, copper, silver, and bluish green), and in aril color (yellow or orange instead of red).

Despite the visual appeal of the colorful, juicy arils, the plants are extremely toxic, only the bird-dispersed arils lacking cyanide-yielding compounds and taxane alkaloids. Some of the latter, especially paclitaxel (sold commercially as taxol), have proven effective in treatment of certain cancers, leading to heavy commercial exploitation of some species and a search for other ways to produce the compounds.

Although the species are very difficult to identify and were sometimes considered varieties of a single species, most cultivated yews belong to only two species (*Taxus baccata* and *T. cuspidata*) and their hybrid (*T.* ×*media*). These two species are probably each other's closest relatives, but relationships within the genus are fairly obscure. There are few visually obvious taxonomic characters, and microscopic features, which have been suggested to fill the gaps, give some results that are difficult to reconcile with other lines of evidence. Despite the pharmaceutical importance of the genus, the available DNA data are rather limited, and the relationships they imply conflict with those suggested by distribution of taxane alkaloids. Both kinds of studies, nonetheless, distinguish most Eurasian species from most North American ones, even if they differ in the implied relationships within and between these continental species groups. Most DNA evidence also favors a monophyletic family Taxaceae with a position within the conifers as sister family to the Cupressaceae. Within Taxaceae, *Taxus* is closest to the similar-looking white cup yew (*Pseudotaxus*), which differs in several obscure ways and one conspicuous one, a white aril rather than the red (or yellow in some cultivars) one found in yews.

The solitary, terminal seed surrounded by a cup-shaped free aril led Florin in 1948 to postulate that the yews are not conifers. He bolstered his argument with fossil specimens from the Jurassic of Yorkshire, about 165 million years old, that now seem more closely related to *Amentotaxus* of the subfamily Amentotaxoideae (T. M. Harris 1979). Secure fossil shoots of *Taxus* extend back only to the Oligocene in Europe, some 30 million years ago, and this relatively recent appearance of the genus accords well with its generally derived features in comparison to other conifers and even to other members of the Taxaceae.

Identification Guide to *Taxus*

The eight species of *Taxus* may be distinguished (with some uncertainty) by the length of the longest needles, the width of the widest needles, and the length of the largest seeds:

Taxus canadensis, needles less than 2 cm by up to 2.5 mm, seeds up to 5 mm

T. brevifolia, needles up to 2 cm by more than 2.5 mm, seeds 6–7 mm

T. baccata, needles 2–3 cm by up to 2.5 mm, seeds 6–7 mm

T. floridana, needles 2–3 cm by up to 2.5 mm, seeds at least 8 mm

T. cuspidata, needles 2–3 cm by more than 2.5 mm, seeds 6–7 mm

T. sumatrana, needles more than 3 cm by up to 2.5 mm, seeds 6–7 mm

T. globosa, needles more than 3 cm by more than 2.5 mm, seeds up to 5 mm

T. wallichiana, needles more than 3 cm by more than 2.5 mm, seeds 6–7 mm

Taxus baccata Linnaeus

COMMON YEW, ENGLISH YEW, EIBE (GERMAN), IF (FRENCH), TASSO (ITALIAN), TEJO (SPANISH)

Tree (or shrub in cultivation) to 20(–29) m tall, with a fluted, short (or, less commonly, unbranched) trunk to 1.5(–3.5) m in diameter. Crown broad, dense, and rounded. Winter bud scales persistent, blunt, not keeled. Needles dark green above (or yellow or variegated in some cultivars), (1–)2–3 cm long, 2–2.5 mm broad, the stomatal bands beneath pale green, tapering gradually to the soft-pointed tip. Plants dioecious. Pollen cones 3–4 mm long. Seeds slightly flattened, (5–)6–7 mm long, conspicuous within the bright red (sometimes yellow) aril (hence the scientific name, Latin for "berrylike"). Most of Europe, northern Africa, western and southern Turkey, Caucasus to southern shore of Caspian Sea. Scattered or in small groves in deciduous or mixed forests, especially near streams or on moist slopes with limestone substrates, in the lowlands in the north but only in mountains in the Mediterranean region and eastwards; (0–)50–1,800(–2,100) m. Zone 6.

Common yew has been cultivated for centuries, and there are numerous cultivars of diverse habits and needle color, length, and orientation. The elastic wood was the premier material for the longbows of the archers (called yeomen after this tree) of England and other parts of medieval Europe, as well as for the bent parts of Windsor chairs. Its durability in contact with soil also made it suitable for fence posts and other outdoor uses. Many groves were sacred in pre-Christian Europe, and trees were later planted or retained in cemeteries. Association with churchyards allows the identification of many individuals as being centuries old. Some of the oldest are estimated to be more than 1,000 years old (or even 3,000 or more), but no such estimate is based on a full ring count because all ancient yews have decayed centers.

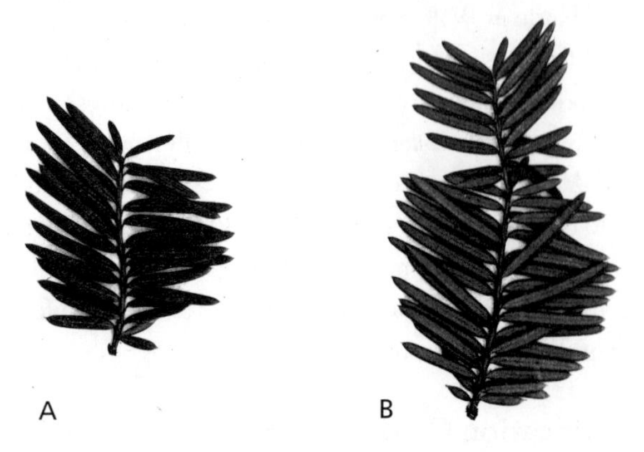

Upper (A) and undersides (B) of twigs of common yew (*Taxus baccata*) with1 or 2 years of growth (B), ×0.5.

Taxus brevifolia T. Nuttall

PACIFIC YEW

Understory shrub, or tree to (15–)25 m tall, with fluted trunk to 0.6(–1.5) m in diameter. Crown conical to dome-shaped, becoming rather open and irregular with age. Winter bud scales persistent or soon shed, narrowly pointed, keeled. Needles yel-

Upper (A) and undersides (B) of twigs of Pacific yew (*Taxus brevifolia*) with 2 years of growth, ×0.5.

lowish green above, 1–2(–3) cm long, 1–3 mm broad, the stomatal bands beneath grayish green, narrowing abruptly to the soft-pointed tip. Plants dioecious. Pollen cones 3–4 mm long. Seeds slightly flattened, 6–8 mm long, conspicuous within the bright red or orange-red aril. Southern Alaska to northern California and southeastern British Columbia to northern Idaho. Understory of moist riverside forests and canyons; 0–2,200 m, depending on latitude and proximity to the sea. Zone 6.

The toughness and durability of the wood made it useful for many kinds of tools, weapons, and implements among the native peoples and later settlers of the region. These uses include fence posts, bows for archery, canoe paddles, bent work, turnery, and bodies for lutes and other stringed instruments. It was, perhaps, overexploited by local craftsmen and more decisively so by the pharmaceutical industry in search of the powerful anticancer agent taxol, obtained in low yield by extracting bark stripped from the trees. The needles are shorter than those of common yew (*Taxus baccata*), hence the scientific name, Latin for "short leaf." The largest known individual in the United States in combined dimensions when measured in 1989 was 16.5 m tall, 1.5 m in diameter, with a spread of 9 m.

Taxus canadensis H. Marshall

CANADA YEW, GROUND HEMLOCK, IF (FRENCH)

Low sprawling shrub, typically 1(–2) m tall, the branches rooting in contact with the soil. Crown low and wide, sparse to dense, with thin, spreading branches turned up at the end. Winter bud scales keeled, blunt at the tip, persistent. Needles dark green to yellowish green above, turning reddish in winter, 1–2(–2.5) cm long, 1–2(–2.5) mm wide, the stomatal bands beneath pale green, narrowing abruptly to the soft-pointed tip. Plants monoecious. Pollen cones 3–4 mm long. Seeds slightly flattened, 4–5 mm long, conspicuous within the bright red aril. Northeastern North America, from southern Labrador and Newfoundland to Minnesota, southeastern Iowa, Tennessee, and Maryland. Understory of many forest types on rich, moist soils; 0–1,500 m. Zone 2.

This species is distinguished from all other yews by its lack of a trunk, lack of microscopic wartlike papillae on the cells of its stomatal bands, and by having pollen cones and seeds on the same plant. Its small stature and straggly appearance preclude its commercial exploitation in forestry and horticulture. It is one parent of the cultivated *Taxus* ×*hunnewelliana,* along with Japanese yew (*T. cuspidata*), a parentage cast in doubt by the first DNA study of the genus but supported by a more recent and more extensive study specifically addressing the issue. Both DNA studies and microscopic leaf features suggest that Canada yew is more closely related to Eurasian species than to the other three North American yews, but these

Mature shrub of Canada yew (*Taxus canadensis*) in nature.

Upper (A, B) and undersides (C) of twigs of Canada yew (*Taxus canadensis*) with 2 years of growth bearing mature and immature seed cones, ×0.5.

relationships are not securely established. The foliage was brewed to make a medicinal tea by various native groups of the region, but such concoctions must have been used with great care because of the strong toxicity of the foliage, particularly when bruised or wilted.

Taxus cuspidata P. Siebold & Zuccarini
JAPANESE YEW, ICHII, ARARAGI (JAPANESE), CHU MOK (KOREAN)

Shrub, or tree to 20 m tall, with fluted trunk to 1 m in diameter. Crown deep and narrow in forest-grown trees, dense to open and irregular, rounded or flat-topped, with numerous slender horizontal or upwardly angled branches. Winter bud scales persistent, pointed and keeled. Needles dark green above, (1–)1.5–2 cm long, 1.5–2(–4.5) mm wide, the stomatal bands beneath yellowish green, narrowing abruptly to the soft-pointed tip (hence the scientific name, from Latin). Plants dioecious. Pollen cones about 3 mm long. Seeds slightly flattened, about 6 mm long, conspicuous within the bright red or yellow aril. Japan, Korea, and adjacent northeastern China and southeastern Russia. Temperate deciduous forests to subalpine forests; 20–2,500 m. Zone 4. Two varieties.

This species is hardier than *Taxus baccata* and together with *T.* ×*media,* its hybrid with that species, largely replaces it in cultivation in colder regions. It may be distinguished from other species by holding its needles in a V-shaped position rather than horizontally. The modest array of available cultivars seem to have been derived primarily from the shrubby variety *nana* rather than from the taller variety *cuspidata.* Cultivar selection has emphasized shorter, yellow needles and compact growth, with some selection for yellow arils.

Taxus cuspidata P. Siebold & Zuccarini **var. *cuspidata***

Tree with needles arranged in two irregular rows on either side of the twigs. Throughout the range of the species in temperate deciduous forests up to 2,000 m.

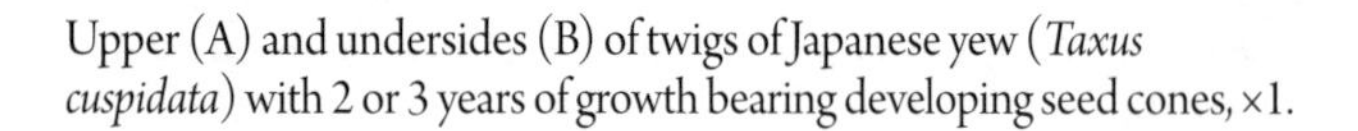

Upper (A) and undersides (B) of twigs of Japanese yew (*Taxus cuspidata*) with 2 or 3 years of growth bearing developing seed cones, ×1.

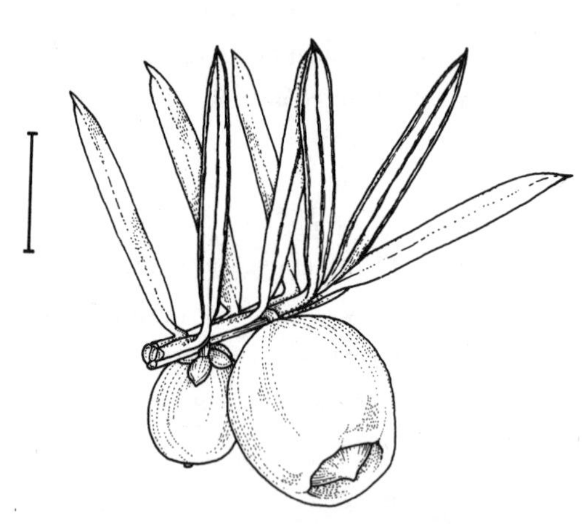

Twig of Japanese yew (*Taxus cuspidata*) with mature seed cones; scale, 1 cm.

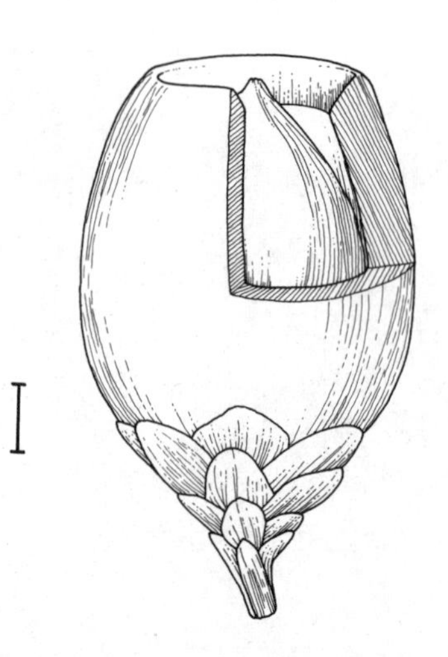

Mature seed cone of Japanese yew (*Taxus cuspidata*) at dispersal with aril partially cut away to show nestled seed; scale, 1 mm.

Taxus cuspidata var. *nana* Rehder

KYARABOKU (JAPANESE)

Spreading shrub to 3 m tall with needles radiating out all around the twigs. Japan Sea side of Honshū in subalpine forest at 1,200–2,500 m.

Taxus floridana A. W. Chapman

FLORIDA YEW

Shrub, or tree to 6(–10) m tall, branching near the base or with a short trunk to 0.2(–0.4) m in diameter. Crown narrowly dome-shaped, open and irregular, with stout or slender, upwardly angled to horizontal branches. Winter bud scales pointed, not keeled, generally shed within a year of shoot emergence. Needles dark green above, (1–)2–2.6 cm long, (1–)1.5–2.2 mm wide, the stomatal band beneath pale green, narrowing gradually to the soft-pointed tip. Plants dioecious. Pollen cones (3–)4–6 mm long. Seeds slightly flattened, about 10 mm long, conspicuous within the dull red aril. Northwestern Florida. Understory of mixed hardwood forests and cedar swamps in moist ravines; 15–30 m. Zone 8.

Florida yew is one of the rarest conifer species. It is confined to slopes and ravines along about a 25-km stretch of land adjoining the eastern side of the upper Apalachicola River and a single Atlantic white cedar (*Chamaecyparis thyoides*) swamp nearby. Its natural range is even more restricted than that of the co-occurring and related stinking cedar (*Torreya taxifolia*), but it has not been decimated by the torreya blight. It is very closely related to Mexican yew (*Taxus globosa*) and may well be treated as part of the same species with additional evidence. It has little apparent relationship with ground hemlock (*T. canadensis*), with which it has sometimes been combined. The largest known individual in combined dimensions when measured in 1986 was 6.1 m tall, 0.2 m in diameter, with a spread of 7.9 m.

Taxus globosa D. F. L. Schlechtendal

MEXICAN YEW, CIPRÉS COLORADO, PINABETE COLORADO (SPANISH)

Shrub, or tree to 20(–40) m tall, with trunk to 0.4 m in diameter. Crown narrowly and irregularly cylindrical, with slender horizontal branches. Winter bud scales blunt, not keeled, partially persistent. Needles olive green above, (1.5–)2–3.5 cm long, 1–3 mm wide, the stomatal bands beneath silvery, narrowing abruptly to the soft-pointed tip. Plants dioecious. Pollen cones about 3 mm long. Seeds somewhat flattened, about 5 mm long, conspicuous within the bright red, nearly spherical aril (hence the scientific name, from Latin). Southern Mexico, central Guatemala, and western Honduras. Cloud forests, often near streams; 2,200–3,000 m. Zone 8.

Mexican yew is the least-known species in the genus. It has a very fragmented distribution with widely separated stands on scattered individual mountain ranges. It appears to be closely related to Florida yew (*Taxus floridana*) based on microscopic leaf structure and DNA studies, but the two are fairly readily distinguishable by the longer needles and smaller pollen cones and seeds of Mexican yew.

Taxus ×*hunnewelliana* Rehder

HUNNEWELL HYBRID YEW

Wide-spreading shrubs (*Taxus canadensis* × *T. cuspidata*) resembling Japanese yew (*T. cuspidata*) in foliage but with lighter green, narrower needles. Zone 4. Although the original Hunnewell yews were raised from seed of ground hemlock (*T. canadensis*) at the Hunnewell Pinetum in Massachusetts, most plants currently available in horticulture must have had Japanese yew as their maternal parent since they contain chloroplast DNA from ground hemlock. (Chloroplast DNA is paternally inherited in conifers, in striking contrast to the almost universal maternal inheritance found in flowering plants.) The minority of plants with a ground hemlock seed origin are a little more like that species in appearance, just as the more common Hunnewell yews are closer morphologically to their maternal parent, Japanese yew.

Taxus ×*media* Rehder

FOUNDATION YEW

Plates 64 and 65

Shrubs or trees (*Taxus baccata* × *T. cuspidata*) closely resembling *T. cuspidata* but with the leaf arrangement often less in a V or even flat, the midrib of the needles less raised above, the bud scales less keeled, and the branchlets remaining green 2 years rather than just 1. Zone 5. Foundation yew is the most widely cultivated yew in North America for foundation plantings, hedges, and specimen shrubs. It is much hardier than common yew (*T. baccata*) and has

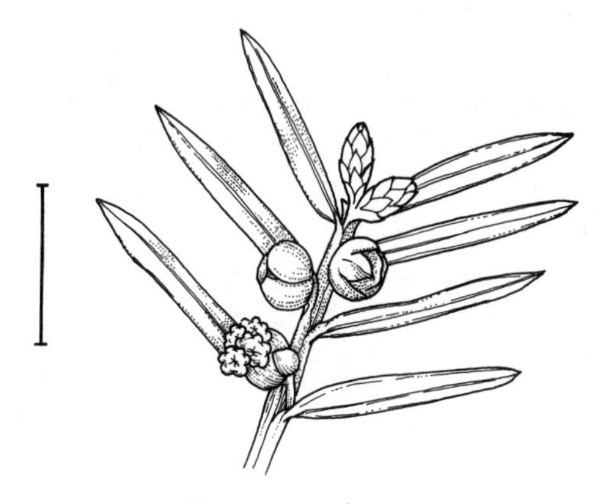

Twig of foundation yew (*Taxus* ×*media*) with emerging pollen cones; scale, 5 mm.

Underside of twig of foundation yew (*Taxus ×media*) with pollen cone buds, ×0.5.

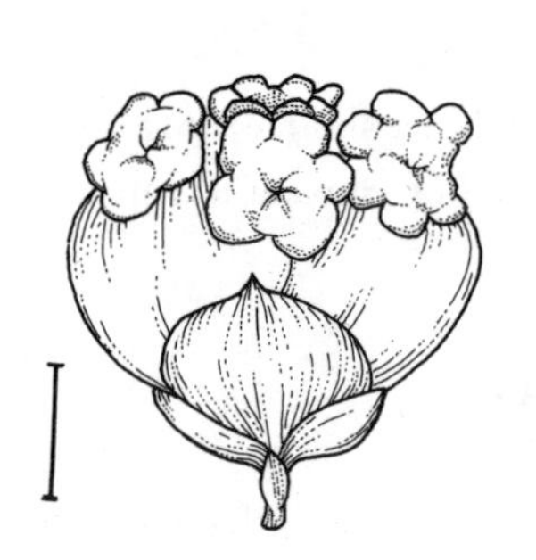

Mature pollen cone of foundation yew (*Taxus ×media*) at time of pollination; scale, 2 mm.

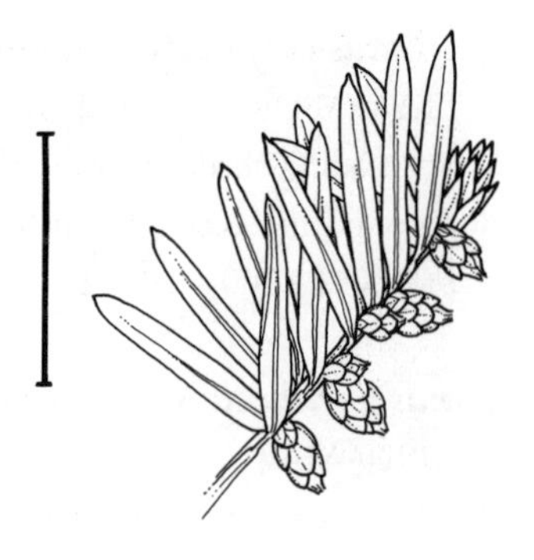

Twig of foundation yew (*Taxus ×media*) with seed cones at time of pollination; scale, 1 cm.

Underside of twig of Himalayan yew (*Taxus wallichiana*) with two growth increments, ×0.5.

spawned a much wider array of cultivars than has Japanese yew (*T. cuspidata*). The numerous existing cultivars are about equally divided between those whose seed parent was common yew and those derived from seed of Japanese yew. This dual origin goes back to the first crosses that gave rise to this hybrid, in about 1900. At what is now the Hunnewell Pinetum, in Massachusetts, seed was collected from an Irish yew (*T. baccata* 'Fastigiata'), while the original plants in the Hicks Nursery in Long Island, New York, were raised from seed collected from a Japanese yew. One seedling of the latter origin was the original plant of *T. ×media* 'Hicksii,' by far the most widely planted hedge yew in North America.

Taxus sumatrana (Miquel) de Laubenfels

CHINESE YEW, HONG DOU SHAN (CHINESE), THANH TÙNG (VIETNAMESE), TAMPINUR BATU (MALAY)

Shrub, or tree to 30(–45) m tall, with fluted trunk to 1(–2) m or more in diameter. Crown dome-shaped, dense, with slender, upwardly angled to horizontal branches. Winter bud scales pointed to blunt, not keeled, soon shed after shoot elongation or some persistent. Needles dark green above, (1–)2–4 cm long, 2–2.5 mm wide, the stomatal bands beneath brownish green, narrowing abruptly or more gradually to the soft-pointed tip. Plants dioecious. Pollen cones 4–6 mm long. Seeds somewhat flattened, 6–7 mm long, conspicuous within the red aril. Central and southern China, Taiwan, Philippines, Indonesia (Sulawesi and Sumatra), Vietnam, and Myanmar. Scattered in the understory or canopy of moist subtropical forests to cloud forests, often along streams or on exposed mossy ridge tops; (100–)500–2,500(–3,500) m. Zone 7. Synonyms: *Taxus celebica* (O. Warburg) H. L. Li, *T. chinensis* (Pilger) Rehder, *T. mairei* (Lemée & H. Léveillé) S. Y. Hu ex Tang S. Liu, *T. speciosa* Florin, *T. wallichiana* var. *chinensis* (Pilger) Florin, *T. wallichiana* var. *mairei* (Lemée & H. Léveillé) L. K. Fu & N. Li.

This species has the most ecologically varied and geographically disjointed distribution of any yew. Although no one has investigated yews from throughout the whole range, those authors who have seen a wider spectrum of material have tended toward recognizing just a single species, while splitters have been more prevalent among those with a more regional focus. A completely convincing taxonomic treatment awaits further study. For example, in China, plants from lower montane forests (sometimes separated as *Taxus mairei*) have longer, curved needles with the margin curled under, while those from cloud forests (sometimes separated as *T. chinensis*) have shorter, straighter needles with a flat margin. Other than the (average) difference in elevation and environment, the two forms have essentially identical ranges, and it is not known to what extent their differences might reflect age- or environment-related influences rather than genetically based taxonomic differences. The plants in the lower montane forests of Sumatra, Sulawesi, and Taiwan are forest giants that yield an excellent timber. The species can never be commercially important, however, because of the scattered distribution of the trees, and their slow growth precludes plantation culture. Unfortunately, their relative rarity and slow growth also mean that these magnificent trees are likely to virtually disappear with continued forest clearance in the region, as they have throughout most of mainland China, and persist only as shrubby remnants that give little hint of their potential.

Taxus wallichiana Zuccarini

HIMALAYAN YEW, BIRMI (KASHMIRI AND OTHERS), THUNER (HINDI), TCHEIRAGULAB (NEPALESE), XI MA LA YA HONG DOU SHAN (CHINESE), KYAUK-TINYE (BURMESE), THANH TÙNG (VIETNAMESE)

Shrub, or tree to 6(–30) m tall, with fluted trunk to 1.5(–3.5) m in diameter. Crown dense, dome-shaped, with numerous thin, upwardly angled branches bearing horizontal or drooping branchlets. Winter bud scales pointed, not strongly keeled, persistent or partially shed. Needles dark glossy green above, (1.5–) 2–4(–4.5) cm long, (1.5–)2–3(–5) mm wide, the stomatal bands beneath paler green, tapering gradually to the long, sometimes stiff-pointed tip. Plants dioecious. Pollen cones (4–)5–8 mm long. Seeds slightly flattened, 6–7 mm long, conspicuous within the bright red aril. Along the entire Himalayan span of ranges from Afghanistan to Yunnan and southwestern Sichuan (China) and in southern Vietnam (Phu Khanh and Lâm Dông). Broad-leaved evergreen forests and conifer forests; 2,300–3,400 m. Zone 8. Synonyms: *Taxus fuana* N. Li & R. Mill, *T. yunnanensis* W. C. Cheng & L. K. Fu.

In the *Flora of China* (Fu et al. 1999e), Himalayan yew was divided into a western Himalayan species (*Taxus fauna*) and an eastern Himalayan species *T. wallichiana*). Furthermore, the widespread Chinese yew was treated as two varieties of the latter rather than belonging with the Malesian *T. sumatrana*, the treatment accepted here. All the yews in this complex, extending from Afghanistan to Taiwan, the Philippines, and Sulawesi (Indonesia), are assigned to just two species here, Himalayan yew and Chinese yew (*T. sumatrana*), admittedly a conservative approach. Some authors recognize even more species than the five taxa accepted in more liberal contemporary treatments. The most effective classification of these plants will remain uncertain until there is thorough study of the complex throughout its range, both in the field and laboratory, including DNA studies. The species name honors Nathaniel Wallich (1786–1854), a Danish botanist who collected and described many Himalayan plants during his tenure as superintendent of the Calcutta Botanic Garden.

Tetraclinis M. T. Masters

ARAR-TREE

Cupressaceae

Evergreen shrubby trees with one to several frequently much branched and crooked trunks clothed with thin, fibrous bark gradually roughening into broken ridges. Branchlets strongly flattened, vertically oriented, brittle at the joints between successive leaves, arranged in flattened sprays because virtually all branching (which is mostly alternate) arises in the axils of lateral leaves. Without definite winter buds. Leaves in alternating pairs; those formed in the transient juvenile phase needlelike; those of adults scalelike, with long attached bases clothing the twig and very short free tips sometimes fringed with tiny hairs. Adult leaves differentiated into facial and lateral pairs, the lateral pairs completely separated by the facial pairs and parallel to them, thus seemingly in whorls of four, like those of *Calocedrus*, with small glands at the tip.

Plants monoecious. Pollen cones single at the ends of short branchlets on a short, slender stalk emerging from a few alternating pairs of bracts that are transitional to the foliage leaves of the supporting branchlets. Individual cones oblong, with five or six alternating pairs of pollen scales. Each scale shield-shaped, the face triangular to pentagonal, with a slender central stalk and four or five pollen sacs. Pollen grains small (25–35 μm in diameter), nearly spherical, almost featureless. Seed cones single at the ends of short branchlets, maturing and gaping open in a single season. Individual cones almost spherical, with two nearly equal alternating pairs of woody seed scales attached at their bases around a common base. Each cone scale composed of an intimately united seed-bearing scale and bract, with a distinct triangular point (of the bract) below the tip, the smaller (upper) pair meeting at their tips, sterile (or with a single abortive seed) and strongly grooved, the larger, more convex pair not touching each other, with (one or) two (or three) seeds. Two additional seeds sitting at the center of the cone and possibly representing an otherwise lost third pair of cone scales (similar to the central plate in seed cones of the related (*Calocedrus*). Seeds oval with two roughly equal wings forming an obtuse angle with and running along the seed body and beyond, and much wider than it, the wings derived from the seed coat. Cotyledons four, each with one vein. Chromosome base number $x = 11$.

Wood very fragrant, relatively heavy and hard, the whitish to yellowish sapwood sharply contrasting with the reddish brown to

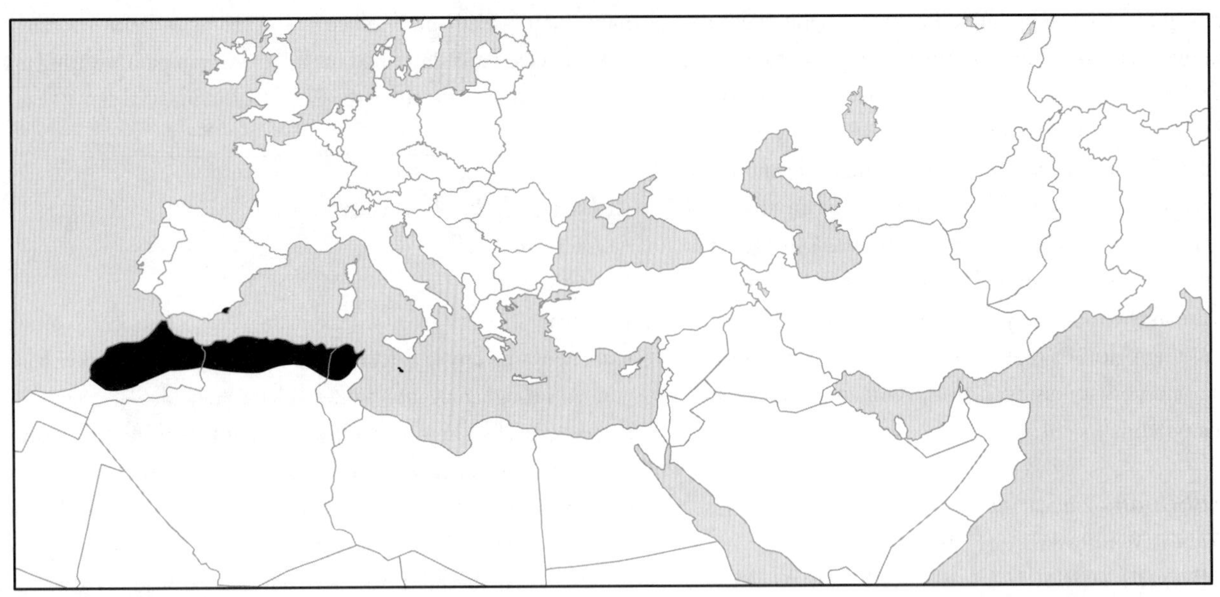

Distribution of *Tetraclinis articulata* in the western Mediterranean of North Africa and southwestern Spain, scale 1 : 55 million

dark chocolate brown or almost black heartwood. Grain very fine and somewhat wavy, with weakly developed growth rings marked by an abrupt transition to one to three rows of smaller but not notably thicker-walled latewood tracheids. Resin canals absent but with a scattering of individual resin parenchyma cells.

Stomates confined to very limited stomatal zones near the tips of the leaves on their inner face and especially lining the grooves between the bases of adjoining lateral and facial leaves. Each stomate sunken beneath and largely hidden by the four to six surrounding subsidiary cells, which are topped by a steep, lobed Florin ring, the remaining epidermal cells in the grooves also often topped by rounded or irregular papillae. Midvein weak, close to the inner edge of the attached leaf base, petering out quickly in the free tip, with little or no accompanying transfusion tissue. Photosynthetic tissue with a single palisade layer inside the epidermis and thin hypodermal layer lining all exposed leaf surfaces, including the attached bases, but displaced from the area of the leaf tips where the centrally located, spherical to oblong resin gland lies directly beneath the epidermis. Spongy mesophyll filling the remainder of the leaf and merging with the stem tissue of the facial leaves.

One species in the western Mediterranean. References: Castroviejo et al. 1986, Farjon 2005b, Kvaček et al. 2000.

Tetraclinis was historically considered the sole northern hemisphere member of the southern hemisphere group of Cupressaceae, with special similarity to *Widdringtonia.* Although the cones superficially resemble those of *Widdringtonia,* details of their structure and development do not. The foliage is more like the northern *Calocedrus,* and leaf pigments and DNA studies also demonstrate a relationship to northern genera, particularly *Calocedrus, Microbiota,* and *Platycladus. Tetraclinis articulata,* while of moderate economic importance, is not terribly ornamental and is scarcely in cultivation outside of its native range and botanical gardens so no cultivar selection has taken place.

The oldest known fossil remains of *Tetraclinis* are from early Eocene (or latest Paleocene) sediments in Europe more than 50 million years old. By the Oligocene, more than 30 million years ago, similar trees were also growing in North America. While they died out in North America during the Miocene, there is a more or less continuous record in Europe into the early Pliocene, some 5 million years ago. There are no records for either Asia or the southern hemisphere, providing further evidence for the northern affinities of the genus.

Tetraclinis articulata (M. H. Vahl) M. T. Masters

ARAR-TREE, SANDARACH, ALERCE (SPANISH)

Shrub, or tree to 15 m tall, trunk to 50 cm in diameter. Bark smooth at first and flaking in small scales or strips, tan to light reddish brown weathering gray, becoming longitudinally furrowed between checkered ridges. Branches dense, ascending. Crown conical, becoming irregular with age. Twigs rather sparsely branching, giving the impression of nearly equal dichotomous forking, reaching about 4–8 mm before forking again (except along the main axes of growth, which have longer intervals). Scale leaves 3–4 mm long, grayish green on both surfaces. Pollen cones 3–5 mm long. Seed cones 8–12 mm across. Seed scales thick and woody, with a grayish waxy coating when young, notably paler inside than out. Seeds light brown, the body 4–6 mm long, about half as wide, the wings kidney-shaped, up to 8 mm wide. Northern Africa from Morocco to western Libya, Malta, southeastern Spain near Cartagena. Dry soils on flats and slopes, 500–1,800 m. Zone 9.

Arar-tree is a rare species in Spain and Malta and uncommon in cultivation where it is susceptible to winter cold and wet. Because of its drought tolerance, it thrives in Mediterranean climates and has become naturalized in South Africa. It is the source of an attractive, fragrant wood and of gum sandarac, a varnish resin. Although formerly used in construction and general carpentry, most exploitation now favors highly figured burls, which are used for carving or turning small objects and as veneers in cabinetry.

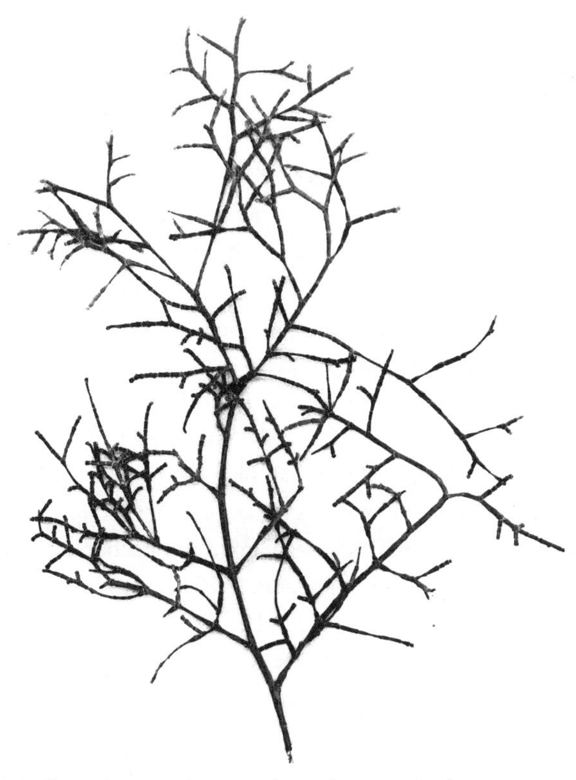

Flattened branch spray of arar-tree (*Tetraclinis articulata*) with several years of growth, ×0.5.

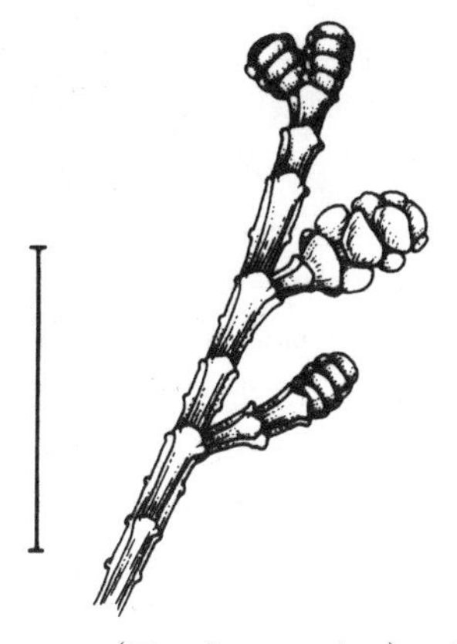

Tip of twig of arar-tree (*Tetraclinis articulata*) with mature and immature pollen cones; scale, 1 cm.

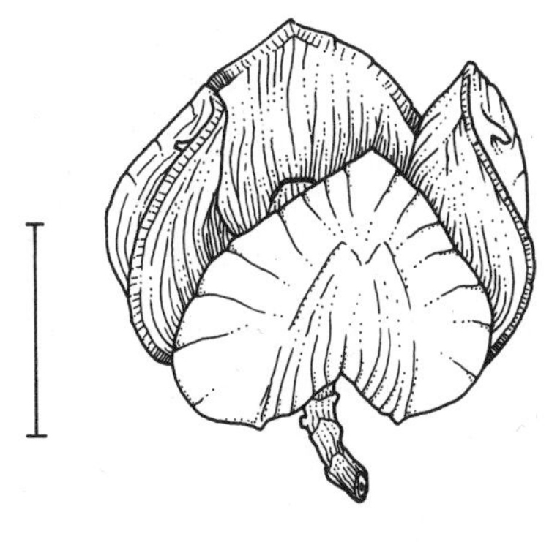

Open seed cone of arar-tree (*Tetraclinis articulata*) after seed dispersal; scale, 1 cm.

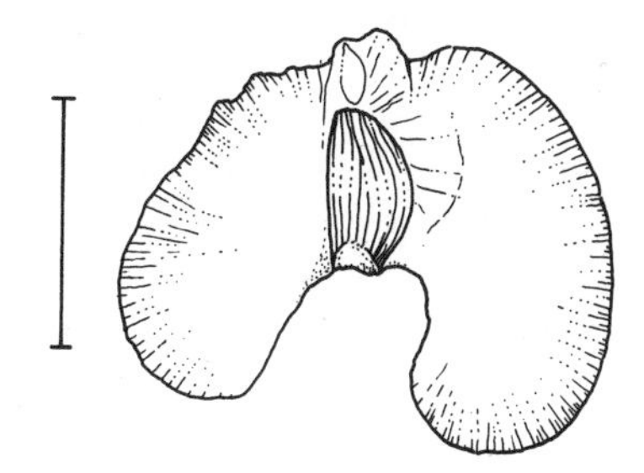

Inner side of dispersed seed of arar-tree (*Tetraclinis articulata*); scale, 5 mm.

Mature, shrubby arar-tree (*Tetraclinis articulata*) in nature.

Thuja Linnaeus

ARBORVITAE, CEDAR

Cupressaceae

Evergreen shrubs to giant trees with one to several trunks clothed with fibrous, furrowed bark peeling in vertical strips and often densely branched from the base with upwardly arcing branches, the lower branches sometimes rooting to form new trunks. Branchlets in flattened, fernlike sprays, held more or less horizontally, those contributing to the permanent framework of branches (long shoots) only weakly differentiated from the more compact, highly branched ones that are shed as intact sprays after a few years (short shoots). Virtually all branching arising from the axils of lateral leaves on alternate sides of the spray or just on the forward (away from the trunk) side, with only a tiny fraction of opposite branching. Tips of branchlets enclosed by arrested ordinary foliage leaves, thus without specialized winter buds. Seedling leaves in alternating quartets, needlelike, standing out from and well spaced on the stem, seedling phase generally short-lived, adult branchlets appearing by the second year. Adult leaves in alternating pairs, scalelike, dense, the bases running down onto and completely clothing the branchlets, with paler (silvery in *Thuja koreana*) stomatal zones beneath, the alternate pairs dissimilar. Lateral pairs keeled, usually not quite touching at their bases, thus separated by the flat successive pairs of facial leaves, which themselves barely overlap, with or without a visible resin pocket on the face.

Plants monoecious. Pollen cones numerous, single at the ends of most short branchlets on those sprays that bear them. Individual cones spherical to slightly oblong, with three to six alternating pairs of pollen scales, each with two to four pollen sacs. Pollen grains small (20–35 μm in diameter), nearly spherical, almost featureless. Seed cones numerous, single at the ends of short branchlets, maturing in a single season and remaining intact as the scales spread. Individual cones oblong, with four or six alternating pairs of thin, woody scales. Each cone scale with the bract almost fully united with the seed scale, completely covering it externally, except for a narrow rim, and ending with a small point below the tip. Middle two or three pairs of scales each usually with two (one to three) seeds. Seeds oval, with two narrow, equal wings running the whole length of the body and a little beyond, the wings derived from the seed coat. Cotyledons two, each with one vein. Chromosome base number $x = 11$.

Wood pleasantly fragrant, light, soft, and brittle but very decay resistant, with a narrow band of nearly white sapwood sharply contrasting with the light yellowish brown or pale brown to reddish brown or dull brown heartwood. Grain very even and fairly fine to somewhat coarse, with well-defined growth rings marked by a gradual to abrupt transition to a narrow to middling, dark band of much smaller, thicker-walled latewood tracheids. Resin canals absent but with sparse individual resin parenchyma cells widely scattered or concentrated in open bands in some growth increments.

Stomatal zones densely packed with stomates, covering most of the downward-facing side of all lateral leaves and broad regions on either side of the midline of only the downward-facing member of each facial pair, the upper side of lateral leaves and the upward-facing facial leaves with stomates largely confined to areas hidden by leaf overlap. Each stomate sunken beneath and partially hidden by the four to six surrounding subsidiary cells (or with a partial second ring of another one or two of them) which are topped by a steep but lobed and incomplete Florin ring. Many other epidermal cells in the stomatal zones also bearing knoblike papillae. Midvein single stranded, close to the twig axis in the attached leaf base and soon ending in the free tip, flanked by small wedges of transfusion

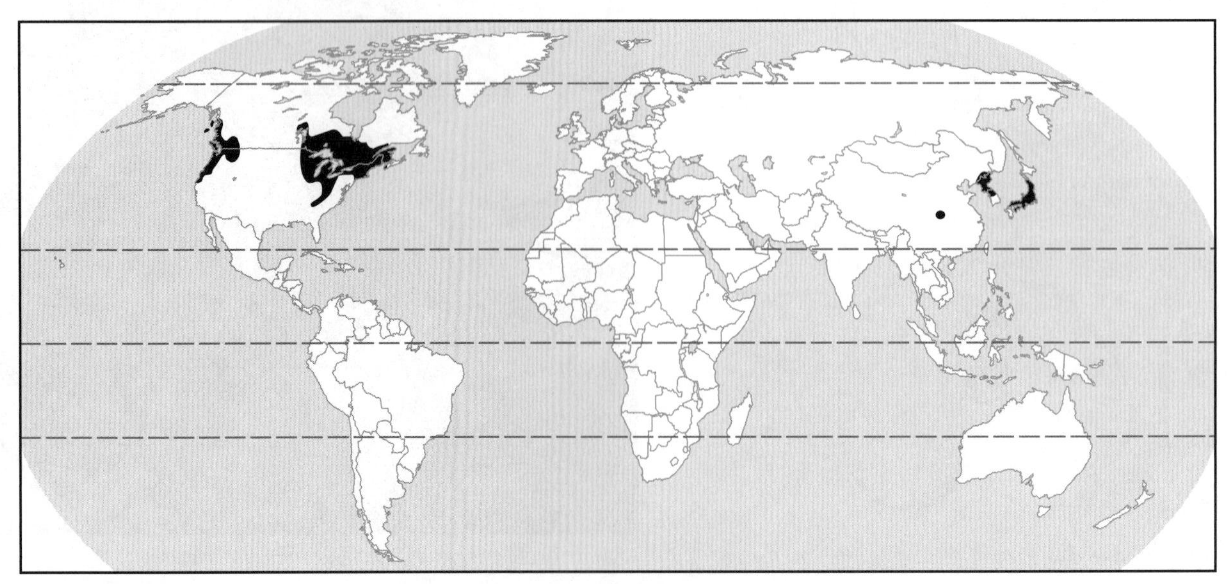

Distribution of *Thuja* in North America and eastern Asia, scale 1 : 200 million

tissue, unaccompanied by resin canals, but sometimes extending to the same level as the resin pocket (when present). Photosynthetic tissue with a well-developed palisade layer only on those leaves or faces covering the upper side of the branchlets below the epidermis and thin accompanying hypodermis, the remaining cells of the spongy mesophyll sometimes somewhat elongated but even then much looser and more open than the palisade parenchyma.

Five species in North America and eastern Asia. References: Chambers 1993, Farjon 2005b, Fu et al. 1999g, McIver and Basinger 1989, Yamazuki 1995, Xiang et al. 2002.

Thuja is an ancient Greek name for a resinous evergreen tree appropriated by Linnaeus for these decidedly non-Greek trees. The genus has been exceptionally productive of cultivars, especially in northern white cedar (*Thuja occidentalis*), with selection for variations in growth rate, size and shape of mature trees (or bushes), and arrangement of foliage sprays as well as for silver and yellow color variants. The foliage may sometimes be hard to distinguish from that of *Chamaecyparis.* This is especially true for permanently juvenile horticultural forms, and some such cultivars are not easily assigned to their proper genus.

The seed cones of *Thuja* are diagnostic and are one of the reasons that *Platycladus orientalis* is no longer included in this genus. Fossil specimens of *Thuja* with both cones and foliage have been found in Paleocene sediments (about 60 million years old) of Spitsbergen Island and Ellesmere Island, both in the arctic, in Eocene sediments (about 50 million years old) of British Columbia, and in Miocene sediments (about 20 million years old) of Japan and the Russian Far East. The oldest forms have more cone scales (eight or nine pairs) than the living species. Foliage resembling that of *Thuja* has an even longer record, extending back to the Cretaceous, but cannot be securely identified with this genus. One cone-bearing shoot from the Cretaceous of Alaska, about 90 million years old, that was recently assigned to the genus appears to be misidentified.

Morphological evidence and DNA studies show that the closest relative of the *Thuja* arborvitaes is the hiba arborvitae (*Thujopsis*) and that these two genera fall at the base of the phylogeny of the northern hemisphere subfamily Cupressoideae, sister to all other genera in the subfamily. This would make a Cretaceous record for *Thuja* (or its common ancestor with *Thujopsis*) phylogenetically plausible so, despite the inadequacy of the Alaskan record just mentioned, the genus may yet be confirmed in Cretaceous sediments.

Identification Guide to *Thuja*

The five species of *Thuja* may be distinguished by whether the facial leaves have a gland, the color of the stomatal areas beneath, and whether the leaves of the main shoots have a long point tight to the twig or a short spreading one:

Thuja occidentalis, facial leaves glandular, stomatal areas yellow-green, leaf point long and tight

T. koraiensis, facial leaves glandular, stomatal areas very white, leaf point short and spreading

T. plicata, facial leaves obscurely glandular, stomatal areas whitish green, leaf point long and tight

T. standishii, facial leaves glandless, stomatal areas grayish white, leaf point short and spreading

T. sutchuensis, facial leaves glandless, stomatal areas greenish white, leaf point short and spreading or tight

Thuja koraiensis T. Nakai

KOREAN ARBORVITAE, NUN CH'UK PAEK (KOREAN), CHAO XIAN YA BAI (CHINESE)

Tree to 10(–20) m tall or spreading shrub to 2 m at high elevations, often persistently shrubby in cultivation, with trunk to 0.3(–0.8) m in diameter. Bark scaly, reddish brown. Crown conical to dome-shaped, dense or more open with slender, widely spreading horizontal to upwardly angled branches bearing ball-like clumps of foliage. Scale leaves 1–3 mm long, dark green above, concave beneath and with very prominent white stomatal zones. Facial leaves with an obscure to prominent resin gland on the midline near the base. Leaves of main branchlets to 15 mm long, with short spreading tips. Pollen cones 1–3 mm long, with three to five pairs of pollen scales, reddish purple. Seed cones (6–)8–10 mm long, with four to six

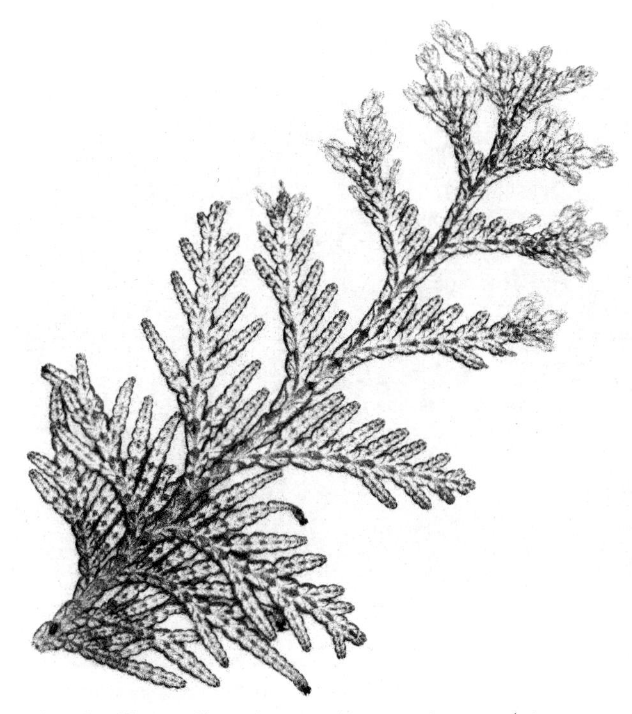

Underside of flattened branch spray of Korean arborvitae (*Thuja koraiensis*) with 3 years of growth, ×1.

pairs of scales, of which the middle two (or three) are fertile, each scale with a triangular point just below the tip. Seeds (one or) two per scale, 4–6 mm long, including the slightly longer wings, each wing slightly narrower than the body. Northern and central Korea and adjacent China (Jilin province). Scattered in mixed forests of valleys and moist mountain slopes; 700–2,000 m. Zone 6.

This tree is very slow to develop a main trunk so it remains shrubby for many years. Like other species of *Thuja,* the crushed foliage is fragrant, described for *T. koraiensis* as smelling like fruitcake. It has, perhaps, the handsomest foliage among the arborvitaes because of the striking contrast between the shiny deep green upper side of the foliage sprays and the extensive and exceptionally silvery white stomatal zones. The foliage is also finer and the sprays more compact than those of the other three commonly cultivated species. Korean arborvitae is rare, especially in China, where it is confined to the Changbai Shan (mountains), but is still cut for its hard, durable wood, valued for use in furniture and construction.

Thuja occidentalis Linnaeus

NORTHERN WHITE CEDAR, ARBORVITAE, CÈDRE BLANC, THUYA OCCIDENTAL (FRENCH)

Tree to 20(–38) m tall in favorable sites, dwarfed and slow growing under harsh conditions, with trunk to 1(–1.8) m in diameter. Bark furrowed, fibrous, reddish brown at first, weathering grayish brown peeling, in longitudinal strips, often spiraling around the trunk. Crown narrowly conical, with closely spaced, short, approximately horizontal branches turned up at the ends. Scale leaves (0.5–)1–3(–5) mm long, yellowish green on both surfaces, with paler stomatal zones. Facial leaves with a prominent gland on the midline near the tip. Leaves of main branchlets to 9 mm with long tips pressed against the twigs. Pollen cones 1–2 mm long, with two to four pairs of pollen scales, reddish brown. Seed cones (6–)8–10(–14) mm long, with four to six pairs of scales, the middle two (or three) pairs fertile, each scale with a tiny blunt point below the tip. Seeds one or two per scale, 4–7 mm long, including the slightly longer wings, each wing slightly narrower than the body. Eastern North America, from Nova Scotia to Manitoba south to North Carolina and Tennessee. Many habitats, from swamps to dry cliff faces, often on calcareous soils; 0–900 m. Zone 2.

Natural stand of northern white cedar (*Thuja occidentalis*), with two marked by arrowheads, mixed with other trees.

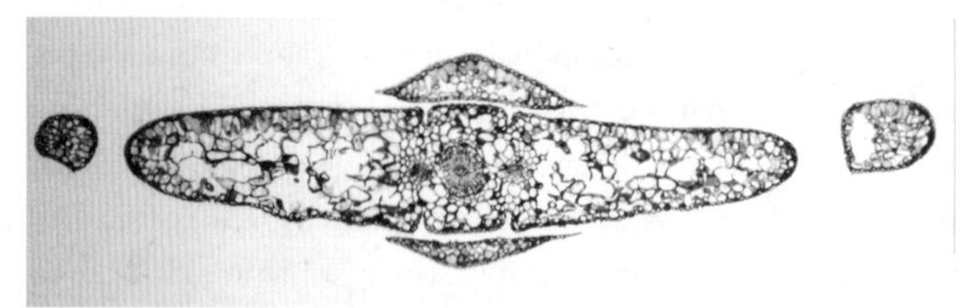

Cross section of preserved, stained, and sectioned twig of northern white cedar (*Thuja occidentalis*) with three crisscross pairs of leaves.

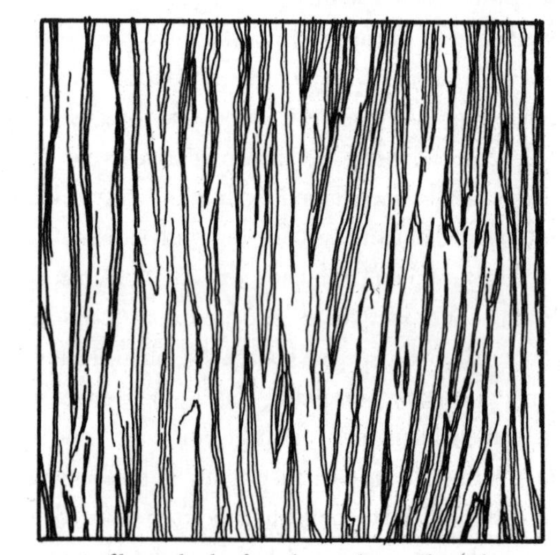

Characteristic fibrous bark of northern white cedar (*Thuja occidentalis*).

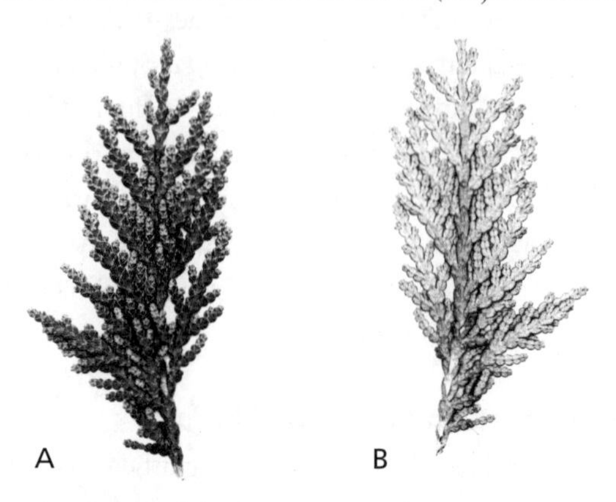

Upper (A) and undersides (B) of branch spray of northern white cedar (*Thuja occidentalis*) with 3 years of growth, ×0.5.

Partially juvenile branch spray of northern white cedar (*Thuja occidentalis* 'Sherwood Moss'), ×1.

Twig of northern white cedar (*Thuja occidentalis*) with closed mature seed cones before seed dispersal; scale, 3 cm.

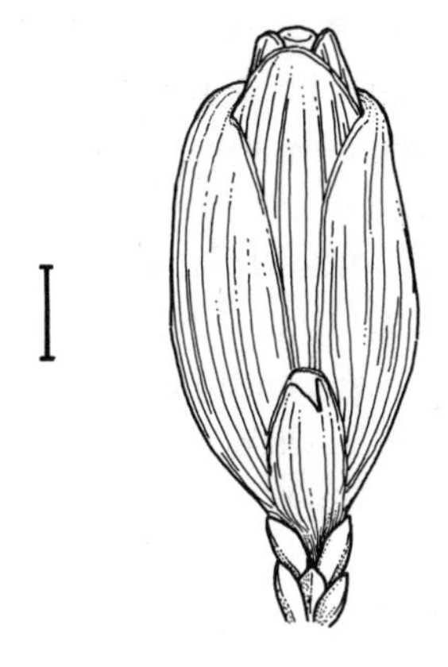

Closed mature seed cone of northern white cedar (*Thuja occidentalis*) before seed dispersal; scale, 2 mm.

Widely planted as a hedge or threshold accent within its native range, *Thuja occidentalis,* like *Chamaecyparis lawsoniana,* has given rise to hundreds of cultivars of varying stature, habit, and foliage form and color. Despite this variability in cultivation, there are no recognized botanical variants, though ecotypic variation is suspected for trees from different habitats. It is among the longest-lived of eastern North American trees, with individuals documented to approach or exceed 1,000 years (even 1,850 years) in several localities under harsh conditions. These dwarfed, gnarled, ancient trees, growing out of cliff faces, like those of the Niagara Escarpment in Ontario, often have living bark and growing branches around only a fraction of their circumference, appearing to barely cling to life.

Northern white cedar had a great number of medicinal uses, both internal and external, among native peoples within its range. Use of a fragrant decoction to fight fevers, in particular, led to its adoption by early European visitors and settlers and to the introduction of the tree to Europe by at least 1558, the earliest known North American tree introduction. During the winter of 1535–1536, Jacques Cartier provided its common name arborvitae (Latin for "tree of life") when the ravages of scurvy among the men of his Quebec expedition were lessened by the vitamin C in the foliage. The decay-resistant wood was historically important for fence posts, palisades, lodges, and canoe frames, and the bark was an important source of cordage as well as being used for the walls of buildings. Although nowhere near as important as formerly, this tree, with its decay-resistant wood, is still a major source of cedar shingles and fence posts. While northern white cedar is generally rather similar to the other North American member of the genus, western red cedar (*Thuja plicata*), the two differ dramatically in heartwood color, distinctly reddish brown in *T. plicata* and pale brown or grayish brown in *T. occidentalis.* The largest known individual in the United States in combined dimensions when measured in 1978 was 34.5 m tall, 1.8 m in diameter, with a spread of 12.8 m. The species name, Latin for "western," contrasts with the other Linnaean species of *Thuja,* now *Platycladus orientalis,* from China.

Thuja plicata D. Don

WESTERN RED CEDAR, CANOE CEDAR, GIANT ARBORVITAE

Tree to 85 m tall under favorable conditions but not more than 50 m in most of its range and occasionally stunted on extreme sites, with trunk to 5(–7) m in diameter. Bark furrowed, fibrous, reddish brown or grayish brown. Crown rigidly conical in youth, becoming irregular with numerous short branches rising from the trunk, the tips hanging in long festoons with age. Scale leaves mostly 1–3 mm long, dark green on the upper side of the sprays, with whitish green

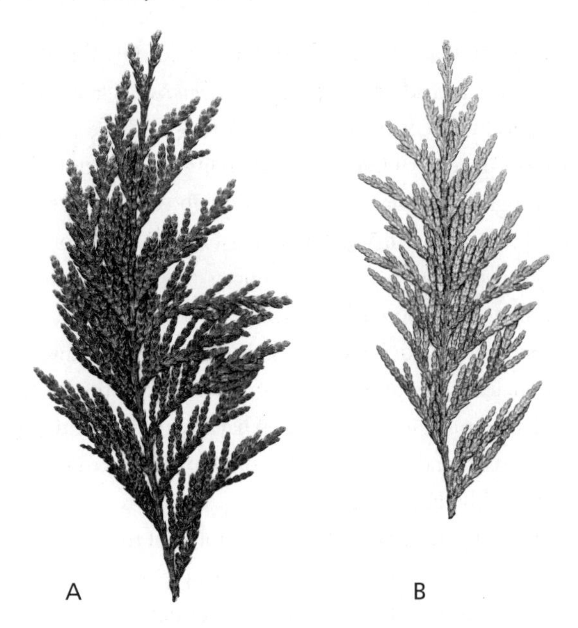

Upper (A) and undersides (B) of mature branch sprays of western red cedar (*Thuja plicata*) with 2 years of growth, ×0.5.

Mature western red cedar (*Thuja plicata*) in nature.

stomatal markings beneath. Facial leaves not evidently glandular. Leaves of main branchlets to 6 mm, with long tips pressed against the twigs. Pollen cones 1–3 mm long, with four to nine pairs of pollen scales, reddish brown. Seed cones 10–18 mm long, with five or six pairs of scales of which the middle two or three are fertile, each scale with a distinct spiny point below the tip. Seeds one or two per scale, 4–7.5 mm long, including the slightly longer wings, each wing about as wide as the body. Western North America from southern Alaska to northern California and in the Rocky Mountains from central Alberta and British Columbia to central Idaho. Mixed coastal and montane conifer forests, usually on soils that remain moist throughout the growing season; 0–1,500(–2,000) m. Zone 6, potentially lower for northern inland provenances.

A tree of great stature, age, and economic and cultural importance in the Pacific coastal forests of North America. The light, decay-resistant wood has many historical and present uses, including dugout canoes of all sizes, totem poles, and framing of beam houses. The most common, and less elegant, contemporary uses of the reddish brown heartwood are for shakes and shingles and all manner of decking. Expansion to an international market has almost depleted commercial old-growth stands, which can contain trees to 800(–2,000?) years old. The largest known living individuals are generally craggy trees, with broken tops, and are thus not the tallest of the species. These broken-topped trees may still exceed 60 m in height, their bell-bottom trunks with diameters to 7 m (one was measured at 16.7 m right at the base), and a crown spread to 16.5 m or more. Although taller or thicker individuals have been known in both Canada and the United States, the largest known living individual in combined dimensions when measured in 1999 was 48.5 m tall, 6.2 m in diameter, with a spread of 13.7 m. In addition to the wood, native peoples made extensive use of the bark for basketry and cordage. An excellent account of the role of western red cedar in native societies in the Pacific Northwest can be found in Stewart (1984). The tree has also succeeded as a plantation species in Europe and New Zealand, where it can be used on land too wet for spruce. The species name refers to the folding of the lateral scale leaves around the base of the facial ones.

Thuja standishii (G. Gordon) Carrière

JAPANESE ARBORVITAE, KUROBE, NEZUKO (JAPANESE)

Tree to 20(–35) m tall, with trunk to 1(–2) m in diameter. Bark furrowed, fibrous, reddish brown, peeling in flakes or longitudinal strips, even in very young plants. Crown conical, with upwardly angled branches bearing branchlets that are less flattened than in other species. Foliage lemon-scented when crushed. Scale leaves 2–3 mm long, dark green above, with grayish white stomatal zones beneath. Facial leaves glandless. Leaves of main branchlets

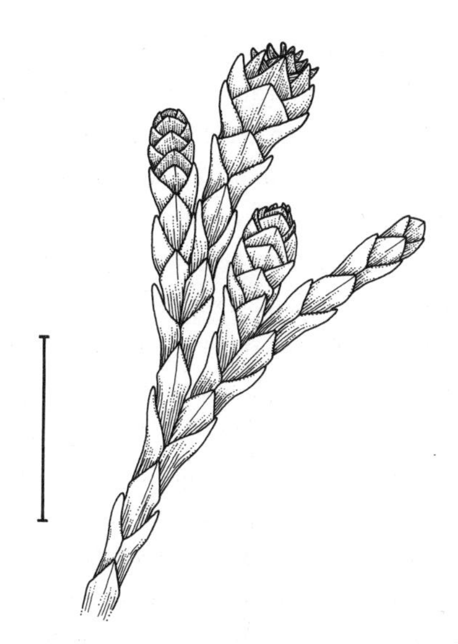

Twig of Japanese arborvitae (*Thuja standishii*) with pollen and seed cones at about the time of pollination; scale, 1 cm.

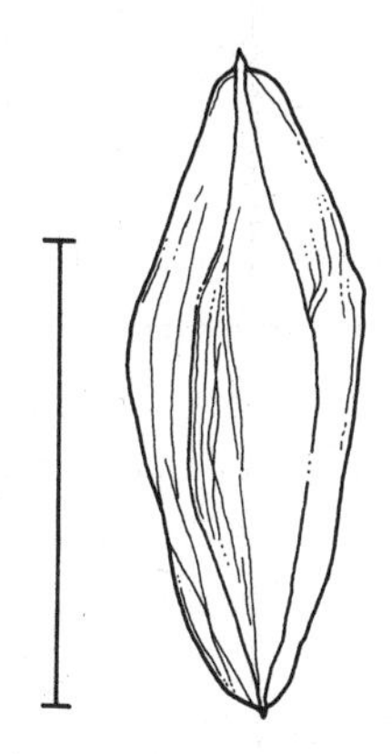

Upper side of dispersed seed of Japanese arborvitae (*Thuja standishii*); scale, 3 mm.

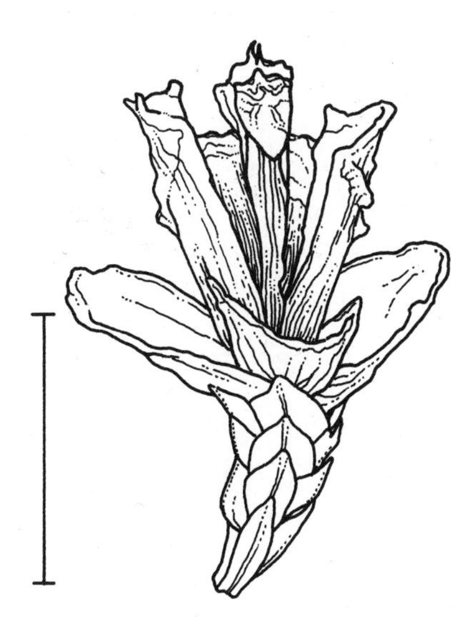

Open seed cone of Japanese arborvitae (*Thuja standishii*) after seed dispersal; scale, 5 mm.

with short, spreading tips. Pollen cones yellow, 1–3 mm long, with four to eight pairs of pollen scales. Seed cones 8–13 mm long, with four to six pairs of scales, the middle two pairs fertile, each scale with a short point below the tip. Seeds one or two (to three) per scale, 5–7 mm long including the slightly longer wings, each wing conspicuously narrower than the body. Restricted to Japan where widespread on Honshū and Shikoku. Moist rocky mountain slopes; (25–)900–1,800(–2,500) m. Zone 7.

Japanese arborvitae is widely cultivated in Japan and elsewhere but rare in nature. In contrast to the North American species in the genus, northern white cedar (*Thuja occidentalis*) and western red cedar (*T. plicata*), little selection of cultivars has been undertaken. While the wood, like that of other species in the genus, is appreciated for shingles and for interior wood paneling, the tree is too rare to be of any present commercial importance. The species name honors John Standish (1814–1875), proprietor of Sunningdale Nursery in Ascot, England, to whom Robert Fortune sent seed in 1860 that he had collected in Tokyo gardens, along with seed of many other species from Japan and China.

Thuja sutchuensis Franchet

SICHUAN ARBORVITAE, YA BAI (CHINESE)

Shrub, or tree to 10(–20) m tall, with trunk to 0.3 m in diameter. Bark bright reddish brown to grayish brown, fibrous, furrowed, peeling in short, curly flakes. Crown narrowly conical, with horizontal to upwardly angled, dense, slender, widely spreading branches bearing horizontal to slightly drooping branchlets. Scale leaves 1.5–4(–9) mm long, shiny bright green, with greenish white stomatal markings beneath. Facial leaves without a surface gland but with an internal resin cavity. Leaves of main branchlets with blunt, spreading or tight tips. Pollen cones 2–3 mm long, with three to five pairs of pollen scales, yellowish brown. Seed cones 5–8 mm long, with four or five pairs of seed scales, of which the middle two or three are fertile, each scale with a bluntly triangular point below the tip. Seeds 3–4 mm long, including the equal-length wings, each wing much narrower than the body. Known only from a single population on the southern side of the Daba Shan (mountains) in Chongqing Municipality, formerly part of northeastern Sichuan (China). Scattered among shrubs or in the understory of mixed deciduous and evergreen hardwood forests on steep, moist slopes and ridge tops, often on limestone-derived soils; (800–)1,000–1,500(–2,100). Zone 7?

Until its rediscovery in 1999, Sichuan arborvitae had not been seen by botanists since 1900, despite repeated searches, and was presumed extinct. Local people, of course, had been exploiting the species all along for applications relying on its durability in the open and in the ground, like shingles and coffins. Essentially all large trees of easy access in the population have been cut, leaving only shrubby individuals and some trees on slopes too steep and rugged to reach. Sichuan arborvitae has the smallest seed cones in the genus. Its largest cones barely reach the length of the smallest fully developed cones of all of the other species. The scientific name is a latinization of Sichuan.

Thujopsis P. Siebold & Zuccarini

HIBA ARBORVITAE

Cupressaceae

Evergreen trees, often bushy when young. Branches almost horizontal or drooping with age from one or more trunks clothed with fibrous, furrowed bark. Branchlets strongly flattened into fernlike sprays with distinct upper and lower surfaces. Without definite winter buds. Leaves in alternating pairs, needlelike in the short-lived seedling and juvenile phase, which gives way during the second and third years to adult branchlet sprays. Adult leaves scalelike, densely clothing the twigs, with a small resin pocket near the tip, dark green with white stomatal patches above and with conspicuous large waxy white stomatal zones beneath. Leaves of branchlets differentiated into lateral and facial pairs. Leaves of each lateral pair very broad and hooked at the tip, completely separated by the facial pair and seemingly inserted at the same level as them, with a thickened green rim surrounding the stomatal zone beneath. The underside member of the facial pair with a thickened green rim running between the lateral leaves at its tip and also with a thickened green keel.

Plants monoecious. Pollen and seed cones relatively few, single at the ends of short lateral shoots, cupped closely around the base by three or four pairs of green bracts transitional to the foliage leaves of the supporting shoot. Pollen cones cylindrical with six to eight alternating pairs of pollen scales, each scale with three or four pollen sacs. Pollen grains small (25–35 μm in diameter), spherical, largely featureless. Seed cones maturing in a single season and then gaping to release the seeds, almost spherical, with four (or five) alternating pairs of seed scales. Cone scales woody, with a large pyramidal knob occupying most of the face and with little outward sign of the boundary between the united outer bract and inner seed scale portions. Lower two pairs of scales with five seeds each, the upper pairs with one to three seeds. Seeds ellipsoid, narrowly two-winged, the wings arising as outgrowths of the seed coat. Cotyledons two, each with one vein. Chromosome base number $x = 11$.

Wood fragrant, light, soft, and brittle but decay resistant, the whitish sapwood not strongly differentiated from the yellowish to light brown heartwood. Grain fine and even with occasional coarse streaks, the obvious growth rings marked by a (usually) gradual transition to a fairly narrow band of smaller but not particularly thicker walled latewood tracheids. Resin canals absent but with numerous individual resin parenchyma cells scattered throughout the growth increment or somewhat concentrated in open bands.

Stomates sunken beneath and about half hidden by the five to seven surrounding subsidiary cells, which are often shared by adjacent stomates. Rim of stomate topped by a complete, lobed Florin ring. Subsidiary cells and other epidermal cells of the stomatal zone also often carrying additional knoblike papillae of raised cuticle. Midvein single-stranded, thin, petering out in the free tip of the leaf, flanked by wedges of transfusion tissue, and often accompanied on its outer side in the attached base by a large resin canal that stops before the embedded resin pocket of the leaf tip. Photosynthetic tissue with a (sometimes partial and ill-defined) palisade layer beneath the epidermis and variably developed hypodermis only on the upper side of the branchlets, the remaining space down to the lower side of the branchlets, including the whole lower facial leaf occupied by very loose spongy mesophyll with a scattering of sclereids.

One species in Japan. References: Farjon 2005b, Gadek et al. 2000, Takahashi et al. 2003, Yamazaki 1995.

While fairly hardy, hiba arborvitae (*Thujopsis dolabrata*), has a high moisture requirement and is successfully cultivated primarily in areas with a humid climate. Unlike many other ornamental Japanese plants, there has been little cultivar selection in its homeland or elsewhere, with just a handful of cultivars displaying variegated foliage or variations in habit and form from dwarf buns to narrower, faster growing columns.

The close relationship between *Thujopsis* and typical arborvitaes (*Thuja*) has been recognized since the sole extant species

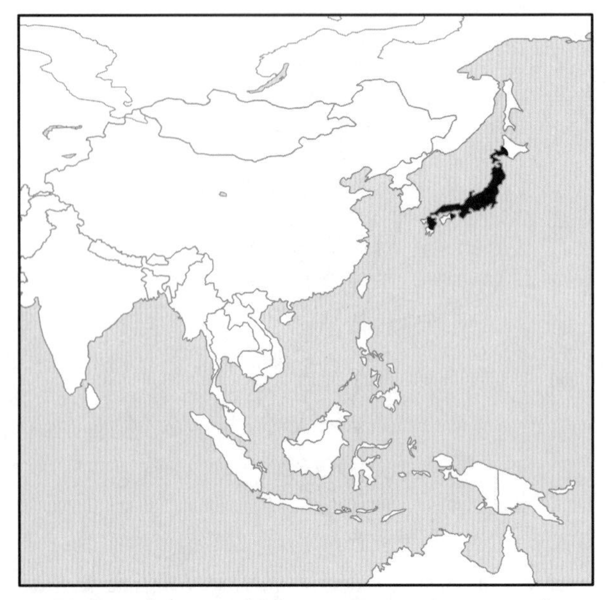

Distribution of *Thujopsis dolabrata* in Japan, scale 1 : 120 million

was first botanically described in 1782 as a species of the latter genus, and this relationship is reflected in the generic name (Greek for "looks like *Thuja*"), not proposed until 60 years later, in 1842. Halfway in between, in 1817, Richard Salisbury was the first to propose a separate generic name for this species, but his sketchy description prevented taxonomists of the time from recognizing what he was talking about, and his name (*Dolophyllum*) was never taken up in the literature. Salisbury's name was finally officially rejected in favor of *Thujopsis* at the Sixteenth International Botanical Congress in Saint Louis in 1999. Although no DNA studies have been specifically designed to assess the relationship between *Thujopsis* and *Thuja*, many broader studies of Cupressaceae included species of these two genera. All studies agree that they are each other's closest living relatives, while most also show that, together, these two genera are the sister group to all the other northern hemisphere members of subfamily Cupressoideae. They are not, in any event, especially closely related to the superficially similar northern incense cedars (*Calocedrus*) or to Oriental arborvitae (*Platycladus orientalis*), which itself is even today still often included as a species of *Thuja*, especially in horticultural references. The hiba arborvitae was even briefly included in *Platycladus* in the mid-19th century, at about the time that it was independently placed in the new separate genus *Thujopsis*.

The known fossil record for *Thujopsis* is limited. Many specimens from Tertiary sediments in Europe and a few from North America were assigned to the genus in the 19th century and the first half of the 20th, but none of these reports seems correct. Instead, the only reliable paleobotanical records for the genus are from Japan, to which the genus is restricted (endemic) today. The oldest of these date to the Miocene, more than 10 million years ago, in Hokkaidō, north of the present natural distributional limits of the genus.

Thujopsis dolabrata (Linnaeus) P. Siebold & Zuccarini

HIBA ARBORVITAE, ASUNARO (JAPANESE)

Tree to 30 m tall, with trunk to 0.6 m in diameter, single or forking at the base or above. Bark reddish brown, shedding in narrow vertical strips. Crown conical, dense, with numerous horizontal branches turned up at the ends and sometimes tilted to the sides. Twigs completely hidden by and deeply grooved between the attached bases of the scale leaves, the texture of the scales on the branchlets unusually reminiscent of reptile skin. Scale leaves 4–5(–8) mm long, shiny dark green on the upper sides of the foliage sprays, with large, bright white waxy patches over the stomatal zones beneath. Pollen cones (3–)5–6 mm long, with four or five pairs of pollen scales, reddish brown with a purple blush. Seed cones 12–15 mm long and wide, somewhat fleshy and covered with a waxy film when young, hardening and darkening with maturity. Seeds 4–5 mm long including the equal-length wings, each wing a little narrower than to as wide as the body. Japan, all four main islands from southern Hokkaidō southward. Rocky open woods, 100–2,000 m. Synonym: *Thujopsis dolabrata* var. *hondae* Makino.

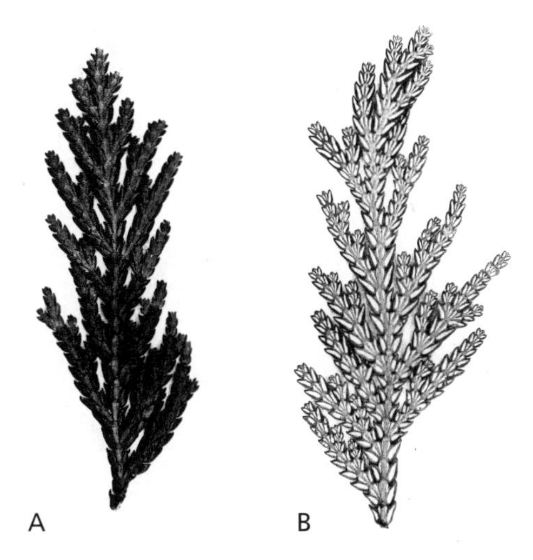

Upper (A) and undersides (B) of branch sprays of hiba arborvitae (*Thujopsis dolabrata*), ×0.5.

Vigorous mature hiba arborvitae (*Thujopsis dolabrata*) in nature.

Open seed cone of hiba arborvitae (*Thujopsis dolabrata*) after seed dispersal; scale, 1 cm.

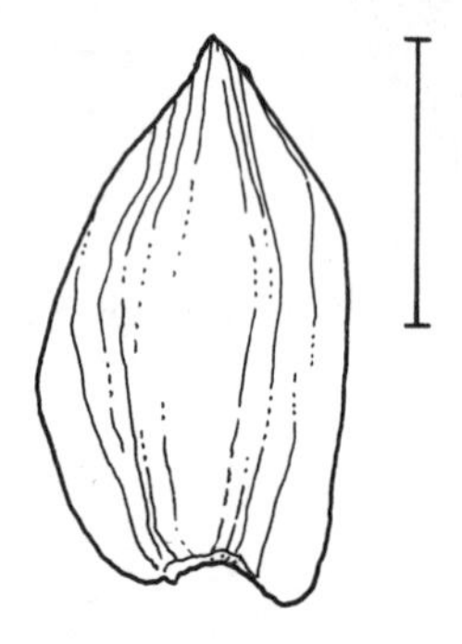

Underside of dispersed seed of hiba arborvitae (*Thujopsis dolabrata*); scale, 2 mm.

Hiba arborvitae is one of the most distinctive of the cupressaceous "cedars" with its large, dark green scale leaves and strongly contrasting white patches. It is often divided into a northern variety *hondae* of Hokkaidō and northern Honshū and a southern variety *dolabrata,* but the two seem scarcely distinct. In addition to its ornamental value, hiba arborvitae is an important timber tree in Japan, with many uses for its wood. In part because of this, there have been several studies of genetic variation in the species.

Torreya G. Arnott

NUTMEG YEW, TORREYA

Taxaceae

Evergreen trees and shrubs with one to few cylindrical to off-center trunks. Bark fibrous, smooth at first, later flaking, and finally peeling in long, thin strips and becoming variously ridged and furrowed. Crown dense and conical in youth, broadening and becoming more irregular with age, initially with regularly spaced tiers of three to seven horizontal to gently rising branches bearing pairs of horizontal to drooping side branchlets. Branchlets all elongate, without distinction into long and short shoots, hairless, remaining green for up to 2 years or turning brown within the first year, completely clothed by and slightly to prominently grooved between the elongate, attached leaf bases. Resting buds well-developed, with about eight crisscross pairs of triangular, brown bud scales that are shed during shoot expansion. Leaves attached in opposite pairs with successive pairs spiraling around the twigs but presented in two ragged rows by bending and twisting of the petioles, needlelike, variously sword-shaped, straight or slightly curved forward, flattened top to bottom, stiff and leathery with a hard, often prickly tip, usually strongly and often pungently aromatic. Midrib usually inconspicuous above, sometimes flanked by grooves or ridges, broadly and shallowly to prominently raised beneath and flanked by flat or slightly depressed whitish green or brownish stomatal bands that are flanked, in turn, by green marginal bands, the leaf edges flat or slightly and narrowly turned down.

Plants dioecious. Pollen cones forming a double row on the lower side of the current growth increments in the axils of ordinary foliage leaves, extending from the lowest pair of leaves to the tip or only along a portion of the branchlet. Each pollen cone with five to eight alternating pairs of tightly packed bud scales on the stalk beneath five to nine alternating whorls of pollen scales and often one terminal one. Each pollen scale with three to five dangling pollen sacs and a ragged upturned tip (and up to seven pollen sacs all around the terminal pollen scale). Pollen grains small (25–35 μm in diameter), roughly spherical but a little squared off and pyramid-shaped, with an ill-defined germination zone at the narrower end, seemingly smooth, but with extremely small, complexly spiny lumps all over the surface. Seed cones forming a short, often interrupted double row on the lower side of reduced or relatively unmodified current growth increments, or sometimes just a single pair. Each seed cone highly reduced, not obviously conelike, without seed scales, consisting of a short reproductive axis with a pair of bracts in the axil of an ordinary foliage leaf. Each of these primary bracts with an axillary fertile shoot bearing two crisscross pairs of bracts at the base of a single upright seed, usually only one seed of the pair maturing, but sometimes with a third seed at the tip of the reproductive axis. Seeds large, the thin fleshy outer seed coat surrounded by and united with a resinous, fleshy aril for most of its length, maturing in two seasons after which the fleshy layers split along one side to release the remainder of the seed within the hard middle stony layer. Cotyledons two, each with one vein. Chromosome base number $x = 11$.

Wood light, variably soft to moderately hard, silky in texture, with a modest contrast between the creamy white to yellowish or light brown sapwood and the yellowish brown to dark brown heartwood. Grain fine and fairly even, with obvious growth rings

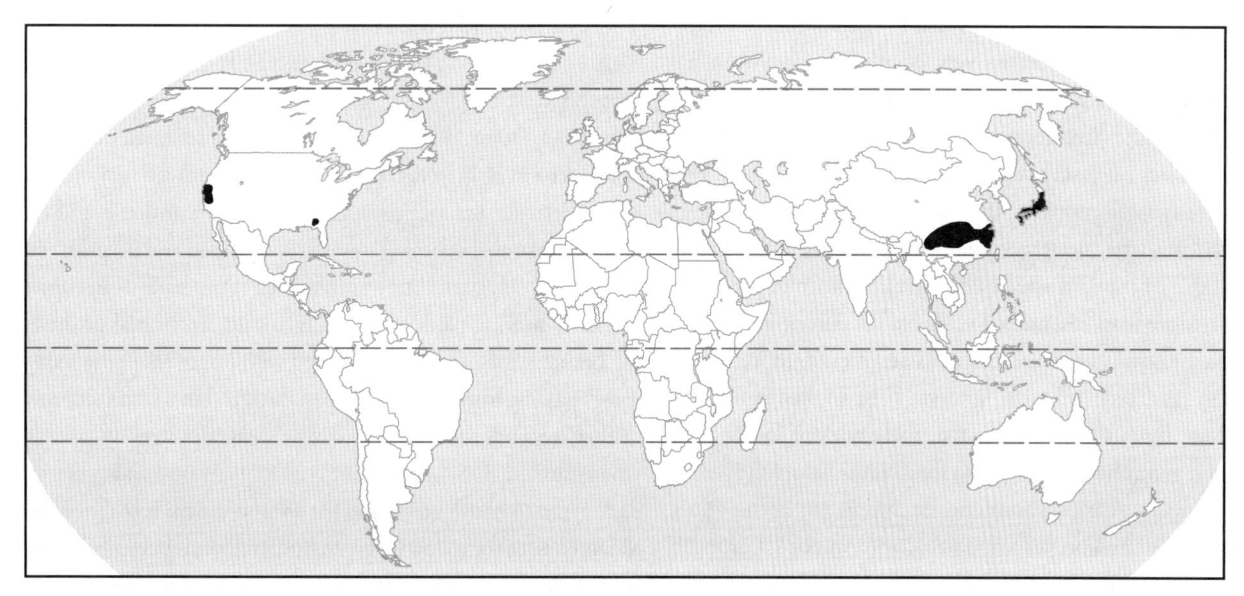

Distribution of *Torreya* in North America, China, and Japan, scale 1 : 200 million

generally marked by a very narrow band of darker latewood. Resin canals absent but with a very sparse scattering of individual resin parenchyma cells, the tracheids (wood cells) with double or triple spiral thickenings, or these too closely spaced to tell.

Stomates densely crowded within the bands so that they often share some of their single circle of 8–10 surrounding subsidiary cells, aligned with the long axis of the leaf but not arranged in distinct lines. Each stomate surrounded by a Florin ring and many other epidermal cells also bearing papillae. Midvein single, surrounded by upper and lower caps of sclerenchyma, with one resin canal embedded among these small, round sclereids immediately beneath the midvein, which is flanked by small bands of transfusion tissue. Photosynthetic tissue with a single or double palisade layer lining the whole upper surface and with horizontal spongy mesophyll filling the blade on either side of the midvein and accompanied by branched sclereids in some species.

Six species in eastern Asia and southern North America. References: Fu et al. 1999e, Hils 1993, Kang and Tang 1995, Li et al. 2001b, Yamazaki 1995.

The wood of most species of *Torreya* is quite durable in contact with soil, so they are frequently used for fence posts, while other timber uses, such as furniture or framing, are much restricted by the generally small size of the trees. Sprouting from cut stumps can ensure a continuing supply of the timber under a coppice system. The large seeds have been roasted and eaten in both eastern Asia and North America, and cultivars were developed in both Japan and China for different seed qualities, including oil content and uses in herbal medicine. A modest number of additional cultivars were selected for dwarf or spreading growth habits and for yellowish foliage. The genus name honors John Torrey (1796–1873), a prominent plant taxonomist at the New York Botanical Garden, who formally named and described California nutmeg about 15 years after the genus was named for him based on the stinking cedar.

The nutmeg yews derived this common name from a modest resemblance of their seeds (visually but not in flavor) to the fruits of nutmeg (*Myristica fragrans*), an unrelated plant. The similarity rests on the splitting husk (aril and fleshy layer of seed coat in nutmeg yew, fruit wall in nutmeg), revealing a hard shell (stony layer of seed coat in nutmeg yew, seed coat in nutmeg) surrounding a convoluted nutritive tissue (female gametophyte in nutmeg yew, endosperm in nutmeg). Nutmeg also has an aril (the mace), but this lies between the husk and the hard shell. The convoluted female gametophyte is quite variable in the different species of *Torreya* and, in some, is quite smooth. Whether it is smooth or wrinkled was used to divide the genus into two botanical sections, but DNA studies show that such a division does not accord with the phylogeny of the species. Instead, the two New World species, one of which has the wrinkles (stinking cedar, *T. taxifolia*) and one of which does not (California nutmeg, *T. californica*), are each other's closest relatives, to the exclusion of the Old World species, which also intermix species with convoluted and smooth gametophytes. Two of the contrasting Asian species, Japanese nutmeg yew (*T. nucifera*), with fairly smooth female gametophytes, and Farges torreya (*T. fargesii*), with the most convoluted female gametophytes in the genus, are essentially identical to each other in one genetic region (the internal transcribed spacer, ITS, region of the ribosome genes) sampled for all species. In fact, all four Asian species are much more similar to one another in their ITS DNA sequences than the two New World species are to each other, suggesting a much more recent divergence.

While even the divergence between the North American and Asian species suggests a separation no longer than about 30 million years ago, the genus appears to be much older. There are fossils with leaf shape and arrangement and epidermal structures just like those of modern species of *Torreya* from mid-Jurassic sediments (more than 160 million years old) in Europe and eastern Siberia and from Lower Cretaceous sediments (more than 100 million years old) in North America. Despite their similarities to extent *Torreya,* it is possible that these ancient fossils are some related genus because there is a long hiatus before the next fossil occurrences. A fairly continuous record began in the Oligocene (about 30 million years ago) in North America, Europe, and eastern Asia. The genus only became extinct in Europe after the Pliocene (less than 2 million years ago) and is still present in North America and eastern Asia.

The bijugate leaf arrangement found in *Torreya,* with spiraling pairs of leaves, is rather rare in plants and, among the conifers, is found elsewhere only in plum yews (*Cephalotaxus*), in another subfamily within the Taxaceae. The closest relatives of the nutmeg yews, as demonstrated by DNA studies and some morphological features, the catkin yews (*Amentotaxus*), with which they share a subfamily, have the more common decussate leaf arrangement, with crisscross leaf pairs forming four rows.

Identification Guide to *Torreya*

The six species of *Torreya* may be distinguished by the length of longer ordinary needles, the fragrance of the crushed foliage, and the contour of the leaves adjacent to the midrib on the upper side:

Torreya nucifera, needles up to 2.5 cm, aromatic, flat adjacent to midrib

T. grandis, needles up to 2.5 cm, unscented, flat or flanked by grooves to about a third of their length adjacent to midrib

T. fargesii, needles 2.5–4 cm, unscented to somewhat aromatic, flanked by grooves to the midpoint or beyond adjacent to midrib

T. taxifolia, needles 2.5–4 cm, sharply and unpleasantly aromatic, flat adjacent to midrib

T. californica, needles more than 4 cm, aromatic, flat adjacent to midrib

T. jackii, needles more than 4 cm, strongly aromatic, flanked by slight ridges adjacent to midrib

Torreya californica J. Torrey

CALIFORNIA NUTMEG, CALIFORNIA TORREYA

Plate 66

Tree to 20(–30) m tall, with trunk to 1(–2) m in diameter, or a shrub under 4 m in chaparral. Bark gray, deeply furrowed between narrow, interlacing, flat-topped ridges. Crown conical, dense, and very regular in youth, becoming more dome-shaped and irregular with age, with closely spaced, upwardly angled to more nearly horizontal branches bearing pairs of upwardly angled, horizontal or drooping branchlets fairly densely clothed with nonoverlapping foliage. Twigs passing from yellowish green through tan to reddish brown in their second, third, or fourth year. Leaves pungently aromatic when crushed, 3–5(–8) cm long, 1.5–2.5(–3) mm wide, very stiff, dark green to bluish green and flat around the midrib above, with indented white or yellowish green stomatal bands beneath. Individual needles straight or slightly curved, widest before the middle, tapering gradually and then more abruptly to the roundly triangular tip with a sharp point 0.5–1.2 mm long, and more abruptly to the rounded base on a very short petiole 2–3 mm long. Pollen cones white or pale yellow, about 7–10 mm long and 3.5–5 mm across. Seeds with aril pale green to whitish green streaked with purple, 2.5–3.5 cm long and 2–3 cm in diameter. Female gametophyte tissue inside the shiny yellowish brown stony layer deeply to only

Underside (and upper) of twig of California nutmeg (*Torreya californica*) with 2 years of growth, ×0.5.

slightly wrinkled. Scattered in the Coast Ranges and on the western slopes of the Sierra Nevada of central California. Scattered or forming small groves in mixed riparian evergreen and deciduous montane forests along streamsides and in chaparral shrublands on slopes; (100–)500–1,800(–2,300) m. Zone 7.

California nutmeg may not be rare, but it is far from common, the widely scattered locations often containing solitary trees or few individuals. Nonetheless, the species has a long history of human uses in California. Like common yew (*Taxus baccata*) in Europe, California nutmeg was used by native people for making bows, reflecting the lightness, elasticity, toughness, and durability of the wood that is fairly common in the yew family. The durability is responsible for most commercial and local usage of the wood for fence posts, bridges, and other outdoor applications, although it is also appreciated in woodworking and furniture making for its pleasant, even color and ability to take a fine finish. Although California nutmeg can reach the largest size among the species of *Torreya*, timber harvesting in the late 19th to mid-20th centuries left few if any stands of large trees. The largest known individual in combined dimensions when measured in 1992 was 29.3 m tall, 2.0 m in diameter, with a spread of 20.7 m. Use of the large seeds for food was to be expected by people for whom agriculture was not the primary means of subsistence. Less obvious uses included stripping of fibers from the roots for basketry and using the sharp needle tips for tattooing. Although California nutmeg has been in cultivation as a handsome, symmetrical evergreen since its botanical description and naming by John Torrey in the 1850s, it is not commonly planted and only a few cultivars have been selected and propagated over the years. Hybrids with Japanese nutmeg yew (*T. nucifera*) have also appeared in gardens but are little cultivated.

Twig of California nutmeg (*Torreya californica*) with mature seed cone; scale, 1 cm.

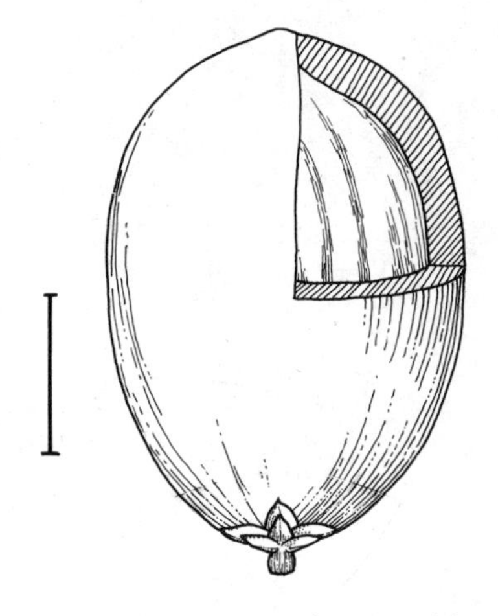

Mature seed cone of California nutmeg (*Torreya californica*) with aril partially cut away to show contained seed; scale, 1 cm.

Torreya fargesii Franchet

FARGES TORREYA, BA SHAN FEI SHU (CHINESE)

Shrub, or tree to 20 m tall, with trunk to 1 m in diameter. Bark light brown and flaking initially, becoming dark gray or grayish brown and deeply furrowed between narrow, irregular ridges. Crown dense, becoming broadly and irregularly dome-shaped with age, with horizontal to strongly downswept branches bearing pairs of horizontal or drooping branchlets densely clothed with slightly separated foliage. Twigs passing through greenish yellow to grayish yellow or yellowish brown in their second and third years. Leaves not especially aromatic when crushed, (1–)2–3.5(–4) cm long, 2–4 mm wide, stiff and slightly drooping, shiny dark green and with grooves flanking the midrib above, with deeply sunken brown stomatal bands beneath. Individual needles straight or curved, widest before the middle, tapering gradually to the narrowly to roundly triangular tip with a sharp point 0.3–1 mm long, and more abruptly to the rounded, often asymmetrical base on a very short petiole 0.5–1 mm long. Pollen cones pale yellow, about 5–6 mm long and 4–5 mm across. Seeds with aril pale green or whitish green with a thin waxy film, 1.5–2.5 cm long and thick. Female gametophyte tissue inside the shiny light brown, stony layer deeply wrinkled. In scattered localities across central China, from Anhui and Jiangxi to southern Shaanxi to northwestern Yunnan. Singly or in small groves in various montane and subalpine forest types along streamsides or on slopes; 1,000–3,400 m. Zone 7. Two varieties.

Farges torreya is the most widely distributed nutmeg yew in China, but known localities are widely separated and the tree is uncommon or rare wherever it occurs. It produces a high-quality timber that, because of its durability in water and in contact with the soil, is used for house and bridge construction as well as for making furniture and hand tools. Because of the quality of the wood, inclusion of the species in reforestation projects in

China has been advocated if not much implemented. As with the rather similar Chinese nutmeg yew (*Torreya grandis*), an oil is extracted from the seeds of this species. Farges torreya is easily distinguishable from all other species by the grooves paralleling the flat midrib for half or more of its length on the upper side of the leaf. Although Farges torreya is similar to Chinese nutmeg yew in so many features that the two species were often confused in the past, DNA studies suggest that it is actually more closely related to Japanese nutmeg yew (*T. nucifera*), which is readily distinguishable from both of these Chinese species by its pungently aromatic crushed foliage and reddish brown third-year branchlets. The plants from northwestern Yunnan are a little different from those from other parts of the range and have been distinguished as a separate species or, as here, a variety. The species name honors Paul Farges (1844–1912), a French priest and botanist who collected the type specimen during his missionary work in Sichuan.

Torreya fargesii Franchet var. *fargesii*

BA SHAN FEI SHU (CHINESE)

Leaves predominantly straight, with grooves flanking the midrib to the midpoint above and with the green marginal bands flanking the stomatal grooves mostly 0.5–1 mm wide. Seeds with the aril mostly 2–2.5 cm long. Throughout the range of the species except in Yunnan. Synonym: *Torreya grandis* var. *fargesii* (Franchet) Silba.

Torreya fargesii var. *yunnanensis* (W. C. Cheng & L. K. Fu) N. Kang

YUN NAN FEI (CHINESE)

Leaves predominantly gently to strongly curved forward, with grooves flanking the midrib to beyond the midpoint above and with the green marginal bands flanking the stomatal grooves mostly 1–1.2 mm wide. Seeds with the aril mostly 2.5–3 cm long. Northwestern Yunnan. Synonyms: *Torreya grandis* var. *yunnanensis* (W. C. Cheng & L. K. Fu) Silba, *T. yunnanensis* W. C. Cheng & L. K. Fu.

Torreya grandis Fortune ex Lindley

CHINESE NUTMEG YEW, FEI SHU (CHINESE)

Tree to 25 m tall, with trunk to 0.5(–2) m in diameter. Bark yellowish brown to grayish brown, becoming dark gray and deeply furrowed between narrow, irregular ridges. Crown dense, becoming broadly conical to broadly dome-shaped with age, with upwardly angled to horizontal branches bearing pairs of horizontal or drooping branchlets densely clothed with slightly separated foliage. Twigs passing through greenish yellow to grayish yellow or yellowish brown in their second and third years. Leaves scarcely scented when crushed, (0.7–)1–2.5(–4.5) cm long, 2–3.5 mm

Large mature Farges torreya (*Torreya fargesii* var. *yunnanensis*) in nature.

Young Chinese nutmeg yew (*Torreya grandis*) in nature.

wide, stiff and slightly drooping, shiny bright to dark green and usually flat around the midrib above or with shallow grooves extending less than halfway and with sunken brown stomatal bands beneath. Individual needles usually straight and nearly parallel-sided, though slightly wider near the base, tapering very gradually and then abruptly to the roundly triangular or rounded tip with a sharp point and abruptly to the broadly rounded, nearly symmetrical base on a very short petiole 0.5–1 mm long. Pollen cones pale yellow, about 7–8 mm long and 4–5 mm across. Seeds with aril dark green to pale purplish brown with a thin, white waxy film, (1.8–)2–4.5(–6) cm long and (1–)1.2–2.5 cm in diameter. Female gametophyte tissue inside the light brown stony layer only slightly wrinkled. In scattered localities across east-central China, from southern Jiangsu to northern Fujian west to northeastern Guizhou. Solitary or forming small groves in various temperate forest types in the mountains and adjacent lowlands; 200–1,400 m. Zone 8. Synonyms: *Torreya grandis* var. *dielsii* H. H. Hu, *T. grandis* var. *jiulongshanensis* Zhi Y. Li, Z. C. Tang & N. Kang, *T. grandis* var. *merrillii* H. H. Hu, *T. grandis* var. *sargentii* H. H. Hu, *T. nucifera* var. *grandis* (Fortune ex Lindley) Pilger.

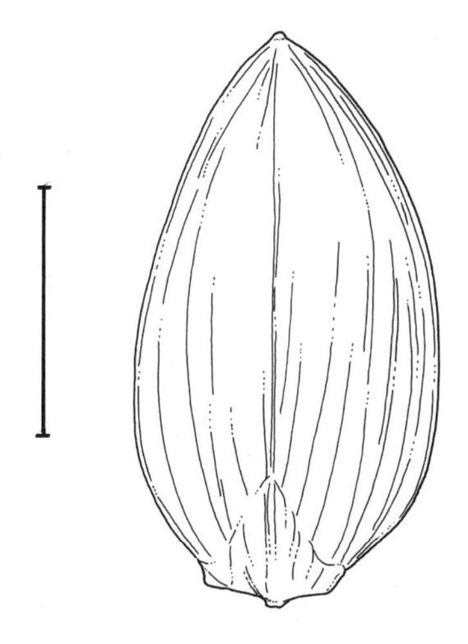

Side view of stony layer of seed of Chinese nutmeg yew (*Torreya grandis*); scale, 1 cm.

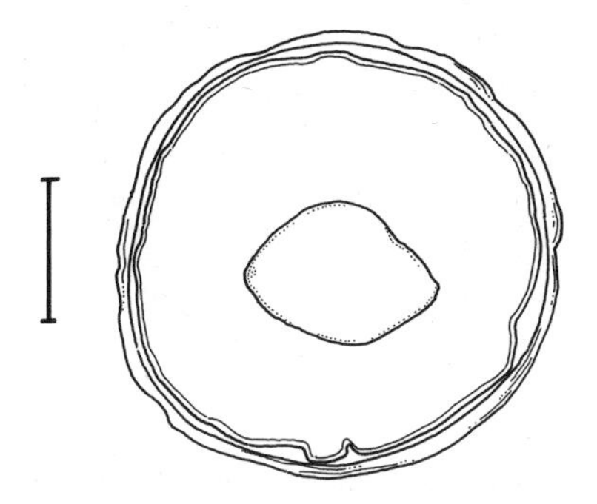

Cross section of seed of Chinese nutmeg yew (*Torreya grandis*); scale, 5 mm.

Chinese nutmeg yew has long been cultivated for its edible seeds (*xiangfei*), from which are also derived an edible oil, a medicine (*feishu*), and commercial torreya oil (from the aril). As a result, several cultivars were selected, based on seed variants, and some of these were also formally described, somewhat inappropriately, as botanical varieties. The large size of the seeds of some of these cultivars is responsible for the species name, from Latin. There is also considerable natural variability, and some individual wild populations, such as one from Jiulong Shan (Mount) in southwestern Zhejiang with exceptionally long leaves to 4.5 cm, were also described as varieties, but there seems to be little justification for doing so, and they are not recognized here, although a few are cited in the synonymy. The timber is of excellent quality and is used for a wide variety of domestic and public applications where its durability, strength, pleasant color and ability to take a fine finish are variously appreciated. Use as a coffin wood, for example, takes advantage of all of these qualities. The range of Chinese nutmeg yew lies generally to the east and south of that of Farges torreya, although the two have some overlap. In the absence of seeds they are not easily distinguishable in their region of overlap and are sometimes treated as a single species. DNA studies, however, mark them as moderately distinct and align Farges torreya with Japanese nutmeg yew (*Torreya nucifera*) rather than with Chinese nutmeg yew.

Torreya jackii W. Y. Chun

LONGLEAF TORREYA, JACK TORREYA, CHANG YE FEI SHU (CHINESE)

Shrub, or tree to 12 m tall, with one or more trunks to 0.2(–0.4) m in diameter. Bark light brown when fresh, weathering ashy gray to dark gray, flaking in thick scales and seldom becoming furrowed. Crown low and broadly dome-shaped, with thin, upwardly angled branches bearing pairs of horizontal or drooping branchlets densely clothed with slightly separated foliage. Twigs passing though greenish brown to shiny reddish brown in their first and second years. Leaves strongly and pleasantly aromatic when crushed, (2.5–)3.5–7(–9) cm long (to 23 cm in juveniles), (2.5–)3–4 mm wide, stiff and drooping, shiny yellowish green, dark green, or bluish green and with low ridges flanking the midrib above, with deeply sunken silvery to grayish white, or later brown, stomatal bands beneath. Individual needles curved away from the twig tip, often twisted, widest about a third of the way from the base, tapering very gradually to the narrowly triangular tip with a sharp, fragile point 0.5–1.5 mm long, and more abruptly to the roundly wedge-shaped, slightly asymmetrical base on a very short, twisted petiole 1–2 mm long. Pollen cones pale yellow, about 6–7 mm long and 4–5 mm across. Seeds with aril reddish yellow with a thin waxy film, 2–3 cm long and 1.5–2.5 cm in diameter. Female gametophyte tissue inside the shiny yellowish brown stony layer deeply wrinkled. With a fairly continuous distribution in southern and western Zhejiang and adjacent northern Fujian and northeastern Jiangxi, eastern China. Scattered in the understory of monsoonal subtropical broad-leaved evergreen forests or secondary woodlands along streamsides and on protected slopes; (250–)400–1,000 m. Zone 8.

Longleaf torreya has the longest needles of any of the nutmeg yews and is also the most consistently shrubby species in the genus. It lives in the (seasonally) driest habitat among the Asian torreyas, and established trunks often die at the tip and are replaced by basal sprouts, resulting in a clump of slender trunks. Even with relatively small trunk dimensions, the fragrant wood is highly prized for household items, and its toughness and strength are exploited in agricultural implements. It is a rare species of conservation concern but grows in fairly rugged country and is somewhat more tolerant of forest disturbance than other species in the genus. The unique (in this genus) twisted, relatively narrow needles may be another drought-resistance feature. Although longleaf torreya stands out in the genus in its morphology and ecology, DNA studies suggest that it is very closely related to Chinese nutmeg yew (*Torreya nucifera*), its closest geographic neighbor. The species name honors its author's graduate supervisor at the Arnold Arboretum in Massachusetts, Canadian-born American botanist John G. Jack (1861–1949).

Torreya nucifera (Linnaeus) P. Siebold & Zuccarini

JAPANESE NUTMEG YEW, KAYA (JAPANESE), PIJA NAMU (KOREAN)

Shrub, or tree to 15(–28) m tall, with trunk to 0.9(–2.0) m in diameter. Bark light reddish brown when fresh, peeling in long, thin flakes, weathering grayish brown and becoming shallowly furrowed between flat-topped ridges. Crown dense, expanding from conical to dome-shaped with age, with thin, gently upwardly angled to horizontal or finally drooping branches bearing pairs of horizontal or drooping branchlets densely clothed with closely spaced to fairly widely separated foliage. Twigs passing from green to shiny reddish brown in their second or third year. Leaves unpleasantly aromatic when crushed, 1.5–2.5(–3.5) cm long, 2–3 (–4) mm wide, stiff and slightly drooping, shiny bright or dark green to bluish green or pale green and flat around the midrib above, with shallowly sunken white to pale yellow stomatal bands beneath. Individual needles straight or slightly curved toward or away from the twig tip, widest before the middle or nearly parallel-sided, tapering very gradually and then abruptly to the roundly triangular tip with a stiff prickle 1–2 mm long, and abruptly to the rounded base on a very short petiole 0.5–1 mm long. Pollen cones greenish yellow, about 7–9 mm long and 4–5 mm across. Seeds with aril dark green, variably blotched or suffused with purple, sometimes very thinly waxy, (1.5–)2–3(–3.5) cm long and (1–)1.5–2 cm thick. Female gametophyte inside the light reddish brown stony layer almost smooth or only slightly wrinkled. Central and southern Japan (Honshū, Shikoku, and Kyūshū) and Cheju (Quelpart I) at the southern tip of Korea. Scattered singly or in small groves (rarely a dominant) in various temperate deciduous or mixed conifer and hardwood forests, often near streamsides or on moist slopes; 0–800 m. Zone 7. Two varieties.

Japanese nutmeg yew is a handsome tree for gardens and is one of the trees planted in temple grounds and around shrines in Japan. A few cultivars have been selected. While the seeds are edible and much appreciated (hence the scientific name, Latin for "nut bearing"), there appears to have been much less selection of seed variants than has occurred in Chinese nutmeg yew (*Torreya grandis*). As with *T. grandis*, an edible oil is also extracted from the seeds. It shares wood qualities with other species in the genus, so it also has dual categories of uses, for example, in furniture, which takes advantage of its attractive color and fine finish, and in water

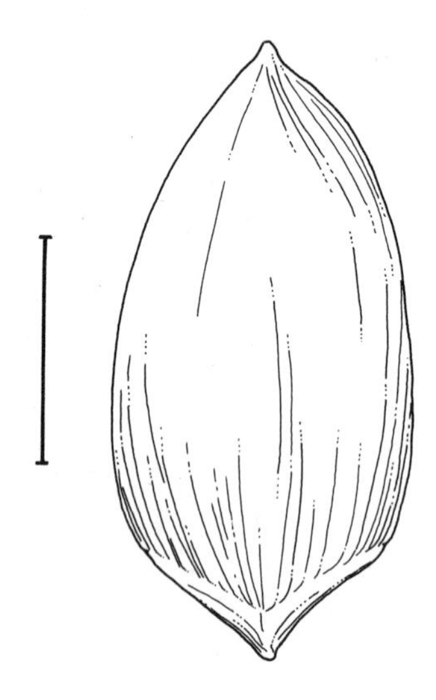

Side view of stony layer of seed of Japanese nutmeg yew (*Torreya nucifera*); scale, 1 cm.

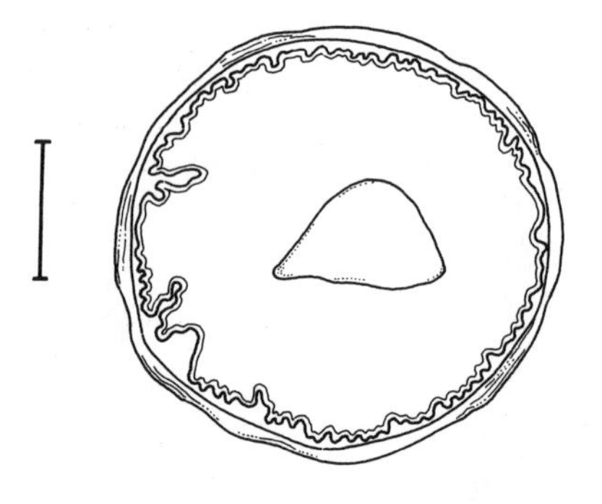

Cross section of seed of Japanese nutmeg yew (*Torreya nucifera*); scale, 5 mm.

buckets, where its durability and fine grain are paramount assets. Plants of exposed mountain slopes of the Sea of Japan side of Honshū are persistently shrubby and are distinguished as a variety. Hybrids with California nutmeg (*T. californica*) have been reported in cultivation.

Torreya nucifera (Linnaeus) P. Siebold & Zuccarini **var. nucifera**

KAYA (JAPANESE)

Tree to 28 m tall, usually with a single main trunk. Second-year branchlets dull reddish brown. Throughout the range of the species but, in Honshū, more prevalent on the Pacific Ocean side from Iwate prefecture southward. Synonyms: *Torreya igaensis* T. Doi & K. Morikawa, *T. macrosperma* Miyoshi ex K. Morikawa, *T. nucifera* var. *igaensis* (T. Doi & K. Morikawa) G. Koidzumi, *T. nucifera* var. *macrosperma* (Miyoshi ex K. Morikawa) Ohwi.

Torreya nucifera var. radicans T. Nakai

CHABO-GAYA (JAPANESE)

Spreading or rising shrub to 3 m high with many trailing branches from the base, rooting in contact with the soil. Second-year branchlets shiny bright red or reddish brown. Sea of Japan side of Honshū from Yamagata prefecture south to Yamaguchi.

Torreya taxifolia G. Arnott

STINKING CEDAR, GOPHER WOOD, FLORIDA TORREYA

Potentially a tree to 12(–18) m tall, with trunk to 0.3(–1) m in diameter, but now mostly reduced in the wild to a sickly shrub less than 1.5 m tall. Bark yellowish brown when fresh, weathering dark grayish brown to almost black, peeling in elongate scales and becoming shallowly furrowed between irregular, narrow, flat-topped ridges. Crown dense to open, conical to narrowly dome-shaped, with slender horizontal branches bearing pairs of horizontal to drooping branchlets densely clothed with slightly separated foliage. Twigs passing through yellowish green to grayish yellow or yellowish brown in their second and third years. Leaves unpleasantly scented when crushed (hence the common name), (1.5–) 2.5–3.5(–4.5) cm long, 2–4 mm wide, shiny bright green and flat or bowed around the midrib above, with fairly flat, grayish green stomatal bands beneath. Individual needles straight or slightly curved, widest near the base or nearly parallel-sided, tapering imperceptibly and then abruptly to the roundly narrowly triangular tip with a stiff, sharp point 0.5–1 mm long, and more abruptly to the rounded, asymmetrical base on a very short petiole 1–2 mm long. Pollen cones pale yellow, 5–8 mm long and 3–5 mm across. Seeds with aril dark green streaked with purple and with a thin waxy film, 2.5–4 cm long and (1.5–)2–3 cm thick. Female gametophyte tissue inside the light reddish brown stony layer deeply wrinkled. Restricted to the vicinity of the Apalachicola River, in the panhandle region of Florida and adjacent Georgia. Scattered as individuals or small groupings in the understory of mixed forests of pine and evergreen and deciduous hardwoods on steep slopes of narrow, moist ravines or on low flatlands; 15–30 m. Zone 7.

Stinking cedar has been in decline at least since the time when it was first formally described botanically and named in 1838. Already a rare tree at that time, it was, as far as known, confined to a strip of ravines and blufflands 35 km long on the eastern side of the Apalachicola River, which it shares with its even more restricted relative, Florida yew (*Taxus floridana*), and to a single population about 10 km west of the river. Cutting for fence posts, livestock grazing, and forest destruction reduced the known stands and removed the largest individuals. A disease swept through all the populations in the 1950s, killing many individuals outright and leaving others weakened and shrunken. Of the fewer than 2,000 (perhaps much fewer) individuals estimated to remain in natural stands, only a handful show no signs of the leaf spots, shoot dieback, and stem cankers that characterize the disease, which is poorly understood. Seed is rarely, if ever, produced by the stunted, perpetually immature shrubs found in the wild populations today. Recovery efforts focus on increasing the stock of cultivated plants with additional genotypes obtained from the native range and plans to reintroduce plants if the disease can ever be controlled. The largest known individual in the United States in combined dimensions when measured in 1972 was a cultivated plant away from the diseased native range 13.7 m tall and 0.9 m in diameter, with a spread of 12.2 m. The foliage is slightly yewlike (hence the scientific name, Latin for "yew leaf"), and certainly more so than other species of *Torreya,* but the leaves are much stiffer and more prickly than those of Florida yew or any other yew species, not to mention their attachment in pairs rather than singly.

Tsuga (Endlicher) Carrière

HEMLOCK

Pinaceae

Evergreen, usually single-trunked trees with furrowed bark, a drooping leader, and horizontal to slightly drooping slender branches. Branchlets thin, flexible, all elongate, without distinction into long and short shoots, with well-developed, but small, scaly, winter buds. Leaves spirally arranged, usually predominantly twisted into a single flat row on each side of the twig (except *Tsuga mertensiana,* in which they angle forward all around the twig), often quite variable in length, with a few shorter ones inverted along the twigs. Individual leaves needlelike, generally broadly sword-shaped, usually flattened and dark green with a shallow central groove above and with a pair of white stomatal bands beneath (except *T. mertensiana,* which has plump needles with stomates on both sides). Leaf blade with a minutely toothed or smooth edge, narrowing abruptly to a short petiole that attaches to a small woody peg running down onto the twig, and with a bluntly pointed, rounded, or notched tip.

Plants monoecious. Pollen cones sparse, single in the axils of leaves, commonly near the tips of first- and second year-twigs, slender-stalked. Individual cones almost spherical, less than 1 cm long, with numerous spirally arranged pollen scales, each scale with two pollen sacs. Pollen grains large to very large (body 55–95 µm in diameter, 60–110 µm overall), usually surrounded by a circular frill (except *Tsuga mertensiana,* which has two air bladders instead). Seed cones hanging at the tips of second-year twigs, maturing in a single season and opening to release the seeds, then falling intact the year after seed release. Each cone spherical to ellipsoid before opening, the bracts tiny to moderate and hidden by the circular to oval thin seed scales. Seeds in pairs, each with a long asymmetric wing derived from the scale. Cotyledons two, each with one vein. Chromosome base number $x = 12$.

Wood soft, odorless, light brown, without a clearly darker heartwood. Grain even, with well-marked growth rings formed by a more or less abrupt transition between small-celled latewood and large-celled earlywood. Vertical resin canals and individual resin cells absent.

Stomates arranged in and oriented with their long axis along definite lines within the stomatal bands. Each stomate sunken beneath and largely hidden by the four surrounding subsidiary cells, which may themselves be sunken beneath the general level of the epidermis. The two subsidiary cells at the ends of the stomate are often shared between adjacent stomates in a line while the two lateral ones are generally not shared between stomates in adjoining rows. Leaf cross section with a single large resin canal touching the epidermis beneath the single-stranded or partially doubled midvein, which is bordered, adjacent to the phloem, by a scattering of transfusion tissue mixed in with other cell types. Photosynthetic tissue with a single well-developed palisade layer directly beneath the upper epidermis without an intervening hypodermis (except sometimes at the leaf edge and along the midline groove) and with spongy mesophyll extending from there down to the lower epidermis with its stomatal bands.

Eight species in eastern and western North America and in southern Asia from India (northern Uttar Pradesh) to northern Vietnam, Taiwan, and Japan. References: Farjon 1990, Fu et al. 1999c, Matsumoto et al. 1995, K. C. Sahni 1990, Swartley 1984, Taylor 1993a, Yamazaki 1995. Synonym: *Hesperopeuce* Lemmon.

Several species of hemlock are important sources of timber, pulp, and (formerly) tannins from the bark. Their graceful habit makes them attractive ornamentals in regions of ample rainfall. Most are quite shade tolerant. Since they require a lot of moisture, they do not grow well in drought-prone areas. Nonetheless, they are moderately commonly cultivated, particularly in the form of one or another of the numerous cultivars, most selected from eastern hemlock (*Tsuga canadensis*). These cultivars are quite varied in overall growth habit, many taking advantage of or exaggerating the natural tendency of the leader and branches to arch gracefully. Others go against the grain in emphasizing stiffly upright growth while still others pull everything into tight buns or show other

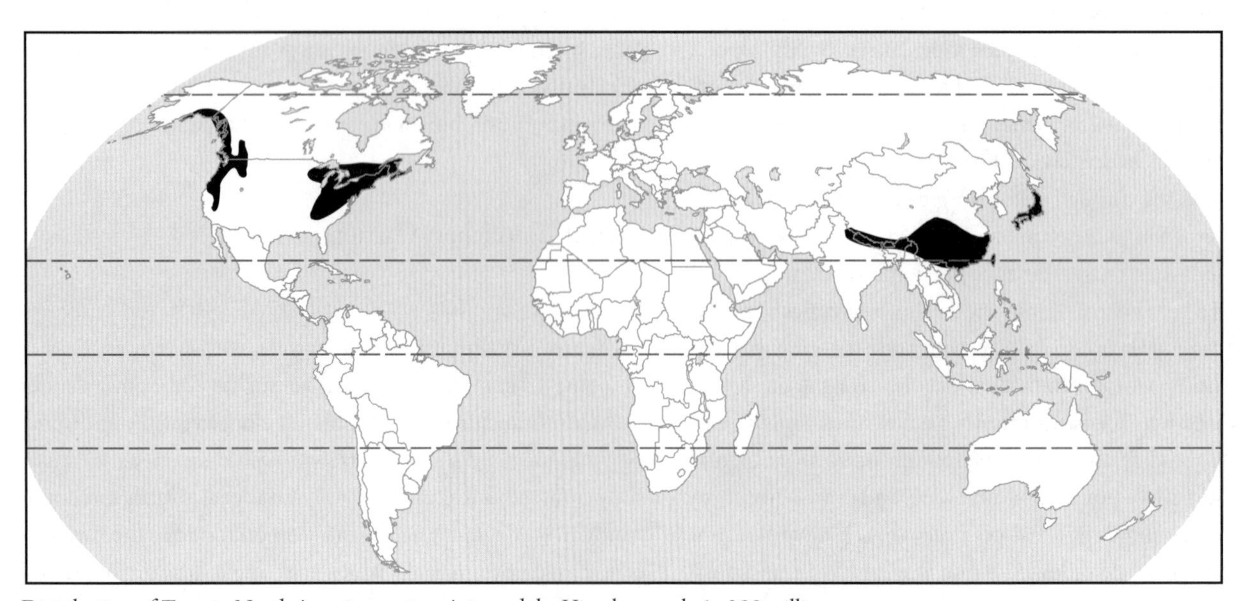

Distribution of *Tsuga* in North America, eastern Asia, and the Himalaya, scale 1 : 200 million

branching irregularities such as twisting or contortion. Cultivar selection has also emphasized variations in needle size, density, and color, including yellows, golds, and creams in the new needles as well as variegation in mature foliage. The name of the genus is the Japanese name for *T. sieboldii.*

With one exception, the species of *Tsuga* are all quite similar to one another. Mountain hemlock (*T. mertensiana*) differs from the others in its leaf form and in having pollen with two bladders. Some taxonomists remove it to its own genus, or botanical section, *Hesperopeuce,* but accumulating DNA studies place it firmly among the other hemlocks. It is especially close to its neighbor, western hemlock (*T. heterophylla*), although the similarities in DNA may be due, in part, to occasional natural hybridization between them. On the other hand, *Cathaya* and *Nothotsuga,* whose species are sometimes included in *Tsuga,* appear amply distinct.

Fossil hemlocks have been found all across the temperate northern hemisphere dating back to the late Cretaceous, perhaps, but most convincingly from the Oligocene, about 30 million years ago. The genus remained in Europe, where it is now absent, until well into the Pleistocene ice ages, as recently as about 300,000 years ago.

Identification Guide to *Tsuga*

Among the eight species of *Tsuga, T. mertensiana* is unique in having plump (not flattened) needles, with stomates on both sides (not just beneath). The remaining seven species may be distinguished by whether the twigs are hairy or not, the length of the longest ordinary needles, whether they are toothed (at least in the upper half) or not, and whether the tips are usually notched or not:

Tsuga sieboldii, twigs hairless, needles up to 2 cm, not toothed, tips notched

T. chinensis, twigs hairless, needles 2–2.5 cm, not toothed, tips notched

T. diversifolia, twigs hairy, needles less than 2 cm, not toothed, tips notched

T. caroliniana, twigs hairy, needles up to 2 cm, not toothed, tips rounded

T. canadensis, twigs hairy, needles less than 2 cm, toothed, tips rounded

T. heterophylla, twigs hairy, needles 2–2.5 cm, toothed, tips rounded

T. dumosa, twigs hairy, needles at least 2.5 cm, toothed, tips pointed

Tsuga canadensis (Linnaeus) Carrière

EASTERN HEMLOCK, CANADA HEMLOCK, PRUCHE (FRENCH)

Tree to 30(–50) m tall, with straight trunk to 1.5 m in diameter. Bark furrowed, scaly, cinnamon brown. Crown conical, shallow with age in dense stands. Twigs downy. Winter buds pointed, 1.5–2.5 mm long. Needles variable in length, 0.5–1.8(–2.0) cm long, tapering from the base to the tip, with tiny teeth along the edge, the tip bluntly pointed or rounded, the white stomatal bands beneath each with five or six lines of stomates. Pollen cones 3–5 mm long, yellow. Seed cones green before maturity, ripening light brown, 1.5–2.5 cm long, opening to 1–1.5 cm wide, the seed scales 8–12 mm long. Seed body 2–4 mm long, the wing 5–8 mm longer. Eastern North America, from Nova Scotia south

Mature eastern hemlock (*Tsuga canadensis*) in nature.

Upper (A) and undersides (B) of twigs of eastern hemlock (*Tsuga canadensis*) with 4 years of growth and closed mature seed cones, ×0.5.

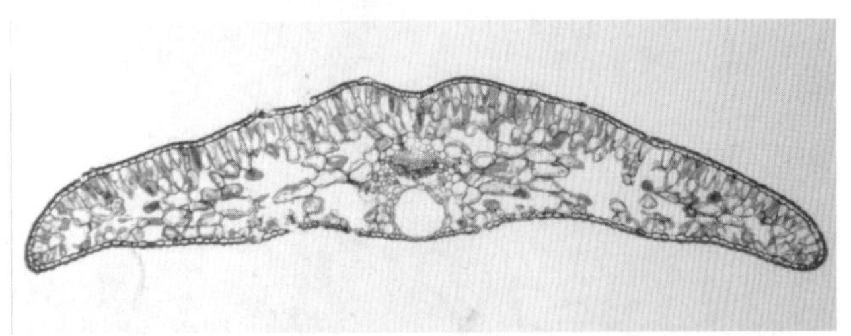

Cross section of preserved, stained, and sectioned needle of eastern hemlock (*Tsuga canadensis*).

to northern Alabama, west to eastern Minnesota. Moist ridges, slopes, and valleys; 600–1,800 m. Zone 5.

Although there are no recognized botanical variants of this common tree, numerous cultivars have been selected. Del Tredici (1983, 1984) traced the history of two of these: 'Minuta' and 'Pendula'. Eastern hemlock was once heavily exploited for tannin obtained from the bark, although the timber was never much valued. Even earlier, about 5,000 years ago, pollen records show a dramatic decline in the abundance of eastern hemlock (the "hemlock crash"), the cause of which has never been fully explained. One of the possible causes for the hemlock crash was a disease or pest, and today eastern hemlocks die in great numbers as a result of feeding by an introduced insect pest, the hemlock woolly adelgid (*Adelges tsugae*), an aphid relative introduced in Virginia in 1951 and gradually expanding its range toward the major range of eastern hemlock in the Great Lakes region. Although this pest can be controlled on ornamental trees once present, its recurrence cannot be prevented, so planting of eastern hemlock and its cultivars should be avoided until the adelgid can be brought under control regionally.

The largest known individual in the United States in combined dimensions when measured in 1998 was 50.3 m tall and 1.6 m in diameter, with a spread of 11.6 m. Roots from adjacent hemlock trees may grow together, forming root grafts that can transport water and nutrients from tree to tree. As a result, a single cut hemlock in a hemlock grove may form a living stump, continuing to add annual growth rings for decades as long as its neighbors stand. The species name reflects Linnaeus's concept of Canada as the northern part of North America north of "Virginia."

Tsuga caroliniana Engelmann

CAROLINA HEMLOCK

Tree to 30 m tall but usually shorter, with straight trunk to 2 m in diameter. Bark furrowed, scaly, brown. Crown conical, more compact than in *Tsuga canadensis.* Twigs thinly hairy. Winter buds pointed, 2–3 mm long. Needles more widely spaced and less strongly two-ranked than in *T. canadensis,* variable in length, (0.5–)1.0–2.0 cm long, parallel-sided, the edge toothless and slightly rolled under, the tip bluntly rounded or slightly notched, the white stomatal bands beneath each with 8–10 lines of stomates. Pollen cones about 5 mm long, yellow. Seed cones green before maturity, ripening light brown, (2–)2.5–4 cm long on stalks 4–5 mm long, opening to 1.5–2.5 cm wide, the seed scales 12–20 mm long. Seed body 5–6 mm long, the wing 8–10 mm longer. Appalachian Mountains from western Virginia to northeastern Georgia. Moist rocky ridges and slopes; 700–1,200 m. Zone 6.

Carolina hemlock is an uncommon tree with a more restricted geographic, elevational, and ecological range than the surrounding eastern hemlock (*Tsuga canadensis*). The majority of its range lies within western North Carolina, from which it takes its name. It is not known to hybridize naturally with its more abundant relative. Like eastern hemlock, Carolina hemlock soon dies (often within 4–6 years) when infested by the introduced hemlock woolly adelgid (*Adelges tsugae*). The largest known individual in combined dimensions when measured in 1999 was 30.2 m tall, 1.3 m in diameter, with a spread of 15.2.

Tsuga chinensis (Franchet) E. Pritzel

CHINESE HEMLOCK, TIE SHAN (CHINESE)

Tree to 50 m tall but usually much shorter, with trunk often forking above 10 m, to 2 m in diameter. Bark blocky, cinnamon brown to gray-brown. Crown broadly conical, flat-topped in age. Twigs with transient hairs in the grooves between the attached leaf bases. Winter buds bluntly rounded, 1–4 mm long. Needles variable in length, (0.6–)1–2.5(–3) cm long, parallel-sided, the edge generally toothless or with a few teeth toward the notched tip, the white stomatal bands beneath each with six to nine lines of stomates. Pollen cones 3–5 mm long, yellow. Seed cones green before maturity, ripening light brown, 1.5–2.8(–4) cm long, opening to (1–)1.5–2.5(–3.5) cm wide, the seed scales 8–14 mm long. Seed body 3.5–4.5 mm long, the wing 3.5–4.5 mm longer. Widely distributed in China from Zhejiang and Guangdong west to southern Gansu and Yunnan, extending just over the border into northwestern Vietnam, and in Taiwan. Humid temperate and montane forests with hardwoods and other conifers; (600–)1,200–3,200 m. Zone 6. Synonyms: *Tsuga chinensis* var. *formosana* (Hayata) H. L. Li & H. Keng, *T. chinensis* var. *oblongisquamata* W. C. Cheng & L. K. Fu, *T. chinensis* var. *patens* (D. Downie) L. K. Fu & N. Li, *T. chinensis* var. *robusta* W. C. Cheng & L. K. Fu, *T. formosana* Hayata, *T. oblongisquamata* (W. C. Cheng & L. K. Fu) L. K. Fu & N. Li, *T. patens* D. Downie, *T. tchekiangensis* Flous.

In the mountainous regions of northwestern Yunnan and vicinity, this species overlaps and hybridizes with the Himalayan hemlock (*Tsuga dumosa*). Here are found a wide range of intermediates in leaf shape and toothing, and hairiness of twigs, several described as species or varieties. The earliest name for these hybrid derivatives is *T.* ×*forrestii*. Even outside the range of overlap there is considerable regional variation in Chinese hemlock, and a number of varieties and segregate species were proposed. There seems little justification for formal taxonomic subdivision within *T. chinensis*, however, as the proposed segregates are merely extremes within a continuum of variation. This is the most widespread hemlock in China, and the tree has many uses for lumber, tannin, resin, and aromatic oils.

Tsuga diversifolia (Maximowicz) M. T. Masters

NORTHERN JAPANESE HEMLOCK, JAPANESE SUBALPINE HEMLOCK, KOMETSUGA (JAPANESE)

Tree to 25 m tall, or shrubby at high elevations in the northern part of its range, with straight or forked trunk to 1 m in diameter. Bark furrowed, scaly, cinnamon brown to gray-brown. Crown rounded, opening up with age. Twigs hairy at first, becoming smooth. Winter buds rounded, 2–3 mm long. Needles variable in length (hence the species name, from the Latin), 0.7–1.5(–1.8) cm long, slightly widening from the base to the notched tip, the edge smooth, the white stomatal bands beneath each with 8–10 lines of stomates. Pollen cones 3–4 mm long, yellow-orange. Seed cones green before maturity, ripening light brown, 1.5–2(–2.5) cm long on stalks 0.5 mm long, opening to 1.5–2 cm wide, the seed scales 7–9(–12) mm long. Seed body 3.5–4.5 mm long, the wing 3.5–4.5 mm longer. Japan, widespread in northern and central Honshū with outliers in Shikoku and Kyūshū. Moist slopes and ridges of mixed montane and subalpine forests; (700–)1,200–2,500 m. Zone 5. Synonym: *Tsuga blaringhemii* Flous.

Of the two endemic hemlock species of Japan, this one occurs at higher elevations. Although they have a broad range of overlap in central Honshū and the southern islands, there are no reports of hybrids between these two closely related species.

Tsuga dumosa (D. Don) A. Eichler

HIMALAYAN HEMLOCK, CHANGA THASI (NEPALI), YUN NAN TIE SHAN (CHINESE)

Tree to 35(–50) m tall, with straight or early forked trunk to 2.7 m in diameter. Bark furrowed, scaly, pinkish brown to gray-brown. Crown broadly conical, flattening with age. Twigs thinly hairy. Winter buds blunt, 2–2.5 mm long. Needles variable in length, 1–2.5(–3.5) cm long, tapering from the base to the pointed tip, with tiny teeth along the edge, the white stomatal bands each with 8–10 lines of stomates. Pollen cones 2.5–5 mm long, yellow. Seed cones green before maturity, ripening light brown, (1.3–)1.8–2.5(–3) cm long, opening to 1.5–2.5(–3) cm wide, the seed scales 10–14 mm long. Seed body 3.5–4.5 mm long, the wing 3.5–4.5 mm longer. Himalaya, from northern Uttar Pradesh (India) to southwestern Sichuan (China) and northern Myanmar. Moist to wet slopes in the monsoon region; (1,700–)2,400–3,000(–3,500) m. Zone 9 (?). Synonyms: *Tsuga calcarea* D. Downie, *T. dumosa* var. *yunnanensis* (Franchet) Silba, T. *yunnanensis* (Franchet) E. Pritzel.

The wood of Himalayan hemlock is used for shingles, carpentry, and construction, and the bark for roofing. This species hybridizes with Chinese hemlock (*Tsuga chinensis*) where the ranges of the two species overlap in western Sichuan and northwestern Yunnan. The hybrids are called *T.* ×*forrestii*. The species name, meaning "bushy," is not terribly appropriate for a large tree but refers to the tendency of the trunk to fork near the base.

Tsuga* ×*forrestii D. Downie

Tree to 30 m tall. A natural hybrid between Chinese hemlock (*Tsuga chinensis*) and Himalayan hemlock (*T. dumosa*) occurring in their region of overlap in the mountains of southwestern Sichuan and northwestern Yunnan. The hybrids are variable but resemble both parents in their numerous shared features. The needles are as long as those of Himalayan hemlock but have a smooth rather than toothed edge as in Chinese hemlock. The

Large mature hybrid hemlock (*Tsuga ×forrestii*) in nature.

Multitrunked mature western hemlock (*Tsuga heterophylla*) in nature.

wood is highly prized and the trees, always uncommon in comparison to their parents, have been decimated by indiscriminate clear-cut logging. Synonym: *T. chinensis* var. *forrestii* (D. Downie) Silba.

Tsuga heterophylla (Rafinesque) C. Sargent

WESTERN HEMLOCK

Tree to 50(–80) m tall, with straight trunk to 2(–3) m in diameter. Bark furrowed, scaly, red-brown to gray-brown. Crown narrowly conical at first, broadening with age. Twigs downy. Winter buds bluntly rounded, 1.5–2 mm long. Needles variable in length (hence the species name, from the Greek), 0.5–2(–3) cm long, parallel-sided, with tiny teeth along the edge, the tip rounded or slightly notched, the white stomatal bands beneath each with 8–10 lines of stomates. Pollen cones 3–5 mm long, red. Seed cones green before maturity, ripening light brown, (1–)1.5–2.5(–3) cm long on stalks 4–6 mm long, opening to 1.5–2.5 cm wide, the seed scales (5–)8–15 mm long. Seed body 3.5–4.5 mm long, the wing 6–8 mm longer. Pacific region of western North America, from central Alaska to northern California and in the northern Rocky Mountain region from northern British Columbia to central Idaho. Wet lowland and montane conifer forests, often as the prime dominant on slopes adjacent to the Pacific Coast; 0–1,500(–1,800) m. Zone 6.

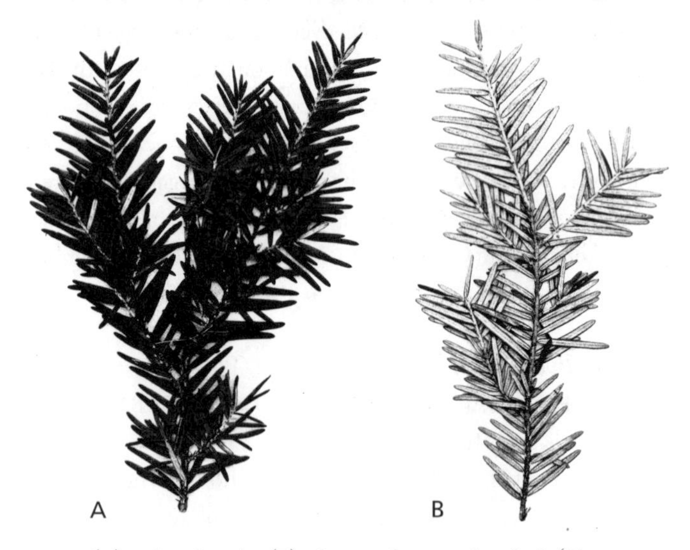

Upper (A) and undersides (B) of twigs of western hemlock (*Tsuga heterophylla*) with 3 years of growth, ×0.5.

This is the largest, most abundant, and most important timber hemlock, producing both good-quality construction lumber and paper pulp. Despite its morphological similarities to other typical hemlocks, DNA studies show that western hemlock is most closely related to the neighboring but morphologically dissimilar mountain hemlock (*Tsuga mertensiana*). Usually living no more than 500 years, exceptional individuals can reach 900 years. Several largest in-

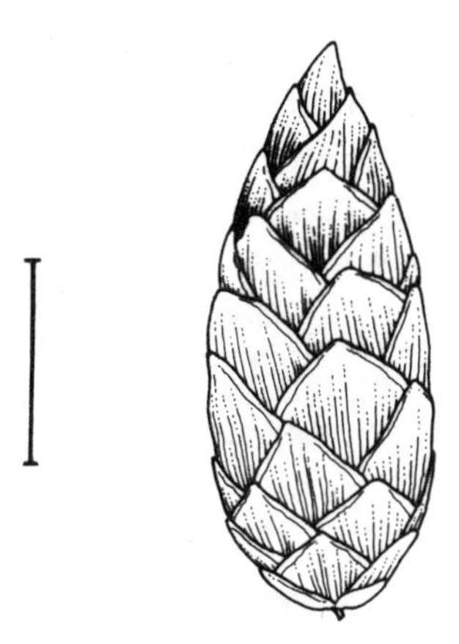

Closed mature seed cone of western hemlock (*Tsuga heterophylla*) before seed dispersal; scale, 1 cm.

dividuals are known, all having similar combined dimensions when measured in the 1980s and 1990s, and vary from 53 to 76 m tall and from 2.2 to 2.8 m in diameter, with a spread of 14.3 to 20.4 m. Although hybrids between western hemlock and the partly sympatric mountain hemlock of higher elevations are found in cultivation under the name *T. ×jeffreyi*, hybridization in nature is very rare.

Tsuga ×jeffreyi (A. Henry) A. Henry

Tree to 30 m tall. A natural and garden hybrid between two apparently dissimilar hemlock species. The trees resemble mountain hemlock (*Tsuga mertensiana*) in their dense branching and radiating needles, but these needles resemble those of western hemlock (*T. heterophylla*) in having a green upper surface and greenish white stomatal bands beneath rather than being waxy bluish green on both surfaces. There are also points of intermediacy, like the length of the needles and of the seed cones and the possession of a few lines of stomates on the upper side of the needles near their tips. This hybrid is very difficult to produce by deliberate crossing, and many seemingly intermediate individuals seen in the wild are more likely aberrant trees of western hemlock at the upper limit of its elevational range. A few individuals have leaf pigments combining those of the parent species and appear to be genuine hybrids. Despite their morphological dissimilarity, the rarity of the hybrids, and the difficulty in producing them, DNA studies suggest that mountain hemlock and western hemlock are each other's closest relatives. Zone 6. Synonym: *Tsuga mertensiana* var. *jeffreyi* (A. Henry) C. K. Schneider.

Tsuga mertensiana (Bongard) Carrière

MOUNTAIN HEMLOCK

Tree to 40(–45) m tall, or shrubby at the alpine tree line, with straight trunk to 1.5(–2) m in diameter. Bark deeply furrowed, scaly, dark red-brown to gray-brown. Crown narrowly conical, broadening somewhat with age. Twigs densely hairy to hairless. Winter buds pointed,

Upper side of twig of mountain hemlock (*Tsuga mertensiana*) with 2 years of growth, ×1.

Open seed cone of mountain hemlock (*Tsuga mertensiana*) after seed dispersal; scale, 1 cm.

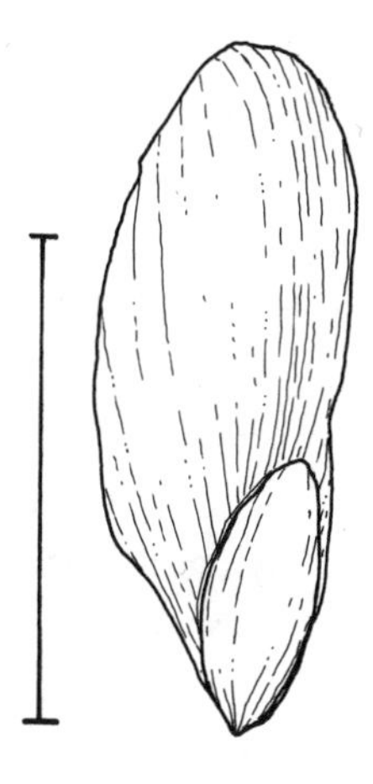

Underside of dispersed seed of mountain hemlock (*Tsuga mertensiana*); scale, 1 cm.

2–3 mm long. Needles plumper than in other hemlocks, spreading all around the twigs rather than largely in two ranks, more even in length, (0.5–)1–2(–3) cm long, parallel sided, with smooth edges, the tip bluntly pointed, waxy gray-green and with stomatal lines above as well as beneath. Pollen cones about 10 mm long, purple maturing yellow. Seed cones purple before maturity, ripening brown, (2–)3–6(–8) cm long, opening to 1.2–2.5(–3.3) cm wide, the seed scales bending backward, 8–13(–18) mm long. Seed body 3–5 mm long, the wing 5–8 mm longer. Pacific region of western North America, from the Kenai Peninsula of Alaska to the southern Sierra Nevada of California and in the Rocky Mountain region from southern British Columbia to northern Idaho and northwestern Montana. Coastal (in the north) and montane forests to the alpine tree line; 0–2,400(–3,350) m. Zone 5. Synonyms: *Tsuga crassifolia* Flous, *T. mertensiana* subsp. *grandicona* Farjon, *Hesperopeuce mertensiana* (Bongard) Rydberg.

Although primarily a tree of cold forests, this species is much more adaptable in cultivation, where it forms a handsome specimen. Because of its lower frequency and greater inaccessibility, it is much less exploited in forestry than is western hemlock (*Tsuga heterophylla*). It rarely hybridizes in the wild with the latter species, but such hybrids are found in cultivation under the name *T. ×jeffreyi*. Mountain hemlock is so unlike the other hemlock species that it was placed in its own genus or even considered a hybrid between a typical hemlock and a spruce (*Picea*). DNA studies place it firmly among the hemlocks, close to western hemlock. The largest known individual in the United States when measured in 1993 was 46.3 m tall, 1.9 m in diameter, with a spread of 12.5 m. The species name honors Karl H. Mertens (1796–1830), who collected the type specimen in southeastern Alaska on a Russian exploring expedition.

Tsuga sieboldii Carrière

SOUTHERN JAPANESE HEMLOCK, TSUGA (JAPANESE), SOLSONG NAMU (KOREAN)

Tree to 25(–30) m tall, with straight or curved trunk to 1(–2.5) m in diameter. Bark furrowed, scaly, gray-brown. Crown sparser than in *Tsuga diversifolia,* broadly conical in youth, broadening and becoming flat-topped with age. Twigs hairless. Winter buds pointed, 2–2.5 mm long. Needles variable in length, 0.7–1.2(–2) cm long, widening gradually toward the notched tip, with smooth edges, the white stomatal bands beneath each with 8–10 lines of stomates. Pollen cones 4–5 mm long, reddish brown maturing yellow. Seed cones green before maturity, ripening light brown, (1.5–)2–2.5(–3) cm long on stalks 2–4 mm long, opening to (1.2–)1.5–2(–2.5) cm wide, the seed scales 8–11 mm long. Seed body 4–5 mm long, the wing 5–6 mm longer. Southern Japan, from southern Honshū to Yaku Shima, and on Ullung Island (Korea). Wet mountain slopes and ridges in mixed forest; (100–)400–1,600 m. Zone 6.

Southern Japanese hemlock occurs at generally lower elevations than the broadly overlapping northern Japanese hemlock (*Tsuga diversifolia*), but the two species are not known to hybridize in the regions of overlap. They are probably not each other's closest relatives, and DNA studies suggest that *T. sieboldii* is closest to Chinese hemlock (*T. chinensis*). Southern Japanese hemlock is the species that gave its name to the genus. Philipp F. von Siebold (1796–1866), coauthor of the second botanical flora of Japan, used the

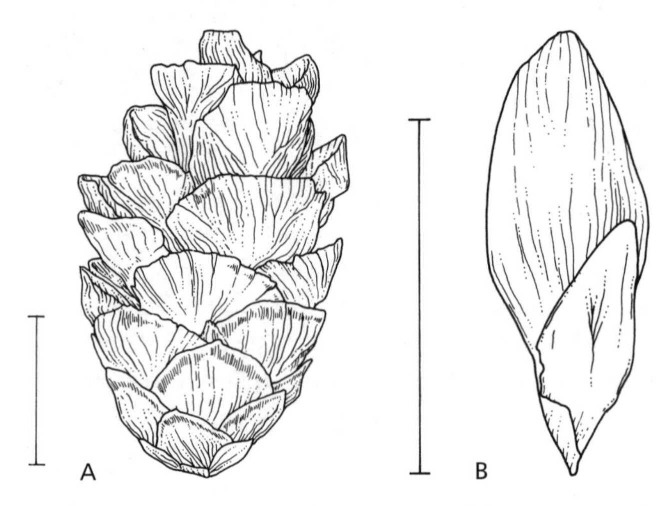

Partially open mature seed cone after seed dispersal (A) and underside of dispersed seed of southern Japanese hemlock (*Tsuga sieboldii*); scale, 1 cm.

Old mature southern Japanese hemlocks (*Tsuga sieboldii*) in nature.

Japanese common name, *tsuga,* when naming it *Abies tsuga.* When Carrière gave the hemlocks their own genus, he used this species' name as the generic name. He then had to find a new name for the species, because botanical nomenclature, unlike zoological nomenclature, does not allow tautonyms, species names that repeat the generic name. Carrière followed a common botanical precedent in coining a replacement name of honoring the author of the name that was being replaced, in this case, von Siebold.

Widdringtonia Endlicher

SOUTH AFRICAN CEDAR

Cupressaceae

Evergreen trees and large shrubs with a dense branching habit. Trunk single or often dividing near the base, clothed with fibrous bark that peels in scales and strips. Twigs all more or less alike, without clear differentiation into long and short shoots, although strong and leading shoots are thicker and have longer leaf spacing than others, branching in three dimensions. Resting buds unspecialized, without distinct bud scales and consisting solely of as yet undeveloped ordinary foliage leaves. Leaves in alternating similar pairs, without differentiation into lateral and facial pairs, or irregularly spiral in youth. Juvenile leaves needlelike, linear, flat, sometimes persisting into, or present as reversions during, the adult phase. Adult leaves scalelike, the attached bases completely clothing the twigs, the free tips pressed against or standing out from the twigs, living 2–4 years.

Plants monoecious. Pollen cones scattered, not stalked, single at the ends of short lateral twigs, seated directly between a pair of foliage leaves without transitional pairs of bracts. Individual cones ellipsoid, with four to seven alternating pairs of pollen scales. Each scale usually with four pollen sacs arrayed along the lower edge of the inverted-heart-shaped, upturned blade. Pollen grains small (20–30 µm in diameter), roughly spherical and with an almost featureless surface. Seed cones single or clustered at the ends of short, stout axillary branchlets, maturing in two or three seasons and then remaining unopened for several years. Individual cones almost spherical before opening, normally with two nearly equal pairs of thick woody scales attached basally around a broad central bump (the columella), the inner (upper) pair meeting in a straight line at the center so that the members of the outer (lower) pair do not touch. Each cone scale flattened or convex externally, the boundary between the fully joined bract and seed scale portions scarcely recognizable (in part because of extreme reduction of the seed scale), with a more or less prominent point or knob near the upper margin and with 1–8(–15) seeds in one or two rows at the base. Seeds angular, two-winged, the wing an outgrowth of the seed coat. Cotyledons two (or three), each with one vein. Chromosome base number $x = 11$.

Wood often powerfully fragrant, light to moderately heavy and soft to moderately hard, highly decay resistant but extremely flammable, shiny, the narrow band of whitish to yellowish sapwood scarcely distinguishable from to sharply contrasting with the yellow or pale reddish to light brown heartwood. Grain even and fine to moderately coarse, with ill-defined to obvious growth rings marked by a gradual transition to a narrow band of smaller but often not notably thicker walled latewood tracheids. Without resin canals but with numerous individual resin parenchyma cells scattered through the growth increments with some concentrated into open bands.

Stomates scattered or in irregular lines across the exposed surface of the leaf tips and in one or a few lines on the hidden face and extending sparingly into the grooves between the attached leaf bases. Each stomate sunken beneath and partly hidden by the (four or) five (or six) surrounding subsidiary cells, which are surmounted by a lobed, complete, steep Florin ring that stands atop the very thick cuticle covering all of the epidermal cells, some of the rest of which have additional, thickened, knoblike papillae. Midvein single, remaining close to the twig axis in the leaf base and soon petering out in the free tip, accompanied toward the outside (lower side) by a large resin canal and often flanked by two smaller ones beyond the large accompanying wedges of transfusion tissue. Photosynthetic tissue with a prominent, partially doubled palisade layer over the whole outer face of the leaves (except beneath the stomates) directly beneath the epidermis without an intervening

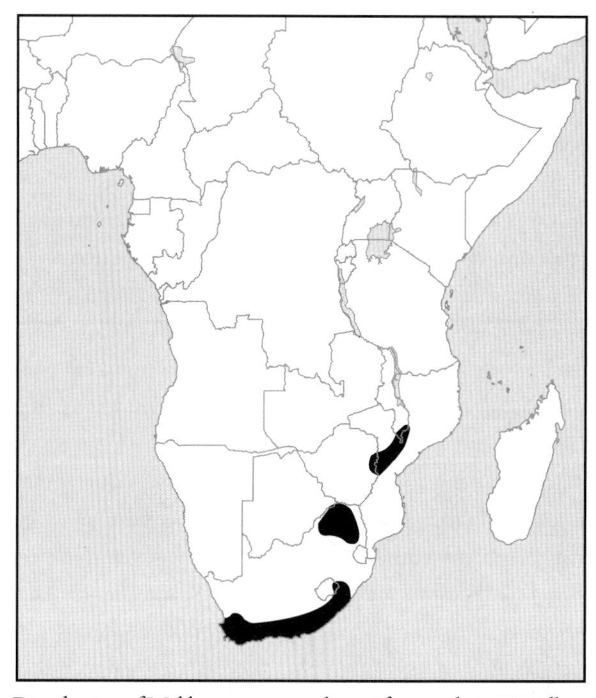

Distribution of *Widdringtonia* in southern Africa, scale 1 : 70 million

hypodermis (or with a partial and incomplete one), the remaining, looser spongy mesophyll dotted with individual large resin cells.

Four species in southern Africa from the Cape of Good Hope, South Africa, to Malawi. References: Chapman 1961, Farjon 2005b, McIver 2001, Marsh 1966, Midgley et al. 1995, Powrie 1972.

Three of the species are very local, while the fourth occupies almost the entire range of the genus. All four occur in fire-prone habitats and require fire for natural regeneration. They also all possess fragrant, highly resinous wood that is resistant to decay but also resistant to paint. The three local ones were important sources of timber before they were depleted, while the single widespread species, *Widdringtonia nodiflora,* rarely reaches commercial size. Since species of the genus are not very hardy and are hardly cultivated outside of botanical gardens and tree collections, it is not surprising that there has been no cultivar selection as yet.

The name of the genus honors British Naval Commander Samuel Edward Widdrington (1787–1856). He had no direct connection to these trees, but in 1842 he published the last of three articles on European *Pinus* and *Abies* species just as Endlicher searched for a new name for the genus because two previously proposed names (one by him to replace an earlier one by Brongniart) had each been scooped in the year before it was proposed for an entirely different plant (1833 versus 1832, and 1841 versus 1840). There may have been a tinge or irony in this choice because Widdrington himself had just changed his own name (in 1840), the two previous articles having been published in 1839 under his birth name, Cook.

Widdringtonia is the only member of the southern hemisphere cypress subfamily Callitroideae native to Africa. It was long thought to be closely related to the superficially similar *Tetraclinis* in North Africa, but a closer look at their structures and DNA studies show that the latter is unrelated and, instead, belongs with other genera in the northern hemisphere subfamily Cupressoideae. Seed cones of *Tetraclinis* lack the central columella that links *Widdringtonia* to the cypress pines (*Callitris*) of Australia and related genera. In fact, *Widdringtonia* is closest to the Tasmanian subalpine shrub *Diselma,* which has similarly constructed but much smaller seed cones less than 5 mm long (versus more than 1 cm, usually much more, in *Widdringtonia*). The other related genera all have leaves and cone scales in threes or fours rather than in pairs. Rare individuals of *Widdringtonia* may also have parts in threes as an aberration, just as opposite-leaved flowering plants sometimes do. Among conifers, several junipers (*Juniperus*), like *Widdringtonia* members of the Cupressaceae, take this variability further and commonly show both conditions.

Peculiarly, given its relationships, there is no fossil record of *Widdringtonia* in the southern hemisphere. Instead, remains attributed to this genus were widely distributed around the North Atlantic (eastern North America, Greenland, and Europe) during the middle of the Cretaceous, from about 95 million years ago, and persisted into the mid-Tertiary (Oligocene, ca. 30 million years ago) in Europe. These fossils are, indeed, very like living *Widdringtonia* species, although the seed cones are generally smaller, so their northern location and age (older than most records of still extant conifer genera), combined with the phylogenetic nesting of the genus firmly among the purely southern genera of subfamily Callitroideae, are puzzling and worthy of further investigation.

Identification Guide to *Widdringtonia*

The four species may be distinguished most easily by the usual width of the open seed cones and whether the rim of the seed scales is strongly warty (irrespective of the condition of the face of the scale):

Widdringtonia whytei, open seed cones 0.8–1.5 cm wide, seed scale rim smooth

W. nodiflora, open seed cones 1.5–2.0 cm wide, seed scale rim smooth

W. schwarzii, open seed cones 1.5–2.0 cm wide, seed scale rim warty

W. cedarbergensis, open seed cones 2.0–2.5 cm wide, seed scale rim warty

Widdringtonia cedarbergensis J. A. Marsh

CLANWILLIAM CEDAR, SEDERBOOM (AFRIKAANS)

Tree to 20 m tall but now rarely more than 7 m in the wild, with trunk to 2 m in diameter, usually branching from near the base. Bark furrowed, beautifully flaking in dark gray on purple-brown. Crown conical at first, spreading broadly with age. Juvenile leaves 1–2 cm long, up to 2 mm wide. Scale leaves 2–4 mm long, the free tip shorter than to as long as the attached base, rounded in cross section. Pollen cones about 2 mm long. Seed cones 2–2.5 cm across, the rim and outer face of the scales both strongly warty, the subapical point well developed. Seeds black, the body 6–10 mm long, ovoid, straight, scarcely winged. South Africa, Cedarberg north of Cape Town. Scattered across rocky outcrops among fire-prone shrublands; 900–1,600 m. Zone 9. Synonym: *Widdringtonia juniperoides* (Linnaeus) Endlicher.

Once an important source of softwood lumber, this species was depleted by overcutting and too frequent fires. It requires fire for seed reproduction but does not resprout after fire, so fires more frequent than the time it takes to start producing cones drastically curtail its reproduction. It appears to be difficult and tender in cultivation despite the heavy frost it experiences during the winter in nature. Older literature uses the name *Widdringtonia juniperoides* for this species, but that name was rejected because no one can determine to which species it originally applied.

Widdringtonia nodiflora (Linnaeus) Powrie

BERG CYPRESS, BERGSIPRES (AFRIKAANS)

Small, shrubby tree exceptionally to 9 m tall but usually 3–6 m, typically with multiple trunks from the base. Bark fibrous, furrowed, gray and peeling in longitudinal strips to reveal fresh red surfaces. Crown narrowly conical at first, spreading with age. Juvenile leaves 1–2 cm long, up to 2 mm wide. Scale leaves 2–3 mm long, the free tip about as long as the attached base, triangular in cross section. Pollen cones about 2 mm long. Seed cones 1.5–2 cm across, the rim of the scales smooth, the face either smooth or warty, the subapical point small. Seeds dark brown, the body 7–10 mm long, oblong, curved, the wings to 4 mm wide. Mountains of southern and southeastern Africa, from Table Mountain, Cape province, South Africa, to Mount Mulanje, southern Malawi and vicinity. Rising above fire-prone shrub lands and grasslands along gullies and slopes; (500–)900–2,100(–2,600) m. Zone 9. Synonyms: *Widdringtonia cupressoides* (Linnaeus) Endlicher, *W. dracomontana* Stapf.

Berg cypress is the shrubbiest species of *Widdringtonia* and the only one capable of resprouting after fires. It occurs with two of the tree species, Willowmore cedar (*W. schwarzii*) and Mulanje cedar (*W. whytei*), but usually at higher elevations and in drier sites. No natural or artificial hybrids among these species are documented, but they are suspected, at least, for the latter species.

Open seed cone of Willowmore cedar (*Widdringtonia schwarzii*) after seed dispersal; scale, 1 cm.

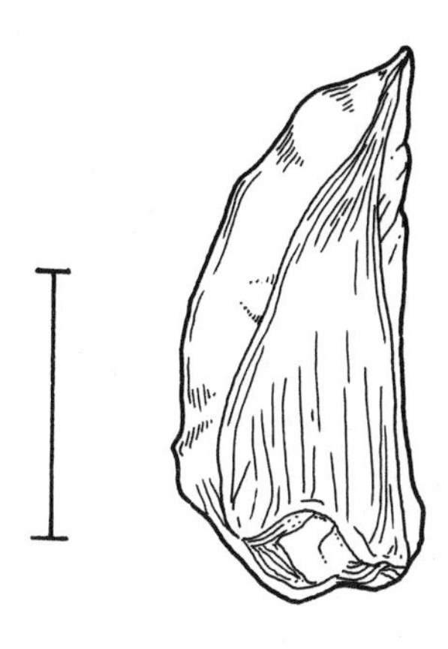

Inner side of dispersed seed of Willowmore cedar (*Widdringtonia schwarzii*); scale, 5 mm.

Widdringtonia schwarzii (Marloth) M. T. Masters

WILLOWMORE CEDAR

Tree to 35 m tall but usually shorter, with trunk to 3 m in diameter. Bark fibrous, flaking, reddish gray. Crown narrowly conical at first, remaining relatively compact with age. Juvenile leaves 1–2 cm long, up to 2 mm wide. Scale leaves (1.5–)2–4(–10) mm long, the free tip often much shorter than the attached base, rounded in cross section. Pollen cones about 2 mm long. Seed cones 1.5–2 cm across, the rim and outer face of the scales both strongly warty, the subapical point moderately developed. Seeds black, the body 5–10 mm long, conical, curved, the larger wing to 3 mm wide, the smaller rudimentary. South Africa, Baviaanskloof and Konga Mountains, Cape province. Rocky, sterile gorges, among fire-prone shrublands; 850–1,200 m. Zone 10?

Like the other two tree species of *Widdringtonia*, the Willowmore cedar yields an attractive, fragrant, decay-resistant wood and has been largely depleted. It is killed by fire without resprouting, although seed regeneration depends on periodic fires.

Widdringtonia whytei Rendle

MULANJE CEDAR

Tree to 42 m tall, with trunk to 2 m in diameter. Bark fibrous, deeply furrowed, longitudinally peeling, reddish brown to gray-brown. Crown conical at first, broadening and flattening with age. Juvenile leaves to 2.5 cm long, up to 2 mm wide. Scale leaves (1.5–)2–3.5(–10) mm long, the free tip shorter than to about as long as the attached base, rounded in cross section. Pollen cones about 2 mm long. Seed cones 0.8–1.5 cm across, the rim of the seed scales smooth and the outer face smooth or lightly warty, the subapical point small. Seeds black, the body 5–9 mm long, oblong, curved, the larger wing to 4 mm wide, the smaller rudimentary. Malawi, Mount Mulanje. Between grassland and closed forest on slopes and in gullies; mostly at 1,600–2,200 m. Zone 9?

There has been taxonomic confusion between this largest species of *Widdringtonia* and the shrubby berg cypress (*W. nodiflora*) because both occur on Mount Mulanje and may hybridize. Plantations on nearby ranges may have such a mixed origin. The

Mulanje cedar is killed by fires and does not resprout like *W. nodiflora,* which occurs at higher elevations and in drier, more exposed sites. Mulanje cedar yields a valuable timber and was severely depleted by a century of overexploitation before plantations began to be established on Mount Mulanje and nearby mountains in Malawi. A small drawing room at the headquarters building of the Royal Commonwealth Society in London was paneled with timbers of this species in the 1950s.

Wollemia W. G. Jones, K. Hill & J. M. Allen

WOLLEMI PINE

Araucariaceae

Evergreen trees with one to many (by natural coppicing) straight, slender trunks clothed with unique bark made up entirely of densely packed bumps. Original branches in distinct tiers, upright at first, then flattening and arching with age until a pollen or seed cone is initiated at their tip after several years of growth increments and they fall after cone maturation and are replaced by new ones growing out from the trunk in irregular positions. Resting buds loosely wrapped by short needle leaves but without specialized bud scales. Leaves opposite or nearly so, needlelike, broadly sword-shaped, flat, without a midrib and with numerous parallel veins, their bases running down onto the stem without an intervening petiole, remaining on the branch until it falls. Juvenile leaves in two horizontal rows, proportionately longer, narrower, and more sharply pointed than the adult leaves. Adult leaves in four rows, with two horizontal rows and two rows forming a V-shaped trough above these, each leaf pair contributing one leaf to a horizontal row and another to the opposite row of the trough.

Plants monoecious. Pollen cones solitary and hanging at the end of branches in the lower portion of the crown. Individual pollen cones not stalked, large, cylindrical, with hundreds of pollen scales, and cupped in a ring of about eight small bracts. Each scale shieldlike, with a large angular boss projecting from the surface and with four to nine elongate pollen sacs. Pollen grains small to medium (30–60 µm in diameter), spherical, minutely bumpy with interspersed larger and smaller granules but otherwise nearly featureless. Seed cones solitary and horizontal to partially upright at the end of branches in the upper portion of the crown after 3–6 years of growth, maturing in 2–3 years and shattering at maturity leaving the bare cone axis and a few basal scales behind. Individual seed cones spherical in outline but the shape obscured by the long, bladelike tips of the hundreds of spirally arranged cone scales, which almost double the apparent cone diameter. Each cone scale with complete fusion of the bract and seed scale portions, broadly winged on either side to yield a rhombic outline overall, contracting abruptly to the free tip that emerges from the broadly and narrowly diamond-shaped external face. Seeds solitary at the midline of the scales, inverted (attached near the far tip and with the original pollination opening pointed toward the cone axis), detaching separately and more or less evenly narrow-winged around the margin. Cotyledons two, each with six to eight veins. Chromosome base number $x = 13$.

Wood odorless, somewhat glossy, soft and medium in weight, with brownish white sapwood not sharply delimited from the light brown heartwood. Grain very fine and even, with clearly defined growth rings marked by a gradual transition to a narrow band of much smaller but not much thicker walled latewood tracheids. Resin canals and individual resin parenchyma cells both absent.

Stomates not in definable bands, in numerous discontinuous lines beneath and absent (juvenile leaves) or in fewer lines (adult leaves) above. Each stomate sunken beneath and partly hidden by the four to six surrounding subsidiary cells, which are themselves sunken beneath the level of the other epidermal cells (and are thus neither shared between neighboring stomates nor topped by a Florin ring). Leaf veins midway between the top and bottom, parallel, more or less evenly spaced across the width, numbering (5–)8–15 in the widest portion, overlying a cap of transfusion tissue, alternating with large resin canals midway between each pair of veins. Photosynthetic tissue with a well-developed palisade of two or three cell layers beneath the epidermis and well-developed associated patches of hypodermal cells above and a less complete one below, the space in between occupied by spongy mesophyll containing scattered small resin canals and special compartmented cells packed with dark material.

One species in southeastern Australia. References: Chambers et al. 1998, Gilmore & Hill, 1997, Jones et al. 1995, Setoguchi et al. 1998, Woodford 2002.

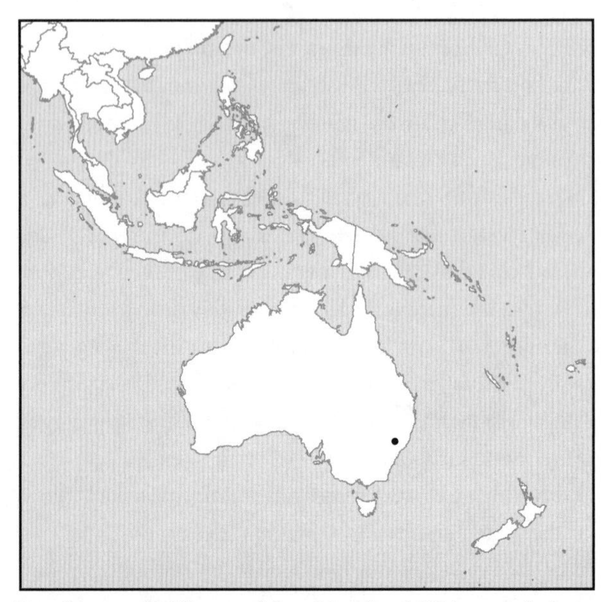

Distribution of *Wollemia nobilis* in New South Wales near Sydney, scale 1 : 119 million

Since the two known stands of the sole living species, *Wollemia nobilis,* in Wollemi National Park (hence the scientific and common names) are about as close to being a single clone as one can get, and are highly homozygous to boot, they have the least natural genetic variation known among the conifers. As a result, we cannot expect much cultivar development once the tree comes into general cultivation, even when it is propagated worldwide, and not just by the massive effort to save the tree undertaken by the Royal Botanic Garden, Sydney. It is now widely available in the form of clonal plantlets produced by tissue culture.

Each pair of living genera in the Araucariaceae shares traits that are absent in the third genus. *Wollemia* shares more such traits with *Araucaria* than it does with *Agathis,* even though two of those shared with the kauris (*Agathis*), opposite leaves (but in a very different arrangement in each genus) and winged seeds attached atop the cone scale rather than embedded within it, are more conspicuous than most similarities with *Araucaria.* Because of the phylogenetic structure revealed by DNA studies, traits shared with *Araucaria,* such as the bristle-tipped cone scales or the special, substance-storing compartmented cells, may be traits inherited from the common ancestor of all three genera. At least, this is a hypothesis that can be explored using a variety of techniques, including genetic, developmental, and paleobotanical studies.

Announcement of the discovery of the sole extant species, *Wollemia nobilis,* in 1995 created an overnight international sensation exceeding, but nowhere near as long lasting, as the one surrounding the announcement of the discovery of *Metasequoia glyptostroboides* in 1948, in part because of different world circumstances, closer proximity to large urban centers (if not easier access), and vastly improved communications. The two genera (one species each) were actually quite similar in scientific importance, both having long, geographically extensive fossil records that required reevaluation, and comparable taxonomic and evolutionary implications within their respective families. DNA studies placing *Agathis* and *Araucaria* together before they join with *Wollemia* suggest that many older araucarian fossils previously identified as *Araucaria* and dating back to the Triassic, more than 200 million years ago, may actually belong to groups ancestral to all three modern genera.

The known fossil record of *Wollemia* itself (or a close relative) extends back to the mid-Cretaceous, some 95–125 million years ago in the Barremian to Cenomanian stages, making it one of the oldest living conifer genera. Juvenile and adult foliage, pollen and seed cones, and seeds are all present in eastern Australian deposits. Related seed cones are also known from the late Cretaceous of New Zealand. Fossil pollen (reasonably securely) presumed to have been produced by ancient *Wollemia* trees (and called *Dilwynites* when dispersed in sediments) was also in Antarctica during the Paleocene, some 60 million years ago. It reached peak abundance during the Paleocene to mid-Eocene (some 60–40 million years ago) in the eastern half of Australia, declined in range and abundance thereafter, and is last recorded as a fossil in Tasmania as recently as 2 million years ago.

Wollemia nobilis W. G. Jones, K. Hill & J. M. Allen

WOLLEMI PINE

Plate 67

Tree to 40 m tall, becoming multitrunked with age through suckering at the base. Bark smooth at first and peeling in small flakes, these later accumulating as dense corky bumps in a layer up to 2 cm thick. Crown narrow. Branches shed after cone production and later replaced by new sprouts from the trunk. Leaves varying in length along the branches, longest near the beginning of each year's growth, but the very first leaves of each flush scale like, about 3 mm long. Juvenile leaves with a pale waxy coating beneath, 2–8 cm long, 2–5 mm wide, with 6–10 veins. Adult leaves 1–4 cm long, 4–8 mm wide, with 9–14 veins. Pollen cones to 11 cm long, 2 cm thick. Seed cones approximately spherical, 5–8 (–12.5) cm in diameter, the clawlike free tips of the bracts 5–10 mm long. Seeds flattened, winged all around, 4–7 mm long, 4–6 mm across, including the wing. Australia in the Blue Mountains of New South Wales. Emergent above warm temperate rain forest in deep-walled canyons protected from fire; 650–800 m. Zone 9?

This rare tree was described from a single grove discovered in Wollemi National Park near Sydney in 1994. Although it generally resembles species of *Agathis* and *Araucaria,* the long-known genera of Araucariaceae, it clearly differs from both in bark, leaf, and seed cone structure. The leaf arrangement of the adult branches is unique among conifers.

Twig of Wollemi pine (*Wollemia nobilis*) with 3 years of growth ending in a slightly immature seed cone, ×0.25.

APPENDIX 1

CONVERSION TABLES

mm	*inches*
1	0.04
2	0.08
3	0.12
4	0.16
5	0.20
6	0.24
7	0.28
8	0.32
9	0.35
10	0.39

cm	*inches*
1	0.4
2	0.8
3	1.2
4	1.6
5	2
6	2.4
7	2.8
8	3.2
9	3.6
10	4
20	8
30	12
40	16
50	20
60	24
70	28
80	32
90	35
100	40

m	*feet*
1	3.3
2	6.6
3	9.8
4	13
5	16
6	20
7	23
8	26
9	30
10	33
20	66
30	98
40	130
50	160
60	200
70	230
80	260
90	300
100	330
200	660
300	980
400	1,300
500	1,600
600	2,000
700	2,300
800	2,600
900	3,000
1000	3,300

km	*miles*
1	0.6
2	1.2
3	1.9
4	2.5
5	3.1
6	3.7
7	4.3
8	5.0
9	5.6
10	6.2
20	12
30	19
40	25
50	31
60	37
70	44
80	50
90	56
100	62
200	120
300	190
400	250
500	310
600	380
700	440
800	500
900	560
1,000	620
2,000	1,200
3,000	1,900
4,000	2,500
5,000	3,100

APPENDIX 2

AUTHORITIES FOR SCIENTIFIC NAMES

Abbreviation	*Full Name*
Abrams	LeRoy Abrams (1874–1956)
R. P. Adams	Robert P. Adams (b. ca. 1945)
Adanson	Michel Adanson (1726–1806)
W. Aiton	William Aiton (1731–1793)
J. M. Allen	Jan M. Allen (fl. 1995)
Andresen	John William Andresen (b. 1925)
F. Antoine	Franz Antoine (1815–1886)
W. Archer bis	William Archer (1820–1874)
J. F. Arnold	Johann Franz Xaver Arnold (fl. 1785)
G. Arnott	George Arnold Walker Arnott (1799–1868)
P. Ascherson	Paul Friedrich August Ascherson (1834–1913)
Ata	Cemil Ata (fl. 1987)
Averyanov	Leonid V. Averyanov (b. 1955)
Axelrod	Daniel Irving Axelrod (1910–1998)
D. Bailey	Dana K. Bailey (b. 1916)
F. M. Bailey	Frederick Manson Bailey (1827–1915)
J. F. Bailey	John Frederick Bailey (1866–1938)
L. H. Bailey	Liberty Hyde Bailey (1858–1954)
E. Bailly	Émile Bailly (1829–1894)
J. G. Baker	John Gilbert Baker (1834–1920)
R. T. Baker	Richard Thomas Baker (1854–1941)
J. Balfour	John Hutton Balfour (1808–1884)
J. Ball	John Ball (1818–1889)
J. Banks	Joseph Banks (1743–1820)
W. Barrett	W. H. G. Barrett (fl. 1962–1972)
Bartel	Jim A. Bartel (fl. 1983)
Batalin	Alexander Theodorowicz Batalin (1847–1896)
Battandier	Jules Aimé Battandier (1848–1922)
Baumann-Bodenheim	Marcel Gustav Baumann-Bodenheim (1920–1996)
J. Beaman	John Homer Beaman (b. 1929)
R. M. Beauchamp	R. Mitchel Beauchamp (fl. 1974)
G. Beck	Günther von Mannagetta und Lërchenau Beck (1856–1931)
Beissner	Ludwig Beissner (1843–1927)
G. Bennett	George Bennett (1804–1893)
Bentham	George Bentham (1800–1884)
Berchtold	Friedrich Graf von Berchtold (1781–1876)
Berthelot	Sabin Berthelot (1794–1880)
A. Bertoloni	Antonio Bertoloni (1775–1869)
M. Bertrand	Marcel C. Bertrand
C. Bessey	Charles Edwin Bessey (1845–1915)
M. Bieberstein	Friedrich August Marschall von Bieberstein (1768–1826)
Bisse	Johannes Bisse (1935–1984)
Bizzarri	Maria Paola Bizzarri (b. 1937)
S. T. Blake	Stanley Thatcher Blake (1910–1959)
C. Blanco	Cenobio E. Blanco
Blume	Carl [Karl] Ludwig von Blume (1796–1862)
A. V. Bobrov	Alexey Vladimir F. Ch. Bobrov (b. 1969)
Boissier	Pierre Edmond Boissier (1810–1885)
B. Boivin	Joseph Robert Bernard Boivin (1916–1985)
Bolaños	Bolaños
Bongard	August [Gustav] Heinrich von Bongard (1786–1839)
Bonpland	Aimé Jaques Alexandre Bonpland (1773–1858)
B. K. Boom	Boudewijn Karel Boom (1903–1980)
Bordères	O. Bordères (-Rey) (fl. 1939–1968)
Borhidi	Attila L. Borhidi (b. 1932)
Boutelje	Julius B. Boutelje (fl. 1954)
N. Britton	Nathaniel Lord Britton (1859–1934)
Ad. Brongniart	Adolphe Théodore Brongniart (1801–1876)
R. Brown	Robert Brown (1773–1858)
S. Brown	Stewardson Brown (1867–1921)
Brügger	Christian Georg Brügger [von Churwalden] (1833–1899)

Abbreviation	*Full Name*
Buchanan-Hamilton	Francis Buchanan-Hamilton (1762–1829)
J. Buchholz	John Theodore Buchholz (1888–1951)
Burdet	Hervé Maurice Burdet (b. 1939)
J. Burgh	Johannes van der Burgh (b. 1937)
Burgsdorff	Friedrich August Ludwig von Burgsdorff (1747–1802)
Businský	Roman Businský (b. 1951)
Bykov	Boris Aleksandrovich Bykov (1910–1990)
M. Caballero	Miguel Caballero Deloya (fl. 1969)
A. Camus	Aimée Antoinette Camus (1879–1965)
Carabia	José Perez Carabia (b. 1910)
Carrière	Élie Abel Carrière (1818–1896)
Carvajal	Servando Carvajal Hernandez (fl. 1981)
Ceballos	Luis Ceballos (Fernández de Córdoba) (1896–1967)
L. J. Čelakovský	Ladislav Josef Čelakovský (1834–1902)
Chamisso	Ludolf Karl Adelbert von Chamisso (1781–1838)
C. C. Chang	Chao Chien Chang (b. 1900)
T. C. Chang	Ting Chien Chang (fl. 1981)
N. Chao	(Liang) Neng Chao (b. 1931)
A. W. Chapman	Alvan (Alvin) Wentworth Chapman (1809–1899)
H. Chapman	H. H. Chapman
Chun G. Chen	Chun Gen Chen (fl. 1981)
Zhi X. Chen	Zhi Xiu Chen (fl. 1989)
W. C. Cheng	Wan Chun Cheng (1904–1983)
A. Chevalier	Auguste Jean Baptiste Chevalier (1873–1956)
S. S. Chien	Sung Shu Chien (1885–1965)
H. Christ	Konrad Hermann Heinrich Christ (1833–1933)
C. D. Chu	Cheng De Chu (b. 1928)
W. Y. Chun	Woon Young Chun (1890–1971)
R. T. Clausen	Robert Theodore Clausen (1911–1981)
Coaz	Johann Wilhelm Fortunat Coaz (1822–1918)
Coincy	Auguste Henri Cornut de Coincy (1837–1903)
Colenso	(John) William Colenso (1811–1899)
Coltman-Rogers	Charles Coltman-Rogers (1854–1929)
R. Compton	Robert Harold Compton (1886–1979)
Coode	Mark James Elgar Coode (b. 1937)
Corbasson	Michel Corbasson (fl. 1968)
Corner	Edred John Henry Corner (1906–1996)
D. Correll	Donovan Stewart Correll (1908–1983)
Cory	Victor Louis Cory (1880–1964)
Côzar	Santiago Sanchez Côzar (fl. 1952)
Craib	William Grant Craib (1882–1993)
W. Critchfield	William Burke Critchfield (1923–1989)
H. Croom	Hardy Bryan Croom (1797–1837)
Csató	J. Csató (1833–1913)

Abbreviation	*Full Name*
Cullen	James Cullen (b. 1936)
A. Cunningham	Allan Cunningham (1791–1839)
Dallimore	William Dallimore (1871–1959)
P. H. Davis	Peter Hadland Davis (1918–1992)
Debeaux	Jean Odon Debeaux (1826–1910)
Debreczy	Zsolt Debreczy (b. 1941)
J. Decaisne	Joseph Decaisne (1807–1882)
A. P. de Candolle	Augustin Pyramus de Candolle (1778–1841)
de Ferré	Yvette de Ferré (fl. 1949)
de Lannoy	de Lannoy (fl. 1863)
de Laubenfels	David John de Laubenfels (b. 1925)
Deppe	Ferdinand Deppe (1794–1861)
Desfontaines	René Louiche (1750–1833)
de Vriese	Willem Hendrik de Vriese (1806–1862)
A. Dietrich	Albert Gottfried Dietrich (1795–1856)
Dode	Louis Albert Dode (1875–1943)
T. Doi	Tōhei Doi (1882–1946)
Domin	Karel Domin (1882–1935)
D. Don	David Don (1799–1841)
G. Don	George Don (1764–1814)
K. Dorman	Keith William Dorman (b. 1910)
D. Douglas	David Douglas (1798–1834)
Doweld	Alexander Borissovitch Doweld (b. 1973)
D. Downie	Dorothy G. Downie (d. 1960)
E. Drake	Emmanuel Drake del Castillo (1855–1904)
Druce	Carl Georg Oscar Druce (1852–1933)
Duhamel	Henri Louis Duhamel du Monceau (1700–1782)
Dummer	Richard Arnold Dummer (1887–1922)
Dunal	Michel Félix Dunal (1789–1856)
S. Dunn	Stephen Troyte Dunn (1868–1938)
Du Roi	Johann Philipp Du Roi (1741–1785)
Du Tour	Du Tour de Salvert (fl. 1803–1815)
Dylis	Nikolai Vladislavovich Dylis (1915–1985)
Eckenwalder	James Emory Eckenwalder (b. 1949)
Eguiluz	Teobaldo Eguiluz Piedra (fl. 1982–1988)
C. G. Ehrenberg	Christian Gottfried Ehrenberg (1795–1876)
A. Eichler	August Wilhelm Eichler (1839–1887)
E. Ekman	Erik Leonard Ekman (1883–1931)
Endlicher	Stephan Friedrich Ladislaws Endlicher (1804–1849)
Engelmann	Georg (George) Engelmann (1809–1884)
S. Z. Fang	S. Z. Fang (fl. 1984)
Farjon	Aljos K. Farjon (b. 1946)
Fassett	Norman Carter Fassett (1900–1954)
B. Fedtschenko	Boris Alexjewitsch (Alexeevich) Fedtschenko (1872–1947)
K. M. Feng	Kuo Mei Feng (1917–2007)
D. K. Ferguson	David Kay Ferguson (b. 1942)

Abbreviation	*Full Name*
Fernald	Merritt Lyndon Fernald (1873–1950)
Fieschi	V. Fieschi (b. ca. 1910)
F. E. L. Fischer	Friedrich Ernst Ludwig von (Fedor Bogdanovic) Fischer (1782–1854)
Fitschen	Jost Fitschen (1869–1947)
Florin	Carl Rudolf Florin (1894–1965)
Flous	Fernande Flous (b. 1908)
Fomin	Aleksandr Vasiljevich Fomin (1869–1935)
Jas. Forbes	James Forbes (1773–1861)
Formánek	Eduard Formánek (1845–1900)
J. G. Forster	Johann Georg Adam Forster (1754–1794)
J. R. Forster	Johann Reinhold Forster (1729–1798)
Fortune	Robert Fortune (1812–1880)
Foxworthy	Frederick William Foxworthy (1877–1950)
Fraas	Carl Nicolaus Fraas (1810–1875)
Franchet	Adrien René Franchet (1834–1900)
Franco	João Manuel Antonio do Amaral Franco (b. 1921)
Frankis	Michael P. Frankis (fl. 1993)
Frémont	John Charles Frémont (1813–1890)
E. M. Fries	Elias Magnus Fries (1794–1878)
L. K. Fu	Li Kuo Fu (b. 1934)
S. H. Fu	Shu Hsia Fu (1916–1986)
J. Gaertner	Joseph Gaertner (1732–1791)
Gandoger	Michel Gandoger (1850–1926)
J. Garden	Joy Garden (Joy Thompson after 1956; b. 1923)
C. A. Gardner	Charles Austin Gardner (1896–1970)
H. Gaussen	Henri Marcel Gaussen (1891–1981)
F. Gay	(Henri Félix) François Gay (1858–1898)
Gáyer	Gyula (Julius) Gáyer (1883–1932)
Georgescu	C. Constantin Georgescu (1898–1968)
L. Gibbs	Lilian Suzette Gibbs (1870–1925)
Glendinning	Robert Glendinning (1805–1862)
Godron	Dominique Alexandre Godron (1807–1880)
Golfari	L. Golfari (fl. 1962)
G. Gordon	George Gordon (1806–1879)
K. Graebner	Karl Otto Robert Peter Paul Graebner (1871–1933)
N. Gray	Netta Elizabeth Gray (1913–1970)
S. F. Gray	Samuel Frederick Gray (1766–1828)
E. Greene	Edward Lee Greene (1843–1915)
Greguss	Pál [Paul] Greguss (1889–1984)
Greuter	Werner Rodolfo Greuter (b. 1938)
Greville	Robert Kaye Greville (1794–1866)
W. Griffith	William Griffith (1810–1845)
A. Grierson	Andrew John Charles Grierson (1929–1990)
Gris	Jean Antoine Arthur Gris (1829–1872)
Griscom	Ludlow Griscom (1890–1959)

Abbreviation	*Full Name*
Grisebach	August Heinrich Rudolf Grisebach (1814–1879)
Guillaumin	André Guillaumin (1885–1974)
Guízar	Enrique Guízar Nozalco (fl. 1995)
Gussone	Giovanni Gussone (1787–1866)
Haenke	Thaddäus (Tadeás) Peregrinus Xaverius Haenke (1761–1817)
M. Hall	Marion Trufant Hall (b. 1920)
J. Haller	John Robert Haller (b. 1930)
Hance	Henry Fletcher Hance (1827–1886)
Handel-Mazzetti	Heinrich R. E. Handel-Mazzetti (1882–1940)
S. H. Hao	Shu Hwa Hao (b. 1931)
D. Harder	Daniel Kenneth Harder (b. 1960)
S. Harrison	Sydney Gerald Harrison (1929–1988)
K. Hartweg	Karl Theodor Hartweg (1812–1871)
Hatusima	Sumihiko Hatusima (b. 1906)
F. Hawksworth	Frank Goode Hawksworth (b. 1926)
Yasaka Hayashi	Yasaka Hayashi (b. 1911)
Hayata	Bunzô Hayata (1874–1934)
Heer	Oswald von Heer (1809–1883)
J. B. Henkel	Johan Baptist Henkel (1815– or 1825–1871)
A. Henry	Augustine Henry (1857–1930)
Heywood	Vernon Hilton Heywood (b. 1927)
P. Hickel	Paul Robert Hickel (1865–1935)
Hiêp	see Nguyên
F. G. Hildebrand	Friedrich Hermann Gustav Hildebrand (1835–1915)
K. Hill	Kenneth D. Hill (b. 1948)
H. Hillier	Sir Harold George Knight Hillier (1905–1985)
C. F. Hochstetter	Christian Ferdinand Friedrich Hochstetter (1787–1860)
W. Hochstetter	Wilhelm Christian Hochstetter (1825–1881)
Höss	Franz Höss (1756–1840)
Holmboe	Jens Holmboe (1880–1943)
Hooibrenk	Danield Hooibrenk (fl. 1848–1861)
J. Hooker	Joseph Dalton Hooker (1817–1911)
W. J. Hooker	William Jackson Hooker (1785–1865)
Hoopes	Josiah Hoopes (1832–1904)
Horsfield	Thomas Horsfield (1773–1859)
J. T. Howell	John Thomas Howell (1903–1994)
K. Hsia	Kuang Cheng Hsia (fl. 1982)
C. F. Hsieh	Chang Fu Hsieh (b. 1947)
Ji R. Hsüeh	Ji Ru [Xue] Hsüeh (b. 1921)
H. H. Hu	Hsen Hsu Hu (1894–1968)
S. Y. Hu	Shiu Ying Hu (b. 1910)
W. L. Huang	Wei Lian Huang (fl. 1984)
Hultén	Oskar Eric Gunnar Hultén (1894–1981)
Humboldt	Friedrich Wilhelm Heinrich Alexander von Humboldt (1769–1859)

Abbreviation	*Full Name*
D. Hunt	David Richard Hunt (b. 1938)
B. Hyland	Bernard Patrick Matthew Hyland (b. 1937)
Ionescu	M. A. Ionescu (b. 1900)
Keis. Ito	Keisuke Ito (1803–1901)
J. Iwata	Jirô Iwata (b. 1909)
A. B. Jackson	Albert Bruce Jackson (1876–1947)
E. James	Edwin James (1797–1861)
Janchen	Erwin Emil Alfred Janchen-Michel von Westland (1882–1970)
Jepson	Willis Linn Jepson (1867–1946)
R. J. Johns	Robert James Johns (b. 1944)
Alb. G. Johnson	Albert G. Johnson (1912–1977)
L. A. Johnson	Lawrence Alexander Sidney Johnson (1925–1997)
I. Johnston	Ivan Murray Johnston (1898–1960)
W. G. Jones	Wyn G. Jones (fl. 1995)
Junghuhn	(Friedrich) Franz Wilhelm Junghuhn (1809–1864)
A. L. Jussieu	Antoine Laurent de Jussieu (1748–1836)
N. Kang	Ning Kang (fl. 1995)
H. Karsten	Gustav Karl Wilhelm Hermann Karsten (1817–1908)
H. Keng	Hsüan Keng (b. 1923)
P. C. Keng	Pai Chieh Keng (b. 1917)
T. Kirk	Thomas Kirk (1828–1898)
Kitamura	Siro Kitamura (1906–2002)
Klotzsch	Johann Friedrich Klotzsch (1805–1860)
J. Knight	Joseph Knight (ca. 1777–1855)
K. Koch	Karl (Carl) Heinrich Emil (Ludwig) Koch (1809–1879)
Koehne	Berhnard Adalbert Emil Koehne (1848–1918)
G. Koidzumi	Gen'ichi [Geniti, Gen-Iti] Koidzumi (1883–1953)
V. Komarov	Vladimir Leontjevich (Leontevich) Komarov (1869–1945)
Koorders	Sijfert Hendrick Koorders (1863–1919)
Kotschy	Carl (Karl) Georg Theodor Kotschy (1813–1866)
M. Koyama	Mitsuo Koyama (1885–1935)
P. Krylov	Porphyru Nikitic Krylov (Krylow) (1850–1931)
C. T. Kuan	Chung Tian Kuan (fl. 1983)
Kuang	Ko Rjen [KoZen] Kuang (1914–1977)
Y. Kudô	Yûshun Kudô (1887–1932)
Kunth	Karl (Carl) Sigismund Kunth (1788–1850)
O. Kuntze	Carl [Karl] Ernst [Eduard] Otto Kuntze (1843–1907)
W. Kurz	Wilhelm Sulpiz Kurz (1834–1878)
Kusaka	Masao Kusaka (b. 1915)
Kuzeneva	Olga Fakinfovna Kuzeneva (1887–1978)

Abbreviation	*Full Name*
Labillardière	Jacques Julien Houtton de Labillardière (1755–1834)
Lacassagne	Marcel Lacassagne (fl. 1929)
Lamarck	Jean Baptiste Antoine Pierre de Monnet de (1744–1829)
A. Lambert	Aylmer Bourke Lambert (1761–1842)
K. M. Lan	Kai Min Lan (fl. 1983)
Lanner	Ronald Martin Lanner (b. 1930)
E. Larsen	Egon Larsen (1928–1968)
L. Laurent	Louis Aimé Alexandre Laurent (1873–1947)
Lauterbach	Carl (Karl) Adolf Georg Lauterbach (1864–1937)
Y. W. Law	Yuh Wu Law (1917–2004)
C. Lawson	Charles Lawson (1794–1873)
P. Lawson	Peter Lawson (d. 1820)
P. Lebreton	P. Lebreton (fl. 1981)
P. Lecomte	Paul Henri Lecomte (1856–1934)
Ledebour	Carl (Karl) Friedrich von Ledebour (1785–1851)
J. G. Lehmann	Johann Georg Christian Lehmann (1792–1860)
Lemée	Albert Marie Victor Lemée (1872–1900)
J. Lemmon	John Gill Lemmon (1832–1908)
H. Léveillé	Augustin Abel Hector Léveillé (1863–1918)
L'Héritier	Charles Louis L'Héritier de Brutelle (1746–1800)
C. X. Li	Cheng Xiang Li (fl. 1995)
H. L. Li	Hui-Lin Li (b. 1911)
N. Li	Nan Li (b. 1963)
X. W. Li	Xing Wen Li (b. 1932)
Zhen Q. Li	Zhen Qing Li (fl. 1989)
Zhi Y. Li	Zhi Yun Li (fl. 1995)
Lindley	John Lindley (1799–1865)
J. Link	Johann Heinrich Friedrich Link (1767–1851)
Linnaeus	Carl Linnaeus (1707–1778)
Linnaeus fil.	Carl Linnaeus (1741–1783)
T. N. Liou	Tchen Ngo Liou (1898–1975)
D. Little	Damon P. Little (fl. 2007)
E. Little	Elbert Luther Little, Jr. (1907–2004)
D. Litvinov	Dimitri Ivanovich Litvinov (1854–1929)
Q. X. Liu	Qi Xing Liu (fl. 1989)
Tang S. Liu	Tang Shui Liu (b. 1911)
C. Loddiges	Conrad (L.) Loddiges (1738–1826)
Lojacano	Michele Lojacano [-Pojero] (1853–1919)
D. Long	David Geoffrey Long (b. 1948)
Loock	E. E. M. Loock (1905–1973)
H. Loret	Henri Loret (1811–1888)
J. C. Loudon	John Claudius Loudon (1783–1843)
Y. J. Lu	Yan Jun Lu (fl. 1992)
Luerssen	Christian Luerssen (1843–1916)

Abbreviation	*Full Name*
C. Lundell	Cyrus Longworth Lundell (b. 1907)
McClelland	John McClelland (1805–1883)
McVaugh	Rogers McVaugh (b. 1909)
Madrigal	Xavier Madrigal Sánchez (fl. 1969)
J. Maire	Réné Charles Joseph Ernest Maire (1878–1949)
Makino	Tomitaro Makino (1862–1957)
Markgraf	Friedrich Markgraf (1897–1987)
Marloth	Hermann Wilhelm Rudolf Marloth (1855–1931)
J. A. Marsh	Judith Anne Marsh (b. 1951)
H. Marshall	Humphrey Marshall (1722–1801)
M. Martínez	Maximino Martínez (1888–1964)
Masamune	Genkei Masamune (1899–1993)
F. Mason	Francis Mason (1799–1874)
H. Mason	Herbert Louis Mason (1896–1994)
M. T. Masters	Maxwell Tylden Masters (1833–1907)
J. Mastrogiuseppe	Joy Dell Mastrogiuseppe (fl. 1980s)
R. Mastrogiuseppe	Ronald J. Mastrogiuseppe (fl. 1980s)
J. Matsumura	Jinzô Matsumura (1856–1928)
G. Mattei	Giovanni Ettore Mattei (1865–1943)
Mattfeld	Johannes Mattfeld (1895–1951)
Matzenko	A. E. Matzenko (b. 1930)
Maximowicz	Carl Johan Maximowicz (1827–1891)
H. Mayr	Heinrich Mayr (1856–1911)
J. Medwedew	Jakob Sergejevitsch Medwedew (1847–1923)
Meijer Drees	E. Meijer Drees (fl. 1938)
Meikle	Robert Desmond Meikle (b. 1923)
Melikian	Alexander P. Melikian (b. 1935)
Melle	Peter Jacobus van Melle (1891–1953)
Melville	Ronald Melville (1903–1985)
C. Menezes	Carlos Azevedo de Menezes (1863–1928)
Merev	Nesime Merev (fl. 1987)
Merriam	Clinton Hart Merriam (1855–1942)
C. Meyer	Carl (Karl) Anton von (Andreevich) Meyer (1795–1855)
A. Michaux	André Michaux (1746–1803)
Miki	Shigeru Miki (1901–1974)
R. Mill	Robert Reid Mill (b. 1950)
P. Miller	Philip Miller (1691–1771)
Miquel	Friedrich Anton Wilhelm Miquel (1811–1871)
Mirbel	Charles Francois Brisseau de Mirbel (1776–1854)
Mirov	Nicholas Tihomitovich Mirov (1893–1980)
A. F. Mitchell	Alan F. Mitchell (1922–1995)
K. Miyabe	Kingo Miyabe (1860–1951)
Miyoshi	Manabu Miyoshi (1861–1939)
S. L. Mo	Sin Li Mo (fl. 1934–1988)
Moench	Conrad Moench (1744–1805)

Abbreviation	*Full Name*
G. Molina	Giovanni Ignazio [Juan Ignacio] Molina (1737–1829)
Molloy	Brian Peter John Molloy (b. 1930)
C. Moore	Charles Moore (1820–1905)
H. Moore	Harold Emery Moore (1917–1980)
P. Morelet	Pierre Marie Arthur Morelet (1809–1892)
K. Morikawa	Kinichi Morikawa (1898–1936)
A. Morrison	Alexander Morrison (1849–1913)
F. J. Mueller	Ferdinand Jacob Heinrich von Mueller (1825–1896)
Muñoz-Schick	Mélica [Elisa] Muñoz-Schick (b. 1941)
A. Murray bis	Andrew Murray (1812–1878)
E. Murray	Edward Murray (fl. 1969)
Nahal	Nahal
T. Nakai	Takenoshin [Takenosin] Nakai (1882–1952)
Narave	Hector Narave Flores (fl. 1989)
C. Nees	Christian Gottfried Daniel Nees von Esenbeck (1776–1858)
F. Neger	Franz (Friedrich) Wilhelm Neger (1869–1923)
J. Nelson	John "Senilis" Nelson (fl. 1860)
H. Neumayer	Hans Neumayer (1887–1945)
T. H. Nguyên	Tiên Hiêp Nguyên (fl. 1980)
Nitzelius	Tor Nitzelius (1914–1999)
T. Nuttall	Thomas Nuttall (1786–1859)
C. Nyman	Carl Frederick Nyman (1820–1893)
Oersted	Anders Sandoe Oersted (1816–1872)
Ohwi	Jisaburo Ohwi (1905–1977)
M. Orr	Matthew Young Orr (ca. 1883–1953)
Ostenfeld	Carl Emil Hansen Ostenfeld (1873–1931)
F. Otto	[Christoph] Friedrich Otto (1783–1856)
C. Page	Christopher Nigel Page (b. 1942)
P. Pallas	Peter (Pyotr) Simon von Pallas (1741–1811)
Pančić	Joseph (Josif) Pančić (1814–1888)
Parfenov	Victor Ivanovich Parfenov (b. 1934)
Parlatore	Filippo Parlatore (1816–1877)
C. Parry	Charles Christopher Parry (1823–1890)
Passini	Marie-Francoise Passini [Robert, Robert-Passini] (fl. 1987–1994)
Patschke	Wilhelm Patschke (b. 1888)
T. F. Patterson	Thomas Frederick Patterson (fl. 1980)
Pavon	José Antonio Pavon (y Jimenéz) (1754–1844)
P. Pérez	P. Pérez (fl. 2002)
Pérez de la Rosa	Jorge A. Pérez de la Rosa (b. 1955)
J. P. Perry	Jesse P. Perry (fl. 1982)
Persoon	Christiaan Hendrik Persoon (1761–1836)
Peyerimhoff	Paul de Peyerimhoff de Fontenelle
K. L. Phan	Ke Lôc Phan (b. 1935)
R. Philippi	Rudolph Amandus [Rodolfo (Rudolf) Amando] Philippi (1808–1904)

Abbreviation	*Full Name*
Pichi Sermolli	Rodolfo Emilio Giuseppe Pichi Sermolli (1912–2005)
Pilger	Robert Knud Friedrich Pilger (1876–1953)
Poeppig	Eduard Friedrich Poeppig (1798–1868)
Poggenburg	Justus Ferdinand Poggenburg (1840–1893)
Poiret	Jean Louis Marie Poiret (1755–1834)
P. Poiteau	Pierre Antoine Poiteau (1766–1854)
Pourtet	J. Pourtet (fl. 1954)
Powrie	Elizabeth Powrie (1925–1977)
C. Presl	Carl [Karl, Carel, Carolus] Borivoj [Boriwog, Boriwag] Presl (1794–1852)
J. Presl	Jan Svatopluk [Swatopluk] Presl (1791–1849)
E. Pritzel	Ernst Georg Pritzel (1875–1946)
Purkyně	Emanuel von Purkyně (1832–1882)
Pursh	Frederick Traugott Pursh (1774–1820)
Quinn	Christopher John Quinn (b. 1936)
Raciborski	Marjan (Maryan, Marian, Maryjan) Raciborski (1863–1917)
I. Rácz	István Rácz (b. 1952)
Rafinesque	Constantine Samuel Rafinesque [-Schmaltz] (1783–1840)
G. Ramírez	Gustavo Ramírez Santiago (fl. 1995)
Ramond	Louis Francois Elisabeth Ramond de Carbonnière (1753–1827)
Raup	Hugh Miller Raup (1901–1995)
E. Regel	Eduard August von Regel (1815–1892)
Rehder	Alfred Rehder (1863–1949)
H. Reichardt	Heinrich Wilhelm Reichardt (1835–1885)
Rendle	Alfred Barton Rendle (1865–1938)
A. Richard	Achille Richard (1794–1852)
L. Richard	Louis Claude Marie Richard (1754–1821)
K. Richter	Karl (Carl) Richter (1855–1891)
H. Ridley	Henry Nicholas Ridley (1855–1956)
Righter	Francis Irving Righter (b. 1897)
Risso	Joseph Antoine Risso (1777–1845)
Rivas Martínez	Salvador Rivas Martínez (b. 1935)
M. Robert	Marie-Francoise Robert [Passini, Robert-Passini] (fl. 1978)
Robert-Passini	Marie-Francoise Robert-Passini [Passini, Robert] (fl. 1981)
Roezl	Benedikt (Benito) Roezl (1824–1885)
B. Robinson	Benjamin Lincoln Robinson (1864–1935)
Rothrock	Joseph Trimble Rothrock (1839–1922)
Rouane	M. L. Rouane
Royle	John Forbes Royle (1798–1858)
Roxburgh	William Roxburgh (1751–1815)

Abbreviation	*Full Name*
F. Ruprecht	Franz Josef (Ivanovich) Ruprecht (1814–1870)
K. Rushforth	Keith D. Rushforth (fl. 1983 etc.)
Rydberg	Per Axel Rydberg (1860–1931)
Rzedowski	Jerzy Rzedowski (b. 1926)
J. Sabine	Joseph Sabine (1770–1837)
K. Sahni	Kailash Chandra Sahni (b. 1921)
R. Salazar	Rosalva Miranda Salazar (fl. 1995)
R. A. Salisbury	Richard Anthony Salisbury (1761–1829)
Salomon	Carl E. Salomon (1829–1899)
Saporta	Louis Charles Joseph Gaston de Saporta (1823–1895)
C. Sargent	Charles Sprague Sargent (1841–1927)
Satake	Yoshisuke Satake (1902–2000)
Schiede	Christian Julius Wilhelm Schiede (1798–1836)
D. F. L. Schlechtendal	Diederich Franz Leonhard von Schlechtendal (1794–1866)
Schlechter	Friedrich Richard Rudolf Schlechter (1872–1925)
Friedr. Schmidt	Friedrich [Karl] Fedor Bogdanovich Schmidt (1832–1908)
P. A. Schmidt	Peter A. Schmidt (b. 1946)
C. K. Schneider	Camillo Karl Schneider (1876–1951)
H. Schott	Heinrich Wilhelm Schott (1794–1865)
K. Schumann	Karl Moritz Schumann (1851–1904)
Schur	Philipp Johan Ferdinand Schur (1799–1878)
O. Schwarz	Otto Karl Anton Schwarz (1900–1983)
F. Schwerin	[Graf] Fritz Kurt Alexander von Schwerin (1856–1934)
Scopoli	Joannes Antonius (Giovanni Antonio) Scopoli (1723–1788)
B. Seemann	Berthold Carl Seemann (1825–1871)
Sénéclauze	A. Sénéclauze (fl. 1868)
Seubert	Moritz August Seubert (1818–1878)
G. R. Shaw	George Russell Shaw (1848–1937)
Shirasawa	Homi (Yasuyoshi) Shirasawa (1868–1947)
P. Siebold	Philipp Franz (Balthasar) von Siebold (1796–1866)
Silba	John Silba (fl. 1984)
Sintenis	Paul Ernest Emil Sintenis (1847–1907)
J. K. Small	John Kunkel Small (1869–1938)
H. G. Smith	Henry George Smith (1852–1924)
J. E. Smith	James Edward Smith (1759–1828)
P. A. Smith	P. A. Smith
W. W. Smith	William Wright Smith (1875–1956)
Solander	Daniel Carl Solander (1733–1782)
Sosnowsky	Dmitrii Ivanovich Sosnowsky (1885–1952)
Spach	Édouard Spach (1801–1879)
F. Späth	Franz Ludwig Späth (1838–1913)

Abbreviation	*Full Name*
K. Sprengel	Curt (Kurt, Curtius) Polycarp Joachim Sprengel (1766–1833)
H. St. John	Harold St. John (1892–1991)
P. Standley	Paul Carpenter Standley (1884–1963)
Stapf	Otto Stapf (1857–1933)
Staszkiewicz	Jerzy Staszkiewicz (b. 1929)
Staunton	George Leonard Staunton (1737–1801)
Sterns	Emerson Ellick Sterns (1846–1926)
Steudel	Ernst Gottlieb von Steudel (1783–1856)
Steven	Christian von Steven (1781–1863)
Steyermark	Julian Alfred Steyermark (1909–1988)
Stockwell	William Palmer Stockwell (1898–1950)
Styles	Brian Thomas Styles (1934–1993)
Sudworth	George Bishop Sudworth (1864–1927)
Sukaczev	Vladimir Nikolajevich Sukaczev (1880–1967)
Suter	Johann Rudolf Suter (1766–1827)
Sig. Suzuki	Sigeyosi Suzuki (1894–1937)
Svoboda	Pravdomil Svoboda (1908–1978)
Swartz	Olof (Peter) Swartz (1760–1880)
R. Sweet	Robert Sweet (1783–1835)
Syrach-Larsen	Carl Syrach-Larsen (b. 1898)
Szafer	Władysław Szafer (1886–1970)
Takenouchi	Makoto Takenouchi (b. 1894)
Takhtajan	Armen Leonovich Takhtajan (b. 1910)
Z. C. Tang	Zhao Cheng Tang (fl. 1995)
Tatewaki	Misao Tatewaki (b. 1899)
R. J. Taylor	Ronald J. Taylor (b. 1932)
M. Tenore	Michele Tenore (1780–1861)
Thivend	Simone Thivend (fl. 1981)
H. H. Thomas	Hugh Hamshaw Thomas (1885–1962)
Joy Thompson	Joy Thompson (Joy Garden before 1956; b. 1923)
Thunberg	Carl Peter Thunberg (1743–1825)
Tieghem	Phillippe Édouard Léon van Tieghem (1839–1941)
Tiong	S. K. K. Tiong (fl. 1984)
J. Torrey	John Torrey (1796–1873)
Trabut	Louis (Charles) Trabut (1853–1929)
Trautvetter	Ernst Rudolf von Trautvetter (1809–1889)
Trew	Christoph Jakob Trew (1695–1769)
Tsiang	Ying Tsiang (1898–1982)
Y. L. Tu	Yu Lin Tu (b. 1941)
Turczaninow	Porphir Kiril Nicolai Stepanowitsch Turczaninow (1796–1863)
Turra	Antonio Turra (1730–1796)
Turrill	William Bertram Turrill (1890–1961)
I. Urban	Ignatz Urban (1848–1931)
Uyeki	Homiki Uyeki (b. 1882)
M. H. Vahl	Martin (Henrichsen) Vahl (1749–1804)
Valeton	Theodoric Valeton (1855–1929)

Abbreviation	*Full Name*
Vasek	Frank Charles Vasek (b. 1927)
G. Vasey	George Vasey (1822–1893)
V. N. Vassiljev	Viktor Nikolayevich Vassiljev (1890–1987)
J. G. Veitch	John Gould Veitch (1839–1870)
Ventenat	Étienne Pierre Ventenat (1757–1808)
Vieillard	Eugène (Deplanche Émile) Vieillard (1819–1896)
Viguié	M.-Th. Viguié (fl. 1960)
E. Villar	Emile (Emílio) Huguet del Villar (1871–1951)
Vipper	Pavel Borisovich Vipper (b. 1920)
Visiani	Roberto de Visiani (1800–1878)
A. Voss	Andreas Voss (1857–1924)
N. Wallich	Nathaniel Wallich (1786–1854)
T. Walter	Thomas Walter (1740–1789)
W. T. Wang	Wen Tsai Wang (b. 1926)
Z. Wang	Zhan (Chan, Chang, Than, Tsang) Wang (1911–2000)
O. Warburg	Otto Warburg (1859–1938)
D. Ward	Daniel Bertram Ward (b. 1928)
Wasscher	Jacob Wasscher (1911–1966)
S. Watson	Sereno Watson (1826–1892)
P. Webb	Philip Barker Webb (1793–1854)
C. White	Cyril Tenison White (1890–1950)
T. Whitmore	Timothy Charles Whitmore (b. 1935)
I. Wiggins	Ira Loren Wiggins (1899–1987)
Wildpret	Wolfredo Wildpret de la Torre (b. 1933)
Willdenow	Carl Ludwig von Willdenow (1765–1812)
E. H. Wilson	Ernest Henry Wilson (1876–1930)
P. Wilson	Percy Wilson (1879–1944)
C. [B.] Wolf	Carl Brandt Wolf (1905–1974)
Woltz	Philippe Woltz (fl. 1986)
Wóycicki	Zygmunt Wóycicki (1871–1941)
C. L. Wu	Chung Luen Wu (fl. 1956)
M. H. Wu	Ming Hsiang Wu (fl. 1980s)
Ji R. Xue	(see Ji R. Hsüeh)
Yo. Yamamoto	Yoshimatsu Yamamoto (1893–1947)
Chun Y. Yang	Chun Yu Yang (b. 1948)
Y. F. Yu	Yong Fu Yu (b. 1967)
X. Y. Yuan	Xiao Ying Yuan (fl. 1985)
Zanoni	Thomas A. Zanoni (b. 1949)
Zavarin	Eugene Zavarin (fl. 1960s–1990s)
Zenari	Silvia Zenari (1896–1956)
F. H. Zhang	F. H. Zhang (fl. 1984)
H. J. Zhang	Han Jie Zhang (fl. 1985)
S. J. Zhang	Shu Jie Zhang (fl. 1995)
Y. C. Zhong	Ye Cong Zhong (fl. 1981)
D. X. Zhou	D. X. Zhou
A. Zippelius	Alexander Zippelius (1797–1828)
Zuccarini	Joseph Gerhard Zuccarini (1797–1848)

APPENDIX 3

CONIFERS WITH DISTINCTIVE FEATURES

Selection lists are provided for conifers with distinctive features or other characteristics as guides for selection and planting. Following those, lists of the most commonly cultivated conifers by region are given.

Conifer Extremes (documented or suggested)

Smallest:
- *Lepidothamnus laxifolius*

Shortest tree form:
- *Cupressus goveniana* var. *pigmaea*
- *Pinus contorta* var. *contorta*
- other hardpan, bog, and white-sand species

Largest tree (height):
- *Sequoia sempervirens*

Largest tree (girth):
- *Taxodium mucronatum*

Largest tree (mass):
- *Sequoiadendron giganteum*

Thinnest bark:
- *Agathis robusta*
- *Amentotaxus argotaenia*
- *Cupressus guadalupensis*
- *Phyllocladus trichomanoides*
- *Pinus bungeana*
- other smooth-barked species

Thickest bark:
- *Sequoiadendron giganteum*

Shortest life span:
- *Actinostrobus arenarius*

Longest life span:
- *Pinus longaeva*

Most frost sensitive:
- *Agathis dammara* subsp. *dammara*
- *Nageia maxima*
- other lowland tropical species

Cold hardiest:
- *Juniperus communis*
- *Microbiota decussata*
- other arctic and alpine tree-line species

Most drought tolerant:
- *Callitris glaucophylla*
- *Cupressus sempervirens* var. *dupreziana*
- *Juniperus phoenicea* subsp. *phoenicea*
- other desert-margin species

Most flood tolerant:
- *Taxodium distichum* var. *imbricarium*

Narrowest ecological tolerance:
- *Retrophyllum minor*

Widest ecological tolerance:
- *Platycladus orientalis*

Smallest natural range:
- *Abies nebrodensis*

Widest natural range:
- *Juniperus communis*

Smallest scale leaves:
- *Cupressus chengiana*
- several others less than 1 mm

Shortest needles:
- *Dacrydium cupressinum*
- *Falcatifolium falciforme*
- *Juniperus brevifolia*
- *Picea orientalis*
- many others less than 1 cm

Largest needles (length):
- *Pinus devoniana*

Largest needles (area):
- *Nageia maxima*

Fewest needles per fascicle (pines):
- *Pinus monophylla*

Most needles per fascicle (pines):
- *Pinus durangensis*

Shortest-lived needles:
- *Larix lyallii*

Longest-lived needles:
- *Araucaria araucana*
- *Pinus longaeva*

Shortest pollen cones:
- *Microbiota decussata*

Longest pollen cones:
Araucaria hunsteinii
Smallest pollen grains:
Metasequoia glyptostroboides
many other Cupressaceae, Podocarpaceae, and Taxaceae
Largest pollen grains:
Abies species
Smallest seed cones:
Microbiota decussata
Largest seed cones (length):
Pinus lambertiana
Largest seed cones (weight):
Araucaria bidwillii
Pinus coulteri, runner-up
Fewest seeds per cone:
Microbiota decussata
many *Juniperus* species
many Podocarpaceae
most Taxaceae (apparently)
Most seeds per cone:
Pinus lambertiana
Fewest seeds per scale:
all Araucariaceae
all Podocarpaceae
most Taxaceae
Most seeds per scale:
Cupressus duclouxiana
Smallest seeds:
Microstrobos niphophilus
Largest seeds:
Araucaria bidwillii

Narrow Spires, Columns, and Cones

Conifers that tend to be narrow exhibit these features especially in youth:

Abies bracteata
Abies lasiocarpa
Abies sibirica
Araucaria bernieri
Araucaria columnaris
Callitris columellaris
Calocedrus decurrens
Chamaecyparis lawsoniana cultivars
Cunninghamia lanceolata
Cupressus gigantea
Cupressus macrocarpa cultivars
Cupressus sempervirens
Juniperus communis cultivars
Juniperus drupacea
Juniperus scopulorum cultivars
Juniperus virginiana
Larix decidua cultivars
Picea abies cultivars
Picea mariana
Picea omorika
Picea schrenkiana
Pinus banksiana
Pinus contorta
Pinus sylvestris cultivars
Taxus baccata cultivars
Thuja occidentalis cultivars
Thuja plicata cultivars
Tsuga mertensiana

Flat-Topped, Widely Spreading, and Lollipops

Conifers with these features are generally more pronounced with age:

Abies beshanzuensis
Abies delavayi
Abies spectabilis
Afrocarpus falcatus
Agathis australis
Agathis ovata
Agathis robusta
Araucaria araucana
Araucaria cunninghamii
Callitris neocaledonica
Cedrus libani
Cephalotaxus harringtonii
Cupressus guadalupensis
Cupressus lusitanica
Cupressus macrocarpa
Dacrycarpus imbricatus
Dacrydium elatum
Juniperus ashei
Juniperus excelsa
Juniperus foetidissima
Juniperus thurifera
Keteleeria fortunei
Metasequoia glyptostroboides
Neocallitropsis pancheri
Pinus coulteri
Pinus densiflora
Pinus halepensis
Pinus luchuensis
Pinus massoniana
Pinus montezumae
Pinus pinaster
Pinus pinea
Pinus sabiniana
Pinus tabuliformis
Pinus thunbergii
Podocarpus brassii
Podocarpus elongatus
Podocarpus latifolius
Podocarpus salignus
Podocarpus totara
Prumnopitys amara
Prumnopitys taxifolia
Pseudolarix amabilis
Pseudotsuga sinensis

Taxodium distichum
Taxodium mucronatum
Taxus brevifolia
Torreya fargesii
Tsuga chinensis
Tsuga sieboldii
Widdringtonia whytei

Weeping Foliage or Branches

Cedrus libani cultivars
Cupressus cashmeriana
Cupressus funebris
Cupressus macrocarpa cultivars
Cupressus nootkatensis
Dacrydium cupressinum
Juniperus communis cultivars
Juniperus flaccida
Juniperus recurva
Juniperus rigida
Juniperus scopulorum cultivars
Lagarostrobos franklinii
Larix decidua cultivars
Larix mastersiana
Picea abies cultivars
Picea brachytyla
Picea breweriana
Picea smithiana
Picea wilsonii
Pinus armandii
Pinus bhutanica
Pinus lumholtzii
Pinus patula
Pinus pinceana
Pinus wallichiana
Podocarpus henkelii
Podocarpus pendulifolius
Podocarpus salicifolius
Pseudotsuga menziesii cultivars
Sequoiadendron giganteum cultivars
Taiwania cryptomerioides
Taxus baccata cultivars
Torreya jackii
Tsuga canadensis cultivars
Tsuga heterophylla cultivars

Ground Covers, Creepers, and Rock Huggers

Cupressus macrocarpa cultivars
Diselma archeri
Juniperus communis cultivars
Juniperus conferta
Juniperus horizontalis
Juniperus sabina
Larix kaempferi cultivars
Lepidothamnus fonkii
Lepidothamnus laxifolius
Microbiota decussata
Microcachrys tetragona
Microstrobos fitzgeraldii
Picea abies cultivars
Pinus banksiana cultivars
Pinus flexilis cultivars
Pinus mugo cultivars
Pinus pumila
Pinus sylvestris cultivars
Podocarpus gnidioides
Podocarpus lawrencei
Podocarpus nivalis
Podocarpus pilgeri
Pseudotsuga menziesii cultivars
Taxus baccata cultivars
Taxus canadensis
Torreya californica cultivars

Colorful or Interesting Bark

Blocky or large-plated bark, or both:
Cedrus libani
Juniperus deppeana
Pinus canariensis
Pinus halepensis
Pinus nigra
Pinus pinaster
Pinus ponderosa
Pseudolarix amabilis

Fibrous, fluted, or deeply furrowed bark, or all three:
Athrotaxis selaginoides
Callitris glaucophylla
Calocedrus decurrens
Chamaecyparis pisifera
Cryptomeria japonica
Cupressus nootkatensis
Cupressus sempervirens
Juniperus occidentalis
Juniperus recurva
Larix decidua
Libocedrus bidwillii
Metasequoia glyptostroboides
Pinus jeffreyi
Podocarpus elatus
Podocarpus totara
Pseudotsuga menziesii
Sequoia sempervirens
Sequoiadendron giganteum
Thuja plicata
Thujopsis dolabrata

Smooth, bumpy, or ringed bark, or all three:
Abies grandis
Abies homolepis
Araucaria araucana

Araucaria bidwillii
Araucaria cunninghamii
Lagarostrobos franklinii
Phyllocladus aspleniifolius
Picea orientalis
Saxegothaea conspicua
Wollemia nobilis

Scaly, flaky, or mottled bark, or all three:
Abies squamata
Agathis robusta
Cephalotaxus fortunei
Cupressus bakeri
Cupressus guadalupensis
Dacrycarpus dacrydioides
Lepidothamnus intermedius
Phyllocladus trichomanoides
Picea omorika
Picea schrenkiana
Picea sitchensis
Pinus bungeana
Pinus contorta
Pinus densiflora
Pinus sylvestris
Prumnopitys taxifolia
Taxus brevifolia

Colorful or Interesting Seed Cones

Berrylike, juicy seed cones (most of these are dioecious, and only the female has seed cones):
Amentotaxus yunnanensis
Juniperus drupacea
Juniperus pinchotii
Microcachrys tetragona
Phyllocladus aspleniifolius
Podocarpus macrophyllus
Podocarpus salignus
Podocarpus totara
Prumnopitys ferruginea
Pseudotaxus chienii
Taxus cuspidata

Colorful dry seed cones (most of these are monoecious, and mostly colorful just before full ripening):
Abies delavayi
Abies koreana
Abies religiosa
Actinostrobus pyramidalis
Cryptomeria japonica
Larix decidua (at pollination)
Pseudotsuga menziesii var. *glauca*

Interesting seed cone structure (most of these are monoecious):
Abies bracteata
Abies procera
Araucaria araucana
Araucaria bidwillii
Callitris preissii
Callitris rhomboidea
Cupressus bakeri
Cupressus duclouxiana
Keteleeria davidiana
Metasequoia glyptostroboides
Picea pungens
Pinus lambertiana
Pinus sabiniana
Pseudolarix amabilis
Sequoiadendron giganteum
Tetraclinis articulata
Thujopsis dolabrata
Widdringtonia schwarzii

Conifers for Extreme Habitats

Seashores:
Araucaria columnaris
Araucaria heterophylla
Callitris columellaris
Cupressus macrocarpa
Juniperus conferta
Pinus contorta
Pinus luchuensis
Pinus pinaster
Pinus pinea
Pinus radiata
Pinus sylvestris
Pinus thunbergii
Podocarpus costalis
Podocarpus polystachyus
Podocarpus spinulosus

Deserts and drylands:
Callitris glaucophylla
Cupressus sempervirens
Juniperus osteosperma
Pinus canariensis
Pinus monophylla

Ski resorts:
Abies fraseri
Abies lasiocarpa
Abies mariesii
Athrotaxis selaginoides
Juniperus communis
Larix lyallii
Larix sibirica
Microcachrys tetragona
Microstrobos niphophilus
Phyllocladus alpinus
Picea engelmannii
Picea rubens

Pinus albicaulis
Pinus aristata
Pinus cembra
Pinus contorta
Pinus culminicola
Pinus flexilis
Pinus hartwegii
Pinus longaeva
Pinus mugo
Pinus pumila
Podocarpus lawrencei
Podocarpus nivalis

Swamps, bogs, and wetlands:
Chamaecyparis thyoides
Dacrycarpus steupii
Dacrydium cornwallianum
Dacrydium pectinatum
Glyptostrobus pensilis
Larix laricina
Lepidothamnus fonkii
Metasequoia glyptostroboides
Nageia maxima
Nageia motleyi
Picea mariana
Podocarpus laubenfelsii
Podocarpus pseudobracteatus
Prumnopitys amara
Retrophyllum minor
Taxodium distichum

Some Commonly Cultivated Conifers by Region

Europe:
Abies koreana
Chamaecyparis lawsoniana
Cryptomeria japonica
Cupressus ×leylandii
Cupressus sempervirens
Juniperus communis
Juniperus virginiana
Larix decidua
Picea abies
Picea glauca
Picea pungens
Picea sitchensis
Pinus mugo
Pinus nigra
Pinus pinea
Pinus radiata
Pinus sylvestris
Platycladus orientalis
Pseudotsuga menziesii
Sequoiadendron giganteum
Taxus baccata
Taxus cuspidata
Thuja occidentalis
Thuja plicata
Tsuga canadensis

Northeastern North America:
Abies concolor
Chamaecyparis pisifera
Cupressus nootkatensis
Juniperus conferta
Juniperus horizontalis
Juniperus ×pfitzeriana
Juniperus scopulorum
Juniperus virginiana
Larix decidua
Larix ×eurolepis
Metasequoia glyptostroboides
Picea abies
Picea glauca
Picea pungens
Pinus mugo
Pinus nigra
Pinus resinosa
Pinus strobus
Pinus sylvestris
Platycladus orientalis
Pseudotsuga menziesii
Taxus cuspidata
Taxus ×media
Thuja occidentalis
Tsuga canadensis

Pacific Northwest:
Abies grandis
Abies lasiocarpa
Araucaria araucana
Calocedrus decurrens
Cedrus deodara
Cedrus libani subsp. *atlantica*
Chamaecyparis lawsoniana
Cryptomeria japonica
Cunninghamia lanceolata
Cupressus ×leylandii
Cupressus nootkatensis
Metasequoia glyptostroboides
Picea abies
Picea pungens
Picea sitchensis
Pinus ponderosa
Pinus thunbergii
Pinus wallichiana
Pseudotsuga menziesii
Sciadopitys verticillata

Sequoia sempervirens
Sequoiadendron giganteum
Taxus cuspidata
Thuja plicata
Tsuga heterophylla

Southwestern United States:
Abies bracteata
Abies concolor
Abies lasiocarpa
Afrocarpus falcatus
Araucaria bidwillii
Araucaria heterophylla
Calocedrus decurrens
Cedrus libani subsp. *atlantica*
Cryptomeria japonica
Cupressus arizonica
Cupressus macrocarpa
Cupressus sempervirens
Juniperus chinensis
Pinus canariensis
Pinus coulteri
Pinus halepensis
Pinus pinea
Pinus ponderosa
Pinus radiata
Pinus thunbergii
Podocarpus macrophyllus
Pseudotsuga menziesii
Sequoia sempervirens
Sequoiadendron giganteum
Taxodium mucronatum

Southeastern United States:
Abies firma
Araucaria araucana
Calocedrus decurrens
Cedrus libani subsp. *libani*
Cephalotaxus harringtonii
Chamaecyparis obtusa
Chamaecyparis pisifera
Chamaecyparis thyoides
Cryptomeria japonica
Cunninghamia lanceolata
Cupressus arizonica
Cupressus ×leylandii
Juniperus chinensis
Juniperus conferta
Juniperus virginiana
Metasequoia glyptostroboides
Nageia nagi
Pinus bungeana
Pinus elliottii
Pinus taeda
Pinus thunbergii
Platycladus orientalis
Podocarpus macrophyllus
Taxodium distichum
Tsuga canadensis

Eastern Asia:
Cedrus deodara
Cephalotaxus fortunei
Cephalotaxus harringtonii
Chamaecyparis obtusa
Chamaecyparis pisifera
Cryptomeria japonica
Cunninghamia lanceolata
Cupressus funebris
Glyptostrobus pensilis
Juniperus chinensis
Juniperus rigida
Keteleeria fortunei
Metasequoia glyptostroboides
Nageia nagi
Pinus bungeana
Pinus densiflora
Pinus parviflora
Platycladus orientalis
Podocarpus macrophyllus
Pseudolarix amabilis
Sciadopitys verticillata
Taxus cuspidata
Thujopsis dolabrata
Torreya grandis
Torreya nucifera

Tropics:
Afrocarpus falcatus
Afrocarpus mannii
Agathis macrophylla
Araucaria angustifolia
Araucaria columnaris
Araucaria heterophylla
Cupressus lusitanica
Dacrydium elatum
Glyptostrobus pensilis
Libocedrus plumosa
Nageia wallichiana
Phyllocladus hypophyllus
Pinus caribaea
Pinus merkusii
Pinus oocarpa
Pinus patula
Platycladus orientalis
Podocarpus elatus
Podocarpus neriifolius
Podocarpus oleifolius

Prumnopitys montana
Taiwania cryptomerioides
Taxodium mucronatum
Widdringtonia nodiflora
Wollemia nobilis

Australia and New Zealand:
Abies nordmanniana
Afrocarpus falcatus
Agathis australis
Agathis robusta
Araucaria bidwillii
Araucaria cunninghamii
Araucaria heterophylla
Callitris columellaris
Callitris glaucophylla
Calocedrus decurrens
Cedrus deodara
Cedrus libani subsp. *atlantica*
Chamaecyparis obtusa
Chamaecyparis pisifera
Cupressus arizonica
Cupressus macrocarpa
Juniperus chinensis
Metasequoia glyptostroboides
Picea abies
Picea pungens
Pinus canariensis
Pinus radiata
Podocarpus elatus
Podocarpus totara
Taxus baccata

Southern South America:
Agathis robusta
Araucaria araucana
Araucaria bidwillii
Araucaria heterophylla
Austrocedrus chilensis
Cedrus deodara
Cedrus libani subsp. *atlantica*
Cryptomeria japonica
Cunninghamia lanceolata
Cupressus arizonica
Cupressus lusitanica
Cupressus macrocarpa
Cupressus sempervirens
Juniperus virginiana
Pinus canariensis
Pinus halepensis
Pinus pinaster
Pinus pinea
Pinus radiata
Pinus strobus
Platycladus orientalis
Podocarpus salignus
Prumnopitys andina
Sequoia sempervirens
Taxodium distichum

APPENDIX 4

NEW NAMES

The following list validates the nomenclatural innovations corresponding to new taxonomic arrangements used in Chapter 8.

Abies lasiocarpa (W. J. Hooker) T. Nuttall var. *bifolia* (A. Murray bis) Eckenwalder, *stat. et comb. nov.* Basionym: *Abies bifolia* A. Murray bis, Proceedings of the Royal Horticultural Society of London 3: 320. 1863.

Amentotaxus argotaenia (Hance) Pilger var. *assamica* (D. K. Ferguson) Eckenwalder, *stat. et comb. nov.* Basionym: *Amentotaxus assamica* D. K. Ferguson, Kew Bulletin 40: 115. 1985.

Cupressus ×*notabilis* (A. F. Mitchell) Eckenwalder, *comb. nov.* Basionym: ×*Cupressocyparis notabilis* A. F. Mitchell, Journal of the Royal Horticultural Society 95: 453. 1970.

Cupressus ×*ovensii* (A. F. Mitchell) Eckenwalder, *comb. nov.* Basionym: ×*Cupressocyparis ovensii* A. F. Mitchell, Journal of the Royal Horticultural Society 95: 454. 1970.

Keteleeria davidiana (M. Bertrand) L. Beissner subsp. *evelyniana* (M. T. Masters) Eckenwalder, *stat. et comb. nov.* Basionym: *Keteleeria evelyniana* M. T. Masters, Gardener's Chronicle, ser. 3, 33: 194. 1903.

Picea brachytyla (Franchet) E. Pritzel var. *farreri* (C. Page & K. Rushforth) Eckenwalder, *stat. et comb. nov.* Basionym: *Picea farreri* C. Page & K. Rushforth, Notes from the Royal Botanic Garden, Edinburgh 38: 130. 1980.

Picea chihuahuana M. Martínez var. *martinezii* (T. F. Patterson) Eckenwalder, *stat. et comb. nov.* Basionym: *Picea martinezii* T. F. Patterson, Sida 13: 131. 1988.

Pinus armandii Franchet var. *fenzeliana* (Handel-Mazzetti) Eckenwalder, *comb. nov.* Basionym: *Pinus fenzeliana* Handel-Mazzetti, Oesterreichische Botanische Zeitschrift 80: 337. 1931.

Pinus culminicola Andresen & J. Beaman var. *bicolor* (E. Little) Eckenwalder, *comb. nov.* Basionym: *Pinus cembroides* var. *bicolor* E. Little, Phytologia 17: 331. 1968.

Pinus culminicola Andresen & J. Beaman var. *remota* (E. Little) Eckenwalder, *comb. nov.* Basionym: *Pinus cembroides* var. *remota* E. Little, Wrightia 3: 183. 1966.

Pinus parviflora P. Siebold & Zuccarini var. *kwangtungensis* (W. Y. Chun & Tsiang) Eckenwalder, *comb. nov.* Basionym: *Pinus kwangtungensis* W. Y. Chun & Tsiang, Sunyatsenia 7: 113. 1948.

Pinus parviflora P. Siebold & Zuccarini var. *wangii* (H. H. Hu & W. C. Cheng) Eckenwalder, *stat. et comb. nov.* Basionym: *Pinus wangii* H. H. Hu & W. C. Cheng, Bullein of the Fan Memorial Institute of Biology, ser. 2, 1: 191. 1948.

Pinus virginiana P. Miller subsp. *clausa* (A. W. Chapman ex Engelmann) Eckenwalder, *comb. nov.* Basionym: *Pinus inops* W. Aiton var. *clausa* A. W. Chapman ex G. Engelmann, Botanical Gazette 2: 125. 1877.

GLOSSARY

In addition to potentially unfamiliar terms used in the text and common terms with special meanings for conifers or plants in general, this glossary includes some terms applied to conifers that are not adopted in this book but that may help the reader in comparing descriptions across different texts. As noted, some of the these additional terms are not really appropriate when applied to conifers. Diagrams in the introduction may help with understanding some terms. Abbreviations used in the present book are also explained here.

abaxial lower side of a leaf (or side branch, usually a horizontal or flattened one), the side facing away from the shoot when the organ formed; in scalelike leaves, the abaxial side is referred to as the dorsal face while in needlelike leaves it is called the ventral face (cf. adaxial)

abietoid referring to the six genera of Pinaceae that together make up the subfamily Abietoideae: *Abies, Cedrus, Keteleeria, Nothotsuga, Pseudolarix,* and *Tsuga* (cf. pinoid)

abscission predetermined shedding of organs, like leaves or shoots; takes place via an abscission zone that develops at the base of an organ to seal off its connections with the parent stem (cf. cladoptosis)

adaxial upper side of a leaf (or side branch, usually a horizontal or flattened one), the side facing the shoot when the organ formed; in scalelike leaves, the adaxial side is referred to as the ventral face while in needlelike leaves it is called the dorsal face (cf. abaxial)

afforestation planting new stands of trees where they were previously absent (cf. reforestation)

air bladder air-filled out-pocketing of the pollen grain wall in most Pinaceae and Podocarpaceae; also called saccus (cf. saccate)

allopatric when two species or varieties have geographic ranges that are entirely separate (cf. sympatric)

alternate see spirally attached

ament see catkin

amphistomatic with stomates on both faces of leaf (cf. epistomatic, hypostomatic)

aneuploid of a chromosome constitution of cells differing by one or two individual chromosomes from its ancestral diploid condition; this usually occurs by the splitting or joining of ancestral chromosomes leading to up or down aneuploidy respectively; aneuploidy is common among Podocarpaceae, in which a chromosome base number $x = 10$ is ancestral but rare elsewhere in the conifers; Douglas fir (*Pseudotsuga menziesii*), but not other *Pseudotsuga* species, is unique within Pinaceae with an aneuploid $x = 13$ compared to the family norm of $x = 12$ (cf. diploid, polyploid)

anther pollen sacs of angiosperms but sometimes inappropriately applied to conifers

apical see distal

apical dominance tendency of a leading shoot to prevent neighboring shoots from growing upright via hormonal signals; the characteristic conical shape of young conifers is due to apical dominance combined with a very regular and rhythmic pattern of growth and branching; some conifers (such as araucarias) have the horizontal growth of lateral (side) branches deeply programmed even without apical dominance so that rooted cuttings taken from such side branches may never turn up at the tips

apomorphic traits altered compared to an ancestral condition in phylogenetic analysis; also called advanced or derived

apophysis external face of the seed scale in *Pinus* at cone maturity, within which the umbo is more or less centered; in hard pines (subgenus *Pinus*) and in soft pines (subgenus *Strobus*) other than those in subsection *Strobus,* it is greatly thickened compared to the thin scale tips of other genera in the family (and of the typical white pines in subsection *Strobus*) (cf. umbo)

appressed of a leaf tip pressed tightly against the twig or neighboring leaf bases (cf. free tip)

arctotertiary pattern of geographic distribution during some or all of the Tertiary (65 million to 1.6 million years ago) in which a species, genus, or family was distributed more or less all the way across the northern continents at middle latitudes; in the original concept, now abandoned, these taxa originated in the arctic and gradually migrated to middle latitudes with cooling climates (cf. holarctic)

aril layer of tissue (usually fleshy) that arises from the stalk of a seed and surrounds at least its lower portion (as in *Phyllocladus* or *Taxus*) and sometimes reaches all the way to its tip (as in *Amentotaxus* and *Torreya*)

axil "armpit" of a leaf (or bract) where its base attaches to the stem; all conifer branches arise from buds in this position, including the seed scales of seed cones; such buds and branches are described as axillary

axillary see axil

axis stem; often applied to the stem running through a pollen or seed cone (cone axis) to which the bracts and scales attach; in true firs (*Abies*), true cedars (*Cedrus*), and araucarias (*Araucaria*) the cone axis persists as a spike or knob sticking up from the branches after the bracts and seed scales have been shed with cone shattering

b. born (used in Appendix 2)

bark complex series of protective tissues that enclose branches and trunks after they initiate secondary (thickening) growth; additions to the bark continue throughout the life of the individual by growth at a cork cambium; bark changes its appearance as the individual ages, and its mature form differs greatly among different conifers, depending on the way that the cork cambiums form, the types of tissues they produce, and which other tissue they capture into the bark

basal see proximal; also, of bracteoles that occur as a pair of very fine, often deciduous, bracts seated just beneath the podocarpium in all species of *Podocarpus* subg. *Foliolatus* but in none of those of subgenus *Podocarpus* (cf. bracteole)

bast secondary phloem, a component of inner bark (cf. phloem)

bijugate underlying leaf arrangement of *Torreya* and *Cephalotaxus* in which opposite pairs of leaves spiral around the shoot like a double spiral staircase

bilobed of an ordinary or modified leaf (such as a bract) expanded forward at the tip on either side of the midrib (cf. emarginate)

bisexual cones or individuals with both pollen and seed organs; bisexual cones are an occasional anomaly in conifers (cf. monoecious, unisexual)

bole branchless portion of a tree trunk underneath the crown (cf. crown)

bordered of pits, which in general are minute openings through the secondary walls of wood cells; adjacent cells develop pits in the same spots on their walls, so they allow enhanced flow of water and dissolved material between connected cells; bordered pits keep a roof of secondary wall with just a narrow opening above the dome-shaped chamber of the pit; these pits, which look like a circle with central hole, help seal off a tracheid if its neighbor becomes filled with air (embolizes)

bract any leaf (though usually smaller than ordinary foliage leaves) connected to or immediately subtending a cone (or inflorescence for flowering plants); bracts that subtend the seed scales in conifer seed cones are referred to as fertile bracts; all others are considered sterile; fertile bracts may also be referred to as bract scales in some conifer literature, particularly descriptions of fossil conifer cones

bract-scale complex see cone scale

bracteole usually applied to a bract subtending a branch within an inflorescence in the flowering plants and sometimes used for conifers when cones are borne on a specialized reproductive branch; used in a special sense for the basal bracteoles of *Podocarpus* (cf. basal)

branchlet slender branch of the current year with little secondary growth; also called twig

bud scale a (more or less) modified leaf surrounding a resting (such as winter) bud; ranging from green and softly needlelike, though smaller than foliage leaves, to brown, hard scales; also called cataphyll (cf. scale leaf)

ca. circa, about

cambium layer of cells that thickens stems and roots by producing new cells to its inside and outside; vascular cambium, a permanent structure, produces wood inwardly and secondary phloem to the outside; cork cambium lies outside the vascular cambium and must be regenerated at intervals because it is crushed by the growing cylinder of wood; its cell products and behavior result in bark formation; when used without a modifier, the term cambium usually refers to the vascular cambium

cataphyll see bud scale

catkin dangling, unisexual, flowering plant inflorescence (hence a compound reproductive structure) associated with wind pollination; sometimes extended to include the simple (or rarely compound) pollen cones of conifers; also called ament

cf. compare

clade single lineage of species consisting of all the descendents of an ancestral species; such a clade is referred to as monophyletic; any clade is also part of larger clades descended from more ancient ancestral species; cladistic taxonomists insist that all taxonomic groups recognized in a classification (except perhaps species) must be clades (cf. paraphyletic)

cladode a simple (unbranched) branch looking like and functioning as a leaf; the double needles of *Sciadopitys* have been interpreted as cladodes (cf. phylloclade)

cladogram branching diagram of relationships between species or other taxa derived by phylogenetic analysis, often involving DNA data (cf. phylogeny)

cladoptosis predetermined shedding (abscission) of whole branches or branchlets; examples include *Wollemia* (Araucariaceae) and *Cunninghamia* (Cupressaceae); dawn redwood (*Metasequoia*) and bald cypress (*Taxodium*) are both annually cladoptosic, shedding all of their deciduous leafy twigs and additional individual leaves on main shoots every autumn

clonal of a patch of seemingly separate plants that is a single genetic individual consisting of separate stems from the ground; can form either by the rooting in of trailing or arching branches, followed by separation from the original stem, or by shoots arising on underground stems or roots; most common in subalpine conifers or those living in bogs, where branches can become buried in peat; redwood (*Sequoia sempervirens*), which sends up suckers around a burned or felled trunk is probably the largest example (cf. genet, ramet)

cone axis see axis

cone scale combination of a seed scale with its bract when these are united in whole or in part; also called bract-scale complex

conspecific belonging to the same species; often used when referring to taxa formerly considered to belong in different species but now merged; the corresponding term describing two or more species belonging to the same genus is congeneric

cordate heart-shaped; may be applied to a whole leaf, scale, or bract or just to the shape of their base

cork cambium see cambium

cortex outer layer of parenchyma cells in young twigs and roots between the central vascular cylinder and the outer epidermis; it is either shed or incorporated into the bark as stems and roots thicken with secondary growth (cf. epidermis, parenchyma, vascular tissue)

cotyledons first leaves produced by a plant embryo within the seed; usually morphologically distinct from the later true leaves, they are still commonly needlelike in conifers; most conifer genera have two cotyledons, but more are found in some families, especially Pinaceae; also called embryonic leaf or seed leaf

crisscross of an arrangement of pairs of leaves in which each opposite pair is placed at right angles to the pairs immediately above and beneath on the shoot; also called decussate

cristate applied to plants with growth distorted so that the branches grow like a cockscomb, with solidly fernlike branchlets

cross-pollination see outcrossing

crown foliage and the branching framework that supports it, including the main trunk (if there is one) above the bole; the crown of sapling conifers and mature shrubs reaches the ground but is well separated from it in mature trees, especially forest-grown ones (cf. bole)

cultivar single genetic variant within a species that is propagated in cultivation and named under the *International Code of Nomenclature of Cultivated Plants;* some selections are from native stands while others have arisen in seed lots or as sports (generally, somatic mutations); thousands of such cultivars have been named but most derive from just a handful of exceptionally variable conifer species; specific cultivars are designated by a capitalized, nonitalicized, non-Latinized (except in older examples) name enclosed in single quotes following the species name or just the generic name

cuticle hard, waxy layer secreted onto the outer surface of all aboveground epidermal cells that makes the epidermis waterproof; consists of cutin, which is chemically related to the sporopollenin that impregnates the outer walls of pollen grains

cutin see cuticle

cutinized covered by a cuticle

decussate with leaves in opposite pairs, and successive pairs along the shoot are at right angles to one another around it (crisscross pairs) so that all leaves fall into four vertical rows (called orthostichies)

decurrent one structure merging with another by running down onto it; in conifers, applied to leaf bases that are attached to the twigs and may completely clothe them

determinate said of a shoot that stops growing, often because its apex is used up in making some structure, like a terminal cone or the needle fascicles of pines; if the leader of a tree is determinate, which happens in some flowering plants and most female cycads but not in any known living conifers, the resulting trunk is sympodial (cf. fascicle, indeterminate, sympodial)

dioecious with pollen- and seed-bearing organs (mostly cones in conifers) on separate individuals; having this sexual distribution is referred to as dioecy (cf. monoecious)

diploid of a chromosome constitution of cells, and the whole plants and animals made of them, with two complete sets of chromosomes in their nucleus (like humans have); in plants, diploidy marks the sporophyte generation that follows the union of sperm and egg (cf. haploid, polyploid, sporophyte)

diploxylon pines see hard pines

disjunct living in widely separated geographic areas, each of which is small compared to the space in between them; the genera *Araucaria, Lepidothamnus,* and *Libocedrus* each have disjunct ranges in southern South America and the southwestern Pacific (cf. endemic)

disseminule a seed and, if present, any structures that are attached to it as it is dispersed, such as the seed wing in *Pinus,* cone scale in *Araucaria,* or aril in *Taxus*

distal positioned relatively toward the tip of a structure (such as a seed cone or growth increment), opposite to the end by which it attaches to its parent axis; also called apical (cf. proximal)

distichous altered leaf (or branch, or both) arrangement in which leaves are brought into two rows on either side of their horizontal supporting stem so that they are roughly parallel to the ground; it may be imposed on either an alternate or an opposite underlying phyllotaxis by twisting or realignment of either the leaves themselves (as in *Taxodium*) or of the internodes between them (as in *Metasequoia*); distichous branching occurs when buds grow out into shoots only if they are in an appropriate position to achieve distichy, as with the lateral leaf pairs but not the facial pairs of *Thuja*

dorsal the backside of a structure such as a leaf but used, rather confusingly, for opposite faces in scalelike leaves compared to needlelike leaves (cf. see abaxial and adaxial)

dorsal gland see gland

earlywood inner (and typically larger) portion of a woody growth ring associated with the initiation and expansion of a growth increment in the foliage; cells are typically larger and with thinner walls than those of the outer latewood

ecotype genetic variant within a species that is physiologically or morphologically (or both) specialized for a particular type of environment within the ecological range of the species, such as (historically, but not at present) upland and swamp ecotypes of tamarack (*Larix laricina*); a species with such ecotypes is said to display ecotypic differentiation, and there must be strong selection for this to occur since conifers typically show little genetic differentiation between populations in selectively neutral traits, like minor enzyme variants (isozymes)

elfin forest dwarf forest at high elevations on ridges of wet tropical mountains where they are also exposed to incessant cold winds; the branches of the trees are often clothed with mosses and other epiphytes; typically occurring in the most exposed portions of cloud forest; this is the habitat for a number of species of *Podocarpus;* also called mossy forest

emarginate notched, as the tip of a leaf (cf. bilobed)

embryonic leaf see cotyledon

endemic found only in a restricted geographic area or environment, since all organisms have a restricted distribution compared to the Earth's land surface area, the term endemic could be applied to any distribution; in practice, however, it is used for a relatively small and coherent area; how small an area is required by convention depends on the rank of the taxonomic group; families can be endemic to an area as large as a continent or even a hemisphere, while genera might be called endemic in an area as large as a continent, and species only in a much smaller area; the only conifer family that is referred to as endemic is Sciadopityaceae, endemic to Japan today, but more widespread in the geologic past; we would say that the genus *Halocarpus* is endemic to New Zealand, but not that the related *Lepidothamnus* is endemic to New Zealand and Chile, because this, while a relatively small area, would not be a coherent one (cf. disjunct, neoendemic, paleoendemic)

endosperm a special nutritive tissue in flowering plant seeds (for example, coconut meat and water); the term is sometimes misapplied to the nutritive tissue of conifer (and other gymnosperm) seeds, which is really the female gametophyte, an entirely different tissue from the angiosperm endosperm

epidermis outer layer of cells originally covering the external surface of all plant organs and usually covered by a waterproof cuticle; epidermal cells, except for the guard cells, are devoid of functional chloroplasts and are not photosynthetic

epigeal germination pattern of seedling establishment in the vast majority of conifers, in which the seed coat is shed and the cotyledons (seed leaves) are raised aboveground and become green and photosynthetic until the first foliage leaves mature (cf. hypogeal)

epimatium highly modified, variously fleshy seed scale of most Podocarpaceae that largely envelops the seed at maturity; its growth during seed and cone development often results in a shift in the direction that the pollination opening (micropyle) of the seed points, from out and up to back in toward the cone axis

epiphyte plant growing on another plant as a physical support; trunks, branches, and older leaves of tropical conifers may be festooned with epiphytes, including lichens, mosses, ferns, and a wide array of flowering plants; the shedding of a heavy burden of epiphytes from the trunk is one reason cited for why many tropical conifers do not have rough, thick bark like most temperate and boreal species but instead maintain smooth, thin bark by shedding of the outer layers

epistomatic with stomates only on the adaxial (usually inner or upper) surface of the leaf; epistomaty is common in scale leaves of dryland conifers (like *Callitris, Cupressus,* and *Juniperus*) or when flattened needles occur on dangling shoots (as in Serbian spruce, *Picea omorika*), in which case the upper side of the leaf is the side toward the ground! In both cases, the stomates are thus protected from heating by direct sunlight, and epistomatic scale leaves also have their stomates sheltered from drying winds (cf. amphistomatic, hypostomatic)

f. forma (plural, formae)

facial leaves of leaf pairs among those Cupressaceae with a crisscross (decussate) leaf arrangement known as various types of cedars (such as *Calocedrus, Libocedrus,* and *Thuja*) in which the branchlets are organized into flattened sprays, the alternating pairs of scale leaves of which are morphologically distinct from one another; facial pairs occupy the upper (adaxial with respect to the parent branch) and lower (abaxial) sides of the branchlet and very rarely support branching from their axillary buds; the abaxial member of the facial pair is sometimes readily distinguishable from its adaxial counterpart because it may have much more prominent waxy patches bearing a much higher density of stomates (cf. lateral leaves)

falcate of a needlelike leaf with the tip curving forward within the plane of the blade somewhat like a sickle; the base of the blade may come out straight from the twig or curve down toward it, producing an overall shallow S shape for the needle

fascicle bundle or tuft, usually bound at the base by a tight clustering of thin scales; needles of pines, uniquely among conifers, are arranged in fascicles, while fascicled pollen cones are somewhat more frequent, such as those of *Amentotaxus* or *Cunninghamia,* and emerge from buds at the tips of twigs

fastigiate narrow, upright crown shape arising when all or most branches grow upright more or less parallel to the trunk in an apparent suppression of apical dominance; two well-known trees exemplify this crown habit, the upright form of Mediterranean cypress (*Cupressus sempervirens*) and a hardwood, the Lombardy poplar (*Populus nigra* 'Italica')

fertile bract see bract

fl. flourished (used in Appendix 2)

Florin ring microscopic, low ridge of thickened cuticle surrounding a stomate; usually discontinuous, with flat gaps between the individual segments because it derives from specially organized, more of less confluent papillae; its function is poorly understood

flower reproductive structure of the flowering plants (angiosperms), with some or all of sepals, petals, stamens, and pistils; the term is sometimes misapplied to the pollen and seed cones (especially the latter) of conifers at the time of pollination, neither of which have any of these organs

free of a leaf tip, that portion of a scale leaf blade that is not pressed against the supporting twig; tips may be appressed, but those that are free are not and may even spread widely away from the twig

fusiform rust an economically destructive disease of hard pines in the southeastern United States caused by the rust fungus *Cronartium fusiforme* Cummings (which is also treated as a variant of another species); it causes swollen galls in and on the branches and trunks that lead to growth reduction, poor wood quality, and death; most severe in young plantations of the two most important plantation species in the region, loblolly and slash pines (*Pinus taeda* and *P. elliottii*), which are rendered almost ungrowable without careful selection of genetic strains and intense management in the areas of highest risk for the disease

gametophyte usually inconspicuous plant generation following meiosis in which all cells have a single (haploid) set of chromosomes in their nuclei; it alternates with the more conspicuous (except in mosses) diploid sporophyte generation; in conifers and other seed-bearing plants, there are separate male and female gametophytes; all cells of the pollen grain belong to the male gametophyte; in contrast, only some tissues of the ovule and seed belong to the female gametophyte, varying with the particular group of seed plants; in conifers, the female gametophyte is made up of one or more egg cells, associated archegonial cells, and the surrounding mass of nutrient-rich tissue that will feed the growing embryo following fertilization; in flowering plants the female gametophyte occurs only in the ovule, disappearing before seed maturation, is typically reduced to just eight cells and nuclei, and the nutritive tissue for the embryo results from a separate fertilization and is called endosperm, a term often misapplied to the nurturing female gametophyte of conifers; female gametophytes of some conifers are eaten as pine nuts or pinyons (cf. sporophyte)

genet original seedling individual in those species, like redwood (*Sequoia sempervirens*), that produce new, physiologically independent individuals by cloning (cf. ramet)

genetic load see selfing

germination (or germinal) papilla single conical, straight or curved outgrowth of the pollen grain in many Cupressaceae, such as *Athrotaxis, Fokienia,* and *Sequoiadendron;* initiates the splitting and shedding of the water-resistant outer wall of the pollen grains as they swell in the pollination droplet sticking out of the tip of an ovule (cf. papilla, pollination droplet)

glabrous without any hairs (trichomes) on the surface (cf. pubescent)

gland structure that secretes substances, such as nectar, oils, or fragrances; many Cupressaceae have conspicuous oil-producing glands (dorsal glands) just below the epidermis on the back (abaxial side) of some or all of their scale leaves (cf. resin pocket)

grass stage early, fire-resistant seedling stage of some pines, such as longleaf pine (*Pinus palustris*), in which the trunk remains essentially at ground level for several years while it builds up thickness (to about 2 cm), resources, and a disproportionately large root system, before quickly elongating until the central shoot apex rises above the level of most ground fires (about 75 cm); during this stage, the needles extend as a dome-shaped tuft, resembling a clump of grass

growth increment that portion of a shoot, including its leaves, produced by a single episode of growth (often in the form of a discrete flush); in temperate and boreal conifers, it usually corresponds to an annual increment and may be marked at its base and apex by bud scales or their scars or by shorter needles, or both, than in the middle of the increment; tropical conifers may or may not show clear growth increments and these, when present, may occur more than once within a year; also applied to a single growth ring of wood

growth ring hollow cylinder of wood (seen as a ring in cross section) marked off from adjoining rings to the inside and outside by changes in density, color, or other attributes; in temperate and boreal conifers, it usually corresponds to a single annual growing season, though extra (false) rings can be produced under certain circumstances and others may be missing entirely or in part; the beginning of growth in the spring results in earlywood with pale, large, thin-walled cells that give way near the end of the growing season to darker, smaller, thicker-walled cells of the latewood; growth rings are marked by the abrupt transition between latewood of one year and earlywood of the next; tropical conifers may display annual growth rings in a seasonal climate (usually with wet and dry seasons), may have rings that reflect active growth without a clear annual cycle, or may effectively lack growth rings entirely if they lack defined resting periods between growth spurts

guard cells pairs of cells within the epidermis that surround the stomatal opening; they look and act like lips that open or close through changes in their cell (turgor) pressure; they are the only epidermal cells with chloroplasts, which help control their function; in many conifers, unlike most angiosperms, the guard cells are sunken in pits (stomatal chambers) below the level of the other epidermal cells

gymnosperm an informal designation for any seed plant, including all conifers, that is not a flowering plant (or angiosperm); once considered a valid taxonomic group contrasting with angiosperms, gymnosperms as a whole (including extinct ones) are now recognized as paraphyletic, although the living groups may actually be monophyletic, an issue as yet left ambiguous by DNA studies; the four extant gymnosperm groups (conifers, cycads, *Ginkgo,* and gnetophytes) are far outnumbered by extinct groups, including seed ferns, cordaites, bennettites, and many others

haploid of a chromosome constitution of cells, and the whole plants and animals made of them, with just a single set of chromosomes in their nucleus; in plants, haploidy marks the gametophyte generation that germinates and matures from spores produced following meiosis in sporangia of the sporophyte (cf. diploid, gametophyte)

haploxylon pines see soft pines

hard pines members of *Pinus* subg. *Pinus,* all of which have the midvein of the needles with two separate vascular strands; most have two or three (but up to eight) needles in fascicles (bundles) wrapped at the base by a persistent sheath of papery cataphylls; the wood, on average, is harder than that of soft pines in subgenus *Strobus;* also called yellow pine, red pine, diploxylon pine

heartwood inner portion of the wood in a stem that no longer has any living cells; conversion of outer sapwood to heartwood takes place after a variable period, most commonly about 8–12 years; the transition may be gradual or abrupt and may be marked by a change in color as various pigments and other chemicals accumulate in the new heartwood

helical see spirally attached

heterozygous when a single gene has two copies in a diploid nucleus that are different from one another at one or more DNA base pairs (each gene variant is called an allele); an individual with many heterozygous genes (loci) is also referred to as heterozygous, and the proportion of heterozygous genes is its heterozygosity; conifers are generally among the most heterozygous diploid organisms, whether plant or animal (cf. homozygous)

hexaploid of a chromosome constitution of cells, and the whole plants and animals made of them, with six complete (three pairs of) sets of chromosomes in their nucleus; hexaploidy, a type of polyploidy, usually arises following hybridization of distinct species and makes the six sets of chromosomes behave as if they were two (diploid), leading to normal meiosis and full fertility, unlike most first generation hybrids with unmatched chromosomes; redwood (*Sequoia sempervirens*) is the only known hexaploid conifer (cf. diploid, polyploid, triploid, tetraploid)

holarctic a geographic distribution all around the northern hemisphere in the arctic, boreal, or north temperate climatic zones, or all three; it need not be completely continuous, just have representation in at least the eastern and western ends of both Eurasia and North America; several large conifer genera, such as *Abies, Juniperus, Picea,* and *Pinus,* have holarctic distributions, but common juniper (*J. communis*) is the only individual conifer species that does (cf. arctotertiary)

homozygous when a single gene has two identical copies in a diploid nucleus (that is, has the same allele at the locus on both homologous chromosome); an individual with many homozygous genes is also referred to as homozygous; red and Torrey pines (*Pinus resinosa* and *P. torreyana*) are examples of the very few highly homozygous conifer species that are thought to have gone through genetic bottlenecks (greatly reduced population sizes) fairly recently, like the one that Wollemi pine (*Wollemia nobilis*) is presently experiencing (cf. heterozygous)

hydric a soil or environment that is waterlogged and hence oxygen-deprived in the root zone for a significant portion of the growing season; bald cypress (*Taxodium distichum*) and Chinese swamp cypress (*Glyptostrobus pensilis*) overcome this with their knees (pneumatophores) that rise above flood levels and allow air down to the roots (cf. mesic, xeric)

hypodermis layer of tissue lying immediately beneath the epidermis of leaves of some conifers that consists of one or more layers of sclereids and adds rigidity to the needle; may be more or less continuous and can have different thicknesses in different areas of the needle (it is often thickest at the outer edge) but always has gaps underlying stomates or stomatal bands; its distribution may be useful in species identification of conifer specimens without cones, as it is in species of *Pinus*

hypogeal germination pattern in a few conifers, like bunya-bunya (*Araucaria bidwillii*), in which the cotyledons remain in the seed coat during germination and the seed coat remains underground so that, eventually, only the plumule (the portion of the seedling above the cotyledons, including the true foliage leaves) emerges aboveground (cf. epigeal)

hypostomatic with stomates only on the abaxial side (usually the underside) of the leaf; most common in conifers with needlelike leaves (cf. amphistomatic, epistomatic)

imbricate overlapping like roofing shingles; in conifers, used for tightly pressed (appressed) scale leaves of many Cupressaceae and Podocarpaceae and for seed scales of seed cones of many conifers, including Pinaceae, Araucariaceae, and some northern hemisphere Cupressaceae, such as *Calocedrus, Cryptomeria, Taiwania,* and *Thuja* (cf. valvate)

inbreeding mating between close relatives, including, in the case of conifers, self-pollination (cf. selfing)

indeterminate a shoot apex continuing to grow and produce organs more or less indefinitely, without being used up in producing a terminal structure that will ultimately be shed or simply die; if the leader of a tree is indeterminate, the resulting trunk is monopodial, as it is under most circumstances in all conifers (cf. determinate, monopodial)

integument protective outer layer of ovule that develops into the seed coat (cf. ovule, seed coat)

International Code of Botanical Nomenclature abbreviated *ICBN* or referred to as the *Code,* publication outlining rules governing scientific names of plants and also including principles, recommendations, and examples; these rules are voted upon by all taxonomists attending special sessions of each International Botanical Congress (held every 6 years); the *Code* is binding in all disputes over scientific names of both living and fossil plants and fungi at all taxonomic ranks and forms the foundation of the *International Code of Nomenclature for Cultivated Plants*

International Code of Nomenclature for Cultivated Plants abbreviated *ICNCP,* publication outlining rules governing the naming of plants in cultivation, including cultivar names; names of the species and genera to which cultivated plants belong must follow the *International Code of Botanical Nomenclature*

internodal elongation lengthening of a shoot between adjacent points of leaf attachment (nodes) until the cells have reached their final size at shoot maturity; the amount of internodal elongation varies greatly among conifers, and often among parts of a single plant; the distinction between long and short shoots in larches (*Larix*) and true cedars (*Cedrus*) is largely a function of whether there is notable internodal elongation (long shoots) or not (short or spur shoots)

internode that part of a shoot lying between two successive points of leaf attachment (nodes) (cf. node)

introgressant an individual of a species with some gene variations (alleles) from another species obtained by initial hybridization with that species followed by repeated backcrossing to the first species; the process of gene transfer is referred to as introgression; it is poorly documented in conifers

isozymes minor variants of an enzyme; often functioning with more or less equivalent efficiency and considered in many cases to be selectively neutral, so that they accumulate through mutations over time; they are easy to assess using standard laboratory procedures (electrophoresis) for certain enzymes and so are commonly used to measure the amount of genetic variation in populations and species; such studies have shown that conifers typically have a great deal of variation in these genes, much more than do people in fact; on the other hand, a few conifer species, like red pine (*Pinus resinosa*), have little (most of their enzymes are monomorphic, with just a single form, and the plants are highly homozygous), just like the cheetah, and such species are assumed to have passed through a period of greatly reduced population size (a population bottleneck) relatively recently so that their stores of variation have not yet been replenished by mutations

juvenile of foliage or leaves; sapling conifers often have leaves that differ in size or relative proportions, or both, from those of mature trees (adult foliage); these juvenile leaves are commonly larger than the adult leaves, sometimes much larger, as they are in many species of yellowwoods (*Podocarpus*); the larger size is to be expected, both because the saplings grow more vigorously than adults and because the saplings reside in the more shady, moister understory while the adult foliage is produced in the exposed, sunny, dry canopy (at least in some conifers)

keel ridge or angle along the back (abaxial face) of a scale leaf or along the length of either side of a needlelike or clawlike leaf; need not be associated directly with the midrib; keels vary from knifelike to rather blunt

lateral leaves leaf pairs in those Cupressaceae with a crisscross (decussate) arrangement of scale leaves in which the branchlets are organized into flattened sprays, alternating pairs of leaves are morphologically distinct from one another; lateral pairs occupy the sides of branchlets, in the plane of branching; they may be wrapped around the branchlet and are often strongly keeled or even greatly expanded (as in *Fokienia* or some species of *Libocedrus*); the two sides of a lateral leaf on either side of the keel (both being part of the abaxial or developmental lower side of the leaf) may look different because the side facing the ground may have waxy stomatal patches lacking on the side facing up; virtually all of the branching that produces the branchlet sprays of these cupressaceous cedars arises from buds in the axils of lateral leaves; the southern hemisphere incense cedars (*Austrocedrus, Libocedrus,* and *Papuacedrus*) often have both members of a pair giving rise to branches, and hence the branchlet sprays have opposite branching; in contrast, northern hemisphere genera (*Calocedrus, Chamaecyparis, Fokienia, Tetraclinis, Thuja,* and *Thujopsis*) very rarely do, and hence their branchlet branching is alternate or pectinate, which is one sided, like teeth on a comb (cf. facial leaves)

latewood outer portion of a growth ring associated with the slowing down and cessation of a growth increment; cells are typically smaller and with thicker walls than those of earlywood (cf. earlywood)

leaf base the part of a leaf that attaches to the parent stem and connects the midvein of the leaf to the stem vascular cylinder; in conifers the leaf base often runs down onto the stem (is decurrent) and the twig may be completely clothed by these leaf bases before they are shed with thickening (secondary) growth and bark replaces them; on the other hand, when scalelike leaves are tightly crowded, the tips of the leaves may completely hide the bases of the ones above them; the base may be jointed (articulated) so that leaf blades may be shed without the base (as in *Picea, Tsuga,* or *Juniperus* subg. *Juniperus*) or there may be no abscission zone and the leaves can only be shed as a whole, including the base; also called leaf cushion

leaf cushion attached (decurrent) leaf base

lenticel warty swelling on young or persistently thin bark filled with mealy cork cells (constituting complementary tissue or filling cells) that provides gas exchange to the inner bark (cf. resin pocket)

long shoot in conifers with morphological and functional specialization of different kinds of shoots (for example, in *Cedrus, Larix, Metasequoia, Pinus,* and *Taxodium*), those shoots that contribute to the branch framework of the tree (or shrub) and that bear short shots; in conifers without such specialization, all shoots look like long shoots, whether they are a permanent part of the framework or more transient (cf. short shoot)

lumen empty (or water-filled in the case of sapwood) part of a wood cell (or other cell types) within the cell wall; latewood tracheids, with their thick walls and small radial dimensions, may have almost no lumen; the lumen is also essentially absent in sclereids

maquis minière, maquis shrubland low, densely twiggy vegetation that may support a few scattered trees or be fringed by them; the term arose around the Mediterranean, where the dominance of shrubs is due to severe summer drought and periodic fires and the associated conifers are species of *Juniperus* and *Pinus;* in New Caledonia, the maquis minière reflects the toxicity of heavy metals, the limited water, and the frequency of fires on soils derived from the mineral-rich serpentine that supports the island's nickel mining industry; here the scrub includes some species of *Podocarpus* and may be overtopped by dwarf kauri (*Agathis ovata*), some species of *Araucaria* and *Callitris,* and chandelier cypress (*Neocallitropsis pancheri*)

mast year year of heavy seed cone and seed production; for many northern conifers, such masting occurs at intervals of only every 3–7 years (varying among cycles and species) with intervening years marked by few cones and little or no viable seed production; an inability of seed predators to increase their population to a level that would demolish the entire seed crop from the base population level of leaner years (via predator saturation) is the most commonly accepted interpretation of this reproductive rhythm

meristem region of cell division responsible for plant growth; apical meristems (shoot and root) are responsible for primary growth (elongation); lateral meristems (vascular and cork cambiums) are responsible for secondary growth (thickening)

mesic soil or environment with adequate moisture throughout the growing season, without severe droughts and with no prolonged periods of waterlogging either (cf. hydric, xeric)

megasporangium part of an ovule within the integument(s) and surrounding the female gametophyte within the megaspore wall; also called nucellus

megasporangiate strobilus see seed cone

mesophyll photosynthetic tissue of a leaf lying between the epidermis (and its associated support and strengthening tissue, if any) and the vascular bundle (and its associated support, strengthening, and extra transport tissue, if any); often differentiated into an upper palisade layer and a lower spongy mesophyll

micropyle opening at the top of the protective jacket (integument) surrounding an ovule through which pollen grains enter; this part of the ovule is called the micropylar end (cf. pollination droplet)

microsporangiate strobilus see pollen cone

microsporangium pollen sac (cf. pollen cone)

microsporophyll pollen scale (cf. pollen cone)

midline an imaginary or real line running down the center of any structure, such as a seed scale or any face of a needle

midrib raised (or even depressed) line running the length of a leaf above (or below, or both) the midvein; generally a manifestation of a mass of support tissue extending out from the vascular bundle to the adjoining epidermis

midvein vascular bundle running the length of a leaf down the center; in most conifers, this is the only vein in the leaf, but in *Nageia* and Araucariaceae, there are several to many parallel veins and no single identifiable midvein while in hard pines (*Pinus* subg. *Pinus*) the midvein consists of two adjoining separate vascular bundles, and in *Sciadopitys* each half of the double needle has its own midvein

molecular phylogeny evolutionary relationships constructed using data from molecular studies, such as amino acid sequences of proteins or DNA sequences from chromosomes in the nucleus, chloroplast, or mitochondria (cf. phylogeny)

monoecious with seed and pollen organs in separate reproductive structures (generally cones in conifers) on the same individual (cf. dioecious)

monophyletic group of species (or higher taxa, like genera or families) that has descended from a common ancestor and that includes all of the descendants of that ancestor, no matter how unlike the others some of them may have become; a stated aim of phylogenetic taxonomy is to ensure that all taxa (taxonomic groups) display this monophyly; we are close to the point where all presently recognized families and genera of conifers are monophyletic, but the subgenera, sections, and subsections often recognized within the larger genera may not yet be so, and some are definitely not (cf. paraphyletic)

monopodial with a trunk that develops by continuing growth of a single, indeterminate shoot apex, the condition found in all conifers (cf. indeterminate, sympodial)

monotypic containing just a single taxon at the next lower (less-inclusive) rank (or even further down the hierarchy); 28 of the 67 genera of conifers presently recognized are monotypic (with just a single species), but only one family, Sciadopityaceae, whose sole genus, *Sciadopitys*, is also monotypic, containing only Japanese umbrella pine (*S. verticillata*); we typically use the term without the qualifier "presently" even if a genus that is monotypic today contains additional extinct fossil species, as do *Cathaya, Pseudolarix, Sequoia,* and *Tetraclinis*

morphogenus an official category recognized in the *International Code of Botanical Nomenclature* (which governs scientific names of plants) for fossil genera (and other taxonomic ranks with the appropriate morpho- prefix) based just on single organs or types of preservation (such as seed cones, pollen grains, wood, or isolated leaves preserved with or without internal structure)

mossy forest see elfin forest

mucronate ending abruptly in a short, straight point (a mucro), as in some leaves or cone scales (cf. subapical)

neoendemic species with a very local distribution because it only recently evolved and has not spread far beyond its place of origin (cf. endemic, paleoendemic)

neotropical pertaining to the New World tropics; contrasts with the paleotropical fauna and flora of the African and Asian tropics

node that point along an axis at which a leaf (whether a normal foliage leaf or some modified version, like a bud scale or bract) is attached (cf. internode)

nomenclature the application of scientific names to plants, animals, and other organisms; scientific names of conifers and other plants are governed by the *International Code of Botanical Nomenclature;* it may not seem this way to the conifer enthusiast upset about name changes in their favorite plants, but the stated goal of the rules is to provide enduring scientific names that everyone around the world will use, and thus provide stability and eliminate confusion (cf. *International Code of Botanical Nomenclature*)

nothosubsp. nothosubspecies

nucellus megasporangium of the ovule, which surrounds the female gametophyte and later embryo and is, in turn, surrounded by the seed coat; it persists as remnants of maternal tissue in the mature seed (cf. megasporangium, ovule)

oleoresin sticky material secreted into the resin canals of conifers by the surrounding cells; contains both smaller, lighter-weight (volatile) oils that evaporate after exposure and the residue of larger, heavier molecules left behind when they do; both components belong to the chemical class of terpenoids made up of different multiples of five-carbon (isoprene) building blocks

opposite arrangement of leaves in which every node (point of leaf attachment) supports two leaves across the twig from each other; branches may also be opposite if the buds in the axils of both leaves grow out (as they often do in *Agathis, Metasequoia,* and *Libocedrus*); quite commonly, however, only one bud of a pair grows out (as in *Microcachrys* and *Thuja*), and then alternate (if there are branches on both sides of the twig) or pectinate (if there are branches on only one side) branching results (cf. bijugate, crisscross, spirally attached, whorl)

outcrossing mating with a genetically distinct individual, particularly one that is genealogically unrelated; in conifers, it is the result of pollen being carried from one individual to the receptive seed cones of another, usually by wind; also called cross pollination (cf. selfing)

ovule immature precursor of the seed, so called until some arbitrary time between fertilization of the egg and maturation of the embryo, when the term immature seed may be used in preference; in conifers, consists of a female gametophyte housing one or more eggs and associated cells (an archegonium) with a maternal megasporangium (nucellus), all surrounded by a single multilayered jacket (integument)

ovuliferous scale seed scale of a seed cone, whether or not it is united with the subtending bract (cf. seed scale)

paleoendemic genus (or species or family) with a very restricted distribution today that is known to have been much more widespread in the geologic past; *Cathaya, Metasequoia, Pseudolarix, Sequoiadendron,* and *Wollemia* are all paleoendemics; also called relictual endemic (cf. endemic, neoendemic)

palisade layer portion of the photosynthetic mesophyll tissue (parenchyma) of a leaf made up of cells that are elongated perpendicular to the epidermis so that in cross section they look like the palisade of a frontier fort; palisade tissue usually occurs on the upper face of a conifer needle, with spongy mesophyll beneath, but it may be absent or occupy both sides of needles that are held vertically or flattened side to side rather than top to bottom, such as *Falcatifolium* or *Retrophyllum;* also called palisade mesophyll, palisade parenchyma, or palisade photosynthetic tissue (cf. mesophyll)

papilla nipplelike protuberance on a surface, typically microscopic as applied to plants; epidermal cells within the stomatal bands of conifer leaves often have one or more papillae (papillate epidermis); many Cupressaceae have pollen grains with a single papilla (cf. germination papilla, stomate)

paraphyletic group of species (or genera, families, etc.) that, while consisting of members all descended from a specified most recent common ancestor, excludes some of the other descendents of that same ancestor, usually because they have achieved specializations not found in taxa retained within the group; paraphyletic groups have been common in traditional taxonomy but are avoided in phylogenetic taxonomy, in which the traditional outcasts are brought back into the main group, which is then monophyletic; the formerly recognized family Taxodiaceae (or redwood family) was paraphyletic because the most recent common ancestor of the nine genera and twelve species assigned to the family was also the ancestor of the much more numerous genera and species of a more narrowly defined (than at present) cypress family Cupressaceae, separated from Taxodiaceae because they typically had scale leaves in an opposite or whorled leaf arrangement unlike that of most taxodiaceous genera; putting the two groups together establishes a monophyletic redwood and cypress family Cupressaceae; exclusion of people from a then-paraphyletic family of great apes including chimpanzees and gorillas is a similar example; paraphyletic groups such as the redwood family Taxodiaceae or the great ape family Pongidae display paraphyly (cf. monophyletic)

parenchyma tissue within plant organs consisting of thin-walled cells; the green photosynthetic mesophyll of leaves is a parenchyma tissue, as are the (usually colorless) cortex and pith of shoots; there are also often parenchyma cells in wood of conifers, in the form of ray parenchyma and individual (or strings of) resin cells or the cells that line resin canals (cf. cortex, mesophyll, pith)

pectinate arranged like the teeth of a comb; usually two-sided with respect to leaves (which are then also distichous, as they are in European silver fir, *Abies alba*) and one-sided with respect to branchlets in the flattened sprays of Cupressaceae like western redcedar (*Thuja plicata*)

peduncle stalk supporting a whole inflorescence (flowering branch) in the angiosperms and, by extension, the stalk of a seed cone in the conifers, itself a reproductive branch system

peltate of a seed cone scale, centrally stalked from a shieldlike outer face

persistent of a needle lasting more than 1 year (the plant then evergreen) but typically applied for unusual persistence such as that found in monkey puzzle tree (*Araucaria araucana*) with leaves green at first but finally turning brown and nonfunctional, persisting for years and widening a little as the stem on which they are borne thickens dramatically through secondary growth

petiole leaf stalk, that portion of the leaf between the base and the blade; flattened needlelike leaves are typically petiolate, even when the petiole is small and indistinct, but plumper needle types may be either petiolate or sessile (lacking a petiole), and scalelike leaves are almost always sessile (cf. sessile)

phenology seasonal timing and sequence of appearance of plant organs; vegetative phenology refers to the formation and bursting of resting buds and the growth of leaves of all types and their supporting shoots; in conifers, reproductive phenology refers to the formation of cones, pollination, seed cone maturation, and seed dispersal; this phenological cycle often differs somewhat between related species and can be governed by seasonal changes in day length and temperature (cf. photoperiodic response)

phloem portion of the vascular tissue that carries products of photosynthesis (photosynthate); it consists of living cells and runs throughout the plant in parallel with the xylem; photosynthate is loaded into the phloem cells (or sieve elements, called sieve cells in conifers) in the leaves and transported from there against the current of water in the xylem; secondary phloem, also called bast, is produced at the same time as the wood and older bast becomes incorporated into the bark as the expanding trunk crushes it (since it lies outside the growing cylinder of wood)

photoperiodic response control of developmental and phenological processes, such as the setting of buds and other preparations for winter dormancy, by day length or by the direction of change in day length, or both (cf. phenology)

photosynthate see phloem

photosynthetic tissue see mesophyll

phylad a single monophyletic lineage, often applied to a group of related species and their extinct relatives within a genus

phylloclade branch system with flattened portions that function instead of leaves as the main photosynthetic organs of plants that have them, while the leaves are then scalelike and inconspicuous; the only conifer with them is *Phyllocladus* (cf. cladode)

phyllotaxis leaf (and bud and branch) arrangement; alternate, opposite, and whorled are the main types, though there are many variations in the final arrangements of leaves after they have been left behind by the extending shoot apex

phylogeny course of evolution in a taxonomic group through geologic time, involving species formation and changes in characters, or a diagram displaying this information, the latter also known as a phylogram or phylogenetic tree (cf. cladogram)

phylogeography a branch of biogeography (distribution of organisms) concerned with tracing past migrations of lineages within species using the distribution and interrelations of molecular markers, such as particular DNA segments

pinnate of a single structure with parts extending out to the sides of a central axis in just two rows, like the photosynthetic segments of the compound phylloclades of *Phyllocladus hypophyllus* or *P. toatoa*

pinoid referring to the five genera of Pinaceae that together make up the subfamily Pinoideae (or subfamilies Laricoideae, Piceoideae, and Pinoideae if the latter is further broken up): *Cathaya, Larix, Picea, Pinus,* and *Pseudotsuga* (cf. abietoid)

pith round, angled, or winged (star-shaped in cross section) central cylinder of parenchyma cells at the core of the vascular tissue in twigs and roots; although pith is not damaged by secondary growth the way the cortex is, it usually dies and rots away as heartwood forms around it and cuts if off from living portions of the trunk (cf. cortex, vascular tissue)

pneumatophore upright root projection allowing air movement to persistently or frequently submerged roots; knees of bald cypress (*Taxodium*) are among the largest known

podocarpium fleshy, often red-skinned structure made up of the few, fused, fertile seed cone bracts (podocarpial bracts) in many Podocarpaceae and serving as a target for dispersal by fruit-eating birds, reptiles, and mammals; mistakenly called a berry, which is a fruit type found in the flowering plants; podocarpial bracts in *Dacrydium* are more numerous than those in *Podocarpus* and end in a free tip that varies in length between species

pollen chamber cavity at the tip of an ovule below the micropyle (opening in the integument) in which pollen grains are trapped before germinating in most conifers (cf. micropyle, ovule, pollen cone, pollination droplet)

pollen cone male reproductive structure borne by conifers; in most conifers it is a simple structure consisting of a central axis to which are attached pollen scales (microsporophylls) in an arrangement that usually agrees with the phyllotaxis of the foliage leaves; each pollen scale bears two or more pollen sacs (microsporangia) in which pollen grains (immature male gametophyte enclosed within the microspore wall) are produced following meiosis; also called microsporangiate strobilus

pollen grain see pollen cone and pollination droplet

pollen sac, pollen scale see pollen cone

pollination droplet droplet (or sometimes flood) of liquid emerging from the opening of the ovule (micropyle) at the time of pollination in most conifers and aiding in the capture of pollen grains and their concentration in the pollen chamber as the droplet withdraws back into the ovule

polyploid of a chromosome constitution of cells and the whole plants and animals made of them with more than two complete sets of chromosomes in their nucleus; polyploidy may follow hybridization or may result from internal doubling; polyploids may have anywhere from three to twelve or more complete sets of chromosomes, though only three (triploid), four (tetraploid), or six (hexaploid) sets are known in conifers; when the number of sets is odd, the plants are usually sterile, while plants with an even number of sets usually have reproductive processes that make the nuclei behave like diploids; polyploidy is common in flowering plants but rare in conifers (cf. aneuploid, diploid, hexaploid, tetraploid, triploid)

prickle sharp, slender protuberance, usually on a seed cone scale in the case of conifers; the position of the prickle helps distinguish seed cones of *Libocedrus* from those of *Austrocedrus* and *Papuacedrus;* many species of *Pinus* have prickles topping off the apopyhyses of their seed scales, and these prickles vary from very slender and fragile (or deciduous), so that you rarely find an old cone with them intact (as in jack pine, *P. banksiana,* or spruce pine, *P. glabra*), to pretty robust, making the cone unpleasant to handle (as in Jeffrey pine, *P. jeffreyi*)

primary growth extension growth, that portion of plant growth produced by the shoot (and root) apices, and the source of all the individual organs (roots, shoots, leaves in all their variations, and cones); once a bit of shoot (along with the organs it carries) has reached its final length behind the shoot apex, primary growth is complete; later further thickening of the shoot is achieved by secondary growth

primordium embryonic leaf

procumbent creeping along the ground, like many branches of creeping juniper (*Juniperus horizontalis*)

propagule seed and any other structures with which it is dispersed, such as the aril of *Taxus* or the podocarpium of *Podocarpus*

proximal positioned relatively toward the point of attachment of a structure to its parent axis; also called basal (cf. distal)

pseudowhorl ring of branches (as in many conifers with spiral phyllotaxis, such as species of *Araucaria, Picea,* and *Pinus*) or leaves (as in *Larix,* needles within the fascicles of *Pinus,* or in *Sciadopitys,* although its needles may actually be branches in origin) concentrated along a short stretch of a branch but not arising at a single node the way a true whorl does (like the three needles in a whorl in *Callitris, Fitzroya,* and many *Juniperus* or the four in *Neocallitropsis*)

pterostegium flap of seed scale tissue wrapped over the top (side away from the scale) of the seed body in true firs (*Abies*) and true cedars (*Cedrus*)

pubescent hairy in general, with many additional technical terms for specific kinds of hairiness; these terms more commonly used in relation to flowering plants than to conifers (cf. glabrous)

ramet any new physiological individuals (or their stems or trunks) arising by cloning of an original genet or any of its clonal descendents, such as each member of a circle of redwoods surrounding their (presumably dead) mother tree (cf. genet)

recurved curled back away from the central stem, like the small, sterile scales at the base of the seed cone of white pine (*Pinus strobus*), the bracts in the seed cone of Rocky Mountain Douglas fir (*Pseudotsuga menziesii* var. *glauca*), or the needles of Min fir (*Abies recurvata*)

red pines see hard pines

reforestation establishment of a replacement forest on the site of one that has been lost due to logging or natural disasters (cf. afforestation)

relict, relictual endemic see paleoendemic

resin blister see resin pocket

resin canal elongate, usually tubelike internal cavity lined with secretory cells and filling up with oleoresin; like vascular bundles, resin canals may branch and display linear continuity through the plant body, sometimes closely paralleling vascular bundles; they are prominent in stems, leaves, and seed cones, often with distributions in these tissues diagnostic of genera and species; resin canals in needles of true firs (*Abies*) and pines (*Pinus*), for example, have been much exploited for help in identifying vegetative specimens; presence or absence of resin canals in wood is very helpful in diagnosing families and genera of conifers in some cases; they are generally considered to have evolved as part of the antiherbivore defenses of conifers and are the primary source of the resins extracted for chemically based taxonomic studies of conifers, since the compounds present vary considerably among species and genera

resin pocket oleoresin-filled cavity associated with a variety of different structures (and usually near the surface) in different conifers: bark (for example, *Abies balsamea,* source of Canada balsam, used scientifically and industrially), leaves (those on the backs of leaves of *Juniperus, Thuja,* and other Cupressaceae are referred to as glands), seeds (for example, some *Chamaecyparis* and *Tsuga* species), and seed scales (for example, bald cypress, *Taxodium distichum*), among others; also called resin blister or resin tubercle

resinous containing oleoresin; resin canals are obviously resinous, but the term is usually applied to larger structures in which the oleoresin is manifested as a piney odor or taste, such as pine wood, or the berrylike seed cones of *Juniperus* or podocarpium of some species of *Podocarpus* (while those of others are not particularly resinous)

resin tubercle see resin pocket

retinosporas cultivars of *Chamaecyparis,* particularly of sawara cypress (*C. pisifera*), with persistently juvenile foliage (and essentially devoid of adult foliage, in fact); it is an obsolete generic name for these forms that disappeared from scientific literature when their nature was finally recognized

revolute curling or rolling under of the edges of more or less flat structures, like the needles of some conifers, such as *Abies delavayi*

riparian, riverine along rivers and their floodplains; species like Chinese swamp cypress (*Glyptostrobus pensilis*) and bald cypress (*Taxodium distichum*) often have riverine distributions in backwater swamps of the floodplains of coastal plain rivers

saccate with air bladders (of pollen grains)

saccus plural, sacci; air bladders of pollen grains

sapwood outer portion of wood that still contains unobstructed tracheids that conduct water and some living cells; this is the portion involved in water transport since the water pathways in the inner heartwood are impeded by accumulation of solids in the tracheids and other interruptions (cf. heartwood, tracheid)

scale leaf either a photosynthetic leaf with a reduced, short, more or less triangular blade (such as those of many Araucariaceae, Cupressaceae, and Podocarpaceae) or a nonphotosynthetic, often papery, modified leaf of various shapes and sizes (such as those whose axillary buds grow out as the needle bundles of *Pinus* or the cladodes of *Sciadopitys*); when the latter make up buds, they are usually called bud scales or cataphylls

sclereids cells with enormously thickened secondary walls that strengthen various plant tissues; sclereids have little or no lumen and so they do not transport water the way tracheids (also thick-walled but not quite so thick) in the xylem do; they may be compactly rounded, irregular in shape, elongated, or variously branched; the hypodermis of conifer needles consists of sclereids, and they may also occur within the mesophyll (those of southern catkin yew, *Amentotaxus yunnanensis,* form cross bands that can be seen when the needles are held up to strong light); they are common in secondary phloem (bast) and, when found there, end up as a component of bark; densely packed, compact sclereids like those in the hypodermis may be called stone cells

sclerenchyma hard or strengthened plant tissue dominated by sclereids, like the seed shell of pine nuts

sclerophyll forest evergreen forest made up of trees and shrubs with relatively small, thick, hardened, drought-resistant leaves, such as is commonly found in regions like California or southwestern Australia with Mediterranean, summer-dry climates, or in other seasonally dry habitats

secondary growth that portion of plant growth leading to thickening of stems, roots, and other parts through production of wood (secondary xylem) and bast (secondary phloem) by the vascular cambium and of various bark components by a cork cambium (both kinds of cambiums referred to as lateral meristems to distinguish them from apical meristems of the root and shoot); secondary growth begins in a plant part after enlargement due to primary growth is complete; it is most obvious in stems but can also be seen in leaves (especially those with a long life) and the central axes of seed cones (where the number of growth rings can help tell how many growing seasons a cone requires to reach maturity, the only technique available for fossilized cones); wood (secondary xylem) is essential for maintaining connections between new (and old) leaves and the roots as primary growth separates them more and more from each other over the years (cf. meristem, primary growth)

secondary xylem wood, the tissue produced by the vascular cambium on its inner side as it expands outward during the secondary growth

sect. section

seed coat outer layers of the seed itself that develop from the corresponding integument(s) of the ovule; in conifer (and other gymnosperm) ovules the single seed coat has three layers: an inner fleshy layer (endotesta) that is usually papery or crushed out of recognizable existence at maturity, a hard stony layer (sclerotesta), made up of sclereids, that is prominent in all conifer seeds, and an outer fleshy layer (sarcotesta) that is usually inconspicuous (and dried out) in most wind-dispersed conifer seeds but well developed and juicy in some animal dispersed ones, like some Podocarpaceae and Taxaceae

seed cone basic female reproductive structure of most gymnosperms which, in conifers, has seeds borne on seed scales in the axils of bracts with which the scales may or may not be united (and is thus compound, unlike the pollen cones); this structure, while recognizable as a "pine cone" in many conifers, including all members of Araucariaceae, Cupressaceae (except *Juniperus*), Pinaceae, and Sciadopityaceae, may also be very unconelike, as those of *Podocarpus,* with their two or three fleshy bracts (podocarpium) and often just one fleshy fertile seed scale; in most Taxaceae, the cone is no longer recognizable, with the seeds seeming to sit by themselves at the end of a stalk; also called megasporangiate strobilus

seed leaves see cotyledons

seed scale portion of the seed cone of conifers to which the seeds are attached; when thoroughly united with the adjacent bract, the two together may be called a cone scale (also called a bract-scale complex); the seed scales may far overtop the bracts and be mostly independent of them, as they are in the Pinaceae, the two structures may contribute nearly equally to a more unified cone scale, as they do in many Araucariaceae and Cupressaceae, or the seed scale may be so reduced that it is just a pad of tissue on one part of the inside of the bract, as it is in *Athrotaxis, Cunninghamia,* and *Taiwania;* also called ovuliferous scale

selfing sexual reproduction by fertilization of an egg by a sperm from the same individual, accomplished in monoecious conifers by movement of pollen from the pollen cones to an ovule in a nearby seed cone; it is also inbreeding, although not strictly selfing, if fertilization involves a close relative; although many conifers are capable of selfing, selfed offspring in most conifers do not survive competition with outcrossed siblings and show various defects relating to genetic load of recessive disadvantageous mutations that are only expressed when homozygous (when two copies are present); also called self-pollination (cf. outcrossing)

self-pollination see selfing

serotinous, serotiny of cones that remain closed on the tree after maturity without releasing their seeds for prolonged periods of years or even decades until opening is triggered by environmental cues, like crown-destroying fires; the frequency of serotiny among conifers is greatest in environments subject to the most frequent and severe fires, which often support serotinous species of cypress (*Cupressus*), cypress pine (*Callitris*), and pine (*Pinus*)

serpentine soil derived from serpentinitic rock contains toxic quantities of heavy metals like nickel that, along with fluctuations between waterlogging and drought, make them very difficult soils for plant growth, often leading to stunted vegetation, like the maquis minière of New Caledonia (which boasts large tracts of such soils); although implying derivation from green, glassy serpentine rocks, the term is also applied more broadly to other ultrabasic soils derived from weathering of similar substrates; while New Caledonia has the richest serpentine conifer flora, a few species favor it elsewhere, such as MacNab cypress (*Cupressus macnabiana*) in California, essentially found on no other soil type

sessile lacking a stalk; said of any structure that attaches to a supporting axis directly, such as needles that lack a petiole or cones without a peduncle (cf. peduncle, petiole)

shoot system aboveground parts of a plant, including trunk, branches, and twigs and any leaves and reproductive structures attached to them; the term can also be used in a more restricted sense that emphasizes patterns of branching to describe such things as the distinction between long and short shoots or the organization of fernlike sprays; contrasts with root system, which lies underground with the exception of such occasional oddball structures as cypress knees (pneumatophores)

short shoot specialized twig not contributing to the permanent branch framework of a tree or shrub; short shoots are borne on the long shoots that collectively build the framework of the crown; the most extreme examples are the determinate needle bundles (fascicles) of pines (*Pinus*), which live only as long as their constituent individual needles; much longer lived are spur shoots of true cedars (*Cedrus*) and larches (*Larix*), which have tufts of needles at the end of a stubby twig that grows for years (and hence is indeterminate) but remains short because there is no internodal elongation between successive leaves; a third kind of short shoot differs from long shoots of the same trees only in being determinate and shed after a definitive period and there is a graduation between these and the situation in which you would say that there is no differentiation of the shoot system into long and short shoots; there is no shortening of an annual increment in these short shoots, which have as much (or almost as much) internodal elongation as in a long shoot so that the needles are well separated from one another and not concentrated into tufts; these shoots may be shed annually, as they are in *Taxodium* and *Metasequoia,* or they may live for several years, as they do in *Wollemia* or *Cunninghamia*

sieve cells, sieve elements see phloem

soft pines members of *Pinus* subg. *Strobus,* all of which have a midvein with a single individual vascular bundle and many of which have longer, more flexible needles than many hard pines (because they have less strengthening tissue); the sheath around the needle fascicles is either shed promptly with needle maturation or curls back to make a little necklace around the base of the needles; the most typical soft pines are the five-needled pines in subsection *Strobi,* such as eastern and western white pines (*P. strobus* and *P. monticola*), and species of the other three subsections in the subgenus less obviously fit the name; also called white pines or haploxylon pines (for a previously current name for the subgenus, emphasizing the midvein)

somatic mutation a genetic change arising within the vegetative tissues of a plant that can be propagated clonally; a major source of new cultivars with variant foliage that are first spotted growing as odd branches (or even just twigs) on a tree or shrub with different characteristics; some of these represent extraneous genetic material introduced by invading bacteria or viruses; also called sports

spirally attached of plant organs (such as leaves, buds, branches, and cone scales) arranged in a shallow, longitudinal coil that rises along the parent axis like a spiral staircase or a loosely to tightly coiled spring; all alternate leaves begin with this arrangement but, at maturity, may continue to stick out all around the branchlet or may bend or twist at their base to achieve other patterns of presentation, such as distichous; also called spiral or helical (cf. opposite, whorl)

spongy mesophyll portion of the photosynthetic tissue in a leaf associated with the stomates; usually lies next to the underside of the leaf; compared to the palisade layer, the cells are about as wide as they are long (though they may have highly irregular, lobed shapes) and there is a great deal of intercellular space between them that is obviously continuous right up to the stomates, promoting transpiration of water and absorption of carbon dioxide; spongy mesophyll usually consists of more layers of cells than the palisade parenchyma, but they usually do not form discrete layers like those that make up the latter

sporophyll any leaf, whether photosynthetic or variously modified, bearing sporangia, the reproductive structures in which meiosis takes place to produce spores; pollen scales of conifer pollen cones are usually sporophylls, though those of yews (*Taxus*) and other Taxaceae are more complicated; so too are the seed cone scales of most cycads and the individual segments of the ovaries of flowering plants; the seed scales of conifer seed cones, however, despite looking like sporophylls, are compound structures equivalent to a whole reproductive dwarf shoot bearing sporophylls rather than being individual sporophylls

sporangiophores unusual pollen scales of Taxaceae, with pollen sacs arranged all around the stalk beneath an umbrella-like (peltate) head rather than just on one side as in other conifers

sporophyte generation that makes up the visible part of familiar green plants (except mosses and liverworts) in which most cells are diploid, with two sets of chromosomes in their nuclei; it alternates with the less conspicuous, nonphotosynthetic (in seed plants like conifers), haploid gametophyte generation that arises following meiosis and is found only within the ovules (female gametophyte) or the pollen grains (male gametophyte)

sport see somatic mutation

spray flattened branch systems of many decussately scale-leaved Cupressaceae (like *Chamaecyparis, Libocedrus,* and *Thuja*) growing in moist, forested environments; branching takes place only in one plane and arises only from the leaf pairs in that plane (lateral leaves) while the pairs at right angles to them on the upper and lower sides of the sprays (facial leaves) generally remain unbranched; the exact patterns of branching, whether alternate or opposite, its frequency (how many lateral leaves not branching in between those that do), and the length of resulting branchlets in different positions all contribute distinctions to the general shape and appearance of the sprays

spur shoot see short shoot, of which it is one type

stamen pollen-bearing organ of the flowering plants corresponding to the pollen scales of the conifers, to which the term is sometimes applied inappropriately (the structure of the two is quite different; cf. anther, pollen scales)

sterigma(ta) small woody peg(s) corresponding to the projecting tip of the otherwise decurrent leaf base to which the needles of spruces (*Picea*) and hemlocks (*Tsuga*) are attached, with or without a petiole as well

sterile bract see bract

stomatal apparatus, bands, zones see stomate

stomate or stoma (plural, stomata), a breathing pore in the epidermis through which water and oxygen are lost and carbon dioxide is absorbed; with their paired, liplike guard cells, stomates look like little mouths, which is what the name means, and like our mouths, they can be opened and shut to control water movement; those on conifer leaves are usually not scattered over the whole surface but are concentrated in discrete patches of various shapes that are helpful taxonomic characters; if the patches extend the length of a needle, they are called stomatal bands, with, typically, one band on either side of the midrib (as in most true firs, *Abies*), otherwise they are just referred to as stomatal zones (common in scale-leaved species); when epidermal cells surrounding the guard cells (subsidiary cells) are differentiated from other epidermal cells, the whole aggregate of stomate and its subsidiaries is referred to as a stomatal apparatus (cf. amphistomatic, epistomatic, and hypostomatic)

stone cell see sclereid

subapical point protuberance, ranging from pimple to prickle, near but not right at the tip of a structure; in conifers, applied most frequently to the cone scales, where the point may be the free tip of the otherwise fused (with the seed scale) bracts, as in *Calocedrus, Thuja,* or *Widdringtonia;* such a free bract tip may occupy many other positions besides subapical, such as right at the tip (*Cunninghamia*), in the middle (*Sequoia*), or nearer to the base (sometimes in *Papuacedrus*) (cf. mucronate)

subg. subgenus

subsect. subsection

subsidiary cells epidermal cells surrounding the guard cells of a stomate that are distinct in shape and arrangement from ordinary epidermal cells; their number, arrangement, shapes, and ornamentation (such as presence of a Florin ring or papillae) vary among extant conifers and have proven very helpful in interpreting fossil conifers with preserved cuticles

subsp. subspecies

subtend to lie immediately beneath, in the position of a leaf with respect to its axillary bud or subsequent branch

successive pair next pair of leaves in a sequence along the same line going out on a twig in a decussate-leaved species, such as the next facial pair in *Chamaecyparis, Libocedrus,* or *Thuja*

support tissue vascular tissue is the main support tissue in conifer stems and in some leaves; outside of the vascular bundle(s), additional support in the leaves may be provided by transfusion tissue and by thick-walled sclereids found in the hypodermis and mesophyll, strengthening them, as they do in the broad, thin leaves of *Podocarpus,* or the slender, elongate needles of *Pinus*

sympatric when two different species or varieties grow or live in the same place; except for the improbable occurrence of long-distance pollination, natural hybridization can only take place in sympatry (cf. allopatric)

sympodial any shoot system in which terminal buds die and are replaced regularly by new leaders that began as axillary buds; a sympodial tree trunk is a compound structure consisting of the products of successive separate determinate shoot apices and leaders covered over by a unifying layer of wood; many pollen and seed cones terminate the branch they are on and the latter, if it continues to grow out, does so by sympodial growth from below the point of attachment of the cone (cf. monopodial, determinate)

taxad member of the family Taxaceae, sometimes used to set these off from conifers by those who believed incorrectly that they had a separate evolutionary history and constituted an independent order, Taxales

taxon plural, taxa; a taxonomic group of any rank in the botanical (or zoological) hierarchy, from forma to kingdom, but typically used when discussing groups of different ranks at the same time

tetracussate leaf arrangement with each whorl of four displaced to the gaps between leaves of the preceding whorl so that the leaves fall into eight ranks (longitudinal rows) overall, each displaced by 45° from its neighboring row; a rare arrangement in conifers found, among others, in *Neocallitropsis pancheri,* cones of *Microcachrys tetragona,* and some shoots of *Callitris macleayana;* this arrangement does not allow production of the flattened sprays of branchlets found in some crisscross-leaved species (cf. crisscross, decussate, tricussate)

tetraploid of a chromosome constitution of cells, and the whole plants and animals made of them, with four sets of chromosomes in their nucleus; Chilean alerce (*Fitzroya cupressoides*) and some strains of Chinese juniper (*Juniperus chinensis*) are among the few documented conifer tetraploids; a type of polyploidy (cf. diploid, hexaploid, triploid, polyploid)

tracheid main type of water-conducting cell in the wood of conifers and other gymnosperms (as well as a few angiosperms); those of the main direction of flow, along the trunk and branches and on into the needles, are elongate, pointed at both ends, hollow because they die before becoming functional, with thick walls compared to parenchyma cells, and with bordered pits as the only water connections between adjacent cells; another kind of tracheary element, vessel elements, the predominant water-transporting cells of the flowering plants, differ, among other things, in removing the walls at the ends of the cells, which are stacked up in columns so that they form continuous, open pipes many cells high; tracheids found in the rays that cross the wood from the center to the bark in some conifers are shorter, more blocky cells than the axial tracheids just described; tracheids are found in primary xylem of leaves, twigs, and branches but are most numerous, by far, in the secondary xylem (the wood) of which they constitute the vast majority of cells; many details of tracheid structure, including types and arrangement of pits, vary among conifers (particularly at the generic level) and thus help in identifying conifer wood in furniture, archeological sites, and the fossil record; transfusion tissue also contains tracheids

transfusion tissue see transport tissue

transpiration movement of water from the soil into the air through the plant body in its xylem, with intake at the roots and loss through the stomates of the leaves

transport tissue the main transport tissues of vascular plants are the vascular tissues (xylem and phloem), but outside the vascular tissue of the midvein, water movement to distant parts of conifer needles is also commonly aided by extra transport (or conducting) cells known as transfusion tissue (and perhaps also by the auxiliary transfusion tissue of uncertain function that is found in *Podocarpus*)

traumatic resin canals, pockets arise in conifer tissues, especially wood, in response to wounding; species that do not normally have resin canals in their wood may produce them following fires, breakage, insect attack, or other traumas

trichome term for all kinds of plant hairs, which are quite varied in shape, size, and function

tricussate leaf arrangement with each whorl of three displaced to the gaps between leaves of the preceding whorl so that leaves fall into six ranks (longitudinal rows) overall, each displaced 60° from its neighboring rows; a minority arrangement in conifers but found in several genera of Cupressaceae, including *Actinostrobus, Callitris, Fitzroya,* and about half the species of *Juniperus;* this arrangement does not allow production of the flattened sprays of branchlets found in some crisscross-leaved species (cf. crisscross, decussate, tetracussate)

triploid of a chromosome constitution of cells, and the whole plants and animals made of them, with three sets of chromosomes in their nucleus; triploids are usually reproductively sterile because of failures in meiosis; a rare type of polyploidy in conifers, triploidy is known in some cultivars of Chinese juniper (*Juniperus chinensis*) (cf. diploid, hexaploid, tetraploid, polyploid)

tubercle wartlike projection on the surface of plant parts, such as the exposed face of the cone scales of many species of *Callitris, Cupressus,* and *Widdringtonia*

umbo central raised area within the apophysis (external face) of seed scales in seed cones of *Pinus* and, by extension with some authors, any projection at the center of the face of a cone scale; the umbo in pines represents the external face of the seed scale at the end of its first year of growth, while it is still relatively small; it often peaks in a prickle and ranges, overall, from almost flat to projecting as a great, curved horn

unisexual a reproductive structure or individual of a single sex; all conifers have unisexual cones, unlike the flowers of angiosperms, which are more often bisexual than unisexual; individual conifers may also be unisexual (and their species dioecious) or they may be bisexual and have separate cones of both sexes (with their species monoecious); for conifers, dioecious species overwhelmingly have female individuals with seeds dispersed by fruit-eating birds and mammals while monoecious species usually have seeds dispersed by wind or by seed-eating and -hoarding birds and mammals (cf. bisexual)

upcurved curled up from a horizontal plane, like the leaves of many firs (*Abies*) on horizontal seed cone bearing branches in the upper crown of the tree

valvate abutting closely without overlapping; in conifers applied primarily to side-to-side seed scales of some southern hemisphere Cupressaceae, such as *Actinostrobus, Callitris, Fitzroya,* and *Widdringtonia,* and to peltate scales of some northern hemisphere members of the same family, such as *Cupressus, Sequoia,* and *Taxodium* (cf. imbricate)

var. variety

vascular cambium see cambium

vascular tissue skeleton of a plant that serves the dual purpose of support and transport; in conifers, as a result of primary growth, it initially consists of individual discrete vascular bundles, usually a single one (the midvein) in their leaves, but a ring of bundles surrounding the pith and making up a vascular cylinder in the new twigs; each vascular bundle contains two distinct tissues, xylem (for transport of water and minerals from roots to leaves) and phloem (for transport of photosynthetic products from leaves to all other living cells); massive increase in stem diameter with secondary growth is also due to accumulation of vascular tissue, primarily secondary xylem, the wood (cf. phloem, primary growth, secondary growth, xylem)

ventral see abaxial and adaxial

water-conducting cell see transport tissue

white pines see soft pines

whorl ring of three or more leaves or their axillary branches around a parent axis attached at the same node so that they were initiated simultaneously on the shoot apex; if, instead, they were initiated in a tight sequence at successive nodes and only seem to be whorled because there has been little internodal elongation as they matures, the resulting ring is referred to as a pseudowhorl; whorled foliage leaves are found only in some Cupressaceae among conifers, where it results in tricussate or tetracussate leaf arrangements (cf. alternate, opposite, tetracussate, tricussate)

wing in conifers, mostly a thin, flat projection on a seed that promotes wind dispersal; general wing types tend to be characteristic of families and those of each genus are varied within those types; species of Pinaceae have a single wing at the far end of the seed that is actually a detached portion of the seed scale while *Sciadopitys* and most Cupressaceae have a pair (or trio) of wings along the sides of the seed body that are an outgrowth of the seed coat itself; seed scales of *Araucaria,* with thin extensions to the sides of a thicker central portion containing the embedded seed are also referred to as winged since they are involved with wind dispersal; the term can be applied as well to frills on the sides of leaves with somewhat less justification

witches'-broom dense, twiggy growth in the crown of a tree or shrub that usually has shorter needles than the tree that bears it; often (but not necessarily) caused by a fungal, bacterial, or viral pathogen, they maintain their distinctive characteristics when propagated vegetatively and are the source of many dwarf conifer cultivars (too many, in the opinion of some conifer enthusiasts)

wood secondary xylem; strengthening and water-transport tissue constituting the bulk of tissue in trunks and branches after their first year; it results from cell divisions in the vascular cambium (cf. cambium, secondary growth, secondary xylem, xylem)

xeric of soils or environments with significant drought stress during a portion of every growing season; vegetation in such environments is often scrubland or woodland with widely scattered trees rather than closed-canopy forest, and most of the conifers growing in them (like most species of *Callitris, Cupressus,* and *Juniperus* subg. *Sabina*) have tightly appressed, scalelike leaves and shoot systems branching in three dimensions rather than arranged in flattened sprays like those of many of their relatives in moister forests (cf. mesic)

xylem water- and dissolved-mineral-conducting portion of the vascular tissue, in conifers, made up mostly of tracheids; see also vascular tissue (cf. phloem, secondary xylem, tracheid, wood)

yellow pines see hard pines

BIBLIOGRAPHY

Because this book is so long, and to enhance readability, the literature consulted is rarely cited within the text (except for selected taxonomic references at the end of each genus description). Instead, books and articles that informed my taxonomic decisions, descriptions, and discussions are gathered here, even when not explicitly cited elsewhere. I hope that this provides access to my sources of published information without encumbering the text. While lengthy, this is not an attempt at an exhaustive bibliography of conifers, and it favors contemporary literature over older works. See Farjon (2005c) for a more extensive listing, including more historical literature.

Adams, J. A., and D. A. Norton. 1991. Soil and vegetation characteristics of some tree windthrow features in a South Westland rimu forest. Journal, Royal Society of New Zealand 21: 33–42.

Adams, R. P. 1973. Reevaluation of the biological status of *Juniperus deppeana* var. *sperryi* Correll. Brittonia 25: 284–289.

Adams, R. P. 1975. Numerical-chemosystematic studies of infraspecific variation in *Juniperus pinchotii* Sudw. Biochemical Systematics and Ecology 3: 71–74.

Adams, R. P. 1977. Chemosystematics: analysis of populational differentiation and variability of ancestral and modern *Juniperus ashei.* Annals of the Missouri Botanical Garden 64: 184–209.

Adams, R. P. 1993. *Juniperus* Linnaeus, pp. 412–420 *in* Flora of North America Editorial Committee, eds., Flora of North America, vol. 2. New York: Oxford University Press.

Adams, R. P. 1994. Geographic variation and systematics of monospermous *Juniperus* (Cupressaceae) from the Chihuahuan Desert based on RAPDs and terpenes. Biochemical Systematics and Ecology 22: 699–710.

Adams, R. P. 1995. Revisionary study of Caribbean species of *Juniperus* (Cupressaceae). Phytologia 78: 134–150.

Adams, R. P., and T. Demeke. 1993. Systematic relationships in *Juniperus* based on random amplified polymorphic DNAs (RAPDs). Taxon 42: 553–571.

Adams, R. P., and J. R. Kistler. 1991. Hybridization between *Juniperus erythrocarpa* Cory and *J. pinchotii* Sudworth in the Chisos Mountains, Texas. Southwestern Naturalist 36: 295–301.

Adams, R., and D. Simmons. 1987. A chemosystematic study of *Callitris* (Cupressaceae) in south-eastern Australia using volatile oils. Australian Forest Research 17: 113–125.

Adams, R. P., E. von Rudloff, T. A. Zanoni, and L. Hogge. 1980. The terpenoids of an ancestral / advanced species pair of *Juniperus.* Biochemical Systematics and Ecology 8: 35–37.

Adams, R. P., T. A. Zanoni, and L. Hogge. 1984. Analyses of the volatile leaf oils of *Juniperus deppeana* and its infraspecific taxa: chemosystematic implications. Biochemical Systematics and Ecology 12: 23–27.

Adams, R. P., J. A. Pérez de la Rosa, and B. M. Charzaro. 1990. The leaf oil of *Juniperus martinezii* Pérez de la Rosa and its taxonomic status. Journal of Essential Oil Research 2: 99–104.

Afzal-Rafii, Z., and R. S. Dodd. 1994. Biometrical variability of foliage and cone characters in *Cupressus bakeri* (Cupressaceae). Plant Systematics and Evolution 192: 151–163.

Aguirre-Planter, E., G. R. Furnier, and L. E. Eguiarte. 2000. Low levels of genetic variation within and high levels of genetic differentiation among populations of species of *Abies* from southern Mexico and Guatemala. American Journal of Botany 87: 362–371.

Ahmed, M., and J. Ogden. 1985. Modern New Zealand tree-ring chronologies. III. *Agathis australis* (Salisb.)—kauri. Tree-Ring Bulletin 45: 11–24.

Aitken, S. N., and W. J. Libby. 1994. Evolution of the pygmy-forest edaphic subspecies of *Pinus contorta* across an ecological staircase. Evolution 48: 1,009–1,019.

Akkemik, Ü. 2003. Tree rings of *Cedrus libani* at the northern boundary of its natural distribution. IAWA Journal 24: 63–73.

Allan, H. H. 1961. Flora of New Zealand, vol. I. Indigenous Tracheophyta. Wellington: R. E. Owen, Government Printer.

Allen, G. S., and J. N. Owens. 1972. The life history of Douglas fir. Ottawa: Environment Canada, Forestry Service.

Allnutt, T. R., P. Thomas, A. C. Newton, and M. F. Gardner. 1998. Genetic variation in *Fitzroya cupressoides* cultivated in the British Isles, assessed using RAPDs. Edinburgh Journal of Botany 55: 329–341.

Allnutt, T. R., A. C. Newton, A. Lara, A. Premoli, J. J. Armesto, R. Vergara, and M. Gardner. 1999. Genetic variation in *Fitzroya cupressoides* (alerce), a threatened South American conifer. Molecular Ecology 8: 975–987.

Allnutt, T. R., J. R. Courtis, M. Gardner, and A. C. Newton. 2001. Genetic variation in wild Chilean and cultivated British populations of *Podocarpus salignus* D. Don (Podocarpaceae). Edinburgh Journal of Botany 58: 459–473.

American Forestry Association. 1996. National register of big trees. American Forests 102: 20–46.

Aplin, R. T., R. C. Cambie, and P. S. Rutledge. 1963. The taxonomic distribution of some diterpene hydrocarbons. Phytochemistry 2: 205–214.

Arcade, A., P. Faivre-Rampart, B. LeGuerroué, L. E. Pâques, and D. Prat. 1996. Heterozygosity and hybrid performance in larch. Theoretical and Applied Genetics 93: 1,274–1,281.

Arno, S. F., and J. R. Habeck. 1972. Ecology of alpine larch (*Larix lyallii* Parl.) in the Pacific Northwest. Ecological Monographs 42: 417–450.

Ash, J. 1983. Tree rings in tropical *Callitris macleayana* F. Muell. Australian Journal of Botany 31: 277–281.

Ash, J. 1985. Growth rings and longevity of *Agathis vitiensis* (Seemann) Benth. & Hook. f. ex Drake in Fiji. Australian Journal of Botany 33: 81–88.

Ash, J. 1986. Growth rings, age, and taxonomy of *Dacrydium* (Podocarpaceae) in Fiji. Australian Journal of Botany 34: 197–205.

Ata, C., and N. Merev. 1987. A new fir taxon in Turkey, Chataldag fir, *Abies ×olcayana* Ata and Merev. Commonwealth Forestry Review 66: 223–238.

Auckland, L. D., J. S. Johnston, H. J. Price, and F. E. Bridgwater. 2001. Stability of nuclear DNA content among divergent and isolated populations of Fraser fir. Canadian Journal of Botany 79: 1,375–1,378.

Aulenback, K. R., and B. A. LePage. 1998. *Taxodium wallisii* sp. nov.: first occurrence of *Taxodium* from the upper Cretaceous. International Journal of Plant Sciences 159: 367–390.

Averyanov, L. V., T. H. Nguyên, K. L. Phan, and V. T. Phan. 2008. The genus *Calocedrus* (Cupressaceae) in the flora of Vietnam. Taiwania 53: 11–22.

Axelrod, D. I. 1976. Evolution of the Santa Lucia fir (*Abies bracteata*) ecosystem. Annals of the Missouri Botanical Garden 63: 24–41.

Axelrod, D. I. 1986. Cenozoic history of some West American pines. Annals of the Missouri Botanical Garden 73: 565–641.

Axelrod, D. I., and T. G. Hill. 1988. *Pinus ×critchfieldii,* a late Pleistocene hybrid pine from coastal southern California. American Journal of Botany 75: 558–569.

Bailey, D. K. 1970. Phytogeography and taxonomy of *Pinus* subsection *Balfourianae.* Annals of the Missouri Botanical Garden 57: 210–249.

Bailey, D. K., and F. G. Hawksworth. 1979. Pinyons of the Chihuahuan Desert region. Phytologia 44: 129–133.

Bailey, D. K., K. Snajberk, and E. Zavarin. 1982. On the question of natural hybridization between *Pinus discolor* and *Pinus cembroides.* Biochemical Systematics and Ecology 10: 111–119.

Baker, R. T., and H. G. Smith. 1910. A research on the pines of Australia. New South Wales Technological Museum, Technical Education Series 16: 1–458.

Ballian, D., R. Longauer, T. Mikić, L. Paule, D. Kajba, and D. Gömöry. 2006. Genetic structure of a rare European conifer, Serbian spruce ((*Picea omorika*) (Panč.) Purk.). Plant Systematics and Evolution 260: 53–63.

Baltunis, B. S., M. S. Greenwood, and T. Eysteinsson. 1998. Hybrid vigor in *Larix:* growth of intra- and interspecific hybrids of *Larix decidua, L. laricina,* and *L. kaempferi* after 5 years. Silvae Genetica 47: 288–293.

Bannan, M. W. 1954. The wood structure of some Arizonan and Californian species of *Cupressus.* Canadian Journal of Botany 32: 285–307.

Barker, P. C. J. 1995. *Phyllocladus aspleniifolius:* phenology, germination, and seedling survival. New Zealand Journal of Botany 33: 325–337.

Barker, P. C. J., and J. B. Kirkpatrick. 1994. *Phyllocladus aspleniifolius:* variability in the population structure, the regeneration niche and dispersion patterns in Tasmanian forests. Australian Journal of Botany 42: 163–190.

Barnes, R. D., and B. T. Styles. 1983. The closed-cone pines of Mexico and Central America. Commonwealth Forestry Review 62: 81–84.

Barres, R. D., J. Burley, G. L. Gibson, and J. P. Garcia de Leon. 1984. Genotype-environment interactions in tropical pines and their effects on the structure of breeding populations. Silvae Genetica 33: 186–198.

Barrett, W. H. 1998. Gymnospermae, pp. 370–391 *in* M. N. Correa, ed., Flora Patagonica, vol. 1. Buenos Aires: Instituto Nacional de Tecnologia Agropecuaria.

Bartel, J. A., R. P. Adams, S. A. James, L. E. Mumba, and R. N. Pandey. 2003. Variation among *Cupressus* species from the western hemisphere based on random amplified polymorphic DNAs. Biochemical Systematics and Ecology 31: 693–702.

Bartholomew, B., D. E. Boufford, and S. A. Spongberg. 1983. *Metasequoia glyptostroboides*—its present status in central China. Journal of the Arnold Arboretum 64: 105–128.

Basinger, J. F. 1981. The vegetative body of *Metasequoia milleri* from the middle Eocene of southern British Columbia. Canadian Journal of Botany 59: 2,379–2,410.

Basinger, J. F. 1984. Seed cones of *Metasequoia milleri* from the middle Eocene of southern British Columbia. Canadian Journal of Botany 62: 281–289.

Beaman, J. H., and R. S. Beaman. 1993. The gymnosperms of Mount Kinabalu. Contributions from the University of Michigan Herbarium 19: 307–340.

Bean, W. J. 1973–1988. Trees and shrubs hardy in the British Isles, ed. 8, vols. 1–4 and supplement, D. L. Clarke, ed. London: John Murray.

Beaulieu, J., and J.-P. Simon. 1995. Mating system in natural populations of eastern white pine in Quebec. Canadian Journal of Forest Research 25: 1,697–1,703.

Bergeron, Y., and D. Gagnon. 1987. Age structure of red pine (*Pinus resinosa* Ait.) at its northern limit in Quebec. Canadian Journal of Forest Research 17: 129–137.

Berry, K. M., N. B. Perry, and R. T. Weavers. 1985. Foliage sequiterpenes of *Dacrydium cupressinum:* identification, variation, and biosynthesis. Phytochemistry 12: 2,893–2,898.

Bevan-Jones, R. 2002. The ancient yew. Macclesfield, U.K.: Windgather Press.

Bidlake, W. R., and R. A. Black. 1989. Vertical distribution of leaf area in *Larix occidentalis:* a comparison of two estimation methods. Canadian Journal of Forest Research 19: 1,131–1,136.

Bieleski, R. L. 1959. Factors affecting growth and distribution of kauri (*Agathis australis* Salisb.). I. Effect of light on the establishment of kauri and of *Phyllocladus trichomanoides* D. Don. Australian Journal of Botany 7: 252–294.

Bigwood, A. J., and R. S. Hill. 1985. Tertiary Araucarian macrofossils from Tasmania. Australian Journal of Botany 33: 645–656.

Biloni, J. S. 1990. Arboles autoctonos argentinos. Buenos Aires: Tipográfica Editora Argentina.

Blackburn, D. T. 1981. Tertiary megafossil flora of Maslin Bay, South Australia: numerical taxonomic study of selected leaves. Alcheringa 5: 9–28.

Bloor, S. J., J. P. Benner, D. Irwin, and P. Boother. 1996. Homoerythrina alkaloids from silver pine, *Lagarostrobos colensoi.* Phytochemistry 41: 801–802.

Bobrov, A. V., and A. P. Melikian. 1998. Specific structures of seed coat in Podocarpaceae Endlicher, 1847 and a possibility of using them in family systematics. Byulleten Moskovskogo Obshchestra Ispytatelei Pirody. Otdel Biologicheskii 103: 56–62. (in Russian)

Bobrov, A. V. F. C., A. P. Melikian, and E. Y. Yembaturova. 1999. Seed morphology, anatomy and ultrastructure of *Phyllocladus* L. C. & A. Rich. ex Mirb. (Phyllocladaceae (Pilg.) Bessey) in connection with generic system and phylogeny. Annals of Botany 83: 601–618.

Boland, D. J., ed. 1984. Forest trees of Australia. East Melbourne: CSIRO Publications.

Boratyńska, K., and M. A. Bobowicz. 2001. *Pinus uncinata* Ramond taxonomy based on needle characters. Plant Systematics and Evolution 227: 183–194.

Bose, M. N., and S. B. Manum. 1990. Mesozoic conifer leaves with '*Sciadopitys*-like' stomatal distribution. A re-evaluation based on fossils from Spitsbergen, Greenland and Baffin Island. Norsk Polarinstitut Skrifter 192: 1–81.

Boutelje, J. B. 1955. The wood anatomy of *Libocedrus* Endl., s. lat., and *Fitzroya* J. D. Hook. Acta Horti Bergiani 17: 177–216.

Bowden R. D., G. T. Geballe, and W. B. Bowden. 1988. Foliar uptake of ^{15}N from simulated cloud water by red spruce (*Picea rubens*) seedlings. Canadian Journal of Forest Research 19: 382–386.

Bowman, D. M. J. S., and S. Harris. 1995. Conifers of Australia's dry forests and open woodlands, pp. 252–270 *in* N. J. Enright & R. S. Hill, eds., Ecology of the southern conifers. Washington, D. C : Smithsonian Institution Press.

Boyd, A. 1990. The Thyra Ø flora: toward an understanding of the climate and vegetation during the early Tertiary in the high arctic. Review of Palaeobotany and Palynology 62: 189–203.

Brodribb, T., and R. S. Hill. 1999. The importance of xylem constraints in the distribution of conifer species. New Phytologist 143: 365–372.

Brody, J. J., R. J. Goldsack, M. Z. Wu, C. J. R. Fookes, and P. I. Forster. 2000. The steam volatile oil of *Wollemia nobilis* and its comparison with other members of the Araucariaceae (*Agathis* and *Araucaria*). Biochemical Systematics and Ecology 28: 563–578.

Brown, P. M., W. D. Shepperd, C. C. Brown, S. A. Mata, and D. L. McClain. 1995. Oldest known Engelmann spruce. US Department of Agriculture Forest Service Research Note RM-RN-534: 1–6.

Brown, R. W. 1936. The genus *Glyptostrobus* in America. Journal, Washington Academy of Sciences 26: 353–357.

Brummitt, R. K. 1987. Report of the Committee for Spermatophyta 33. Taxon 36: 734–739. [*Agathis dammara*]

Brummitt, R. K. 2008. Report of the nomenclature committee for vascular plants: 59. Taxon 56: 1,289–1,296. [*Xanthocyparis*]

Brunsfeld, S. J., P. S. Soltis, D. E. Soltis, P. A. Gadek, C. J. Quinn, D. D. Strenge, and T. A. Ranker. 1994. Phylogenetic relationships among genera of Taxodiaceae and Cupressaceae: evidence from *rbc*L sequences. Systematic Botany 19: 253–262.

Buchholz, J. T. 1939. The generic segregation of the sequoias. American Journal of Botany 26: 535–538.

Buchholz, J. T. 1951. A flat-leaved pine from Annam, Indochina. American Journal of Botany 38: 245–252.

Buchholz, J. T., and N. E. Gray. 1947. A Fijian *Acmopyle.* Journal of the Arnold Arboretum 28: 141–143.

Buchholz, J. T., and N. E. Gray. 1948a. A taxonomic revision of *Podocarpus.* I. The sections of the genus and their subdivisions with special reference to leaf anatomy. Journal of the Arnold Arboretum 29: 49–63.

Buchholz, J. T., and N. E. Gray. 1948b. A taxonomic revision of *Podocarpus.* II. The American species of *Podocarpus:* section *Stachycarpus.* Journal of the Arnold Arboretum 29: 64–76.

Buchholz, J. T., and N. E. Gray. 1948c. A taxonomic revision of *Podocarpus.* IV. The American species of section *Eupodocarpus* subsections C and D. Journal of the Arnold Arboretum 29: 123–151.

Buckley, B., J. Ogden, J. Palmer, A. Fowler, and J. Salinger. 2000. Dendroclimatic interpretation of tree-rings in *Agathis australis* (kauri). 1. Climate correlation functions and master chronology. Journal, Royal Society of New Zealand 30: 263–276.

Burke, J. G. 1975. Human use of the California nutmeg tree *Torreya californica* and other members of the genus. Economic Botany 29: 127–139.

Burley, J. 1976. Genetic systems and genetic conservation of tropical pines. Linnean Society Symposium Series 2: 85–100.

Burley, J., and P. M. Burrows. 1971. Multivariate analysis of variation in needles among provenances of *Pinus kesiya* Royle ex Gordon (syn. *P. khasya* Royle; *P. insularis* Endlicher). Silvae Genetica 21: 69–77.

Burns, B. R., and M. C. Smale. 1990. Changes in structure and composition over fifteen years in a secondary kauri (*Agathis australis*)–tanekaha (*Phyllocladus trichomanoides*) forest stand, Coromandel Peninsula, New Zealand. New Zealand Journal of Botany 28: 141–158.

Burrows, C. J. 1996. Radiocarbon dates for Holocene fires and associated events, Canterbury, New Zealand. New Zealand Journal of Botany 34: 111–121.

Burrows, G. E. 1999a. Wollemi pine (*Wollemia nobilis,* Araucariaceae) possesses the same unusual leaf axil anatomy as other investigated members of the family. Australian Journal of Botany 47: 61–68.

Burrows, G. E. 1999b. Leaf anatomy of Wollemi pine (*Wollemia nobilis,* Araucariaceae). Australian Journal of Botany 47: 795–806.

Bush, E. W., and M. F. Doyle. 1997. Taxonomic description of *Acmopyle sahniana* (Podocarpaceae): additions, revisions, discussion. Harvard Papers in Botany 2: 229–233.

Calamassi, R., S. R. Puglisi, and G. G. Vendramin. 1988. Genetic variation in morphological and anatomical needle characteristics in *Pinus brutia* Ten. Silvae Genetica 37: 199–206.

Cambie, R. C., and S. M. Bocks. 1966. A *p*-diphenol oxidase from gymnosperms. Phytochemistry 5: 391–396.

Cambie, R. C., and B. F. Cain. 1960. Bark extractives of *Dacrydium cupressinum* Soland. New Zealand Journal of Science 3: 121–126.

Cambie, R. C., R. E. Cox, K. C. Croft, and D. Sidwell. 1983. Phenolic diterpenoids of some podocarps. Phytochemistry 22: 1,163–1,166.

Cambie, R. C., R. E. Cox, and D. Sidwell. 1984. Phenolic diterpenoids of *Podocarpus ferrugineus* and other podocarps. Phytochemistry 23: 333–336.

Cambie, R. C., D. Sidwell, and C. K. Cheong. 1983. Extractives of *Decussocarpus wallichianus.* Journal of Natural Products 47: 562.

Camus, A. 1914. Les cyprès (genre *Cupressus*). Paris: Paul Lechevalier.

Cantril, D. J. 1991. Broad leafed coniferous foliage from the lower Cretaceous Otway Group, southeastern Australia. Alcheringa 15: 177–190.

Carder, A. C. 1995. Forest giants of the world past and present. Markham, Ontario: Fitzhenry & Whiteside.

Carlson, C. E. 1994. Germination and early growth of western larch (*Larix occidentalis*), alpine larch (*Larix lyallii*), and their reciprocal hybrids. Canadian Journal of Forest Research 24: 911–916.

Carlson, C. E., and L. J. Theroux. 1993. Cone and seed morphology of western larch (*Larix occidentalis*), alpine larch (*Larix lyallii*), and their hybrids. Canadian Journal of Forest Research 23: 1,264–1,269.

Carlson, C. E., R. G. Cates, and S. C. Spencer. 1991. Foliar terpenes of a putative hybrid swarm (*Larix occidentalis* × *Larix lyallii*) in western Montana. Canadian Journal of Forest Research 21: 876–881.

Carman, R. M., and R. A. Marty. 1970. Diterpenoids. XXIV. A survey of the *Agathis* species of north Queensland. Two new resin acids. Australian Journal of Chemistry 23: 1,457–1,464.

Carpenter, R. J., and M. Pole. 1995. Eocene plant fossils from the Lefroy and Cowan paleodrainages, Western Australia. Australian Systematic Botany 8: 1,107–1,154.

Carvajal, S., and R. McVaugh. 1992. *Pinus* L., pp. 32–100 *in* R. McVaugh and W. R. Anderson, eds., Flora Novo-Galiciano, vol. 17. Ann Arbor, Michigan: University of Michigan Herbarium.

Castroviejo, S., M. Laínz, G. López González, P. Montserrat, F. Muñoz Garmendia, J. Paiva, and L. Villar, eds. 1986. Flora Iberica, vol. 1. Lycopodiaceae–Papaveraceae. Madrid: Real Jardín Botánico.

Chadwick, L. C., and R. A. Keen. 1976. A study of the genus *Taxus.* Ohio Agricultural Research and Development Center Research Bulletin 1086: 1–57.

Chalk, L., J. Burtt Davy, and H. E. Desch. 1932. Some East African Coniferae and Leguminosae. Oxford, Clarendon Press.

Chalk, L., M. M. Chattaway, J. Burtt Davy, F. S. Laughton, and M. H. Scott. 1935. Fifteen South African high forest timber trees. Oxford, Clarendon Press.

Chambers, K. L. 1993. *Thuja* Linnaeus, pp. 410–411 *in* Flora of North America Editorial Committee, eds., Flora of North America, vol. 2. New York: Oxford University Press.

Chambers, T. C., A. N. Drinnan, and S. McLoughlin. 1998. Some morphological features of Wollemi pine (*Wollemia nobilis:* Araucariaceae) and their comparison to Cretaceous plant fossils. International Journal of Plant Sciences 159: 160–171.

Chaney, R. W. 1951. A revision of fossil *Sequoia* and *Taxodium* in western North America based on the recent discovery of *Metasequoia.* Transactions, American Philosophical Society, New Series 40: 169–263.

Chapman, J. D. 1961. Some notes on the taxonomy, distribution, ecology and economic importance of *Widdringtonia,* with particular reference to *W. whytei.* Kirkia 1: 138–154.

Chater, A. O. 1993. *Abies* Miller, pp. 37–38 *in* T. G. Tutin, N. A. Burges, A. O. Chater, J. R. Edmondson, V. H. Heywood, D. M. Moore, D. H. Valentine, S. M. Walters, and D. A. Webb, eds., Flora Europaea, vol. 1, ed. 2. Cambridge: Cambridge University Press.

Cheliak, W. M., J. Wang, and J. A. Pitel. 1988. Population structure and genic diversity in tamarack, *Larix laricina* (Du Roi) K. Koch. Canadian Journal of Forest Research 18: 1,318–1,324.

Chen, L. L. 1986. The anatomy of the bark of *Agathis* in New Zealand. IAWA Bulletin, n.s. 7: 229–241.

Chen, Z. K., and F. H. Wang. 1990. On the embryology and relationship of the Cephalotaxaceae and Taxaceae. Cathaya 2: 41–52.

Chen, Z. K., J. H. Zhang, and F. Zhou. 1995. The ovule structure and development of female gametophyte in *Cathaya* (Pinaceae). Cathaya 7: 165–176.

Cheng, Y., R. G. Nicolson, K. Tripp, and S. M. Chaw. 2000. Phylogeny of Taxaceae and Cephalotaxaceae genera inferred from chloroplast *mat*K gene and nuclear rDNA ITS region. Molecular Phylogenetics and Evolution 14: 353–365.

Chernavskaya, M. M., H. D. Grissino-Mayer, A. N. Kreuke, and A. V. Pushin. 1999. *Pinus tropicalis* growth responses to seasonal precipitation changes in western Cuba, pp. 185–190 *in* R. Wimmer and R. E. Vetter, eds., Tree ring analysis. New York: CABI Publishing.

Cherrier, J. F. 1981. Les kaoris de Nouvelle-Calédonie. Revue Forestière Française 33: 373–382.

Ching, K. K. 1959. Hybridization between Douglas-fir and bigcone Douglas-fir. Forest Science 5: 246–254.

Christiansen, H. 1972. On the development of pollen and the fertilization mechanisms of *Larix* and *Pseudotsuga menziesii.* Silvae Genetica 21: 166–174.

Christensen, K. I. 1987. Taxonomic revision of the *Pinus mugo* complex and *P.* ×*rhaetica* (*P. mugo* × *sylvestris*) (Pinaceae). Nordic Journal of Botany 7: 383–408.

Christensen, K. I. 1997. Gymnospermae, pp. 1–17 *in* A. Strid and K. Tan, eds., Flora Hellenica, vol. 1. Königstein: Koeltz Scientific Books.

Christensen, K. I., and G. H. Dar. 1996. A morphometric analysis of spontaneous and artificial hybrids of *Pinus mugo* × *sylvestris* (Pinaceae). Nordic Journal of Botany 17: 77–86.

Chu, C. D. 1981. A brief introduction to the Chinese species of the genus *Pseudotsuga.* Davidsonia 12: 15–17.

Chu, K. L., and W. S. Cooper. 1950. An ecological reconnaissance in the native home of *Metasequoia glyptostroboides.* Ecology 31: 260–278.

Chuan, T. I., and W. W. L. Hu. 1963. Study of *Amentotaxus argotaenia* (Hance) Pilger. Botanical Bulletin, Academia Sinica 4: 10–14.

Chun, W. Y. 1922. Chinese economic trees. Shanghai: Commercial Press.

Chun, W. Y., and K. Z. Kuang. 1958. A new genus of Pinaceae *Cathaya* Chun et Kuang gen. nov., from southern and western China. Botanicheskii Zhurnal 43: 461–470. (in Russian and Latin with English summary)

Clarkson, R. B., and D. E. Fairbrothers. 1970. A serological and electrophoretic investigation of eastern North American *Abies* (Pinaceae). Taxon 19: 720–727.

Clifford, H. T., and J. Constantine. 1980. Ferns, fern allies, and conifers of Australia. St. Lucia: University of Queensland Press.

Clifton, N. C. 1990. New Zealand timbers. Christchurch: G. P. Books.

Clout, M. N., and J. A. V. Tilley. 1992. Germination of miro (*Prumnopitys ferruginea*) seeds after consumption by New Zealand pigeons (*Hemiphaga novaeseelandiae*). New Zealand Journal of Botany 30: 25–28.

Colenutt, M. E., and B. H. Luckman. 1995. The dendrochronological characteristics of alpine larch. Canadian Journal of Forest Research 23: 777–789.

Collins, D., R. R. Mill, and M. Möller. 2003. Species separation of *Taxus baccata, T. canadensis,* and *T. cuspidata* (Taxaceae) and origins of their reputed hybrids inferred from RAPD and cpDNA data. American Journal of Botany 90: 175–182.

Conkle, M. T., G. Schiller, and C. Grunwald. 1988. Electrophoretic analysis of diversity and phylogeny of *Pinus brutia* and closely related taxa. Systematic Botany 13: 411–424.

Connor, K. F., and R. M. Lanner. 1987. The architectural significance of interfoliar branches in *Pinus* subsection *Balfourianae.* Canadian Journal of Forest Research 17: 269–272.

Conran, J. G., G. M. Wood, P. G. Martin, J. M. Dowd, C. J. Quinn, P. A. Gadek, and R. A. Price. 2000. Generic relationships within and between the gymnosperm families Podocarpaceae and Phyllocladaceae based on an analysis of the chloroplast gene *rbc*L. Australian Journal of Botany 48: 715–724.

Contreras-Medina, R., and I. Luna Vega. 2002. On the distribution of gymnosperm genera, their areas of endemism and cladistic biogeography. Australian Systematic Botany 15: 193–203.

Contreras-Medina, R., I. Luna Vega, and J. J. Morrone. 1999. Biogeographic analysis of the genera of Cycadales and Coniferales (Gymnospermae): a panbiogeographic approach. Biogeographica 75: 163–176.

Coode, M. J. E., and J. Cullen. 1965. Gymnospermae, pp. 67–85 *in* P. H. Davis, ed., Flora of Turkey and the East Aegean Islands, vol. 1. Edinburgh: Edinburgh University Press.

Cookson, I. C. 1953. The identification of the sporomorph *Phyllocladidites* with *Dacrydium* and its distribution in southern Tertiary deposits. Australian Journal of Botany 1: 64–70.

Cookson, I. C., and S. L. Duigan. 1951. Tertiary Araucariaceae from south-eastern Australia, with notes on living species. Australian Journal of Scientific Research, series B 4: 415–449.

Cookson, I. C., and K. M. Pike. 1953a. The Tertiary occurrence and distribution of *Podocarpus* (section *Dacrycarpus*) in Australia and Tasmania. Australian Journal of Botany 1: 71–82.

Cookson, I. C., and K. M. Pike. 1953b. A contribution to the Tertiary occurrence of the genus *Dacrydium* in the Australian region. Australian Journal of Botany 1: 474–484.

Cookson, I. C., and K. M. Pike. 1954. The fossil occurrence of *Phyllocladus* and two other podocarpaceous types in Australia. Australian Journal of Botany 2: 60–68.

Cool, L. G., and E. Zavarin. 1992. Terpene variability of mainland *Pinus radiata.* Biochemical Systematics and Ecology 20: 133–144.

Cool, L. G., Z. L. Hu, and E. Zavarin. 1998. Foliage terpenoids of Chinese *Cupressus* species. Biochemical Systematics and Ecology 26: 899–913.

Cool, L. G., A. B. Power, and E. Zavarin. 1991. Variability of foliage terpenes of *Fitzroya cupressoides.* Biochemical Systematics and Ecology 19: 421–432.

Cope, E. A. 1986. Native and cultivated conifers of northwestern North America. Ithaca: Cornell University Press.

Cope, E. A. 1992. Pinophyta (Gymnosperms) of New York State. New York State Museum Bulletin 483: 1–80.

Cope, E. A., 1998. Taxaceae: the genera and cultivated species. Botanical Review 64: 291–322.

Correll, D. S., and H. B. Correll. 1982. Flora of the Bahama Archipelago. Vaduz, Liechtenstein: J. Cramer.

Covas, G. 1995. Podocarpaceae, Araucariaceae, and Cupressaceae, pp. 1–23 *in* A. T. Hunziker, ed., Flora fánerogamica Argentina, fasc. 4. Córdoba, Argentina: Pro Flora Conicet.

Coyne, J. F., and W. B. Critchfield. 1974. Identity and terpene composition of Honduran pines attacked by the bark beetle *Dendroctonus frontalis* (Scolytidae). Turrialba 24: 327–331.

Cranwell, L. M. 1940. Pollen grains of New Zealand conifers. New Zealand Journal of Science and Technology 22: 1B–17B.

Critchfield, W. B. 1957. Geographic variation in *Pinus contorta.* Maria Moors Cabot Foundation Publication 3.

Critchfield, W. B. 1967. Crossability and relationships of the closed-cone pines. Silvae Genetica 16: 89–97.

Critchfield, W. B. 1975. Interspecific hybridization in *Pinus:* a summary review. Proceedings, Canadian Tree Improvement Association 14(2): 99–105.

Critchfield, W. B. 1977. Hybridization of foxtail and bristlecone pines. Madroño 24: 193–212.

Critchfield, W. B. 1984. Crossability and relationships of Washoe pine. Madroño 31: 144–170.

Critchfield, W. B. 1986. Hybridization and classification of the white pines (*Pinus* section *Strobus*). Taxon 35: 647–656.

Critchfield, W. B. 1988. Hybridization of the California firs. Forest Science 34: 139–151.

Critchfield, W. B., and B. B. Kinloch. 1986. Sugar pine and its hybrids. Silvae Genetica 35: 138–145.

Critchfield, W. B., and E. L. Little, Jr. 1966. Geographic distribution of the pines of the world. US Department of Agriculture Forest Service Miscellaneous Publication 991: 1–96.

Crowden, R. K., and M. J. Grubb. 1971. Anthocyanins from five species of the Podocarpaceae. Phytochemistry 10: 2,821–2,822.

Cullen, P. J., and J. B. Kirkpatrick 1988. The ecology of *Athrotaxis* D. Don (Taxodiaceae). II. The distributions and ecological differentiation of *A. cupressoides* and *A. selaginoides.* Australian Journal of Botany. 36: 561–573.

Curtis, W. M. 1956. The student's flora of Tasmania, part 1. Hobart: Government Printer.

Curtis, B. A., P. D. Tyson, and T. G. J. Dyer. 1978. Dendrochronological age determination of *Podocarpus falcatus.* South African Journal of Science 74: 92–95.

Dallimore, W., A. B. Jackson, and S. G. Harrison. 1966. A handbook of Coniferae and Ginkgoaceae, ed. 4. London: Edward Arnold.

Darrow, W. K., and T. A. Zanoni. 1990. Hispaniolan pine (*Pinus occidentalis* Swartz), a little known sub-tropical pine of economic potential. Commonwealth Forestry Review 69: 133–146, 259–271.

Dasgupta, B., B. A. Burke, and K. L. Stuart. 1981. Biflavonoids, norditerpenes and a nortriterpene from *Podocarpus urbanii.* Phytochemistry 20: 153–156.

Daubenmire, R. 1968. Some geographic variations in *Picea sitchensis* and their ecologic interpretation. Canadian Journal of Botany 46: 787–798.

Daubenmire, R. 1972. On the relation between *Picea pungens* and *Picea engelmannii* in the Rocky Mountains. Canadian Journal of Botany 50: 733–742.

Daubenmire, R. 1974. Taxonomic and ecologic relationships between *Picea glauca* and *Picea engelmannii.* Canadian Journal of Botany 52: 1,545–1,560.

Davies, B. J., I. E. W. O'Brien, and B. G. Murray. 1997. Karyotypes, chromosome bands and genome size variation in New Zealand endemic gymnosperms. Plant Systematics and Evolution 208: 169–185.

Debreczy, Z., and I. Rácz. 1995. New species and varieties of conifers from Mexico. Phytologia 78: 217–243.

Debreczy, Z., and I. Rácz. 2000. Fenyök a föld Körül [conifers around the world]. Budapest, Dendrológiai Alapítváng.

de Ferré, Y., M. L. Rouane, and P. Woltz. 1977. Systematique et anatomie comparée des feuilles de Taxaceae, Podocarpaceae, Cupressaceae de Nouvelle-Calédonie. Cahiers du Pacifique 20: 241–266.

de Laubenfels, D. J. 1953. The external morphology of coniferous leaves. Phytomorphology 3: 1–20.

de Laubenfels, D. J. 1959. Parasitic conifer found in New Caledonia. Science 130: 97. [*Parasitaxus ustus*]

de Laubenfels, D. J. 1965. The relationships of *Fitzroya cupressoides* (Molina) Johnston and *Diselma archeri* J. D. Hooker based on morphological considerations. Phytomorphology 15: 414–419.

de Laubenfels, D. J. 1969. A revision of the Malesian and Pacific rainforest conifers, I. Podocarpaceae, in part. Journal of the Arnold Arboretum 50: 274–369.

de Laubenfels, D. J. 1972. Gymnospermes, pp. 1–168 *in* Aubréville and J.-F. Leroy, eds., Flore de la Nouvelle Calédonie et dépendances 4. Paris: Muséum National d'Histoire Naturelle.

de Laubenfels, D. J. 1972. Podocarpaceae, pp. 9–22 *in* J.-F. Leroy, ed., Flore de Madagascar et des Comores, Gymnospermes. Paris: Muséum National d'Histoire Naturelle.

de Laubenfels, D. J. 1978a. The genus *Prumnopitys* (Podocarpaceae) in Malesia. Blumea 24: 189–190.

de Laubenfels, D. J. 1978b. The Moluccan dammars (*Agathis,* Araucariaceae). Blumea 24: 499–504.

de Laubenfels. D. J. 1978c. The taxonomy of Philippine Coniferae and Taxaceae. Kalikasan 7: 117–152.

de Laubenfels, D. J. 1979. The species of *Agathis* (Araucariaceae) of Borneo. Blumea 25: 531–541.

de Laubenfels, D. J. 1982. Podocarpaceae. Flora de Venezuela 11(2): 7–41.

de Laubenfels, D. J. 1984. Un nuevo *Podocarpus* (Podocarpaceae) de la Española. Moscosoa 3: 149–150.

de Laubenfels, D. J. 1985. A taxonomic revision of the genus *Podocarpus.* Blumea 30: 251–278.

de Laubenfels, D. J. 1987. Revision of the genus *Nageia* (Podocarpaceae). Blumea 12: 209–211.

de Laubenfels, D. J. 1988. Coniferales. Flora Malesiana, ser. 1, 10: 337–453.

de Laubenfels, D. J. 1990. The Podocarpaceae of Costa Rica. Brenesia 33: 119–121.

de Laubenfels, D. J. 1991. Las Podocar[p]aceas del Peru. Boletin de Lima 73: 57–60.

de Laubenfels, D. J. 1992. *Podocarpus acuminatus* (Podocarpaceae), a new species from South America. Novon 2: 329.

de Laubenfels, D. J. 2003. A new species of *Podocarpus* from the Maquis of New Caledonia. New Zealand Journal of Botany 41: 715–718.

de Laubenfels, D. J., and J. Silba. 1987. The *Agathis* of Espiritu Santo (Araucariaceae, New Hebrides). Phytologia 61: 448–452.

de Laubenfels, D. J., and J. Silba. 1988a. Notes on Asian-Pacific Podocarpaceae, I (*Podocarpus*). Phytologia 64: 290–292.

de Laubenfels, D. J., and J. Silba. 1988b. Notes on Asian and trans-Pacific Podocarpaceae, II. Phytologia 65: 329–332.

Del Fueyo, G. M., and S. Archangelsky. 2005. A new araucarian pollen cone with in situ *Cyclusphaera* Elsik from the Aptian of Patagonia, Argentina. Cretaceous Research 26: 757–768.

Delgado, P., D. Piñero, A. Chaos, N. Pérez-Nasser, and E. R. Alvarez-Buylla. 1999. High population differentiation and genetic variation in the endangered Mexican pine *Pinus rzedowskii* (Pinaceae). American Journal of Botany 86: 669–676.

Del Tredici, P. 1983. A giant among the dwarfs: the mystery of Sargent's weeping hemlock. Little Compton, Rhode Island: Theophrastus Press.

Del Tredici, P. 1984. St. George among the pygmies: the story of *Tsuga canadensis* 'Minuta'. Little Compton, Rhode Island: Theophrastus Press.

Dennis, J. V. 1988. The great cypress swamps. Baton Rouge: Louisiana State University Press. [*Taxodium distichum*]

den Ouden, P., and B. K. Boom. 1965. Manual of cultivated conifers hardy in the cold- and warm-temperate zone. The Hague: Martinus Nijhoff.

den Outer, R. W., and E. Toes. 1974. The secondary phloem of *Amentotaxus*. Journal of the Arnold Arboretum 55: 119–122.

Dettmann, M. E., and D. M. Jarzen. 2000. Pollen of extant *Wollemia* (Wollemi pine) and comparisons with pollen of other extant and fossil Araucariaceae, pp. 187–203 *in* M. M. Harley, C. M. Morton, and S. Blackmore, eds., Pollen and spores: morphology and biology. Richmond, Surrey: Royal Botanic Gardens, Kew.

Devall, M. S., B. R. Parresol, and J. J. Armesto. 1998. Dendroecological analysis of a *Fitzroya cupressoides* and a *Nothofagus nitida* stand in the Cordillera Pelada, Chile. Forest Ecology and Management 108: 135–145.

DeVerno, L. L., P. J. Charest, and L. Bonen. 1993. Inheritance of mitochondrial DNA in the conifer *Larix*. Theoretical and Applied Genetics 86: 383–388.

Dial, S. C., W. T. Batson, and R. Stalter. 1976. Some ecological and morphological observations of *Pinus glabra* Walter. Castanea 41: 361–377.

Dodd, M. C., and J. van Staden. 1981. Germination and viability studies on the seeds of *Podocarpus henkelii* Stapf. South African Journal of Science 77: 171–174.

Dodd, R. S., and Z. A. Rafii. 1995. Ecogeographic variation in seed fatty acids of *Austrocedrus chilensis*. Biochemical Systematics and Ecology 23: 825–833.

Dong, J., and D. B. Wagner. 1993. Taxonomic and population differentiation of mitochondrial diversity in *Pinus banksiana* and *Pinus contorta*. Theoretical and Applied Genetics 86: 573–578.

Dorado, O., G. Avila, D. M. Arias, R. Ramírez, D. Salinas, and G. Valladares. 1996. The Arbol del Tule (*Taxodium mucronatum* Ten.) is a single genetic individual. Madroño 43: 445–452.

Doyle, J. 1945. Developmental lines in pollination mechanisms in the Coniferales. Scientific Proceedings, Royal Dublin Society 24: 43–62.

Doyle, J., and W. J. Looby. 1939. Embryogeny in *Saxegothaea* in relation to other podocarps. Scientific Proceedings, Royal Dublin Society 22: 127–147.

Doyle, J., and M. O'Leary. 1935. Pollination in *Saxegothaea*. Scientific Proceedings, Royal Dublin Society 21: 175–179.

Doyle, M. F. 1998. Gymnosperms of the SW Pacific—I. Fiji. Endemic and indigenous species: changes in nomenclature, key, annotated checklist, and discussion. Harvard Papers in Botany 5: 101–106.

Dungey, E. S. 2001. Pine hybrids—a review of their use, performance, and genetics. Forest ecology and management 148: 243–258.

Dunwiddie, P. W. 1979. Dendrochronological studies of indigenous New Zealand trees. New Zealand Journal of Botany 17: 251–266.

Dvorak, W. S., and R. H. Raymond. 1991. The taxonomic status of closely related closed cone pines in Mexico and Central America. New Forests 4: 291–307.

Echt, C. S., G. G. Vendramin, C. D. Nelson, and P. Marquardt. 1999. Microsatellite DNA as shared genetic markers among conifer species. Canadian Journal of Forest Research 29: 365–371.

Eckenwalder, J. E. 1976. Re-evaluation of Cupressaceae and Taxodiaceae: a proposed merger. Madroño 23; 237–256.

Eckenwalder, J. E. 1993. *Cupressus*, pp. 405–408 *in* Flora of North America Editorial Committee, eds., Flora of North America, vol. 2. New York: Oxford University Press.

Ecroyd, C. E. 1982. Biological flora of New Zealand. 8. *Agathis australis* (D. Don) Lindl. (Araucariaceae) kauri. New Zealand Journal of Botany 20: 17–36.

Edwards-Burke, M. A., J. L. Hamrick, and R. A. Price. 1997. Frequency and direction of hybridization in sympatric populations of *Pinus taeda* and *P. echinata* (Pinaceae). American Journal of Botany 84: 879–886.

Eguiluz-Piedra, T. 1984. Geographic variation in needles, cones and seeds of *Pinus tecunumanii* in Guatemala. Silvae Genetica 33: 72–78.

Eguiluz-Piedra, T. 1986. Taxonomic relationships of *Pinus tecunumanii* from Guatemala. Commonwealth Forestry Review 65: 303–313.

El-Kassaby, Y. A., A. M. Colangeli, and O. Sziklai. 1983. A numerical analysis of karyotypes in the genus *Pseudotsuga*. Canadian Journal of Botany 61: 536–544.

Engelmark, O. 1984. Forest fires in the Muddus National Park (northern Sweden) during the past 600 years. Canadian Journal of Botany 62: 893–898.

Engstrom, F. B., and D. H. Mann. 1991. Fire ecology of red pine (*Pinus resinosa*) in northern Vermont, U.S.A. Canadian Journal of Forest Research 21: 882–889.

Enright, N. J., and D. Goldblum. 1998. Stand structure of the emergent conifer *Agathis ovata* in forest and maquis, province Sud, New Caledonia. Journal of Biogeography 25: 641–648.

Enright, N. J., and R. S. Hill, eds. 1995. Ecology of the southern conifers. Washington, D.C.: Smithsonian Institution Press.

Erdtman, G. 1957. Pollen and spore morphology / plant taxonomy, vol. 2. Stockholm: Almquist & Wiksell.

Erdtman, G. 1965. Pollen and spore morphology / plant taxonomy, vol. 3. Stockholm: Almquist & Wiksell.

Erize, F. 1997a. El nuevo libro del arbol, vol. 1. Especies forestales de la Argentina occidental. Buenos Aires: El Ateneo.

Erize, F. 1997b. El nuevo libro del arbol, vol. 2. Especies forestales de la Argentina oriental. Buenos Aires: El Ateneo.

Erize, F. 2000. El nuevo libro del arbol, vol. 3. Especies exóticas de uso ornamental. Buenos Aires: El Ateneo.

Ernst, S. G., J. W. Hanover, and D. E. Keathley. 1990. Assessment of natural interspecific hybridization of blue and Engelmann spruce in southwestern Colorado. Canadian Journal of Botany 68: 1,489–1,496.

Ewers, F. W. 1982. Secondary growth in needle leaves of *Pinus longaeva* (bristlecone pine) and other conifers: quantitative data. American Journal of Botany 69: 1,552–1,559.

Ewers, F. W., and R. Schmid. 1981. Longevity of needle fascicles in *Pinus longaeva* (bristlecone pine) and other North American pines. Oecologia 51: 107–115.

Fady, B., M. Arbez, and A. Marpeau. 1992. Geographic variability of terpene composition in *Abies cephalonica* Loudon and *Abies* species around the Aegean: hypotheses for their possible phylogeny from the Miocene. Trees 6: 162–171.

Farjon, A. 1989. A second revision of the genus *Keteleeria* Carrière. Notes, Royal Botanic Garden, Edinburgh 46: 81–99.

Farjon, A. 1990. Pinaceae. Königstein: Koeltz Scientific Books.

Farjon, A. 1992. The taxonomy of multiseed junipers (*Juniperus* sect. *Sabina*) in southwest Asia and East Africa. Edinburgh Journal of Botany 49: 251–283.

Farjon, A. 1993. Nomenclature of the Mexican cypress or "cedar of Goa," *Cupressus lusitanica* Mill. (Cupressaceae). Taxon 42: 81–84.

Farjon, A. 1994. *Cupressus cashmeriana.* Kew Magazine 11: 156–166.

Farjon, A. 1996. Biodiversity of *Pinus* (Pinaceae) in Mexico: speciation and palaeo-endemism. Botanical Journal, Linnean Society 121: 365–384.

Farjon, A. 2001. World checklist and bibliography of conifers, ed. 2. Richmond, Surrey: Royal Botanic Gardens, Kew.

Farjon, A. 2005a. Pines, drawings and descriptions of the genus *Pinus,* ed. 2. Leiden: Brill.

Farjon, A. 2005b. A monograph of Cupressaceae and *Sciadopitys.* Richmond, Surrey: Royal Botanic Gardens, Kew.

Farjon, A. 2005c. A bibliography of conifers, ed. 2. Richmond, Surrey: Royal Botanic Gardens, Kew.

Farjon, A., and C. N. Page, compilers. 1999. Conifers: status survey and conservation action plan. Cambridge, U.K.: International Union for the Conservation of Nature.

Farjon, A., and K. D. Rushforth. 1989. A classification of *Abies* Miller (Pinaceae). Notes, Royal Botanic Garden, Edinburgh 46: 59–79.

Farjon, A., and B. T. Styles. 1997. *Pinus* (Pinaceae). Flora Neotropica, monograph 75. New York: New York Botanical Garden.

Farjon, A., T. H. Nguyên, D. K. Harder, K. L. Phan, and L. Averyanov. 2002. A new genus and species in Cupressaceae (Coniferales) from northern Vietnam, *Xanthocyparis vietnamensis.* Novon 12: 179–189.

Farnsworth, D. H., G. E. Gatherum, J. J. Jokela, H. B. Kriebel, D. T. Lester, C. Merritt, S. S. Pauley, R. A. Read, R. L. Saidak, and J. W. Wright. 1972. Geographic variation in Japanese larch in north central United States plantations. Silvae Genetica 21: 139–147.

Farrar, J. L. 1995. Trees in Canada. Ottawa: Canadian Forest Service.

Fedorov, A. A. 1999. Flora of Russia, the European part and bordering regions, vol. 1. Rotterdam: A. A. Balkema.

Ferguson, D. K. 1967. On the phytogeography of Coniferales in the European Cenozoic. Palaeogeography, Palaeoclimatology, Palaeoecology 3: 73–110.

Ferguson, D. K. 1978. Some current research on fossil and recent taxads. Review of Palaeobotany and Palynology 26: 213–226.

Ferguson, D. K. 1984. A new species of *Amentotaxus* (Taxaceae) from northeastern India. Kew Bulletin 40: 115–119.

Ferguson, D. K. 1989. On Vietnamese *Amentotaxus* (Taxaceae). Adansonia, sér. 4, 11: 315–318.

Ferguson, D. K. 1992. A Kräusel legacy: advances in our knowledge of the taxonomy and evolution in *Amentotaxus.* Courier Forschungs-Institut Senckenberg 147: 255–285.

Ferguson, D. K., H. Jähnichen, and K. L. Alvin. 1978. *Amentotaxus* Pilger from the European Tertiary. Feddes Repertorium 89: 379–410.

Fernandez de la Reguera, P. A., J. Burley, and F. H. C. Marriott. 1988. Putative hybridization between *P. caribaea* Morelet and *P. oocarpa* Schiede: a canonical approach. Silvae Genetica 37: 88–93.

Fins, L., and W. J. Libby. 1982. Population variation in *Sequoiadendron:* seed and seedling studies, vegetative propagation, and isozyme variation. Silvae Genetica 31: 102–110.

Fins, L., and L. W. Seeb. 1986. Genetic variation in allozymes of western larch. Canadian Journal of Forest Research 16: 1,013–1,018.

Florin, R. 1930. Die Koniferegattung *Libocedrus* Endl. in Ostasien. Svensk Botanisk Tidskrift. 24: 117–131.

Florin, R. 1940a. Die heutige und frühere Verbreitung der Koniferengattung *Acmopyle* Pilger. Svensk Botanisk Tidskrift 34: 117–140.

Florin, R. 1940b. Notes on the past geographical distribution of the genus *Amentotaxus* Pilger (Coniferales). Svensk Botanisk Tidskrift 34: 162–165.

Florin, R. 1940c. The Tertiary fossil conifers of south Chile and their phytogeographical significance. Kongliga Svenska Vetenskapsakademiens Handlingar, ser. 3, 19(2): 1–107.

Florin, R. 1948a. On the morphology and relationships of the Taxaceae. Botanical Gazette 110: 31–39.

Florin, R. 1948b. On *Nothotaxus;* a new genus of the Taxaceae from eastern China. Acta Horti Bergiani. 14: 385–395.

Florin, R. 1952. On *Metasequoia,* living and fossil. Botaniska Notiser 105: 1–29.

Florin, R. 1958. Notes on the systematics of the Podocarpaceae. Acta Horti Bergiani 17: 403–411.

Florin, R. 1963. The distribution of conifer and taxad genera in time and space. Acta Horti Bergiani 20: 121–312.

Florin, R., and J. B. Boutelje. 1954. External morphology and epidermal structure of leaves in the genus *Libocedrus,* s. lat. Acta Horti Bergiani 17: 7–37.

Flous, F. 1936. Révision du genre *Keteleeria.* Bulletin, Societé d'Histoire Naturelle de Toulouse 70: 273–348.

Foo, L. Y. 1987. Phenylpropanoid derivatives of catechin, epicatechin and phylloflavan from *Phyllocladus trichomanoides.* Phytochemistry 26: 2,825–2,830.

Foo, L. Y. 1989. Flavanocoumarins and flavanophenylpropanoids from *Phyllocladus trichomanoides.* Phytochemistry 28: 2,477–2,481.

Forrest, G. I. 1987. A rangewide comparison of outlying and central lodgepole pine populations based on oleoresin monoterpene analysis. Biochemical Systematics and Ecology 15: 19–30.

Fowler, A., J. Palmer, J, Salinger, and J. Ogden. 2000. Dendroclimatic interpretation of tree-rings in *Agathis australis* (kauri): 2. Evidence of a significant relationship with ENSO. Journal, Royal Society of New Zealand 30: 277–292.

Fowler, D. P. 1983. The hybrid black × Sitka spruce, implications to phylogeny of the genus *Picea.* Canadian Journal of Forest Research 13: 108–115.

Franco, J. do A. 1945. A *Cupressus lusitanica* Miller, notas acerca da sua história e sistemática. Agros 28: 3–87.

Franco, J. do A. 1962. Taxonomy of the common juniper. Boletim, Sociedade Broteriana, ser. 2, 36: 101–120.

Franco, J. do A. 1969. On Himalyan-Chinese cypresses. Portugaliae Acta Biologica (series B) 9: 183–195.

Franco, J. do A. 1993a. *Picea* A. Dietr., pp. 39–40 *in* T. G. Tutin, N. A. Burges, A. O. Chater, J. R. Edmondson, V. H. Heywood, D. M. Moore, D. H. Valentine, S. M. Walters, and D. A. Webb, eds., Flora Europaea, vol. 1, ed. 2. Cambridge: Cambridge University Press.

Franco, J. do A. 1993b. *Larix* Miller, p. 40 *in* T. G. Tutin, N. A. Burges, A. O. Chater, J. R. Edmondson, V. H. Heywood, D. M. Moore, D. H. Valentine, S. M. Walters, and D. A. Webb, eds., Flora Europaea, vol. 1, ed. 2. Cambridge: Cambridge University Press.

Frankis, M. P. 1988. Generic inter-relationships in Pinaceae. Notes, Royal Botanic Garden, Edinburgh 45: 527–548.

Franklin, D. A. 1968. Biological flora of New Zealand, 3. *Dacrydium cupressinum* Lamb. (Podocarpaceae) rimu. New Zealand Journal of Botany 6: 493–513.

Franklin, D. A. 1969. Growth rings in rimu from south Westland terrace forest. New Zealand Journal of Botany 7: 177–188.

Freer, P. C., ed. 1910. The study of Manila Copal. Philippine Journal of Science A 5: 171–227. [*Agathis*]

Fryer, J. H. 1987. Agreement between patterns of morphological variability and isozyme band phenotypes in pitch pine. Silvae Genetica 36: 199–206.

Fu, D. Z. 1992. Nageiaceae—a new gymnosperm family. Acta Phytotaxonomica Sinica 30: 515–528.

Fu, L. K., editor. 1992. China plant red data book. Beijing: Science Press.

Fu, L. K., N. Li, and T. S. Elias. 1999a. *Picea* A. Dietrich, pp. 25–32 *in* Z. Y. Wu and P. H. Raven, eds., Flora of China, vol. 4. St Louis: Missouri Botanical Garden Press.

Fu, L. K., N. Li, and T. S. Elias. 1999b. *Abies* Miller, pp. 44–52 *in* Z. Y. Wu and P. H. Raven, eds., Flora of China, vol. 4. St Louis: Missouri Botanical Garden Press.

Fu, L. K., N. Li, and R. R. Mill. 1999c. Pinaceae, pp. 11–52 *in* Z. Y. Wu and P. H. Raven, eds., Flora of China, vol. 4. St Louis: Missouri Botanical Garden Press.

Fu, L. K., N. Li, and R. R. Mill. 1999d. Cephalotaxaceae, pp. 85–88 *in* Z. Y. Wu and P. H. Raven, eds., Flora of China, vol. 4. St Louis: Missouri Botanical Garden Press.

Fu, L. K., N. Li, and R. R. Mill. 1999e. Taxaceae, pp. 89–96 *in* Z. Y. Wu and P. H. Raven, eds., Flora of China, vol. 4. St. Louis: Missouri Botanical Garden Press.

Fu, L. K., Y. Li, and R. R. Mill. 1999f. Podocarpaceae, pp. 78–84 *in* Z. Y. Wu and P. H. Raven, eds., Flora of China, vol. 4. St Louis: Missouri Botanical Garden Press.

Fu, L. K., Y. F. Yu, and A. Farjon. 1999g. Cupressaceae, pp. 62–77 *in* Z. Y. Wu and P. H. Raven, eds., Flora of China, vol. 4. St Louis: Missouri Botanical Garden Press.

Fu, L. K., Y. F. Yu, and R. R. Mill. 1999h. Taxodiaceae, pp. 54–61 *in* Z. Y. Wu and P. H. Raven, eds., Flora of China, vol. 4. St Louis: Missouri Botanical Garden Press.

Fuller, E. H. 1976. *Metasequoia*—fossil and living—an initial thirty-year (1941–1970) annotated and indexed bibliography with an historical introduction. Botanical Review 42: 215–284.

Fung, L. E. 1994. A literature review of *Cunninghamia lanceolata*. Commonwealth Forestry Review 73: 172–192.

Furman, T. E. 1970. The nodular mycorrhizae of *Podocarpus rospigliosii*. American Journal of Botany 57: 910–915.

Furnier, G. R., P. Knowles, M. A. Clyde, and B. P. Dancik. 1987. Effects of avian seed dispersal on the genetic structure of whitebark pine populations. Evolution 41: 607–612.

Furnier, G. R., M. Stine, C. A. Mohn, and M. A. Clyde. 1991. Geographic patterns of variation in allozymes and height growth in white spruce. Canadian Journal of Forest Research 21: 707–712.

Gadek, P. A., and C. J. Quinn. 1981. Biflavones of *Dacrydium* sensu lato. Phytochemistry 20: 677–681.

Gadek, P. A., and C. J. Quinn. 1985. Biflavones of the subfamily Cupressoideae, Cupressaceae. Phytochemistry 24: 267–272.

Gadek, P. A., D. L. Alpers, M. H. Heslewood, and C. J. Quinn. 2000. Relationships within Cupressaceae sensu lato: a combined morphological and molecular approach. American Journal of Botany 87: 1,044–1,057.

Garden, J. 1957. A Revision of the genus *Callitris* Vent. Contributions, New South Wales National Herbarium 2: 363–392.

Gardner, C. A. 1964. Contributiones florae Australiae Occidentalis XIII Cupressaceae. Journal, Royal Society of Western Australia 47: 54. [*Actinostrobus arenarius*]

Gardner, M., and P. Thomas. 1996. The conifer conservation programme. The New Plantsman 3: 5–21.

Gauquelin, T., V. Bertaudière-Montès, W. Badri, and N. Montès. 2002. Sex ratio and sexual dimorphism in mountain dioecious thuriferous juniper (*Juniperus thurifera* L., Cupressaceae). Botanical Journal, Linnean Society 138: 237–244.

Gaussen, H., V. H. Heywood, and A. O. Chater. 1993. *Pinus* L., pp. 40–44 *in* T. G. Tutin, N. A. Burges, A. O. Chater, J. R. Edmondson, V. H. Heywood, D. M. Moore, D. H. Valentine, S. M. Walters, and D. A. Webb, eds., Flora Europaea, vol. 1, ed. 2. Cambridge: Cambridge University Press.

Ge, S., D. Y. Hong, H. Q. Wang, Z. Y. Liu, and C. M. Zhang. 1998. Population genetic structure of an endangered conifer, *Cathaya argyrophylla* (Pinaceae). International Journal of Plant Sciences 159: 351–357.

Gelbart, G., and P. von Aderkas. 2002. Ovular secretions as part of pollination mechanisms in conifers. Annals of Forest Science 59: 345–357.

Gernandt, D. S., and A. Liston. 1999. Internal transcribed spacer region evolution in *Larix* and *Pseudotsuga* (Pinaceae). American Journal of Botany 86: 711–723.

Gernandt, D. S., A. Liston, and D. Piñero. 2001. Variation in the nrDNA ITS of *Pinus* subsection *Cembroides:* implications for molecular systematic studies of pine species complexes. Molecular Phylogenetics and Evolution 21: 449–467.

Gernandt, D. S., G. Geada L., S. Ortiz G., and A. Liston. 2005. Phylogeny and classification of *Pinus.* Taxon 54: 29–42.

Gibson, J. P., and J. L. Hamrick. 1990. Genetic diversity and structure in *Pinus pungens* (table mountain pine) populations. Canadian Journal of Forest Research 21: 635–642.

Gilmore, S., and K. D. Hill. 1997. Relationships of the Wollemi pine (*Wollemia nobilis*) and a molecular phylogeny of the Araucariaceae. Telopea 7: 275–291.

Givnish, T. J. 1980. Ecological constraints on the evolution of breeding systems in seed plants: dioecy and dispersal in gymnosperms. Evolution 34: 959–972.

Glidewell, S. M., M. Möller, G. Duncan, R. R. Mill, D. Masson, and B. Williamson. 2002. NMR imaging as a tool for noninvasive taxonomy: comparison of female cones of two Podocarpaceae. New Phytologist 154: 197–207.

Godet, J.-D. 1988. Collins photographic key to the trees of Britain and northern Europe. London: Collins.

Godfrey, R. K 1988. Trees, shrubs, and woody vines of northern Florida and adjacent Georgia and Alabama. Athens: University of Georgia Press.

Goggans, J. F., and C. E. Posey. 1968. Variation in seeds and ovulate cones of some species and varieties of *Cupressus.* Auburn University Agricultural Experiment Station Circular 160: 1–23.

Goncharenko, G. G., V. E. Padutov, and A. E. Silin. 1993. Allozyme variation in natural populations of Eurasian pines. I & II. Silvae Genetica 42: 237–253.

Goncharenko, G. G., A. E. Silin, and V. E. Padutov. 1995. Intra- and interspecific genetic differentiation in closely related pines from *Pinus* subsection *Sylvestres* (Pinaceae) in the former Soviet Union. Plant Systematics and Evolution 194: 39–54.

Gooch, N. L. 1992. Two new species of *Pseudolarix* Gordon (Pinaceae) from the middle Eocene of the Pacific Northwest. PaleoBios 14: 13–19.

Gordon, A. G. 1968. Ecology of *Picea chihuahuana* Martínez. Ecology 49: 880–896.

Gordon, A. G. 1976. The taxonomy and genetics of *Picea rubens* and its relationship to *Picea mariana.* Canadian Journal of Botany 54: 781–813.

Gordon, A. G. 1990. Crossability in the genus *Picea* with special emphasis on the Mexican species. Proceedings, Joint Meeting of the Western Forest Genetics Association and IUFRO Working Parties S2.02–05, 06, 12, and 14.

Gorman S. W., R. D. Teasdale, and C. A. Cullis. 1992. Structure and organization of the 5S rRNA genes (5S DNA) in *Pinus radiata* (Pinaceae). Plant Systematics and Evolution 183: 223–234.

Govindaraju, D. R., D. B. Wagner, G. P. Smith, and B. P. Dancik. 1988. Chloroplast DNA variation within individual trees of a *Pinus banksiana–Pinus contorta* sympatric region. Canadian Journal of Forest Research 18: 1,347–1,350.

Govindaraju, D. R., P. Lewis, and C. Cullis. 1992. Phylogenetic analysis of pines using ribosomal DNA restriction fragment length polymorphisms. Plant Systematics and Evolution 179: 141–153.

Gower, S. T., and J. H. Richards. 1990. Larches: deciduous conifers in an evergreen world. BioScience 40: 818–826.

Gower, S. T., C. C. Grier, D. J. Vogt, and K. A. Vogt. 1987. Allometric relations of deciduous (*Larix occidentalis*) and evergreen conifers (*Pinus contorta* and *Pseudotsuga menziesii*) of the Cascade Mountains in central Washington. Canadian Journal of Forest Research 17: 630–634.

Graham, G. C., R. J. Henry, I. D. Godwin, and D. G. Nikles. 1996. Phylogenetic position of hoop pine (*Araucaria cunninghamii*). Australian Systematic Botany 9: 893–902.

Gray, N. E. 1953a. A taxonomic revision of *Podocarpus.* VII. The African species of *Podocarpus:* section *Afrocarpus.* Journal of the Arnold Arboretum 34: 67–76.

Gray, N. E. 1953b. A taxonomic revision of *Podocarpus.* VIII. The African species of section *Eupodocarpus,* subsections A and E. Journal of the Arnold Arboretum 34: 163–175.

Gray, N. E. 1955. A taxonomic revision of *Podocarpus.* IX. The South Pacific species of section *Eupodocarpus,* subsection F. Journal of the Arnold Arboretum 36: 199–206.

Gray, N. E. 1956. A taxonomic revision of *Podocarpus.* X. The South Pacific species of section *Eupodocarpus,* subsection D. Journal of the Arnold Arboretum 37: 160–172.

Gray, N. E. 1958. A taxonomic revision of *Podocarpus.* XI. The South Pacific species of section *Podocarpus,* subsection B. Journal of the Arnold Arboretum 39: 424–477.

Gray, N. E. 1960. A taxonomic revision of *Podocarpus.* XII. Section *Microcarpus.* Journal of the Arnold Arboretum 41: 36–39.

Gray, N. E. 1962. A taxonomic revision of *Podocarpus:* XIII. Section *Polypodiopsis* in the South Pacific. Journal of the Arnold Arboretum 43: 67–79.

Gray, N. E., and J. T. Buchholz. 1948. A taxonomic revision of *Podocarpus,* III. The American species of *Podocarpus* section *Polypodiopsis.* Journal of the Arnold Arboretum 29: 117–122.

Gray, N. E., and J. T. Buchholz. 1951a. A taxonomic revision of *Podocarpus.* V. The South Pacific species of *Podocarpus:* section *Stachycarpus.* Journal of the Arnold Arboretum 32: 82–92.

Gray, N. E., and J. T. Buchholz. 1951b. A taxonomic revision of *Podocarpus.* VI. The South Pacific species of *Podocarpus:* section *Sundacarpus.* Journal of the Arnold Arboretum 32: 93–97.

Greenwood, D. R. 1987. Early Tertiary Podocarpaceae from the Eocene Anglesea locality, Victoria, Australia. Australian Journal of Botany 35: 111–133.

Greenwood, D. R., and J. F. Basinger. 1994. The paleoecology of high-latitude Eocene swamp forests from Axel Heiberg Island, Canadian high arctic. Review of Palaeobotany and Palynology 81: 83–97.

Gregor, H.-J. 1979. Fruktifikationen der Gattung *Cephalotaxus* Siebold & Zuccarini aus dem Tertiär Europas und Japans. Feddes Repertorium 90: 1–10.

Greguss, P. 1955. Identification of living gymnosperms on the basis of xylotomy. Budapest: Akadémiai Kiadó.

Greuter, W., H. M. Burdet, and G. Long. 1984. Med-checklist, vol. 1. Geneva: Conservatoire et Jardin Botaniques.

Grierson, A. J. C., D. G. Long, and C. N. Page. 1980. Notes relating to the flora of Bhutan: III. *Pinus bhutanica:* a new 5-needle pine from Bhutan and India. Notes, Royal Botanic Garden, Edinburgh 38: 297–310.

Gullberg, U., R. Yazdani, D. Rudin, and N. Ryman. 1985. Allozyme variation in Scots pines (*Pinus sylvestris* L.) in Sweden. Silvae Genetica 34: 193–201.

Guries, R. P., and F. T. Ledig. 1982. Genetic diversity and population structure in pitch pine (*Pinus rigida* Mill.). Evolution 36: 387–402.

Haase, P. 1992. Isozyme variation and genetic relationships in *Phyllocladus trichomanoides* and *P. alpinus* (Podocarpaceae). New Zealand Journal of Botany 30: 359–363.

Habeck, J. R., and T. W. Weaver. 1969. A chemosystematic analysis of some hybrid spruce (*Picea*) populations in Montana. Canadian Journal of Botany 47: 1,565–1,570.

Hair, J. B. 1963. Cytogeographical relationships of the southern podocarps. *In* J. L. Gressitt, ed., Pacific Basin biogeography. Honolulu: Bishop Museum Press.

Hall, G. S. 1966. Age distribution of needles in red pine crowns. Forest Science 12: 369–371.

Hall, J. P., and I. R. Brown. 1977. Embryo development and yield of seed in *Larix*. Silvae Genetica 26: 77–84.

Hall, M. T. 1971. A new species of *Juniperus* from Mexico. Fieldiana, Botany. 34: 45–53.

Haller, J. R. 1959. Factors affecting the distribution of ponderosa and Jeffrey pines in California. Madroño 15: 65–96.

Haller, J. R. 1986. The taxonomy and relationships of the mainland and island populations of *Pinus torreyana* (Pinaceae). Systematic Botany 11: 39–50.

Hanover, J. W., and R. C. Wilkinson. 1969. A new hybrid between blue spruce and white spruce. Canadian Journal of Botany 47: 1,693–1,700.

Hanson, L. 2001. Chromosome number, karyotype and DNA C-value of the Wollemi pine (*Wollemia nobilis*, Araucariaceae). Botanical Journal, Linnean Society 135: 271–274.

Harden, G. J., and J. Thompson. 1990. Coniferopsida. *In* G. J. Harden, ed., Flora of New South Wales, vol. 1. Kensington, Australia: New South Wales University Press.

Haridasan, K. 1988. *Amentotaxus* (Taxaceae)—a rare gymnosperm from Arunachal Pradesh. Indian Forester 114: 868–870.

Harlow, W. M. 1931. The identification of the pines of the United States, native and introduced, by needle structure. New York State College of Forestry Bulletin 4(2a).

Harris, L. J., J. H. Borden, H. D. Pierce, Jr., and A. C. Oehlschlager. 1983. Cortical resin monoterpenes in Sitka spruce and resistance to the white pine weevil, *Pissodes strobi* (Coleoptera: Curculionidae). Canadian Journal of Forest Research 13: 350–352.

Harris, S., and J. B. Kirkpatrick. 1991. The distribution, dynamics and ecological differentiation of *Callitris* species of Tasmania. Australian Journal of Botany 39: 187–202.

Harris, T. M. 1976. Two neglected aspects of fossil conifers. American Journal of Botany 63: 902–910.

Harris, T. M. 1979. The Yorkshire Jurassic flora, V. Coniferales. London: Trustees of the British Museum (Natural History).

Hattemer, H. H. 1968–1969. Versuche zur geographischen Variation bei der japanischen Lärche. Silvae Genetica 17: 186–192; 18: 1–23.

Hawkins, B. J., and G. B. Sweet. 1989. Genetic variation in rimu—an investigation using isozyme analysis. New Zealand Journal of Botany 27: 83–90.

Hawkins, B. J., G. B. Sweet, D. H. Greer, and D. O. Bergin. 1991. Genetic variation in the frost hardiness of *Podocarpus totara*. New Zealand Journal of Botany 29: 455–458.

Hawley, G. J., and D. H. DeHayes. 1985. Hybridization among several North America firs. I. Crossability. Canadian Journal of Forest Research 15: 42–49.

Hawley, G. J., and D. H. DeHayes. 1994. Genetic diversity and population structure of red spruce (*Picea rubens*). Canadian Journal of Botany 72: 1,778–1,786.

Hayman, A. R., and R. T. Weavers. 1990. Terpenes of foliage oils from *Halocarpus bidwillii*. Phytochemistry 29: 3,157–3,162.

Heady, R. D., J. G. Banks, and P. D. Evans. 2002. Wood anatomy of Wollemi pine (*Wollemia nobilis*, Araucariaceae). IAWA Journal 23: 339–357.

Henry, A., and M. McIntyre. 1926. The swamp cypresses, *Glyptostrobus* of China and *Taxodium* of America, with notes on allied genera. Proceedings, Royal Irish Academy 37B: 90–116.

Herbert, J., P. M. Hollingsworth, M. F. Gardner, R. R. Mill, P. I. Thomas, and T. Jaffré. 2002. Conservation genetics and phylogenetics of New Caledonian *Retrophyllum* (Podocarpaceae) species. New Zealand Journal of Botany 40: 175–188.

Hermann, R. K. 1982. The genus *Pseudotsuga:* historical records and nomenclature. Oregon State University, College of Forestry, Forest Research Laboratory Special Publication 2a: 1–29.

Hermann, R. K. 1985. The genus *Pseudotsuga:* ancestral history and past distribution. Oregon State University, College of Forestry, Forest Research Laboratory Special Publication 2b: 1–32.

Hewes, J. J. 1981. Redwoods—the world's largest trees. Bison Books Limited, London.

Heywood, V. H., D. M. Moore, D. H. Valentine, S. M. Walters, and D. A. Webb, eds. 1964. Flora Europaea, vol. 1, Psilotaceae to Platanaceae. Cambridge: Cambridge University Press.

Hill, K. D. 1997. Architecture of the Wollemi pine (*Wollemia nobilis*, Araucariaceae), a unique combination of model and reiteration. Australian Journal of Botany 45: 817–826.

Hill, K. D. 1998. Pinophyta, pp. 545–596 *in* P. M. McCarthy, ed., Flora of Australia, vol. 48. Ferns, gymnosperms, and allied groups. Melbourne: CSIRO Publications.

Hill, R. S. 1989. New species of *Phyllocladus* (Podocarpaceae) macrofossils from southeastern Australia. Alcheringa 13: 193–208.

Hill, R. S. 1990. *Araucaria* (Araucariaceae) species from Australian Tertiary sediments—a micromorphological study. Australian Systematic Botany 3: 203–220.

Hill, R. S. 1995. Conifer origin, evolution and diversification in the southern hemisphere, pp. 10–29 *in* N. J. Enright and R. S. Hill, eds., Ecology of the southern conifers. Washington, D.C.: Smithsonian Institution Press.

Hill, R. S., and A. J. Bigwood. 1987. Tertiary gymnosperms from Tasmania: Araucariaceae. Alcheringa 11: 325–335.

Hill, R. S., and T. J. Brodribb. 1999. Southern conifers in time and space. Australian Journal of Botany 47: 639–696.

Hill, R. S., and R. J. Carpenter. 1989. Tertiary gymnosperms from Tasmania: Cupressaceae. Alcheringa 13: 89–102.

Hill, R. S., and R. J. Carpenter. 1991. Evolution of *Acmopyle* and *Dacrycarpus* (Podocarpaceae) foliage as inferred from macrofossils in south-eastern Australia. Australian Systematic Botany 4: 449–479.

Hill, R. S., and D. C. Christophel. 2001. Two new species of *Dacrydium* (Podocarpaceae) based on vegetative fossils from middle Eocene sediments at Nelly Creek, South Australia. Australian Systematic Botany 14: 193–205.

Hill, R. S., and M. K. Macphail. 1985. A fossil flora from rafted Plio-Pleistocene mudstones at Regatta Point, Tasmania. Australian Journal of Botany 33: 497–517.

Hill, R. S., and H. E. Merrifield. 1993. An early Tertiary macroflora from West Dale, southwestern Australia. Alcheringa 17: 285–326.

Hill, R. S., and M. S. Pole. 1992. Leaf and shoot morphology of extant *Afrocarpus, Nageia* and *Retrophyllum* (Podocarpaceae) species, and species with similar leaf arrangement, from Tertiary sediments in Australia. Australian Journal of Botany 5: 337–358.

Hill, R. S., and L. J. Scriven. 1999. *Falcatifolium* (Podocarpaceae) macrofossils from Paleogene sediments in south-eastern Australia: a reassessment. Australian Systematic Botany 11: 711–720.

Hill, R. S., and S. S. Whang. 1996. A new species of *Fitzroya* (Cupressaceae) from Oligocene sediments in north-western Tasmania. Australian Systematic Botany 9: 867–875.

Hill, R. S., and S. S. Whang. 2000. *Dacrycarpus* (Podocarpaceae) macrofossils from Miocene sediments at Elands, eastern Australia. Australian Systematic Botany 13: 395–408.

Hill, R. S., G. J. Jordan, and R. J. Carpenter. 1993. Taxodiaceous macrofossils from Tertiary and Quaternary sediments in Tasmania. Australian Systematic Botany 6: 237–249.

Hills, L. V., and R. T. Ogilvie. 1970. *Picea banksii* n. sp. Beaufort Formation (Tertiary), northwestern Banks Island, arctic Canada. Canadian Journal of Botany 48: 457–464.

Hils, M. W. 1993. Taxaceae Gray, pp. 423–427 *in* Flora of North America Editorial Committee, eds., Flora of North America, vol. 2. New York: Oxford University Press.

Ho, C. C. 1950. Leaf anatomy of Chinese species of *Podocarpus.* Botanical Bulletin of Academia Sinica 3: 146–150.

Hoffman J., A. E. 1994. Flora silvestre de Chile, zona araucana, ed. 3. Santiago de Chile: Ediciones Fundacion Claudio Gay.

Holloway, J. T. 1937. Ovule anatomy and development and embryogeny in *Phyllocladus alpinus* (Hook.) and in *P. glaucus* (Carr.). Transactions, Royal Society of New Zealand 67: 149–165.

Hong, Y. P., V. D. Hipkins, and S. H. Strauss. 1993a. Chloroplast DNA diversity among trees, populations and species in the California closed-cone pines (*Pinus radiata, Pinus muricata* and *Pinus attenuata*). Genetics 135: 1,187–1,196.

Hong, Y. P., A. B. Krupkin, and S. H. Strauss. 1993b. Chloroplast DNA transgresses species boundaries and evolves at variable rates in the California closed-cone pines (*Pinus radiata, P. muricata* and *P. attenuata*). Molecular Phylogenetics and Evolution 2: 322–329.

Hu, H. H., and W. Y. Chun. 1927–1937. Icones plantarum sinicarum. Shanghai: Commercial Press.

Hu, Y. S., and F. H. Wang. 1984. Anatomical studies of *Cathaya* (Pinaceae). American Journal of Botany 71: 727–735.

Hunt, R. S. 1993. *Abies* Miller, pp. 354–362 *in* Flora of North America Editorial Committee, eds., Flora of North America, vol. 2. New York: Oxford University Press.

Hyland, B. P. M. 1978. A revision of the genus *Agathis* in Australia. Brunonia 1: 103–115.

Isoda, K., T. Brodribb, and S. Shiraishi. 2000. Hybrid origin of *Athrotaxis laxifolia* (Taxodiaceae) confirmed by random amplified polymorphic DNA analysis. Australian Journal of Botany 48: 753–758.

Jackson, A. B., and W. Dallimore. 1926. A new hybrid conifer. Bulletin of Miscellaneous Information for 1926, Royal Botanic Gardens, Kew. 113–115. [*Cupressus* ×*leylandii*]

Jackson, T., and C. Weng. 1999. Late Quaternary extinction of a tree species in eastern North American. Proceedings, National Academy of Sciences of the USA 96: 13,847–13,852. [*Picea critchfieldii*]

Jacobs, B. F., C. R. Werth, and S. I. Guttman. 1984. Genetic relationships in *Abies* (fir) of eastern United States: an electrophoretic study. Canadian Journal of Botany 62: 609–616.

Jähnichen, H. 1990. New records of the conifer *Amentotaxus gladifolia* (Ludwig) Ferguson, Jähnichen & Alvin, 1978, from the Polish and Czechoslovakian Tertiary and its recognition in Canada, North America, and Europe. Tertiary Research 12: 69–80.

Jaffré, T. 1995. Distribution and ecology of the conifers of New Caledonia, pp. 171–196 *in* N. J. Enright and R. S. Hill, eds., Ecology of the southern conifers. Washington, D.C.: Smithsonian Institution Press.

Jaffré, T., J.-M. Veillon, and J.-F. Cherrier. 1987. Sur la présence de deux Cupressaceae, *Neocallitropsis pancheri* (Carr.) Laubenf. et *Libocedrus austrocaledonica* Brongn. & Gris dans le massif Paéoua et localités nouvelles de gymnospermes en Nouvelle-Calédonie. Bulletin, Museum National de Histoire Naturel de Paris, sér. 4, 9 section B: 273–288.

Jagel, A., and Th. Stützel. 2001. Zur Abgrenzung von *Chamaecyparis* Spach und *Cupressus* L. (Cupressaceae) und die systematische Stellung von *Cupressus nootkatensis* D. Don [= *Chamaecyparis nootkatensis* (D. Don) Spach]. Feddes Repertorium 112: 179–229.

Jain, K. K. 1975. A taxonomic revision of the Himalayan firs. Indian Forester 101: 199–204.

Jain, K. K. 1976. Introgressive hybridization in the west Himalayan silver firs. Silvae Genetica 25: 107–109.

Jain, K. K. 1977. Morphology of female strobilus in *Podocarpus neriifolius.* Phytomorphology 27: 215–224.

Jardon, Y., L. Filion, and C. Cloutier. 1994. Tree-ring evidence for endemicity of the larch sawfly in North American. Canadian Journal of Forest Research 24: 742–747.

Jørgensen, S., J. L. Hamrick, and P. V. Wells. 2002. Regional patterns of genetic diversity in *Pinus flexilis* (Pinaceae) reveal complex species history. American Journal of Botany 89: 792–800.

Jones, W. G., K. D. Hill, and J. M. Allen. 1995. *Wollemia nobilis,* a new living Australian genus and species in the Araucariaceae. Telopea 6:173–176.

Johnson, A. H., E. R. Cook, and T. G. Siccama. 1988. Climate and red spruce growth and decline in the northern Appalachians. Proceedings, National Academy of Sciences of the USA 85: 5,369–5,373.

Jordan, G. J. 1995. Extinct conifers and conifer diversity in the early Pleistocene of western Tasmania. Review of Palaeobotany and Palynology 84: 375–387.

Jordan, G. J. 1997. Evidence of Pleistocene plant extinction and diversity from Regatta Point, western Tasmania, Australia. Botanical Journal, Linnean Society 123: 45–71.

Joyce, D. G. 1987. Adaptive variation in cold hardiness of eastern larch, *Larix laricina,* in northern Ontario. Canadian Journal of Forest Research 18: 85–89.

Joyner, K. L., X. R. Wang, J. S. Johnston, H. J. Price, and C. G. Williams. 2001. DNA content for Asian pines parallels New World relatives. Canadian Journal of Botany 79: 192–196.

Jules, E. S., M. J. Kauffman, W. D. Ritts, and A. L. Carroll. 2002. Spread of an invasive pathogen over a variable landscape: a nonnative root rot on Port Orford Cedar. Ecology 83: 3,167–3,181.

Kärkkäinen, K., V. Koski, and O. Savolainen. 1996. Geographical variation in the inbreeding depression of Scots pine. Evolution 50: 111–119.

Kang, N., and Z-X. Tang. 1995. Studies on the taxonomy of the genus *Torreya*. Bulletin of Botanical Research 15:349–362. [Chinese with English abstract]

Karalamangala, R. P., and D. L. Nickrent. 1989. An electrophoretic study of representatives of subgenus *Diploxylon* of *Pinus*. Canadian Journal of Botany 67: 1,750–1,759.

Kaundun, S. S., P. Lebreton, and B. Fady. 1998a. Geographical variability of *Pinus halepensis* Mill. as revealed by foliar flavonoids. Biochemical Systematics and Ecology 26: 83–96.

Kaundun, S. S., P. Lebreton, and B. Fady. 1998b. Genetic variation in the needle flavonoid composition of *Pinus brutia* var. *brutia* populations. Biochemical Systematics and Ecology 26: 485–494.

Kelch, D. G. 1997. The phylogeny of Podocarpaceae based on morphological evidence. Systematic Botany 22: 113–131.

Kelch, D. G. 1998. Phylogeny of Podocarpaceae: comparison of evidence from morphology and 18S rDNA. American Journal of Botany 85: 986–996.

Kelch, D. G. 2002. Phylogenetic assessment of the monotypic genera *Sundacarpus* and *Manoao* (Coniferales: Podocarpaceae) utilising evidence from 18S rDNA sequences. Australian Systematic Botany 15: 29–35.

Kellison, R. C., and B. J. Zobel. 1974. Genetics of Virginia pine. USDA Forest Service Research Paper WO-21: 1–10.

Keng, H. 1963. Aspects of morphology of *Phyllocladus hypophyllus*. Annals of Botany, n.s. 27: 69–78.

Keng, H. 1969. Aspects of morphology of *Amentotaxus formosana* with a note on the taxonomic position of the genus. Journal of the Arnold Arboretum 50: 432–446.

Keng, H. 1973. On the family Phyllocladaceae. Taiwania 18: 142–145.

Keng, H. 1974. The phylloclade of *Phyllocladus* and its possible bearing on the branch systems of progymnosperms. Annals of Botany 38: 757–764.

Keng, H. 1978. The genus *Phyllocladus* (Phyllocladaceae). Journal of the Arnold Arboretum 59: 249–273.

Keng, H. 1980. A new interpretation of the compound strobilar structures of cordaites and conifers. Reinwardtia 9: 377–384.

Keng, H., and E. L. Little, Jr. 1961. Needle characteristics of hybrid pines. Silvae Genetica 10: 131–146.

Keng, R.-S. L. 1979. Comparative anatomy of the species of *Phyllocladus* (Coniferae), a preliminary study. Quarterly Journal, Taiwan Museum 32: 221–305.

Khalil, M. A. K. 1987. Genetic variation in red spruce (*Picea rubens* Sarg.). Silvae Genetica 36: 164–171.

Kirkpatrick, J. 1997. Alpine Tasmania. Melbourne: Oxford University Press.

Klaus, W. 1989. Mediterranean pines and their history. Plant Systematics and Evolution 162: 133–163.

Knapp, M., R. M. Mudaliar, D. Havell, S. J. Wagstaff, and P. J. Lockhart. 2007. The drowning of New Zealand and the problem of *Agathis*. Systematic Biology 56: 862–870.

Köpke, E., L. J. Musselman, and D. J. de Laubenfels. 1981. Studies on the anatomy of *Parasitaxus ustus* and its root connections. Phytomorphology 31: 85–92.

Komarov, V. L. 1986. Coniferales, pp. 101–154 *in* V. L. Komarov, ed., Flora of the USSR, vol. 1. Dehra Dun: Bishen Singh Mahendra Pal Singh. (translation of 1934 original in Russian)

Konnert, M., and F. Bergmann. 1995. The geographical distribution of genetic variation of silver fir (*Abies alba*, Pinaceae) in relation to its migration history. Plant Systematics and Evolution 196: 19–30.

Kormutak, A., B. Vookova, A. Gajdosova, and A. Salaj. 1992. Hybridological relationships between *Pinus nigra* Arn., *Pinus thunbergii* Parl. and *Pinus tabulaeformis* Carrière. Silvae Genetica 41: 228–234.

Korn, R. W. 2001. Analysis of shoot apical organization in six species of the Cupressaceae based on chimeric behavior. American Journal of Botany 88: 1,945–1,952.

Korn, R. W. 2002. Chimeric patterns in *Juniperus chinensis* 'Torulosa Variegata' (Cupressaceae) expressed during leaf and stem formation. American Journal of Botany 89: 758–765.

Korol, L., A. Madmony, Y. Riov, and G. Schiller. 1995. *Pinus halepensis* × *Pinus brutia* subsp. *brutia* hybrids? Identification using morphological and biochemical traits. Silvae Genetica 44: 186–190.

Korol, L., G. Shklar, and G. Schiller. 2002. Diversity among circum-Mediterranean populations of Aleppo pine and differentiation from Brutia pine in their isoenzymes: additional results. Silvae Genetica 51: 35–41.

Kossuth, S. V., and G. H. Fechner. 1973. Incompatibility between *Picea pungens* Engelm. and *Picea engelmannii* Parry. Forest Science 19: 50–60.

Kovar-Eder, J., and Z. Kvacek. 1995. Der Nachweis eines fertilen Zweiges von *Tetraclinis brachyodon* (Brongniart) Mai et Walther aus Radoboj, Kroatien (Mittel-Miozän). Flora 190: 261–264.

Kral, R. 1993. *Pinus* Linnaeus, pp. 373–398 *in* Flora of North America Editorial Committee, eds., Flora of North America, vol. 2. New York: Oxford University Press.

Krasnoborov, I. M., ed. 2000. Flora of Siberia, vol. 1. Lycopodiaceae–Hydrocharitaceae. Enfield, New Hampshire: Science Publishers.

Krassilov, V. A. 1974. *Podocarpus* from the upper Cretaceous of eastern Asia and its bearing on the theory of conifer evolution. Palaeontology 17: 365–370.

Kriebel, H. B., and D. P. Fowler. 1965. Variability in needle characteristics of soft pine species and hybrids. Silvae Genetica 14: 73–76.

Krüssmann, G. 1985. Manual of cultivated conifers. Transl. M. E. Epp. Portland, Oregon: Timber Press.

Krupkin, A. B., A. Liston, and S. H. Strauss. 1996. Phylogenetic analysis of the hard pines (*Pinus* subgenus *Pinus*, Pinaceae) from chloroplast DNA restriction site analysis. American Journal of Botany 83: 489–498.

Kummerow, Jochen. 1966. Vegetative propagation of *Pinus radiata* by means of needle fascicles. Forest Science 12: 391–398.

Kvaček, Z., S. R. Manchester, and H. E. Schorn. 2000. Cones, seeds, and foliage of *Tetraclinis salicornioides* (Cupressaceae) from the Oligocene and Miocene of western North America: a geographic extension of the European Tertiary species. International Journal of Plant Science 161: 331–344.

Laderman, A. D., ed. 1998. Coastally restricted forests. New York: Oxford University Press. [*Chamaecyparis*]

Lang, K. J. 1994. *Abies alba* Mill.: differentiation of provenances and provenance groups by the monoterpene patterns in the cortex resin of twigs. Biochemical Systematics and Ecology 22: 53–63.

Lanner, R. M. 1974a. Natural hybridization between *Pinus edulis* and *Pinus monophylla* in the American Southwest. Silvae Genetica 23: 108–116.

Lanner, R. M. 1974b. A new pine from Baja California and the hybrid origin of *Pinus quadrifolia*. Southwestern Naturalist 19: 75–95.

Lanner, R. M. 1982. Adaptations of whitebark pine for dispersal by Clark's nutcracker. Canadian Journal of Forest Research 12: 391–402.

Lanner, R. M. 1999. Conifers of California. Los Olivos, California: Cachuma Press.

Lanner, R. M., and A. M. Phillips III. 1992. Natural hybridization and introgression of pinyon pines in northwestern Arizona. International Journal of Plant Sciences 153: 250–257.

Lanner, R. M., and T. R. van Devender. 1974. Morphology of pinyon pine needles from fossil packrat middens in Arizona. Forest Science 20: 207–211.

Lara, A., and R. Villalba. 1993. A 3620-year temperature record from *Fitzroya cupressoides* tree rings in southern South America. Science 260: 1,104–1,106.

La Roi, G. H., and J. R. Dugle. 1968. A systematic and genecological study of *Picea glauca* and *P. engelmannii*, using paper chromatograms of needle extracts. Canadian Journal of Botany 46: 649–687.

Latta, R. G., Y. B. Linhart, D. Fleck, and M. Elliot. 1998. Direct and indirect estimates of seed versus pollen movement within a population of ponderosa pine. Evolution 52: 61–67.

Lauranson-Broyer, J., and P. Lebreton. 1993. Flavonoids and morphological traits of needles as markers of natural hybridization between *Pinus uncinata* Ram. and *Pinus sylvestris* L. Biochemical Systematics and Ecology 21: 241–247.

Lauria, F. 1991. Taxonomy, systematics, and phylogeny of *Pinus*, subsection *Ponderosae* Loudon (Pinaceae). Alternative concepts. Linzer Biologischer Beiträge 23: 129–202.

Lawrence, L., R. Bartschot, E. Zavarin, and J. R. Griffin. 1975. Natural hybridization of *Cupressus sargentii* and *C. macnabiana* and the composition of the derived essential oils. Biochemical Systematics and Ecology 2: 113–119.

Ledig, F. T., and M. T. Conkle. 1983. Gene diversity and genetic structure in a narrow endemic, Torrey pine (*Pinus torreyana* Parry ex Carr.). Evolution 37: 79–85.

Ledig, F. T., V. Jacob-Cervantes, P. D. Hodgskiss, and T. Eguiluz-Piedra. 1997. Recent evolution and divergence among populations of a rare Mexican endemic, Chihuahua spruce, following Holocene climatic warming. Evolution 51: 1,815–1,827.

Ledig, F. T., M. T. Conkle, B. Bermejo-Velazquez, T. Eguiluz-Piedra, P. D. Hodgskiss, D. R. Johnson, and W. S. Dvorak. 1999. Evidence for an extreme bottleneck in a rare Mexican pinyon: genetic diversity, disequilibrium, and the mating system in *Pinus maximartinezii*. Evolution 53: 91–99.

Ledig, F. T., M. Mápula-Larreta, B. Bermejo-Velázquez, V. Reyes-Hernández, C. Flores-López, and M. A. Capó-Arteaga. 2000. Locations of endangered spruce populations in México and the demography of *Picea chihuahuana*. Madroño 47: 71–88.

Ledig, F. T., M. A. Capó-Arteaga, P. D. Hodgskiss, H. Sbay, C. Flores-López, M. T. Conkle, and B. Bermejo-Velázquez. 2001. Genetic diversity and the mating system of a rare Mexican piñon, *Pinus pinceana*, and a comparison with *Pinus maximartinezii* (Pinaceae). American Journal of Botany 88: 1,977–1,987.

Lee, C. H. 1968. Geographic variation in European black pine. Silvae Genetica 17: 165–172.

Lee, C. L. 1952. The anatomy and ontogeny of the leaf of *Dacrydium taxoides*. American Journal of Botany 39: 393–398.

Lee, S. C. 1962. Taiwan red- and yellow-cypress and their conservation. Taiwania 8: 1–15.

Lee, T. B. 1979. Illustrated flora of Korea. Seoul: Hyangmunsa.

Leistner, O. A. 1966. Podocarpaceae. Flora of Southern Africa 1: 34–41.

Leistner, O. A., G. F. Smith, and H. F. Glen. 1995. Notes on *Podocarpus* in southern Africa and Madagascar. Bothalia 25: 233–245.

Leopold, D. J., W. C. McComb, and R. N. Muller. 1998. Trees of the central hardwood forests of North America. Portland, Oregon: Timber Press.

LePage, B. A. 2003. A new species of *Thuja* (Cupressaceae) from the late Cretaceous of Alaska: implications of being evergreen in a polar environment. American Journal of Botany 90: 167–174.

LePage, B. A., and J. F. Basinger. 1991. A new species of *Larix* (Pinaceae) from the early Tertiary of Axel Heiberg Island, arctic Canada. Review of Palaeobotany and Palynology 70: 89–111.

LePage, B. A., and J. F. Basinger. 1995a. The evolutionary history of the genus *Larix* (Pinaceae). US Department of Agriculture Forest Service General Technical Report INT-319: 19–29.

LePage, B. A., and J. F. Basinger. 1995b. Evolutionary history of the genus *Pseudolarix* Gordon (Pinaceae). International Journal of Plant Sciences 156 (6): 910–950.

Lewandowski, A. 1997. Genetic relationships between European and Siberian larch, *Larix* spp. (Pinaceae), studied by allozymes. Is the Polish larch a hybrid between these two species? Plant Systematics and Evolution 204: 65–73.

Lewandowski, A., and L. Meinartowicz. 1990. Inheritance of allozymes in *Larix decidua* Mill. Silvae Genetica 39: 184–188.

Li, H. L. 1952. The genus *Amentotaxus*. Journal of the Arnold Arboretum 33: 192–198.

Li, H. L. 1953. A reclassification of *Libocedrus* and the Cupressaceae. Journal of the Arnold Arboretum 34: 17–34.

Li, H. L. 1962. A new species of *Chamaecyparis*. Morris Arboretum Bulletin 13: 43–46.

Li, H. L. 1963. Woody flora of Taiwan. Narberth, Pennsylvania: Livingston Publishing Company.

Li, H. L. 1968. The golden larch, *Pseudolarix amabilis*. Morris Arboretum Bulletin 19: 19–25.

Li, H. L., and H. Keng. 1994. Gymnospermae, *in* T.-C. Huang, ed., Flora of Taiwan, ed. 2. Taipei: Editorial Committee of the Flora of Taiwan.

Li, J. H., C. C. Davis, P. del Tredici, and M. J. Donoghue. 2001a. Phylogeny and biogeography of *Taxus* (Taxaceae) inferred from sequences of the internal transcribed spacer region of nuclear ribosomal DNA. Harvard Papers in Botany 6: 267–274.

Li, J. H., C. C. Davis, M. J. Donoghue, S. Kelley, and P. del Tredici. 2001b. Phylogenetic relationships of *Torreya* (Taxaceae) inferred from sequences of nuclear ribosomal DNA ITS region. Harvard Papers in Botany 6: 275–281.

Li, J. H., D. L. Zhang, and M. J. Donoghue. 2003. Phylogeny and biogeography of *Chamaecyparis* (Cupressaceae) inferred from DNA sequences of the nuclear ribosomal ITS region. Rhodora 105: 106–117.

Li, N., and L. K. Fu. 1997. Notes on gymnosperms I. Taxonomic treatment of some Chinese conifers. Novon 7: 261–264.

Lidholm, J., and P. Gustafsson. 1991. The chloroplast genome of the gymnosperm *Pinus contorta:* a physical map and a complete collection of overlapping clones. Current Genetics 20: 161–166.

Lin, J. X., Y. S. Hu, and F. H. Wang. 1995. Wood and bark anatomy of *Nothotsuga* (Pinaceae). Annals of the Missouri Botanical Garden 82: 603–609.

Lindley, J. 1851. Notices of certain ornamental plants lately introduced into England. Journal, Horticultural Society of London 6: 258–273. [*Fitzroya, Saxegothaea,* some Araucariaceae]

Linhart, Y. B., J. B. Mitton, K. B. Sturgeon, and M. L. Davis. 1981. Genetic variation in space and time in a population of ponderosa pine. Heredity 46: 407–426.

Linnaeus, C. 1753. Species plantarum, vol. 2. Stockholm: Laurentius Salvius.

Lipscomb, B. 1993. *Pseudotsuga* Carrière, pp. 365–366 *in* Flora of North America Editorial Committee, eds., Flora of North America, vol. 2. New York: Oxford University Press.

Liston, A., M. Parker-Defeniks, J. V. Syring, A. Willyard, and R. Cronn, 2007. Interspecific phylogenetic analysis enhances intraspecific phylogeographic inference: a case study in *Pinus lambertiana.* Molecular Ecology 16: 3,926–3,937.

Little, D. P. 2006. Evolution and circumscription of the true cypresses (Cupressaceae: *Cupressus*). Systematic Botany 31: 461–480.

Little, E. L., Jr. 1966. Varietal transfers in *Cupressus* and *Chamaecyparis.* Madroño 18: 161–167.

Little, E. L., Jr., and W. B. Critchfield. 1969. Subdivisions of the genus *Pinus* (pines). US Department of Agriculture Forest Service Miscellaneous Publication 1144: 1–51.

Little, E. L., Jr., and F. I. Righter. 1965. Botanical descriptions of forty artificial pine hybrids. US Department of Agriculture Forest Service Technical Bulletin 1345: 1–47.

Little, E. L., Jr., and R. G. Skolmen. 1989. Common forest trees of Hawaii native and introduced). US Department of Agriculture Agricultural Handbook 679. Washington, D.C.: Government Printing Office.

Liu, T. S. 1971. A monograph of the genus *Abies.* Taipei: Department of Forestry, National Taiwan University.

Liu, T. S., and H. J. Su. 1983. Biosystematic studies on *Taiwania* and numerical evaluations of the systematics of Taxodiaceae. Taiwan Museum Special Publication 2: 1–113.

Lockhart, L. A. 1990a. Chemotaxonomic relationships within the Central American closed-cone pines. Silvae Genetica 39: 173–184.

Lockhart, L. A. 1990b. The xylem resin terpene composition of *Pinus greggii* Engelm. and *Pinus pringlei* Shaw. Silvae Genetica 39: 198–201.

Long, D. G. 1980. Notes relating to the flora of Bhutan: IV. The weeping cypress, *Cupressus corneyana* Carr. Notes, Royal Botanic Garden, Edinburgh 38: 311–314.

López, G. G., K. Kamiya, and K. Harada. 2002. Phylogenetic relationships of *Diploxylon* pines (subgenus *Pinus*) based on plastid sequence data. International Journal of Plant Sciences 163: 737–747.

Lorimer, S. D., and R. T. Weavers. 1987. Foliage sesquiterpenes and diterpenes of *Podocarpus spicatus.* Phytochemistry 26: 3,207–3,215.

Lowry, J. B. 1972. Anthocyanins of the Podocarpaceae. Phytochemistry 11: 725–731.

Macbride, J. F. 1936. Flora of Peru, part 1. Chicago: Field Museum of Natural History.

McCarter, P. S., and J. S. Birks. 1985. *Pinus patula* subsp. *tecunumanii:* the application of numerical techniques to some problems of its taxonomy. Commonwealth Forestry Review 64: 117–132.

McClelland, B. R., S. S. Frissell, W. C. Fischer, and C. H. Halvorson. 1979. Habitat management for hole-nesting birds in forests of western larch and Douglas-fir. Journal of Forestry 77: 480–483.

McClintock, E. 1986. The identity of *Cupressus lusitanica* in Mexico. International Dendrology Society Yearbook 1986: 96–99.

McComb, A. L. 1955. The European larch: its races, site requirements and characteristics. Forest Science 1: 298–318.

McIver. E. E. 1992. Fossil *Fokienia* (Cupressaceae) from the Paleocene of Alberta, Canada. Canadian Journal of Botany 70: 742–749.

McIver, E. E. 1994. An early *Chamaecyparis* (Cupressaceae) from the late Cretaceous of Vancouver Island, British Columbia, Canada. Canadian Journal of Botany 72: 1,787–1,796.

McIver, E. E. 2001. Cretaceous *Widdringtonia* Endl. (Cupressaceae) from North America. International Journal of Plant Sciences 162: 937–961.

McIver, E. E., and K. R. Aulenback. 1994. Morphology and relationships of *Mesocyparis umbonata* sp. nov.: fossil Cupressaceae from the late Cretaceous of Alberta, Canada. Canadian Journal of Botany 72: 273–295.

McIver, E. E., and J. F. Basinger. 1989. The morphology and relationships of *Thuja polaris* sp. nov. (Cupressaceae) from the early Tertiary, Ellesmere Island, arctic Canada. Canadian Journal of Botany 67: 1,903–1,915.

McIver, E. E., and J. F. Basinger. 1990. Fossil seed cones of *Fokienia* (Cupressaceae) from the Paleocene Ravenscrag Formation of Saskatchewan, Canada. Canadian Journal of Botany 68: 1,609–1,618.

McLea, W. L. 1996. The late-Quaternary pollen records of south-east Nelson, South Island, New Zealand. New Zealand Journal of Botany 34: 523–538.

McMillan, C. 1952. The third locality for *Cupressus abramsiana* Wolf. Madroño 11: 189–194.

McVaugh, R. 1992. Flora Novo-Galiciano, vol. 17. Gymnosperms and pteridophytes. Ann Arbor, Michigan: University of Michigan Herbarium.

Mai, D. H. 1987. Neue Arten nach Fruchten und Samen aus dem Tertiär von Nordwestsachsen und der Lausitz. Feddes Repertorium 98: 105–126.

Maley, J., and P. Brenac. 1998. Vegetation dynamics, palaeoenvironments and climatic changes in the forests of western Cameroon during the last 28,000 years B.P. Review of Palaeobotany and Palynology 99: 157–187.

Malusa, J. 1992. Phylogeny and biogeography of the pinyon pines (*Pinus* subsect. *Cembroides*). Systematic Botany 17: 42–66.

Manley, S. A. M., and F. T. Ledig. 1979. Photosynthesis in black and red spruce and their hybrid derivatives: ecological isolation and hybrid adaptive inferiority. Canadian Journal of Botany 57: 305–314.

Markgraf, V., J. P. Bradbury, and J. R. Busby. 1986. Paleoclimates in southwestern Tasmania during the last 13,000 years. Palaios 1: 368–380.

Markham, K. R., C. Vilain, and B. P. J. Molloy. 1985a. Uniformity and distinctness of *Phyllocladus* as evidenced by flavonoid accumulation. Phytochemistry 24: 2,607–2,609.

Markham, K. R., R. F. Webby, L. A. Whitehouse, B. P. J. Molloy, C. Vilain, and R. Mues. 1985b. Support from flavonoid glycoside distribution for the division of *Podocarpus* in New Zealand. New Zealand Journal of Botany 23: 1–13.

Markham, K. R., R. F. Webby, B. P. J. Molloy, and C. Vilain. 1989. Support from flavonoid glycoside distribution for the division of *Dacrydium* sensu lato. New Zealand Journal of Botany 27: 1–11.

Marsh, J. A. 1966. Cupressaceae. Flora of Southern Africa 1: 43–48.

Marshall, K. A., and D. B. Neale. 1992. The inheritance of mitochondrial DNA in Douglas-fir (*Pseudotsuga menziesii*). Canadian Journal of Forest Research 22: 73–75.

Martin, H. A. 1997. The use of ecological tolerances for the reconstruction of Tertiary palaeoclimates. Australian Journal of Botany 45: 475–492.

Martin, P. C. 1950. A morphological comparison of *Biota* and *Thuja*. Proceedings, Pennsylvania Academy of Science 24: 65–112.

Martín, R. A. 1993. Podocarpaceae S. Endlicher, pp. 641–645 *in* T. J. Killeen, E. García E., and S. G. Beck, eds., Guía de arboles de Bolivia. La Paz: Herbario Nacional de Bolivia.

Martínez, M. 1963. Las pináceas mexicanas, ed. 3. Mexico, D.F.: Universidad Nacional Autónoma de México.

Masson, D., M. Glidewell, M. Möller, R. R. Mill, B. Williamson, and R. M. Bateman. 2001. Non-destructive examination of herbarium material for taxonomic studies using NMR imaging. Edinburgh Journal of Botany 58: 1–14.

Matos, J. A. 1995. *Pinus hartwegii* and *P. rudis:* a critical assessment. Systematic Botany 20: 6–21.

Matsumoto, M., T. Ohsawa, and M. Nishida. 1995. *Tsuga shimokawaensis,* a new species of premineralized conifer leaves from the middle Miocene Shimokawa Group, Hokkaido, Japan. Journal of Plant Research 108: 417–428.

Medan, D., and R. D. Tortosa. 1981. Anatomy of the root and stem nodules of *Dacrydium fonckii* (Phil.) Florin (Podocarpaceae). Botanical Journal, Linnean Society 83: 85–91.

Meijer Drees, E. 1940. The genus *Agathis* in Malesia. Bulletin, Jardin Botanique de Buitenzorg, sér. 3, 16: 455–474.

Meikle, R. D. 1977. Flora of Cyprus, vol. 1. Richmond, Surrey: Royal Botanic Gardens, Kew.

Melikian, A. P., and A. V. F. Ch. Bobrov. 2000. Morphology of female reproductive structures and an attempt of the construction of phylogenetic system of orders Podocarpales, Cephalotaxales and Taxales. Botanicheskii Zhurnal 85(7): 50–68. (in Russian)

Melnikova, A. B., and A. N. Makhinov. 2004. On the record of *Microbiota decussata* (Cupressaceae) at uncommonly low altitude. Botanicheskii Zhurnal 89: 1,470–1,472. (in Russian)

Melville, R. 1955. The podocarps of East Africa. Kew Bulletin 9: 563–574.

Michaux, J., J.-P. Suc, and J.-L. Vernet. 1979. Climatic inference from the history of the Taxodiaceae during the Pliocene and the early Pleistocene in western Europe. Review of Palaeobotany and Palynology 27: 185–191.

Michener, D. C. 1993. *Chamaecyparis* Spach, pp. 408–410 *in* Flora of North America Editorial Committee, eds., Flora of North America, vol. 2. New York: Oxford University Press.

Midgley, J. J., W. J. Bond, and C. J. Geldenhuys. 1995. The ecology of southern African conifers, pp. 64–80 *in* N. J. Enright and R. S. Hill, eds., Ecology of the southern conifers. Washington, D.C.: Smithsonian Institution Press.

Miki, S., and S. Hikita. 1951. Probable chromosome number of fossil *Sequoia* and *Metasequoia* in Japan. Science 113: 3–4.

Mikkola, L. 1969. Observations on interspecific sterility in *Picea.* Annales Botanici Fennici 6: 285–339.

Mill, R. R. 1999a. A new combination in *Nageia* (Podocarpaceae). Novon 9: 77–78.

Mill, R. R. 1999b. A new species of *Larix* (Pinaceae) from southeast Tibet and other nomenclatural notes on Chinese *Larix.* Novon 9: 79–82.

Mill, R. R. 2001. A new sectional combination in *Nageia* Gaertn. (Podocarpaceae). Edinburgh Journal of Botany 58: 499–501.

Mill, R. R., and C. J. Quinn. 2001. *Prumnopitys andina* reinstated as the correct name for "lleuque," the Chilean conifer recently renamed *P. spicata* (Podocarpaceae). Taxon 50: 1,143–1,154.

Mill, R. R., M. Möller, F. Christie, S. M. Glidewell, D. Masson, and B. Williamson. 2001. Morphology, anatomy and ontogeny of female cones of *Acmopyle pancheri* (Brongn. & Gris) Pilg. (Podocarpaceae). Annals of Botany 88: 55–67.

Millar, C. I. 1986. The Californian closed cone pines (subsection *Oocarpae* Little and Critchfield): a taxonomic history and review. Taxon 35: 657–670.

Millar, C. I. 1993. Impact of the Eocene on the evolution of *Pinus* L. Annals of the Missouri Botanical Garden 80: 471–498.

Millar, C. I. 1998. Early evolution of pines, pp. 69–91 *in* D. M. Richardson, ed., Ecology and biogeography of *Pinus.* Cambridge University Press.

Millar, C. I., S. H. Strauss, M. T. Conkle, and R. D. Westfall. 1988. Allozyme differentiation and biosystematics of the Californian closed-cone pines (*Pinus* subsect. *Oocarpae*). Systematic Botany 13: 351–370.

Miller, C. N., Jr. 1976. Early evolution in the Pinaceae. Review of Palaeobotany and Palynology 21: 101–117.

Miller, C. N., Jr. 1989. A new species of *Picea* based on silicified seed cones from the Oligocene of Washington. American Journal of Botany 76: 747–754.

Miller, C. N., Jr. 1992. Structurally preserved cones of *Pinus* from the neogene of Idaho and Oregon. International Journal of Plant Sciences 153: 147–154.

Miller, C. N., Jr., and J. M. Malinky. 1986. Seed cones of *Pinus* from the late Cretaceous of New Jersey, U.S.A. Review of Palaeobotany and Palynology 46: 257–272.

Miller, H. J. 1973. The wood of *Amentotaxus*. Journal of the Arnold Arboretum 54: 111–119.

Miller, R. W. 1980. A brief survey of *Taxus* alkaloids and other taxane derivatives. Journal of Natural Products 43: 425–437.

Mills, J. S. 1973. Diterpenes of *Larix* oleoresins. Phytochemistry 12: 2,407–2,412.

Milton, J. B., R. G. Latta, and G. E. Rehfeldt. 1997. The pattern of inbreeding in Washoe pine and survival of inbred progeny under optimal environmental conditions. Silvae Genetica 46: 215–219.

Mirams, R. V. 1957. Aspects of the natural regeneration of the kauri (*Agathis australis* Salisb.) Transactions, Royal Society of New Zealand 84: 661–680.

Miranda, V., and M. Chaphekar. 1980. SEM study of the inner periclinal surface of leaf cuticles in the family Pinaceae. Botanical Journal, Linnean Society 81: 61–78.

Mirov, N. T. 1953. Taxonomy and chemistry of the white pines. Madroño 12: 81–89.

Mirov, N. T. 1967. The genus *Pinus*. New York: Ronald Press.

Mirov, N. T., E. Zavarin, and K. Snajberk. 1965. Chemical composition of the turpentines of some eastern Mediterranean pines in relation to their classification. Phytochemistry 5: 97–102.

Mirov, N. T., E. Zavarin, K. Snajberk, and K. Costello. 1966. Further studies of turpentine composition of *Pinus muricata* in relation to its taxonomy. Phytochemistry 5: 343–355.

Mitchell, A. F. 1972. Conifers in the British Isles. Forestry Commission Booklet 33. London: HMSO.

Mitsopoulos, D. J., and C. P. Panetsos. 1987. Origin of variation in fir forests of Greece. Silvae Genetica 36: 1–15.

Mitton, J. B. 1992. The dynamic mating system of conifers. New Forests 6: 197–216.

Mitton, J. B., M. C. Grant, and A. M. Yoshino. 1998. Variation in allozymes and stomatal size in pinyon (*Pinus edulis*, Pinaceae), associated with soil moisture. American Journal of Botany 85: 1,262–1,265.

Mitton, J. B., B. R. Kreiser, and R. G. Latta. 2000. Glacial refugia of limber pine (*Pinus flexilis* James) inferred from the population structure of mitochondrial DNA. Molecular Ecology 9: 91–97.

Moar, N. T. 1955. Adventitious root-shoots of *Dacrydium colensoi* Hook. in Westland, South Island, New Zealand. New Zealand Journal of Science and Technology 37A: 207–213.

Möller, M., R. R. Mill, M. Glidewell, D. Masson, B. Williamson, and R. M. Bateman. 2000. Comparative biology of pollination mechanisms in *Acmopyle pancheri* and *Phyllocladus hypophyllus* (Podocarpaceae s.l.) Annals of Botany 86: 149–158.

Moles, A. T., and D. R. Drake. 1999. Postdispersal seed predation on eleven large-seeded species from the New Zealand flora: a preliminary study in secondary forest. New Zealand Journal of Botany 37: 679–685.

Molina R., A. 1964. Coníferas de Honduras. Ceiba 10: 5–21.

Molina-Freaner, F., P. Delgado, D. Piñero, N. Perez-Nasser, and E. Alvarez-Buylla. 2001. Do rare pines need different conservation strategies? Evidence from three Mexican species. Canadian Journal of Botany 79: 131–138.

Molloy, B. P. J. 1995. *Manoao* (Podocarpaceae), a new monotypic conifer genus endemic to New Zealand. New Zealand Journal of Botany 33: 183–201.

Molloy, B. P. J. 1996. A new species name in *Phyllocladus* (Phyllocladaceae) from New Zealand. New Zealand Journal of Botany 34: 287–297.

Molloy, B. P. J., and K. R. Markham. 1999. A contribution to the taxonomy of *Phyllocladus* (Phyllocladaceae) from the distribution of key flavonoids. New Zealand Journal of Botany 37: 375–382.

Molloy, B. P. J., and M. Muñoz-Schick. 1999. The correct name for the Chilean conifer lleuque (Podocarpaceae). New Zealand Journal of Botany 37: 189–193.

Moore, H. E., Jr. 1965. *Chrysolarix*, a new name for the golden larch. Baileya 13: 131–134.

Moore, H. E., Jr. 1973. *Chrysolarix* renounced—a comedy of restoration. Taxon 22: 587–589.

Moore, P. R., and R. Wallace. 2000. Petrified wood from the Miocene volcanic sequences of Coromandel Peninsula, northern New Zealand. Journal, Royal Society of New Zealand 30: 115–130.

Moran, G. F., D. Smith, J. C. Bell, and R. Appels. 1992. The 5S RNA gene in *Pinus radiata* and the spacer region as a probe for relationships between *Pinus* species. Plant Systematics and Evolution 183: 209–221.

Moriguchi, Y., A. Matsumoto, M. Saito, Y. Tsumura, and H. Taira. 2001. DNA analysis of clonal structure of an old growth isolated forest of *Cryptomeria japonica* in snowy region. Canadian Journal of Forest Research 31: 377–383.

Morris, R. W., W. B. Critchfield, and D. P. Fowler. 1980. The putative Austrian × red pine hybrid: a test of paternity based on allelic variation at enzyme-specifying loci. Silvae Genetica 29: 93–100.

Mosseler, A., D. J. Innes, and B. A. Roberts. 1991. Lack of allozyme variation in disjunct Newfoundland populations of red pine (*Pinus resinosa*). Canadian Journal of Forest Research 21: 525–528.

Moulalis, D., C. Bassiotis, and D. Mitsopoulos. 1976. Controlled pollinations among pine species in Greece. Silvae Genetica 25: 95–107.

Murphy, J. O. 1993. Dendrochronological investigation of *Halocarpus biformis* (pink pine) from a site at the West Arm of Lake Manapouri in New Zealand. Australian Systematic Botany 6: 481–489.

Myers, O., Jr., and F. H. Bormann. 1963. Phenotypic variation in *Abies balsamea* in response to altitudinal and geographic gradients. Ecology 44: 429–431.

Neale, D. B., and R. R. Sederoff. 1989. Paternal inheritance of chloroplast DNA and maternal inheritance of mitochondrial DNA in loblolly pine. Theoretical and Applied Genetics 77: 212–216.

Neale, D. B., N. C. Wheeler, and R. W. Allard. 1986. Paternal inheritance of chloroplast DNA in Douglas-fir. Canadian Journal of Forest Research 16: 1,152–1,154.

Nelson, C. D., W. L. Nance, and D. B. Wagner. 1994. Chloroplast DNA variation within and among taxonomic varieties of *Pinus caribaea* and *Pinus elliottii.* Canadian Journal of Forest Research 24: 424–426.

Nguyên, T. H., and J. E. Vidal. 1996. Gymnospermae, pp. 1–168 *in* Ph. Morat, ed., Flore du Cambodge de Laos et du Viêtnam, vol. 28. Paris: Muséum National d'Histoire Naturelle.

Nicholson, R. G. 1986. Collecting rare conifers in North Africa. Arnoldia 46: 20–29.

Niebling, C. R., and M. T. Conkle. 1990. Diversity of Washoe pine and comparisons with allozymes of ponderosa pine trees. Canadian Journal of Forest Research 20: 298–308.

Niemann, G. J. and H. H. van Genderen. 1980. Chemical relationships between Pinaceae. Biochemical Systematics and Ecology 8: 237–240.

Nikolić, D., and N. Tucić. 1983. Isoenzyme variation within and among populations of European black pine (*Pinus nigra* Arnold). Silvae Genetica 32: 80–89.

Nkongolo, K. K. 1996. Chromosome analysis and DNA homology in three *Picea* species, *P. mariana, P. rubens,* and *P. glauca* (Pinaceae). Plant Systematics and Evolution 203: 27–40.

Nkongolo, K. K., and K. Klimaszewska. 1995. Cytological and molecular relationships between *Larix decidua, L. leptolepis* and *Larix* ×*eurolepis:* identification of species-specific chromosomes and synchronization of mitotic cells. Theoretical and Applied Genetics 90: 827–834.

Norin, T. 1972. Some aspects of the chemistry of the order Pinales. Phytochemistry 11: 1,231–1,242.

Norton, D. A. 1991. Seedling and sapling distribution patterns in a coastal podocarp forest, Hokitika Ecological District, New Zealand. New Zealand Journal of Botany 29: 463–466.

Norton, D. A., J. W. Herbert, and A. E. Beveridge. 1988. The ecology of *Dacrydium cupressinum:* a review. New Zealand Journal of Botany 26: 37–62.

Offler, C. E. 1984. Extant and fossil Coniferales of Australia and New Guinea. Part 1: a study of the external morphology of the vegetative shoots of the extant species. Palaeontographica Abt. B, 193: 18–120.

Offord, C. A., C. L. Porter, P. F. Meagher, and G. Errington. 1999. Sexual reproduction and early plant growth of the Wollemi pine (*Wollemia nobilis*), a rare and threatened Australian conifer. Annals of Botany 84: 1–9.

Ogden, J. 1978. Investigations of the dendrochronology of the genus *Athrotaxis* D. Don (Taxodiaceae) in Tasmania. Tree-Ring Bulletin 38: 1–13.

Ogden, J., A. Wilson, C. Hendy, and R. M. Newnham. 1992. The late Quaternary history of kauri (*Agathis australis*) in New Zealand and its climatic significance. Journal of Biogeography 19: 611–622.

Ogilvie, R. T., and E. von Rudloff. 1968. Chemosystematic studies in the genus *Picea* (Pinaceae). IV. The introgression of white and Engelmann spruce as found along the Bow River. Canadian Journal of Botany 46: 901–908.

Ohsawa, T. 1994. Anatomy and relationships of petrified seed cones of the Cupressaceae, Taxodiaceae, and Sciadopityaceae. Journal of Plant Research 107: 503–512.

Ohtsu, H., R. Tanaka, T. Michida, T. Shingu, and S. Matsunaga. 1998. Tetracyclic triterpenes and other constituents from the leaves and bark of *Larix kaempferi.* Phytochemistry 49: 1,761–1,768.

O'Reilly, G. J., W. H. Parker, and W. M. Cheliak. 1985. Isozyme differentiation of upland and lowland *Picea mariana* stands in northern Ontario. Silvae Genetica 34: 214–221.

Orr, M. Y. 1933. Plantae Chinenses Forrestianae: Coniferae. Notes, Royal Botanic Garden, Edinburgh 18: 119–158.

Orr, M. Y. 1944. The leaf anatomy of *Podocarpus.* Transactions, Botanical Society of Edinburgh 34: 1–54.

Ortiz, P. L., M. Arista, and S. Talavera. 2002. Sex ratio and reproductive effort in the dioecious *Juniperus communis* subsp. *alpina* (Suter) Čelak. (Cupressaceae) along an altitudinal gradient. Annals of Botany 89: 205–211.

Ostenfeld, C. H., and C. S. Larsen. 1930. The species of the genus *Larix* and their geographical distribution. Biologiske Meddelelser 9(2): 1–107.

Owens, J. N., G. L. Catalano, S. J. Morris, and J. Aitken-Christie. 1995a. The reproductive biology of kauri (*Agathis australis*). I. Pollination and prefertilization development. International Journal of Plant Sciences 156: 257–269.

Owens, J. N., G. L. Catalano, S. J. Morris, and J. Aitken-Christie. 1995b. The reproductive biology of kauri (*Agathis australis*). II. Male gametes, fertilization, and cytoplasmic inheritance. International Journal of Plant Sciences 156: 404–416.

Owens, J. N., G. L. Catalano, S. J. Morris, and J. Aitken-Christie. 1995c. The reproductive biology of kauri (*Agathis australis*). III. Proembryogeny and early embryogeny. International Journal of Plant Sciences 156: 793–806.

Owens, J. N., G. L. Catalano, and J. Aitken-Christie. 1997. The reproductive biology of kauri (*Agathis australis*). IV. Late embryogeny, histochemistry, cone and seed morphology. International Journal of Plant Sciences 158: 395–407.

Page, C. N. 1974. Morphology and affinities of *Pinus canariensis.* Notes, Royal Botanic Garden, Edinburgh 33: 317–323.

Page, C. N. 1980. Leaf micromorphology in *Agathis* and its taxonomic implications. Plant Systematics and Evolution 135: 71–79.

Page, C. N. 1988. New and maintained genera in the conifer families Podocarpaceae and Pinaceae. Notes, Royal Botanic Garden, Edinburgh 45: 377–395.

Palmer, E., and N. Pitman. 1972. Trees of southern Africa, vol. 1. Cape Town: A. A. Balkema.

Pandey, P. C., B. Singh, and P. S. Rehill. 1983. Windthrow in deodar. Commonwealth Forestry Review 62: 37–39.

Panetsos, C. P. 1975. Natural hybridization between *Pinus halepensis* and *Pinus brutia* in Greece. Silvae Genetica 24: 163–168.

Panetsos, K. P., A. Christou and A. Scaltsoyiannes. 1992. First analysis on allozyme variation in cedar species (*Cedrus* sp.). Silvae Genetica 41: 339–341.

Pant, D. D., and N. Basu. 1977. A comparative study of the leaves of *Cathaya argyrophylla* Chun & Kuang and three species of *Keteleeria* Carrière. Botanical Journal, Linnean Society 75: 271–282.

Park, Y. S., and D. P. Fowler. 1982. Effects of inbreeding and genetic variances in a natural population of tamarack (*Larix laricina* (Du Roi) K. Koch) in eastern Canada. Silvae Genetica 31: 21–26.

Parker, K. C., and J. L. Hamrick. 1996. Genetic variation in sand pine (*Pinus clausa*). Canadian Journal of Forest Research 26: 244–254.

Parker, K. C., J. L. Hamrick, A. J. Parker, and E. A. Stacy. 1997. Allozyme diversity in *Pinus virginiana* (Pinaceae): intraspecific and interspecific comparisons. American Journal of Botany 84: 1,372–1,382.

Parker, W. H. 1993. *Larix* Miller, pp. 366–368 *in* Flora of North America Editorial Committee, eds., Flora of North America, vol. 2. New York: Oxford University Press.

Parker, W. H., and T. A. Dickinson. 1990. Range-wide morphology and anatomical variation in *Larix laricina.* Canadian Journal of Botany 68: 832–840.

Parker, W. H., and D. G. McLachlan. 1978. Morphological variation in white and black spruce: investigation of natural hybridization between *Picea glauca* and *P. mariana.* Canadian Journal of Botany 56: 2,512–2,520.

Parker, W. H., J. Maze, and G. E. Bradfield. 1981. Implications of morphological and anatomical variation in *Abies balsamea* and *A. lasiocarpa* (Pinaceae) from western Canada. American Journal of Botany 68: 843–854.

Parker, W. H., J. Maze, F. E. Bennett, T. A. Cleveland, and D. G. McLachlan. 1984. Needle flavonoid variation in *Abies balsamea* and *A. lasiocarpa* from western Canada. Taxon 33: 1–12.

Passini, M.-F., and N. Pinel. 1989. Ecology and distribution of *Pinus lagunae* in the Sierra de la Laguna, Baja California Sur, Mexico. Madroño 36: 84–92.

Patel, R. N. 1967. Wood anatomy of Podocarpaceae indigenous to New Zealand, 1. *Dacrydium.* New Zealand Journal of Botany 5: 171–184.

Patel, R. N. 1968a. Wood anatomy of Podocarpaceae indigenous to New Zealand, 3. *Phyllocladus.* New Zealand Journal of Botany 6: 3–8.

Patel, R. N. 1968b. Wood anatomy of Cupressaceae and Araucariaceae indigenous to New Zealand. New Zealand Journal of Botany 6: 9–18.

Pattanavibool, R., P. von Aderkas, A. Hanhijärvi, L. K. Simola, and J. M. Bonga. 1995. Diploidization in megagametophyte-derived cultures of the gymnosperm *Larix decidua.* Theoretical and Applied Genetics 90: 671–674.

Pauw, C. A., and H. P. Linder. 1997. Tropical African cedars (*Widdringtonia,* Cupressaceae): systematics, ecology, and conservation status. Botanical Journal, Linnean Society 123: 297–319.

Pederick, L. A. 1968. Chromosome inversions in *Pinus radiata.* Silvae Genetica 17: 22–26.

Pederick, L. A. 1970. Chromosome relationships between *Pinus* species. Silvae Genetica 19: 171–180.

Perez de la Rosa, J., S. A. Harris, and A. Farjon. 1995. Noncoding chloroplast DNA variation in Mexican pines. Theoretical and Applied Genetics 91: 1,101–1,106.

Perry, J. P., Jr. 1991. The pines of Mexico and Central America. Portland, Oregon: Timber Press.

Perry, N. B., and R. T. Weavers. 1985a. Intraspecific variation of foliage diterpenes of *Dacrydium cupressinum.* Phytochemistry 24: 2,233–2,237.

Perry, N. B., and R. T. Weavers. 1985b. Foliage diterpenes of *Dacrydium intermedium:* identification, variation and biosynthesis. Phytochemistry 24: 2,899–2,904.

Perry, N. B., M. H. Benn, L. M. Foster, A. Routledge, and R. T. Weavers. 1996. The glycosidic precursor of (*Z*)-5-ethylidene-2(5*H*)-furanone in *Halocarpus biformis* juvenile foliage. Phytochemistry 42: 453–459.

Peteet, D. M. 1991. Postglacial migration history of lodgepole pine near Yakutat, Alaska. Canadian Journal of Botany 69: 786–796.

Philipson, W. R., and B. P. J. Molloy. 1990. Seedling, shoot, and adult morphology of New Zealand conifers. The genera *Dacrycarpus, Podocarpus, Dacrydium,* and *Prumnopitys.* New Zealand Journal of Botany 28: 73–84.

Phillips, E. W. J. 1941. The identification of coniferous woods by their microscopic structure. Journal of the Linnean Society, Botany 52: 259–320.

Phipps, C. J., J. M. Osborn, and R. A. Stockey. 1995. *Pinus* pollen cones from the middle Eocene Princeton Chert (Allenby Formation) of British Columbia, Canada. International Journal of Plant Sciences 156: 117–124.

Pilger, R. 1903. Taxaceae. Das Pflanzenreich 18: 1–124. (in German)

Pilger, R. 1926. Klasse Coniferae, pp. 121–403 *in* A. Engler and R. Prantl, eds., Die natürlichen Pflanzenfamilien, ed. 2, vol. 3. Leipzig: Verlag von Wilhelm Engelmann. (in German)

Playford, G., and M. E. Dettmann. 1978. Pollen of *Dacrydium franklinii* Hook. f. and comparable early Tertiary microfossils. Pollen et Spores 20: 513–534.

Pocknall, D. T. 1981a. Pollen morphology of the New Zealand species of *Dacrydium* Solander, *Podocarpus* L'Héritier, and *Dacrycarpus* Endlicher (Podocarpaceae). New Zealand Journal of Botany 19: 67–95.

Pocknall, D. T. 1981b. Pollen morphology of *Phyllocladus* L. C. et A. Rich. New Zealand Journal of Botany 19: 259–266.

Pocknall, D. T. 1981c. Pollen morphology of the New Zealand species of *Libocedrus* Endlicher (Cupressaceae) and *Agathis* Salisbury (Araucariaceae). New Zealand Journal of Botany 19: 267–272.

Pole, M. S. 1992a. Eocene vegetation from Hasties, north-eastern Tasmania. Australian Systematic Botany 5: 431–475.

Pole, M. S. 1992b. Early Miocene flora of the Manuherikia Group, New Zealand. 2. Conifers. Journal, Royal Society of New Zealand 22: 287–302.

Pole, M. S. 1993. Miocene broad-leaved *Podocarpus* from Foulden Hills, New Zealand. Alcheringa 17: 173–177.

Pole, M. S. 1994. An Eocene macroflora from the Taratu Formation at Livingstone, North Otago, New Zealand. Australian Journal of Botany 42: 341–367.

Pole, M. S. 1997a. Miocene conifers from the Manuherikia Group, New Zealand. Journal, Royal Society of New Zealand 27: 355–370.

Pole, M. S. 1997b. Paleocene plant macrofossils from Kakahu, South Canterbury, New Zealand. Journal, Royal Society of New Zealand 27: 371–400.

Pole, M. S. 1998. Paleocene gymnosperms from Mount Somers, New Zealand. Journal, Royal Society of New Zealand 28: 375–403.

Pollack, J. C., and B. P. Dancik. 1985. Monoterpene and morphological variation and hybridization of *Pinus contorta* and *P. banksiana* in Alberta. Canadian Journal of Botany 63: 201–210.

Posey, C. E., and J. F. Goggans. 1967. Observations on species of cypress indigenous to the United States. Auburn University Agricultural Experiment Station Circular 153: 1–19.

Powell, G. R. 1992. Patterns of leaf size and morphology in relation to shoot length in *Tsuga canadensis*. Trees 7: 59–66.

Powrie, E. 1972. The typification of *Brunia nodiflora* L. Journal of South African Botany 38: 301–304. [*Widdringtonia*]

Preest, D. S. 1967. A note on the dispersal characteristics of the seed of the New Zealand podocarps and beeches and their biogeographical significance, *in* J. L. Gressitt, ed., Pacific Basin biogeography. Honolulu: Bishop Museum Press.

Premoli, A. C., T. Kitzberger, and T. T. Veblen. 2000. Isozyme variation and recent biogeographical history of the long-lived conifer *Fitzroya cupressoides*. Journal of Biogeography 27: 251–260.

Price, R. A. 1990. The genera of Taxaceae in the southeastern United States. Journal of the Arnold Arboretum 71: 69–91.

Price, R. A., and J. M. Lowenstein. 1989. An immunological comparison of the Sciadopityaceae, Taxodiaceae, and Cupressaceae. Systematic Botany 14: 141–149.

Price, R. A., A. Liston, and S. H. Strauss. 1998. Phylogeny and systematics of *Pinus*, pp. 49–68 *in* D. M. Richardson, ed., Ecology and biogeography of *Pinus*. Cambridge: Cambridge University Press.

Prus-Glowacki, W., J. Szweykowski, and R. Novak. 1985. Serotaxonomical investigation of the European pine species. Silvae Genetica 34: 162–170.

Qian, T., R. A. Ennos, and T. Helgason. 1995. Genetic relationships among larch species based on analysis of restriction fragment variation for chloroplast DNA. Canadian Journal of Forest Research 25: 1,197–1,202.

Quinn, C. J. 1965. Gametophyte development and embryogeny in the Podocarpaceae. II. *Dacrydium laxifolium*. Phytomorphology 15: 37–45.

Quinn, C. J. 1970. Generic boundaries in the Podocarpaceae. Proceedings, Linnean Society of New South Wales 94: 166–172.

Quinn, C. J. 1982. Taxonomy of *Dacrydium*. Australian Journal of Botany 30: 311–320.

Quinn, C. J. 1986. Embryogeny in *Phyllocladus*. New Zealand Journal of Botany 24: 575–579.

Quinn, C. J. 1987. The Phyllocladaceae Keng—a critique. Taxon 36: 559–565.

Quinn, C. J., and P. Gadek. 1981. Biflavones of *Dacrydium* sensu lato. Phytochemistry 20: 677–681.

Quinn, C. J., and J. A. Rattenbury. 1972. Structural hybridity in New Zealand *Dacrydium*. New Zealand Journal of Botany 10: 427–436.

Quinn, C. J., R. A. Price, and P. A. Gadek. 2002. Familial concepts and relationships in the conifers based on *rbc*L and *mat*K sequence comparisons. Kew Bulletin 57: 513–531.

Rafii, Z., L. G. Cool, R. Jonas, and E. Zavarin. 1992a. Chemical diversity in *Cupressus bakeri*, 1. megagametophyte fatty acids. Biochemical Systematics and Ecology 20: 25–30.

Rafii, Z., L. G. Cool, and E. Zavarin. 1992b. Variability of foliar mono- and sequiterpenoids of *Cupressus bakeri*. Biochemical Systematics and Ecology 20: 123–131.

Rafii, Z., R. S. Dodd, and E. Zavarin. 1996. Genetic diversity in foliar terpenoids among natural populations of European black pine. Biochemical Systematics and Ecology 24: 325–339.

Raubeson. L. A., and R. K. Jansen. 1992. A rare chloroplast-DNA structural mutation is shared by all conifers. Biochemical Systematics and Ecology 20: 17–24.

Rehder, A., and E. H. Wilson. 1916. Pinaceae, pp. 10–62 *in* C. S. Sargent, ed., Plantae Wilsonianae, vol. 2. Cambridge, Massachusetts: Harvard University Press.

Rehfeldt, G. E. 1982. Differentiation of *Larix occidentalis* populations from the northern Rocky Mountains. Silvae Genetica 31: 13–19.

Rehfeldt, G. E. 1988. Ecological genetics of *Pinus contorta* from the Rocky Mountains (USA): a synthesis. Silvae Genetica 37: 131–135.

Rehfeldt, G. E. 1990a. Genecology of *Larix laricina* (Du Roi) K. Koch in Wisconsin. I. Patterns of natural variation. Silvae Genetica 39: 9–16.

Rehfeldt, G. E. 1990b. Genetic differentiation among populations of *Pinus ponderosa* from the upper Colorado River basin. Botanical Gazette 151: 125–137.

Rehfeldt, G. E. 1992. Breeding strategies for *Larix occidentalis:* adaptations to the biotic and abiotic environment in relation to improving growth. Canadian Journal of Forest Research 22: 5–13.

Rehfeldt, G. E. 1999a. Systematics and genetic structure of *Ponderosae* taxa (Pinaceae) inhabiting the mountain islands of the southwest. American Journal of Botany 86: 741–752.

Rehfeldt, G. E. 1999b. Systematics and genetic structure of Washoe pine: applications in conservation genetics. Silvae Genetica 48: 167–173.

Richardson, D. M., editor. 1998. Ecology and biogeography of *Pinus*. Cambridge: Cambridge University Press.

Riedl, H. 1966. Pinaceae. Flora Iranica 14: 1–9.

Riedl, H. 1968. Cupressaceae. Flora Iranica 50: 1–10.

Robison, C. R. 1977. *Pinus triphylla* and *Pinus quinquefolia* from the upper Cretaceous of Massachusetts. American Journal of Botany 64: 726–732.

Roche, L. 1969. A genecological study of the genus *Picea* in British Columbia. New Phytologist 68: 505–554.

Rodríguez, R., and M. Quezada. 1995. Gymnospermae. *In* C. Maticorena and R. Rodríguez, eds., Flora de Chile, vol. 1. Concepción: Universidad de Concepción.

Rodríguez, R., R., O. Matthei S., and M. Quezada M. 1983. Flora arbórea de Chile. Concepción: Universidad de Concepcion Press.

Rogers, D. L., C. I. Millar, and R. D. Westfall. 1999. Fine-scale genetic structure of whitebark pine (*Pinus albicaulis*): associations with watershed and growth form. Evolution 53: 74–90.

Romero, A., M. Luna, E. Garcia, and M. F. Passini. 2000. Phenetic analysis of the Mexican midland pinyon pines, *Pinus cembroides* and *Pinus johannis.* Botanical Journal, Linnean Society 133: 181–194.

Rothwell, G. W., and J. F. Basinger. 1979. *Metasequoia milleri* n. sp., anatomically preserved pollen cones from the middle Eocene (Allenby Formation) of British Columbia. Canadian Journal of Botany 57: 958–970.

Rowell, R. J. 1996. *Ornamental conifers for Australian gardens.* Sydney: University of New South Wales Press.

Rowett, A. I. 1992. Dispersed cuticular floras of South Australian Tertiary coalfields, part 2: Lochiel. Transactions, Royal Society of South Australia 116: 95–107.

Ruby, J. L., and J. W. Wright. 1976. A revised classification of geographical varieties in Scots pine. Silvae Genetica 25: 169–175.

Rudolph, T. D., N. C. Wheeler, and N. K. Dhir. 1986. Cone clusters in jack pine. Canadian Journal of Forest Research 16: 1,180–1,184.

Rundel, P. W. 1972. An annotated checklist of the groves of *Sequoiadendron giganteum* in the Sierra Nevada, California. Madroño 21: 319–328.

Runions, C. J., and J. N. Owens. 1998. Evidence of pre-zygotic self-incompatibility in a conifer, pp. 255–264 *in* S. J. Owens and P. J. Rudall, eds., Reproductive biology. Richmond, Surrey: Royal Botanic Gardens, Kew.

Rushforth, K. D. 1989. Two new species of *Abies* (Pinaceae) from western Mexico. Notes, Royal Botanic Garden, Edinburgh 46: 101–109.

Rushforth, K., R. P. Adams, M. Zhong, X.-q. Ma, and R. N. Pandey. 2003. Variation among *Cupressus* species from the eastern hemisphere based on random amplified polymorphic DNAs (RAPDs). Biochemical Systematics and Ecology 31: 17–24.

Šaden-Krehula, M., M. Tajić, and D. Kolbah. 1979. Sex hormones and corticosteroids in pollen of *Pinus nigra.* Phytochemistry 18: 345–346.

Sahni, B. 1920. On the structure and affinities of *Acmopyle pancheri* Pilger. Philosophical Transactions, Royal Society of London, series B, 210: 253–310.

Sahni, B., and A. K. Mitra. 1927. Notes on the anatomy of some New Zealand species of *Dacrydium.* Annals of Botany 91: 75–89.

Sahni, K. C. 1990. Gymnosperms of India and adjacent countries. Dehra Dun: Bishen Singh Mahendra Pal Singh.

Said, C., M. Villar, and P. Zandonella. 1991. Ovule receptivity and pollen viability in Japanese larch (*Larix leptolepis* Gord.). Silvae Genetica 40: 1–6.

Saiki, K. 1996. *Pinus mutoi* (Pinaceae), a new species of permineralized seed cone from the Upper Cretaceous of Hokkaido, Japan. American Journal of Botany 83: 1,630–1,636.

Salazar, R. 1983. Genetic variation in needles of *Pinus caribaea* var. *hondurensis* Barr. et Golf. from natural stands. Silvae Genetica 32: 52–59.

Salmon, J. T. 1980. The native trees of New Zealand. Wellington: A. H. & A. W. Reed.

Salter, J., B. G. Murray, and J. E. Braggins. 2002. Wettable and unsinkable: the hydodynamics of saccate pollen grains in relation to the pollination mechanism in the two New Zealand species of *Prumnopitys* Phil. (Podocarpaceae). Annals of Botany 89: 133–144. 2002.

Sargent, C. S. 1898. The silva of North America, vol. 12, Coniferae. Boston: Houghton Mifflin.

Sax, H. J. 1932. Chromosome pairing in *Larix* species. Journal of the Arnold Arboretum 13: 368–374.

Saylor, L. C. 1964. Karyotype analysis of *Pinus*—group *Lariciones.* Silvae Genetica 13: 165–170.

Saylor, L. C. 1972. Karyotype analysis of the genus *Pinus*—subgenus *Pinus.* Silvae Genetica 21: 155–163

Saylor, L. C. 1983. Karyotype analysis of the genus *Pinus*—subgenus *Strobus.* Silvae Genetica 32: 119–124.

Saylor, L. C., and R. L. Koenig. 1967. The slash × sand pine hybrid. Silvae Genetica 16: 134–138.

Schaefer, P. R., and J. W. Hanover. 1990. An investigation of sympatric populations of blue and Engelmann spruces in Scotch Creek drainage, Colorado. Silvae Genetica 39: 72–81.

Schiller, G., and A. Genizi. 1993. An attempt to identify the origin of *Pinus brutia* Ten. plantations in Israel by needle resin composition. Silvae Genetica 42: 63–68.

Schiller, G., and C. Grunwald. 1987. Resin monoterpenes in range-wide provenance trials of *Pinus halepensis* Mill. in Israel. Silvae Genetica 36: 109–114.

Schiller, G., M. T. Conkle, and C. Grunwald. 1986. Local differentiation among Mediterranean populations of Aleppo pine in their isoenzymes. Silvae Genetica 35: 11–19.

Schlarbaum, S. E., and T. Tsuchiya. 1984a. Cytotaxonomy and phylogeny in certain species of Taxodiaceae. Plant Systematics and Evolution 147: 29–54.

Schlarbaum, S. E., and T. Tsuchiya. 1984b. A chromosome study of coast redwood, *Sequoia sempervirens* (D. Don) Endl. Silvae Genetica 33: 56–62.

Schlarbaum, S. E., L. C. Johnson, and T. Tsuchiya 1983. Chromosome studies of *Metasequoia glyptostroboides* and *Taxodium distichum.* Botanical Gazette 144: 559–565.

Schlarbaum, S. E., T. Tsuchiya, and L. C. Johnson. 1984. The chromosomes and relationships of *Metasequoia* and *Sequoia* (Taxodiaceae): an update. Journal of the Arnold Arboretum 65: 251–254.

Schmid, M. 1981. Fleurs et plantes de Nouvelle-Caledonie. Papeete: Éditions du Pacifique.

Schmidt, P. A. 1989. Beitrag zur Systematik und Evolution der Gattung *Picea* A. Dietr. Flora 182: 435–461.

Schmidt, W. C. 1995. Around the world with *Larix:* an introduction. US Department of Agriculture Forest Service General Technical Report INT-319: 6–18.

Schmidt, W. C., and K. J. McDonald, eds. 1995. Ecology and management of *Larix* forests: a look ahead. US Department of Agriculture Forest Service General Technical Report INT-319: 1–521.

Schmidtling, R. C. 1984. Planting south of origin increases flowering in shortleaf (*Pinus echinata* Mill.) and Virginia pine (*P. virginiana* Mill.). Silvae Genetica 33: 140–144.

Schoettle, A. W., and S. G. Rochelle. 2000. Morphological variation of *Pinus flexilis* (Pinaceae), a bird-dispersed pine, across a range of elevations. American Journal of Botany 87: 1,797–1,806.

Schoonraad, E., and H. P. van der Schijff. 1974. Anatomy of leaves of the genus *Podocarpus* in South Africa. Phytomorphology 24: 75–85.

Schoonraad, E., and H. P. van der Schijff. 1975. Distribution and some interesting morphological aspects of the South African Podocarpaceae. Boissiera 24: 135–144.

Schorn, H. E. 1994. A preliminary discussion of fossil larches (*Larix*, Pinaceae) from the arctic. Quaternary International 22–23: 173–183.

Schorn, H. E., and W. C. Wehr. 1996. The conifer flora from the Eocene uplands at Republic, Washington. Washington Geology 24: 22–24.

Schulman, E. 1958. Bristlecone pine, oldest known living thing. National Geographic, March 1958: 355–372.

Schwartz, M. W., and S. M. Hermann. 1993. The continuing population decline of *Torreya taxifolia* Arn. Bulletin, Torrey Botanical Club 120: 275–286.

Schwartz, M. W., and S. M. Hermann. 1999. Is slow growth of the endangered *Torreya taxifolia* (Arn.) normal? Journal, Torrey Botanical Society 126: 307–312.

Schwarz, O., and H. Weide. 1962. Systematische Revision der Gattung *Sequoia* Endl. Feddes Repertorium 66: 159–192.

Semerikov, V. L., and M. Lascoux. 2003. Nuclear and cytoplasmic variation within and between Eurasian *Larix* (Pinaceae) species. American Journal of Botany 90: 1,113–1,123.

Semerikov, V. L., H. Zhang, M. Sun, and M. Lascoux. 2003. Conflicting phylogenies of *Larix* (Pinaceae) based on cytoplasmic and nuclear DNA. Molecular Phylogenetics and Evolution 27: 173–184.

Serre, F. 1978. The dendroclimatological value of the European larch (*Larix decidua* Mill.) in the French Maritime Alps. Tree-Ring Bulletin 38: 25–34.

Setoguchi, H., T. A. Osawa, J.-C. Pintaud, T. Jaffré, and J.-M. Veillon. 1998. Phylogenetic relationships within Araucariaceae based on rbcL gene sequences. American Journal of Botany 85: 1,507–1,516.

Seward, A. C. 1906. The Araucarieae, recent and extinct. Philosophical Transactions, Royal Society of London, series B, 198: 305–411.

Shah, A., D. Z. Li, M. Möller, L. M. Gao, M. L. Hollingsworth, and M. Gibby. 2008. Delimitation of *Taxus fauna* Nan Li & R. R. Mill (Taxaceae) based on morphological and molecular data. Taxon 57: 211–222.

Shapcott, A., M. J. Brown, J. B. Kirkpatrick, and J. B. Reid. 1995. Stand structure, reproductive activity and sex expression in Huon pine (*Lagarostrobos franklinii* (Hook. L. [sic]) Quinn.) Journal of Biogeography 22: 1,035–1,045.

Shaw, G. R. 1909. The pines of Mexico. Publications of the Arnold Arboretum 1: 1–30.

Shaw, G. R. 1914. The genus *Pinus*. Publications of the Arnold Arboretum 5: 1–96.

Shea, K. L., and G. R. Furnier. 2002. Genetic variation and population structure in central and isolated populations of balsam fir, *Abies balsamea* (Pinaceae). American Journal of Botany 89: 783–791.

Shimizu, Y., M. Ando, and F. Sakai. 2002. Clonal structure of natural populations of *Cryptomeria japonica* growing at different positions on slopes, detected using RAPD markers. Biochemical Systematics and Ecology 30: 733–748.

Shin, D. I., G. K. Podila, Y. H. Huang, and D. F. Karnosky. 1994. Transgenic larch expressing genes for herbicide and insect resistance. Canadian Journal of Forest Research 24: 2,059–2,067.

Sigurgeirsson, A., and A. E. Szmidt. 1993. Phylogenetic and biogeographic implications of chloroplast DNA variation in *Picea*. Nordic Journal of Botany 13: 233–246.

Silba, J. 1981. Revised generic concepts of *Cupressus* L. (Cupressaceae). Phytologia 49: 390–399.

Silba, J. 1983. Addendum to a revision of *Cupressus* L. (Cupressaceae). Phytologia 52: 349–361.

Silba, J. 1984. An international census of the Coniferae, I. Phytologia Memoirs 7: 1–79.

Silba, J. 1985a. The infraspecific taxonomy of *Pinus culminicola* Andr. et Beam. (Pinaceae). Phytologia 56: 489–491.

Silba, J. 1985b. A supplement to the international census of the Coniferae, I. Phytologia 58: 365–370.

Silba, J. 1986. Encyclopedia Coniferae. Phytologia Memoirs 8: 1–217.

Silba, J. 1987. Nomenclature of the weeping Himalayan cypress (*Cupressus*, Cupressaceae). Phytologia 64: 78–80.

Silba, J. 1990. A supplement to the international census of the Coniferae, II. Phytologia 68: 7–78.

Silba, J. 1994. The trans-Pacific relationship of *Cupressus* in India and North America. Journal, International Conifer Preservation Society 1: 1–28.

Sillet, S. C., J. C. Spickler, and R. van Pelt. 2000. Crown structure of the world's second largest tree. Madroño 47: 127–133.

Sinclair, W. T., R. R. Mill, M. F. Gardner, P. Woltz, T. Jaffré, J. Preston, M. L. Hollingsworth, A. Ponge, and M. Möller. 2002. Evolutionary relationships of the New Caledonian heterotrophic conifer, *Parasitaxus usta* (Podocarpaceae), inferred from chloroplast *trn*L–F intron / spacer and nuclear rDNA ITS2 sequences. Plant Systematics and Evolution 233: 79–104.

Sivak, J. 1976. Nouvelles espèces du genre *Cathaya* d'après leurs grains de pollen dans le Tertiaire du sud de la France. Pollen et Spores 18: 243–288.

Smith, A. C. 1979. Flora Vitiensis nova: a new flora of Fiji (spermatophytes only), vol. I. Lawai, Hawaii: Pacific Tropical Botanical Garden.

Smith, R. M., R. A. Marty, and C. F. Peters. 1981. The diterpene acids in the bled resins of three Pacific kauri, *Agathis vitiensis, A. lanceolata* and *A. macrophylla*. Phytochemistry 20: 2,205–2,207.

Smouse, P. E., and L. C. Saylor. 1973. Studies of the *Pinus rigida–serotina* complex I. a study of geographic variation. Annals of the Missouri Botanical Garden 60: 174–191.

Snajberk, K., and E. Zavarin. 1986. Monoterpenoid differentiation in relation to the morphology of *Pinus remota*. Biochemical Systematics and Ecology 14: 155–163.

Snajberk, K., E. Zavarin, and R. Debry. 1982. Terpenoid and morphological variability of *Pinus quadrifolia* and the natural hybridization with *P. monophylla* in the San Jacinto Mountains of California. Biochemical Systematics and Ecology 10: 121–132.

Sniezko, R. A., and L. J. Mullin. 1987. Taxonomic implications of bush pig damage and basal shoots in *Pinus tecunumanii.* Commonwealth Forestry Review 66: 313–316.

Stahlhut, R., G. Park, R. Petersen, W. Ma, and P. Hylands. 1999. The occurrence of the anti-cancer diterpene taxol in *Podocarpus gracilior* Pilger (Podocarpaceae). Biochemical Systematics and Ecology 27: 613–622.

Stainton, A. 1988. Flowers of the Himalaya: a supplement. Oxford: Oxford University Press.

Stairs, G. R. 1968. Monoterpene composition in *Larix.* Silvae Genetica 17: 182–186.

Standley, P. C., and J. A. Steyermark. 1958. Flora of Guatemala. Fieldiana: Botany 24, part 1.

Stead, J. W., and B. T. Styles. 1984. Studies of Central American pines: a revision of the "*pseudostrobus*" group (Pinaceae). Botanical Journal, Linnean Society 89: 249–275.

Stebbins, G. L. 1948. The chromosomes and relationships of *Sequoia* and *Metasequoia.* Science 108: 95–98.

Stefanović, S., M. Jager, J. Deutsch, J. Broutin, and M. Masselot. 1998. Phylogenetic relationships of conifers inferred from partial 28S rRNA gene sequences. American Journal of Botany 85: 688–697.

Steinhoff, R. J., and J. W. Andresen. 1971. Geographic variation in *Pinus flexilis* and *Pinus strobiformis* and its bearing on their taxonomic status. Silvae Genetica 20: 159–167.

Stephenson, N. L. 2000. Estimated ages of some large giant sequoias: General Sherman keeps getting younger. Madroño 47: 61–67.

Stevenson, D. W. 1991. Podocarpaceae Endlicher (1847), pp. 23–30 *in* A. R. A. Görts-van Rijn, ed., Flora of the Guianas, fasc. 9. Köeltz Scientific Books.

Stewart, H. 1984. Cedar, tree of life to the Northwest Coast Indians. Toronto: Douglas and McIntyre.

Stine, M., B. B. Sears, and D. E. Keathley. 1989. Inheritance of plastids in interspecific hybrids of blue spruce and white spruce. Theoretical and Applied Genetics 78: 768–774.

Stockey, R. A. 1982. The Araucariaceae: an evolutionary perspective. Review of Palaeobotany and Palynology 37: 133–154.

Stockey, R. A. 1994. Mesozoic Araucariaceae: morphology and systematic relationships. Journal of Plant Research 107: 493–502.

Stockey, R. A., and I. J. Atkinson. 1993. Cuticle micromorphology of *Agathis* Salisbury. International Journal of Plant Sciences 154: 187–225.

Stockey, R. A., and B. J. Frevel. 1997. Cuticle micromorphology of *Prumnopitys* Philippi (Podocarpaceae). International Journal of Plant Sciences 158: 198–221.

Stockey, R. A., and H. Ko. 1986. Cuticle micromorphology of *Araucaria* de Jussieu. Botanical Gazette 147: 508–548.

Stockey, R. A., and H. Ko. 1988. Cuticle micromorphology of some New Caledonian podocarps. Botanical Gazette 149: 240–252.

Stockey, R. A., and H. Ko. 1990. Cuticle micromorphology of *Dacrydium* (Podocarpaceae) from New Caledonia. Botanical Gazette 151: 138–149.

Stockey, R. A., and M. Nishida. 1986. *Pinus haboroensis* sp. nov. and the affinities of permineralized leaves from the upper Cretaceous of Japan. Canadian Journal of Botany 64: 1,856–1,866.

Stockey, R. A., and Y. Ueda. 1986. Permineralized pinaceous leaves from the upper Cretaceous of Hokkaido. American Journal of Botany 73: 1,157–1,162.

Stockey, R. A., H. Ko, and P. Woltz. 1992. Cuticle micromorphology of *Falcatifolium* de Laubenfels (Podocarpaceae). International Journal of Plant Sciences 153: 589–601.

Stockey, R. A., H. Ko, and P. Woltz. 1995. Cuticle micromorphology of *Parasitaxus* de Laubenfels (Podocarpaceae). International Journal of Plant Sciences 156: 723–730.

Stockey, R. A., B. J. Frevel, and P. Woltz. 1998. Cuticle micromorphology of *Podocarpus* subgenus *Podocarpus,* section *Scytopodium* (Podocarpaceae) of Madagascar and South Africa. International Journal of Plant Sciences 159: 923–940.

Stone, E. L. 1974. The communal root system of red pine: growth of girdled trees. Forest Science 20: 294–305.

Strauss, S. H., and A. H. Doerksen. 1990. Restriction fragment analysis of pine phylogeny. Evolution 44: 1,081–1,096.

Strauss, S. H., J. D. Palmer, G. T. Howe, and A. H. Doerksen. 1988. Chloroplast genomes of two conifers lack a large inverted repeat and are extensively rearranged. Proceedings, National Academy of Sciences of the USA 85: 3,898–3,902.

Strauss, S. H., A. H. Doerksen, and J. R. Byrne. 1990. Evolutionary relationships of Douglas-fir and its relatives (genus *Pseudotsuga*) from DNA restriction fragment analysis. Canadian Journal of Botany 68: 1,502–1,510.

Strauss, S. H., Y. P. Hong, and V. D. Hipkins. 1993. High levels of population differentiation for mitochondrial DNA haplotypes in *Pinus radiata, muricata,* and *attenuata.* Theoretical and Applied Genetics 86: 605–611.

Strobel, G. A., W. M. Hess, J. Y. Li, E. Ford, J. Sears, R. S. Sidhu, and B. Summerell. 1997. *Pestalotiopsis guepinii,* a taxol-producing endophyte of the Wollemi pine, *Wollemia nobilis.* Australian Journal of Botany 45: 1,073–1,082.

Stützel, T., and I. Röwekamp. 1999. Female reproductive structures in Taxales. Flora 194: 145–157.

Sudworth, G. B. 1908. Forest trees of the Pacific slope. Washington, D.C.: US Department of Agriculture Forest Service. [reprint New York: Dover Publications]

Sudworth, G. B. 1915. The cypress and juniper trees of the Rocky Mountain Region. US Department of Agriculture Bulletin 207: 1–36.

Sudworth, G. B. 1916. The spruce and balsam fir trees of the Rocky Mountain region. US Department of Agriculture Bulletin 327: 1–43.

Süss, H., and E. Velitzelos. 2000. Zwei neue fossile Hölzer der Formgattung *Podocarpoxylon* Gotham aus tertiären Schichten der Insel Lesbos, Greichenland. Feddes Reppertorium 111: 135–149.

Sullivan, J. J., C. J. Burrows, and J. S. Dugdale. 1995. Insect predation of seeds of native New Zealand woody plants in some central South Island localities. New Zealand Journal of Botany 33: 355–364.

Sutton, B. C. S., D. J. Flanagan, J. R. Gawley, C. H. Newton, D. T. Lester, and Y. A. El-Kassaby. 1991. Inheritance of chloroplast and mitochondrial DNA in *Picea* and composition of hybrids from introgression zones. Theoretical and Applied Genetics 82: 242–248.

Swartley, J. C. 1984. The cultivated hemlocks. Revised by H. J. Welch. Portland, Oregon: Timber Press.

Swetnam, T. W. 1993. Fire history and climate change in giant sequoia groves. Science 262: 885–889.

Syring, J., K. Farrell, R. Businký, R. Cronn, and A. Liston. 2007. Widespread genealogical nonmonophly in species of *Pinus* subgenus *Strobus*. Systematic Biology 56: 163–181.

Szmidt, A. E., and X. R. Wang. 1993. Molecular systematics and genetic differentiation of *Pinus sylvestris* (L.) and *P. densiflora* (Sieb. et Zucc.). Theoretical and Applied Genetics 86: 159–165.

Szmidt, A. E., A. Torsten, and J.-E. Hällgren. 1987. Paternal inheritance of chloroplast DNA in *Larix*. Plant Molecular Biology 9: 59–64.

Takahashi, K., S. Nagahama, T. Nakashima, and H. Suenaga. 2003. Chemotaxonomy of leaf oil constituents of *Thujopsis dolabrata* Sieb. et Zucc.—analysis of acidic extracts. Biochemical Systematics and Ecology 31: 723–738.

Takano, I., I. Yasuda, M. Nishijima, Y. Hitotsuyanagi, K. Takeya, and H. Itokawa. 1996. Alkaloids from *Cephalotaxus harringtonia*. Phytochemistry 43: 299–303.

Takaso, T., and J. N. Owens. 1995. Pollination drop and microdrop secretions in *Cedrus*. International Journal of Plant Sciences 156: 640–649.

Takaso, T., and P. B. Tomlinson. 1989a. Aspects of cones and ovule ontogeny in *Cryptomeria* (Taxodiaceae). American Journal of Botany 76: 692–705.

Takaso, T., and P. B. Tomlinson. 1989b. Cone and ovule development in *Callitris* (Cupressaceae—Callitroideae). Botanical Gazette 150: 378–390.

Takaso, T., and P. B. Tomlinson. 1990. Cone and ovule ontogeny in *Taxodium* and *Glyptostrobus* (Taxodiaceae—Coniferales). American Journal of Botany 77: 1,209–1,221.

Takaso, T., and P. B. Tomlinson. 1991. Cone and ovule development in *Sciadopitys*. (Taxodiaceae—Coniferales). American Journal of Botany 78: 417–428.

Tanaka, R., H. Ohtsu, and S. Matsunaga. 1997. Abietane diterpene acids and other constituents from the leaves of *Larix kaempferi*. Phytochemistry 46: 1,051–1,057.

Taylor, R. J. 1972. The relationship and origin of *Tsuga heterophylla* and *Tsuga mertensiana* based on phytochemical and morphological interpretations. American Journal of Botany 59: 149–157.

Taylor, R. J. 1993a. *Tsuga* (Endlicher) Carrière, pp. 362–365 *in* Flora of North America Editorial Committee, eds., Flora of North America, vol. 2. New York: Oxford University Press.

Taylor, R. J. 1993b. *Picea* A. Dietrich, pp. 369–373 *in* Flora of North America Editorial Committee, eds., Flora of North America, vol. 2. New York: Oxford University Press.

Taylor, R. J., and T. F. Patterson. 1980. Biosystematics of Mexican spruce species and populations. Taxon 29: 421–440.

Taylor, R. L., and S. Taylor. 1980. *Tsuga mertensiana* in British Columbia. Davidsonia 11: 78–84.

Taylor, R. L., and S. Taylor. 1981. *Taxus brevifolia* in British Columbia. Davidsonia 12: 89–94.

Taylor, R. J., T. F. Patterson, and R. J. Harrod. 1994. Systematics of Mexican spruce—revisited. Systematic Botany 19: 47–59.

Tengnér, J. 1965. *Dacrydium*—anatomy and taxonomy. Botaniska Notiser 118: 450–452.

Tengnér, J. 1967. Anatomy and taxonomy in the Podocarpaceae. Botaniska Notiser 120: 504–506.

Thielges, B. A. 1969. A chromatographic investigation of interspecific relationships in *Pinus* (subsection *Sylvestres*). American Journal of Botany 56: 406–409.

Thielges, B. A. 1972a. A chromatographic study of foliage polyphenols in pine hybrids (subsection *Sylvestres*). Silvae Genetica 21: 109–114.

Thielges, B. A. 1972b. Intraspecific variation in foliage polyphenols of *Pinus* (subsection *Sylvestres*). Silvae Genetica 21: 114–119.

Thompson, J., and L. A. S. Johnson. 1986. *Callitris glaucophylla*, Australia's "white cypress pine"—a new name for an old species. Telopea 2: 731–736.

Thomson, R. B. 1909a. On the pollen of *Microcachrys tertragona*. Botanical Gazette 47: 26–29.

Thomson, R. B. 1909b. The megasporophyll of *Saxegothaea* and *Microcachrys*. Botanical Gazette 47: 345–354.

Tidwell, W. D., L. R. Parker, and V. K. Folkman. 1986. *Pinuxylon woolardii* sp. nov., a new petrified taxon of Pinaceae from the Miocene basalts of eastern Oregon. American Journal of Botany 73: 1,517–1,524.

Toda, R., and S. Mikami. 1976. The provenance trials of Japanese larch established in Japan and the tentative achievements. Silvae Genetica 25: 209–216.

Tolmachev, A. I., J. G. Packer, and G. C. D. Griffiths. 1995. Flora of the Russian arctic, vol. 1. Edmonton: University of Alberta Press.

Tomlinson, P. B. 1991. Pollen scavenging. National Geographic Research and Exploration 7: 188–195.

Tomlinson, P. B. 1992. Aspects of cone morphology and development in Podocarpaceae (Coniferales). International Journal of Plant Sciences 153: 572–588.

Tomlinson, P. B. 1994. Functional morphology of saccate pollen in conifers with special reference to Podocarpaceae. International Journal of Plant Sciences 155: 699–715.

Tomlinson, P. B. 2000. Structural features of saccate pollen types in relation to their functions, pp. 147–162 *in* M. M. Harley, C. M. Morton, and S. Blackmore, eds., Pollen and spores: morphology and biology. Richmond, Surrey: Royal Botanic Gardens, Kew.

Tomlinson, P. B., and T. Takaso. 1998. Hydrodynamics of pollen capture in conifers, pp. 265–275 *in* S. J. Owens and P. J. Rudall, eds., Reproductive biology. Richmond, Surrey: Royal Botanic Gardens, Kew.

Tomlinson, P. B., and E. H. Zacharias. 2001. Phyllotaxis, phenology and architecture in *Cephalotaxus*, *Torreya*, and *Amentotaxus* (Coniferales). Botanical Journal, Linnean Society 135: 215–228.

Tomlinson, P. B., T. Takaso, and J. A. Rattenbury. 1989a. Cone and ovule ontogeny in *Phyllocladus* (Podocarpaceae). Botanical Journal, Linnean Society 99: 209–221.

Tomlinson, P. B., T. Takaso, and J. A. Rattenbury. 1989b. Developmental shoot morphology in *Phyllocladus* (Podocarpaceae). Botanical Journal, Linnean Society 99: 223–248.

Tomlinson, P. B., J. E. Braggins, and J. A. Rattenbury. 1991. Pollination drop in relation to cone morphology in Podocarpaceae: a novel reproductive mechanism. American Journal of Botany 78: 1,289–1,303.

Tomlinson, P. B., T. Takaso, and E. K. Cameron. 1993. Cone development in *Libocedrus* (Cupressaceae)—phenological and morphological aspects. American Journal of Botany 80: 649–659.

Tomlinson, P. B., J. E. Braggins, and J. A. Rattenbury. 1997. Contrasted pollen capture mechanisms in Phyllocladaceae and certain Podocarpaceae (Coniferales). American Journal of Botany 84: 214–233.

Torres-Romero, J. H. 1988. Podocarpaceae, *in* P. Pinto and G. Lozano, eds., Flora de Colombia, Monograph 5. Bogota: Universidad Nacional de Colombia.

Townrow, J. A. 1965. Notes on some Tasmanian pines. I. Some lower Tertiary podocarps. Papers and Proceedings, Royal Society of Tasmania 99: 87–107.

Tripp, K. E. 1995. *Cephalotaxus:* the plum yews. Arnoldia 55: 24–39.

Tsudzuki, J., K. Nakashima, T. Tsudzuki, J. Hiratsuka, M. Shibata, T. Wakasugi, and M. Sugiura. 1992. Chloroplast DNA of black pine retains a residual inverted repeat lacking rRNA genes: nucleotide sequences of *trn*Q, *trn*K, *psb*A, *trn*I and *trn*H and the absence of *rps16*. Molecular and General Genetics 236: 206–214.

Tsukada, M. 1982. *Cryptomeria japonica:* glacial refugia and late glacial and postglacial migration. Ecology 63: 1,091–1,105.

Tsumura, Y., and Y. Suyama. 1998. Differentiation of mitochondrial polymorphisms in populations of five Japanese *Abies* species. Evolution 52: 1,031–1,042.

Tsumura, Y., Y. Suyama, and K. Yoshimura. 2000. Chloroplast DNA inversion polymorphism in populations of *Abies* and *Tsuga*. Molecular Biology and Evolution 17: 1,302–1,312.

Van der Burgh, J. 1973. Hölzer der niederrheinischen Braunkohlenformation, 2. Hölzer der Braunkohlengruben "Maria Theresia" zu Herzogenrath, "Zukunft West" zu Eschweiler und "Victor" (Zülpich Mitte) zu Zülpich. Nebst einer systematisch-anatomischen Bearbeitung der Gattung *Pinus* L. Review of Palaeobotany and Palynology 15: 73–275.

van der Burgt, X. M. 1997. Determination of the age of *Pinus occidentalis* in La Celestina, Dominican Republic, by the use of growth rings. IAWA [International Association of Wood Anatomists] Journal, vol. 18: 139–146.

van Gelderen, D. M., and J. R. P. van Hoey Smith. 1996. Conifers: the illustrated encyclopedia, vols. 1–2. Portland, Oregon: Timber Press.

van Pelt, R. 1996. Champion trees of Washington state. Seattle: University of Washington Press.

van Rozendaal, E. L. M., S. J. J. Kurstjens, T. A. van Beek, and R. G. van den Berg. 1999. Chemotaxonomy of *Taxus*. Phytochemistry 52: 427–433.

Vasilieva, G. V. 1972. Materials on the comparative anatomy of leaf in species *Agathis* Salisb. (Araucariaceae). Botanicheskii Zhurnal 57: 108–118. (in Russian)

Vaucher, H. 2003. Tree bark: a color guide. Transl. J. E. Eckenwalder. Portland, Oregon: Timber Press.

Veblen, T. T., B. R. Burns, T. Kitzberger, A. Lara, and R. Villalba. 1995. The ecology of the conifers of southern South America, pp. 120–155 *in* N. J. Enright and R. S. Hill, eds., Ecology of the southern conifers. Washington, D.C.: Smithsonian Institution Press.

Veillon, J. P. 1962. Coniferas autoctonas de Venezuela—Los Podocarpus. Mérida, Venezuela: Universidad de los Andes.

Veldkamp, J. F., and D. J. de Laubenfels. 1984. Proposal to reject *Pinus dammara* (Araucariaceae). Taxon 33: 337–347.

Venning, J. 1986. Cupressaceae, pp. 105–108 *in* J. P. Jessop and H. R. Toelken, eds., Flora of South Australia 1. Adelaide: South Australian Government Printing Division.

Vidaković, M. 1957. Investigations on the intermediate type between the Austrian and the Scots pine. Silvae Genetica 6: 12–19.

Vidaković, M. 1991. Conifers: morphology and variation. Zagreb, Graficki Zavod Hrvatske.

Villar, M., R. B. Knox, and C. Dumas. 1984. Effective pollination period and nature of pollen-collecting apparatus in the gymnosperm, *Larix leptolepis*. Annals of Botany 53: 279–284.

Volney, W. J. A., and H. F. Cerezke. 1992. The phenology of white spruce and the spruce budworm in northern Alberta. Canadian Journal of Forest Research 22: 198–205.

von Aderkas, P., K. Klimaszewska, and J. M. Bonga. 1990. Diploid and haploid embryogenesis in *Larix leptolepis, L. decidua,* and their reciprocal hybrids. Canadian Journal of Forest Research 20: 9–14.

von Schantz, M., and S. Juvonen. 1966. Chemotaxonomische Untersuchungen in der Gattung *Picea*. Acta Botanica Fennica 73: 1–51.

Wagner, D. B., J. Dong, M. R. Carlson, and A. D. Yanchuk. 1991. Paternal leakage of mitochondrial DNA in *Pinus*. Theoretical and Applied Genetics 82: 510–514.

Walker, E. H. 1976. Flora of Okinawa and the southern Ryukyu Islands. Washington, D.C.: Smithsonian Institution Press.

Wang, X. Q., and Y. Q. Shu. 2000. Chloroplast *mat*K gene phylogeny of Taxaceae and Cephalotaxaceae with additional reference to the systematic position of *Nageia*. Acta Phytotaxonomica Sinica 38: 201–210. (in Chinese)

Wang, X. R., and A. E. Szmidt. 1993. Chloroplast DNA phylogeny of Asian *Pinus* species (Pinaceae). Plant Systematics and Evolution 188: 197–211.

Wang, X. R., and A. E. Szmidt. 1994. Hybridization and chloroplast DNA variation in a *Pinus* species complex from Asia. Evolution 48: 1,020–1,031.

Wang, X. Q., Y. Han, and D. Y. Hong. 1998. A molecular systematic study of *Cathaya,* a relic genus of the Pinaceae in China. Plant Systematics and Evolution 213: 165–172.

Wang, X. R., Y. Tsumura, H. Yoshimaru, K. Nagasaka, and A. E. Szmidt. 1999. Phylogenetic relationships of Eurasian pines (*Pinus,* Pinaceae) based on chloroplast *rbc*L, *mat*K, *rpl20–rps18* spacer, and *trn*V intron sequences. American Journal of Botany 86: 1,742–1,753.

Wang, X. Q., D. C. Tank, and T. Sang. 2000. Phylogeny and divergence times in Pinaceae: evidence from three genomes. Molecular Biology and Evolution 17: 773–781.

Wang, Z. M., and S. E. Macdonald. 1992. Peatland and upland black spruce populations in Alberta, Canada: isozyme variation and seed germination ecology. Silvae Genetica 41: 117–122.

Wardle, P. 1968. Engelmann spruce (*Picea engelmannii*) at its upper limits on the Front Range, Colorado. Ecology 49: 483–495.

Wardle, P. 1969. Biological flora of New Zealand. 4. *Phyllocladus alpinus* Hook. f. (Podocarpaceae) mountain toatoa, celery pine. New Zealand Journal of Botany 7: 76–95.

Wardle, P., and M. C. Coleman. 1992. Evidence for rising upper limits of four native New Zealand forest trees. New Zealand Journal of Botany 30: 303–314.

Waring, R. H., W. H. Emmingham, and S. W. Running. 1975. Environmental limits of an endemic spruce, *Picea breweriana*. Canadian Journal of Botany 53: 1,599–1,613.

Wasscher, J. 1941. The genus *Podocarpus* in the Netherlands Indies. Blumea 4: 359–481.

Watano, Y., M. Imazu, and T. Shimizu. 1995. Chloroplast DNA typing by PCR-SSCP in the *Pinus pumila–P. parviflora* var. *pentaphylla* complex (Pinaceae). Journal of Plant Research 108: 493–499.

Watson, F. D. 1985. The nomenclature of pondcypress and baldcypress (Taxodiaceae). Taxon 34: 506–509.

Watson, F. D. 1993. *Sequoia* Endlicher, *Sequoiadendron* Buchholz, *Taxodium* Richard, pp. 401–404 *in* Flora of North America Editorial Committee, eds., Flora of North America, vol. 2. New York: Oxford University Press.

Webby, R. F., K. R. Markham, and B. P. J. Molloy. 1987. The characterisation of New Zealand *Podocarpus* hybrids using flavonoid markers. New Zealand Journal of Botany 25: 355–366.

Weissmann, Von G., and S. Reck. 1987. Identifizierung von Hybridlärchen mit Hilfe chemischer Merkmale. Silvae Genetica 36: 60–64.

Welch, H. J. 1991. The conifer manual, vol. 1. Dordrecht: Kluwer Academic Publishers. [*Abies* to *Phyllocladus*]

Wells, J. A. 1972. Ecology of *Podocarpus hallii* in central Otago, New Zealand. New Zealand Journal of Botany 10: 399–426.

Wells, O. O. 1964. Geographic variation in ponderosa pine. I. The ecotypes and their distribution. Silvae Genetica 13: 89–103.

Wells, P. M., and R. S. Hill. 1989a. Leaf morphology of the imbricate-leaved Podocarpaceae. Australian Systematic Botany 2: 369–386.

Wells, P. M., and R. S. Hill. 1989b. Fossil imbricate-leaved Podocarpaceae from Tertiary sediments in Tasmania. Australian Systematic Botany 2: 387–423.

Wheatley, J. I. 1992. A guide to the common trees of Vanuatu. Port Vila, Vanuatu: Department of Forestry.

Wheeler, N. C., R. P. Guries, and D. M. O'Malley. 1983. Biosystematics of the genus *Pinus*, subsection *Contortae*. Biochemical Systematics and Ecology 11: 333–340.

White, E. E. 1983. Genetic variation in resin canal frequency and relationship to terpene production in foliage of *Pinus contorta*. Silvae Genetica 33: 79–84.

White, F., F. Dowsett-Lemaire, and J. D. Chapman. 2001. Evergreen forest flora of Malawi. Richmond, Surrey: Royal Botanic Gardens, Kew.

Whitmore, T. C. 1966. Guide to the forests of the British Solomon Islands. Oxford: Oxford University Press.

Whitmore, T. C. 1977. A first look at *Agathis*. Tropical Forestry Papers 11: 1–54.

Whitmore, T. C. 1980. A monograph of *Agathis*. Plant Systematics and Evolution 135: 41–69.

Wieland, G. R. 1935. The Cerro Cuadrado petrified forest. Washington, D.C.: Carnegie Institution of Washington.

Wiggins, I. L. 1980. Flora of Baja California. Stanford, California: Stanford University Press.

Wilde, M. H. 1944. A new interpretation of coniferous cones. I. Podocarpaceae (*Podocarpus*). Annals of Botany 12: 311–326.

Wilde, M. H. 1975. A new interpretation of microsporangiate cones in Cephalotaxaceae and Taxaceae. Phytomorphology 25: 435–450.

Wilde, M. H., and A. J. Eames. 1952. The ovule and "seed" of *Araucaria bidwillii* with discussion of the taxonomy of the genus. II. Taxonomy. Annals of Botany 16: 27–47.

Wilson, E. H. 1916. The conifers and taxads of Japan. Publications of the Arnold Arboretum 8: 1–91.

Wilson, E. H. 1926. The taxads and conifers of Yunnan. Journal of the Arnold Arboretum 7: 37–68.

Wilson, H., and T. Galloway. 1993. Small-leaved shrubs of New Zealand. Christchurch, Manuka Press.

Wilson, V. R., and J. N. Owens. 1999. The reproductive biology of totara (*Podocarpus totara*) (Podocarpaceae). Annals of Botany 83: 401–411.

Wittlake, E. B. 1975. The androstrobilus of *Glyptostrobus nordenskioldi* (Heer) Brown. American Midland Naturalist 94: 215–223.

Wolf, C. B. 1948. Taxonomic and distributional studies of the New World cypresses. Aliso 1: 1–250.

Woltz, P., and M. L. Rouane. 1980. A propos du *Microcachrys tetragona* Hook. f. (Podocarpacées) et de l'evolution vasculaire de la plantule. Bulletin, Société Botanique de France. Lettres Botaniques 127: 151–158.

Woltz, P., R. A. Stockey, M. Gondran, and J.-F. Cherrier. 1994. Interspecific parasitism in the gymnosperms: unpublished data on two endemic New Caledonian Podocarpaceae using scanning electron microscopy. Acta Botanica Gallica 141: 731–746.

Woodford, J. 2002. The Wollemi pine. Melbourne: Text Publishing.

Wright, H. E. 1968. The roles of pine and spruce in the forest history on Minnesota and adjacent areas. Ecology 49: 937–955.

Wright, J. W. 1955. Species crossability in spruce in relation to distribution and taxonomy. Forest Science 1: 319–349.

Wright, J. W., and W. J. Gabriel. 1958. Species hybridization in the hard pines, series *Sylvestres*. Silvae Genetica 7: 109–115.

Wright, J. W., W. A. Lemmien, and D. S. Canavera. 1969. Abundant natural hybridization between Austrian and Japanese red pines in southern Michigan. Forest Science 15: 269–274.

Wright, J. W., R. A. Read, D. T. Lester, C. Merritt, and C. Mohn. 1972. Geographic variation in red pine. Silvae Genetica 21: 205–210.

Xiang, Q. P., and A. Farjon. 2003. Cuticle morphology of a newly discovered conifer, *Xanthocyparis vietnamensis* (Cupressaceae), and a comparison with some of its nearest relatives. Botanical Journal, Linnean Society 143: 315–322.

Xiang, Q. P., A. Farjon, Z. Y. Li, L. K. Fu, and Z. Y. Liu. 2002. *Thuja sutchuensis*: a rediscovered species of the Cupressaceae. Botanical Journal, Linnean Society 139: 305–310.

Xie, C. Y., B. P. Dancik, and F. C. Yeh. 1992. Genetic structure of *Thuja orientalis.* Biochemical Systematics and Ecology 20: 433–441.

Xiong, L., N. Okado, T. Fujiwara, S. Ohta, and J. G. Palmer. 1998. Chronology development and climate response analysis of different New Zealand pink pine (*Halocarpus biformis*) tree-ring parameters. Canadian Journal of Forest Research 28: 566–573.

Yamazaki, T. 1995. Gymnospermae, pp. 261–287 *in* K. Iwatsuki, T. Yamazaki, D. E. Boufford, and H. Ohba, eds., Flora of Japan, vol. 1. Tokyo: Kodansha.

Yazvenko, S. B., and D. J. Rapport. 1997. The history of ponderosa pine pathology—implications for management. Journal of Forestry Dec 1997: 16–20.

Yeatman, C. W. 1967. Biogeography of jack pine. Canadian Journal of Botany 45: 2,201–2,211.

Yeaton, R. I. 1978. Some ecological aspects of reproduction in the genus *Pinus* L. Bulletin, Torrey Botanical Club 105: 306–311.

Yeh, F. C., and J. T. Arnott. 1986. Electrophoretic and morphological differentiation of *Picea sitchensis, Picea glauca,* and their hybrids. Canadian Journal of Forest Research 16: 791–798.

Ying, B. P., I. Kubo, Chairul, T. Matsumoto, and Y. Hayashi. 1990. Congeners of norditerpene dilactones from *Podocarpus nagi.* Phytochemistry 29: 3,953–3,955.

Ying, L., and E. K. Morgenstern. 1991. The population structure of *Larix laricina* in New Brunswick, Canada. Silvae Genetica 40: 180–184.

Ying, T. S., Y. L. Zhang, and D. E. Boufford. 1993. The endemic genera of seed plants of China. Beijing: Science Press.

Yu, B. and J. F. Zeng, eds. 1992. China plant red data book: rare and endangered plants, vol. 1. Beijing: Science Press.

Zanoni, T. A. 1978. The American junipers of the section *Sabina* (*Juniperus,* Cupressaceae)—a century later. Phytologia 38: 433–454.

Zanoni, T. A. 1980. Notes on *Cupressus* in Mexico. Boletin, Sociedad Botanica de Mexico 39: 128–133.

Zanoni, T. A., and R. P. Adams. 1979. The genus *Juniperus* (Cupressaceae) in Mexico and Guatemala: synonymy, key, and distributions of the taxa. Boletin, Sociedad Botanica de Mexico 38: 83–121.

Zavarin, E., and K. Snajberk. 1986a. Monoterpenoid differentiation in relation to the morphology of *Pinus discolor* and *Pinus johannis.* Biochemical Systematics and Ecology 14: 1–11.

Zavarin, E., and K. Snajberk. 1986b. Monoterpenoid differentiation in relation to the morphology of *Pinus remota.* Biochemical Systematics and Ecology 14: 155–163.

Zavarin, E., and K. Snajberk. 1987. Monoterpene differentiation in relation to the morphology of *Pinus culminicola, Pinus nelsonii, Pinus pinceana,* and *Pinus maximartinezii.* Biochemical Systematics and Ecology 15: 307–312.

Zavarin, E., N. T. Mirov, and K. Snajberk. 1966. Turpentine chemistry and taxonomy of three pines of southeastern Asia. Phytochemistry 5: 91–96.

Zavarin, E., W. Hathaway, T. Reichert, and Y. B. Linhart. 1967. Chemotaxonomic study of *Pinus torreyana* Parry turpentine. Phytochemistry 6: 1,019–1,023.

Zavarin, E., L. Lawrence, and M. C. Thomas. 1971. Compositional variations of leaf monoterpenes in *Cupressus macrocarpa, C. pygmaea, C. goveniana, C. abramsiana* and *C. sargentii.* Phytochemistry 10: 379–393.

Zavarin, E., K. Snajberk, D. K. Bailey, and E. C. Rockwell. 1982. Variability in essential oils and needle resin canals of *Pinus longaeva* from eastern California and western Nevada in relation to other members of subsection *Balfourianae.* Biochemical Systematics and Ecology 10: 11–20.

Zavarin, E., K. Snajberk, and L. Cool. 1989. Monoterpenoid differentiation in relation to the morphology of *Pinus edulis.* Biochemical Systematics and Ecology 17: 271–282.

Zavarin, E., K. Snajberk, and L. Cool. 1990. Chemical differentiation in relation to the morphology of the single-needle pinyons. Biochemical Systematics and Ecology 18: 125–137.

Zavarin, E., Z. Rafii, L. G. Cool, and K. Snajberk. 1991. Geographic monoterpene variability of *Pinus albicaulis.* Biochemical Systematics and Ecology 19: 147–156.

Zhang, X., and J. S. States. 1991. Selective herbivory of ponderosa pine by Abert squirrels: a re-examination of the role of terpenes. Biochemical Systematics and Ecology 19: 111–115.

Zobel, D. B. 1998. *Chamaecyparis* forests, a comparative analysis, pp. 39–53 *in* A. D. Laderman, ed., Coastally restricted forests. New York: Oxford University Press.

Zohary, M. 1966. Flora Palaestina, part 1. Equisetaceae to Moringaceae. Jerusalem: Israel Academy of Science and Humanities.

INDEX

When more than one entry is cited, the principal one is boldfaced.